THEORY OF DIFFERENTIAL EQUATIONS IN ENGINEERING AND MECHANICS

THEORY OF DIFFERENTIAL EQUATIONS IN ENGINEERING AND MECHANICS

K.T. CHAU

CRC Press
Taylor & Francis Group
Boca Raton London New York

CRC Press is an imprint of the
Taylor & Francis Group, an **informa** business

CRC Press
Taylor & Francis Group
6000 Broken Sound Parkway NW, Suite 300
Boca Raton, FL 33487-2742

Printed on acid-free paper
Version Date: 20170811

International Standard Book Number-13: 978-1-138-74813-2 (Paperback)
International Standard Book Number-13: 978-1-4987-6778-1 (Hardback)

Library of Congress Cataloging-in-Publication Data

Names: Chau, K. T., author.
Title: Theory of differential equations in engineering and mechanics / by
K.T. Chau.
Description: Boca Raton : CRC Press, [2017] | Includes bibliographical
references and index.
Identifiers: LCCN 2017001029| ISBN 9781138748132 (pbk. : alk. paper) | ISBN
9781498767781 (hardback : acid-free paper) | ISBN 9781498767798 (ebook :
acid-free paper)
Subjects: LCSH: Differential equations.
Classification: LCC QA371 .C445 2017 | DDC 515/.35--dc23
LC record available at https://lccn.loc.gov/2017001029

Visit the Taylor & Francis Web site at
http://www.taylorandfrancis.com

and the CRC Press Web site at
http://www.crcpress.com

Printed and bound in the United States of America by
Edwards Brothers Malloy on sustainably sourced paper

To

My wife Lim, son Magnum, and daughter Jaquelee

and

my mother, father, sisters, and brothers

CONTENTS

PREFACE

THE AUTHOR

PREFACE

Without mathematics, there is not much we can do to progress in our knowledge and technologies in mechanics, engineering, and science. In recent years, this statement has become applicable to business (such as the mathematical modeling and prediction of stock and currency exchange markets), or even social sciences (such as statistical analysis of data). The study of differential equations has been a very important branch of mathematics or of applied mathematics since the time of Leibniz, Bernoulli and Euler, and this is the main subject of the present book. Its evolution ties closely with engineering and scientific applications in the 18th century. In these two-book sequels, we focus on discussion of the theories and applications of differential equations in engineering and mechanics. This volume focuses on the theory, and the next one is on applications.

In college, I took two "engineering mathematics" courses, one from the Mathematics Department taught by a mathematician, and the other from the Civil Engineering Department taught by Dr. Yan Sze Kwan, an MIT PhD graduate majoring in physics. Although I loved both the textbooks that they adopted for the Advanced Engineering Mathematics course (one used Kreyszig, 1979 and the other used Wylie, 1975), I enjoyed the teaching from the physicist much more than from the pure mathematician. Their emphases are different: the mathematician covered no background, physical meaning or applications, whilst the physicist emphasized application and usefulness of the technique, without loss of rigor. The syllabi of these subjects did cover a wide range of topics (although sometimes the coverage was very superficial) that is much broader than what we teach to our students today. For example, I learned vectors, matrices, vector calculus, Gauss and Stokes Theorems, statistics and probability, complex variables including the residue theorem and conformal mapping, special functions, calculus of variations, Laplace transform, and differential equations. For the differential equation part, the teaching mainly covered ordinary differential equations but not much on partial differential equations (except for the wave equation), let alone the classification of second order PDE as hyperbolic, parabolic and elliptic, integral transforms, integral equations, etc.

Once I entered graduate school, I found that a lot of the mathematical techniques in solving differential equations were employed in various graduate courses in mechanics and engineering, like wave propagation in solids, continuum and solid mechanics, fracture mechanics, and wave hydrodynamics. Although my mathematics education at college exposed me to a wide range of topics, I still had to pick up various subjects in applied mathematics by self-learning or by sitting in on courses that I did not have the luxury to take, such as perturbation theory, asymptotic analysis, Fourier series expansion, integral transforms, boundary integral equations, Green's function method, integro-differential equations, eigenfunction expansions, variational principle, and weighted residue methods. However, there was no single book covering my needs. After my master's degree studies in structural engineering from the Asian Institute of Technology and with the encouragement of Professor Pisidhi Karasudhi, I decided to pursue a PhD degree in Theoretical and Applied Mechanics (TAM) from Northwestern University (NU) under the supervision of Professor John Rudnicki. The TAM program at Northwestern University requires students to take three mathematics courses before graduation. I formally took a series of three one-quarter courses on Differential Equations on Mathematical Physics I, II & III offered by Professor William L. Kath. Even though the course covered extensively on partial differential equations that included Laplace equation (or the potential theory), wave equations, Green's functions and distribution theory, I found that it remains essential for me to continue self-learning on different mathematical topics in solving differential equations for my research work as well as for understanding technical papers by others.

The most commonly raised question by educators and students alike is how much engineering mathematics is enough for an engineer or a scientist. There is a continuous debate on what needs to be covered for engineering and mechanics students. But one thing for sure is that you need to read a lot of books in mathematics for which the target audiences are not engineers. Some books are by far too theoretical and jammed with pages of axioms and theorems without talking about applications, and others covered a lot of applications but the topics are not advanced enough and specific enough for engineers and mechanicians. For example, most textbooks on differential equations contain a chapter on numerical methods, yet after learning that, you still have no idea on what the Newton-Raphson method, Newmark beta method and Wilson theta method are. They are the standards in solving structural dynamics problems. Another example is that nearly all textbooks on partial differential equations discuss solely the solutions of the wave equation, Laplace equation and heat equation, but engineers and mechanicians are typically required to obtain the solutions of biharmonic equations, the Poisson equation, Helmholtz equation, or nonhomogeneous wave equation. There is a clear gap between the traditional topics covered on PDEs and what engineers and mechanicians need to learn. The second volume of Selvadurai's (2000) two-volume book series entitled *Partial Differential Equations in Mechanics* focuses on biharmonic and Poisson equations and is clearly motivated by such deficiency.

A few years back, I was given the opportunity to teach Engineering Analysis and Computation for undergraduate civil engineering students. I find that there is no single book that can cover what I want to teach, with focus on engineering and mechanics applications. About one sixth of the materials in this two-book series resulted from preparation of the lecture notes for the course. The remaining content is clearly beyond the undergraduate level, and targets graduate students and researchers.

When I was a graduate student, I was tempted to ask those "big shots" with superb mathematical skills what "mathematics books" they studied when they were young. Nowadays, many of my graduate students are asking me the same question. They seem to believe that there exists a single "secret book" that I studied seriously when I was young. To be frank, there is no such book. You have to learn your way step by step. With this in mind, the present book, hopefully, can partially satisfy some of my students. If the present book will be the first book being recommended to students by professors as their "secret book," my time in preparing the present book would not be wasted. One main feature of the present book is that detailed step-by-step analyses and proofs are given such that most college students with basic training in engineering mathematics can follow through many advanced topics. Hopefully, seemingly difficult and advanced topics will be made accessible to everyone. Following the advice of Morris Kline (1977), whenever possible, we will introduce a topic through its historical background, intuitive reason, physical motivations, potential applications, and relevance to the real world. Interesting stories of mathematicians and their problems will be given (either in the text or in the biography section). Illustrative examples and workable problems are included for each topic, hoping to provoke readers to read, learn, think, and ask.

The main purpose of these two companion-volume books on differential equations is to provide readers with most of the essential topics on differential equations that one is likely to encounter in solving engineering and mechanics problems. The target students are those studying engineering and mechanics, but science students may find the present work useful. However, substantial amount of material in these book volumes will form useful reference materials for researchers and practitioners in dealing with differential equations. There are twenty-eight chapters in these two volumes, fifteen for volume one and thirteen for volume two.

Volume one consists of fifteen chapters and focuses on reviewing the essential mathematical techniques in solving differential equations. Some of these chapters discuss classical techniques for undergraduate students, some are mainly for graduate students, and there are also some exclusively for researchers. We are going more for the breath rather than the depth in our coverage, hoping that students can find and learn the essential mathematical skills in a more comprehensive single volume. Many of the topics covered by these chapters can easily evolve into an independent book. Chapter 1 provides an overall summary of mathematical preliminaries, Chapter 2 gives an overall view on differential equations, Chapter 3 deals with ODEs, Chapter 4 considers series solutions for second order ODEs, and Chapter 5 discusses systems of first order ODEs. These four chapters can easily fill up lectures for an undergraduate course on "introduction to ordinary differential equations" or more. Chapter 6 covers first order PDEs, which is normally not covered in engineering courses, Chapter 7 considers higher order PDEs, Chapter 8 summarizes the idea of Green's function method, Chapter 9 deals with the classical topics of wave, diffusion and potential equations in mathematical physics, and Chapter 10 discusses solution of boundary value problems in terms of eigenfunction expansions. These four chapters can form a graduate course on "partial differential equations in engineering and mechanics." Chapter 11 covers integral and integro-differential equations, Chapter 12 serves an introduction to the asymptotic expansion and perturbation, Chapter 13 deals with calculus of variations, Chapter 14 summarizes the concept of variational methods, and finally Chapter 15 examines numerical methods. Each of the six chapters can be a graduate course on its own right. These chapters are expected to provide general knowledge on more advanced topics in engineering and mechanics. Readers interested in numerical analyses probably should study Chapters 13–15 together. Chapter 12 is an important introduction to the domain of studying nonlinear differential equations.

Volume two consists of thirteen chapters focused more on applications of differential equations on variable topics. The choices of this topic are shaped by my former background and interests in engineering and in mechanics.

The unfailing and continuous support from my wife Lim, my son Magnum, and my daughter Jaquelee is what keeps me going when I face difficult times.

Regular training with the PolyU swimming team and interactions with swimmers from masters swimming allows me to refresh my mind and my body from being drained by this ambitious book project.

After I finished my first book *Analytic Methods of Geomechanics*, I told myself that I was not going to write another one because it is just too demanding both physically and mentally. However, most first-time marathon finishers will swear that they will never run another marathon again in their lives right after the race. It is just too demanding. Yet many return to marathon training after only a few months. Marathon running is an addictive behavior. As a former marathoner, I know too well this feeling and here I go again.

This book project was encouraged by Mr. Tony Moore, a senior editor of civil engineering at CRC Press (imprint of Taylor & Francis). The expert assistance from Production Editor Michele Dimont and Editorial Assistant Scott Oakley is highly appreciated. A special thank you goes to my former undergraduate students, who through their comments shaped some of the contents of this book.

K.T. Chau
Hong Kong

THE AUTHOR

Professor K.T. Chau, Ph.D., is the Chair Professor of Geotechnical Engineering of the Department of Civil and Environmental Engineering at the Hong Kong Polytechnic University. He obtained his honors diploma with distinction from Hong Kong Baptist College (Hong Kong); his master's of engineering in structural engineering from the Asian Institute of Technology (Thailand), where he was also awarded the Tim Kendall Memorial Prize (an academic prize for the best graduating student); his Ph.D. in Theoretical and Applied Mechanics from Northwestern University (USA); and an Executive Certificate from the Graduate School of Business of Stanford University.

Dr. Chau worked as a full-time tutor/demonstrator/technician at Hong Kong Baptist College (1984–1985), as a research associate at the Asian Institute of Technology (summer of 1987), a research assistant at Northwestern University (1987–1991), and as a post-doctoral fellow at Northwestern University (1991–1992). At Hong Kong Polytechnic University (PolyU), he has served as a lecturer, an assistant professor, an associate professor, a full professor, and a chair professor since 1992. At PolyU, he was the Associate Dean (Research and Development) of the Faculty of Construction and Environment, the Associated Head of the Department of Civil and Structural Engineering, the Chairman of the Appeals and Grievance Committee, the Alternate Chairman of the Academic Appeals Committee, and the Alternate Chairman of the University Staffing Committee.

Dr. Chau is a fellow of the Hong Kong Institution of Engineers (HKIE), the past Chairman of the Geomechanics Committee (2005–2010) of the Applied Mechanics Division (AMD) of ASME, and the past Chairman of the Elasticity Committee (2010–2013) of the Engineering Mechanics Institute (EMI) of ASCE. He is a recipient of the Distinguished Young Scholar Award of the National Natural Science Foundation, China (2003); the France-Hong Kong Joint Research Scheme (2003–2004) of RGC of Hong Kong; and the Young Professor Overseas Placement Scheme of PolyU. He was a recipient of the Excellent Teaching Award of the Department of Civil and Environmental Engineering (2014). He is a past President of the Hong Kong Society of Theoretical and Applied Mechanics (2004–2006) after serving as member-at-large and Vice President. He also served as a Scientific Advisor of the Hong Kong Observatory of HKSAR Government, an RGC Engineering Panel member of the HKSAR Government for 7 consecutive years, and served as the Vice President of the Hong Kong Institute of Science. He held visiting positions at Harvard University (USA), Kyoto University (Japan), Polytech-Lille (France), Shandong University (China), Taiyuan University of Technology (China), the Rock Mechanics Research Center of CSIRO (Australia), and the University of Calgary (Canada).

Dr. Chau's research interests have included geomechanics and geohazards, including bifurcation and stability theories in geomaterials, rock mechanics, fracture and damage mechanics in brittle rocks, three-dimensional elasticity, earthquake engineering and mechanics, landslides and debris flows, tsunami and storm surges, and rockfalls and dynamic impacts, seismic pounding, vulnerability of tall buildings with transfer systems, and shaking table tests. He is the author of more than 100 journal papers and 200 conference publications. His first book entitled *Analytic Methods in Geomechanics* was published by CRC Press in 2013.

In his leisure time, he enjoys swimming and takes part in master swimming competitions. He is the Honorable Manager of the Hong Kong Polytechnic University Swimming Team. Since 2001, he has competed in Hong Kong Masters Games, the Hong Kong Territory-wise Age-Group Swimming Competition, the Hong Kong Amateur Swimming Association (HKASA) Masters Swimming Championships, and District Swimming Meets of the Leisure and Cultural Services Department (LCSD). He has also participated in international masters swimming competitions, including the Macau Masters Swimming Championship, the Singapore National Masters Swimming, Standard Chartered Asia Pacific Masters Swim Meet, Wisdom-Act International Swimming Championship (Taiwan), Standard Chartered Singapore Masters Swim 2007, Japan Masters Long Distance Swim Meet 2008 (Aichi Meet and Machida Meet), Japan Short Course Masters Swimming Championship 2009 (Kyoto), Marblehead Sprint Classics (USA), The Masters Games Hamilton (New Zealand), Hawaii Senior Olympics, National China Masters Swimming Championships, the Third Annual Hawaii International Masters Swim Meet, and the Fifth Penang Invitational Masters Swimming Championship (Malaysia). By 2007, he had competed in all long-course FINA events (i.e., 50 m, 100 m, 200 m, 400 m, 800 m and 1500 m freestyle; 50 m, 100 m and 200 m butterfly; 50 m, 100 m and 200 m breaststroke; 50 m, 100 m and 200 m backstroke; and 200 m and 400 m individual medley).

He also enjoys jogging and has completed four full marathons, including the Hong Kong International Marathon, the Chicago Oldstyle Marathon, and the China Coast Marathon with a personal best time of 3 hours, 34 minutes, and 15 seconds.

CHAPTER ONE

Mathematical Preliminaries

1.1 INTRODUCTION

The main focuses of the present book are the application on differential equations in engineering and in mechanics, and on the associated solution techniques used in solving these equations. Readers are expected to have general education and training in engineering mathematics, including differentiation and integration, vector analysis, matrix operation, and basics in algebra and trigonometry. Indeed, engineering and science students would have taken engineering mathematics before they embark on taking a course on differential equations. It is, however, impossible to cover all preliminary mathematics in a single chapter before we continue on our discussion on differential equations. This chapter will not only cover some elementary topics that we need for later development in this book, but also touch upon some less familiar, yet important, topics.

We will in this chapter review binomial theorem, differentiation, integration, Jacobian, complex variables, Euler's polar form for complex variables, elementary functions like circular functions (e.g., sine and cosine) and hyperbolic functions (e.g., sinh and cosh), differentiation of complex functions, integration using the Cauchy integral formula and residue theorem, Gauss and Kelvin-Stokes theorems, series expansions (including Taylor series expansion, Maclaurin series expansion, Laurent's series expansion), and vector calculus. Most of these topics are typically covered in the first two years of engineering mathematics courses. In addition, we have added some more advanced topics that are unlikely being covered in elementary engineering mathematics courses, and they are the Frullani-Cauchy integral, Ramanujan's master theorem, Ramanujan's integral theorem, Darboux's formula, Mittag-Leffler's expansion, Borel's theorem, and tensor analysis. Both differential and integral forms of gradient, divergence, and curl are discussed. Helmholtz's representation theorem is also discussed. Some of these advanced topics can be skipped in the first reading but they form the essential basis for more advanced analysis in applied mathematics and in engineering mechanics.

Before we summarize some essential mathematical backgrounds needed for later analysis of differential equations, we recall here the definitions of algebraic functions and transcendental functions. If $w(x)$ is a solution of the following polynomials,

$$P_0(x)w^n + P_1(x)w^{n-1} + ... + P_{n-1}(x)w + P_n(x) = 0 \tag{1.1}$$

where $P_0 \neq 0$, $P_1(x)$,..., $P_n(x)$ are polynomials of x and n is positive integer, $w = f(x)$ is called an algebraic function. On the other hand, any function which cannot be expressed as a solution of (1.1) is called a transcendental function. Transcendental functions include exponential functions, trigonometric functions (or circular functions such as sine and cosine), hyperbolic functions (such as sinh and cosh), and logarithmic functions.

1.2 BINOMIAL THEOREM

Typically, the binomial theorem has been covered in high school or in a first-year college mathematics course. It is, however, so useful that we will employ it repeatedly in our analyses on differential equations. To start with, we all learn in high school that the square of a sum can be expanded like

$$(a+b)^2 = (a+b)(a+b) = a^2 + 2ab + b^2 \tag{1.2}$$

where a and b are any real numbers. This expansion can easily be generalized to other integer (whole number) powers as

$$\begin{aligned}
&(a+b)^0 = 1 \\
&(a+b)^1 = a+b \\
&(a+b)^2 = a^2 + 2ab + b^2 \\
&(a+b)^3 = a^3 + 3a^2b + 3ab^2 + b^3 \\
&(a+b)^4 = a^4 + 4a^3b + 6a^2b^2 + 4ab^3 + b^4 \\
&\ldots
\end{aligned} \tag{1.3}$$

Clearly, there is a pattern in the coefficient of this expansion as the value of the integer power increases. It can be checked in a straightforward manner that the coefficients can be put in a triangular pattern as shown Figure 1.1.

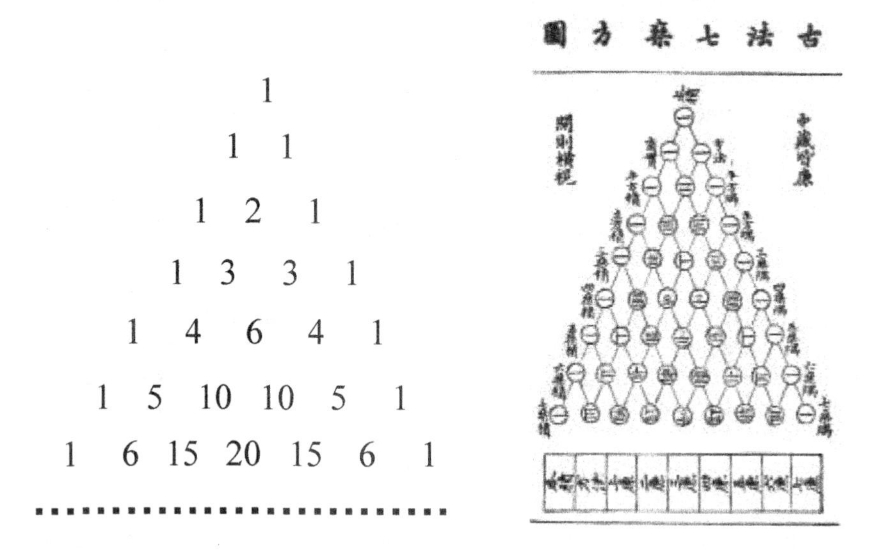

Figure 1.1 The Pascal triangle or Yang Hui/Jia Xian triangle (the figure on the right is adopted from Zhu, 1303)

The triangle shown in Figure 1.1 expressed in terms of coefficients of power expansions is normally called the Pascal triangle in western literature and was discovered by Pascal in 1654, but it later was discovered that it has been known to other mathematicians well before Pascal. For example, it was known to German mathematician Petrus Apianus in 1527 and to Yang Hui in 1261 (his last name is Yang). The triangle on the right of Figure 1.1 is the extraction from Zhu's book in 1303 (Zhu, 1303). In China, the term "Yang Hui triangle" gets its popularity because of the excellent book by Hua Loo Keng (Hua, 1956), the most influential mathematician in modern China (see biography section at the end of this book). In its introduction, Hua (1956) remarked that Yang Hui mentioned that this triangle was reported earlier in a book by Jia Xian (last name is Jia). Based on the fact that Jia's book is lost today, Hua decided to called it the Yang Hui triangle in his book and even use it as the title of his excellent book. Being the authority in mathematics in China, it was not challenged. This use or misuse of terminology by Hua (1956) has been recently criticized by historians. Thus, in China, both the names of "Yang Hui triangle" and "Jia Xian triangle" have been used in the literature.

Mathematically, the binomial theorem can be expressed as:

$$(x+h)^n = x^n + nx^{n-1}h + \frac{n(n-1)}{2}x^{n-2}h^2 + \cdots + nxh^{n-1} + h^n \tag{1.4}$$

Although the Jia Xian triangle has been known since at least AD 1200, the binomial formula given in (1.4) was believed discovered by Isaac Newton. The coefficient on the right hand side of (1.4) is called the binomial coefficient. It can be defined as

$$C_n^r = \frac{n(n-1)...(n-r+1)}{r!} = \frac{n!}{r!(n-r)!} \tag{1.5}$$

so that

$$(x+h)^n = x^n + C_1^n x^{n-1}h + C_2^n x^{n-2}h^2 + \cdots + C_{n-1}^n xh^{n-1} + C_n^n h^n \tag{1.6}$$

In (1.5), the factorial function $n!$ has been used (i.e., $5! = 5\times4\times3\times2\times1 = 120$). This binomial coefficient is found extremely useful in probability analysis, which however is out of the scope of the present book. For example, the binomial coefficient given in (1.5) can be interpreted as the number of different ways of taking r objects from n different objects. Taking 1 object at a time (or r =1), we can clearly see that there are exactly n different ways. Putting r = 1 in the right hand side of (1.5) gives exactly n (note that $1! = 1$). It is clear from Figure 1.1 that the Pascal triangle or Jia Xian triangle can be generated by two adjacent terms to form the binomial coefficient of the next row as shown in Figure 1.2. This summing rule can also be proved analytically in Example 1.1.

Example 1.1

$$\begin{aligned} C_{n-1}^{r-1} + C_{n-1}^r &= \frac{(n-1)!}{(r-1)!(n-r)!} + \frac{(n-1)!}{r!(n-1-r)!} \\ &= \frac{(n-1)!}{r!(n-r)!}[r + (n-r)] = \frac{n!}{r!(n-r)!} = C_n^r \end{aligned} \tag{1.7}$$

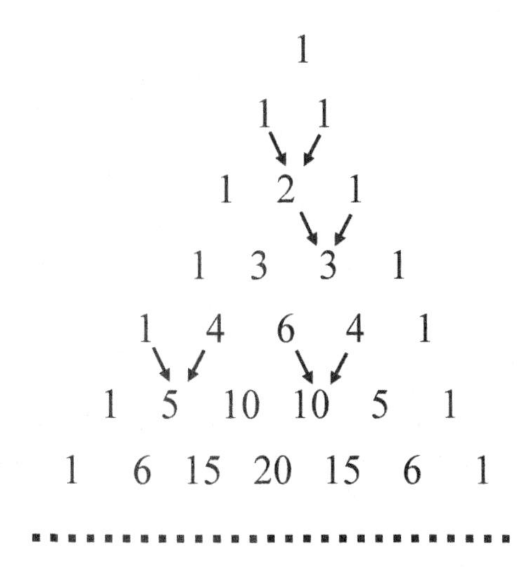

Figure 1.2 Summing rule in generating the coefficients on next row

Example 1.2 illustrates the series expansion of the function e^x with e being the base of natural logarithms (i.e., $e = 2.71828\ 18284\ 59045\ 23536\ 0287...$). The definition of e will be given in Example 1.4.

Example 1.2

$$\begin{aligned} e^x &= \lim_{n\to\infty}(1+\frac{x}{n})^n = \lim_{n\to\infty}[1+\frac{n}{1!}(\frac{x}{n})+\frac{n(n-1)}{2!}(\frac{x}{n})^2+\frac{n(n-1)(n-2)}{3!}(\frac{x}{n})^3+...] \\ &= \lim_{n\to\infty}[1+\frac{x}{1!}+\frac{(1-1/n)x^2}{2!}+\frac{(1-1/n)(1-2/n)x^3}{3!}+...] \\ &= 1+\frac{x}{1!}+\frac{x^2}{2!}+\frac{x^3}{3!}+... \end{aligned} \tag{1.8}$$

Actually (1.8) can be viewed as a Taylor series expansion, and this will be covered later in Section 1.13.2.

In fact, if we set $x = 1$ in (1.8), we also have the definition of the e. This was first considered by Leonard Euler, who was the most prolific researcher in the history of mathematics. There are 886 publications of Euler listed in the website of the Mathematics Department at Dartmouth University. Interested readers can download them from http://www.math.dartmouth.edu/~euler/tour/tour00.html. We will see later in this book that Euler's name will be mentioned repeatedly. Euler is actually one of the main pioneers of differential equations and nearly all of his theories on differential equations are related to the solving of real problems. Most of the notations that we use today were proposed by Euler, such as π, e, i ($=(-1)^{1/2}$), $f(x)$, $\sin x$ and $\cos x$ (Havil, 2003). Laplace once said "Read Euler, read Euler, he is the master of us all" (Dunham, 1999).

We will next review the essence of calculus, which consists of differentiation and integration. They are the bread and butter of differential equations.

1.3 DIFFERENTIATION

To the general public, calculus is believed to be invented by Isaac Newton. In fact, it was invented independently by British physicist and mathematician Isaac Newton and by German mathematician and diplomat Gottfried Wilhelm Leibniz. The currently adopted notation for differentiation and integration are all due to Leibniz. In a sense, we owe more to Leibniz than to Newton. The following disputed story between Newton and Leibniz was told by the most popular physicist of our time, Professor Stephen W. Hawking (Hawking, 1988) in the appendix of his celebrated book *A Brief History of Time*.

> Isaac Newton was not a pleasant man. His relations with other academics were notorious, with most of his later life spent embroiled in heated disputes....
>
> A more serious dispute arose with the German philosopher Gottfried Leibniz. Both Leibniz and Newton had independently developed a branch of mathematics called calculus, which underlies most of modern physics. Although we now know that Newton discovered calculus years before Leibniz, he published his work much later. A major row ensued over who had been first, with scientists vigorously defending both contenders. It is remarkable, however, that most of the articles appearing in defence of Newton were originally written by his own hand—and only published in the name of friends! As the row grew, Leibniz made the mistake of appealing to the Royal Society to resolve the dispute. Newton, as president, appointed an "impartial" committee to investigate, coincidentally consisting entirely of Newton's friends! But that was not all: Newton then wrote the committee report himself and had the Royal Society publish it, officially accusing Leibniz of plagiarism. Still unsatisfied, he then wrote an anonymous review of the report in the Royal Society's own periodical. Following the death of Leibniz, Newton is reported to have declared that he had taken great satisfaction in "breaking Leibniz's heart." (pp. 181–182, Hawking, 1988)

In fact, the initial relation between Newton and Leibniz is not as Hawking (1988) described above. The following extraction from Newton's first edition of *Principia* (1687) gives another side of the story:

> In letters which went between me and that most excellent geometer, G.W. Leibniz, 10 years ago, when I signified I was in the knowledge of a method of determining maxima and minima, of drawing tangents, and the like ... that most distinguished man wrote back that he had also fallen on a method of the same kind, and communicated his method which hardly different from mine, except in his form and words and symbols. (Gjertsen, 1986)

However, in the third edition of *Principia* (1726), which was published 10 years after the death of Leibniz, the above reference to Leibniz work was removed. Probably, we all should learn something from this notorious story.

We now return to the technical side. Differentiation of a function $f(x)$ can be defined as a limit process of looking the ratio of the change of f versus the change of x:

$$\frac{df}{dx} = f'(x) = \lim_{h \to 0} \frac{f(x+h) - f(x)}{h} \tag{1.9}$$

If we plot f versus x, differentiation of f with respect to x is actually the slope of the curve. Figure 1.3 illustrates this geometric meaning of differentiation. It is also clear from Figure 1.3 that when $f'(x) = 0$, the point x is either a maximum or a minimum. At the maximum point, clearly the slope of the curve is changing from positive to negative whereas the slope at the maximum point is zero. Thus, we must have $f''(x) < 0$ at a maximum point, and a similar argument shows that we must have $f''(x) > 0$ at a minimum point. Because of these, differentiation is useful in maximizing or minimizing certain functions. In addition, for a derivative to exist, the curve shown in Figure 1.3 must be continuous and smooth. In mathematical terms, we require that the same differentiation is obtained at point a whether we approach the point a from the left or from the right

$$\lim_{x \to a-\varepsilon} f'(x) = \lim_{x \to a+\varepsilon} f'(x) \tag{1.10}$$

where ε is a small number.

In Figure 1.3, we also demonstrated the mean value theorem for differentiation. In particular, if $f(x)$ is continuous and differentiable in the interval $(a, a+h)$, there is a value $a+\theta h$ $(0 < \theta < 1)$ such that

$$f(a+h) = f(a) + hf'(a+\theta h) \tag{1.11}$$

Before we discuss more rules of differentiation, let us consider a very important formula related to taking a limit. It is called L'Hôpital's rule. In particular, when $f(x)/g(x)$ tends to the indeterminate form of 0/0 or ∞/∞ as the limit of $x \to a$, then we can evaluate the limit as:

$$\lim_{x \to a} \frac{f(x)}{g(x)} = \lim_{x \to a} \frac{f'(x)}{g'(x)} \tag{1.12}$$

If the right hand side of (1.12) remains as in an indeterminate form of 0/0 or ∞/∞, we can reapply (1.12) repeatedly. Note that a finite limit of any indeterminate form may exist or may not exist. The so-called L'Hôpital's rule is actually discovered by Johann Bernoulli (also known as John Bernoulli), but his student Gillaume Francois Antoine de L'Hôpital published this rule in his first textbook on calculus in 1696. As a result, this was mistaken as the result by L'Hôpital (Maor, 1994).

1.3.1 General Formulas

The most important formula for differentiation is probably the power law rule, which is defined as

$$\frac{d}{dx}\left(x^n\right) = nx^{n-1} \tag{1.13}$$

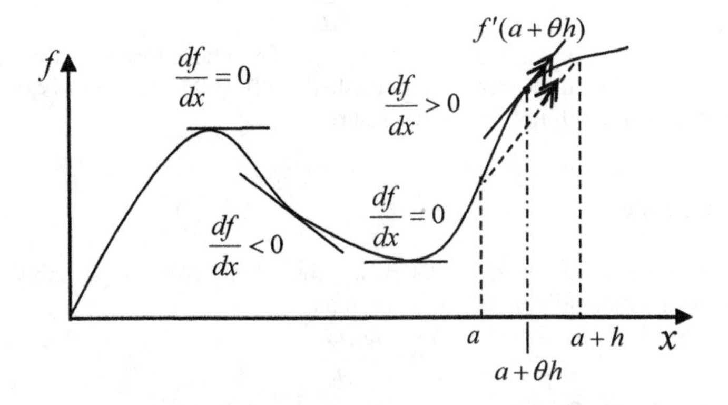

Figure 1.3 Geometric interpretation of differentiation

To see the proof, we employ the binomial theorem given in (1.4). In particular, we can expand the following limit by using the binomial theorem:

$$\lim_{h\to 0}\frac{(x+h)^n - x^n}{h} = \lim_{h\to 0}\frac{1}{h}[x^n + nx^{n-1}h + \frac{n(n-1)}{2}x^{n-2}h^2 + \cdots + nxh^{n-1} + h^n - x^n]$$
$$= \lim_{h\to 0}\frac{1}{h}[nx^{n-1}h + \frac{n(n-1)}{2}x^{n-2}h^2 + \cdots + nxh^{n-1} + h^n]$$
$$= nx^{n-1} + \lim_{h\to 0}[\frac{n(n-1)}{2}x^{n-2}h + \cdots + nxh^{n-2} + h^{n-1}] = nx^{n-1} \tag{1.14}$$

Without going into the details, we report the following commonly used formulas of differentiation:

Constant rule:

$$\frac{d}{dx}(c) = 0 \tag{1.15}$$

Sum rule:

$$\frac{d}{dx}(u+v) = \frac{du}{dx} + \frac{dv}{dx} \tag{1.16}$$

Product rule:

$$\frac{d}{dx}(uv) = u\frac{dv}{dx} + v\frac{du}{dx} \tag{1.17}$$

Quotient rule:

$$\frac{d}{dx}\left(\frac{u}{v}\right) = \frac{v\frac{du}{dx} - u\frac{dv}{dx}}{v^2} \tag{1.18}$$

Constant multiplier rule:

$$\frac{d}{dx}(cu) = c\frac{du}{dx} \tag{1.19}$$

where c is a constant, u *and* v are both function of x. These formulas were known to Leibniz in 1675, but they were not published until 1684. The proofs of them are straightforward and will not be reported here.

1.3.2 Chain Rule

Among different rules of differentiation, the chain rule is probably the most powerful. Mathematically, it can be written as

$$\frac{dy}{dx} = \frac{dy}{du}\frac{du}{dx} \tag{1.20}$$

where $y(u)$ and $u(x)$. This can be proved by noting that

$$\frac{dy}{du} = \lim_{\Delta u \to 0} \frac{y(u+\Delta u) - y(u)}{\Delta u}, \quad \frac{du}{dx} = \lim_{\Delta x \to 0} \frac{u(x+\Delta x) - u(x)}{\Delta x} \tag{1.21}$$

Then, the differentiation on the left of (1.20) can be written as:

$$\begin{aligned} \frac{dy\{u(x)\}}{dx} &= \lim_{\Delta x \to 0} \frac{y\{u(x+\Delta x)\} - y\{u\}}{\Delta x} \\ &= \lim_{\Delta x \to 0} \frac{y\{u(x+\Delta x)\} - y\{u(x)\}}{u(x+\Delta x) - u(x)} [\frac{u(x+\Delta x) - u(x)}{\Delta x}] \\ &= \lim_{\Delta u \to 0} \frac{y\{u+\Delta u\} - y\{u\}}{\Delta u} \lim_{\Delta x \to 0} \frac{u(x+\Delta x) - u(x)}{\Delta x} \\ &= \frac{dy}{du}\frac{du}{dx} \end{aligned} \tag{1.22}$$

In deriving (1.22), we note by definition that $\Delta u = u(x+\Delta x) - u(x)$, and $\Delta u \to 0$ as $\Delta x \to 0$. This completes the proof of (1.20).

The mastery of the chain rule is essential for later analysis of differential equations. For example, we can apply the chain rule to the power rule of an arbitrary $u(x)$ as (putting $y(u) = u^n$ in (1.13)):

$$\frac{d}{dx}(u^n) = \frac{d}{du}(u^n)\frac{du}{dx} = nu^{n-1}\frac{du}{dx} \tag{1.23}$$

Differentiation of elementary functions forms the basis for the analysis of differential equations. The commonly used differentiation formulas for circular or trigonometric functions are

$$\frac{d}{dx}\sin(u) = \cos u\frac{du}{dx}, \quad \frac{d}{dx}\cos(u) = -\sin u\frac{du}{dx}, \quad \frac{d}{dx}\tan(u) = \sec^2 u\frac{du}{dx}$$

$$\frac{d}{dx}\sin^{-1} u = \frac{1}{\sqrt{1-u^2}}\frac{du}{dx}, \quad \frac{d}{dx}\cos^{-1} u = -\frac{1}{\sqrt{1-u^2}}\frac{du}{dx}, \quad \frac{d}{dx}\tan^{-1} u = \frac{1}{1+u^2}\frac{du}{dx} \tag{1.24}$$

The differentiation formulas for hyperbolic functions are

$$\frac{d}{dx}(\sinh u)=\cosh u\frac{du}{dx},\quad \frac{d}{dx}(\cosh u)=\sinh u\frac{du}{dx},\quad \frac{d}{dx}(\tanh u)=\operatorname{sech}^2 u\frac{du}{dx}$$

$$\frac{d}{dx}(\sinh^{-1}u)=\frac{1}{\sqrt{1+u^2}}\frac{du}{dx},\quad \frac{d}{dx}(\cosh^{-1}u)=\frac{1}{\sqrt{u^2-1}}\frac{du}{dx},$$

$$\frac{d}{dx}(\tanh^{-1}u)=\frac{1}{1-u^2}\frac{du}{dx} \tag{1.25}$$

Differentiation of logarithms and exponential functions are

$$\frac{d}{dx}\ln u=\frac{1}{u}\frac{du}{dx},\quad \frac{d}{dx}\log_a u=\frac{1}{u\ln a}\frac{du}{dx},\quad \frac{d}{dx}\left(a^u\right)=a^u\ln a\frac{du}{dx},$$

$$\frac{d}{dx}e^u=e^u\frac{du}{dx} \tag{1.26}$$

1.3.3 Leibniz Theorem on n-th Order Differentiation

The Leibniz rule for taking the n-th differentiation of a product of two functions f and g is:

$$\frac{d^n}{dx^n}(fg)=\sum_{k=0}^{n}C_n^k\frac{d^k f}{dx^k}\frac{d^{n-k}g}{dx^{n-k}} \tag{1.27}$$

where the binomial coefficient appears in (1.27) has been defined in (1.5). For example, if $n = 4$, we have

$$\frac{d^4}{dx^4}(fg)=\frac{d^4 f}{dx^4}g+4\frac{d^3 f}{dx^3}\frac{dg}{dx}+6\frac{d^2 f}{dx^2}\frac{d^2 g}{dx^2}+4\frac{df}{dx}\frac{d^3 g}{dx^3}+f\frac{d^4 g}{dx^4} \tag{1.28}$$

This formula is useful in dealing with higher order differentiation. For small n, the validity of (1.27) can be checked easily. The proof of (1.27) for general n can be done by using mathematical induction. For $n = 1$, we have

$$\frac{d}{dx}(fg)=C_1^0 g\frac{df}{dx}+C_1^1 f\frac{dg}{dx}=\frac{1!}{0!(1-0)!}g\frac{df}{dx}+\frac{1!}{1!(1-1)!}f\frac{dg}{dx}=g\frac{df}{dx}+f\frac{dg}{dx} \tag{1.29}$$

where we have used (1.5). The validity for $n = 1$ is established. Next, we assume (1.27) is true for the case of $n = k$. That is, we have

$$\frac{d^k}{dx^k}(fg)=C_k^0\frac{d^k f}{dx^k}g+C_k^1\frac{d^{k-1}f}{dx^{k-1}}\frac{dg}{dx}+C_k^2\frac{d^{k-2}f}{dx^{k-2}}\frac{d^2 g}{dx^2}$$

$$+\dots+C_k^{k-1}\frac{df}{dx}\frac{d^{k-1}g}{dx^{k-1}}+C_k^k f\frac{d^k g}{dx^k} \tag{1.30}$$

Taking differentiation with respect to x once more time, we obtain

$$\frac{d^{k+1}}{dx^{k+1}}(fg) = C_k^0(\frac{d^{k+1}f}{dx^{k+1}}g + \frac{d^k f}{dx^k}\frac{dg}{dx}) + C_k^1(\frac{d^k f}{dx^k}\frac{dg}{dx} + \frac{d^{k-1}f}{dx^{k-1}}\frac{d^2 g}{dx^2})$$

$$+C_k^2(\frac{d^{k-2}f}{dx^{k-2}}\frac{d^3 g}{dx^3} + \frac{d^{k-1}f}{dx^{k-1}}\frac{d^2 g}{dx^2})$$

$$+...+C_k^{k-1}(\frac{d^2 f}{dx^2}\frac{d^{k-1}g}{dx^{k-1}} + \frac{df}{dx}\frac{d^k g}{dx^k}) + C_k^k(f\frac{d^{k+1}g}{dx^{k+1}} + \frac{df}{dx}\frac{d^k g}{dx^k}) \quad (1.31)$$

$$= C_k^0\frac{d^{k+1}f}{dx^{k+1}}g + (C_k^1 + C_k^0)\frac{d^k f}{dx^k}\frac{dg}{dx} + (C_k^1 + C_k^2)\frac{d^{k-1}f}{dx^{k-1}}\frac{d^2 g}{dx^2} + ...$$

$$+(C_k^{k-1} + C_k^k)\frac{df}{dx}\frac{d^k g}{dx^k} + C_k^k f\frac{d^{k+1}g}{dx^{k+1}}$$

By employing the definition of the binomial coefficient given in (1.5) and the result established in Example 1.1, we can easily show that

$$C_k^0 = C_{k+1}^0 = C_k^k = C_{k+1}^{k+1} = 1, \quad C_k^1 + C_k^0 = C_k^1 + 1 = k+1 = C_{k+1}^1,$$
$$C_k^1 + C_k^2 = C_{k+1}^1, \quad C_k^{k-1} + C_k^k = C_k^{k-1} + 1 = k+1 = C_{k+1}^k \quad (1.32)$$

Thus, substitution of (1.32) into (1.31) gives

$$\frac{d^{k+1}}{dx^{k+1}}(fg) = C_{k+1}^0\frac{d^{k+1}f}{dx^{k+1}}g + C_{k+1}^1\frac{d^k f}{dx^k}\frac{dg}{dx} + C_{k+1}^2\frac{d^{k-1}f}{dx^{k-1}}\frac{d^2 g}{dx^2} + ...$$
$$+C_{k+1}^k\frac{df}{dx}\frac{d^k g}{dx^k} + C_{k+1}^{k+1}f\frac{d^{k+1}g}{dx^{k+1}} \quad (1.33)$$

This agrees with (1.27) for the case of $n = k+1$. Therefore, if (1.27) is true for the case of k, (1.27) is also true for $k+1$. By mathematical induction, (1.27) is valid for all values of n starting from $n = 1$.

1.3.4 Leibniz Rule of Differentiation for Integral

Another important formula by Leibniz is the differentiation of integral:

$$\frac{d}{dx}\int_{g(x)}^{h(x)} f(x,\xi)d\xi = \int_{g(x)}^{h(x)}\frac{df(x,\xi)}{dx}d\xi + f[x,h(x)]\frac{dh(x)}{dx} - f[x,g(x)]\frac{dg(x)}{dx} \quad (1.34)$$

where both lower and upper limits are functions of x. The formal definition of integration will be deferred to a later section. To prove (1.27), we first write the integral as

$$\phi(x) = \int_{g(x)}^{h(x)} f(x,\xi)d\xi \quad (1.35)$$

The change of this integral with respect to an increment of Δx is considered:

$$\Delta\phi(x) = \phi(x+\Delta x) - \phi(x) = \int_{g(x+\Delta x)}^{h(x+\Delta x)} f(x+\Delta x,\xi)d\xi - \int_{g(x)}^{h(x)} f(x,\xi)d\xi \quad (1.36)$$

The first integral on the right hand side of (1.36) can be split into three integrals as

$$\Delta\phi(x)=\int_{g(x+\Delta x)}^{g(x)} f(x+\Delta x,\xi)d\xi+\int_{g(x)}^{h(x)} f(x+\Delta x,\xi)d\xi$$
$$+\int_{h(x)}^{h(x+\Delta x)} f(x+\Delta x,\xi)d\xi-\int_{g(x)}^{h(x)} f(x,\xi)d\xi \tag{1.37}$$

The second and fourth integrals on the right hand side have the same limits and are grouped together, and the limits of the first integral on the right hand side are reversed. Thus, (1.37) can be rewritten as:

$$\Delta\phi(x)=\int_{g(x)}^{h(x)} [f(x+\Delta x,\xi)-f(x,\xi)]d\xi+\int_{h(x)}^{h(x+\Delta x)} f(x+\Delta x,\xi)d\xi$$
$$-\int_{g(x)}^{g(x+\Delta x)} f(x+\Delta x,\xi)d\xi \tag{1.38}$$

The first term on the right of (1.38) can be expressed as:

$$\int_{g(x)}^{h(x)} [f(x+\Delta x,\xi)-f(x,\xi)]d\xi=\Delta x\int_{g(x)}^{h(x)} \frac{[f(x+\Delta x,\xi)-f(x,\xi)]}{\Delta x}d\xi\,; \tag{1.39}$$

whereas, by the mean value theorem for integration (Spiegel, 1963), we can express the last two integrals on the right of (1.38) as

$$\int_{h(x)}^{h(x+\Delta x)} f(x+\Delta x,\xi)d\xi=f(x+\Delta x,\xi_1)[h(x+\Delta x)-h(x)] \tag{1.40}$$

$$\int_{g(x)}^{g(x+\Delta x)} f(x+\Delta x,\xi)d\xi=f(x+\Delta x,\xi_2)[g(x+\Delta x)-g(x)] \tag{1.41}$$

where $h(x)<\xi_1<h(x+\Delta x)$, and $g(x)<\xi_2<g(x+\Delta x)$. Substituting (1.39)–(1.41) into (1.38) and dividing the whole expression by Δx yields

$$\frac{\Delta\phi(x)}{\Delta x}=\int_{g(x)}^{h(x)} \frac{[f(x+\Delta x,\xi)-f(x,\xi)]}{\Delta x}d\xi+f(x+\Delta x,\xi_1)\frac{\Delta h}{\Delta x}-f(x+\Delta x,\xi_2)\frac{\Delta g}{\Delta x} \tag{1.42}$$

Finally, we can take the limit that Δx goes to zero to give

$$\frac{d\phi(x)}{dx}=\lim_{\Delta x\to 0}\frac{\Delta\phi(x)}{\Delta x}=\int_{g(x)}^{h(x)} \frac{\partial f(x,\xi)}{\partial x}d\xi+f(x,h(x))\frac{dh}{dx}-f(x,g(x))\frac{dg}{dx} \tag{1.43}$$

which is the Leibniz formula given in (1.34). This completes the proof of (1.34). This formula will be used repeatedly in the later part of this book.

1.3.5 Partial Derivative

When a function f depends on more than one variable (say x and y), differentiation has to be modified as partial differentiation. In particular, two partial derivatives are defined as, depending on differentiating with respect to x or y

$$\frac{\partial f}{\partial x}=\lim_{\Delta x\to 0}\frac{f(x+\Delta x,y)-f(x,y)}{\Delta x}=(\frac{\partial f}{\partial x})_{y=const.}=f_{,x}$$
$$\frac{\partial f}{\partial y}=\lim_{\Delta y\to 0}\frac{f(x,y+\Delta y)-f(x,y)}{\Delta y}=(\frac{\partial f}{\partial y})_{x=const.}=f_{,y} \tag{1.44}$$

Note that the partial differentiation with respect to the variable x or y is represented by the *comma-subscript convention* (see the last terms in (1.44)). When we take the partial differentiation with respect to x, we will keep y constant. Similarly, in taking partial differentiation with respect to y, we keep x as constant. This idea can easily be extended to the partial differentiation of function with three or more variables (say, x, y, and z) derivative. The symbol ∂ is not a Greek letter, and it was invented by Legendre and has been adopted since (Cajori, 1993). Some people pronounce ∂ as "partial differentiation," some call it "partial d," but it was introduced as a rounded "d" so as to distinguish it from the normal "d." Therefore, it seems easier to call it as "round".

An important consequence of partial differentiation lies in the calculation of the total change of a function or its total differential, which is defined as:

$$df = \frac{\partial f}{\partial x}dx + \frac{\partial f}{\partial y}dy \tag{1.45}$$

This fundamental result will repeatedly be used in the analysis of differential equations (e.g., the exactness of 1st ODE and solving 1st order PDE). To show the validity of (1.45), we first note that the total change of f when x increases to $x+\Delta x$ and y increases to $y+\Delta y$ is

$$\begin{aligned} \Delta f &= f(x+\Delta x, y+\Delta y) - f(x,y) \\ &= f(x+\Delta x, y+\Delta y) - f(x, y+\Delta y) + f(x, y+\Delta y) - f(x,y) \\ &= [\frac{f(x+\Delta x, y+\Delta y) - f(x, y+\Delta y)}{\Delta x}]\Delta x + [\frac{f(x, y+\Delta y) - f(x,y)}{\Delta y}]\Delta y \end{aligned} \tag{1.46}$$

$$\begin{aligned} df &= \lim_{\Delta x \to 0}[\frac{f(x+\Delta x, y+\Delta y) - f(x, y+\Delta y)}{\Delta x}]\Delta x + \lim_{\Delta y \to 0}[\frac{f(x, y+\Delta y) - f(x,y)}{\Delta y}]\Delta y \\ &= \frac{\partial f}{\partial x}dx + \frac{\partial f}{\partial y}dy \end{aligned} \tag{1.47}$$

This completes the proof of (1.45). When f depends on n variables (x_i, $i = 1,2,..., n$), the total differential becomes

$$df = \frac{\partial f}{\partial x_1}dx_1 + \frac{\partial f}{\partial x_2}dx_2 + ... + \frac{\partial f}{\partial x_n}dx_n \tag{1.48}$$

The chain rule discussed in Section 1.3.2 has to be revised for the case of multiple variables; and, in view of (1.45) we have:

$$\frac{df}{dt} = \frac{\partial f}{\partial x}\frac{dx}{dt} + \frac{\partial f}{\partial y}\frac{dy}{dt} \tag{1.49}$$

where f is a function of both x and y, and both x and y are functions of t. Equation (1.46) is of particular importance in fluid mechanics and it is the basis of the so-called convective derivative or material derivative. We will return to this in Section 2.8.2 of Chapter 2.

Example 1.3 Consider the following partial derivative of a product of two functions u and v:

$$\frac{\partial^2}{\partial x_1 \partial x_2}(uv) = \frac{\partial}{\partial x_1}(u\frac{\partial v}{\partial x_2} + v\frac{\partial u}{\partial x_2})$$
$$= \frac{\partial u}{\partial x_1}\frac{\partial v}{\partial x_2} + u\frac{\partial^2 v}{\partial x_1 \partial x_2} + \frac{\partial v}{\partial x_1}\frac{\partial u}{\partial x_2} + v\frac{\partial^2 u}{\partial x_1 \partial x_2} \quad (1.50)$$

The geometric meaning of partial derivative $\partial f/\partial x$ and $\partial f/\partial y$ are demonstrated in Figure 1.4. They are the corresponding slopes for function $f(x,a)$ and $f(b,y)$.

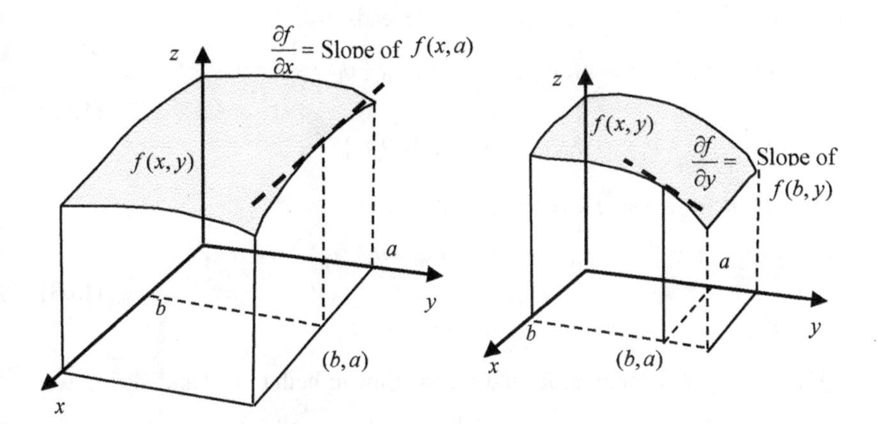

Figure 1.4 Geometric meaning of $\partial f/\partial x$ and $\partial f/\partial y$

1.3.6 Commutative Rule for Partial Derivatives

If a function has a continuous and smooth second derivative, and the second derivative is commutative as:

$$\frac{\partial^2 f}{\partial x \partial y} = \frac{\partial^2 f}{\partial y \partial x} \quad (1.51)$$

This is known as Clairaut theorem. To prove this, let us consider the following function:

$$G = f(x_0 + h, y_0 + k) - f(x_0, y_0 + k) - f(x_0 + h, y_0) + f(x_0, y_0) \quad (1.52)$$

Next, let us further assume that

$$\phi(x, y) = f(x + h, y) - f(x, y) \quad (1.53)$$

$$\psi(x, y) = f(x, y + k) - f(x, y) \quad (1.54)$$

With the definition given in (1.53), we can consider x being x_0 and y being y_0 or y_0+k

$$\phi(x_0, y_0 + k) = f(x_0 + h, y_0 + k) - f(x_0, y_0 + k) \quad (1.55)$$

$$\phi(x_0, y_0) = f(x_0 + h, y_0) - f(x_0, y_0) \quad (1.56)$$

Then, (1.52) can be rewritten as

$$G=\phi(x_0,y_0+k)-\phi(x_0,y_0) \tag{1.57}$$

Similarly, using the definition of (1.54) we have

$$\psi(x_0,y_0+k)=f(x_0+h,y_0+k)-f(x_0+h,y_0) \tag{1.58}$$

$$\psi(x_0,y_0)=f(x_0,y_0+k)-f(x_0,y_0) \tag{1.59}$$

Thus, alternatively the function G defined in (1.52) can be expressed as

$$G=\psi(x_0+h,y_0)-\psi(x_0,y_0) \tag{1.60}$$

Recall the mean value theorem that

$$\frac{f(b)-f(a)}{b-a}=f'(\xi) \quad a<\xi<b \tag{1.61}$$

Application of this mean value theorem to (1.57) leads to

$$\begin{aligned} G&=\phi(x_0,y_0+k)-\phi(x_0,y_0)=k\frac{\partial\phi}{\partial y}(x_0,y_0+\theta_1 k) \quad 0<\theta_1<1 \\ &=k\{\frac{\partial f}{\partial y}(x_0+h,y_0+\theta_1 k)-\frac{\partial f}{\partial y}(x_0,y_0+\theta_1 k)\} \end{aligned} \tag{1.62}$$

Application of this mean value theorem to (1.60) gives

$$\begin{aligned} G&=\psi(x_0+h,y_0)-\psi(x_0,y_0)=h\frac{\partial\psi}{\partial x}(x_0+\theta_2 h,y_0) \quad 0<\theta_2<1 \\ &=h\{\frac{\partial f}{\partial x}(x_0+\theta_2 h,y_0+k)-\frac{\partial f}{\partial x}(x_0+\theta_2 h,y_0)\} \end{aligned} \tag{1.63}$$

Applying the mean value theorem for the second time to both (1.62) and (1.63), we obtain

$$G=hk\frac{\partial^2 f}{\partial x\partial y}(x_0+\theta_3 h,y_0+\theta_1 k) \quad 0<\theta_3<1 \tag{1.64}$$

$$G=kh\frac{\partial^2 f}{\partial y\partial x}(x_0+\theta_2 h,y_0+\theta_4 k) \quad 0<\theta_4<1 \tag{1.65}$$

Thus, we obtain

$$\frac{\partial^2 f}{\partial y\partial x}(x_0+\theta_2 h,y_0+\theta_4 k)=\frac{\partial^2 f}{\partial x\partial y}(x_0+\theta_3 h,y_0+\theta_1 k) \tag{1.66}$$

Consideration of the limiting case that $h\to 0$ and $k\to 0$ gives

$$\frac{\partial^2 f}{\partial y\partial x}(x_0,y_0)=\frac{\partial^2 f}{\partial x\partial y}(x_0,y_0) \tag{1.67}$$

Thus, the order of partial differentiation can be reversed if the second derivative is a smooth function of x and y at point (x_0, y_0).

1.4 INTEGRATION

When a function $f(x)$ is plotted against x as shown in Figure 1.5(a), the area under the curve f can be approximated by summing the areas of n columns. Mathematically, we can write

$$A = [f(x_1)\Delta x + f(x_2)\Delta x + ... + f(x_n)\Delta x] \tag{1.68}$$

The actual area under the curve can be written as an integration if we take the limit of n approaching infinity

$$\int_a^b f(x)dx = \lim_{n\to\infty} \sum_{i=1}^{n} f(x_i)\Delta x \tag{1.69}$$

If the integration is not integrated along the x axis from a to b but instead along a curve C in space or in a plane as shown in Figure 1.5(b), we have the contour integral. A curve C is depicted in Figure 1.5(b) from Point A to Point B. The direction of a curve C can be defined from Point A to Point B in Figure 1.5(b) or vice versa. Similar to (1.69), the line integration along curve C is defined as a sum

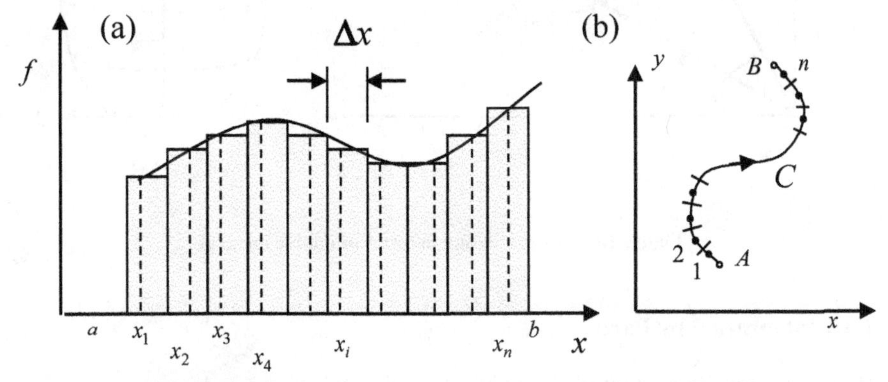

Figure 1.5 Integration as a sum of areas and a sum of lengths

$$\int_C f(x,y)ds = \lim_{n\to\infty} \sum_{i=1}^{n} f(x_i, y_i)\Delta s \tag{1.70}$$

where the length of each segment along the curve C is Δs, and point (x_i, y_i) is the point at the center of the segment i. The value of the line integral will be negative if we integrate from Point B to Point A. Formula (1.70) can be extended easily to 3-D space. If the curve C forms a closed contour (Point A overlaps with Point B), the integral sign will be changed to

$$\oint f(x,y)ds = \lim_{n\to\infty} \sum_{i=1}^{n} f(x_i, y_i)\Delta s \tag{1.71}$$

When a function f depends on more than one variable, we can extend the idea of a single integration to multiple integrations. For example, using the polar coordinate shown in Figure 1.6(a), we can define area integration as:

$$\iint_R f(r,\theta)dA = \int_\alpha^\beta \int_{g_1(\theta)}^{g_2(\theta)} f(r,\theta)dr d\theta \tag{1.72}$$

Similarly, we can extend the area integration to volume integration, which is formed by the revolution of a curve $y(x)$ about the x axis as shown in Figure 1.6(b) or about the y axis as shown in Figure 1.6(c).

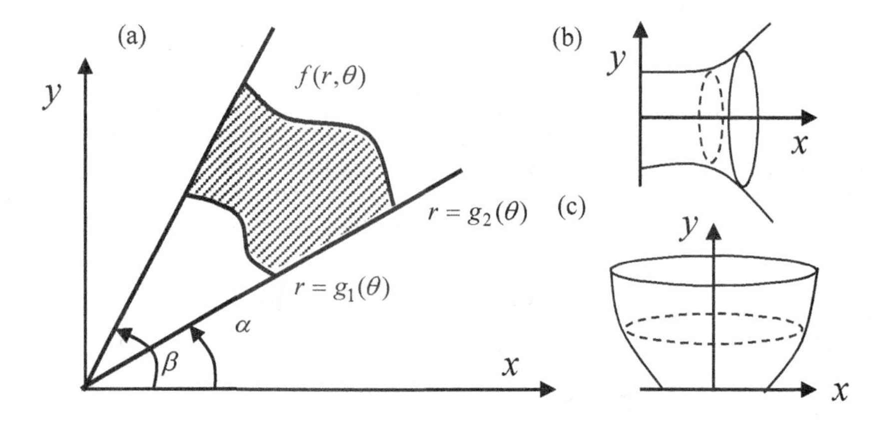

Figure 1.6 Area evaluation in terms of double integral

1.4.1 Integration by Parts

One of the most important formulas of integration is integration by parts:

$$\int u\, dv = uv - \int v\, du \tag{1.73}$$

This technique will be used repeatedly in solving differential equations, and the mastery of this technique is essential. This formula can be generalized to the case of an integrand involving the n-th derivative of a function f multiplying another function g:

$$\int f^{(n)} g\, dx = f^{(n-1)} g - f^{(n-2)} g' + f^{(n-3)} g'' - ...(-1)^n \int f g^{(n)} dx \tag{1.74}$$

When upper and lower limits of integration are given, these formulas should be modified accordingly by assigning the limits to the non-integral parts in (1.73) and (1.74).

1.4.2 General Rules of Integration

There are a lot of mathematical handbooks on integration (e.g., Gradshteyn and Ryzhik, 1980; Zwillinger, 2012; Spiegel, 1968). Carrying out integration may not be an easy matter and, in general, integration is much more difficult than differentiation. Instead of summarizing all known integration formulas here, we only report some useful formulas that we would encounter repeatedly in solving differential equations

$$\int u^n dx = \frac{u^{n+1}}{n+1} \tag{1.75}$$

$$\int \frac{1}{u} du = \ln|u|, \quad \int e^u du = e^u \tag{1.76}$$

$$\int a^u du = \int e^{u \ln a} du = \frac{e^{u \ln a}}{\ln a} = \frac{a^u}{\ln a} \tag{1.77}$$

In addition, we report two general rules for carrying out integration

$$\int f(ax)dx = \frac{1}{a}\int f(u)du \tag{1.78}$$

$$\int F\{f(x)\}dx = \int F(u)\frac{dx}{du}du = \int \frac{F(u)}{f'(x)}du \tag{1.79}$$

1.4.3 Some Transformation Rules

There are some well-established rules for integrands containing certain groups of arguments. Here are some of them and their associated change of variables should be used:

$$\int F(ax+b)dx = \frac{1}{a}\int F(u)du, \quad u = ax+b \tag{1.80}$$

$$\int F(\sqrt{ax+b})dx = \frac{2}{a}\int uF(u)du, \quad u = \sqrt{ax+b} \tag{1.81}$$

$$\int F(\sqrt[n]{ax+b})dx = \frac{n}{a}\int u^{n-1}F(u)du, \quad u = \sqrt[n]{ax+b} \tag{1.82}$$

$$\int F(\sqrt{a^2-x^2})dx = a\int F(a\cos u)\cos u du, \quad x = a\sin u \tag{1.83}$$

$$\int F(\sqrt{a^2+x^2})dx = a\int F(a\sec u)\sec^2 u du, \quad x = a\tan u \tag{1.84}$$

$$\int F(\sqrt{x^2-a^2})dx = a\int F(a\tan u)\sec u \tan u du, \quad x = a\sec u \tag{1.85}$$

$$\int F(e^{ax})dx = \frac{1}{a}\int \frac{F(u)}{u}du, \quad u = e^{ax} \tag{1.86}$$

$$\int F(\ln x)dx = \int F(u)e^u du, \quad u = \ln x \tag{1.87}$$

$$\int F(\sin^{-1}\frac{x}{a})dx = a\int F(u)\cos u du, \quad u = \sin^{-1}\frac{x}{a} \tag{1.88}$$

$$\int F(\sin x, \cos x)dx = 2\int F(\frac{2u}{1+u^2}, \frac{1-u^2}{1+u^2})\frac{du}{1+u^2}, \quad u = \tan\frac{x}{2} \tag{1.89}$$

Note, however, that there is no guarantee that these changes of variables will always work. It depends on the particular functional form of F involved. In Section 1.7.4, we will see that Cauchy's integral formula for complex variables is a very powerful technique even for integrating real functions. Another less known technique is Ramanujan's master theorem and the related Ramanujan's integral theorem, and these will also be discussed in Sections 1.9 and 1.10.

1.4.4 Mean Value Theorem

Similar to the mean value theorem in differentiation given in (1.11), the mean value theorem also exists in integration. We have actually used the mean value theorem in the proof of the Leibniz rule of differentiation for the integrals in (1.40) and (1.41) in Section 1.3.4. In particular, there exists a point c within the upper and lower limits such that:

$$\int_a^b f(x)dx = (b-a)f(c), \quad a < c < b \tag{1.90}$$

$$\int_a^b f(x)g(x)dx = f(c)\int_a^b g(x)dx, \quad a < c < b \tag{1.91}$$

where $f(x)$ is a less rapid changing function comparing to g(x). This idea can be extended to double integrations. Similarly, the mean value theorem for the double integral can be established as

$$\iint_R f(r,\theta)dA = Af(r_0,\theta_0) \tag{1.92}$$

where the point (r_0, θ_0) lies within the domain of integration.

1.4.5 Improper Integral

When the integrand is not defined at its end point of integration, the integration can be modified as

$$\int_a^b f(x)dx = \lim_{\varepsilon\to 0}\int_{a+\varepsilon}^b f(x)dx \tag{1.93}$$

where $f(a)$ tends to infinity. The integral given in (1.93) must exist in order that the integral is well defined. Another typical improper integral involves infinity as the upper limit

$$\int_a^\infty f(x)dx = \lim_{b\to\infty}\int_a^b f(x)dx \tag{1.94}$$

Again, the limit on the right hand side of (1.94) must exist in order for (1.94) to be well defined.

1.4.6 Laplace/Gauss Integral

Let us consider an integral that we often encounter in engineering applications. The integral was known as the Laplace integral or Gauss integral. It is defined as

$$I = \int_0^\infty e^{-x^2} dx \tag{1.95}$$

Note that (1.95) is an improper integral in the sense defined in the last section. As is normally done in engineering, we will omit the limit process but simply assume the limit exists. To evaluate (1.95), we will first note that the integrand in (1.95) is an even function with respect to $x = 0$. Thus, we can extend the integral to

$$I = \frac{1}{2}\int_{-\infty}^{\infty} e^{-x^2} dx \tag{1.96}$$

Next, a clever mathematical trick is proposed. We can expand the integration to an area integration of an infinite domain as

$$I^2 = \frac{1}{2}\int_{-\infty}^{\infty} e^{-y^2} dy \frac{1}{2}\int_{-\infty}^{\infty} e^{-x^2} dx = \frac{1}{4}\int_{-\infty}^{\infty}\int_{-\infty}^{\infty} e^{-(x^2+y^2)} dxdy \tag{1.97}$$

This integral can easily be evaluated by using the polar form, as shown in Figure 1.7.

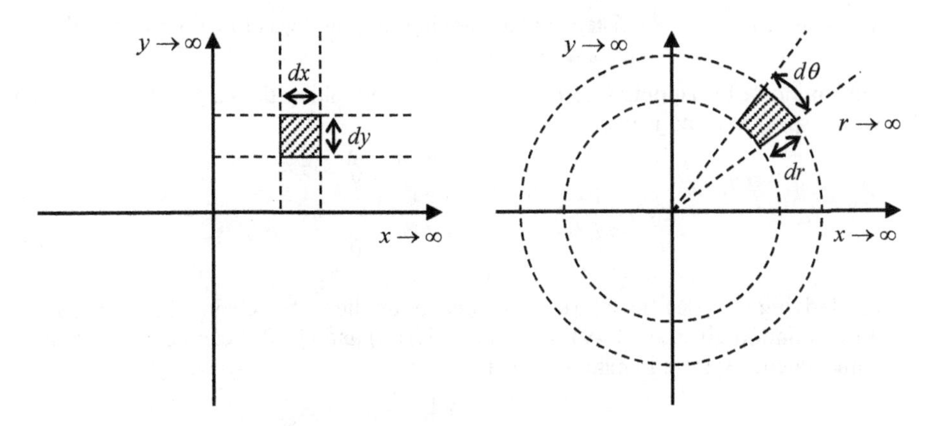

Figure 1.7 Area evaluation in terms of the double integral

In particular, we assume

$$x = r\cos\theta, \quad y = r\sin\theta, \quad r^2 = x^2 + y^2 \tag{1.98}$$

The limits of integration for $-\infty < x,\ y < \infty$ become $0 < r < \infty$ and $0 < \theta < 2\pi$. The integral actually represents an area integral. From geometric consideration given in Figure 1.7, we must have

$$dA = dxdy = rd\theta dr \tag{1.99}$$

Thus, the integral becomes

$$I^2 = \frac{1}{4}\int_{-\infty}^{\infty}\int_{-\infty}^{\infty} e^{-(x^2+y^2)} dxdy = \int_0^{2\pi}\int_0^{\infty} e^{-r^2} rd\theta dr \tag{1.100}$$

The integration can be conducted separately for θ and r as

$$I^2 = \frac{1}{4}[\theta]_0^{2\pi}(-\frac{1}{2})\int_0^{\infty} e^{-r^2} d(-r^2) = \frac{\pi}{4} \tag{1.101}$$

Finally, taking the square root of (1.101) gives

$$\int_0^{\infty} e^{-x^2} dx = \frac{\sqrt{\pi}}{2} \tag{1.102}$$

This result will be used repeatedly in the later part of this book.

1.5 JACOBIAN

In (1.100) of the last section, we have seen that when a change of variables is applied to an area integral, an additional factor needs to multiply the increment of the new variables. In general, we can cast the integral as

$$\iint f(x,y)dxdy = \iint f(\xi,\eta)|J|d\xi d\eta \tag{1.103}$$

where $|J|$ is the additional factor. The symbol $|J|$ is normally used to represent this factor. The factor J was derived by Jacobi and is referred to as Jacobian. It is an important quantity for any mapping of a function of two variables.

Following Hardy (1944), let us consider the following function in terms of u and v:

$$\phi(u,v) = 0 \tag{1.104}$$

We assume there is an existence of a mapping for u and v as functions of x and y:

$$u = u(x,y), \quad v = v(x,y) \tag{1.105}$$

Substitution of (mapping (1.105) into (1.104) and taking the partial differentiation with respect to x and y gives

$$\frac{\partial \phi}{\partial x} = \frac{\partial \phi}{\partial u}\frac{\partial u}{\partial x} + \frac{\partial \phi}{\partial v}\frac{\partial v}{\partial x} = 0 \tag{1.106}$$

$$\frac{\partial \phi}{\partial y} = \frac{\partial \phi}{\partial u}\frac{\partial u}{\partial y} + \frac{\partial \phi}{\partial v}\frac{\partial v}{\partial y} = 0 \tag{1.107}$$

In deriving (1.106) to (1.107), we have applied the chain rule for partial differentiation given in (1.49). Equations (1.106) and (1.107) can be rewritten as a homogeneous system in matrix form as

$$\begin{bmatrix} \dfrac{\partial u}{\partial x} & \dfrac{\partial v}{\partial x} \\ \dfrac{\partial u}{\partial y} & \dfrac{\partial v}{\partial y} \end{bmatrix} \begin{Bmatrix} \dfrac{\partial \phi}{\partial u} \\ \dfrac{\partial \phi}{\partial v} \end{Bmatrix} = \begin{Bmatrix} 0 \\ 0 \end{Bmatrix} \tag{1.108}$$

If the partial differentiations of ϕ with respect to u and v exist, the determinant of the square matrix in (1.108) must be zero. The determinant is called Jacobian and is defined as:

$$J = \det \begin{vmatrix} \dfrac{\partial u}{\partial x} & \dfrac{\partial v}{\partial x} \\ \dfrac{\partial v}{\partial x} & \dfrac{\partial v}{\partial y} \end{vmatrix} = \frac{\partial u}{\partial x}\frac{\partial v}{\partial y} - \frac{\partial v}{\partial x}\frac{\partial v}{\partial x} = u_x v_y - v_x u_y \tag{1.109}$$

If (1.104) exists, it means that u can be expressed in terms of v (or vice versa). That is, u and v are not independent. In a slightly different way of saying this, if J = 0, $\partial\phi/\partial u$ and $\partial\phi/\partial v$ are nonzero. Then, consequently (1.104) exists. Conversely, if $J \neq 0$, $\partial\phi/\partial u$ and $\partial\phi/\partial v$ are zeros. Then, u and v are independent, and consequently the mapping given by (1.105) exists. Therefore, for a mapping to exist, we must have its Jacobian defined in (1.109) not equal to zero.

We now return to (1.103) and see why the Jacobian is involved. Referring to Figure 1.8, we consider a small diagonal element $\boldsymbol{dr}$ as

$$\boldsymbol{r} = x\boldsymbol{e}_1 + y\boldsymbol{e}_2, \quad \boldsymbol{dr} = \boldsymbol{e}_1 dx + \boldsymbol{e}_2 dy \tag{1.110}$$

Now consider a mapping:

$$x = x(\xi,\eta), \quad y = y(\xi,\eta) \tag{1.111}$$

Therefore, (1.110) becomes

$$\boldsymbol{r} = x(\xi,\eta)\boldsymbol{e}_1 + y(\xi,\eta)\boldsymbol{e}_2, \quad d\boldsymbol{r} = \frac{\partial \boldsymbol{r}}{\partial \xi}d\xi + \frac{\partial \boldsymbol{r}}{\partial \eta}d\eta \tag{1.112}$$

The two vectors on the right hand side of (1.112) are physically shown in Figure 1.8 as the components of $d\boldsymbol{r}$ in the ξ-η plane. They can be determined using the first equation of (1.110) as

$$\frac{\partial \boldsymbol{r}}{\partial \xi} = \frac{\partial x}{\partial \xi}\boldsymbol{e}_1 + \frac{\partial y}{\partial \xi}\boldsymbol{e}_2, \quad \frac{\partial \boldsymbol{r}}{\partial \eta} = \frac{\partial x}{\partial \eta}\boldsymbol{e}_1 + \frac{\partial y}{\partial \eta}\boldsymbol{e}_2 \tag{1.113}$$

The small rectangular element with diagonal $d\boldsymbol{r}$ in the x-y plane in Figure 1.8 is

$$d\boldsymbol{A} = (dx\boldsymbol{e}_1)\times(dx\boldsymbol{e}_1) = dxdy\boldsymbol{e}_3 \tag{1.114}$$

whereas the area of the curvilinear element with diagonal $d\boldsymbol{r}$ in the ξ-η plane is

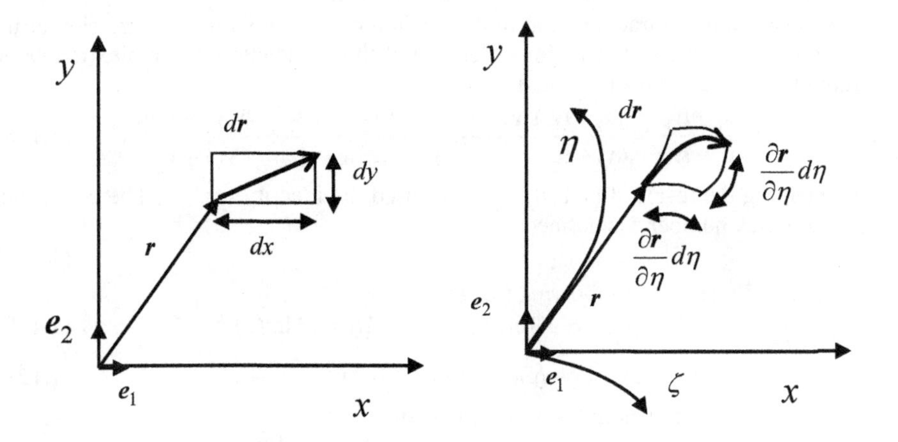

Figure 1.8 The mapping of a small vector *dr*

$$\begin{aligned} d\boldsymbol{A} &= \frac{\partial \boldsymbol{r}}{\partial \xi}d\xi \times \frac{\partial \boldsymbol{r}}{\partial \eta}d\eta = (\frac{\partial x}{\partial \xi}\boldsymbol{e}_1 + \frac{\partial y}{\partial \xi}\boldsymbol{e}_2)\times(\frac{\partial x}{\partial \eta}\boldsymbol{e}_1 + \frac{\partial y}{\partial \eta}\boldsymbol{e}_2)d\xi d\eta \\ &= (\frac{\partial x}{\partial \xi}\frac{\partial y}{\partial \eta} - \frac{\partial y}{\partial \xi}\frac{\partial x}{\partial \eta})d\xi d\eta \boldsymbol{e}_3 = |J|\, d\xi d\eta \boldsymbol{e}_3 \end{aligned} \tag{1.115}$$

These areas must be equal before and after the mapping, thus we have demonstrated the validity of (1.103).

Although formula (1.103) was given in nearly all textbooks on engineering mathematics (e.g., Kreyszig, 1979; Wylie, 1975), its proof is seldom given.

1.6 COMPLEX VARIABLES AND EULER'S FORMULA

The analysis of complex variables is an important area in applied mathematics. It is not possible to review such a huge topic in a chapter, let alone a subsection like this. Nevertheless, for later analysis we need to review some essential ideas here.

The idea of introducing imaginary number can date back to 1545, when Italian mathematician Gerolamo Cardano considered the solution of the cubic equation. All real numbers can be considered as special cases of complex numbers. Setting the imaginary part of any complex numbers to zero, we get real numbers. When both the real and imaginary parts of a complex number are replaced by changing variables, we have a complex variable

$$z = x + iy, \quad i = \sqrt{-1} \tag{1.116}$$

where x = Re(z) and y = Im(z) are the real and imaginary parts of z. When a function depends on the complex variable z, we have a complex function

$$f(z) = u(x, y) + iv(x, y) \tag{1.117}$$

The summation, subtraction, and multiplication of complex numbers are similar to that for real numbers. The major difference of the arithmetic for complex numbers from that for real number lies in division:

$$\frac{z_1}{z_2} = \frac{x_1 + iy_1}{x_2 + iy_2} = \frac{(x_1 + iy_1)(x_2 - iy_2)}{(x_2 + iy_2)(x_2 - iy_2)} = \frac{x_1 x_2 + y_1 y_2}{x_2^2 + y_2^2} + i\frac{y_1 x_2 - x_1 y_2}{x_2^2 + y_2^2} \tag{1.118}$$

In obtaining the result of (1.118), we have used the identity $i^2 = -1$. The conjugate of a complex number z is defined as:

$$\bar{z} = x - iy \tag{1.119}$$

We can easily show the following identities

$$\bar{z} + z = 2x = 2\,\mathrm{Re}(z), \quad z - \bar{z} = 2iy = i2\,\mathrm{Im}(z) \tag{1.120}$$

$$|z|^2 = z\bar{z} = (x + iy)(x - iy) = x^2 - i^2 y^2 = x^2 + y^2 \tag{1.121}$$

The $|z|$ is called the modulus of a complex number.

The polar form of a complex number is defined in Figure 1.9, and is given as

$$z = r\cos\theta + ir\sin\theta = re^{i\theta} \tag{1.122}$$

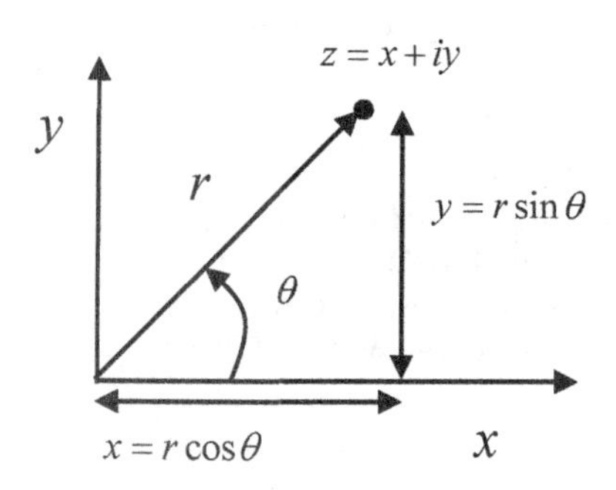

Figure 1.9 Polar form of a complex number

where $r = |z|$ and θ is the argument of the complex number. Equation (1.122) is known as Euler formula. By applying the Euler polar form, we immediately obtain DeMoivre's Formula

$$z^n = r^n e^{in\theta} = r^n(\cos n\theta + i\sin n\theta) \tag{1.123}$$

Let us consider the special case that $\theta = \pi$ in (1.122), we obtain

$$e^{i\pi} = \cos\pi + i\sin\pi = -1 \tag{1.124}$$

Rearranging (1.124) gives the well-known Euler's formula

$$e^{i\pi} + 1 = 0 \tag{1.125}$$

Many mathematicians and scientists called this the most beautiful formula in mathematics. It links five fundamental constants in mathematics: the base of natural logarithms (or Euler's number), imaginary constant, ratio of circumference to diameter of a circle pi, one, and zero; otherwise, these constants are seemingly unrelated. Nobel prize laureate in physics, Richard Feynman, called this the most remarkable formula in mathematics (in the cover page of Nahin, 2006). Among these five constants, π and *e* are the most fundamental and appear in all branches of mathematics and engineering applications. Appendix F compiles some important formulas for π, whereas the definition for *e* is given next.

Example 1.4 Let us consider the origin of the base of natural logarithms *e*. If a sum of money *P* is deposited at a bank, and the bank offered an annual interest rate of *r*, thus, at the end of one year, we gain an interest of *Pr* or the total return sum *S* as

$$S = P(1+r) \tag{1.126}$$

Then, what happens if we take out the money from the bank at the end of one-half year and redeposit it into the bank with the new initial sum. Mathematically, the return sum at the end of one year becomes

$$S = [P(1+\frac{r}{2})](1+\frac{r}{2}) = P(1+\frac{r}{2})^2 \tag{1.127}$$

This formula is actually the compound interest formula that we should have learned in high school. Clearly, the sum given in (1.127) is more than that of (1.126). If after each $1/n$ year, we take out the money and redeposit it again, when we do this *n* times in a year, we ultimately get

$$S = [P(1+\frac{r}{n})...](1+\frac{r}{n}) = P(1+\frac{r}{n})^n \tag{1.128}$$

One may ask the ultimate question, what would be the largest gain in a year, if we decrease the deposit period to an infinitely small period and redeposit again and again. Thus, mathematically, we have the sum at the end of one year as

$$S = P\lim_{n\to\infty}(1+\frac{r}{n})^n = Pe^r \tag{1.129}$$

The last of (1.129) gives the definition of *e*, and this gives birth to natural logarithms. Clearly, we have

$$e = \lim_{n\to\infty}(1+\frac{1}{n})^n = 2.71828... \tag{1.130}$$

The origin of *e* is from the deposit and interest problem, yet we will see that it is of central importance to mathematics as well as to the solution of differential equations.

The amazing properties of Euler's formula do not stop at (1.125). Let us consider another example of (1.122).

Example 1.5 What is i to the power i?

Solution: Consider the special case that $\theta = \pi/2 + 2n\pi$ in (1.122); we find

$$e^{i(\pi/2+2n\pi)} = \cos(\frac{\pi}{2}+2n\pi)+i\sin(\frac{\pi}{2}+2n\pi)=i \tag{1.131}$$

Then, let us consider the amazing form of i to the power i

$$i^i = e^{i\times i(\pi/2+2n\pi)} = e^{-(\pi/2+2n\pi)} \tag{1.132}$$

Surprisingly, i to the power i is actually real. There are infinite answers to it, the "simplest" one being $e^{-\pi/2}$ or 0.207879576... Some authors also called this the principal value among the infinite solutions. This amazing result was obtained by Euler in 1746. It has been reported unofficially that some mathematics departments will ask this question at the interview of potential candidates for their bachelor degree program. In the problem section at the end of the chapter, we will consider more amazing formulas resulting from (1.122).

1.7 ANALYTIC FUNCTION

Let us now consider the differentiation of a complex variable. In analogy to the definition given for a real variable, we can define

$$f'(z) = \frac{df(z)}{dz} = \lim_{\Delta z\to 0}\frac{f(z+\Delta z)-f(z)}{\Delta z} \tag{1.133}$$

If this derivative exists, we call this complex function analytic. Actually, it is just the complex counterpart of a real function being differentiable. Therefore, let us reiterate that, in essence, analytic is just another word for differentiable in the complex variable context. We have learned that a function must be smooth for it being differentiable (or left hand limit equals right hand limit shown in (1.10)) and, similarly, we also have certain conditions that a complex function needs to satisfy in order to be analytic. This required condition is called the Cauchy-Riemann relations and is considered next.

1.7.1 Cauchy-Riemann Relations

Let us assume that a complex function can be written in terms of a real part function and a complex part function as given in (1.117). Then, the complex and its incremental change due to the change of z are

$$\begin{aligned} f(z) &= u(x,y)+iv(x,y), \\ f(z+\Delta z) &= u(x+\Delta x, y+\Delta y)+iv(x+\Delta x, y+\Delta y) \end{aligned} \tag{1.134}$$

Substitution of (1.134) into (1.133) yields

$$f'(z) = \lim_{\Delta z \to 0} \frac{f(z+\Delta z) - f(z)}{\Delta z} = \lim_{\Delta z \to 0} \frac{u(x+\Delta x, y+\Delta y) - u(x,y) + i[v(x+\Delta x, y+\Delta y) - v(x,y)]}{\Delta x + i\Delta y} \tag{1.135}$$

Since $\Delta z = \Delta x + i\Delta y$, we have two independent paths of taking the limit. First, we can take $\Delta y \to 0$, then $\Delta x \to 0$:

$$f'(z) = \lim_{\Delta x \to 0} \frac{u(x+\Delta x, y) - u(x,y) + i[v(x+\Delta x, y) - v(x,y)]}{\Delta x} = \frac{\partial u}{\partial x} + i\frac{\partial v}{\partial x} \tag{1.136}$$

Secondly, we can take $\Delta x \to 0$, then $\Delta y \to 0$.

$$f'(z) = \lim_{\Delta y \to 0} \frac{u(x, y+\Delta y) - u(x,y) + i[v(x, y+\Delta y) - v(x,y)]}{i\Delta y} = \frac{1}{i}\frac{\partial u}{\partial y} + \frac{\partial v}{\partial y} = \frac{\partial v}{\partial y} - i\frac{\partial u}{\partial y} \tag{1.137}$$

The resulting limits by following these paths must be the same if the differentiation is unique. Equating the real and imaginary parts of (1.136) and (1.137) gives a pair of equations:

$$\frac{\partial u}{\partial x} = \frac{\partial v}{\partial y}, \quad \frac{\partial v}{\partial x} = -\frac{\partial u}{\partial y} \tag{1.138}$$

These are the Cauchy-Riemann relations.

Example 1.6 Check whether the following complex function is analytic:

$$f(z) = z^2 \tag{1.139}$$

Solution: Expansion of the square term gives

$$f(z) = (x+iy)^2 = x^2 + 2ixy - y^2 = u + iy \tag{1.140}$$

Thus, u and v are

$$u = x^2 - y^2, \quad v = 2xy \tag{1.141}$$

Substitution of (1.141) into (1.138) gives

$$\frac{\partial u}{\partial x} = \frac{\partial v}{\partial y} = 2x, \quad \frac{\partial v}{\partial x} = -\frac{\partial u}{\partial y} = 2y \tag{1.142}$$

Thus, (1.139) is analytic or the differentiation of (1.139) exists.

Example 1.7 Check whether the following complex function is analytic:

$$f(z) = \bar{z}^2 \tag{1.143}$$

Solution: Expansion of (1.143) gives

$$f(z) = (x-iy)^2 = x^2 - 2ixy - y^2 = u + iy \tag{1.144}$$

Comparing the real and imaginary parts, we obtain

$$u = x^2 - y^2, \quad v = -2xy \tag{1.145}$$

Substitution of (1.145) into (1.138) gives

$$\frac{\partial u}{\partial x} = 2x, \quad \frac{\partial v}{\partial y} = -2x, \quad \frac{\partial v}{\partial x} = -2y, \quad \frac{\partial u}{\partial y} = -2y \tag{1.146}$$

Thus, (1.143) is not analytic.

Without going into details, we simply claim here that all functions involving only z are analytic, whereas any function containing the conjugate of z is not analytic. In addition, if u and v depends on Im(z) or Re(z), we can show that its resulting f is not analytic (see Problem 1.11).

1.7.2 Liouville Theorem

Suppose that for all z in the entire complex plane: (i) $f(z)$ is analytic and (ii) $f(z)$ is bounded (i.e., $|f(z)| < M$ for some constant M). Then, $f(z)$ must be a constant. In layman's language, we can rephrase this as "if $f(z)$ is analytic and bounded everywhere, $f(z)$ is a constant." This theorem was actually derived by Cauchy in 1844. The term "Liouville theorem" was coined by Borchardt in 1880, who learned about this theorem in one of Liouville's lectures in 1847 and unfortunately named it after Liouville. Some mathematicians called it the Cauchy-Liouville theorem. There are 16 concepts and theorems associated with Cauchy's name, and this is the most among mathematicians (see biography at the back of this book).

1.7.3 Cauchy-Goursat Theorem

We now review the most important formula for complex variable—the Cauchy integral formula. First, let us consider an associated result called Green's lemma or Green's theorem in the plane. Figure 1.10(a) shows a plane domain R with boundary C.

Green's lemma can be stated as

$$\iint_R \left(\frac{\partial Q}{\partial x} - \frac{\partial P}{\partial y}\right) dxdy = \oint_C (Pdx + Qdy) \tag{1.147}$$

To prove this, we first consider the left hand side of (1.147). As shown in Figure 1.10, the lowest point of the curve C is y_1, whereas the highest point of C is y_2 and the arcs 123 and 143 are single value functions represented by $g_1(y)$ and $g_2(y)$. The first term on the left of (1.147) can be reduced to

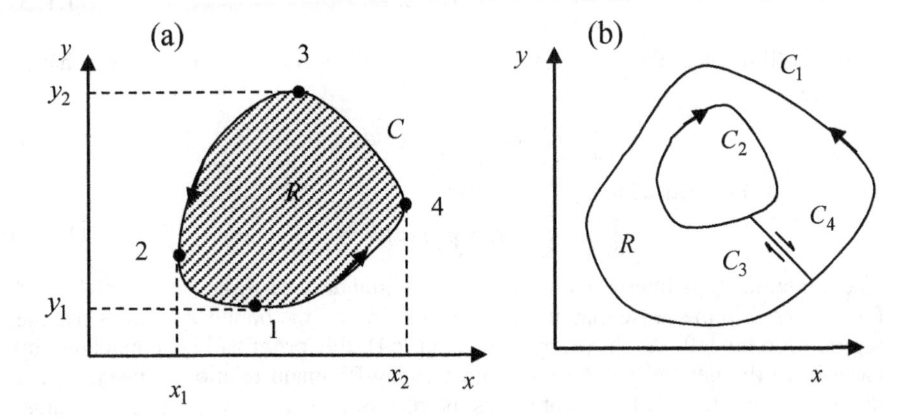

Figure 1.10 A plane region and its boundary

$$
\begin{aligned}
\iint_R \left(\frac{\partial Q}{\partial x}\right) dxdy &= \int_{y_1}^{y_2}\int_{g_1(y)}^{g_2(y)} \frac{\partial Q}{\partial x} dxdy = \int_{y_1}^{y_2} \{Q[g_2(y)] - Q[g_1(y)]\}\, dy \\
&= \int_{y_1}^{y_2} Q[g_2(y)]dy - \int_{y_1}^{y_2} Q[g_1(y)]dy \\
&= \int_{y_1}^{y_2} Q[g_2(y)]dy + \int_{y_2}^{y_1} Q[g_1(y)]dy = \oint_C Qdy
\end{aligned} \tag{1.148}
$$

Similarly, the second term can be simplified as

$$
\begin{aligned}
\iint_R \left(\frac{\partial P}{\partial y}\right) dxdy &= \int_{x_1}^{x_2}\int_{f_1(x)}^{f_2(x)} \frac{\partial P}{\partial y} dxdy = \int_{x_1}^{x_2} \{P[f_2(x)] - P[f_1(x)]\}\, dx \\
&= -\int_{x_2}^{x_1} P[f_2(x)]dx - \int_{x_1}^{x_2} P[f_1(x)]dx = -\oint_C Pdx
\end{aligned} \tag{1.149}
$$

Combining (1.148) and (1.149) gives (1.147).

Now consider the closed contour integration of an analytic function

$$
\oint_C f(z)dz = \oint_C (u+iv)(dx+idy) = \oint_C (udx - vdy) + i\oint_C (vdx + udy) \tag{1.150}
$$

Application of Green's lemma given in (1.147) to (1.150) yields

$$
\oint_C f(z)dz = \iint_R \left[-\frac{\partial v}{\partial x} - \frac{\partial u}{\partial y}\right] dxdy + i\iint_R \left[\frac{\partial u}{\partial x} - \frac{\partial v}{\partial y}\right] dxdy \tag{1.151}
$$

By virtue of the Cauchy-Riemann relation, we arrive at the well-known Cauchy integral theorem as:

$$
\oint_C f(z)dz = 0 \tag{1.152}
$$

which is true when $f'(z)$ is continuous. Figure 1.10(b) shows the modification of the proof to a multi-connected region. We first construct a cross-cut (dotted line in Figure 1.10(b)) such that the domain becomes simply connected as Figure 1.10(a). The closed contour integral of an analytic function becomes

$$\oint_{C_1+C_3+C_2+C_4} f(z)dz = 0 \tag{1.153}$$

However, line integrals for curves C_3 and C_4 are the negative of each other. That is, we have

$$\int_{C_3} f(z)dz = -\int_{C_4} f(z)dz \tag{1.154}$$

Thus, (1.152) is reduced to

$$\oint_{C_1+C_2} f(z)dz = \oint_C f(z)dz = 0 \tag{1.155}$$

where contour C is interpreted as a complete boundary of R, consisting of C_1 and C_2, traversed in the sense that an observer walking on the boundary always has the region R on his left. As shown by Spiegel (1964), this proof had been extended by Goursat to the general situation that no Cauchy-Riemann relation is needed (i.e., the condition that $f'(z)$ is continuous is relaxed). This is the Cauchy-Goursat theorem.

1.7.4 Cauchy Integral Formula

If $f(z)$ is analytic within and on the boundary C of a simply connected region R whose boundary C is sectionally smooth and if z_0 is any point in the interior of R, then

$$f(z_0) = \frac{1}{2\pi i} \oint_C \frac{f(z)dz}{z - z_0} \tag{1.156}$$

where C is integrated in positive sense.

To prove this Cauchy integral formula, we note first that $f(z)/(z-z_0)$ is analytic everywhere in R except at point z_0. Thus, similar to Figure 1.10, we can construct a circle centered at z_0 with radius ρ (see Figure 1.11). Then, a cross-cut can be made to make it simply connected. Eventually, following the argument used in the last section or applying the Cauchy-Goursat theorem given in the last section, we obtain

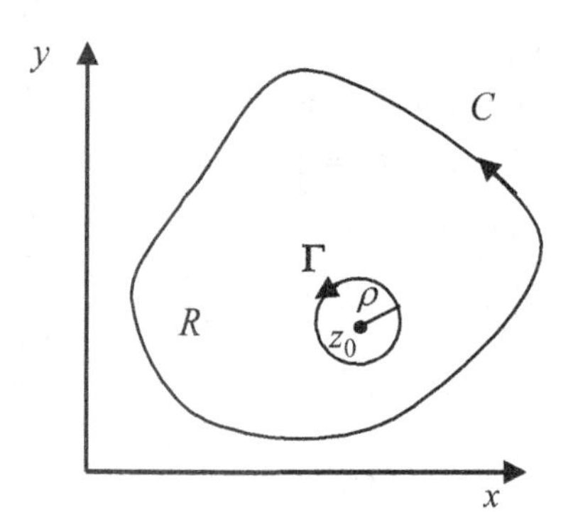

Figure 1.11 An analytic region for *f(z)* containing a point z_0

$$\oint_C \frac{f(z)dz}{z-z_0} = \oint_\Gamma \frac{f(z)dz}{z-z_0} \tag{1.157}$$

with $\rho \to 0$. Note also that we have set both C and Γ contours counter-clockwise, which is different from that of Figure 1.10(b). On the circular contour Γ, the following change of variable can be applied

$$z - z_0 = \rho e^{i\theta}, \quad dz = i\rho e^{i\theta} d\theta \tag{1.158}$$

The right hand side of (1.157) becomes

$$\begin{aligned} \oint_\Gamma \frac{f(z)dz}{z-z_0} &= \int_0^{2\pi} \frac{f(z_0+\rho e^{i\theta})\rho e^{i\theta} id\theta}{\rho e^{i\theta}} = \lim_{\rho \to 0} \int_0^{2\pi} f(z_0+\rho e^{i\theta}) id\theta \\ &= f(z_0)\int_0^{2\pi} id\theta = 2\pi i f(z_0) \end{aligned} \tag{1.159}$$

Substitution of (1.159) into (1.157), we obtain the Cauchy integral formula

$$f(z_0) = \frac{1}{2\pi i}\oint_C \frac{f(z)dz}{z-z_0} \tag{1.160}$$

The proof is now completed.

Equation (1.156) is actually the zero order of the more general Cauchy Integral formula, which is given by

$$f^{(n)}(z_0) = \frac{n!}{2\pi i}\oint_C \frac{f(z)dz}{(z-z_0)^{n+1}} \tag{1.161}$$

Mathematical induction will be used to prove this general formula. The zero order of (1.161) has just been proved. Next, suppose that (1.161) is valid for $n = k$ that

$$f^{(k)}(z_0) = \frac{k!}{2\pi i}\oint_C \frac{f(z)dz}{(z-z_0)^{k+1}} \tag{1.162}$$

Taking differentiation of (1.162) one more time, we obtain

$$\begin{aligned} f^{(k+1)}(z_0) &= \lim_{\Delta z_0 \to 0} \frac{f^{(k)}(z_0+\Delta z_0) - f^{(k)}(z_0)}{\Delta z_0} \\ &= \frac{k!}{2\pi i} \lim_{\Delta z_0 \to 0} (\frac{1}{\Delta z_0}) \left[\oint_C \frac{f(z)dz}{(z-z_0-\Delta z_0)^{k+1}} - \oint_C \frac{f(z)dz}{(z-z_0)^{k+1}} \right] \\ &= \frac{k!}{2\pi i} \lim_{\Delta z_0 \to 0} (\frac{1}{\Delta z_0}) \left[\oint_C [\frac{1}{(z-z_0-\Delta z_0)^{k+1}} - \frac{1}{(z-z_0)^{k+1}}] f(z)dz \right] \end{aligned} \tag{1.163}$$

The square bracket term inside the integration sign can be simplified as

$$\begin{aligned} L &= \frac{1}{(z-z_0-\Delta z_0)^{k+1}} - \frac{1}{(z-z_0)^{k+1}} = \frac{(z-z_0)^{k+1} - (z-z_0-\Delta z_0)^{k+1}}{(z-z_0-\Delta z_0)^{k+1}(z-z_0)^{k+1}} \\ &= \frac{(z-z_0)^{k+1}\left\{1-(1-\Delta z_0/(z-z_0))^{k+1}\right\}}{(z-z_0-\Delta z_0)^{k+1}(z-z_0)^{k+1}} \end{aligned} \tag{1.164}$$

Then, we can apply binomial theorem discussed in Section 1.2 to get

$$L = \frac{1}{(z - z_0 - \Delta z_0)^{k+1}} \left\{ 1 - [1 - \frac{\Delta z_0}{z - z_0}(k+1) + \frac{(\Delta z_0)^2}{(z - z_0)^2} \frac{k(k+1)}{2} + ...] \right\} \tag{1.165}$$

Back substitution of (1.165) into (1.163) we have

$$f^{(k+1)}(z_0) = \frac{k!}{2\pi i} \lim_{\Delta z_0 \to 0} \oint_C f(z) dz [\frac{k+1}{(z - z_0)(z - z_0 - \Delta z_0)^{k+1}} - O(\Delta z_0)] \tag{1.166}$$

Finally, we have

$$f^{(k+1)}(z_0) = \frac{(k+1)!}{2\pi i} \oint_C \frac{f(z) dz}{(z - z_0)^{k+2}} \tag{1.167}$$

Therefore, (1.161) is also true for $n = k+1$ if it is true for $n = k$. The proof is completed.

The Cauchy integral formula is remarkable because it shows that if a complex function $f(z)$ is known on the simple closed curve C, then the value of the function and all its derivatives can be found at all points inside C. It also implied that if a function has a first derivative (i.e., analytic) in a region R, all its higher derivatives exist in R. This is a major difference between a complex variable and real variable.

1.7.5 Residue Theorem

Finally, we will review the most powerful technique in complex variables—the residue theorem. Let us first consider the Laurent theorem about a singular point a (i.e., the function is not analytic at point a). If $f(z)$ is analytic inside and on the boundary of the ring-shaped region R bounded by two concentric circles C_1 and C_2 with center at a and respective radii r_1 and r_2, then for all z in R

$$f(z) = \frac{a_{-n}}{(z-a)^n} + \frac{a_{-n+1}}{(z-a)^{n-1}} + ... + \frac{a_{-1}}{(z-a)} + a_0 + a_1(z-a) + a_2(z-a)^2 + ...$$
$$= \sum_{k=-\infty}^{\infty} a_k (z-a)^k \tag{1.168}$$

where

$$a_k = \frac{1}{2\pi i} \oint_C \frac{f(\zeta)}{(\zeta - a)^{k+1}} d\zeta, \quad k = 0, \pm 1, \pm 2, ... \tag{1.169}$$

The negative power terms of $(z-a)$ are the principal part, from which the singularity comes, and the positive power terms of $(z-a)$ are the analytic parts of the expansion. If the principal parts have infinite terms, the point a is an essential singularity. Note also that Taylor series expansion contains only positive power, and hence it is only a special case of the Laurent expansion.

We will not provide a rigorous proof of the Laurent theorem, but we only sketch the main ideas. Applying the Cauchy integral formula given in the last section, we can write

$$f(z) = \frac{1}{2\pi i} \oint_{C_1} \frac{f(w)}{(w-z)} dw - \frac{1}{2\pi i} \oint_{C_2} \frac{f(w)}{(w-z)} dw \tag{1.170}$$

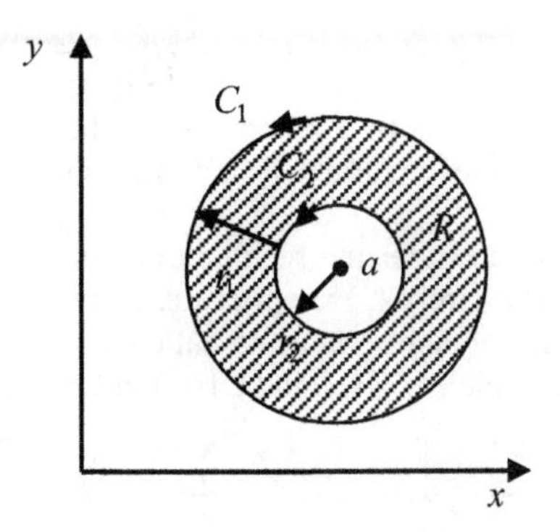

Figure 1.12 Annular region around a point *a*

First, we can expand $1/(w-z)$ as

$$\begin{aligned}\frac{1}{w-z} &= \frac{1}{(w-a)[1-(z-a)/(w-a)]} \\ &= \frac{1}{(w-a)}\{1+\frac{z-a}{w-a}+...+(\frac{z-a}{w-a})^{n-1}+(\frac{z-a}{w-a})^{n}+...\} \\ &= \frac{1}{(w-a)}+\frac{z-a}{(w-a)^2}+...+\frac{(z-a)^{n-1}}{(w-a)^n}+(\frac{z-a}{w-a})^n\frac{1}{w-a}+...\end{aligned} \tag{1.171}$$

We have applied Taylor series expansion in obtaining the result in (1.171). Similarly, we can also expand $-1/(w-z)$ as

$$\begin{aligned}-\frac{1}{w-z} &= \frac{1}{(z-a)[1-(w-a)/(z-a)]} \\ &= \frac{1}{(z-a)}\{1+\frac{w-a}{z-a}+...+(\frac{w-a}{z-a})^{n-1}+(\frac{w-a}{z-a})^{n}+...\} \\ &= \frac{1}{(z-a)}+\frac{w-a}{(z-a)^2}+...+\frac{(w-a)^{n-1}}{(z-a)^n}+(\frac{w-a}{z-a})^n\frac{1}{z-a}+...\end{aligned} \tag{1.172}$$

Substitution of (1.171) into the first term on the right of (1.170) gives

$$\frac{1}{2\pi i}\oint_{C_1}\frac{f(w)}{(w-z)}dw = a_0+a_1(z-a)+...+a_{n-1}(z-a)^{n-1}+U_n \tag{1.173}$$

where

$$a_k = \frac{1}{2\pi i}\oint_{C_1}\frac{f(w)dw}{(w-a)^{k+1}}, \qquad U_n = \frac{1}{2\pi i}\oint_{C_1}(\frac{z-a}{w-a})^n\frac{f(w)dw}{w-a} \tag{1.174}$$

with $k = 1,2,...n-1$. Substitution of (1.172) into the second term on the right of (1.170) gives

$$-\frac{1}{2\pi i}\oint_{C_2}\frac{f(w)}{w-z}dw = \frac{a_{-1}}{z-a}+\frac{a_{-2}}{(z-a)^2}+...+\frac{a_{-n}}{(z-a)^n}+V_n \tag{1.175}$$

where

$$a_{-k} = \frac{1}{2\pi i}\oint_{C_2}(w-a)^{k-1}f(w)dw, \quad V_n = \frac{1}{2\pi i}\oint_{C_2}(\frac{w-a}{z-a})^n \frac{f(w)dw}{z-a} \tag{1.176}$$

with $k = 1,2,...n$. As $n \to \infty$, both U_n and V_n can be shown approaching zero provided the value of the line integral $\int_C f(z)dz$ is bounded (Spiegel, 1964). This completes the proof.

We are now ready to consider the residue theorem. Let $f(z)$ be single valued and analytic inside and on a circle C except at the point $z = a$, which is chosen as the center of C (i.e., a is a singular point or so-called a pole). We now consider the closed line integral of $f(z)$ which is given in (1.168) in Laurent series:

$$\oint_C f(z)dz = \sum_{n=-\infty}^{\infty} a_n \oint_C (z-a)^n dz + \sum_{n=1}^{\infty} a_{-n} \oint_C \frac{dz}{(z-a)^n} \tag{1.177}$$

We drop all positive power terms because all positive power terms are analytic and by the Cauchy-Goursat theorem, the closed contour integral must be zero. The remaining job is to evaluate the negative powers. Use the result of the Cauchy-Goursat theorem again or refer to (1.157); we have

$$\oint_C \frac{dz}{(z-a)^n} = \oint_\Gamma \frac{dz}{(z-a)^n} \tag{1.178}$$

where the contour Γ is given in Figure 1.11. When $n = 1$, we can again adopt the change of variables given in (1.158) and obtain

$$\oint_C \frac{dz}{z-z_0} = \oint_\Gamma \frac{dz}{z-z_0} = \int_0^{2\pi} \frac{\rho e^{i\theta} id\theta}{\rho e^{i\theta}} = \int_0^{2\pi} id\theta = 2\pi i \tag{1.179}$$

For $n > 1$, we use the same change of variables given in (1.158) and obtain

$$\begin{aligned}\oint_C \frac{dz}{(z-z_0)^n} &= \oint_\Gamma \frac{dz}{(z-z_0)^n} = \int_0^{2\pi} \frac{\rho e^{i\theta} id\theta}{\rho^n e^{in\theta}} = \frac{i}{\rho^{n-1}} \int_0^{2\pi} e^{(1-n)i\theta} d\theta \\ &= \frac{1}{(1-n)\rho^{n-1}}[e^{(1-n)2\pi i} - 1] = 0\end{aligned} \tag{1.180}$$

Substitution of (1.179) and (1.180) into (1.177) gives

$$\oint_C f(z)dz = \sum_{n=1}^{\infty} a_{-n} \oint_C \frac{dz}{(z-a)^n} = 2\pi i a_{-1} \tag{1.181}$$

where a_{-1} is called the residue of $f(z)$ at $z = a$.

Thus, we can find the residue as

$$a_{-1} = \lim_{z\to a}(z-a)f(z) \tag{1.182}$$

When $z = a$ is a singularity of order k, we find the residue by the following formula

$$a_{-1} = \lim_{z\to a} \frac{1}{(k-1)!} \frac{d^{k-1}}{dz^{k-1}} \{(z-a)^k f(z)\} \tag{1.183}$$

The formula can readily be verified by the Cauchy integral formula. Let us consider that a function $f(z)$ has a pole of order m at $z = a$, then we can write it as

$$f(z) = \frac{F(z)}{(z-a)^m} \tag{1.184}$$

where $F(z)$ is analytic inside and on C, and $f(a) \neq 0$. Then by Cauchy's integral formula given in (1.161):

$$\frac{1}{2\pi i}\oint_C f(z)dz = \frac{1}{2\pi i}\oint_C \frac{F(z)dz}{(z-a)^m} = \frac{F^{(m-1)}(a)}{(m-1)!}$$
$$= \lim_{z\to a}\frac{1}{(m-1)!}\frac{d^{m-1}}{dz^{m-1}}\{(z-a)^m f(z)\} \tag{1.185}$$

Combining (1.185) and (1.181) gives (1.183) and this completes the proof. If there is more than one singular point within contour C as shown in Figure 1.13, we have

$$\oint_C f(z)dz = 2\pi i(a_{-1} + b_{-1} + c_{-1} + ...) \tag{1.186}$$

The residue theorem will be used later in Section 1.7.7.

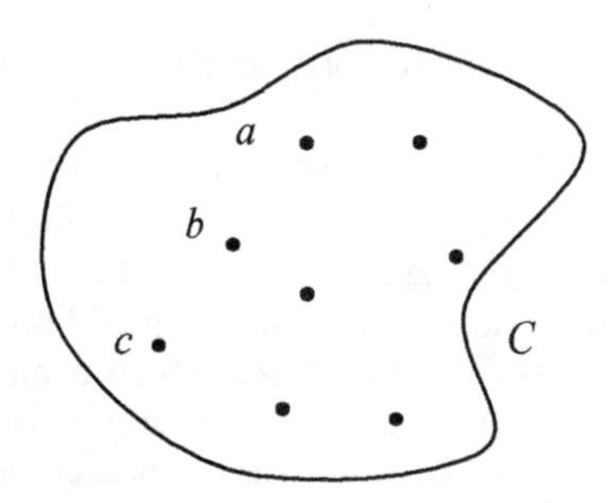

Figure 1.13 Annular region around points *a*, *b*, and *c*.

1.7.6 Branch Point and Branch Cut

One major problem with complex analysis is the multi-value properties of complex functions around a certain point, which is called a branch point. Let us consider the function that

$$w = z^{1/n}, \quad z = \rho e^{i\theta} \tag{1.187}$$

where n is a positive integer. Use the polar form given in the second equation of (1.187) we have

$$w = \rho^{1/n} e^{i\theta/n} \tag{1.188}$$

If we go around the origin starting from Point A in Figure 1.14, following a complete counterclockwise circuit, we have

$$w = \rho^{1/n} e^{i(\theta+2\pi)/n} = \rho^{1/n} e^{i\theta/n} e^{i2\pi/n} \tag{1.189}$$

which is different from (1.188). Thus, we end up with a different value for the same Point A. However, if we go around the origin n times, we should obtain the same value of w as in (1.188). We can describe one branch of the multi-valued function $z^{1/n}$ for $0 \le \theta < 2\pi$ as long as we do not exceed 2π, and similarly we can find the other branches of the solution for larger values of θ. Grouping all the answers, we obtain n branches of solution

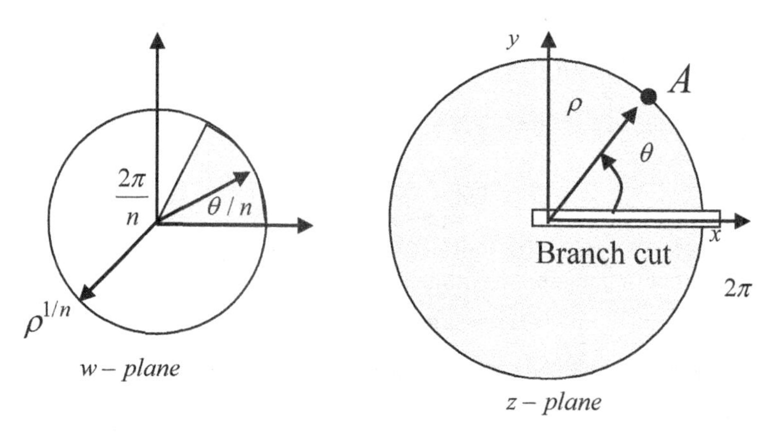

Figure 1.14 A function of $w = z^{1/n}$

$$0 \le z < 2\pi, \quad 2\pi \le z < 4\pi, ..., (4n-2)\pi \le z < 4n\pi \tag{1.190}$$

Each branch is single valued, but to keep the function single valued, we must set up an artificial barrier that joins the origin and infinity. This artificial barrier is called a branch line or a branch cut, and this barrier should not be crossed for each branch of solutions. For this case, the origin, for which a multi-valued function appears when we go around this point a complete circuit, is called a branch point. As illustrated in Figure 1.14, we have chosen the branch cut along the positive x-axis, but actually any other branch cut connecting the origin and infinity along any orientation can be selected.

1.7.7 Titchmarsh's Contour Integral

In this section, we consider the following integral:

$$\int_0^\infty \frac{x^{p-1}}{1+x} dx = \frac{\pi}{\sin p\pi}, \quad 0 < p < 1 \tag{1.191}$$

To get this result, we will first generalize the integral in (1.191) to a complex variable with the contour integral shown in Figure 1.15:

$$\oint_C \frac{z^{p-1}}{1+z} dz, \quad 0 < p < 1 \tag{1.192}$$

The contour shown in Figure 1.15 was proposed by Titchmarsh. The contour ABCDEGHJA is a closed integral containing the pole or singular point at $z = -1$.

On the numerator, z^{p-1} is a multi-valued function, as we discussed in the last section. That is, when we make a circuit going around the origin, the argument will increase by an amount of 2π such that the value of z^{p-1} changes its value. Thus, the branch point is at $z = 0$. By using the residue theorem, the residue of the integral in (1.192) is

$$\lim_{z \to -1} (1+z) \frac{z^{p-1}}{1+z} = (e^{\pi i})^{p-1} = e^{(p-1)\pi i} \tag{1.193}$$

On the left hand side of (1.192), we can divide the contour path given in Figure 1.15 into

$$\oint_C \frac{z^{p-1}}{1+z}dz = \int_{AB} ...+ \int_{BCDEG} ...+ \int_{GH} ...+ \int_{HJA} ... = 2\pi i e^{(p-1)\pi i} \tag{1.194}$$

We have substituted (1.193) into (1.192) to get the last of (1.194). The contours given in (1.194) can be written as

$$\int_r^R \frac{x^{p-1}}{1+x}dx + \int_0^{2\pi} \frac{(\mathrm{R}e^{i\theta})^{p-1} iR^{i\theta} d\theta}{1+\mathrm{R}e^{i\theta}} + \int_R^r \frac{(xe^{2\pi i})^{p-1}dx}{1+xe^{2\pi i}} + \int_{2\pi}^0 \frac{(re^{i\theta})^{p-1} ir^{i\theta} d\theta}{1+re^{i\theta}} \tag{1.195}$$

in which we have used $dz=iR^{i\theta}d\theta$ and $z = xe^{2\pi i}$ along GH (i.e., $\theta = 2\pi$ along GH). Note that the Titchmarsh contour excludes the branch point and a branch cut is formed along the positive x-axis. As long as the pole $z = -1$ is kept within the closed contour ABCDEGHJA, we can deform the closed contour arbitrarily such that the integral remains the same. Consequently, we can let $r \to 0$ and $R \to \infty$ such that

$$\int_0^{2\pi} \frac{(\mathrm{R}e^{i\theta})^{p-1} iR^{i\theta} d\theta}{1+\mathrm{R}e^{i\theta}} \to 0 \quad \text{as} \quad R \to \infty, p<1 \tag{1.196}$$

$$\int_{2\pi}^0 \frac{(re^{i\theta})^{p-1} ir^{i\theta} d\theta}{1+re^{i\theta}} \to 0 \text{ as } r \to 0, p>0 \tag{1.197}$$

Thus, (1.194) becomes

$$\int_0^\infty \frac{x^{p-1}}{1+x}dx + \int_\infty^0 \frac{e^{2\pi(p-1)i} x^{p-1} dx}{1+x} = 2\pi i e^{(p-1)\pi i} \tag{1.198}$$

Or equivalently, we have

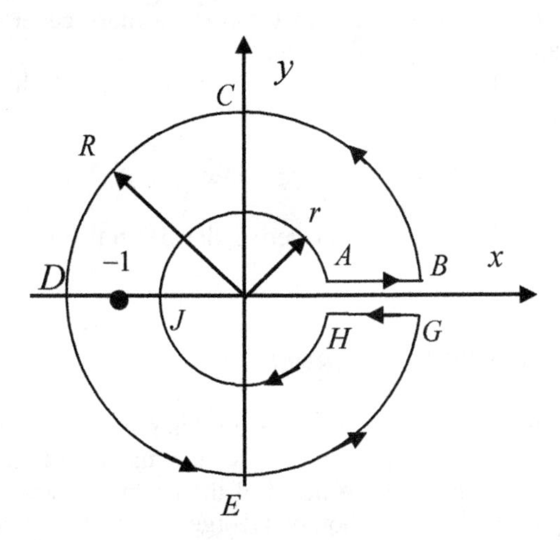

Figure 1.15 Titchmarsh contour for integral given in (1.194)

$$(1-e^{2\pi(p-1)i})\int_0^\infty \frac{x^{p-1}}{1+x}dx = 2\pi i e^{(p-1)\pi i} \tag{1.199}$$

Rearranging (1.199), we get

$$\begin{aligned}\int_0^\infty \frac{x^{p-1}}{1+x}dx &= \frac{2\pi i e^{(p-1)\pi i}}{(1-e^{2\pi(p-1)i})} = \frac{2\pi i}{e^{-(p-1)\pi i} - e^{\pi(p-1)i}} = \frac{2\pi i}{[e^{\pi i}e^{-p\pi i} - e^{-\pi i}e^{\pi p i}]} \\ &= \frac{2\pi i}{e^{\pi p i} - e^{-p\pi i}} = \frac{\pi}{\sin p\pi}\end{aligned} \tag{1.200}$$

Finally, we complete the proof for (1.191). This formula will be used later relating to the discussion of the gamma function in Chapter 4.

1.8 FRULLANI-CAUCHY INTEGRAL

The following integral was first mentioned in a letter by Italian mathematician Frullani in 1821 and was later published in 1828

$$\int_0^\infty \frac{f(ax)-f(bx)}{x}dx = f(0)\ln\frac{b}{a} \tag{1.201}$$

This so-called Frullani's integral is, however, exact for the case that $f(\infty)$ vanishes. Cauchy in 1823 and again in 1827 provided the following exact form as:

$$\int_0^\infty \frac{f(ax)-f(bx)}{x}dx = \{f(0)-f(\infty)\}\ln\frac{b}{a} \tag{1.202}$$

provided that the derivative of $f(x)$ is continuous and the integral converges. The integral given in (1.202) is commonly known as the Frullani-Cauchy integral. In a sense, the Frullani integral given in (1.201) is correct only for the special case that $f(\infty) = 0$. We will see in Section 1.10 that there is a more general integral than (1.202) called Ramanujan's integral theorem.

Consider a special case that $f(x) = e^{-x}$ and $a = 1$, we have an integral representation of $\ln(x)$:

$$\int_0^\infty \frac{e^{-\xi}-e^{-x\xi}}{\xi}d\xi = \ln x \tag{1.203}$$

This result could be employed later to derive the integral representation of the digamma function in Chapter 4.

1.9 RAMANUJAN MASTER THEOREM

The Ramanujan master theorem is considered in this section and this result could be useful to derive the integral representation for the digamma function. Although nearly all mathematicians should have heard of the legendary story of Ramanujan and the re-discovery of his lost notebook by George Andrews in 1976 in one of the boxes left behind by G.N. Watson. The coverage of Ramanujan's discovery in mathematics textbooks, especially in engineering mathematics, is non-existent. In this section, we will show how his so-called master theorem leads to down-to-earth results that we have used regularly in engineering mathematics. Ramanujan was a self-taught Indian mathematical genius, and his story started with his correspondence with Hardy. Upon invitation of Professor G.H. Hardy of Cambridge University, Ramanujan spent five years in England before he returned to India in 1919. Unfortunately, he passed away in 1920 at the age of thirty-two (Kanigel, 1991). Ramanujan's brief biography is given at the end of this book. Ramanujan also published a number of formulas for generating accurate results for π. Some of these amazing formulas are reported in Appendix F.

Ramanujan considered the following improper integral (Berndt, 1985):

$$I \equiv \int_0^\infty x^{n-1} F(x) dx \tag{1.204}$$

where n is not necessarily an integer. Ramanujan asserted that this integral I can be evaluated if $F(x)$ is assumed to be expandable in the Maclaurin series. In particular, it is assumed that

$$F(x) = \sum_{k=0}^{\infty} \frac{\varphi(k)(-x)^k}{k!} \tag{1.205}$$

where $\varphi(k)$ is the coefficient of the Maclaurin series expansion of $F(x)$. Then, Ramanujan found that

$$I = \Gamma(n)\varphi(-n) \tag{1.206}$$

where $\Gamma(x)$ is the gamma function (Abramowitz and Stegun, 1964). Conversely, if the integral is known, the coefficients of the Maclaurin series can be found. This was called Ramanujan's master theorem by Berndt (1985). The proof of this theorem involves our knowledge of the gamma function and Laplace transform, and this knowledge is yet to be covered in this chapter. Thus, it will not be given here but is reported in Appendix C1. Readers can refer to the proof in the Appendix after we introduce the gamma function and Laplace transform in later chapters. Ramanujan's master theorem was reported in the first quarterly report of Ramanujan submitted to Presidence College in Madras, which admitted Ramanujan upon the recommendation of Hardy. It was published before Ramanujan's 5-year visit to England.

1.10 RAMANUJAN INTEGRAL THEOREM

A natural consequence of Ramanujan's master theorem is Ramanujan's integral theorem, which can be obtained for the limiting case of $n \to 0$ in (1.204). In particular, Ramanujan's integral theorem states that

$$\int_0^\infty \frac{\{f(ax) - g(bx)\}}{x} dx = \{f(0) - f(\infty)\} \left[\ln(\frac{b}{a}) + \frac{d}{ds}\left(\ln[\frac{v(s)}{u(s)}] \right)_{s=0} \right] \tag{1.207}$$

provided that $f(0) = g(0)$ and $f(\infty) = g(\infty)$ with $a, b > 0$. In addition, functions u and v are defined as the coefficients of the Maclaurin series of functions f and g as:

$$f(x) - f(\infty) = \sum_{k=0}^{\infty} \frac{u(k)(-x)^k}{k!}, \quad g(x) - g(\infty) = \sum_{k=0}^{\infty} \frac{v(k)(-x)^k}{k!} \tag{1.208}$$

This is a power extension of Frullani's integral theorem given in (1.201). The proof of (1.207) is given in Appendix C2. For the special case that $f = g$, we have the last term in (1.207) vanishing and giving

$$\int_0^\infty \frac{\{f(ax)-f(bx)\}}{x}dx = \{f(0)-f(\infty)\}\ln(\frac{b}{a}) \tag{A.209}$$

which is the Frullani-Cauchy integral given in (1.202). Thus, Frullani-Cauchy's integral is a special case of Ramanujan's integral theorem. Ramanujan's integral theorem is a highly original and powerful result. Here is an example.

Example 1.8 Prove the following infinite integration formula of the Bessel function by Weber (see p. 391 Watson, 1944) by using Ramanujan's master theorem:

$$\int_0^\infty \frac{J_\nu(t)}{t^{\nu-\mu+1}}dt = \frac{\Gamma(\frac{1}{2}\mu)}{2^{\nu-\mu+1}\Gamma(\nu-\frac{1}{2}\mu+1)} \tag{1.210}$$

Solution: By using the definition of the Bessel function (Watson, 1944), the left hand side can be written as:

$$I = \int_0^\infty t^{-\nu+\mu-1}J_\nu(t)dt = \int_0^\infty t^{-\nu+\mu-1}\sum_{k=0}^\infty \frac{(-1)^k}{k!\Gamma(\nu+k+1)}(\frac{t}{2})^{2k+\nu}dt \tag{1.211}$$

By applying a change of variable $t = y^\alpha$ to (1.211), we obtain:

$$I = \int_0^\infty y^{\mu\alpha-1}\sum_{k=0}^\infty \frac{(-1)^k}{k!\Gamma(\nu+k+1)}(\frac{\alpha}{2^{2k+\nu}})y^{2k\alpha}dy \tag{1.212}$$

In order to apply Ramanujan's master theorem, we set $\alpha = 1/2$ and obtain

$$I = \int_0^\infty y^{\mu/2-1}\sum_{k=0}^\infty \frac{(-y)^k}{k!\Gamma(\nu+k+1)}(\frac{1}{2^{2k+\nu+1}})dy = \int_0^\infty y^{\mu/2-1}\sum_{k=0}^\infty \frac{\varphi(k)(-y)^k}{k!}dy \tag{1.213}$$

where

$$\varphi(k) = \frac{1}{2^{2k+\nu+1}\Gamma(\nu+k+1)} \tag{1.214}$$

Ramanujan's master theorem gives

$$I = \int_0^\infty y^{\mu/2-1}\sum_{k=0}^\infty \frac{\varphi(k)(-y)^k}{k!}dy = \Gamma(\frac{\mu}{2})\varphi(-\frac{\mu}{2}) = \frac{\Gamma(\frac{\mu}{2})}{2^{\nu-\mu+1}\Gamma(\nu-\frac{\mu}{2}+1)} \tag{1.215}$$

This completes the proof of the Weber infinite integral for the Bessel function.

1.11 CIRCULAR FUNCTIONS

Trigonometric functions are also known as circular functions. As we all learned in high school, sine, cosine, and tangent are ratios of the sides of a right angle triangle. In this chapter, we will look at sine and cosine from the viewpoint of differential equations. More specifically, let us consider the first differential equation of this book:

$$\frac{d^2y}{dx^2} + \lambda^2 y = 0 \tag{1.216}$$

Let us assume that the solution of $y(x)$ can be expressed in exponent function with an unknown constant α:

$$y(x) = e^{\alpha x} \tag{1.217}$$

Substitution of (1.217) in (1.216), we get the so-called characteristics equation for the differential equation:

$$(\alpha^2 + \lambda^2)e^{\alpha x} = 0 \tag{1.218}$$

The roots for α are

$$\alpha = \pm i\lambda \tag{1.219}$$

Thus, the solution for y(x) becomes

$$y(x) = C_1 e^{i\lambda x} + \bar{C}_1 e^{-i\lambda x} \tag{1.220}$$

Note that the constant C_1 is complex and since (1.216) is real, we have $y(x)$ real. Recalling that a complex number adding to its complex conjugate is real, we must set the unknown constant for the second term as the complex conjugate of C_1. Applying the Euler formula, we have

$$e^{i\lambda x} = \cos x + i \sin x\,, \quad e^{-i\lambda x} = \cos x - i \sin x \tag{1.221}$$

Substitution of (1.221) into (1.220) leads to the standard solution form for (1.216):

$$y(x) = D_1 \cos \lambda x + D_2 \sin \lambda x \tag{1.222}$$

The validity of (1.222) can be shown easily by substituting (1.222) into (1.216). The unknown constants D_1 and D_2 need to be determined by appropriate boundary conditions of $y(x)$. The reader should bear in mind the solution form of (1.222) because we will encounter this differential equation many more times in this book. Sine and cosine are the only functions for which their second derivatives are equal to the negative values of themselves.

1.12 HYPERBOLIC FUNCTIONS

Hyperbolic functions appear naturally in our daily lives. The shape of a hanging chain, a spider web, or a hanging power line (catenary problem solved by Huygens), optimum shapes of arches (like the Gateway Arch at St. Louis, Missouri, USA or the largest vault Taq-i Kisra in Iraq), and the optimum shape of a soap sheet between two rings (catenoid problem solved by Euler). Apparently, it links naturally to the deformed shape of a flexible body under gravity. It also appears naturally in the solutions of waves in solids, the motion of falling objects with air resistance, and the solution of the temperature distribution in cooling solids. Its importance in physics, engineering, and applied mathematics should not be undermined. They are also regarded as the most commonly encountered elementary functions, probably only second to circular functions (e.g., sine, cosine etc.). Even the “M” logo of McDonald’s is also a double-inverted catenary. By definition, hyperbolic functions are defined as

$$\sinh x = \frac{e^x - e^{-x}}{2}, \quad \cosh x = \frac{e^x + e^{-x}}{2}, \quad \tanh x = \frac{e^x - e^{-x}}{e^x + e^{-x}} \tag{1.223}$$

$$\tanh x = \frac{\sinh x}{\cosh x}, \quad \operatorname{csch} x = \frac{1}{\sinh x}, \quad \operatorname{sech} x = \frac{1}{\cosh x}, \quad \coth x = \frac{\cosh x}{\sinh x} \tag{1.224}$$

The close resemblance of circular and hyperbolic functions is obvious. These functions were first defined by J.H. Lambert, who proved the irrationality of π and was known to contribute to map projection and stereonet projection related to his works on cosmology. A brief biography is given at the end of this book. It is straightforward to show that

$$\sinh(-x) = -\sinh x, \quad \cosh(-x) = \cosh x \tag{1.225}$$

$$\cosh^2 x - \sinh^2 x = 1, \quad 1 - \tanh^2 x = \operatorname{sech}^2 x, \quad \coth^2 x - \operatorname{csch}^2 x = 1 \tag{1.226}$$

$$\sinh(x+y) = \sinh x \cosh y + \cosh x \sinh y \tag{1.227}$$

$$\cosh(x+y) = \cosh x \cosh y + \sinh x \sinh y \tag{1.228}$$

These will be left as exercises for the readers.

Physically, the argument of the circular functions equals twice the gray area of the unit circle on the left of Figure 1.16, whereas the argument of the hyperbolic functions equals twice the gray area on the right of Figure 1.16. The curves of the circle and hyperbola are given respectively by:

$$x^2 + y^2 = 1, \quad x^2 - y^2 = 1 \tag{1.229}$$

The gray area of the unit circle is clearly proportional to θ or

$$A_\theta = \pi(1)^2 \frac{\theta}{2\pi} = \frac{\theta}{2} \tag{1.230}$$

The proof for the argument of the hyperbolic functions is less obvious. We can rewrite the gray area as the difference of ΔOPB and sector APB. The area of triangle OPB is clearly

$$A_{OPB} = \frac{1}{2} \cosh\theta \sinh\theta \tag{1.231}$$

The area of the curved sector can be integrated as

$$A_{ABP} = \int_1^{\cosh\theta} y dx = \int_1^{\cosh\theta} \sqrt{x^2 - 1} dx \tag{1.232}$$

We can make a change of variables $x = \cosh u$ and obtain

$$A_{ABP} = \int_0^\theta \sqrt{\cosh^2 u - 1} \sinh u du = \int_0^\theta \sinh^2 u du \tag{1.233}$$

By virtue of the definition of (1.223), we have

$$\begin{aligned} A_{ABP} &= \int_0^\theta \left(\frac{e^u - e^{-u}}{2}\right)^2 du = \int_0^\theta \left(\frac{e^{2u} - 2 + e^{-2u}}{4}\right) du = \int_0^\theta \left(\frac{\cosh 2u - 1}{2}\right) du \\ &= \frac{1}{4}\sinh 2\theta - \frac{\theta}{2}, \end{aligned} \tag{1.234}$$

by noting from (1.231) that

$$A_{OPB} = \frac{1}{2}\cosh\theta\sinh\theta = \frac{1}{8}(e^\theta + e^{-\theta})(e^\theta - e^{-\theta}) = \frac{1}{8}(e^{2\theta} - e^{-2\theta}) = \frac{1}{4}\sinh 2\theta \tag{1.235}$$

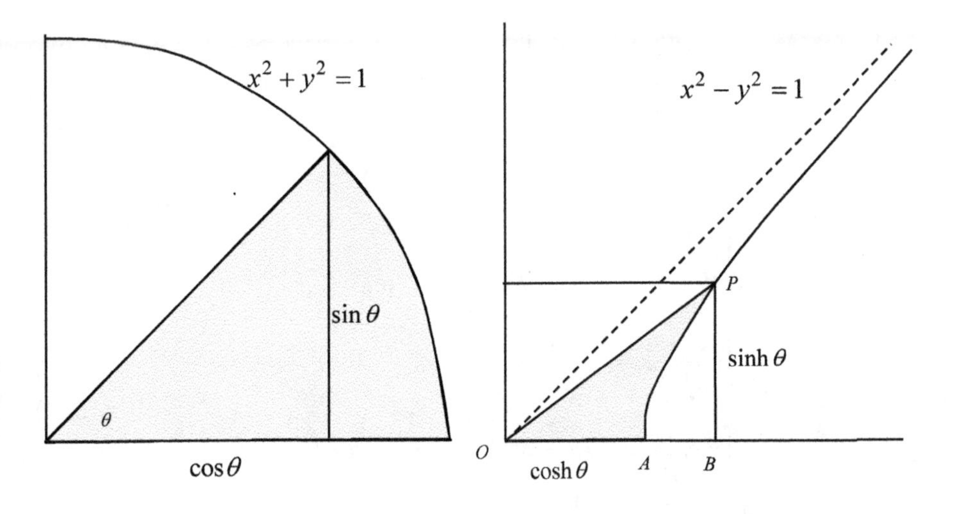

Figure 1.16 Physical meaning of the argument of circular and hyperbolic functions

This can also be obtained from (1.227).

Therefore, we finally have

$$A_{OPB} - A_{APB} = \frac{1}{2}\theta \tag{1.236}$$

Again, the gray area equals half of the argument.

We here only report some of the commonly used formulas of differentiation of hyperbolic functions without proof:

$$\frac{d}{dx}\sinh x = \cosh x, \quad \frac{d}{dx}\cosh x = \sinh x \tag{1.237}$$

$$\frac{d}{dx}\tanh x = \text{sech}^2 x, \quad \frac{d}{dx}\coth x = -\text{csch}^2 x \tag{1.238}$$

$$\frac{d}{dx}\text{sech} x = -\text{sech} x \tanh x, \quad \frac{d}{dx}\text{csch} x = -\text{csch}\, x \coth x \tag{1.239}$$

$$\frac{d}{dx}\sinh^{-1} x = \frac{1}{\sqrt{1+x^2}}, \quad \frac{d}{dx}\cosh^{-1} x = \frac{\pm 1}{\sqrt{x^2-1}} \tag{1.240}$$

where the plus sign is for $\cosh^{-1}x > 0$ and minus when $\cosh^{-1}x < 0$. In addition, we have

$$\frac{d}{dx}\tanh^{-1} x = \frac{1}{1-x^2}, \quad |x|<1 \quad \frac{d}{dx}\coth^{-1} x = -\frac{1}{x^2-1}, \quad |x|>1 \tag{1.241}$$

$$\frac{d}{dx}\text{csch}^{-1} x = -\frac{1}{|x|\sqrt{1+x^2}}, \quad \frac{d}{dx}\text{sech}^{-1} x = \frac{\mp 1}{x\sqrt{1-x^2}} \tag{1.242}$$

where the minus sign is for $\text{sech}^{-1}x > 0$ and plus when $\text{sech}^{-1}x < 0$. Some useful formulas of integration are

$$\int \frac{du}{\sqrt{a^2+u^2}} = \sinh^{-1}\frac{u}{a}+C$$

$$\int \frac{du}{\sqrt{u^2-a^2}} = \cosh^{-1}\frac{u}{a}+C \tag{1.243}$$

$$\int \frac{du}{a^2-u^2} = \begin{Bmatrix} \frac{1}{a}\tanh^{-1}\frac{u}{a}+C & \text{if } \mathrm{u}^2 < a^2 \\ \frac{1}{a}\coth^{-1}\frac{u}{a}+C & \text{if } \mathrm{u}^2 > a^2 \end{Bmatrix}$$

$$\int \frac{du}{u\sqrt{a^2-u^2}} = \frac{-1}{a}\operatorname{sech}^{-1}\frac{u}{a}+C \qquad 0<u<a$$

$$\int \frac{du}{u\sqrt{u^2+a^2}} = \frac{-1}{a}\operatorname{csch}^{-1}\left|\frac{u}{a}\right|+C \qquad u \neq 0 \tag{1.244}$$

The series expansions of $\sinh x$ and $\cosh x$ are

$$\sinh x = x+\frac{x^3}{3!}+\frac{x^5}{5!}+..., \quad \cosh x = 1+\frac{x^2}{2!}+\frac{x^4}{4!}+... \tag{1.245}$$

These series expansions can be obtained by Taylor series expansions, which will be considered in Section 1.13.2.

Equation (1.237) suggests that the general solutions of the following second order ordinary differential equation are sinh and cosh functions:

$$\frac{d^2y}{dx^2}-\lambda^2 y=0 \tag{1.246}$$

Let us assume that the solution of $y(x)$ can be expressed in an exponential function with an unknown constant α:

$$y(x)=e^{\alpha x} \tag{1.247}$$

Substitution of (1.247) in (1.246) gives the so-called characteristics equation for the differential equation:

$$(\alpha^2-\lambda^2)e^{\alpha x}=0 \tag{1.248}$$

The roots for α are

$$\alpha = \pm\lambda \tag{1.249}$$

Thus, the solution for y(x) becomes

$$y(x)=C_1 e^{\lambda x}+C_2 e^{-\lambda x} \tag{1.250}$$

Note from (1.223) that

$$e^{\lambda x}=\cosh\lambda x+\sinh\lambda x, \quad e^{-\lambda x}=\cosh\lambda x-\sinh\lambda x \tag{1.251}$$

Substitution of (1.251) into (1.250) leads to the standard solution form of (1.246):

$$y(x)=D_1\sinh\lambda x+D_2\cosh\lambda x \tag{1.252}$$

The unknown constants D_1 and D_2 need to be determined by appropriate boundary conditions of $y(x)$. This is the second most commonly encountered ODE in this book. Readers should familiar themselves with this differential equation.

1.13 SERIES EXPANSIONS

In various areas of applied mathematics, functions are often expanded around certain points and their behaviors are considered around these selected points. Asymptotic expansion and perturbation are typical examples of series expansion (see Chapter 12). In this section, we will consider series expansion for some well-behaved functions.

1.13.1 Darboux's Formula

We first consider the most general series expansion of a function, the so-called Darboux's formula. According to Whittaker and Watson (1927), this formula was derived by Darboux in 1876, who was an expert in applying differential geometry to differential equations (see biography). Darboux's formula relies on the following identity of the differentiation of a product of a power series of $(z-a)$ of order m, a polynomial $\phi(t)$ of degree n, and an analytic function f(t) at all points from a to z

$$\begin{aligned}&\frac{d}{dt}\sum_{m=1}^{n}\left\{(-1)^m (z-a)^m \phi^{(n-m)}(t) f^{(m)}[a+t(z-a)]\right\} \\ &\quad = -(z-a)\phi^{(n)}(t) f'[a+t(z-a)] + (-1)^n (z-a)^{n+1}\phi(t) f^{(n+1)}[a+t(z-a)]\end{aligned} \tag{1.253}$$

where the superscript bracket "(m)" indicates m-th differentiation of the function. This identity is remarkable in the sense that the differentiation of a series of n-terms will result in only two terms on the right hand side. To demonstrate the validity of (1.253), let us consider $n = 1$ such that the left hand side is

$$\begin{aligned}&\frac{d}{dt}\{-(z-a)\phi(t) f'[a+t(z-a)]\} \\ &\quad = -(z-a)\phi'(t) f'[a+t(z-a)] - (z-a)^2 \phi(t) f''[a+t(z-a)]\end{aligned} \tag{1.254}$$

The right hand side of (1.253) is

$$RHS = -(z-a)\phi'(t) f'[a+t(z-a)] - (z-a)^2 \phi(t) f''[a+t(z-a)] \tag{1.255}$$

Thus, the validity for $n = 1$ is demonstrated. For $n = 2$, the left hand side of (1.253) is

$$\begin{aligned}&\frac{d}{dt}\left\{-(z-a)\phi'(t) f'[a+t(z-a)] + (z-a)^2 \phi(t) f''[a+t(z-a)\right\} \\ &= -(z-a)\phi''(t) f'[a+t(z-a)] - (z-a)^2 \phi'(t) f''[a+t(z-a)] \\ &\quad +(z-a)^2 \phi'(t) f''[a+t(z-a)] + (z-a)^3 \phi(t) f^{(3)}[a+t(z-a)] \\ &= -(z-a)\phi''(t) f'[a+t(z-a)] + (z-a)^3 \phi(t) f^{(3)}[a+t(z-a)]\end{aligned} \tag{1.256}$$

Note that the middle two terms cancel out one and another, and thus only two terms were left and they are equal to the right hand side of (1.253). Thus, the validity of (1.253) for $n = 2$ is demonstrated. This procedure can be easily extended to $n = k$; we have the left hand side of (1.253) as

$$\frac{d}{dt}\{-(z-a)\phi^{(k-1)}(t)f'[a+t(z-a)]+(z-a)^2\phi^{(k-2)}(t)f''[a+t(z-a)]+...$$
$$+(-1)^{k-1}(z-a)^{k-1}\phi^{(1)}(t)f^{(k-1)}[a+t(z-a)]$$
$$+(-1)^{k}(z-a)^{k}\phi(t)f^{(k)}[a+t(z-a)]\}$$
$$=-(z-a)\phi^{(k)}(t)f'[a+t(z-a)]-(z-a)^2\phi^{(k-1)}(t)f^{(2)}[a+t(z-a)]$$
$$+(z-a)^2\phi^{(k-1)}(t)f''[a+t(z-a)]+(z-a)^3\phi^{(k-2)}(t)f^{(3)}[a+t(z-a)]+...$$
$$+(-1)^{k-1}(z-a)^{k-1}\phi^{(2)}(t)f^{(k-1)}[a+t(z-a)]$$
$$+(-1)^{k-1}(z-a)^{k}\phi^{(1)}(t)f^{(k)}[a+t(z-a)]$$
$$+(-1)^{k}(z-a)^{k}\phi^{(1)}(t)f^{(k)}[a+t(z-a)]+(-1)^{k}(z-a)^{k+1}\phi(t)f^{(k+1)}[a+t(z-a)]$$
$$=-(z-a)\phi^{(k)}(t)f'[a+t(z-a)]+(-1)^{k}(z-a)^{k+1}\phi(t)f^{(k+1)}[a+t(z-a)] \tag{1.257}$$

Note that all intermediate terms will cancel one another except for the first and the last terms. The last of (1.257) is the right hand side of (1.253). Thus, the validity for $n = k$ is demonstrated.

We now rearrange (1.253) to get

$$(z-a)\phi^{(n)}(t)f'[a+t(z-a)]$$
$$=-\frac{d}{dt}\sum_{m=1}^{n}\{(-1)^m(z-a)^m\phi^{(n-m)}(t)f^{(m)}[a+t(z-a)]\} \tag{1.258}$$
$$+(-1)^n(z-a)^{n+1}\phi(t)f^{(n+1)}[a+t(z-a)]$$

Recall that $\phi(t)$ is a polynomial of degree n, and we have

$$\phi^{(n)}(t)=\phi^{(n)}(0)=c \tag{1.259}$$

where c is a constant. Note that

$$(z-a)f'[a+t(z-a)]=\frac{d}{dt}f[a+t(z-a)] \tag{1.260}$$

We then integrate (1.258) from 0 to 1 to get

$$\phi^{(n)}(0)\int_0^1\frac{d}{dt}f[a+t(z-a)]dt=-\int_0^1\frac{d}{dt}\sum_{m=1}^{n}\{(-1)^m(z-a)^m\phi^{(n-m)}(t)f^{(m)}[a+t(z-a)]\}dt$$
$$+(-1)^n(z-a)^{n+1}\int_0^1\phi(t)f^{(n+1)}[a+t(z-a)]dt \tag{1.261}$$

The first integral on the right hand side of (1.261) can be evaluated by reversing the order of integration and summation such that

$$-\int_0^1 \frac{d}{dt}\sum_{m=1}^{n}\left\{(-1)^m (z-a)^m \phi^{(n-m)}(t) f^{(m)}[a+t(z-a)]\right\}dt$$
$$=\sum_{m=1}^{n}(-1)^{m-1}(z-a)^m\{\phi^{(n-m)}(1)f^{(m)}(z)-\phi^{(n-m)}(0)f^{(m)}(a)\} \tag{1.262}$$

Substitution of (1.262) into the first integral on the right hand side of (1.261) and reversing of the order of integration and summation gives

$$\phi^{(n)}(0)\{f(z)-f(a)\}=\sum_{m=1}^{n}(-1)^{m-1}(z-a)^m\{\phi^{(n-m)}(1)f^{(m)}(z)-\phi^{(n-m)}(0)f^{(m)}(a)\}$$
$$+(-1)^n(z-a)^{n+1}\int_0^1 \phi(t)f^{(n+1)}[a+t(z-a)]dt \tag{1.263}$$

This is Darboux's formula of the series expansion of function $f(z)$.

1.13.2 Taylor Series Expansion

Two different proofs for Taylor series expansion formula are given here.

Proof 1

We now consider a special case of Darboux's formula given in the last section that $\phi(t) = (t-1)^n$. Consequently, we have

$$\phi^{(1)}(t)=n(t-1)^{n-1},$$
$$\phi^{(2)}(t)=n(n-1)(t-1)^{n-2}, \tag{1.264}$$
$$\phi^{(n)}(t)=\phi^{(n)}(0)=n!$$

$$\phi^{(n-m)}(t)=n...(m+1)(t-1)^m,$$
$$\phi^{(n-m)}(0)=n(n-1)...(m+1)(-1)^m=\frac{n!}{m!}(-1)^m, \tag{1.265}$$
$$\phi^{(n-m)}(1)=0$$

Substitution of (1.264) and (1.265) into (1.263) results in

$$n!\{f(z)-f(a)\}=-\sum_{m=1}^{n}(-1)^{m-1}(z-a)^m\frac{n!}{m!}(-1)^m f^{(m)}(a)$$
$$+(-1)^n(z-a)^{n+1}\int_0^1 \phi(t)f^{(n+1)}[a+t(z-a)]dt \tag{1.266}$$

Rearranging (1.266) gives

$$f(z)=f(a)+\sum_{m=1}^{n}\frac{f^{(m)}(a)}{m!}(z-a)^m$$
$$+\frac{(-1)^n(z-a)^{n+1}}{n!}\int_0^1 \phi(t)f^{(n+1)}[a+t(z-a)]dt \tag{1.267}$$

Now consider the case that $\phi(t)$ is an infinite series or $n \to \infty$. Since f is an analytic function, all its higher derivatives exist such that $f^{(n+1)}[a+t(z-a)]$ is finite. The term $(-1)^n(z-a)^{n+1}$ is an oscillating function and remains finite compared to $n \to \infty$. Thus, the last term will approach zero as $n \to \infty$. We finally obtain

$$f(z) = f(a) + \sum_{m=1}^{\infty} \frac{f^{(m)}(a)}{m!}(z-a)^m \tag{1.268}$$

This is the Taylor series expansion which was derived by Brook Taylor in 1715, which is a special case of Darboux's formula of function expansion. Note that the last integral in (1.267) actually gives the error in the Taylor series expansion if a finite number of terms is used.

Proof 2

We can also prove the Taylor series expansion formula by starting with Cauchy's integral formula. Recall from Section 1.7.4 that any analytic function $f(z)$ can be written as

$$f(z) = \frac{1}{2\pi i} \oint_C \frac{f(\zeta)d\zeta}{\zeta - z} \tag{1.269}$$

where C is integrated in the positive sense. We recognize that (1.269) can first be rearranged as

$$\begin{aligned} f(z) &= \frac{1}{2\pi i} \oint_C \frac{f(\zeta)d\zeta}{\zeta - z} = \frac{1}{2\pi i} \oint_C \frac{f(\zeta)d\zeta}{(\zeta - z_0) - (z - z_0)} \\ &= \frac{1}{2\pi i} \oint_C \frac{f(\zeta)d\zeta}{(\zeta - z_0)[1 - (z - z_0)/(\zeta - z_0)]} = \frac{1}{2\pi i} \oint_C \sum_{n=0}^{\infty} \frac{(z - z_0)^n f(\zeta)d\zeta}{(\zeta - z_0)^{n+1}} \end{aligned} \tag{1.270}$$

In obtaining (1.270), we have applied the following expansion

$$\frac{1}{1-t} = 1 + t + t^2 + \ldots = \sum_{n=0}^{\infty} t^n \tag{1.271}$$

This expansion can be obtained as a special case of the binomial theorem by setting $n = -1$ in (1.4). We now interchange the order of summation and integral to obtain

$$f(z) = \sum_{n=0}^{\infty} \left[\frac{1}{2\pi i} \oint_C \frac{f(\zeta)d\zeta}{(\zeta - z_0)^{n+1}} \right] (z - z_0)^n = \sum_{n=0}^{\infty} \frac{f^{(n)}(z_0)}{n!}(z - z_0)^n \tag{1.272}$$

The last of (1.272) is obtained by employing the general Cauchy's integral formula given in (1.167). This completes another proof of Taylor series expansion formula.

1.13.3 Maclaurin Series Expansion

When we set $a = 0$ in (1.268), we obtain the so-called Maclaurin series expansion:

$$f(z) = f(0) + \sum_{m=1}^{\infty} \frac{f^{(m)}(0)}{m!} z^m \tag{1.273}$$

This formula was actually derived by Stirling in 1717 but was published by Maclaurin in 1742.

1.13.4 Laurent's Series Expansion

One main assumption is that we have tacitly assumed the Taylor or Maclaurin series expansions are analytic, and only the positive power of the expansion exists. A very important theorem was published in 1843 by Laurent who considered that the point of expansion a is not analytic and in such case, by virtue of (1.168) or Cauchy's integral formula, we can expand a function f as

$$f(z) = \sum_{n=0}^{\infty} a_n (z-a)^n + \sum_{n=1}^{\infty} \frac{a_{-n}}{(z-a)^n} \tag{1.274}$$

where

$$a_n = \frac{1}{2\pi i} \oint_{C_1} \frac{f(w)}{(w-a)^{n+1}} dw, \quad a_{-n} = \frac{1}{2\pi i} \oint_{C_2} \frac{f(w)}{(w-a)^{-n+1}} dw \tag{1.275}$$

The contours C_1 and C_2 are given in Figure 1.17. This theorem was actually contained in a paper by Weierstrass in 1841, but it was not published until 1894. The proof of this theorem has already been given in Section 1.7.5. If the principal part (second summation in (1.274)) of the Laurent's series is zero, Taylor series expansion is recovered.

1.13.5 Lagrange's Theorem

In the previous section, we have seen that functions can be expanded in a power series, either by Laurent's series or by Taylor series. A more general theorem was derived by Lagrange in 1770, which deals with the expansion of a function in terms of another function. Let both $f(z)$ and $\phi(z)$ be analytic functions and expansions can be done about a point a as:

$$f(\zeta) = f(a) + \sum_{n=1}^{\infty} \frac{t^n}{n!} \frac{d^{n-1}}{da^{n-1}} [f'(a)\{\phi(a)\}^n] \tag{1.276}$$

where

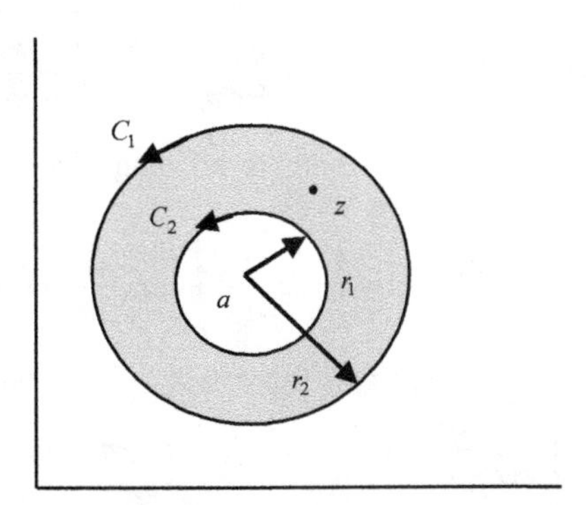

Figure 1.17 The function f is analytic within the annular ring between r_1 and r_2 considered in Laurent's series expansion

$$\zeta = a + t\phi(\zeta) \tag{1.277}$$

A special form of (1.276) is

$$\zeta = a + \sum_{n=1}^{\infty} \frac{t^n}{n!} \frac{d^{n-1}}{da^{n-1}} [\{\phi(a)\}^n] \tag{1.278}$$

The proof of this theorem can be found in Whittaker and Watson (1927) and will not be given here.

1.13.6 Mittag-Leffler Expansion

Some functions can be expanded in rational functions by applying the residue theorem. Such expansion is known as Mittag-Leffler's expansion, which was derived by Swedish mathematician G. Mittag-Leffler in 1880. Although Mittag-Leffler is not particularly famous among applied mathematicians and engineers, it is probably because of Mittag-Leffler that we do not have a Nobel Prize in mathematics today. This story is given in the biography section at the back of this book.

Let us consider the following identity:

$$\frac{1}{z-x} = \frac{1}{z} + \frac{x}{z(z-x)} \tag{1.279}$$

which can be shown in a straightforward manner. Next consider a function $f(z)$ containing singular points at a_1, a_2, ... with corresponding residues b_1, b_2, ... and choose a sequence of circles C_m of radius R_m with center at the origin, such that it does not pass through the poles and $f(z)$ is bounded on the circle c_m. Multiplying (1.279) with $f(z)$ and considering the contour integral along c_m, we obtain

$$\frac{1}{2\pi i}\int_{c_m} \frac{f(z)dz}{z-x} = \frac{1}{2\pi i}\int_{c_m} \frac{f(z)dz}{z} + \frac{x}{2\pi i}\int_{c_m} \frac{f(z)dz}{z(z-x)} \tag{1.280}$$

Applying the residue theorem to the left hand side and to the first term on the right, we obtain

$$f(x) + \sum_{r=1}^{m} \frac{b_r}{a_r - x} = f(0) + \sum_{r=1}^{m} \frac{b_r}{a_r} + \frac{x}{2\pi i}\int_{c_m} \frac{f(z)dz}{z(z-x)} \tag{1.281}$$

in which we have assumed that there is no singularity at $z = 0$ for $f(z)$. Now consider the limit that $m \to \infty$; we have

$$f(x) = f(0) + \sum_{n=1}^{\infty} b_n \{\frac{1}{x-a_n} + \frac{1}{a_n}\} + \lim_{m\to\infty} \frac{x}{2\pi i}\int_{c_m} \frac{f(z)dz}{z(z-x)} \tag{1.282}$$

Noting that $f(z)$ is analytic (i.e., $f(z)$ is finite as $R_m \to \infty$) and x is finite, we have

$$\int_{c_m} \frac{f(z)dz}{z(z-x)} = O(\frac{1}{R_m}) \to 0, \quad as \quad m \to \infty \tag{1.283}$$

where $O(1/R_m)$ implies that (1.283) is of the order of $1/R_m$. Therefore for a function with infinite poles, we have

$$f(x) = f(0) + \sum_{n=1}^{\infty} b_n \{\frac{1}{x-a_n} + \frac{1}{a_n}\} \tag{1.284}$$

This is the so-called Mittag-Leffler expansion in terms of rational functions, which was derived by Mittag-Leffler in 1884.

Example 1.9 Show the validity of the following expansion by using Mittag-Leffler's expansion:

$$\text{cosec} z = \frac{1}{z} + \sum_{n=-\infty}^{\infty} (-1)^n \left\{\frac{1}{z-n\pi} - \frac{1}{n\pi}\right\} \tag{1.285}$$

Solution: Let us consider $f(z)$ as

$$f(z) = \text{cosec} z - \frac{1}{z} \tag{1.286}$$

By noting that cosec z =1/sin z, we have infinite poles at $z = n\pi$ for n = 1,2,3,... For this function we can consider a circular contour on which $|z|=(n+1/2)\pi$. The residue at singularity z = 0 is

$$b_0 = a_{-1} = \lim_{z\to 0}\{zf(z)\} = \lim_{z\to 0}\{z(\frac{1}{\sin z} - \frac{1}{z})\} = \lim_{z\to 0}\{\frac{z-\sin z}{\sin z}\} \tag{1.287}$$

The last limit of (1.287) is the indeterminate form of 0/0, and thus we can apply L'Hôpital's rule to get

$$b_0 = \lim_{z\to 0}\left\{\frac{\frac{d}{dz}(z-\sin z)}{\frac{d}{dz}(\sin z)}\right\} = \lim_{z\to 0}\left\{\frac{1-\cos z}{\cos z}\right\} = 0 \tag{1.288}$$

Again by applying L'Hôpital's rule, the residue at singularities z = $\pm n\pi$ with n = 1,2,3,... can be evaluated as

$$b_n = \lim_{z\to n\pi}\left\{(z-n\pi)\frac{z-\sin z}{z\sin z}\right\} = \lim_{z\to n\pi}\left\{\frac{z-\sin z+(z-n\pi)(1-\cos z)}{\sin z + z\cos z}\right\} \tag{1.289}$$
$$= (-1)^n$$

Substitution of (1.288) and (1.289) into (1.284) and by noting that $f(0) = 0$ gives

$$\text{cosec} z = \frac{1}{z} + \sum_{n=-\infty}^{\infty} (-1)^n \left\{\frac{1}{z-n\pi} - \frac{1}{n\pi}\right\} \tag{1.290}$$

This form can be rewritten in a slightly different form by grouping the $-n$ and $+n$ terms as

$$\frac{1}{z-n\pi} - \frac{1}{n\pi} + \frac{1}{z+n\pi} + \frac{1}{n\pi} = \frac{2z}{(z-n\pi)(z+n\pi)} = \frac{2z}{(z^2-n^2\pi^2)} \tag{1.291}$$

By virtue of (1.291), the first few terms of (1.290) can be written as

$$\text{cosec} z = \frac{1}{z} + 2z\left\{\frac{-1}{z^2-\pi^2} + \frac{1}{z^2-4\pi^2} - \frac{1}{z^2-9\pi^2} + \ldots\right\} \tag{1.292}$$

This formula agrees with that given by Spiegel (1964).

Following a similar procedure, one can show the validity of the following expansions using Mittag-Leffler's expansion (Spiegel, 1964):

$$\sec z = \pi \left\{ \frac{1}{(\pi/2)^2 - z^2} - \frac{3}{(3\pi/2)^2 - z^2} + \frac{5}{(5\pi/2)^2 - z^2} + \dots \right\} \tag{1.293}$$

$$\tan z = 2z \left\{ \frac{1}{(\pi/2)^2 - z^2} + \frac{1}{(3\pi/2)^2 - z^2} + \frac{1}{(5\pi/2)^2 - z^2} + \dots \right\} \tag{1.294}$$

$$\cot z = \frac{1}{z} + 2z \left\{ \frac{1}{z^2 - \pi^2} + \frac{1}{z^2 - 4\pi^2} + \frac{1}{z^2 - 9\pi^2} + \dots \right\} \tag{1.295}$$

$$\operatorname{csch} z = \frac{1}{z} - 2z \left\{ \frac{1}{z^2 + \pi^2} - \frac{1}{z^2 + 4\pi^2} + \frac{1}{z^2 + 9\pi^2} + \dots \right\} \tag{1.296}$$

$$\operatorname{sech} z = \pi \left\{ \frac{1}{(\pi/2)^2 + z^2} - \frac{3}{(3\pi/2)^2 + z^2} + \frac{5}{(5\pi/2)^2 + z^2} - \dots \right\} \tag{1.297}$$

$$\tanh z = 2z \left\{ \frac{1}{(\pi/2)^2 + z^2} + \frac{1}{(3\pi/2)^2 + z^2} + \frac{1}{(5\pi/2)^2 + z^2} + \dots \right\} \tag{1.298}$$

$$\coth z = \frac{1}{z} + 2z \left\{ \frac{1}{z^2 + \pi^2} + \frac{1}{z^2 + 4\pi^2} + \frac{1}{z^2 + 9\pi^2} + \dots \right\} \tag{1.299}$$

$$\frac{\pi^2}{ab} \coth \pi a \coth \pi b = \sum_{m=-\infty}^{\infty} \sum_{n=-\infty}^{\infty} \frac{1}{(m^2 + a^2)(n^2 + b^2)} \tag{1.300}$$

1.13.7 Borel's Theorem

In this section, we introduce Borel's theorem. If a function $\phi(z)$ is expandable in an infinite series as

$$\phi(z) = \sum_{n=0}^{\infty} \frac{a_n z^n}{n!}, \tag{1.301}$$

which is called Borel's function, then the following integral can be evaluated as

$$f(z) = \int_0^{\infty} e^{-t} \phi(zt) dt = \sum_{n=0}^{\infty} a_n z^n \tag{1.302}$$

This is Borel's theorem, which can be proved easily by reversing the order of integration and summation and by observing the definition of the gamma function. In particular, we can substitute (1.301) into (1.302) and reverse the order of integration and summation:

$$f(z)=\int_0^{\infty}e^{-t}\phi(zt)dt=\int_0^{\infty}e^{-t}\sum_{n=0}^{\infty}\frac{a_n(zt)^n}{n!}dt=\sum_{n=0}^{\infty}\frac{a_nz^n}{n!}\int_0^{\infty}e^{-t}t^n dt$$
$$=\sum_{n=0}^{\infty}\frac{a_nz^n}{n!}\Gamma(n+1)=\sum_{n=0}^{\infty}a_nz^n \tag{1.303}$$

where $\Gamma(z)$ is Euler's gamma function and is defined as

$$\Gamma(n+1)=\int_0^{\infty}e^{-t}t^n dt \tag{1.304}$$

Note that for n being an integer, we have $\Gamma(n+1) = n!$. In view of the importance of the gamma function in applied mathematical analysis, the properties of the gamma function will be discussed in more detail in Chapter 4. This completes the proof of Borel's theorem. Comparison of Borel's theorem and Ramanujan's master theorem given in Section 1.9 shows that there is a close resemblance between these two theorems. The idea behind their proofs is exactly the same.

1.14 FUNCTIONS AS INFINITE PRODUCT

In this section, we will employ Mittag-Leffler's expansion formula to expand certain types of functions as infinite products. Let us consider that a function $f(z)$ has simple zeros at points a_1, a_2, ..., and in addition

$$\lim_{n\to\infty}|a_n|\to\infty \tag{1.305}$$

Other than this, $f(z)$ is assumed to be analytic for all values of z. Thus, $f'(z)$ must also be analytic. Consequently, $f'(z)/f(z)$ can have singularities at the points a_1, a_2, ... Use Taylor series expansion for $f(z)$ such that

$$f(z)=(z-a_r)f'(a_r)+\frac{(z-a_r)^2}{2}f''(a_r)+... \tag{1.306}$$

Note that we have used $f(a_r) = 0$ in obtaining (1.306). Then, the differentiation of (1.306) gives

$$f'(z)=f'(a_r)+(z-a_r)f''(a_r)+... \tag{1.307}$$

Then, we have

$$\frac{f'(z)}{f(z)}=\frac{\{f'(a_r)+(z-a_r)f''(a_r)+...\}}{(z-a_r)\{f'(a_r)+\frac{(z-a_r)}{2}f''(a_r)+...\}}. \tag{1.308}$$

The residue for the pole $z = a_r$ can be evaluated as

$$a_{-1}=\lim_{z\to a_r}(z-a_r)\frac{\{f'(a_r)+(z-a_r)f''(a_r)+...\}}{(z-a_r)\{f'(a_r)+\frac{(z-a_r)}{2}f''(a_r)+...\}}=1 \tag{1.309}$$

Let $f(0)\neq 0$; we can apply Mittag-Leffler's expansion to $f'(z)/f(z)$

$$\frac{f'(z)}{f(z)}=\frac{f'(0)}{f(0)}+\sum_{n=1}^{\infty}\{\frac{1}{z-a_n}+\frac{1}{a_n}\} \tag{1.310}$$

We can integrate both sides with respect to z as

$$\int \frac{1}{f(z)}\frac{df(z)}{dz}dz = \ln f(z) = \frac{f'(0)}{f(0)}z + \sum_{n=1}^{\infty}\{\ln(z-a_n)+\frac{z}{a_n}\}+C_1 \tag{1.311}$$

This can be inverted as

$$f(z) = Ce^{\frac{f'(0)}{f(0)}z+\sum_{n=1}^{\infty}\{\ln(z-a_n)+\frac{z}{a_n}\}} = Ce^{\frac{f'(0)}{f(0)}z}\prod_{n=1}^{\infty}(z-a_n)e^{z/a_n} \tag{1.312}$$

where the product function is defined as

$$\prod_{n=1}^{m} a_n = a_1a_2a_3a_4\cdots a_m \tag{1.313}$$

Equation (1.312) can further be simplified by grouping constant terms as

$$f(z) = \left\{-C\prod_{n=1}^{\infty}a_n\right\}e^{\frac{f'(0)}{f(0)}z}\prod_{n=1}^{\infty}(1-\frac{z}{a_n})e^{z/a_n} = \bar{C}e^{\frac{f'(0)}{f(0)}z}\prod_{n=1}^{\infty}\{(1-\frac{z}{a_n})e^{z/a_n}\} \tag{1.314}$$

Note that $f(0) = \bar{C}$, and thus (1.314) can be

$$f(z) = f(0)e^{\frac{f'(0)}{f(0)}z}\prod_{n=1}^{\infty}\{(1-\frac{z}{a_n})e^{z/a_n}\} \tag{1.315}$$

This formula was reported in Section 7.5 of Whittaker and Watson (1927) and is a direct consequence of Mittag-Leffler's expansion formula.

Example 1.10 Show the validity of the following expansion:

$$\frac{\sin z}{z} = \prod_{n=1}^{\infty}\left\{(1-\frac{z}{n\pi})e^{\frac{z}{n\pi}}\right\}\left\{(1+\frac{z}{n\pi})e^{-\frac{z}{n\pi}}\right\} \tag{1.316}$$

Solution: Let us consider $f(z) = \sin z/z$, and the zeros of $f(z)$ are $z = \pm n\pi$. Thus, the limits are

$$\lim_{z\to 0} f(z) = \lim_{z\to 0}\frac{\sin z}{z} = \lim_{z\to 0}\frac{\cos z}{1} = 1 \tag{1.317}$$

$$\lim_{z\to 0} f'(z) = \lim_{z\to 0}(-\frac{\sin z}{z^2}+\frac{\cos z}{z}) = \lim_{z\to 0}(\frac{-\sin z + z\cos z}{z^2}) = \lim_{z\to 0}\left\{-\frac{\sin z}{2}\right\} = 0 \tag{1.318}$$

Grouping terms for $z = -n\pi$ and $z = +n\pi$, we obtain from (1.315)

$$\begin{aligned}\frac{\sin z}{z} &= \prod_{n=1}^{\infty}\left\{(1-\frac{z}{n\pi})e^{\frac{z}{n\pi}}\right\}\left\{(1+\frac{z}{n\pi})e^{-\frac{z}{n\pi}}\right\} \\ &= (1-\frac{z^2}{\pi^2})(1-\frac{z^2}{4\pi^2})\cdots\end{aligned} \tag{1.319}$$

This completes the proof.

1.15 VECTOR CALCULUS

The vector calculus considered here can be considered as a special case of tensor calculus, which was developed by Ricci and Levi-Civita and motivated by a paper by Riemann in 1854 on Riemann geometry. Subsequent development on tensor calculus was made by Weyl, Eddington, and Schouten. Other main contributors of vector analysis include Gibbs, Heaviside, Föppl, Wilson, and many others (e.g., Crowe, 1993).

Let us consider vector calculus using the example of a moving particle in space. The position vector of the particle is a function of time t and is represented by

$$\boldsymbol{r}(t) = x_1(t)\boldsymbol{e}_1 + x_2(t)\boldsymbol{e}_2 + x_3(t)\boldsymbol{e}_3 \tag{1.320}$$

The time derivative of this position vector is the velocity vector $\boldsymbol{v}$

$$\boldsymbol{v} = \frac{d\boldsymbol{r}}{dt} = \frac{dx_1}{dt}\boldsymbol{e}_1 + \frac{dx_2}{dt}\boldsymbol{e}_2 + \frac{dx_3}{dt}\boldsymbol{e}_3 \tag{1.321}$$

If a particle moves along a curve C and the arc length along curve C is s, that is, $\boldsymbol{r} = \boldsymbol{r}$ (s), we can use chain rule to rewrite the velocity vector as

$$\boldsymbol{v} = \frac{d\boldsymbol{r}}{dt} = \frac{d\boldsymbol{r}}{ds}\frac{ds}{dt} = \boldsymbol{r}'\frac{ds}{dt} \tag{1.322}$$

where $\boldsymbol{r}'$ is the unit vector along the curve C of the moving particle and $ds/dt = v$ is the speed of the particle along the curve. Let us call the unit vector along $\boldsymbol{v}$ as $\boldsymbol{T}$. The acceleration of the particle is

$$\boldsymbol{a} = \frac{d\boldsymbol{v}}{dt} = \frac{d}{dt}(\boldsymbol{T}\frac{ds}{dt}) = \frac{d\boldsymbol{T}}{ds}(\frac{ds}{dt})^2 + \boldsymbol{T}\frac{d^2 s}{dt^2} = \frac{\boldsymbol{N}}{\rho}(\frac{ds}{dt})^2 + \boldsymbol{T}\frac{d^2 s}{dt^2} \tag{1.323}$$

The direction of vector $\boldsymbol{T}$ is parallel to velocity $\boldsymbol{v}$ as suggested by (1.323) and $\boldsymbol{N}$ is a unit vector perpendicular to $\boldsymbol{T}$ or $\boldsymbol{v}$ as shown in Figure 1.18. The last of (1.323) can be proved as

$$\left|\frac{d\boldsymbol{T}}{ds}\right| = \lim_{\Delta s \to 0}\left|\frac{\Delta \boldsymbol{T}}{\Delta s}\right| = \lim_{\Delta s \to 0}\left|\frac{\Delta \theta}{\Delta s}\right| = \frac{1}{\rho} \tag{1.324}$$

where ρ is the radius of curvature. The change of $\boldsymbol{T}$ can only be in the direction of the magnitude of a unit vector fixed as unity. Thus, $|\Delta \boldsymbol{T}| = |\Delta\theta|$ is the radius of the circle shown in Figure 1.18 is fixed at 1. Equation (1.323) can be rewritten as

$$\boldsymbol{a} = \frac{dv}{dt}\boldsymbol{T} + \frac{v^2}{\rho}\boldsymbol{N} \tag{1.325}$$

Therefore, there are two components of acceleration, one along the tangential direction and the other along the direction perpendicular to tangential direction. The second term is clearly the centrifugal force of circular motion (French, 1971).

1.15.1 Gradient

More generally, let us consider a scalar function $\phi = \phi(x_1, x_2, x_3)$, which is a function of position, such that the total differential of it is

$$d\phi = \frac{\partial \phi}{\partial x_1} dx_1 + \frac{\partial \phi}{\partial x_2} dx_2 + \frac{\partial \phi}{\partial x_3} dx_3 \tag{1.326}$$

This is a scalar, but the structural form of (1.326) suggests that it can be viewed as the dot product of two vectors as:

$$d\phi = \nabla\phi \cdot d\boldsymbol{r} = (\frac{\partial \phi}{\partial x_1} \boldsymbol{e}_1 + \frac{\partial \phi}{\partial x_2} \boldsymbol{e}_2 + \frac{\partial \phi}{\partial x_3} \boldsymbol{e}_3) \cdot (dx_1 \boldsymbol{e}_1 + dx_2 \boldsymbol{e}_2 + dx_3 \boldsymbol{e}_3) \tag{1.327}$$

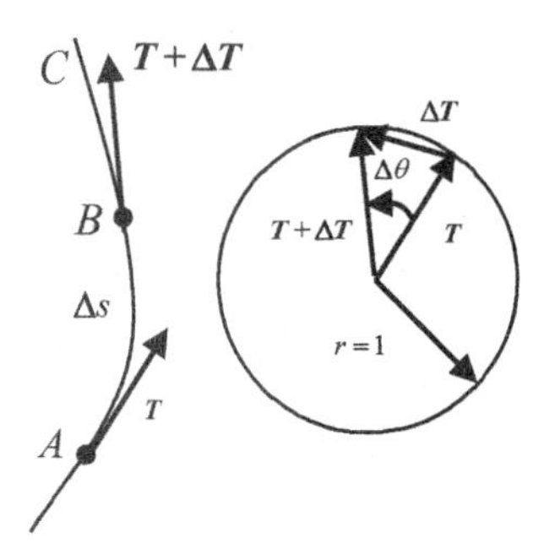

Figure 1.18 The change in the direction of the unit vector along the moving path *C*

where $\boldsymbol{e}_i$ (i = 1,2,3) are the base vectors of a particular coordinate system for which we want to consider the changes of the function ϕ with respect to the variables x_i. In (1.327), we have used the vector differential operator that is expressed as ∇. It is commonly called "del" or "nabla" and in Cartesian coordinates it can be written as

$$\text{grad}\phi = \nabla\phi = \boldsymbol{e}_1 \frac{\partial \phi}{\partial x_1} + \boldsymbol{e}_2 \frac{\partial \phi}{\partial x_2} + \boldsymbol{e}_3 \frac{\partial \phi}{\partial x_3} \tag{1.328}$$

In physics, it is sometimes referred to as the Hamilton operator as this symbol was introduced by Sir Hamilton (we will talk about his Hamilton's Principle in Section 14.5). His brief biography is given at the back of this book. The term$\nabla\phi$ is also termed the gradient of a scalar function ϕ. Physically, the gradient of a scalar can be interpreted as the normal to the surface defined by $\phi(x_1, x_2, x_3) = C_1$, as shown in Figure 1.19. It is also the direction of the greatest rate of change of the function ϕ. By considering the change of ϕ along the path C, we can rewrite (1.327) as

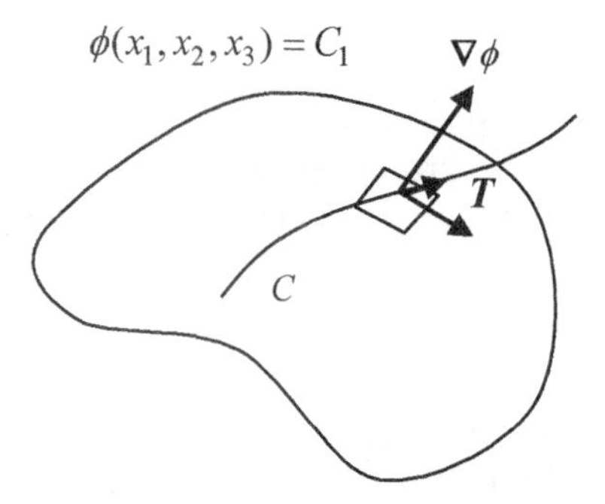

Figure 1.19 Physical meaning of the gradient of a scalar function ϕ

$$\frac{d\phi}{ds} = \nabla\phi \cdot \frac{d\boldsymbol{r}}{ds} = \nabla\phi \cdot \boldsymbol{T} \tag{1.329}$$

Thus, $d\phi/ds$ is the value of $\nabla\phi$ projected along the direction of $\boldsymbol{T}$. If the direction of $\boldsymbol{T}$ is along a direction perpendicular to the normal of the surface $\phi(x_1, x_2, x_3) = C_1$, we have $d\phi/ds = 0$.

1.15.2 Divergence

Since ∇ is a vector operator, its operation on a vector can be expressed through either a dot product or a cross product. Indeed, the dot product of ∇ on a vector function $\boldsymbol{v}$ is called divergence (a term introduced by Heaviside) or div of $\boldsymbol{v}$ as:

$$\text{div}\ \boldsymbol{v} = \nabla \bullet \boldsymbol{v} = \frac{\partial v_1}{\partial x_1} + \frac{\partial v_2}{\partial x_2} + \frac{\partial v_3}{\partial x_3} \tag{1.330}$$

The physical meaning of divergence is best understood in terms of fluid flow. For example, a control volume shown in Figure 1.20 for a flowing fluid can be used to derive the condition of continuity. The outflow subtracts the inflow along the x_2 direction and gives:

$$[\rho v_2 + \frac{\partial(\rho v_2)}{\partial x_2} dx_2] dx_1 dx_3 - \rho v_2 dx_1 dx_3 = \frac{\partial(\rho v_2)}{\partial x_2} dx_1 dx_2 dx_3 \tag{1.331}$$

Similarly, the net outflow along the x_1 and x_3 directions are

$$\frac{\partial(\rho v_1)}{\partial x_1} dx_1 dx_2 dx_3, \quad \frac{\partial(\rho v_3)}{\partial x_3} dx_1 dx_2 dx_3 \tag{1.332}$$

The net outflow per unit volume from all three independent directions is

$$\frac{\partial(\rho v_1)}{\partial x_1} + \frac{\partial(\rho v_2)}{\partial x_2} + \frac{\partial(\rho v_3)}{\partial x_3} = \nabla \bullet (\rho \boldsymbol{v}) = \frac{\partial \rho}{\partial t} \tag{1.333}$$

Physically, the left hand side is the amount of "diverging" fluid from the control volume and the left hand side is the change of density to accommodate such changes. By now, the physical meaning of so-called divergence is clear. If density of the fluid does not change, we have zero divergence or

$$\nabla \bullet \boldsymbol{v} = 0 \tag{1.334}$$

Similarly, we can also consider the divergence of electric or magnetic flux through a control volume.

1.15.3 Curl

The cross product between ∇ and a vector function $\boldsymbol{v}$ can be expressed as the so-called curl or rot as

$$\text{curl } \boldsymbol{v} = \nabla \times \boldsymbol{v} = \begin{vmatrix} \boldsymbol{e}_1 & \boldsymbol{e}_2 & \boldsymbol{e}_3 \\ \dfrac{\partial}{\partial x_1} & \dfrac{\partial}{\partial x_2} & \dfrac{\partial}{\partial x_3} \\ v_1 & v_2 & v_3 \end{vmatrix} = (\frac{\partial v_3}{\partial x_2} - \frac{\partial v_2}{\partial x_3})\boldsymbol{e}_1 + (\frac{\partial v_1}{\partial x_3} - \frac{\partial v_3}{\partial x_1})\boldsymbol{e}_2 + (\frac{\partial v_2}{\partial x_1} - \frac{\partial v_1}{\partial x_2})\boldsymbol{e}_3 \tag{1.335}$$

Physically, the curl of a rotating fluid is related to the circulation or the strength of a vortex. If the curl of $\boldsymbol{v}$ is zero, the flow is irrotational (i.e., no vortex can be formed).

1.15.4 Physical Meaning of Gradient, Divergence and Curl

Alternatively, we can also consider the integral form of the definitions of gradient, divergence, and curl as:

$$\nabla \phi = \lim_{\Delta V \to 0} \frac{1}{\Delta V} \oint_{\partial V} \phi d\boldsymbol{S} \tag{1.336}$$

$$\nabla \bullet \boldsymbol{v} = \lim_{\Delta V \to 0} \frac{1}{\Delta V} \oint_{\partial V} \boldsymbol{v} \bullet d\boldsymbol{S} \tag{1.337}$$

$$[\nabla \times \boldsymbol{v}(\boldsymbol{r})] \bullet \boldsymbol{n} = \lim_{\Delta S \to 0} \frac{1}{\Delta S} \oint \boldsymbol{v} \bullet d\boldsymbol{r} \tag{1.338}$$

The surface $d\boldsymbol{S}$ is the surface pointing outwardly from the volume ΔV. The point of evaluation for $\nabla\phi$ and $\nabla \cdot \boldsymbol{v}$ is within the interior of the small volume ΔV. The integration in (1.336) to (1.338) is carried out over a closed surface or around a closed path as indicated by a small circle through the integral sign. The integral definition of divergence given in (1.337) provides another physical meaning that divergence is a measure of the strength of sources or sinks within the element ΔV.

The integral form of the divergence can be interpreted from Figure 1.20 again by defining the flow (or flux) through a control volume ΔV

$$\Phi = \int_S (\rho \boldsymbol{v}) \bullet d\boldsymbol{S} \tag{1.339}$$

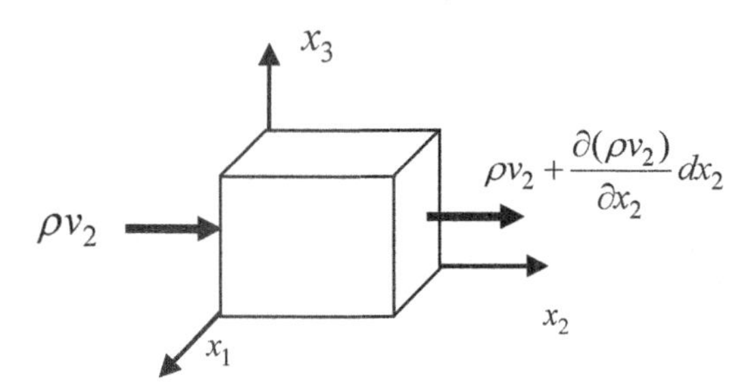

Figure 1.20 Interpretation of divergence through the fluid flow problem

However, by (1.333) the total flow per unit volume through the element is

$$\frac{\partial(\rho v_1)}{\partial x_1}+\frac{\partial(\rho v_2)}{\partial x_2}+\frac{\partial(\rho v_3)}{\partial x_3}=\nabla \bullet (\rho \boldsymbol{v}) \tag{1.340}$$

Balancing the flow per unit volume, we have

$$\lim_{\Delta V \to 0} \frac{1}{\Delta V} \oint_{\partial V} (\rho \boldsymbol{v}) \bullet d\boldsymbol{S} = \nabla \bullet (\rho \boldsymbol{v}) \tag{1.341}$$

For constant density, (1.341) is the same as (1.334).

The physical meaning of (1.338) can be interpreted through Figure 1.21. Let us consider a rotating fluid element *ABCD* as shown in Figure 1.21. The tangential velocities on the sides of the element are shown. If the fluid element *ABCD* is rotating, a resultant peripheral velocity exists around the element counter-clockwise. However, the center of rotation is not known; it has been proposed that rotation be expressed in terms of the sum of the products of circulating velocity and distance around the contour. This sum is called "circulation" Γ and is defined:

$$\Gamma = \oint \boldsymbol{v} \bullet d\boldsymbol{r} \tag{1.342}$$

Taking the integration counter-clockwise as positive, we can evaluate the closed contour integral as the summation

$$\begin{aligned}\Gamma_{ABCD} &= v_1 dx_1 + (v_2 + \frac{\partial v_2}{\partial x_1} dx_1) dx_2 - (v_1 + \frac{\partial v_1}{\partial x_2} dx_2) dx_1 - v_2 dx_2 \\ &= (\frac{\partial v_2}{\partial x_1} - \frac{\partial v_1}{\partial x_2}) dx_1 dx_2 = (\nabla \times \boldsymbol{v})_3 \cdot (d\boldsymbol{S})_3\end{aligned} \tag{1.343}$$

This expression is true for the configuration shown in Figure 1.21 (i.e., for a constant value of x_3), but it can be readily extended to the cases of more general non-planar and non-rectangular contour. The circulation can be generalized to

$$\Gamma = (\nabla \times \boldsymbol{v}) \cdot (d\boldsymbol{S}) = (\nabla \times \boldsymbol{v}) \cdot \boldsymbol{n} dS \doteq (\nabla \times \boldsymbol{v}) \cdot \boldsymbol{n} \Delta S \tag{1.344}$$

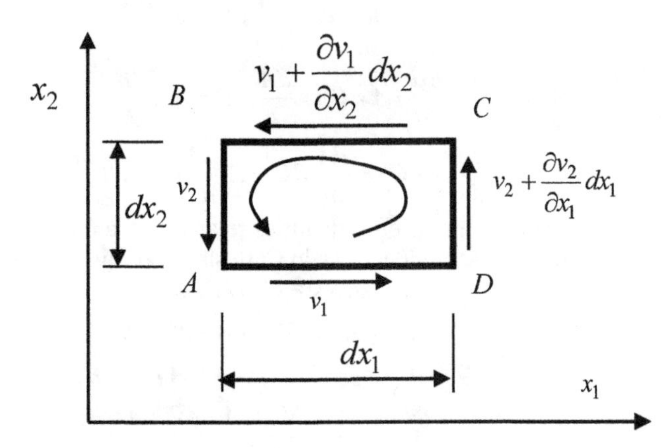

Figure 1.21 The circulation for a rotating fluid ABCD

Note that since Γ is a scalar and thus independent of the actual directions of the velocity and normal vector, therefore it is also independent of the choice of the coordinate system. Equating (1.344) and (1.342), we obtain

$$(\nabla \times \boldsymbol{v}) \cdot \boldsymbol{n} = \lim_{\Delta S \to 0} \frac{1}{\Delta S} \oint \boldsymbol{v} \bullet d\boldsymbol{r} \tag{1.345}$$

This completes the proof for (1.338).

1.15.5 Vector Identities

A number of identities of vector calculus will be found useful and are reported here.

The two most important formulas in vector calculus are probably

$$\nabla \times (\nabla \phi) = 0 \tag{1.346}$$

$$\nabla \bullet (\nabla \times \boldsymbol{A}) = 0 \tag{1.347}$$

Because of these identities, we can see later that any vector can be represented by an irrotational part and a solenoidal part. This is Helmholtz's representation theorem or Helmholtz's decomposition theorem for vectors as discussed by Chau (2013) and will be presented in Section 1.16. If the curl of a vector vanishes or mathematically

$$\nabla \times \boldsymbol{A} = \boldsymbol{0}, \tag{1.348}$$

it is called irrotational and it has an important consequence on fluid mechanics. There will be no effect of viscosity. If the divergence of a vector vanishes or mathematically

$$\nabla \bullet \boldsymbol{A} = 0, \tag{1.349}$$

it is called solenoidal (a term that came from electromagnetism on wire coil) or divergence-free. More discussions on this will be given in the next section.

Some basic formulas of the time differentiation of vectors are

$$\frac{d}{dt}(\boldsymbol{A} \bullet \boldsymbol{B}) = \frac{d\boldsymbol{A}}{dt} \bullet \boldsymbol{B} + \boldsymbol{A} \bullet \frac{d\boldsymbol{B}}{dt} \tag{1.350}$$

$$\frac{d}{dt}(\boldsymbol{A} \times \boldsymbol{B}) = \frac{d\boldsymbol{A}}{dt} \times \boldsymbol{B} + \boldsymbol{A} \times \frac{d\boldsymbol{B}}{dt} \tag{1.351}$$

$$\frac{d}{dt}(\boldsymbol{ABC}) = \frac{d\boldsymbol{A}}{dt} \boldsymbol{BC} + \boldsymbol{A} \frac{d\boldsymbol{B}}{dt} \boldsymbol{C} + \boldsymbol{AB} \frac{d\boldsymbol{C}}{dt} \tag{1.352}$$

There are many important and useful identities related to the properties of gradient, curl, and divergence, and some identities in Cartesian coordinates are

$$\nabla \bullet (\nabla \phi \times \nabla \psi) = 0 \tag{1.353}$$

$$\nabla \bullet (\phi \boldsymbol{A}) = (\nabla \phi) \bullet \boldsymbol{A} + \phi \nabla \bullet \boldsymbol{A} \tag{1.354}$$

$$\nabla \bullet (\boldsymbol{A} \times \boldsymbol{B}) = (\nabla \times \boldsymbol{A}) \bullet \boldsymbol{B} - \boldsymbol{A} \bullet (\nabla \times \boldsymbol{B}) \tag{1.355}$$

$$\nabla \times (\phi \boldsymbol{A}) = \nabla \phi \times \boldsymbol{A} + \phi \nabla \times \boldsymbol{A} \tag{1.356}$$

$$\nabla \bullet \nabla \phi = \nabla^2 \phi \tag{1.357}$$

$$\nabla^2 (\nabla \phi) = \nabla (\nabla^2 \phi) \tag{1.358}$$

$$\nabla \bullet (\nabla^2 \boldsymbol{A}) = \nabla^2 (\nabla \bullet \boldsymbol{A}) \tag{1.359}$$

$$\nabla \times (\nabla^2 \boldsymbol{A}) = \nabla^2 (\nabla \times \boldsymbol{A}) \tag{1.360}$$

$$\nabla(\boldsymbol{A} \cdot \boldsymbol{B}) = \boldsymbol{A} \cdot \nabla \boldsymbol{B} + \boldsymbol{B} \cdot \nabla \boldsymbol{A} + \boldsymbol{A} \times (\nabla \times \boldsymbol{B}) + \boldsymbol{B} \times (\nabla \times \boldsymbol{A}) \tag{1.361}$$

$$\nabla \times (\nabla \times \boldsymbol{A}) = \nabla(\nabla \bullet \boldsymbol{A}) - \nabla^2 \boldsymbol{A} \tag{1.362}$$

$$\nabla(\phi\psi) = \phi \nabla \psi + \psi \nabla \phi \tag{1.363}$$

$$\nabla(\phi / \psi) = (\psi \nabla \phi - \phi \nabla \psi) / \psi^2 \tag{1.364}$$

$$\nabla(\phi^n) = n\phi^{n-1} \nabla \phi \tag{1.365}$$

$$\nabla^2 (\phi\psi) = \phi \nabla^2 \psi + 2(\nabla \phi) \bullet (\nabla \psi) + \psi \nabla^2 \phi \tag{1.366}$$

$$\nabla \bullet (\phi \nabla \psi) = \phi \nabla^2 \psi + \nabla \phi \bullet \nabla \psi \tag{1.367}$$

$$\nabla^2 (\boldsymbol{A} \bullet \boldsymbol{r}) = 2\nabla \bullet \boldsymbol{A} + \boldsymbol{r} \bullet \nabla^2 \boldsymbol{A} \tag{1.368}$$

$$\nabla^2 (\phi \boldsymbol{r}) = 2\nabla \phi + \boldsymbol{r} \nabla^2 \phi \tag{1.369}$$

$$\nabla(\boldsymbol{A} \bullet \boldsymbol{r}) = \boldsymbol{A} + (\nabla \boldsymbol{A}) \bullet \boldsymbol{r} \tag{1.370}$$

where ϕ, ψ are scalar functions of position $\boldsymbol{x}$, $\boldsymbol{A}$, $\boldsymbol{B}$ are vector functions of position $\boldsymbol{x}$, and $\boldsymbol{r}$ is the position vector. There are also a number of important integral identities in vector calculus and some of them are

$$\oint d\boldsymbol{S} = \boldsymbol{0} \tag{1.371}$$

$$\int_V \nabla \phi dV = \oint_{\partial V} \phi d\boldsymbol{S} \tag{1.372}$$

$$\int_V (\phi \nabla^2 \psi + \nabla \phi \cdot \nabla \psi) dV = \oint_{\partial V} \phi \nabla \psi \cdot d\boldsymbol{S} \tag{1.373}$$

$$\int_V (\phi \nabla^2 \psi - \psi \nabla^2 \phi) dV = \oint_{\partial V} (\phi \nabla \psi - \psi \nabla \phi) \cdot d\boldsymbol{S} \tag{1.374}$$

$$\int_V \nabla \cdot \boldsymbol{A} dV = \int_{\partial V} \boldsymbol{A} \cdot d\boldsymbol{S} \tag{1.375}$$

$$\int_S (\nabla \times \boldsymbol{A}) \cdot d\boldsymbol{S} = \int_C \boldsymbol{A} \cdot d\boldsymbol{r} \tag{1.376}$$

Green's first and second identities are given by (1.373) and (1.374), which will be discussed again in Chapter 8, whereas Gauss's divergence theorem and the Kelvin-Stoke's theorem are given in (1.375) and (1.376). They will be discussed in Sections 1.17 and 1.18.

1.16 HELMHOLTZ REPRESENTATION THEOREM

The Helmholtz representation theorem for any vector is a natural consequence of (1.346) and (1.347). Mathematically, it can be stated as (Chau, 2013)

$$\boldsymbol{v} = \nabla \phi + \nabla \times \boldsymbol{A} \tag{1.377}$$

The first term on the right hand side of (1.377) is called the irrotational field of a vector, whereas the second of the right hand side of (1.377) is called the solenoidal field of a vector. The function ϕ is called the scalar potential of a vector, whereas the vector function $\boldsymbol{A}$ is called the vector potential of a vector. In fluid mechanics, $\nabla\times\boldsymbol{A}$ is called the vorticity of $\boldsymbol{A}$, whereas $\nabla\phi$ is called the gradient field describing the irrotational fluid. Whenever a vector can be represented by the first term of (1.377) only, the field is called conservative (such as a gravitational or electrostatic field).

By identity (1.346), the first term is evidently irrotational, and by identity (1.347) the second term is evidently solenoidal because

$$\nabla\times(\nabla\phi)=0,\quad \nabla\cdot(\nabla\times\boldsymbol{A})=0 \tag{1.378}$$

Rearranging (1.377), we have

$$\boldsymbol{v}-\nabla\phi=\nabla\times\boldsymbol{A} \tag{1.379}$$

Taking the divergence of (1.379) gives

$$\nabla\cdot(\boldsymbol{v}-\nabla\phi)=\nabla\cdot(\nabla\times\boldsymbol{A})=0 \tag{1.380}$$

Thus, we can express the divergence of $\boldsymbol{v}$ as

$$\nabla\cdot\boldsymbol{v}=\nabla^2\phi \tag{1.381}$$

Similarly, rearranging (1.377), we have

$$\boldsymbol{v}-\nabla\times\boldsymbol{A}=\nabla\phi \tag{1.382}$$

Taking the curl of (1.382) gives

$$\nabla\times(\boldsymbol{v}-\nabla\times\boldsymbol{A})=\nabla\times(\nabla\phi)=0 \tag{1.383}$$

Consequently, the following identity is obtained for the curl of vector $\boldsymbol{v}$

$$\nabla\times\boldsymbol{v}=\nabla\times\nabla\times\boldsymbol{A}=\nabla(\nabla\cdot\boldsymbol{A})-\nabla^2\boldsymbol{A} \tag{1.384}$$

The last of (1.384) is a result of the application of identity (1.362). As discussed by Chau (2013), we have the freedom of setting one constraint for the four unknown potentials (one scalar potential and three components of a vector potential). When $\nabla\times\boldsymbol{A}$ decreases with distance r more rapidly than $1/r$, the term $\nabla(\nabla\cdot\boldsymbol{A})$ vanishes. Thus, without loss of generality we have

$$\nabla\times\boldsymbol{v}=-\nabla^2\boldsymbol{A} \tag{1.385}$$

In conclusion, the divergence of $\boldsymbol{v}$ is expressible in a scalar potential and the curl is expressible in a vector potential (see (1.377) and (1.378)).

1.17 GAUSS DIVERGENCE THEOREM

The Gauss divergence theorem can be visualized by using the integral form of divergence given in (1.337). In particular, consider a finite volume V in Figure 1.22 being subdivided into a number of smaller volumes, and its finite closed surface S is subdivided into a number of smaller closed surfaces. By applying (1.337) to each of the small closed surfaces we have

$$\int_V \nabla\bullet(\rho\boldsymbol{v})dV=\sum_i\Big(\lim_{\Delta V_i\to 0}\frac{1}{\Delta V_i}\oint_{\partial V_i}(\rho\boldsymbol{v})\bullet d\boldsymbol{S}\Big)=\oint_{\partial V}(\rho\boldsymbol{v})\bullet\boldsymbol{n}dS \tag{1.386}$$

This is the so-called Gauss divergence theorem and it transforms the volume integral to the surface integral or vice versa. This equation also applies to the case where $\boldsymbol{v}$ is an n-th order tensor (e.g., see Segel, 1987). The mathematical proof for the divergence theorem can be found in standard textbooks for engineering mathematics (Kreyszig, 1979), which somewhat differs from the current approach using a physical argument.

In electrostatics, the electric field intensity $\boldsymbol{E}$ is related to the charge density ρ as

$$\nabla \bullet \boldsymbol{E} = \rho(\boldsymbol{r}) / \varepsilon_0 \tag{1.387}$$

where ε_0 is the electric permittivity of free space. Applying the Gauss divergence theorem, we have

$$\oint_{\partial V} \boldsymbol{E} \bullet \boldsymbol{n} dS = \frac{1}{\varepsilon_0} \int_V \rho(\boldsymbol{r}) dV = \frac{Q}{\varepsilon_0} \tag{1.388}$$

This is the Gauss law of electrostatics and it states that the total flux of electric field intensity coming out across a closed surface is proportional to the total charge Q enclosed by the closed surface S.

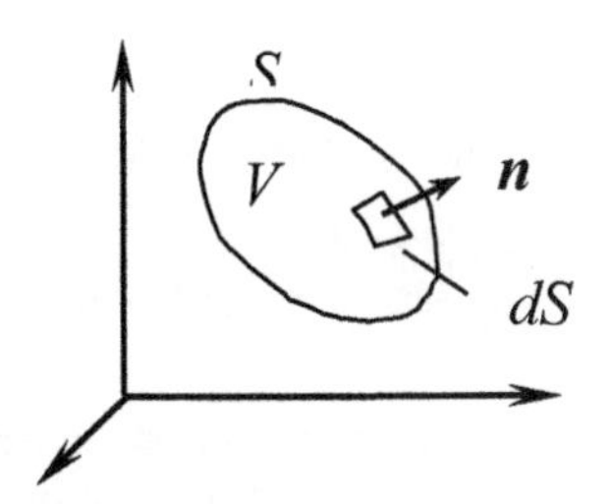

Figure 1.22 The domain for the divergence theorem

1.18 KELVIN-STOKES THEOREM

Let S be a smooth surface bounded by a simple closed curve C, which does not intersect itself as shown in Figure 1.23. Similar to the argument used in deriving the Gauss divergence theorem, we can subdivide a closed loop C into a number of smaller closed loops C_i and the surface S into a number of small surfaces S_i. By applying the integral form of the curl given in (1.338) we have

$$\int_S [\nabla \times \boldsymbol{v}(\boldsymbol{r})] \bullet \boldsymbol{n} dS = \sum_i \left(\lim_{\Delta S_i \to 0} \frac{1}{\Delta S_i} \oint_{C_i} \boldsymbol{v} \bullet d\boldsymbol{r} \right) = \oint_C \boldsymbol{v} \bullet d\boldsymbol{r} \tag{1.389}$$

This is the so-called Kelvin-Stokes theorem.

In electromagnetism, Ampere's law relates the magnetic induction $\boldsymbol{B}$ to the current density $\boldsymbol{J}$ by

$$\nabla \times \boldsymbol{B} = \mu_0 \boldsymbol{J} \tag{1.390}$$

where μ_0 is the permeability of the free space. Applying the Kelvin-Stokes' theorem, we have

$$\oint_C \boldsymbol{B} \bullet d\boldsymbol{r} = \mu_0 \oint_C \boldsymbol{J} \boldsymbol{I} \bullet d\boldsymbol{r} = \mu_0 \boldsymbol{I} \tag{1.391}$$

This states that magnetic induction is proportional to the total current passing through the surface enclosed by C.

As discussed by Chau (2013), Lord Kelvin played a fundamental role in its development, so it is also known as the Kelvin–Stokes theorem (see biography of G.G. Stokes).

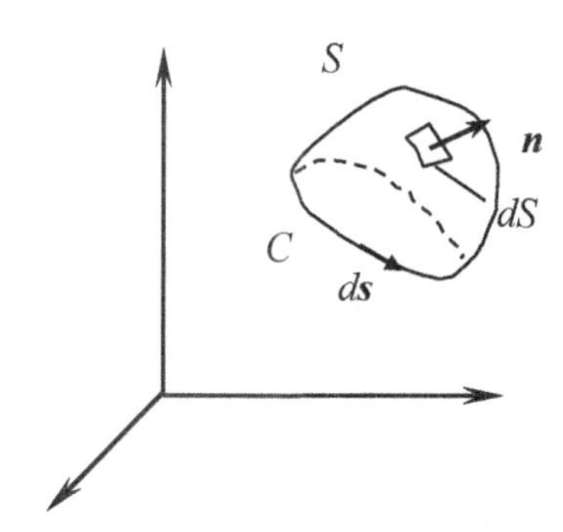

Figure 1.23 The domain for the Stokes theorem

1.19 VECTORS AND TENSORS

Certain quantities, like temperature and pressure, are independent of direction and are known as scalars. Displacement, velocity, acceleration, and force are quantities that depend on direction, and they are called vectors. The term vector was coined by Hamilton in the nineteenth century. In the three-dimensional domain, a vector has three physical components, as shown in Figure 1.24. For example, a vector $\boldsymbol{v}$ can be expressed as a sum of three quantities and each one is the projected value along certain directions and these directions are indicated by their base vectors $\boldsymbol{e}_i$ (i = 1,2,3). In particular, $\boldsymbol{v}$ can be expressed as

$$\boldsymbol{v} = v_1\boldsymbol{e}_1 + v_2\boldsymbol{e}_2 + v_3\boldsymbol{e}_3 = \sum_{i=1}^{3} v_i\boldsymbol{e}_i = v_i\boldsymbol{e}_i \tag{1.392}$$

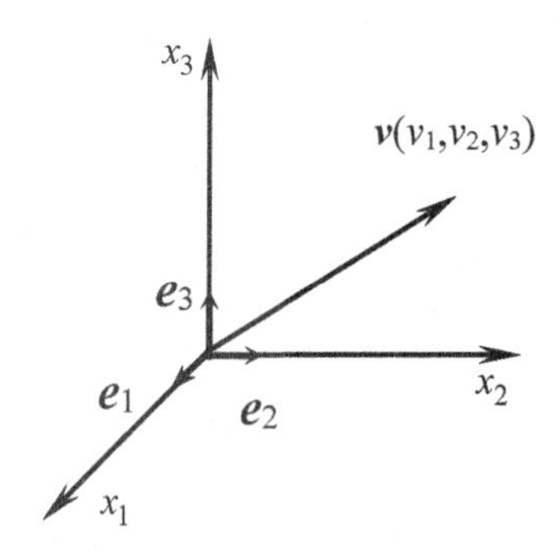

Figure 1.24 A vector in Cartesian coordinates

where v_i is also called the indicial notation or index notation of a vector. The physical component of the vector is denoted as v_i; physically, it is the length of the vector $\boldsymbol{v}$ projected along the i-th coordinate of the system. In the last part of (1.392), the summation sign Σ is usually neglected; thus, repeated indices imply summation automatically. This is usually referred to as the Einstein convention for summation. The index i becomes a dummy index (i.e., it can be replaced arbitrarily by j, k, etc.), and it is no longer a free index (i.e., i cannot be set to 1, 2, or 3 arbitrarily). Instead of using physical components, vectors can be denoted by the *symbolic or Gibbs* notation (i.e., $\boldsymbol{v}$).

Certain quantities like stress and strain depend on two directions; first the direction of the surface where stress and strain are measured, and second the direction of the stress and strain acting on this surface. For example, stress can be represented by

$$\boldsymbol{\sigma} = \sum_{i=1}^{3}\sum_{j=1}^{3} \sigma_{ij}\boldsymbol{e}_i\boldsymbol{e}_j = \sigma_{ij}\boldsymbol{e}_i\boldsymbol{e}_j \tag{1.393}$$

where the indicial form is written in terms of Cartesian coordinates. Again, the Einstein convention is used in the last of (1.393). Or more explicitly, we have

$$\begin{aligned} \boldsymbol{\sigma} = &\sigma_{11}\boldsymbol{e}_1\boldsymbol{e}_1 + \sigma_{22}\boldsymbol{e}_2\boldsymbol{e}_2 + \sigma_{33}\boldsymbol{e}_3\boldsymbol{e}_3 + \sigma_{13}\boldsymbol{e}_1\boldsymbol{e}_3 + \sigma_{31}\boldsymbol{e}_3\boldsymbol{e}_1 + \sigma_{12}\boldsymbol{e}_1\boldsymbol{e}_2 \\ &+ \sigma_{21}\boldsymbol{e}_2\boldsymbol{e}_1 + \sigma_{23}\boldsymbol{e}_2\boldsymbol{e}_3 + \sigma_{32}\boldsymbol{e}_3\boldsymbol{e}_2 \end{aligned} \tag{1.394}$$

The symbolic form $\boldsymbol{\sigma}$ is most general and is independent of any coordinate system, whereas the indicial form σ_{ij} is the physical component corresponding to a particular Cartesian coordinate system. Therefore, second order tensors can be written in terms of two vectors side-by-side called *dyads,* such as $\boldsymbol{e}_1\boldsymbol{e}_3$. Note that dyads are in general not commutative, i.e., $\boldsymbol{e}_1\boldsymbol{e}_3 \neq \boldsymbol{e}_3\boldsymbol{e}_1$. This dyadic form was proposed by Gibbs in the 1880s. Gibbs was the first USA PhD graduate in engineering in 1863 from Yale. There are also tensors of order higher than second. For example, a typical fourth order tensor is the stiffness tensor

$$\boldsymbol{C} = \sum_{i=1}^{3}\sum_{j=1}^{3}\sum_{k=1}^{3}\sum_{l=1}^{3} C_{ijkl}\boldsymbol{e}_i\boldsymbol{e}_j\boldsymbol{e}_k\boldsymbol{e}_l = C_{ijkl}\boldsymbol{e}_i\boldsymbol{e}_j\boldsymbol{e}_k\boldsymbol{e}_l \tag{1.395}$$

In explicit terms, there are 81 terms (= 3×3×3×3), but we are not going to write them out here. The scalar can be considered as a zero-th order tensor, a vector is a first order tensor, stress is a second order tensor, and so on. In other words, all physical quantities can be considered as tensors. This is like saying that all numbers are complex numbers, since real numbers are just complex numbers with zero imaginary parts. The tensor is a very powerful concept and its analysis allows very tedious steps to be simplified.

To see the power of tensor analysis, let us note that the dot product between two base vectors can be expressed as

$$\boldsymbol{e}_i \cdot \boldsymbol{e}_j = \delta_{ij} = \begin{cases} 1 & i = j \\ 0 & i \neq j \end{cases} \tag{1.396}$$

where $i, j = 1, 2, 3$; and δ_{ij} is called the *Kronecker delta* function. Therefore, if we write

$$\boldsymbol{u} = u_1\boldsymbol{e}_1 + u_2\boldsymbol{e}_2 + u_3\boldsymbol{e}_3\,, \quad \boldsymbol{v} = v_1\boldsymbol{e}_1 + v_2\boldsymbol{e}_2 + v_3\boldsymbol{e}_3\,, \tag{1.397}$$

the dot product between $\boldsymbol{u}$ and $\boldsymbol{v}$ is clearly

$$\boldsymbol{u}\cdot\boldsymbol{v} = (u_i\mathbf{e}_i)\cdot(u_j\mathbf{e}_j) = u_iu_j\delta_{ij} = u_iv_i = u_1v_1 + u_2v_2 + u_3v_3 \tag{1.398}$$

Again, Einstein's notation for summing over all possible i and j is implied. The use of the Kronecker delta function suggests that

$$u_iu_j\delta_{ij} = u_iv_i = u_jv_j \tag{1.399}$$

If $i \neq j$, we must have zero values. Thus, we must substitute i into j or j into i, otherwise this quantity is zero. In addition, there is no free index (index only appears once) on both sides of (1.399) because both i and j appear twice or are repeated (recall from the Einstein convention that a repeated index means summation). The rule of this simple "substitution" allows a very efficient and effective way to carry out a dot product using tensor analysis.

Another important operation between tensors is the double dot product between two tensors. Hooke's law will be used to demonstrate the double dot product, and it can be expressed as

$$\boldsymbol{\sigma} = \boldsymbol{C} : \boldsymbol{\varepsilon} \tag{1.400}$$

Using the symbolic form, we have

$$\begin{aligned} \boldsymbol{\sigma} = \boldsymbol{C} : \boldsymbol{\varepsilon} &= (C_{ijkl}\boldsymbol{e}_i\boldsymbol{e}_j\boldsymbol{e}_k\boldsymbol{e}_l) : (\varepsilon_{mn}\boldsymbol{e}_m\boldsymbol{e}_n) = C_{ijkl}\varepsilon_{mn}\boldsymbol{e}_i\boldsymbol{e}_j(\boldsymbol{e}_k \bullet \boldsymbol{e}_m)(\boldsymbol{e}_l \bullet \boldsymbol{e}_n) \\ &= C_{ijkl}\varepsilon_{mn}\delta_{km}\delta_{ln}\boldsymbol{e}_i\boldsymbol{e}_j = C_{ijkl}\varepsilon_{kl}\boldsymbol{e}_i\boldsymbol{e}_j \end{aligned} \tag{1.401}$$

Note that the result is a second order tensor with a pair of base vectors for each component. Note that any index in a tensor equation cannot be reused, and the index only appears once or twice and cannot be more than two. Mathematically, fourth order tensor can be considered as a mapping function between two second order tensors.

Equation (1.401) can also be written using the component form (instead of dyadic or polyadic forms) as

$$\sigma_{ij} = C_{ijkl}\varepsilon_{kl} \tag{1.402}$$

The cross product between base vectors can also be conducted using algebra of a symbolic operation similar to the Kronecker delta in the dot product. In particular, cross products between any combinations of two base vectors are:

$$\begin{aligned} &\boldsymbol{e}_1 \times \boldsymbol{e}_2 = \boldsymbol{e}_3, \quad \boldsymbol{e}_2 \times \boldsymbol{e}_3 = \boldsymbol{e}_1, \quad \boldsymbol{e}_3 \times \boldsymbol{e}_1 = \boldsymbol{e}_2, \quad \boldsymbol{e}_1 \times \boldsymbol{e}_1 = 0, \quad \boldsymbol{e}_2 \times \boldsymbol{e}_2 = 0, \\ &\boldsymbol{e}_2 \times \boldsymbol{e}_1 = -\boldsymbol{e}_3, \quad \boldsymbol{e}_3 \times \boldsymbol{e}_2 = -\boldsymbol{e}_1, \quad \boldsymbol{e}_1 \times \boldsymbol{e}_3 = -\boldsymbol{e}_2, \quad \boldsymbol{e}_3 \times \boldsymbol{e}_3 = 0 \end{aligned} \tag{1.403}$$

By applying (1.403), the cross product between two vectors is

$$\begin{aligned} \boldsymbol{u} \times \boldsymbol{v} &= (u_1\boldsymbol{e}_1 + u_2\boldsymbol{e}_2 + u_3\boldsymbol{e}_3) \times (v_1\boldsymbol{e}_1 + v_2\boldsymbol{e}_2 + v_3\boldsymbol{e}_3) \\ &= (u_2v_3 - u_3v_2)\boldsymbol{e}_1 + (u_3v_1 - u_1v_3)\boldsymbol{e}_2 + (u_1v_2 - u_2v_1)\boldsymbol{e}_3 \end{aligned} \tag{1.404}$$

The cross product is sometimes easier to remember using the following expansion of determinant:

$$\boldsymbol{u} \times \boldsymbol{v} = \begin{vmatrix} \boldsymbol{e}_1 & \boldsymbol{e}_2 & \boldsymbol{e}_3 \\ u_1 & u_2 & u_3 \\ v_1 & v_2 & v_3 \end{vmatrix} = (u_2v_3 - u_3v_2)\boldsymbol{e}_1 + (u_3v_1 - u_1v_3)\boldsymbol{e}_2 + (u_1v_2 - u_2v_1)\boldsymbol{e}_3 \tag{1.405}$$

Now, we attempt to generalize the cross product analysis using algebra. Consider the special case of (1.405) that

$$\boldsymbol{e}_1 \times \boldsymbol{e}_2 = \begin{vmatrix} \boldsymbol{e}_1 & \boldsymbol{e}_2 & \boldsymbol{e}_3 \\ 1 & 0 & 0 \\ 0 & 1 & 0 \end{vmatrix} = e_{121}\boldsymbol{e}_1 + e_{122}\boldsymbol{e}_2 + e_{123}\boldsymbol{e}_3 = \sum_{k=1}^{3} e_{12k}\boldsymbol{e}_k = e_{12k}\boldsymbol{e}_k = \boldsymbol{e}_3 \quad (1.406)$$

where e_{ijk} (in general, i, j, k = 1,2,3) is a symbol to represent the different components of the cross product for base vectors. This was introduced by Levi-Civita and thus is known as the Levi-Civita symbol or permutation symbol. Since the cross product of two vectors must itself be a vector with three components, clearly, we obtain from (1.406)

$$e_{121} = 0, \quad e_{122} = 0, \quad e_{123} = 1 \quad (1.407)$$

Consider another special case of (1.405) that

$$\boldsymbol{e}_3 \times \boldsymbol{e}_2 = \begin{vmatrix} \boldsymbol{e}_1 & \boldsymbol{e}_2 & \boldsymbol{e}_3 \\ 0 & 0 & 1 \\ 0 & 1 & 0 \end{vmatrix} = e_{321}\boldsymbol{e}_1 + e_{322}\boldsymbol{e}_2 + e_{323}\boldsymbol{e}_3 = \sum_{k=1}^{3} e_{32k}\boldsymbol{e}_k = e_{32k}\boldsymbol{e}_k = -\boldsymbol{e}_1 \quad (1.408)$$

Similarly, we must have the following values for the permutation symbols as

$$e_{321} = -1, \quad e_{322} = 0, \quad e_{323} = 0 \quad (1.409)$$

Let us consider another example of (1.405)

$$\boldsymbol{e}_3 \times \boldsymbol{e}_3 = \begin{vmatrix} \boldsymbol{e}_1 & \boldsymbol{e}_2 & \boldsymbol{e}_3 \\ 0 & 0 & 1 \\ 0 & 0 & 1 \end{vmatrix} = e_{331}\boldsymbol{e}_1 + e_{332}\boldsymbol{e}_2 + e_{333}\boldsymbol{e}_3 = \sum_{k=1}^{3} e_{33k}\boldsymbol{e}_k = e_{33k}\boldsymbol{e}_k = 0 \quad (1.410)$$

Comparison of components on (1.410) gives

$$e_{331} = 0, \quad e_{332} = 0, \quad e_{333} = 0 \quad (1.411)$$

Repeating this procedure for the rest of the combination of cross product between base vectors, we get

$$\begin{aligned} &e_{231} = 1, \quad e_{232} = 0, \quad e_{233} = 0, \quad e_{311} = 0, \quad e_{312} = 1, \quad e_{313} = 0, \\ &e_{111} = 0, \quad e_{112} = 0, \quad e_{113} = 0, \quad e_{221} = 0, \quad e_{222} = 0, \quad e_{223} = 0, \\ &e_{211} = 0, \quad e_{212} = 0, \quad e_{213} = -1, \quad e_{131} = 0, \quad e_{132} = -1, \quad e_{133} = 0, \end{aligned} \quad (1.412)$$

By now, we have obtained all 27 components of the permutation symbols. We observe that

$$e_{123} = e_{231} = e_{312} = 1, \quad e_{132} = e_{213} = e_{321} = -1 \quad (1.413)$$

Note that only nonzero values for e_{ijk} with indices $i \neq j \neq k$, whereas all other components with repeated indices are zeros. We can summarize the values of the permutation tensor in a cyclic sequence shown in Figure 1.25. The magnitude of e_{ijk} is either -1, $+1$, or 0, which is determined by the following rules: (1) e_{ijk} equals 0 if any two indices are equal; (2) e_{ijk} equals $+1$ when i, j, k are 1, 2, 3 or an even permutation of 1, 2, 3; and (3) e_{ijk} equals -1 when i, j, k are 3, 2, 1, or an odd permutation of 1, 2, 3. As shown in Figure 1.25, if the different subscripts follow an even or clockwise permutation, we have the values being 1; whereas for anti-clockwise or odd permutation, we have the values being -1. We also find that moving subscripts in a cyclic order (say the first subscript is moved to the last while all other indices shift forward) yields the same value. However, if the positions of any two indices are reversed, we will have the original value being multiplied by

-1. This permutation tensor was proposed by Italian mathematician Levi-Civita. In summary, we have

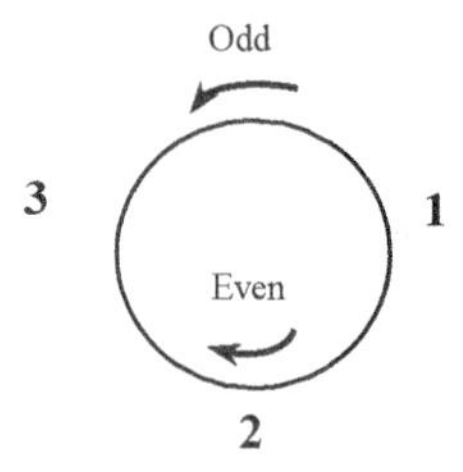

Figure 1.25 The odd and even permutations for 1, 2, and 3 in the Levi-Civita symbol

$$e_{ijk} = e_{jki} = e_{kij} = -e_{jik} = -e_{kji} = -e_{ikj} \tag{1.414}$$

By now, we can rewrite the cross product of any two base vectors as

$$\boldsymbol{e}_i \times \boldsymbol{e}_j = e_{kij} \boldsymbol{e}_k = e_{ijk} \boldsymbol{e}_k \tag{1.415}$$

This formula can now be applied to the cross product of any two vectors

$$\boldsymbol{w} = \boldsymbol{u} \times \boldsymbol{v} = u_i \boldsymbol{e}_i \times v_j \boldsymbol{e}_j = u_i v_j e_{kij} \boldsymbol{e}_k \tag{1.416}$$

In terms of tensor notation, the physical components w_i can be given by

$$w_i = e_{ijk} u_j v_k \tag{1.417}$$

In addition, the permutation tensor also finds application to the calculation of the determinant in tensor form:

$$\det\left|A_{ij}\right| = e_{ijk} A_{i1} A_{j2} A_{k3} \tag{1.418}$$

More amazing formulas for e_{ijk} are given Problems 1.37-1.41.

1.20 e-δ IDENTITY

The following *e-δ identity* has been found extremely useful in tensor analysis:

$$e_{ijk} e_{irs} = \delta_{jr} \delta_{ks} - \delta_{js} \delta_{kr} \tag{1.419}$$

There is more than one way to prove this formula. The straightforward but tedious way is to exhaust all possibilities of j, k, r, and s. Instead, here we will present a less obvious approach.

First, we will establish the following identity of the delta function and determinant

$$e_{pqs} e_{mnr} = \begin{vmatrix} \delta_{mp} & \delta_{mq} & \delta_{ms} \\ \delta_{np} & \delta_{nq} & \delta_{ns} \\ \delta_{rp} & \delta_{rq} & \delta_{rs} \end{vmatrix} \tag{1.420}$$

Consider the determinant of A_{ij} to be given as

$$\det A = \begin{vmatrix} A_{11} & A_{12} & A_{13} \\ A_{21} & A_{22} & A_{23} \\ A_{31} & A_{32} & A_{33} \end{vmatrix} \tag{1.421}$$

When we interchange any two rows or any two columns of the determinant, the sign of the determinant changes sign. That is,

$$\begin{vmatrix} A_{21} & A_{22} & A_{23} \\ A_{11} & A_{12} & A_{13} \\ A_{31} & A_{32} & A_{33} \end{vmatrix} = \begin{vmatrix} A_{12} & A_{11} & A_{13} \\ A_{22} & A_{21} & A_{23} \\ A_{32} & A_{31} & A_{33} \end{vmatrix} = -\det A \tag{1.422}$$

For an arbitrary number of row changes, we can write it as

$$\begin{vmatrix} A_{m1} & A_{m2} & A_{m3} \\ A_{n1} & A_{n2} & A_{n3} \\ A_{r1} & A_{r2} & A_{r3} \end{vmatrix} = e_{mnr} \det A \tag{1.423}$$

Similarly, for an arbitrary number of column changes, we can write it as

$$\begin{vmatrix} A_{1p} & A_{1q} & A_{1s} \\ A_{2p} & A_{2q} & A_{2s} \\ A_{3p} & A_{3q} & A_{3s} \end{vmatrix} = e_{pqs} \det A \tag{1.424}$$

Thus, for an arbitrary number of row and column change sequences, we can combine

$$\begin{vmatrix} A_{mp} & A_{mq} & A_{ms} \\ A_{np} & A_{nq} & A_{ns} \\ A_{rp} & A_{rq} & A_{rs} \end{vmatrix} = e_{pqs} e_{mnr} \det A \tag{1.425}$$

Take the special case that $A_{ij} = \delta_{ij}$, and $\det A = 1$; we have (1.420). Next, (1.420) can be expanded as

$$\begin{aligned} e_{pqs} e_{mnr} = {} & \delta_{mp}(\delta_{nq}\delta_{rs} - \delta_{ns}\delta_{rq}) - \delta_{mq}(\delta_{np}\delta_{rs} - \delta_{ns}\delta_{rp}) \\ & + \delta_{ms}(\delta_{np}\delta_{rq} - \delta_{nq}\delta_{rp}) \end{aligned} \tag{1.426}$$

Now we can set $s = r$ in (1.426) to get

$$\begin{aligned} e_{pqs} e_{mns} = {} & \delta_{mp}(\delta_{nq}\delta_{ss} - \delta_{ns}\delta_{sq}) - \delta_{mq}(\delta_{np}\delta_{ss} - \delta_{ns}\delta_{sp}) \\ & + \delta_{ms}(\delta_{np}\delta_{sq} - \delta_{nq}\delta_{sp}) \\ = {} & 3\delta_{mp}\delta_{nq} - \delta_{mp}\delta_{nq} - 3\delta_{mq}\delta_{np} + \delta_{mq}\delta_{np} + \delta_{mq}\delta_{np} - \delta_{nq}\delta_{mp} \\ = {} & (\delta_{mp}\delta_{nq} - \delta_{mq}\delta_{np}) \end{aligned} \tag{1.427}$$

Thus, the e-δ identity is established.

1.21 TENSOR ANALYSIS IN CARTESIAN COORDINATES

The formulas of expressing the dot product by the delta function, of expressing the cross product by permutation tensor, and of e-δ identity are useful tools for tensor analysis. We will demonstrate the validity of (1.362) as follows:

$$(\nabla \times \boldsymbol{A}) = (\boldsymbol{e}_j \frac{\partial}{\partial x_j}) \times (A_k \boldsymbol{e}_k) = (\boldsymbol{e}_j \times \boldsymbol{e}_k) \frac{\partial A_k}{\partial x_j} = e_{ijk} A_{k,j} \boldsymbol{e}_i \tag{1.428}$$

$$\nabla \times (\nabla \times \boldsymbol{A}) = (\boldsymbol{e}_n \frac{\partial}{\partial x_n}) \times (e_{ijk} A_{k,j} \boldsymbol{e}_i) = (\boldsymbol{e}_n \times \boldsymbol{e}_i) e_{ijk} A_{k,jn} = e_{mni} e_{ijk} A_{k,jn} \boldsymbol{e}_m \tag{1.429}$$

Now, we apply the e-δ identity

$$\begin{aligned}\nabla\times(\nabla\times A) &= e_{mni}e_{ijk}A_{k,jn}\boldsymbol{e}_m = (\delta_{mj}\delta_{nk}-\delta_{mk}\delta_{nj})A_{k,jn}\boldsymbol{e}_m \\ &= (\delta_{mj}\delta_{nk}A_{k,jn}-\delta_{mk}\delta_{nj}A_{k,jn})\boldsymbol{e}_m \\ &= (A_{n,mn}-A_{m,nn})\boldsymbol{e}_m\end{aligned} \tag{1.430}$$

On the right hand side, we have

$$\nabla\bullet A = (\boldsymbol{e}_j\frac{\partial}{\partial x_j})\bullet(A_k\boldsymbol{e}_k) = (\boldsymbol{e}_j\bullet\boldsymbol{e}_k)A_{k,j} = \delta_{jk}A_{k,j} = A_{k,k} = A_{n,n} \tag{1.431}$$

$$\nabla(\nabla\bullet A) = \boldsymbol{e}_j\frac{\partial A_{n,n}}{\partial x_j} = A_{n,nm}\boldsymbol{e}_m \tag{1.432}$$

$$\nabla^2 A = (\boldsymbol{e}_j\frac{\partial}{\partial x_j})\bullet(\boldsymbol{e}_k\frac{\partial}{\partial x_k})A_m\boldsymbol{e}_m = \delta_{jk}A_{m,jk}\boldsymbol{e}_m = A_{m,jj}\boldsymbol{e}_m \tag{1.433}$$

Summing (1.433) from (1.432), we have the right hand side of (1.362) equal to

$$\nabla(\nabla\bullet A)-\nabla^2 A = A_{n,nm}\boldsymbol{e}_m - A_{m,jj}\boldsymbol{e}_m = (A_{n,nm}-A_{m,nn})\boldsymbol{e}_m = \nabla\times(\nabla\times A) \tag{1.434}$$

The last of (1.434) is obtained by using the result of (1.430). Note also that the order of partial differentiations is interchangeable, that is,

$$A_{n,nm} = A_{n,mn} \tag{1.435}$$

Similarly, other vector identities given in (1.353) to (1.370) can also be proved similar to the above tensor analysis.

1.22 TENSOR ANALYSIS IN CYLINDRICAL COORDINATES

Tensor analysis in polar cylindrical coordinates is more complicated than those in Cartesian coordinates. The main reason is that the base vector is no longer fixed in the space. Any position vector $\boldsymbol{r}$ in a Cartesian coordinate system can be written in terms of a cylindrical coordinate system (ρ, ϕ, z) as shown in Figure 1.26:

$$\boldsymbol{r} = x_1\boldsymbol{e}_1 + x_2\boldsymbol{e}_2 + x_3\boldsymbol{e}_3 = \rho\cos\phi\,\boldsymbol{e}_1 + \rho\sin\phi\,\boldsymbol{e}_2 + z\boldsymbol{e}_3 \tag{1.436}$$

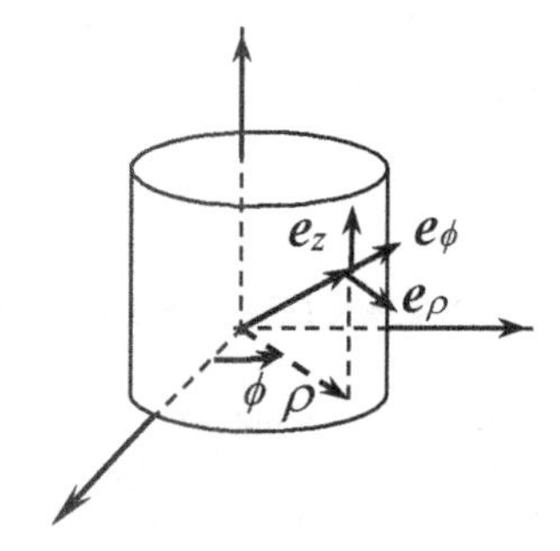

Figure 1.26 Cylindrical coordinates

The new set of base vectors in cylindrical coordinates is defined by

$$e_\alpha = \frac{1}{h_\alpha}\frac{\partial \boldsymbol{r}}{\partial q_\alpha} \tag{1.437}$$

where

$$h_\alpha = \left|\frac{\partial \boldsymbol{r}}{\partial q_\alpha}\right| = \sqrt{\left(\frac{\partial x_1}{\partial q_\alpha}\right)^2 + \left(\frac{\partial x_2}{\partial q_\alpha}\right)^2 + \left(\frac{\partial x_3}{\partial q_\alpha}\right)^2} \tag{1.438}$$

and q_α equals ρ, ϕ, or z. In particular, we have for cylindrical coordinates

$$h_\rho = \left|\frac{\partial \boldsymbol{r}}{\partial \rho}\right| = \sqrt{\left(\frac{\partial x_1}{\partial \rho}\right)^2 + \left(\frac{\partial x_2}{\partial \rho}\right)^2 + \left(\frac{\partial x_3}{\partial \rho}\right)^2} = \sqrt{\cos^2\phi + \sin^2\phi} = 1 \tag{1.439}$$

$$h_\phi = \left|\frac{\partial \boldsymbol{r}}{\partial \phi}\right| = \sqrt{\left(\frac{\partial x_1}{\partial \phi}\right)^2 + \left(\frac{\partial x_2}{\partial \phi}\right)^2 + \left(\frac{\partial x_3}{\partial \phi}\right)^2} = \sqrt{\rho^2\cos^2\phi + \rho^2\sin^2\phi} = \rho \tag{1.440}$$

$$h_z = \left|\frac{\partial \boldsymbol{r}}{\partial z}\right| = \sqrt{\left(\frac{\partial x_1}{\partial z}\right)^2 + \left(\frac{\partial x_2}{\partial z}\right)^2 + \left(\frac{\partial x_3}{\partial z}\right)^2} = 1 \tag{1.441}$$

Thus, substitution of (1.439) to (1.441) into (1.437) gives

$$\begin{aligned} &\boldsymbol{e}_\rho = \frac{1}{h_\rho}\frac{\partial \boldsymbol{r}}{\partial \rho} = \cos\phi \boldsymbol{e}_1 + \sin\phi \boldsymbol{e}_2, \quad \boldsymbol{e}_\phi = \frac{1}{h_\phi}\frac{\partial \boldsymbol{r}}{\partial \phi} = -\sin\phi \boldsymbol{e}_1 + \cos\phi \boldsymbol{e}_2 \\ &\boldsymbol{e}_z = \frac{1}{h_z}\frac{\partial \mathbf{r}}{\partial z} = \boldsymbol{e}_3 \end{aligned} \tag{1.442}$$

It is obvious that we must have the following identities:

$$\frac{\partial \boldsymbol{e}_\rho}{\partial \phi} = \frac{\partial}{\partial \phi}(\cos\phi \boldsymbol{e}_1 + \sin\phi \boldsymbol{e}_2) = -\sin\phi \boldsymbol{e}_1 + \cos\phi \boldsymbol{e}_2 = \boldsymbol{e}_\phi \tag{1.443}$$

$$\frac{\partial \boldsymbol{e}_\phi}{\partial \phi} = \frac{\partial}{\partial \phi}(-\sin\phi \boldsymbol{e}_1 + \cos\phi \boldsymbol{e}_2) = -(\cos\phi \boldsymbol{e}_1 + \sin\phi \boldsymbol{e}_2) = -\boldsymbol{e}_r \tag{1.444}$$

while all other derivatives of the base vectors vanish. For example, the displacement gradient tensor can be formulated in dyadic notation as

$$\begin{aligned} \nabla \boldsymbol{u} &= (\boldsymbol{e}_\rho \frac{\partial}{\partial r} + \boldsymbol{e}_\phi \frac{\partial}{\rho \partial \phi} + \boldsymbol{e}_z \frac{\partial}{\partial z})(u_\rho \boldsymbol{e}_\rho + u_\phi \boldsymbol{e}_\phi + u_z \boldsymbol{e}_z) \\ &= \frac{\partial u_\rho}{\partial \rho}\boldsymbol{e}_\rho \boldsymbol{e}_\rho + \frac{1}{\rho}(u_\rho + \frac{\partial u_\phi}{\partial \phi})\boldsymbol{e}_\phi \boldsymbol{e}_\phi + \frac{\partial u_z}{\partial z}\boldsymbol{e}_z \boldsymbol{e}_z + \frac{1}{\rho}\frac{\partial u_z}{\partial \phi}\boldsymbol{e}_\phi \boldsymbol{e}_z + \frac{\partial u_\phi}{\partial z}\boldsymbol{e}_z \boldsymbol{e}_\phi \\ &+ \frac{\partial u_\rho}{\partial z}\boldsymbol{e}_z \boldsymbol{e}_\rho + \frac{\partial u_z}{\partial r}\boldsymbol{e}_\rho \boldsymbol{e}_z + \frac{\partial u_\phi}{\partial r}\boldsymbol{e}_\rho \boldsymbol{e}_\phi + \frac{1}{\rho}(\frac{\partial u_\rho}{\partial \phi} - u_\phi)\boldsymbol{e}_\phi \boldsymbol{e}_\rho \end{aligned} \tag{1.445}$$

In obtaining the above equation, we have already used the coordinate variation of the base vectors obtained in (1.443) and (1.444). For example, the derivative taken with respect to ϕ is

$$\boldsymbol{e}_\phi \frac{\partial}{\rho \partial \phi}(u_\rho \boldsymbol{e}_\rho + u_\phi \boldsymbol{e}_\phi + u_z \boldsymbol{e}_z) = \boldsymbol{e}_\phi \frac{\partial}{\rho \partial \phi}(u_\rho \boldsymbol{e}_\rho) + \boldsymbol{e}_\phi \frac{\partial}{\rho \partial \phi}(u_\phi \boldsymbol{e}_\phi) + \boldsymbol{e}_\phi \frac{\partial}{\rho \partial \phi} u_z \boldsymbol{e}_z$$

$$= \boldsymbol{e}_\phi \frac{\partial u_\rho}{\rho \partial \phi} \boldsymbol{e}_\rho + \boldsymbol{e}_\phi \frac{u_\rho}{\rho} \frac{\partial \boldsymbol{e}_\rho}{\partial \phi} + \boldsymbol{e}_\phi \frac{\partial u_\phi}{\rho \partial \phi} \boldsymbol{e}_\phi + \boldsymbol{e}_\phi \frac{u_\phi}{\rho} \frac{\partial \boldsymbol{e}_\phi}{\partial \phi} + \boldsymbol{e}_\phi \boldsymbol{e}_z \frac{\partial u_z}{\rho \partial \phi} \tag{1.446}$$

$$= \boldsymbol{e}_\phi \boldsymbol{e}_\rho \frac{\partial u_\rho}{\rho \partial \phi} + \boldsymbol{e}_\phi \boldsymbol{e}_\phi \frac{u_\rho}{\rho} + \boldsymbol{e}_\phi \boldsymbol{e}_\phi \frac{\partial u_\phi}{\rho \partial \phi} - \boldsymbol{e}_\phi \boldsymbol{e}_\rho \frac{u_\phi}{\rho} + \boldsymbol{e}_\phi \boldsymbol{e}_z \frac{\partial u_z}{\rho \partial \phi}$$

This result explains why two extra terms appear in the bracket terms on the right hand side of (1.445), and they are the results of the direction change of base vectors with respect to the change in variable ϕ. Clearly, $\nabla \boldsymbol{u}$ is a second order tensor since every component is accomplished by two base vectors. This result can readily be used to obtain the strain tensor, which is defined as

$$\varepsilon = \frac{1}{2}(\nabla \boldsymbol{u} + \boldsymbol{u} \nabla) \tag{1.447}$$

where $\boldsymbol{u}\nabla = (\nabla \boldsymbol{u})^{T}$ or the transpose of $\nabla \boldsymbol{u}$. Substitution of (1.445) into (1.447) yields the following physical components in cylindrical coordinates

$$\varepsilon_{zz} = \frac{\partial u_z}{\partial z}, \quad \varepsilon_{\rho\rho} = \frac{\partial u_\rho}{\partial \rho}, \quad \varepsilon_{\phi\phi} = \frac{u_\rho}{\rho} + \frac{1}{\rho}\frac{\partial u_\phi}{\partial \phi} \tag{1.448}$$

$$\varepsilon_{\rho\phi} = \frac{1}{2}\left(\frac{\partial u_\phi}{\partial \rho} + \frac{1}{\rho}\frac{\partial u_\rho}{\partial \phi} - \frac{u_\phi}{\rho}\right), \quad \varepsilon_{\phi z} = \frac{1}{2}\left(\frac{1}{\rho}\frac{\partial u_z}{\partial \phi} + \frac{\partial u_\phi}{\partial z}\right),$$
$$\varepsilon_{\rho z} = \frac{1}{2}\left(\frac{\partial u_\rho}{\partial z} + \frac{\partial u_z}{\partial \rho}\right) \tag{1.449}$$

These equations are the same as those obtained by Timoshenko and Goodier (1982), starting from the kinematics of compatibility in deformations. Thus, the tensor equation (1.447) provides a concise and elegant form for the strain-displacement relation, and, more importantly, it is independent of any coordinate system.

Similarly, the following identities in cylindrical coordinates can be obtained:

$$\nabla \bullet \boldsymbol{u} = \frac{\partial u_\rho}{\partial \rho} + \frac{1}{\rho} u_\rho + \frac{1}{\rho}\frac{\partial u_\phi}{\partial \phi} + \frac{\partial u_z}{\partial z} \tag{1.450}$$

$$\nabla \times \boldsymbol{u} = \boldsymbol{e}_\rho\left(\frac{1}{\rho}\frac{\partial u_z}{\partial \phi} - \frac{\partial u_\phi}{\partial z}\right) + \boldsymbol{e}_\phi\left(\frac{\partial u_\rho}{\partial z} - \frac{\partial u_z}{\partial \rho}\right) + \boldsymbol{e}_z\left(\frac{\partial u_\phi}{\partial \rho} + \frac{u_\phi}{\rho} - \frac{1}{\rho}\frac{\partial u_\rho}{\partial \phi}\right) \tag{1.451}$$

$$\nabla^2 f = \nabla \bullet \nabla f = \frac{\partial^2 f}{\partial \rho^2} + \frac{1}{\rho}\frac{\partial f}{\partial \rho} + \frac{1}{\rho^2}\frac{\partial^2 f}{\partial \phi^2} + \frac{\partial^2 f}{\partial z^2} \tag{1.452}$$

$$\nabla^2 \boldsymbol{u} = \boldsymbol{e}_\rho\left(\nabla^2 u_\rho - \frac{2}{\rho^2}\frac{\partial u_\rho}{\partial \phi} - \frac{u_\rho}{\rho^2}\right) + \boldsymbol{e}_\phi\left(\nabla^2 u_\phi + \frac{2}{\rho^2}\frac{\partial u_\rho}{\partial \phi} - \frac{u_\phi}{\rho^2}\right) + \boldsymbol{e}_z \nabla^2 u_z \tag{1.453}$$

The three components of the equilibrium equations, $\nabla \cdot \boldsymbol{\sigma} = 0$, can be written explicitly as

$$\frac{\partial \sigma_{\rho\rho}}{\partial \rho} + \frac{\partial \sigma_{z\rho}}{\partial z} + \frac{1}{\rho}\frac{\partial \sigma_{\rho\phi}}{\partial \phi} + \frac{\sigma_{\rho\rho} - \sigma_{\phi\phi}}{\rho} = 0 \tag{1.454}$$

$$\frac{1}{\rho}\frac{\partial}{\partial \rho}(\rho\sigma_{\rho z})+\frac{\partial \sigma_{zz}}{\partial \rho}+\frac{1}{\rho}\frac{\partial \sigma_{z\phi}}{\partial \phi}=0 \tag{1.455}$$

$$\frac{1}{\rho}\frac{\partial \sigma_{\phi\phi}}{\partial \phi}+\frac{\partial \sigma_{\rho\phi}}{\partial \rho}+\frac{\partial \sigma_{z\phi}}{\partial z}+2\frac{\sigma_{\rho\phi}}{\rho}=0 \tag{1.456}$$

These equations are the same as those obtained by Timoshenko and Goodier (1982) by considering force equilibriums along ρ-, ϕ-, and z-directions for an infinitesimally small element. However, the present approach is more efficient and systematic in obtaining the component form of the equilibrium equations in terms of stresses.

1.23 TENSOR ANALYSIS IN SPHERICAL COORDINATES

The development of this section follows closely the discussion in the previous section. Any position vector $\boldsymbol{r}$ in a Cartesian coordinate system can be written in terms of a polar spherical coordinate system (r, φ, θ) as shown in Figure 1.27:

$$\boldsymbol{r}=x_1\boldsymbol{e}_1+x_2\boldsymbol{e}_2+x_3\boldsymbol{e}_3=\rho\sin\theta\cos\varphi\,\boldsymbol{e}_1+\rho\sin\theta\sin\varphi\,\boldsymbol{e}_2+\rho\cos\theta\,\boldsymbol{e}_3 \tag{1.457}$$

For spherical coordinates, q_α equals either ρ, φ, or θ. Thus,

$$\begin{aligned} h_\rho&=\left|\frac{\partial \boldsymbol{r}}{\partial \rho}\right|=\sqrt{\left(\frac{\partial x_1}{\partial \rho}\right)^2+\left(\frac{\partial x_2}{\partial \rho}\right)^2+\left(\frac{\partial x_3}{\partial \rho}\right)^2}\\ &=\sqrt{\sin^2\theta\cos^2\phi+\sin^2\theta\sin^2\phi+\cos^2\theta}=1 \end{aligned} \tag{1.458}$$

$$\begin{aligned} h_\varphi&=\left|\frac{\partial \boldsymbol{r}}{\partial \varphi}\right|=\sqrt{\left(\frac{\partial x_1}{\partial \varphi}\right)^2+\left(\frac{\partial x_2}{\partial \varphi}\right)^2+\left(\frac{\partial x_3}{\partial \varphi}\right)^2}\\ &=\sqrt{\rho^2\sin^2\theta\sin^2\varphi+\rho^2\sin^2\theta\cos^2\varphi}=\rho\sin\theta \end{aligned} \tag{1.459}$$

$$\begin{aligned} h_\theta&=\left|\frac{\partial \boldsymbol{r}}{\partial \theta}\right|=\sqrt{\left(\frac{\partial x_1}{\partial \theta}\right)^2+\left(\frac{\partial x_2}{\partial \theta}\right)^2+\left(\frac{\partial x_3}{\partial \theta}\right)^2}\\ &=\sqrt{\rho^2\cos^2\theta\cos^2\varphi+\rho^2\cos^2\theta\sin^2\varphi+\rho^2\sin^2\theta}=\rho \end{aligned} \tag{1.460}$$

Substitution of (1.458) to (1.460) into (1.437) results in the following base vectors

$$\begin{aligned} \boldsymbol{e}_\rho&=(\sin\theta\cos\varphi)\boldsymbol{e}_1+(\sin\theta\sin\varphi)\boldsymbol{e}_2+\cos\theta\,\boldsymbol{e}_3\\ \boldsymbol{e}_\theta&=(\cos\theta\cos\varphi)\boldsymbol{e}_1+(\cos\theta\cos\varphi)\boldsymbol{e}_2-\sin\theta\,\boldsymbol{e}_3\\ \boldsymbol{e}_\varphi&=-\sin\varphi\,\boldsymbol{e}_1+\cos\varphi\,\boldsymbol{e}_2 \end{aligned} \tag{1.461}$$

The variation of base vectors along coordinate directions is more complicated than that for cylindrical coordinates; in particular, the following nonzero terms are obtained:

$$\begin{aligned} &\frac{\partial \boldsymbol{e}_\rho}{\partial \theta}=\boldsymbol{e}_\theta,\quad \frac{\partial \boldsymbol{e}_\theta}{\partial \theta}=-\boldsymbol{e}_\rho,\quad \frac{\partial \boldsymbol{e}_\theta}{\partial \varphi}=\cos\theta\,\boldsymbol{e}_\varphi,\\ &\frac{\partial \boldsymbol{e}_\varphi}{\partial \varphi}=-\sin\theta\,\boldsymbol{e}_\rho-\cos\theta\,\boldsymbol{e}_\theta,\quad \frac{\partial \boldsymbol{e}_\rho}{\partial \varphi}=\sin\theta\,\boldsymbol{e}_\varphi \end{aligned} \tag{1.462}$$

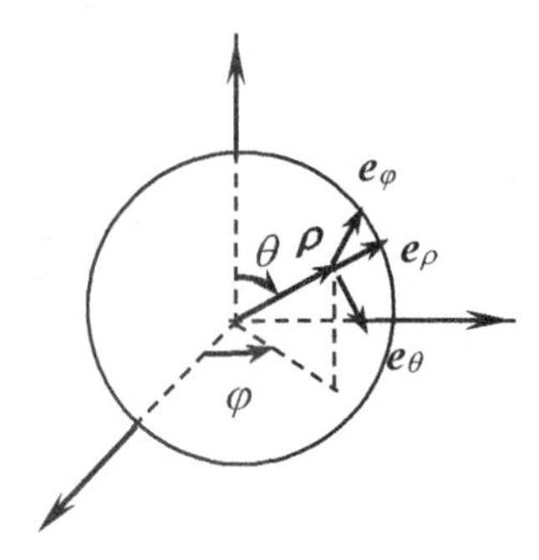

Figure 1.27 Spherical coordinates

while all other derivatives of the base vectors vanish. The differential operator in spherical polar coordinates is

$$\nabla = \left(\boldsymbol{e}_\rho \frac{\partial}{\partial \rho} + \boldsymbol{e}_\theta \frac{1}{\rho}\frac{\partial}{\partial \theta} + \boldsymbol{e}_\varphi \frac{1}{\rho \sin\theta}\frac{\partial}{\partial \varphi} \right) \tag{1.463}$$

Without showing the details here, we quote the following physical components for the strain-displacement by specializing (1.447) to the spherical coordinates:

$$\varepsilon_{\varphi\varphi} = \frac{1}{\rho\sin\theta}\frac{\partial u_\varphi}{\partial \varphi} + \frac{u_\rho}{\rho} + \frac{u_\theta}{\rho}\cot\theta, \quad \varepsilon_{\rho\rho} = \frac{\partial u_\rho}{\partial \rho}, \quad \varepsilon_{\theta\theta} = \frac{u_\rho}{\rho} + \frac{1}{\rho}\frac{\partial u_\theta}{\partial \theta}, \tag{1.464}$$

$$\varepsilon_{\rho\theta} = \frac{1}{2}\left(\frac{\partial u_\theta}{\partial \rho} + \frac{1}{\rho}\frac{\partial u_\rho}{\partial \theta} - \frac{u_\theta}{\rho}\right), \quad \varepsilon_{\rho\varphi} = \frac{1}{2}\left(\frac{1}{\rho\sin\theta}\frac{\partial u_\rho}{\partial \varphi} + \frac{\partial u_\varphi}{\partial \rho} - \frac{u_\varphi}{\rho}\right), \tag{1.465}$$

$$\varepsilon_{\theta\varphi} = \frac{1}{2}\left(\frac{1}{\rho\sin\theta}\frac{\partial u_\theta}{\partial \varphi} + \frac{1}{\rho}\frac{\partial u_\varphi}{\partial \theta} - \frac{u_\varphi}{\rho}\cot\theta\right) \tag{1.466}$$

Again these results are the same as those obtained by considering the kinematics of compatibility in deformations considered in Timoshenko and Goodier (1982).

The following identities can be obtained in polar coordinates as

$$\nabla\cdot \boldsymbol{u} = \frac{\partial u_\rho}{\partial \rho} + \frac{2}{\rho}u_\rho + \frac{1}{\rho}\left(\frac{\partial u_\theta}{\partial \theta} + \cot\theta u_\theta\right) + \frac{1}{\rho\sin\theta}\frac{\partial u_\varphi}{\partial \varphi} \tag{1.467}$$

$$\nabla\times \boldsymbol{u} = \mathbf{e}_\rho \frac{1}{\rho}\left(\frac{\partial u_\varphi}{\partial \theta} + \cot\theta u_\varphi - \frac{1}{\sin\theta}\frac{\partial u_\theta}{\partial \varphi}\right) + \mathbf{e}_\theta\left(\frac{1}{\rho\sin\theta}\frac{\partial u_\rho}{\partial \varphi} - \frac{\partial u_\varphi}{\partial \rho} - \frac{u_\varphi}{\rho}\right)$$
$$+\mathbf{e}_\varphi\left(\frac{\partial u_\theta}{\partial \rho} + \frac{u_\theta}{\rho} - \frac{1}{\rho}\frac{\partial u_\rho}{\partial \theta}\right) \tag{1.468}$$

$$\nabla^2 f = \frac{1}{\rho^2}\frac{\partial}{\partial \rho}\left(\rho^2 \frac{\partial f}{\partial \rho}\right) + \frac{1}{\rho^2\sin\theta}\frac{\partial}{\partial \theta}\left(\sin\theta\frac{\partial f}{\partial \theta}\right) + \frac{1}{\rho^2\sin^2\theta}\frac{\partial^2 f}{\partial \varphi^2} \tag{1.469}$$

$$\nabla^2 \boldsymbol{u} = \boldsymbol{e}_\rho \left(\nabla^2 u_\rho - \frac{2u_\rho}{\rho^2} - \frac{2\cot\theta}{\rho^2} u_\theta - \frac{2}{\rho^2}\frac{\partial u_\theta}{\partial\theta} - \frac{2}{\rho^2 \sin\theta}\frac{\partial u_\varphi}{\partial\varphi}\right)$$
$$+\boldsymbol{e}_\theta \left(\nabla^2 u_\theta + \frac{2}{\rho^2}\frac{\partial u_\rho}{\partial\theta} - \frac{u_\theta}{\rho^2 \sin^2\theta} - \frac{2\cot\theta}{\rho^2\sin\theta}\frac{\partial u_\varphi}{\partial\varphi}\right) \quad (1.470)$$
$$+\boldsymbol{e}_\varphi \left(\nabla^2 u_\varphi + \frac{2}{\rho^2\sin\theta}\frac{\partial u_\rho}{\partial\varphi} - \frac{u_\varphi}{\rho^2 \sin^2\theta} + \frac{2\cot\theta}{\rho^2\sin\theta}\frac{\partial u_\theta}{\partial\varphi}\right)$$

In spherical coordinates, the three components of the equilibrium equations, $\nabla \cdot \boldsymbol{\sigma} = 0$, can be written explicitly as

$$\frac{\partial\sigma_{\rho\rho}}{\partial\rho} + \frac{1}{\rho}\frac{\partial\sigma_{\rho\theta}}{\partial\theta} + \frac{1}{\rho\sin\theta}\frac{\partial\sigma_{\rho\varphi}}{\partial\varphi} + \frac{1}{\rho}(2\sigma_{\rho\rho} - \sigma_{\theta\theta} - \sigma_{\varphi\varphi} + \sigma_{\rho\theta}\cot\theta) = 0 \quad (1.471)$$

$$\frac{\partial\sigma_{\rho\theta}}{\partial\rho} + \frac{1}{\rho}\frac{\partial\sigma_{\theta\theta}}{\partial\theta} + \frac{1}{\rho\sin\theta}\frac{\partial\sigma_{\theta\varphi}}{\partial\varphi} + \frac{1}{\rho}[3\sigma_{\rho\theta} + (\sigma_{\theta\theta} - \sigma_{\varphi\varphi})\cot\theta] = 0 \quad (1.472)$$

$$\frac{\partial\sigma_{\rho\varphi}}{\partial\rho} + \frac{1}{\rho}\frac{\partial\sigma_{\theta\varphi}}{\partial\theta} + \frac{1}{\rho\sin\theta}\frac{\partial\sigma_{\varphi\varphi}}{\partial\varphi} + \frac{1}{\rho}(3\sigma_{\rho\varphi} + 2\sigma_{\theta\varphi}\cot\theta) = 0 \quad (1.473)$$

Readers are advised to work out the details in obtaining these equations.

1.24 SUMMARY AND FURTHER READING

In this chapter, we have summarized a number of important theorems and formulas that will be useful for our later determination of the solution of differential equations in engineering and mechanics. Due to the space limitation, we only sketch their proofs and demonstrate their applications through examples. For elementary mathematical analysis, we recommend the books by Hardy (1944) and Tranter (1957). For more discussions on classical methods and analyses, we refer readers to the advanced textbook by Whittaker and Watson (1927). Stories of two Nobel Prize winners should be mentioned here. Lars Onsager, 1968 Nobel Prize winner in chemistry, worked out all "difficult problems" in Whittaker and Watson (1927) when he was a teenager, and this laid down the mathematical skill for his latter achievements. In 1933, Onsager was awarded the Sterling and Gibbs Fellow at Yale University, but an embarrassing situation occurred, when it was discovered that he had never received a PhD. His colleagues advised him to try for a Yale PhD, and Onsager wrote a thesis on *Solutions of the Mathieu Equation of Period 4π and certain related functions*. Nobody in the chemistry and physics could understand it and it was sent to Prof. E. Hille of Mathematics Department, who was an expert in the area. Prof. Hille was so impressed that he suggested to Prof. Hill (head of Chemistry Department) that the Mathematics Department at Yale would be happy to recommend him for PhD. Not wishing to be upstaged, the Chemistry Department awarded Onsager a PhD in Chemistry. Mathieu equation was covered in Chapter 19 of Whittaker and Watson (1927). The second story is on Subrahmanyan Chandrasekhar (1983 Nobel Prize winner in physics). During his 1930 voyage to England, Chandrasekhar derived his celebrated "Chrandrasekhar limit" for white dwarf stars, and studied Whittaker and Watson (1927) seriously on board. The mastery of this "old" book is not easy but clearly would be helpful in understanding more advanced methods of mathematical analyses.

Spiegel (1963) summarized a lot of essential results in calculus. For the analysis of complex variables, Spiegel (1964) covered a lot of classical results in a concise manner, and it is a good reference book to start with. Other references on complex variables include Watson (1914), Silverman (1974), Copson (1935), and Forsyth (1893). For results related to Ramanujan's master theorem, *Ramanujan's Lost Notebooks* by Berndt (1985, 1989) are the authorities. For the discussion on tensor analysis, we refer to Segel (1987), Wong (1991), and Chau (2013).

1.25 PROBLEMS

Problem 1.1 Use the binomial theorem to find the following sum of binomial coefficients:

$$S = 1 + C_n^1 + C_n^2 + \cdots + C_n^r + \ldots + C_n^{n-1} + 1 \tag{1.474}$$

Ans: $S = 2^n$

Problem 1.2 Find the following sum

$$S_1 = 1 - C_n^1 + C_n^2 + \ldots + (-1)^{n-1} C_n^{n-1} + (-1)^n \tag{1.475}$$

Ans: $S_1 = 0$

Problem 1.3 Use integration by parts to prove the formula

$$\int_0^\infty x^{2n} e^{-x^2} dx = \frac{(2n-1)!!}{2^{n+1}} \sqrt{\pi} \tag{1.476}$$

where n is an integer and $(2n-1)!! = 1 \cdot 3 \cdot 5 \cdots (2n-1)$.

Problem 1.4 Prove the formula

$$\int_0^\infty x^{2n+1} e^{-x^2} dx = \frac{n!}{2} \tag{1.477}$$

where n is an integer.

Problem 1.5 Show that the Jacobian for the change of variables given in Section 1.4.6 is indeed r.

Problem 1.6 Find the value of the following number:

$$\sqrt[i]{i} \tag{1.478}$$

Ans: Infinite answers and the principal one is 4.810477381…

Problem 1.7 Show that

$$\cos i = \frac{e + 1/e}{2}, \quad \sin i = \frac{1/e - e}{2i}, \tag{1.479}$$

Problem 1.8 Find the value of the following number:

$$i^{i \cos i} \tag{1.480}$$

Ans: Infinite answers and the principal one is 0.08858...

Problem 1.9 Find the value of the following number:

$$i^{\sin i} \tag{1.481}$$

Ans: Infinite answers and the principal one is 0.15787...

Problem 1.10 Find the roots of

$$\sqrt[3]{1} \tag{1.482}$$

Ans: 1, $(-1+i\sqrt{3})/2$, and $(-1-i\sqrt{3})/2$

Problem 1.11 Check whether the following complex function is analytic:

$$f(z) = F_1[\mathrm{Re}(z)] + iF_2[\mathrm{Im}(z)] \tag{1.483}$$

Ans: No

Problem 1.12 Show the following identities:

$$\cosh(iy) = \cos y, \quad \sinh(iy) = i \sin y \tag{1.484}$$

Problem 1.13 Show that:

$$z^{\zeta} = e^{\xi \ln \sigma - \eta(\psi + 2m\pi)} \{\cos[\eta \ln \sigma + \xi(\psi + 2m\pi)] + i \sin[\eta \ln \sigma + \xi(\psi + 2m\pi)]\} \tag{1.485}$$

where

$$\zeta = \xi + i\eta, \quad z = \sigma(\cos\psi + i \sin\psi) \tag{1.486}$$

and m is an integer.

Problem 1.14 Use the result of Problem 1.13 to show that:

$$x^i + x^{-i} = 2\cos(\ln x)\cosh[2\pi(n+m)] - 2i \sin(\ln x)\sinh[2\pi(n+m)] \tag{1.487}$$

where x is real and m and n are integers.

Problem 1.15 Show that the modulus of the result in Problem 1.14 is

$$\left|x^i + x^{-i}\right| = \sqrt{2}\{\cos(2\ln x) + \cosh[4\pi(n+m)]\}^{1/2} \tag{1.488}$$

where x is real and m and n are integers.

Problem 1.16 Find the real part of

$$\mathrm{Re}\{i^{\ln(1-i)}\} = e^{(4k+1)\pi^2/8} \cos\{\frac{1}{4}(4k+1)\pi \ln 2\} \tag{1.489}$$

where k is an integer.

Problem 1.17 Find the value of

$$2\ln\left(\frac{1-i}{1+i}\right)^i \tag{1.490}$$

Ans: There are infinite solutions, and the simplest one of it is π

Note that this is the definition of π given by Wronski! No wonder he is interested in metaphysics (see biography section)!

Problem 1.18 Show by chain rule that if $\phi = \phi(u)$ and $u = u(x_1, x_2, x_3)$, the gradient of ϕ is

$$\nabla\phi = \frac{d\phi}{du}\nabla u \tag{1.491}$$

Problem 1.19 If A is any vector function, show that

$$\frac{d}{dt}(\boldsymbol{A}\times\frac{d\boldsymbol{A}}{dt}) = \boldsymbol{A}\times\frac{d^2\boldsymbol{A}}{dt^2} \tag{1.492}$$

Problem 1.20 Find the time derivative of

$$\boldsymbol{A}\cdot(\frac{d\boldsymbol{A}}{dt}\times\frac{d^2\boldsymbol{A}}{dt^2}) \tag{1.493}$$

Ans: $\boldsymbol{A}\cdot\frac{d\boldsymbol{A}}{dt}\times\frac{d^3\boldsymbol{A}}{dt^3}$

Problem 1.21 Show the validity of the formula

$$\nabla\bullet(\phi\nabla\psi) - \nabla\bullet(\psi\nabla\phi) = \phi\nabla^2\psi - \psi\nabla^2\phi \tag{1.494}$$

Problem 1.22 Show the validity of the formula

$$\boldsymbol{v}\times(\nabla\times\boldsymbol{v}) = \frac{1}{2}\nabla(\boldsymbol{v}\cdot\boldsymbol{v}) - (\boldsymbol{v}\cdot\nabla)\boldsymbol{v} \tag{1.495}$$

Problem 1.23 Evaluate the divergence and curl of the position vector $\boldsymbol{r}$

$$\nabla\times\boldsymbol{r}, \quad \nabla\cdot\boldsymbol{r} \tag{1.496}$$

Ans: 0 and 3

Problem 1.24 Show that

$$\nabla^2(\frac{1}{r}) = 0 \tag{1.497}$$

Hint: Chau (2013) called $1/r$ the granddaddy of all solutions of Laplace equations.

Problem 1.25 Find the value of n such that

$$\nabla^2 r^n = 0 \tag{1.498}$$

Hint: More than 1 answer

Problem 1.26 For spherical coordinates, prove the following identities

$$\frac{\partial}{\partial\theta}(\boldsymbol{e}_\rho\boldsymbol{e}_\rho) = \boldsymbol{e}_\rho\boldsymbol{e}_\theta + \boldsymbol{e}_\theta\boldsymbol{e}_\rho \tag{1.499}$$

$$\frac{\partial}{\partial\theta}(\boldsymbol{e}_\rho\boldsymbol{e}_\theta) = -\boldsymbol{e}_\rho\boldsymbol{e}_\rho + \boldsymbol{e}_\theta\boldsymbol{e}_\theta \tag{1.500}$$

$$\frac{\partial}{\partial \theta}(\boldsymbol{e}_{\varphi}\boldsymbol{e}_{\theta}) = -\boldsymbol{e}_{\varphi}\boldsymbol{e}_{\rho} \tag{1.501}$$

$$\frac{\partial}{\partial \varphi}(\boldsymbol{e}_{\rho}\boldsymbol{e}_{\rho}) = (\boldsymbol{e}_{\varphi}\boldsymbol{e}_{\rho} + \boldsymbol{e}_{\rho}\boldsymbol{e}_{\varphi})\sin\theta \tag{1.502}$$

$$\frac{\partial}{\partial \varphi}(\boldsymbol{e}_{\rho}\boldsymbol{e}_{\theta}) = \boldsymbol{e}_{\varphi}\boldsymbol{e}_{\theta}\sin\theta + \boldsymbol{e}_{\rho}\boldsymbol{e}_{\varphi}\cos\theta \tag{1.503}$$

$$\frac{\partial}{\partial \varphi}(\boldsymbol{e}_{\rho}\boldsymbol{e}_{\varphi}) = \boldsymbol{e}_{\varphi}\boldsymbol{e}_{\varphi}\sin\theta - \boldsymbol{e}_{\rho}(\boldsymbol{e}_{\rho}\sin\theta + \boldsymbol{e}_{\theta}\cos\theta) \tag{1.504}$$

$$\frac{\partial}{\partial \varphi}(\boldsymbol{e}_{\theta}\boldsymbol{e}_{\theta}) = (\boldsymbol{e}_{\varphi}\boldsymbol{e}_{\theta} + \boldsymbol{e}_{\theta}\boldsymbol{e}_{\varphi})\cos\theta \tag{1.505}$$

$$\frac{\partial}{\partial \varphi}(\boldsymbol{e}_{\varphi}\boldsymbol{e}_{\rho}) = (\boldsymbol{e}_{\varphi}\boldsymbol{e}_{\varphi} - \boldsymbol{e}_{\rho}\boldsymbol{e}_{\rho})\sin\theta - \boldsymbol{e}_{\theta}\boldsymbol{e}_{\rho}\cos \tag{1.506}$$

Problem 1.27 For cylindrical coordinates, prove the following identity

$$\begin{aligned}\nabla(\nabla\cdot\boldsymbol{u}) = {}& \boldsymbol{e}_{\rho}(\frac{\partial^2 u_{\rho}}{\partial \rho^2} + \frac{1}{\rho}\frac{\partial u_{\rho}}{\partial \rho} - \frac{u_{\rho}}{\rho^2} + \frac{1}{\rho}\frac{\partial^2 u_{\phi}}{\partial \rho \partial \phi} - \frac{1}{\rho^2}\frac{\partial u_{\phi}}{\partial \phi} + \frac{\partial^2 u_z}{\partial \rho \partial z}) \\ & + \boldsymbol{e}_{\phi}(\frac{1}{\rho}\frac{\partial^2 u_z}{\partial z \partial \phi} + \frac{1}{\rho^2}\frac{\partial^2 u_{\phi}}{\partial \phi^2} + \frac{1}{\rho}\frac{\partial^2 u_{\rho}}{\partial \rho \partial \phi} + \frac{1}{\rho^2}\frac{\partial u_{\rho}}{\partial \phi}) \\ & + \boldsymbol{e}_z(\frac{\partial^2 u_{\rho}}{\partial z \partial \rho} + \frac{1}{\rho}\frac{\partial u_{\rho}}{\partial z} + \frac{1}{\rho}\frac{\partial^2 u_{\phi}}{\partial z \partial \phi} + \frac{\partial^2 u_z}{\partial z^2})\end{aligned} \tag{1.507}$$

Problem 1.28 For spherical coordinates, prove the following identity

$$\begin{aligned}\nabla(\nabla\cdot\boldsymbol{u}) = {}& \boldsymbol{e}_{\rho}(\frac{\partial^2 u_{\rho}}{\partial \rho^2} + \frac{2}{\rho}\frac{\partial u_{\rho}}{\partial \rho} - \frac{2u_{\rho}}{\rho^2} + \frac{1}{\rho}\frac{\partial^2 u_{\theta}}{\partial \rho \partial \theta} - \frac{1}{\rho^2}\frac{\partial u_{\theta}}{\partial \theta} + \frac{\cot\theta}{\rho}\frac{\partial u_{\theta}}{\partial \rho} \\ & - \frac{\cot\theta}{\rho^2}u_{\theta} + \frac{1}{\rho\sin\theta}\frac{\partial^2 u_{\varphi}}{\partial \rho \partial \varphi} - \frac{1}{\rho^2\sin\theta}\frac{\partial u_{\varphi}}{\partial \varphi}) \\ & + \boldsymbol{e}_{\theta}(\frac{1}{\rho}\frac{\partial^2 u_{\rho}}{\partial \theta \partial \rho} + \frac{2}{\rho^2}\frac{\partial u_{\rho}}{\partial \theta} + \frac{1}{\rho^2}\frac{\partial^2 u_{\theta}}{\partial \theta^2} + \frac{\cot\theta}{\rho^2}\frac{\partial u_{\theta}}{\partial \theta} - \frac{u_{\theta}}{\rho^2\sin^2\theta} \\ & + \frac{1}{\rho^2\sin\theta}\frac{\partial^2 u_{\varphi}}{\partial \theta \partial \varphi} - \frac{\cot\theta}{\rho^2\sin\theta}\frac{\partial u_{\varphi}}{\partial \varphi}) \\ & + \boldsymbol{e}_{\varphi}(\frac{1}{\rho\sin\theta}\frac{\partial^2 u_{\rho}}{\partial \varphi \partial \rho} + \frac{2}{\rho^2\sin\theta}\frac{\partial u_{\rho}}{\partial \varphi} + \frac{1}{\rho^2\sin\theta}\frac{\partial^2 u_{\theta}}{\partial \theta \partial \varphi} \\ & + \frac{\cot\theta}{\rho^2\sin\theta}\frac{\partial u_{\theta}}{\partial \varphi} + \frac{1}{\rho^2\sin^2\theta}\frac{\partial^2 u_{\varphi}}{\partial \varphi^2})\end{aligned} \tag{1.508}$$

Problem 1.29 For cylindrical coordinates, prove the following identity

$$\nabla\times(\nabla\times\boldsymbol{u}) = \boldsymbol{e}_\rho\left(\frac{1}{\rho}\frac{\partial^2 u_\phi}{\partial\rho\partial\phi}+\frac{1}{\rho^2}\frac{\partial u_\phi}{\partial\phi}-\frac{1}{\rho^2}\frac{\partial^2 u_\rho}{\partial\phi^2}-\frac{\partial^2 u_\rho}{\partial z^2}+\frac{\partial^2 u_z}{\partial\rho\partial z}\right)$$
$$+\boldsymbol{e}_\phi\left(\frac{1}{\rho}\frac{\partial^2 u_z}{\partial z\partial\phi}-\frac{\partial^2 u_\phi}{\partial z^2}-\frac{\partial^2 u_\phi}{\partial\rho^2}-\frac{1}{\rho}\frac{\partial u_\phi}{\partial\rho}+\frac{u_\phi}{\rho^2}-\frac{1}{\rho^2}\frac{\partial u_\rho}{\partial\phi}+\frac{1}{\rho}\frac{\partial^2 u_\rho}{\partial\rho\partial\phi}\right) \quad (1.509)$$
$$+\boldsymbol{e}_z\left(\frac{\partial^2 u_\rho}{\partial z\partial\rho}+\frac{1}{\rho}\frac{\partial u_\rho}{\partial z}-\frac{\partial^2 u_z}{\partial\rho^2}-\frac{1}{\rho}\frac{\partial u_z}{\partial\rho}-\frac{1}{\rho^2}\frac{\partial^2 u_z}{\partial\phi^2}+\frac{1}{\rho}\frac{\partial^2 u_\phi}{\partial z\partial\phi}\right)$$

Problem 1.30 For spherical coordinates, prove the following identity

$$\nabla\times(\nabla\times\boldsymbol{u}) = \boldsymbol{e}_\rho\left(\frac{1}{\rho}\frac{\partial^2 u_\theta}{\partial\rho\partial\theta}+\frac{1}{\rho^2}\frac{\partial u_\theta}{\partial\theta}-\frac{1}{\rho^2}\frac{\partial^2 u_\rho}{\partial\theta^2}+\frac{\cot\theta}{\rho}\frac{\partial u_\theta}{\partial\rho}+\frac{\cot\theta}{\rho^2}u_\theta\right.$$
$$\left.-\frac{\cot\theta}{\rho^2}\frac{\partial u_\rho}{\partial\theta}-\frac{1}{\rho^2\sin^2\theta}\frac{\partial^2 u_\rho}{\partial\varphi^2}+\frac{1}{\rho\sin\theta}\frac{\partial^2 u_\varphi}{\partial\rho\partial\varphi}+\frac{1}{\rho^2\sin\theta}\frac{\partial u_\varphi}{\partial\varphi}\right)$$
$$+\boldsymbol{e}_\theta\left(\frac{1}{\rho^2\sin\theta}\frac{\partial^2 u_\varphi}{\partial\theta\partial\varphi}+\frac{\cot\theta}{\rho^2\sin\theta}\frac{\partial u_\varphi}{\partial\varphi}-\frac{1}{\rho^2\sin^2\theta}\frac{\partial^2 u_\theta}{\partial\varphi^2}-\frac{\partial^2 u_\theta}{\partial\rho^2}\right.$$
$$\left.-\frac{2}{\rho}\frac{\partial u_\theta}{\partial\rho}+\frac{1}{\rho}\frac{\partial^2 u_\rho}{\partial\rho\partial\theta}\right) \quad (1.510)$$
$$+\boldsymbol{e}_\varphi\left(\frac{1}{\rho\sin\theta}\frac{\partial^2 u_\rho}{\partial\rho\partial\varphi}-\frac{\partial^2 u_\varphi}{\partial\rho^2}-\frac{2}{\rho}\frac{\partial u_\varphi}{\partial\rho}-\frac{1}{\rho^2}\frac{\partial^2 u_\varphi}{\partial\theta^2}-\frac{\cot\theta}{\rho^2}\frac{\partial u_\varphi}{\partial\theta}\right.$$
$$\left.+\frac{u_\varphi}{\rho^2\sin^2\theta}+\frac{1}{\rho^2\sin\theta}\frac{\partial^2 u_\theta}{\partial\theta\partial\varphi}-\frac{\cot\theta}{\rho^2\sin\theta}\frac{\partial u_\theta}{\partial\varphi}\right)$$

Problem 1.31 Show the validity of the following equations

$$\mathrm{Re}[(1+i)^{1+i}] = e^{\ln\sqrt{2}-\pi/4-2n\pi}\cos(\ln\sqrt{2}+\pi/4+2n\pi) \quad (1.511)$$
$$\mathrm{Im}[(1+i)^{1+i}] = e^{\ln\sqrt{2}-\pi/4-2n\pi}\sin(\ln\sqrt{2}+\pi/4+2n\pi) \quad (1.512)$$

Problem 1.32 Find the value of

$$(2^i+2^{-i})/2 \quad (1.513)$$

Ans: 0.769238901...

Problem 1.33 Find the value of

$$(-1)^{\sin i} \quad (1.514)$$

Ans: $e^{-\pi\sinh(1)}$

Problem 1.34 Prove the following identity

$$-e=\sum_{n=0}^{\infty}\frac{(1+i\pi)^n}{n!} \tag{1.515}$$

Problem 1.35 Prove the following identity

$$\cos\theta+\cos 3\theta+...+\cos(2n-1)\theta=\frac{\sin 2n\theta}{2\sin\theta} \tag{1.516}$$

Hint: Use Euler's formula and note the following identity:

$$x^{2n}-y^{2n}=(x-y)(x^{2n-1}+x^{2n-2}y+x^{2n-3}y^2+...+xy^{2n-2}+y^{2n-1}) \tag{1.517}$$

Problem 1.36 Solve for z in the following equation

$$\cos(z)=2 \tag{1.518}$$

Ans: $z=i\ln(2\pm\sqrt{3})$

Hint: $\cos(ix)=\cosh x$

Problem 1.37 Show the following tensor identity for the permutation tensor:

$$e_{mjk}e_{njk}=2\delta_{mn} \tag{1.519}$$

Hint: Use the δ-*e* identity.

Problem 1.38 Show the following tensor identity for the permutation tensor:

$$e_{kmj}e_{kln}e_{npq}=\delta_{ml}e_{jpq}-\delta_{jl}e_{mpq} \tag{1.520}$$

Hint: Use the δ-*e* identity.

Problem 1.39 Show the following tensor identity for the permutation tensor:

$$e_{kmj}e_{kln}e_{npm}=e_{jpl} \tag{1.521}$$

Hint: Use the δ-*e* identity.

Problem 1.40 Show the following tensor identity for the permutation tensor:

$$e_{mjk}e_{kln}e_{npq}e_{qim}=\delta_{ij}\delta_{pl}+\delta_{jl}\delta_{pi} \tag{1.522}$$

Hint: Use the δ-*e* identity.

Problem 1.41 Show the following vector identity:

$$e_{mjk}e_{kln}e_{npq}e_{qim}=\delta_{ij}\delta_{pl}+\delta_{jl}\delta_{pi} \tag{1.523}$$

Hint: Use the δ-*e* identity.

Problem 1.42 Show the following vector identity:

$$\nabla\times(\boldsymbol{u}\times\boldsymbol{v})=\boldsymbol{u}(\nabla\cdot\boldsymbol{v})+(\boldsymbol{v}\cdot\nabla)\boldsymbol{u}-[\boldsymbol{v}(\nabla\cdot\boldsymbol{u})+(\boldsymbol{u}\cdot\nabla)\boldsymbol{v}] \tag{1.524}$$

Problem 1.43 Show the following vector identity:

$$\nabla(\boldsymbol{u}\cdot\boldsymbol{v})=(\boldsymbol{v}\cdot\nabla)\boldsymbol{u}+(\boldsymbol{u}\cdot\nabla)\boldsymbol{v}+\boldsymbol{v}\times(\nabla\times\boldsymbol{u})+\boldsymbol{u}\times(\nabla\times\boldsymbol{v}) \tag{1.525}$$

Problem 1.44 Show the following infinite product for the sine function:

$$\sin x = x(1-\frac{x^2}{\pi^2})(1-\frac{x^2}{2^2\pi^2})(1-\frac{x^2}{3^2\pi^2})\cdots \tag{1.526}$$

Hint: Recall from Taylor series expansion that

$$\sin x = x-\frac{x^3}{3!}+\frac{x^5}{5!}-\frac{x^7}{7!}+... \tag{1.527}$$

Can you factorize the infinite polynomials as products of their infinite roots? Note also that this result agrees with formula (1.316) given in Example 1.10, but in this problem, without using Mittag-Leffler's expansion formula.

Problem 1.45 Following the similar idea used in Problem 1.44, show the following infinite product for the cosine function:

$$\cos x = (1-\frac{4x^2}{\pi^2})(1-\frac{4x^2}{3^2\pi^2})(1-\frac{4x^2}{5^2\pi^2})\cdots \tag{1.528}$$

Hint: Recall from Taylor series expansion that

$$\cos x = 1-\frac{x^2}{2!}+\frac{x^4}{4!}-\frac{x^6}{6!}+... \tag{1.529}$$

Can you factorize the infinite polynomials as a product of their infinite roots?

Problem 1.46 This problem considers the Basel problem. Following Problem 1.44, we can rewrite (1.526) and (1.527) as

$$\frac{\sin x}{x} = (1-\frac{x^2}{\pi^2})(1-\frac{x^2}{2^2\pi^2})(1-\frac{x^2}{3^2\pi^2})\cdots \tag{1.530}$$

$$\frac{\sin x}{x} = 1-\frac{x^2}{3!}+\frac{x^4}{5!}-\frac{x^6}{7!}+... \tag{1.531}$$

(i) Show that the coefficient for the x^2 term on the right hand side of (1.530) is

$$-(\frac{1}{\pi^2}+\frac{1}{2^2\pi^2}+\frac{1}{3^2\pi^2}+...)=-\frac{1}{\pi^2}\sum_{n=1}^{\infty}\frac{1}{n^2} \tag{1.532}$$

(ii) Compare this coefficient for the x^2 term on the right hand side of (1.531) and prove that

$$\sum_{n=1}^{\infty}\frac{1}{n^2}=\frac{1}{1^2}+\frac{1}{2^2}+\frac{1}{3^2}+\cdots+\frac{1}{n^2}+...=\frac{\pi^2}{6} \tag{1.533}$$

The quest for the exact sum of (1.533) was first posed by Mengoli in 1644 and again by Jacob Bernoulli in 1689 to the boarder mathematical community. This is known as the Basel problem because of its association with the Bernoulli family who lived in Basel, Switzerland. This result was first obtained by Euler in 1734 at the age of twenty-eight. The closed form sum in the last part of (1.533) brought Euler instant fame in the mathematical world. This result also led to his later analysis of the Euler-Riemann zeta function and the well-known Riemann hypothesis (see Problem 4.56 in Chapter 4).

Problem 1.47 Show the following series expansion of sine integral $Si(x)$ defined as (Abramowitz and Stegun, 1964):

$$Si(x) = \int \frac{\sin x}{x} dx = \sum_{n=0}^{\infty} \frac{(-1)^n x^{2n+1}}{(2n+1)(2n+1)!} \tag{1.534}$$

Hint: Recall (1.531).

CHAPTER TWO

Introduction to Differential Equations

2.1 INTRODUCTION

Most of the textbooks on the history of mathematics either did not cover or only briefly mention the discovery of calculus, let alone the historical development of differential equations. In terms of the coverage of the development of differential equations, the three-volume series by Kline (1972) *Mathematical Thought from Ancient to Modern Times* are the best. In addition to mathematical thought, a lot of mathematical details were covered. The application background of differential equations was also covered in detail.

The origin of differential equations naturally arises from the study of calculus, discovered independently by British physicist and mathematician Newton and German mathematician and diplomat Leibniz. The dispute on the priority of the discovery of calculus between Newton and Leibniz is briefly covered in the biography section. Our present usage of symbols and terminology owes mainly to Leibniz. Leibniz also discovered the method of separation of variables and the general theory of homogeneous type for first order ordinary differential equations (ODEs). Through correspondence between Leibniz and the Bernoulli brothers, many problems of differential equations were solved. The brachistochrone problem occupies a central place in the development of differential equations, and it was solved by Jacob and Johann Bernoulli, Newton, Leibniz, and L'Hôpital. Daniel Bernoulli, son of Johann Bernoulli, derived partial differential equations for fluid mechanics, whereas Leonhard Euler, a student of Johann Bernoulli, derived and applied differential equations for mechanics analysis. The method of integrating factors for ODEs was also discovered by Euler, and he was responsible for the first systematic solution technique for solving differential equations. Lagrange, an Italian-born French mathematician, contributed significantly to the development of theory for the solutions of differential equations, particular for the particular solution for nonhomogeneous ODEs and first order partial differential equations (PDEs). Laplace used differential equations to model the motions of celestial bodies, and the Laplace equation is one of the most important second order PDEs. Considering the vibrations of strings, D'Alembert considered the first solution of wave equations, another important second order PDE. Many others made important contributions to differential equations, including Riccati, Clairaut, Fourier, Airy, Jacobi, Poisson, Bessel, Legendre, Hermite, Hankel, Monge, Charpit, Cauchy, Pfaff, Gauss, Goursat, Ampere, Green, Helmholtz, Riemann, Tricomi, Courant, Robin, Dirichlet, Fredholm, Fuchs, Liouville, Sturm, Painlevé, and many others.

2.2 TOTAL AND PARTIAL DERIVATIVES

If a function $u(x)$ depends on only one variable, the change of the function with respect to the change on the variable is denoted as

$$\frac{du}{dx} \tag{2.1}$$

This notation was proposed by Leibniz and "*d*" for differentiation or derivative. However, when a function depends on more than one variables, say $u(x_1, x_2, ..., x_n)$, as discussed in Chapter 1, the change in function u with respect to the change of only one variable is called partial differentiation. Our current adopted notation for partial differentiation was proposed by Legendre as the "partial *d*" or "∂" (Cajori, 1993)

$$\frac{\partial u(x_1, x_2, ..., x_n)}{\partial x_i} \tag{2.2}$$

where $i = 1,2,\ldots,n$ and x_1, ..., x_n are called independent variables. Since "∂" also symbolizes a rounded d, it is sometimes pronounced as "round d" or simply "round."

2.3 ORDER OF DIFFERENTIAL EQUATIONS

The unknown function u is also called the dependent variable. In general, a differential equation for an unknown u can be written symbolically as:

$$F(x_1,\cdots,x_n,u,\frac{\partial u}{\partial x_1},\cdots,\frac{\partial u}{\partial x_n},\cdots,\frac{\partial^m u}{\partial x_1^{m_1}\partial x_2^{m_2}\cdots\partial x_n^{m_n}})=0 \tag{2.3}$$

where m_1, m_2, ..., m_n are integers. This is clearly a partial differential equation or PDE. The order of the differential equation is determined by the highest derivative term. For example, the order of the equation given in (2.3) is

$$m = m_1 + m_2 + \cdots + m_n \tag{2.4}$$

For the case of single variable x, (2.3) can be simplified to:

$$F_1(x,u,\frac{du}{dx},\cdots,\frac{d^m u}{dx^m}) = 0 \tag{2.5}$$

Clearly, the order of the differential equation is m.

2.4 NONLINEAR VERSUS LINEAR

When a differential equation is linear, all terms of the unknown or its derivatives in the differential equations can only appear linearly. For example, the following is a linear ODE of u

$$\frac{d^n u}{dx^n} + a_1(x)\frac{d^{n-1}u}{dx^{n-1}} + \cdots + a_n(x)u = f(x) \tag{2.6}$$

This ODE is linear no matter how nonlinear the functions a_1,..., a_n depend on the variable x. Whether the differential equation is linear or nonlinear, it all depends on the nonlinearity of u and has nothing to do with the coefficient functions a_1,...,

a_n, which are not functions of the unknown u. All terms of u or its derivatives appear in (2.6) in their first order or first degree. In general, if a differential equation is not linear, it is nonlinear. However, in more advanced books dealing with nonlinear differential equations, there are two kinds of nonlinearity: (i) quasi-linear differential equations, in which nonlinearity of u appears in the coefficient functions in $a_1,\ldots,a_n$ in (2.6); and (ii) nonlinear equations in which nonlinearity occurs in the highest order derivative term (Kevorkian and Cole, 1981). Clearly, the first kind of nonlinearity is called quasi-linear because its solution technique is simpler than those for the second type of nonlinearity.

For example, the following second order partial differential equation is quasi-linear because there is no nonlinearity in its highest derivative (second order):

$$\begin{aligned} &A_{11}(x_1,x_2,u,\frac{\partial u}{\partial x_1},\frac{\partial u}{\partial x_2})\frac{\partial^2 u}{\partial x_1^2}+A_{12}(x_1,x_2,u,\frac{\partial u}{\partial x_1},\frac{\partial u}{\partial x_2})\frac{\partial^2 u}{\partial x_1 \partial x_2} \\ &+A_{22}(x_1,x_2,u,\frac{\partial u}{\partial x_1},\frac{\partial u}{\partial x_2})\frac{\partial^2 u}{\partial x_2^2}=F(x_1,x_2,u\frac{\partial u}{\partial x_1},\frac{\partial u}{\partial x_2}) \end{aligned} \tag{2.7}$$

For first order partial differential equations, the following form is quasi-linear

$$A_1(x_1,x_2,u)\frac{\partial u}{\partial x_1}+A_2(x_1,x_2,u)\frac{\partial u}{\partial x_2}=B(x_1,x_2,u) \tag{2.8}$$

Most of the established methods of solutions for differential equations are for linear ODEs or PDEs. Indeed, most of the differential equations to be discussed in this book are linear. Although some methods for solving nonlinear differential equations approximately have been developed (such as the perturbation method to be discussed in Chapter 12), in general, there are no well-accepted methods to deal with nonlinear differential equations. Quite often we have to rely on numerical methods to get approximate solutions for nonlinear differential equations. Exact solutions for nonlinear ODEs or PDEs exist only for certain special forms of simple nonlinear equations. Certain types of nonlinear ODEs can be transformed to linear ODEs and are solvable in known functions. The most notable examples are the first order Riccati equation and the Bernoulli equation. In considering a problem posed by Picard for second order nonlinear ODEs, Painlevé in 1900 identified six equation types that can be transformed to linear ODEs and can be solved analytically. Their solutions are called Painlevé transcendents and these equations contain only movable singularities as poles (or the so-called Painlevé property) (see Section 4.14). It was discovered that Painlevé equations arise as reductions of soliton equations, which are solvable by the inverse scattering technique (IST) (Ablowitz and Clarkson, 1991). See more details in Chapter 4. Painlevé was a student of Flex Klein, and served as French Prime Minister twice (see biography section). Therefore, our ability of classifying whether a differential equation is linear or nonlinear is crucial before we even try to look for a solution.

One major problem in solving nonlinear PDEs or ODEs is that nonlinear differential equations may have more than one solution at some points of the independent variables. That is, there is no guarantee of having a unique solution at a certain domain of the problem. Typical examples include buckling of a long bar under compression (mathematically it was called Euler's buckling because of his contribution to this problem), necking of a long steel bar under tension, and barrelling of a short soil specimen under confined compression. All these physical phenomena of a uniquely and homogeneously deforming material yielding suddenly to a non-uniformly deforming pattern can be modeled mathematically as bifurcation problems of nonlinear differential equations. If the specimen continues to deform uniformly, the solution is called the equilibrium state. If the specimen deforms into a non-uniform state, the solution is called a bifurcated state or a bifurcation (since a unique solution bifurcates into two or more solutions at that point). When such non-unique solutions exist, the point in the domain is called a bifurcation point. If there is more than one solution in the solution domain, there is a related issue of which of these bifurcated solutions or the equilibrium state is stable (or it would appear in nature). The issue of dynamic stability needs to be considered mathematically (see the chapter on nonlinear buckling in volume two of this book series). This problem of bifurcation can occur repeatedly, that is, the bifurcated state can further bifurcate into a secondary bifurcation, and then to a tertiary bifurcation, and so on. When repeated bifurcation occurs, it is called a cascading bifurcation. It is believed that the onset of such complicated cascading bifurcations will lead to a "nearly" unpredictable solution or chaotic behavior of the solutions (or simply called chaos). In 1975, a Los Alamos scientist, Mitchell Feigenbaum, when solving a differential equation using the difference equation, discovered that period doubling of a nonlinear iterative solution has a universal pattern for a large class of nonlinear equations when chaotic solutions occur (the so-called Feigenbaum number of 4.6692 for predicting the occurrence of period doubling). This discovery leads to the new hope and belief that the so-called chaos is actually predictable, as opposed to the original definition of chaos. This view was popularized by James Gleick (1988) by his book *Chaos: Making of a New Science*. Unfortunately, not much progress has been made so far.

In the present book, we mainly deal with linear differential equations. But some simple nonlinear differential equations will be covered whenever exact solutions for them can be found.

2.5 PDE VERSUS ODE

The main difference between ordinary differential equations (ODEs) and partial differential equations (PDEs) lies on the number of independent variables. If an unknown u depends on only one variable, the resulting differential equation is an ODE, otherwise, it is a PDE. The method of solution for ODEs is more well established (which will be covered in the next few chapters) whereas PDEs are often converted to ODEs using the method of separation of variables. For example, we can use separation of variables to convert the following PDE called the Laplace equation into an ODE:

$$\frac{\partial^2 u}{\partial x^2}+\frac{\partial^2 u}{\partial y^2}=0 \tag{2.9}$$

In particular, we assume

$$u(x,y)=X(x)Y(y) \tag{2.10}$$

Substitution of (2.10) into (2.9) yields

$$\frac{\partial^2(XY)}{\partial x^2}+\frac{\partial^2(XY)}{\partial y^2}=\frac{d^2X(x)}{dx^2}Y(y)+X(x)\frac{d^2Y(y)}{dy^2}=0 \tag{2.11}$$

Note that the round ∂ is replaced by d since X is only a function of x, whereas Y is only a function of y. Rearranging (2.11) gives

$$\frac{1}{X(x)}\frac{d^2X(x)}{dx^2}=-\frac{1}{Y(y)}\frac{d^2Y(y)}{dy^2}=\lambda \tag{2.12}$$

where X and Y are clearly nonzero (otherwise, there is no solution for u). Since the first term is only a function x whereas the second term is only a function of y, both of these terms must be constant. Thus, λ must also be a constant. Whether λ is positive or negative, it depends on the boundary condition. Later discussions in Chapters 7 and 9 on separation of variables will further tackle this issue. Nevertheless, (2.12) actually gives both the governing equations (ODEs) for functions X and Y:

$$\frac{d^2X}{dx^2}-\lambda X=0 \tag{2.13}$$

$$\frac{d^2Y}{dy^2}+\lambda Y=0 \tag{2.14}$$

From our discussion in Sections 1.11 and 1.12, one should immediately realize that the general solutions for (2.13) and (2.14) are respectively

$$X=C_1\sinh\sqrt{\lambda}x+C_2\cosh\sqrt{\lambda}x \tag{2.15}$$

$$Y=C_3\sin\sqrt{\lambda}y+C_4\cos\sqrt{\lambda}y \tag{2.16}$$

where the unknown constants need to be determined by boundary conditions. For certain given boundary conditions, λ need to prescribe as negative (or $\lambda=-|\lambda|$), and thus the general solutions for X and Y will be reversed:

$$Y=C_1\sinh\sqrt{|\lambda|}y+C_2\cosh\sqrt{|\lambda|}y \tag{2.17}$$

$$X=C_3\sin\sqrt{|\lambda|}x+C_4\cos\sqrt{|\lambda|}x \tag{2.18}$$

For this particular example, we see that solving one PDE (i.e., (2.9)) is equivalent to solving two ODEs (i.e., (2.15) and (2.16)) if separation of variables is applicable. Unfortunately, as we will demonstrate in a later chapter, separation of variables does not work for all PDEs.

2.6 NONHOMOGENEOUS VERSUS HOMOGENEOUS

Recall from equation (2.6) that there is a term $f(x)$ on the right hand side which is independent of the unknown function u. Whenever such term exists, the differential equation is called nonhomogeneous and $f(x)$ is called the nonhomogeneous term. For example, we can rewrite (2.6) as

$$\frac{d^nu}{dx^n}+a_1(x)\frac{d^{n-1}u}{dx^{n-1}}+\cdots+a_n(x)u+g(x)=0 \tag{2.19}$$

Clearly, this is the same as (2.6) if we realize that $f=-g$. Thus, the nonhomogeneous term can be on the left or on the right; it will not change the fact that it is nonhomogeneous. As will be seen in the next chapter, the general solution of nonhomogeneous differential equations must be the sum of the solution of the corresponding homogeneous equation (i.e., by setting f in (2.6) or g in (2.19) to zero) and the particular solution satisfying the nonhomogeneous equation. Mathematically, we have

$$u(x) = u_h(x) + u_p(x) \tag{2.20}$$

In other words, if the differential equation is nonhomogeneous, we have to solve the problem twice. Therefore, it is important to classify the type of differential equation before trying to solve for its solution.

Example 2.1 Classify the following differential equations

$$\frac{dy}{dx} = 2x + 1 \tag{2.21}$$

Solution: This is a linear first order nonhomogeneous ODE. The term $2x+1$ makes it nonhomogeneous. The highest derivative term is first order. There is only one variable, so it is an ordinary differential equation. Note that this first order ODE can be integrated directly as

$$y = \int (2x+1)dx + C = x^2 + x + C \tag{2.22}$$

This is called a separable ODE and will be further discussed in Section 3.2.1.

Example 2.2 Classify the following differential equation

$$(y - x^2)dy - 5ydx = 0 \tag{2.23}$$

Solution: First of all, we can divide the whole equation by dx to get

$$(y - x^2)\frac{dy}{dx} - 5y = 0 \tag{2.24}$$

The coefficient for dy/dx is a function of y, thus it is a nonlinear differential equations, or more precisely, quasi-linear. It is a homogeneous first order ordinary differential equation because there is no term in the equation that is independent of y.

Example 2.3 Classify the following differential equation

$$\frac{d^2 y}{dt^2} + t^2 y \left(\frac{dy}{dt}\right)^3 + y = 0 \tag{2.25}$$

Solution: There are both first and second derivative terms, but the highest order derivative term always controls. Thus, it is of second order. The differential equation is nonlinear because of the nonlinearly in the second term, but is homogeneous as all terms are functions of the unknown y. There is only one variable and "d" is used instead of "∂," so it is an ODE.

Example 2.4 Classify the following differential equation

$$\frac{d^4x}{dt^4}+5\frac{d^2x}{dt^2}+3x=\sin t \tag{2.26}$$

Solution: Again the highest derivative term controls, and thus it is a fourth order ODE, not second order. In contrast to the last example, here x is the unknown, not the independent variable (t is the independent variable). The term $\sin t$ on the right is independent of the unknown x and thus is the nonhomogeneous term. Thus, it is a nonhomogeneous ODE. All x and its derivative terms appear linearly, and thus it is a linear ODE.

Example 2.5 Classify the following differential equation

$$\frac{\partial^2 u}{\partial x^2}+\frac{\partial^2 u}{\partial y^2}+x+y-uz=0 \tag{2.27}$$

Solution: The highest derivative is of second order. This is clearly a PDE as u is a function of x, y, and z. All terms involving u and its derivatives are linear, and thus it is a linear PDE. The term $x+y$ is independent of the unknown u and thus it is a nonhomogeneous term. Therefore, (2.27) is a linear nonhomogeneous PDE of second order.

2.7 SOME SIGNIFCIANT NONLINEAR PDES

In reality, nearly all real systems are nonlinear in nature. It just happens that most of these physically real systems are well behaved and can be modeled adequately by a linear model most of the time, except at certain given parameters that satisfy the bifurcation condition. Many significant physical phenomena cannot be explained by a linear model. In this section, we will list some of them.

The pendulum equation is one of the first nonlinear differential equations that attracted the attention of applied mathematicians. The horizontal force equilibrium for the mass m shown in Figure 2.1 is

$$m\frac{d^2x}{dt^2}+F_T\sin\theta=0 \tag{2.28}$$

where θ is the angular rotation and is measured in radians. For small-amplitude oscillations, the vertical acceleration of the mass can be neglected. For such case, the tension in the spring can be related to the weight of the pendulum

$$F_T\cos\theta=mg \tag{2.29}$$

where g is the gravitational constant (9.81 m/s^2). For a small value of θ, we can further assume the following approximations:

$$x=L\sin\theta\approx L\theta,\quad \cos\theta\approx 1,\quad F_T\approx mg \tag{2.30}$$

where L is the length of the pendulum shown in Figure 2.1. Substitution of (2.29) and (2.30) into (2.28) yields the following differential equation for the angular rotation of the pendulum

$$\frac{d^2\theta}{dt^2}+\frac{g}{L}\sin\theta=0 \tag{2.31}$$

If the amplitude of pendulum motion is small, we can approximate $\sin\theta \approx \theta$. As expected, the linearized equation becomes

$$\frac{d^2\theta}{dt^2}+\omega^2\theta=0 \tag{2.32}$$

where $\omega^2 = g/L$. This is the governing equation for sine and cosine as we discussed in Section 1.11, and thus we have

$$\theta = C_1\sin\omega t + C_2\cos\omega t \tag{2.33}$$

Physically, ω is the natural circular frequency of the swinging motion. It is also clear that there is no damping built into the model given in (2.28). In reality, there is always air resistance or internal friction in the pendulum against the swinging motion. Nevertheless, neglecting air resistance and assuming small amplitude of pendulum motion, we have the natural period of oscillations as

$$T = 2\pi\sqrt{\frac{L}{g}} \tag{2.34}$$

There are at least two major significances associated with this simple result of pendulum motion. In the eighteenth century, Newton used the observed changes in the period of oscillation of a pendulum at various places on the earth's surface and deduced that the gravitational constant is not a constant. He further deduced that the earth bulges at the equator. This period of oscillation is an intrinsic property of the earth. For example, the period of oscillations of the same pendulum on Mars would be different because of the difference in the gravitation field on Mars.

On the other hand, the strength of gravity on any planet clearly relates to the total mass of the planet which, in turn, depends on the size of the planet. Because of this, it has been proposed that the "standard length" on earth should be related to the size or gravity of the earth. It was proposed that the standard length meter can be defined as the length L of a pendulum that gives a fundamental period of 2 seconds. Thus, rearranging,

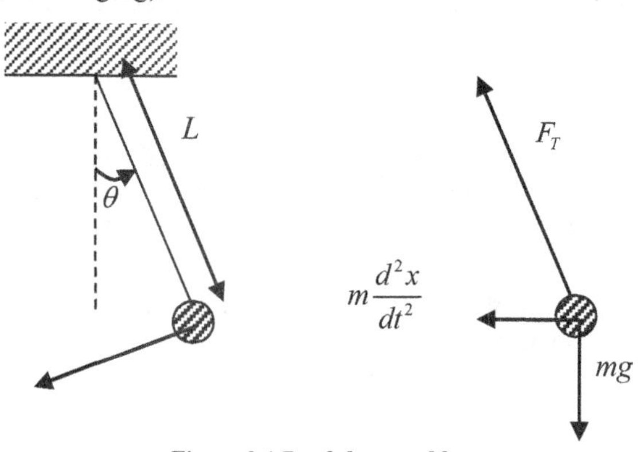

Figure 2.1 Pendulum problem

$$L = g(\frac{T}{2\pi})^2 \approx 9.821(\frac{2}{2\pi})^2 \approx 0.995075344m \tag{2.35}$$

However, the currently adopted definition of a meter is taken as one ten-millionth of the distance between the equator and the pole.

$$L = (\frac{2\pi R}{4})\frac{1}{10000000} \approx (\frac{2\pi \times 6366000}{4})\frac{1}{10000000} = 0.999968941m \tag{2.36}$$

where R is the radius of the Earth.

Example 2.6 Solve the nonlinear pendulum equation given in (2.28) with the boundary condition that the maximum angular rotation equals ω when $d\theta/dt = 0$.

Solution: The first and also the most crucial step is to recognize the first integral (or integrate the equation once) to get

$$\frac{1}{2}(\frac{d\theta}{dt})^2 - \frac{g}{L}\cos\theta = C \tag{2.37}$$

where C is an arbitrary constant. If we differentiate (2.37) once, we have

$$(\frac{d\theta}{dt})\frac{d^2\theta}{dt^2} + \frac{g}{L}\sin\theta(\frac{d\theta}{dt}) = \frac{d\theta}{dt}(\frac{d^2\theta}{dt^2} + \frac{g}{L}\sin\theta) = 0 \tag{2.38}$$

Since $d\theta/dt$ cannot be zero (otherwise there will be no oscillation), we must have

$$\frac{d^2\theta}{dt^2} + \frac{g}{L}\sin\theta = 0 \tag{2.39}$$

This is precisely (2.28). Next, we can impose the boundary condition to (2.37) to find C as

$$C = -\frac{g}{L}\cos\omega \tag{2.40}$$

Substitution of (2.40) into (2.37) and rearranging the resulting equation gives

$$\frac{d\theta}{dt} = \sqrt{\frac{2g}{L}}\sqrt{(\cos\theta - \cos\omega)} \tag{2.41}$$

Clearly, all functions of θ can be put on one side to yield (this is so-called separable and will be discussed in the next chapter)

$$dt = \sqrt{\frac{L}{2g}}\frac{d\theta}{\sqrt{(\cos\theta - \cos\omega)}} \tag{2.42}$$

The following change of variables is then assumed

$$\cos\theta = 1 - 2k^2\sin^2\phi, \quad k = \sin\frac{\omega}{2} \tag{2.43}$$

Recall the double angle formula for cosine; we have

$$1 - \cos\omega = 2\sin^2(\frac{\omega}{2}) \tag{2.44}$$

Using this formula, we have

$$\cos\theta - \cos\omega = 2k^2\cos^2\phi,$$
$$\sin\theta = 2k\sin\phi\sqrt{1-k^2\sin^2\phi}, \tag{2.45}$$
$$\sin\theta d\theta = 4k^2\sin\phi\cos\phi d\phi$$

Substitution of (2.45) into (2.42) leads to

$$dt = \sqrt{\frac{L}{g}}\frac{d\phi}{\sqrt{1-k^2\sin^2\phi}} \tag{2.46}$$

For $\theta = 0$, (2.43) shows that $\cos\theta = 1$ and thus $\phi = 0$. If the angle rotation increases to θ_0, ϕ_0 can be evaluated from

$$\sin^2\phi_0 = \frac{1-\cos\theta_0}{2k^2} = \frac{\sin^2(\theta_0/2)}{k^2} \tag{2.47}$$

Therefore, we have

$$\phi_0 = \sin^{-1}[\frac{\sin(\theta_0/2)}{k}] \tag{2.48}$$

Integrating (2.46) from 0 to ϕ_0, we have

$$t = \sqrt{\frac{L}{g}}\int_0^{\phi_0}\frac{d\phi}{\sqrt{1-k^2\sin^2\phi}} = \sqrt{\frac{L}{g}}F(\phi_0,k) \tag{2.49}$$

where $F(\phi,k)$ is the elliptic integral of the first kind with $k < 1$ (Abramowitz and Stegun, 1964). This integral cannot be evaluated in terms of any known functions, and thus, a numerical table of this integral has been evaluated. The elliptic integral appears naturally in many problems in engineering and science, including the elliptic motions of celestial bodies. This function was studied extensively by A. Legendre. To find the period of pendulum oscillations, we note that when $\theta = \omega$, the angular velocity is $d\theta/dt = 0$ and thus $\phi = \pi/2$. Consequently, the period T becomes

$$T = 4\sqrt{\frac{L}{g}}\int_0^{\pi/2}\frac{d\phi}{\sqrt{1-k^2\sin^2\phi}} = 4\sqrt{\frac{L}{g}}F(\pi/2,k) = 4\sqrt{\frac{L}{g}}K(k) \tag{2.50}$$

where $K(k)$ is the complete elliptic integral of the first kind (Abramowitz and Stegun, 1964).

For the general solution of rotation at any time, we have

$$t = \sqrt{\frac{L}{g}}\int_0^{\phi}\frac{d\psi}{\sqrt{1-k^2\sin^2\psi}} \tag{2.51}$$

Finally, the angular rotation can be evaluated from (2.51) by adopting Jacobi's elliptic functions, which was proposed independently by C.G.J. Jacobi and N.H. Abel in 1827. In particular, we can write

$$u = \int_0^{\phi}\frac{d\psi}{\sqrt{1-k^2\sin^2\psi}} \tag{2.52}$$

We can solve for ϕ from (2.52) and denote its solution symbolically as

$$\mathrm{sn}(u,k) = \sin\phi, \quad \mathrm{cn}(u,k) = \cos\phi \tag{2.53}$$

where sn is called the Jacobi elliptic sine and cn is called the Jacobi elliptic cosine (Abramowitz and Stegun, 1964). In a sense, you can consider the Jacobi elliptic functions are the solution of the integral (2.52) with given u and k. Elliptic integral and Jacobi's elliptic functions are among the very first functions specially defined to solve nonlinear differential equations. They are considered as "special functions" (versus the standard functions of circular and hyperbolic functions) and the essential results for studying nonlinear differential equations. Applying (2.52) and (2.53) to (2.51), we obtain

$$\sin\phi = \mathrm{sn}(t\sqrt{\frac{g}{L}}, k) = \frac{1}{k}\sin(\frac{\theta}{2}) \tag{2.54}$$

The last of (2.54) is a consequence of using the first equation of (2.43). Finally, we have the exact solution for the rotation θ as

$$\theta(t) = 2\sin^{-1}\left[k\,\mathrm{sn}(t\sqrt{\frac{g}{L}}, k)\right], \quad k = \sin(\frac{\omega}{2}) \tag{2.55}$$

This is one of the very first nonlinear differential equation for which an exact solution can be found. But, in general, this is not possible for most nonlinear differential equations.

More mathematical properties of the Jacobi elliptic function can be found in Appendix D.

Another nonlinear differential equation used to model the oscillations of current in an electric circuit connected to a tunnel diode with nonlinear damping is called the van der Pol equation, which was named after Dutch electrical engineer Balthasar van der Pol (1889–1959). Mathematically, the van der Pol equation reads as

$$\frac{d^2y}{dt^2} - \mu(1-y^2)\frac{dy}{dt} + y = 0 \tag{2.56}$$

Although the nonlinear damping term is only of second order and looks innocent, but it leads to chaotic behavior of the oscillations.

In gas dynamics and traffic flow, a one-dimensional model called Burgers' equation is found useful

$$\frac{\partial u}{\partial t} + u\frac{\partial u}{\partial x} = \nu\frac{\partial^2 u}{\partial x^2} \tag{2.57}$$

This equation was studied by Burgers in 1948. We will see later that this is just a special case of the Navier-Stokes equation for fluid flows (see Section 2.8.2).

Example 2.7 Show that Burgers equation can be transformed into a linear diffusion equation of a function ϕ by the Cole-Hopf transformation defined by

$$u = \psi_x = -2\nu\frac{\phi_x}{\phi} \tag{2.58}$$

Solution: The first and the most crucial step is to rewrite (2.57) into a form of conservation law:

$$u_t + (\frac{1}{2}u^2 - \nu u_x)_x = 0 \tag{2.59}$$

Independently, Hopf in 1950 and Cole in 1951 introduced the following system

$$u = \psi_x \tag{2.60}$$

$$\psi_t = \nu u_x - \frac{1}{2}u^2 \tag{2.61}$$

It is straightforward to show that (2.60) and (2.61) is equivalent to (2.59). Next, they introduced the following function ϕ

$$\psi = -2\nu \ln \phi \tag{2.62}$$

Differentiation of (2.62) with respect to both x and t gives respectively

$$u = \psi_x = -2\nu \frac{\phi_x}{\phi}, \quad \psi_t = -2\nu \frac{\phi_t}{\phi} \tag{2.63}$$

Differentiating u once with respect to x gives

$$u_x = \psi_{xx} = -2\nu \left\{ \frac{\phi_{xx}}{\phi} - (\frac{\phi_x}{\phi})^2 \right\} \tag{2.64}$$

Substitution of the first equation of (2.63) and (2.64) into (2.61) gives

$$\begin{aligned} \psi_t &= \nu \psi_{xx} - \frac{1}{2}\psi_x^2 = -2\nu^2 \left[\frac{\phi_{xx}}{\phi} - (\frac{\phi_x}{\phi})^2 \right] - \frac{1}{2} 4\nu^2 (\frac{\phi_x}{\phi})^2 \\ &= -2\nu^2 \frac{\phi_{xx}}{\phi} = -2\nu \frac{\phi_t}{\phi} \end{aligned} \tag{2.65}$$

Recognizing the second of (2.63), we get the last of (2.65). Thus, the last two of (2.65) can be reduced to

$$\nu \phi_{xx} = \phi_t \tag{2.66}$$

This is the linear diffusion equation, which will be discussed later in Chapter 9. This leads to a much deeper question: What kind of nonlinear PDE can be transformed into a linear PDE? As discussed in Section 2.4, Painlevé equations provide a partial answer for the case of the second order ODEs (see Section 4.14).

Another important nonlinear PDE is the KdV equation:

$$\frac{\partial u}{\partial t} = \frac{\partial^3 u}{\partial x^3} + u \frac{\partial u}{\partial x} \tag{2.67}$$

Korteweg and his PhD student de Vries studied the solitary wave observed in Scotland by civil engineer, John Scott Russell (1808–1882), using this nonlinear equation. The two terms on the right hand side of (2.67) are competing terms; the dispersion term tries to diffuse out the wave amplitude while the nonlinear convective term tries to build amplitude nonlinearly. When there is a delicate balance between these two terms, the amplitude of such a wave will not decay with propagation and it is called a soliton. Scientists believe that the "giant red" eye on Jupiter is a soliton, rogue or freak wave observed in oceans is a soliton, and a cloud system called "morning glory" in Australia is also a soliton. A soliton is not a phenomenon restricted to waves in fluids, and it has been observed in many different disciplines of science and engineering. For example, Alan Hodgkin won the Nobel Prize in medicine in 1963 when he found that the nerve pulse in squid behaves as a soliton. This equation can be solved by a new analytic technique for nonlinear differential equations called the inverse scattering technique (IST). In a sense, the IST is similar to the Fourier transform technique for linear differential equations, and is a major advance in solving soliton problems (Ablowitz and Clarkson, 1991). It is also related to Painlevé equations.

Another differential equation closely related to KdV is the Boussinesq equation:

$$\frac{\partial^2 u}{\partial t^2} - \frac{\partial^2 u}{\partial x^2} - \frac{\partial^4 u}{\partial x^4} + 3\frac{\partial^2 (u^2)}{\partial x^2} = 0 \tag{2.68}$$

It was named after French physicist and mathematician J. Boussinesq (1842–1929). The first two terms form a "typical" 1-D wave equation, the third term is the fourth order dispersive term, and the last term is the nonlinear term. In a sense, this closely resembles the KdV equation, with the third and fourth terms competing one against another. Again, soliton type solutions can be obtained for the Boussinesq equation. However, it should be noted that there are many different versions of the so-called Boussinesq equation in the literature, although somewhat similar to (2.68).

The Duffing equation is a nonlinear differential equation that models nonlinear oscillations of a pendulum subjected to friction and driven by a period force:

$$\frac{\partial^2 y}{\partial t^2} + y + \varepsilon y^3 = 0 \tag{2.69}$$

George Duffing (1861–1944) published a small book on this equation in 1918 and it has been investigated by many others since then. The period of oscillations not only depends on ε, but also on the amplitude of the motion.

The sine-Gordon equation was studied by Walter Gordon (1893–1939) and is another equation closely related to the soliton:

$$\frac{\partial^2 u}{\partial t^2} + \frac{\partial^2 u}{\partial y^2} \pm \sin u = 0 \tag{2.70}$$

This equation can again be solved by using the IST.

A nonlinear differential equation called the Thomas-Fermi equation, named after British physicist L.H. Thomas and 1938 Nobel Prize recipient E. Fermi, was formulated in modeling the distribution of electrons in a heavy atom. This theory was proposed independently by Thomas and Fermi in 1927. The Thomas-Fermi reads (Davis, 1962)

$$\frac{d^2 y}{dx^2} = \frac{1}{\sqrt{x}} y^{3/2} \tag{2.71}$$

This equation is of fundamental importance in quantum mechanics, and has been studied and improved by many famous physicists. The unknown y is proportional to the potential energy of an atom (the sum of electrostatic and chemical potentials) or equivalently proportional to 2/3 of the power of the electron density, and the variable x is a normalized distance from the center of an atom. This equation is for a neutral atom with spherical symmetry in electron density.

Example 2.7 Give an approximate solution for the Thomas-Fermi equation with the following boundary conditions:

$$y(0) = 1, \quad y(x) \to 0, \quad \text{as } x \to \infty \tag{2.72}$$

Solution: The following approximate solution was proposed by A. Sommerfeld in 1932 (see Davis, 1962). We will mention Sommerfeld again in later chapters and his legendary story of being nominated 81 times for the Nobel Prize (see Appendix B). First, the following change of variables for (2.71) is applied

$$x=\frac{1}{t}, \quad y=\frac{w}{t} \tag{2.73}$$

Thus, we have

$$dx=-\frac{1}{t^2}dt \tag{2.74}$$

$$\frac{dy}{dx}=\frac{1}{t}\frac{dw}{dx}-\frac{w}{t^2}\frac{dt}{dx}=-t\frac{dw}{dt}+w \tag{2.75}$$

The last of (2.75) is a result of the substitution of (2.74). Differentiating (2.75) one more time with respect to x, we get

$$\begin{aligned}\frac{d^2y}{dx^2}&=-\frac{dt}{dx}\frac{dw}{dx}-t\frac{d^2w}{dxdt}+\frac{dw}{dx}\\&=t^2\frac{dw}{dt}+t^3\frac{d^2w}{dt^2}-t^2\frac{dw}{dt}=t^3\frac{d^2w}{dt^2}\end{aligned} \tag{2.76}$$

The right hand side of (2.71) becomes

$$y^{3/2}x^{-1/2}=\sqrt{t}\frac{w^{3/2}}{t^{3/2}}=\frac{1}{t}w^{3/2} \tag{2.77}$$

Equating (2.76) and (2.77), we obtain a governing equation for w

$$t^4\frac{d^2w}{dt^2}=w^{3/2} \tag{2.78}$$

The boundary conditions given by (2.72) become

$$w \sim t, \quad \text{as } t\to\infty, \quad w(0)=0 \tag{2.79}$$

The degree of nonlinearity of (2.78) is actually similar to that of (2.71). First of all, Thomas in 1927 recognized that the following is a particular solution of (2.78)

$$w_1=144t^4 \tag{2.80}$$

Differentiating (2.80) twice leads to

$$\frac{d^2w}{dt^2}=1728t^2 \tag{2.81}$$

On the other hand, substitution of (2.80) into the right hand side of (2.78) results in

$$w^{3/2}=(144)^{3/2}t^{4(3/2)}=12^3t^6=1728t^6 \tag{2.82}$$

Evidently, (2.78) is satisfied by (2.80). In addition, the second boundary condition is satisfied. However, the first boundary condition remains to be satisfied. We must modify the solution given in (2.80). We assume the following form

$$w=w_1(1+\alpha t^\lambda) \tag{2.83}$$

Substitution of (2.83) into the left hand side of (2.78) yields

$$\begin{aligned}t^4\frac{d^2w}{dt^2}&=t^4\frac{d^2w_1}{dt^2}(1+\alpha t^\lambda)+2\frac{dw_1}{dt}\alpha\lambda t^{\lambda+3}+w_1\alpha\lambda(\lambda-1)t^{\lambda+2}\\&=1728t^6(1+\alpha t^\lambda)+1152\alpha\lambda t^{\lambda+6}+144\alpha\lambda(\lambda-1)t^{\lambda+6}\end{aligned} \tag{2.84}$$

This can be further simplified to

$$
\begin{aligned}
t^4 \frac{d^2 w}{dt^2} &= 1728t^6 \left\{1 + \alpha t^\lambda + \frac{2}{3}\alpha\lambda t^\lambda + \frac{1}{12}\alpha\lambda(\lambda-1)t^\lambda\right\} \\
&= 1728t^6 \left\{1 + \alpha t^\lambda [1 + \frac{2}{3}\lambda + \frac{1}{12}\lambda(\lambda-1)]\right\} \\
&= 1728t^6 \left\{1 + \frac{1}{12}\alpha t^\lambda (3+\lambda)(4+\lambda)\right\}
\end{aligned}
\tag{2.85}
$$

Substitution of (2.83) into the right hand side of (2.78) leads to

$$
\begin{aligned}
w^{3/2} &= w_1^{3/2}(1+\alpha t^\lambda)^{3/2} = (144)^{3/2} t^{4(3/2)} (1+\alpha t^\lambda)^{3/2} \\
&= 1728t^6 \left\{1 + \frac{3}{2}\alpha t^\lambda + \frac{3}{8}\alpha^2 t^{2\lambda} - \ldots\right\}
\end{aligned}
\tag{2.86}
$$

The last expansion of (2.86) is a result of Binomial series expansion (see Section 1.2 of Chapter 1). Clearly, (2.85) cannot be set equal to (2.86). However, when $t \to 0$, higher order terms of t on the right hand side of (2.86) can be neglected. Thus, we can get an approximation as

$$
\frac{1}{12}(3+\lambda)(4+\lambda) = \frac{3}{2} \tag{2.87}
$$

Or equivalently, we have

$$
\lambda^2 + 7\lambda - 6 = 0 \tag{2.88}
$$

The two roots of this quadratic equation are

$$
\lambda_1 = \frac{-7+\sqrt{73}}{2} = 0.772, \quad \lambda_2 = \frac{-7-\sqrt{73}}{2} = -7.772 \tag{2.89}
$$

We obtain an approximation of (2.78) if these values of λ are chosen. But yet we have not satisfied the first boundary condition given in (2.79). Note also that we have not arrived at any value for α. We further extend the approximation to a more general form (compare (2.83))

$$
w = w_1(1+\beta t^\lambda)^n \tag{2.90}
$$

We seek to choose β and n properly such that the first boundary condition can be satisfied. Substitution of (2.80) into (2.90) into gives

$$
w = w_1(1+\beta t^\lambda)^n = [144t^3(1+\beta t^\lambda)^n]t \tag{2.91}
$$

For $t \to \infty$, we need to satisfy the first equation of (2.79) so that it results in

$$
144t^3(1+\beta t^\lambda)^n = 144t^3\beta^n t^{\lambda n}(1+\frac{1}{\beta t^\lambda})^n \sim 1 \tag{2.92}
$$

For large t, we can set

$$
3 + n\lambda = 0, \quad 144\beta^n = 1 \tag{2.93}
$$

The first equation of (2.93) will ensure that (2.92) is proportional to t^0 (i.e., independent of t), and the second of (2.93) will ensure the magnitude of (2.92) is one as $t \to \infty$. Recall from (2.88) that

$$
\lambda^2 + 7\lambda - 6 = (\lambda-\lambda_1)(\lambda-\lambda_2) = \lambda^2 - (\lambda_1+\lambda_2) + \lambda_1\lambda_2 \tag{2.94}
$$

Thus, we must have

$$
\lambda_1\lambda_2 = -6 \tag{2.95}
$$

We can set

$$\lambda = \lambda_1, \quad n = -\frac{3}{\lambda_1} = \frac{\lambda_2}{2}, \quad \beta = \frac{1}{144^{1/n}} \tag{2.96}$$

where λ_1 and λ_2 have been obtained from (2.89). Substitution of (2.80) and (2.96) into (2.90) results in

$$w = 144t^4\{1 + \frac{1}{144^{-\lambda_1/3}} t^{\lambda_1}\}^{\lambda_2/2} \tag{2.97}$$

We can now back substitute (2.97) into (2.73)

$$y = \frac{144}{x^3}\{1 + \frac{1}{144^{-\lambda_1/3}}(\frac{1}{x})^{\lambda_1}\}^{\lambda_1/2} = \frac{144}{x^3}\{1 + (\frac{144}{x^3})^{\lambda_1/3}\}^{\lambda_2/2} \tag{2.98}$$

This can be written as

$$y = y_1(x)\{1 + [y_1(x)]^{\lambda_1/3}\}^{\lambda_2/2}, \qquad y_1(x) = \frac{144}{x^3} \tag{2.99}$$

As a double check, we see that as $x \to 0$ we have $y_1 \to \infty$ and thus

$$y = y_1(x)\{1 + [y_1(x)]^{\lambda_1/3}\}^{\lambda_2/2} \approx y_1(x)[y_1(x)]^{\lambda_1\lambda_2/6} = y_1(x)[y_1(x)]^{-1} = 1 \tag{2.100}$$

Therefore, the first boundary condition in (2.72) is satisfied.

The solution given in Example 2.7 is found in reasonably good agreement with numerical integration of (2.71). We can see from this example that one needs extraordinary mathematical skill and insight to find an approximation to the nonlinear differential equation. Unlike the pendulum problem, no exact solution is found. In addition, such ingenious solution is only an approximation, and the solution procedure is not straightforward and cannot be generalized easily to other nonlinear differential equations. Because of this difficulty, the method of solution for nonlinear equations is less developed.

2.8 SYSTEMS OF DIFFERENTIAL EQUATIONS

In this section, we will report a number of systems of differential equations that we commonly encounter in engineering and mechanics. They include Maxwell's equations, Navier-Stokes equation, and equations of motion of waves in solids.

2.8.1 Maxwell Equations

For electrodynamics, the magnetic field vector $\boldsymbol{B}$ and electric field vector $\boldsymbol{E}$ are coupled through the set of Maxwell equations:

$$\begin{gathered} \nabla \cdot \boldsymbol{E} = \frac{\rho}{\varepsilon}, \quad \nabla \cdot \boldsymbol{B} = 0, \\ \nabla \times \boldsymbol{E} = -\frac{\partial \boldsymbol{B}}{\partial t}, \quad \nabla \times \boldsymbol{B} = \mu(\boldsymbol{J} + \varepsilon \frac{\partial \boldsymbol{E}}{\partial t}) \end{gathered} \tag{2.101}$$

where ε is the permittivity of the material, μ is the permeability of the material, ρ is the charge density, and $\boldsymbol{J}$ is the electric current. The first, second, third, and fourth equations of (2.101) are the Gauss law for electricity (relation between electric field and electric charge), the Gauss law for magnetism (non-existence of magnetic charge), the Faraday law (electric field induced by changing magnetic flux) and the Ampere law (magnetic field induced from electric current). This set of Maxwell equations unified all magnetic and electric effect in terms of electrodynamics. Note also that the Maxwell equations can be expressed in integral forms. However, we will not pursue such possibility in the present book, as our focus here is on the study of differential equations.

For the special case of a vacuum with no electric charge ($\rho = 0$) and no electric current ($\boldsymbol{J} = 0$), electromagnetic waves can be specialized as:

$$\nabla \cdot \boldsymbol{E} = 0, \quad \nabla \cdot \boldsymbol{B} = 0,$$
$$\nabla \times \boldsymbol{E} = -\frac{\partial \boldsymbol{B}}{\partial t}, \quad \nabla \times \boldsymbol{B} = \frac{1}{c^2}\frac{\partial \boldsymbol{E}}{\partial t} \tag{2.102}$$

where

$$c^2 = \frac{1}{\mu_0 \varepsilon_0} \tag{2.103}$$

Here c is the speed of light in a vacuum and μ_0 and ε_0 are the permeability and permittivity in a vacuum.

Taking the curl of the third of (2.102), we have

$$\begin{aligned}\nabla \times (\nabla \times \boldsymbol{E}) &= \nabla(\nabla \cdot \boldsymbol{E}) - \nabla^2 \boldsymbol{E} = -\nabla^2 \boldsymbol{E} \\ &= -\frac{\partial (\nabla \times \boldsymbol{B})}{\partial t}\end{aligned} \tag{2.104}$$

where we have used the Gauss law of electricity that the divergence of $\boldsymbol{E}$ is zero. Take the time derivative of the fourth of (2.102) yields

$$\frac{\partial}{\partial t}(\nabla \times \boldsymbol{B}) = \frac{1}{c^2}\frac{\partial^2 \boldsymbol{E}}{\partial t^2} \tag{2.105}$$

Combining (2.104) and (2.105) gives

$$\nabla^2 \boldsymbol{E} = \frac{1}{c^2}\frac{\partial^2 \boldsymbol{E}}{\partial t^2} \tag{2.106}$$

This is a vector wave equation for the electric field $\boldsymbol{E}$. Similarly, taking the curl of the fourth equation of (2.102) gives

$$\begin{aligned}\nabla \times (\nabla \times \boldsymbol{B}) &= \nabla(\nabla \cdot \boldsymbol{B}) - \nabla^2 \boldsymbol{B} = -\nabla^2 \boldsymbol{B} \\ &= \frac{1}{c^2}\frac{\partial (\nabla \times \boldsymbol{E})}{\partial t}\end{aligned} \tag{2.107}$$

On the other hand, the time derivative of the third of (2.102) gives

$$\frac{\partial}{\partial t}(\nabla \times \boldsymbol{E}) = -\frac{\partial^2 \boldsymbol{B}}{\partial t^2} \tag{2.108}$$

Combining (2.107) and (2.108) results in another vector wave equation for $\boldsymbol{B}$

$$\nabla^2 \boldsymbol{B} = \frac{1}{c^2}\frac{\partial^2 \boldsymbol{B}}{\partial t^2} \tag{2.109}$$

In conclusion, for the case of a vacuum, both the electric field and magnetic field propagate as waves, called electromagnetic waves, and their speed of propagation is the speed of light in vacuum ($c \approx 2.99792458\times10^8$m/s). In addition, if the electric and magnetic fields are independent of time, we have

$$\nabla^2 \boldsymbol{B} = 0, \quad \nabla^2 \boldsymbol{E} = 0 \tag{2.110}$$

That is, electrostatics is governed by the Laplace equation. This also suggests the existence of an electric potential and a magnetic potential. Since the electric and magnetic fields do not change with time, physically the Laplace equation models the equilibrium problem of an electric field.

Without going into the details, we report that for the case of a non-vacuum with charge and current, we can express $\boldsymbol{B}$ and $\boldsymbol{E}$ in terms of a vector potential and a scalar potential as

$$\boldsymbol{E} = -\nabla\varphi - \frac{\partial \boldsymbol{A}}{\partial t}, \quad \boldsymbol{B} = \nabla\times \boldsymbol{A} \tag{2.111}$$

The resulting governing equations for $\boldsymbol{A}$ and φ are

$$\nabla^2 \boldsymbol{A} - \mu\varepsilon\frac{\partial^2 \boldsymbol{A}}{\partial t^2} = -\mu \boldsymbol{J} \tag{2.112}$$

$$\nabla^2 \varphi - \mu\varepsilon\frac{\partial^2 \varphi}{\partial t^2} = -\frac{\rho}{\varepsilon}, \tag{2.113}$$

provided that the following condition, called the Lorenz gauge, is assumed

$$\nabla\cdot \boldsymbol{A} = -\mu\varepsilon\frac{\partial \varphi}{\partial t} \tag{2.114}$$

This Lorenz gauge is only one choice of the gauge freedoms and it is not unique. The Lorenz gauge is named after Danish mathematician L.V. Lorenz (1829-1891). Note that (2.112) and (2.113) are nonhomogeneous vector and scalar wave equations. Therefore, nonhomogeneous wave equations occupy a central place in the solution of electrodynamics or Maxwell equations. However, the nonhomogeneous wave equation is not covered in most textbooks on differential equations. We will discuss it in Section 9.3 and the solution is given in (9.207).

Another choice of gauge is the so-called Coulomb gauge (Zangwill, 2013) which is

$$\nabla\cdot \boldsymbol{A} = 0 \tag{2.115}$$

It is named after French physicist Charles-Augustin de Coulomb (1736–1806), who made contributions on electrostatic force and friction. The resulting governing equations for $\boldsymbol{A}$ and φ become

$$\nabla^2 \boldsymbol{A} - \mu\varepsilon\frac{\partial^2 \boldsymbol{A}}{\partial t^2} = -\mu \boldsymbol{J} + \mu\varepsilon\nabla(\frac{\partial \varphi}{\partial t}) \tag{2.116}$$

$$\nabla^2 \varphi = -\frac{\rho}{\varepsilon} \tag{2.117}$$

This gauge choice is popular because we can solve the Poisson equation for φ given in (2.117) before solving (2.116).

In a later chapter in the second book of this series, we will discuss these gauge theories and the mathematical theory of Maxwell equations in more detail.

2.8.2 Navier-Stokes Equations

In deriving the Navier-Stokes equations, we need to use the concept of a control volume versus a point mass considered for rigid body dynamics. The idea of control volume is depicted in Figure 2.2. In classical mechanics, the dynamics of a rigid body is referred as the Lagrangian formulation in which the displacement vector is referred to as a fixed point in a rigid body. We say that the displacement vector is formulated in Lagrangian coordinates. For fluid mechanics problems, if control volume is used, the displacement vector is referred to as a fixed point in space (i.e., does not follow a fluid particle). We say that the displacement vector is formulated in Eulerian coordinates because the control volume formulation was first proposed by Euler.

First of all, we assume that there exists a flow field described by the flow velocity field $\boldsymbol{u}$, which is a function of a three-dimensional position (x, y, z) and time t. Let us consider the total differential of $\boldsymbol{u}$ with respect to its dependent variables,

$$d\boldsymbol{u} = \frac{\partial \boldsymbol{u}}{\partial t} dt + \frac{\partial \boldsymbol{u}}{\partial x} dx + \frac{\partial \boldsymbol{u}}{\partial y} dy + \frac{\partial \boldsymbol{u}}{\partial z} dz \tag{2.118}$$

Dividing the whole expression by dt, we have the acceleration of the flow

$$\boldsymbol{a} = \frac{D\boldsymbol{u}}{Dt} = \frac{\partial \boldsymbol{u}}{\partial t} + \frac{\partial \boldsymbol{u}}{\partial x}\frac{dx}{dt} + \frac{\partial \boldsymbol{u}}{\partial y}\frac{dy}{dt} + \frac{\partial \boldsymbol{u}}{\partial z}\frac{dz}{dt} = \frac{\partial \boldsymbol{u}}{\partial t} + u\frac{\partial \boldsymbol{u}}{\partial x} + v\frac{\partial \boldsymbol{u}}{\partial y} + w\frac{\partial \boldsymbol{u}}{\partial z} \tag{2.119}$$

where the physical components of the velocity $\boldsymbol{u}$ are denoted as u, v, and w. Note that instead of d we have employed the notation D. The differential operator D/Dt is called the material time derivative or the advective time derivative in contrast to the ordinary time derivative d/dt. In terms of tensor notation, (2119) reads as:

$$\boldsymbol{a} = \frac{D\boldsymbol{u}}{Dt} = \frac{\partial \boldsymbol{u}}{\partial t} + \boldsymbol{u} \cdot \nabla \boldsymbol{u} \tag{2.120}$$

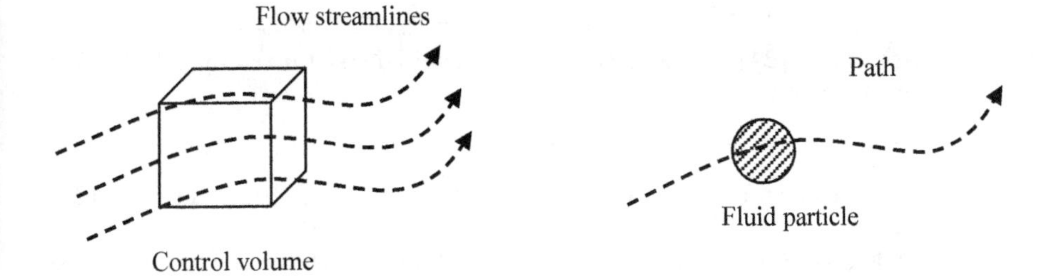

Figure 2.2 Particle and control volume descriptions of a fluid flow

where the first term is a result of unsteady flow and the second term is the acceleration induced by fluid flowing in and out of the control volume. Physically, the acceleration calculated in (2.119) is experienced by somebody who moves locally with the fluid in the control volume. Since the second term is resulting from fluid flowing in and out, it is normally called the convective term. The main nonlinearity in Navier-Stokes equations actually comes from this term. Alternatively, the material time derivative can also be expressed as

$$\boldsymbol{a}=\frac{D\boldsymbol{u}}{Dt}=\frac{\partial \boldsymbol{u}}{\partial t}+\boldsymbol{u}\cdot\nabla\boldsymbol{u}=\frac{\partial \boldsymbol{u}}{\partial t}+\nabla(\frac{|\boldsymbol{u}|^2}{2})-\boldsymbol{u}\times\nabla\times\boldsymbol{u} \tag{2.121}$$

This result is illustrated in the following example.

Example 2.8 Prove the following vector identity:

$$\frac{1}{2}\nabla|\boldsymbol{u}|^2=\boldsymbol{u}\times\nabla\times\boldsymbol{u}+(\boldsymbol{u}\cdot\nabla)\boldsymbol{u} \tag{2.122}$$

Solution: To simplify the notation, let us rewrite the magnitude of $\boldsymbol{u}$ as

$$u=|\boldsymbol{u}|=\sqrt{\boldsymbol{u}\cdot\boldsymbol{u}}=(u_k u_k)^{1/2} \tag{2.123}$$

In terms of the Cartesian tensor, the left hand side of (2.122) is

$$\begin{aligned}\frac{1}{2}\nabla|\boldsymbol{u}|^2&=\frac{1}{2}\boldsymbol{e}_i\frac{\partial u^2}{\partial x_i}=\frac{1}{2}\boldsymbol{e}_i 2u\frac{\partial u}{\partial x_i}\\&=\boldsymbol{e}_i u\frac{1}{2}\frac{1}{(u_k u_k)^{1/2}}[u_k\frac{\partial u_k}{\partial x_i}+\frac{\partial u_k}{\partial x_i}u_k]=u_k\frac{\partial u_k}{\partial x_i}\boldsymbol{e}_i=u_k u_{k,i}\boldsymbol{e}_i\end{aligned} \tag{2.124}$$

On the other hand, we have (recall tensor analysis in Section 1.19 of Chapter 1)

$$\begin{aligned}\boldsymbol{u}\times\nabla\times\boldsymbol{u}&=\boldsymbol{u}\times(e_{ijk}u_{k,j}\boldsymbol{e}_i)=e_{nmi}e_{ijk}u_m u_{k,j}\boldsymbol{e}_n\\&=(\delta_{nj}\delta_{mk}-\delta_{nk}\delta_{mj})u_m u_{k,j}\boldsymbol{e}_n\\&=(u_k u_{k,n}-u_k u_{n,k})\boldsymbol{e}_n\end{aligned} \tag{2.125}$$

and

$$(\boldsymbol{u}\cdot\nabla)\boldsymbol{u}=(u_i\boldsymbol{e}_i\cdot\boldsymbol{e}_j\frac{\partial}{\partial x_j})u_k\boldsymbol{e}_k=\delta_{ij}u_i\frac{\partial}{\partial x_j}u_k\boldsymbol{e}_k=u_i u_{k,i}\boldsymbol{e}_k=u_k u_{n,k}\boldsymbol{e}_n \tag{2.126}$$

Thus, adding (2.125) and (2.126) we have

$$\boldsymbol{u}\times\nabla\times\boldsymbol{u}+(\boldsymbol{u}\cdot\nabla)\boldsymbol{u}=(u_k u_{k,n}-u_k u_{n,k}+u_k u_{n,k})\boldsymbol{e}_n=u_k u_{k,n}\boldsymbol{e}_n \tag{2.127}$$

Comparison of (2.124) and (2.127) yields the required result given in (2.122). This completes the proof.

With the proved identity given in (2.122), the last convective term given in (2.121) is obtained.

We now consider the continuity of fluid within any control volume. The mass flux through the closed surface of the control volume adding to the change of mass with time must be zero:

$$\oint_{\partial V}\rho\boldsymbol{u}\cdot d\mathbf{A}+\frac{\partial}{\partial t}\int_V\rho dV=0 \tag{2.128}$$

By applying the Gauss theorem to the first integral we have

$$\int_V\nabla\cdot(\rho\boldsymbol{u})dV+\frac{\partial}{\partial t}\int_V\rho dV=\int_V[\nabla\cdot(\rho\boldsymbol{u})+\frac{\partial\rho}{\partial t}]dV=0 \tag{2.129}$$

Since the choice of control volume is arbitrary, we must require the integrand itself to vanish or

$$\nabla \cdot (\rho \boldsymbol{u}) + \frac{\partial \rho}{\partial t} = 0 \tag{2.130}$$

This is called the continuity condition of fluid flow. Note that this continuity equation is true for all fluids, whether viscous or not. If the density of the fluid does not change with time, we have

$$\nabla \cdot \boldsymbol{u} = 0 \tag{2.131}$$

This is called the incompressibility of fluid. When this assumption is valid, the fluid flow is called incompressible flow. We will see that incompressible flow also leads to the Laplace equation, indicating the existence of a potential (see Section 9.7 for details). Therefore, incompressible flow is also called potential flow.

We will now look at force equilibrium in fluid flow. The force acting on a control volume of fluid can be considered as the divergence of a stress tensor:

$$\rho(\frac{\partial \boldsymbol{u}}{\partial t} + \boldsymbol{u} \cdot \nabla \boldsymbol{u}) = \nabla \cdot \boldsymbol{\sigma} \tag{2.132}$$

Physically, the change of stress along spatial coordinates (the right hand side of (2.132)) leads to the unbalancing of force on the control volume and this force will lead to acceleration of the control volume (the left hand side of (1.132)). For fluids, the stress tensor can be subdivided into two contributions, a pressure term and a viscous stress term, as

$$\boldsymbol{\sigma} = -p\boldsymbol{I} + \boldsymbol{\sigma}' \tag{2.133}$$

where $\boldsymbol{I}$ is the unit second order tensor, p is the fluid pressure at a fixed point, and $\boldsymbol{\sigma}'$ is called the viscous stress tensor. In terms of dyadic notation, the unit tensor is

$$\boldsymbol{I} = \delta_{ij} \boldsymbol{e}_i \boldsymbol{e}_j \tag{2.134}$$

The viscous stress models the irreversible "viscous" transfer of momentum whereas the pressure term is for reversible transfer of momentum. Viscosity also leads to dissipation of energy through a frictional type of heat loss, and this is why it is irreversible.

Naturally, the stress within the fluid depends on how fast the fluid is flowing. A good first order of approximation is that the viscous stress tensor is proportional to the gradient of the velocity vector (Landau and Lifshitz, 1987):

$$\boldsymbol{\sigma}' = \eta[(\nabla \boldsymbol{u})^T + \nabla \boldsymbol{u} - \frac{2}{3}\boldsymbol{I}(\nabla \cdot \boldsymbol{u})] + \zeta \boldsymbol{I}(\nabla \cdot \boldsymbol{u}) \tag{2.135}$$

where η and ζ are the coefficients of viscosity and second viscosity. At first look, the viscosity term looks odd and there is a common factor overlapping with the second viscosity term. The choice of the first term actually reflects that the viscosity is only related to the deviatoric terms (or non-axial terms) and physically viscous irreversible transfer of momentum only relates to the rate of shear deformation and is independent of the rate of volumetric change. To see this, we can take the trace of (2.135) such that

$$tr(\boldsymbol{\sigma}') = \eta[2u_{k,k} - \frac{2}{3}\delta_{kk}u_{j,j}] + \zeta\delta_{kk}u_{j,j} = 3\zeta u_{j,j} \tag{2.136}$$

This clearly reflects that the viscosity η relates to the rate of shearing deformation only. We can rearrange (2.135) rewriting the viscous stress as

$$\boldsymbol{\sigma}' = \eta[(\nabla \boldsymbol{u})^T + \nabla \boldsymbol{u}] + \lambda \boldsymbol{I}(\nabla \cdot \boldsymbol{u}) \tag{2.137}$$

where

$$\zeta = \lambda + \frac{2}{3}\eta \tag{2.138}$$

Physically, both η and ζ must be positive or

$$\zeta > 0, \quad \eta > 0 \tag{2.139}$$

The first inequality of (2.139) reflects the fact that dissipation leads to a decrease in mechanical energy (instead of increase); whereas the second term of (2.139) reflects the fact that irreversible processes of internal friction lead to increase in entropy, a term in thermodynamics associated with the change from less probable state to a more probable state. The value of ζ reflects how fast or slow the relaxation time is to restore equilibrium. Normally, the order of magnitude of viscosity and second viscosity are comparable (i.e., $\zeta \approx \eta$). Under certain circumstances, we may have $\zeta >> \eta$ and in such case there will be a change of volume and in turn a change in density. Although we normally assume ζ is a constant, in reality ζ reflects the rate of compression versus the relaxation time. For example, for sound waves in fluids, ζ should depend on frequency (how fast the compression and tension are applied). When this happens, the waves in fluid are called dispersive.

Substitution of (2.134) and (2.133) into (1.132) leads to

$$\rho(\frac{\partial \boldsymbol{u}}{\partial t} + \boldsymbol{u} \cdot \nabla \boldsymbol{u}) = -\nabla p + \nabla \cdot \left\{ \eta[(\nabla \boldsymbol{u})^T + \nabla \boldsymbol{u} - \frac{2}{3}\boldsymbol{I}(\nabla \cdot \boldsymbol{u})] \right\} + \nabla[\zeta(\nabla \cdot \boldsymbol{u})] \tag{2.140}$$

This is the most general Navier-Stokes equation for viscous fluids. This can be rewritten in a more compact form by noting the following identity:

$$\nabla \cdot \left\{ \eta(\nabla \boldsymbol{u})^T \right\} = \nabla \times [\eta(\nabla \times \boldsymbol{u})] + \nabla \cdot [\eta(\nabla \boldsymbol{u})] \tag{2.141}$$

This is proved in the following example.

Example 2.9 Prove the following vector identity:

$$\nabla \cdot \left\{ \eta(\nabla \boldsymbol{u})^T \right\} = \nabla \times [\eta(\nabla \times \boldsymbol{u})] + \nabla \cdot [\eta(\nabla \boldsymbol{u})] \tag{2.142}$$

Solution: Let us consider the first term on right hand side of (2.142)

$$\begin{aligned} \nabla \times [\eta(\nabla \times \boldsymbol{u})] &= \nabla \times (\eta e_{ijk} u_{k,j} \boldsymbol{e}_i) = e_{nmi} e_{ijk} (\eta u_{k,j})_m \boldsymbol{e}_n = e_{inm} e_{ijk} (\eta u_{k,j})_m \boldsymbol{e}_n \\ &= (\delta_{nj}\delta_{mk} - \delta_{nk}\delta_{mj})(\eta u_{k,j})_m \boldsymbol{e}_n = (\eta u_{m,n})_m \boldsymbol{e}_n - (\eta u_{n,m})_m \boldsymbol{e}_n \\ &= \nabla \cdot \left\{ \eta(\nabla \boldsymbol{u})^T \right\} - \nabla \cdot [\eta(\nabla \boldsymbol{u})] \end{aligned} \tag{2.143}$$

Rearranging (2.143) gives

$$\nabla \cdot \left\{ \eta(\nabla \boldsymbol{u})^T \right\} = \nabla \times [\eta(\nabla \times \boldsymbol{u})] + \nabla \cdot [\eta(\nabla \boldsymbol{u})] \tag{2.144}$$

This completes the proof.

With the identity given in (2.141) in hand, the last two terms of (2.140) can be rewritten as

$$\nabla\cdot\left\{\eta[(\nabla\boldsymbol{u})^T+\nabla\boldsymbol{u}-\frac{2}{3}\boldsymbol{I}(\nabla\cdot\boldsymbol{u})]\right\}+\nabla[\zeta(\nabla\cdot\boldsymbol{u})]$$

$$=\nabla\times[\eta(\nabla\times\boldsymbol{u})]+\nabla\cdot[\eta(\nabla\boldsymbol{u})]+\nabla\cdot[\eta(\nabla\boldsymbol{u})]-\frac{2}{3}\nabla[\eta(\nabla\cdot\boldsymbol{u})]+\nabla[\zeta(\nabla\cdot\boldsymbol{u})] \quad (2.145)$$

$$=\nabla\times[\eta(\nabla\times\boldsymbol{u})]+2\nabla\cdot[\eta(\nabla\boldsymbol{u})]+\nabla\{(\zeta-\frac{2}{3}\eta)(\nabla\cdot\boldsymbol{u})\}$$

Finally, combining (2.145) and (2.140) we have another form of Navier-Stokes equations

$$\rho(\frac{\partial\boldsymbol{u}}{\partial t}+\boldsymbol{u}\cdot\nabla\boldsymbol{u})=-\nabla p+\nabla\times[\eta(\nabla\times\boldsymbol{u})]+2\nabla\cdot[\eta(\nabla\boldsymbol{u})]+\nabla\{(\zeta-\frac{2}{3}\eta)(\nabla\cdot\boldsymbol{u})\} \quad (2.146)$$

Note, however, that this equation differs from (3.48) of Hughes and Brighton (1967) and from (1.33) of Hughes and Gaylord (1964). There are clear typos in both of them. To the best of our knowledge, this correct form has not been reported in the literature.

In general, both ζ and η can be functions of pressure and temperature. If both ζ and η are independent of pressure and temperature, we have

$$\rho(\frac{\partial\boldsymbol{u}}{\partial t}+\boldsymbol{u}\cdot\nabla\boldsymbol{u})=-\nabla p+\eta\nabla^2\boldsymbol{u}+(\zeta+\frac{1}{3}\eta)\nabla(\nabla\cdot\boldsymbol{u}) \quad (2.147)$$

For incompressible flow, by virtue of (2.131) we have

$$\rho(\frac{\partial\boldsymbol{u}}{\partial t}+\boldsymbol{u}\cdot\nabla\boldsymbol{u})=-\nabla p+\eta\nabla^2\boldsymbol{u} \quad (2.148)$$

Let us reiterate that all nonlinearity of the Navier-Stokes equation comes from the convective term. In fact, the KdV equation, Burgers equation, and Boussinesq equation are special cases of Navier-Stokes equations. For example, for 1-D flow with no pressure gradient applied, we can replace $\boldsymbol{u}$ by u and

$$\frac{\partial u}{\partial t}+u\frac{\partial u}{\partial x}=\frac{\eta}{\rho}\frac{\partial^2 u}{\partial x^2}=\nu\frac{\partial^2 u}{\partial x^2} \quad (2.149)$$

where

$$\nu=\frac{\eta}{\rho} \quad (2.150)$$

is the kinematic viscosity (η is also known as dynamic viscosity). Evidently, (2.150) is the Burgers equation given in (2.57). At 20°C, the values of ν for water and air are 0.010 and 0.150 cm^2/sec. Thus, water is 15 times more viscous than air, and this is the reason why it is so hard to swim fast in water.

In closing, we should mention that the Clay Mathematics Institute offered US $1 million for anyone making “substantial process toward a mathematical theory which will unlock the secrets hidden in the Navier-Stokes equations.” Indeed, this nonlinear differential equation leads to all phenomena that we observe in fluids every day, including waves behind our boats, and turbulent currents following a modern flight jet. For example, there is no satisfactory theory for turbulent flow yet and much remains to be done mathematically on Navier-Stokes equations.

2.8.3 Elastodynamics

Following the discussion by Chau (2013), for dynamics problem for solids, we consider the force equilibrium of a solid of volume V as

$$\int_V \boldsymbol{F} dV + \int_S \boldsymbol{n} \cdot \boldsymbol{\sigma} dS = \frac{\partial}{\partial t} \int_V \rho \frac{\partial \boldsymbol{u}}{\partial t} dV \tag{2.151}$$

where $\boldsymbol{F}$, $\boldsymbol{\sigma}$, $\boldsymbol{u}$, $\boldsymbol{n}$ and ρ are the body force vector, stress tensor, displacement vector, normal vector pointing outside the volume, and density of the solid (see Figure 2.3).

Applying the Gauss theorem to the second term on the left of (2.151), we have

$$\int_V (\nabla \cdot \boldsymbol{\sigma} + \boldsymbol{F}) dV = \int_V \rho \frac{\partial^2 \boldsymbol{u}}{\partial t^2} dV \tag{2.152}$$

Since the body is taken arbitrarily, we must have

$$\nabla \cdot \boldsymbol{\sigma} + \boldsymbol{F} = \rho \frac{\partial^2 \boldsymbol{u}}{\partial t^2} \tag{2.153}$$

For isotropic elastic solids, Hooke's law only involves two independent moduli λ and μ, which are called the Lamé constants. In tensor notation, it is

$$\boldsymbol{\sigma} = \lambda tr(\boldsymbol{\varepsilon}) \boldsymbol{I} + 2\mu \boldsymbol{\varepsilon} \tag{2.154}$$

For small deformation and small strain, the strain-displacement relation is defined (Chau, 2013)

$$\boldsymbol{\varepsilon} = \tfrac{1}{2}[\nabla \boldsymbol{u} + (\nabla \boldsymbol{u})^T] \tag{2.155}$$

Substitution of (2.155) into (2.154) gives

$$\boldsymbol{\sigma} = \lambda \boldsymbol{I}(\nabla \cdot \boldsymbol{u}) + \mu[\nabla \boldsymbol{u} + (\nabla \boldsymbol{u})^T] \tag{2.156}$$

Then, substitution of (2.156) into (2.153) yields

$$\lambda \nabla(\nabla \cdot \boldsymbol{u}) + \mu[\nabla \cdot \nabla \boldsymbol{u} + \nabla \cdot (\nabla \boldsymbol{u})^T] + \boldsymbol{F} = \rho \frac{\partial^2 \boldsymbol{u}}{\partial t^2} \tag{2.157}$$

Note that

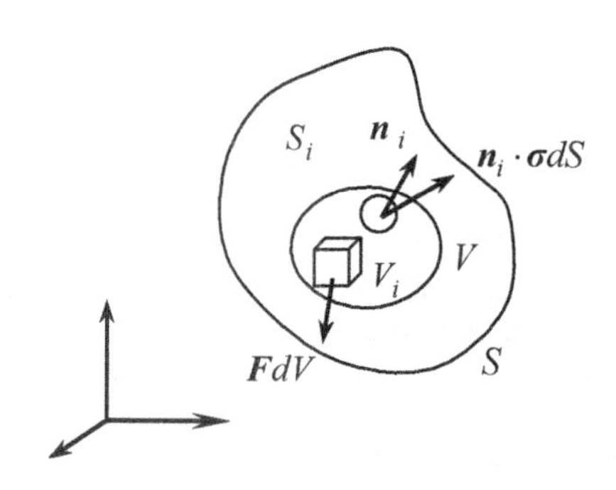

Figure 2.3 Force equilibrium of a solid of volume V and surface S

$$\nabla(\nabla\cdot\boldsymbol{u})=\nabla(\boldsymbol{e}_i\frac{\partial}{\partial x_i}\cdot u_j\boldsymbol{e}_j)=\boldsymbol{e}_k\frac{\partial}{\partial x_k}(\delta_{ij}u_{j,i})=\boldsymbol{e}_k\frac{\partial}{\partial x_k}(u_{j,j})=u_{j,ji}\boldsymbol{e}_i \quad (2.158)$$

$$\begin{aligned}\nabla\cdot(\nabla\boldsymbol{u})^T&=\nabla\cdot(\boldsymbol{e}_i\frac{\partial}{\partial x_i}u_j\boldsymbol{e}_j)^T=\boldsymbol{e}_k\frac{\partial}{\partial x_k}\cdot(u_{j,i}\boldsymbol{e}_i\boldsymbol{e}_j)^T=\boldsymbol{e}_k\frac{\partial}{\partial x_k}\cdot(u_{j,i}\boldsymbol{e}_j\boldsymbol{e}_i)\\&=\delta_{kj}u_{j,ik}\boldsymbol{e}_i=u_{j,ij}\boldsymbol{e}_i=u_{j,ji}\boldsymbol{e}_i\end{aligned} \quad (2.159)$$

Therefore, we can group these terms in (2.157) as

$$(\lambda+\mu)\nabla(\nabla\cdot\boldsymbol{u})+\mu\nabla\cdot\nabla\boldsymbol{u}+\boldsymbol{F}=\rho\frac{\partial^2\boldsymbol{u}}{\partial t^2} \quad (2.160)$$

This is the equation of motion for an elastic isotropic solid. By virtue of vector identity (1.362) of Chapter 1, we can rewrite (2.160) as

$$(\lambda+2\mu)\nabla\nabla\cdot\boldsymbol{u}-\mu\nabla\times(\nabla\times\boldsymbol{u})+\boldsymbol{F}=\rho\frac{\partial^2\boldsymbol{u}}{\partial t^2} \quad (2.161)$$

Expressed in terms of compressional and shear wave speeds, (2.161) becomes

$$c_d^2\nabla\nabla\cdot\boldsymbol{u}-c_s^2\nabla\times(\nabla\times\boldsymbol{u})+\frac{1}{\rho}\boldsymbol{F}=\frac{\partial^2\boldsymbol{u}}{\partial t^2} \quad (2.162)$$

where

$$c_d=\sqrt{\frac{\lambda+2\mu}{\rho}},\quad c_s=\sqrt{\frac{\mu}{\rho}} \quad (2.163)$$

Let us consider the divergence of (2.162) as

$$c_d^2\nabla^2(\nabla\cdot\boldsymbol{u})+\frac{1}{\rho}\nabla\cdot\boldsymbol{F}=\frac{\partial^2(\nabla\cdot\boldsymbol{u})}{\partial t^2} \quad (2.164)$$

This is a nonhomogeneous wave equation for dilatation $\nabla\cdot\boldsymbol{u}$. In obtaining (2.164), we have used the vector identity given in (1.347) setting $\boldsymbol{A}=\nabla\times\boldsymbol{u}$ as

$$\nabla\cdot(\nabla\times A)=0 \quad (2.165)$$

As discussed in Chapter 1, all vectors can be decomposed into an irrotational term and a solenoid term according to Helmholtz theorem (see Section 1.16 of Chapter 1). Let us decompose the displacement vector as:

$$\boldsymbol{u}=\nabla\phi+\nabla\times\psi \quad (2.166)$$

Taking the divergence of $\boldsymbol{u}$ gives

$$\nabla\cdot\boldsymbol{u}=\nabla^2\phi+\nabla\cdot(\nabla\times\psi)=\nabla^2\phi \quad (2.167)$$

Substitution of (2.167) into (2.164) yields

$$\nabla^2[c_d^2\nabla^2\phi-\frac{\partial^2\phi}{\partial t^2}]=-\frac{1}{\rho}\nabla\cdot\boldsymbol{F} \quad (2.168)$$

This is a nonhomogeneous differential equation for the potential ϕ. Physically, (2.167) shows that $\nabla^2\phi$ is the dilatation. Therefore, (2.168) is also called the dilatation wave equation, and it governs the motion of the so-called P-waves.

On the other hand, if we take the curl of (2.162), we arrive at

$$c_d^2\nabla\times\nabla(\nabla\cdot\boldsymbol{u})-c_s^2\nabla\times[\nabla\times(\nabla\times\boldsymbol{u})]+\frac{1}{\rho}\nabla\times\boldsymbol{F}=\frac{\partial^2\nabla\times\boldsymbol{u}}{\partial t^2} \quad (2.169)$$

The first term vanishes by virtue of vector identity (1.347) while the second term can be expanded by using (1.362) as

$$\nabla\times[\nabla\times(\nabla\times\boldsymbol{u})]=\nabla[\nabla\cdot(\nabla\times\boldsymbol{u})]-\nabla^2(\nabla\times\boldsymbol{u})=-\nabla^2(\nabla\times\boldsymbol{u}) \tag{2.170}$$

The last result follows from (1.347) or (2.165). Substitution of (2.170) into (2.169) gives

$$c_s^2\nabla^2(\nabla\times\boldsymbol{u})-\frac{\partial^2\nabla\times\boldsymbol{u}}{\partial t^2}=-\frac{1}{\rho}\nabla\times\boldsymbol{F} \tag{2.171}$$

This is a nonhomogeneous vector wave equation for the rotation tensor $\nabla\times\boldsymbol{u}/2$. Taking the curl of (2.166) we obtain

$$\begin{aligned}\nabla\times\boldsymbol{u}&=\nabla\times\nabla\phi+\nabla\times\nabla\times\boldsymbol{\psi}=\nabla(\nabla\cdot\boldsymbol{\psi})-\nabla^2\boldsymbol{\psi}\\&=-\nabla^2\boldsymbol{\psi}\end{aligned} \tag{2.172}$$

In terms of vector potential, (2.172) can be written as

$$\nabla^2[c_s^2\nabla^2\boldsymbol{\psi}-\frac{\partial^2\boldsymbol{\psi}}{\partial t^2}]=\frac{1}{\rho}\nabla\times\boldsymbol{F} \tag{2.173}$$

This is a nonhomogeneous vector differential equation for the vector potential $\boldsymbol{\psi}$.

If there is no body force, we have

$$c_d^2\nabla^2\phi=\frac{\partial^2\phi}{\partial t^2}+h \tag{2.174}$$

$$c_s^2\nabla^2\boldsymbol{\psi}=\frac{\partial^2\boldsymbol{\psi}}{\partial t^2}+\boldsymbol{H} \tag{2.175}$$

where h and $\boldsymbol{H}$ are arbitrary harmonic functions. However, they are normally neglected. That is,

$$c_d^2\nabla^2\phi=\frac{\partial^2\phi}{\partial t^2},\quad c_s^2\nabla^2\boldsymbol{\psi}=\frac{\partial^2\boldsymbol{\psi}}{\partial t^2} \tag{2.176}$$

In summary, in terms of dilatation and rotation tensors the elastodynamics problems are governed by the following nonhomogeneous wave equations:

$$c_d^2\nabla^2(\nabla\cdot\boldsymbol{u})-\frac{\partial^2(\nabla\cdot\boldsymbol{u})}{\partial t^2}=-\frac{1}{\rho}\nabla\cdot\boldsymbol{F} \tag{2.177}$$

$$c_s^2\nabla^2(\nabla\times\boldsymbol{u})-\frac{\partial^2\nabla\times\boldsymbol{u}}{\partial t^2}=-\frac{1}{\rho}\nabla\times\boldsymbol{F} \tag{2.178}$$

Physically, the wave equations given in (2.177) and (2.178) closely resemble the scalar and vector wave equations given in (2.112) and (2.113) as Maxwell equations. Therefore, elastodynamics and electrodynamics are mathematically the same problems. The mathematical technique for one problem is applicable to the other. To the best of our knowledge, this resemblance has not been mentioned in any textbook on differential equations. The solution of the nonhomogeneous wave equation is discussed in Section 9.3 and is given in (9.207).

2.8.4 Three-Dimensional Elasticity

Let us look at the special case that the displacement field is independent of time.

$$(\lambda+\mu)\nabla(\nabla\cdot\boldsymbol{u})+\mu\nabla^2\boldsymbol{u}+\boldsymbol{F}=0 \tag{2.179}$$

If the Helmholtz theorem is used, we can specialize (2.168) and (2.173)

$$\nabla^2\nabla^2\phi=\nabla^4\phi=-\frac{1}{\lambda+2\mu}\nabla\cdot\boldsymbol{F} \tag{2.180}$$

$$\nabla^2\nabla^2\psi=\nabla^4\psi=\frac{1}{\mu}\nabla\times\boldsymbol{F} \tag{2.181}$$

These are nonhomogeneous scalar and vector biharmonic equations for ϕ and ψ.

It is also possible to introduce a single vector potential called the Galerkin vector $\boldsymbol{G}$

$$\boldsymbol{u}=2(1-\nu)\nabla^2\boldsymbol{G}-\nabla\nabla\bullet\boldsymbol{G} \tag{2.182}$$

Taking the Laplacian of (2.182) gives

$$\nabla^2\boldsymbol{u}=2(1-\nu)\nabla^2\nabla^2\boldsymbol{G}-\nabla^2(\nabla\nabla\bullet\boldsymbol{G}) \tag{2.183}$$

Similarly, we consider the gradient of the divergence of $\boldsymbol{u}$ as

$$\begin{aligned}\nabla\nabla\bullet\boldsymbol{u}&=2(1-\nu)\nabla\nabla\bullet(\nabla^2\boldsymbol{G})-\nabla\nabla\bullet(\nabla\nabla\bullet\boldsymbol{G})\\&=2(1-\nu)\nabla^2(\nabla\nabla\bullet\boldsymbol{G})-\nabla^2(\nabla\nabla\bullet\boldsymbol{G})\\&=(1-2\nu)\nabla^2(\nabla\nabla\bullet\boldsymbol{G})\end{aligned} \tag{2.184}$$

Substitution of (2.183) and (2.184) into (2.179) yields

$$(\lambda+\mu)(1-2\nu)\nabla^2(\nabla\nabla\bullet\boldsymbol{G})+2\mu(1-\nu)\nabla^2\nabla^2\boldsymbol{G}-\mu\nabla^2(\nabla\nabla\bullet\boldsymbol{G})+\boldsymbol{F}=0 \tag{2.185}$$

Simplification of (2.185) results in

$$[\lambda-2\nu(\lambda+\mu)]\nabla^2(\nabla\nabla\bullet\boldsymbol{G})+2\mu(1-\nu)\nabla^2\nabla^2\boldsymbol{G}+\boldsymbol{F}=0 \tag{2.186}$$

However, from the result of elasticity, we know that Poisson's ratio ν can be related to Lamé constants as (Chau, 2013)

$$2\nu=\frac{\lambda}{\lambda+\mu} \tag{2.187}$$

Substitution (2.187) into (2.186) gives

$$\nabla^2\nabla^2\boldsymbol{G}=\nabla^4\boldsymbol{G}=-\frac{1}{2\mu(1-\nu)}\boldsymbol{F} \tag{2.188}$$

This is a nonhomogeneous biharmonic equation for vector potential $\boldsymbol{G}$. For the special case of zero body force, we have biharmonic equations

$$\nabla^4\boldsymbol{G}=0 \tag{2.189}$$

Thus, we see that the biharmonic equation is of fundamental importance for three-dimensional elasticity. Yet, most textbooks on differential equations for engineers and scientists do not cover the biharmonic equation. Further discussions of the biharmonic equation will be given in Sections 7.9, 8.13.4 and 14.12.

2.9 SUMMARY AND FURTHER READING

Although the main focus of this book is to discuss the mathematical techniques in solving differential equations and their applications in engineering and mechanics, it is essential that we should know what kind of information is available even before we attempt to solve them. We should be familiar with handbooks on differential equations. A good start is to refer to the handbook by Zwillinger (1997), which summarizes different procedures for different types of differential equations with examples. There are also a lot of useful and handy comments on different techniques. For the most comprehensive and up-to-date handbooks on differential equations, we recommend the series of handbooks by Polyanin (2001), Polyanin and Zaitsev (2002, 2003), and Polyanin et al. (2002). For integral equations, which are closely related to differential equations and the focus of Chapter 12, one should consult the handbook of Polyanin and Manzhirov (2008). However, there is not much discussion on the method of analyses, but instead only the final solution is given if they are available. It is a good idea to check the handbooks of Polyanin (2001), Polyanin and Zaitsev (2002, 2003), and Polyanin et al. (2002) for your differential equation at hand before even thinking about solving it by yourself.

Regarding textbooks on differential equations, there are hundreds of available textbooks on differential equations but most of them were written by mathematicians, not much by engineers. A number of very good books are recommended here, but it should be warned that some of them are not easy to read. The series of books by Forsyth (1890, 1893, 1900, 1902, 1906, 1918, 1956) remain classic although there are some obsolete topics. Some topics in these references are not readily comprehensible. Other good books include Airy (1873), Craig (1889), Bateman (1918, 1944), Whittaker and Watson (1927), Sommerfeld (1949), Erdelyi (1953), Ayres (1952), Ince (1956), Myint-U (1987), Myint-U and Debnath (1987), Zill (1993), Zachmanoglou and Thoe (1986), and Zill and Cullen (2005).

For the history of differential equations, the three volume book series by Kline (1972) is a must read. Struik (1987) also provides some useful information about differential equations.

A number of handbooks of mathematical functions should be available if you want to learn differential equation seriously. The number one handbook is Abramowitz and Stegun (1964), and Olver et al. (2010) is considered an updated version of Abramowitz and Stegun (1964). In terms of handbooks on integration, we recommend Gradshteyn and Ryzhik (1980). A handy handbook by Spiegel (1964) provides a quick reference.

2.10 PROBLEMS

Problem 2.1 Consider a fluid with uniform rotation with angular velocity $\boldsymbol{\Omega}$; the velocity field $\boldsymbol{v}$ is defined as

$$\boldsymbol{v} = \boldsymbol{\Omega} \times \boldsymbol{r} \tag{2.190}$$

Show that such a uniform rotational field will not produce any viscous stress $\boldsymbol{\sigma}'$. That is, there will no energy loss for such uniform rotation. This result is somehow consistent with Kelvin's theorem on the conservation of circulation. Actually, conservation of circulation in fluids was first established by Helmholtz in 1858 for incompressible flows, but was extended to compressible flows by Kelvin in 1869.

Problem 2.2 Classify the following system of differential equations for plate buckling

$$\nabla^4 F = Eh[(\frac{\partial^2 w}{\partial x \partial y})^2 - \frac{\partial^2 w}{\partial x^2}\frac{\partial^2 w}{\partial y^2}]$$
$$D\nabla^4 w = p + (\frac{\partial^2 F}{\partial y^2}\frac{\partial^2 w}{\partial x^2} + \frac{\partial^2 F}{\partial x^2}\frac{\partial^2 w}{\partial y^2} - 2\frac{\partial^2 F}{\partial x \partial y}\frac{\partial^2 w}{\partial x \partial y}) \tag{2.191}$$

where w is the deflection of the plate and F is the Airy stress function for the stress in the plate. This is called the von Karman-Foppl equation. It models the buckling of plates and was proposed by von Karman and Foppl independently in 1910 and 1907 respectively.

Problem 2.3 Classify the following system of differential equations for shallow shell buckling

$$\nabla^8 w - \nabla^4 (N_{xx}^0 \frac{\partial^2 w}{\partial x^2} + 2N_{xy}^0 \frac{\partial^2 w}{\partial x \partial y} + N_{yy}^0 \frac{\partial^2 w}{\partial y^2}) + \frac{Eh}{R^2}\frac{\partial^4 w}{\partial x^4} = 0 \tag{2.192}$$

where the initial stress state in the shells is denoted by a superscript "0." This is the linearized Donnell equation for shallow shells proposed by Donnell in 1934.

Problem 2.4 Classify the following differential equation for cylindrical shell with nonuniform wall thickness shallow shell buckling

$$\frac{1}{x}\frac{d^2}{dx^2}(x^3 \frac{d^2 w}{dx^2}) + \frac{12(1-\nu^2)}{\alpha^2 a^2} w = -\frac{12(1-\nu^2)}{E\alpha^3}\gamma_w \left(1 - \frac{x_0}{x}\right) \tag{2.193}$$

Where the thickness is $h = \alpha x$ and a is the radius of the shell. The depth of the water is defined as $d-x_0$ and d is the height of the cylindrical tank. This theory was derived by Reissner in 1908.

Problem 2.5 Classify the following differential equation for a circular ice plate supported on water (Nevel, 1959):

$$\nabla^2 \nabla^2 w + w = \frac{q}{k} \tag{2.194}$$

Its solution was given by Wyman (1950) in terms of Kelvin's functions, which will be discussed in Section 4.9. This problem relates to landing of aircraft on ice sheets (Assur, 1959).

Problem 2.6 Classify the following differential equation for vibrations of circular plate:

$$\frac{1}{r}\frac{\partial}{\partial r}\{r \frac{\partial}{\partial r}[\frac{1}{r}\frac{\partial}{\partial r}(r \frac{\partial u}{\partial r})]\} + \frac{1}{b^4}\frac{\partial^2 u}{\partial t^2} = 0 \tag{2.195}$$

where b is a constant. The axisymmetric boundary value problem was considered comprehensively by Reid (1962) for simply supported, fixed support and free support.

Problem 2.7 Classify the following differential equation for the bending theory of cylindrical shells (Timoshenko and Woinowsky-Krieger, 1959):

$$\nabla^2\nabla^2\nabla^2\nabla^2 F + \frac{(1-\nu^2)}{c^2}\frac{\partial^4 F}{\partial \xi^4} = 0 \tag{2.196}$$

where F is Vlasov's stress function, ξ is a normalized distance measured from the center, ν is Poisson's ratio, and c is defined as:

$$c^2 = \frac{h^2}{12a^2} \tag{2.197}$$

The thickness of the shell is h and the radius of the cylindrical shell is a.

Problem 2.8 Classify the following Monge-Ampere equation, which is a differential equation encountered in differential geometry, gas dynamics, and meteorology:

$$(\psi_{\eta\xi})^2 - \psi_{\eta\eta}\psi_{\xi\xi} = f(\eta,\xi) \tag{2.198}$$

Problem 2.9 Classify the following differential equation for shells with surface of revolution under axisymmetric bending (the unknown U is related to shear force and the variable is φ):

$$LL(U) + \mu^4 U = 0, \quad \mu^4 = \frac{Eh}{D} - \frac{\nu^2}{r_1^2} \tag{2.199}$$

$$L(..) = \frac{r_2}{r_1^2}\frac{d^2(..)}{d\varphi^2} + \frac{1}{r_1}\left[\frac{d}{d\varphi}\left(\frac{r_2}{r_1}\right) + \frac{r_2}{r_1}\cot\varphi\right]\frac{d(..)}{d\varphi} - \frac{r_1\cot^2\varphi}{r_2 r_1}(..) \tag{2.200}$$

where E = Young's modulus, ν = Poisson's ratio, h = thickness, D [= $Eh^3/(1-\nu^2)$], and r_1 and r_2 are parameters controlling the shape of the shells.

Problem 2.10 Classify the following differential equation for a mode III dynamic crack branching problem:

$$\frac{\partial^2 W}{\partial s^2} - \coth s\frac{\partial W}{\partial s} + \frac{\partial^2 W}{\partial \theta^2} + \frac{1}{4}W = 0 \tag{2.201}$$

CHAPTER THREE

Ordinary Differential Equations

3.1 INTRODUCTION

The techniques for solving ordinary differential equations (ODEs) are of fundamental importance in the theory of differential equations. Most of the techniques to be discussed in this chapter were developed by founders of differential equations, including Leibniz, Euler, and Bernoulli. The mastery of these techniques is essential for our later chapters in solving partial differential equations (PDEs). In later chapters, we will see that a PDE is normally converted to a number of ODEs by assuming a technique called "separation of variables." For the two-dimensional case, separation of variables leads to two ODEs; and for the three-dimensional case, separation of variables leads to three ODEs. Thus, solving PDE becomes solving ODEs. Therefore, the solution of ODEs is of fundamental importance in solving PDEs.

For first order ODEs, the topics to be covered in the present chapter include separable equations, homogeneous type equations, exact ODEs, integrable condition, integrating factors, the Stokes method for homogeneous type, the Jacobi method, the Euler method, standard linear form, the Bernoulli equation, the Riccati equation, integration by differentiation, the Clairaut equation, singular solution, the Lagrange equation, factorization of nonlinear form, and Taylor series expansion. For second order ODEs with constant coefficients, the solution methods for a differential equation (DE) with non-zero nonhomogeneous terms include undetermined coefficients, variations of parameters, and operator factors.

For second order ODE with non-constant coefficients, we discuss the Euler equation, Laplace type equation, Liouville problem, Mainardi approach for Liouville problem, and finally the Liouville transformation. Both homogeneous and nonhomogeneous forms are considered. Another general transformation (differing from Liouville transformation) and its associated concept of invariants of ODEs are discussed.

For higher order ODEs, we consider the Euler equation of order n, adjoint differential equation of an n-th order ODE, Sarrus method, rule of transformation, and homogeneous equation. For nonhomogeneous equations, we have extended the method of undetermined coefficients and variation of parameters to n-th order ODEs. The technique of reduction of an n-th order ODE to a lower order ODE is discussed. The exact condition for an n-th order ODE is presented. The idea of factorization of ODEs is presented. A symbolic method for solving nonhomogeneous ODEs is discussed. The special technique of removal of the second highest derivative is also covered in the context of the n-th order ODE. The general idea of reduction of order for autonomous n-th order ODEs is summarized.

3.2 FIRST ORDER ODE

The most general form of first order ordinary differential equation (ODE) is

$$F(\frac{dy}{dx}, y, x) = 0 \tag{3.1}$$

Assuming that dy/dx can be solved from (3.1), we may express the first order ODE as

$$\frac{dy}{dx} = F_1(x, y) \tag{3.2}$$

Note however that, in general, (3.2) may not be obtainable by solving (3.1).

3.2.1 Separable ODE

If the function on the right hand side of (3.2) can be separated as a product of two functions, one is a function of x only and the other is a function of y only, we get

$$\frac{dy}{dx} = f(x)\phi(y) \tag{3.3}$$

Thus, we can put all functions of y on the left hand side with dy and all other functions on the right hand side with dx as:

$$\frac{dy}{\phi(y)} = f(x)dx \tag{3.4}$$

Integration can be carried out independently on both sides of (3.4):

$$\int \frac{dy}{\phi(y)} = \int f(x)dx + c \tag{3.5}$$

Therefore, the solution of the differential equation is obtained by integration only. Alternatively, we can rewrite first order ODE as

$$M(x, y)dx + N(x, y)dy = 0 \tag{3.6}$$

The condition of being separable implies M being a function of x only and N being a function of y only. That is, we have

$$M(x)dx = -N(y)dy \tag{3.7}$$

Thus, the solution of the first order ODE can be obtained by integrating on both sides separately as:

$$\int M(x)dx = -\int N(y)dy + C \tag{3.8}$$

In other words, the first order ODE can be solved by integration alone if it is separable. Thus, a separable first order ODE can be considered as the simplest type of first order ODE.

Example 3.1 Find the general solution of the following first order ODE

$$\frac{dy}{dx} = x^2\left(y^2 + 1\right) \tag{3.9}$$

Solution: The given ODE is clearly separable and can be rewritten as

$$\frac{dy}{y^2+1} = x^2 dx \tag{3.10}$$

The integration on the left hand side can be conducted by using the change of variables as

$$y = \tan\theta, \quad dy = \sec^2\theta d\theta \tag{3.11}$$

Recalling the trigonometric identity that

$$\tan^2\theta + 1 = \sec^2\theta \tag{3.12}$$

we find

$$\begin{aligned} \int \frac{dy}{y^2+1} &= \int \frac{\sec^2\theta d\theta}{\tan^2\theta+1} = \int d\theta = \theta \\ &= \tan^{-1} y \end{aligned} \tag{3.13}$$

Consequently, the solution can be expressed as

$$\tan^{-1} y = \frac{1}{3}x^3 + C \tag{3.14}$$

Alternatively, y can be written as:

$$y = \tan(\frac{1}{3}x^3 + C) \tag{3.15}$$

Example 3.2 Find the general solution of the following first order ODE

$$\frac{dy}{dx} = \frac{x^2}{\left(1-y^2\right)} \tag{3.16}$$

Solution: This ODE is clearly separable and can be arranged as:

$$\left(1-y^2\right)dy = x^2 dx \tag{3.17}$$

Integrating on both sides, we have the solution as

$$-\frac{x^2}{3} + \left(y - \frac{y^3}{3}\right) = C \tag{3.18}$$

Note that the solution of the first order ODE has only one unknown constant, and it must be used to fit the boundary value prescribed for the ODE. Let us assume that

$$\frac{dH_1(x)}{dx} = M(x), \quad \frac{dH_2(y)}{dy} = N(y) \tag{3.19}$$

Substitution of (3.19) into (3.8) gives:

$$H_1(x) + H_2(y) = C \tag{3.20}$$

Now suppose that we know the value of $y = y_0$ for $x = x_0$ as:

$$y(x_0) = y_0 \tag{3.21}$$

Application of this boundary condition into (3.20) gives the following value of C as

$$H_1(x_0)+H_2(y_0)=C \tag{3.22}$$

Back substitution of this value of constant into (3.20) gives

$$H_1(x)+H_2(y)-H_1(x_0)-H_2(y_0)=0 \tag{3.23}$$

Note that now the solution contains no unknown constant. This solution can also be expressed in terms of the original given function M and N by noting that

$$H_1(x)-H_1(x_0)=\int_{x_0}^{x} M(\xi)d\xi \tag{3.24}$$

$$H_2(y)-H_2(y_0)=\int_{y_0}^{y} N(\xi)d\xi \tag{3.25}$$

Therefore, the final solution can also be written as

$$\int_{x_0}^{x} M(\xi)d\xi+\int_{y_0}^{y} N(\xi)d\xi=0 \tag{3.26}$$

Note again that the boundary condition given in (3.21) has been used in prescribing the lower limit of the integration.

Example 3.3 Find the general solution of the following first order ODE

$$\frac{dy}{dx}=\frac{3x^2+4x+2}{2(y-1)}, \quad y(0)=-1 \tag{3.27}$$

Solution: This equation is separable and can be integrated with respect to y on the left hand side and with respect to x on the right hand side as:

$$2(y-1)dy=(3x^2+4x+2)dx \tag{3.28}$$

Integration on both sides gives

$$y^2-2y=x^3+2x^2+2x+C \tag{3.29}$$

Applying the boundary condition at $x=0$ we have

$$(-1)^2-2(-1)=0^3+2\times 0^2+2\times 0+C \tag{3.30}$$

This gives the unknown constant C

$$C=3 \tag{3.31}$$

Substitution of (3.31) into (3.29) results in the final solution

$$y^2-2y=x^3+2x^2+2x+3 \tag{3.32}$$

This is a quadratic equation in y and its solution is

$$y=\frac{-B\pm\sqrt{B^2-4AC}}{2A} \tag{3.33}$$

where A, B and C are coefficients of the quadratic equation. However, we have to be very careful in picking the final solution. It can be shown that the solution is (instead of taking the positive sign in front of the square root):

$$y=1-\sqrt{x^3+2x^2+2x+4} \tag{3.34}$$

To check this, we have to back-substitute (3.34) into the boundary condition:

$$y(0)=1-2=-1 \tag{3.35}$$

If we take the positive sign, we can see that the boundary condition is not satisfied:

$$y(0) = 1 + 2 = 3 \neq -1 \tag{3.36}$$

The process of taking the square root creates an extra solution that does not satisfy the boundary condition. Therefore, we have to be extra careful when we deal with the square root of the quadratic equations.

3.2.2 Homogeneous Equation

We now consider a special form of first order ODE that can be readily transformed into a separable form and thus can be solved easily in principle (provided that we know how to conduct the subsequent integration). This type of ODE is called the homogeneous type. However, we must emphasize that this homogeneous type first order ODE should be not confused with the so-called homogeneous ODE in the classification discussed in Chapter 2, in which all terms in the differential equation involve the unknown of the equation.

The functional form on the right hand side of (3.2) is only a function of y/x. That is, whenever x and y appear, their appearance must be in the form of y/x. In other words, we have

$$\frac{dy}{dx} = g(\frac{y}{x}) \tag{3.37}$$

where g can be any arbitrary function of y/x. To solve (3.37), we apply the following change of variables:

$$u = \frac{y}{x}, \quad or \quad y = ux \tag{3.38}$$

Differentiation of the second of (3.38) gives

$$\frac{dy}{dx} = x\frac{du}{dx} + u = g(u) \tag{3.39}$$

The last two terms of (3.39) can be rearranged as:

$$\frac{du}{dx} = \frac{g(u) - u}{x} \tag{3.40}$$

This is clearly separable and can be integrated as:

$$\int \frac{du}{g(u) - u} = \ln x + C \tag{3.41}$$

This provides the implicit solution of the differential equation. If integration can be carried out explicitly, we can obtain the solution in closed form by substituting this result into (3.38). This depends on the given function $g(u)$.

Example 3.4 Find the general solution of the following first order ODE

$$x\frac{dy}{dx} - 2\sqrt{xy} = y \qquad (x > 0) \tag{3.42}$$

Solution: This is a linear first order ODE of homogeneous type

$$\frac{dy}{dx}=2\sqrt{\frac{y}{x}}+\frac{y}{x} \tag{3.43}$$

Adopting the following change of variables

$$u=\frac{y}{x}, \tag{3.44}$$

we obtain

$$\frac{dy}{dx}=x\frac{du}{dx}+u=2\sqrt{u}+u \tag{3.45}$$

Comparison of the last equation of (3.45) gives

$$\frac{du}{2\sqrt{u}}=\frac{dx}{x} \tag{3.46}$$

Integrating both sides independently, we find

$$\sqrt{u}=\ln x+c \tag{3.47}$$

Thus, u becomes

$$u=[\ln x+c]^2 \tag{3.48}$$

Finally, back-substitution of (3.48) into (3.44) gives the solution as

$$y=x[\ln x+c]^2 \tag{3.49}$$

For the case that M and N are given as

$$Mdx+Ndy=0 \tag{3.50}$$

$$M=x^n\phi(\frac{y}{x}), \quad N=x^n\psi(\frac{y}{x}) \tag{3.51}$$

we have

$$\phi(\frac{y}{x})dx+\psi(\frac{y}{x})dy=0 \tag{3.52}$$

This is obviously a homogeneous type and we use the following change of variables

$$y=vx, \quad dy=vdx+xdv \tag{3.53}$$

The first order ODE becomes

$$\phi(v)dx+\psi(v)(vdx+xdv)=0 \tag{3.54}$$

Grouping the differential terms gives

$$\{\phi(v)+v\psi(v)\}dx+\psi(v)xdv=0 \tag{3.55}$$

$$\frac{dx}{x}+\frac{\psi(v)dv}{\phi(v)+v\psi(v)}=0 \tag{3.56}$$

$$x=C\exp\{-\int\frac{\psi(v)dv}{\phi(v)+v\psi(v)}\} \tag{3.57}$$

Back-substitution of the definition of v after integration gives the final solution.

3.2.3 Rational Polynomials

Another form of the first order ODE, which is closely related to the homogeneous type first order ODE, is given as:

$$\frac{dy}{dx}=\frac{a_1x+b_1y+c_1}{a_2x+b_2y+c_2} \tag{3.58}$$

Both the denominator and numerator are given as a linear function of x and y. There are three possible scenarios associated with (3.58).

Case 1: The simplest scenario is that the constant terms in both numerator and denominator are zero:

$$c_1=c_2=0 \tag{3.59}$$

Dividing all terms in the denominator and numerator by x, we obtain a homogenous type of ODE as:

$$\frac{dy}{dx}=\frac{a_1x+b_1y}{a_2x+b_2y}=\frac{a_1+b_1\dfrac{y}{x}}{a_2+b_2\dfrac{y}{x}}=g(\frac{y}{x}) \tag{3.60}$$

As discussed in the last section, we make the standard change of variables given in (3.38) to make this ODE separable as:

$$\frac{dy}{dx}=x\frac{du}{dx}+u=g(u) \tag{3.61}$$

$$\frac{du}{dx}=\frac{g(u)-u}{x}=\frac{a_1+(b_1-a_2)u-b_2u^2}{(a_2+b_2u)x} \tag{3.62}$$

Therefore, we can integrate it as

$$\int\frac{(a_2+b_2u)du}{a_1+(b_1-a_2)u-b_2u^2}=\int\frac{dx}{x}=\ln x+C \tag{3.63}$$

where C is an integration constant.

There are three possible forms of the integration of (3.63), depending on the values of a_1, b_1, a_2, and b_2. To integrate it, let us first note that the following integration formulas will be useful to integrate (3.63) (Formulas 14.66, 14.265, and 14.266 of Spiegel, 1968):

$$\int\frac{dx}{ax^2+bx+c}=\begin{cases}\dfrac{2}{\sqrt{4ac-b^2}}\tan^{-1}\left(\dfrac{2ax+b}{\sqrt{4ac-b^2}}\right) & b^2<4ac\\[2ex] -\dfrac{2}{2ax+b} & b^2=4ac\\[2ex] \dfrac{1}{\sqrt{b^2-4ac}}\ln\left(\dfrac{2ax+b-\sqrt{b^2-4ac}}{2ax+b+\sqrt{b^2-4ac}}\right) & b^2>4ac\end{cases} \tag{3.64}$$

$$\int\frac{xdx}{ax^2+bx+c}=\frac{1}{2a}\ln(ax^2+bx+c)-\frac{b}{2a}\int\frac{dx}{ax^2+bx+c} \tag{3.65}$$

Applying these integration formulas, for $(a_2-b_1)^2+4a_1b_2>0$, we have

$$\ln x + C = -\frac{1}{2}\ln[b_2(\frac{y}{x})^2 + (a_2 - b_1)\frac{y}{x} - a_1]$$
$$-\frac{a_2 + b_1}{2\sqrt{(a_2 - b_1)^2 + 4a_1b_2}}\ln\left\{\frac{2b_2(y/x) + (a_2 - b_1) - \sqrt{(a_2 - b_1)^2 + 4a_1b_2}}{2b_2(y/x) + (a_2 - b_1) + \sqrt{(a_2 - b_1)^2 + 4a_1b_2}}\right\} \quad (3.66)$$

For $(a_2 - b_1)^2 + 4a_1b_2 = 0$, we have

$$\ln x + C = \frac{a_2 + b_1}{2b_2(y/x) + a_2 - b_1} - \ln[(\frac{y}{x}) + \frac{a_2 - b_1}{2b_2}] \quad (3.67)$$

For $(a_2 - b_1)^2 + 4a_1b_2 < 0$, we have

$$\ln x + C = -\frac{1}{2}\ln[b_2(\frac{y}{x})^2 + (a_2 - b_1)\frac{y}{x} - a_1]$$
$$-\frac{a_2 + b_1}{\sqrt{-(a_2 - b_1)^2 - 4a_1b_2}}\tan^{-1}\left\{\frac{2b_2(y/x) + (a_2 - b_1)}{\sqrt{-(a_2 - b_1)^2 - 4a_1b_2}}\right\} \quad (3.68)$$

Case 2: The second scenario corresponds to the case that the following determinant formed by the coefficients of x and y in the numerator and denominator equals zero. That is,

$$\begin{vmatrix} a_1 & a_2 \\ b_1 & b_2 \end{vmatrix} = 0 \quad (3.69)$$

Alternatively, this can be expressed as:

$$a_1b_2 - a_2b_1 = 0 \quad (3.70)$$

Thus, a_1 and b_1 can be expressed in term of a_2 and b_2 as:

$$\frac{a_1}{a_2} = \frac{b_1}{b_2} = k \quad (3.71)$$

where k is a constant. Substitution of (3.71) into (3.58) gives

$$\frac{dy}{dx} = \frac{a_1x + b_1y + c_1}{a_2x + b_2y + c_2} = \frac{k(a_2x + b_2y) + c_1}{a_2x + b_2y + c_2} = f(a_2x + b_2y) \quad (3.72)$$

Note the fact that x and y only appear as a functional form of u defined as

$$u = a_2x + b_2y \quad (3.73)$$

Naturally, we can adopt u as the change of variables. Differentiation of (3.73) gives

$$\frac{du}{dx} = a_2 + b_2\frac{dy}{dx} = a_2 + b_2f(u) \quad (3.74)$$

This renders (3.74) separable and it can be integrated as:

$$\int \frac{du}{a_2 + b_2f(u)} = x + C \quad (3.75)$$

This can be integrated easily to get:

$$\frac{1}{a_2 + b_2k}\left\{a_2x + b_2y + \frac{b_2(c_2k - c_1)}{a_2 + b_2k}\ln[a_2x + b_2y + \frac{a_2c_2 + b_2c_1}{a_2 + b_2k}]\right\} = x + C \quad (3.76)$$

Case 3: The third and last scenario is for the case that the following determinant formed by the coefficients a_1, b_1, a_2, and b_2 is nonzero. That is,

$$\begin{vmatrix} a_1 & a_2 \\ b_1 & b_2 \end{vmatrix} \neq 0 \quad c_1 \neq 0 \ \text{and} \quad c_2 \neq 0 \tag{3.77}$$

To make this ODE homogeneous, we can use a change of variables to remove the constants c_1 and c_2, such that the mathematical form of (3.60) can be recovered. To find the change of variables, we first formulate a system of two equations of straight lines by setting the numerator and denominator of (3.58) to zero:

$$\begin{cases} a_1 x + b_1 y + c_1 = 0 \\ a_2 x + b_2 y + c_2 = 0 \end{cases} \tag{3.78}$$

Let the solution of this system be $x = \alpha$ and $y = \beta$. A set of new variables can be defined as

$$\begin{cases} X = x - \alpha \\ Y = y - \beta \end{cases} \tag{3.79}$$

Alternatively, this can be rewritten as:

$$\begin{cases} x = X + \alpha \\ y = Y + \beta \end{cases} \tag{3.80}$$

Substitution of (3.80) into the numerator and denominator of (3.58) yields

$$\begin{aligned} a_1 x + b_1 y + c_1 &= a_1(X+\alpha) + b_1(Y+\beta) + c_1 \\ &= a_1 X + b_1 Y + (a_1\alpha + b_1\beta + c_1) \\ &= a_1 X + b_1 Y \end{aligned} \tag{3.81}$$

$$\begin{aligned} a_2 x + b_2 y + c_2 &= a_2(X+\alpha) + b_2(Y+\beta) + c_2 \\ &= a_2 X + +b_2 Y + (a_2\alpha + b_2\beta + c_2) \\ &= a_2 X + +b_2 Y \end{aligned} \tag{3.82}$$

Therefore, (3.58) can be reduced to a homogeneous form in terms of the new variables X and Y as:

$$\frac{dY}{dX} = \frac{a_1 X + b_1 Y}{a_2 X + b_2 Y} \tag{3.83}$$

Obviously, the solution given in (3.66) to (3.68) remains valid for (3.83) if the following substitutions are made:

$$x \leftarrow x - \alpha, \quad y \leftarrow y - \beta. \tag{3.84}$$

Example 3.5 Find the general solution of the following first order ODE

$$\frac{dy}{dx} = \frac{x + y - 1}{x - y + 3} \tag{3.85}$$

Solution: The determinant of the factor of (3.77) is clearly non-zero. This corresponds to Case 3 discussed above.

By setting the numerator and denominator to zeros, we have

$$\begin{cases} x+y-1=0 \\ x-y+3=0 \end{cases} \tag{3.86}$$

Addition and subtraction between these equations give the following solutions

$$x=-1, \quad y=2, \tag{3.87}$$

That is, we can set the shift of coordinates using

$$\alpha=-1, \quad \beta=2 \tag{3.88}$$

The new variables X and Y can be defined as

$$X=x+1, \quad Y=y-2 \tag{3.89}$$

Substitution of (3.89) into (3.85) gives

$$\frac{dY}{dX}=\frac{dy}{dx}=\frac{x+y-1}{x-y+3}=\frac{X+Y}{X-Y}=\frac{1+\dfrac{Y}{X}}{1-\dfrac{Y}{X}} \tag{3.90}$$

This is clearly a homogeneous form and we can assume the standard change of variable as

$$u=\frac{Y}{X} \tag{3.91}$$

With this change of variable, (3.91) is converted to the following separable form:

$$X\frac{du}{dX}=\frac{1+u^2}{1-u} \tag{3.92}$$

Putting all terms containing u on the left and grouping all terms containing X to the right, we have

$$\frac{(1-u)du}{1+u^2}=\frac{dX}{X} \tag{3.93}$$

Integrating both sides, we obtain

$$\tan^{-1}u-\frac{1}{2}\ln(1+u^2)=\ln|X|+c \tag{3.94}$$

Back substitution of (3.91) into (3.94) gives

$$\tan^{-1}u=\tan^{-1}(\frac{Y}{X})=\ln[\sqrt{(1+u^2)}X]+c=\ln[\sqrt{(X^2+Y^2)}]+c \tag{3.95}$$

Finally, substitution of (3.89) into (3.95) yields the solution in the original unknown and variable y and x:

$$\tan^{-1}(\frac{y-2}{x+1})=\ln\sqrt{(x+1)^2+(y-2)^2}+c \tag{3.96}$$

This is the final solution in implicit form and we would not attempt to solve for y explicitly.

3.2.4 Integrable Condition

For non-separable first order ODEs, it is very tempting to integrate with respect to x while treating y as constant and similarly for integration with respect to y. Whether we can do that is the subject of this section. Euler showed that only certain forms of first order ODEs could be integrated directly on both sides even though the equation is not separable.

Consider the total differential of a function u and by definition we have

$$du(x,y) = \frac{\partial u}{\partial x} dx + \frac{\partial u}{\partial y} dy \tag{3.97}$$

Suppose that u equals a constant:

$$u(x,y) = c \tag{3.98}$$

Thus, we arrive at the following differential form of an ODE:

$$du(x,y) = \frac{\partial u(x,y)}{\partial x} dx + \frac{\partial u(x,y)}{\partial y} dy = 0 \tag{3.99}$$

Recasting this into a standard form of ODE using M and N gives

$$\frac{\partial u(x,y)}{\partial x} dx + \frac{\partial u(x,y)}{\partial y} dy = M(x,y)dx + N(x,y)dy = 0 \tag{3.100}$$

This equation reveals that if the differential form given in the second of (3.100) is integrable, there must exist a function u such that

$$\frac{\partial u}{\partial x} = M(x,y), \quad \frac{\partial u}{\partial y} = N(x,y) \tag{3.101}$$

Differentiation of M with respect to y gives

$$\frac{\partial M}{\partial y} = \frac{\partial^2 u}{\partial y \partial x} \tag{3.102}$$

On the other hand, differentiation of N with respect to x gives

$$\frac{\partial N}{\partial x} = \frac{\partial^2 u}{\partial x \partial y} \tag{3.103}$$

Recalling the fact that the order of differentiation can be reversed, we must require

$$\frac{\partial^2 u}{\partial y \partial x} = \frac{\partial^2 u}{\partial x \partial y} \tag{3.104}$$

Thus, the integrable condition is obtained:

$$\frac{\partial M(x,y)}{\partial y} = \frac{\partial N(x,y)}{\partial x} \tag{3.105}$$

If and only if (3.105) is satisfied by functions M and N of a given ODE, we can integrate the equation using the following procedure. We can start with the definition for M that (alternatively we can start with that of N):

$$\frac{\partial u}{\partial x} = M(x,y) \tag{3.106}$$

Integrating with respect to x by treating the variable y in M as constant, we get

$$u(x,y) = \int M(x,y)dx + \phi(y) \tag{3.107}$$

Note that we have to add an arbitrary function of y in the process (instead of a constant) because of the definition of partial differentiation. Next, differentiation of u just obtained in (3.107) with respect to y leads to

$$\frac{\partial u}{\partial y} = \frac{\partial}{\partial y}\int M(x,y)dx + \frac{d\phi(y)}{dy} = N \tag{3.108}$$

The last of (3.108) is obtained by virtue of the definition of N. Rearranging (3.108) gives an equation for finding the arbitrary function ϕ:

$$\frac{d\phi(y)}{dy} = N - \frac{\partial}{\partial y}\int M(x,y)dx \tag{3.109}$$

Integrating both sides with respect to y, we get

$$\phi(y) = \int\left[N - \frac{\partial}{\partial y}\int M(x,y)dx\right]dy \tag{3.110}$$

Note that we can treat x in the bracket of (3.110) as constant. Finally, back substitution of (3.110) into (3.107) gives

$$u(x,y) = \int M(x,y)dx + \int[N - \frac{\partial}{\partial y}\int M(x,y)dx]dy = c \tag{3.111}$$

We should emphasize that the whole procedure works because we start with $u(x,y) = c$ Therefore, it is of utmost importance to set u to a constant as the last of (3.111). Otherwise, we still do not obtain any solution, but just define a function u. Most students will make this careless mistake when they first learn this method.

Example 3.6 Find the general solution of the following first order ODE

$$(e^x + y)dx + (x - 2\sin y)dy = 0 \tag{3.112}$$

Solution: Identifying M and N, we have

$$M(x,y) = e^x + y, N(x,y) = x - 2\sin y \tag{3.113}$$

It is straightforward to see that the integrable condition in (3.105) is satisfied:

$$\frac{\partial M(x,y)}{\partial y} = 1 = \frac{\partial N(x,y)}{\partial x} \tag{3.114}$$

We can integrate directly by assuming the existence of a function u as

$$\frac{\partial u}{\partial x} = e^x + y, \tag{3.115}$$

$$\frac{\partial u}{\partial y} = x - 2\sin y, \tag{3.116}$$

Integration of (3.115) with respect to x gives

$$u(x,y) = \int(e^x + y)dx + \phi(y) = e^x + yx + \phi(y) \tag{3.117}$$

Note that we have treated y as a constant during the integration as:

$$\int M(x,y)dx = e^x + xy \tag{3.118}$$

Differentiating with respect to x and equating it to N, we find

$$\frac{\partial u}{\partial y} = \frac{\partial}{\partial y}\int M(x,y)dx + \frac{d\phi(y)}{dy} = N = x - 2\sin y \tag{3.119}$$

Taking partial differentiation of (3.118) gives

$$\frac{\partial}{\partial y}\int M(x,y)dx = x \tag{3.120}$$

Finally, we obtain a differential equation for ϕ from (3.119)

$$x + \frac{d\phi(y)}{dy} = x - 2\sin y \tag{3.121}$$

Canceling the function x on both sides, we obtain

$$\frac{d\phi(y)}{dy} = -2\sin y \tag{3.122}$$

We want to emphasize that if the function of x in (3.121) does not cancel out on both sides, we must have made a careless mistake. Finally, integration of (3.122) gives

$$\phi(y) = 2\cos y, \tag{3.123}$$

Putting ϕ into (3.117), we obtain the required function u

$$u(x,y) = e^x + yx + \phi(y) = e^x + yx + 2\cos y \tag{3.124}$$

As remarked earlier, the most important step is to set u to a constant to get the final solution:

$$e^x + yx + 2\cos y = c \tag{3.125}$$

3.2.5 Integrating Factor

If the integrable condition is not satisfied, we can make it integrable by multiplying an integrating factor. This was first discovered by Euler. To illustrate the idea, we can recast the left hand side of the ODE as

$$Mdx + Ndy = \frac{1}{2}\{(Mx+Ny)(\frac{dx}{x}+\frac{dy}{y}) + (Mx-Ny)(\frac{dx}{x}-\frac{dy}{y})\} \tag{3.126}$$

The validity of this identity can be demonstrated by expanding the right hand side. Note also that we have an exact integral for both brackets on the right hand side as

$$\frac{dx}{x}+\frac{dy}{y} = d\ln(xy), \quad \frac{dx}{x}-\frac{dy}{y} = d\ln(\frac{x}{y}) \tag{3.127}$$

Both of these are exact integrals, thus, providing a simple way to determine the integrating factor. However, this method is not covered in most of the textbooks on differential equations.

We now look at a few special cases of (3.126).

3.2.5.1 Case 1: Mx+Ny=0

For this case, we can rearrange the differential equation (3.126) as

$$\frac{Mdx + Ndy}{(Mx - Ny)} = \frac{1}{2} d\ln(\frac{x}{y}) \tag{3.128}$$

Since the right hand side is an exact integral, the integrating factor for such case is thus

$$\frac{1}{Mx - Ny} \tag{3.129}$$

This is also an integrating factor for the following special forms of M and N:

$$M = F_1(xy)y, \quad N = F_2(xy)x \tag{3.130}$$

To prove this, we can rearrange this equation as

$$\begin{aligned}\frac{Mdx + Ndy}{(Mx - Ny)} &= \frac{1}{2}\{\frac{Mx + Ny}{Mx - Ny} d\ln(xy) + d\ln(\frac{x}{y})\} \\ &= \frac{1}{2}\{\frac{F_1(xy) + F_2(xy)}{F_1(xy) - F_2(xy)} d\ln(xy) + d\ln(\frac{x}{y})\}\end{aligned} \tag{3.131}$$

We see the first term on the right of (3.131) is a function of xy only, and thus the right hand side is again an exact integral. This completes the proof.

3.2.5.2 Case 2: $Mx-Ny=0$

For this case, to find the integrating factor we can rearrange (3.126) as

$$\frac{Mdx + Ndy}{Mx + Ny} = \frac{1}{2} d\ln(xy) \tag{3.132}$$

Since the right hand side is an exact integral, the integrating factor for this case is

$$\frac{1}{Mx + Ny} \tag{3.133}$$

3.2.5.3 Case 3: $Mx+Ny\neq 0$ & $Mx-Ny\neq 0$

When both of these groups are not zeros, the integrating factor is

$$\frac{1}{Mx + Ny} \tag{3.134}$$

If $Mdx+Ndy = 0$ is homogeneous, the integrating factor is

$$\frac{1}{Mx - Ny} \tag{3.135}$$

Also recall that if the differential equation can be expressed as $F_1(xy)ydx + F_2(xy)xdy = 0$, we have shown that the integrating factor is given by (3.135).

3.2.5.4 Stokes Method for Homogeneous Equation

If M and N are homogeneous functions of x and y of the degree n, we have

$$M = x^n \phi(v), \quad N = x^n \psi(v), \quad v = y/x \tag{3.136}$$

Subsequently, the differential equation can be written as

$$\begin{aligned}Mdx + Ndy &= x^n \phi(v)dx + x^n \psi(v)dy \\ &= x^n \{\phi(v) + v\psi(v)\}dx + x^{n+1}\psi(v)dv\end{aligned} \tag{3.137}$$

Case 1: $\phi(v)+v\psi(v)=0$ (or $Mx+Ny=0$)

For this case, we have

$$Mdx+Ndy=x^{n+1}\psi(v)dv \tag{3.138}$$

$$\frac{Mdx+Ndy}{x^{n+1}}=\psi(v)dv \tag{3.139}$$

The right hand side is an exact integral, and thus the left hand side is also exact. Consequently, the integrating factor for case 1 is

$$\frac{1}{x^{n+1}} \tag{3.140}$$

Case 2: $\phi(v)+v\psi(v)\neq 0$ (or $Mx+Ny\neq 0$)

Using (3.137), we have

$$\frac{Mdx+Ndy}{x^{n+1}\{\phi(v)+v\psi(v)\}}=\frac{dx}{x}+\frac{\psi(v)dv}{\phi(v)+v\psi(v)} \tag{3.141}$$

Both terms on the right are exact differentials and, in turn, the right hand side is also exact

$$\frac{Mdx+Ndy}{Mx+Ny}=\frac{dx}{x}+\frac{\psi(v)dv}{\phi(v)+v\psi(v)} \tag{3.142}$$

Thus, the integrating factor is

$$\frac{1}{Mx+Ny} \tag{3.143}$$

This result obtained by Stokes is more general than the solution discussed in the previous section.

Note, however, that in Section 3.2.2 we have shown that (3.136) is homogeneous. Therefore, we can actually solve it without finding the integrating factor.

3.2.5.5 Differential Equation for Integrating factor

We now consider the general case of finding an integrating factor μ. If a first order ODE is not exact, we can make it exact by multiplying a function called the integrating factor:

$$\mu M(x,y)dx+\mu N(x,y)dy=0 \tag{3.144}$$

This is actually the definition of an integrating factor. Since (3.144) becomes exact after multiplying by μ, we must have the exact condition being satisfied:

$$\frac{\partial\mu(x,y)M(x,y)}{\partial y}=\frac{\partial\mu(x,y)N(x,y)}{\partial x} \tag{3.145}$$

Differentiation and rearrangement of (3.145) gives

$$N\frac{\partial\mu}{\partial x}-M\frac{\partial\mu}{\partial y}=(\frac{\partial M}{\partial y}-\frac{\partial N}{\partial x})\mu \tag{3.146}$$

This is a first order PDE and is very difficult to solve for arbitrary functions M and N. The commonly adopted approach is to make certain assumptions regarding the functional form of the integrating factor, and check whether the assumed functional form is valid. In a sense, it is a trial and error approach. In particular, we assume the integrating factor can be expressed in terms of ζ, which is a function of x and y:

$$\mu(x,y)=\mu(\zeta), \quad \zeta=\zeta(x,y) \tag{3.147}$$

Substituting (3.417) into (3.146) gives

$$N\frac{d\mu}{d\zeta}\frac{\partial\zeta}{\partial x}-M\frac{d\mu}{d\zeta}\frac{\partial\zeta}{\partial y}=(\frac{\partial M}{\partial y}-\frac{\partial N}{\partial x})\mu \tag{3.148}$$

Rearranging (3.148), we obtain

$$\frac{d\mu}{\mu}=\frac{(\frac{\partial M}{\partial y}-\frac{\partial N}{\partial x})}{N\frac{\partial\zeta}{\partial x}-M\frac{\partial\zeta}{\partial y}}d\zeta \tag{3.149}$$

We now observe that if the function on the right of (3.149) is a function of ζ only, we can immediately integrate both sides and result in the integrating factor. If this is the case, we can first rewrite it as

$$\frac{d\mu}{\mu}=\frac{(\frac{\partial M}{\partial y}-\frac{\partial N}{\partial x})}{N\frac{\partial\zeta}{\partial x}-M\frac{\partial\zeta}{\partial y}}d\zeta=\chi(\zeta)d\zeta \tag{3.150}$$

Integration leads to the following result for μ

$$\mu=e^{\int\chi(\zeta)d\zeta}, \quad \chi(\zeta)=\frac{(\frac{\partial M}{\partial y}-\frac{\partial N}{\partial x})}{N\frac{\partial\zeta}{\partial x}-M\frac{\partial\zeta}{\partial y}} \tag{3.151}$$

Note that we did not add an integration constant in (3.151) because we can always multiply an integrating factor by a constant to get another integrating factor. In fact, there are infinite integrating factors for the same ODE. However, if χ is not a function of ζ, we cannot solve it this way. If $\zeta=x$, we have to check whether the following function is expressible in x only

$$\chi=\frac{(\frac{\partial M}{\partial y}-\frac{\partial N}{\partial x})}{N} \tag{3.152}$$

If $\zeta=y$, we have to check whether the following function is expressible in y only

$$\chi=\frac{(\frac{\partial M}{\partial y}-\frac{\partial N}{\partial x})}{-M} \tag{3.153}$$

There are, however, infinite possibilities for the structural form of ζ. Therefore, this method should be our last resort when all other methods fail to apply. The following example illustrates another choice of ζ.

Example 3.7 Find the integrating factor for the following first order ODE

$$(3x+\frac{6}{y})dx+(\frac{x^2}{y}+3\frac{y}{x})dy=0 \tag{3.154}$$

Solution: We can identify

$$M=(3x+\frac{6}{y}),\quad N=(\frac{x^2}{y}+3\frac{y}{x}) \tag{3.155}$$

It is straightforward to see that it is not exact.

$$\frac{\partial M}{\partial y}-\frac{\partial N}{\partial x}=-\frac{6}{y^2}-2\frac{x}{y}+3\frac{y}{x^2} \tag{3.156}$$

To find the integrating factor, we assume

$$\mu(x,y)=\mu(xy) \tag{3.157}$$

That is, we have $\zeta = xy$. We have

$$N\frac{\partial \zeta}{\partial x}-M\frac{\partial \zeta}{\partial y}=(\frac{x^2}{y}+3\frac{y}{x})y-(3x+\frac{6}{y})x=-2x^2+3\frac{y^2}{x}-\frac{6x}{y}$$
$$=xy(-\frac{6}{y^2}-2\frac{x}{y}+3\frac{y}{x^2}) \tag{3.158}$$

Differentiating (3.160) with respect to y gives

$$\chi(\zeta)=\frac{(\frac{\partial M}{\partial y}-\frac{\partial N}{\partial x})}{N\frac{\partial \zeta}{\partial x}-M\frac{\partial \zeta}{\partial y}}=\frac{1}{\zeta}=\frac{1}{xy} \tag{3.159}$$

Substitution of (3.158) into (3.151) gives

$$\mu=e^{\int \frac{1}{\zeta}d\zeta}=e^{\ln\zeta}=\zeta=xy \tag{3.160}$$

The original ODE becomes exact by using (3.157) as

$$\mu Mdx+\mu Ndx=(3x^2y+6x)dx+(x^3+3y^2)dy=0 \tag{3.161}$$

Applying the technique discussed in Section 3.2.4, we have

$$\frac{\partial u}{\partial x}=\mu M=3x^2y+6x \tag{3.162}$$

Integrating with respect to x, we find

$$u=x^3y+3x^2+\phi(y) \tag{3.163}$$

Differentiating (3.163) with respect to y gives

$$\frac{\partial u}{\partial y}=\mu N=x^3+3y^2=x^3+\phi'(y) \tag{3.164}$$

This leads to a first order ODE for the unknown function ϕ:

$$\frac{d\phi(y)}{dy}=3y^2 \tag{3.165}$$

Integrating and substituting the result back into (3.163), we obtain the final solution as

$$u = x^3 y + 3x^2 + y^3 = C \tag{3.166}$$

3.2.5.6 *Integrating Factors by Inspection*

Certain first order ODEs allow one to guess the corresponding integrating factor. Here are some examples. A particular example is that a first order ODE contains the following group:

$$xdy - ydx \tag{3.167}$$

The following Table 3.1 gives six possible integrating factors for (3.167) that we can use.

Table 3.1 System of integrating factors by inspection

Integrating factors	Total differential	Applicable ODE form
$\frac{1}{x^2}$	$\frac{xdy - ydx}{x^2} = d(\frac{y}{x})$	$xdy - ydx + f(x)dx = 0$
$\frac{1}{y^2}$	$\frac{xdy - ydx}{y^2} = -d(\frac{x}{y})$	$xdy - ydx + f(y)dy = 0$
$\frac{1}{xy}$	$\frac{xdy - ydx}{yx} = d[\ln(\frac{y}{x})]$	$xdy - ydx + f(xy)(xdy + ydx) = 0$
$\frac{1}{x^2 + y^2}$	$\frac{xdy - ydx}{y^2 + x^2} = d[\tan^{-1}(\frac{y}{x})]$	$xdy - ydx + f(x^2 + y^2)(xdy + ydx) = 0$
$\frac{1}{x^2 - y^2}$	$\frac{xdy - ydx}{x^2 - y^2} = \frac{1}{2}d[\ln(\frac{x + y}{x - y})]$	$xdy - ydx + f(x^2 - y^2)(xdy - ydx) = 0$

The validity of the second column of Table 3.1 can be established easily, and this will be left for readers to show (see Problem 3.36). The following examples illustrate this method of inspection.

Example 3.8 Find the integrating factor for the following first order ODE

$$(y^2 - xy)dx + x^2 dy = 0 \tag{3.168}$$

Solution: First, we rearrange the ODE as

$$y^2 dx - x(ydx - xdy) = 0 \tag{3.169}$$

The functional form in (3.167) appears. Thus, we find that the integrating factor is $1/(xy^2)$, and using this we have

$$\frac{dx}{x} - \frac{ydx - xdy}{y^2} = 0 \tag{3.170}$$

Using the result in the second row of Table 3.1, we have

$$\frac{dx}{x} - d\left(\frac{x}{y}\right) = 0 \tag{3.171}$$

Thus, the solution of the ODE given in (3.168) is

$$\ln x - \frac{x}{y} = C \tag{3.172}$$

where C is an arbitrary constant.

Example 3.9 Find the integrating factor for the following first order ODE

$$(x+y)dx - (x-y)dy = 0 \tag{3.173}$$

Solution: First, we rearrange the ODE as

$$xdx + ydy + (ydx - xdy) = 0 \tag{3.174}$$

The structural group given in (3.167) again appears. By inspection, we go for the fourth row of Table 3.1 and the integrating factor is $1/(x^2+y^2)$, and using this we have

$$\frac{1}{2}\frac{d(x^2+y^2)}{x^2+y^2} + \frac{ydx - xdy}{x^2+y^2} = 0 \tag{3.175}$$

Integration gives the solution of the ODE as

$$\frac{1}{2}\ln(x^2+y^2) - \tan^{-1}\left(\frac{y}{x}\right) = C \tag{3.176}$$

where C is an arbitrary constant.

3.2.6 Standard Linearized Form

The most general first order "linear" ODE can be written as

$$\frac{dy}{dx} = p(x)y + Q(x) \tag{3.177}$$

If we set $Q(x) = 0$, (3.177) can instantly be integrated since it becomes separable. The solution is:

$$y = ce^{\int p(x)dx} \tag{3.178}$$

To solve (3.177), we can set the constant as a function of x in (3.178). This idea originated from Euler and is called variation of parameters. Thus, we assume the solution form (3.177) as

$$y = c(x)e^{\int p(x)dx} \tag{3.179}$$

Differentiation of (3.179) gives

$$\frac{dy}{dx} = \frac{dc(x)}{dx}e^{\int p(x)dx} + c(x)p(x)e^{\int p(x)dx} = \frac{dc(x)}{dx}e^{\int p(x)dx} + p(x)y \tag{3.180}$$

Substitution of (3.180) into (3.177) leads to a differential equation for $c(x)$

$$\frac{dc(x)}{dx} = Q(x)e^{-\int p(x)dx} \tag{3.181}$$

This can be readily integrated to give

$$c(x) = \int Q(x)e^{-\int p(x)dx} dx + \bar{C} \tag{3.182}$$

Substitution of (3.182) into (3.179) gives the final solution as

$$y = e^{\int p(x)dx} (\int Q(x)e^{-\int p(x)dx} dx + \bar{C}) \tag{3.183}$$

We will see in a later section that many first order ODEs can be converted to the standard linearized form given in (3.177).

Another approach in solving (3.177) is integrating factor approach discussed in Section 3.2.6. In particular, we can multiply (3.177) by μ:

$$\mu(x)\frac{dy}{dx} - \mu(x)p(x)y = \mu(x)Q(x) \tag{3.184}$$

Next, we note the following identity:

$$\frac{d\mu y}{dx} = \mu\frac{dy}{dx} + y\frac{d\mu}{dx} \tag{3.185}$$

Comparison of the left hand side of (3.184) with (3.185) gives the following condition, if the equation is exact:

$$\frac{d\mu}{dx} = -\mu(x)p(x) \tag{3.186}$$

This is clearly separable, and we have the integrating factor as

$$\mu = e^{-\int p(x)dx} \tag{3.187}$$

Returning to (3.184) and using the integrating factor obtained in (3.189), it can be simplified as

$$\frac{d\mu y}{dx} = \mu Q \tag{3.188}$$

Integration of (3.188) with respect to x gives

$$\mu y = \int \mu Q dx + c \tag{3.189}$$

This gives the final solution:

$$y = \frac{1}{\mu}[\int \mu Q dx + c] = e^{\int p(x)dx} [\int Q e^{-\int p(x)dx} dx + c] \tag{3.190}$$

This is, of course, the same as (3.183). Once we can convert any first order ODE to the standard form in (3.177), we can solve the ODE. Note again that there is one unknown constant in the solution of first order ODEs before we imposed any boundary condition.

3.2.7 Bernoulli Equation

We now consider a nonlinear first order ODE of the form:

$$\frac{dy}{dx} = p(x)y + Q(x)y^n \tag{3.191}$$

This nonlinear ODE was first considered by Jacob Bernoulli and for this equation, the following change of variable can be proposed:

$$z = y^{1-n} \tag{3.192}$$

Differentiation of (3.192) with respect to x gives

$$\frac{dz}{dx} = (1-n)y^{-n}\frac{dy}{dx} \tag{3.193}$$

Substitution of (3.193) into (3.191) gives the following ODE for z

$$\frac{dz}{dx} = (1-n)P(x)z + (1-n)Q(x) \tag{3.194}$$

This becomes the standard linear first order ODE that we have discussed in the last section. Thus, we know how to solve it. We now illustrate the technique in the next example.

Example 3.10 Find the general solution of the following first order ODE

$$\frac{dy}{dx} = \frac{y}{2x} + \frac{x^2}{2y} \tag{3.195}$$

Solution: This is the Bernoulli equation with the following identifications:

$$n = -1, \quad z = y^2 \tag{3.196}$$

Using this change of variables suggested in (3.196), we obtain

$$\frac{dz}{dx} = 2y\frac{dy}{dx} \tag{3.197}$$

Substitution of (3.196) and (3.197) into (3.195) gives

$$\frac{dz}{dx} = \frac{1}{x}z + x^2 \tag{3.198}$$

In the context of the standard linearized ODE, we have

$$p(x) = \frac{1}{x}, \quad Q(x) = x^2 \tag{3.199}$$

It can be solved as in Section 3.2.6 to give

$$z = e^{\int \frac{1}{x}dx}\left(\int x^2 e^{-\int \frac{1}{x}dx} dx + c\right) = cx + \frac{1}{2}x^3 \tag{3.200}$$

Back substitution of the second of (3.196) into (3.200) gives the final solution for y as:

$$y^2 = cx + \frac{1}{2}x^3 \tag{3.201}$$

Note that (3.200) is not our final solution since z does not appear in our original equation. Thus, we must use the second equation of (3.196) to convert z to y. StudentsThis final step that students normally forget.

3.2.8 Riccati Equation

As summarized by Watson (1944), the Riccati equation first appeared in a paper of John Bernoulli in 1694, who however never received the credit that he deserved. Riccati's paper on the same differential equation was published in 1724. The generalized Riccati equation is given as

$$\frac{dy}{dt} = q_1(t) + q_2(t)y + q_3(t)y^2 \tag{3.202}$$

This Riccati equation is not easy to solve, in general. Euler in 1763 found that the general solution can be determined if one of the particular solution y_1 is known and can be written as

$$y(t) = y_1(t) + \frac{1}{v(t)} \tag{3.203}$$

Substitution of (3.203) into the left hand side of (3.202) gives

$$LHS = \frac{dy}{dt} = \frac{dy_1}{dt} - \frac{1}{v^2}\frac{dv}{dt} = q_1 + q_2 y_1 + q_3 y_1^2 - \frac{1}{v^2}\frac{dv}{dt} \tag{3.204}$$

On the other hand, substituting the following value of y^2

$$y^2 = y_1^2 + 2\frac{y_1}{v} + \frac{1}{v^2} \tag{3.205}$$

into the right hand side of (3.202) gives

$$RHS = \frac{dy}{dt} = q_1 + q_2(y_1 + \frac{1}{v}) + q_3(y_1^2 + \frac{2y_1}{v} + \frac{1}{v^2}) \tag{3.206}$$

Equating (3.204) and (3.206), we obtain

$$\frac{dv}{dt} = -(q_2 + 2q_3 y_1)v - q_3 \tag{3.207}$$

This is in the form of the most general first order ODE discussed in Section 3.2.6 and, therefore, can be integrated exactly.

Example 3.11 Find the general solution of the following first order ODE

$$\frac{dy}{dt} = -\frac{2}{t} + \frac{1}{t}y + \frac{1}{t}y^2 \tag{3.208}$$

Solution: We can easily show that a particular solution of this Riccati equation is

$$y_1 = 1 \tag{3.209}$$

We apply the following change of variables recommended by Euler

$$y = 1 + \frac{1}{v} \tag{3.210}$$

Differentiation of (3.210) and substitution of the result into (3.208) gives

$$\frac{dv}{dt} = -(q_2 + 2q_3 y_1)v - q_3 = -(\frac{3}{t})v - \frac{1}{t} \tag{3.211}$$

This converts the Riccati equation given in (3.208) into the standard linearized form discussed in Section 3.2.6. Thus, it can be solved using the standard approach.

In the years between 1697 and 1704, John Bernoulli found a change of variables to transform the Riccati equation into a linear second order ODE. In particular, we first rewrite the generalized Riccati equation as

$$\frac{dy}{dx}+\phi y+\psi y^2+\chi=0 \tag{3.212}$$

The following change of variables was communicated to Leibniz by John Bernoulli:

$$y=\frac{1}{\psi u}\frac{du}{dx}=\frac{u'}{\psi u} \tag{3.213}$$

Differentiating (3.213) with respect to x, we see that

$$y'=\frac{\psi u u'-u'(\psi' u+\psi u')}{\psi^2 u^2} \tag{3.214}$$

Substitution of (3.214) into (3.212) leads to the following linear second order ODE:

$$u''+(\phi-\frac{\psi'}{\psi})u'+\chi\psi u=0 \tag{3.215}$$

Therefore, the nonlinear first order Riccati equation becomes a linear second order ODE. Actually, only certain types of nonlinear ODEs can be converted to linear ODEs. The Riccati equation is one of these special types of nonlinear ODEs. As shown in Section 4.14 in Chapter 4, a nonlinear ODE that has poles as its only movable singularities can be converted to a linear ODE. Although, in general, this linear second order ODE is easier to solve than the nonlinear ODE, its solution may not be easy to obtain. In the next example, we will consider a special form of the Riccati equation that can be solved readily.

Example 3.12 Find the general solution of the following first order ODE

$$\frac{dy}{dt}=-\frac{y}{t}-1-y^2 \tag{3.216}$$

Solution: First, we can rewrite it as

$$\frac{dy}{dt}+\frac{y}{t}+y^2+1=0 \tag{3.217}$$

Use the change of variables suggested by John Bernoulli

$$y=\frac{u'}{u} \tag{3.218}$$

Substitution of (3.218) into (3.217) gives

$$u''+\frac{1}{t}u'+u=0 \tag{3.219}$$

This is a standard linear second order ODE, which is called the Bessel function of zero order. The solution is the Bessel function of the first kind and of the second kind of zero order (J_0 and Y_0):

$$u = AJ_0(t) + BY_0(t) \tag{3.220}$$

More detailed discussions of Bessel functions will be given in Chapter 4. In fact, there is a close relation between the Riccati equation and the Bessel equation as discussed by Watson (1944).

We now consider a special form of the Riccati equation that allows an exact solution to be found. In particular, we start from the original equation considered by Riccati as

$$\frac{du}{dx} + bu^2 = cx^m \tag{3.221}$$

Use a change of variables of

$$u = \frac{y}{x} \tag{3.222}$$

Differentiation of (3.222) gives

$$\frac{du}{dx} = \frac{1}{x}\frac{dy}{dx} - \frac{y}{x^2} \tag{3.223}$$

Substitution of (3.223) into (3.221) results in

$$x\frac{dy}{dx} - y + by^2 = cx^{m+2} \tag{3.224}$$

This can be generalized to the following form

$$x\frac{dy}{dx} - ay + by^2 = cx^n \tag{3.225}$$

This form is, however, a special form of (3.202) or (3.212).

In the following discussion, we will consider a special case of $n = 2a$, which allows an exact solution to be obtained. In particular, we now consider

$$x\frac{dy}{dx} - ay + by^2 = cx^{2a} \tag{3.226}$$

A change of variables can be proposed as

$$y = x^a v \tag{3.227}$$

With this change of variables, it is straightforward to find

$$\frac{dy}{dx} = ax^{a-1}v + x^a\frac{dv}{dx} \tag{3.228}$$

Substitution of (3.228) into (3.226) arrives at the following separable ODE:

$$x^{1-a}\frac{dv}{dx} + bv^2 = c \tag{3.229}$$

Note that the linear order term is removed by this change of variables, and it becomes separable:

$$\frac{dv}{b(c/b - v^2)} = \frac{dx}{x^{1-a}} \tag{3.230}$$

If b and c are of equal signs, we can use a partial fraction to find

$$\frac{1}{(c/b-v^2)} = \frac{1}{(\sqrt{c/b}-v)(\sqrt{c/b}+v)} = \frac{1}{2}\sqrt{\frac{b}{c}}\{\frac{1}{\sqrt{c/b}-v} + \frac{1}{\sqrt{c/b}+v}\} \tag{3.231}$$

Therefore, integration on both sides gives

$$\frac{1}{2b}(\frac{b}{c})^{1/2}\ln\{\frac{\sqrt{c/b}+v}{\sqrt{c/b}-v}\} = \frac{1}{a}x^a + \bar{C} \tag{3.232}$$

Rearranging this equation yields

$$\frac{\sqrt{c/b}+v}{\sqrt{c/b}-v} = Ce^{\frac{2\sqrt{bc}x^a}{a}} \tag{3.233}$$

Solving for v, we obtain

$$v = \sqrt{\frac{c}{b}}\left\{\frac{Ce^{\frac{2\sqrt{bc}x^a}{a}} - 1}{1+Ce^{\frac{2\sqrt{bc}x^a}{a}}}\right\} \tag{3.234}$$

Finally, substitution of (3.234) into (3.227) gives the final solution of the particular form of Riccati equation given in (3.226) as:

$$y = x^a\sqrt{\frac{c}{b}}\left\{\frac{C\exp(\frac{2\sqrt{bc}x^a}{a}) - 1}{1+C\exp(\frac{2\sqrt{bc}x^a}{a})}\right\} \tag{3.235}$$

As expected, we only have one unknown constant.

For the case that b and c are of unequal signs, we can integrate (3.230) in terms of a tangent function by introducing the following change of variables

$$v = \sqrt{-c/b}\tan\theta \tag{3.236}$$

That is, we have

$$\frac{dv}{-b(-c/b-v^2)} = \frac{(-c/b)^{1/2}\sec^2\theta d\theta}{-b(-c/b)(1+\tan^2\theta)} = -\frac{d\theta}{-b(-c/b)^{1/2}} \tag{3.237}$$

Therefore, integration of both sides (3.237) and (3.230) gives

$$-\frac{\theta}{-b(-c/b)^{1/2}} = \frac{1}{a}x^a + \bar{C} \tag{3.238}$$

Therefore, integration of both sides results in

$$\theta = \tan^{-1}\{\frac{v}{\sqrt{-c/b}}\} = -b\sqrt{-\frac{c}{b}}\{\frac{1}{a}x^a + \bar{C}\} \tag{3.239}$$

Rearranging (3.239) and substituting (3.227) into the result, we obtain the final solution as

$$y = \sqrt{-c/b}\,x^a\tan\{C - \frac{(-bc)^{1/2}x^a}{a}\} \tag{3.240}$$

Another specific form of Riccati equation encountered in viscous incompressible flows will be considered in Problem 3.14.

Example 3.13 A special form of Riccati equation was found equivalent to the following Ramanujan differential equation:

$$q\frac{dP}{dq} = \frac{1}{12}(P^2 - Q), \quad q\frac{dQ}{dq} = \frac{1}{3}(PQ - R), \quad q\frac{dR}{dq} = \frac{1}{2}(PR - Q^2), \tag{3.241}$$

This system relates to number theory studied by Ramanujan. As shown in Hill et al. (2007), this is equivalent to solving the following system after applying an appropriate change of variables:

$$(u^2 - 1)\frac{dv}{du} = \frac{1}{6}(v^2 - 2uv + 1) \tag{3.242}$$

$$\frac{dw}{du} = -\frac{w}{3}\frac{(uv-1)}{v(u^2-1)} - \frac{2}{u^2-1} \tag{3.243}$$

Once v is obtained by solving (3.242), then w can be determined by solving (3.243). Our focus is to solve the Riccati equation given in (3.242).

Solution: To solve (3.242), we apply the change of variables proposed by Bernoulli in (3.213):

$$v(u) = -\frac{6(u^2-1)}{X}\frac{dX}{du} \tag{3.244}$$

Differentiation of (3.244) gives

$$\frac{dv}{du} = -\frac{12u}{X}\frac{dX}{du} + \frac{6(u^2-1)}{X^2}\left(\frac{dX}{du}\right)^2 - \frac{6(u^2-1)}{X}\frac{d^2X}{du^2} \tag{3.245}$$

Substitution of (3.245) into (3.242) gives

$$(u^2-1)^2\frac{d^2X}{du^2} + \frac{7u}{3}(u^2-1)\frac{dX}{du} + \frac{X}{36} = 0 \tag{3.246}$$

We now apply another round of change of variables

$$w = \frac{u}{(u^2-1)^{1/2}} \tag{3.247}$$

Using this change of variables, we obtain

$$\frac{dX}{du} = -\frac{1}{(u^2-1)^{3/2}}\frac{dX}{dw} \tag{3.248}$$

$$\frac{d^2X}{du^2} = \frac{3u}{(u^2-1)^{5/2}}\frac{dX}{dw} + \frac{1}{(u^2-1)^3}\frac{d^2X}{dw^2} \tag{3.249}$$

Substitution of (3.248) and (3.249) into (3.246) arrives at

$$(w^2-1)\frac{d^2X}{dw^2} + \frac{2}{3}w\frac{dX}{dw} + \frac{X}{36} = 0 \tag{3.250}$$

Finally, we impose the following shift of variables:

$$\gamma = (w+1)/2 \tag{3.251}$$

In terms of this new variable, we have

$$\gamma(1-\gamma)\frac{d^2X}{d\gamma^2}+(\frac{1}{3}-\frac{2\gamma}{3})\frac{dX}{d\gamma}-\frac{X}{36}=0 \tag{3.252}$$

This equation is a special form of hypergeometric equation, which was first discovered by Gauss as:

$$x(1-x)\frac{d^2y}{dx^2}+\{c-(a+b+1)x\}\frac{dy}{dx}-aby=0 \tag{3.253}$$

The solution of (3.253) can be written as

$$y=AF(a,b;c;x)+Bx^{1-c}F(a-c+1,b-c+1;2-c;x) \tag{3.254}$$

where F is the hypergeometric series defined as:

$$y=1+\frac{a\cdot b}{1\cdot c}x+\frac{a(a+1)b}{1\cdot 2\cdot c(c+1)}x^2+... \tag{3.255}$$

Detailed discussion of this hypergeometric series will be given in Chapter 4. Comparison of (3.252) and (3.253) gives the solution as:

$$X=AF(-\frac{1}{6},-\frac{1}{6};\frac{1}{3};\gamma)+B\gamma^{2/3}F(\frac{1}{2},\frac{1}{2};\frac{5}{3};\gamma) \tag{3.256}$$

Back substitution into the original unknown and variables gives the final solution as:

$$v=\frac{3}{(u^2-1)^{1/2}}\psi(\frac{u}{2(u^2-1)^{1/2}}+\frac{1}{2}) \tag{3.257}$$

where

$$\psi(\gamma)=\frac{F'(-\frac{1}{6},-\frac{1}{6};\frac{1}{3};\gamma)+C\gamma^{2/3}F'(\frac{1}{2},\frac{1}{2};\frac{5}{3};\gamma)+\frac{2}{3}C\gamma^{-1/3}F'(\frac{1}{2},\frac{1}{2};\frac{5}{3};\gamma)}{F(-\frac{1}{6},-\frac{1}{6};\frac{1}{3};\gamma)+C\gamma^{2/3}F(\frac{1}{2},\frac{1}{2};\frac{5}{3};\gamma)} \tag{3.258}$$

where the superimposed prime means differentiation with respect to γ. Therefore, a certain form of Riccati equation can be expressed in terms of hypergeometric series. Another particular form of Riccati equation that can be solved in terms of hypergeometric series is given in Problem 4.26 of Chapter 4.

3.2.9 Jacobi Method

Jacobi considered the following first order ODE:

$$(A+A'x+A''y)(xdy-ydx)-(B+B'x+B''y)dy+(C+C'x+C''y)dx=0 \tag{3.259}$$

More specifically, Jacobi showed that (3.259) can be converted to the Riccati equation. Allow the following change of variables

$$x=\xi+\alpha,\quad y=\eta+\beta \tag{3.260}$$

Substitution of these into the differential equation given in (3.259) gives

$$\begin{aligned}&[A+A'(\xi+\alpha)+A''(\eta+\beta)][(\xi+\alpha)d\eta-(\eta+\beta)d\xi]\\&-[B+B'(\xi+\alpha)+B''(\eta+\beta)]d\eta+[C+C'(\xi+\alpha)+C''(\eta+\beta)]d\xi=0\end{aligned} \tag{3.261}$$

This ODE can be reduced to a form

$$(a\xi + a'\eta)(\xi d\eta - \eta d\xi) - (b\xi + b'\eta)d\eta + (c\xi + c'\eta)d\xi = 0 \tag{3.262}$$

provided that

$$\alpha[A + A'\alpha + A''\beta] - [B + B'\alpha + B''\beta] = 0 \tag{3.263}$$

$$-\beta[A + A'\alpha + A''\beta] + [C + C'\alpha + C''\beta] = 0 \tag{3.264}$$

$$a = A', \quad a' = A'' \tag{3.265}$$

$$b = B' - A - 2A'\alpha - \beta A'', \quad b' = B'' - \alpha A'' \tag{3.266}$$

$$c = C' - A'\beta, \quad c' = C'' - A - A'\alpha - 2\beta A'' \tag{3.267}$$

Equations (3.263) and (3.264) require

$$A + A'\alpha + A''\beta = \frac{C + C'\alpha + C''\beta}{\beta} = \frac{B + B'\alpha + B''\beta}{\alpha} = \lambda \tag{3.268}$$

In a sense, λ is like a characteristics value and depends on the given differential equation. These can be rearranged as

$$\begin{aligned} A - \lambda + A'\alpha + A''\beta = 0 \\ B + (B' - \lambda)\alpha + B''\beta = 0 \\ C + C'\alpha + (C'' - \lambda)\beta = 0 \end{aligned} \tag{3.269}$$

Elimination of both α and β gives

$$\begin{aligned} (A - \lambda)(B' - \lambda)(C'' - \lambda) - B''C'(A - \lambda) - A''C(B' - \lambda) \\ - A'B(C'' - \lambda) + A'B''C + A''BC' = 0 \end{aligned} \tag{3.270}$$

This is a third order equation for λ and a closed form root is in general not possible. The value of λ needs to be evaluated numerically. Then, α and β can be determined accordingly by back substitution of λ into (3.269). Obviously, (3.262) can be simplified by using the following change of variables

$$v = \frac{\eta}{\xi} \tag{3.271}$$

$$dv = \frac{\xi d\eta - \mu d\xi}{\xi^2}, \quad d\eta = v d\xi + \xi dv \tag{3.272}$$

Substitution of (3.271) and (3.272) into (3.262) gives

$$[(c + c'v) - v(b + b'v)]\frac{d\xi}{dv} - (b + b'v)\xi + (a + a'v)\xi^2 = 0 \tag{3.273}$$

This is the Riccati equation and its solution technique has been discussed in the previous section. Therefore, ODE of the form given in (3.259) can be converted to the Riccati equation as demonstrated by Jacobi.

3.2.10 Integration by Differentiation

In this section, we will introduce a concept that sounds contradictory at first sight. Actually, there are certain types of differential equations that can be solved by differentiation. Yes, we carry integration by differentiation. Consider a general form of first order ODE

$$y = F(x, p), \quad p = \frac{dy}{dx} \tag{3.274}$$

This type of ODE can be solved by differentiation. Differentiation of (3.274) gives

$$\frac{dy}{dx} = \frac{\partial F(x,p)}{\partial x} + \frac{\partial F(x,p)}{\partial p}\frac{dp}{dx} \tag{3.275}$$

Substitution of (3.275) into (3.274) leads to

$$p = \phi(x, p, \frac{dp}{dx}) \tag{3.276}$$

If we can integrate (3.276) once, we can get

$$p = f(x) + c \tag{3.277}$$

Back substitution of (3.277) into the original ODE, we get symbolically

$$G_1(x, y, c) = 0 \tag{3.278}$$

Of course, in reality the workability of this technique depends on whether we can integrate (3.276).

Another form of first order ODE that can be solved by integration by differentiation is

$$x = F(y, p), \quad p = \frac{dy}{dx} \tag{3.279}$$

Differentiation of (3.279) with respect to y gives

$$\frac{dx}{dy} = F_y(y, p) + F_p(y, p)\frac{dp}{dy} \tag{3.280}$$

This equation can be rewritten as

$$\frac{1}{p} = f(y, p, \frac{dp}{dy}) \tag{3.281}$$

In general, we can integrate (3.281) to obtain

$$p = g(y) + c \tag{3.282}$$

Back substitution of (3.282) into the original ODE, we obtain the solution as

$$G_2(x, y, c) = 0 \tag{3.283}$$

A special form of the above forms of first order ODE is that both x and y appear linearly in the functional form of

$$x\phi(p) + y\psi(p) = \chi(p) \tag{3.284}$$

Rewriting (3.284), we have

$$y = -x\frac{\phi(p)}{\psi(p)} + \frac{\chi(p)}{\psi(p)} = x\varphi_1(p) + \zeta_1(p) \tag{3.285}$$

Alternatively, we can also rewrite (3.285) as

$$x = -y\frac{\psi(p)}{\phi(p)} + \frac{\chi(p)}{\phi(p)} = y\varphi_2(p) + \zeta_2(p) \tag{3.286}$$

Differentiation of (3.285) gives

$$p = \varphi_1(p) + \{x\varphi_1'(p) + \zeta_1'(p)\}\frac{dp}{dx} \tag{3.287}$$

Rearrangement of (3.287) leads to

$$\{p-\varphi_1(p)\}\frac{dx}{dp}-x\varphi_1'(p)=\zeta_1'(p) \tag{3.288}$$

Dividing through by the bracket term gives

$$\frac{dx}{dp}-x\frac{\varphi_1'(p)}{p-\varphi_1(p)}=\frac{\zeta_1'(p)}{p-\varphi_1(p)} \tag{3.289}$$

This is the most general linear first order differential equation that we have considered earlier in Section 3.2.6. Thus, the solution can be solved in terms of integration and expressed symbolically as:

$$x=g(p) \tag{3.290}$$

Suppose that we can invert this equation to give:

$$p=g_2(x) \tag{3.291}$$

Substitution of this into the original differential equation (3.284) gives the final solution.

Example 3.14 Find the general solution of the following first order ODE

$$x+yp=ap^2 \tag{3.292}$$

Solution: *Method 1*: Differentiating this equation with respect to x gives

$$1+p^2+y\frac{dp}{dx}=2ap\frac{dp}{dx} \tag{3.293}$$

However, the original ODE can be used to find:

$$y=\frac{ap^2-x}{p} \tag{3.294}$$

Substitution of (3.294) into (3.293) gives

$$1+p^2+(\frac{ap^2-x}{p})\frac{dp}{dx}=2ap\frac{dp}{dx} \tag{3.295}$$

Inverting this equation, we obtain a general linear ODE for unknown x:

$$\frac{dx}{dp}-\frac{x}{p(1+p^2)}=\frac{ap}{1+p^2} \tag{3.296}$$

Integration of this equation gives

$$x=\frac{p}{\sqrt{1+p^2}}[c+a\ln\{p+\sqrt{1+p^2}\}] \tag{3.297}$$

The original differential equation can also be used to give the value of p:

$$p=\frac{y\pm\sqrt{y^2+4ax}}{2a} \tag{3.298}$$

The final solution is obtained by substituting (3.298) into (3.292).

Method 2: Differentiating this equation with respect to y gives

$$\frac{dx}{dy}+p+y\frac{dp}{dy}=2ap\frac{dp}{dy} \tag{3.299}$$

This is a general linear ODE for unknown y with variable p. The solution of this is

$$y=\frac{1}{\sqrt{1+p^2}}[c+ap\sqrt{1+p^2}-a\ln\{p+\sqrt{1+p^2}\}] \tag{3.300}$$

Again, the final solution is obtained by substituting (3.298) into (3.300).

3.2.11 Clairaut Equation

The Clairaut equation is a special case of (3.284) that can be solved using integration by differentiation. In particular, the Clairaut equation is recovered as a special case if we set

$$\psi(p)=1, \quad \phi(p)=-p, \quad \chi(p)=f(p) \tag{3.301}$$

That is, the Clairaut equation is

$$y=x\frac{dy}{dx}+f(\frac{dy}{dx}) \tag{3.302}$$

Differentiation of (3.302) with respect to x gives

$$\frac{dy}{dx}=\frac{dy}{dx}+x\frac{d^2y}{dx^2}+f'(\frac{dy}{dx})\frac{d^2y}{dx^2} \tag{3.303}$$

This can be factorized as

$$[x+f'(\frac{dy}{dx})]\frac{d^2y}{dx^2}=0 \tag{3.304}$$

Therefore, this solution can be found by setting either factor in (3.304) to zero:

$$\frac{d^2y}{dx^2}=0, \quad or \quad [x+f'(\frac{dy}{dx})]=0 \tag{3.305}$$

Taking the first equation in (3.305), we have

$$\frac{dy}{dx}=C \tag{3.306}$$

However, we should not integrate this again to get y because this will lead to another unknown constant. Instead, we should substitute (3.306) into (3.302) directly to get

$$y(x)=Cx+f(C) \tag{3.307}$$

As expected, we only have one unknown constant C. The solution of the second of (3.305) does not lead to a solution of the original differential equation, but instead yields a so-called singular solution, which actually gives the envelope of the family of parametric solutions given in (3.307):

$$x=-f'(\frac{dy}{dx}) \tag{3.308}$$

More discussion on this singular solution is next.

3.2.12 Singular Solution

Recall from the last section that there are two solutions for the Clairaut equation because of differentiation. However, one of them is called a singular solution and is not the actual solution of the Clairaut equation, but instead it gives the envelope of the actual solutions with a different constant C. The singular solution had been studied extensively and theoretically by many mathematicians, including Leibniz in 1694, Taylor in 1715, Clairaut in 1734, Euler in 1756, Laplace, De Morgan, Lagrange, and Cauchy in 1772. It is known that exact differential equation does not admit a singular solution. The singular solution only appears in nonlinear first order ODEs, just like the Clairaut equation discussed in the last section. We will briefly summarize its determination here.

Not all first order ODEs will have a singular solution. If a first order ODE is given as

$$\phi(x, y, p) = 0, \tag{3.309}$$

there is no singular solution if the equation is linear in p. If the solution of the differential equation is

$$f(x, y, c) = 0, \tag{3.310}$$

the partial derivative of this with respect to x is

$$\frac{\partial f}{\partial x} + \frac{\partial f}{\partial y}\frac{\partial y}{\partial x} = 0 \tag{3.311}$$

A technique is called c-discriminant (Forsyth, 1956) if it assumes that c is a function of x such that:

$$\frac{\partial f}{\partial x} + \frac{\partial f}{\partial y}\frac{\partial y}{\partial x} + \frac{\partial f}{\partial c}\frac{dc}{dx} = 0 \tag{3.312}$$

To convert back to the (3.311), we need to set

$$\frac{\partial f}{\partial c}\frac{dc}{dx} = 0 \tag{3.313}$$

This requires

$$\frac{\partial f}{\partial c} = 0 \tag{3.314}$$

This and the solution (3.310) together provide a system to determine the singular solution by eliminating c from:

$$f(x, y, c) = 0, \quad \frac{\partial f}{\partial c} = 0 \tag{3.315}$$

However, this topic does not have obvious applications in science and engineering, becomes obsolete, and is not covered in most textbooks in engineering mathematics. Our coverage of it will end here.

3.2.13 Lagrange Equation

A slight extension of the Clairaut equation is called the Lagrange equation. It can be expressed in the following general form (Zwillinger, 1997; Piaggio, 1920; Forsyth, 1956)

$$y = xf(p) + \phi(p), \quad p = \frac{dy}{dx} \tag{3.316}$$

Note that if $f(p) = p$, the Clairaut equation is recovered as a special case. Differentiation of the Lagrange equation yields

$$p = f(p) + [xf'(p) + \phi'(p)]\frac{dp}{dx} \tag{3.317}$$

This remains a nonlinear first order ODE. However, if we reverse the roles of x and p and we arrive at a linear ODE for x with p being the variable:

$$\frac{dx}{dp} + x\frac{f'(p)}{f(p) - p} = \frac{\phi'(p)}{p - f(p)} \tag{3.318}$$

The solution of this can be found exactly. Let the solution be expressed symbolically in the following form:

$$F(x, p, c) = 0 \tag{3.319}$$

We can solve for p and substitute this solution of p into (3.316) to obtain the final solution.

3.2.14 Factorization of Nonlinear Form

Let us consider a nonlinear first order ODE in the form

$$(\frac{dy}{dx})^n + P_1(\frac{dy}{dx})^{n-1} + P_2(\frac{dy}{dx})^{n-2} + ... + P_n = 0 \tag{3.320}$$

where P_k (k = 1,2,...,n) are, in general, functions of x and y. Suppose that (3.320) can be factorized as:

$$(\frac{dy}{dx} - p_1)(\frac{dy}{dx} - p_2)\cdots(\frac{dy}{dx} - p_n) = 0 \tag{3.321}$$

where p_k (k = 1,2,...,n) are functions of x and y. The solution can be considered as

$$(\frac{dy}{dx} - p_1) = 0, \quad (\frac{dy}{dx} - p_2) = 0, \cdots, \quad (\frac{dy}{dx} - p_n) = 0 \tag{3.322}$$

Let the corresponding solutions of these first order ODEs be:

$$u_1(x,y) - c_1 = 0, \quad u_2(x,y) - c_2 = 0, \cdots, \quad u_n(x,y) - c_n = 0 \tag{3.323}$$

The solution of (3.320) can be formed as a product of any two or more of these solutions, for example

$$[u_1(x,y) - c][u_2(x,y) - c]\cdots[u_n(x,y) - c] = 0 \tag{3.324}$$

A number of examples are used to illustrate this technique.

Example 3.15 Find the general solution of the following first order nonlinear ODE

$$(\frac{dy}{dx})^2 - a^2 y^2 = 0 \quad (3.325)$$

Solution: Factorization of (3.325) gives

$$(\frac{dy}{dx} - ay)(\frac{dy}{dx} + ay) = 0 \quad (3.326)$$

Setting both of these brackets to zero results in 2 equations:

$$\ln y - ax - c_1 = 0, \quad \ln y + ax - c_2 = 0 \quad (3.327)$$

The solution is then given as

$$(\ln y - ax - c)(\ln y + ax - c) = 0 \quad (3.328)$$

or

$$(y - ce^{ax})(y - ce^{-ax}) = 0 \quad (3.329)$$

Note again that we have set the constants equal in both equations of (3.327). Readers are advised to check the validity of this solution.

To visualize the power of factorization, let us consider the following nonlinear ODE of the first order with non-constant coefficients:

$$(\frac{dy}{dx})^2 - 3x\frac{dy}{dx} + 2x^2 = 0 \quad (3.330)$$

It is straightforward to show that it is equivalent to

$$(\frac{dy}{dx} - 2x)(\frac{dy}{dx} - x) = 0 \quad (3.331)$$

By setting the first factor to zero, we have

$$(\frac{dy}{dx} - 2x) = 0 \quad (3.332)$$

This is a separable first order ODE and can be readily integrated to give

$$y = x^2 + c_1 \quad (3.333)$$

Similarly, setting the second factor to zero results in the following solution

$$y = \frac{x^2}{2} + c_2 \quad (3.334)$$

The final solution is

$$(y - x^2 - c)(y - \frac{x^2}{2} - c) = 0 \quad (3.335)$$

Since the ODE is of first order, we only need one unknown constant c.

Example 3.16 Find the general solution of the following nonlinear first order ODE

$$(\frac{dy}{dx})^2 = 4x^2 \quad (3.336)$$

Solution: Factorization gives

$$(\frac{dy}{dx} - 2x)(\frac{dy}{dx} + 2x) = 0 \quad (3.337)$$

The solutions for the first and second factors are respectively

$$y = x^2 + c_1, \quad y = -x^2 + c_2 \tag{3.338}$$

The final solution is then

$$(y - x^2 - c)(y + x^2 - c) = 0 \tag{3.339}$$

Or equivalently, we have

$$(y - c)^2 - x^4 = 0 \tag{3.340}$$

3.2.15 Solution by Taylor Series Expansion

If a boundary condition is given, we can find an approximation of the solution by Taylor series expansion. First, we can rewrite a first order ODE as

$$\frac{dy}{dx} = -\frac{M}{N} = f_1(x, y) \tag{3.341}$$

Taking differentiation with respect to x, we find

$$\frac{d^2 y}{dx^2} = \frac{\partial f_1}{\partial x} + \frac{\partial f_1}{\partial y}\frac{dy}{dx} = \frac{\partial f_1}{\partial x} + \frac{\partial f_1}{\partial y} f_1 = f_2(x, y) \tag{3.342}$$

$$\frac{d^3 y}{dx^3} = \frac{\partial f_2}{\partial x} + \frac{\partial f_2}{\partial y}\frac{dy}{dx} = \frac{\partial f_2}{\partial x} + \frac{\partial f_2}{\partial y} f_1 = f_3(x, y) \tag{3.343}$$

$$\frac{d^n y}{dx^n} = \frac{\partial f_{n-1}}{\partial x} + \frac{\partial f_{n-1}}{\partial y}\frac{dy}{dx} = \frac{\partial f_{n-1}}{\partial x} + \frac{\partial f_{n-1}}{\partial y} f_1 = f_n(x, y) \tag{3.344}$$

Using Taylor series expansion about a point x_0, we get

$$y = y_0 + f_1(x_0, y_0)(x - x_0) + f_2(x_0, y_0)\frac{(x - x_0)^2}{2!} + \ldots \tag{3.345}$$

Suppose that the boundary condition is given as

$$y(x_0 = 0) = y_0 = c \tag{3.346}$$

An approximation of the solution is

$$y(x) = c + f_1(0, c)x + f_2(0, c)\frac{x^2}{2!} + \ldots \tag{3.347}$$

When the number of terms goes to infinity, the solution approaches an exact solution. In reality, depending on the explicit form of f_1, the process of differentiation may lead to very tedious calculation. Nevertheless, this approach can give a fast first order approximation of the result.

3.3 SECOND ORDER ODE

In this section, we consider second order ODEs. The most general case can be expressed as

$$\frac{d^2y}{dt^2} = f(t, y, \frac{dy}{dt}) \tag{3.348}$$

If we assume the second order ODE is linear, we can express it as

$$y'' + p(t)y' + q(t)y = g(t) \tag{3.349}$$

Because of the nonhomogeneous term g, we must write the solution as a sum of two solutions. That is, the homogeneous solution plus the particular solution:

$$y(t) = y_h(t) + y_p(t) \tag{3.350}$$

The proof of this is deferred to Section 3.3.2. The procedure of finding the particular solution will be discussed in Sections 3.3.3 and 3.3.4.

For the case of a homogeneous differential equation, suppose that we can find the homogeneous solution as:

$$y = C_1 y_1(x) + C_2 y_2(x) \tag{3.351}$$

Whether a boundary value problem can be solved, we have to study the so-called Wronskian. In particular, let us assume that the boundary conditions are given as:

$$y(t_0) = y_0, \; y'(t_0) = y_* \tag{3.352}$$

Substitution of (3.351) into (3.352) gives

$$\begin{aligned} C_1 y_1(t_0) + C_2 y_2(t_0) &= y_0 \\ C_1 y_1'(t_0) + C_2 y_2'(t_0) &= y_* \end{aligned} \tag{3.353}$$

This system provides two equations for two unknowns, and using Cramer's rule we can solve for the unknown constants:

$$c_1 = \frac{\begin{vmatrix} y_0 & y_2(t_0) \\ y^* & y_2'(t_0) \end{vmatrix}}{W}, \quad c_2 = \frac{\begin{vmatrix} y_1(t_0) & y_0 \\ y_1'(t_0) & y^* \end{vmatrix}}{W} \tag{3.354}$$

where W is called the Wronskian and is defined as:

$$W = \begin{vmatrix} y_1(t_0) & y_2(t_0) \\ y_1'(t_0) & y_2'(t_0) \end{vmatrix} = y_1(t_0)y_2'(t_0) - y_1'(t_0)y_2(t_0) \tag{3.355}$$

For the system to be solvable, we must require that the Wronskian is non-zero. That is,

$$W(y_1, y_2)(t_0) \neq 0 \tag{3.356}$$

3.3.1 ODE with Constant Coefficients

Let us consider the simplest case of constant coefficients without a nonhomogeneous term. That is, p and q in (3.349) are both constants:

$$y'' + py' + qy = 0 \tag{3.357}$$

For constant coefficients, it can be proved that the solution must be in the form of an exponential function as:

$$y = e^{rx} \tag{3.358}$$

Substitution of (3.358) into (3.357) gives the following equation:

$$r^2 + pr + q = 0\,. \tag{3.359}$$

Since our assumed solution should not be zero (if it is zero, the solution becomes a trivial solution), (3.359) must be zero. This is also the characteristic equation of the ODE given in (3.357). There are two roots of this quadratic equation:

$$r_{1,2} = \frac{-p \pm \sqrt{p^2 - 4q}}{2}, \quad \Delta = p^2 - 4q \tag{3.360}$$

These characteristic roots can be real, repeated, or complex, depending on the values of Δ. The three scenarios are:

Case 1: $\Delta > 0$

For this case, we have two distinct real roots:

$$r_1 = \frac{-p + \sqrt{p^2 - 4q}}{2}, \quad r_2 = \frac{-p - \sqrt{p^2 - 4q}}{2} \tag{3.361}$$

The corresponding independent solutions are:

$$y_1 = e^{r_1 x}, \quad y_2 = e^{r_2 x} \tag{3.362}$$

The general solutions are then linear combinations of these solutions

$$y_h = C_1 e^{r_1 x} + C_2 e^{r_2 x} \tag{3.363}$$

The unknown constants need to satisfy the corresponding boundary condition (but this will be done after we obtain the particular solution first).

Case 2: $\Delta = 0$

In this case, since $\Delta = 0$ we must have two roots that are the same. Thus, the repeated roots are

$$r_1 = r_2 = -\frac{p}{2} \tag{3.364}$$

The first solution is obvious, while the second independent solution cannot be the same as the first one. To find the second one, we use Euler's approach of variation of parameter (i.e., replacing the constant by a function of x). Or equivalently, we can use the well-known theorem (Forsyth, 1956) that the second independent solution must be in the form of an unknown function multiplying the first known solution. In either case, we have two independent solutions as:

$$y_1 = e^{r_1 x}, \quad y_2 = u(x) e^{r_1 x} \tag{3.365}$$

To find the function u, we note that the differentiations of y_2 are:

$$y_2' = (u r_1 + u') e^{r_1 x} \tag{3.366}$$

$$y_2'' = (u r_1^2 + 2 r_1 u' + u'') e^{r_1 x} \tag{3.367}$$

Substitution of these results into (3.357) gives

$$u'' + (2r_1 + p)u' + (r_1^2 + p r_1 + q)u = u'' = 0 \tag{3.368}$$

Both bracket terms are zeros, and we end up with the last term in (3.368). Thus, we have

$$u = Cx \tag{3.369}$$

With the second independent solution, we can now have two independent solutions as:

$$y_h = (C_1 + C_2 x)e^{r_1 x} \tag{3.370}$$

Although the present analysis is for second order only, a similar result is also obtained for the case of a higher order ODE with roots of higher multiplicity.

Case 3: $\Delta < 0$

Finally, for $\Delta < 0$ the two roots are complex conjugate pairs. Thus, the characteristic roots are:

$$r_{1,2} = \frac{-p \pm \sqrt{\Delta}}{2}, \quad \Delta = p^2 - 4q \tag{3.371}$$

Thus, expressing the real and imaginary parts of it as α and β we get

$$r_1 = \alpha + i\beta,, \quad r_2 = \alpha - i\beta, \tag{3.372}$$

The two independent solutions are

$$y_1 = e^{(\alpha + i\beta)x}, \quad y_2 = e^{(\alpha - i\beta)x} \tag{3.373}$$

However, (3.357) is real but the solutions given in (3.373) are complex. We can take the real part and the imaginary part as the solutions:

$$y_3 = \frac{1}{2}(y_1 + y_2) = \cos \beta x e^{\alpha x}, \quad y_4 = \frac{1}{2i}(y_1 - y_2) = \sin \beta x e^{\alpha x} \tag{3.374}$$

Finally, the homogeneous solutions are

$$y_h = e^{\alpha x}(C_1 \cos \beta x + C_2 \sin \beta x) \tag{3.375}$$

We will illustrate the solution in the following examples.

Example 3.17 Find the general solution of the following second order ODE

$$y'' - 2y' - 3y = 0 \tag{3.376}$$

Solution: For an ODE with constant coefficients, we can assume the solution as an exponential function as:

$$y = e^{rx} \tag{3.377}$$

Substitution of (3.377) into (3.376) gives the following characteristic equation

$$r^2 - 2r - 3 = 0 \tag{3.378}$$

Two real distinct roots are found, and we have Case 1. The two distinct roots are

$$r_1 = -1, \quad r_2 = 3 \tag{3.379}$$

Finally, the general solution is

$$y = C_1 e^{-x} + C_2 e^{3x} \tag{3.380}$$

Example 3.18 Find the general solution of the following second order ODE with given initial condition

$$\frac{d^2 s}{dt^2} + 2\frac{ds}{dt} + s = 0 \tag{3.381}$$

$$s\big|_{t=0} = 4, \quad s'\big|_{t=0} = -2 \tag{3.382}$$

Solution: Again, an exponential solution is expected:

$$s = e^{rt} \tag{3.383}$$

The corresponding characteristic equation is

$$r^2 + 2r + 1 = 0 \tag{3.384}$$

It is straightforward to see that the roots are equal:

$$r_1 = r_2 = -1, \tag{3.385}$$

Thus, the corresponding solution is

$$s = (C_1 + C_2 t)e^{-t} \tag{3.386}$$

Differentiating this with respect to t, we obtain

$$s' = C_2 e^{-t} - (C_1 + C_2 t)e^{-t} \tag{3.387}$$

Applying the two initial conditions, we get two equations for two unknowns:

$$s\big|_{t=0} = 4 = C_1, \quad s'\big|_{t=0} = -2 = C_2 - C_1 \tag{3.388}$$

The unknown constants are solved as

$$C_1 = 4, C_2 = 2 \tag{3.389}$$

Finally, the solution is

$$s = (4 + 2t)e^{-t} \tag{3.390}$$

Example 3.19 Find the general solution of the following second order ODE

$$y'' - 2y' + 5y = 0 \tag{3.391}$$

Solution: Finally, this example illustrates the case of a complex conjugate pair of roots. Assuming exponential function

$$y = e^{rx}, \tag{3.392}$$

we get the following characteristic equation

$$r^2 - 2r + 5 = 0, \tag{3.393}$$

The corresponding pair of complex conjugate roots is

$$r_{1,2} = 1 \pm 2i, \tag{3.394}$$

The general solution becomes

$$y = e^x (C_1 \cos 2x + C_2 \sin 2x) \tag{3.395}$$

3.3.2 Nonhomogeneous ODE

It was discovered by Lagrange that the general solution of a nonhomogeneous ODE consists of two parts, namely the homogeneous solution (solution of the differential equation without the nonhomogeneous term) plus a particular solution of the nonhomogeneous ODE. Mathematically speaking, for a linear second ODE,

$$y'' + P(x)y' + Q(x)y = f(x) \tag{3.396}$$

the general solution is

$$y(x) = y_h(x) + y_p(x) \tag{3.397}$$

where y_h and y_p are the homogeneous solutions that satisfy (3.396) with $f(x) = 0$ and the particular solution satisfies (3.396) with nonzero $f(x)$. To show this, we first substitute (3.397) into the left hand side of (3.396) as

$$\begin{aligned} y'' + P(x)y' + Q(x)y &= (y_h'' + y_p'') + P(x)(y_h' + y_p') + Q(x)(y_h + y_p) \\ &= [y_p'' + P(x)y_p' + Q(x)y_p] + [y_h'' + P(x)y_h' + Q(x)y_h] \\ &= f(x) + 0 = f(x) \end{aligned} \tag{3.398}$$

The final result is obtained by observing that the term containing y_p equals $f(x)$ and the term containing y_p equals zero.

Therefore, it is of utmost importance that classification in terms of homogeneous or nonhomogeneous is made properly otherwise, the solution is automatically incorrect.

In the next two sections, we will discuss the method of undetermined coefficient and the method of variation of parameters in obtaining the particular solution.

3.3.3 Undetermined Coefficient

For the following discussions, we will restrict to the cases of constant coefficient, that is

$$y'' + py' + qy = f(x) \tag{3.399}$$

where p and q are constants. Consider the case that the nonhomogeneous term is a product of polynomials and an exponential function

$$f(x) = e^{\lambda x} P_m(x) \tag{3.400}$$

where P_m is defined as

$$P_m(x) = a_0 x^m + a_1 x^{m-1} + \ldots + a_n \tag{3.401}$$

Since the exponential function cannot be altered or killed by differentiation, we must assume that the particular solution must also be proportional to the same exponential function. Thus, we assume the particular solution as:

$$y = Q(x)e^{\lambda x} \tag{3.402}$$

Differentiation of this assumed solution form gives

$$y' = Q'e^{\lambda x} + \lambda Q e^{\lambda x} \tag{3.403}$$

$$y'' = Q''e^{\lambda x} + 2\lambda Q'e^{\lambda x} + \lambda^2 Q e^{\lambda x} \tag{3.404}$$

Substitution of (3.403) and (3.404) into (3.399) leads to

$$Q''(x) + (2\lambda + p)Q'(x) + (\lambda^2 + p\lambda + q)Q(x) = P_m(x) \tag{3.405}$$

If we look closely, we discover that the bracket terms in (3.405) are the characteristic equation of the homogeneous ODE. Thus, it becomes very important for us to check whether the exponential power λ of the nonhomogeneous term is the same as the characteristic roots of the homogeneous solution given in (3.359).

<u>Case 1</u>: λ is not a characteristic root

For this case, none of the bracket terms in (3.405) vanishes or:

$$\lambda^2 + p\lambda + q \neq 0, \quad 2\lambda + p \neq 0 \tag{3.406}$$

Thus, the highest power term of Q on the right hand side of (3.405) must be the same as polynomials P_m:

$$Q(x) = Q_m(x) \tag{3.407}$$

More specifically, we expect the particular solution is

$$y_p = Q_m(x)e^{\lambda x} \tag{3.408}$$

where Q_m is

$$Q_m(x) = b_0 x^m + b_1 x^{m-1} + ... + b_n \tag{3.409}$$

Substitution of (3.409) into (3.399) gives

$$\begin{aligned}&[m(m-1)b_0 x^{m-2} + ... + 2b_{m-2}] + (2\lambda + p)[mb_0 x^{m-1} + ... + b_{m-1}] \\ &+(\lambda^2 + p\lambda + q)[b_0 x^m + b_1 x^{m-1} + ... + b_m] = a_0 x^m + a_1 x^{m-1} + ... + a_m\end{aligned} \tag{3.410}$$

Matching coefficients for different powers of x, we find m equations in ascending order as:

$$\begin{aligned}&(\lambda^2 + p\lambda + q)b_0 = a_0 \\ &(2\lambda + p)mb_0 + (\lambda^2 + p\lambda + q)b_1 = a_1 \\ &\vdots \\ &2b_{m-2} + (2\lambda + p)b_{m-1} + (\lambda^2 + p\lambda + q)b_m = a_m\end{aligned} \tag{3.411}$$

Once this system of equations is solved, we obtain the particular solution. Therefore, Case 1 is purely a matter of matching terms on both sides.

Case 2: λ is a simple characteristic root

If the exponential power λ matches the characteristic roots of the homogeneous differential equation (3.360), we have

$$\lambda^2 + p\lambda + q = 0, \quad 2\lambda + p \neq 0 \tag{3.412}$$

Thus, (3.405) is reduced to

$$Q''(x) + (2\lambda + p)Q'(x) = P_m(x) \tag{3.413}$$

The highest order term of the power series on the left is Q' (i.e., the first derivative of Q) which must have power of order m in order to match the highest order term on the right hand side. This suggests that we should choose Q as

$$Q(x) = xQ_m(x) \tag{3.414}$$

Consequently, for Case 2 (simple root matching) we have the assumed particular solution in the following form:

$$y_p = xQ_m(x)e^{\lambda x} \tag{3.415}$$

Case 3: λ is a double characteristic root

Finally, for Case 3, power λ matches with the double or repeated characteristic roots of the homogeneous equation, and we have

$$\lambda^2 + p\lambda + q = 0, \quad 2\lambda + p = 0 \tag{3.416}$$

Thus, (3.405) is simplified to

$$Q''(x) = P_m(x) \tag{3.417}$$

Similar to the earlier discussion given in Case 2, we have to raise the power of the polynomials by two in order to match the highest power on both sides of (3.413):

$$Q(x) = x^2 Q_m(x) \tag{3.418}$$

Therefore, we expect the particular solution to be expressed as:

$$y_p = x^2 Q_m(x) e^{\lambda x} \tag{3.419}$$

In summary, we can combine all three different scenarios into a simple formula:

$$y_p = x^k e^{\lambda x} Q_m(x) \tag{3.420}$$

where k equals zero, one, or two for the case of no match of the root (zero matching), match a simple root (one matching), or match the double root (two matching). Let us illustrate again by example.

Example 3.20 Find the particular solution for the special case of a polynomial $P_m = A$

$$y'' + py' + qy = Ae^{\lambda x} \tag{3.421}$$

Solution: The nonhomogeneous term is simply an exponential constant and thus the polynomial is of order zero:

$$Q''(x) + (2\lambda + p)Q'(x) + (\lambda^2 + p\lambda + q)Q(x) = P_m(x) = A \tag{3.422}$$

Case 1: λ is a not characteristic root

For this simple case, we have a constant term for our polynomials Q_m

$$Q = c_0 \tag{3.423}$$

Obviously, we have

$$Q''(x) = Q'(x) = 0 \tag{3.424}$$

Balancing terms on both sides, we have

$$(\lambda^2 + p\lambda + q)Q = (\lambda^2 + p\lambda + q)c_0 = A \tag{3.425}$$

Then, we have the unknown c_0 as:

$$c_0 = \frac{A}{(\lambda^2 + p\lambda + q)} \tag{3.426}$$

Finally, the particular solution is

$$y_p = \frac{A}{\lambda^2 + p\lambda + q} e^{\lambda x} \tag{3.427}$$

Since A is given in (3.421), there is no unknown constant in (3.427).

Case 2: λ is a simple characteristic root

For this case, we have

$$\lambda^2 + p\lambda + q = 0, \quad 2\lambda + p \neq 0 \tag{3.428}$$

Thus, for $k = 1$ we have to raise the order by one:

$$Q = c_1 x \tag{3.429}$$

Differentiation gives

$$Q''(x) = 0, \quad Q'(x) = c_1 \tag{3.430}$$

Finally, we can match all terms on both sides to get

$$Q''(x) + (2\lambda + p)Q'(x) = (2\lambda + p)c_1 = A \tag{3.431}$$

This gives the constant as:

$$c_1 = \frac{A}{2\lambda + p} \tag{3.432}$$

The final particular solution becomes:

$$y_p = \frac{A}{2\lambda + p} x e^{\lambda x} \tag{3.433}$$

Case 3: λ is a double characteristic root

Finally, if λ matches the double characteristic root of the homogeneous ODE, we have

$$\lambda^2 + p\lambda + q = 0, \quad 2\lambda + p = 0 \tag{3.434}$$

For $k = 2$, we have to raise the order by two:

$$Q = c_2 x^2 \tag{3.435}$$

The ODE (3.422) reduces to

$$Q'' = 2c_2 = A \tag{3.436}$$

The unknown constant is

$$c_2 = \frac{A}{2} \tag{3.437}$$

The final particular solution becomes:

$$y_p = \frac{A}{2} x^2 e^{\lambda x} \tag{3.438}$$

In summary, we have the following solution depending on λ:

$$y_p = \begin{cases} \dfrac{A}{\lambda^2 + p\lambda + q} e^{\lambda x} & \lambda \text{ is not characteristic root} \\ \dfrac{A}{2\lambda + p} x e^{\lambda x} & \lambda \text{ is a simple characteristic root} \\ \dfrac{A}{2} x^2 e^{\lambda x} & \lambda \text{ is a double characteristic root} \end{cases} \tag{3.439}$$

Example 3.21 Find the general solution for the following nonhomogeneous second order ODE

$$y''-2y'-3y=3x+1 \tag{3.440}$$

Solution: Note that the nonhomogeneous term is only a first order power series. We have to solve for the characteristic root of the corresponding ODE (i.e., setting the right hand side zero):

$$y''-2y'-3y=0 \tag{3.441}$$

Assuming the exponential solution form leads to the following characteristic equation:

$$r^2-2r-3=0 \tag{3.442}$$

Two distinct real roots are obtained:

$$r_1=-1, \quad r_2=3 \tag{3.443}$$

This is Case 1, and we can assume the particular solution as:

$$y_p=b_0x+b_1 \tag{3.444}$$

Substitution of (3.444) into (3.441) yields

$$-3b_0x-2b_0-3b_1=3x+1 \tag{3.445}$$

Comparing terms on both sides, we obtain

$$b_0=-1, \quad b_1=\frac{1}{3} \tag{3.446}$$

The particular solution is

$$y_p=-x+\frac{1}{3} \tag{3.447}$$

The general solution is then the sum of homogeneous solution and the particular solution:

$$y=C_1e^{-x}+C_2e^{3x}-x+\frac{1}{3} \tag{3.448}$$

Example 3.22 Find the general solution of the following second order nonhomogeneous ODE

$$y''-3y'+2y=xe^{2x} \tag{3.449}$$

Solution: Let us consider the homogeneous equation of (3.449):

$$y''-3y'+2y=0 \tag{3.450}$$

Since the coefficients of this differential equation are constant, the solution is of exponential form. Substitution of a solution form of e^{rx} leads to

$$r^2-3r+2=0 \tag{3.451}$$

The roots of this characteristic equation are

$$r_1=1, \; r_2=2 \tag{3.452}$$

Thus, the general solution of the homogeneous equation is

$$y=C_1e^{x}+C_2e^{2x} \tag{3.453}$$

The nonhomogeneous term of (3.449) is a first order polynomial and thus, a polynomial of the same order as

$$Q_m(x)=Ax+B \tag{3.454}$$

However, we have $\lambda=r_2$ thus we have to assume

$$y_p = x(Ax + B)e^{2x} \tag{3.455}$$

Substitution of this into (3.449) gives

$$2Ax + B + 2A = x \tag{3.456}$$

Matching coefficients on both sides gives

$$A = \frac{1}{2}, \quad B = -1 \tag{3.457}$$

Finally, the general solution is obtained

$$y = C_1 e^x + C_2 e^{2x} + x(\frac{1}{2}x - 1)e^{2x} \tag{3.458}$$

Let us now consider that the nonhomogeneous term is a product of polynomials, an exponential function, and a circular function:

$$f(x) = e^{\lambda x}[P_l \cos \omega x + P_m \sin \omega x] \tag{3.459}$$

One simple way to deal with the cosine and sine terms is to replace them by exponential functions using Euler's formula discussed in Chapter 1:

$$f(x) = e^{\lambda x}[P_l \frac{e^{i\omega x} + e^{-i\omega x}}{2} + P_m \frac{e^{i\omega x} - e^{-i\omega x}}{2i}] \tag{3.460}$$

Grouping similar terms we can rewrite (3.460) as

$$\begin{aligned} f(x) &= (\frac{P_l}{2} + \frac{P_m}{2i})e^{(\lambda + i\omega)x} + (\frac{P_l}{2} - \frac{P_m}{2i})e^{(\lambda - i\omega)x} \\ &= P(x)e^{(\lambda + i\omega)x} + \bar{P}(x)e^{(\lambda - i\omega)x} \end{aligned} \tag{3.461}$$

Allowing for complex constants for the polynomials, we get a particular solution for the first term on the right of (3.461)

$$y'' + py' + qy = P(x)e^{(\lambda + i\omega)x}, \quad \bar{y}_1 = x^k Q_m e^{(\lambda + i\omega)x} \tag{3.462}$$

Similarly, for the second term on the right hand side of (3.461) we have

$$y'' + py' + qy = \bar{P}(x)e^{(\lambda - i\omega)x}, \quad \bar{y}_2 = x^k \bar{Q}_m e^{(\lambda - i\omega)x} \tag{3.463}$$

Finally, we can combine these particular solutions to get the particular solution for (3.459)

$$\begin{aligned} y_p &= x^k e^{\lambda x}[Q_m e^{i\omega x} + \bar{Q}_m e^{-i\omega x}] \\ &= x^k e^{\lambda x}[R_m^{(1)}(x)\cos \omega x + R_m^{(2)}(x)\sin \omega x], \end{aligned} \tag{3.464}$$

Example 3.23 Find the general solution for the following nonhomogeneous ODE

$$y'' + y = 4\sin x \tag{3.465}$$

Solution: Solving for the homogeneous case of (3.465) we obtain the following characteristic equation and its corresponding roots:

$$\lambda^2 = -1, \quad \lambda = \pm i \tag{3.466}$$

The homogeneous solution is

$$y = C_1 \cos x + C_2 \sin x \tag{3.467}$$

Using the approach just discussed, we can first rewrite (3.465) as the imaginary part of the following ODE:

$$y''+y=4e^{ix} \tag{3.468}$$

Thus, the nonhomogeneous term matches the characteristic roots (i.e., $\pm i$). We have to raise the order of power by one:

$$y_p = Axe^{ix} \tag{3.469}$$

Differentiation of (3.469) gives

$$y_p' = Ae^{ix} + Axie^{ix} \tag{3.470}$$

$$y_p'' = 2iAe^{ix} - Axe^{ix} \tag{3.471}$$

Substitution of these results into (3.468) results in

$$2Ai = 4 \tag{3.472}$$

This can be solved to give

$$A = -2i \tag{3.473}$$

Finally, our particular solution for (3.468) is

$$y_p = -2ixe^{ix} = 2x\sin x - (2x\cos x)i \tag{3.474}$$

Comparing (4.465) and (4.468), we find that taking the imaginary part of the right hand side of (3.468) gives the right hand side of (4.465). Thus, to get the particular solution of (4.465) we also take the imaginary part of (3.474):

$$y_p = -2x\cos x \tag{3.475}$$

Adding the particular solution to the homogeneous solution gives the final solution as

$$y = C_1\cos x + C_2\sin x - 2x\cos x \tag{3.476}$$

The undetermined coefficient approach discussed here can be considered as a lucky guess method. For any nonhomogeneous terms other than those considered in this section, we cannot assume the proper form of the particular solution. In the next section, we will discuss a much more powerful technique—the method of variation of parameters.

3.3.4 Variation of Parameters

For first order ODEs, we have already learned the method of variation of parameters by Euler. The same technique was used by Lagrange to derive the particular solution of a second order ODE. In particular, consider the following second order ODE with non-constant coefficients:

$$y'' + p(t)y' + q(t)y = g(t) \tag{3.477}$$

Assume that we can solve for the homogeneous equation of (3.477)

$$y = C_1y_1(t) + C_2y_2(t) \tag{3.478}$$

Using the idea of variation of parameters, Lagrange assumed that the particular solution for (3.478) could be expressed as:

$$y(t)=u_1(t)y_1(t)+u_2(t)y_2(t) \tag{3.479}$$

Differentiation of (3.479) gives

$$y'(t)=u_1'(t)y_1(t)+u_1(t)y_1'(t)+u_2'(t)y_2(t)+u_2(t)y_2'(t) \tag{3.480}$$

Lagrange observed that there are two unknown functions that need to be found in (3.479). That means we need two conditions for two unknowns. He then decided to choose the first condition before he continued to get the second derivative. In particular, we set the following condition:

$$u_1'(t)y_1(t)+u_2'(t)y_2(t)=0 \tag{3.481}$$

In doing so, we have reduced (3.480) to

$$y'(t)=u_1(t)y_1'(t)+u_2(t)y_2'(t) \tag{3.482}$$

Differentiation of (3.482) one more time leads to

$$y''(t)=u_1'(t)y_1'(t)+u_1(t)y_1''(t)+u_2'(t)y_2'(t)+u_2(t)y_2''(t) \tag{3.483}$$

Substitution of (3.482) and (3.483) into (3.481) gives

$$u_1'y_1'+u_1y_1''+u_2'y_2'+u_2y_2''+p(u_1y_1'+u_2y_2')+q(u_1y_1+u_2y_2)=g \tag{3.484}$$

Grouping terms, we get

$$\begin{aligned}&u_1[y_1''+py_1'+qy_1]+u_2[y_2''+py_2'+qy_2]+u_1'y_1'+u_2'y_2'\\&=u_1'y_1'+u_2'y_2'=g\end{aligned} \tag{3.485}$$

Note that the square bracket terms in (3.485) are zeros because both y_1 and y_2 are the solutions of a homogeneous equation. Thus, we have two equations for two unknowns:

$$u_1'y_1'+u_2'y_2'=g, \quad u_1'y_1+u_2'y_2=0 \tag{3.486}$$

Using Cramer's rule, we find the solutions as:

$$u_1'(t)=-\frac{y_2(t)g(t)}{W(y_1,y_2)(t)}, \quad u_2'(t)=\frac{y_1(t)g(t)}{W(y_1,y_2)(t)} \tag{3.487}$$

where W is the Wronskian. Integrating both with respect to t, we find

$$u_1(t)=-\int\frac{y_2(t)g(t)}{W(y_1,y_2)(t)}dt, \quad u_2(t)=\int\frac{y_1(t)g(t)}{W(y_1,y_2)(t)}dt \tag{3.488}$$

Back substitution of these results into (3.479) gives

$$y_p(t)=-y_1(t)\int\frac{y_2(t)g(t)}{W(y_1,y_2)(t)}dt+y_2(t)\int\frac{y_1(t)g(t)}{W(y_1,y_2)(t)}dt \tag{3.489}$$

Finally adding this to the homogeneous solution, we obtain

$$y=C_1y_1(x)+C_2y_2(x)+y_p(t) \tag{3.490}$$

Note that this formula obtained by Lagrange is much more general than the undetermined coefficient method discussed in the last section because it is applicable to ODEs with non-constant coefficients. In addition, it is valid for all functions g as long as we can find the integration. However, first of all we must have the homogeneous solutions of the ODE (i.e., y_1 and y_2). Even for linear ODEs, the homogeneous solution may not be easy to find for the case of non-constant coefficients. Secondly, even if a homogeneous solution is available, integration given in (3.488) is no easy task. Very likely, a closed form solution is not possible, and we may have to use numerical technique to conduct the integration.

3.3.5 Operator Factors

The following method of factorization of differential operators was considered by Cayley in 1886 and reported in Ince (1956). Let us consider a second order ODE in the following form:

$$\frac{dy^2}{dx^2}+2p(x)\frac{dy}{dx}+q(x)y=0 \tag{3.491}$$

Next, we assume that it can be factorized in the following form:

$$[\frac{d}{dx}+\alpha_2(x)][\frac{d}{dx}+\alpha_1(x)]y=0 \tag{3.492}$$

This equation can be rewritten as

$$\{\frac{d^2}{dx^2}+(\alpha_1+\alpha_2)\frac{d}{dx}+(\alpha_1\alpha_2+\alpha_1')\}y=0 \tag{3.493}$$

Note that the expansion of the differential operators in (3.492) to the form (3.493) involves differentiation and thus cannot be treated as simple expansion of algebraic factors. If (3.493) and (3.491) are the same, we must have the following identities

$$2p=\alpha_1+\alpha_2, \quad q=\alpha_1\alpha_2+\alpha_1' \tag{3.494}$$

However, we can only say that one of the solutions of (3.491) is

$$[\frac{d}{dx}+\alpha_1(x)]y=0 \tag{3.495}$$

Since it is in a separable form, one of the solutions of (3.491) is

$$y=Ce^{-\int\alpha_1(x)dx}. \tag{3.496}$$

We consider a special case that (3.491) can also be factorized as

$$[\frac{d}{dx}+\alpha_1(x)][\frac{d}{dx}+\alpha_2(x)]y=0 \tag{3.497}$$

In other words, we are looking for the condition that this factorization is commutative. Similarly, (3.497) can be expanded to get

$$\{\frac{d^2}{dx^2}+(\alpha_1+\alpha_2)\frac{d}{dx}+(\alpha_1\alpha_2+\alpha_2')\}y=0 \tag{3.498}$$

Thus, we must have

$$q=\alpha_1\alpha_2+\alpha_1'=\alpha_1\alpha_2+\alpha_2' \tag{3.499}$$

Equality of (3.474) and (3.499) can only be realized by setting

$$\alpha_1'=\alpha_2' \tag{3.500}$$

This implies that the factorized function can only differ by a constant:

$$\alpha_1=\alpha_2+A \tag{3.501}$$

Thus, we can factorize the second order ODE as

$$[\frac{d}{dx}+\alpha_1(x)][\frac{d}{dx}+\alpha_1(x)+A]y=(P+A)Py=0 \tag{3.502}$$

where the differential operator P is defined as

$$Py = [\frac{d}{dx} + \alpha_1(x)]y = \frac{dy}{dx} + \alpha_1(x)y \tag{3.503}$$

That is, the factorization is commutative

$$[\frac{d}{dx} + \alpha_1(x)][\frac{d}{dx} + \alpha_1(x) + A]y = [\frac{d}{dx} + \alpha_1(x) + A][\frac{d}{dx} + \alpha_1(x)]y = 0 \tag{3.504}$$

For such cases, the final solution is

$$y = C_1 e^{-\int \alpha_1(x)dx} + C_2 e^{-\int \alpha_1(x)dx - Ax} \tag{3.505}$$

Let us consider a special case that allows the factorization to be commutative

$$\frac{dy^2}{dx^2} + 2p(x)\frac{dy}{dx} + [p^2(x) + p'(x) - a^2]y = 0 \tag{3.506}$$

It is straightforward to show that it can be factorized as:

$$[\frac{d}{dx} + p(x) - a][\frac{d}{dx} + p(x) + a]y = 0 \tag{3.507}$$

The corresponding solution becomes

$$y = C_1 e^{-\int [p(x)+a]dx} + C_2 e^{-\int [p(x)-a]dx} \tag{3.508}$$

Example 3.24 Find the solution of the following ODE by factorization

$$\frac{dy^2}{dx^2} + 2x^2 \frac{dy}{dx} + [x^4 + 2x - 1]y = 0 \tag{3.509}$$

Solution: Recall (3.506) that

$$\frac{dy^2}{dx^2} + 2p(x)\frac{dy}{dx} + [p^2(x) + p'(x) - a^2]y = 0 \tag{3.510}$$

Comparison of (3.510) and (3.509) gives

$$p(x) = x^2, \quad a = 1 \tag{3.511}$$

Thus, the ODE can be factorized as

$$[\frac{d}{dx} + x^2 - 1][\frac{d}{dx} + x^2 + 1]y = [\frac{d}{dx} + x^2 + 1][\frac{d}{dx} + x^2 - 1]y = 0 \tag{3.512}$$

The two independent solutions can be obtained by solving the following ODEs:

$$[\frac{d}{dx} + x^2 - 1]y_1 = 0, \quad [\frac{d}{dx} + x^2 + 1]y_2 = 0, \tag{3.513}$$

Finally, the solution is

$$y_1 = C_1 \exp[x(1 - \frac{x^2}{3})] + C_2 \exp[-x(1 + \frac{x^2}{3})] \tag{3.514}$$

3.3.6 Reduction to First Order

Second order ODEs can be reduced to first order by using various means. If the right hand side is only a function of x:

$$\frac{d^2y}{dx^2} = f(x), \tag{3.515}$$

we can just integrate it once. This is the simplest type of second order ODE. If the unknown y does not appear explicitly on the right hand side as:

$$\frac{d^2y}{dx^2} = f(x, \frac{dy}{dx}), \tag{3.516}$$

the following change of variable can be applied:

$$p(x) = \frac{dy}{dx} \tag{3.517}$$

Then, the resulting ODE becomes first order and p becomes the new unknown:

$$\frac{dp}{dx} = f(x, p) \tag{3.518}$$

Clearly, it will not work if y appears explicitly. Another widely encountered situation is that the variable does not appear in the ODE as:

$$\frac{d^2y}{dx^2} = f(y, \frac{dy}{dx}) \tag{3.519}$$

For such cases, we can still use the change of variables given in (3.517) but we observe that we can convert differentiation with respect to x to become differentiation with respect to y as:

$$\frac{d^2y}{dx^2} = \frac{dp}{dx} = \frac{dp}{dy}\frac{dy}{dx} = p\frac{dp}{dy} = f(y, p) \tag{3.520}$$

The last equation of (3.520) is an ODE with p being the unknown and y being the variable. This type of differential equation is called the autonomous type because its coefficients do not change with x. Higher order autonomous ODE will be discussed in Section 3.5.14.

Example 3.25 This example considers the escape velocity of a rocket of mass m firing from Earth's surface

$$m\frac{d^2h}{dt^2} = -\frac{GMm}{h^2} \tag{3.521}$$

$$h\big|_{t=0} = R, \quad \left.\frac{dh}{dt}\right|_{t=0} = v_0 \tag{3.522}$$

where h is the elevation of the rocket measured from the center of the Earth, G is the universal gravitational constant, and M is the mass of the Earth.

Solution: Note that (3.521) is autonomous or the time variable does not appear explicitly in the ODE. We observe that

$$\frac{d^2h}{dt^2} = \frac{dv}{dh}(\frac{dh}{dt}) = v\frac{dv}{dh} \tag{3.523}$$

where v is the velocity. Substitution of (3.523) into (3.521) gives

$$v\frac{dv}{dh} = -\frac{GM}{h^2} \tag{3.524}$$

We have effectively reduced the equation to first order and clearly, it is separable

$$v\,dv = -\frac{GM}{h^2}dh \tag{3.525}$$

Integration of both sides gives the following relation between v and h:

$$\frac{1}{2}v^2 = \frac{GM}{h} + C \tag{3.526}$$

Substitution of the initial condition given in (3.522) gives

$$C = \frac{1}{2}v_0^2 - \frac{GM}{R} \tag{3.527}$$

With this constant, we can rewrite the solution in (3.526) as

$$\frac{1}{2}v^2 = \frac{1}{2}v_0^2 + GM\left(\frac{1}{h} - \frac{1}{R}\right) \tag{3.528}$$

If the rocket is going to escape from the gravitational pull of the Earth, we will have a nonzero v even at $h \to \infty$:

$$\lim_{h\to+\infty}\frac{1}{2}v^2 = \frac{1}{2}v_0^2 - GM\frac{1}{R} \tag{3.529}$$

This is equivalent of requiring the initial velocity to satisfy the following condition

$$v_0 \geq \sqrt{\frac{2GM}{R}} \tag{3.530}$$

At the surface of the earth, the gravitational pull equals the weight of the rocket:

$$\frac{GMm}{R^2} = mg \quad (g = 9.81\text{m/s}^2) \tag{3.531}$$

Thus, GM can be found in terms of R and h as

$$GM = R^2 g \tag{3.532}$$

Finally, substitution of (3.532) into (3.530) gives

$$v_0 \geq \sqrt{2Rg} = \sqrt{2\times 63\times 10^5 \times 9.81} \approx 11.2\times 10^3\,(\text{m/s}) \tag{3.533}$$

The escape velocity is about 11.2 km/s, which is a very fast initial velocity and is difficult to achieve. In reality, the rocket will gradually accelerate from launch.

3.4 SECOND ORDER ODE WITH NONCONSTANT COEFFICIENTS

In general, a second order ODE with non-constant coefficients is not easy to solve. We will see that when the Bessel equation is considered in Chapter 4. In this section, we will consider the condition that an ODE with non-constant coefficients can be transformed into one with constant coefficients. One of the most classical examples is the so-called Euler equation.

3.4.1 Euler Equation

The Euler equation can be considered as the simplest type of ODE with non-constant coefficients. The coefficient is in a power of x and its order is exactly the same as the order of differential terms as:

$$x^2 y'' + pxy' + qy = f(x) \tag{3.534}$$

Euler discovered that the following change of variables

$$x = e^t \quad \text{or} \quad \ln x = t \tag{3.535}$$

can be used to convert it to an ODE with constant coefficients. Differentiation of the second equation of (3.535) gives

$$dx = xdt \tag{3.536}$$

In particular, we can apply the chain rule to get

$$\frac{dy}{dx} = \frac{dy}{dt}\frac{dt}{dx} = \frac{1}{x}\frac{dy}{dt} \tag{3.537}$$

Similarly, we can find the second derivative as

$$\begin{aligned} \frac{d^2y}{dx^2} &= \frac{d}{dx}(\frac{1}{x})\frac{dy}{dt} + \frac{1}{x}\frac{d}{dx}(\frac{dy}{dt}) = -(\frac{1}{x^2})\frac{dy}{dt} + \frac{1}{x}\frac{d}{xdt}(\frac{dy}{dt}) \\ &= -(\frac{1}{x^2})\frac{dy}{dt} + \frac{1}{x^2}(\frac{d^2y}{dt^2}) \end{aligned} \tag{3.538}$$

Substitution of (3.537) and (3.538) into (3.534) leads to

$$\frac{d^2y}{dt^2} + (p-1)\frac{dy}{dt} + qy = f(e^t) \tag{3.539}$$

This is a second order ODE with a constant coefficient, which has been considered in Section 3.3.3. Thus, we know how to solve it.

3.4.2 Transformation to Constant Coefficient

In this section, we will consider the condition that an ODE with non-constant can be transformed into one with constant coefficients. Let us consider the following ODE

$$a(x)\frac{d^2u}{dx^2} + b(x)\frac{du}{dx} + c(x)u = f(x) \tag{3.540}$$

Consider a change of variables as:

$$t = \varphi(x) \tag{3.541}$$

Differentiation of u with respect to the new variable defined in (3.541) results in

$$\frac{du}{dx} = \frac{du}{dt}\frac{dt}{dx} = \varphi'(x)\frac{du}{dt} \tag{3.542}$$

$$\frac{d^2u}{dx^2} = \varphi''(x)\frac{du}{dt} + [\varphi'(x)]^2\frac{d^2u}{dt^2} \tag{3.543}$$

Substitution of (3.542) and (3.543) into (3.540) gives

$$a(\varphi')^2 \frac{d^2u}{dt^2} + (a\varphi'' + b\varphi')\frac{du}{dt} + cu = f \tag{3.544}$$

We now impose the following conditions for φ:

$$a(\varphi')^2 = \lambda c, \quad (a\varphi'' + b\varphi') = \mu c \tag{3.545}$$

where λ and μ are constants. Thus, all coefficients become constant:

$$\lambda \frac{d^2u}{dt^2} + \mu \frac{du}{dt} + u = \frac{f}{c} \tag{3.546}$$

Substitution of the first equation of (3.545) into the second equation of (3.545) leads to

$$a\frac{d}{dx}(\sqrt{\frac{\lambda c}{a}}) + b\sqrt{\frac{\lambda c}{a}} = \mu c \tag{3.547}$$

This gives the condition that a, b, and c need to satisfy if it can be converted to a constant coefficient ODE. This can be further simplified to a more compact form

$$\frac{(a\dfrac{dc}{dx} - c\dfrac{da}{dx} + 2bc)^2}{ac^3} = \frac{4\mu^2}{\lambda} \tag{3.548}$$

Note that the right hand side is a constant. If (3.548) is satisfied, the first equation of (3.545) gives the required change of variables:

$$\varphi' = \sqrt{\frac{\lambda c}{a}}, \quad \text{or} \quad \varphi = \sqrt{\lambda}\int \sqrt{\frac{c}{a}} dx \tag{3.549}$$

Example 3.26 Let us consider the classic non-constant ODE of Euler type as an example:

$$px^2 \frac{d^2u}{dx^2} + qx\frac{du}{dx} + ru = f(x) \tag{3.550}$$

Solve the Euler type ODE using the result of this section.

Solution: Thus, from (3.540) we have

$$a = px^2, \quad b = qx, \quad c = r \tag{3.551}$$

Substitution of these values into (3.549) gives

$$\varphi' = \frac{1}{x}\sqrt{\frac{\lambda r}{p}} \tag{3.552}$$

It is natural to choose

$$\lambda = \frac{p}{r} \tag{3.553}$$

The change of variables becomes

$$\varphi = \ln x = t\text{, or} \quad x = e^t \tag{3.554}$$

This of course agrees with Euler's approach of solving this equation. Substitution of (3.554) into (3.548) gives

$$\mu = \frac{q-p}{r} \tag{3.555}$$

Finally, using (3.553) and (3.555) in (3.546) we have the following constant coefficient ODE

$$p\frac{d^2u}{dt^2} + (q-p)\frac{du}{dt} + ru = \frac{rf}{c} \tag{3.556}$$

This can be solved by assuming the standard exponential form. Thus, we can see that you do not need to be Euler to figure out the appropriate change of variables for Euler's equation. The systematic approach presented in this section shows you how to propose the proper change of variables.

3.4.3 Laplace Type

The following ODE is known as the Laplace type and was considered by Weiler in 1856, Schlomilch in 1879, Pochhammer in 1891, and Bolza in 1893. In particular, Laplace's type equation is

$$(a+lx)\frac{d^2y}{dx^2} + (b+mx)\frac{dy}{dx} + (c+nx)y = 0 \tag{3.557}$$

For non-zero l, we can let

$$\xi = a + lx \tag{3.558}$$

It is straightforward to show that (3.557) can be written as:

$$\frac{d^2y}{d\xi^2} + 2(\frac{h}{\xi} + f)\frac{dy}{d\xi} + (\frac{2q}{\xi} + r)y = 0 \tag{3.559}$$

For $l = 0$, (3.557) can be rewritten as:

$$\frac{d^2y}{dx^2} + 2(h + fx)\frac{dy}{dx} + (2qx + r)y = 0 \tag{3.560}$$

Both of these can be considered as special cases of the following more general ODE:

$$\frac{d^2y}{dx^2} + 2(\frac{h}{x} + f)\frac{dy}{dx} + (\frac{p}{x^2} + \frac{2q}{x} + r)y = 0 \tag{3.561}$$

$$\frac{d^2y}{dx^2} + 2(h + fx)\frac{dy}{dx} + (px^2 + 2qx + r)y = 0 \tag{3.562}$$

Note that we have added extra terms in the last bracket terms in (3.561) and (3.562). We will focus our discussion for these two ODEs given in (3.561) and (3.562), since the Laplace type ODE is only a special case of them (i.e., $p = 0$). We first examine a particular change of variables as:

$$x = \kappa\xi, \quad y = \nu\xi^{\lambda}e^{\mu\xi}\zeta \tag{3.563}$$

We can see that

$$\frac{dy}{dx} = \frac{dy}{d\xi}\frac{d\xi}{dx} = \frac{1}{\kappa}\frac{dy}{d\xi}, \quad \frac{d^2y}{dx^2} = \frac{1}{\kappa^2}\frac{d^2y}{d\xi^2} \tag{3.564}$$

$$\frac{dy}{d\xi}=v\xi^{\lambda}e^{\mu\xi}\{\frac{d\zeta}{d\xi}+(\mu+\frac{\lambda}{\xi})\zeta\} \tag{3.565}$$

$$\frac{d^2y}{d\xi^2}=v\xi^{\lambda}e^{\mu\xi}\{\frac{d^2\zeta}{d\xi^2}+2(\mu+\frac{\lambda}{\xi})\frac{d\zeta}{d\xi}+[\mu^2+\frac{2\lambda\mu}{\xi}+\frac{\lambda(\lambda-1)}{\xi^2}]\zeta\} \tag{3.566}$$

Substitution of (3.564) to (3.566) into (3.561) gives

$$\frac{d^2\zeta}{d\xi^2}+2(\frac{h'}{\xi}+f')\frac{d\zeta}{d\xi}+(\frac{p'}{\xi^2}+\frac{2q'}{\xi}+r')\zeta=0 \tag{3.567}$$

where

$$\begin{aligned} &h'=\lambda+h, \quad f'=\mu+\kappa f, \quad r'=r\kappa^2+2\kappa\mu f+\mu^2, \\ &p'=\lambda(\lambda-1)+2h\lambda+p, \quad q'=\lambda\mu+\kappa\lambda f+h\mu+\kappa q \end{aligned} \tag{3.568}$$

It can be shown that there are invariants for some combinations of these constants such that:

$$p'+h'-h'^2=p+h-h^2=A \tag{3.569}$$

$$q'-h'f'=\kappa(q-hf)=\kappa B \tag{3.570}$$

$$r'-f'^2=\kappa^2(r-f^2)=\kappa^2 C \tag{3.571}$$

These can be proved directly by using the definitions given in (3.568). We can also define another invariant as:

$$I=C/B^2 \tag{3.572}$$

Now, we want to remove the first order derivative term by setting $h'=f'=0$ and this implies

$$\lambda=-h, \quad \mu=-\kappa f \tag{3.573}$$

Thus, we propose the change of variables (compare (3.563))

$$y=z^{-h}e^{-\kappa fz}v(z) \tag{3.574}$$

From (3.569) to (3.571), we have the special cases of

$$p'=A \tag{3.575}$$

$$q'=\kappa B \tag{3.576}$$

$$r'=\kappa^2 C \tag{3.577}$$

Therefore, with these special values, we have the following normalized form of

$$\frac{d^2v}{dz^2}+(\frac{A}{z^2}+\frac{2\kappa B}{z}+\kappa^2 C)v=0 \tag{3.578}$$

There are a number of scenarios for the invariants (with $\kappa=1/B$ in (3.563)):

Case I: $B\neq 0$, $C\neq 0$

$$\frac{d^2v}{dz^2}+(\frac{A}{z^2}+\frac{2}{z}+I)v=0 \tag{3.579}$$

Case II: $B\neq 0$, $C=0$

$$\frac{d^2v}{dz^2}+(\frac{A}{z^2}+\frac{2}{z})v=0 \tag{3.580}$$

Case III: $B=0$, $C\neq 0$

$$\frac{d^2v}{dz^2}+(\frac{A}{z^2}+I)v=0 \tag{3.581}$$

Case IV: $B = 0$, $C= 0$

$$\frac{d^2v}{dz^2}+\frac{A}{z^2}v=0 \tag{3.582}$$

where

$$A=p+h-h^2, \quad B=q-hf, \quad C=r-f^2, \quad I=C/B^2 \tag{3.583}$$

Note that a special form of (3.578) is the Whittaker equation (Whittaker and Watson, 1927: Abramowitz and Stegun, 1964):

$$\frac{d^2W}{dz^2}+\{-\frac{1}{4}+\frac{k}{z}+\frac{1/4-m^2}{z^2}\}W=0 \tag{3.584}$$

To recover the Laplace type ODE given in (3.557) as a special form given in (3.578), we can make the following definitions:

$$2h=\frac{bl-ma}{l^2}, \quad 2f=\frac{m}{l^2}, \quad p=0, \quad 2q=\frac{cl-na}{l^3}, \quad r=\frac{n}{l^3} \tag{3.585}$$

To reduce (3.567) to (3.559), we can choose λ according to the following equation:

$$p'=\lambda(\lambda-1)+2h\lambda+p=0 \tag{3.586}$$

A special form of second order ODE of (3.561) is called the Bessel equation, which is one of the most important second order ODEs in physics and engineering, namely

$$\frac{d^2y}{dx^2}+\frac{1}{x}\frac{dy}{dx}+(1-\frac{n^2}{x^2})y=0 \tag{3.587}$$

which is obtained by setting

$$2h=1, \quad f=0, \quad p=-n^2, \quad q=0, \quad r=1 \tag{3.588}$$

Thus, for the Bessel equation, we can find λ by (3.586) as

$$\lambda^2-n^2=0 \tag{3.589}$$

Thus, making use of the following change of variables (see (3.574))

$$y=x^n z \tag{3.590}$$

we have the derivatives as

$$\frac{dy}{dx}=nx^{n-1}z+x^n\frac{dz}{dx} \tag{3.591}$$

$$\frac{d^2y}{dx^2}=n(n-1)x^{n-2}z+2nx^{n-1}\frac{dz}{dx}+x^n\frac{d^2z}{dx^2} \tag{3.592}$$

With these derivatives ready, it is easy to show that the Bessel equation given in (3.587) can be written as

$$\frac{d^2z}{dx^2}+(\frac{2n+1}{x})\frac{dz}{dx}+z=0 \tag{3.593}$$

This provides another mathematical form of the Bessel equation and is also known as Weiler's canonical form of Laplace type ODE. Thus, the Bessel equation is a special form of (3.561). In conclusion, (3.561) can be converted to one of the canonical forms given in (3.579) to (3.582).

Similarly, we can also convert (3.562) to three canonical forms. In doing so, we observe that the following change of variables can be used

$$x = \kappa\xi + \lambda \tag{3.594}$$

$$y = e^{\mu\xi + \frac{1}{2}\nu\xi^2} \zeta \tag{3.595}$$

Differentiation of y with respect to x can be changed to differentiation with respect to ξ by the chain rule:

$$\frac{dy}{dx} = \frac{dy}{d\xi}\frac{d\xi}{dx} = \frac{1}{\kappa}\frac{dy}{d\xi}, \quad \frac{d^2y}{dx^2} = \frac{1}{\kappa^2}\frac{d^2y}{d\xi^2} \tag{3.596}$$

$$\frac{dy}{d\xi} = e^{\mu\xi + \frac{1}{2}\nu\xi^2} \{(\mu + \nu\xi)\zeta + \frac{d\zeta}{d\xi}\} \tag{3.597}$$

$$\frac{d^2y}{d\xi^2} = e^{\mu\xi + \frac{1}{2}\nu\xi^2} \{[(\mu + \nu\xi)^2 + \nu]\zeta + 2(\mu + \nu\xi)\frac{d\zeta}{d\xi} + \frac{d^2\zeta}{d\xi^2}\} \tag{3.598}$$

With these results, (3.562) is transformed to

$$\frac{d^2\zeta}{d\xi^2} + 2(h' + f'\xi)\frac{d\zeta}{d\xi} + (p'\xi^2 + 2q'\xi + r')\zeta = 0 \tag{3.599}$$

$$\begin{aligned} h' &= \mu + \kappa h + f\kappa\lambda \\ f' &= \nu + \kappa^2 f \\ p' &= \nu^2 + 2\kappa^2\nu f + p\kappa^4 \\ q' &= \mu\nu + \kappa\nu(h + f\lambda) + \kappa^2\mu f + \kappa^3(\lambda p + q) \\ r' &= \kappa^2(p\lambda^2 + 2q\lambda + \lambda) + 2\kappa\mu(h + f\lambda) + \mu^2 + \nu \end{aligned} \tag{3.600}$$

Similar to the former case, there exist invariants for these coefficients. In particular, we can easily show that

$$P' = p' - f'^2 = \kappa^4(p - f^2) = \kappa^4 P \tag{3.601}$$

$$Q' = q' - h'f' = \kappa^3(\lambda P + Q) \tag{3.602}$$

$$R' = r' - h'^2 - f' = \kappa^2[R + \lambda^2 P + 2\lambda Q] \tag{3.603}$$

$$P'\xi^2 + 2Q'\xi + R' = \kappa^2[Px^2 + 2Qx + R] \tag{3.604}$$

$$S' = P'R' - Q'^2 = \kappa^6(PR - Q^2) = \kappa^6 S \tag{3.605}$$

To remove the first derivative term, we set the following special case of (3.599)

$$h' = 0, \quad f' = 0 \tag{3.606}$$

For this case, it is observed that

$$P' = p', \quad Q' = q', \quad R' = r' \tag{3.607}$$

Consequently, the mathematical structure of (3.599) is reduced to

$$\frac{d^2\zeta}{d\xi^2} + (P'\xi^2 + 2Q'\xi + R')\zeta = 0 \tag{3.608}$$

Now we can use the following change of variables to simplify (3.608):

$$\xi = \alpha z + \beta \tag{3.609}$$

The result is

$$\frac{d^2\zeta}{dz^2}+[P'\alpha^4 z^2+2\alpha^3(P'\beta+Q')z+\alpha^2(P'\beta^2+2Q'\beta+R')]\zeta=0 \quad (3.610)$$

Comparing (3.608) and (3.610), we see that the invariant forms given in (3.601) to (3.603) reappear.

Case I: $P\neq 0$,

For this case, we can choose $\beta=-Q'/P'$, $\alpha^4=1/P'$ such that

$$\frac{d^2\zeta}{dz^2}+[z^2+\sqrt{I}]\zeta=0 \quad (3.611)$$

where

$$I=\frac{S'^2}{P'^3} \quad (3.612)$$

This equation can be converted to a confluent hypergeometric equation, but the details will be given later in this section.

Case II: $P'=0$, $S\neq 0$,

By choosing $\beta=-R'/(2Q')$, $\alpha^3=-1/(6Q')$, we get

$$\frac{d^2\zeta}{dz^2}-\frac{1}{3}z\zeta=0 \quad (3.613)$$

This equation is known as the Scherk-Lobatto equation. However, by assuming a change of variables of

$$z=3^{1/3}x \quad (3.614)$$

we have the Airy equation as a result:

$$\frac{d^2\zeta}{dx^2}-x\zeta=0 \quad (3.615)$$

The solution of it is the Airy functions of the first and second kinds (Stegun and Abramowitz, 1964):

$$\zeta=C_1 Ai(x)+C_2 Bi(x) \quad (3.616)$$

Case III: $P'=0$, $S'=0$,

In this case, we assume $\alpha^2=1/R'$ such that

$$\frac{d^2\zeta}{dz^2}+\zeta=0 \quad (3.617)$$

This is the well-known harmonic equation with solution

$$\zeta=C_1\sin x+C_2\cos x \quad (3.618)$$

Case IV: $P'=0$, $S'=0$, $R'=0$

This is the simplest case, and thus we have

$$\frac{d^2\zeta}{dz^2} = 0 \tag{3.619}$$

Thus, the solution is

$$\zeta = Az + B \tag{3.620}$$

We can see that (3.561) and (3.562) are very general. The special cases of them include the Whittaker equation in (3.584), the Bessel equation in (3.587), and the Airy equation in (3.615).

3.4.4 Solution as Confluent Hypergeometric Functions

In this section, we show that the Laplace type ODE is equivalent to the confluent hypergeometric equation, which is a special case of the hypergeometric equation. The consideration of the solutions of these functions will be postponed to Chapter 4 (see Section 4.12) on series solution.

We now return to (3.557) and will show that it can always be transformed into the following equation, called the confluent hypergeometric or Kummer equation (Abramowitz and Stegun, 1964):

$$s\frac{d^2u}{ds^2} + (\gamma - s)\frac{du}{ds} - \alpha u = 0 \tag{3.621}$$

Its solutions can be expressed as Kummer's function and are given as

$$u = C_1\Phi(\alpha, \gamma; s) + s^{1-\gamma}C_2\Phi(\alpha - \gamma + 1, 2 - \gamma; s) \tag{3.622}$$

where the Humbert's symbol $\Phi(\alpha,\gamma;s)$ is defined by

$$\Phi(\alpha, \gamma; s) = 1 + \sum_{k=1}^{\infty} \frac{(\alpha)_k s^k}{(\gamma)_k k!} \tag{3.623}$$

and the following Pochhammer symbol has been used:

$$(\alpha)_k = \alpha(\alpha+1)\cdots(\alpha+k-1), \quad (\alpha)_0 = 1 \tag{3.624}$$

It was defined by Pochhammer in 1890. The detailed discussion of such solution will be presented in Chapter 4 on series solutions.

<u>Case 1</u>: $l \neq 0$, $m \neq 0$

Let us assume a change of variables:

$$y = e^{\lambda x}\eta, \tag{3.625}$$

then (3.557) can be written as

$$(a + lx)\frac{d^2\eta}{dx^2} + [2\lambda a + b + x(2\lambda l + m)]\frac{d\eta}{dx} + [a\lambda^2 + b\lambda + c + x(l\lambda^2 + m\lambda + n)]\eta = 0 \tag{3.626}$$

We can use the following equation to determine λ

$$l\lambda^2 + m\lambda + n = 0 \tag{3.627}$$

Thus, (3.626) can be expressed as

$$(a+lx)\frac{d^2\eta}{dx^2}+(h+kx)\frac{d\eta}{dx}+j\eta=0 \tag{3.628}$$

where

$$h=2\lambda a+b, \quad k=2\lambda l+m, \quad j=a\lambda^2+b\lambda+c \tag{3.629}$$

Adopting another change of variables

$$a+lx=\nu\xi \tag{3.630}$$

we can convert (3.628) to

$$\xi\frac{d^2\eta}{d\xi^2}+(\frac{lh-ak}{l^2}+\frac{k\nu}{l^2}\xi)\frac{d\eta}{d\xi}+\frac{j\nu}{l^2}\eta=0 \tag{3.631}$$

We can further choose ν as

$$\xi\frac{d^2\eta}{d\xi^2}+(\gamma-\xi)\frac{d\eta}{d\xi}-\alpha\eta=0 \tag{3.632}$$

where

$$\gamma=\frac{ak-lh}{k\nu}, \quad \alpha=\frac{j}{k} \tag{3.633}$$

If there are equal roots for

$$l\lambda^2+m\lambda+n=0 \tag{3.634}$$

we have the root as:

$$2l\lambda=-m \tag{3.635}$$

Equation (3.626) can be simplified as

$$(a+lx)\frac{d^2\eta}{dx^2}+h\frac{d\eta}{dx}+k\eta=0 \tag{3.636}$$

Adopting a change of variables of

$$a+lx=\nu\xi^2 \tag{3.637}$$

gives

$$\frac{d\eta}{dx}=\frac{l}{2\nu\xi}\frac{d\eta}{d\xi} \tag{3.638}$$

$$\frac{d^2\eta}{dx^2}=\frac{l^2}{4\nu^2\xi^2}(\frac{d^2\eta}{d\xi^2}-\frac{1}{\xi}\frac{d\eta}{d\xi}) \tag{3.639}$$

Substitution of (3.638) and (3.639) into (3.636) gives

$$\xi\frac{d^2\eta}{d\xi^2}+(\frac{2h}{l}-1)\frac{d\eta}{d\xi}+\frac{4\nu k}{l^2}\xi\eta=0 \tag{3.640}$$

To reduce this equation further, we use another change of variables

$$\eta=e^{\frac{1}{2}\xi}u \tag{3.641}$$

Equation (3.640) can then be written as

$$\xi\frac{d^2u}{d\xi^2}+(\frac{2h}{l}-1+\xi)\frac{du}{d\xi}+[\frac{h}{l}-\frac{1}{2}+(\frac{4\nu k}{l^2}+\frac{1}{4})\xi]u=0 \tag{3.642}$$

We now make the identifications for the parameters of this change of variables:

$$16\nu k + l^2 = 0, \quad \xi = -s \tag{3.643}$$

Hence, (3.642) becomes

$$s\frac{d^2u}{ds^2} + (\frac{2h}{l} - 1 - s)\frac{du}{ds} - [\frac{h}{l} - \frac{1}{2}]u = 0 \tag{3.644}$$

This can be recast as

$$s\frac{d^2u}{ds^2} + (\gamma - s)\frac{du}{ds} - \alpha u = 0 \tag{3.645}$$

where

$$\gamma = \frac{2h}{l} - 1, \quad \alpha = \frac{h}{l} - \frac{1}{2} \tag{3.646}$$

This is clearly the confluent hypergeometric equation given in (3.621).

Case 2: $l = 0$, $m \neq 0$

For this case, we introduce a change of variables

$$y = \eta e^{-\frac{nx}{m}}, \tag{3.647}$$

then (3.557) can be written as

$$\frac{d^2\eta}{dx^2} + (h + kx)\frac{d\eta}{dx} + g\eta = 0 \tag{3.648}$$

where

$$h = \frac{b}{a} - 2\frac{n}{m}, \quad k = \frac{m}{a}, \quad g = \frac{c}{a} - \frac{bn}{am} + \frac{n^2}{m^2} \tag{3.649}$$

Then, we assume another change of variables:

$$h + kx = \sqrt{2k\xi} \tag{3.650}$$

Hence, (3.648) becomes

$$\xi\frac{d^2\eta}{d\xi^2} + (\xi + \frac{1}{2})\frac{d\eta}{d\xi} + \omega\eta = 0 \tag{3.651}$$

where

$$\omega = \frac{g}{2k} = \frac{c}{2m} - \frac{bn}{2m^2} + \frac{an^2}{2m^3} \tag{3.652}$$

Finally, let $\xi = -s$, and we obtain

$$s\frac{d^2\eta}{ds^2} + (\frac{1}{2} - s)\frac{d\eta}{ds} - \omega\eta = 0 \tag{3.653}$$

By identifying that γ = 1/2 and ω = α, it is clear that it is a special form of the confluent hypergeometric equation given in (3.621).

Case 3: $l = 0$, $m = 0$

For this case, the following change of variables can be used:

$$y = \eta e^{-\frac{bx}{2a}}, \tag{3.654}$$

then (3.557) can be written as

$$\frac{d^2\eta}{dx^2} + (h + kx)\eta = 0 \tag{3.655}$$

where

$$h = \frac{c}{a} - \frac{b^2}{4a^2}, \quad k = \frac{n}{a} \tag{3.656}$$

It can further be converted to a simpler form by

$$h + kx = \nu\xi^{2/3} \tag{3.657}$$

The resulting equation becomes

$$\xi\frac{d^2\eta}{d\xi^2} + \frac{1}{3}\frac{d\eta}{d\xi} + \lambda\xi\eta = 0 \tag{3.658}$$

where

$$\lambda = \frac{4\nu^3}{9k^3} \tag{3.659}$$

We further let

$$\eta = e^{\rho\xi}u \tag{3.660}$$

Consequently, (3.658) can be transformed to

$$\xi\frac{d^2u}{d\xi^2} + (2\rho\xi + \frac{1}{3})\frac{du}{d\xi} + [(\rho^2 + \lambda)\xi + \frac{1}{3}\rho]u = 0 \tag{3.661}$$

To recover the mathematical form of the confluent hypergeometric equation, we can select the following parameters in the change of variables:

$$2\rho = -1, \quad \rho^2 + \lambda = 0 \tag{3.662}$$

Consequently, we obtain the following special form of the confluent hypergeometric equation:

$$\xi\frac{d^2u}{d\xi^2} + (\frac{1}{3} - \xi)\frac{du}{d\xi} - \frac{1}{6}u = 0 \tag{3.663}$$

Therefore, we conclude that the Laplace type of second order ODE given in (3.557) can always be transformed into a confluent hypergeometric equation, and thus can always be solved in terms of Kummer's functions.

3.4.5 Liouville Problem

The following nonlinear second order ODE was considered by Liouville:

$$\frac{d^2y}{dx^2} + f(x)\frac{dy}{dx} + F(y)(\frac{dy}{dx})^2 = 0 \tag{3.664}$$

To solve this equation, we first drop the nonlinear term to get

$$\frac{d^2y}{dx^2} + f(x)\frac{dy}{dx} = 0 \tag{3.665}$$

This can be rewritten as

$$p = \frac{dy}{dx}, \quad \frac{dp}{dx} + f(x)p = 0 \tag{3.666}$$

It is separable and the solution is

$$p = \frac{dy}{dx} = ce^{-\int f(x)dx} \tag{3.667}$$

Now, we follow a similar idea of variation of parameters introduced by Euler:

$$\frac{dy}{dx} = c(y)e^{-\int f(x)dx} \tag{3.668}$$

Note, however, that Euler's variation of parameters assumes C as a function of x instead of y. Differentiation of (3.668) gives

$$\begin{aligned} \frac{d^2y}{dx^2} &= [\frac{dc}{dy}\frac{dy}{dx} - cf(x)]e^{-\int f(x)dx} = \frac{1}{c}\frac{dy}{dx}\{\frac{dc}{dy}\frac{dy}{dx} - cf(x)\} \\ &= \frac{1}{c}\frac{dc}{dy}(\frac{dy}{dx})^2 - f(x)\frac{dy}{dx} \end{aligned} \tag{3.669}$$

Substitution of (3.669) into (3.664) gives

$$\frac{1}{c}\frac{dc}{dy}(\frac{dy}{dx})^2 - f(x)\frac{dy}{dx} + f(x)\frac{dy}{dx} + F(y)(\frac{dy}{dx})^2 = 0 \tag{3.670}$$

The cancellation of the middle terms leads to

$$(\frac{dy}{dx})^2\{\frac{1}{c}\frac{dc}{dy} + F(y)\} = 0 \tag{3.671}$$

Since $dy/dx \neq 0$, we have

$$\frac{dc}{c} = -F(y)dy \tag{3.672}$$

The unknown function can be evaluated as

$$c = Ae^{-\int F(y)dy} \tag{3.673}$$

Back substitution of (3.673) into (3.668) gives

$$\frac{dy}{dx} = Ae^{-\int F(y)dy}e^{-\int f(x)dx} \tag{3.674}$$

This first order ODE is separable and the result is

$$\int e^{\int F(y)dy}dy = A\int e^{-\int f(x)dx}dx + B \tag{3.675}$$

This is the solution of the Liouville problem.

3.4.6 Mainardi Approach for Liouville Problem

Mainardi provides a different approach to solve the Liouville problem. In particular, we can divide through the (3.664) by dy/dx as

$$\frac{1}{\frac{dy}{dx}}\frac{d^2y}{dx^2}+f(x)+F(y)(\frac{dy}{dx})=0 \tag{3.676}$$

Note that the first and last terms in (3.676) can be rewritten as:

$$\frac{d}{dx}\{\ln(\frac{dy}{dx})\}=\frac{1}{\frac{dy}{dx}}\frac{d^2y}{dx^2} \tag{3.677}$$

$$\frac{d}{dx}\{\int F(y)dy\}=\frac{d}{dy}\{\int F(y)dy\}\frac{dy}{dx}=F(y)(\frac{dy}{dx}) \tag{3.678}$$

In view of (3.677) and (3.678), we find (3.676) becoming

$$\frac{d}{dx}\{\ln(\frac{dy}{dx})\}+\frac{d}{dx}\{\int F(y)dy\}=-f(x) \tag{3.679}$$

Integration of (3.679) with respect to x gives

$$\ln(\frac{dy}{dx})+\int F(y)dy=-\int f(x)dx+C \tag{3.680}$$

This can be expressed as

$$\frac{dy}{dx}=\bar{C}e^{-\int F(y)dy}e^{-\int f(x)dx} \tag{3.681}$$

Equation (3.681) is separable and can be integrated as

$$\frac{dy}{e^{-\int F(y)dy}}=\bar{C}e^{-\int f(x)dx}dx \tag{3.682}$$

$$\int e^{\int F(y)dy}dy=\bar{C}\int e^{-\int f(x)dx}dx+C \tag{3.683}$$

This solution is exactly the same as that given in Section 3.4.5.

3.4.7 Liouville Transformation

Consider a second order ODE with non-constant coefficients

$$y''+p(x)y'+q(x)y=0 \tag{3.684}$$

According to Piaggio (1920), the first order derivative term can be removed from (3.684) by using the following change of variables

$$y(x)=W(x)\exp\{-\frac{1}{2}\int_0^x p(\xi)d\xi\} \tag{3.685}$$

$$y'(x)=W'(x)\exp\{-\frac{1}{2}\int_0^x p(\xi)d\xi\}-\frac{W(x)}{2}p(x)\exp\{-\frac{1}{2}\int_0^x p(\xi)d\xi\} \tag{3.686}$$

$$y''(x) = W''(x)\exp\{-\frac{1}{2}\int_0^x p(\xi)d\xi\} - W'(x)p(x)\exp\{-\frac{1}{2}\int_0^x p(\xi)d\xi\}$$
$$+\frac{1}{4}p^2(x)W(x)\exp\{-\frac{1}{2}\int_0^x p(\xi)d\xi\} - \frac{1}{2}W(x)p'(x)\exp\{-\frac{1}{2}\int_0^x p(\xi)d\xi\} \tag{3.687}$$

Substitution of these results into (3.684) gives

$$W''(x) + Q(x)W(x) = 0, \quad Q(x) = q - \frac{1}{4}p^2 - \frac{1}{2}p' \tag{3.688}$$

where

$$W''(x) = \frac{d^2W(x)}{dx^2} \tag{3.689}$$

The term $Q(x)$ is called the invariant of the second order ODE. That is, if the invariants of two different ODEs are the same, these ODEs are actually equivalent (see Problem 3.25). Further discussion of the invariant will be given in Section 3.4.8. The ODE given in (3.688) is the intermediate form or so-called normal form of these different ODEs. Note that the invariant of the adjoint problem of an ODE is the same as the original invariant (see Problem 3.29). The existence of invariants in ODEs was discovered by Laguerre. This invariant allows transformation between ODEs. In 1873, Lie group transformation was used in studying whether an ODE can be integrated. All hodograph transformation, Legendre transformation, and Riccati transformation can be derived from Lie Group method. This topic is, however, out of the scope of the present study.

We further note that (3.688) can be simplified by using Liouville transformation or the Liouville-Green transformation

$$W(x) = \sqrt{\dot{x}}w(t), \quad t = t(x) \tag{3.690}$$

where

$$\sqrt{\dot{x}} = \sqrt{\frac{dx}{dt}} \tag{3.691}$$

$$\dot{w} = \frac{dw}{dt} = \frac{d}{dt}(\frac{1}{\sqrt{\dot{x}}})W + W'\sqrt{\dot{x}} \tag{3.692}$$

$$\ddot{w} = \frac{d\dot{w}}{dt} = \frac{d^2}{dt^2}(\frac{1}{\sqrt{\dot{x}}})W + \frac{d}{dt}(\frac{1}{\sqrt{\dot{x}}})W'\dot{x} + W''\sqrt{\dot{x}}\dot{x} + W'\frac{d}{dt}(\sqrt{\dot{x}}) \tag{3.693}$$

It is straightforward to prove that

$$\frac{d}{dt}(\frac{1}{\sqrt{\dot{x}}})\dot{x} = -\frac{d}{dt}(\sqrt{\dot{x}}) \tag{3.694}$$

Thus, (3.693) can be simplified as

$$\ddot{w} = \frac{d^2}{dt^2}(\frac{1}{\sqrt{\dot{x}}})W + W''\sqrt{\dot{x}}\dot{x} \tag{3.695}$$

Substitution of (3.690) and (3.688) into (3.695) gives

$$\ddot{w} = \sqrt{\dot{x}}\frac{d^2}{dt^2}(\frac{1}{\sqrt{\dot{x}}})w - \dot{x}^2 Q(x)w \tag{3.696}$$

Finally, the differential equation becomes

$$\ddot{w} - \psi(x)w = 0 \tag{3.697}$$

where

$$\psi(x) = -\dot{x}^2 Q(x) + \sqrt{\dot{x}}\frac{d^2}{dt^2}(\frac{1}{\sqrt{\dot{x}}}) \tag{3.698}$$

The second term on the right hand side of (3.698) can be rewritten in terms of the Schwarzian derivative. That is

$$\frac{d}{dt}(\frac{1}{\sqrt{\dot{x}}}) = -\frac{1}{2}\frac{\ddot{x}}{\dot{x}^{3/2}} \tag{3.699}$$

$$\frac{d^2}{dt^2}(\frac{1}{\sqrt{\dot{x}}}) = \frac{3}{4}\frac{\ddot{x}^2}{\dot{x}^{5/2}} - \frac{1}{2}\frac{\dddot{x}}{\dot{x}^{3/2}} \tag{3.700}$$

$$\sqrt{\dot{x}}\frac{d^2}{dt^2}(\frac{1}{\sqrt{\dot{x}}}) = \frac{3}{4}\frac{\ddot{x}^2}{\dot{x}^2} - \frac{1}{2}(\frac{\dddot{x}}{\dot{x}}) = -\frac{1}{2}\{x,t\} \tag{3.701}$$

where the Schwarzian derivative is defined as

$$\{x,t\} = \frac{\dddot{x}}{\dot{x}} - \frac{3}{2}(\frac{\ddot{x}}{\dot{x}})^2 \tag{3.702}$$

The term "Schwarzian derivative" was coined by A. Cayley in honor of Schwarz. However, this derivative had been studied implicitly by Lagrange and Jacobi, and explicitly by Kummer. Whether the resultant differential equation is easier to solve depends on the function $Q(x)$.

Further discussion on Liouville transformation can be found in Temme (1996) and Zwillinger (1997).

3.4.8 Transformation and Invariants of ODE

To consider the invariants of ODEs, we will consider general transformation. Consider again a general second order ODE with non-constant coefficients as:

$$L(u) = a(x)\frac{d^2u}{dx^2} + b(x)\frac{du}{dx} + c(x)u = f(x) \tag{3.703}$$

We assume that the unknown function u can be considered as a product of two functions with one of them at our disposal:

$$u = z(x)w(x) \tag{3.704}$$

Differentiation of (3.704) and substitution of its result into (3.703) results in

$$aw\frac{d^2z}{dx^2} + (2a\frac{dw}{dx} + bw)\frac{dz}{dx} + (a\frac{d^2w}{dx^2} + b\frac{dw}{dx} + cw)z = f(x) \tag{3.705}$$

<u>Case 1</u>:

The function w is at our disposal, and we select w that

$$a\frac{d^2w}{dx^2}+b\frac{dw}{dx}+cw=0 \tag{3.706}$$

Note, however, that this is precisely the homogeneous equation of (3.703). Thus, w is the homogeneous solution of (3.703). But, in general, the determination of $w(x)$ may not be a straightforward task. In a sense, (3.704) can be interpreted as a variation of parameters. With (3.706), (3.705) can be written as

$$\frac{d^2z}{dx^2}+(\frac{2}{w}\frac{dw}{dx}+\frac{b}{a})\frac{dz}{dx}=\frac{f(x)}{aw} \tag{3.707}$$

Naturally, we assume a change of variable as

$$\frac{dz}{dx}=\zeta \tag{3.708}$$

This reduces (3.707) to a first order ODE

$$\frac{d\zeta}{dx}+(\frac{2}{w}\frac{dw}{dx}+\frac{b}{a})\zeta=\frac{f(x)}{aw} \tag{3.709}$$

The homogeneous solution of this ODE (i.e., solution of (3.709) with $f = 0$) is

$$\zeta_h=C\frac{1}{w^2}e^{-\int\frac{b}{a}dx} \tag{3.710}$$

Application of variation of parameters leads to the following solution form:

$$\zeta=C(x)\frac{1}{w^2}e^{-\int\frac{b}{a}dx} \tag{3.711}$$

Substitution of (3.711) into (3.709) gives a differential equation of $C(x)$

$$\frac{dC}{dx}=\frac{fw}{a}e^{\int\frac{b}{a}dx} \tag{3.712}$$

Thus, we have

$$C=\int\frac{fw}{a}e^{\int\frac{b}{a}dx}dx+C_1 \tag{3.713}$$

Thus, the solution for ζ is

$$\frac{dz}{dx}=\zeta=\frac{1}{w^2}e^{-\int\frac{b}{a}dx}[\int\frac{fw}{a}e^{\int\frac{b}{a}dx}dx+C_1] \tag{3.714}$$

Finally, integration of (3.714) gives

$$z=\int\frac{1}{w^2(x)g(x)}[\int_x\frac{f(\xi)w(\xi)}{a(\xi)}g(\xi)d\xi+C_1]dx+C_2 \tag{3.715}$$

where $g(x) = e^{\int(b/a)dx}$. Again, the determination of $w(x)$ may not be a straightforward task.

Case 2:

Instead of removing the linear order term of z, but rather the first derivative term in (3.705), we have

$$2a\frac{dw}{dx}+bw=0 \tag{3.716}$$

That is, we have

$$\frac{dw}{dx}=-\frac{bw}{2a} \tag{3.717}$$

$$\frac{d^2w}{dx^2}=-\frac{1}{2}(\frac{w}{a}\frac{db}{dx}-\frac{wb^2}{2a^2}-\frac{bw}{a^2}\frac{da}{dx}) \tag{3.718}$$

Substitution of (3.717) and (3.716) into (3.705) gives

$$\frac{d^2z}{dx^2}+Iz=\frac{f}{aw} \tag{3.719}$$

where

$$I=\frac{c}{a}-\frac{b^2}{4a^2}-\frac{1}{2a}\frac{db}{dx}+\frac{b}{2a^2}\frac{da}{dx} \tag{3.720}$$

We find that if we assume $u = zv$ instead of $u = zw$, we arrive at exactly the same equation as (3.720) except for v instead of w:

$$\frac{d^2z}{dx^2}+Iz=\frac{f}{av} \tag{3.721}$$

where the value of I is the same as that given in (3.720). For different ODE of the form:

$$a'\frac{d^2z}{dx^2}+b'\frac{dz}{dx}+c'z=g(x) \tag{3.722}$$

If it can be transformed to the form (3.720) by $u = zv$:

$$\frac{d^2z}{dx^2}+I'z=\frac{g}{av} \tag{3.723}$$

we must have

$$\lambda a'=av, \quad \lambda b'=2a\frac{dv}{dx}+bv, \quad \lambda c'=a\frac{d^2v}{dx^2}+b\frac{dv}{dx}+cv \tag{3.724}$$

where λ is a function of x only. Thus, we have

$$I'=\frac{c'}{a'}-\frac{b'^2}{4a'^2}-\frac{1}{2a'}\frac{db'}{dx}+\frac{b'}{2a'^2}\frac{da'}{dx}=I=\frac{c}{a}-\frac{b^2}{4a^2}-\frac{1}{2a}\frac{db}{dx}+\frac{b}{2a^2}\frac{da}{dx} \tag{3.725}$$

Therefore, I is called an invariant.

Case 3:

Finally, we consider a self-adjoint second order ODE:

$$\frac{d}{dx}(p\frac{du}{dx})+qu=F(x) \tag{3.726}$$

Thus, we have

$$p\frac{d^2u}{dx^2}+\frac{dp}{dx}\frac{du}{dx}+qu=F(x) \tag{3.727}$$

To put this into the standard form, we can let

$$\nu a = p, \quad \nu b = \frac{dp}{dx}, \quad \nu c = q, \quad \nu f = F(x) \tag{3.728}$$

The first two equations of (3.728) give a differential equation for ν

$$\nu b = \frac{d(\nu a)}{dx} \tag{3.729}$$

This can be recast as

$$\frac{b}{a}dx = \frac{d(\nu a)}{\nu a} \tag{3.730}$$

Integration gives

$$\nu = \frac{1}{a}e^{\int \frac{b}{a}dx} \tag{3.731}$$

To remove the middle term in (3.727), we can assume a new variable as

$$t = \int \frac{dx}{p} \tag{3.732}$$

Then,

$$\frac{du}{dx} = \frac{du}{dt}\frac{dt}{dx} = \frac{1}{p}\frac{du}{dt} \tag{3.733}$$

$$\frac{d}{dx}(p\frac{du}{dx}) = \frac{d}{dx}(\frac{du}{dt}) = \frac{1}{p}\frac{d^2u}{dt^2} \tag{3.734}$$

We deduce from (3.734) that (3.726) becomes

$$\frac{d^2u}{dt^2} + pqu = pF \tag{3.735}$$

where pq can be rewritten as

$$pq = \frac{c}{a}e^{2\int \frac{b}{a}dx} \tag{3.736}$$

This resulted from (3.728) and (3.731).

3.5 HIGHER ORDER ODE

The general form of linear ODE of order n can be given as

$$L[y] = \frac{d^n y}{dt^n} + p_1(t)\frac{d^{n-1}y}{dt^{n-1}} + \cdots + p_{n-1}(t)\frac{dy}{dt} + p_n(t)y = g(t) \tag{3.737}$$

The corresponding boundary conditions are

$$y(t_0) = y_0, \; y'(t_0) = y_0', \ldots, y^{(n-1)}(t_0) = y_0^{(n-1)} \tag{3.738}$$

Without going into mathematical proof, we simply recall that we have learned that there are two independent solutions for second order ODE. Actually, this observation is also true for any order. Thus, the general solution for the homogeneous form of (3.737) can be expressed as

$$y(t) = c_1 y_1(t) + c_2 y_2(t) + \cdots + c_n y_n(t) \tag{3.739}$$

To show the validity of this solution, we can substitute (3.739) into the homogeneous form of (3.737) to give

$$\begin{aligned} & c_1[\frac{d^n y_1}{dt^n} + p_1(t)\frac{d^{n-1} y_1}{dt^{n-1}} + \cdots + p_{n-1}(t)\frac{dy_1}{dt} + p_n(t)y_1] \\ & + c_2[\frac{d^n y_2}{dt^n} + p_1(t)\frac{d^{n-1} y_2}{dt^{n-1}} + \cdots + p_{n-1}(t)\frac{dy_2}{dt} + p_n(t)y_2] \\ & + \ldots \\ & + c_n[\frac{d^n y_n}{dt^n} + p_1(t)\frac{d^{n-1} y_n}{dt^{n-1}} + \cdots + p_{n-1}(t)\frac{dy_n}{dt} + p_n(t)y_n] = 0 \end{aligned} \tag{3.740}$$

Each square bracket is zero since y_1, y_2, ... and y_n are solutions of the homogeneous equation.

Note again that a second order ODE possesses two unknown constants. Analogously, there must be n unknown constants for an n-th order ODE, as shown in (3.740). Consequently, for an n-th order ODE being well-posed we must need n properly prescribed boundary conditions (see (3.738)). Otherwise, the differential equation is not solvable.

The next question is whether the boundary conditions are given properly, or whether the unknown constants can be determined uniquely by the prescribed boundary conditions. The idea of the Wronskian for second order ODEs can be readily adapted to n-th ODEs. Application of (3.739) into (3.738) gives a system of equations as:

$$\begin{aligned} c_1 y_1(t_0) + \cdots + c_n y_n(t_0) &= y_0 \\ c_1 y_1'(t_0) + \cdots + c_n y_n'(t_0) &= y_0' \\ &\vdots \\ c_1 y_1^{(n-1)}(t_0) + \cdots + c_n y_n^{(n-1)}(t_0) &= y_0^{(n-1)} \end{aligned} \tag{3.741}$$

Clearly, for the system to have a unique solution, we must have the Wronskian be nonzero:

$$W(y_1, y_2, \ldots, y_n)(t_0) = \begin{vmatrix} y_1(t_0) & y_2(t_0) & \cdots & y_n(t_0) \\ y_1'(t_0) & y_2'(t_0) & \cdots & y_n'(t_0) \\ \vdots & \vdots & \ddots & \vdots \\ y_1^{(n-1)}(t_0) & y_2^{(n-1)}(t_0) & \cdots & y_n^{(n-1)}(t_0) \end{vmatrix} \neq 0 \tag{3.742}$$

We first consider the homogeneous ODE that

$$y^{(n)} + P_1 y^{(n-1)} + \cdots + P_{n-1} y' + P_n y = 0 \tag{3.743}$$

As before, we seek an exponential function as the solution for the case of constant coefficients:

$$y = e^{rx} \tag{3.744}$$

Substitution of (3.744) into (3.743) gives the following characteristic equation:

$$r^n + P_1 r^{n-1} + \cdots + P_{n-1} r + P_n = 0 \tag{3.745}$$

It is a mathematical theorem by Gauss established in 1799 that an n-th order polynomial has n roots, and thus (3.745) allows a factorization as

$$Z(r) = (r - r_1)(r - r_2)\cdots(r - r_n) = 0 \tag{3.746}$$

Suppose that all the roots are real and that the first k roots are equal. Thus, we have the following factorization:

$$Z(r) = (r - r_1)^k (r - r_{k+1})(r - r_{k+2})\cdots(r - r_n) = 0 \tag{3.747}$$

We also said that the real root r_1 is of order k. The k independent solutions corresponding to the repeated root of order k can be expressed as:

$$y(x) = (C_0 + C_1 x + \cdots + C_{k-1} x^{k-1}) e^{rx} \tag{3.748}$$

The proof for the case of the third order root will be given in the following example. For the case with complex conjugate roots (i.e., $r = \alpha \pm i\beta$) of order k, we have the k independent solutions as:

$$\begin{aligned} y(x) = [&(C_0 + C_1 x + \cdots + C_{k-1} x^{k-1}) \cos \beta x \\ &+ (D_0 + D_1 x + \cdots + D_{k-1} x^{k-1}) \sin \beta x] e^{\alpha x} \end{aligned} \tag{3.749}$$

For the case of a simple root, we have the solution form as

$$y(x) = C_1 e^{r_1 x} \tag{3.750}$$

The general solution can be expressed as (derived by D'Alembert)

$$y = C_1 y_1 + C_2 y_2 + \cdots + C_n y_n + Y(x) \tag{3.751}$$

Example 3.27 Show that for the case of a triple root, the 3rd order ODE and its general solution are

$$y''' + py'' + qy' + sy = 0 \tag{3.752}$$

$$y = (C_1 + C_2 t + C_3 t^2) e^{rt} \tag{3.753}$$

where r is the root of the following characteristic equation

$$r^3 + pr^2 + qr + s = 0 \tag{3.754}$$

Solution: For the case of constant coefficients, we seek an exponential solution in the form:

$$y(t) = e^{rt} \tag{3.755}$$

Thus, we have

$$(r^3 + pr^2 + qr + s) e^{rt} = 0 \tag{3.756}$$

Since e^{rt} cannot be zero (otherwise we have the trivial solution), we arrive at the characteristic equation for r. For the case of triple roots, we mean that the characteristic equation can be factorized as

$$r^3 + pr^2 + qr + s = (r - r_0)^3 = 0 \tag{3.757}$$

On the other hand, we expand the last term in (3.757) as

$$r^3 - 3r_0 r^2 + 3r_0^2 r - r_0^3 = 0 \tag{3.758}$$

Comparison of (3.757) and (3.758) yields the following values of p, q, and s as:

$$p = -3r_0, \quad q = 3r_0^2, \quad s = -r_0^3 \tag{3.759}$$

Therefore, the first obvious solution is

$$y = e^{r_0 t} \tag{3.760}$$

To find the other independent solutions for the case of triple roots, similar to the double root case, we seek a solution in the form:

$$y = u(t)e^{rt} \tag{3.761}$$

where r satisfies (3.757). Differentiation of (3.761) gives

$$y''' = (u''' + 3ru'' + 3r^2u' + r^3u)e^{rt} \tag{3.762}$$

$$y'' = (u'' + 2ru' + r^2u)e^{rt} \tag{3.763}$$

$$y' = (u' + ru)e^{rt} \tag{3.764}$$

Substitution of (3.762) into (3.764) into (3.752) gives

$$u''' + (3r + p)u'' + (3r^2 + 2pr + q)u' + (r^3 + pr^2 + qr + s)u = 0 \tag{3.765}$$

Substitution of (3.757) into (3.765) gives

$$u''' + 3(r - r_0)u'' + 3(r - r_0)^2 u' + (r - r_0)^3 u = 0 \tag{3.766}$$

Therefore, all terms become zero except the first terms, and we have, after integrating it three times:

$$u(t) = (C_1 + C_2 t + C_3 t^2) \tag{3.767}$$

Combining (3.767) and (3.761), we obtain the required result

$$y = (C_1 + C_2 t + C_3 t^2)e^{rt} \tag{3.768}$$

3.5.1 Euler Equation of Order n

The definition of the Euler equation for second order given in Section 3.4.1 can easily be extended to a higher order. In particular, the Euler equation for the n-th order is

$$x^n \frac{d^n y}{dx^n} + a_1 x^{n-1} \frac{d^{n-1} y}{dx^{n-1}} + \dots + a_{n-1} x \frac{dy}{dx} + a_n y = f(x) \tag{3.769}$$

Note that the coefficient of each derivative term is a power function x and its degree equals the order of differentiation of the same term. Again, we can adopt the following change of variables:

$$x = \mathrm{e}^t, \quad y(x) = Y(t) \tag{3.770}$$

Taking the total differential, we find

$$dx = e^t dt, \quad \frac{dt}{dx} = \frac{1}{x} \tag{3.771}$$

Then, applying chain rule of differentiation, we have the first derivative of y as

$$\frac{dy}{dx} = \frac{dY}{dt}\frac{dt}{dx} = \frac{1}{x}\frac{dY}{dt} \tag{3.772}$$

Similarly, we can apply the chain rule of differentiation one more time to get

$$\frac{d^2y}{dx^2}=\frac{d}{dx}(\frac{1}{x}\frac{dY}{dt})=\frac{1}{x}\frac{d^2Y}{dt^2}\frac{dt}{dx}-\frac{1}{x^2}\frac{dY}{dt}$$
$$=\frac{1}{x^2}\frac{d^2Y}{dt^2}-\frac{1}{x^2}\frac{dY}{dt}=\frac{1}{x^2}\frac{d}{dt}(\frac{d}{dt}-1)Y \quad (3.773)$$

Note that the last of (3.773) rearranges the differentiation in a factorized fashion. This step is very important, and this will become obvious when we continue the process of differentiation. In particular, applying the chain rule of differentiation for the third time gives

$$\frac{d^3y}{dx^3}=\frac{1}{x^3}\frac{d^3Y}{dt^3}-\frac{3}{x^3}\frac{d^2Y}{dt^2}+\frac{2}{x^3}\frac{dY}{dt}=\frac{1}{x^3}\frac{d}{dt}\{\frac{d^2}{dt^2}-3\frac{d}{dt}+2\}Y$$
$$=\frac{1}{x^3}\frac{d}{dt}(\frac{d}{dt}-1)(\frac{d}{dt}-2)Y \quad (3.774)$$

More importantly, we observe that the result can be factorized in an orderly fashion. To double check, we consider the fourth order differentiation of y and the result is

$$\frac{d^4y}{dx^4}=\frac{1}{x^4}\frac{d^4Y}{dt^4}-\frac{6}{x^4}\frac{d^3Y}{dt^3}+\frac{11}{x^4}\frac{d^2Y}{dt^2}-\frac{6}{x^4}\frac{dY}{dt}$$
$$=\frac{1}{x^4}\frac{d}{dt}(\frac{d}{dt}-1)(\frac{d}{dt}-2)(\frac{d}{dt}-3)Y \quad (3.775)$$

Therefore, it becomes obvious that the change of variable for higher differentiation can be expressed in a compact form. In fact, one can show that for the k-th derivative term, we have

$$\frac{d^ky}{dx^k}=\frac{1}{x^k}\frac{d}{dt}(\frac{d}{dt}-1)...(\frac{d}{dt}-k+1)Y \quad (3.776)$$

Using the general formula in (3.776), the original ODE becomes

$$\frac{d}{dt}(\frac{d}{dt}-1)...(\frac{d}{dt}-n+1)Y+a_1\frac{d}{dt}(\frac{d}{dt}-1)...(\frac{d}{dt}-n+2)Y$$
$$+...+a_{n-1}\frac{dY}{dt}+a_nY=F(t) \quad (3.777)$$

It is obvious that this is an n-th order ODE with constant coefficients. Thus, the solution must be expressed as an exponential function and the corresponding characteristics equation is

$$r(r-1)...(r-n+1)+a_1r(r-1)...(r-n+2)+...+a_{n-1}r+a_n=0 \quad (3.778)$$

It is obvious that this is an n-th order ODE with constant coefficients.

Example 3.28 Show that the following generalized Euler equation can also be solved by using Euler's method. This is also known as Lagrange problem.

$$(ax+b)^n\frac{d^ny}{dx^n}+A_1(ax+b)^{n-1}\frac{d^{n-1}y}{dx^{n-1}}+...+A_{n-1}(ax+b)\frac{dy}{dx}+A_ny=0 \quad (3.779)$$

Solution: Using Euler's approach, we can assume

$$ax+b=e^t, \quad adx=e^tdt \quad (3.780)$$

Then, the first and second order derivative terms can be rewritten as

$$\frac{\mathrm{d}y}{\mathrm{d}x} = \frac{a}{e^t}\frac{dy}{dt} \tag{3.781}$$

$$\begin{aligned}\frac{\mathrm{d}^2 y}{\mathrm{d}x^2} &= \frac{a}{e^t}\frac{d}{dt}\left(\frac{a}{e^t}\frac{dy}{dt}\right) = \frac{a^2}{e^t}\left(-e^{-t}\frac{dy}{dt} + e^{-t}\frac{d^2y}{dt^2}\right) \\ &= \frac{a^2}{e^{2t}}\left(\frac{d^2y}{dt^2} - \frac{dy}{dt}\right)\end{aligned} \tag{3.782}$$

Note that all derivatives of y with respect to t appear linearly on the right hand side of (3.781) and (3.782). To show that this statement is true for all higher derivatives, we use mathematical induction. From (3.781), the statement is true for the case of first order, and we assume that it is true for the k-th order:

$$\frac{\mathrm{d}^k y}{\mathrm{d}x^k} = \frac{a^k}{e^{kt}}P_k \tag{3.783}$$

where P_k is a linear function of derivative of y. That is,

$$P_k = P_k\left(\frac{dy}{dt}, \frac{d^2y}{dt^2}, ..., \frac{d^ky}{dt^k}\right) \tag{3.784}$$

with all derivatives, appears linearly. Differentiation of (3.783) once with respect to x gives

$$\begin{aligned}\frac{\mathrm{d}^{k+1} y}{\mathrm{d}x^{k+1}} &= \frac{a}{e^t}\frac{d}{dt}\left(\frac{a^k}{e^{kt}}P_k\right) \\ &= \frac{a^{k+1}}{e^t}\left\{-ke^{-kt}P_k + e^{-kt}\frac{dP_k}{dt}\right\} \\ &= \frac{a^{k+1}}{e^{(k+1)t}}\left(-P_k + \frac{dP_k}{dt}\right) = \frac{a^{k+1}}{e^{(k+1)t}}P_{k+1}\end{aligned} \tag{3.785}$$

It is clear that P_{k+1} is a linear function of y's derivatives. This completes the proof.

Generalizing Euler's analysis for (3.779) gives

$$y = C_1(ax+b)^{r_1} + C_2(ax+b)^{r_2} + ... + C_n(ax+b)^{r_n} \tag{3.786}$$

where r_i (i =1,2,3,...,n) are the characteristic roots of

$$r(r-1)...(r-n+1) + A_1 r(r-1)...(r-n+2) + ... + A_{n-1}r + A_n = 0 \tag{3.787}$$

If there is a multiple root of order m, the solution becomes

$$\begin{aligned}y &= (ax+b)^{r_1}\{C_0 + C_1\ln(ax+b) + ... + C_{m-1}[\ln(ax+b)]^{m-1}\} \\ &\quad + C_{m+1}(ax+b)^{r_2} + ... + C_n(ax+b)^{r_{n-m}}\end{aligned} \tag{3.788}$$

3.5.2 Adjoint ODE

One general way to solve an n-th order ODE is to consider its adjoint ODE. Physically, the adjoint ODE is the governing equation of the integrating factor of the origin problem. A particularly important special case is that when the adjoint problem is the same as the original problem (differential equation as well as boundary conditions), it is called self-adjoint.

Consider the following n-th order linear ODE

$$L(u) = p_0(x)\frac{d^n u}{dx^n} + p_1(x)\frac{d^{n-1}u}{dx^{n-1}} + \cdots + p_{n-1}(x)\frac{du}{dx} + p_n(x)u = 0 \quad (3.789)$$

First, we note the product rule of differentiation that

$$\frac{d}{dx}\{UV\} = U\frac{dV}{dx} + V\frac{dU}{dx} \quad (3.790)$$

Rewrite differentiation as a power in bracket form, we have

$$VU^{(1)} = \frac{d}{dx}\{UV\} - UV^{(1)} \quad (3.791)$$

We are going to see that this formula can be generalized to higher order easily. Consider the following identity

$$\frac{d}{dx}\left\{U^{(1)}V - UV^{(1)}\right\} = U^{(2)}V + U^{(1)}V^{(1)} - U^{(1)}V^{(1)} - UV^{(2)} = U^{(2)}V - UV^{(2)} \quad (3.792)$$

Note that all immediate terms canceled successively. This can be rewritten as (3.790)

$$VU^{(2)} = \frac{d}{dx}\left\{U^{(1)}V - UV^{(1)}\right\} + UV^{(2)} \quad (3.793)$$

Following the same procedure, it is straightforward to see that

$$VU^{(3)} = \frac{d}{dx}\left\{U^{(2)}V - U^{(1)}V^{(1)} + UV^{(2)}\right\} - UV^{(3)} \quad (3.794)$$

In fact, we have the following general form for any order k

$$VU^{(k)} = \frac{d}{dx}\left\{U^{(k-1)}V - U^{(k-2)}V^{(1)} + U^{(k-3)}V^{(2)} - \dots + (-1)^{(k-1)}UV^{(k-1)}\right\} + (-1)^k UV^{(k)} \quad (3.795)$$

Now, we consider the following function:

$$vL(u) = vp_0(x)\frac{d^n u}{dx^n} + vp_1(x)\frac{d^{n-1}u}{dx^{n-1}} + \cdots + vp_{n-1}(x)\frac{du}{dx} + vp_n(x)u \quad (3.796)$$

Applying (3.795) to each term on the right of (3.796), we find

$$\begin{aligned} vp_0(x)\frac{d^n u}{dx^n} &= \frac{d}{dx}\left\{u^{(n-1)}p_0 v - u^{(n-2)}(p_0 v)' + \dots + (-1)^{n-1}u(p_0 v)^{(n-1)}\right\} \\ &\quad + (-1)^n u\frac{d^n(p_0 v)}{dx^n} \end{aligned} \quad (3.797)$$

$$\begin{aligned} vp_1(x)\frac{d^{n-1}u}{dx^{n-1}} &= \frac{d}{dx}\left\{u^{(n-2)}p_1 v - u^{(n-3)}(p_1 v)' + \dots + (-1)^{n-2}u(p_1 v)^{(n-2)}\right\} \\ &\quad + (-1)^{n-1}u\frac{d^{n-1}(p_1 v)}{dx^{n-1}} \end{aligned} \quad (3.798)$$

$$vp_{n-2}(x)\frac{d^2 u}{dx^2} = \frac{d}{dx}\{u'p_{n-2}v - u(p_{n-2}v)'\} + u\frac{d^2(p_{n-2}v)}{dx^2} \quad (3.799)$$

$$vp_{n-1}(x)\frac{du}{dx}=\frac{d}{dx}\{up_{n-1}v\}-u\frac{d(p_{n-1}v)}{dx} \tag{3.800}$$

$$vp_n u = up_n v \tag{3.801}$$

Substitution of (3.797) to (3.801) into (3.796) gives

$$\begin{aligned} vL(u) &= \frac{d}{dx}\left\{u^{(n-1)}p_0v-u^{(n-2)}(p_0v)'+...+(-1)^{n-1}u(p_0v)^{(n-1)}\right\} \\ &+\frac{d}{dx}\left\{u^{(n-2)}p_1v-u^{(n-3)}(p_1v)'+...+(-1)^{n-2}u(p_1v)^{(n-2)}\right\} \\ &+... \\ &+\frac{d}{dx}\{u'p_{n-2}v-u(p_{n-2}v)'\} \\ &+\frac{d}{dx}\{up_{n-1}v\}+u\bar{L}(v) \end{aligned} \tag{3.802}$$

where the adjoint operator is defined as

$$\bar{L}(v)=(-1)^n\frac{d^n(p_0v)}{dx^n}+(-1)^{n-1}\frac{d^{n-1}(p_1v)}{dx^{n-1}}+\cdots-\frac{d(p_{n-1}v)}{dx}+p_nv \tag{3.803}$$

Equation (3.802) can further be simplified to the following form:

$$vL(u)-u\bar{L}(v)=\frac{d}{dx}\{P(u,v)\} \tag{3.804}$$

where $P(u,v)$ can be identified from (3.802) as the summation of all terms under the differentiation sign d/dx. If we want to solve the original ODE such that u satisfies (3.789) and v satisfies the adjoint ODE

$$\bar{L}(v)=0, \tag{3.805}$$

then, the left hand side of (3.804) equals zero, and we have

$$\frac{d}{dx}\{P(u,v)\}=0 \tag{3.806}$$

Integration of (3.806) yields

$$P(u,v)=c \tag{3.807}$$

If we can solve the adjoint ODE given in (3.805), we will have n independent solutions. Substitution of each of these solutions into (3.807) results in n algebraic equations for n unknown u, u', u'', ..., and $u^{(n-1)}$ as:

$$P(u,v_1)=c_1,\ P(u,v_2)=c_2,\ ...,\ P(u,v_n)=c_n \tag{3.808}$$

If we solve for u, the solution can be expressed in terms of n unknown constants as:

$$u=u(v_1,v_2,...,v_n;c_1,c_2,...,c_n) \tag{3.809}$$

An important particular case is that the associated adjoint ODE is the same as the original ODE. Most of the important ODEs found in physics, science, and mechanics are of the adjoint type. Mathematically, the self-adjoint ODE is defined as

$$L(u)=\bar{L}(u) \tag{3.810}$$

Note that the adjoint of the adjoint of an ODE is the original ODE.

Example 3.29 This example considers a second order ODE with circular functions as its coefficients

$$\sin x\frac{d^2u}{dx^2}+2\cos x\frac{du}{dx}+(\sin x)u=0 \tag{3.811}$$

Solution: This is a linear first order homogeneous ODE. For n = 2, (3.789) and (3.805) become

$$L(u)=p_0(x)\frac{d^2u}{dx^2}+p_1(x)\frac{du}{dx}+p_2(x)u=0 \tag{3.812}$$

$$\bar{L}(v)=\frac{d^2(p_0v)}{dx^2}-\frac{d(p_1v)}{dx}+p_2v=0 \tag{3.813}$$

The adjoint ODE given in (3.813) can be written explicitly as

$$\bar{L}(v)=p_0v''+(2p_0'-p_1)v'+(p_0''-p_1'+p_2)v=0 \tag{3.814}$$

Comparison of (3.813) and (3.814) gives

$$p_0(x)=\sin x,\quad p_1(x)=2\cos x,\quad p_2(x)=\sin x \tag{3.815}$$

Substitution of (3.815) into (3.814) gives

$$(\sin x)v''+(2\sin x)v=0 \tag{3.816}$$

This can be further simplified as

$$v''+2v=0 \tag{3.817}$$

The general solution of (3.817) is

$$v=C_1\sin\sqrt{2}x+C_2\cos\sqrt{2}x \tag{3.818}$$

For the second order ODE, the function P can be defined as:

$$vL(u)-u\bar{L}(v)=\frac{d}{dx}\{P(u,v)\}=0 \tag{3.819}$$

$$P(u,v)=u'p_0v-u(p_0v)'+up_1v=c \tag{3.820}$$

The two independent functions are

$$v_1=\sin\beta x=\sin\sqrt{2}x,\quad v_2=\cos\beta x=\cos\sqrt{2}x \tag{3.821}$$

The simultaneous equations given in (3.808) are

$$\begin{aligned}P(u,v_1)&=u'p_0v_1-u(p_0v_1)'+up_1v_1\\&=u'\sin x\sin\beta x+u(\cos x\sin\beta x-\beta\sin x\cos\beta x)=C_1\end{aligned} \tag{3.822}$$

$$\begin{aligned}P(u,v_2)&=u'p_0v_2-u(p_0v_2)'+up_1v_2\\&=u'\sin x\cos\beta x+u(\cos x\cos\beta x+\beta\sin x\sin\beta x)=C_2\end{aligned} \tag{3.823}$$

That is, we have two equations for two unknowns, and the solution for u is

$$u=\frac{1}{\sqrt{2}\sin x}\left\{-C_1\cos(\sqrt{2}x)+C_2\sin(\sqrt{2}x)\right\} \tag{3.824}$$

Note that if the adjoint ODE cannot be solved, the present approach does not work. Therefore, obviously only very special kinds of ODEs can be solved using this approach.

3.5.3 Sarrus Method

The Sarrus method allows one to reduce the order of a second order ODE to become a first order ODE. Let us illustrate the method by using a particular example:

$$y+3x\frac{dy}{dx}+2y(\frac{dy}{dx})^3+(x^2+2y^2\frac{dy}{dx})\frac{d^2y}{dx^2}=0 \tag{3.825}$$

Note in (3.825) that the highest order differential (i.e., d^2y/dx^2) must be of the first degree (i.e., linear in power) because it is resulted from differentiation. It is a necessary condition that it can be integrated once by exact differential. If the highest order in the ODE does not appear linearly, the Sarrus method is inapplicable. Let us assume that the following exact differential exists

$$dU=\{y+3x\frac{dy}{dx}+2y(\frac{dy}{dx})^3+(x^2+2y^2\frac{dy}{dx})\frac{d^2y}{dx^2}\}dx \tag{3.826}$$

As usual, let us rewrite the first derivative as

$$\frac{dy}{dx}=p \tag{3.827}$$

$$\begin{aligned} dU &= \{y+3xp+2yp^3+(x^2+2y^2p)\frac{dp}{dx}\}dx \\ &= (y+3xp+2yp^3)dx+(x^2+2y^2p)dp \end{aligned} \tag{3.828}$$

Let us assume that

$$dU_1=(x^2+2y^2p)dp \tag{3.829}$$

Integrating once gives

$$U_1=x^2p+y^2p^2=x^2\frac{dy}{dx}+y^2(\frac{dy}{dx})^2 \tag{3.830}$$

Now take the total differential of U_1 again, but this time with respect to dx.

$$dU_1=\{2x\frac{dy}{dx}+2y(\frac{dy}{dx})^3+(x^2+2y^2\frac{dy}{dx})\frac{d^2y}{dx^2}\}dx \tag{3.831}$$

Subtraction of (3.831) from (3.828) gives

$$dU-dU_1=(y+x\frac{dy}{dx})dx \tag{3.832}$$

The first derivative term of dy/dx only appears once on the right hand side, and this is a consequence of an exact differential. The left hand side is clearly exact. Integrating once more, we have

$$U-U_1=xy \tag{3.833}$$

Thus,

$$U=U_1+xy=x^2\frac{dy}{dx}+y^2(\frac{dy}{dx})^2+xy \tag{3.834}$$

Note that $dU=0$ is our differential equation and this implies that $U=C$. Therefore, we have the first integral of the second ODE as

$$x^2\frac{dy}{dx}+y^2(\frac{dy}{dx})^2+xy=C \tag{3.835}$$

Unfortunately, this equation is clearly not exact and we cannot proceed using the Sarrus method. Therefore, the Sarrus method is not a very general method. That is, once the resulting equation is not linear in its highest order, we cannot reapply the Sarrus method again.

3.5.4 Rule of Transformation

We showed earlier that sometimes it is easier to solve a differential equation if the roles of variable and dependent function are reversed. This idea will reappear in our later discussion of PDEs (e.g., hodograph transformation discussed in Section 6.15 and in Section 7.7.2).

Interchange of variable: The interchange of variables is defined as

$$\frac{dy}{dx} = \frac{1}{dx/dy} \tag{3.836}$$

With this transformation, we have the second derivative as:

$$\begin{aligned} \frac{d^2y}{dx^2} &= \frac{d}{dx}\left(\frac{dy}{dx}\right) = \frac{d}{dy}\left(\frac{dy}{dx}\right)\frac{dy}{dx} = \frac{d}{dy}\left(\frac{1}{dx/dy}\right)\frac{1}{(dx/dy)} \\ &= -\frac{1}{(dx/dy)^2}\frac{d^2x}{dy^2}\frac{1}{(dx/dy)} = -\frac{d^2x/dy^2}{(dx/dy)^3} \end{aligned} \tag{3.837}$$

Similarly, the third derivative can be evaluated similarly as

$$\begin{aligned} \frac{d^3y}{dx^3} &= \frac{d}{dx}\left(\frac{d^2y}{dx^2}\right) = \frac{d}{dy}\left\{-\frac{d^2x/dy^2}{(dx/dy)^3}\right\}\frac{dy}{dx} \\ &= \left\{-\frac{d^3x/dy^3}{(dx/dy)^3} + \frac{3(d^2x/dy^2)^2}{(dx/dy)^4}\right\}\left(\frac{1}{dx/dy}\right) \\ &= \frac{-(d^3x/dy^3)(dx/dy) + 3(d^2x/dy^2)^2}{(dx/dy)^5} \end{aligned} \tag{3.838}$$

Higher derivatives can also be derived accordingly. Let us consider an example.

Example 3.30 Solve the following third order nonlinear ODE by interchange of variables

$$3\left(\frac{d^2y}{dx^2}\right)^2 - \frac{dy}{dx}\frac{d^3y}{dx^3} - \frac{d^2y}{dx^2}\left(\frac{dy}{dx}\right)^2 = 0 \tag{3.839}$$

Solution: Application of the formulas given in (3.836) to (3.839) gives

$$\begin{aligned} &3\left\{-\frac{d^2x/dy^2}{(dx/dy)^3}\right\}^2 - \frac{1}{dx/dy}\frac{-(d^3x/dy^3)(dx/dy) + 3(d^2x/dy^2)^2}{(dx/dy)^5} \\ &\quad + \frac{d^2x/dy^2}{(dx/dy)^3}\frac{1}{(dx/dy)^2} = 0 \end{aligned} \tag{3.840}$$

This is a linear first order nonhomogeneous ODE.

$$3\{\frac{d^2x/dy^2}{(dx/dy)^3}\}^2+\frac{d^3x}{dy^3}\frac{1}{(dx/dy)^5}-3(\frac{d^2x}{dy^2})^2\frac{1}{(dx/dy)^6}+\frac{d^2x}{dy^2}\frac{1}{(dx/dy)^5}=0 \quad (3.841)$$

The first and third terms cancel out and the equation is simplified to

$$\frac{d^3x}{dy^3}+\frac{d^2x}{dy^2}=0 \quad (3.842)$$

To solve (3.842), we can assume a reduction of order by

$$z=\frac{d^2x}{dy^2} \quad (3.843)$$

This technique of reduction of order will be discussed in more detail in Section 3.5.8. Using (3.843), (3.842) becomes

$$\frac{dz}{dy}+z=0 \quad (3.844)$$

Integrating this by using separation of variables, we have

$$z=\frac{d^2x}{dy^2}=C_0e^{-y} \quad (3.845)$$

Thus, the final solution is

$$x=C_0e^{-y}+C_1y+C_2 \quad (3.846)$$

Change of dependent variable: Consider y a function of x or $y(x)$, and we propose to change the unknown from y to z:

$$y=\phi(z) \quad (3.847)$$

Applying the chain rule, we find the first, second, and third derivatives as

$$\frac{dy}{dx}=\frac{dy}{dz}\frac{dz}{dx}=\phi'(z)\frac{dz}{dx} \quad (3.848)$$

$$\frac{d^2y}{dx^2}=\frac{d}{dx}(\frac{dy}{dx})=\frac{d}{dx}\phi'(z)\frac{dz}{dx}+\phi'(z)\frac{d^2z}{dx^2}=\phi''(z)(\frac{dz}{dx})^2+\phi'(z)\frac{d^2z}{dx^2} \quad (3.849)$$

$$\begin{aligned}\frac{d^3y}{dx^3}&=\frac{d}{dx}(\frac{d^2y}{dx^2})\\&=\phi'''(z)(\frac{dz}{dx})^3+2\phi''(z)(\frac{dz}{dx})\frac{d^2z}{dx^2}+\phi''(z)\frac{d^2z}{dx^2}\frac{dz}{dx}+\phi'(z)\frac{d^3z}{dx^3} \quad (3.850)\\&=\phi'''(z)(\frac{dz}{dx})^3+3\phi''(z)(\frac{dz}{dx})\frac{d^2z}{dx^2}+\phi'(z)\frac{d^3z}{dx^3}\end{aligned}$$

Higher derivatives can be found by following a similar procedure. The dependent variable is changed from y to z. The most difficult part is the identification of an appropriate function $\phi(z)$. The application of this change of dependent variable is considered through the following example.

Example 3.31 Solve the following second order nonlinear ODE by changing the dependent variable

$$(1+y^2)\frac{d^2y}{dx^2}-(2y-1)(\frac{dy}{dx})^2+3x(1+y^2)\frac{dy}{dx}=0 \tag{3.851}$$

Solution: We consider the following change of variable

$$y=\tan z \tag{3.852}$$

Note that

$$\frac{dy}{dx}=\sec^2 z\frac{dz}{dx}, \quad 1+\tan^2 z=\sec^2 z \tag{3.853}$$

$$\frac{d^2y}{dx^2}=2\sec^2 z\tan z(\frac{dz}{dx})^2+\sec^2 z\frac{d^2z}{dx^2} \tag{3.854}$$

Substitutions of (3.853) and (3.854) into (3.851) gives

$$\frac{d^2z}{dx^2}+(\frac{dz}{dx})^2+3x\frac{dz}{dx}=0 \tag{3.855}$$

The use of reduction of order gives

$$\frac{d\phi}{dx}+\phi^2+3x\phi=0 \tag{3.856}$$

where

$$\phi=\frac{dz}{dx} \tag{3.857}$$

This is the Bernoulli equation and can be solved analytically as discussed in Section 3.2.7.

Change of independent variable: Consider y as a function of x or $y(x)$, and we propose to change the unknown from x to z:

$$x=\phi(z) \tag{3.858}$$

The first, second, and third derivatives can be evaluated as

$$\frac{dy}{dx}=\frac{dy}{dz}\frac{dz}{dx}=\frac{dy}{dx}\frac{1}{\phi'(z)} \tag{3.859}$$

$$\begin{aligned}\frac{d^2y}{dx^2}&=\frac{d}{dx}(\frac{dy}{dx})=\frac{d}{dz}\{\frac{dy}{dz}\frac{1}{\phi'(z)}\}\frac{1}{dx/dz}\\&=\frac{d^2y}{dz^2}\frac{1}{\{\phi'(z)\}^2}-\frac{dy/dz}{\{\phi'(z)\}^2}\frac{\phi''(z)}{\phi'(z)}\\&=\frac{1}{\{\phi'(z)\}^2}\frac{d^2y}{dz^2}-\frac{\phi''(z)}{\{\phi'(z)\}^3}\frac{dy}{dz}\end{aligned} \tag{3.860}$$

$$\frac{d^3y}{dx^3}=\frac{d}{dx}(\frac{d^2y}{dx^2})=\frac{d}{dz}\{\frac{1}{\{\phi'(z)\}^2}\frac{d^2y}{dz^2}-\frac{\phi''(z)}{\{\phi'(z)\}^3}\frac{dy}{dz}\}\frac{1}{dx/dz}$$
$$=\frac{1}{\{\phi'(z)\}^3}\frac{d^3y}{dz^3}-3\frac{\phi''(z)}{\{\phi'(z)\}^4}\frac{d^2y}{dz^2}-\frac{\phi'(z)\phi''(z)-3\{\phi''(z)\}^2}{\{\phi'(z)\}^5}\frac{dy}{dz} \quad (3.861)$$

Again, the most difficult part is finding an appropriate function $\phi(z)$. The following example illustrates this method.

Example 3.32 Consider the second order ODE

$$\frac{d^2y}{dx^2}-\frac{x}{1-x^2}\frac{dy}{dx}+\frac{y}{1-x^2}=0 \quad (3.862)$$

Solve this differential equation by using the following change of variables

$$x=\cos z \quad (3.863)$$

Solution: The derivative can be evaluated using the chain rule as:

$$\frac{dy}{dx}=\frac{dy}{dz}\frac{dz}{dx}=-\operatorname{cosec} z\frac{dy}{dz} \quad (3.864)$$

$$\frac{d^2y}{dx^2}=\frac{d}{dx}(\frac{dy}{dx})=\frac{d}{dz}(-\operatorname{cosec} z\frac{dy}{dz})\frac{1}{(dx/dz)}$$
$$=-\operatorname{cosec}^2 z\cot z\frac{dy}{dz}+\operatorname{cosec}^2 z\frac{d^2y}{dz^2} \quad (3.865)$$

Note also that

$$\frac{x}{1-x^2}=\frac{\cos z}{1-\cos^2 z}=\frac{\cos z}{\sin^2 z}=\operatorname{cosec} z\cot z \quad (3.866)$$

Substitution of (3.864) to (3.866) into (3.862) yields

$$\frac{d^2y}{dx^2}+y=0 \quad (3.867)$$

This is a harmonic equation and the solution is

$$y=A\cos z+B\sin z \quad (3.868)$$

Thus, the final solution is

$$y=Ax+B\sqrt{1-x^2} \quad (3.869)$$

3.5.5 Homogeneous Equation

We have learned that certain kinds of first order ODEs are called homogeneous and they can be solved by a standard change of variables to make it becoming separable. The idea of homogeneous can be extended to consider higher order ODEs. The following discussion follows from Bateman (1918).

In this approach, we assign a number system to each quantity of a differential equation as:

Table 3.2 Number system used in the homogeneous system

Variables	x	y	y'	y''	y'''	$\cdots$	x^2	y^2	y'^2
Weights	m	n	$n-m$	$n-2m$	$n-3m$	$\cdots$	$2m$	$2n$	$2(n-m)$

The differential equation is called "homogeneous" if the sum of the numbers for each term in the differential system is the same when m and n are chosen properly. Sometimes, a differential equation is homogeneous for all values of m and n, but more often, there is only one value of the ratio n/m for which the differential equation is homogeneous. The ratio n/m is called the grade of the equation, denoted by p.

Consider the following differential equation as an example:

$$x^2(\frac{dy}{dx})^2 + x^2 y\frac{d^2y}{dx^2} - y^2 = 0 \tag{3.870}$$

According to the weighting system given in Table 3.2, the sum of the number of the first term is $2m+2n-2m = 2n$, that for the second term is $2m+n+(n-2m) = 2n$ and that for the last one is simply $2n$. Thus, the sum of the numbers for each term is the same regardless of the values of m and n. Consider another example that

$$x^4\frac{d^2y}{dx^2} - (x^3 + 6xy)\frac{dy}{dx} - 5y^2 = 0 \tag{3.871}$$

The middle term actually consists of two different sums of the numbers. The sum of the number of the first term is $4m+n-2m = 2m+n$, that of the second term is $3m+n-m = 2m+n$, that of the third terms is $m+n+n-m = 2n$, and that of the fourth term is $2n$. Thus, for the differential equation to be homogeneous, we require that

$$2m+n = 2n, \quad \text{or } 2m = n \tag{3.872}$$

Therefore, the grade of the equation is two (i.e., $p=2$).

When $m \neq 0$, the following change of variables can be applied for all values of m and n:

$$y = x^p\xi, \quad \frac{dy}{dx} = x^{p-1}\eta \tag{3.873}$$

where ξ and η are the new independent variable and the new unknown. Differentiating the first of (3.873) gives

$$\frac{dy}{dx} = px^{p-1}\xi + x^p\frac{d\xi}{dx} \tag{3.874}$$

The last of (3.874) can be obtained by putting the second definition of (3.873) into the left hand side of (3.874). This results in

$$\frac{d\xi}{dx} = \frac{1}{x}(\eta - p\xi) \tag{3.875}$$

Note from the chain rule and (3.875) that

$$\frac{d\eta}{dx} = \frac{d\eta}{d\xi}\frac{d\xi}{dx} = \frac{1}{x}(\eta - p\xi)\frac{d\eta}{d\xi} \tag{3.876}$$

Differentiating the second equation of (3.873) gives

$$\frac{d^2y}{dx^2}=(p-1)x^{p-2}\eta+x^{p-1}\frac{d\eta}{dx} \tag{3.877}$$

Combining (3.877) and (3.876) gives

$$\frac{d^2y}{dx^2}=x^{p-2}[(p-1)\eta+(\eta-p\xi)\frac{d\eta}{d\xi}] \tag{3.878}$$

Differentiation of (3.878) with respect to x one more time gives

$$\begin{aligned}\frac{d^3y}{dx^3}&=(p-2)x^{p-3}[(p-1)\eta+(\eta-p\xi)\frac{d\eta}{d\xi}]\\&+x^{p-2}\{(p-1)\frac{d\eta}{dx}+(\frac{d\eta}{dx}-p\frac{d\xi}{dx})\frac{d\eta}{d\xi}+(\eta-p\xi)\frac{d}{dx}(\frac{d\eta}{d\xi})\}\end{aligned} \tag{3.879}$$

Substitution of (3.876) into (3.879) leads to

$$\begin{aligned}\frac{d^3y}{dx^3}&=x^{p-3}\{(p-1)(p-2)\eta+(p-3)(\eta-p\xi)\frac{d\eta}{d\xi}\\&+(\eta-p\xi)(\frac{d\eta}{d\xi})^2+(\eta-p\xi)^2\frac{d^2\eta}{d\xi^2}\}\end{aligned} \tag{3.880}$$

Similarly, higher order derivatives can be found. In general, the derivatives on the right of these equations are one order lower than those on the left before transformation. Thus, substituting these results into the original differential equation, the order of the original differential equation will be reduced one order after the change of variables.

If the original equation is homogeneous for all values of m and n, it must have the form

$$F\left\{y,x\frac{dy}{dx},x^2\frac{d^2y}{dx^2},x^3\frac{d^3y}{dx^3}\dots\right\}=0 \tag{3.881}$$

Note that (3.870) is another example of this form. If such a differential equation is linear, it is clearly of Euler type. If the equation is second order, transformation of (3.873) will convert it to first order and it is solvable by the technique of first order homogeneous type. If the equation is of third order, it would be converted into second order of grade 1 (i.e., $p = n/m = 1$). For the case that $m = 0$, we can use a substitution of

$$y=e^{\int udx} \tag{3.882}$$

The method will be illustrated in the next example.

Example 3.33 Consider the second order ODE

$$x^2y\frac{d^2y}{dx^2}=x^2(\frac{dy}{dx})^2-y^2 \tag{3.883}$$

Solution: As shown earlier in (3.870), the equation is homogeneous as the total weights of all terms are $2n$ and thus it is homogeneous for all values of m and n. Substitution of (3.878) and (3.873) into (3.883) gives

$$x^2(x^p\xi)x^{p-2}[(p-1)\eta+(\eta-p\xi)\frac{d\eta}{d\xi}]=x^2(x^{2p-2}\eta^2)-x^{2p}\xi^2 \quad (3.884)$$

Simplification leads to

$$\frac{d\eta}{d\xi}=\frac{\eta}{\xi}+(\frac{\eta-\xi}{\eta-p\xi}) \quad (3.885)$$

Since (3.885) is homogeneous for all values of m and n, we can set p = 1 for simplicity. Integration of (3.885) gives

$$\eta=(\ln\xi+C_1)\xi \quad (3.886)$$

Back substitution of the definition given in (3.873) gives

$$\frac{dy}{dx}=\frac{y}{x}[\ln(\frac{y}{x})+C_1] \quad (3.887)$$

As expected, it is the homogeneous type of first order ODE. Thus, we can use the standard change of variables as

$$\frac{y}{x}=u \quad (3.888)$$

Using this change of variables and integrating, we obtain the final result as:

$$y=xe^{C_1+C_2x} \quad (3.889)$$

3.5.6 Undetermined Coefficient for Nonhomogeneous ODE

The method of undetermined coefficients discussed for second order ODEs can be easily extended to the case of higher order ODEs. All earlier discussions given in Section 3.3.3 remain valid here. That is, by examining the mathematical form of the nonhomogeneous term, we can propose a particular solution form that can provide terms that can match those given in the nonhomogeneous terms, one by one. For example, a nonhomogeneous term in polynomials must be matched by assuming the solution in polynomials. We will proceed by considering the specific examples:

Example 3.34 Consider a third order ODE with the nonhomogeneous term given in an exponential function as

$$y'''-3y''+3y'-y=4e^t \quad (3.890)$$

Solution: Note that this is a linear ODE with constant coefficients. Thus, we should try for an exponential solution as for the homogeneous ODE (i.e., ignoring the term on the right hand side of (3.890) as

$$y(t)=e^{rt} \quad (3.891)$$

Substitution of (3.891) into (3.890) gives the following characteristic equation:

$$r^3-3r^2+3r-1=0 \quad (3.892)$$

Recalling the binomial theorem, we recognize that it can be factorized as

$$(r-1)^3 = 0 \tag{3.893}$$

Thus, the solution can be expressed as

$$y_h(t) = c_1 e^t + c_2 t e^t + c_3 t^2 e^t \tag{3.894}$$

Since the nonhomogeneous term on the right hand side of (3.890) matches the triple root of the homogeneous ODE, we need to add t^3 to the trial particular function as

$$y_p(t) = At^3 e^t \tag{3.895}$$

Differentiation of (3.895) gives

$$y_p' = 3At^2 e^t + At^3 e^t \tag{3.896}$$

$$y_p'' = 6Ate^t + 6At^2 e^t + At^3 e^t \tag{3.897}$$

$$y_p''' = 6Ae^t + 18Ate^t + 9At^2 e^t + At^3 e^t \tag{3.898}$$

Substitution of these results into (3.890) gives

$$6Ae^t = 4e^t \tag{3.899}$$

Thus, the unknown constant is

$$A = \frac{2}{3} \tag{3.900}$$

Finally, the particular solution becomes

$$y_p = \frac{2}{3} t^3 e^t \tag{3.901}$$

Adding the homogeneous solution and the particular solution, we obtain

$$y(t) = c_1 e^t + c_2 t e^t + c_3 t^2 e^t + \frac{2}{3} t^3 e^t \tag{3.902}$$

Example 3.35 Consider a fourth order ODE with nonhomogeneous terms given in terms of a circular function as

$$y^{(4)} + 2y'' + y = 3\sin t - 5\cos t \tag{3.903}$$

Solution: The homogeneous solution is considered by seeking

$$y(t) = e^{rt} \tag{3.904}$$

The corresponding characteristic equation is

$$r^4 + 2r^2 + 1 = (r^2 + 1)(r^2 + 1) = 0 \tag{3.905}$$

The roots of r are i and $-i$ and both are double roots. The homogeneous solution is

$$y_h(t) = c_1 \cos t + c_2 \sin t + c_3 t \cos t + c_4 t \sin t \tag{3.906}$$

As expected, there are four unknown constants. For the particular solution, we observe that the nonhomogeneous terms do match the characteristic roots and thus we seek a particular solution in the form

$$Y(t) = t^2 (A \sin t + B \cos t) \tag{3.907}$$

Differentiation of (3.907) gives

$$Y'' = (4At + 2B - Bt^2)\cos t + (2A - 4Bt - At^2)\sin t \tag{3.908}$$

$$Y''' = (-12B - 8At + Bt^2)\cos t + (-12A + 8Bt - At^2)\sin t \tag{3.909}$$

Substitution of (3.908) and (3.909) into (3.903) gives

$$-8A\sin t - 8B\cos t = 3\sin t - 5\cos t \tag{3.910}$$

Matching the coefficients of sine and cosine on both sides gives two equations for A and B and their solutions are:

$$A = -3/8, \quad B = 5/8 \tag{3.911}$$

Therefore, the particular solution is

$$Y(t) = -\frac{3}{8}t^2 \sin t + \frac{5}{8}t^2 \cos t \tag{3.912}$$

Summation of the homogeneous solution and particular solution leads to

$$\begin{aligned} y(t) &= y_h(t) + Y(t) \\ &= c_1 \cos t + c_2 \sin t + c_3 t \cos t + c_4 t \sin t - \frac{3}{8}t^2 \sin t + \frac{5}{8}t^2 \cos t \end{aligned} \tag{3.913}$$

3.5.7 Variation of Parameters

The idea of variation of parameters was introduced by Euler but its application to higher order ODEs was done by Lagrange. In particular, we consider the following linear n-th order ODE with a nonhomogeneous term:

$$y^{(n)} + p_1(t)y^{(n-1)} + \cdots + p_{n-1}(t)y' + p_n(t)y = g(t) \tag{3.914}$$

Let us assume that the homogeneous solution is known:

$$y_h(t) = c_1 y_1(t) + c_2 y_2(t) + \cdots + c_n y_n(t) \tag{3.915}$$

Using variation of parameters, we assume the particular solution for (3.914) as

$$y_p(t) = u_1(t)y_1(t) + u_2(t)y_2(t) + \cdots + u_n(t)y_n(t) \tag{3.916}$$

Differentiation of (3.914) gives

$$y_p{}'(t) = (u_1'y_1 + u_2'y_2 + \cdots + u_n'y_n) + (u_1y_1' + u_2y_2' + \cdots + u_ny_n') \tag{3.917}$$

Following our proof of the second order, Lagrange set the following group to zero:

$$u_1'y_1 + u_2'y_2 + \cdots + u_n'y_n = 0 \tag{3.918}$$

This is the first equation governing the first derivative of the unknown functions u_i. Differentiation of (3.917) with condition (3.918) gives

$$y_p{}'' = (u_1'y_1' + u_2'y_2' + \cdots + u_n'y_n') + (u_1y_1'' + u_2y_2'' + \cdots + u_ny_n'') \tag{3.919}$$

We again set the first bracket term on the right hand side of (3.885) to zero

$$u_1'y_1' + u_2'y_2' + \cdots + u_n'y_n' = 0 \tag{3.920}$$

This is the second equation governing the first derivative of the unknown functions u_i. We can repeat this differentiation procedure. In summary, up to differentiation of $n-1$ times we have the following relations:

$$u_1'y_1^{(k-1)} + u_2'y_2^{(k-1)} + \cdots + u_n'y_n^{(k-1)} = 0, \quad k = 1, \ldots, n-2 \tag{3.921}$$

$$y_p^{(k)} = u_1 y_1^{(k)} + \cdots + u_n y_n^{(k)}, \quad k = 0, 1, \ldots, n-1 \tag{3.922}$$

For the n-th differentiation, we have

$$y_p^{(n)} = \left(u_1' y_1^{(n-1)} + \cdots + u_n' y_n^{(n-1)} \right) + \left(u_1 y_1^{(n)} + \cdots + u_n y_n^{(n)} \right) \tag{3.923}$$

Substitution of (3.922) and (3.923) into (3.914) gives

$$\begin{aligned} &\left(u_1' y_1^{(n-1)} + \cdots + u_n' y_n^{(n-1)} \right) + \left(u_1 y_1^{(n)} + \cdots + u_n y_n^{(n)} \right) \\ &+ p_1(t) \left(u_1 y_1^{(n-1)} + \cdots + u_n y_n^{(n-1)} \right) + \cdots \\ &+ p_{n-1}(t) \left(u_1 y_1' + \cdots + u_n y_n' \right) + p_n(t) \left(u_1 y_1 + \cdots + u_n y_n \right) = g(t) \end{aligned} \tag{3.924}$$

Grouping all terms of u_i, we get

$$\begin{aligned} &u_1 \left(y_1^{(n)} + p_1 y_1^{(n-1)} + \cdots + p_n y_1 \right) + u_2 \left(y_2^{(n)} + p_1 y_2^{(n-1)} + \cdots + p_n y_2 \right) \\ &+ \ldots + u_n \left(y_n^{(n)} + p_1 y_n^{(n-1)} + \cdots + p_n y_n \right) + \left(u_1' y_1^{(n-1)} + \cdots + u_n' y_n^{(n-1)} \right) = g(t) \end{aligned} \tag{3.925}$$

Since y_k ($k = 1,2,\ldots,n$) are the homogeneous solutions of (3.914), we must have all the bracket terms in (3.925) identically zero, except for the last one on the left hand side of (3.925). This provides the last equation for the first derivative of u_i as:

$$u_1' y_1^{(n-1)} + \cdots + u_n' y_n^{(n-1)} = g \tag{3.926}$$

In summary, we have n equations for n unknowns:

$$\begin{aligned} u_1' y_1 + \cdots + u_n' y_n &= 0 \\ u_1' y_1' + \cdots + u_n' y_n' &= 0 \\ &\vdots \\ u_1' y_1^{(n-1)} + \cdots + u_n' y_n^{(n-1)} &= g \end{aligned} \tag{3.927}$$

The system can be expressed in matrix form:

$$\begin{bmatrix} y_1 & \cdots & y_n \\ \vdots & \cdots & \vdots \\ y_1^{(n-1)} & \cdots & y_n^{(n-1)} \end{bmatrix} \begin{Bmatrix} u_1' \\ \vdots \\ u_n' \end{Bmatrix} = \begin{Bmatrix} 0 \\ \vdots \\ 1 \end{Bmatrix} g \tag{3.928}$$

The solution for each unknown can be solved by Cramer's rule as:

$$u_k'(t) = \frac{g(t) W_k(t)}{W(t)}, \quad \text{where } W(t) = W(y_1, \ldots, y_n)(t) \tag{3.929}$$

Integration of (3.929) gives the unknown functions defined in (3.916)

$$u_k(t) = \int_{t_0}^{t} \frac{g(s) W_k(s)}{W(s)} ds, \quad k = 1, \ldots, n \tag{3.930}$$

Finally, the particular solution is obtained as

$$y_p(t) = \sum_{k=1}^{n} \left[\int_{t_0}^{t} \frac{g(s) W_k(s)}{W(s)} ds \right] y_k(t) \tag{3.931}$$

The homogeneous solution (3.915) can be added to this particular solution to obtain the general solution for (3.914).

Example 3.36 Consider the following third order ODE with its homogeneous solutions given:

$$y''' - y'' - y' + y = g(t),$$
$$y_1(t) = e^t,\ y_2(t) = te^t,\ y_3(t) = e^{-t} \tag{3.932}$$

Solution: We now apply Lagrange's variation of parameters technique. For the present case $n = 3$, we have the particular solution as:

$$y_p(t) = \sum_{k=1}^{3} \left[\int_{t_0}^{t} \frac{g(s)W_k(s)}{W(s)} ds \right] y_k(t) \tag{3.933}$$

Recall the definition of the Wronskian

$$W(y_1, y_2, \dots, y_n)(t_0) = \begin{vmatrix} y_1(t_0) & y_2(t_0) & \cdots & y_n(t_0) \\ y_1'(t_0) & y_2'(t_0) & \cdots & y_n'(t_0) \\ \vdots & \vdots & \ddots & \vdots \\ y_1^{(n-1)}(t_0) & y_2^{(n-1)}(t_0) & \cdots & y_n^{(n-1)}(t_0) \end{vmatrix} \neq 0 \tag{3.934}$$

Substitution of the homogeneous solutions into (3.934) gives

$$W(t) = \begin{vmatrix} e^t & te^t & e^{-t} \\ e^t & (t+1)e^t & -e^{-t} \\ e^t & (t+2)e^t & e^{-t} \end{vmatrix} = 4e^t \tag{3.935}$$

Replacing the first column by (0,0,1) we have

$$W_1(t) = \begin{vmatrix} 0 & te^t & e^{-t} \\ 0 & (t+1)e^t & -e^{-t} \\ 1 & (t+2)e^t & e^{-t} \end{vmatrix} = -2t - 1 \tag{3.936}$$

Similarly, we can find W_2 and W_3 as

$$W_2(t) = \begin{vmatrix} e^t & 0 & e^{-t} \\ e^t & 0 & -e^{-t} \\ e^t & 1 & e^{-t} \end{vmatrix} = 2 \tag{3.937}$$

$$W_3(t) = \begin{vmatrix} e^t & te^t & 0 \\ e^t & (t+1)e^t & 0 \\ e^t & (t+2)e^t & 1 \end{vmatrix} = e^{2t} \tag{3.938}$$

Substitution of these results into (3.933) gives

$$y_p(t)=\sum_{k=1}^{3}\left[\int_{t_0}^{t}\frac{g(s)W_k(s)}{W(s)}ds\right]y_k(t)$$
$$=e^t\int_{t_0}^{t}\frac{g(s)(-2s-1)}{4e^s}ds+te^t\int_{t_0}^{t}\frac{g(s)2}{4e^s}ds+e^{-t}\int_{t_0}^{t}\frac{g(s)e^{2s}}{4e^s}ds \quad (3.939)$$
$$=\frac{1}{4}\int_{t_0}^{t}\left[e^{t-s}\left(-1+2(t-s)\right)+e^{-(t-s)}\right]g(s)ds$$

The final solution given in (3.939) is a function of g and integration can be conducted once this nonhomogeneous term is given.

3.5.8 Reduction to Lower Order

We now consider a special form of n-th order ODE that all derivatives of order lower than k do not appear in the ODE. Mathematically, we have

$$F(t,x^{(k)},x^{(k+1)},\cdots,x^{(n)})=0 \quad (3.940)$$

where x is the unknown function and t is the variable. Naturally, this suggests a change of variables that the lowest order derivative term is the new unknown as:

$$x^{(k)}=y \quad (3.941)$$

It is clear that now the new differential equation is of order n–k and appears as

$$F(t,y,y',\cdots,y^{(n-k)})=0 \quad (3.942)$$

Suppose that the solution of this k-th ODE can be solved, and the solution is written symbolically as:

$$y=\phi(t,c_1,\cdots,c_{n-k}) \quad (3.943)$$

Then, back substitution of (3.394) into (3.941) gives a k-th order ODE:

$$x^{(k)}=\phi(t,c_1,\cdots,c_{n-k}) \quad (3.944)$$

Note, however, that now the right hand side is only a function of t. Therefore, it can be solved by integrating the right hand side k times as:

$$x=\psi(t,c_1,\cdots,c_n) \quad (3.945)$$

The following example will illustrate its usefulness.

Example 3.37 Solve the following fifth order ODE:

$$\frac{d^5x}{dt^5}-\frac{1}{t}\frac{d^4x}{dt^4}=0 \quad (3.946)$$

Solution: This is a linear first order nonhomogeneous ODE.

$$\frac{d^4x}{dt^4}=y \quad (3.947)$$

$$\frac{dy}{dt}-\frac{1}{t}y=0 \quad (3.948)$$

This can be integrated readily to get

$$y = ct \tag{3.949}$$

Back substitution of the definition of y gives

$$\frac{d^4x}{dt^4} = ct \tag{3.950}$$

Finally, integration of (3.950) yields the solution

$$x = c_1t^5 + c_2t^3 + c_3t^2 + c_4t + c_5 \tag{3.951}$$

Example 3.38 Solve the following nonlinear ODE

$$x\frac{d^2x}{dt^2} - (\frac{dx}{dt})^2 = 0 \tag{3.952}$$

Solution: We can reduce the order by assuming

$$\frac{dx}{dt} = y \tag{3.953}$$

Substitution of (3.953) into (3.952) gives

$$(x\frac{dy}{dx} - y)y = 0 \tag{3.954}$$

There are two solutions for y and they are

$$y = 0, \quad \frac{dy}{dx} = \frac{y}{x} \tag{3.955}$$

The solution of the second equation of (3.955) is

$$y = c_1x \tag{3.956}$$

Back substitution of (3.953) into (3.956) yields

$$\frac{dx}{dt} = c_1x \tag{3.957}$$

Integration gives the final solution as

$$x = c_2e^{c_1t} \tag{3.958}$$

Another type of reduction of order for an n-th ODE will be considered in Section 3.5.14 for the case of autonomous differential equations (i.e., variable does not appear explicitly in the ODE).

3.5.9 Exact Condition

Consider the most general form of linear ODE of order n

$$p_0(x)y^{(n)} + p_1(x)y^{(n-1)} + ... + p_n(x)y = f(x) \tag{3.959}$$

The adjoint of this n-th order ODE is

$$(-1)^n \frac{d^n(p_0 v)}{dx^n} + (-1)^{n-1} \frac{d^{n-1}(p_1 v)}{dx^{n-1}} + ... - \frac{d(p_{n-1} v)}{dx} + p_n v = 0 \tag{3.960}$$

This adjoint problem has been discussed in Section 3.5.2. The adjoint ODE actually is the governing equation for the integrating factor v. Thus, if an n-th order ODE is exact we must have $v = 1$, and with this information we have the following condition for the coefficients to satisfy

$$(-1)^n \frac{d^n p_0}{dx^n} + (-1)^{n-1} \frac{d^{n-1} p_1}{dx^{n-1}} + ... - \frac{dp_{n-1}}{dx} + p_n = 0 \tag{3.961}$$

This is the condition for an n-th order ODE to be exact.

3.5.10 Factorization of ODE

The factorization technique considered here closely relates to the symbolic methods reported in some textbooks on differential equations. This symbolic technique was developed by Boole 1859 and Lobatto in 1837. Consider an n-th order ODE with constant coefficients

$$(\frac{d^n}{dx^n} + A_1 \frac{d^{n-1}}{dx^{n-1}} + A_2 \frac{d^{n-2}}{dx^{n-2}} + ... + A_n)u = f(x) \tag{3.962}$$

For the homogeneous case, assuming an exponential function, we arrive at a characteristic equation

$$r^n + A_1 r^{n-1} + A_2 r^{n-2} + ... + A_n = 0 \tag{3.963}$$

We can factorize (3.962) as

$$(\frac{d}{dx} - a_1)(\frac{d}{dx} - a_1)\cdots(\frac{d}{dx} - a_n)u = f(x) \tag{3.964}$$

where $a_i = 1,2,..., n$ are the roots of (3.963). The homogeneous solution is

$$u = C_1 e^{a_1 x} + C_2 e^{a_2 x} + ... + C_n e^{a_n x} \tag{3.965}$$

To consider the particular solution, we first consider the case with only one differential operator:

$$(\frac{d}{dx} - a)u = f(x) \tag{3.966}$$

This is the most general linear form of first order ODE discussed in Section 3.2.6, and its solution is

$$u = e^{ax} \int e^{-ax} f dx \tag{3.967}$$

This can be solved symbolically as the inverse of the differential operator

$$u = (\frac{d}{dx} - a)^{-1} f = e^{ax} \int e^{-ax} f dx \tag{3.968}$$

For the homogeneous case that $f = 0$, we can define

$$u = (\frac{d}{dx} - a)^{-1} 0 = Ce^{ax} \tag{3.969}$$

For the cases of two factorized operators, we considered the following second order ODE

$$\frac{d^2u}{dx^2}-(a+b)\frac{du}{dx}+abu=f \tag{3.970}$$

This can be factorized easily as

$$(\frac{d}{dx}-a)(\frac{d}{dx}-b)u=f(x) \tag{3.971}$$

The solution of u can be given symbolically as

$$u=(\frac{d}{dx}-b)^{-1}(\frac{d}{dx}-a)^{-1}f=\{(\frac{d}{dx}-a)(\frac{d}{dx}-b)\}^{-1}f \tag{3.972}$$

To solve this, in principle, we can apply (3.967) successively to (3.972). Alternatively, a symbolic approach based on rational function decomposition was developed by Gregory. That is, the solution can be written such that the inverse process can be applied to each term of (3.964) one-by-one:

$$\begin{aligned} u&=\{(\frac{d}{dx}-a_1)(\frac{d}{dx}-a_2)\cdots(\frac{d}{dx}-a_n)\}^{-1}f \\ &=\left\{N_1(\frac{d}{dx}-a_1)^{-1}+N_2(\frac{d}{dx}-a_2)^{-1}+...+N_n(\frac{d}{dx}-a_n)^{-1}\right\}f \end{aligned} \tag{3.973}$$

where N_i $(i = 1,2,..., n)$ can be found by algebraic means of partial fraction. If r roots are equal to a and the rest are distinct, we have

$$\begin{aligned} u&=\{(\frac{d}{dx}-a)^r(\frac{d}{dx}-a_{r+1})\cdots(\frac{d}{dx}-a_n)\}^{-1}f \\ &=\{N_1(\frac{d}{dx}-a)^{-r}+N_2(\frac{d}{dx}-a)^{-r+1}+... \\ &\quad +N_r(\frac{d}{dx}-a)^{-1}+N_{r+1}(\frac{d}{dx}-a_{r+1})^{-1}+...+N_n(\frac{d}{dx}-a_n)^{-1}\}f \end{aligned} \tag{3.974}$$

For the case of repeated roots, by applying (3.967) r times we have

$$(\frac{d}{dx}-a)^{-r}f=e^{ax}\int\cdots\int e^{-ax}f dx\cdots dx \tag{3.975}$$

Using (3.971) and (3.972) as an illustration, we assume $u = e^{mx}$ and this leads to the following considerations:

$$\frac{1}{(m-a)(m-b)}=\frac{N_1}{(m-a)}+\frac{N_2}{(m-b)}=\frac{(m-b)N_1+(m-a)N_2}{(m-a)(m-b)} \tag{3.976}$$

Equating the constant and m order term, we have

$$N_1+N_2=0, \quad bN_1+aN_2=-1 \tag{3.977}$$

The solutions of this set of equations are

$$N_1=\frac{1}{a-b}, \quad N_2=-\frac{1}{a-b} \tag{3.978}$$

To see the validity of this algebraic approach, we can consider the actual differential operator:

$$\{(\frac{d}{dx}-a)(\frac{d}{dx}-b)\}^{-1}f = N_1(\frac{d}{dx}-a)^{-1}f + N_2(\frac{d}{dx}-b)^{-1}f \tag{3.979}$$

Applying the original differential operator on both sides of (3.979), the left hand side is f and the right hand side becomes

$$\begin{aligned} RHS &= N_1(\frac{d}{dx}-a)(\frac{d}{dx}-b)(\frac{d}{dx}-a)^{-1}f + N_2(\frac{d}{dx}-a)(\frac{d}{dx}-b)(\frac{d}{dx}-b)^{-1}f \\ &= N_1(\frac{d}{dx}-b)f + N_2(\frac{d}{dx}-a)f = N_1\frac{df}{dx} - N_1 bf + N_2\frac{df}{dx} - N_2 af \\ &= (N_1+N_2)\frac{df}{dx} - (N_1 b + N_2 a)f \end{aligned} \tag{3.980}$$

To get back the original nonhomogeneous function, we must set

$$N_1 + N_2 = 0, \quad bN_1 + aN_2 = -1 \tag{3.981}$$

These are precisely what we got in the partial fraction analysis for (3.977). This is the reason why we can replace the inverse differential operator by algebraic analysis.

Return to our second order derivative problem given in (3.971)

$$\begin{aligned} u &= \{(\frac{d}{dx}-a)(\frac{d}{dx}-b)\}^{-1}f = \frac{1}{a-b}\{(\frac{d}{dx}-a)^{-1}f - (\frac{d}{dx}-b)^{-1}f\} \\ &= \frac{1}{a-b}\{e^{ax}\int e^{-ax} f dx - e^{bx}\int e^{-bx} f dx\} \end{aligned} \tag{3.982}$$

This completes the procedure of symbolic analysis for the factorized form.

Example 3.39 Find the particular solution of the following fourth order differential equation using the factorization and symbolic technique

$$\frac{d^4y}{dx^4} + 4\frac{d^3y}{dx^3} + 3\frac{d^2y}{dx^2} - 4\frac{dy}{dx} - 4y = f(x) \tag{3.983}$$

Solution: The characteristic equation of (3.983) is

$$m^4 + 4m^3 + 3m^2 - 4m - 4 = 0 \tag{3.984}$$

This can be factorized as

$$(m+2)^2(m-1)(m+1) = 0 \tag{3.985}$$

According to (3.964), the differential equation can be factorized as

$$(\frac{dy}{dx}+2)^2(\frac{dy}{dx}+1)(\frac{dy}{dx}-1)y = f(x) \tag{3.986}$$

Applying symbolic analysis, we have

$$y = \left\{(\frac{dy}{dx}+2)^2(\frac{dy}{dx}+1)(\frac{dy}{dx}-1)\right\}^{-1} f(x) \tag{3.987}$$

The associated algebraic analysis using partial fractions is

$$\frac{1}{(m+2)^2(m+1)(m-1)} = \frac{4m+11}{9(m+2)^2} - \frac{1}{2(m+1)} + \frac{1}{18(m-1)} \tag{3.988}$$

Therefore, the final solution is given as

$$y = \frac{1}{9}(4\frac{d}{dx}+11)e^{-2x}\iint e^{2x} f dx dx - \frac{1}{2}e^{-x}\int e^{x} f dx + \frac{1}{18}e^{x}\int e^{-x} f dx \quad (3.989)$$

For the case of non-constant coefficients, the symbolic analysis can be extended to the case of an arbitrary differential operator instead of *d/dx*. In particular, we can write

$$(\Pi^n + A_1\Pi^{n-1} + A_2\Pi^{n-2} + ... + A_n)u = f(x) \quad (3.990)$$

where the operator Π satisfies the following rules:

$$\Pi(au) = a\Pi(u),$$
$$\Pi(u+v) = \Pi(u) + \Pi(v), \quad (3.991)$$
$$\Pi^m \Pi^n(u) = \Pi^{m+n}(u)$$

Then, the solution can be expressed as

$$u = (\Pi^n + A_1\Pi^{n-1} + A_2\Pi^{n-2} + ... + A_n)^{-1} f(x) \quad (3.992)$$

Using symbolic method, the solution can be rewritten as

$$u = N_1(\Pi - a_1)^{-1} f + N_2(\Pi - a_2)^{-1} f + ... + N_1(\Pi - a_n)^{-1} f \quad (3.993)$$

This method is demonstrated in the following example.

Example 3.40 Solve the following second order ODE with non-constant coefficients by factorization:

$$\frac{d^2u}{dx^2} - (2x+1)\frac{du}{dx} + (x^2 + x - 1)u = 0 \quad (3.994)$$

Solution: Let us define the operator Π as:

$$\Pi u = \frac{du}{dx} - \alpha(x)u, \quad (\Pi - 1)u = \frac{du}{dx} - \alpha(x)u - u \quad (3.995)$$

We now consider that

$$\begin{aligned} \Pi(\Pi-1)(u) &= (\frac{d}{dx} - \alpha)(\frac{du}{dx} - \alpha u - u) \\ &= \frac{d^2u}{dx^2} - (2\alpha+1)\frac{du}{dx} + (\alpha^2 + \alpha - \alpha')u \end{aligned} \quad (3.996)$$

Therefore, we recognize that if $\alpha = x$, we recover (3.994). Therefore, we have

$$u = (\Pi - 1)^{-1}0 - \Pi^{-1}0 \quad (3.997)$$

The corresponding equations are

$$(\Pi - 1)y = \frac{dy}{dx} - (x+1)y = 0, \quad (\Pi)z = \frac{dz}{dx} - xz = 0 \quad (3.998)$$

The solution is

$$u = C_1 e^{(x+1)^2/2} + C_2 e^{x^2/2} \quad (3.999)$$

3.5.11 Symbolic Method for Nonhomogeneous ODE

This idea of using the algebraic symbolic method to differential calculus can be traced back to the time of Leibniz when he discovered the analogy between his n-th differentiation formula for two functions and the binomial theorem (see Eq. (1.27)). Subsequently, in 1774 Lagrange also found an algebraic analogy between Taylor series expansion and an exponential function of differential operator:

$$u(x+h)=\sum_{n=0}^{\infty}\frac{h^n}{n!}\frac{d^n u}{dx^n},\quad \Delta h(u)=(e^{h\frac{d}{dx}}-1)u \tag{3.1000}$$

Cauchy reported the symbolic calculus by Brisson in 1821 and 1823 (both papers have been lost). Then, symbolic methods branched into the British school (with G. Boole, Gregory, de Morgan, Carmichael, etc.) and the French school (with Arbogast, Francais, Cauchy, Laplace, etc.). The application of symbolic algebra to differential equations with variable coefficients was made by Boole in 1844, and the approach was summarized in a book by Carmichael in 1855.

In particular, differentiation can be defined as:

$$D=\frac{d}{dx},\quad D^2=\frac{d^2}{dx^2},\quad D^3=\frac{d^3}{dx^3},\cdots,D^n=\frac{d^n}{dx^n} \tag{3.1001}$$

The use of the symbol D gives this method another name, the so-called "D-operator method." Using this notation, we observe that

$$De^{ax}=ae^{ax},\quad D^2e^{ax}=a^2e^{ax},\ldots,\quad D^n=a^ne^{ax} \tag{3.1002}$$

Let us define a general differential operator as:

$$\begin{aligned}F(D)e^{ax}&=(p_0D^n+p_1D^{n-1}+\ldots+p_{n-1}D+p_n)e^{ax}\\&=e^{ax}F(a)\end{aligned} \tag{3.1003}$$

Therefore, differentiation of e^{ax} with respect to x is done by setting D in F to a. Next, we consider the differentiation of another functional form:

$$D(e^{ax}V)=D(e^{ax})V+e^{ax}D(V) \tag{3.1004}$$

where V is a function of x. Similarly, higher derivatives can be evaluated as

$$D^2(e^{ax}V)=D^2(e^{ax})V+2D(e^{ax})D(V)+e^{ax}D^2(V) \tag{3.1005}$$

$$D^3(e^{ax}V)=D^3(e^{ax})V+3D^2(e^{ax})D(V)+3D(e^{ax})D^2(V)+e^{ax}D^3(V) \tag{3.1006}$$

More generally, we can use the Leibniz formula in (1.27) as:

$$\begin{aligned}D^n(e^{ax}V)&=D^n(e^{ax})V+nD^{n-1}(e^{ax})D(V)+\frac{1}{2}n(n-1)D^{n-2}(e^{ax})D^2(V)\\&+\ldots+e^{ax}D^n(V)\end{aligned} \tag{3.1007}$$

Carrying out the differentiation with respect to the exponential function, we find

$$\begin{aligned}D^n(e^{ax}V)&=a^ne^{ax}V+na^{n-1}e^{ax}D(V)+\frac{1}{2}n(n-1)a^{n-2}e^{ax}D^2(V)+\ldots+e^{ax}D^n(V)\\&=e^{ax}(a^n+na^{n-1}D+\frac{1}{2}n(n-1)a^{n-2}D^2+\ldots+D^n)V\end{aligned} \tag{3.1008}$$

Therefore, we have

$$D^n(e^{ax}V) = e^{ax}(D+a)^n V \tag{3.1009}$$

The exponential function appears to shift the differential operator from D to $D+a$. The result just obtained in (3.1009) can be used in the following differential operator:

$$\begin{aligned} F(D)(e^{ax}V) &= (p_0 D^n + p_1 D^{n-1} + ... + p_{n-1} D + p_n) e^{ax} V \\ &= e^{ax}\{p_0(D+a)^n + p_1(D+a)^{n-1} + ... + p_{n-1}(D+a) + p_n\}V \\ &= e^{ax} F(D+a)V \end{aligned} \tag{3.1010}$$

This analysis can be extended to other functions as (the readers are encouraged to verify these themselves):

$$F(D^2)\cos ax = F(-a^2)\cos ax \tag{3.1011}$$

$$F(D^2)\sin ax = F(-a^2)\sin ax \tag{3.1012}$$

$$(D-a)^2\{e^{ax}V\} = e^{ax}(D-a+a)^2 V = e^{ax} D^2 V \tag{3.1013}$$

Consider a special form of (3.1010) as

$$\begin{aligned} (D-a)^p \phi(D)\{\frac{e^{ax}}{\phi(a)}\frac{x^p}{p!}\} &= \phi(D)\left[(D-a)^p\{\frac{e^{ax}}{\phi(a)}\frac{x^p}{p!}\}\right] \\ &= \phi(D)[\frac{e^{ax}}{\phi(a)} D^p\{\frac{x^p}{p!}\}] \\ &= \phi(D)[\frac{e^{ax}}{\phi(a)}] = e^{ax} \end{aligned} \tag{3.1014}$$

Note that we have used (3.1010) in getting (3.1014). We now introduce the strangest notation of this symbolic approach. In particular, it is obvious that

$$Dx = 1 \tag{3.1015}$$

We introduce the algebraic form of writing its inverse as

$$x = \frac{1}{D}(1) \tag{3.1016}$$

Recall that D is a differential operator, but not a coefficient. However, using the symbolic approach, we just divide through by the operator D as we would do for algebra. Similarly, we can extend the idea to the following form

$$D^p\{\frac{x^p}{p!}\} = \frac{p!}{p!} = 1 \tag{3.1017}$$

Thus, algebraically we can write:

$$\frac{x^p}{p!} = \frac{1}{D^p}(1) \tag{3.1018}$$

Let us now proceed to consider the symbolic method in solving a nonhomogeneous ODE as

$$F(D)y = f(x) \tag{3.1019}$$

We now propose to write the particular solution of (3.1019) as

$$y = \frac{1}{F(D)} f(x) \tag{3.1020}$$

Note that we have treated the operator D as if it were an ordinary algebraic quantity. In this approach, we will follow all plausible algebraic operations and then check the final result by differentiation.

Case (i) For $f(x) = e^{ax}$, (3.1003) suggests that

$$\frac{1}{F(D)} e^{ax} = \frac{1}{F(a)} e^{ax} \tag{3.1021}$$

We can verify this by differentiating (3.1021) as

$$F(D)\{\frac{1}{F(a)} e^{ax}\} = \frac{e^{ax} F(a)}{F(a)} = e^{ax} \tag{3.1022}$$

This confirms our usage of symbolic algebra for solving (3.1019).

Case (ii) For $f(x) = e^{ax}$ and $F(D) = (D-a)^p \phi(D)$, the symbolic method suggests that

$$\frac{1}{F(D)} e^{ax} = \frac{1}{(D-a)^p \phi(D)} e^{ax} = \frac{1}{(D-a)^p} \{\frac{e^{ax}}{\phi(a)}\} = \{\frac{e^{ax}}{\phi(a)}\} \frac{1}{(D)^p}$$
$$= \{\frac{e^{ax}}{\phi(a)} \frac{x^p}{p!}\} \tag{3.1023}$$

In obtaining the last of (3.1023), we have taken $1/D$ as integration as defined in (3.1018). To verify this result, we can differentiate the result by using (3.1017):

$$(D-a)^p \phi(D)\{\frac{e^{ax}}{\phi(a)} \frac{x^p}{p!}\} = e^{ax} \tag{3.1024}$$

Thus, the result is verified. We will illustrate the method by the following examples.

Example 3.41 Solve the following second order ODE

$$(D+3)^2 y = 50e^{2x} \tag{3.1025}$$

Solution: The particular solution is

$$y_p = \frac{1}{(D+3)^2} 50e^{2x} = \frac{1}{(2+3)^2} 50e^{2x} = 2e^{2x} \tag{3.1026}$$

The last of (3.1026) is suggested by the result of (3.1021). It is easy to verify the validity of (3.1026). Adding the homogeneous solution gives

$$y = (A+Bx)e^{-3x} + 2e^{2x} \tag{3.1027}$$

Example 3.42 Solve the following second order ODE

$$(D-2)^2 y = 50e^{2x} \tag{3.1028}$$

Solution: The particular solution is

$$y_p = \frac{1}{(D-2)^2} 50e^{2x} \tag{3.1029}$$

For this case, we cannot apply (3.1021) or directly substitute $D = 2$ because it will lead to infinity. Instead, we can apply (3.1013) and (3.1018) to get

$$y_p = \frac{1}{(D-2)^2} 50e^{2x} = 50e^{2x} \frac{1}{D^2} = 50e^{2x} (\frac{1}{2} x^2) = 25x^2 e^{2x} \tag{3.1030}$$

It is easy to verify the validity of (3.1030). Adding the homogeneous solution gives

$$y = (A + Bx)e^{2x} + 25x^2 e^{2x} \tag{3.1031}$$

For this case, the nonhomogeneous term matches the repeated root of the homogeneous ODE. Thus, similar to the method of undetermined coefficients that we discussed earlier, special treatment is needed in obtaining our particular solution in (3.1030).

Example 3.43 Find the general solution of the following second order ODE

$$(D^2 + 3D + 2)y = \cos 2x \tag{3.1032}$$

Solution: Using the symbolic method and employing (3.1011), we have

$$y_p = \frac{1}{(D^2 + 3D + 2)} \cos 2x = \frac{1}{(-4 + 3D + 2)} \cos 2x = \frac{1}{(3D - 2)} \cos 2x \tag{3.1033}$$

Note that we can only substitute the value of D^2 but not D (see (3.1011)). Next, (3.1033) can be simplified as

$$\begin{aligned} y_p &= \frac{3D + 2}{(9D^2 - 4)} \cos 2x = \frac{3D + 2}{(-9 \times 4 - 4)} \cos 2x \\ &= -\frac{1}{40} (3D \cos 2x + 2 \cos 2x) \\ &= \frac{1}{20} (3 \sin 2x - \cos 2x) \end{aligned} \tag{3.1034}$$

Note that we can employ (3.1011) again in (3.1034).

Example 3.44 Find the particular solution of the following third order ODE

$$(D^3 + 6D^2 + 11D + 6)y = 2 \sin 3x \tag{3.1035}$$

Solution: Using the symbolic method and employing (3.1012), we have

$$\begin{aligned} y_p &= \frac{1}{(D^3 + 6D^2 + 11D + 6)} 2 \sin 3x = \frac{2}{(-9D - 54 + 11D + 6)} \sin 3x \\ &= \frac{1}{(D - 24)} \sin 3x \end{aligned} \tag{3.1036}$$

Again, note that we can only substitute the value of D^2 but not D. Thus, we proceed like the last example to get

$$y_p = \frac{1}{(D-24)}\sin 3x = \frac{D+24}{(D^2-576)}\sin 3x$$
$$= \frac{D+24}{-9-576}\sin 3x = -\frac{1}{585}(3\cos 3x + 24\sin 3x) \qquad (3.1037)$$
$$= -\frac{1}{195}(\cos 3x + 8\sin 3x)$$

If the nonhomogeneous term is given as a power of x, we should expand the "inverted" differential operator as:

$$F(D)y = x^m \qquad (3.1038)$$

$$y_p = \frac{1}{F(D)}x^m = (a_0 + a_1 D + a_2 D^2 + ...)x^m \qquad (3.1039)$$

Let us illustrate this with examples.

Example 3.45 Solve the following third order ODE by the symbolic method

$$(D^2+4)y = x^2 \qquad (3.1040)$$

Solution: Using the symbolic method and employing (3.1039), we have

$$y_p = \frac{1}{(D^2+4)}x^2 = \frac{1}{4}\frac{1}{(1+\frac{1}{4}D^2)}x^2 = \frac{1}{4}(1-\frac{1}{4}D^2 + \frac{1}{16}D^4 + ...)x^2 \qquad (3.1041)$$
$$= \frac{1}{4}(x^2 - \frac{1}{2})$$

Therefore, the general solution becomes

$$y = y_h + y_p = A\cos 2x + B\sin 2x + \frac{1}{4}(x^2 - \frac{1}{2}) \qquad (3.1042)$$

We can see that a higher order of the Taylor series expansion is not needed since the nonhomogeneous term is only up to second degree in power. For power series of higher order, we need to retain more terms in the series expansion of the differential operator D.

Example 3.46 Solve the following third order ODE by the symbolic method

$$(D^2 - 4D + 3)y = x^3 \qquad (3.1043)$$

Solution: Using the symbolic method and employing (3.973), we have

$$y_p = \frac{1}{(D^2-4D+3)}x^3 = \frac{1}{2}(\frac{1}{1-D} - \frac{1}{3-D})x^3$$

$$= \frac{1}{2}\{(1+D+D^2+D^3+D^4+...) - \frac{1}{3}(1+\frac{D}{3}+\frac{D^2}{9}+\frac{D^3}{27}+\frac{D^4}{81}+...)\}x^3 \quad (3.1044)$$

$$= \frac{1}{3}x^3 + \frac{4}{3}x^2 + \frac{26}{9}x + \frac{80}{27}$$

Therefore, the general solution becomes

$$y = y_h + y_p = Ae^x + Be^{3x} + \frac{1}{3}x^3 + \frac{4}{3}x^2 + \frac{26}{9}x + \frac{80}{27} \quad (3.1045)$$

As illustrated, more terms are needed in Taylor series expansion for this example.

The symbolic method introduced here is also known as the "inverse differential operator," which is more efficient for finding a particular solution. If the order of a differential equation is higher than third order, the traditional technique of undetermined coefficients becomes tedious, whereas the symbolic method becomes very effective. Table 3.3 compiles some typical formulas for the symbolic method, and the differential operator $F(D)$ in Table 3.3 is defined in (3.1010).

Table 3.3 Some formulas of symbolic method

No.	ODE	g	Remark
1	$F(D)g = e^{ax}$	$g = \frac{1}{F(a)}e^{ax}$	(3.1021)
2	$F(D)g = e^{ax}Q(x)$	$g = e^{ax}\frac{1}{F(D+a)}Q(x)$	(3.1013)
3	$F(D-a)g = Q(x)$	$g = e^{ax}\int e^{-ax}Q(x)dx$	(3.967)
4	$F(D^2)g = \cos ax$	$g = \frac{1}{F(-a^2)}\cos ax$	(3.1011)
5	$F(D^2)g = \sin ax$	$g = \frac{1}{F(-a^2)}\sin ax$	(3.1012)
6	$F(D^2+a^2)g = \cos ax$	$g = \frac{x\sin ax}{2a}$	
7	$F(D^2+a^2)g = \sin ax$	$g = -\frac{x\cos ax}{2a}$	
8	$F(D-a)^n g = Q(x)$	$g = e^{ax}\iint\cdots\int e^{-ax}Q(x)dx\cdots dxdx$	

3.5.12 Removal of the Second Highest Derivative

A linear n-th order ODE can be written as

$$y^{(n)} + a_1(x)y^{(n-1)} + ... + a_{n-1}(x)y' + a_n(x)y = f(x) \tag{3.1046}$$

Let y be expressed as a product of two functions

$$y(x) = y_0(x)u(x) \tag{3.1047}$$

Applying the Leibniz rule of differentiation, we have

$$y' = y_0'u + y_0u' \tag{3.1048}$$

$$y'' = y_0''u + 2y_0'u' + y_0u'' \tag{3.1049}$$

$$y^{(n-2)} = y_0u^{(n-2)} + (n-2)y_0'u^{(n-3)} + ... \tag{3.1050}$$

$$y^{(n-1)} = y_0u^{(n-1)} + (n-1)y_0'u^{(n-2)} + ... \tag{3.1051}$$

$$y^{(n)} = y_0u^{(n)} + ny_0'u^{(n-1)} + \frac{n(n-1)}{2}y_0''u^{(n-2)} + ... \tag{3.1052}$$

Substitution of (3.1047) to (3.1052) into (3.1046) and collection of all terms up to $u^{(n-2)}$ gives

$$y_0u^{(n)} + (ny_0' + a_1y_0)u^{(n-1)} + [\frac{n(n-1)}{2}y_0'' + (n-1)a_1y_0' + a_2y_0]u^{(n-2)} + ... = f(x) \tag{3.1053}$$

To remove the second term of (3.1053), we must set its coefficient to zero

$$\frac{y_0'}{y_0} = -\frac{a_1}{n} \tag{3.1054}$$

This is separable and it can be rewritten as

$$\frac{dy_0}{y_0} = -\frac{a_1}{n}dx \tag{3.1055}$$

$$\ln y_0 = \exp[-\frac{1}{n}\int a_1 dx + c] \tag{3.1056}$$

$$y_0 = C\exp[-\frac{1}{n}\int a_1 dx] \tag{3.1057}$$

Therefore, the second highest derivative term can be removed by using the following transformation

$$y = u(x)\exp[-\frac{1}{n}\int a_1(x)dx] \tag{3.1058}$$

Example 3.47 Solve the following non-constant coefficient ODE by removing the second highest derivative term

$$\frac{dy^2}{dx^2} + 2\frac{dy}{dx} + [1 + \frac{2}{(1+3x)^2}]y = 0 \tag{3.1059}$$

Solution: We see that for this particular ODE

$$a_1 = 2 \tag{3.1060}$$

Thus, we have the following transformation

$$y = ue^{-x} \tag{3.1061}$$

The derivative terms of (3.1061) are

$$y' = (-u + u')e^{-x} \tag{3.1062}$$

$$y'' = (u - 2u' + u'')e^{-x} \tag{3.1063}$$

Substitution of (3.1061) and (3.1063) into (3.1059) gives a simplified ODE

$$u'' + \frac{2}{(1+3x)^2} u = 0 \tag{3.1064}$$

This equation can actually be considered as an extended Euler equation (see Example 3.28). The following change of variables can be assumed

$$1 + 3x = e^t, \quad or \quad t = \ln(1+3x) \tag{3.1065}$$

Thus, we have

$$\frac{dt}{dx} = \frac{3}{1+3x} = \frac{3}{e^t}, \quad u(x) = U(t) \tag{3.1066}$$

$$u' = \frac{du}{dx} = \frac{dU}{dt}\frac{dt}{dx} = \frac{3}{e^t}\frac{dU}{dt} \tag{3.1067}$$

$$\begin{aligned} u'' &= \frac{d^2u}{dx^2} = \frac{d}{dt}\{\frac{3}{e^t}\frac{dU}{dt}\}\frac{dt}{dx} = \frac{3}{e^t}\{-\frac{3}{e^t}\frac{dU}{dt} + \frac{3}{e^t}\frac{d^2U}{dt^2}\} \\ &= \frac{1}{e^{2t}}\left\{9\frac{d^2U}{dt^2} - 9\frac{dU}{dt}\right\} \end{aligned} \tag{3.1068}$$

Finally, the differential equation is converted to an ODE with constant coefficients

$$9\frac{d^2U}{dt^2} - 9\frac{dU}{dt} + 2U = 0 \tag{3.1069}$$

Following the standard procedure of assuming that an exponential solution leads to the characteristic equation

$$9\lambda^2 - 9\lambda + 2 = 0 \tag{3.1070}$$

$$(\lambda - \frac{1}{3})(\lambda - \frac{2}{3}) = 0, \tag{3.1071}$$

the solution of U is

$$U = Ae^{t/3} + Be^{2t/3} \tag{3.1072}$$

Substitution of this result into the definition of change of variables given in (3.1065) and (3.1061) gives the final solution

$$u = [A(1+3x)^{1/3} + B(1+3x)^{2/3}] \tag{3.1073}$$

$$y = [A(1+3x)^{1/3} + B(1+3x)^{2/3}]e^{-x} \tag{3.1074}$$

3.5.13 Particular Forms

In this section, we will summarize some particular forms of higher order ODEs that can be solved easily. The simplest ODE of order n is

$$\frac{d^n y}{dx^n} = f(x) \tag{3.1075}$$

This ODE can be solved by applying direct integration n times as:

$$\begin{aligned} y = &\iiint \cdots \int f(x_n) dx_n dx_{n-1} \cdots dx_2 dx \\ &+ c_1 x^{n-1} + c_2 x^{n-2} + \ldots + c_n \end{aligned} \tag{3.1076}$$

Another simple form of an n-th ODE is

$$\frac{d^n y}{dx^n} = f(y) \tag{3.1077}$$

The integration of this differential equation is less obvious. Direct integration can only be applied for the case n=1, 2. For n = 1, we have

$$\frac{dy}{dx} = f(y) \tag{3.1078}$$

This is a special form of separable ODE, and thus the solution can be determined as

$$x = \int \frac{dy}{f(y)} + C \tag{3.1079}$$

For the case of n = 2, the ODE given (3.1077) becomes

$$\frac{d^2 y}{dx^2} = f(y) \tag{3.1080}$$

Multiplying both sides by $2(dy/dx)$, we have

$$2\frac{dy}{dx}\frac{d^2 y}{dx^2} = 2f(y)\frac{dy}{dx} \tag{3.1081}$$

Note that the left hand side can be recognized as

$$2\frac{dy}{dx}\frac{d^2 y}{dx^2} = \frac{d}{dx}\left(\frac{dy}{dx}\right)^2 = 2f(y)\frac{dy}{dx} \tag{3.1082}$$

This technique actually closely relates to the evaluation of energy in physical problems. It is a very powerful mathematical technique and will be used again in later chapters. Thus, we can integrate the last one as

$$\left(\frac{dy}{dx}\right)^2 = 2\int f(y)dy + C_1 = \varphi(y) + C_1 \tag{3.1083}$$

The first derivative becomes

$$\frac{dy}{dx} = \pm\sqrt{\varphi(y) + C_1} \tag{3.1084}$$

Integrating one more time, we have

$$x = \pm \int \frac{dy}{\sqrt{\varphi(y) + C_1}} + C_2 \tag{3.1085}$$

3.5.14 Autonomous *n*-th Order ODE

In this essay, the n-th order ODE is independent of the variable x or more specifically x does not appear in the ODE except in the derivative terms. That is,

$$F(y, \frac{dy}{dx}, ..., \frac{d^n y}{dx^n}) = 0 \tag{3.1086}$$

Since none of the coefficients change with x, we call such an ODE autonomous. The functional form is independent of x. We can apply a standard change of variables as

$$\frac{dy}{dx} = p \tag{3.1087}$$

Differentiation of (3.1087) gives

$$\frac{d^2 y}{dx^2} = \frac{dp}{dx} = \frac{dp}{dy}\frac{dy}{dx} = p\frac{dp}{dy} \tag{3.1088}$$

$$\frac{d^3 y}{dx^3} = \frac{d}{dx}(\frac{d^2 y}{dx^2}) = \frac{d}{dy}(p\frac{dp}{dy})\frac{dy}{dx} = p^2\frac{d^2 p}{dy^2} + p(\frac{dp}{dy})^2 \tag{3.1089}$$

$$\begin{aligned}\frac{d^4 y}{dx^4} &= \frac{d}{dx}(\frac{d^3 y}{dx^3}) = \frac{d}{dy}[p^2\frac{d^2 p}{dy^2} + p(\frac{dp}{dy})^2]p \\ &= p^3\frac{d^3 p}{dy^3} + p(\frac{dp}{dy})^3 + 4p^2(\frac{dp}{dy})\frac{d^2 p}{dy^2}\end{aligned} \tag{3.1090}$$

$$\begin{aligned}\frac{d^5 y}{dx^5} &= \frac{d}{dx}(\frac{d^4 y}{dx^4}) = \frac{d}{dy}[p^3\frac{d^3 p}{dy^3} + p(\frac{dp}{dy})^3 + 4p^2(\frac{dp}{dy})\frac{d^2 p}{dy^2}]p \\ &= p^4\frac{d^4 p}{dy^4} + 7p^3\frac{dp}{dy}\frac{d^3 p}{dy^3} + 11p^2(\frac{dp}{dy})^2\frac{d^2 p}{dy^2} + 4p^3(\frac{d^2 p}{dy^2})^2 + p(\frac{dp}{dy})^4\end{aligned} \tag{3.1091}$$

All higher derivatives can be evaluated using a similar procedure. However, it does not appear to have a general form for $d^n y/dx^n$. However, it is clear that

$$\frac{d^n y}{dx^n} = f(p, \frac{dp}{dy}, ..., \frac{d^{n-1} p}{dy^{n-1}}) \tag{3.1092}$$

Therefore, the original ODE becomes symbolically

$$F(y, p, \frac{dp}{dy}, ..., \frac{d^{n-1} p}{dy^{n-1}}) = 0 \tag{3.1093}$$

The original unknown y now becomes a variable, and its first derivative p becomes the new unknown. Now the variable y appears explicitly in the differential equation. Therefore, the procedure used in (3.1087) can no longer be used. In addition, even when the original ODE (3.1092) is linear, a change of variable given in (3.1087) will make it highly nonlinear. Therefore, this technique is more useful if the highest order is second, and in such case, the resulting nonlinear first order ODE is more likely to be solved.

3.6 SUMMARY AND FURTHER READING

The methods for solving ODEs have occupied the minds of great mathematicians in the last three hundred years. Many of the techniques covered in the chapter were developed by Bernoulli, Euler, Fourier, Kirchhoff, Lagrange, D'Alembert, Jacobi, Laplace, Poisson, Legendre, Helmholtz, Gauss, Kummer, Clairaut, Riccati, Bessel, Boole, Hankel, Fuchs, Cauchy, Riemann, Picard, Frobenius, Stokes, Liouville, Monge, Ampere, Darboux, Goursat, Bateman, Airy, Sarrus, Kelvin, Forsyth, and many others. Yet, there remain many unsolved ODEs, especially nonlinear ODEs or ODEs with non-constant coefficients. Many naive looking ODEs require the application of substantial mathematical skills and insights to obtain their solutions. In this chapter, we only summarize some of the most notable techniques in obtaining the solutions of ODE. These skills are essential before we discuss the solution techniques for partial differential equation (PDEs). Very often, when we solve a PDE, we convert the PDE into a number of ODEs (such as the separation of variables). There are a number of handbooks on differential equations. The most comprehensive and newest ones are a series of handbooks by Polyanin and co-authors (Polyanin, 2001, Polyanin and Zaitsev, 2002 and 2003). The handbook by Zwillinger (1997) covers a number of different techniques and provides insights in solving different ODEs as well as PDEs. We suggest that readers identify and classify the ODE at hand. It is advisable to look it up in a handbook before trying to solve the differential equations by other techniques (such as a series solution technique to be covered in Chapter 4). It is quite common that an exotic differential equation can easily be transformed into one that has been solved, if appropriate change of variables is applied. Indeed, in this chapter we have repeatedly demonstrated the power of change of variables.

Many techniques covered in this chapter are not covered in most modern textbooks on differential equations. We have referred extensively to more classic textbooks on differential equations, such as Boole (1865), Forysth (1890, 1900, 1902, 1906, 1956), Goursat (1917), Bateman (1918), Piaggio (1920), Ince (1956), Sommerfeld (1949), Erdelyi (1951), and Sneddon (1957). We recommend that readers refer to these books for more classical techniques in solving ODEs.

3.7 PROBLEMS

Problem 3.1 Find the solution of the following first order ODE

$$\left\{y+\sqrt{(x^2+y^2)}\right\}dx-xdy=0 \tag{3.1094}$$

Ans:

$$x^2=c\{y+\sqrt{x^2+y^2}\} \tag{3.1095}$$

Problem 3.2 Find the solution of the following nonlinear first order ODE

$$(\frac{dy}{dx})^2=a^2-y^2 \tag{3.1096}$$

Hint: Use factorization.

Ans:

$$(x-c)^2-(\sin^{-1}\frac{y}{a})^2=0 \tag{3.1097}$$

Problem 3.3 Find the solution of the following second order ODE

$$x^2\frac{d^2y}{dx^2}-2x\frac{dy}{dx}+(x^2+2)y=0 \tag{3.1098}$$

Hint: Use removal of the second highest derivative.
Ans:

$$y=Ax\cos x+Bx\sin x \tag{3.1099}$$

Problem 3.4 Find the solution of the following second order ODE

$$\frac{d^2y}{dx^2}+\frac{2}{x}\frac{dy}{dx}-\frac{2}{(1+x)^2}y=0 \tag{3.1100}$$

Hint: Use removal of the second highest derivative.
Ans:

$$y=A\frac{1}{x(1+x)}+B\frac{(1+x)^2}{x} \tag{3.1101}$$

Problem 3.5 Find the integrating factor of the following first order ODE

$$F_1(xy)ydx+F_2(xy)xdy=0 \tag{3.1102}$$

Ans:

$$\mu=\frac{1}{xy\{F_1(xy)-F_2(xy)\}} \tag{3.1103}$$

Problem 3.6 Use the integrating factor to solve the following first order ODE

$$(x^2y^2+1)ydx+(x^2y^2-1)xdy=0 \tag{3.1104}$$

Ans:

$$\frac{1}{2}x^2y^2+\ln(\frac{x}{y})=c \tag{3.1105}$$

Problem 3.7 Use the integrating factor to solve the following first order ODE

$$(2x^3y^2-y)dx+(2x^2y^3-x)dy=0 \tag{3.1106}$$

Ans:

$$x^2+y^2+\frac{1}{xy}=c \tag{3.1107}$$

Problem 3.8 Find the integrating factor of the following first order ODE

$$(\frac{1}{y}+\sec\frac{y}{x})dx-\frac{x}{y^2}dy=0 \tag{3.1108}$$

Ans:

$$\mu = \cos(\frac{y}{x}) \tag{3.1109}$$

Problem 3.9 Find the general solution of the following n-th order ODE with the n-th repeated root, r_0:

$$p_n y^{(n)} + ... + p_3 y''' + p_2 y'' + p_1 y' + p_0 y = 0 \tag{3.1110}$$

Ans:

$$y = (C_0 + C_1 t + ... + C_n t^{n-1}) e^{r_0 t} \tag{3.1111}$$

where r satisfies the following characteristic equation

$$p_n r^n + p_{n-1} r^{n-1} + ... + p_1 r + p_0 = (r - r_0)^n = 0 \tag{3.1112}$$

Problem 3.10 By differentiating the solution obtained in Example 3.15, show that it is indeed the solution of the given ODE.

Problem 3.11 In solving a particular form of Navier-Stoke Equation for the case of viscous incompressible fluid flow, one arrives at the following form of Riccati equation in polar form (Sedov, 1993)

$$\frac{d\varphi}{d\theta} - \frac{1}{2}\varphi^2 - \varphi \cot\theta = 0 \tag{3.1113}$$

Show that by applying the following change of variables

$$\varphi = -\frac{2}{u}\frac{du}{d\theta} \tag{3.1114}$$

we can convert (3.1113) into the following second order ODE:

$$\frac{d^2 u}{d\theta^2} - \cot\theta \frac{du}{d\theta} = 0 \tag{3.1115}$$

Problem 3.12 Show that we can apply the following change of variables

$$\frac{du}{d\theta} = v \tag{3.1116}$$

to solve (3.1115) obtained in Problem 3.11 and thus obtain the solution for φ.

Ans:

$$\varphi = \frac{2\sin\theta}{A + \cos\theta} \tag{3.1117}$$

where A is an unknown constant.

Problem 3.13 Repeat the solution procedure for solving (3.1115) by using another change of variables

$$\mu = \cos^2(\frac{\theta}{2}) \tag{3.1118}$$

Ans:

$$\varphi = \frac{2\sin\theta}{A + \cos\theta} \tag{3.1119}$$

where A is an unknown constant.

Problem 3.14 Show that the solution for the following particular form of Riccati equation encountered in viscous compressible flow

$$(\frac{d\varphi}{d\theta}-\frac{1}{2}\varphi^2)\sin^2\theta-\varphi\sin\theta\cos\theta=\frac{1}{2}\cos 2\theta+1 \tag{3.1120}$$

is

$$\varphi=-[\cot\theta+\coth(\frac{\theta+\theta_0}{2})] \tag{3.1121}$$

Hint: See Chapter 3 of Sedov (1993).

Problem 3.15 Solve the following nonlinear first order ODE by differentiation:

$$y\sqrt{1+(\frac{dy}{dx})^2}=f(x+y\frac{dy}{dx}) \tag{3.1122}$$

(i) In particular, first consider that

$$x+y\frac{dy}{dx}=a, \quad y\sqrt{1+(\frac{dy}{dx})^2}=b \tag{3.1123}$$

Show that differentiation of both of these with respect to x leads to the following ODE:

$$1+(\frac{dy}{dx})^2+y\frac{d^2y}{dx^2}=0 \tag{3.1124}$$

(ii) Next, show that differentiation of (3.1122) also leads to (3.1124).

(iii) Finally, show that the solution of (3.1122) is

$$y^2+(a-x)^2=\{f(a)\}^2 \tag{3.1125}$$

where a is defined in (3.1123).

Problem 3.16 Find the differential equation for the singular solution of (3.1122).

Ans:

$$\frac{\frac{dy}{dx}}{\sqrt{1+(\frac{dy}{dx})^2}}-f'(x+y\frac{dy}{dx})=0 \tag{3.1126}$$

Problem 3.17 Find the solution of the following ODE

$$y=x\frac{dy}{dx}+m\frac{dx}{dy} \tag{3.1127}$$

Ans:

$$y=cx+\frac{m}{c} \tag{3.1128}$$

Problem 3.18 Find the solution of the following ODE

$$a\frac{d^2y}{dx^2}\frac{d^3y}{dx^3}=\sqrt{1+(\frac{d^2y}{dx^2})^2} \tag{3.1129}$$

Hint: Assume the following change of variables:

$$\frac{d^2y}{dx^2}=z \tag{3.1130}$$

Ans:

$$y=\iint\sqrt{(\frac{x-c}{a})^2-1}dxdx+c_1x+c_2 \tag{3.1131}$$

Problem 3.19 Find the solution of the following ODE

$$a^2\frac{d^4y}{dx^4}=\frac{d^2y}{dx^2} \tag{3.1132}$$

Ans:

$$y=c_1e^{x/a}+c_2e^{-x/a}+c_3x+c_4 \tag{3.1133}$$

Problem 3.20 Find the solution of the following ODE

$$ye^{xy}dx+xe^{xy}dy=0 \tag{3.1134}$$

Ans:

$$e^{xy}=c \tag{3.1135}$$

Problem 3.21 Find the solution of the following ODE

$$(x-2xy+e^y)dx+(y-x^2+xe^y)dy=0 \tag{3.1136}$$

Ans:

$$x^2+y^2-2x^2y+2xe^y=c \tag{3.1137}$$

Problem 3.22 Apply the Liouville transformation discussed in Section 3.4.7 to the following second order ODE:

$$\frac{d^2y}{dx^2}-\frac{1}{x^{1/2}}\frac{dy}{dx}+\frac{1}{4x^2}(x+x^{1/2}-8)y=0 \tag{3.1138}$$

(i) Use the following transformation

$$y=W\exp\{\frac{1}{2}\int\frac{dx}{x^{1/2}}\} \tag{3.1139}$$

Show that the governing equation for v is

$$\frac{d^2W}{dx^2}-\frac{2}{x^2}W=0 \tag{3.1140}$$

(ii) Apply a change of variables $x=e^t$, and show that the answer of (3.1138) is

$$y=\exp(x^{1/2})\{C_1x^2+\frac{C_2}{x}\} \tag{3.1141}$$

Problem 3.23 Apply Liouville transformation discussed in Section 3.4.7 to the following second order ODE:

$$(1-x^2)\frac{d^2z}{dx^2}+(1-3x)\frac{dz}{dx}+kz=0 \tag{3.1142}$$

Find the governing equation for W defined in (3.685).

Ans:

$$\frac{d^2W}{dx^2}-\frac{(1+x)(1-3x)}{4(1-x^2)^2}W=0 \tag{3.1143}$$

Problem 3.24 Apply Liouville transformation discussed in Section 3.4.7 to the following second order ODE:

$$(1-x^2)\frac{d^2\zeta}{dx^2}-(1+x)\frac{d\zeta}{dx}+(k+1)\zeta=0 \tag{3.1144}$$

Find the governing equation for W defined in (3.685).

Ans:

$$\frac{d^2W}{dx^2}-\frac{(1+x)(1-3x)}{4(1-x^2)^2}W=0 \tag{3.1145}$$

Problem 3.25 Consider two different ODEs as:

$$\frac{d^2z}{dx^2}+P(x)\frac{dz}{dx}+Q(x)z=0 \tag{3.1146}$$

$$\frac{d^2\zeta}{dx^2}+P_1(x)\frac{d\zeta}{dx}+Q_1(x)\zeta=0 \tag{3.1147}$$

(i) By observing the results from Problems 3.23 and 3.24, show that if the invariant I is the same for different ODEs, these ODEs are equivalent. More specifically, show that the invariants for them are the same if

$$I=Q_1-\frac{1}{2}\frac{dP_1}{dx}-\frac{1}{4}P_1^2=Q-\frac{1}{2}\frac{dP}{dx}-\frac{1}{4}P^2 \tag{3.1148}$$

(ii) Prove the following relation between z and ζ:

$$\zeta=z\exp\{\frac{1}{2}\int(P-P_1)dx \tag{3.1149}$$

Problem 3.26 Show that z in Problem 3.23 and ζ in Problem 3.24 are related by

$$z=\zeta(1+x) \tag{3.1150}$$

Hint: Use the result of Problem 3.25.

Problem 3.27 It will be shown in Chapter 4 that the Bessel equation is

$$\frac{d^2z}{dx^2}+\frac{1}{x}\frac{dz}{dx}+(1-\frac{n^2}{x^2})z=0 \tag{3.1151}$$

Its solutions are the Bessel function of the first and second kinds:

$$z = AJ_n(x) + BY_n(x) \tag{3.1152}$$

Find the solution of the following ODE in terms of Bessel functions

$$\frac{d^2y}{dx^2} + \alpha\frac{dy}{dx} + (1 + \frac{\alpha^2}{4} + \frac{1}{4x^2} - \frac{n^2}{x^2})y = 0 \tag{3.1153}$$

Hint: Use the result of Problem 3.25.

Ans:

$$y = (\frac{x}{e^{\alpha x}})^{1/2}[AJ_n(x) + BY_n(x)] \tag{3.1154}$$

Problem 3.28 It has been shown that

$$\frac{d^2y}{dx^2} + P(x)\frac{dy}{dx} + Q(x)y = 0 \tag{3.1155}$$

can be converted to

$$\frac{d^2v}{dx^2} + I(x)v = 0 \tag{3.1156}$$

where I is defined in (3.1148). If the solutions of this second order ODE are v_1 and v_2, and their ratio is $s = v_1/v_2$, show that

$$\frac{s'''}{s'} - \frac{3}{2}(\frac{s''}{s'})^2 = \{s, x\} = 2I \tag{3.1157}$$

where $\{s,x\}$ is the Schwarzian derivative, which is defined in (3.702).

Problem 3.29 Show that the invariant I of the adjoint ODE of

$$\frac{d^2y}{dx^2} + P(x)\frac{dy}{dx} + Q(x)y = 0 \tag{3.1158}$$

is the same as the invariant of the original ODE.

Problem 3.30 Use the symbolic method to find the particular solution of the following linear ODE

$$[\phi(D^2) + D\psi(D^2)]y = P\cos ax + Q\sin ax \tag{3.1159}$$

where ϕ and ψ are arbitrary linear differential operators.

Ans:

$$y_p = \frac{\phi(-a^2)(P\cos ax + Q\sin ax) + a\psi(-a^2)(P\sin ax - Q\cos ax)}{\{\phi(-a^2)\}^2 + a^2\{\psi(-a^2)\}^2} \tag{3.1160}$$

Problem 3.31 Find the integrating factor by inspection and solve the following ODE

$$(x^2 + y^2)dx - 2xydy = 0 \tag{3.1161}$$

Ans:

$$x^2 - y^2 = Cx \tag{3.1162}$$

Problem 3.32 Find the integrating factor by inspection and solve the following ODE

$$(x^2 - y^2)dx + 2xydy = 0 \tag{3.1163}$$

Ans:

$$\ln x + \frac{y^2}{x} = C \tag{3.1164}$$

Problem 3.33 Find the integrating factor by inspection and solve the following ODE

$$xdy - ydx = (x^2 + y^2)dx \tag{3.1165}$$

Ans:

$$\tan^{-1}(\frac{y}{x}) = x + C \tag{3.1166}$$

Problem 3.34 Find the integrating factor by inspection and solve the following ODE

$$\frac{xdy - ydx}{\sqrt{x^2 - y^2}} = xdy \tag{3.1167}$$

Ans:

$$\sin^{-1}(\frac{y}{x}) = y + C \tag{3.1168}$$

Problem 3.35 Solve the following ODE

$$xdx + ydy + ydx - xdy = 0 \tag{3.1169}$$

Hint: Use polar form.
Ans:

$$\ln\sqrt{x^2 + y^2} - \tan^{-1}(\frac{y}{x}) = C \tag{3.1170}$$

Problem 3.36 Show the validity of column 2 in Table 3.1.

Problem 3.37 Show that (3.300) is the solution of (3.3299).

Problem 3.38 Employ the technique discussed in Section 3.5.14 to solve

$$y\frac{d^2y}{dx^2} - (\frac{dy}{dx})^2 - y^2\frac{dy}{dx} = 0 \tag{3.1171}$$

Ans:

$$\frac{y}{y + C_1} = C_2 e^{C_1 x} \tag{3.1172}$$

Problem 3.39 Solve the following ODE

$$yf_1(xy)dx + xf_2(xy)dy = 0 \tag{3.1173}$$

Ans:

$$\int \frac{f_2(v)dv}{v[f_1(v) - f_2(v)]} + \ln x + C = 0, \quad v = xy \tag{3.1174}$$

Problem 3.40 It was reported in Example 3.13 that Ramanujan's differential equations given in (3.241) can be converted to (3.242) and (3.243). This problem provides the detail of the derivation.

(i) Applying the following change of variables

$$q = e^{-y}, \tag{3.1175}$$

show that (3.241) can be rewritten as

$$\frac{dP}{dy} = -\frac{1}{12}(P^2 - Q), \quad \frac{dQ}{dy} = -\frac{1}{3}(PQ - R), \quad \frac{dR}{dy} = -\frac{1}{2}(PR - Q^2) \tag{3.1176}$$

(ii) Apply the following stretching transformation

$$y_1 = e^{\varepsilon} y, \quad P_1 = e^{-\varepsilon} P, \quad Q_1 = e^{-2\varepsilon} Q, \quad R_1 = e^{-3\varepsilon} R, \tag{3.1177}$$

show that (3.1176) can further be reduced to

$$\frac{dP_1}{dy_1} = -\frac{1}{12}(P_1^2 - Q_1), \quad \frac{dQ_1}{dy_1} = -\frac{1}{3}(P_1Q_1 - R_1), \quad \frac{dR_1}{dy_1} = -\frac{1}{2}(P_1R_1 - Q_1^2) \tag{3.1178}$$

Comparing (3.1176) and (3.1178), we see that the ODEs are invariants under the stretching transformation given in (3.1177). This information allows us to apply a scaling transformation to be considered next.

(iii) Apply the following scaling transformation

$$u = \frac{R}{Q^{3/2}}, \quad v = \frac{Q^{1/2}}{P}, \quad w = y_1 Q^{1/2}, \tag{3.1179}$$

derive the following system of ODEs for u,v, and w:

$$y_1\frac{du}{dy_1} = -\frac{w}{2}(u^2 - 1), \quad y_1\frac{dv}{dy_1} = -\frac{w}{12}(v^2 - 2uv + 1), \quad y_1\frac{dv}{dy_1} = w\{1 + \frac{w}{6}(u - \frac{1}{v})\} \tag{3.1180}$$

(iv) Finally, prove the following ODEs given in (3.242) and (3.243)

$$(u^2 - 1)\frac{dv}{du} = \frac{1}{6}(v^2 - 2uv + 1) \tag{3.1181}$$

$$\frac{dw}{du} = -\frac{w}{3}\frac{(uv - 1)}{v(u^2 - 1)} - \frac{2}{u^2 - 1} \tag{3.1182}$$

CHAPTER FOUR

Series Solutions of Second Order ODEs

4.1 INTRODUCTION

Series solutions for second order ODEs with non-constant coefficients occupy an essential place in the development of the solution techniques for differential equations. Physically, many phenomena in sciences and engineering can be modeled by second order ODEs. The difficult part of solving many second order ODEs relates to the existence of regular singular points and irregular singular points in these ODEs. Fuchs, in 1866 and 1868, initiated study of the regular and irregular singular points of linear second order differential equations with non-constant coefficients. This problem had been considered by a number of mathematicians including Hermite, Jordan, Hadamard, Darboux, Poincare, Frobenius, Goursat, Thomé, and Painlevé. For example, the Bessel equation, which was found useful in modeling many physical phenomena, has a regular singular point at the origin of the variable. A technique called the Frobenius series has been found useful in solving it. Many special functions were defined as the solutions for these second order ODEs with non-constant coefficients. The solutions for the Bessel equation are called Bessel functions of the first and second kinds. In this chapter, we will consider some classical differential equations that fall into this category, including the Bessel equation, modified Bessel equation, Legendre equation, associated Legendre equation, hypergeometric equation, and generalized hypergeometric equation.

Before we devote ourselves to the series solutions, a thorough introduction to the gamma function will be given, which is essential for the later part of the chapter in obtaining solutions of these important differential equations.

4.2 GAMMA FUNCTION

The birth of the gamma function is related to the factorial function, which is defined as

$$n! = n(n-1)\cdots 3\cdot 2\cdot 1 \tag{4.1}$$

In addition, it is defined that

$$0! = 1 \tag{4.2}$$

We will see why we have such a strange definition for the zero factorial shortly. The gamma function was proposed by Euler. In 1729, at the age of 22, Euler wrote a letter to Christian Goldbach (1690-1764), a German living in Moscow, to describe the idea of extending the factorial function to the non-integer case. Euler proposed (Davis, 1959)

$$\left[\left(\frac{2}{1}\right)^n \frac{1}{n+1}\right]\left[\left(\frac{3}{2}\right)^n \frac{2}{n+2}\right]\left[\left(\frac{4}{3}\right)^n \frac{3}{n+3}\right]\ldots = \lim_{m\to\infty} \frac{m!(m+1)^{n+1}}{(n+1)(n+2)\cdots(n+1+m)} = n! \tag{4.3}$$

A proof of (4.3) will be given in Section 4.2.2. This definition leads to an amazing formula of π in terms of infinite product (see Problem 4.55). The motivation for studying (4.3) is that a^m is true not only for integer m but also for the non-integer case, and Euler said to the extreme, that the factorial function defined in (4.1) can also allow the non-integer case. For example, 2! = 2 and 3! = 6, Euler expected that 2.5! would be somewhere between 2 and 6. If we take the first four terms of the infinite product given in (4.3), we have $2.5! \approx 2.3828....$ As expected, it is between 2 and 6. Euler later proposed the gamma function in terms of an infinite integral, which will be discussed in the next section. The analysis of gamma functions is by no means simple but occupies a central place in mathematical analysis, and its properties have been investigated by many great mathematicians, including Euler, Legendre, Gauss, Weierstrass, Dirichlet, Binet, Hankel, Stirling, Pochhammer, and Hadamard. It also forms the basis for studying many functions of series solutions, like Bessel functions and hypergeometric functions.

In the next few sections, we will summarize some essential properties of the gamma function. We will see that the gamma function becomes infinity at zero and negative integers. To remedy this, French mathematician Hadamard proposed an alternative factorial function, which will be discussed in Section 4.3.

4.2.1 Euler's Integral Definition

Instead of using (4.3), Euler discovered that the following integral did give the required definition of the factorial for the case of integer $z = n$, with n being an integer:

$$\Gamma(z) = \int_0^\infty t^{z-1} e^{-t} dt \tag{4.4}$$

A plot of this gamma function is given in Figure 4.1. First, we rewrite (4.4) in terms of argument n+1 and apply integration by parts as:

$$\Gamma(n+1) = \int_0^\infty t^n e^{-t} dt = -t^n e^{-t}\Big|_0^\infty + n\int_0^\infty t^{n-1} e^{-t} dt = -t^n e^{-t}\Big|_0^\infty + n\Gamma(n) \tag{4.5}$$

Note that the upper limit of the first term on the right hand side is in a form of ∞/∞, whereas the lower limit is 0. Then, L'Hôpital's rule can be applied repeatedly to reduce the numerator to a constant and thus the upper limit is also zero:

$$-t^n e^{-t}\Big|_0^\infty = -\lim_{t\to\infty} \frac{t^n}{e^t} + 0 = -\lim_{t\to\infty} \frac{nt^{n-1}}{e^t} = \cdots = -\lim_{t\to\infty} \frac{n!}{e^t} = 0 \tag{4.6}$$

In short, we have a recursive formula:

$$\Gamma(n+1) = n\Gamma(n) \tag{4.7}$$

Applying (4.7) repeatedly, we finally get the factorial definition:

$$\Gamma(n+1) = n(n-1)(n-2)...1 \cdot \Gamma(1) = n! \tag{4.8}$$

The last part of (4.8) resulted from the fact that $\Gamma(1) = 1$, which will be shown in the following example.

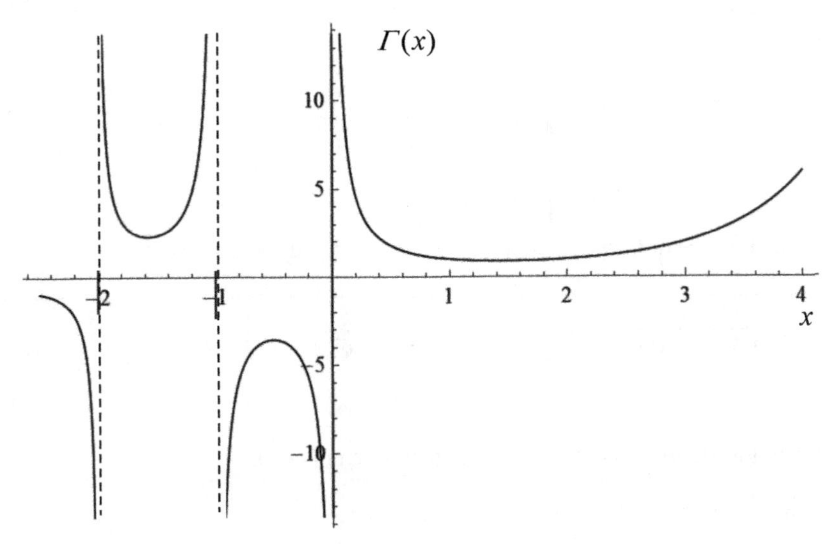

Figure 4.1. Plots of gamma function

Example 4.1 Using Euler's definition for the gamma function given in (4.4), show that 0! = 1.

Solution: Set $n = 0$ into the following relation between the gamma function and the factorial function for the case of integer argument:

$$\Gamma(n+1) = n! \tag{4.9}$$

Thus, we have

$$\Gamma(1) = 0! \tag{4.10}$$

Using Euler's integral definition, we obtain

$$\Gamma(1) = \int_0^{\infty} e^{-t} dt = -e^{-t}\Big|_0^{\infty} = -(0-1) = 1 \tag{4.11}$$

Substitution of (4.11) into (4.10) yields

$$0! = 1 \tag{4.12}$$

This provides the definition 0! = 1 that we reported earlier. This is a definition given to high school students without proof, and now we see why it is so.

Example 4.2 Find the value of $\Gamma(1/2)$

Solution: By substituting $z = 1/2$ in (4.4), we have

$$\Gamma(\frac{1}{2}) = \int_0^\infty t^{-1/2} e^{-t} dt \tag{4.13}$$

Application of the following change of variables

$$t = y^2, \quad dt = 2y dy \tag{4.14}$$

gives

$$\begin{aligned} \Gamma(\frac{1}{2}) &= \int_0^\infty e^{-y^2} \frac{1}{y} 2y dy = 2\int_0^\infty e^{-y^2} dy \\ &= \int_{-\infty}^\infty e^{-y^2} dy = \sqrt{\pi} \end{aligned} \tag{4.15}$$

This is the Laplace or Gauss integral that we encountered and proved in Chapter 1.

Example 4.3 The original integral definition of Euler in 1729 is actually given as

$$\Gamma(z) = \int_0^1 (\ln \frac{1}{y})^{z-1} dy \tag{4.16}$$

Show that this definition is equivalent to that given in (4.4).

Solution: Apply the following change of variables,

$$t = \ln(1/y), \text{ or } y = e^{-t} \tag{4.17}$$

Taking the total differential on both sides, we have

$$dt = -\frac{1}{y} dy, \quad \text{or} \quad dy = -e^{-t} dt \tag{4.18}$$

Noting from (4.17) that $y = 1$ for $t = 0$ and $y = 0$ for $t \to \infty$, we can rewrite the integral given in (4.16) as

$$\Gamma(z) = -\int_\infty^0 t^{z-1} e^{-t} dt = \int_0^\infty e^{-t} t^{z-1} dt \tag{4.19}$$

This is the same as (4.4) given above.

4.2.2 Euler's Factorial Form

We now show that the gamma function can be expressed in terms of a factorial function, which was introduced by Euler in 1729 and has been cited early in (4.3). First, we rewrite the integral (4.4) by using the definition of the exponential function defined in Example 1.2 in Chapter 1:

$$\Gamma(z) = \lim_{n\to\infty} \int_0^n t^{z-1} e^{-t} dt = \lim_{n\to\infty} \int_0^n t^{z-1} (1 - \frac{t}{n})^n dt \tag{4.20}$$

We now apply the following change of variables

$$t = n\beta \tag{4.21}$$

Substitution of (4.21) into (4.20) gives

$$\Gamma(z) = \lim_{n\to\infty} \int_0^1 n^{z-1} \beta^{z-1} (1-\beta)^n n d\beta = \lim_{n\to\infty} n^z \int_0^1 \beta^{z-1} (1-\beta)^n d\beta \tag{4.22}$$

Integration by parts reduces the integral to

$$\Gamma(z) == \lim_{n\to\infty} n^z [\frac{\beta^z (1-\beta)^n}{z}\bigg|_0^1 - \frac{1}{z}\int_0^1 \beta^z (1-\beta)^{n-1} n(-d\beta)]$$
$$= \lim_{n\to\infty} n^z [\frac{n}{z}\int_0^1 \beta^z (1-\beta)^{n-1} d\beta] \tag{4.23}$$

Comparison of (4.22) and (4.23) gives a recursive formula as

$$\int_0^1 \beta^{z-1} (1-\beta)^n d\beta = \frac{n}{z}\int_0^1 \beta^z (1-\beta)^{n-1} d\beta \tag{4.24}$$

Repeating the process of integration by parts or applying (4.24) repeatedly, we obtain another form of the gamma function

$$\Gamma(z) = \lim_{n\to\infty} n^z [\frac{n(n-1)}{z(z+1)}\int_0^1 \beta^{z+1} (1-\beta)^{n-2} d\beta]$$
$$= \lim_{n\to\infty} n^z [\frac{n(n-1)(n-2)...1}{z(z+1)(z+2)...(z+n-1)}\int_0^1 \beta^{z+n-1} (1-\beta)^{n-n} d\beta] \tag{4.25}$$
$$= \lim_{n\to\infty} [\frac{n^z n!}{z(z+1)(z+2)...(z+n)}]$$

In obtaining the last result in (4.25), we have used the following obvious result:

$$\int_0^1 \beta^{z+n-1} d\beta = \frac{1}{z+n} \tag{4.26}$$

The result given in (4.25) provides another definition of the gamma function in terms of a limit of the factorial function. The result given in (4.25) is the same as (4.3) as $n \to \infty$. This result will be used in the next section as an immediate step to show Weierstrass's product definition.

4.2.3 Weierstrass's Product Definition

Euler also proposed another form of the gamma function in terms of infinite products as:

$$\frac{1}{\Gamma(z)} = z e^{\gamma z} \prod_{n=1}^{\infty} \left\{ (1+\frac{z}{n}) e^{-z/n} \right\} \tag{4.27}$$

where $\gamma \approx 0.5772157$ is the Euler constant. This definition was proposed by Euler in 1729 in his original letter to Goldbach, but many of the properties of this product definition were considered by Weierstrass. Indeed, Weierstrass preferred this definition to Euler's integral definition given in (4.4). In the literature, it was referred as Weierstrass's canonical form of definition (e.g., Whittaker and Watson, 1927).

To show that this definition is the same as the integral definition of Euler given in (4.4), we first recall the definition of Euler's constant:

$$\gamma = \lim_{m\to\infty}\left(\sum_{k=1}^{m}\frac{1}{k}-\ln m\right)\cong 0.5772157 \tag{4.28}$$

A review of the Euler constant is given in Appendix E at the back of this book. Let us rewrite the infinite product as a limit

$$\frac{1}{\Gamma(z)} = \lim_{n\to\infty} z e^{\gamma z}\prod_{k=1}^{n}\left\{(1+\frac{z}{k})e^{-z/k}\right\} = \lim_{n\to\infty}\frac{1}{\Gamma_n(z)} \tag{4.29}$$

Rearranging this equation results in

$$z\Gamma_n(z) = e^{\sum_{k=1}^{n}-\frac{z}{k}+z\ln n}\prod_{k=1}^{n}\left\{(1+\frac{z}{k})^{-1}e^{z/k}\right\} = e^{z\ln n}\prod_{k=1}^{n}\left\{(1+\frac{z}{k})^{-1}\right\} \tag{4.30}$$

The last part of this result is obtained by recognizing the following identity

$$\exp(\sum_{k=1}^{n}-\frac{z}{k}) = \exp(-\frac{z}{1}-\frac{z}{2}+\ldots-\frac{z}{n}) = \exp(-\frac{z}{1})\exp(-\frac{z}{2})\cdots\exp(-\frac{z}{n}) = \prod_{k=1}^{n}e^{-z/k} \tag{4.31}$$

In this equation, it is important to recognize that the exponential function of a sum equals the product of the individual exponential functions.

To further simplify (4.30), we want to establish another identity

$$\ln n = [\ln 2-\ln 1]+[\ln 3-\ln 2]+\ldots+[\ln n-\ln(n-1)]+[\ln(n+1)-\ln n]+\ln(\frac{n}{n+1})$$

$$= -\ln 1+\ln n = \ln n = \sum_{k=1}^{n}\ln(\frac{k+1}{k})+\ln(\frac{n}{n+1}) \tag{4.32}$$

Substitution of (4.32) into (4.30) gives

$$\begin{aligned} z\Gamma_n(z) &= e^{\sum_{k=1}^{n} z\ln(1+\frac{1}{k})+z\ln(\frac{n}{n+1})}\prod_{k=1}^{n}\left\{(1+\frac{z}{k})^{-1}\right\} \\ &= (\frac{n}{n+1})^z\prod_{k=1}^{n}(1+\frac{1}{k})^z\prod_{k=1}^{n}(1+\frac{z}{k})^{-1} \\ &= (\frac{n}{n+1})^z\prod_{k=1}^{n}\frac{(1+k)^z}{k^z}(\frac{k}{k+z}) \\ &= (\frac{n}{n+1})^z\frac{2^z3^z\ldots n^z(n+1)^z}{1^z2^z\ldots n^z}\frac{n!}{(z+1)(z+2)\ldots(z+n)} \\ &= \frac{n^z n!}{(z+1)(z+2)\ldots(z+n)} \end{aligned} \tag{4.33}$$

Substitution of (4.33) into (4.29) gives

$$\Gamma(z) = \lim_{n\to\infty}[\Gamma_n(z)] = \lim_{n\to\infty}[\frac{n^z n!}{z(z+1)(z+2)\ldots(z+n)}] \tag{4.34}$$

This is obviously the same as (4.25) obtained in the last section. Consequently, the equivalence of (4.27) and (4.4) is established through the immediate result of the factorial form derived in (4.25).

4.2.4 Reflection Formula

An important result of the gamma function is called the reflection formula and it is given as:

$$\Gamma(1-z)\Gamma(z)=\frac{\pi}{\sin \pi z} \tag{4.35}$$

This formula is sometimes known as the complement formula. It is a very powerful identity but its proof is not simple.

To prove (4.35), we start the left side as

$$\Gamma(z)\Gamma(1-z)=\int_0^\infty t^{z-1}e^{-t}dt\int_0^\infty s^{(1-z)-1}e^{-s}ds=\int_0^\infty\int_0^\infty e^{-(t+s)}s^{-z}t^{z-1}dsdt \tag{4.36}$$

In writing (4.36), we recognize that both t and s are dummy variables of the definite integral of the gamma function. Now apply a change of variables of

$$u=s+t, \quad v=\frac{t}{s} \tag{4.37}$$

Using the second equation of (4.37), we can rewrite u as

$$u=s(1+v), \quad or \quad s=\frac{u}{1+v} \tag{4.38}$$

Subsequently, t becomes

$$t=sv=\frac{uv}{1+v} \tag{4.39}$$

Now s and t are completely in terms of u and v. The physical or geometrical interpretation of (4.37) can be depicted in Figure 4.2. Every radial line corresponds to a specific value of v, with $v = 0$ for horizontal (i.e., $t = 0$) and $v \to \infty$ for vertical (i.e., $s = 0$). The new variable u is the length measured along the radial lines for various values of v. Therefore, for $v = 0$ we have $u = s$ and $v \to \infty$ we have $u = t$. In summary, we find

$$s=\frac{u}{1+v}, \quad t=sv=\frac{uv}{1+v} \tag{4.40}$$

For the double integration in (4.36), we need to consider the Jacobian of the mapping [see (1.109) of Chapter 1]:

$$dtds=\left|\frac{\partial(t,s)}{\partial(u,v)}\right|dudv \tag{4.41}$$

By using (4.38) and (4.39), we obtain the Jacobian as

$$\frac{\partial(t,s)}{\partial(u,v)} = \begin{vmatrix} \frac{\partial t}{\partial u} & \frac{\partial t}{\partial v} \\ \frac{\partial s}{\partial u} & \frac{\partial s}{\partial v} \end{vmatrix} = \frac{\partial t}{\partial u}\frac{\partial s}{\partial v} - \frac{\partial t}{\partial v}\frac{\partial s}{\partial u}$$

$$= \frac{v}{1+v}\frac{(-u)}{(1+v)^2} - \left(\frac{u}{1+v} - \frac{vu}{(1+v)^2}\right)\frac{1}{1+v} \tag{4.42}$$

$$= \frac{1}{(1+v)^3}[-uv - u(1+v) + uv] = \frac{-u}{(1+v)^2}$$

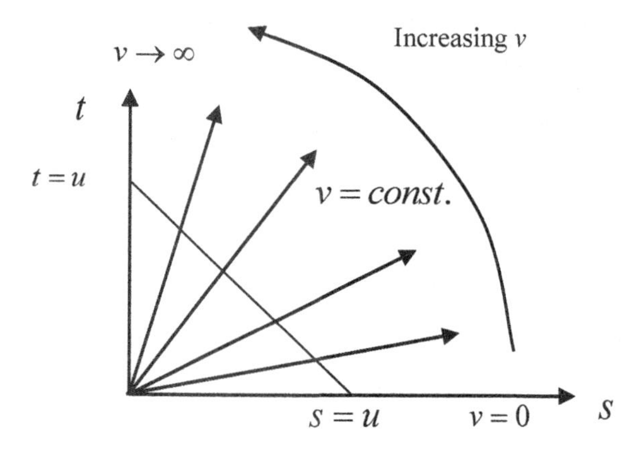

Figure 4.2 Mapping from *s*-*t* space to *u*-*v* space

Substitution of (4.42) into (4.41) leads to

$$dtds = \left|\frac{\partial(t,s)}{\partial(u,v)}\right| dudv = \frac{u}{(1+v)^2} dudv \tag{4.43}$$

In addition, it is straightforward to see that every point in the quarter plane s-t is uniquely defined by u and v. In particular, the ranges of these variables are

$$\begin{aligned} &s: 0 \to \infty, \quad t: 0 \to \infty, \\ &u = s + t: 0 \to \infty, \quad v = t/s: 0 \to \infty \end{aligned} \tag{4.44}$$

The ranges of the new variables are $0 < u < \infty$ and $0 < v < \infty$. Finally, (4.36) can be expressed as

$$\Gamma(z)\Gamma(1-z) = \int_0^\infty \int_0^\infty e^{-u} v^{z-1} \left(\frac{1+v}{u}\right) \frac{u}{(1+v)^2} dudv \tag{4.45}$$

Expressing in terms of the new variables, using (4.43), we have (4.45) becoming

$$\Gamma(z)\Gamma(1-z)=\int_0^\infty\int_0^\infty e^{-u}v^{z-1}\frac{dudv}{1+v}=\int_0^\infty e^{-u}du\int_0^\infty\frac{v^{z-1}dv}{1+v}$$
$$=\int_0^\infty\frac{v^{z-1}dv}{1+v}=\frac{\pi}{\sin z\pi} \tag{4.46}$$

The last result of (4.46) is a result of Titchmarsh's contour integral derived in Section 1.7.7 in Chapter 1. The above proof was outlined in Lebedev (1972) without giving detail. A simpler proof by Euler of (4.35) is given in Problem 4.58 using infinite product form of sine function. This formula is also found useful in evaluating certain Mellin transform (see Problem 11.6 in Chapter 11). The reflection formula is closely related to the singularity of the gamma function. Note from the reflection formula that there are poles at z = 0, −1, −2, ... The poles coincide with all the zeros of the periodic sine function (with a period of 2). For example, from the recursive formula given in (4.7) we have

$$\Gamma(z)=\frac{\Gamma(1+z)}{z} \tag{4.47}$$

For z = 0, we have

$$\Gamma(0)=\frac{\Gamma(1)}{0}\to\infty \tag{4.48}$$

Thus, Γ(z) is singular at $z = 0$. We can extend this idea further for $z = -n$

$$\Gamma(-n)=\frac{\Gamma(1-n)}{-n}=\frac{1}{(-n)}\frac{1}{(-n+1)}\Gamma(-n+2)$$
$$=\frac{(-1)^2}{n(n-1)}\Gamma(-n+2)=\frac{(-1)^n}{n!}\Gamma(0)\to\infty \tag{4.49}$$

The last result is a consequence of repeated applications of the recursive formula given in (4.7). From this, we can observe that there is a simple pole at $z = 0, -1, -2$, ... For complex z, the residue at the pole z = 0 is:

$$\operatorname*{Res}_{z\to 0}\Gamma(z)=\lim_{z\to 0}z\Gamma(z)=\lim_{z\to 0}\Gamma(z+1)=\Gamma(1)=1 \tag{4.50}$$

The residue of the negative integer is

$$\operatorname*{Res}_{z\to -n}\Gamma(z)=\lim_{z\to -n}(z+n)\Gamma(z)=\lim_{\zeta\to 0}\zeta\Gamma(\zeta-n)=\lim_{\zeta\to 0}\frac{\zeta}{\zeta-n}\Gamma(\zeta-n+1)$$
$$=\lim_{\zeta\to 0}\frac{\zeta}{(\zeta-n)(\zeta-n+1)\cdots(\zeta-1)}\Gamma(\zeta) \tag{4.51}$$
$$=\frac{(-1)^n}{n!}\lim_{\zeta\to 0}\zeta\,\Gamma(\zeta)=\frac{(-1)^n}{n!}$$

Alternatively, we can investigate the singularity of the gamma function as

$$\Gamma(z) = \int_0^\infty e^{-t} t^{z-1} dt = \int_0^1 e^{-t} t^{z-1} dt + \int_1^\infty e^{-t} t^{z-1} dt$$
$$= \int_1^\infty e^{-t} t^{z-1} dt + \int_0^1 \sum_{n=0}^\infty \frac{(-1)^n}{n!} t^{n+z-1} dt \tag{4.52}$$
$$= \int_1^\infty e^{-t} t^{z-1} dt + \sum_{n=0}^\infty \frac{(-1)^n}{n!} \frac{1}{z+n}$$

In obtaining (4.52), we have substituted the power series of the exponential function inside the integration. The residue is clearly the same as that found in (4.51). The first integral is analytic or finite whereas the second summation contains all the poles at $n = 0, -1, -2, \ldots$ This is called the Mittag-Leffler expansion of the gamma function and is also known as Prym's decomposition. Mittag-Leffler played a crucial role in the final decision of "not" including mathematics in the Nobel Prize. This story is reported in the bibliography section. This singularity of the gamma function leads to the search for other factorial functions that contain no singularity. One such choice is called the Hadamard factorial function and will be examined in more detail in Section 4.3.

4.2.5 Recurrence Formula

The recurrence formula is given by

$$\Gamma(z+1) = z\Gamma(z) \tag{4.53}$$

This is essentially the same as (4.7) for integer argument. This can be proved easily by integration by parts and L'Hôpital's rule.

$$\Gamma(z+1) = \int_0^\infty t^z e^{-t} dt = -t^z e^{-t}\Big|_0^\infty + z\int_0^\infty t^{z-1} e^{-t} dt$$
$$= -\lim_{t\to\infty} \frac{z(z-1)\cdots t^{z-m}}{e^t} + 0 + z\int_0^\infty t^{z-1} e^{-t} dt \tag{4.54}$$
$$= z\Gamma(z)$$

We have applied L'Hôpital's rule m times with $m > z$. When $z = n$ is an integer, we have the factorial function $\Gamma(n+1) = n!$ as shown in Example 4.1. This is the most basic property and explains why the gamma function is an extension of the factorial function to the non-integer case.

4.2.6 Legendre Duplication Formula

The following duplication formula was derived by Legendre

$$\Gamma(2z) = \frac{1}{\sqrt{2\pi}} 2^{2z-1/2} \Gamma(z)\Gamma(z+\frac{1}{2}) \tag{4.55}$$

To prove this, we consider the following term contained in the right hand side of (4.55):

$$\begin{aligned} I &= 2^{2z-1}\Gamma(z)\Gamma(z+\frac{1}{2}) = 2^{2z-1}\int_0^\infty e^{-t}t^{z-1}dt\int_0^\infty e^{-s}s^{z-1/2}ds \\ &= 2^{2z-1}\int_0^\infty\int_0^\infty e^{-(s+t)}t^{z-1}s^{z-1/2}dtds \end{aligned} \tag{4.56}$$

Applying a change of variables as

$$\alpha = \sqrt{s}, \quad \beta = \sqrt{t} \tag{4.57}$$

and taking the differential on both sides, we obtain

$$d\alpha = \frac{1}{2\sqrt{s}}ds, \quad d\beta = \frac{1}{2\sqrt{t}}dt \tag{4.58}$$

With these new variables, the double integration becomes

$$\begin{aligned} I &= 2^{2z-1}\int_0^\infty\int_0^\infty e^{-(\alpha^2+\beta^2)}\beta^{2z-2}\alpha^{2z-1}4\alpha\beta d\alpha d\beta \\ &= 2^{2z+1}\int_0^\infty\int_0^\infty e^{-(\alpha^2+\beta^2)}(\alpha\beta)^{2z-1}\alpha d\alpha d\beta \end{aligned} \tag{4.59}$$

On the other hand, we can independently apply another change of variables as

$$\alpha = \sqrt{t}, \quad \beta = \sqrt{s} \tag{4.60}$$

Following a similar procedure, it is straightforward to see that

$$\begin{aligned} I &= 2^{2z-1}\int_0^\infty\int_0^\infty e^{-(\alpha^2+\beta^2)}\alpha^{2z-2}\beta^{2z-1}4\alpha\beta d\alpha d\beta \\ &= 2^{2z+1}\int_0^\infty\int_0^\infty e^{-(\alpha^2+\beta^2)}(\alpha\beta)^{2z-1}\beta d\alpha d\beta \end{aligned} \tag{4.61}$$

Addition of these two independent results given in (4.59) and (4.61) leads to

$$2I = 2^{2z+1}\int_0^\infty\int_0^\infty e^{-(\alpha^2+\beta^2)}(\alpha\beta)^{2z-1}(\alpha+\beta)d\alpha d\beta \tag{4.62}$$

Because the integrand is symmetric with respect to both α and β, we have the integration as symmetric with respect to $\alpha = \beta$ as shown in Figure 4.3. In particular, we can express the double integral for the first quadrant (i.e., $0 < \alpha < \infty$ and $0 < \beta < \infty$) as twice the shaded area I_2 shown in Figure 4.3 (i.e., $0 < \alpha < \beta$, and $0 < \beta < \infty$):

$$\begin{aligned} I &= 2\int_0^\infty\int_0^\infty e^{-(\alpha^2+\beta^2)}(2\alpha\beta)^{2z-1}(\alpha+\beta)d\alpha d\beta \\ &= 4\int_0^\infty\int_0^\beta e^{-(\alpha^2+\beta^2)}(2\alpha\beta)^{2z-1}(\alpha+\beta)d\alpha d\beta = 4I_2 \end{aligned} \tag{4.63}$$

Note that I, which is symmetric with respect to α and β, can be expressed in terms of I_2, which is shown in Figure 4.3. Apply another change of variables

$$u = \alpha^2 + \beta^2, \quad v = 2\alpha\beta \tag{4.64}$$

Thus, we find that

$$u + v = (\alpha+\beta)^2 \tag{4.65}$$

Inversely, we can rewrite (4.65) as

$$\alpha + \beta = \sqrt{u+v} \tag{4.66}$$

The domain integration can be expressed in terms of the Jacobian of the mapping function as

$$dudv = \frac{\partial(u,v)}{\partial(\alpha,\beta)} d\alpha d\beta \tag{4.67}$$

Or, inversely, it can be expressed as

$$d\alpha d\beta = \frac{1}{\dfrac{\partial(u,v)}{\partial(\alpha,\beta)}} dudv \tag{4.68}$$

Substitution of (4.64) into (4.67) gives

$$\begin{aligned} \frac{\partial(u,v)}{\partial(\alpha,\beta)} &= \begin{vmatrix} \dfrac{\partial u}{\partial \alpha} & \dfrac{\partial u}{\partial \beta} \\ \dfrac{\partial v}{\partial \alpha} & \dfrac{\partial v}{\partial \beta} \end{vmatrix} = \begin{vmatrix} 2\alpha & 2\beta \\ 2\beta & 2\alpha \end{vmatrix} = 4(\alpha^2 - \beta^2) \\ &= 4(\alpha+\beta)(\alpha-\beta) = 4\sqrt{u+v}\sqrt{u-v} \end{aligned} \tag{4.69}$$

The domain of integration for

$$\alpha : 0 \to \beta, \quad \beta : 0 \to \infty, \tag{4.70}$$

is converted to:

$$u : 0 \to \infty, \quad v : 0 \to \infty \tag{4.71}$$

Thus, the integral I becomes

$$\begin{aligned} I &= 4\int_0^\infty \int_0^\infty e^{-u} v^{2z-1} \sqrt{u+v} \frac{dudv}{4\sqrt{u+v}\sqrt{u-v}} \\ &= \int_0^\infty \int_0^\infty e^{-u} v^{2z-1} \frac{dudv}{\sqrt{u-v}} \\ &= \int_0^\infty v^{2z-1} e^{-v} \int_0^\infty \frac{e^{-(u-v)} d(u-v)}{\sqrt{u-v}} dv \end{aligned} \tag{4.72}$$

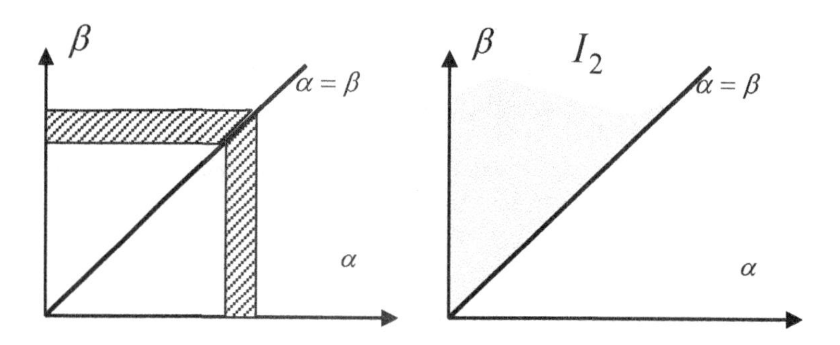

Figure 4.3 The domain of symmetry and the physical domain for I_2

Clearly, a change of variables for the inner integral can be used:

$$u - v = w^2, \quad d(u-v) = 2wdw \tag{4.73}$$

The inner integral is thus integrable as

$$\int_0^\infty \frac{e^{-(u-v)}d(u-v)}{\sqrt{u-v}} = \int_0^\infty \frac{e^{-w^2}}{w} 2wdw$$
$$= 2\int_0^\infty e^{-w^2} dw = \sqrt{\pi} \tag{4.74}$$

The last result is obtained by using the Laplace or Gauss integral considered in Chapter 1. With the result obtained in (4.74), the remaining integration in (4.72) is precisely $\Gamma(2z)$. Finally, we have derived.

$$\sqrt{\pi}\Gamma(2z) = 2^{2z-1}\Gamma(z)\Gamma(z+\frac{1}{2}) \tag{4.75}$$

Dividing through by $\pi^{1/2}$, we finally obtain the duplication formula obtained by Legendre. A more general multiplication formula was derived by Gauss

$$\Gamma(nz) = (2\pi)^{(1-n)/2} n^{nz-1/2}\Gamma(z)\Gamma(z+\frac{1}{n})\Gamma(z+\frac{2}{n})\cdots\Gamma(z+\frac{n-1}{n}) \tag{4.76}$$

For $n = 2$, the Legendre duplication formula is recovered. The proof of this formula is more advanced as it involves the Stirling formula to be discussed in Section 4.2.8, and thus the proof of the Gauss multiplication formula will be postponed until Problem 4.28.

Example 4.4 Prove the following identity

$$\Gamma(1-x) = -\frac{x\Gamma(\frac{1}{2}-\frac{x}{2})\Gamma(-\frac{x}{2})}{\sqrt{\pi}2^{x+1}} \tag{4.77}$$

Solution: Set $z = -x/2$ in Legendre's duplication formula given in (4.55) to get

$$-x\Gamma(-x) = -\frac{1}{\sqrt{\pi}}x2^{-x-1}\Gamma(-\frac{x}{2})\Gamma(\frac{1}{2}-\frac{x}{2}) = \Gamma(1-x) \tag{4.78}$$

Rearranging (4.78), we obtain

$$\Gamma(1-x) = -\frac{x}{\sqrt{\pi}2^{x+1}}\Gamma(-\frac{x}{2})\Gamma(\frac{1}{2}-\frac{x}{2}) \tag{4.79}$$

This is the required result. This result will be used later in deriving the recursive formula of Hadamard factorial function in Section 4.3.

4.2.7 Digamma or Psi Function

Literally, digamma function means differentiation of the gamma function. It is defined as the differentiation of logarithm of gamma function, instead of the gamma function itself:

$$\psi(z) = \frac{d}{dz}\ln\Gamma(z) = \frac{\Gamma'(z)}{\Gamma(z)} \tag{4.80}$$

A plot of the digamma function is given in Figure 4.4. Because the normal symbol for the digamma function is the Greek letter psi or ψ, it is also sometimes called the psi function. The digamma function is a special function in its own right. There are a lot properties associated with the digamma function. It will be shown in Section 4.6.2 on the Bessel equation that the digamma function appears naturally in the definition of the Bessel function of the second kind or $Y_\nu(z)$.

Some essential properties of the digamma function are reported here. First, the recurrence formula of the digamma function is

$$\psi(z+1) = \frac{1}{z} + \psi(z) \tag{4.81}$$

This can be proved easily based on the recurrence formula of the gamma function:

$$\Gamma(1+z) = z\Gamma(z) \tag{4.82}$$

Differentiation of (4.82) with respect to z gives

$$\Gamma'(1+z) = \Gamma(z) + z\Gamma'(z) \tag{4.83}$$

where $\Gamma'(1+z)$ means the differentiation of $\Gamma(1+z)$ with respect to the argument (i.e., $1+z$ in this case). Dividing the whole expression by $\Gamma(1+z)$, we get

$$\frac{\Gamma'(1+z)}{\Gamma(1+z)} = \frac{\Gamma(z)}{\Gamma(1+z)} + \frac{z\Gamma'(z)}{\Gamma(1+z)} \tag{4.84}$$

Recalling that $\Gamma(1+z) = z\Gamma(z)$, we can further simplify (4.84) as

$$\frac{\Gamma'(1+z)}{\Gamma(1+z)} = \frac{1}{z} + \frac{\Gamma'(z)}{\Gamma(z)} \tag{4.85}$$

By virtue of the definition given in (4.80), we immediately obtain

$$\psi(z+1) = \frac{1}{z} + \psi(z) \tag{4.86}$$

This completes the proof.

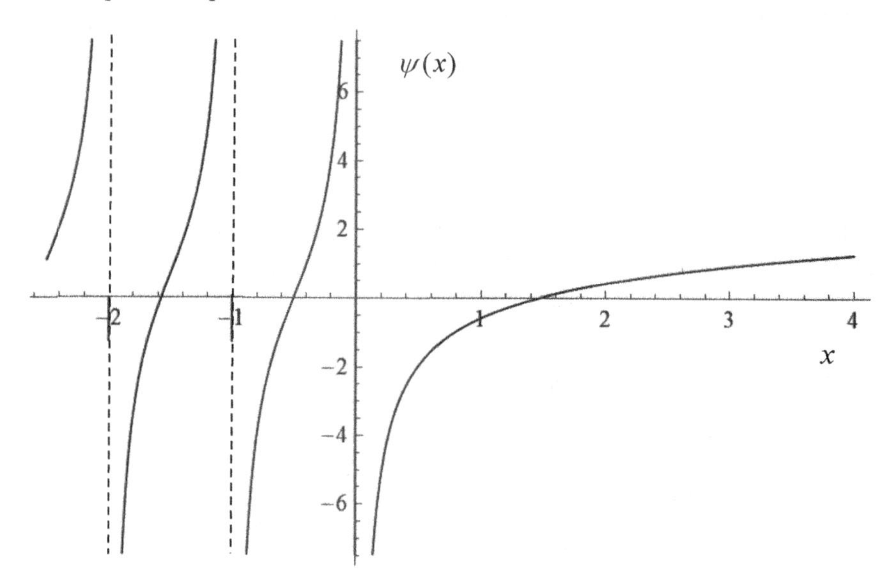

Figure 4.4 Plots of digamma function

Another important identity is the duplication formula of the digamma function (Formula 6.3.8 of Abramowitz and Stegun, 1964)

$$\psi(2z) = \frac{1}{2}\psi(z) + \frac{1}{2}\psi(z+\frac{1}{2}) + \ln 2 \tag{4.87}$$

To prove (4.87), we start with the Legendre duplication formula:

$$\Gamma(2z) = \frac{1}{\sqrt{2\pi}} 2^{2z-1/2}\Gamma(z)\Gamma(z+\frac{1}{2}) \tag{4.88}$$

Differentiation of (4.88) with respect to z gives

$$\begin{aligned}2\Gamma'(2z) = \frac{1}{\sqrt{2\pi}}\{&(\ln 2)2^{2z-1/2}\Gamma(z)\Gamma(z+\frac{1}{2})\\ &+2^{2z-1/2}\Gamma'(z)\Gamma(z+\frac{1}{2}) + 2^{2z-1/2}\Gamma(z)\Gamma'(z+\frac{1}{2})\}\end{aligned} \tag{4.89}$$

Dividing through the whole expression by $\Gamma(2z)$ and recalling the duplication formula in (4.88) again, we obtain

$$2\frac{\Gamma'(2z)}{\Gamma(2z)} = \frac{\sqrt{2\pi}}{\sqrt{2\pi}}\frac{2^{2z-1/2}}{2^{2z-1/2}}\{2\ln 2 + \frac{\Gamma'(z)}{\Gamma(z)} + \frac{\Gamma'(z+1/2)}{\Gamma(z+1/2)}\} \tag{4.90}$$

Using the definition of the digamma function, we finally get

$$2\psi(2z) = 2\ln 2 + \psi(z) + \psi(z+1/2) \tag{4.91}$$

This completes the proof of (4.87).

The reflection formula of the digamma function is (Formula 8.365.8 of Gradshteyn and Ryzhik, 1980; (1.3.4) of Lebedev, 1972)

$$\psi(1-z) = \psi(z) + \pi\cot\pi z \tag{4.92}$$

To prove this reflection formula, we start with the reflection formula of the gamma function (see (4.35))

$$\Gamma(z)\Gamma(1-z) = \frac{\pi}{\sin z\pi} \tag{4.93}$$

Differentiation of both sides with respect to z gives

$$\frac{d}{dz}\{\Gamma(z)\Gamma(1-z)\} = -\Gamma(z)\Gamma'(1-z) + \Gamma'(z)\Gamma(1-z) = -\pi^2\frac{\cot z\pi}{\sin z\pi} \tag{4.94}$$

The last of (4.94) can be rearranged to get

$$\Gamma(z)\Gamma(1-z)\left\{-\frac{\Gamma'(1-z)}{\Gamma(1-z)} + \frac{\Gamma'(z)}{\Gamma(z)}\right\} = -\pi^2\frac{\cot z\pi}{\sin z\pi} \tag{4.95}$$

Reapplying the reflection formula in (4.93) to the left hand side, we obtain

$$\frac{\pi}{\sin z\pi}\left\{-\psi(1-z) + \psi(z)\right\} = -\pi^2\frac{\cot z\pi}{\sin z\pi} \tag{4.96}$$

Canceling the common factor on both sides, we arrive at

$$\left\{-\psi(1-z) + \psi(z)\right\} = -\pi\cot z\pi \tag{4.97}$$

This of course agrees with (4.92).

Some special formulas for the digamma function are given here without proof:

$$\ln \Gamma(z) = \int_0^z \psi(\zeta) d\zeta \tag{4.98}$$

Dirichlet formula

$$\psi(z) = \int_0^\infty \left[e^{-x} - \frac{1}{(1+x)^z} \right] \frac{1}{x} dx \tag{4.99}$$

Gauss formula

$$\psi(z) = \int_0^\infty \left[\frac{e^{-x}}{x} - \frac{e^{-xz}}{1-e^{-x}} \right] dx \tag{4.100}$$

$$\psi(z) = -\gamma - \frac{1}{z} + z \sum_{k=1}^{\infty} \frac{1}{k(z+k)} \tag{4.101}$$

$$\psi(z) = -\gamma + (z-1) \sum_{k=0}^{\infty} \frac{1}{(k+1)(z+k)} \tag{4.102}$$

$$\psi(n+\frac{1}{2}) = -\gamma - 2\ln 2 + \sum_{k=1}^{n} \frac{2}{2k-1} \tag{4.103}$$

$$\psi(1) = -\gamma \tag{4.104}$$

Another important formula of the digamma function ψ will be given in (4.301) and will be proved there.

Example 4.5 Prove the following identity

$$\psi(\frac{1}{2}+z) = \psi(\frac{1}{2}-z) + \pi \tan \pi z \tag{4.105}$$

Solution: We start with the following identity for the gamma function (see proof in Problem 4.1):

$$\Gamma(\frac{1}{2}+z)\Gamma(\frac{1}{2}-z) = \frac{\pi}{\cos \pi z} \tag{4.106}$$

Taking the natural logarithm of (4.106) gives

$$\ln \Gamma(\frac{1}{2}+z) + \ln \Gamma(\frac{1}{2}-z) = \ln \pi - \ln \cos \pi z \tag{4.107}$$

Differentiation with respect to z gives

$$\psi(\frac{1}{2}+z) - \psi(\frac{1}{2}-z) = \pi \tan \pi z \tag{4.108}$$

Rearranging this gives the identity given in (4.105).

4.2.8 Stirling Formula

For a large argument, asymptotic expansion of the gamma function was derived by Stirling in 1730, which is given by

$$n! \sim \sqrt{2\pi} n^{n+1/2} e^{-n} \tag{4.109}$$

for $n \to \infty$. To derive this formula, we recall for $x > 0$ that

$$\Gamma(x+1) = \int_0^\infty t^x e^{-t} dt \tag{4.110}$$

We can introduce a new variable as

$$t = sx, \quad dt = xds \tag{4.111}$$

Substitution of (4.111) into (4.110) gives

$$\begin{aligned}\Gamma(x+1) &= \int_0^\infty (sx)^x e^{-sx} xds = x^{x+1} \int_0^\infty s^x e^{-sx} ds \\ &= x^{x+1} \int_0^\infty e^{x(-s+\ln s)} ds = x^{x+1} \int_0^\infty \exp[x(\phi(s)]ds\end{aligned} \tag{4.112}$$

where

$$\phi(s) = -s + \ln s \tag{4.113}$$

We can examine the asymptotic behavior of the integral for large values of x (i.e., $x \to \infty$) by a method similar to the Laplace method for the case of real $\phi(s)$. This method is closely related to the so-called Riemann method of steepest descents (or Debye's Saddle point method) if $\phi(s)$ is complex, or the Stoke's and Kelvin's method of stationary phase if $\phi(s)$ is purely imaginary (Erdelyi, 1956). More details on these methods can be found in Bleistein and Handelsman (1986) and Chapter 12.

First, we need to find the maximum or critical point of the function $\phi(s)$ by considering:

$$\phi'(s) = -1 + \frac{1}{s} = 0 \tag{4.114}$$

We now consider a Taylor series expansion of $\phi(s)$ around the critical point $s = 1$ (the solution of s in (4.114)) as

$$\begin{aligned}\phi(s) &= \phi(1) + (s-1)\phi'(1) + \frac{1}{2}(s-1)^2 \phi''(1) + ... \\ &= -1 - \frac{1}{2}(s-1)^2 + \frac{1}{3}(s-1)^3 + ...\end{aligned} \tag{4.115}$$

Note that

$$\phi'(1) = 0, \quad \phi''(1) = -1 < 0 \tag{4.116}$$

and this confirms that the critical point is indeed a maximum. Now we apply another round of change of variables to (4.112) as:

$$u = s - 1, \quad du = ds \tag{4.117}$$

$$\Gamma(x+1) = x^{x+1} \int_{-1}^\infty \exp[x(-1 - \frac{u^2}{2} + ...)]du \approx x^{x+1} e^{-x} \int_{-1}^\infty \exp(-\frac{xu^2}{2})du \tag{4.118}$$

Since we are expanding $\phi(s)$ about $s = 1$ or $u \to 0$, all higher order terms can be neglected as a first approximation. In addition, we are interested in the asymptotic case that $x \to \infty$, and the main contribution to the integration comes from the region around the maximum value of function $\phi(s)$ around $s = 1$. Let us consider another change of variables:

$$\frac{xu^2}{2} = y^2, \quad du = \frac{2y}{xu} dy = \sqrt{\frac{2}{x}} dy \tag{4.119}$$

Thus, the integral in (4.118) becomes:

$$\int_{-1}^{\infty} \exp(-\frac{xu^2}{2}) du = \sqrt{\frac{2}{x}} \int_{-\sqrt{x/2}}^{\infty} \exp(-y^2) dy \tag{4.120}$$

This integral can readily be integrated if we recall from Abramowitz and Stegun (1964) that an error function can be defined as

$$\frac{2}{\sqrt{\pi}} \int_{z}^{\infty} \exp(-t^2) dt = 1 - erf(z), \quad erf(-z) = -erf(z) \tag{4.121}$$

Finally, we have

$$\int_{-1}^{\infty} \exp(-\frac{xu^2}{2}) du = \sqrt{\frac{\pi}{2x}} \{1 + erf(\sqrt{\frac{x}{2}})\} \tag{4.122}$$

Substitution of (4.122) into (4.118) gives

$$\Gamma(x+1) \approx x^{x+1/2} e^{-x} \sqrt{\frac{\pi}{2}} \{1 + erf(\sqrt{x/2})\} \tag{4.123}$$

As $x \to \infty$, an error function can be approximated by (Abramowitz and Stegun, 1964)

$$erf(\infty) = \frac{2}{\sqrt{\pi}} \int_{0}^{\infty} \exp(-t^2) dt = \frac{2}{\sqrt{\pi}} \frac{\sqrt{\pi}}{2} = 1 \tag{4.124}$$

In obtaining (4.124), we have used the Laplace or Gauss integral discussed in Chapter 1. Finally, we obtain the following approximation

$$\Gamma(x+1) \approx \sqrt{2\pi} x^{x+1/2} e^{-x} \tag{4.125}$$

When $x = n$, we arrive at the Stirling formula given in (4.109). In fact, the Stirling formula can also be proved using the Laplace method but the proof is quite lengthy and will not be presented here.

If more terms are retained in the Taylor series expansion given in (4.115), more terms in the asymptotic expansion can be obtained as :

$$\begin{aligned} \Gamma(z+1) \approx \sqrt{2\pi} z^{z+1/2} e^{-z} \{ 1 + \frac{1}{12z} + \frac{1}{288z^2} - \frac{139}{51840z^3} - \frac{571}{2488320z^4} \\ + \frac{163879}{209018880z^5} + \frac{5246819}{75246796800z^6} - \frac{534703531}{902961561600z^7} + \ldots \} \end{aligned} \tag{4.126}$$

This eight-term series was obtained by Stirling in 1730. Another completely different form of asymptotic expansion of the gamma function was obtained by the Indian prodigy Ramanujan

$$n! \approx \sqrt{\pi} (\frac{n}{e})^n \sqrt[6]{8n^3 + 4n^2 + n + \frac{1}{30}} \tag{4.127}$$

This is completely different the from Stirling formula given in (4.126). However, if we retain only the first term inside the square root of (4.127), we obtain precisely the first term of (4.126) or (4.109). The biography of Ramanujan is given at the back of this book.

4.2.9 Important Formulas

To conclude our introduction to Euler's gamma function, some important identities of the gamma function are reported here without proof:

$$\Gamma(\frac{1}{3})\Gamma(\frac{2}{3}) = \frac{2\pi}{\sqrt{3}} \tag{4.128}$$

$$\frac{\Gamma(z)}{\Gamma(z-n)} = (-1)^n \frac{\Gamma(-z+n+1)}{\Gamma(-z+1)} \tag{4.129}$$

$$\frac{\Gamma(-z+n)}{\Gamma(-z)} = (-1)^n \frac{\Gamma(z+1)}{\Gamma(z-n+1)} \tag{4.130}$$

$$\Gamma(\frac{1}{3}) = 2.6789385347... \tag{4.131}$$

$$\Gamma(z) = p^z \int_0^\infty t^{z-1} e^{-pt} dt \tag{4.132}$$

$$\frac{1}{\Gamma(z)} = z \lim_{n\to\infty} \left\{ \frac{1}{n^z} \prod_{k=1}^{n} (1+\frac{z}{k}) \right\} \tag{4.133}$$

Euler's interpolation formula

$$\lim_{n\to\infty} \frac{\Gamma(n+z)}{n^z \Gamma(n)} = 1 \tag{4.134}$$

Wendel limit

$$\lim_{n\to\infty} \left[x^{b-a} \frac{\Gamma(x+a)}{\Gamma(x+b)} \right] = 1 \tag{4.135}$$

Legendre formula

$$\psi(z) = -\gamma + \int_0^1 \frac{t^{z-1}-1}{t-1} dt \tag{4.136}$$

Binet's first formula

$$\ln \Gamma(x) = (x-\frac{1}{2}) \ln x - x + \ln \sqrt{2\pi} + \int_0^\infty (\frac{1}{e^t - 1} - \frac{1}{t} + \frac{1}{2}) \frac{e^{-xt}}{t} dt \tag{4.137}$$

Binet's second formula

$$\ln \Gamma(x) = (x-\frac{1}{2}) \ln x - x + \ln \sqrt{2\pi} + 2\int_0^\infty \frac{\tan^{-1}(t/x)}{e^{2\pi t} - 1} dt \tag{4.138}$$

Mittag-Leffler expansion (this has been proved in (4.52))

$$\Gamma(z) = \sum_{n=0}^{\infty} \frac{(-1)^n}{n!(n+z)} + \int_1^\infty e^{-t} t^{z-1} dt \tag{4.139}$$

In recent years, some physical problems involve the study of complex arguments of the gamma function. However, I am not aware of its application in engineering and mechanics yet. Nevertheless, as a general introduction, some cases of complex argument are reported here without proof (in fact, their proofs are straightforward).

If the argument of the gamma function is complex, we have the following identities

$$\Gamma(1+iy) = iy\Gamma(iy) \tag{4.140}$$

$$\Gamma(iy)\Gamma(-iy) = \frac{\pi}{y\sinh \pi y} \tag{4.141}$$

$$\Gamma(\frac{1}{2}+iy)\Gamma(\frac{1}{2}-iy) = \frac{\pi}{\cosh \pi y} \tag{4.142}$$

$$\Gamma(1+iy)\Gamma(1-iy) = \frac{\pi y}{\sinh \pi y} \tag{4.143}$$

$$\mathrm{Re}\{\ln\Gamma(iy)\} = \frac{1}{2}\ln(\frac{\pi}{y\sinh \pi y}) \tag{4.144}$$

See Problems 4.20, 4.21, and 4.24 for two more formulas for the gamma function with a complex argument. There are also special functions which are closely related to the gamma function, including the incomplete gamma function (e.g., Lebedev, 1972), multiple gamma function, Barnes G-function (or double gamma function), and beta function. But these are outside the scope of the present book. Thus, the studies of the gamma function are by no means simple.

4.3 HADAMARD FACTORIAL FUNCTION

One major problem with the definition of the gamma function by Euler is that it possesses an infinite number of singularities at $n = 0, -1, -2, -3, \ldots$ In fact, there is more than one way to define the factorial function that coincides with the factorial for the limiting case of the integer. An interesting choice was proposed in 1894 by French mathematician J. Hadamard (Davis, 1959) and the following definition of the Hadamard factorial function was reported in Question 46 of Chapter 12 of Whittaker and Watson (1927) (without mentioning the name of Hadamard):

$$H(x) = \sqrt{\pi}(V\frac{dU}{dx} - U\frac{dV}{dx}) \tag{4.145}$$

where U and V are defined as:

$$U(x) = \frac{2^{x/2}}{\Gamma(1-\frac{x}{2})}, \quad V(x) = \frac{2^{x/2}}{\Gamma(\frac{1}{2}-\frac{x}{2})} \tag{4.146}$$

To the best of my knowledge, Hadamard's factorial function $H(x)$ has not been covered in any textbook. Some of the identities to be covered in this section are new and have not been published before.

First, we will show the following recursive formula of the Hadamard factorial function

$$H(x+1) = xH(x) + \frac{1}{\Gamma(1-x)} \tag{4.147}$$

Note the definition of the digamma function and the following result

$$\frac{d}{dx}2^{x/2} = (\frac{1}{2}\ln 2)2^{x/2} \tag{4.148}$$

We obtain

$$\frac{dU}{dx} = \frac{2^{x/2}}{2\Gamma(1-\frac{x}{2})}[\ln 2 + \psi(1-\frac{x}{2})] \tag{4.149}$$

$$\frac{dV}{dx} = \frac{2^{x/2}}{2\Gamma(\frac{1}{2}-\frac{x}{2})}[\ln 2 + \psi(\frac{1}{2}-\frac{x}{2})] \tag{4.150}$$

Substitution of (4.149) and (4.150) into (4.145) results in

$$H(x) = \frac{\sqrt{\pi}2^x}{2\Gamma(1-\frac{x}{2})\Gamma(\frac{1}{2}-\frac{x}{2})}[\psi(1-\frac{x}{2}) - \psi(\frac{1}{2}-\frac{x}{2})] \tag{4.151}$$

Replacing x by $x+1$, we obtain

$$\begin{aligned} H(x+1) &= \frac{\sqrt{\pi}2^{x+1}}{2\Gamma(1-\frac{x+1}{2})\Gamma(\frac{1}{2}-\frac{x+1}{2})}[\psi(1-\frac{x+1}{2}) - \psi(\frac{1}{2}-\frac{x+1}{2})] \\ &= \frac{\sqrt{\pi}2^x}{\Gamma(\frac{1}{2}-\frac{x}{2})\Gamma(-\frac{x}{2})}[\psi(\frac{1}{2}-\frac{x}{2}) - \psi(-\frac{x}{2})] \end{aligned} \tag{4.152}$$

On the other hand,

$$\begin{aligned} xH(x) &= \frac{\sqrt{\pi}2^{x+1}}{2\Gamma(1-\frac{x}{2})\Gamma(\frac{1}{2}-\frac{x}{2})}[\psi(1-\frac{x}{2}) - \psi(\frac{1}{2}-\frac{x}{2})] \\ &= -\frac{\sqrt{\pi}2^x}{\Gamma(-\frac{x}{2})\Gamma(\frac{1}{2}-\frac{x}{2})}[-\frac{2}{x} + \psi(-\frac{x}{2}) - \psi(\frac{1}{2}-\frac{x}{2})] \end{aligned} \tag{4.153}$$

We have used the following identities in arriving at the last equation of (4.153)

$$\Gamma(1-\frac{x}{2}) = -\frac{x}{2}\Gamma(-\frac{x}{2}), \quad \psi(1-\frac{x}{2}) = -\frac{2}{x} + \psi(-\frac{x}{2}) \tag{4.154}$$

Next, we consider

$$\begin{aligned}
xH(x)+\frac{1}{\Gamma(1-x)} &= \frac{\sqrt{\pi}2^x}{\Gamma(-\frac{x}{2})\Gamma(\frac{1}{2}-\frac{x}{2})}[\frac{2}{x}-\psi(-\frac{x}{2})+\psi(\frac{1}{2}-\frac{x}{2})]+\frac{1}{\Gamma(1-x)} \\
&= -\frac{1}{\Gamma(1-x)}+\frac{\sqrt{\pi}2^x}{\Gamma(-\frac{x}{2})\Gamma(\frac{1}{2}-\frac{x}{2})}[-\psi(-\frac{x}{2})+\psi(\frac{1}{2}-\frac{x}{2})]+\frac{1}{\Gamma(1-x)} \\
&= \frac{\sqrt{\pi}2^x}{\Gamma(-\frac{x}{2})\Gamma(\frac{1}{2}-\frac{x}{2})}[-\psi(-\frac{x}{2})+\psi(\frac{1}{2}-\frac{x}{2})] \\
&= H(x+1)
\end{aligned} \tag{4.155}$$

In obtaining the second line of (4.155), we have applied the result of Example 4.4. This is the required result.

Consider the special case that $x \to n$ and (4.155) becomes

$$H(n+1) = nH(n)+\frac{1}{\Gamma(1-n)} \tag{4.156}$$

For integer n (= 1,2,...), the gamma function $\Gamma(1-n)$ is unbounded or

$$\frac{1}{\Gamma(1-k)} \to 0, \quad k = 1,2,3,... \tag{4.157}$$

Therefore, for positive integers we must have

$$H(n+1) = nH(n) = n(n-1)\cdots 1\cdot H(1) = \Gamma(n+1) \tag{4.158}$$

where $H(1)$ = 1 (see result of Problem 4.29). Clearly, the Hadamard factorial function is another choice that extends the factorial function to the case of the non-integer argument.

One major problem of Euler's gamma function is that the gamma function becomes unbounded at zero and all negative integers. We will examine whether the Hadamard factorial function will be finite-valued. Note first that the Hadamard factorial function given in (4.151) can be rewritten as

$$H(x) = \frac{\sqrt{\pi}2^x}{2}\left\{\frac{\psi(1-\frac{x}{2})}{\Gamma(1-\frac{x}{2})\Gamma(\frac{1}{2}-\frac{x}{2})}-\frac{\psi(\frac{1}{2}-\frac{x}{2})}{\Gamma(1-\frac{x}{2})\Gamma(\frac{1}{2}-\frac{x}{2})}\right\} \tag{4.159}$$

The only problems clearly arises from gamma and digamma functions. We know that both gamma and digamma functions are finite except at 0, −1, −2, ... (all negative integers). Thus, we only need to investigate the behavior of $H(x)$ at those arguments where Γ and ψ becomes unbounded.

Note that

$$\Gamma(\frac{1}{2}-\frac{x}{2}) \to \infty, \quad \psi(\frac{1}{2}-\frac{x}{2}) \to \infty, \quad x = 1,3,5,7,... \tag{4.160}$$

$$\Gamma(1-\frac{x}{2}) \to \infty, \quad \psi(1-\frac{x}{2}) \to \infty, \quad x = 2,4,6,8,... \tag{4.161}$$

Therefore, we only need to study the behavior of the following terms:

$$\chi_1(\frac{1}{2}-\frac{x}{2}) = \left\{ \frac{\psi(\frac{1}{2}-\frac{x}{2})}{\Gamma(\frac{1}{2}-\frac{x}{2})} \right\}, \quad \text{for} \quad x = 1,3,5,7,... \tag{4.162}$$

and

$$\chi_2(1-\frac{x}{2}) = \left\{ \frac{\psi(1-\frac{x}{2})}{\Gamma(1-\frac{x}{2})} \right\}, \quad \text{for} \quad x = 2,4,6,8,... \tag{4.163}$$

Both χ_1 and χ_2 have the indeterminate form of ∞/∞ as x approaches odd and even integers respectively, but we will show that both of them have a finite limit as x approaches an integer. Consider the reflection formula for the gamma function given in (4.35) with $z = x/2+1/2$; we have

$$\Gamma(\frac{1}{2}-\frac{x}{2}) = \Gamma[1-(\frac{1}{2}+\frac{x}{2})] = \frac{\pi}{\Gamma(\frac{1}{2}+\frac{x}{2})\sin\pi x} \tag{4.164}$$

Similarly, using the reflection formula for the digamma function given in (4.92) with $z = x/2+1/2$, we obtain

$$\psi(\frac{1}{2}-\frac{x}{2}) = \psi[1-(\frac{1}{2}+\frac{x}{2})] = \psi(\frac{1}{2}+\frac{x}{2}) + \pi\cot\pi x \tag{4.165}$$

Substitution of (4.164) and (4.165) into (4.162) yields

$$\begin{aligned}\chi_1(\frac{1}{2}-\frac{x}{2}) &= \frac{\psi(\frac{1}{2}-\frac{x}{2})}{\Gamma(\frac{1}{2}-\frac{x}{2})} = \frac{\Gamma(\frac{1}{2}+\frac{x}{2})\sin\pi x}{\pi}[\psi(\frac{1}{2}+\frac{x}{2}) + \pi\cot\pi x] \\ &= \frac{1}{\pi}\psi(\frac{1}{2}+\frac{x}{2})\Gamma(\frac{1}{2}+\frac{x}{2})\sin\pi x + \cos\pi x\Gamma(\frac{1}{2}+\frac{x}{2})\end{aligned} \tag{4.166}$$

For odd integers, we have the following limiting value

$$\lim_{x\to 2m+1}\chi_1(x) = \lim_{x\to 2m+1}\left\{ \frac{\psi(\frac{1}{2}-\frac{x}{2})}{\Gamma(\frac{1}{2}-\frac{x}{2})} \right\} = (-1)^{2m+1}\Gamma(1+m) \tag{4.167}$$

This is finite for all m = 0,1,2,... Similarly, we find

$$\begin{aligned}\chi_2(1-\frac{x}{2}) &= \frac{\psi(1-\frac{x}{2})}{\Gamma(1-\frac{x}{2})} = \frac{1}{\pi}\left\{ \psi(\frac{x}{2}) + \pi\cot\pi x \right\}\sin\pi x\Gamma(\frac{x}{2}) \\ &= \frac{1}{\pi}\psi(\frac{x}{2})\sin\pi x\Gamma(\frac{x}{2}) + \cos\pi x\Gamma(\frac{x}{2})\end{aligned} \tag{4.168}$$

$$\lim_{x\to 2m} \chi_2(1-\frac{x}{2}) = \lim_{x\to 2m} \left\{ \frac{\psi(1-\frac{x}{2})}{\Gamma(1-\frac{x}{2})} \right\} = (-1)^{2m}\Gamma(m) = \Gamma(m) \tag{4.169}$$

Thus, the Hadamard factorial function is finite for all values of x. More specifically, the Hadamard factorial function of the integer argument becomes

$$\begin{aligned} H(x) &= -\frac{\sqrt{\pi}\,2^{2m+1}}{2} \frac{\Gamma(1+m)}{\Gamma(\frac{1}{2}-m)}, \quad x = 2m+1 \\ &= \frac{\sqrt{\pi}\,2^{2n}}{2} \frac{\Gamma(n)}{\Gamma(1/2-n)}, \quad x = 2n \end{aligned} \tag{4.170}$$

where m = 0,1,2,... and n = 1,2,... It can be shown that the Hadamard factorial function can also be defined in the following form (Davis, 1959):

$$H(x) = \frac{1}{\Gamma(1-x)} \frac{d}{dx} \ln \left\{ \frac{\Gamma(\frac{1-x}{2})}{\Gamma(1-\frac{x}{2})} \right\} \tag{4.171}$$

To prove this, consider the differentiation term on the right of (4.171) as

$$\begin{aligned} \frac{d}{dx} \ln \left\{ \frac{\Gamma(\frac{1-x}{2})}{\Gamma(1-\frac{x}{2})} \right\} &= \frac{d}{dx} \left\{ \ln\Gamma(\frac{1-x}{2}) - \ln\Gamma(1-\frac{x}{2}) \right\} \\ &= -\frac{1}{2} \left\{ \psi(\frac{1-x}{2}) - \psi(1-\frac{x}{2}) \right\} \end{aligned} \tag{4.172}$$

By noting that

$$\Gamma(1-x) = -x\Gamma(-x) \tag{4.173}$$

we have

$$\frac{1}{\Gamma(1-x)} \frac{d}{dx} \ln \left\{ \frac{\Gamma(\frac{1-x}{2})}{\Gamma(1-\frac{x}{2})} \right\} = \frac{1}{2x\Gamma(-x)} \left\{ \psi(\frac{1-x}{2}) - \psi(1-\frac{x}{2}) \right\} \tag{4.174}$$

Next, we observe that for $z = -x/2$, the Legendre duplication formula given in (4.35) becomes

$$\Gamma(-x) = \frac{1}{\sqrt{2\pi}} 2^{-x-1/2} \Gamma(-\frac{x}{2}) \Gamma(\frac{1}{2}-\frac{x}{2}) \tag{4.175}$$

Substitution of (4.175) into (4.174) yields the RHS of (4.174) as

$$RHS = \frac{2^x \sqrt{\pi}}{x\Gamma(-\frac{x}{2})\Gamma(\frac{1}{2}-\frac{x}{2})} \left\{ \psi(\frac{1-x}{2}) - \psi(1-\frac{x}{2}) \right\} \tag{4.176}$$

On the other hand, Legendre's duplication formula gives

$$\Gamma(1-\frac{x}{2}) = -\frac{x}{2}\Gamma(-\frac{x}{2}) \tag{4.177}$$

Using this result,

$$RHS = \frac{2^x\sqrt{\pi}}{2\Gamma(1-\frac{x}{2})\Gamma(\frac{1}{2}-\frac{x}{2})}\left\{\psi(1-\frac{x}{2}) - \psi(\frac{1-x}{2})\right\} = H(x)\,. \tag{4.178}$$

This proves (4.171), which is given in both Whittaker and Watson (1927) and Davis (1959).

The intermediate result of (4.174) gives a slightly different definition for the Hadamard factorial function

$$H(x) = \frac{1}{2x\Gamma(-x)}\left\{\psi(\frac{1-x}{2}) - \psi(1-\frac{x}{2})\right\} \tag{4.179}$$

Therefore, $H(x)$ can be defined equivalently by (4.151), (4.159), (4.171), and (4.179).

4.3.1 Recurrence Formula

The recurrence formula for the Hadamard factorial function can be found in Alzer (2009):

$$H(x+1) = xH(x) + \frac{1}{\Gamma(1-x)} \tag{4.180}$$

This recurrence formula has been proved in the last section (see (4.147)). We have used this result to see why it leads to the factorial function type for the integer argument (see (4.158)).

4.3.2 Reflection Formula

Before we consider the reflection formula for the Hadamard factorial function, we first report the following Luschny formula for the Hadamard gamma function (Luschny, 2006; Alzer, 2009)

$$H(x) = \Gamma(x)\left\{1 + \frac{\sin \pi x}{2\pi}[\psi(\frac{x}{2}) - \psi(\frac{x}{2}+\frac{1}{2})]\right\} \tag{4.181}$$

The proof of this identity is given in the next example.

Example 4.6 Prove the Luschny formula for the Hadamard gamma function

$$H(x) = \Gamma(x)\left\{1 + \frac{\sin \pi x}{2\pi}[\psi(\frac{x}{2}) - \psi(\frac{x}{2}+\frac{1}{2})]\right\} \tag{4.182}$$

Solution: First, the reflection formula for the gamma function can be written as

$$\frac{\sin \pi x}{\pi} = \frac{1}{\Gamma(x)\Gamma(1-x)} \tag{4.183}$$

Substitution of (4.183) into the right hand side of (4.182) gives

$$RHS = \Gamma(x)\left\{1+\frac{1}{2\Gamma(x)\Gamma(1-x)}[\psi(\frac{x}{2})-\psi(\frac{x}{2}+\frac{1}{2})]\right\} \tag{4.184}$$

The reflection formula for the digamma function gives

$$\psi(\frac{x}{2}) = \psi(1-\frac{x}{2}) - \frac{\pi}{\tan(\pi x/2)} \tag{4.185}$$

Recall from the result of Problem 4.2 that

$$\psi(\frac{1}{2}+z) = \psi(\frac{1}{2}-z) + \pi \tan \pi z \tag{4.186}$$

Putting $z = x/2$ into (4.186) gives

$$\psi(\frac{1}{2}+\frac{x}{2}) = \psi(\frac{1}{2}-\frac{x}{2}) + \pi \tan(\frac{\pi x}{2}) \tag{4.187}$$

Subtracting (4.187) from (1.185) gives

$$\begin{aligned}
\psi(\frac{x}{2})-\psi(\frac{x}{2}+\frac{1}{2}) &= \psi(1-\frac{x}{2})-\psi(\frac{1}{2}-\frac{x}{2})-\pi\left\{\frac{1}{\tan(\frac{\pi x}{2})}+\tan(\frac{\pi x}{2})\right\} \\
&= \psi(1-\frac{x}{2})-\psi(\frac{1}{2}-\frac{x}{2})-\frac{\pi}{\tan(\frac{\pi x}{2})}\left\{1+\tan^2(\frac{\pi x}{2})\right\} \\
&= \psi(1-\frac{x}{2})-\psi(\frac{1}{2}-\frac{x}{2})-\frac{2\pi}{\sin \pi x}
\end{aligned} \tag{4.188}$$

Putting this result into (4.184), we obtain

$$\begin{aligned}
RHS &= \Gamma(x)+\frac{1}{2\Gamma(1-x)}[\psi(1-\frac{x}{2})-\psi(\frac{1}{2}-\frac{x}{2})]-\frac{\pi}{\Gamma(1-x)\sin \pi x} \\
&= \frac{1}{2\Gamma(1-x)}[\psi(1-\frac{x}{2})-\psi(\frac{1}{2}-\frac{x}{2})] = H(x)
\end{aligned} \tag{4.189}$$

By virtue of the reflection formula of the gamma function given in (4.93), we get the final result of (4.182).

We now return to consider the reflection formula. First, we rewrite the Luschny formula as

$$H(x) = \Gamma(x)P(x) \tag{4.190}$$

$$P(x) = 1-g(x)\frac{\sin \pi x}{\pi}, \quad g(x) = \frac{x}{2}[\psi(\frac{x}{2}+\frac{1}{2})-\psi(\frac{x}{2})] \tag{4.191}$$

Next, we consider the Hadamard function using (4.171)

$$H(1-x) = \frac{1}{\Gamma(1-1+x)}\frac{d}{dx}\ln\left\{\frac{\Gamma(\frac{1-1+x}{2})}{\Gamma(1-\frac{1-x}{2})}\right\} = \frac{1}{\Gamma(x)}\frac{d}{dx}\ln\left\{\frac{\Gamma(\frac{x}{2})}{\Gamma(\frac{1}{2}+\frac{x}{2})}\right\} \tag{4.192}$$

Multiplying (4.190) and (4.192), we have

$$H(x)H(1-x) = \frac{1}{\Gamma(x)\Gamma(1-x)} \frac{d}{dx} \ln\left\{ \frac{\Gamma(\frac{x}{2})}{\Gamma(\frac{1}{2}+\frac{x}{2})} \right\} \frac{d}{dx} \ln\left\{ \frac{\Gamma(\frac{1-x}{2})}{\Gamma(1-\frac{x}{2})} \right\} \tag{4.193}$$

$$= \frac{\sin \pi x}{4\pi} [\psi(1-\frac{x}{2}) - \psi(\frac{1-x}{2})][\psi(\frac{x}{2}) - \psi(\frac{1}{2}+\frac{x}{2})]$$

The final expression of (4.193) results from differentiation as well as using the reflection formula (4.35) for the gamma function. We note that

$$g(1-x) = \frac{(1-x)}{2}[\psi(\frac{1-x}{2}+\frac{1}{2}) - \psi(\frac{1-x}{2})] = \frac{(1-x)}{2}[\psi(1-\frac{x}{2}) - \psi(\frac{1-x}{2})] \tag{4.194}$$

Now, it is clear that the right hand side of (4.193) can be expressed in term of $g(x)$ defined in (4.191) as

$$H(x)H(1-x) = -\frac{\sin \pi x}{\pi x(1-x)} g(x)g(1-x) \tag{4.195}$$

We call this the reflection formula for the Hadamard factorial function; it is a new formula for the Hadamard function that has not been reported before. This formula bears similarity with the reflection formula of Euler's gamma function given in (4.35).

4.4 HARMONIC EQUATION

With our basic knowledge of the gamma function and factorial function, we return to series solution of second order ODEs. First, we consider the following second order ODE

$$y'' + \alpha(x)y' + \beta(x)y = 0 \tag{4.196}$$

where $\alpha(x)$ and $\beta(x)$ are analytic functions of x and thus can be expanded in a Taylor series expansion. That is, they are single-valued and possess derivatives of all orders. Thus, the solution of y exists in series form around a regular point a

$$y = \sum_{r=0}^{\infty} c_r (x-a)^r \tag{4.197}$$

The term "regular point" was coined by German mathematician Thomé, who was a student of Weierstrass, in 1873. We will start the series solution technique for the simplest case, that is, the harmonic equation with the solution being sine and cosine. We have introduced the following differential equation (called the harmonic equation) in Chapter 1:

$$\frac{d^2 y}{dx^2} + y = 0 \tag{4.198}$$

We seek a series solution for (4.198) as

$$y = \sum_{r=0}^{\infty} c_r x^r \tag{4.199}$$

$$y'' = \sum_{r=0}^{\infty} r(r-1)c_r x^{r-2} \tag{4.200}$$

The harmonic equation (4.198) requires

$$\sum_{r=0}^{\infty} r(r-1)c_r x^{r-2} + \sum_{r=0}^{\infty} c_r x^r = 0 \tag{4.201}$$

The index of the infinite series is then shifted by $r = k-2$ and consequently (4.201) becomes

$$\sum_{k=-2}^{\infty} (k+2)(k+1)c_{k+2} x^k + \sum_{k=0}^{\infty} c_k x^k = 0 \tag{4.202}$$

However, the first two terms of the first series are evidently zero, so (4.202) can be simplified as

$$(k+2)(k+1)c_{k+2} + c_k = 0 \tag{4.203}$$

Thus,

$$c_{k+2} = -\frac{c_k}{(k+2)(k+1)} \tag{4.204}$$

Considering the even terms (i.e., $k+2 = 2m$), the recurrence formula becomes

$$c_{2m} = -\frac{c_{2m-2}}{(2m-1)(2m)} \tag{4.205}$$

We can reapply the recurrence formula (4.205) to the coefficient c_{2m-2} on the right of (4.205) to get

$$c_{2m} = \frac{(-1)^2 c_{2m-4}}{(2m-1)(2m-3)(2m)(2m-2)} \tag{4.206}$$

Repeat the application of the recurrence formula (4.205) m times and the following formula is obtained

$$c_{2m} = \frac{(-1)^m c_0}{(2m)!} \tag{4.207}$$

For odd terms in the series solution, we set $k+2 = 2m+1$ in (4.205)

$$c_{2m+1} = -\frac{c_{2m-1}}{(2m+1)(2m)} \tag{4.208}$$

Similarly, reapplying (4.208) m times, we get

$$c_{2m+1} = \frac{(-1)^m c_1}{(2m+1)!} \tag{4.209}$$

Back substitution of these coefficients into the series solution, we get

$$y = \sum_{r=0}^{\infty} \frac{(-1)^m c_0}{(2m)!} x^{2m} + \sum_{r=0}^{\infty} \frac{(-1)^m c_1}{(2m+1)!} x^{2m+1}$$

$$= c_0 \left\{1 - \frac{x^2}{2!} + \frac{x^4}{4!} + ...\right\} + c_1 \left\{x - \frac{x^3}{3!} + \frac{x^5}{5!} + ...\right\} \tag{4.210}$$

$$= c_0 \cos x + c_1 \sin x$$

This result, of course, agrees with the conclusion that we got in Chapter 1 that the general solution for the harmonic equation is sine and cosine.

4.5 FUCHSIAN ODE WITH REGULAR SINGULAR POINTS

Before we consider the Bessel equation, let us consider the series solution of a more general second order ordinary differential equation

$$(x-a)^2 y'' + (x-a)\alpha(x) y' + \beta(x) y = 0 \tag{4.211}$$

where $\alpha(x)$ and $\beta(x)$ are analytic functions of x and thus can be expanded in a Taylor series expansion. Rearranging (4.211), we get

$$y'' + \frac{\alpha(x)}{(x-a)} y' + \frac{\beta(x)}{(x-a)^2} y = 0 \tag{4.212}$$

This type of differential equation is called a Fuchsian type ODE. Clearly, this differential equation is singular at $x = a$. However, if the singularity does not concur with (4.212), the technique discussed in this chapter does not apply. The point $x = a$ is called a regular singular point, and this term was coined by Thomé in 1873, as remarked earlier. For such equations, a standard procedure of series solution exists, which is called the Frobenius series.

By the assumption of $\alpha(x)$ and $\beta(x)$ being analytic, we can expand them in

$$\alpha(x) = \sum_{r=0}^{\infty} p_r (x-a)^r, \quad \beta(x) = \sum_{r=0}^{\infty} q_r (x-a)^r \tag{4.213}$$

And the series solution is assumed in the following form with an extra index ρ as

$$y(x) = \sum_{r=0}^{\infty} c_r (x-a)^{\rho+r} \tag{4.214}$$

This is called the Frobenius series. This extra index ρ allows us to satisfy the governing equation, and thus it depends on the type of ODE.

$$\sum_{r=0}^{\infty} c_r (\rho+r)(\rho+r-1)(x-a)^{\rho+r} + \sum_{s=0}^{\infty} p_s (x-a)^s \sum_{r=0}^{\infty} c_r (\rho+r)(x-a)^{\rho+r}$$
$$+ \sum_{s=0}^{\infty} q_s (x-a)^s \sum_{r=0}^{\infty} c_r (x-a)^{\rho+r} = 0 \tag{4.215}$$

Collecting the coefficient for $(x-a)^\rho$ and we find the indicial equation:

$$\rho^2 + (p_0 - 1)\rho + q_0 = 0 \tag{4.216}$$

if $c_0 \neq 0$. Clearly, there are two roots for the index ρ, namely ρ_1 and ρ_2. The term indicial equation was coined by Cayley. Collecting the coefficients for $(x-a)^{\rho+r}$ leads to

$$c_r(\rho+r)(\rho+r-1)+\sum_{s=0}^{\infty}[p_s(\rho+r-s)+q_s]c_{r-s}=0 \tag{4.217}$$

Grouping the first term in the infinite sum with the first term of (4.217), we obtain

$$c_r[(\rho+r)(\rho+r-1)+p_0(\rho+r)+q_0]+\sum_{s=1}^{\infty}[p_s(\rho+r-1)+q_s]c_{r-s}=0 \tag{4.218}$$

This recurrence formula allows us to determine the coefficient c_r.

In general, there are two independent solutions for the second order ODE (4.212)

$$y_1(x)=\sum_{r=0}^{\infty}c_r(x-a)^{\rho_1+r} \tag{4.219}$$

$$y_2(x)=\sum_{r=0}^{\infty}c_r^*(x-a)^{\rho_2+r} \tag{4.220}$$

where c_r is obtained from (4.218) with $\rho = \rho_1$ and c_r* is obtained from (4.218) with $\rho = \rho_2$ with $c_0 \neq 0$ and $c_0^* \neq 0$. However, special consideration is needed when the roots for ρ are not distinct or when their difference equals an integer. In summary, there are three scenarios depending on the roots of ρ.

Case 1: $\rho_1-\rho_2 \neq 0$, and $\rho_1-\rho_2 \neq$ integer

$$y_1(x)=\sum_{r=0}^{\infty}c_r(x-a)^{\rho_1+r},\quad y_2(x)=\sum_{r=0}^{\infty}c_r^*(x-a)^{\rho_2+r} \tag{4.221}$$

with $c_0 \neq 0$ and $c_0^* \neq 0$.

Case 2: $\rho_1=\rho_2$

$$y_1(x)=\sum_{r=0}^{\infty}c_r(x-a)^{\rho_1+r} \tag{4.222}$$

$$y_2(x)=y_1\ln(x-a)+(x-a)^{\rho_1}\sum_{r=1}^{\infty}b_{r_1}(x-a)^r \tag{4.223}$$

Case 3: $\rho_2=\rho_1-n$ (n = integer)

$$y_1(x)=\sum_{r=0}^{\infty}c_r(x-a)^{\rho_1+r} \tag{4.224}$$

$$y_2(x)=g_n y_1\ln(x-a)+(x-a)^{\rho_2}\sum_{r=0}^{\infty}b_r(x-a)^r \tag{4.225}$$

where g_n is the coefficient of x_n of the expansion of the following term

$$\frac{x^{n+1}}{\{y_1(x-a)\}^2}\exp\left[-\int_0^x p(u)du\right] \tag{4.226}$$

The proof of this result was outlined in Section 10.15 of Copson (1935) and will be proved in detail here. Note that a similar result was given in Sneddon (1956) but there are various typos and the results are given without proof.

If $\rho_1-\rho_2 = n$ and n is an integer or zero, then for $n \neq 0$, we have case 3, otherwise it is case 2. Therefore, cases 3 and 2 can be considered simultaneously. It is a well-known theorem in differential equations that if a solution of an n-th order differential equation is known, we can depress the order of the differential equation to $n-1$ (e.g., Forsyth, 1956). Using this idea, we first rewrite (4.212) as

$$y'' + p(x)y' + q(x)y = 0 \tag{4.227}$$

Then, we define a new unknown v as:

$$y = y_1(x)v(x) \tag{4.228}$$

where $y_1(x)$ is the known solution for exponent ρ_1 and the purpose is to find the unknown function $v(x)$. Then, we have

$$\frac{dy}{dx} = y_1'v + y_1v' \tag{4.229}$$

$$\frac{d^2y}{dx^2} = y_1''v + 2y_1'v' + y_1v'' \tag{4.230}$$

Substitution of these results into (4.227) gives

$$(y_1'' + py_1' + qy_1)v + y_1v'' + (2y_1' + py_1)v' = 0 \tag{4.231}$$

The first term in the bracket is clearly zero since $y_1(x)$ is the solution of (4.227). Thus, we have

$$v'' + (p + 2\frac{y_1'}{y_1})v' = 0 \tag{4.232}$$

Clearly, from Section 3.5.8 we can let

$$u = v' \tag{4.233}$$

such that

$$u' + (p + 2\frac{y_1'}{y_1})u = 0 \tag{4.234}$$

This first order ODE can be solved exactly because it is separable:

$$\frac{du}{u} + (p + 2\frac{y_1'}{y_1})dx = 0 \tag{4.235}$$

$$\ln u = -\int (p + 2\frac{y_1'}{y_1})dx + C_1 \tag{4.236}$$

The final solution is

$$u = B\exp[-\int (p + 2\frac{y_1'}{y_1})dx] \tag{4.237}$$

Let us simplify the integrand as

$$-\int(p+2\frac{y_1'}{y_1})dx=-\int pdx-2\int\frac{y_1'}{y_1}dx$$
$$=-\int pdx-2\ln y_1 \tag{4.238}$$
$$=-\int pdx+\ln\{\frac{1}{y_1^2}\}$$

Back substituting of (4.237) and (4.238) into (4.233) and conducting integration with respect to the argument, we obtain

$$v(x)=A+B\int^x\frac{1}{\{y_1(\xi)\}^2}\exp[-\int^{\xi}p(\zeta)d\zeta]d\xi \tag{4.239}$$

Substitution of (4.239) into (4.228) yields the second solution

$$y(x)=y_1(x)\int^x\frac{1}{\{y_1(\xi)\}^2}\exp[-\int^{\xi}p(\zeta)d\zeta]d\xi \tag{4.240}$$

In obtaining the above solution, we have dropped the integration constants.

To further simplify this solution, we recall the characteristic equation from (4.216)

$$\rho^2+(p_0-1)\rho+q_0=0 \tag{4.241}$$

We now consider the special case that $\rho_2=\rho_1-n$ into (4.241) gives

$$(\rho_1-n)^2+(p_0-1)(\rho_1-n)+q_0=0 \tag{4.242}$$

But, we have from (4.241) that

$$q_0=-\rho_1^2-(p_0-1)\rho_1 \tag{4.243}$$

Substitution of (4.243) into (4.242) gives

$$p_0=n+1-2\rho_1 \tag{4.244}$$

Let us consider the Frobenius series for the first solution and recall the Laurent's series expansion of $p(x)$:

$$y_1=x^{\rho_1}(a_0+a_1x+...) \tag{4.245}$$

$$xp(x)=\sum_{r=0}^{\infty}p_rx^r,\quad p(x)=\frac{p_0}{x}+p_1+p_2x+... \tag{4.246}$$

By virtue of (4.245) and (4.246), the integrand of (4.240) becomes

$$I=\frac{1}{\{y_1(x)\}^2}\exp[-\int^x p(x)dx]$$
$$=\frac{1}{x^{2\rho_1}\{a_0+a_1x+...\}^2}\exp\{\int^x[\frac{2\rho_1-1-n}{x}-p_1-p_2x-...]dx\} \tag{4.247}$$
$$=\frac{\exp\{\ln x^{2\rho_1-1-n}\}}{x^{2\rho_1}\{a_0+a_1x+...\}^2}\exp\{-\int^x(p_1+p_2x+...)dx\}$$

Therefore, the integrand can be simplified as

$$I = \frac{x^{-1-n}}{\{a_0 + a_1 x + ...\}^2} \exp\{-\int_x (p_1 + p_2 x + ...)dx\} = x^{-1-n} g(x) \tag{4.248}$$

We have

$$g(0) = \frac{1}{a_0^2} \tag{4.249}$$

For $a_0 \neq 0$, $g(x)$ must be regular and thus $g(x)$ allows a Taylor series expansion as:

$$g(x) = \sum_{m=0}^{\infty} g_m x^m \tag{4.250}$$

Thus, we have

$$\begin{aligned} y(x) &= y_1(x) \int_x x^{-1-n} \sum_{m=0}^{\infty} g_m x^m dx = y_1(x) \int_x \sum_{m=0}^{\infty} g_m x^{m-1-n} dx \\ &= y_1(x) \int_x \{\sum_{m=0}^{n-1} g_m x^{m-1-n} + g_n \frac{1}{x} + \sum_{m=n+1}^{\infty} g_m x^{m-1-n}\} dx \end{aligned} \tag{4.251}$$

This can be integrated readily as:

$$y(x) = y_1(x) \{\sum_{m=0}^{n-1} \frac{g_m x^{m-n}}{m-n} + g_n \ln x + \sum_{m=n+1}^{\infty} \frac{g_m x^{m-n}}{m-n}\} \tag{4.252}$$

When $n = 0$ (i.e., case 2), we have

$$y(x) = g_o y_1(x) \ln x + x^{\rho_1} \sum_{k=0}^{\infty} a_k x^k \sum_{m=1}^{\infty} \frac{g_m x^m}{m} \tag{4.253}$$

Let $i = m-1$; we have

$$\sum_{m=1}^{\infty} \frac{g_m x^m}{m} = \sum_{i=0}^{\infty} \frac{g_{i+1} x^{i+1}}{i+1} = x \sum_{i=0}^{\infty} \frac{g_{i+1} x^i}{i+1} \tag{4.254}$$

Thus, the two summations can be combined as one

$$y(x) = g_o y_1(x) \ln x + x^{\rho_1 + 1} \sum_{m=0}^{\infty} b_m x^m \tag{4.255}$$

Since $g_0 = g(0) \neq 0$ as given in (4.249), we can divide (4.255) through by g_0:

$$y_2(x) = y_1(x) \ln x + x^{\rho_1} \sum_{m=1}^{\infty} \bar{b}_m x^m \tag{4.256}$$

When $n \neq 0$, we have

$$x^{\rho_1} \sum_{k=0}^{\infty} a_k x^k \sum_{m=0, m\neq n}^{\infty} \frac{g_m x^{m-n}}{m-n} = x^{\rho_1 - n} \sum_{j=0}^{\infty} c_j x^j = x^{\rho_2} \sum_{j=0}^{\infty} c_j x^j \tag{4.257}$$

With this, the final solution is of the form:

$$y(x) = g_n y_1(x) \ln x + x^{\rho_2} \sum_{m=0}^{\infty} c_m x^m \tag{4.258}$$

where g_n may be zero and therefore, we cannot scale it like (4.256). This completes the proof for the case of $a = 0$ in (4.221) to (4.226), and this analysis can be easily extended to the case of nonzero a.

4.6 BESSEL EQUATION

We will now consider the following Bessel equation

$$x^2 y'' + xy' + (x^2 - \nu^2) y = 0 \tag{4.259}$$

where ν is called the order of the Bessel equation. Clearly, it falls into our class of ODEs with a regular singular point at $x = 0$. The zero order Bessel equation was first considered and solved by Daniel Bernoulli in 1733 when he considered the problem of vibrations of chains. The first order Bessel equation was derived by Bernoulli when he extended chain vibrations to non-uniform sections. Euler, in 1739, considered a chain with a weight proportional to x_n and arrived at the Bessel function of general order. In 1764, Euler also found that the Bessel function is related to vibrations of a stretched membrane. Note that the Bessel function with an imaginary argument will lead to a modified Bessel function and it will be considered in Section 4.6.4.

Another popular form of the Bessel equation is

$$\frac{d^2 Z_\nu}{d\rho^2} + \frac{1}{\rho}\frac{dZ_\nu}{d\rho} + \left(1 - \frac{\nu^2}{\rho^2}\right) Z_\nu = 0 \tag{4.260}$$

The subscript ν is used to highlight the fact that the solution is a function of ν.

4.6.1 Frobenius Series for Non-Integer Order

A solution in terms of the Frobenius series is sought

$$Z_\nu = \sum_{r=0}^{\infty} a_r \rho^{\lambda + r} = \rho^\lambda (a_0 + a_1\rho + a_2\rho^2 + \ldots + a_k \rho^k + \ldots) \tag{4.261}$$

Differentiation of this series solution once and twice leads respectively to

$$\frac{1}{\rho}\frac{dZ_\nu}{d\rho} = \lambda \rho^{\lambda-2}(a_0 + a_1\rho + a_2\rho^2 + \ldots + a_k\rho^k + \ldots) + \rho^{\lambda-1}(a_1 + 2a_2\rho + \ldots + ka_k\rho^{k-1} + \ldots) \tag{4.262}$$

$$\begin{aligned}\frac{d^2 Z_\nu}{d\rho^2} &= \lambda(\lambda-1)\rho^{\lambda-2}(a_0 + a_1\rho + a_2\rho^2 + \ldots + a_k\rho^k + \ldots) \\ &+ 2\lambda\rho^{\lambda-1}(a_1 + 2a_2\rho + \ldots + ka_k\rho^{k-1} + \ldots) \\ &+ \rho^\lambda(2a_2 + 6a_3\rho + \ldots + k(k-1)a_k\rho^{k-2} + \ldots)\end{aligned} \tag{4.263}$$

$$-\frac{\nu^2}{\rho^2} Z_\nu = -\nu^2 \rho^{\lambda-2}(a_0 + a_1\rho + a_2\rho^2 + \ldots + a_k\rho^k + \ldots) \tag{4.264}$$

Substituting these into (4.260) and collecting the coefficients for $\rho^{\lambda-2}$ gives the following indicial equation for λ

$$[\lambda(\lambda-1)+\lambda-\nu^2]a_0=0 \tag{4.265}$$

For nonzero a_0, two roots for the indices in the Frobenius series are obtained

$$\lambda=\pm\nu \tag{4.266}$$

Equating the coefficients for $\rho^{\lambda+k-2}$ on both sides of the equation gives

$$[\lambda(\lambda-1)+2\lambda k+k(k-1)+\lambda+k-\nu^2]a_k+a_{k-2}=0 \tag{4.267}$$

Using the indicial value $\lambda=\nu$ obtained in (4.266) gives the recurrence formula for the coefficient

$$k(2\lambda+k)a_k+a_{k-2}=0 \tag{4.268}$$

In view of (4.266), it can be simplified as

$$a_k=-\frac{a_{k-2}}{k(2\lambda+k)} \tag{4.269}$$

For even k (i.e., $k=2m$), the coefficient of the series solution becomes

$$\begin{aligned} a_{2m} &= -\frac{a_{2m-2}}{2m(2\lambda+2m)}=-\frac{a_{2m-2}}{4m(\lambda+m)}=-\frac{1}{4m(\lambda+m)}\frac{(-1)a_{2m-4}}{(2m-2)(2\lambda+2m-2)} \\ &= \frac{(-1)^2}{2^4}\frac{a_{2m-4}}{m(m-1)(\lambda+m)(\lambda+m-1)} \\ &= \frac{(-1)^m}{2^{2m}}\frac{a_0}{m!(\lambda+m)(\lambda+m-1)\cdots(\lambda+1)} \end{aligned} \tag{4.270}$$

The initial constants a_0 and a_1 are arbitrary constants, and thus without loss of generality can be set to

$$a_0=\frac{1}{2^\nu}[\frac{1}{\Gamma(\lambda+1)}], \quad a_1=0 \tag{4.271}$$

$$a_{2m}=\frac{(-1)^m}{2^{2m+\nu}}\frac{1}{m!\Gamma(\lambda+m+1)}, \quad a_{2m+1}=0 \tag{4.272}$$

With these coefficients, the series solution becomes

$$Z_\nu=\sum_{m=0}^{\infty}\frac{(-1)^m}{m!\Gamma(\nu+m+1)}(\frac{\rho}{2})^{2m+\nu}=J_\nu(\rho) \tag{4.273}$$

This gives the solution as $J_\nu(\rho)$, the Bessel function of the first kind of order ν, which is plotted in Figure 4.5. In principle, we can also retain odd order terms instead of even order terms, but the resulting function is not the Bessel function. For the second solution, we can simply make the substitution of $\lambda=-\nu$

$$J_{-\nu}(\rho)=\sum_{m=0}^{\infty}\frac{(-1)^m}{m!\Gamma(-\nu+m+1)}(\frac{\rho}{2})^{2m-\nu} \tag{4.274}$$

Therefore, the general solution of the Bessel equation is

$$Z_\nu=AJ_\nu(\rho)+BJ_{-\nu}(\rho) \tag{4.275}$$

provided that $\nu\neq$ integer. However, this solution is rather limited and is normally not used because ν is an integer for many physical problems.

4.6.2 Bessel Function of Second Kind for Integer Order

The Bessel function of the negative order reported in the last section breaks down when the order is an integer, and thus (4.275) is not the most general solution.

To illustrate the problem, we see that the recurrence equation becomes for $\nu = n$

$$(k^2 - 2kn)a_k + a_{k-2} = 0 \tag{4.276}$$

Since we have assumed that k is even (or $k = 2m$, where $m = 1,2,3,...$), (4.276) can be written as

$$(4m^2 - 4mn)a_k + a_{k-2} = 0 \tag{4.277}$$

The solution is for k being an infinite series (or $m = 1,2,..., \infty$). Therefore, no matter what is the given order in the ODE, m will match n sooner or later. The recurrence formula of the series solution breaks down when $m = n$.

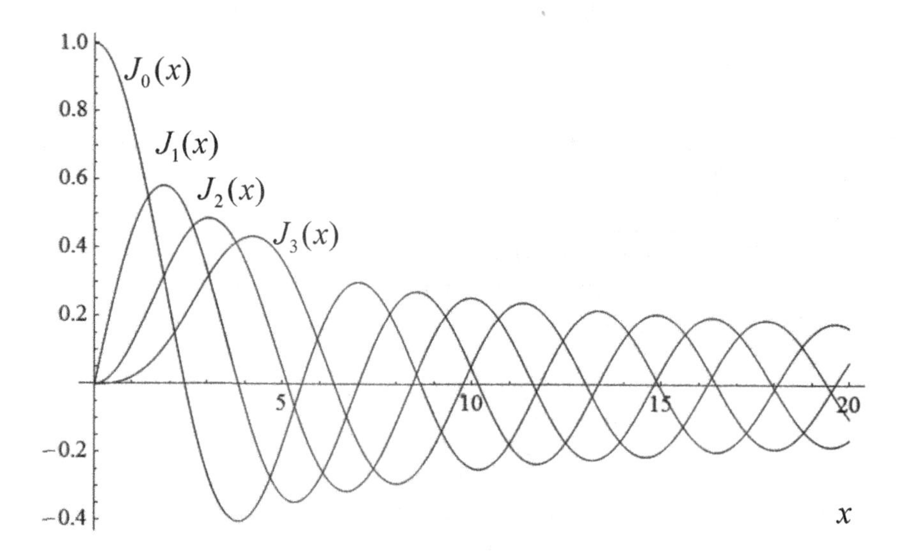

Figure 4.5 Bessel functions of the first kind

In fact, for the case of integer order (i.e., $\nu = n$), J_{-n} is not independent of J_n. In particular, they satisfy the following identity

$$J_{-n}(\rho) = (-1)^n J_n(\rho) \tag{4.278}$$

To prove this, we first note that the gamma function in the Bessel function becomes a factorial function for the case of integers. That is,

$$J_n(\rho) = \sum_{m=0}^{\infty} \frac{(-1)^m}{m!\Gamma(n+m+1)} (\frac{\rho}{2})^{2m+n} = \sum_{m=0}^{\infty} \frac{(-1)^m}{m!(n+m)!} (\frac{\rho}{2})^{2m+n} \tag{4.279}$$

Similarly, the Bessel function of negative order is

$$J_{-n}(\rho) = \sum_{m=0}^{\infty} \frac{(-1)^m}{m!\Gamma(m-n+1)} (\frac{\rho}{2})^{2m-n} \tag{4.280}$$

No matter what the initial value of the integer n is, we encounter the unbounded cases of the gamma function when $m < n$. More specially, we have

$$J_{-n}(\rho) \sim \frac{1}{\Gamma(m-n+1)} \to \frac{1}{\infty} = 0, \quad m < n \tag{4.281}$$

Effectively, we can drop all terms with $m < n$ since they are zeros. The solution (4.280) becomes

$$J_{-n}(\rho) = \sum_{m=n}^{\infty} \frac{(-1)^m}{m!\Gamma(m-n+1)} (\frac{\rho}{2})^{2m-n} \tag{4.282}$$

Since the series is infinite, we can shift back the lower index of summation to zero by setting $k = m-n$:

$$\begin{aligned} J_{-n}(\rho) &= \sum_{k=0}^{\infty} \frac{(-1)^{k+n}}{(k+n)!\Gamma(k+1)} (\frac{\rho}{2})^{2k+n} = (-1)^n \sum_{k=0}^{\infty} \frac{(-1)^k}{(k+n)!k!} (\frac{\rho}{2})^{2k+n} \\ &= (-1)^n \sum_{m=0}^{\infty} \frac{(-1)^m}{(m+n)!m!} (\frac{\rho}{2})^{2m+n} = (-1)^n J_n(\rho) \end{aligned} \tag{4.283}$$

This is (4.278) given above.

Our main job is now to search for an independent solution for the Bessel function for the case of integer order. There is no unique choice in this process. Recall that even the choice of J_ν is not unique (we have selected a particular value of a_0 such that the definition of Bessel function becomes more compact). The following definition by Weber and Schlafli is widely accepted (Watson, 1944):

$$Y_n(\rho) = \lim_{\nu \to n} \frac{J_\nu(\rho)\cos\nu\pi - J_{-\nu}(\rho)}{\sin\nu\pi} \tag{4.284}$$

where Y_n is called the Bessel function of the second kind of order n. When the integer limit is taken, both numerator and denominator approach zero. Since the limit is of the form 0/0, L'Hôpital's rule can be applied to (4.284) to arrive at

$$Y_n(\rho) = \lim_{\nu \to n} \frac{\dfrac{\partial J_\nu(\rho)}{\partial \nu}\cos\nu\pi - \pi\sin\nu\pi J_\nu(\rho) - \dfrac{\partial J_{-\nu}(\rho)}{\partial \nu}}{\pi\cos\nu\pi} \tag{4.285}$$

Note that

$$\cos n\pi = (-1)^n, \quad \sin n\pi = 0 \tag{4.286}$$

In view of these limiting values, (4.285) becomes

$$Y_n(\rho) = \frac{1}{\pi} \lim_{\nu \to n} \{\frac{\partial J_\nu(\rho)}{\partial \nu} + (-1)^{n+1} \frac{\partial J_{-\nu}(\rho)}{\partial \nu}\} \tag{4.287}$$

Note that we have applied the limit partially for sine and cosine functions only. Our next step is to find the derivative of J_ν with respect to the order ν. In particular, the power is first rewritten as:

$$(\frac{\rho}{2})^{2m\pm\nu} = e^{\ln(\rho/2)^{2m\pm\nu}} = e^{(2m\pm\nu)\ln(\rho/2)} \tag{4.288}$$

Then, the differentiation can be evaluated as

$$\frac{\partial J_\nu(\rho)}{\partial \nu} = \frac{\partial}{\partial \nu}\sum_{m=0}^{\infty}(-1)^m \frac{1}{m!\Gamma(\nu+m+1)} e^{(2m+\nu)\ln(\rho/2)}$$
$$= J_\nu(\rho)\ln(\rho/2) - \sum_{m=0}^{\infty}(-1)^m \frac{\psi(\nu+m+1)}{m!\Gamma(\nu+m+1)}(\rho/2)^{2m+\nu} \tag{4.289}$$

Similarly, we have

$$\frac{\partial J_{-\nu}(\rho)}{\partial \nu} = -J_{-\nu}(\rho)\ln(\rho/2) + \sum_{m=0}^{\infty}(-1)^m \frac{\psi(-\nu+m+1)}{m!\Gamma(-\nu+m+1)}(\rho/2)^{2m-\nu} \tag{4.290}$$

Now, we can take the limiting case as $\nu \to n$

$$[\frac{\partial J_\nu(\rho)}{\partial \nu}]_{\nu\to n} = J_n(\rho)\ln(\rho/2) - \sum_{m=0}^{\infty}(-1)^m \frac{\psi(n+m+1)}{m!\Gamma(n+m+1)}(\rho/2)^{2m+n} \tag{4.291}$$

Both ψ and Γ are finite and it is clear that for the positive order no special treatment is needed. For the negative order, we have

$$[\frac{\partial J_{-\nu}(\rho)}{\partial \nu}]_{\nu\to n} = -J_{-n}(\rho)\ln(\rho/2) + \sum_{m=0}^{n-1}(-1)^m [\frac{\psi(-\nu+m+1)}{m!\Gamma(-\nu+m+1)}]_{\nu\to n}(\rho/2)^{2m-n}$$
$$+\sum_{m=n}^{\infty}(-1)^m [\frac{\psi(-\nu+m+1)}{m!\Gamma(-\nu+m+1)}]_{\nu\to n}(\rho/2)^{2m-n} \tag{4.292}$$

Both the gamma function and the digamma function in (4.292) approach infinity for $m < n$. Thus, a special treatment is needed for the first sum on the right hand side of (4.292). Our objective is to evaluate the ratio of them given in the square bracket. Applying the reflection formulas for both digamma and gamma functions, we have

$$\psi(-\nu+m+1) = \psi[1-(\nu-m)] = \psi(\nu-m) + \pi\cot[\pi(\nu-m)] \tag{4.293}$$

$$\Gamma(-\nu+m+1) = \Gamma(1-(\nu-m)) = \frac{\pi}{\sin\pi(\nu-m)}\frac{1}{\Gamma(\nu-m)} \tag{4.294}$$

Dividing (4.293) by (4.294) gives

$$[\frac{\psi(-\nu+m+1)}{\Gamma(-\nu+m+1)}] = \frac{1}{\pi}[\psi(\nu-m)+\pi\cot[\pi(\nu-m)]]\sin\pi(\nu-m)\Gamma(\nu-m)$$
$$= \frac{\psi(\nu-m)\sin\pi(\nu-m)\Gamma(\nu-m)}{\pi} + \cos\pi(\nu-m)\Gamma(\nu-m) \tag{4.295}$$

Note that

$$\sin\pi(n-m) = 0 \tag{4.296}$$

$$[\frac{\psi(-\nu+m+1)}{\Gamma(-\nu+m+1)}]_{\nu\to n} = \Gamma(n-m)\cos(n-m)\pi = (-1)^{n-m}(n-m-1)! \tag{4.297}$$

Therefore, the limit between digamma and gamma functions exists and is finite for the first summation term on the right of (4.292). In addition, we want to shift the summation index of the last term in (4.292) back to starting from 0, instead of n. In particular, we can shift the index of summation by assuming $m = n+k$:

$$\begin{aligned}
&\sum_{m=n}^{\infty}(-1)^m [\frac{\psi(-\nu+m+1)}{m!\Gamma(-\nu+m+1)}]_{\nu\to n}(\frac{\rho}{2})^{2m-n} \\
&= (-1)^n \sum_{k=0}^{\infty}(-1)^k \frac{1}{(n+k)!}\frac{\psi(k+1)}{\Gamma(k+1)}](\frac{\rho}{2})^{2k+n} \\
&= (-1)^n \sum_{m=0}^{\infty}(-1)^m \frac{1}{(n+m)!}\frac{\psi(m+1)}{\Gamma(m+1)}](\frac{\rho}{2})^{2m+n} \\
&= (-1)^n \sum_{m=0}^{\infty}(-1)^m \frac{\psi(m+1)}{m!(n+m)!}(\frac{\rho}{2})^{2m+n}
\end{aligned} \tag{4.298}$$

Back substitution of these results into (4.292) we have

$$\begin{aligned}
[\frac{\partial J_{-\nu}(\rho)}{\partial \nu}]_{\nu\to n} &= (-1)^{n+1} J_n(\rho)\ln(\rho/2) + (-1)^n \sum_{m=0}^{n-1}\frac{(n-m-1)!}{m!}(\rho/2)^{2m-n} \\
&+(-1)^n \sum_{m=0}^{\infty}(-1)^m \frac{\psi(m+1)}{m!(n+m)!}(\rho/2)^{2m+n}
\end{aligned} \tag{4.299}$$

Finally, the second solution for the Bessel equation is obtained by substituting (4.299) and (4.291) into (4.287)

$$\begin{aligned}
Y_n(\rho) &= \frac{2}{\pi} J_n(\rho)\ln(\rho/2) - \frac{1}{\pi}\sum_{m=0}^{n-1}\frac{(n-m-1)!}{m!}(\rho/2)^{2m-n} \\
&-\frac{1}{\pi}\sum_{m=0}^{\infty}(-1)^m \frac{(\rho/2)^{2m+n}}{m!(n+m)!}\{\psi(n+m+1)+\psi(m+1)\}
\end{aligned} \tag{4.300}$$

Instead of expressing in terms of digamma functions, whose evaluation requires computer programs or numerical tables, we can express the digamma function in finite series as

$$\psi(m+1) = -\gamma + \sum_{k=1}^{m}\frac{1}{k} \tag{4.301}$$

To see the validity of this equation, we start with Weierstrass's canonical form given in (4.27)

$$\frac{1}{\Gamma(z)} = z e^{\gamma z}\prod_{n-1}^{\infty}\left\{(1+\frac{z}{n})e^{-z/n}\right\} \tag{4.302}$$

Inversion of (4.302) gives the following form of the gamma function

$$\Gamma(z) = \frac{1}{z e^{\gamma z}\prod_{n-1}^{\infty}\left\{(1+\frac{z}{n})e^{-z/n}\right\}} \tag{4.303}$$

Taking the logarithm of the gamma function given in (4.303), we obtain

$$\ln\Gamma(z) = -\ln z - \gamma z - \sum_{n=1}^{\infty} \ln[(1+\frac{z}{n})e^{-z/n}] = -\ln z - \gamma z - \sum_{n=1}^{\infty}[\ln(1+\frac{z}{n}) - \frac{z}{n}] \quad (4.304)$$

Differentiation of (4.304) gives

$$\frac{d}{dz}\ln\Gamma(z) = \frac{\Gamma'(z)}{\Gamma(z)} = \psi(z) = -\frac{1}{z} - \gamma - \sum_{n=1}^{\infty}[\frac{1}{n+z} - \frac{1}{n}] \quad (4.305)$$

Next, we want to evaluate the special value of $\psi(1)$

$$\begin{aligned}\psi(1) &= -1 - \gamma - \sum_{n=1}^{\infty}[\frac{1}{n+1} - \frac{1}{n}] = -\gamma - \sum_{n=0}^{\infty}[\frac{1}{n+1}] - \sum_{n=1}^{\infty}\frac{1}{n} \\ &= -\gamma - \sum_{n=0}^{\infty}\frac{1}{n+1} - \sum_{k=0}^{\infty}\frac{1}{k+1} = -\gamma\end{aligned} \quad (4.306)$$

Note that we have shifted the index of summation using $n = k+1$ for the last sum. Finally, we recall the recurrence formula:

$$\begin{aligned}\psi(n+1) &= \frac{1}{n} + \psi(n) = \frac{1}{n} + \frac{1}{n-1} + \psi(n-1) = \frac{1}{n} + \frac{1}{n-1} + ... + 1 + \psi(1) \\ &= -\gamma + \sum_{k=1}^{n}\frac{1}{k}\end{aligned} \quad (4.307)$$

To obtain the final line in (4.307), we have substituted the result obtained in (4.306). This gives the required series expansion for the digamma function given in (4.301).

Finally, with substitution of (4.307) into (4.300) we get the commonly seen definition of the Bessel function of the second kind or Y_n as

$$\begin{aligned}Y_n(\rho) &= -\frac{1}{\pi}\sum_{m=0,n>0}^{n-1}\frac{(n-m-1)!}{m!}(\rho/2)^{2m-n} \\ &+ \frac{2}{\pi}\sum_{m=0}^{\infty}(-1)^m\frac{(\rho/2)^{2m+n}}{m!(n+m)!}[\ln(\frac{\rho}{2}) + \gamma - \frac{1}{2}(\sum_{k=1}^{n+m}\frac{1}{k} + \sum_{k=1}^{m}\frac{1}{k})]\end{aligned} \quad (4.308)$$

where the Euler or Euler-Mascheroni constant γ is defined as

$$\gamma = \lim_{m\to\infty}\left(\sum_{k=1}^{m}\frac{1}{k} - \ln m\right) \cong 0.5772157 \quad (4.309)$$

More discussion of Euler constant is given in Appendix E. It was Mascheroni who extended Euler's 16 digit calculation of γ to 32 digits in 1790 (although wrong) and gave it the symbol "gamma". Figure 4.6 plots the Bessel function of the second kind. Finally, we obtain the general solution, which is valid for both integer and non-integer order

$$Z_\nu = AJ_\nu(\rho) + BY_\nu(\rho) \quad (4.310)$$

The Bessel function is considered one of the most commonly encountered functions other than circular and hyperbolic functions. It appears naturally in many branches of science and engineering. As summarized by Watson (1944), it is intimately related to the Riccati equation. The most comprehensive authority of the Bessel function remains the classic book by Watson (1944). Some people regarded it as the best mathematics book of all time. A brief biography of Watson is given in the back of this book.

To end this section, we record the first few terms in the series form of the Bessel functions as:

$$J_0(x) = 1 - \frac{1}{(1!)^2}(\frac{1}{2}x)^2 + \frac{1}{(2!)^2}(\frac{1}{2}x)^4 - \frac{1}{(2!)^2}(\frac{1}{2}x)^6 + ... \quad (4.311)$$

$$J_0(x) = 1 - \frac{x^2}{2^2} + \frac{x^4}{2^2 \cdot 4^2} - \frac{x^6}{2^2 \cdot 4^2 \cdot 6^2} + ...$$
$$= 1 - \frac{x^2}{4} + \frac{x^4}{64} - \frac{x^6}{2304} + ... \quad (4.312)$$

$$J_1(x) = \frac{1}{2}x[1 - \frac{1}{1!2!}(\frac{1}{2}x)^2 + \frac{1}{2!3!}(\frac{1}{2}x)^4 - \frac{1}{3!4!}(\frac{1}{2}x)^6 + ...] \quad (4.313)$$

$$J_2(x) = (\frac{1}{2})^2[1 - \frac{1}{1!3!}(\frac{1}{2}x)^2 + \frac{1}{2!4!}(\frac{1}{2}x)^4 - + ...] \quad (4.314)$$

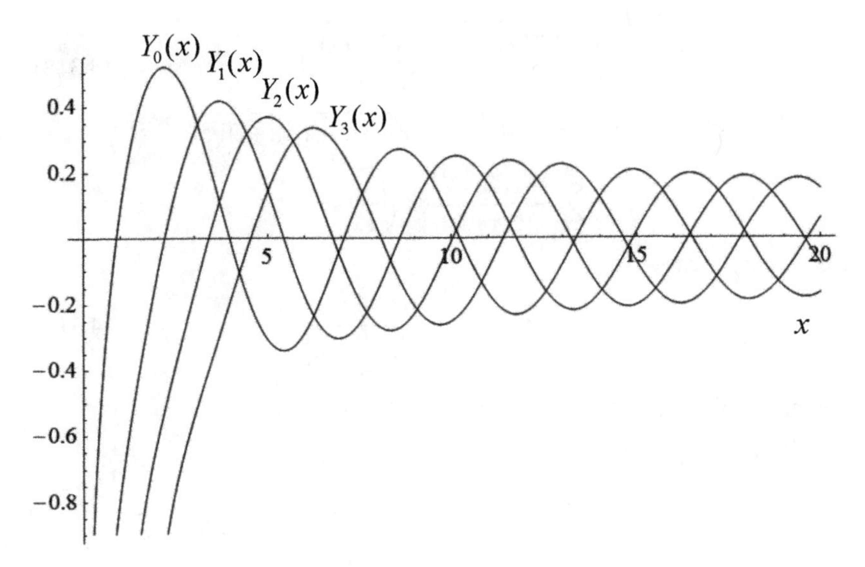

Figure 4.6 Bessel functions of the second kind Y_n

$$Y_0(x) = \frac{2}{\pi}[\ln(\frac{x}{2}) + \gamma]J_0(x)$$
$$+ \frac{2}{\pi}[\frac{x^2}{2^2} - \frac{x^4}{2^2 \cdot 4^2}(1 + \frac{1}{2}) + \frac{x^6}{2^2 \cdot 4^2 \cdot 6^2}(1 + \frac{1}{2} + \frac{1}{3}) - ...] \quad (4.315)$$

4.6.3 Circular Functions and Bessel Functions of Half Order

The Bessel function of half order can be expressed in terms of circular functions. Let us consider the case that ν = 1/2:

$$J_{1/2}(x)=\sum_{m=0}^{\infty}\frac{(-1)^m x^{2m+1/2}}{2^{2m+1/2}m!\Gamma(\frac{3}{2}+m)} \tag{4.316}$$

The gamma function can be evaluated as

$$\begin{aligned}\Gamma(\frac{3}{2}+m)&=(\frac{1}{2}+m)\Gamma(\frac{1}{2}+m)=(\frac{1}{2}+m)(m-\frac{1}{2})\Gamma(m-\frac{1}{2})\\&=(\frac{1}{2}+m)(m-\frac{1}{2})\cdots\frac{1}{2}\Gamma(\frac{1}{2})=\frac{(2m+1)(2m-1)\cdots3\cdot1}{2^{m+1}}\sqrt{\pi}\end{aligned} \tag{4.317}$$

Substitution of (4.317) into (4.316) gives

$$\begin{aligned}J_{1/2}(x)&=\sum_{m=0}^{\infty}\frac{(-1)^m x^{2m+1/2}2^{m+1}}{2^{2m+1/2}m!1\cdot3\cdot5\cdots(2m+1)\sqrt{\pi}}\\&=\sqrt{\frac{2}{\pi x}}\sum_{m=0}^{\infty}\frac{(-1)^m x^{2m+1}}{2^m m!1\cdot3\cdot5\cdots(2m+1)}\\&=\sqrt{\frac{2}{\pi x}}\sum_{m=0}^{\infty}\frac{(-1)^m x^{2m+1}}{(2\cdot4\cdot6\cdots2m)[1\cdot3\cdot5\cdots(2m+1)]}\\&=\sqrt{\frac{2}{\pi x}}\sum_{m=0}^{\infty}\frac{(-1)^m x^{2m+1}}{(2m+1)!}=\sqrt{\frac{2}{\pi x}}\sin x\end{aligned} \tag{4.318}$$

Let us consider the case that ν = −1/2

$$J_{-1/2}(x)=\sum_{m=0}^{\infty}\frac{(-1)^m x^{2m-1/2}}{2^{2m-1/2}m!\Gamma(\frac{1}{2}+m)} \tag{4.319}$$

The gamma function can be evaluated as

$$\begin{aligned}\Gamma(\frac{1}{2}+m)&=(m-\frac{1}{2})\Gamma(m-\frac{1}{2})=(m-\frac{1}{2})(m-\frac{3}{2})\Gamma(m-\frac{3}{2})\\&=(m-\frac{1}{2})(m-\frac{3}{2})\cdots\frac{3}{2}\frac{1}{2}\Gamma(\frac{1}{2})=\frac{(2m-1)(2m-3)\cdots3\cdot1}{2^m}\sqrt{\pi}\end{aligned} \tag{4.320}$$

Substitution of (4.320) into (4.319) gives

$$J_{-1/2}(x)=\sum_{m=0}^{\infty}\frac{(-1)^m(x/2)^{2m-1/2}2^m}{m!1\cdot3\cdot5\cdots(2m-1)\sqrt{\pi}}$$
$$=\sqrt{\frac{2}{\pi x}}\sum_{m=0}^{\infty}\frac{(-1)^m x^{2m}}{2^m(1\cdot3\cdot5\cdots m)[1\cdot3\cdot5\cdots(2m-1)]}$$
$$=\sqrt{\frac{2}{\pi x}}\sum_{m=0}^{\infty}\frac{(-1)^m x^{2m}}{(2\cdot4\cdot6\cdots2m)[1\cdot3\cdot5\cdots(2m-1)]} \tag{4.321}$$
$$=\sqrt{\frac{2}{\pi x}}\sum_{m=0}^{\infty}\frac{(-1)^m x^{2m}}{(2m)!}=\sqrt{\frac{2}{\pi x}}\cos x$$

Therefore, the sine is related to $J_{1/2}$ and the cosine is related to $J_{-1/2}$. The following example provides another proof of the functional form of the two identities.

Example 4.7 Show that the solution of the following Bessel equation

$$x^2y''+xy'+(x^2-\frac{1}{4})y=0 \tag{4.322}$$

can be expressed as

$$y=A\frac{1}{\sqrt{x}}\cos x+B\frac{1}{\sqrt{x}}\sin x \tag{4.323}$$

Solution: Assume the following change of variables

$$y=\frac{1}{\sqrt{x}}w(x) \tag{4.324}$$

Differentiation of y can be expressed in terms of w as

$$y'=\left\{-\frac{1}{2}x^{-3/2}w(x)+\frac{1}{\sqrt{x}}w'\right\} \tag{4.325}$$

$$y''=\left\{\frac{3}{4}x^{-5/2}w-x^{-3/2}w'+\frac{1}{\sqrt{x}}w''\right\} \tag{4.326}$$

Substitution of (4.326) and (4.325) into (4.322) results in

$$w''+w'=0 \tag{4.327}$$

The general solution is of course sine and cosine

$$w=A\cos x+B\sin x \tag{4.328}$$

Thus, we have

$$y=A\frac{1}{\sqrt{x}}\cos x+B\frac{1}{\sqrt{x}}\sin x \tag{4.329}$$

On the other hand, the solution of (4.322) is Bessel functions of the order 1/2:

$$y=C_1J_{-1/2}(x)+C_2J_{1/2}(x) \tag{4.330}$$

This agrees with the results obtained in (4.318) and (4.321).

4.6.4 Modified Bessel Function

When the argument of a Bessel function becomes purely imaginary, we can make the following substitution: $\rho \leftarrow i\rho$ or $\rho \leftarrow -i\rho$. The Bessel equation becomes the so-called modified Bessel equation

$$\frac{d^2 Z_\nu}{d\rho^2}+\frac{1}{\rho}\frac{dZ_\nu}{d\rho}-\left(1+\frac{\nu^2}{\rho^2}\right)Z_\nu=0 \tag{4.331}$$

Its solution is evidently $J_\nu(i\rho)$. Since (4.331) is a homogeneous differential equation, any constant multiplying the Bessel function of the imaginary argument is also a solution of the modified Bessel function. Thus, we can define its solution as:

$$\begin{aligned} I_\nu(\rho) &= i^{-\nu} J_\nu(i\rho), \quad (-\pi < \arg\rho \le \pi/2) \\ &= i^{\nu} J_\nu(-i\rho), \quad (\pi/2 < \arg\rho \le \pi) \end{aligned} \tag{4.332}$$

This is called the modified Bessel function of the first kind, and it was defined by Basset in 1889. Substitution of $\rho \leftarrow i\rho$ into the series definition of the Bessel function gives

$$I_\nu(\rho)=\sum_{m=0}^{\infty}\frac{(\rho/2)^{2m+\nu}}{m!\Gamma(\nu+m+1)} \tag{4.333}$$

which is plotted in Figure 4.7. Thus, the general solution can be expressed as:

$$Z_\nu(\rho)=AI_\nu(\rho)+BI_{-\nu}(\rho) \tag{4.334}$$

For the case of negative integer order, similar to the case of the Bessel function of integer order, these two solutions are no longer independent. In fact, we can show that

$$I_n(\rho)=I_{-n}(\rho) \tag{4.335}$$

where n is an integer.

To show this, we first note that

$$I_n(\rho)=\sum_{m=0}^{\infty}\frac{(\rho/2)^{2m+n}}{m!\Gamma(n+m+1)}=\sum_{m=0}^{\infty}\frac{(\rho/2)^{2m+n}}{m!(n+m)!} \tag{4.336}$$

The modified Bessel function of negative order becomes

$$I_{-n}(\rho)=\sum_{m=0}^{\infty}\frac{(\rho/2)^{2m-n}}{m!\Gamma(m-n+1)} \tag{4.337}$$

As we have shown before, the gamma function $\Gamma(k)$ becomes infinite for $k = 0, -1, -2, \ldots$ Thus, all terms with $m < n$ vanish in the summation, or we have

$$I_{-n}(\rho)=\sum_{m=n}^{\infty}\frac{(\rho/2)^{2m-n}}{m!\Gamma(m-n+1)} \tag{4.338}$$

Now, we can make the following change of summation index:

$$k=m-n \tag{4.339}$$

Then, (4.488) can be rewritten as

$$I_{-n}(\rho)=\sum_{k=0}^{\infty}\frac{(\rho/2)^{2k+n}}{(k+n)!\Gamma(k+1)}=\sum_{k=0}^{\infty}\frac{(\rho/2)^{2k+n}}{(k+n)!k!}=I_n(\rho) \tag{4.340}$$

This completes the proof.

For integer order, we therefore need to define another independent solution. The following second independent solution is called the modified Bessel function of the second kind (Abramowitz and Stegun, 1964):

$$K_\nu(\rho)=\frac{\pi}{2}\frac{I_{-\nu}(\rho)-I_\nu(\rho)}{\sin\nu\pi} \tag{4.341}$$

which was defined by MacDonald in 1899 and therefore it is also known as the MacDonald function. For integer $\nu = n$, the definition of K_n was proposed by Basset in 1889. The modified Bessel function of the second kind is plotted in Figure 4.8. Applying L'Hôpital's rule to (4.341) we obtain

$$K_n(\rho)=\frac{(-1)^n}{2}\{\frac{\partial I_{-\nu}(\rho)}{\partial \nu}-\frac{\partial J_\nu(\rho)}{\partial \nu}\}_{\nu\to n} \tag{4.342}$$

Following a similar step in obtaining Y_n we have

$$\begin{aligned}\frac{\partial I_\nu(\rho)}{\partial \nu}&=\frac{\partial}{\partial \nu}\sum_{m=0}^{\infty}\frac{1}{m!\Gamma(\nu+m+1)}e^{(2m+\nu)\ln(\rho/2)}\\&=I_\nu(\rho)\ln(\rho/2)-\sum_{m=0}^{\infty}\frac{1}{m!}\frac{\psi(\nu+m+1)}{\Gamma(\nu+m+1)}(\rho/2)^{2m+\nu}\end{aligned} \tag{4.343}$$

Similarly, we have

$$\frac{\partial I_{-\nu}(\rho)}{\partial \nu}=-I_{-\nu}(\rho)\ln(\rho/2)+\sum_{m=0}^{\infty}\frac{1}{m!}\frac{\psi(-\nu+m+1)}{\Gamma(-\nu+m+1)}(\rho/2)^{2m-\nu} \tag{4.344}$$

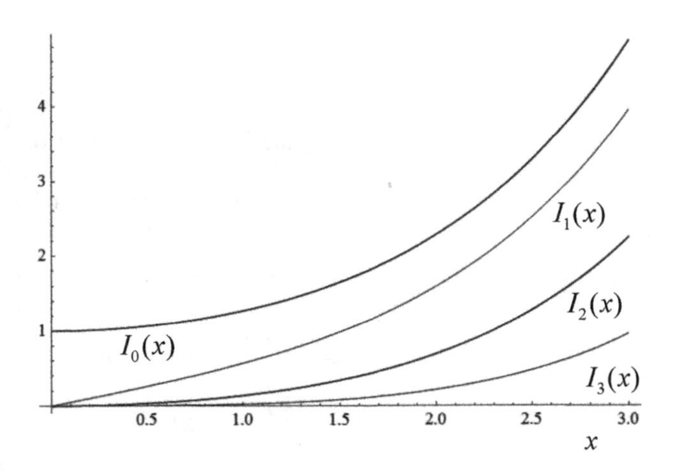

Figure 4.7 Modified Bessel functions of the first kind I_n

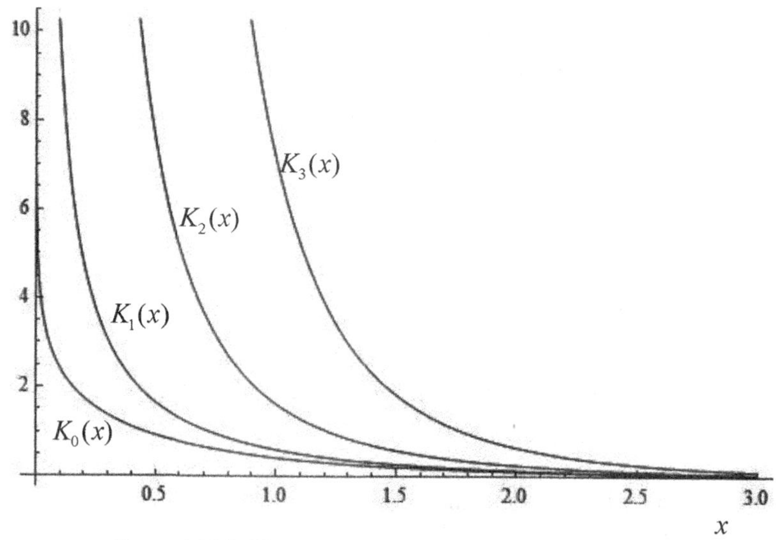

Figure 4.8 Modified Bessel functions of the second kind K_n

Now, we can take the limiting case as $\nu \to n$

$$[\frac{\partial I_\nu(\rho)}{\partial \nu}]_{\nu\to n} = I_n(\rho)\ln(\rho/2) - \sum_{m=0}^{\infty}\frac{\psi(n+m+1)}{m!(n+m)!}(\rho/2)^{2m+n} \tag{4.345}$$

It is clear that for the positive order no special treatment is needed, as the right hand side of (4.345) is finite. For the negative order, we have

$$[\frac{\partial I_{-\nu}(\rho)}{\partial \nu}]_{\nu\to n} = -I_{-n}(\rho)\ln(\rho/2) + \sum_{m=0}^{n-1}\frac{1}{m!}[\frac{\psi(-\nu+m+1)}{\Gamma(-\nu+m+1)}]_{\nu\to n}(\rho/2)^{2m-n}$$
$$+\sum_{m=n}^{\infty}\frac{1}{m!}[\frac{\psi(-\nu+m+1)}{\Gamma(-\nu+m+1)}]_{\nu\to n}(\rho/2)^{2m-n} \tag{4.346}$$

Our next objective is to find the limit in the square bracket in the first summation term in (4.346). Both the gamma function and the digamma function in (4.346) approach infinity for $m < n$. Thus, we have an indeterminate form of ∞/∞. By applying the reflection formula of gamma and digamma functions, we have:

$$\begin{aligned}[\frac{\psi(-\nu+m+1)}{\Gamma(-\nu+m+1)}]_{\nu\to n} &= \frac{1}{\pi}[\{\psi(\nu-m)+\pi\cot[\pi(\nu-m)]\}\sin\pi(\nu-m)\Gamma(\nu-m)]_{\nu\to n} \\ &= \left[\frac{\psi(\nu-m)\sin\pi(\nu-m)\Gamma(\nu-m)}{\pi} + \cos\pi(\nu-m)\Gamma(\nu-m)\right]_{\nu\to n} \\ &= (-1)^{n-m}(n-m-1)!\end{aligned} \tag{4.347}$$

Similar to the calculation for (4.298), we have

$$\sum_{m=n}^{\infty}\frac{1}{m!}\Big[\frac{\psi(-\nu+m+1)}{\Gamma(-\nu+m+1)}\Big]_{\nu\to n}(\frac{\rho}{2})^{2m-n}=\sum_{k=0}^{\infty}\frac{1}{(n+k)!}\frac{\psi(k+1)}{\Gamma(k+1)}\Big](\frac{\rho}{2})^{2k+n}$$
$$=\sum_{m=0}^{\infty}\frac{\psi(m+1)}{m!(n+m)!}(\frac{\rho}{2})^{2m+n} \tag{4.348}$$

Back substitution of these results into (4.346) yields

$$\Big[\frac{\partial I_{-\nu}(\rho)}{\partial \nu}\Big]_{\nu\to n}=-I_n(\rho)\ln(\rho/2)+\sum_{m=0}^{n-1}(-1)^{n-m}\frac{(n-m-1)!}{m!}(\rho/2)^{2m-n}$$
$$+\sum_{m=0}^{\infty}\frac{\psi(m+1)}{m!(n+m)!}(\rho/2)^{2m+n} \tag{4.349}$$

The second solution is called the modified Bessel equation of the second kind

$$K_n(\rho)=(-1)^{n+1}I_n(\rho)\ln(\rho/2)+\frac{1}{2}\sum_{m=0}^{n-1}(-1)^m\frac{(n-m-1)!}{m!}(\rho/2)^{2m-n}$$
$$+\frac{(-1)^n}{2}\sum_{m=0}^{\infty}\frac{(\rho/2)^{2m+n}}{m!(n+m)!}\{\psi(n+m+1)+\psi(m+1)\} \tag{4.350}$$

Finally, by using the series expansion for digamma given in (4.301) we obtain

$$K_n(\rho)=\frac{1}{2}\sum_{m=0,n>0}^{n-1}(-1)^m\frac{(n-m-1)!}{m!}(\rho/2)^{2m-n}$$
$$+(-1)^{n+1}\sum_{m=0}^{\infty}\frac{(\rho/2)^{2m+n}}{m!(n+m)!}\Big[\ln(\frac{\rho}{2})+\gamma-\frac{1}{2}(\sum_{k=1}^{n+m}\frac{1}{k}+\sum_{k=1}^{m}\frac{1}{k})\Big] \tag{4.351}$$

where Euler's constant γ (≈ 0.5772157).

Thus, the general solution of (4.331) is

$$Z_\nu(\rho)=AI_\nu(\rho)+BK_\nu(\rho) \tag{4.352}$$

To summarize, we simply report the first few terms of the modified Bessel functions as

$$I_0(x)=1+\frac{1}{(1!)^2}(\frac{1}{2}x)^2+\frac{1}{(2!)^2}(\frac{1}{2}x)^4+\frac{1}{(2!)^2}(\frac{1}{2}x)^6+... \tag{4.353}$$

$$I_1(x)=\frac{1}{2}x[1+\frac{1}{1!2!}(\frac{1}{2}x)^2+\frac{1}{2!3!}(\frac{1}{2}x)^4+\frac{1}{3!4!}(\frac{1}{2}x)^6+...] \tag{4.354}$$

$$K_0(x)=-(\ln\frac{x}{2}+\gamma)I_0(x)+$$
$$+\frac{1}{(1!)^2}(\frac{1}{2}x)^2+(1+\frac{1}{2})\frac{1}{(2!)^2}(\frac{1}{2}x)^4+\frac{1}{(3!)^2}(\frac{1}{2}x)^6+... \tag{4.355}$$

$$K_1(x) = \frac{1}{x} + (\ln\frac{x}{2} + \gamma)I_1(x) - \frac{x}{2}[\frac{1}{2} + (1 + \frac{1}{2}\cdot\frac{1}{2})\frac{1}{1!2!}(\frac{1}{2}x)^2$$
$$+(1 + \frac{1}{2} + \frac{1}{2}\cdot\frac{1}{3})\frac{1}{2!3!}(\frac{1}{2}x)^4 + (1 + \frac{1}{2} + \frac{1}{3} + \frac{1}{2} + \frac{1}{2}\cdot\frac{1}{4})\frac{1}{3!4!}(\frac{1}{2}x)^6 + + ...] \tag{4.356}$$

We have shown that Bessel functions of half integer order can be related to circular functions (i.e., sine and cosine). We will consider the half integer order of modified functions here.

$$I_{1/2}(\rho) = \sum_{m=0}^{\infty} \frac{(\rho/2)^{2m+1/2}}{m!\Gamma(m+1/2+1)} = \sqrt{\frac{2}{\pi\rho}} \sum_{m=0}^{\infty} \frac{(\rho/2)^{2m+1}}{m!\Gamma(m+3/2)}\sqrt{\pi} \tag{4.357}$$

By using (4.317), the gamma function in (4.357) can be rewritten as:

$$\begin{aligned} & m!\Gamma(m+3/2)2^{2m+1} \\ & = [2^m m(m-1)\cdots 2\cdot 1]\left\{\frac{2^{m+1}(2m+1)(2m-1)(2m-3)\cdots 1\cdot\sqrt{\pi}}{2^{m+1}}\right\} \\ & = [2m(2m-2)\cdots 4\cdot 2][(2m+1)(2m-1)\cdots 3\cdot 1\cdot\sqrt{\pi}] \\ & = (2m+1)!\sqrt{\pi} \end{aligned} \tag{4.358}$$

Therefore, (4.357) can be written as

$$I_{1/2}(\rho) = \sqrt{\frac{2}{\pi\rho}} \sum_{m=0}^{\infty} \frac{\rho^{2m+1}}{(2m+1)!} = \sqrt{\frac{2}{\pi\rho}}\{\rho + \frac{\rho^3}{3!} + \frac{\rho^5}{5!} + ...\} = \sqrt{\frac{2}{\pi\rho}}\sinh\rho \tag{4.359}$$

Similarly, the modified Bessel function of $-1/2$ can be considered following the same procedure:

$$I_{-1/2}(\rho) = \sum_{m=0}^{\infty} \frac{(\rho/2)^{2m-1/2}}{m!\Gamma(m-1/2+1)} = \sqrt{\frac{2}{\pi\rho}} \sum_{m=0}^{\infty} \frac{(\rho/2)^{2m}}{m!\Gamma(m+1/2)}\sqrt{\pi} \tag{4.360}$$

By using (4.320), the gamma function in (4.360) can be rewritten as:

$$\begin{aligned} & m!\Gamma(m+1/2)2^{2m} \\ & = [2^m m(m-1)\cdots 2\cdot 1][(m-\frac{1}{2})(m-\frac{3}{2})(2m-\frac{5}{2})\cdots\frac{3}{2}\cdot\frac{1}{2}\Gamma(\frac{1}{2})]2^m \\ & = [2m(2m-2)\cdots 4\cdot 2][(2m-1)(2m-3)\cdots 3\cdot 1\cdot\sqrt{\pi}] \\ & = (2m)!\sqrt{\pi} \end{aligned} \tag{4.361}$$

Therefore, (4.360) can be written as

$$I_{-1/2}(\rho) = \sqrt{\frac{2}{\pi\rho}} \sum_{m=0}^{\infty} \frac{\rho^{2m}}{(2m)!} = \sqrt{\frac{2}{\pi\rho}}\{1 + \frac{\rho^2}{2!} + \frac{\rho^4}{4!} + ...\} = \sqrt{\frac{2}{\pi\rho}}\cosh\rho \tag{4.362}$$

In summary, the modified Bessel function of half integer order can be expressed in terms of hyperbolic functions. In the next section, we will consider the integral representation of the Bessel function.

4.6.5 Helmholtz Equation and Integral Representation

To consider the integral representation of the Bessel function of the first kind, we start from the 2-D wave equation

$$\frac{\partial^2 w}{\partial x^2}+\frac{\partial^2 w}{\partial y^2}-\frac{1}{c}\frac{\partial^2 w}{\partial t^2}=0 \tag{4.363}$$

We seek a periodic waves or harmonic waves of the form

$$w=u(x,y)e^{i\omega t} \tag{4.364}$$

Substitution of (4.364) into (4.363) gives the following Helmholtz equation

$$\frac{\partial^2 u}{\partial x^2}+\frac{\partial^2 u}{\partial y^2}+k^2 u=0 \tag{4.365}$$

where

$$k^2=\frac{\omega^2}{c} \tag{4.366}$$

We seek a solution of the form

$$u=Ae^{i(ax+by)} \tag{4.367}$$

Substitution of (4.367) into (4.365) gives

$$a^2+b^2=k^2 \tag{4.368}$$

Thus, it is natural to assume a new parameter α such that

$$a=k\cos\alpha, \quad b=k\sin\alpha \tag{4.369}$$

This suggests a change of variables in polar form:

$$x=r\cos\varphi, \quad y=r\sin\varphi \tag{4.370}$$

Combining these, we obtain the following solution form of (4.367)

$$u=Ae^{ikr(\cos\alpha\cos\varphi+\sin\alpha\sin\varphi)}=Ae^{ikr\cos(\varphi-\alpha)} \tag{4.371}$$

For the special case of $\alpha=\varphi$, we have the case of a plane wave propagating in the direction of $\varphi=\alpha$ (or in the form of a one-dimensional wave). The change of variables given in (4.370) results in

$$r=(x^2+y^2)^{1/2}, \quad \frac{\partial r}{\partial x}=\frac{x}{(x^2+y^2)^{1/2}}=\cos\varphi, \quad \frac{\partial r}{\partial y}=\sin\varphi$$
$$\varphi=\tan^{-1}(y/x), \quad \frac{\partial \varphi}{\partial x}=\frac{-y/x^2}{1+(y/x)^2}=\frac{-y}{x^2+y^2}=\frac{-\sin\varphi}{r}, \quad \frac{\partial \varphi}{\partial y}=\frac{\cos\varphi}{r} \tag{4.372}$$

Application of (4.372) and the chain rule of partial differentiation leads to

$$\frac{\partial u}{\partial x}=\frac{\partial r}{\partial x}\frac{\partial u}{\partial r}+\frac{\partial \varphi}{\partial x}\frac{\partial u}{\partial \varphi}=\cos\varphi\frac{\partial u}{\partial r}-\frac{\sin\varphi}{r}\frac{\partial u}{\partial \varphi} \tag{4.373}$$

Similarly, we also have

$$\frac{\partial u}{\partial y}=\sin\varphi\frac{\partial u}{\partial r}+\frac{\cos\varphi}{r}\frac{\partial u}{\partial \varphi} \tag{4.374}$$

Application of (4.373) twice, we have

$$\frac{\partial^2 u}{\partial x^2} = \frac{\partial}{\partial x}(\frac{\partial u}{\partial x})$$
$$= \cos\varphi \frac{\partial}{\partial r}[\cos\varphi \frac{\partial u}{\partial r} - \frac{\sin\varphi}{r}\frac{\partial u}{\partial \varphi}] - \frac{\sin\varphi}{r}\frac{\partial}{\partial \varphi}[\cos\varphi \frac{\partial u}{\partial r} - \frac{\sin\varphi}{r}\frac{\partial u}{\partial \varphi}] \tag{4.375}$$
$$= \frac{\partial^2 u}{\partial r^2}\cos^2\varphi + \sin^2\varphi(\frac{1}{r}\frac{\partial u}{\partial \varphi} + \frac{1}{r^2}\frac{\partial^2 u}{\partial \varphi^2}) + 2\sin\varphi\cos\varphi(\frac{1}{r^2}\frac{\partial u}{\partial \varphi} - \frac{1}{r}\frac{\partial^2 u}{\partial \varphi \partial r})$$

Similarly, application of (4.374) twice leads to

$$\frac{\partial^2 u}{\partial y^2} = \frac{\partial}{\partial y}(\frac{\partial u}{\partial y})$$
$$= \sin\varphi \frac{\partial}{\partial r}[\sin\varphi \frac{\partial u}{\partial r} + \frac{\cos\varphi}{r}\frac{\partial u}{\partial \varphi}] + \frac{\cos\varphi}{r}\frac{\partial}{\partial \varphi}[\sin\varphi \frac{\partial u}{\partial r} + \frac{\cos\varphi}{r}\frac{\partial u}{\partial \varphi}] \tag{4.376}$$
$$= \frac{\partial^2 u}{\partial r^2}\sin^2\varphi + \cos^2\varphi(\frac{1}{r}\frac{\partial u}{\partial \varphi} + \frac{1}{r^2}\frac{\partial^2 u}{\partial \varphi^2}) - 2\sin\varphi\cos\varphi(\frac{1}{r^2}\frac{\partial u}{\partial \varphi} - \frac{1}{r}\frac{\partial^2 u}{\partial \varphi \partial r})$$

Finally, combining (4.375) and (4.376), we obtain the polar form of the Helmholtz equation

$$\frac{\partial^2 u}{\partial x^2} + \frac{\partial^2 u}{\partial y^2} + k^2 u = \frac{\partial^2 u}{\partial r^2} + \frac{1}{r}\frac{\partial u}{\partial r} + \frac{1}{r^2}\frac{\partial^2 u}{\partial \varphi^2} + k^2 u = 0 \tag{4.377}$$

Applying another round of change of variables of $\rho = kr$, we have

$$\frac{\partial^2 u}{\partial \rho^2} + \frac{1}{\rho}\frac{\partial u}{\partial \rho} + \frac{1}{\rho^2}\frac{\partial^2 u}{\partial \varphi^2} + u = 0 \tag{4.378}$$

Next, we seek an angular dependence of the form:

$$u = Z_n(\rho)e^{in\varphi} \tag{4.379}$$

Substitution of (4.379) into (4.378) gives

$$\frac{d^2 u}{d\rho^2} + \frac{1}{\rho}\frac{du}{d\rho} + (1 - \frac{n^2}{\rho^2})u = 0 \tag{4.380}$$

This is of course the Bessel equation that we discussed earlier. If the boundedness of the solution of origin is enforced, the solution of u can only be written as:

$$u = AJ_n(\rho)e^{in\varphi} \tag{4.381}$$

Now we can rewrite (4.371) by assuming

$$A = C_n e^{in\alpha} \tag{4.382}$$

and integrate the solution from β to γ as

$$u = C_n \int_\beta^\gamma e^{i\rho\cos(\varphi-\alpha)} e^{in\alpha} d\alpha \tag{4.383}$$

Physically, this corresponds to a bundle of waves with directions varying from $\alpha = \beta$ to $\alpha = \gamma$.

We apply the following change of variables:

$$\alpha = w + \varphi \tag{4.384}$$

to give

$$u = C_n e^{in\varphi} \int_{w_0}^{w_1} e^{i\rho\cos w} e^{inw} dw \tag{4.385}$$

where

$$w_0 = \beta - \varphi, \quad w_1 = \gamma - \varphi \tag{4.386}$$

With this change of variables, the angular dependence of φ appears explicitly in (4.385) and it closely resembles the exact solution that we got from (4.381). The only problem is that the limit of integration cannot be dependent on the angular variable φ. Thus, we must remove the φ dependence from w_0 and w_1. This is probably the most crucial step in this analysis. This can only be achieved if we allow both β and γ to tend to infinity. Or, we have

$$w_0 \to \infty, \quad w_1 \to \infty \tag{4.387}$$

Thus, the issue becomes the study of the convergence of the integral given in (4.385) as β and γ → ∞. Since physically the wave will decay to zero at infinity, we must have a converging solution. In other words, we need to make u remain finite as $w \to \infty$. This can be done if we allow w to be a complex number. That is,

$$w = p + iq \tag{4.388}$$

We note that

$$\begin{aligned} \cos w &= \cos(p+iq) = \cos p \cos(iq) - \sin p \sin(iq) \\ &= \cos p \cosh q - i \sin p \sinh q \end{aligned} \tag{4.389}$$

In obtaining the above equation, we have used the following identities

$$\cos(iq) = \cosh q, \quad \sin(iq) = i \sinh q \tag{4.390}$$

These identities can be obtained readily by noting Euler's formula that

$$e^{i\theta} = \cos\theta + i\sin\theta, \quad e^{-i\theta} = \cos\theta - i\sin\theta \tag{4.391}$$

Allowing θ be purely imaginary (or θ $=i\phi$), we get

$$e^{-\phi} = \cos(i\phi) + i\sin(i\phi), \quad e^{\phi} = \cos(i\phi) - i\sin(i\phi) \tag{4.392}$$

Adding these equations gives

$$\cosh\phi = \frac{e^{\phi} + e^{-\phi}}{2} = \frac{\cos(i\phi) - i\sin(i\phi) + \cos(i\phi) + i\sin(i\phi)}{2} = \cos(i\phi) \tag{4.393}$$

Similarly, subtracting them gives

$$\sinh\phi = \frac{e^{\phi} - e^{-\phi}}{2} = \frac{\cos(i\phi) - i\sin(i\phi) - \cos(i\phi) - i\sin(i\phi)}{2} = -i\sin(i\phi) \tag{4.394}$$

Thus, for real ρ in (4.385) we must have

$$\text{Re}(i\cos w) = \sin p \sinh q < 0 \tag{4.395}$$

Referring to the complex plane shown in Figure 4.9, we are searching a path along which (4.385) will converge to zero as $w \to \infty$.

For the upper w-plane, we have $q > 0$ and we need

$$\sin p < 0, \qquad -\pi < p < 0 \tag{4.396}$$

For the lower w-plane, we have $q < 0$ and we need

$$\sin p > 0, \qquad 0 < p < \pi \tag{4.397}$$

For $p > \pi$ and $p < -\pi$, we can also establish the converging zone according. The regions for which the passage for w_0 and w_1 being infinity is permissible are shaded in Figure 4.9. For the case that ρ is complex, the regions shown in Figure 4.9 only shift horizontally, but we will not consider this possibility here (e.g., see Sommerfeld, 1949).

We now consider the special case that

$$w_0 = a + i\infty \to \infty, \quad w_1 = b + i\infty \to \infty \tag{4.398}$$

where

$$-\pi < a < 0, \quad \pi < b < 2\pi \tag{4.399}$$

Let us take the constant c_n in (4.385) as

$$C_n = \frac{1}{2\pi} e^{-in\pi/2} \tag{4.400}$$

Comparison of (4.381) and (4.385) yields the following result

$$J_n(\rho) = \frac{1}{2\pi} \int_W e^{i\rho\cos w} e^{in(w-\pi/2)} dw \tag{4.401}$$

where W is the rectangular contour ABCD shown in Figure 4.9 such that

$$-\pi + i\infty \to -\pi \to \pi \to \pi + i\infty \tag{4.402}$$

Note that this contour integral over a complex path W has a great advantage over the real representation in that it is not limited to only integral values of n but remains valid for arbitrary values of n. Consider the path BC and use a change of variables of

$$\beta = w - \pi / 2 \tag{4.403}$$

This β axis is shown in Figure 4.9. Thus, we have

$$\begin{aligned} \cos w &= \cos(\beta + \pi / 2) = \cos\beta\cos(\pi / 2) - \sin\beta\sin(\pi / 2) \\ &= -\sin\beta \end{aligned} \tag{4.404}$$

With this change of variables, the integration path BC of $-\pi/2 < w < 3\pi/2$ is mapped to becoming $-\pi < \beta < \pi$. The integral in (4.401) along path BC becomes

$$J_n(\rho)\big|_{BC} = \frac{1}{2\pi} \int_{-\pi}^{\pi} e^{i(n\beta - \rho\sin\beta)} d\beta \tag{4.405}$$

For the path AB, we use a change of variables of

$$\beta = -\pi + i\gamma \tag{4.406}$$

Thus, we have

$$\begin{aligned} \sin\beta &= \sin(-\pi + i\gamma) = -\sin\pi\cos(i\gamma) + \sin(i\gamma)\cos\pi \\ &= -\sin(i\gamma) \end{aligned} \tag{4.407}$$

For points A and B, we have

$$A: \beta \to -\pi + i\infty \quad (\gamma \to \infty); \qquad B: \beta \to -\pi \quad (\gamma \to 0) \tag{4.408}$$

The integral in (4.401) along path AB becomes

$$\begin{aligned} J_n(\rho)\big|_{AB} &= \frac{i}{2\pi} \int_{\infty}^{0} e^{in(-\pi + i\gamma) + i\rho\sin(i\gamma)} d\gamma = \frac{-i}{2\pi} e^{-in\pi} \int_0^{\infty} e^{-n\gamma + i\rho\sin(i\gamma)} d\gamma \\ &= \frac{-i}{2\pi} e^{-in\pi} \int_0^{\infty} e^{-(\gamma n + \rho\sinh\gamma)} d\gamma \end{aligned} \tag{4.409}$$

For the path CD, we use another change of variables of

$$\beta = \pi + i\gamma \tag{4.410}$$

Thus, we have

$$\begin{aligned} \sin\beta &= \sin(\pi + i\gamma) = \sin\pi\cos(i\gamma) + \sin(i\gamma)\cos\pi \\ &= -i\sinh\gamma \end{aligned} \tag{4.411}$$

For points D and C, we have

$$D: \beta \to \pi + i\infty \quad (\gamma \to \infty); \qquad C: \beta \to \pi \quad (\gamma \to 0) \tag{4.412}$$

The integral in (4.401) along path CD becomes

$$J_n(\rho)\big|_{CD} = \frac{i}{2\pi}\int_0^{\infty} e^{in(\pi+i\gamma)-\rho\sinh\gamma}d\gamma = \frac{i}{2\pi}e^{in\pi}\int_0^{\infty} e^{-(\gamma n+\rho\sinh\gamma)}d\gamma \qquad (4.413)$$

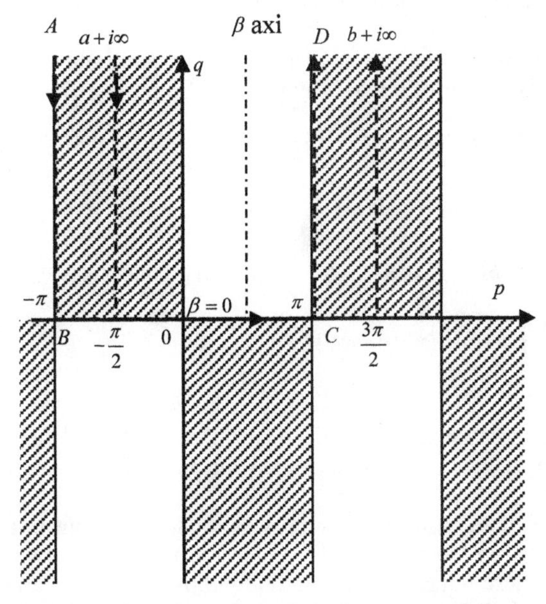

Figure 4.9 Permissible regions in the complex plane for contours giving converging integral (Sommerfeld's contour)

Adding all these results gives the following integral representation of the Bessel function of the first kind:

$$\begin{aligned} J_n(\rho) &= \frac{1}{2\pi}\int_W e^{i\rho\cos w}e^{in(w-\pi/2)}dw = J_n(\rho)\big|_{AB} + J_n(\rho)\big|_{BC} + J_n(\rho)\big|_{CD} \\ &= \frac{1}{2\pi}\int_{-\pi}^{\pi}[\cos(n\beta-\rho\sin\beta)+i\sin(n\beta-\rho\sin\beta)]d\beta - \frac{\sin n\pi}{\pi}\int_0^{\infty} e^{-(\gamma n+\rho\sinh\gamma)}d\gamma \\ &= \frac{1}{2\pi}\int_{-\pi}^{\pi}\cos(n\beta-\rho\sin\beta)d\beta - \frac{\sin n\pi}{\pi}\int_0^{\infty} e^{-(\gamma n+\rho\sinh\gamma)}d\gamma \\ &= \frac{1}{\pi}\int_0^{\pi}\cos(n\beta-\rho\sin\beta)d\beta - \frac{\sin n\pi}{\pi}\int_0^{\infty} e^{-(\gamma n+\rho\sinh\gamma)}d\gamma \end{aligned} \qquad (4.414)$$

where n may not be an integer. The imaginary part of (4.414) vanishes because the sine function is an odd function from $-\pi$ to π and the limit of the cosine term is rewritten from $-\pi$ to π to 0 to π by recognizing the even properties of the cosine function. This is Schlafli's generalized Bessel integral obtained in 1873 (Watson, 1944). However, the current proof follows from that of Sommerfeld (1949) instead of from Watson (1944). Other integral representations were reported by Watson (1944).

For the case of integer order n, we have $\sin(n\pi) = 0$ and thus

$$J_n(\rho) = \frac{1}{\pi}\int_0^{\pi} \cos(n\beta - \rho\sin\beta)d\beta \tag{4.415}$$

This is the well-known integral representation of the Bessel function of the first kind.

Let us demonstrate the power of this integral representation. For the Bessel function of negative order of integer $-n$, we have from (4.415)

$$\begin{aligned}
J_{-n}(\rho) &= \frac{1}{\pi}\int_0^{\pi} \cos(n\beta + \rho\sin\beta)d\beta = \frac{1}{\pi}\int_0^{\pi} \cos[n(\pi-\alpha) + \rho\sin(\pi-\alpha)]d\alpha \\
&= \frac{1}{\pi}\int_0^{\pi} \cos[n\pi - n\alpha + \rho\sin\alpha]d\alpha \\
&= \frac{1}{\pi}\int_0^{\pi} \cos[-n\alpha + \rho\sin\alpha]\cos(n\pi) - \sin[-n\alpha + \rho\sin\alpha]\sin(n\pi)d\alpha \\
&= (-1)^n \frac{1}{\pi}\int_0^{\pi} \cos[-n\alpha + \rho\sin\alpha]d\alpha \\
&= (-1)^n J_n(\rho)
\end{aligned} \tag{4.416}$$

This, of course, agrees with our earlier result that the Bessel function of the first kind of negative integer order is dependent on that of the positive integer order given in (4.278).

This type of integral representation was found important in the investigation of light diffraction problems and this is also the prime objective of Sommerfeld when he studied these integrals. Sommerfeld published in 1894 an important paper on diffraction of light by a screen that improved on the results from Fresnel, Kirchhoff, and Poincare on short wavelength limits. His results were confirmed by experiments for large and small diffraction angles. This piece of work brought considerable fame to Sommerfeld. In wave propagation problems, Sommerfeld's radiation condition is of profound importance.

We will say a little more about the legendary story of Prof. Arnold Sommerfeld here. According to Crawford (2001), Sommerfeld was the king of Nobel Prize nominations and had been nominated to receive the Nobel Prize for a record of 81 times over a span of 34 years from 1917 to 1950. On average, he received 2.38 nominations per year in the 34 years. Clearly, he should receive a Guinness World Record certificate for this. On April 26 1951, Sommerfeld was run over by a car when he was playing with his grandkids. Unfortunately, he never received the prize. Ironically, he himself had made two nominations, one to Einstein and one to Planck, and both of them of course received the Nobel Prize. Even more ironically, he had taught 7 PhD or post-doctoral students, who eventually received the Nobel Prize. These recipients include W. Heisenberg (physics, 1932), H. Bethe (physics, 1967), W. Pauli (physics, 1945), P. Debye (chemistry, 1936), L. Pauling (chemistry, 1954), M. von Laue (physics, 1914) and I.I. Rabi (physics, 1944). It was reported that Einstein once told Sommerfeld: "What I especially admire about you is that you have, as it were, pounded out of the soil such a large number of young talents."

His series of textbooks, *Theoretical Lectures on Physics*, made significant impacts on the new generation of scientists and on the development of physics. His books include: *Mechanics (Theoretical Lectures on Physics Vol. 1)*, *Mechanics of Deformable Bodies (Theoretical Lectures on Physics Vol. 2)*, *Electrodynamics*

(Theoretical Lectures on Physics Vol. 3), Optics (Theoretical Lectures on Physics Vol. 4), Mathematical Theory of Diffraction, Differential Equations in Physics (Vol. 6), Atomic Structures and Spectral Lines, and *The Theory of Top Volume III.*

4.7 LOMMEL DIFFERENTIAL EQUATION

The Lommel differential equation is a special type of nonhomogeneous Bessel function:

$$x^2 y'' + xy' + \left(x^2 - \nu^2\right) y = x^{\mu+1} \tag{4.417}$$

The solutions of this equation are the Lommel functions

$$s_{\mu,\nu}(x) = \frac{\pi}{2}\left[Y_\nu(x)\int_0^x s^\mu J_\nu(s)ds - J_\nu(x)\int_0^x s^\mu Y_\nu(s)ds \right] \tag{4.418}$$

$$S_{\mu,\nu}(x) = s_{\mu,\nu}(x) - \frac{2^{\mu-1}\Gamma(\frac{1+\mu+\nu}{2})}{\pi\Gamma(\frac{\nu-\mu}{2})}\{J_\nu(x) - \cos[\pi(\mu-\nu)/2]Y_\nu(x)\} \tag{4.419}$$

There are other types of nonhomogeneous Bessel equations, including the Anger differential equation and the Weber differential equation. Details can be found in Abramowitz and Stegun (1964). Lommel did research on meteorology, light and physical optics. He was the PhD advisor of J. Stark, Nobel Prize winner in physics in 1919.

4.8 HANKEL FUNCTION

We have given the solutions of the Bessel equation as J_ν and Y_ν. Actually, the general solutions to the Bessel function can also appear in different forms. For example, N. Nielsen in 1902 defined the following functions, which are now known as Hankel functions of the first and second kinds or Bessel functions of the third kind,

$$H_\nu^{(1)}(x) = J_\nu(x) + iY_\nu(x), \quad H_\nu^{(2)}(x) = J_\nu(x) - iY_\nu(x) \tag{4.420}$$

The symbol H was chosen by Nielsen to honor the contribution of Hankel on the integral representation and asymptotic expansions of Bessel functions.

Recall from (4.284) that

$$Y_\nu(\rho) = \frac{J_\nu(\rho)\cos\nu\pi - J_{-\nu}(\rho)}{\sin\nu\pi} \tag{4.421}$$

Substitution of (4.421) into the first equation of (4.420) gives

$$
\begin{aligned}
H_\nu^{(1)}(x) &= J_\nu(x) + i[\frac{J_\nu(x)\cos\nu\pi - J_{-\nu}(x)}{\sin\nu\pi}] \\
&= \frac{J_\nu(x)\sin\nu\pi + i[J_\nu(x)\cos\nu\pi - J_{-\nu}(x)]}{\sin\nu\pi} \\
&= -i\{\frac{J_{-\nu}(x) - J_\nu(x)(\cos\nu\pi - i\sin\nu\pi)}{\sin\nu\pi}\} \\
&= \frac{J_{-\nu}(x) - e^{-\nu\pi i}J_\nu(x)}{i\sin\nu\pi}
\end{aligned}
\tag{4.422}
$$

Similarly, the second equation of (4.420) can be expressed as

$$
H_\nu^{(2)}(x) = \frac{J_{-\nu}(x) - e^{\nu\pi i}J_\nu(x)}{-i\sin\nu\pi} \tag{4.423}
$$

Inverting these two equations, we can also express the Bessel function of the first kind in terms of the Hankel functions as

$$
J_\nu(x) = \frac{H_\nu^{(1)}(x) + H_\nu^{(2)}(x)}{2}, \quad J_{-\nu}(x) = \frac{e^{\nu\pi i}H_\nu^{(1)}(x) + e^{-\nu\pi i}H_\nu^{(2)}(x)}{2} \tag{4.424}
$$

A slightly different form of these formulas can be obtained by recognizing the following identity:

$$
e^{\pm\nu\pi i} = e^{\pm 2\nu(\pi i/2)} = i^{\pm 2\nu} \tag{4.425}
$$

Using this identity, we have

$$
H_\nu^{(1)}(x) = \frac{i[J_\nu(x)i^{-2\nu} - J_{-\nu}(x)]}{\sin\nu\pi}, \quad H_\nu^{(2)}(x) = -\frac{i[J_\nu(x)i^{2\nu} - J_{-\nu}(x)]}{\sin\nu\pi} \tag{4.426}
$$

Analogously, relations between Hankel functions and Bessel functions closely resemble the relation between exponential functions and sine and cosine functions as illustrated in Table 4.1.

The analogy is more vivid if we consider the asymptotic expansion of Bessel and Hankel functions:

$$
J_\nu(x) \approx \sqrt{\frac{2}{\pi x}}\cos(x - \frac{\nu\pi}{2} - \frac{\pi}{4}), \quad Y_\nu(x) \approx \sqrt{\frac{2}{\pi x}}\sin(x - \frac{\nu\pi}{2} - \frac{\pi}{4}) \tag{4.427}
$$

$$
H_\nu^{(1)}(x) \approx \sqrt{\frac{2}{\pi x}}\exp(x - \frac{\nu\pi}{2} - \frac{\pi}{4}), \quad H_\nu^{(2)}(x) \approx \sqrt{\frac{2}{\pi x}}\exp[-(x - \frac{\nu\pi}{2} - \frac{\pi}{4})] \tag{4.428}
$$

With this analogy in mind, it is not difficult to visualize the importance of Hankel functions. In fact, for physical problems of wave propagation, the role of exponential functions and circular functions in Cartesian coordinates is actually reflected by Hankel and Bessel functions in polar cylindrical coordinates. Therefore, Hankel functions are commonly encountered in wave propagation problems (see Section 9.2.3 in Chapter 9).

Since Bessel functions of imaginary argument can be expressed as modified Bessel functions, clearly we can do the same for Hankel functions of imaginary argument. In particular, by substituting x by ix into the first equation of (4.426), we have

Table 4.1. Analogy between circular functions and Bessel functions

Function type	Analogy between functions	
	Exponential function	Hankel functions
1	e^{ix}	$H_\nu^{(1)}(x)$
2	e^{-ix}	$H_\nu^{(2)}(x)$
3	$\cos x$	$J_\nu(x)$
4	$\sin x$	$Y_\nu(x)$
5	$e^{x} = \cos x + i \sin x$	$H_\nu^{(1)} = J_\nu(x) + iY_\nu(x)$
6	$e^{-ix} = \cos x - i \sin x$	$H_\nu^{(1)} = J_\nu(x) - iY_\nu(x)$

$$H_\nu^{(1)}(ix) = \frac{i}{\sin \nu\pi}[J_\nu(ix)i^{-2\nu} - J_{-\nu}(ix)] \tag{4.429}$$

For $-\pi < \arg x \le \pi/2$, recall that the definition of a modified Bessel function is

$$I_\nu(x) = i^{-\nu} J_\nu(ix), \quad I_{-\nu}(x) = i^{\nu} J_{-\nu}(ix), \tag{4.430}$$

Substitution of (4.430) into (4.429) yields

$$\begin{aligned} H_\nu^{(1)}(ix) &= \frac{i^{1-\nu}}{\sin \nu\pi}[I_\nu(x) - I_{-\nu}(x)] = \frac{i^{-(1+\nu)}}{\sin \nu\pi}[I_{-\nu}(x) - I_\nu(x)] \\ &= \frac{2}{\pi} i^{-(1+\nu)} \frac{\pi}{2} \frac{[I_{-\nu}(x) - I_\nu(x)]}{\sin \nu\pi} = \frac{2}{\pi} i^{-(1+\nu)} K_\nu(x) \end{aligned} \tag{4.431}$$

The last result immediately follows from the definition of the modified Bessel of the second kind given in (4.341). Inversely, we can substitute x by $-ix$ into the left hand side of (4.431) to give

$$H_\nu^{(1)}(x) = \frac{2}{\pi} i^{-(1+\nu)} K_\nu(-ix), \quad (-\pi/2 < \arg x \le \pi) \tag{4.432}$$

Following the same procedure, it is straightforward to show that for the case of $-\pi/2 < \arg x \le \pi$, we have

$$H_\nu^{(2)}(-ix) = \frac{2}{\pi} i^{1+\nu} \frac{\pi}{2} \frac{[I_{-\nu}(x) - I_\nu(x)]}{\sin \nu\pi} = \frac{2}{\pi} i^{1+\nu} K_\nu(x) \tag{4.433}$$

Again, substitution of x by ix into the left hand side of (4.433) results in

$$H_\nu^{(2)}(x) = \frac{2}{\pi} i^{1+\nu} K_\nu(ix), \quad (-\pi < \arg x \le \pi/2) \tag{4.434}$$

This provides relations between Hankel functions and modified Bessel functions of imaginary argument.

4.9 KELVIN FUNCTIONS

A kind of special function closely related to the Bessel function is called the Kelvin functions. They were proposed by Lord Kelvin in 1889 when he considered certain electrical problems. Actually, Kelvin functions not only appear in electrical problems that Kelvin encountered, they also emerged naturally in the bending of a cylindrical shell, in the problems of circular plates on elastic foundations, and in symmetrical bending of shallow spherical shells. The Kelvin equation of order ν is

$$x^2 y'' + xy' - (ix^2 + \nu^2) y = 0 \tag{4.435}$$

The general solutions of this equation can be written as special forms of Bessel or modified Bessel functions:

$$y = C_1 J_\nu (i\sqrt{i}x) + C_2 Y_\nu (i\sqrt{i}x) \tag{4.436}$$

$$y = C_1 I_\nu (\sqrt{i}x) + C_2 K_\nu (\sqrt{i}x) \tag{4.437}$$

However, these modified Bessel functions are complex and the real and imaginary parts of these functions are called Kelvin functions, which are defined as:

$$J_\nu (xe^{i3\pi/4}) = J_\nu (i\sqrt{i}x) = i^\nu I_\nu (\sqrt{i}x) = \mathrm{ber}_\nu (x) + i\mathrm{bei}_\nu (x) \tag{4.438}$$

$$e^{-\nu\pi i/2} K_\nu (xe^{\pi i/4}) = i^{-\nu} K_\nu (\sqrt{i}x) = \mathrm{ker}_\nu (x) + i\mathrm{kei}_\nu (x) \tag{4.439}$$

Note that these Kelvin functions are real. For the case of zero orders, we have

$$I_0 (\sqrt{i}x) = \mathrm{ber}(x) + i\mathrm{bei}(x) \tag{4.440}$$

$$K_0 (\sqrt{i}x) = \mathrm{ker}(x) + i\mathrm{kei}(x) \tag{4.441}$$

It is customary not to write the subscript "0" for the case of zero order, because only the zero order was actually proposed by Kelvin in 1889. Kelvin proposed the name of ber(x) and bei(x) because they closely resemble the roles of circular functions cos(x) and sin(x). Actually ker(x) and kei(x) were defined by Russell in 1909 while all higher order Kelvin functions were proposed by Whitehead in 1911 (Watson, 1944). With the Kelvin functions, the solution of (4.435) becomes

$$y = A[\mathrm{ber}_\nu (x) + i\mathrm{bei}_\nu (x)] + B[\mathrm{ker}_\nu (x) + i\mathrm{kei}_\nu (x)] \tag{4.442}$$

Kelvin functions $\mathrm{ber}_\nu(x)$ and $\mathrm{bei}_\nu(x)$ are plotted in Figures 4.10 and 4.11 respectively. For the case of moving loads on a circular ice plate on water, the following equation is encountered (Wyman, 1950; Nevel, 1959; Assur, 1959):

$$\nabla^2\nabla^2 \zeta + \zeta = \left(\frac{d^2}{d\bar{r}^2} + \frac{1}{\bar{r}}\frac{d}{d\bar{r}} + i\right)\left(\frac{d^2}{d\bar{r}^2} + \frac{1}{\bar{r}}\frac{d}{d\bar{r}} - i\right)\zeta = 0 \tag{4.443}$$

where

$$\bar{r} = r\left(\frac{\gamma_w}{D}\right)^{1/4} \tag{4.444}$$

and γ_w is the unit weight of water and D is the bending stiffness of the plate. The solution of (4.443) is

$$\zeta = c_1 \mathrm{ber}(\bar{r}) + c_2 \mathrm{bei}(\bar{r}) + c_3 \mathrm{ker}(\bar{r}) + c_4 \mathrm{kei}(\bar{r}) \tag{4.445}$$

A similar equation is also obtained for the case of bending of spherical shallow shells (see Timoshenko and Woinowsky-Krieger, 1959).

Equation (4.443) is not normally named in the literature, and we call it the Kelvin equation, as its solutions are Kelvin functions. The role of the Kelvin equation versus the biharmonic equation is similar to the role of the Helmholtz equation versus the Laplace equation. This is illustrated in Table 4.2. We can see that there is a close resemblance of the Laplace and Helmholtz equations versus the biharmonic and Kelvin equations.

In series form, Kelvin functions of the first kind are defined as

$$\mathrm{ber}(x) = 1 - \frac{(x/2)^4}{2!^2} + \frac{(x/2)^8}{4!^2} - ... \tag{4.446}$$

$$\mathrm{bei}(x) = (\frac{x}{2})^2 - \frac{(x/2)^6}{3!^2} + \frac{(x/2)^{10}}{5!^2} - ... \tag{4.447}$$

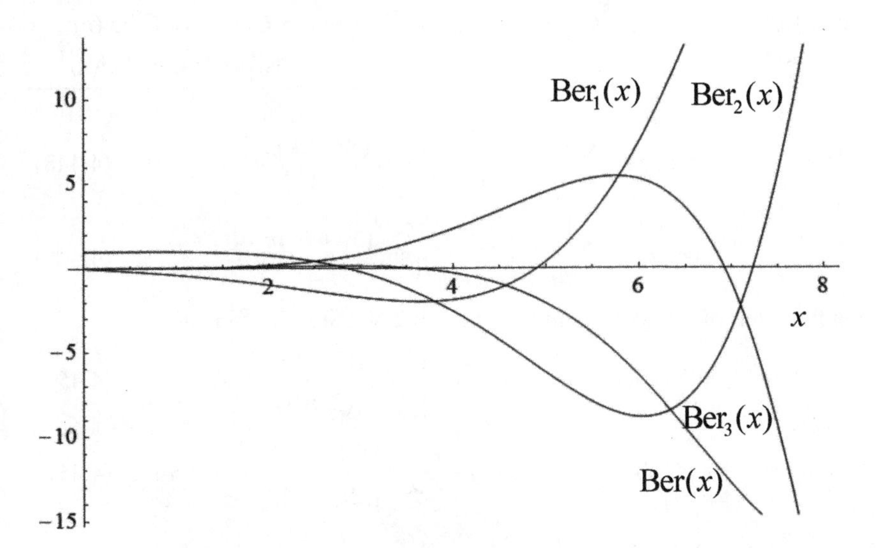

Figure 4.10 Kelvin function ber$_\nu$

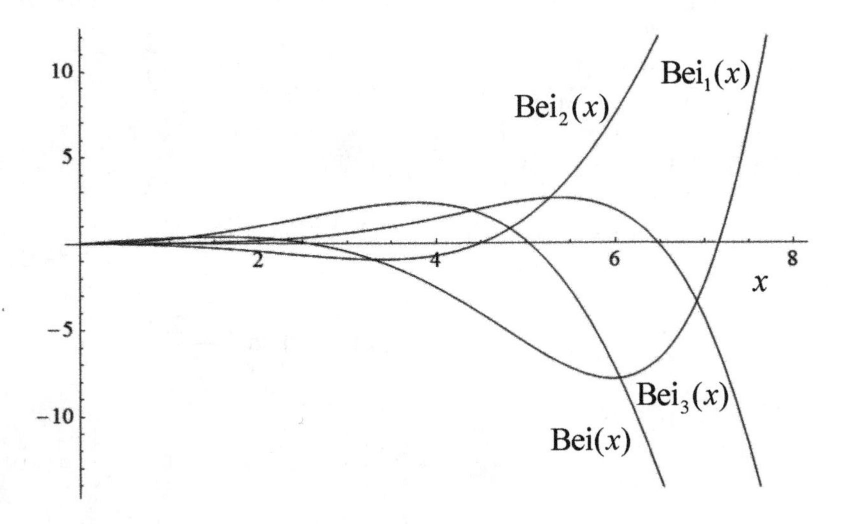

Figure 4.11 Kelvin function bei$_\nu$

Table 4.2. Analogy between Laplace and biharmonic equations

Function type	Analogy between Laplace and Biharmonic equations	
	Laplace type equations	Biharmonic type equations
Basic equation	$\nabla^2\zeta = 0$ (Laplace)	$\nabla^2\nabla^2\zeta = 0$ (biharmonic)
Additional term	$\nabla^2\zeta + \zeta = 0$ (Helmholtz)	$\nabla^2\nabla^2\zeta + \zeta = 0$ (Kelvin)
Polar form solution	$\zeta = C_1 J_0(r) + C_1 Y_0(r)$	$\zeta = C_1\text{ber}(r) + C_2\text{bei}(r) + C_3\text{ker}(r) + C_4\text{kei}(r)$

$$\text{ber}_\nu(x) = \sum_{k=0}^{\infty} \frac{(x/2)^{2k+\nu}}{k!\Gamma(\nu+k+1)} \cos\frac{(3\nu+2k)\pi}{4} \tag{4.448}$$

$$\text{bei}_\nu(x) = \sum_{k=0}^{\infty} \frac{(x/2)^{2k+\nu}}{k!\Gamma(\nu+k+1)} \sin\frac{(3\nu+2k)\pi}{4} \tag{4.449}$$

Kelvin functions of the second kind can be expressed in series form as:

$$\text{ker}(x) = -\ln(x/2)\text{ber}(x) + \frac{\pi}{4}\text{bei}(x) + \sum_{m=0}^{\infty} \frac{(-1)^m (x/2)^{4m}}{[(2m)!]^2} \psi(2m+1) \tag{4.450}$$

$$\text{kei}(x) = -\ln(x/2)\text{bei}(x) - \frac{\pi}{4}\text{ber}(x) + \sum_{m=0}^{\infty} \frac{(-1)^m (x/2)^{4m+2}}{[(2m+1)!]^2} \psi(2m+2) \tag{4.451}$$

$$\begin{aligned} \text{ker}_n(x) &= -\ln(x/2)\text{ber}_n(x) + \frac{\pi}{4}\text{bei}_n(x) \\ &+ \frac{1}{2}\sum_{k=0}^{n-1} \frac{(n-k-1)!(x/2)^{2k-n}}{k!} \cos\frac{(3n+2k)\pi}{4} \\ &+ \frac{1}{2}\sum_{k=0}^{\infty} \frac{(x/2)^{n+2k}}{(n+k)!k!} \{\psi(k+1) + \psi(n+k+1)\} \cos\frac{(3n+2k)\pi}{4} \end{aligned} \tag{4.452}$$

$$\begin{aligned} \text{kei}_n(x) &= -\{\ln(x/2) + \gamma\}\text{bei}_n(x) - \frac{\pi}{4}\text{ber}_n(x) \\ &- \frac{1}{2}\sum_{k=0}^{n-1} \frac{(n-k-1)!(x/2)^{2k-n}}{k!} \sin\frac{(3n+2k)\pi}{4} \\ &+ \frac{1}{2}\sum_{k=0}^{\infty} \frac{(x/2)^{n+2k}}{(n+k)!k!} \{\psi(k+1) + \psi(n+k+1)\} \sin\frac{(3n+2k)\pi}{4} \end{aligned} \tag{4.453}$$

where the digamma function has been defined (4.307) as

$$\psi(n+1) = -\gamma + \sum_{k=1}^{n} \frac{1}{k} \tag{4.454}$$

We can use the Kelvin equation to derive the formula for the second differentiation of the Kelvin function. For example, the zero order of (4.435) is

$$x^2y''+xy'-ix^2y=0 \tag{4.455}$$

One of its solution is

$$y=\text{ber}(x)+i\text{bei}(x) \tag{4.456}$$

Substitution of (4.456) into (4.455) gives

$$x^2\text{ber}''(x)+x\text{ber}'(x)+x^2\text{bei}(x)+i[x^2\text{bei}''(x)+x\text{bei}'(x)-x^2\text{ber}(x)]=0 \tag{4.457}$$

Setting the real part and imaginary part of (4.457) to zeros gives

$$\text{ber}''(x)=-\frac{1}{x}\text{ber}'(x)-\text{bei}(x) \tag{4.458}$$

$$\text{bei}''(x)=\text{ber}(x)-\frac{1}{x}\text{bei}'(x) \tag{4.459}$$

Another solution of (4.455) is

$$y=\text{ker}(x)+i\text{kei}(x) \tag{4.460}$$

Following a similar procedure in getting (4.458) and (4.459), we obtain

$$\text{ker}''(x)=-\frac{1}{x}\text{ker}'(x)-\text{kei}(x) \tag{4.461}$$

$$\text{kei}''(x)=\text{ker}(x)-\frac{1}{x}\text{kei}'(x) \tag{4.462}$$

Differentiating (4.459) once more gives

$$\text{bei}'''(x)=\text{ber}'(x)+\frac{1}{x^2}\text{bei}'(x)-\frac{1}{x}\text{bei}''(x) \tag{4.463}$$

Substitution of (4.459) into (4.463) gives

$$\text{bei}'''(x)=\text{ber}'(x)+\frac{2}{x^2}\text{bei}'(x)-\frac{1}{x}\text{ber}(x) \tag{4.464}$$

Similarly, we have the following identities for other Kelvin functions

$$\text{ber}'''(x)=-\text{bei}'(x)+\frac{2}{x^2}\text{ber}'(x)+\frac{1}{x}\text{bei}(x) \tag{4.465}$$

$$\text{ker}'''(x)=-\text{kei}'(x)+\frac{2}{x^2}\text{ker}'(x)+\frac{1}{x}\text{kei}(x) \tag{4.466}$$

$$\text{kei}'''(x)=\text{ker}'(x)+\frac{2}{x^2}\text{kei}'(x)-\frac{1}{x}\text{ker}(x) \tag{4.467}$$

Next, we will consider some simple formulas of integrals involving Kelvin functions. More specifically, we differentiate the following term:

$$\begin{aligned}\frac{d}{dx}[x\text{bei}'(x)]&=\text{bei}'(x)+x\text{bei}''(x)\\&=\text{bei}'(x)+x[\text{ber}(x)-\frac{1}{x}\text{bei}'(x)]=x\text{ber}(x)\end{aligned} \tag{4.468}$$

Integrating both side, we have

$$\int x\text{ber}(x)dx=x\text{bei}'(x) \tag{4.469}$$

Similarly, one can easily obtain the following formulas

$$\int x\text{bei}(x)dx = -x\text{ber}'(x) \tag{4.470}$$

$$\int x\text{ker}(x)dx = x\text{kei}'(x) \tag{4.471}$$

$$\int x\text{kei}(x)dx = -x\text{ker}'(x) \tag{4.472}$$

Note also the following recurrence relations for differentiation (Abramowitz and Stegun, 1964)

$$\text{ber}'(x) = \frac{1}{\sqrt{2}}[\text{ber}_1(x) + \text{bei}_1(x)] \tag{4.473}$$

$$\text{bei}'(x) = \frac{1}{\sqrt{2}}[-\text{ber}_1(x) + \text{bei}_1(x)] \tag{4.474}$$

$$\text{ker}'(x) = \frac{1}{\sqrt{2}}[\text{ker}_1(x) + \text{kei}_1(x)] \tag{4.475}$$

$$\text{kei}'(x) = \frac{1}{\sqrt{2}}[-\text{ker}_1(x) + \text{kei}_1(x)] \tag{4.476}$$

Expansion for the cross product can be found in Abramowitz and Stegun (1964):

$$\text{ber}_\nu^2 + \text{bei}_\nu^2 = (x/2)^{2\nu} \sum_{k=0}^{\infty} \frac{1}{\Gamma(\nu+k+1)\Gamma(\nu+2k+1)} \frac{(x^2/4)^{2k}}{k!} \tag{4.477}$$

For the special case of $\nu = 0$, we have (p. 82 of Watson, 1944)

$$\text{ber}^2(\text{x}) + \text{bei}^2(x) = 1 + \frac{(x/2)^4}{2!} + \frac{(x/2)^8}{4 \cdot 4!} + \frac{(x/2)^{12}}{6^2 \cdot 6!} + \frac{(x/2)^{16}}{8^2 \cdot 9!} + \ldots \tag{4.478}$$

More extensive formulas about Kelvin functions can be found in Abramowitz and Stegun (1964).

In closing, we report the following limiting values of Kelvin functions at zero argument:

$$\begin{aligned} &\text{ber}(0) = 1, \ \text{bei}(0) = 0, \text{ker}(0) \to \infty, \ \text{kei}(0) = -\pi/4, \\ &\text{ber}_1(0) = 0, \ \text{bei}_1(0) = 0, \text{ker}_1(0) \to -\infty, \ \text{kei}_1(0) \to -\infty \end{aligned} \tag{4.479}$$

In predicting the onset of diffuse mode bifurcations of thick-walled hollow cylinders of geomaterials, Chau and Choi (1998) found that the evaluations of Bessel functions of the first and second kinds of complex arguments are necessary. However, Fortran subroutine for such calculations is not readily available in standard textbooks (e.g., Press at el., 1992). To check the accuracy of their Fortran subroutine, it was found that numerical tables of Kelvin functions given in Abramowitz and Stegun (1964) could be used.

4.10 LEGENDRE EQUATION

The Legendre equation is resulted from the spherical form of the Helmholtz equation. The solution of the Legendre equation is called the Legendre polynomials and it is closely related to spherical harmonics, which is the basic eigenfunction expansion of spherical coordinates. Legendre polynomials were first proposed by

Legendre in 1785 when he expand Newton's gravitational potential in series expansion. Laplace in 1782 also studied them in relation to the gravitational field due to a spherical planet. But, later mathematicians, like Jacobi, Dirichlet and Heine agreed that credit should go to Legendre. First, we recall the Helmholtz equation in spherical coordinates

$$\nabla^2\psi + k^2\psi = 0 \tag{4.480}$$

In terms of spherical coordinates, we can write the Helmholtz equation as

$$\frac{1}{r^2}\frac{\partial}{\partial r}(r^2\frac{\partial\psi}{\partial r}) + \frac{1}{r^2\sin\theta}\frac{\partial}{\partial\theta}(\sin\theta\frac{\partial\psi}{\partial\theta}) + \frac{1}{r^2\sin^2\theta}\frac{\partial^2\psi}{\partial\phi^2} + k^2\psi = 0 \tag{4.481}$$

Let us assume the following separation of variables

$$\psi(r,\theta,\phi) = R(r)\Theta(\theta)\Phi(\phi) \tag{4.482}$$

Substitution of (4.482) into (4.481) gives

$$\frac{\sin^2\theta}{R}\frac{d}{dr}(r^2\frac{dR}{dr}) + \frac{\sin\theta}{\Theta}\frac{d}{d\theta}(\sin\theta\frac{d\Theta}{d\theta}) + k^2r^2\sin^2\theta = -\frac{1}{\Phi}\frac{d^2\Phi}{d\phi^2} = m^2 \tag{4.483}$$

Then we have

$$\frac{d^2\Phi}{d\phi^2} + m^2\Phi = 0 \tag{4.484}$$

The solution of (4.484) is of course

$$\Phi = A\cos m\phi + B\sin m\phi \tag{4.485}$$

Rearranging (4.483) we can rewrite

$$\frac{1}{\Theta\sin\theta}\frac{d}{d\theta}(\sin\theta\frac{d\Theta}{d\theta}) - \frac{m^2}{\sin^2\theta} = -\frac{1}{R}\frac{d}{dr}(r^2\frac{dR}{dr}) - k^2r^2 = -\beta^2 \tag{4.486}$$

Thus, the theta-dependent function can be expressed as

$$\frac{1}{\sin\theta}\frac{d}{d\theta}(\sin\theta\frac{d\Theta}{d\theta}) + (\beta^2 - \frac{m^2}{\sin^2\theta})\Theta = 0 \tag{4.487}$$

Applying the following change of variables,

$$x = \cos\theta, \quad P(x) = \Theta(\theta) \tag{4.488}$$

we get

$$\frac{d}{dx}[(1-x^2)\frac{dP}{dx}] + (\beta^2 - \frac{m^2}{1-x^2})P = 0 \tag{4.489}$$

Further simplification of this equation by assuming $\beta^2 = n(n+1)$ leads to

$$(1-x^2)\frac{d^2P}{dx^2} - 2x\frac{dP}{dx} + [n(n+1) - \frac{m^2}{1-x^2}]P = 0 \tag{4.490}$$

This is the associated Legendre equation. For the special case of $m = 0$ (i.e., ψ is independent of ϕ), we have the following Legendre equation:

$$(1-x^2)\frac{d^2P}{dx^2} - 2x\frac{dP}{dx} + n(n+1)P = 0 \tag{4.491}$$

We see that the Legendre equation results from the separation of variables of the Helmholtz equation.

4.10.1 Series Solution

Let us consider the solution in terms of infinite series as:

$$P = \sum_{m=0}^{\infty} a_m x^m \tag{4.492}$$

Differentiation of this proposed form gives

$$\frac{dP}{dx} = \sum_{m=1}^{\infty} m a_m x^{m-1}, \quad \frac{d^2 P}{dx^2} = \sum_{m=2}^{\infty} m(m-1) a_m x^{m-2} \tag{4.493}$$

Substitution of (4.492) and (4.493) into (4.491) gives

$$(1-x^2)\sum_{m=2}^{\infty} m(m-1) a_m x^{m-2} - 2x \sum_{m=1}^{\infty} m a_m x^{m-1} + n(n+1)\sum_{m=0}^{\infty} a_m x^m = 0 \tag{4.494}$$

Multiplying the non-constant coefficients into the series gives

$$\sum_{m=2}^{\infty} m(m-1) a_m x^{m-2} - \sum_{m=2}^{\infty} m(m-1) a_m x^m - \sum_{m=1}^{\infty} 2m a_m x^m + \sum_{m=0}^{\infty} n(n+1) a_m x^m = 0 \tag{4.495}$$

As the summation is for infinite terms, we can always shift the summation to start from $m = 0$. More specifically, we can assume $k = m+2$ in the first sum and write the series as:

$$\sum_{m=0}^{\infty} (m+2)(m+1) a_{m+2} x^m - \sum_{m=2}^{\infty} m(m-1) a_m x^m - \sum_{m=1}^{\infty} 2m a_m x^m + \sum_{m=0}^{\infty} n(n+1) a_m x^m = 0 \tag{4.496}$$

Writing out the first few terms explicitly, we have

$$2a_2 + 6a_3 x - 2a_1 x + n(n+1)a_0 + n(n+1)a_1 x$$
$$+\sum_{m=2}^{\infty} [(m+2)(m+1)a_{m+2} - m(m-1)a_m - 2m a_m + n(n+1)a_m] x^m = 0 \tag{4.497}$$

Thus, setting the coefficients of the zero, first, and general m-order terms to zero, we obtain

$$2a_2 + n(n+1)a_0 = 0 \tag{4.498}$$

$$6a_3 - 2a_1 + n(n+1)a_1 = 0 \tag{4.499}$$

$$(m+2)(m+1)a_{m+2} - [m(m-1) + 2m - n(n+1)]a_m = 0 \tag{4.500}$$

These three equations give

$$a_2 = -\frac{n(n+1)}{2} a_0 \tag{4.501}$$

$$a_3 = -\frac{(n+2)(n-1)}{6} a_1 \tag{4.502}$$

$$a_{m+2} = \frac{(m-n)(m+n+1)}{(m+2)(m+1)} a_m \tag{4.503}$$

Therefore, for $m = 2$ we have

$$a_4 = \frac{(2-n)(2+n+1)}{12} a_2 = -\frac{(2-n)(2+n+1)n(n+1)}{4!} a_0 \tag{4.504}$$

For $m = 3$, we have

$$a_5 = \frac{(3-n)(4+n)}{20} a_3 = -\frac{(3-n)(4+n)(n-1)(n+2)}{5!} a_1 \tag{4.505}$$

Using (4.503) to (4.505), we can express all terms using only two constants:

$$P(x) = a_0 y_1 + a_1 y_2 \tag{4.506}$$

where

$$y_1 = 1 - \frac{n(n+1)}{2!} x^2 + \frac{(n-2)n(n+1)(n+3)}{4!} x^4 - ... \tag{4.507}$$

$$y_2 = x - \frac{(n-1)(n+2)}{3!} x^3 + \frac{(n-3)(n-1)(n+2)(n+4)}{5!} x^5 - ... \tag{4.508}$$

This provides the general solution for the Legendre equation. From (4.500), if n is an integer, we have $a_{m+2} = 0$, for $m = n, n+2, ...$ That is, the series becomes finite and is of order n (for both even and odd). We can work backward to express all terms in terms of a_n. In particular, we first rewrite (4.500) as:

$$a_m = -\frac{(m+2)(m+1)}{(n-m)(m+n+1)} a_{m+2}, \quad m \le n-2 \tag{4.509}$$

Letting $m = n-2$ in (4.509) gives

$$a_{n-2} = -\frac{n(n-1)}{2(2n-1)} a_n \tag{4.510}$$

Applying this expression twice, we get

$$a_{n-4} = -\frac{(n-2)(n-3)}{4(2n-3)} a_{n-2} = (-1)^2 \frac{n(n-1)(n-2)(n-3)}{2 \cdot 4 \cdot (2n-1)(2n-3)} a_n \tag{4.511}$$

Application of the recursive formula m times results in

$$a_{n-2m} = (-1)^m \frac{n(n-1)(n-2)(n-3)\cdots(n-2m+1)}{2 \cdot 4 \cdots 2m(2n-1)(2n-3)\cdots(2n-2m+1)} a_n \tag{4.512}$$

With this recursive formula in mind, the function $P(x)$ becomes

$$P(x) = a_n x^n + a_{n-2} x^{n-2} + a_{n-4} x^{n-4} + ... + a_0 \tag{4.513}$$

for n even and

$$P(x) = a_n x^n + a_{n-2} x^{n-2} + a_{n-4} x^{n-4} + ... + a_1 x \tag{4.514}$$

for n odd. These expressions can be rewritten as a single expression:

$$P(x) = a_n \sum_{m=0}^{M} (-1)^m \frac{n(n-1)(n-2)(n-3)\cdots(n-2m+1)}{2 \cdot 4 \cdots 2m(2n-1)(2n-3)\cdots(2n-2m+1)} x^{n-2m} \tag{4.515}$$

where

$$\begin{aligned} M &= n/2, && n \text{ even} \\ &= (n-1)/2, && n \text{ odd} \end{aligned} \tag{4.516}$$

To further simplify this form, we first derive the following identities:

$$\begin{aligned} & n(n-1)(n-2)\cdots(n-2m+1) \\ & = n(n-1)(n-2)\cdots(n-2m+1)\frac{(n-2m)(n-2m-1)\cdots 3\cdot 2\cdot 1}{(n-2m)\cdots 3\cdot 2\cdot 1} \\ & = \frac{n!}{(n-2m)!} \end{aligned} \tag{4.517}$$

$$2\cdot 4\cdots 2m = 2^m m! \tag{4.518}$$

$$\begin{aligned} & (2n-1)(2n-3)\cdots(2n-2m+1) \\ & = \frac{2n(2n-1)(2n-2)(2n-3)\cdots(2n-2m+1)}{2n(2n-2)((2n-4)\cdots(2n-2m+2)}\frac{(2n-2m)!}{(2n-2m)!} \\ & = \frac{(2n)!}{2^m n(n-1)(n-2)\cdots(n-m+1)(2n-2m)!} \\ & = \frac{(2n)!}{2^m (2n-2m)! n(n-1)(n-2)\cdots(n-m+1)}\frac{(n-m)!}{(n-m)!} \\ & = \frac{(2n)!(n-m)!}{2^m (2n-2m)! n!} \end{aligned} \tag{4.519}$$

Substitution of (4.517)–(4.519) into (4.515) arrives at

$$\begin{aligned} P(x) &= a_n \sum_{m=0}^{M} \frac{(-1)^m n! 2^m (2n-2m)! n!}{(n-2m)! 2^m m! (2n)! (n-m)!} x^{n-2m} \\ &= a_n \sum_{m=0}^{M} \frac{(-1)^m (n!)^2 (2n-2m)!}{m!(n-2m)!(n-m)!(2n)!} x^{n-2m} \end{aligned} \tag{4.520}$$

Since a_n is an arbitrary constant, we can choose any value for it. For the following choice

$$a_n = \frac{(2n)!}{2^n (n!)^2}, \tag{4.521}$$

we can express the solution as

$$P_n(x) = \sum_{m=0}^{M} \frac{(-1)^m (2n-2m)!}{2^n m!(n-m)!(n-2m)!} x^{n-2m} \tag{4.522}$$

This is the Legendre polynomial for the case of integer n. The Legendre polynomial given by (4.522) is finite for all values of $-1 \le x \le 1$. For n being a non-integer, we have $P_n(x) \to \infty$ as $x \to \pm 1$. The first few terms of the Legendre polynomials are:

$$P_0(x) = 1,$$
$$P_1(x) = x,$$
$$P_2(x) = \frac{1}{2}(3x^2 - 1), \tag{4.523}$$
$$P_3(x) = \frac{1}{2}(5x^3 - 3x),$$
$$P_4(x) = \frac{1}{8}(35x^4 - 30x^2 + 3)$$

The first five Legendre polynomials (except zero order) are plotted in Figure 4.12. Since (4.491) is a linear second order ODE, we need to find another independent solution in additional to the Legendre polynomials obtained in (4.522). To do so, let us assume the other independent solution in the form:

$$P = A(x)P_n(x) \tag{4.524}$$

where $A(x)$ is the unknown function to be found. Again, (4.524) results from a standard theorem stated in Forsyth (1956). Substitution of (4.524) into (4.191) leads to

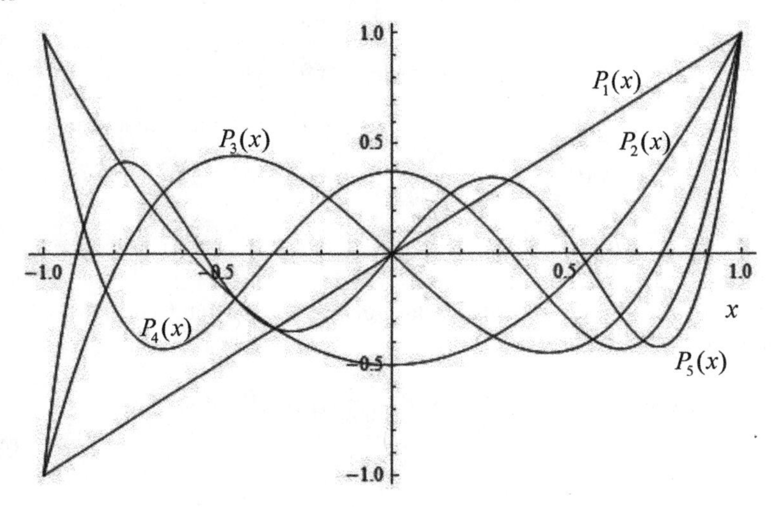

Figure 4.12 Legendre polynomials of the first kind P_n

$$(1-x^2)(AP_n'' + 2A'P_n' + A''P_n) - 2x(AP_n' + A'P_n) + n(n+1)AP_n = 0 \tag{4.525}$$

This can be regrouped as

$$A\{(1-x^2)P_n'' - 2xP_n' + n(n+1)P_n\} + (1-x^2)(2A'P_n' + A''P_n) - 2xA'P_n = 0 \tag{4.526}$$

The first term is evidently zero as the Legendre polynomial is the solution of (4.491), and thus we are left with

$$2\frac{P_n'}{P_n} + \frac{A''}{A'} - \frac{2x}{1-x^2} = 0 \tag{4.527}$$

This can be rewritten by assuming $u = A'$

$$2\frac{dP_n}{P_n}+\frac{du}{u}-\frac{2xdx}{1-x^2}=0 \tag{4.528}$$

Integrating term by term, we have

$$\ln P_n^2+\ln u+\ln(1-x^2)=C_1 \tag{4.529}$$

This can be written as

$$u=\frac{dA}{dx}=\frac{C}{(1-x^2)P_n^2} \tag{4.530}$$

Thus, integration leads to the result for A and the final solution results in Legendre polynomials of the second kind:

$$Q_n(x)=P_n\int_0^x\frac{dx}{(1-x^2)P_n^2} \tag{4.531}$$

As examples, we demonstrate the first few terms of the Legendre polynomials of the second kind. For $P_0(x) = 1$, the Legendre polynomials of the second kind of zero order are

$$\begin{aligned}Q_0(x)&=\int_0^x\frac{dx}{1-x^2}=\frac{1}{2}\int_0^x(\frac{1}{1+x}+\frac{1}{1-x})dx\\&=\frac{1}{2}[\ln(1+x)-\ln(1-x)]=\frac{1}{2}\ln(\frac{1+x}{1-x})\end{aligned} \tag{4.532}$$

For $P_1(x) = x$, the Legendre polynomials of the second kind of first order are

$$\begin{aligned}Q_1(x)&=x\int^x\frac{dx}{(1-x^2)x^2}=x\int^x(\frac{1}{1-x^2}+\frac{1}{x^2})dx\\&=\frac{x}{2}\ln(\frac{1+x}{1-x})-1\end{aligned} \tag{4.533}$$

More generally, Legendre polynomials of the second kind can be evaluated as (Abramowitz and Stegun, 1964)

$$\begin{aligned}Q_n(x)&=\frac{1}{2}P_n(x)\ln(\frac{1+x}{1-x})-\sum_{m=1}^{n}\frac{1}{m}P_{m-1}(x)P_{n-m}(x)\quad |x|<1\\&=\frac{1}{2}P_n(x)\ln(\frac{x+1}{x-1})-\sum_{m=1}^{n}\frac{1}{m}P_{m-1}(x)P_{n-m}(x)\quad |x|>1\end{aligned} \tag{4.534}$$

where $P_{-m}(x) = 0$. Thus, we have

$$Q_2(x)=P_2(x)Q_0(x)-\frac{3}{2}x=\frac{1}{4}(3x^2-1)\ln(\frac{1+x}{1-x})-\frac{3}{2}x \tag{4.535}$$

$$Q_3(x)=P_3(x)Q_0(x)-\frac{5}{2}x^2+\frac{2}{3}=\frac{1}{4}(5x^3-3x)\ln(\frac{1+x}{1-x})-\frac{5}{2}x^2+\frac{2}{3} \tag{4.536}$$

Thus, the general solution of (4.491) becomes

$$P(x)=AP_n(x)+BQ_n(x) \tag{4.537}$$

Note, however, that $Q_n(1)\to\infty$ and for solid spheres containing the poles (i.e., $\theta=0,\pi$ or $x=\pm1$ at poles) we normally require $B = 0$ based on physical grounds.

4.10.2 Rodrigues Formula

Actually, Legendre polynomials can also be generated by using the Rodrigues formula:

$$P_n(x) = \frac{1}{2^n n!} \frac{d^n}{dx^n} (x^2 - 1)^n \tag{4.538}$$

The formula was discovered by Rodrigues in 1816, and rediscovered by Ivory in 1824 and Jacobi in 1827. The current name was given by Heine in 1878 in his book. To prove this identity, we can start with the binominal theorem for the function under the differentiation sign:

$$(x^2 - 1)^n = \sum_{m=0}^{n} \frac{(-1)^m n!}{m!(n-m)!} x^{2n-2m} = \sum_{m=0}^{n} C_m^n (-1)^m x^{2n-2m} \tag{4.539}$$

Substitution of (4.539) into the left hand side of (4.538) yields

$$\frac{1}{2^n n!} \frac{d^n}{dx^n} (x^2 - 1)^n = \frac{1}{2^n n!} \sum_{m=0}^{n} \frac{(-1)^m n!}{m!(n-m)!} \frac{d^n}{dx^n} x^{2n-2m} \tag{4.540}$$

The differentiation term on the right hand side is zero if $2n-2m < n$ and thus we must have $n < 2m$ or $n/2 < m$ for even n and $(n-1)/2 < m$ for odd n. Note also that

$$\frac{d^n}{dx^n} x^p = p(p-1)(p-2)\cdots(p-n+1)x^{p-n} = \frac{p!}{(p-n)!} x^{p-n} \tag{4.541}$$

Thus, differentiation of (4.540) becomes

$$\frac{d^n}{dx^n} x^{2n-2m} = \frac{(2n-2m)!}{(n-2m)!} x^{n-2m} \tag{4.542}$$

Back substitution of (4.542) into (4.540) gives

$$\frac{1}{2^n n!} \frac{d^n}{dx^n} (x^2 - 1)^n = \sum_{m=0}^{M} \frac{(-1)^m (2n-2m)!}{2^n m!(n-m)!(n-2m)!} x^{n-2m} = P_n(x) \tag{4.543}$$

where M is defined in (4.516). This completes the proof of the Rodrigues formula. Legendre polynomials are of fundamental importance in the analysis of spherical coordinates, and are the marrow of spherical harmonics. They also provide the basis of eigenfunctions for spheres.

4.11 ASSOCIATED LEGENDRE EQUATION

For the case that $m \neq 0$ in (4.490), it is the associated Legendre equation:

$$(1-x^2)\frac{d^2W}{dx^2} - 2x\frac{dW}{dx} + [n(n+1) - \frac{m^2}{1-x^2}]W = 0 \tag{4.544}$$

Let us start from the Legendre function (i.e., $m = 0$) that

$$(1-x^2)\frac{d^2V}{dx^2} - 2x\frac{dV}{dx} + n(n+1)V = 0 \tag{4.545}$$

Differentiating this equation m times, we have

$$\frac{d^m}{dx^m}[(1-x^2)\frac{d^2V}{dx^2}]-\frac{d^m}{dx^m}(2x\frac{dV}{dx})+n(n+1)\frac{d^mV}{dx^m}=0 \tag{4.546}$$

Recalling the Leibniz rule for differentiation, we have

$$\frac{d^m(uv)}{dx^m}=\sum_{r=0}^{m}C_r^m\frac{d^ru}{dx^r}\frac{d^{m-r}v}{dx^{m-r}}=\sum_{r=0}^{m}\frac{m!}{(m-r)!r!}\frac{d^ru}{dx^r}\frac{d^{m-r}v}{dx^{m-r}} \tag{4.547}$$

Only a finite number of terms remain in (4.546) by recognizing u and v as

$$u=(1-x^2),\quad v=\frac{d^2V}{dx^2} \tag{4.548}$$

Thus, we have

$$\frac{d^m}{dx^m}[(1-x^2)\frac{d^2V}{dx^2}]=(1-x^2)\frac{d^{m+2}V}{dx^{m+2}}-\frac{m}{1!}\frac{d^{m+1}V}{dx^{m+1}}2x-\frac{m(m-1)}{2!}\frac{d^mV}{dx^m}2 \tag{4.549}$$

$$\frac{d^m}{dx^m}(2x\frac{dV}{dx})=\frac{d^{m+1}V}{dx^{m+1}}2x+2\frac{m}{1!}\frac{d^mV}{dx^m} \tag{4.550}$$

Finally, we have

$$(1-x^2)\frac{d^2U}{dx^2}-2x(m+1)\frac{dU}{dx}+[n(n+1)-m(m+1)]U=0 \tag{4.551}$$

where

$$U=\frac{d^mV}{dx^m}=\frac{d^mP_n(x)}{dx^m} \tag{4.552}$$

The last part of (4.552) is obtained by recognizing that the solution of (4.545) is the Legendre polynomial.

Let us now introduce the following change of variables

$$W=(1-x^2)^{m/2}U \tag{4.553}$$

Thus, the derivative terms of (4.552) can be expressed as:

$$\frac{dU}{dx}=mx(1-x^2)^{-m/2-1}W+(1-x^2)\frac{dW}{dx} \tag{4.554}$$

$$\frac{d^2U}{dx^2}=mx(1-x^2)^{-m/2-2}W[(1-x^2)+(m+2)x^2]+2mx(1-x^2)^{-m/2-1}\frac{dW}{dx}$$
$$+(1-x^2)^{-m/2}\frac{d^2W}{dx^2} \tag{4.555}$$

Substitution of (4.554) and (4.555) into (4.551) gives

$$(1-x^2)\frac{d^2W}{dx^2}-2x\frac{dW}{dx}+[n(n+1)-\frac{m^2}{1-x^2}]W=0 \tag{4.556}$$

This is precisely our associated Legendre equation. Thus, finally we have the solution of (4.556) as the associated Legendre polynomial:

$$P_n^m(x)=W=(1-x^2)^{m/2}U=(1-x^2)^{m/2}\frac{d^mP_n(x)}{dx^m} \tag{4.557}$$

This associated Legendre function of the first kind is also known as Ferrers functions, proposed by British mathematician Norman Ferrers (Olver, 2012).

Clearly, the zero order associated polynomials (i.e., $m = 0$) become the Legendre polynomials. It is straightforward to see that this procedure is equally applicable to the associated Legendre function of the second kind, and thus we have

$$Q_n^m(x) = W = (1-x^2)^{m/2}U = (1-x^2)^{m/2}\frac{d^m Q_n(x)}{dx^m} \tag{4.558}$$

Combining these two solutions, we have the general solution for the associated Legendre equation as

$$W(x) = AP_n^m(x) + BQ_n^m(x) \tag{4.559}$$

Similar to the observations for Legendre polynomials, for solid spheres containing the poles (i.e., $\theta = 0, \pi$ or $x = \pm 1$ at poles) the associated Legendre function of the second kind becomes infinite. Thus, for such problems we normally require $B = 0$ based on physical grounds.

4.12 HYPERGEOMETRIC FUNCTION

The hypergeometric series or functions had been studied by mathematicians before its governing equation was known (or called the hypergeometric equation). It has been considered by Wallis, Euler, Gauss, Kummer, and Riemann. Its application is related to the analyses of a weightless cable containing point masses, of particle physics, and of fluctuation in electric circuit. The term hypergeometric series was coined by J.F. Pfaff (1765–1825), who was the advisor of Gauss. Unlike Bessel functions, the hypergeometric function is one of the topics of special functions that has been commonly left out in the syllabus of engineering mathematics. In the bending theory of a shell having a surface of revolution subjected to axisymmetric loadings, the hypergeometric equation and function appear naturally (e.g., Chapter 16 of Timoshenko and Woinowsky-Krieger, 1959). However, such topics are not normally covered even in graduate courses in "the theory of plates and shells" because the hypergeometric function is considered too complicated for graduate students or even for engineering professors. In fact, if we do not go too deep into the mathematical theory of hypergeometric function, it is not so formidable. In this section, we will illustrate that it is just another series solution with three parameters and one variable. If one tries to find the analytic solution of a complicated integral using the software Mathematica, it is very likely that Mathematica's solution will involve the hypergeometric function. Therefore, one should be familiar with this special function.

4.12.1 Frobenius Series Solution

The following hypergeometric equation was discovered by Gauss:

$$z(1-z)y'' + [\gamma - (\alpha+\beta+1)z]y' - \alpha\beta y = 0 \tag{4.560}$$

where α, β, and γ are parameters of the equation. Near $z = 0$, let us seek a solution in terms of the Frobenius series

$$y = z^{\lambda}(a_0 + a_1 z + a_2 z^2 + ... + a_k z^k + ...) \tag{4.561}$$

Thus, we have

$$y' = \lambda a_0 z^{\lambda-1} + (\lambda+1)a_1 z^{\lambda} + (\lambda+2)a_2 z^{\lambda+1} + ... + (\lambda+k)a_k z^{k+\lambda-1} + ... \quad (4.562)$$

$$\begin{aligned} y'' &= \lambda(\lambda-1)a_0 z^{\lambda-2} + \lambda(\lambda+1)a_1 z^{\lambda-1} + (\lambda+2)(\lambda+1)a_2 z^{\lambda} \\ &\quad + ... + (\lambda+k)(k+\lambda-1)a_k z^{k+\lambda-2} + ... \end{aligned} \quad (4.563)$$

Substitution of these into (4.560) results in

$$\begin{aligned} & z(1-z)[\lambda(\lambda-1)a_0 z^{\lambda-2} + \lambda(\lambda+1)a_1 z^{\lambda-1} + (\lambda+2)(\lambda+1)a_2 z^{\lambda} \\ & + ... + (\lambda+k)(k+\lambda-1)a_k z^{k+\lambda-2} + ...] \\ & + [\gamma - (\alpha+\beta+1)z]\{\lambda a_0 z^{\lambda-1} + (\lambda+1)a_1 z^{\lambda} + (\lambda+2)a_2 z^{\lambda+1} \\ & + ... + (\lambda+k)a_k z^{k+\lambda-1} + ...\} \\ & - \alpha\beta(a_0 z^{\lambda} + a_1 z^{\lambda+1} + a_2 z^{\lambda+2} + ... + a_k z^{\lambda+k} + ...) = 0 \end{aligned} \quad (4.564)$$

Collecting the coefficients for $z^{\lambda-1}$, we have

$$\lambda(\lambda-1+\gamma) = 0 \quad (4.565)$$

There are two solutions for the index λ

$$\lambda = 0, \quad \lambda = 1-\gamma \quad (4.566)$$

Collecting the coefficients for $z^{\lambda+k}$, we have

$$\begin{aligned} & [(\lambda+k+1)(\lambda+k) + \gamma(\lambda+k+1)]a_{k+1} \\ & = [(\lambda+k)(\lambda+k-1) + (\alpha+\beta+1)(\lambda+k) + \alpha\beta]a_k \end{aligned} \quad (4.567)$$

For the case of $\lambda = 0$, the recursive formula for the coefficients is

$$a_{k+1} = \frac{(\alpha+k)(\beta+k)}{(k+1)(k+\gamma)} a_k \quad (4.568)$$

We further set $a_0 = 1$, and the first solution is Gauss hypergeometric series

$$y = y_1 = F(\alpha,\beta,\gamma,z) = 1 + \frac{\alpha\beta}{1\cdot\gamma} z + \frac{\alpha(\alpha+1)\beta(\beta+1)}{1\cdot 2\cdot\gamma(\gamma+1)} z^2 + ... \quad (4.569)$$

where $F(\alpha,\beta,\gamma,z)$ is called the hypergeometric series or hypergeometric function, and the first three arguments indicate the indices of the ODE and the last one is the variable. This hypergeometric function can be defined in a more compact form:

$$F(\alpha,\beta,\gamma,z) = \sum_{k=0}^{\infty} \frac{(\alpha)_k (\beta)_k}{(\gamma)_k k!} z^k \quad (4.570)$$

where the Pochhammer symbol is defined as

$$(\alpha)_k = \alpha(\alpha+1)\cdots(\alpha+k-1) = \frac{\Gamma(\alpha+k)}{\Gamma(\alpha)} \quad (4.571)$$

It was proposed by Prussian mathematician Pochhammer, and its name was coined by Appell in 1880. Thus, we can define the hypergeometric series in terms of the gamma function

$$F(\alpha,\beta,\gamma,z) = \frac{\Gamma(\gamma)}{\Gamma(\alpha)\Gamma(\beta)} \sum_{k=0}^{\infty} \frac{\Gamma(\alpha+k)\Gamma(\beta+k)}{\Gamma(\gamma+k)} \frac{z^k}{k!} \quad (4.572)$$

For the second root of λ in (4.566), $\lambda = 1-\gamma$, we have

$$y = y_2 = z^{1-\gamma} F(\alpha-\gamma+1, \beta-\gamma+1, 2-\gamma, z) \tag{4.573}$$

Therefore, the general solution of the hypergeometric equation is

$$y = AF(\alpha,\beta,\gamma,z) + Bz^{1-\gamma} F(\alpha-\gamma+1, \beta-\gamma+1, 2-\gamma, z) \tag{4.574}$$

This solution is only valid for

$$-1 < z < 1, \quad \gamma-(\alpha+\beta) > -1 \tag{4.575}$$

Note that the differentiation of the hypergeometric series is

$$\frac{d}{dz} F(\alpha,\beta,\gamma,z) = \frac{\alpha\beta}{\gamma} F(\alpha+1, \beta+1, \gamma+1, z) \tag{4.576}$$

To show this, can find that

$$F(\alpha+1,\beta+1,\gamma+1,z) = 1 + \frac{(\alpha+1)(\beta+1)}{\gamma+1} z + \frac{(\alpha+1)(\alpha+2)(\beta+1)(\beta+2)}{1\cdot 2\cdot(\gamma+1)(\gamma+2)} z^2 + \dots \tag{4.577}$$

$$\begin{aligned} \frac{\alpha\beta}{\gamma} F(\alpha+1,\beta+1,\gamma+1,z) = & \frac{\alpha\beta}{\gamma} + \frac{\alpha(\alpha+1)\beta(\beta+1)}{\gamma(\gamma+1)} z \\ & + \frac{\alpha(\alpha+1)(\alpha+2)\beta(\beta+1)(\beta+2)}{1\cdot 2\cdot \gamma(\gamma+1)(\gamma+2)} z^2 + \dots \end{aligned} \tag{4.578}$$

The function on the right hand side is precisely the derivative of the hypergeometric series. This completes the proof.

4.12.2 Confluent Hypergeometric Function

A related function is called the confluent hypergeometric function or Kummer function, which is defined as:

$$M(\alpha,\gamma,\rho) = 1 + \frac{\alpha}{\gamma}\frac{\rho}{1!} + \frac{\alpha(\alpha+1)}{\gamma(\gamma+1)}\frac{\rho}{2!} + \dots = \sum_{n=0}^{\infty} \frac{(\alpha)_n \rho^n}{(\gamma)_n n!} \tag{4.579}$$

which was introduced by Kummer in 1837. It can be obtained as a special case of the hypergeometric function that

$$\beta \to \infty, \quad z \to 0, \quad \beta z \to \rho \tag{4.580}$$

Using the following change of variables

$$z = \frac{\rho}{\beta} \tag{4.581}$$

we have

$$\frac{dy}{dz} = \beta \frac{dy}{d\rho}, \quad \frac{d^2y}{dz^2} = \beta^2 \frac{d^2y}{d\rho^2} \tag{4.582}$$

Substitution of these into (4.560) gives

$$\frac{\rho}{\beta}(1-\frac{\rho}{\beta})\beta^2 \frac{d^2y}{d\rho^2} + [\gamma - (\alpha+\beta+1)\frac{\rho}{\beta}]\beta \frac{dy}{d\rho} - \alpha\beta y = 0 \tag{4.583}$$

This can further be simplified as

$$\rho(\beta-\rho)\frac{d^2y}{d\rho^2}+[\gamma\beta-(\alpha+\beta+1)\rho]\frac{dy}{d\rho}-\alpha\beta y=0 \tag{4.584}$$

Finally, we can take the limit $\beta \to \infty$ and we obtain the following confluent hypergeometric equation or Kummer's equation

$$\rho\frac{d^2y}{d\rho^2}+(\gamma-\rho)\frac{dy}{d\rho}-\alpha y=0 \tag{4.585}$$

The second independent solution of this second ODE was introduced by Tricomi in 1947

$$\Phi(\alpha,\gamma,\rho)=\frac{\Gamma(1-\gamma)}{\Gamma(\alpha-\gamma+1)}M(\alpha,\gamma,\rho)+\frac{\Gamma(\gamma-1)}{\Gamma(\alpha)}\rho^{1-\gamma}M(\alpha-\gamma+1,2-\gamma,\rho) \tag{4.586}$$

which is also known as the Tricomi function. The hypergeometric series also gives the solution of Riemann's P-equation or the Papperitz equation (see chapter 10 of Whittaker and Watson, 1927).

The appearance of the solution form in (4.574) is not unique. If the range of z is not constrained by (4.575), the solution of the hypergeometric equation will appear in different forms. The next section will present the classification of solutions proposed by Kummer.

4.12.3 Kummer's Classification of Hypergeometric Series

Actually, there are three singular points at z = 0, 1, ∞ for the hypergeometric equation. The solution near singular points z = 1, and $z \to \infty$ will appear differently. Near each of the three singular points, there are always two linearly independent solutions. There are many different ways to express these solutions. The following Kummer classification is presented by Goursat in his thesis and was translated to English in Chapter 7 of Craig's (1889) book. For the solution near the point z = 0, we have two characteristic roots (as given in the last section), and the situation is the same for indicial equations at the other two singular points. The indicial equations at z= 1 and $z = \infty$ are respectively:

$$\mu(\mu-\gamma+\alpha+\beta)=0 \tag{4.587}$$

$$r(r-1)-r(\alpha+\beta-1)+\alpha\beta=0 \tag{4.588}$$

Similar to (4.565), these equations are also quadratic and thus there exist two solutions. This is summarized in Table 4.3 below. To show the validity of (4.588), we introduce the following change of variables

$$z=\frac{1}{u} \tag{4.589}$$

Table 4.3. Roots of the indicial equation near different singular points

Singular point	Roots of indicial equation	
0	$\lambda_1 = 0$	$\lambda_2 = 1-\gamma$
1	$\mu_1 = 0$	$\mu_2 = \gamma-\alpha-\beta$
∞	$r_1 = \alpha$	$r_2 = \beta$

Note that the singular point $z \to \infty$ is mapped to $u = 0$. The derivatives of the new variables are:

$$\frac{dy}{dz} = \frac{dy}{du}\frac{du}{dz} = -u^2 \frac{dy}{du} \tag{4.590}$$

$$\frac{d^2y}{dz^2} = \frac{d}{du}(-u^2 \frac{dy}{du})\frac{du}{dz} = 2u^3 \frac{dy}{du} + u^4 \frac{d^2y}{du^2} \tag{4.591}$$

Substitution of these derivatives into the hypergeometric equation (4.560) gives

$$u^2(u-1)\frac{d^2y}{du^2} + u[(\alpha+\beta-1)-(\gamma-2)u]\frac{dy}{du} - \alpha\beta y = 0 \tag{4.592}$$

Thus, a power series in u about $u = 0$ corresponds to the singular point $z \to \infty$. Near $u = 0$, let us seek a solution in terms of the Frobenius series

$$y = u^r(a_0 + a_1 u + a_2 u^2 + ... + a_k u^k + ...) \tag{4.593}$$

Thus, we have

$$y' = ra_0 u^{r-1} + (r+1)a_1 u^r + (r+2)a_2 u^{r+1} + ... + (r+k)a_k u^{k+r-1} + ... \tag{4.594}$$

$$\begin{aligned} y'' &= r(r-1)a_0 u^{r-2} + r(r+1)a_1 u^{r-1} + (r+2)(r+1)a_2 u^r \\ &+ ... + (r+k)(r+k-1)a_k u^{k+r-2} + ... \end{aligned} \tag{4.595}$$

Substitution of these into (4.592) yields

$$\begin{aligned} &u^2(u-1)[r(r-1)a_0 u^{r-2} + r(r+1)a_1 u^{r-1} + (r+2)(r+1)a_2 u^r \\ &+ ... + (r+k)(k+r-1)a_k u^{k+r-2} + ...] \\ &+u[(\alpha+\beta-1)-(\gamma-2)u]\{ra_0 u^{r-1} + (r+1)a_1 u^r + (r+2)a_2 u^{r+1} \\ &+ ... + (r+k)a_k u^{k+r-1} + ...\} \\ &-\alpha\beta(a_0 u^r + a_1 u^{r+1} + a_2 u^{r+2} + ... + a_k u^{r+k} + ...) = 0 \end{aligned} \tag{4.596}$$

Collecting the coefficient of the u^r term, we obtain for nonzero a_0

$$r(r-1) - r(\alpha+\beta-1) + \alpha\beta = 0 \tag{4.597}$$

This is precisely (4.588). The two roots are $r = \alpha, \beta$. For the case of $r = \alpha$, we have

$$a_{k+1} = \frac{(\alpha+k)(1+k)}{\beta(k-1) - \alpha(k+1) - (k-1)(k+1)} a_k \tag{4.598}$$

To show the validity of (4.587), we introduce the following change of variables

$$z = 1 - u \tag{4.599}$$

Note that the singular point $z = 1$ is mapped to $u = 0$. The hypergeometric equation becomes

$$u(1-u)\frac{d^2y}{du^2} + [(\alpha+\beta+1-\gamma) - u(\alpha+\beta+1)]\frac{dy}{du} - \alpha\beta y = 0 \tag{4.600}$$

Near $u = 0$, let us seek a solution in terms of the Frobenius series

$$y = u^\mu(a_0 + a_1 u + a_2 u^2 + ... + a_k u^k + ...) \tag{4.601}$$

Thus, we have

$$y' = \mu a_0 u^{\mu-1} + (\mu+1)a_1 u^{\mu} + (\mu+2)a_2 u^{\mu+1} + ... + (\mu+k)a_k u^{k+\mu-1} + ... \quad (4.602)$$

$$y'' = \mu(\mu-1)a_0 u^{\mu-2} + \mu(\mu+1)a_1 u^{\mu-1} + (\mu+2)(\mu+1)a_2 u^{\mu} + ... + (\mu+k)(k+\mu-1)a_k u^{k+\mu-2} + ... \quad (4.603)$$

Substitution of these into (4.600) leads to

$$\begin{aligned} & u(1-u)[\mu(\mu-1)a_0 u^{\mu-2} + \mu(\mu+1)a_1 u^{\mu-1} + (\mu+2)(\mu+1)a_2 u^{\mu} \\ & + ... + (\mu+k)(k+\mu-1)a_k u^{k+\mu-2} + ...] \\ & + [(\alpha+\beta-\gamma+1) - u(\alpha+\beta+1)]\{\mu a_0 u^{\mu-1} + (\mu+1)a_1 u^{\mu} + (\mu+2)a_2 u^{\mu+1} \\ & + ... + (\mu+k)a_k u^{k+\mu-1} + ...\} \\ & - \alpha\beta(a_0 u^{\mu} + a_1 u^{\mu+1} + a_2 u^{\mu+2} + ... + a_k u^{\mu+k} + ...) = 0 \end{aligned} \quad (4.604)$$

Collecting the coefficients of the u^r term, we obtain for nonzero a_0

$$\mu(\mu+\alpha+\beta-\gamma) = 0 \quad (4.605)$$

This is (4.587). The two roots are $\mu = 0$ and $\gamma-\alpha-\beta$. For the case of $\mu = 0$, we have

$$a_{k+1} = \frac{k(k+\alpha+\beta)+\alpha\beta}{(k+1)(k+\beta+\alpha-\gamma+1)} a_k \quad (4.606)$$

In general, the solution forms near the singular points can be expressed in the solutions summarized in Table 4.4.

Table 4.4. General solution of hypergeometric functions near different singular points

Singular point	General solution form of Frobenius series
0	$y = C_1 z^{\lambda_1} U_1 + C_2 z^{\lambda_2} U_2$
1	$y = C_1(1-z)^{\mu_1} V_1 + C_2(1-z)^{\mu_2} V_2$
∞	$y = C_1 z^{-r_1} W_1 + C_2 z^{-r_2} W_2$

Therefore, there are six types of solution forms, and they are U_1, U_2, V_1, V_2, W_1, and W_2, which, we have the following functional forms

$$U_1, U_2 = U_1(z), U_2(z) \quad (4.607)$$

$$V_1, V_2 = V_1(z-1), V_2(z-1) \quad (4.608)$$

$$W_1, W_2 = W_1(1/z), W_2(1/z) \quad (4.609)$$

They can be expressed in terms of hypergeometric functions. The hypergeometric equation can be rewritten in different forms, depending on whether the solution is being sought near z = 0, 1 or ∞. There are six different ways of transforming an ODE to the standard form by the following transformation:

$$x = \frac{au+b}{cu+d} \quad (4.610)$$

Mapping of this form is also called bilinear transformation. More specifically, we can transform the differential equation to different forms by using the mapping shown in Table 4.5.

Table 4.5. Six mappings that can rewrite the differential equations in different forms near 3 singular points

Mapping	Value of u		
	$z=0$	$z=1$	$z=\infty$
$z=u$	0	1	∞
$z=u/(u-1)$	0	∞	1
$z=1-u$	1	0	∞
$z=(u-1)/u$	1	∞	0
$z=1/(1-u)$	∞	0	1
$z=1/u$	∞	1	0

The third and sixth mappings have been applied earlier to obtain the hypergeometric equations near the singular at $x \to \infty$ and $x = 1$. For the solution near each singular point, we can express the solution as:

$$y = z^{\lambda}(1-z)^{\mu} f(z) \tag{4.611}$$

where $f(z)$ is the hypergeometric function of certain parameters. Since there are two roots for both λ and μ, four ways of expressing the solution exist. Each solution of these different forms can be represented by six different forms (see (4.407) to (4.409)). According to this system, we have 24 solution forms. This is called Kummer's 24 solutions for hypergeometric equations. The results are summarized in Table 4.6 for the solutions near the singular point $z = 0$, in Table 4.7 for the solutions near the singular point $z = 1$, and in Table 4.8 for the solutions near the singular point $z \to \infty$.

The first four Kummer solutions given in Table 4.6 are equivalent whereas the fifth to eighth are equal. That is why they were grouped into solutions y_1 and y_2. Near the point $z = 0$, the general solution of the hypergeometric equation is

$$y = Ay_1(z) + By_2(z) \tag{4.612}$$

Altogether we have six groups of solutions with four each as shown in Tables 4.6–4.8. Thus, a total of 24 solutions can be obtained.

Near the point $z = 1$, the general solution of the hypergeometric equation is

$$y = Ay_3(z) + By_4(z) \tag{4.613}$$

Near the point $z \to \infty$, the general solution of the hypergeometric equation is

$$y = Ay_5(z) + By_6(z) \tag{4.614}$$

It can be shown that all solutions of y_3, y_4, y_5, and y_6 are dependent functions of y_1 and y_2 (Craig, 1889).

4.12.4 Hypergeometric Series versus Other Functions

The hypergeometric series is considered as one of the most general functions that can embrace many functions, which can be expressed in terms of infinite series with rational coefficients, as its special case. Here are some examples:

$$F(-n, \beta, \beta, -x) = (1+x)^n \tag{4.615}$$

$$\lim_{n\to\infty} F(1,n,1,x/n) = e^x \tag{4.616}$$

$$F(1,\beta,\beta,x) = 1/(1-x) \tag{4.617}$$

$$F(1,1,2,-z) = \frac{\ln(1+z)}{z} \tag{4.618}$$

$$\lim_{n,m\to\infty} F(m,n,1/2,z^2/(4mn)) = \cosh z \tag{4.619}$$

$$\lim_{n,m\to\infty} F(m,n,3/2,z^2/(4mn)) = \frac{\sinh z}{z} \tag{4.620}$$

$$\lim_{n,m\to\infty} F(m,n,3/2,-z^2/(4mn)) = \frac{\sin z}{z} \tag{4.621}$$

Table 4.6. General solution of hypergeometric functions near $z = 0$

Number	Solution	Type	region
1	$F(\alpha,\beta,\gamma,z)$	y_1	$z=0$
2	$(1-z)^{\gamma-\alpha-\beta}F(\gamma-\alpha,\gamma-\beta,\gamma,z)$		
3	$(1-z)^{-\alpha}F(\alpha,\gamma-\beta,\gamma,z/(z-1))$		
4	$(1-z)^{-\alpha}F(\beta,\gamma-\alpha,\gamma,z/(z-1))$		
5	$z^{1-\gamma}F(\alpha-\gamma+1,\beta-\gamma+1,2-\gamma,z)$	y_2	$z=0$
6	$z^{1-\gamma}(1-z)^{\gamma-\alpha-\beta}F(1-\alpha,1-\beta,2-\gamma,z)$		
7	$z^{1-\gamma}(1-z)^{\gamma-\alpha-1}F(\alpha-\gamma+1,1-\beta,2-\gamma,z/(z-1))$		
8	$z^{1-\gamma}(1-z)^{\gamma-\beta-1}F(\beta-\gamma+1,1-\alpha,2-\gamma,z/(z-1))$		

Table 4.7. General solution of hypergeometric functions near $z = 1$

Number	Solution	Type	region
1	$F(\alpha,\beta,\alpha+\beta-\gamma+1,1-z)$	y_3	$z=1$
2	$z^{1-\gamma}F(\alpha-\gamma+1,\beta-\gamma+1,\alpha+\beta-\gamma+1,1-z)$		
3	$z^{-\alpha}F(\alpha,\alpha-\gamma+1,\alpha+\beta-\gamma+1,(z-1)/z)$		
4	$z^{-\beta}F(\beta,\beta-\gamma+1,\alpha+\beta-\gamma+1,(z-1)/z)$		
5	$(1-z)^{\gamma-\alpha-\beta}F(\gamma-\alpha,\gamma-\beta,\gamma-\alpha-\beta+1,1-z)$	y_4	$z=1$
6	$(1-z)^{\gamma-\alpha-\beta}z^{1-\gamma}F(1-\alpha,1-\beta,\gamma-\alpha-\beta+1,1-z)$		
7	$(1-z)^{\gamma-\alpha-\beta}z^{\alpha-\gamma}F(1-\alpha,\gamma-\alpha,\gamma-\alpha-\beta+1,(1-z)/z)$		
8	$(1-z)^{\gamma-\alpha-\beta}z^{\beta-\gamma}F(1-\beta,\gamma-\beta,\gamma-\alpha-\beta+1,(1-z)/z)$		

Table 4.8. General solution of hypergeometric functions near $z \to \infty$

Number	Solution	Type	region
1	$z^{-\alpha}F(\alpha,\alpha-\gamma+1,\alpha-\beta+1,1/z)$	y_5	$z \to \infty$
2	$z^{-\alpha}(1-1/z)^{\gamma-\alpha-\beta}F(1-\beta,\gamma-\beta,\alpha-\beta-\gamma+1,1/z)$		
3	$z^{-\alpha}(1-1/z)^{-\alpha}F(\alpha,\gamma-\beta,\alpha-\beta+1,1/(1-z))$		
4	$z^{-\alpha}(1-1/z)^{\gamma-\alpha-1}F(\alpha-\gamma+1,1-\beta,\alpha-\beta+1,1/(1-z))$		
5	$z^{-\beta}F(\beta,\beta-\gamma+1,\beta-\alpha+1,1/z)$	y_6	$z \to \infty$
6	$z^{-\beta}(1-1/z)^{\gamma-\alpha-\beta}F(1-\alpha,\gamma-\alpha,\beta-\alpha+1,1/z)$		
7	$z^{-\beta}(1-1/z)^{-\beta}F(\beta,\gamma-\alpha,\beta-\alpha+1,1/(1-z))$		
8	$z^{-\beta}(1-1/z)^{\gamma-\beta-1}F(\beta-\gamma+1,1-\alpha,\beta-\alpha+1,1/(1-z))$		

$$\lim_{n,m\to\infty} F(m,n,1/2,-z^2/(4mn)) = \cos z \tag{4.622}$$

$$F(-n,n+1,1,(1-x)/2) = P_n(x) \tag{4.623}$$

$$F(m-n,m+n+1,m+1,(1-x)/2) = \frac{(n-m)!2^m m!}{(n+m)!(1-x^2)^{m/2}} P_n^m(x) \tag{4.624}$$

$$F(-n,n,1/2,(1-x)/2) = T_n(x) \tag{4.625}$$

$$F(-n+1,n+1,3/2,(1-x)/2) = \frac{U_n(x)}{\sqrt{1-x^2}} \tag{4.626}$$

where $T_n(x)$ and $U_n(x)$ are Chebyshev polynomials of the first and second kinds (Abramowitz and Stegun, 1964).

4.13 GENERALIZED HYPERGEOMETRIC EQUATION

A related function is called the generalized hypergeometric function and is defined as

$${}_mF_n(\alpha_1,\alpha_2,...\alpha_m;\beta_1,\beta_2,...\beta_n;x) = \sum_{r=0}^{\infty} \frac{(\alpha_1)_r(\alpha_2)_r\cdots(\alpha_m)_r x^r}{(\beta_1)_r(\beta_2)_r\cdots(\beta_n)_r r!} \tag{4.627}$$

which satisfies the following generalized hypergeometric equation:

$$\{x\frac{d}{dx}(x\frac{d}{dx}+\beta_1-1)\cdots(x\frac{d}{dx}+\beta_n-1)-x(x\frac{d}{dx}+\alpha_1)\cdots(x\frac{d}{dx}+\alpha_m)\}y=0 \tag{4.628}$$

This ODE can be recast into product form as

$$\left\{D\prod_{j=1}^{n}(D+\beta_j-1)-x\prod_{j=1}^{m}(D+\alpha_j)\right\}y(x)=0 \tag{4.629}$$

where the differential operator D is the Euler derivative and is defined as:

$$D = x\frac{d}{dx} \tag{4.630}$$

Note the following special property of this operator

$$Dx^n = x\frac{d}{dx}(x^n) = nx^n \tag{4.631}$$

We now assume a power series for y in (4.629)

$$y(x) = \sum_{k=0}^{\infty} c_k x^k \tag{4.632}$$

Substitution of (4.632) into (4.629) we have

$$\sum_{k=0}^{\infty} \left\{ k\prod_{j=1}^{n}(k+\beta_j-1)c_k - \prod_{j=1}^{m}(k+\alpha_j)c_{k-1} \right\} x^k = 0 \tag{4.633}$$

Thus, the recursive formula for the two successive constant coefficients is

$$c_k = \frac{\prod_{j=1}^{m}(k+\alpha_j)}{k\prod_{j=1}^{n}(k+\beta_j-1)} c_{k-1} \tag{4.634}$$

This recursive formula made it possible to formulate the series solution without much difficulty. Reapplying this recursive formula n times, we have

$$c_k = \frac{\prod_{j=1}^{m}(k+\alpha_j)\prod_{j=1}^{m}(k-1+\alpha_j)\cdots\prod_{j=1}^{m}(1+\alpha_j)}{k(k-1)\cdots 1\prod_{j=1}^{n}(k+\beta_j-1)\prod_{j=1}^{n}(k+\beta_j-2)\ldots\prod_{j=1}^{n}\beta_j} c_0 \tag{4.635}$$

Reshuffling the terms and setting $c_0 = 1$, we have the very compact form of c_k in terms of the Pochhammer symbol as

$$c_k = \frac{\prod_{i=1}^{m}(\alpha_i)_k}{k!\prod_{j=1}^{n}(\beta_i)_k} \tag{4.636}$$

Finally, substitution of (4.636) into (4.632) leads to

$$y(x) = \sum_{k=0}^{\infty} \frac{\prod_{i=1}^{m}(\alpha_i)_k}{k!\prod_{j=1}^{n}(\beta_i)_k} x^k = \sum_{r=0}^{\infty} \frac{(\alpha_1)_r(\alpha_2)_r\cdots(\alpha_m)_r x^r}{(\beta_1)_r(\beta_2)_r\cdots(\beta_n)_r r!} \tag{4.637}$$

$$= {}_m F_n(\alpha_1,\alpha_2,\ldots\alpha_m;\beta_1,\beta_2,\ldots\beta_n;x)$$

The compact notation given in (4.637) was introduced by Pochhammer in 1890 whereas the notation of ${}_mF_n$ with one subscript before and one after F was introduced by Barnes in 1907. This verifies that the solution of ODE (4.628) is indeed the generalized hypergeometric function given in (4.627).

For the special case of $m = 2$ and $n = 1$, we have the hypergeometric function

$$\begin{aligned} {}_2F_1(\alpha_1,\alpha_2;\beta_1;x) &= \sum_{r=0}^{\infty}\frac{(\alpha_1)_r(\alpha_2)_r x^r}{(\beta_1)_r r!} = \sum_{r=0}^{\infty}\frac{\Gamma(\alpha_1+r)\Gamma(\alpha_2+r)\Gamma(\beta_1)x^r}{\Gamma(\alpha_1)\Gamma(\alpha_2)\Gamma(\beta_1+r)r!} \\ &= F(\alpha_1,\alpha_2,\beta_1,x) \end{aligned} \tag{4.638}$$

For the special case of $m = 1$ and $n = 1$, we have the confluent hypergeometric function or Kummer's function $M(\alpha,\beta,\gamma)$ being recovered

$$\begin{aligned} {}_1F_1(\alpha_1;\beta_1;x) &= \sum_{r=0}^{\infty}\frac{(\alpha_1)_r x^r}{(\beta_1)_r r!} = \sum_{r=0}^{\infty}\frac{\Gamma(\alpha_1+r)\Gamma(\beta_1)x^r}{\Gamma(\alpha_1)\Gamma(\beta_1+r)r!} \\ &= M(\alpha_1,\beta_1,x) \end{aligned} \tag{4.639}$$

There are also a number of important functions that can be expressed in terms of the generalized hypergeometric function

$${}_1F_1(n+1/2,2n+1,2ix) = (\frac{2}{x})^n n! e^{ix} J_n(x) \tag{4.640}$$

$${}_1F_1(-n,1/2,x^2) = (-1)^n \frac{n!}{(2n)!} H_{2n}(x) \tag{4.641}$$

$${}_1F_1(-n,3/2,x^2) = (-1)^n \frac{n!}{2(2n+1)!} H_{2n+1}(x) \tag{4.642}$$

where $H_n(x)$ is the Hermite polynomial (Abramowitz and Stegun, 1964). Note there is symmetry between α and β for ${}_2F_1$ that we obtain

$${}_2F_1(\alpha,\beta,\gamma,x) = {}_1F_2(\alpha,\beta,\gamma,x) = \sum_{r=0}^{\infty}\frac{(\alpha)_r(\beta)_r x^r}{(\gamma)_r r!} \tag{4.643}$$

4.14 MOVABLE SINGULARITIES AND PAINLEVE EQUATIONS

In Section 3.2.8, we have seen that the Riccati equation can be converted to a linear ODE. Fuchs realized that the Riccati equation only has movable poles. Whenever a nonlinear ODE that has a pole as its only movable singularity, we say that this type of ODE is the Painlevé type. More importantly, it was found that only this kind of nonlinear ODE with movable singularities can be transformed to a linear ODE, just like the Riccati equation. By definition, a singularity is called a movable singularity if its location depends on the initial conditions of the ODE. That is, the location of the singularity is not fixed solely by the coefficients of the ODE.

To illustrate the idea of the movable simple pole, we consider the following ODE:

$$\frac{dw}{dz} = w^2, \quad w(z_0 = 0) = w_0 \tag{4.644}$$

It can be shown that the solution of (4.644) is

$$w = -\frac{1}{z - z_0} \tag{4.645}$$

We see that the location of the singularity is at z_0, which is a function of the boundary condition w_0 and is not fixed by the ODE alone.

The Painlevé type of nonlinear ODEs are known to be linearized and perhaps can be solved exactly. Indeed, Painlevé conducted an extensive analysis on what kind of second order nonlinear ODEs can be convertible to linear ODEs, like the Riccati equation. In particular, Painlevé investigated the following second order ODE:

$$\frac{d^2 w}{dz^2} = F(\frac{dw}{dz}, w, z) \tag{4.646}$$

He examined the second order ODE of this form that only has poles as moving singularities. He found a total of fifty ODEs, and all of them can be reduced to (a) linear ODEs, (b) Riccati equations, (c) equations satisfied by elliptic functions, and (d) six "new" equations. Painlevé discovered that these six equations are not reducible to "known" differential equations. These are called Painlevé transcendents:

P_{I}:

$$\frac{d^2 w}{dz^2} = 6w^2 + z \tag{4.647}$$

P_{II}:

$$\frac{d^2 w}{dz^2} = 2w^3 + zw + a \tag{4.648}$$

P_{III}:

$$\frac{d^2 w}{dz^2} = \frac{(w')^2}{w} - \frac{w'}{z} + \frac{(aw^2 + b)}{z} + cw^3 + \frac{d}{w} \tag{4.649}$$

P_{IV}:

$$\frac{d^2 w}{dz^2} = \frac{(w')^2}{2w} + \frac{3w^3}{2} + 4zw^2 + 2(z^2 - a)w + \frac{b}{w} \tag{4.650}$$

P_{V}:

$$\frac{d^2 w}{dz^2} = (\frac{1}{2w} + \frac{1}{w-1})(w')^2 - \frac{w'}{z} + \frac{(w-1)^2}{z^2}(aw + \frac{b}{w}) + \frac{cw}{z} + \frac{dw(w+1)}{w-1} \tag{4.651}$$

P_{VI}:

$$\begin{aligned} \frac{d^2 w}{dz^2} &= \frac{1}{2}(\frac{1}{w} + \frac{1}{w-1} + \frac{1}{w-z})(w')^2 - (\frac{1}{z} + \frac{1}{z-1} + \frac{1}{w-z})w' \\ &+ \frac{w(w-1)(w-2)}{z^2(z-1)^2}[a + \frac{bz}{w^2} + \frac{c(z-1)}{(w-1)^2} + \frac{dz(z-1)}{(w-z)^2}] \end{aligned} \tag{4.652}$$

where $w' = dw/dz$ and a, b, c, and d are constants. These equations have been shown to be convertible to linear integral equations. But such transformations involve complicated complex analysis and are out of the scope of this chapter. Recent research shows that some very important nonlinear PDE can be reduced to Painlevé transcendents. They include the KdV equation, Sine-Gordon equation, and Boussinesq equation, and these are all soliton types of equations. If any PDE can be reduced to Painlevé transcendents, we say that it satisfies the Painlevé property.

More importantly, they can be solved by using "inverse scattering transform" and their solutions are of the soliton type (Ablowitz and Clarkson, 1991).

4.15 SUMMARY AND FURTHER READING

The series solution method is a major topic in the theory of ODE. The fact that a solution of a second ODE can be expressed in terms of an infinite series allows the evaluation of the solutions. Many so-called "special" functions are actually a particular type of series solution of ODE, including the Bessel functions, modified Bessel functions, Kelvin functions, Legendre polynomials, associated Legendre polynomials, hypergeometric functions, confluent hypergeometric functions, and general hypergeometric functions discussed here. The study of ODEs with regular singular points leads to the investigation of the Fuchsian type ODE. Among these special functions, the general hypergeometric function appears to be the most general and powerful, and it covers nearly all of the special functions as its special case. If you try to do complicated integration by a symbolic manipulation program such as Maple and Mathematica, it is very likely that you could get an analytic result in terms of some sorts of general hypergeometric functions. This is one of the major reasons that it should be covered in a chapter on series solutions. More discussion on hypergeometric functions can be found in Craig (1889), Bateman (1918), Piaggio (1920), Copson (1935), Poole (1936), Erdelyi (1953), Spiegel (1968), Lebedev et al. (1965), and Lebedev (1972).

The investigation of special functions probably started with Bernoulli and Euler in 1700s. They include elliptic integrals and Bessel functions. We have given an introduction on elliptic integrals in Chapter 2 when we discussed the pendulum problem. Further information on Jacobi's elliptic integral is also given in Appendix A. Euler introduced gamma functions as a non-integer continuation of factorial and studied elliptic integrals related to pendulums, and Bessel functions related to vibrations of circular drums. Nearly all of his investigations are driven by everyday applications. Related to celestial mechanics and potential theory, Legendre polynomials emerged. There are also other special functions in terms of polynomials, such as Hermite polynomials, Laguerre polynomials, Chebyshev polynomials, and Jacobi polynomials. Because of the orthogonal properties of these polynomials, they are important as the basis of eigenfunction expansion (see Chapter 10). Some of them relate to the theory of probabilities, quantum mechanics and wave scattering theory.

One of the best coverages of special functions remains the classic book by Erdelyi (1953) (i.e., the *Higher Transcendental Functions* of the Bateman manuscript project). Other textbooks on special functions include Sneddon (1961), Lebedev (1972), and Bell (1968). For Bessel functions, the number one authority is still the classic book by Watson (1944). Serious readers should consult these books. The number one reference book on special functions is Abramowitz and Stegun (1964) and its updated version by Olver et al. (2010). It was reported by Biosvert and Lozier (2001) that Abramowitz and Stegun (1964) was cited more than 2000 times in 2009 alone by journal articles. In view of the importance of the use of special functions in engineering and industry, the National Institute of Standards and Technology (NIST) also published an online version called the

Digital Library of Mathematical Functions, which provides a link to software for evaluating many of the known special functions (http://dlmf.nist.gov/software/). The handbook of Gradshteyn and Ryzhik (1980) also compiled many results of integrations related to special functions.

4.16 PROBLEMS

Problem 4.1 Prove the following reflection formula for the gamma function

$$\Gamma(\frac{1}{2}+z)\Gamma(\frac{1}{2}-z)=\frac{\pi}{\cos\pi z} \tag{4.653}$$

Hint: Use appropriate substitution of the argument into the original reflection formula.

Problem 4.2 Prove the following reflection formula for the digamma function

$$\psi(\frac{1}{2}+z)=\psi(\frac{1}{2}-z)+\pi\tan\pi z \tag{4.654}$$

Hint: Use the result of Problem 4.1.

Problem 4.3 Prove the following identity for the gamma function

$$\Gamma(\frac{1}{4}+n)\Gamma(\frac{3}{4}-n)=\frac{\sqrt{2}\pi}{(-1)^n} \tag{4.655}$$

where n is an integer.
Hint: Use the reflection formula.

Problem 4.4 Prove the following identity for the digamma function

$$\psi(\frac{1}{4}+x)-\psi(\frac{3}{4}-x)=-\pi(\frac{1-\tan\pi x}{1+\tan\pi x}) \tag{4.656}$$

where n is an integer.
Hint: Use the reflection formula.

Problem 4.5 Show that

$$\psi(\frac{3}{4}-n)=\psi(\frac{1}{4}+n)+\pi \tag{4.657}$$

where n is an integer.

Problem 4.6 Show that

$$\frac{\Gamma(z+1)}{\Gamma(z-n)}=z(z-1)\cdots(z-n) \tag{4.658}$$

where n is an integer.

Problem 4.7 Show that

$$\Gamma(-\frac{1}{2})=-2\sqrt{\pi} \tag{4.659}$$

Hint: Use the recurrence formula.

Problem 4.8 Show that

$$\Gamma(-\frac{5}{2}) = -\frac{8\sqrt{\pi}}{15} \tag{4.660}$$

Hint: Use the recurrence formula.

Problem 4.9 Show that

$$\psi(\frac{3}{2}) = -\gamma - 2\ln 2 + 2 \tag{4.661}$$

Hint: Use the recurrence formula.

Problem 4.10 Show that

$$\psi(\frac{1}{2}) = -\gamma - 2\ln 2 \tag{4.662}$$

Hint: Use the recurrence formula.

Problem 4.11 Show that

$$\prod_{n=1}^{8} \Gamma(\frac{n}{3}) = \frac{640}{3^6}(\frac{\pi}{\sqrt{3}})^3 \tag{4.663}$$

Hint: Use the multiplication formula of Gauss with $m = 3$.

Problem 4.12 Prove the following identity for the n-th derivative of the digamma function

$$\psi^{(n)}(z+1) = \psi^{(n)}(z) + \frac{(-1)^n n!}{z^{n+1}} \tag{4.664}$$

Hint: Use the recurrence formula.

Problem 4.13 Prove the following identity for the gamma function

$$\Gamma(z) = \frac{1}{z}\prod_{n=1}^{\infty} \frac{(1+1/n)^z}{(1+z/n)} \tag{4.665}$$

Hint: Try to show that the right hand side equals the factorial definition given in Section 4.2.2.

Problem 4.14 Prove the following identity for the gamma function

$$\Gamma(z)\Gamma(-z) = -\frac{\pi}{z\sin \pi z} \tag{4.666}$$

Problem 4.15 Prove the following identity

$$\Gamma(\frac{1}{2} - \frac{x}{2}) = \frac{\pi}{\Gamma(\frac{1}{2} + \frac{x}{2})\cos(\frac{\pi x}{2})} \tag{4.667}$$

Hint: Use the reflection formula for the gamma function.

Problem 4.16 Show that U and V defined in (4.146) can be expressed as

$$U = \frac{2^{x/2}}{\pi}\sin(\frac{\pi x}{2})\Gamma(\frac{x}{2}), \quad V = \frac{2^{x/2}}{\pi}\cos(\frac{\pi x}{2})\Gamma(\frac{1}{2}+\frac{x}{2}) \tag{4.668}$$

Hint: Use the reflection formula for the gamma function and the result of Problem 4.15.

Problem 4.17 Use the results obtained in Section 4.3 to prove the following formulas for the integer argument of the Hadamard factorial function:

$$H(2m) = \frac{\sqrt{\pi}2^{2m}}{2}\frac{\Gamma(m)}{\Gamma(\frac{1}{2}-m)}, \quad H(2m+1) = -\frac{\sqrt{\pi}2^{2m+1}}{2}\frac{\Gamma(m+1)}{\Gamma(\frac{1}{2}-m)}, \tag{4.669}$$

where $m = 1,2,3,...$

Problem 4.18 Prove the following formula of the Hadamard factorial function:

$$H(x)H(-x) = \frac{P(x)}{x^2} + \frac{\sin \pi x}{\pi x}[\frac{g(x)g(1-x)}{x(1-x)}] \tag{4.670}$$

where $P(x)$ and $g(x)$ were defined in (4.191).

Problem 4.19 Prove the following formula:

$$\Gamma(1+x)\Gamma(1-x) = g(x) + g(-x) - 1 \tag{4.671}$$

where $g(x)$ were defined in (4.191).

Problem 4.20 Prove the following formula:

$$\Gamma(\frac{1}{4}+iy)\Gamma(\frac{3}{4}-iy) = \frac{\sqrt{2}\pi}{\cosh \pi y + i \sinh \pi y} \tag{4.672}$$

Problem 4.21 Prove the following asymptotic formula as $y \to \infty$

$$\mathrm{Re}[\ln \Gamma(iy)] \sim \frac{1}{2}\ln(2\pi) - \frac{1}{2}\ln y - \frac{1}{2}\pi y \tag{4.673}$$

Problem 4.22 Prove the following formula

$$\Gamma(m+\frac{1}{2}) = \frac{\sqrt{\pi}(2m)!}{2^{2m}m!} \tag{4.674}$$

Hint: Use the Legendre duplication formula.

Problem 4.23 Prove the following formula

$$\Gamma(m+\frac{1}{2}) = \frac{\sqrt{\pi}(2m-1)!!}{2^m} \tag{4.675}$$

where

$$(2m-1)!! = 1 \cdot 3 \cdot 5 \cdots (2m-3)(2m-1) \tag{4.676}$$

Problem 4.24 Prove the following formulas

$$\int_0^\infty \sin(y\ln t)e^{-t}dt = \frac{y}{2}\{\Gamma(iy)+\Gamma(-iy)\} \tag{4.677}$$

$$\int_0^\infty \sinh(y\ln t)e^{-t}dt = \frac{y}{2}\{\Gamma(y)+\Gamma(-y)\} \tag{4.678}$$

Problem 4.25 Use the definition of the Pochhammer symbol to show that
(i)

$$(-n)_r = (n+r)!/n! \tag{4.679}$$

(ii)

$$(1)_r = r! \tag{4.680}$$

Problem 4.26 Show that the following hypergeometric equation can be converted to a Riccati type equation

$$t(t-1)\frac{d^2w}{dt^2}+[\gamma-(\alpha+\beta+1)t]\frac{dw}{dx}-\alpha\beta w = 0 \tag{4.681}$$

by

$$w = e^{\int_0^t S(\alpha,\beta,\xi)d\xi} \tag{4.682}$$

Ans:

$$\frac{dS}{dt}+S^2+\left[\frac{\gamma-(\alpha+\beta+1)t}{t(t-1)}\right]S-\frac{\alpha\beta}{t(t-1)} = 0 \tag{4.683}$$

Problem 4.27 Prove that the error function discussed in Section 4.2.8 can be expressed as:

$$erf(z) = \frac{2}{\sqrt{\pi}}\int_0^z \exp(-t^2)dt \tag{4.684}$$

Problem 4.28 Prove the Gauss multiplication formula given in (4.76):

$$\Gamma(nz) = (2\pi)^{(1-n)/2}n^{nz-1/2}\Gamma(z)\Gamma(z+\frac{1}{n})\Gamma(z+\frac{2}{n})\cdots\Gamma(z+\frac{n-1}{n}) \tag{4.685}$$

(i) First show that this is equal to

$$\Gamma(nz) = (2\pi)^{(1-n)/2}n^{nz-1/2}\prod_{k=0}^{n-1}\Gamma(z+\frac{k}{n}) \tag{4.686}$$

(ii) Use the recursive formula given in (4.53) and Euler's factorial form (4.25) to show that

$$\Gamma(z+\frac{k}{n}) = \lim_{m\to\infty}\frac{m!m^{z+k/n-1}}{\left(z+\frac{k}{n}\right)\left(z+\frac{k}{n}+1\right)\cdots\left(z+\frac{k}{n}-1+m\right)} \tag{4.687}$$

(iii) Use the Stirling formula given in (4.109) to show that

$$\Gamma(z+\frac{k}{n}) = \lim_{m\to\infty}\frac{\sqrt{2\pi}(mn/e)^m m^{z+k/n-1/2}}{(nz+k)(nz+k+n)\cdots(nz+k-n+mn)} \tag{4.688}$$

(iv) Take the product function of (4.686) from $k = 0$ to $n-1$, and change the index by replacing mn with n to show

$$\prod_{k=0}^{n-1}\Gamma(z+\frac{k}{n}) = \lim_{m\to\infty}\frac{(\sqrt{2\pi})^n (m/e)^m m^{nz-1/2} n^{1/2-nz}}{(nz+k)(nz+k+n)\cdots(nz+k-n+mn)} \tag{4.689}$$

(v) Reapply the Stirling formula given in (4.109) and Euler's factorial form given in (4.25) to show

$$\prod_{k=0}^{n-1}\Gamma(z+\frac{k}{n}) = (2\pi)^{(n-1)/2} n^{1/2-nz}\Gamma(nz) \tag{4.690}$$

(vi) Finally, prove (4.685) by using (4.690).

Problem 4.29 Using (4.167) and (4.159) to show that

$$H(1) = 1 \tag{4.691}$$

Problem 4.30 Solve the following ODE by assuming a change of variable of $z = x^2$

$$\frac{d^2u}{dx^2}+\frac{1}{x}\frac{du}{dx}+4(x^2-\frac{\nu^2}{x^2})u = 0 \tag{4.692}$$

Ans:

$$u = AJ_\nu(x^2)+BY_\nu(x^2) \tag{4.693}$$

Problem 4.31 Solve the following ODE by assuming a change of variable of $z = x^{1/2}$

$$\frac{d^2u}{dx^2}+\frac{1}{x}\frac{du}{dx}+\frac{1}{4x}(1-\frac{\nu^2}{x})u = 0 \tag{4.694}$$

Ans:

$$u = AJ_\nu(\sqrt{x})+BY_\nu(\sqrt{x}) \tag{4.695}$$

Problem 4.32 Solve the following ODE by assuming a change of variable of $z = \beta x^\alpha$

$$\frac{d^2u}{dx^2}+\frac{1}{x}\frac{du}{dx}+[(\beta\alpha x^{\alpha-1})^2-\frac{\nu^2\alpha^2}{x^2}]u = 0 \tag{4.696}$$

Ans:

$$u = AJ_\nu(\beta x^\alpha)+BY_\nu(\beta x^\alpha) \tag{4.697}$$

Problem 4.33 Solve the following ODE by assuming two change of variables:
(i) $Z = x^{\nu/2}u$
(ii) $\zeta = x^{1/2}$

$$\frac{d^2u}{dx^2}+\frac{1-\nu}{x}\frac{du}{dx}+\frac{1}{4x}u = 0 \tag{4.698}$$

Ans:

$$u = x^{\nu/2}[AJ_\nu(\sqrt{x})+BY_\nu(\sqrt{x})] \tag{4.699}$$

Problem 4.34 Solve the following ODE by assuming two change of variables:

(i) $Z = x^{\beta\nu-\alpha}u$
(ii) $\zeta = \gamma x^{\beta}$

$$\frac{d^2u}{dx^2}+\frac{2\alpha-2\beta\nu+1}{x}\frac{du}{dx}+[(\beta\gamma x^{\beta-1})^2+\frac{\alpha(\alpha-2\beta\nu)}{x^2}]u=0 \tag{4.700}$$

Ans:

$$u = x^{\beta\nu-\alpha}[AJ_\nu(\gamma x^\beta)+BY_\nu(\gamma x^\beta)] \tag{4.701}$$

Problem 4.35 Solve the following ODE by assuming two change of variables:
(i) $Z = \sqrt{x}u$
(ii) $y = \beta x$

$$\frac{d^2u}{dx^2}+(\beta^2-\frac{4\nu^2-1}{4x^2})u=0 \tag{4.702}$$

Ans:

$$u = \sqrt{x}[AJ_\nu(\beta x)+BY_\nu(\beta x)] \tag{4.703}$$

Problem 4.36 Solve the following ODE by assuming two change of variables:
(i) $Z = x^{\alpha}u$
(ii) $\zeta = \beta x^{\gamma}$

$$\frac{d^2u}{dx^2}+\frac{1-2\alpha}{x}\frac{du}{dx}+[(\beta\gamma x^{\gamma-1})^2+\frac{\alpha^2-\nu^2\gamma^2}{x^2}]u=0 \tag{4.704}$$

Ans:

$$u = x^{\alpha}[AJ_\nu(\beta x^\gamma)+BY_\nu(\beta x^\gamma)] \tag{4.705}$$

Problem 4.37 Solve the following ODE by assuming two change of variables:
(i) $Z = \sqrt{x}u$
(ii) $y = \beta x^{1/2}$

$$\frac{d^2u}{dx^2}+(\frac{\beta^2}{4x}-\frac{\nu^2-1}{4x^2})u=0 \tag{4.706}$$

Ans:

$$u = \sqrt{x}[AJ_\nu(\beta\sqrt{x})+BY_\nu(\beta\sqrt{x})] \tag{4.707}$$

Problem 4.38 Solve the following ODE by assuming two change of variables:
(i) $Z = \sqrt{x}u$
(ii) $y = \gamma x^{\beta}$

$$\frac{d^2u}{dx^2}+(\beta\gamma x^{\beta-1})^2u=0 \tag{4.708}$$

Ans:

$$u = \sqrt{x}[AJ_{1/(2\beta)}(\gamma x^\beta)+BY_{1/(2\beta)}(\gamma x^\beta)] \tag{4.709}$$

Problem 4.39 Solve the following ODE by assuming the following change of variables: $y = \beta e^{x}$

$$\frac{d^2u}{dx^2}+(\beta^2 e^{2x}-\nu^2)u=0 \tag{4.710}$$

Ans:

$$u = AJ_\nu(\beta e^x)+BY_\nu(\beta e^x) \tag{4.711}$$

Problem 4.40 Solve the following Airy equation in terms of modified Bessel functions of the first and second kinds by assuming two change of variables:
(i) $Z=\sqrt{x}u$
(ii) $\zeta=(2/3)x^{3/2}$

$$\frac{d^2u}{dx^2}-xu=0 \tag{4.712}$$

Ans:

$$u=\sqrt{x}[AI_{1/3}(\frac{2}{3}x^{3/2})+BK_{1/3}(\frac{2}{3}x^{3/2})] \tag{4.713}$$

Problem 4.41 Solve the following equation by assuming two change of variables:
(i) $Z=\sqrt{x}u$
(ii) $\zeta=(2/3)x^{3/2}$

$$\frac{d^2u}{dx^2}+xu=0 \tag{4.714}$$

Ans:

$$u=\sqrt{x}[AJ_{1/3}(\frac{2}{3}x^{3/2})+BY_{1/3}(\frac{2}{3}x^{3/2})] \tag{4.715}$$

Problem 4.42 Solve the following equation by assuming two change of variables:
(i) $Z=\sqrt{x}u$
(ii) $\zeta=(4/3)x^{3/4}$

$$\frac{d^2u}{dx^2}+\frac{u}{\sqrt{x}}=0 \tag{4.716}$$

Ans:

$$u=\sqrt{x}[AJ_{2/3}(\frac{4}{3}x^{3/4})+BY_{2/3}(\frac{4}{3}x^{3/4})] \tag{4.717}$$

Problem 4.43 Solve the following equation by assuming two change of variables:
(i) $Z=\sqrt{x}u$
(ii) $\zeta=(4/3)x^{3/4}$

$$\frac{d^2u}{dx^2}-\frac{u}{\sqrt{x}}=0 \tag{4.718}$$

Ans:

$$u=\sqrt{x}[AI_{2/3}(\frac{4}{3}x^{3/4})+BK_{2/3}(\frac{4}{3}x^{3/4})] \tag{4.719}$$

Problem 4.44 Solve the following equation

$$\frac{d^2u}{dx^2}+\frac{e^{2/x}-\nu^2}{x^4}u=0 \tag{4.720}$$

Ans:

$$u=x[AJ_\nu(e^{1/x})+BY_\nu(e^{1/x})] \tag{4.721}$$

Problem 4.45 Solve the following equation by assuming a change of variables: $u=\sec x\, Z(x)$

$$\frac{d^2u}{dx^2}+(\frac{1}{x}-2\tan x)\frac{du}{dx}-(\frac{\nu^2}{x^2}+\frac{\tan x}{x})u=0 \tag{4.722}$$

Ans:

$$u=\sec x[AJ_\nu(x)+BY_\nu(x)] \tag{4.723}$$

Problem 4.46 Solve the following equation by assuming a change of variables: $u=\csc x\, Z(x)$

$$\frac{d^2u}{dx^2}+(\frac{1}{x}+2\cot x)\frac{du}{dx}-(\frac{\nu^2}{x^2}-\frac{\cot x}{x})u=0 \tag{4.724}$$

Ans:

$$u=\csc x[AJ_\nu(x)+BY_\nu(x)] \tag{4.725}$$

Problem 4.47 Solve the following equation

$$\begin{aligned}&\frac{d^2u}{dx^2}-[\frac{g''(x)}{g'(x)}+(2\nu-1)\frac{g'(x)}{g(x)}+2\frac{f'(x)}{f(x)}]\frac{du}{dx}+\{[\frac{g''(x)}{g'(x)}+(2\nu-1)\frac{g'(x)}{g(x)}\\&+2\frac{f'(x)}{f(x)}]\frac{f'(x)}{f(x)}-\frac{f''(x)}{f(x)}+[g'(x)]^2\}u=0\end{aligned} \tag{4.726}$$

Ans:

$$u=f(x)[g(x)]^\nu[AJ_\nu(g(x))+BY_\nu(g(x))] \tag{4.727}$$

Problem 4.48 Solve the following equation

$$\begin{aligned}&\frac{d^2u}{dx^2}-\frac{f'(x)}{f(x)}\frac{du}{dx}+\{\frac{3}{4}[\frac{f'(x)}{f(x)}]^2-\frac{1}{2}\frac{f''(x)}{f(x)}-\frac{3}{4}[\frac{g''(x)}{g'(x)}]^2+\frac{1}{2}\frac{g'''(x)}{g'(x)}\\&+[g^2(x)-\nu^2+\frac{1}{4}][\frac{g'(x)}{g(x)}]^2\}u=0\end{aligned} \tag{4.728}$$

Ans:

$$u=\sqrt{\frac{f(x)g(x)}{g'(x)}}[AJ_\nu(g(x))+BY_\nu(g(x))] \tag{4.729}$$

Problem 4.49 Solve the following equation

$$\frac{d^2u}{dx^2}+\{\frac{1}{2}\frac{f'''(x)}{f'(x)}-\frac{3}{4}[\frac{f''(x)}{f'(x)}]^2+[f^2(x)-\nu^2+\frac{1}{4}][\frac{f'(x)}{f(x)}]^2\}u=0 \tag{4.730}$$

Ans:

$$u = \sqrt{\frac{f(x)}{f'(x)}}[AJ_\nu(f(x)) + BY_\nu(f(x))] \tag{4.731}$$

Problem 4.50 Solve the following 2nd order ODE with non-constant coefficients

$$\frac{d^2u}{dx^2} - [\frac{f''(x)}{f'(x)} + (2\mu - 1)\frac{f'(x)}{f(x)}]\frac{du}{dx} + [\mu^2 - \nu^2 + f^2(x)][\frac{f'(x)}{f(x)}]^2 u = 0 \tag{4.732}$$

Ans:

$$u = [f(x)]^\mu [AJ_\nu(f(x)) + BY_\nu(f(x))] \tag{4.733}$$

Problem 4.51 Solve the following 4-th order ODE

$$\frac{d^4u}{dx^4} + \frac{2}{x}\frac{d^3u}{dx^3} - \frac{2\nu^2+1}{x^2}\frac{d^2u}{dx^2} + \frac{2\nu^2+1}{x^3}\frac{du}{dx} + [\frac{\nu^2(\nu^2-4)}{x^4} - 1]u = 0 \tag{4.734}$$

Ans:

$$u = AJ_\nu(x) + BY_\nu(x) + CI_\nu(x) + DK_\nu(x) \tag{4.735}$$

Problem 4.52 Find the general solution of the following Airy equation in terms of Bessel functions:

$$\frac{d^2u}{dx^2} + x\nu^2 u = 0 \tag{4.736}$$

Hint: By assuming

$$u = \sqrt{x} Z(\frac{2}{3}\nu x^{3/2}) \tag{4.737}$$

Ans:

$$u = \sqrt{x}\{AJ_{1/3}(\frac{2}{3}\nu x^{3/2}) + BY_{1/3}(\frac{2}{3}\nu x^{3/2})\} \tag{4.738}$$

Problem 4.53 Prove the validity of (4.465) to (4.467).

Problem 4.54 Prove the validity of (4.470) to (4.472).

Problem 4.55 Prove the validity of the following identity:

$$\frac{\sqrt{\pi}}{2} = (\frac{2\cdot 2}{1\cdot 3})(\frac{4\cdot 4}{3\cdot 5})(\frac{6\cdot 6}{5\cdot 7})(\frac{8\cdot 8}{7\cdot 9})\cdots \tag{4.739}$$

Hint: Refer to (4.3) and choose a proper value of n.

Problem 4.56 Prove the validity of the following identity:

$$\psi(x) = -\gamma + \sum_{n=1}^{\infty}\{\frac{1}{n} - \frac{1}{n+x-1}\} \tag{4.740}$$

Problem 4.57 Prove the validity of the following identity:

$$\zeta(x)=\frac{1}{\Gamma(x)}\int_0^\infty \frac{u^{x-1}}{e^u-1}du \tag{4.741}$$

where $\zeta(x)$ is the Euler-Riemann zeta function defined as

$$\zeta(x)=\sum_{n=1}^{\infty}\frac{1}{n^x} \tag{4.742}$$

Hints: Apply the change of variables $t = nu$ to the definition of gamma function given in (4.4) and sum n from 1 to ∞, the reverse order of integration and summation, and finally sum the infinite series inside the integral by geometric series. For real x, (4.742) was studied by Euler in 1737 and was extended to complex x by Riemann in 1857. Note that this Euler-Riemann zeta function is of fundamental significance to prime number distribution and is linked to the celebrated problem of the Riemann hypothesis (e.g., see Havil, 2003; Sabbagh, 2003). In short, the hypothesis links to the distribution of primes (e.g., 2,3,5,7,11,13,17,..., 15,485,863, ... are primes) for large integers to the tnontrivial roots of the Euler-Riemann zeta function of the complex argument. The Clay Mathematics Institute set US$ 1 million as the reward for proving or disproving this Riemann hypothesis as one of the Millennium Problems.

Problem 4.58 Rederive the following reflection formula (i.e., (4.35)) by answering the following sub-problems:

$$\Gamma(1-z)\Gamma(z)=\frac{\pi}{\sin\pi z} \tag{4.743}$$

(i) Apply the recurrence formula of the gamma function (4.53) and the Weierstrass canonical form of the gamma function (4.302) to prove the following identity

$$\frac{1}{\Gamma(z)}\frac{1}{\Gamma(-z)}=z\prod_{n=1}^{\infty}\left\{(1-\frac{z^2}{n^2})\right\} \tag{4.744}$$

(ii) Expand $\sin x$ in terms of infinite series, observe the infinite roots of $\sin x$ being $n\pi$, and show that

$$\frac{\sin x}{x}=(1-\frac{x^2}{\pi^2})(1-\frac{x^2}{4\pi^2})(1-\frac{x^2}{9\pi^2})..=\prod_{n=1}^{\infty}\left\{(1-\frac{x^2}{n^2\pi^2})\right\} \tag{4.745}$$

(iii) Combine the results in parts (i) and (ii) to derive (4.743).

Note: This proof was given by Euler in 1739 and was reported in Whittaker and Watson (1927). This proof is simpler than that given in Section 4.2.4.

CHAPTER FIVE

Systems of First Order Differential Equations

5.1 INTRODUCTION

In this chapter, we consider a system of coupled first order ODEs. This is a relatively modern topic, which evolves into the present form with the invention and popularization of the computer and the extensive use of numerical analysis after the Second World War in the 1940s. For example, it is not covered in the classic textbooks of Forsyth in the early twentieth century. Most of the modern computer software for solving ODEs, especially nonlinear ODEs, are formulated as a system of first order ODE. It is because all systems of differential equations, regardless of the order and size, can be recast in the standard form of a system of first order ODEs. It turns out that the solution of such systems involves solving the matrix eigenvalue problem. This chapter mainly deals with the linear system of ODEs for which analytic solutions can be obtained. For nonlinear ODEs, a more advanced perturbation method for studying the stability of the evolving systems of ODEs focuses mainly on the first order ODE system.

In particular, we will show that all ODE systems can be recast into the following standard form:

$$\begin{aligned} x_1' &= F_1(t, x_1, x_2, \dots x_n) \\ x_2' &= F_2(t, x_1, x_2, \dots x_n) \\ &\vdots \\ x_n' &= F_n(t, x_1, x_2, \dots x_n) \end{aligned} \tag{5.1}$$

where x_i $(i = 1,2,\dots,n)$ are the unknown functions and t is the variable. If the functional form of F_i is nonlinear, numerical analysis is normally used. For example, the Runge-Kutta method can be used to integrate this nonlinear system and its discussion will be given in Chapter 15. However, nonlinear systems of ODEs may have more than one solution at certain values of the variable t. That is, no unique solution can be guaranteed for a nonlinear system. The solutions may sometimes evolve in an unpredictable chaotic manner.

This branch of mathematics has evolved into a major branch of applied mathematics, such as the theory of catastrophes, application of topologies, and chaos theory in solving nonlinear ODEs. In 1961, MIT meteorologist Edward Lorenz studied the chaotic behavior of the convective weather system and found that there is a strange attractor for chaotic solutions of the nonlinear ODE system. The shape of the strange attractor of Lorenz closely resembles a butterfly. The term "butterfly effect" emerges and becomes a fashionable term. It suggests the notion that a butterfly flipping its wings will induce small uncertainties in the initial condition that make error cascades upward through a chain of turbulent features in a nonlinear system. It is "exaggerated" that storm systems on the other side of the globe months later are a consequence of this butterfly effect. The bottom line is that a small error in the initial condition can make a nonlinear system totally unpredictable. This contradicts the so-called "Laplace hypothesis" that the world is predictable if all physical laws and initial conditions are known. In 1966, Thom proposed a theory of catastrophe that includes the study of the geometry and topology of chaos structures. He believed that complicated chaotic structure at singularities (such as the butterfly-shaped Lorenz attractor) can be unfolded using studies of differentiable manifolds. Terms like cusp, bifurcation, and jump were introduced. In 1975, a Los Alamos based scientist, Mitchell Feigenbaum, made a major breakthrough in modern nonlinear analysis. He studied Robert May's "period-doubling" of biology populations. May discovered that chaotic solutions emerge through a series of bifurcations, and at each bifurcation point the period doubles and one cycle splits into two cycles, and at the next bifurcation from two cycles to four cycles, and so on. This splitting makes a fascinating pattern. Feigenbaum discovered that the splitting appears to come at a faster and faster rate, but there is a scaling that the rate of splitting is a constant. To his amazement, no matter how different the nonlinear system that he started with, he arrived at the same or a "universal" convergence rate. The number is approximately 4.669201609103 and now this number is called the Feigenbaum number. More importantly, he discovered an order within the so-called chaos. This sheds light on the hope that there is order within chaos. In other words, chaotic solution is predictable, and it is just our limited knowledge or ignorance of patterns in the so-called chaos. This also leads to the revitalization of scaling studies of fractal dimension, the Julia set and the Mandelbrot set. Unfortunately, there has not been another major breakthrough since Feigenbaum's discovery.

Returning to linear systems of ODEs in this chapter, we will show how systems of ODE of arbitrary order and with an arbitrary number of unknowns are transformed into a system of first order ODEs. We will also summarize the uncoupling of ODEs for multiple unknowns involving higher derivatives originated by Jacobi and Chrystal. The determinacy of ODE systems and its relation to the number of arbitrary constants in the solution are considered. The solution of ODE system with constant coefficients will be presented in the context of a matrix eigenvalue problem. The solution form is classified into the cases of distinct eigenvalue, repeated eigenvalues for Hermitian and non-Hermitian matrices, and complex conjugate pairs of eigenvalues.

Systems of differential equations have been formulated by Euler, D'Alembert, Lagrange, and Laplace. The development of the matrix method for solving a system of first order ODEs can trace back to the time of Laplace, Lagrange and Cauchy, when they considered the eigenvalue problems and formulating problem in celestial mechanics. They were also concerned with the stability of the solution of differential equations. Lagrange successfully dealt with the case of repeated eigenvalues, but it was Laplace who recognized the importance of the symmetric matrix on eigenvalues. However, Laplace, Lagrange, and Cauchy were not aware of the relation between eigenvalue problems and solving systems of first order ODE. It was J.C.F. Sturm, a student of Fourier and co-founder of the Sturm-Liouville problem, who pointed out to Cauchy in 1828 that the eigenvalue problem is related to solving the system of first order ODEs. Later contributors to the development of the matrix method for systems of ODEs include Cayley, Weierstrass, Dirichlet, Jordan, and Frobenius, among others. As summarized by Hawkins (1975a-b, 1977), the development of matrix theory is, in fact, closely related to the solution of systems of first order ODEs.

5.2 REDUCTION OF N-TH ORDER ODE TO SYSTEM OF EQUATIONS

Consider an n-th order ODE given symbolically as

$$y^{(n)} = F\left(t, y, y', y'', \ldots, y^{(n-1)}\right) \tag{5.2}$$

This ODE can be recast as a system of n ODEs. In particular, we make the following identifications:

$$x_1 = y,\ x_2 = y',\ x_3 = y'',\ \ldots, x_n = y^{(n-1)} \tag{5.3}$$

where the new unknowns are the original unknown y and its higher derivatives up to order $n-1$. Therefore, one unknown is split into n unknowns, and the first $n-1$ definitions in (5.3) actually also provide the first $n-1$ ODEs of the system:

$$\begin{aligned} x_1' &= x_2 \\ x_2' &= x_3 \\ &\vdots \\ x_{n-1}' &= x_n \\ x_n' &= F(t, x_1, x_2, \ldots x_n) \end{aligned} \tag{5.4}$$

Thus, (5.4) is the equivalent n ODE system for the single n-th order ODE given in (5.2).

Let us consider a practical problem of two oscillators connected through a string and a dashpot (or a mechanical damper) given in Figure 5.1.

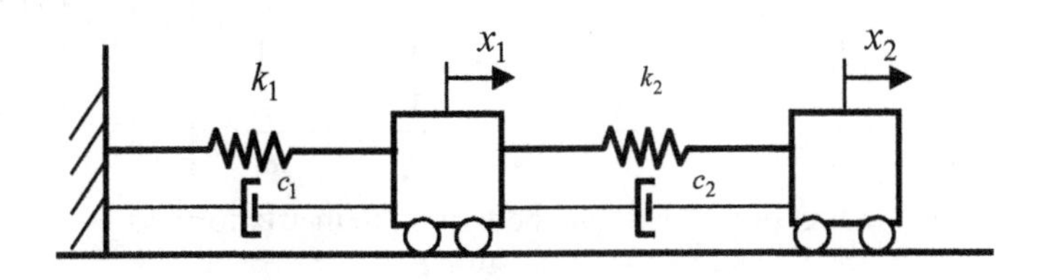

Figure 5.1 Two connected movable masses as a system of oscillators

Newton's second law of force equilibrium can be applied to formulate the equations of motion for the two masses. In this formulation, we will neglect the frictions (both static and kinematic or dynamic) between the base of the mass and the horizontal ground that the two masses rest on. The damping forces are assumed proportional to the velocity of the damper. We have the following equations of motion:

$$\begin{aligned} m_1 \frac{d^2 x_1}{dt^2} &= k_2(x_2 - x_1) - k_1 x_1 + c_2 \frac{d(x_2 - x_1)}{dt} - c_1 \frac{dx_1}{dt} + F_1(t) \\ &= -(k_1 + k_2)x_1 + k_2 x_2 - (c_1 + c_2)\frac{dx_1}{dt} + c_2 \frac{dx_2}{dt} + F_1(t) \\ m_2 \frac{d^2 x_2}{dt^2} &= -k_2(x_2 - x_1) - c_2 \frac{d(x_2 - x_1)}{dt} + F_2(t) \\ &= k_2 x_1 - k_2 x_2 + c_2 \frac{dx_1}{dt} - c_2 \frac{dx_2}{dt} + F_2(t) \end{aligned} \tag{5.5}$$

Similar to our earlier discussion, we can redefine the unknown as:

$$y_1 = x_1, \quad y_2 = \frac{dx_1}{dt} \tag{5.6}$$

$$y_3 = x_2, \quad y_4 = \frac{dx_2}{dt} \tag{5.7}$$

Using these definitions, (5.4) and (5.5) can be recast as a system of first order ODEs

$$\begin{aligned} y_1' &= y_2 \\ y_2' &= \frac{1}{m_1}\left[-(k_1 + k_2)y_1 + k_2 y_3 - (c_1 + c_2)y_2 + c_2 y_4 + F_1(t)\right] \\ y_3' &= y_4 \\ y_4' &= \frac{1}{m_2}\left[k_2 y_1 - k_2 y_3 + c_2 y_2 - c_2 y_4 + F_2(t)\right] \end{aligned} \tag{5.8}$$

Putting (5.8) in matrix form, we have

$$\frac{d}{dt}\begin{Bmatrix} y_1 \\ y_2 \\ y_3 \\ y_4 \end{Bmatrix} = \begin{bmatrix} 0 & 1 & 0 & 0 \\ -\frac{1}{m_1}(k_1 + k_2) & -\frac{1}{m_1}(c_1 + c_2) & \frac{1}{m_1}k_2 & \frac{1}{m_1}c_2 \\ 0 & 0 & 0 & 1 \\ \frac{1}{m_2}k_2 & \frac{1}{m_2}c_2 & -\frac{1}{m_1}k_2 & -\frac{1}{m_1}c_2 \end{bmatrix} \begin{Bmatrix} y_1 \\ y_2 \\ y_3 \\ y_4 \end{Bmatrix} + \begin{Bmatrix} 0 \\ F_1(t) \\ 0 \\ F_2(t) \end{Bmatrix} \tag{5.9}$$

Symbolically, the first order ODE can be written in matrix form as

$$y' = \boldsymbol{A}y + \boldsymbol{F} \tag{5.10}$$

Therefore, it is clear that all coupled ODEs can be expressed as a system of first order ODEs. This is particularly important if the ODE is nonlinear and an exact solution cannot be found. Nowadays, nearly all computer software for numerical analyses deals exclusively with systems of first order ODEs (e.g., Press et al., 1992). One of the most popular methods is the fourth order Rung-Kutta method, which will be discussed in more detail in Chapter 15.

5.3 ANALYTIC SOLUTION BY ELIMINATION

To illustrate the idea of solving a system of equations by elimination, we consider the case of two linear coupled ODEs as:

$$\begin{aligned} F_{11}(D)x + F_{12}(D)y &= f_1 \\ F_{21}(D)x + F_{22}(D)y &= f_2 \end{aligned} \tag{5.11}$$

where F_{ij} are differential operators defined as:

$$F_{11}(D)x = (a_0 D^k + a_1 D^{k-1} + ... + a_k)x \tag{5.12}$$

$$F_{12}(D)y = (b_0 D^k + b_1 D^{k-1} + ... + b_k)y \tag{5.13}$$

Similarly, F_{21} and F_{22} can be defined. We will restrict our discussions to ODEs with constant coefficients, but we will not restrict ourselves to first order, which is the focus for the rest of this chapter. Differentiation with respect to t is denoted symbolically by D as:

$$D^k y = \frac{d^k y}{dt^k} \tag{5.14}$$

To eliminate x from (5.11), we can first take differential operator F_{11} on the second equation of (5.11) and take differential operator F_{21} on the first of (5.11). Then, the difference of these two resulting equations gives:

$$(F_{21}F_{12} - F_{11}F_{22})y = F_{21}f_1 - F_{11}f_2 \tag{5.15}$$

Similarly, we can also eliminate y to give

$$(F_{21}F_{12} - F_{11}F_{22})x = -F_{22}f_1 + F_{12}f_2 \tag{5.16}$$

where

$$\begin{vmatrix} F_{11} & F_{12} \\ F_{21} & F_{22} \end{vmatrix} = -(F_{21}F_{12} - F_{11}F_{22}) \neq 0 \tag{5.17}$$

This determinant is called the operational determinant (Edwards and Penney, 2005), the characteristic determinant (Ince, 1956), or simply the determinant of the system (Tenenbaum and Pollard, 1963). If the determinant on (5.17) is zero, the system may have either no solution or infinite solutions. Thus, (5.11) can be uncoupled to give two ODE for x and y separately. Note that the operators on the left of (5.15) and (5.16) are the same. Both of them are nonhomogeneous linear ODEs and thus can be solved by the technique covered in an earlier chapter.

However, the resulting uncoupled ODE is of higher order, and thus the elimination process induces a number of unnecessary and undesirable arbitrary constants. They need to be eliminated by substituting these solutions back into the original ODE system. In other words, the uncoupled ODE is not equivalent to the original system. The following examples illustrate this problem.

Example 5.1 A system of two second order ODEs is given as

$$\begin{aligned} (\frac{d^2}{dt^2} - 2)x - \frac{d^2 y}{dt^2} &= -2\sin\sqrt{2}t \\ (\frac{d^2}{dt^2} + 2)x + \frac{d^2 y}{dt^2} &= 0 \end{aligned} \tag{5.18}$$

Find the solution of the system.

Solution: Adding these two ODEs gives

$$\frac{d^2x}{dt^2} = -\sin\sqrt{2}t \tag{5.19}$$

Integrating twice, we obtain

$$x = \frac{1}{2}\sin\sqrt{2}t + A_1 t + A_2 \tag{5.20}$$

Substitution of (5.20) into the first equation of (5.18) gives

$$\frac{d^2y}{dt^2} = -2A_1 t - 2A_2 \tag{5.21}$$

Direct integration on both sides yields the solution of y

$$y = -\frac{1}{3}A_1 t^3 - A_2 t^2 + B_1 t + B_2 \tag{5.22}$$

Since both x and y are second order in the system of (5.18), it is expected that there should be four unknown constants.

Example 5.2 A system of two second order ODEs is given as

$$\begin{aligned} &(\frac{d^2}{dt^2} - 2)x + \frac{d^2y}{dt^2} = -2\sin\sqrt{2}t \\ &(\frac{d^2}{dt^2} + 2)x + \frac{d^2y}{dt^2} = 0 \end{aligned} \tag{5.23}$$

Find the solution of the system.

Solution: Subtracting these two ODEs gives

$$x = \frac{1}{2}\sin\sqrt{2}t \tag{5.24}$$

Substitution of this result into (5.23) leads to

$$\frac{d^2y}{dt^2} = 0 \tag{5.25}$$

Substitution of (5.25) into the first equation of (5.23) gives

$$y = B_1 t + B_2 \tag{5.26}$$

Thus, there are only two unknown constants. Thus, the corresponding boundary conditions must be given as

$$y = y_0, \quad \frac{dy}{dt} = y_1, \quad t = t_0 \tag{5.27}$$

Note in this case that we only have two unknown constants instead of four as obtained in Example 5.1. We will discuss this peculiar result in the next section.

Example 5.3 A system of two second order ODEs is given as

$$(\frac{d^2}{dt^2}-2)x+(\frac{d^2}{dt^2}-4)y=-4\sin\sqrt{2}t$$
$$(\frac{d^2}{dt^2}+2)x+\frac{d^2y}{dt^2}=0 \tag{5.28}$$

Find the solution of the system.

Solution: Subtracting these two ODEs gives

$$x+y=\sin\sqrt{2}t \tag{5.29}$$

Differentiation of (5.29) leads to

$$\frac{d^2(x+y)}{dt^2}=-2\sin\sqrt{2}t \tag{5.30}$$

Substitution of (5.30) into the second equation of (5.28) gives

$$x=\sin\sqrt{2}t \tag{5.31}$$

Consequently, we must have

$$y=0 \tag{5.32}$$

We have just obtained another peculiar result that there is no arbitrary unknown constant in the solution. Or, we cannot impose any initial condition. Again, we will discuss this in the next section.

Example 5.4 A system of two second order ODEs is given as

$$(\frac{d^2}{dt^2}-1)x+(\frac{d^2}{dt^2}-\frac{d}{dt})y=-2\sin t$$
$$(\frac{d^2}{dt^2}+\frac{d}{dt})x+\frac{d^2y}{dt^2}=0 \tag{5.33}$$

Find the solution of the system.

Solution: Subtracting the first of (5.33) from the second of (5.33) gives

$$(\frac{d}{dt}+1)x+\frac{dy}{dt}=2\sin t \tag{5.34}$$

Differentiation of (5.34) with respect to t gives

$$(\frac{d^2}{dt^2}+\frac{d}{dt})x+\frac{d^2y}{dt^2}=2\cos t \tag{5.35}$$

This ODE is however inconsistent with the second equation of (5.33). Thus, this inconsistency leads to no solution.

Example 5.5 A system of three equations is

$$(\frac{d^2}{dt^2}-2\frac{d}{dt}+3)x+(\frac{d}{dt}-1)y+\frac{dz}{dt}=0$$
$$(3\frac{d}{dt}+1)x-3\frac{dy}{dt}-\frac{dz}{dt}=1 \tag{5.36}$$
$$2x-2y-z=-4$$

Find the solution of the system.

Solution: Differentiation of the last equation of (5.36) results in

$$\frac{dz}{dt}=2\frac{dx}{dt}-2\frac{dy}{dt} \tag{5.37}$$

Substitution of (5.37) into the first two of (5.36) leads to a system of ODEs for x and y only:

$$(\frac{d^2}{dt^2}+3)x-(\frac{d}{dt}+1)y=0$$
$$(\frac{d}{dt}+1)x-\frac{dy}{dt}=1 \tag{5.38}$$

Elimination of y from (5.38) gives

$$(\frac{d^3}{dt^3}-\frac{d^2}{dt^2}+\frac{d}{dt}-1)x=-1 \tag{5.39}$$

Assuming an exponent solution for the homogeneous ODE of (5.39), we obtain

$$(\lambda^3-\lambda^2+\lambda-1)=(\lambda-1)(\lambda^2+1)=0 \tag{5.40}$$

Thus, the homogeneous solution is

$$x_h=C_1e^t+C_2\cos t+C_3\sin t \tag{5.41}$$

It is straightforward to see that the particular solution for (5.39) is

$$x_p=1 \tag{5.42}$$

Therefore, we have

$$x=1+C_1e^t+C_2\cos t+C_3\sin t \tag{5.43}$$

Substitution of this result into the second equation of (5.38) leads to

$$\frac{dy}{dt}=(\frac{d}{dt}+1)x-1$$
$$=(C_3+C_2)\cos t+(C_3-C_2)\sin t+2C_1e^t \tag{5.44}$$

Integration of (5.44) gives

$$y=(C_3+C_2)\sin t-(C_3-C_2)\cos t+2C_1e^t+C_4 \tag{5.45}$$

Substitution of these solutions of x and y into the first equation of (5.38) leads to

$$C_4=3 \tag{5.46}$$

Finally, substitution of (5.43) and (5.45) into the third equation of (5.36) gives the solution of z

$$z=-2C_2\sin t+2C_3\cos t-2C_1e^t \tag{5.47}$$

This gives the complete solution for the system of ODEs given in (5.36) and note that there are only three unknown constants.

It can be seen from Example 5.5 that one of the constants needs to be fixed by the original ODE. Instead of back substitution into the original system, we can also make the transformed ODE system equivalent to the original one by using a multiplier system during the elimination procedure (Ince, 1956). As remarked by Ince (1956), this topic was first considered by Jacobi in 1865 and was refined by Chrystal in 1895.

Let us consider two systems of coupled ODEs of n unknowns y_i ($i=1,...,n$)

$$U_r = F_{r1}(D)y_1 + F_{r2}(D)y_2 + ... + F_{rn}(D)y_n - f_r(t) = 0 \tag{5.48}$$

$$V_r = G_{r1}(D)y_1 + G_{r2}(D)y_2 + ... + G_{rn}(D)y_n - g_r(t) = 0 \tag{5.49}$$

where $r=1,2,...,n$ and F_{ij} and G_{ij} are polynomials in terms of differential operator D and with constant coefficients. For the case of constant coefficient polynomials of D, it is straightforward to see that these operators are commutative, associative, and distributive (Tenenbaum and Pollard, 1963), but the proofs of them will not be covered here.

Now, we are looking for the condition that these two ODE systems are equivalent (i.e., the solutions of (5.49) are also solution of (5.48) and vice versa). If every solution of U satisfies V, we have

$$\begin{aligned} V_1 &= T_{11}(D)U_1 + ... + T_{1n}(D)U_n \\ &... \\ V_n &= T_{n1}(D)U_1 + ... + T_{nn}(D)U_n \end{aligned} \tag{5.50}$$

where T_{ij} are polynomial operators. It is clear that if U satisfies (5.48), V defined in (5.50) would also satisfy (5.49). If (5.50) is written in matrix form, we must have

$$\begin{Bmatrix} V_1 \\ \vdots \\ V_n \end{Bmatrix} = \begin{pmatrix} T_{11} & \cdots & T_{1n} \\ \vdots & \cdots & \vdots \\ T_{n1} & & T_{nn} \end{pmatrix} \begin{Bmatrix} U_1 \\ \vdots \\ U_n \end{Bmatrix} \tag{5.51}$$

$$\Delta = \begin{vmatrix} T_{11} & \cdots & T_{1n} \\ \vdots & \cdots & \vdots \\ T_{n1} & & T_{nn} \end{vmatrix} \neq 0 \tag{5.52}$$

The determinant is nonzero since all V are independent. Ince (1956) called T_{ij} the multiplier system. If every solution of V satisfies U, we also have, by the theory of the determinant (Ince, 1956),

$$\begin{aligned} \Delta U_1 &= \Delta_{11}(D)V_1 + ... + \Delta_{1n}(D)V_n \\ &... \\ \Delta U_n &= \Delta_{n1}(D)V_1 + ... + \Delta_{nn}(D)V_n \end{aligned} \tag{5.53}$$

where

$$\frac{1}{\Delta} = \begin{vmatrix} \Delta_{11} & \cdots & \Delta_{1n} \\ \vdots & \cdots & \vdots \\ \Delta_{n1} & \cdots & \Delta_{nn} \end{vmatrix} \neq 0 \tag{5.54}$$

In fact, (5.53) can be regarded as the inversion of the system of ODEs given in (5.51). If V satisfies (5.49), we have

$$\Delta U_1 = 0, ..., \Delta U_n = 0 \tag{5.55}$$

Thus, if Δ is a constant, every solution of $V = 0$ satisfies $U = 0$. In other words, Δ defined in (5.54) cannot be a function of D. This is the condition in which (5.48) and (5.49) are equivalent.

A natural question would be how to find the multiplier system. The first row of it is naturally chosen by our aim of eliminating one of the unknowns of the system. The other rows must be found in such a way that the resulting determinant (5.52) will become a constant. We will illustrate more specifically the elimination process for the first two equations of (5.48). Written explicitly, these equations are

$$U_1 = F_{11}(D)y_1 + F_{12}(D)y_2 + ... + F_{1n}(D)y_n - f_1(t) = 0 \tag{5.56}$$

$$U_2 = F_{21}(D)y_1 + F_{22}(D)y_2 + ... + F_{2n}(D)y_n - f_2(t) = 0 \tag{5.57}$$

Let us assume that y_1 can be eliminated from (5.56) and (5.57) to get

$$LU_1 + MU_2 = 0 \tag{5.58}$$

$$L^* U_1 + M^* U_2 = 0 \tag{5.59}$$

In matrix form, this equivalent pair of ODEs becomes

$$\begin{bmatrix} L & M \\ L^* & M^* \end{bmatrix} \begin{Bmatrix} U_1 \\ U_2 \end{Bmatrix} = \begin{Bmatrix} 0 \\ 0 \end{Bmatrix} \tag{5.60}$$

Since both U_1 and U_2 are identical zero, we have the determinant of the operator coefficient being nonzero:

$$LM^* - ML^* = \text{constant} \tag{5.61}$$

Assume further that there is a common polynomial factor Γ between F_{11} and F_{21} such that

$$F_{11}(D) = \Gamma\Phi, \quad F_{21}(D) = \Gamma\Psi, \tag{5.62}$$

Clearly, to eliminate y_1 we should let

$$L = \Psi, \quad M = -\Phi \tag{5.63}$$

where L and M are relatively prime with respect to D. Equation (5.61) provides a condition to find the last two operators of the multiplier system L^* and M^*, which were called adjoint operators (Golomb and Shanks, 1965). However, it can easily be confused with the adjoint ODE and we will not use such terminology in this book.

We now illustrate this technique of the multiplier system.

Example 5.6 Use the method of the multiplier system to solve the following ODE system

$$\begin{aligned} (D+1)y_1 + D^2 y_2 + (D+1)y_3 &= 0 \\ (D-1)y_1 + Dy_2 + (D-1)y_3 &= 0 \\ y_1 + y_2 + Dy_3 &= 0 \end{aligned} \tag{5.64}$$

Solution: To eliminate y_2 from the second and third equations of (5.64), we can sum the result of multiplying -1 to the second equation of (5.64) and the result of applying D to the third equation of (5.64). That is, we have $L = -1$ and $M = 0$. The associated multiplier system can be found by inspection as:

$$\begin{pmatrix} L & M \\ L^* & M^* \end{pmatrix} = \begin{pmatrix} -1 & D \\ -1 & D-1 \end{pmatrix} \tag{5.65}$$

It is straightforward to show that its determinant is 1 (although any constant will do, we have chosen to use 1), and thus (5.61) is satisfied. Thus, we have found L^* and M^* in (5.65) by inspection. The resulting ODE becomes

$$\begin{aligned} (D+1)y_1 + D^2 y_2 + (D+1)y_3 &= 0 \\ y_1 + (D^2 - D + 1)y_3 &= 0 \\ -y_2 + (D^2 - 2D + 1)y_3 &= 0 \end{aligned} \tag{5.66}$$

To eliminate y_1 from the first and second equations of (5.64), we can use another multiplier system:

$$\begin{pmatrix} -1 & D+1 \\ -1 & D \end{pmatrix} \tag{5.67}$$

The first row of (5.67) is the pairs of operators need in eliminating y_1 and the second row is the corresponding operators to make the determinant of (5.67) being 1 (see (5.61)) and it is obtained by inspection. The new system of ODEs becomes

$$\begin{aligned} -y_1 - D^2 y_2 + (D^3 - D^2 - 1)y_3 &= 0 \\ -D^2 y_2 + (D^3 - D)y_3 &= 0 \\ -y_2 + (D^2 - 2D + 1)y_3 &= 0 \end{aligned} \tag{5.68}$$

Finally, we eliminate y_2 from the second and third equations of (5.68) by using the following multiplier system:

$$\begin{pmatrix} 0 & 1 \\ -1 & D^2 \end{pmatrix} \tag{5.69}$$

Eventually, we have

$$\begin{aligned} -y_1 - D^2 y_2 + (D^3 - D^2 - 1)y_3 &= 0 \\ -y_2 + (D^2 - 2D + 1)y_3 &= 0 \\ (D^4 - 3D^3 + D^2 + D)y_3 &= 0 \end{aligned} \tag{5.70}$$

The last of (5.70) is uncoupled and can be solved by exponential function (i.e., $y_3 = e^{\lambda t}$). The corresponding characteristics equation for y_3 is

$$\lambda^4 - 3\lambda^3 + \lambda^2 + \lambda = 0 \tag{5.71}$$

Without going through the details, one can show that it can be factorized as

$$\lambda(\lambda - 1)(\lambda - 1 - \sqrt{2})(\lambda - 1 + \sqrt{2}) = 0 \tag{5.72}$$

The general solution can be expressed as

$$y_3 = C_1 + C_2 e^t + C_3 e^{(1+\sqrt{2})t} + C_4 e^{(1-\sqrt{2})t} \tag{5.73}$$

Substitution of (5.73) into the rest of (5.70) yields

$$y_2 = C_1 + (3+\sqrt{2})C_3 e^{(1+\sqrt{2})t} + (3-\sqrt{2})C_4 e^{(1-\sqrt{2})t} \tag{5.74}$$

$$y_1 = -C_1 - C_2 e^t - 2(5+3\sqrt{2})C_3 e^{(1+\sqrt{2})t} - 2(5-3\sqrt{2})C_4 e^{(1-\sqrt{2})t} \tag{5.75}$$

This completes the full solution.

To understand the scenarios that we have observed from Examples 5.1–5.6, we will consider the determinacy of system of ODEs in the next section.

5.4 DETERMINACY OF SYSTEM OF EQUATIONS

Let us consider the following system of ODEs symbolically:

$$\begin{aligned} P_1(D)x+Q_1(D)y+R_1(D)z&=F_1\\ P_2(D)x+Q_2(D)y+R_2(D)z&=F_2\\ P_3(D)x+Q_3(D)y+R_3(D)z&=F_3 \end{aligned} \tag{5.76}$$

where P_i, Q_i and R_i are differential operators. This system of coupled ODEs can be written in matrix form as:

$$\begin{bmatrix} P_1(D) & Q_1(D) & R_1(D)\\ P_2(D) & Q_2(D) & R_2(D)\\ P_3(D) & Q_3(D) & R_3(D) \end{bmatrix}\begin{Bmatrix} x\\ y\\ z \end{Bmatrix}=\begin{Bmatrix} F_1\\ F_2\\ F_3 \end{Bmatrix} \tag{5.77}$$

$$[A(D)]\{x\}=\{F\} \text{ or } \boldsymbol{A}(D)\boldsymbol{x}=\boldsymbol{F} \tag{5.78}$$

Let the determinant of the matrix of differential operators be $\Delta(D)$ = det$|\boldsymbol{A}(D)|$, where D is defined in (5.14). The system of ODEs is determinate (i.e., solution can be found) if $\Delta(D) \neq 0$ and the system is indeterminate if $\Delta(D) = 0$. The degree in D of the determinant $\Delta(D)$ indicates the number of arbitrary unknown constants involved in the solution. Therefore, it is necessary to check the determinacy of the system before attempting to solve it.

Let us check the determinacy of each of the examples considered in the last section. The $\Delta(D)$ of Example 5.1 is

$$\Delta(D)=\begin{vmatrix} D^2-2 & -D^2\\ D^2+2 & D^2 \end{vmatrix}=2D^4 \tag{5.79}$$

Thus, $\Delta(D) \neq 0$ and this system is determinate. The degree of D is four and we have four unknown constants. This agrees with the result in Example 5.1. The determinant for the system given in Example 5.2 is

$$\Delta(D)=\begin{vmatrix} D^2-2 & D^2\\ D^2+2 & D^2 \end{vmatrix}=-4D^2 \tag{5.80}$$

Hence, it is determinate and there are two unknown constants, and it is what we found in Example 5.2. The determinacy of the system given in Example 5.3 can be determined as:

$$\Delta(D)=\begin{vmatrix} D^2-2 & D^2-4\\ D^2+2 & D^2 \end{vmatrix}=8 \tag{5.81}$$

Thus, there is no unknown constant like we found in Example 5.3. For Example 5.4, we have $\Delta(D)$

$$\Delta(D)=\begin{vmatrix} D^2-1 & D^2-D \\ D^2+D & D^2 \end{vmatrix}=0 \tag{5.82}$$

Therefore, it is indeterminate and no solution can be found. Finally, the determinacy of the system in Example 5.5 is

$$\Delta(D)=\begin{vmatrix} D^2-2D+3 & D-1 & D \\ 3D+1 & -3D & D \\ 2 & -2 & -1 \end{vmatrix}=D^3-D^2+D-1 \tag{5.83}$$

There are three unknown constants for the system of ODEs, and this agrees with our results. Therefore, it is advisable to check the determinacy of the system before we actually solve it.

Finally, the characteristics determinant of Example 5.6 becomes

$$\Delta(D)=\begin{vmatrix} D+1 & D^2 & D+1 \\ D-1 & D & D-1 \\ 1 & 1 & D \end{vmatrix}=-D^4+3D^3-D^2+D \tag{5.84}$$

This suggests four unknown constants and this agrees with the result of Example 5.6.

5.5 REVIEW ON MATRIX

It will be shown that the solution of the system of first order ODEs involves solving the eigenvalue problem in a matrix. Some fundamental matrix results will be summarized here. Consider an $m\times n$ matrix $\boldsymbol{A}$

$$\boldsymbol{A}=\begin{pmatrix} a_{11} & a_{12} & \cdots & a_{1n} \\ a_{21} & a_{22} & \cdots & a_{2n} \\ \vdots & \vdots & \ddots & \vdots \\ a_{m1} & a_{m2} & \cdots & a_{mn} \end{pmatrix} \tag{5.85}$$

The matrix is sometimes denoted as $\boldsymbol{A} = (a_{ij})$. The transpose of it is $\boldsymbol{A}^T = (a_{ji})$ or more explicitly:

$$\boldsymbol{A}^T=\begin{pmatrix} a_{11} & a_{21} & \cdots & a_{m1} \\ a_{12} & a_{22} & \cdots & a_{m2} \\ \vdots & \vdots & \ddots & \vdots \\ a_{1n} & a_{2n} & \cdots & a_{mn} \end{pmatrix} \tag{5.86}$$

That is, the first column becomes the first row of the transpose matrix and similarly for other rows. A zero matrix is defined as $\boldsymbol{0}=(0)$ or all entries in the matrix are zeros. If two matrices are the same, all elements must be identical:

$$\boldsymbol{A}=\boldsymbol{B},\quad a_{ij}=b_{ij} \tag{5.87}$$

Matrix addition is defined as:

$$\boldsymbol{A}+\boldsymbol{B}=\boldsymbol{C},\quad (c_{ij})=(a_{ij}+b_{ij}) \tag{5.88}$$

Scalar multiplication is defined as:

$$k\mathbf{A} = (ka_{ij}) \tag{5.89}$$

Matrix multiplication is defined as $\boldsymbol{AB} = \boldsymbol{C}$, or

$$c_{ij} = \sum_{k=1}^{n} a_{ik} b_{kj} \tag{5.90}$$

where matrix $\boldsymbol{A}$ is $m \times n$, matrix $\boldsymbol{B}$ is $n \times r$, and the resulting matrix $\boldsymbol{C}$ is $m \times r$. In other words, the i-th row multiplying the j-th column yields the element of c_{ij}. In general, matrix multiplication is not commutative, that is $\boldsymbol{AB} \neq \boldsymbol{BA}$. In addition, we may also have $\boldsymbol{AB} = \mathbf{0}$ even though $\boldsymbol{A} \neq \mathbf{0}$ and $\boldsymbol{B} \neq \mathbf{0}$.

Example 5.7 Show that the multiplication of the following matrices is not commutative:

$$\boldsymbol{A} = \begin{pmatrix} 1 & 2 \\ 3 & 4 \end{pmatrix}, \quad \boldsymbol{B} = \begin{pmatrix} 1 & 3 \\ 2 & 4 \end{pmatrix} \tag{5.91}$$

Solution: Applying the rule of matrix multiplication given in (5.91), we get

$$\boldsymbol{AB} = \begin{pmatrix} 1 & 2 \\ 3 & 4 \end{pmatrix}\begin{pmatrix} 1 & 3 \\ 2 & 4 \end{pmatrix} = \begin{pmatrix} 1\times1+2\times2 & 1\times3+2\times4 \\ 3\times1+4\times2 & 3\times3+4\times4 \end{pmatrix} = \begin{pmatrix} 5 & 11 \\ 11 & 25 \end{pmatrix} \tag{5.92}$$

On the other hand,

$$\boldsymbol{BA} = \begin{pmatrix} 1 & 3 \\ 2 & 4 \end{pmatrix}\begin{pmatrix} 1 & 2 \\ 3 & 4 \end{pmatrix} = \begin{pmatrix} 1\times1+3\times3 & 1\times2+3\times4 \\ 2\times1+4\times3 & 2\times2+4\times4 \end{pmatrix} = \begin{pmatrix} 10 & 14 \\ 14 & 20 \end{pmatrix} \tag{5.93}$$

From this example, it is clear that matrix multiplication is, in general, not commutative.

A unit matrix can be defined as:

$$\boldsymbol{I} = \begin{pmatrix} 1 & 0 & \cdots & 0 \\ 0 & 1 & \cdots & 0 \\ \vdots & \vdots & \ddots & \vdots \\ 0 & 0 & \cdots & 1 \end{pmatrix} \tag{5.94}$$

In addition, it is straightforward to illustrate that for any square matrix $\boldsymbol{A}$

$$\boldsymbol{IA} = \boldsymbol{AI} = \boldsymbol{A} \tag{5.95}$$

Then, the inverse of a matrix $\boldsymbol{A}$ can be defined as

$$\boldsymbol{AA}^{-1} = \boldsymbol{A}^{-1}\boldsymbol{A} = \boldsymbol{I} \tag{5.96}$$

The inverse exists as long as the determinant of $\boldsymbol{A}$ is nonzero. We also have

$$(\boldsymbol{A}^{-1})^{-1} = \boldsymbol{A} \tag{5.97}$$

The dot product between vectors can be expressed in terms of matrix multiplication of a row vector (or an $n\times1$ matrix) with a column matrix (or a $1\times n$ matrix) as:

$$\mathbf{x}^T\mathbf{y} = \sum_{i=1}^{n} x_i y_i \tag{5.98}$$

More generally, if the elements in the vectors are complex, we can define an inner product (which is an extension of the dot product to the complex case):

$$(\mathbf{x},\mathbf{y}) = \mathbf{x}^T\overline{\mathbf{y}} = \sum_{i=1}^{n} x_i \overline{y}_i \tag{5.99}$$

which is in essence the same as the dot product except for taking the complex conjugate of vector y. The dot product of a vector and itself forms the square of the length of the vector. If the dot product or inner product of two vectors (real or complex) equals zero, these vectors are called orthogonal. The word "orthogonal" is equivalent to the term perpendicular the case of 2-D or 3-D vectors (i.e., vector with 2 or 3 elements). The adjective perpendicular is not very meaningful, if we go beyond 3-D cases. This is because perpendicularity is normally visualized in a geometric sense that cannot be applied to a 4-D or higher dimensional space.

As a side story, we should also mention that when the great mathematician Hilbert tried solving eigenvalue problems of integral equations in 1901, he inaugurated the study of spectrum theory (i.e., the spectrum of eigenvalues and eigenfunctions of integral equations). Hilbert also showed that differential equations could be converted to integral equations. His studies prompted other mathematicians, like Schmidt, Riesz, Volterra, Fischer, Lebesgue, Frechet, and Banach, to introduce and develop the concept of the more abstract analysis of functional space or vector space. Actually, the inner product defined in (5.99) is one of the major tools used in functional analysis. It has been shown that the differential operator needs a different definition of inner product such that the resulting eigenfunctions (in terms of vector) are orthogonal. Nowadays, the Hilbert space and Banach space concepts become very important in numerical analysis of complex differential equations. However, these topics are considered abstract for most engineers and are out of the scope of the present book.

5.5.1 Hermitian Matrix

For real matrices, if $A = A^T$, it is called symmetric. For matrices with complex elements, if

$$\mathbf{A} = \overline{\mathbf{A}}^T \tag{5.100}$$

it is called self-adjoint. The bar over matrix A indicates that all elements in the matrix are replaced by their complex conjugates. Self-adjoint matrices are also known as Hermitian matrices, named after French mathematician Hermite. In particular, if a matrix is Hermitian, all eigenvalues of the matrix are real and all eigenvectors are independent. For Hermitian matrices with an eigenvalue of algebraic multiplicity m, it is possible to choose m mutually orthogonal eigenvectors. This is important in the sense that the solution form in terms of eigenvectors may depend on whether A *is* Hermitian. This will be discussed shortly in this chapter.

There is also a close resemblance between the eigenvalue problem of the Hermitian matrix and the eigenfunction expansion of a homogeneous boundary value problem of ODEs. In fact, if an ODE is self-adjoint (note that this terminology is the same as the one used in (5.100) for matrix), the eigenvalues of the boundary value problem are also real and all corresponding eigenfunctions are orthogonal.

5.5.2 Eigenvalue Problem

The eigenvalue of a matrix is of profound importance as its applications include the calculation of vibration frequency of mechanical systems or solids, of principal stresses in solids, and of dynamic stability of systems. The prefix eigen- is adopted from the German "eigen" for "self." Matrix multiplication can be considered as a function evaluation or mapping. For example, if a matrix A is multiplied by a vector x and the result is another vector y:

$$\mathbf{A}x = y \tag{5.101}$$

It is a linear function transformation between vector spaces. The eigenvalue problem is related to a transformation such that a vector x is mapped to itself except for a scalar multiple λ, which is called the eigenvalue of the matrix. That is,

$$\mathbf{A}x = \lambda x \tag{5.102}$$

It turns out that it is, in general, not possible to find such a system except for some particular values of λ. The number of the eigenvalue depends on the dimension of the matrix A. Rearranging (5.102) we have

$$Ax - \lambda x = (A - \lambda I)x = \mathbf{0} \tag{5.103}$$

If vector x is not zero, the determinant of the matrix in the bracket must be zero

$$\det(A - \lambda I) = 0 \tag{5.104}$$

The solution of this algebraic equation gives the required so-called eigenvalue λ of the matrix A. The corresponding vector x is called the eigenvector.

Example 5.8 Find the eigenvalue and vector of the following matrix:

$$A = \begin{pmatrix} 3 & -1 \\ 4 & -2 \end{pmatrix} \tag{5.105}$$

Solution: The eigenvalue of this matrix A can be evaluated as:

$$\det(\mathbf{A} - \lambda\mathbf{I}) = \det\left[\begin{pmatrix} 3 & -1 \\ 4 & -2 \end{pmatrix} - \lambda\begin{pmatrix} 1 & 0 \\ 0 & 1 \end{pmatrix}\right] = \det\begin{pmatrix} 3-\lambda & -1 \\ 4 & -2-\lambda \end{pmatrix} \tag{5.106}$$

$$= \lambda^2 - \lambda - 2 = (\lambda - 2)(\lambda + 1)$$

It is straightforward to see that a 2×2 matrix leads to an algebraic equation of second order. Thus, we have two eigenvalues as

$$\lambda = 2, \ \lambda = -1 \tag{5.107}$$

For $\lambda = 2$, applying (5.103) we have

$$\begin{pmatrix} 3-2 & -1 \\ 4 & -2-2 \end{pmatrix}\begin{pmatrix} x_1 \\ x_2 \end{pmatrix} = \begin{pmatrix} 0 \\ 0 \end{pmatrix} \tag{5.108}$$

This matrix equation implies

$$x_1 = x_2 \tag{5.109}$$

Thus, the eigenvector for $\lambda = 2$ is

$$\mathbf{x}^{(1)} = \begin{pmatrix} x_2 \\ x_2 \end{pmatrix} = c\begin{pmatrix} 1 \\ 1 \end{pmatrix} = \begin{pmatrix} 1 \\ 1 \end{pmatrix} \tag{5.110}$$

where c is an arbitrary constant. In the last part of (5.106), we have set $c = 1$ to get the eigenvector, but .other values of c can also be used.

For $\lambda = -1$, the eigenvector can be calculated as

$$\begin{pmatrix} 3+1 & -1 \\ 4 & -2+1 \end{pmatrix}\begin{pmatrix} x_1 \\ x_2 \end{pmatrix} = \begin{pmatrix} 0 \\ 0 \end{pmatrix} \Leftrightarrow \begin{pmatrix} 4 & -1 \\ 4 & -1 \end{pmatrix}\begin{pmatrix} x_1 \\ x_2 \end{pmatrix} = \begin{pmatrix} 0 \\ 0 \end{pmatrix} \tag{5.111}$$

This implies that

$$x_2 = 4x_1 \tag{5.112}$$

This gives

$$\mathbf{x}^{(2)} = \begin{pmatrix} x_1 \\ 4x_1 \end{pmatrix} = c\begin{pmatrix} 1 \\ 4 \end{pmatrix} = \begin{pmatrix} 1 \\ 4 \end{pmatrix} \tag{5.113}$$

Again, we have set $c = 1$ to get the last eigenvector. Clearly, the eigenvector is not unique.

5.5.3 Differentiation of Matrix

More generally, the elements of a matrix can be a function of the variable:

$$A(t) = \begin{pmatrix} a_{11}(t) & a_{12}(t) & \cdots & a_{1n}(t) \\ a_{21}(t) & a_{22}(t) & \cdots & a_{2n}(t) \\ \vdots & \vdots & \ddots & \vdots \\ a_{m1}(t) & a_{m2}(t) & \cdots & a_{mn}(t) \end{pmatrix} \tag{5.114}$$

Therefore, sometimes it is desirable to consider the calculus on matrices or vectors. Some essential formulas are reported here:

$$\frac{d\mathbf{A}}{dt} = \left(\frac{da_{ij}}{dt}\right), \quad \int_a^b \mathbf{A}(t)dt = \left(\int_a^b a_{ij}(t)dt\right) \tag{5.115}$$

For a constant matrix $\mathbf{C}$ and a constant c, we have this differentiation rule

$$\frac{d(\mathbf{CA})}{dt} = \mathbf{C}\frac{d\mathbf{A}}{dt} \tag{5.116}$$

$$\frac{d(c\mathbf{A})}{dt} = c\frac{d\mathbf{A}}{dt} \tag{5.117}$$

Similarly, we have the following distributive rules of differentiation:

$$\frac{d(\mathbf{A}+\mathbf{B})}{dt} = \frac{d\mathbf{A}}{dt} + \frac{d\mathbf{B}}{dt} \tag{5.118}$$

$$\frac{d(\mathbf{AB})}{dt} = \left(\frac{d\mathbf{A}}{dt}\right)\mathbf{B} + \mathbf{A}\left(\frac{d\mathbf{B}}{dt}\right) \tag{5.119}$$

5.6 HOMOGENEOUS SYSTEM WITH CONSTANT COEFFICIENTS

A system of n linear first order ODEs can be expressed as:

$$\begin{aligned}
x_1' &= a_{11}(t)x_1 + a_{12}(t)x_2 + \ldots + a_{1n}(t)x_n + f_1(t) \\
x_2' &= a_{21}(t)x_1 + a_{22}(t)x_2 + \ldots + a_{2n}(t)x_n + f_2(t) \\
&\vdots \\
x_n' &= a_{n1}(t)x_1 + a_{n2}(t)x_2 + \ldots + a_{nn}(t)x_n + f_n(t)
\end{aligned} \tag{5.120}$$

where f is the nonhomogeneous term. This system can be expressed in matrix form as

$$\frac{d\mathbf{X}}{dt} = \mathbf{AX} + \mathbf{F} \quad \text{or } \mathbf{X}' = \mathbf{AX} + \mathbf{F} \tag{5.121}$$

where

$$\mathbf{X} = \begin{bmatrix} x_1(t) \\ x_2(t) \\ \vdots \\ \vdots \\ x_n(t) \end{bmatrix}, \quad \mathbf{A} = \begin{bmatrix} a_{11}(t) & a_{12}(t) & \cdots & \cdots & a_{1n}(t) \\ a_{21}(t) & a_{22}(t) & \cdots & \cdots & a_{2n}(t) \\ \vdots & \vdots & \ddots & & \vdots \\ \vdots & \vdots & & \ddots & \vdots \\ a_{n1}(t) & a_{n2}(t) & \cdots & \cdots & a_{nn}(t) \end{bmatrix}, \quad \mathbf{F} = \begin{bmatrix} f_1(t) \\ f_2(t) \\ \vdots \\ \vdots \\ f_n(t) \end{bmatrix} \tag{5.122}$$

If $\boldsymbol{F} = 0$, the system is homogeneous. If matrix $\boldsymbol{A} \neq \boldsymbol{A}(t)$, the system is called autonomous. The initial condition of this system can be given as

$$\mathbf{X}(t_0) = \begin{bmatrix} x_1(t_0) \\ x_2(t_0) \\ \vdots \\ \vdots \\ x_n(t_0) \end{bmatrix} = \mathbf{X}_0 = \begin{bmatrix} r_1 \\ r_2 \\ \vdots \\ \vdots \\ r_n \end{bmatrix} \tag{5.123}$$

If $\boldsymbol{A}$ and $\boldsymbol{F}$ are continuous in an interval containing point t_0, the initial value problem has a unique solution in this interval. For a $n \times n$ matrix $\boldsymbol{A}$ and $\boldsymbol{F} = 0$, there exists n independent solution vectors $\boldsymbol{X}_i$ ($i = 1,2,\ldots,n$) and the general solution for (5.121) with $\boldsymbol{F} = 0$ is given by

$$\mathbf{X} = c_1\mathbf{X}_1 + c_2\mathbf{X}_2 + \cdots\cdots + c_n\mathbf{X}_n \tag{5.124}$$

where c_1, ..., c_n are arbitrary constants. The specific form of $\boldsymbol{X}_i$ will be discussed later. Note that there are n unknown constants in the general solution. This set of independent solution vectors forms a fundamental set of solutions of the system.

Similar to our discussion on ODEs, we can use the Wronskian to check their independence:

$$W(\mathbf{X}_1, \mathbf{X}_2, \cdots\cdots \mathbf{X}_n) = \det \begin{bmatrix} x_{11}(t) & x_{12}(t) & \cdots & \cdots & x_{1n}(t) \\ x_{21}(t) & x_{22}(t) & \cdots & \cdots & x_{2n}(t) \\ \vdots & \vdots & \ddots & & \vdots \\ \vdots & \vdots & & \ddots & \vdots \\ x_{n1}(t) & x_{n2}(t) & \cdots & \cdots & x_{nn}(t) \end{bmatrix} \tag{5.125}$$

where

$$\mathbf{X_1} = \begin{bmatrix} x_{11}(t) \\ x_{21}(t) \\ \vdots \\ \vdots \\ x_{n1}(t) \end{bmatrix} \quad \mathbf{X_2} = \begin{bmatrix} x_{12}(t) \\ x_{22}(t) \\ \vdots \\ \vdots \\ x_{n2}(t) \end{bmatrix} \quad \cdots\cdots \quad \mathbf{X_n} = \begin{bmatrix} x_{1n}(t) \\ x_{2n}(t) \\ \vdots \\ \vdots \\ x_{nn}(t) \end{bmatrix} \tag{5.126}$$

If these solutions are independent, the Wronskian must be nonzero, or $W(\mathbf{X}_1, \mathbf{X}_2, \ldots, \mathbf{X}_n) \neq 0$.

Example 5.9 Check whether the following system

$$\frac{dx}{dt} = x + 3y$$
$$\frac{dy}{dt} = 5x + 3y \tag{5.127}$$

has the following two independent solutions

$$\mathbf{X_1} = \begin{bmatrix} e^{-2t} \\ -e^{-2t} \end{bmatrix} = \begin{bmatrix} 1 \\ -1 \end{bmatrix} e^{-2t}, \quad \mathbf{X_2} = \begin{bmatrix} 3e^{6t} \\ 5e^{6t} \end{bmatrix} = \begin{bmatrix} 3 \\ 5 \end{bmatrix} e^{6t} \tag{5.128}$$

Solution: First , we rewrite the system in matrix form as

$$X' = AX = \begin{bmatrix} 1 & 3 \\ 5 & 3 \end{bmatrix} X, \quad \mathbf{X} = \begin{bmatrix} x \\ y \end{bmatrix} \tag{5.129}$$

Differentiation of the first solution in (5.128) gives

$$\mathbf{X}_1' = \begin{bmatrix} -2e^{-2t} \\ 2e^{-2t} \end{bmatrix} \tag{5.130}$$

On the other hand, we have

$$\mathbf{AX_1} = \begin{bmatrix} 1 & 3 \\ 5 & 3 \end{bmatrix} \begin{bmatrix} e^{-2t} \\ -e^{-2t} \end{bmatrix} = \begin{bmatrix} -2e^{-2t} \\ 2e^{-2t} \end{bmatrix} \tag{5.131}$$

Comparison of (5.130) and (5.131) shows that the first vector given in (5.128) is indeed a solution of the system.

Similarly, the differentiation of the second vector in (5.128) is

$$\mathbf{X}_2' = \begin{bmatrix} 18e^{6t} \\ 30e^{6t} \end{bmatrix} \tag{5.132}$$

The right hand side of the ODE system is

$$\mathbf{AX_2} = \begin{bmatrix} 1 & 3 \\ 5 & 3 \end{bmatrix} \begin{bmatrix} 3e^{6t} \\ 5e^{6t} \end{bmatrix} = \begin{bmatrix} 18e^{6t} \\ 30e^{6t} \end{bmatrix} \tag{5.133}$$

This again shows the validity of the second solution vector in (5.128). To check the independence, we can form the Wronskian as

$$W(\mathbf{X}_1, \mathbf{X}_2) = \begin{vmatrix} e^{-2t} & 3e^{6t} \\ -e^{-2t} & 5e^{6t} \end{vmatrix} = \begin{vmatrix} 1 & 3 \\ -1 & 5 \end{vmatrix} e^{-2t+6t} = 8e^{4t} \neq 0 \tag{5.134}$$

Therefore, the independence of these solutions is demonstrated. Thus, they do form the fundamental solution set.

Now let us consider the special case of an ODE system with a constant coefficient and no nonhomogeneous term $\boldsymbol{F}$ in (5.121). We seek a solution of exponential form

$$\mathbf{X} = \begin{bmatrix} k_1 e^{\lambda t} \\ k_2 e^{\lambda t} \\ \vdots \\ \vdots \\ k_n e^{\lambda t} \end{bmatrix} = \begin{bmatrix} k_1 \\ k_2 \\ \vdots \\ \vdots \\ k_n \end{bmatrix} e^{\lambda t} = \mathbf{K} e^{\lambda t} \tag{5.135}$$

Differentiation of (5.135) results in

$$\mathbf{X}' = \begin{bmatrix} k_1 \lambda e^{\lambda t} \\ k_2 \lambda e^{\lambda t} \\ \vdots \\ \vdots \\ k_n \lambda e^{\lambda t} \end{bmatrix} = \mathbf{K} \lambda e^{\lambda t} \tag{5.136}$$

Substitution of (5.135) and (5.136) into (5.121) gives

$$\mathbf{K} \lambda e^{\lambda t} = \mathbf{A} \mathbf{K} e^{\lambda t} \tag{5.137}$$

Since the solution should be true for all values of variable t, we can cancel the exponential function on both sides of (5.137). Putting all terms on the left, we obtain an eigenvalue problem as:

$$(\mathbf{A} - \lambda \mathbf{I})\mathbf{K} = 0 \tag{5.138}$$

Since $\boldsymbol{K}$ is not zero, we must require

$$\det(\mathbf{A} - \lambda \mathbf{I}) = 0 \tag{5.139}$$

Thus, solving a system of ODEs reduces to solving the matrix eigenvalue problem. The general solution form depends on what kind of roots we get for the eigenvalue. We will consider all scenarios one by one.

5.6.1 Case 1: Distinct Eigenvalues

If all eigenvalues are real and distinct, the general solution is

$$\mathbf{X} = c_1 \mathbf{K}_1 e^{\lambda_1 t} + c_2 \mathbf{K}_2 e^{\lambda_2 t} + c_3 \mathbf{K}_3 e^{\lambda_3 t} + \cdots\cdots + c_n \mathbf{K}_n e^{\lambda_n t} \tag{5.140}$$

where vector $\boldsymbol{K}$ is the corresponding eigenvector for the eigenvalue λ. This is the simplest scenario.

5.6.2 Case 2: Repeated Eigenvalues

If one of the eigenvalue λ_m has a multiplicity of m whilst all other eigenvalues are distinct and real, the solution can be expressed in two different forms depending on whether A is Hermitian.

Case 2.1: If A is Hermitian, we would be able to find m independent eigenvectors for the same eigenvalue. The general solution is

$$\mathbf{X} = c_1\mathbf{K}_1 e^{\lambda_m t} + ... + c_m \mathbf{K}_m e^{\lambda_m t} + c_{m+1}\mathbf{K}_{m+1} e^{\lambda_{m+1} t} + \cdots + c_n \mathbf{K}_n e^{\lambda_n t} \tag{5.141}$$

where vectors $\boldsymbol{K}_i$ (i = 1,2,...,m) are the independent eigenvectors corresponding to repeated eigenvalue λ_m.

Case 2.2: If A is not Hermitian, the general solution is

$$\mathbf{X} = c_1\mathbf{X}_{\mathbf{a},1} + c_2\mathbf{X}_{\mathbf{a},2} + ... + c_m\mathbf{X}_{\mathbf{a},m} + c_{m+1}\mathbf{K}_{m+1} e^{\lambda_{m+1} t} + \cdots\cdots + c_n\mathbf{K}_n e^{\lambda_n t} \tag{5.142}$$

where solution vectors for the repeated eigenvalues are

$$\begin{aligned} \mathbf{X}_{\mathbf{a},1} &= \mathbf{K}_{\mathbf{a},1} e^{\lambda_a t} \\ \mathbf{X}_{\mathbf{a},2} &= \mathbf{K}_{\mathbf{a},1} t\, e^{\lambda_a t} + \mathbf{K}_{\mathbf{a},2}\, e^{\lambda_a t} \\ &\;\;\vdots \\ \mathbf{X}_{\mathbf{a,m}} &= \mathbf{K}_{\mathbf{a},1} \frac{t^{m-1}}{(m-1)!} e^{\lambda_a t} + \mathbf{K}_{\mathbf{a},2} \frac{t^{m-2}}{(m-2)!} e^{\lambda_a t} + \cdots\cdots + \mathbf{K}_{\mathbf{a,m}}\, e^{\lambda_a t} \end{aligned} \tag{5.143}$$

The vectors $\boldsymbol{K}$ are calculated by

$$\begin{aligned} (\mathbf{A} - \lambda_a \mathbf{I})\mathbf{K}_{\mathbf{a},1} &= 0 \\ (\mathbf{A} - \lambda_a \mathbf{I})\mathbf{K}_{\mathbf{a},2} &= \mathbf{K}_{\mathbf{a},1} \\ &\;\;\vdots \\ (\mathbf{A} - \lambda_a \mathbf{I})\mathbf{K}_{\mathbf{a,m}} &= \mathbf{K}_{\mathbf{a,m-1}} \end{aligned} \tag{5.144}$$

If we have two or more repeated eigenvalues for the matrix A, we have to add more solutions for the repeated eigenvalues similar to that given in (5.142).

Proof: The proof of this formula will be sketched here. For any p =1,2,...,m, we have the solution as

$$\mathbf{X}_{\mathbf{a},p} = \mathbf{K}_{\mathbf{a},1} \frac{t^{p-1}}{(p-1)!} e^{\lambda_a t} + \mathbf{K}_{\mathbf{a},2} \frac{t^{p-2}}{(p-2)!} e^{\lambda_a t} + \cdots\cdots + \mathbf{K}_{\mathbf{a},p}\, e^{\lambda_a t} \tag{5.145}$$

Differentiation of (5.145) gives

$$\begin{aligned} \mathbf{X}'_{\mathbf{a,p}} &= \lambda_a \mathbf{K}_{\mathbf{a},1} \frac{t^{p-1}}{(p-1)!} e^{\lambda_a t} + (\mathbf{K}_{\mathbf{a},1} + \lambda_a \mathbf{K}_{\mathbf{a},2}) \frac{t^{p-2}}{(p-2)!} e^{\lambda_a t} \\ &+ (\mathbf{K}_{\mathbf{a},2} + \lambda_a \mathbf{K}_{\mathbf{a},3}) \frac{t^{p-3}}{(p-3)!}\, e^{\lambda_a t} \\ &+ \cdots\cdots + (\mathbf{K}_{\mathbf{a,p\text{-}2}} + \lambda_a \mathbf{K}_{\mathbf{a,p\text{-}1}}) \frac{t^1}{1!} e^{\lambda_a t} + (\mathbf{K}_{\mathbf{a,p\text{-}1}} + \lambda_a \mathbf{K}_{\mathbf{a,p}}) e^{\lambda_a t} \end{aligned} \tag{5.146}$$

Substitution of (5.145) and (5.146) into the ODE system gives

$$\mathbf{A}\mathbf{X}_{a,p}-\mathbf{X}'_{a,p}$$
$$=(\mathbf{A}-\lambda_a \boldsymbol{I})\mathbf{K}_{\mathbf{a,1}}\frac{t^{p-1}}{(p-1)!}e^{\lambda_a t}+[(\mathbf{A}-\lambda_a \boldsymbol{I})\mathbf{K}_{\mathbf{a,2}}-\mathbf{K}_{\mathbf{a,1}}]\frac{t^{p-2}}{(p-2)!}e^{\lambda_a t}+...=0 \quad (5.147)$$

The function of t is in general nonzero, and therefore, their coefficients must be zero:

$$\begin{aligned}(\mathbf{A}-\lambda_a\mathbf{I})\mathbf{K}_{\mathbf{a,1}}&=0\\ (\mathbf{A}-\lambda_a\mathbf{I})\mathbf{K}_{\mathbf{a,2}}&=\mathbf{K}_{\mathbf{a,1}}\\ &\dots\\ (\mathbf{A}-\lambda_a\mathbf{I})\mathbf{K}_{\mathbf{a,m}}&=\mathbf{K}_{\mathbf{a,m-1}}\end{aligned} \quad (5.148)$$

which is (5.144). This completes the proof.

5.6.3 Case 3: Complex Eigenvalues

If there is a pair of complex eigenvalues, the two complex eigenvalues must be in conjugate form. Let us assume they are

$$\lambda_1=\alpha+i\beta, \quad \lambda_2=\alpha-i\beta \quad (5.149)$$

And their corresponding eigenvalues are

$$\boldsymbol{K}_1=\boldsymbol{B}_1+i\boldsymbol{B}_2, \quad \boldsymbol{K}_2=\boldsymbol{B}_1-i\boldsymbol{B}_2 \quad (5.150)$$

The general solution is

$$\mathbf{X}=c_1\mathbf{X}_{\mathbf{a}}+c_2\mathbf{X}_b+c_3\mathbf{K}_3 e^{\lambda_3 t}+...+c_n\mathbf{K}_{\mathbf{n}}e^{\lambda_n t} \quad (5.151)$$

where

$$\begin{aligned}\mathbf{X}_{\mathbf{a}}&=[\mathbf{B}_1\cos\beta t-\mathbf{B}_2\sin\beta t]e^{\alpha t},\\ \mathbf{X}_{\mathbf{b}}&=[\mathbf{B}_2\cos\beta t+\mathbf{B}_1\sin\beta t]e^{\alpha t}\end{aligned} \quad (5.152)$$

If there is more than one pair of complex conjugate eigenvalues, we can add another pair of eigenvectors accordingly.

Proof: Let a pair of complex eigenvalues of the real matrix $\boldsymbol{A}$ be $\lambda_1=\alpha+i\beta$ and $\lambda_2=\alpha-i\beta$, and further assume that the eigenvector for λ_1 is:

$$\boldsymbol{K}_1=\boldsymbol{B}_1+i\boldsymbol{B}_2 \quad (5.153)$$

By definition, we must have $\boldsymbol{K}_1$ satisfy

$$\mathbf{A}\boldsymbol{K}_1=\lambda_1\boldsymbol{K}_1 \quad (5.154)$$

Taking the complex conjugate of (5.154), we obtain

$$\bar{\mathbf{A}}\bar{\boldsymbol{K}}_1=\bar{\lambda}_1\bar{\boldsymbol{K}}_1 \quad (5.155)$$

However, $\boldsymbol{A}$ is real and the complex conjugate of λ_1 is λ_2 and (5.155) becomes

$$\mathbf{A}\bar{\boldsymbol{K}}_1=\lambda_2\bar{\boldsymbol{K}}_1 \quad (5.156)$$

By definition again, we must have the eigenvector of λ_2 satisfy

$$\mathbf{A}\boldsymbol{K}_2=\lambda_2\boldsymbol{K}_2 \quad (5.157)$$

Comparison of (5.156) and (5.157) shows that

$$K_2 = \bar{K}_1 = B_1 - iB_2 \tag{5.158}$$

Thus, the general solution due to the complex conjugate pair is:

$$\begin{aligned} X &= c_a(\mathbf{B_1} + i\mathbf{B_2})e^{(\alpha+i\beta)t} + c_b(\mathbf{B_1} - i\mathbf{B_2})e^{(\alpha-i\beta)t} \\ &= c_a e^{\alpha t}(\mathbf{B_1} + i\mathbf{B_2})(\cos\beta t + i\sin\beta t) + c_b e^{\alpha t}(\mathbf{B_1} - i\mathbf{B_2})(\cos\beta t - i\sin\beta t) \\ &= (c_a + c_b)e^{\alpha t}(\mathbf{B_1}\cos\beta t - \mathbf{B_2}\sin\beta t) + i(c_a - c_b)e^{\alpha t}(\mathbf{B_1}\sin\beta t + \mathbf{B_2}\cos\beta t) \end{aligned} \tag{5.159}$$

Since we start with a real ODE system, we should expect the solution to be real. This can be done easily by choosing c_b as a complex conjugate of c_a. Thus, we can express the solution as

$$X = c_1 e^{\alpha t}(\mathbf{B_1}\cos\beta t - \mathbf{B_2}\sin\beta t) + c_2 e^{\alpha t}(\mathbf{B_1}\sin\beta t + \mathbf{B_2}\cos\beta t) \tag{5.160}$$

where c_1 and c_2 are real constants. This completes the proof of (1.151).

Example 5.10 Find the general solution of the following system of ODEs

$$\begin{aligned} \frac{dx}{dt} &= 2x + 3y \\ \frac{dy}{dt} &= 2x + y \end{aligned} \tag{5.161}$$

Solution: The matrix form of the system is

$$X' = AX = \begin{bmatrix} 2 & 3 \\ 2 & 1 \end{bmatrix} X, \quad \mathbf{X} = \begin{bmatrix} x \\ y \end{bmatrix} \tag{5.162}$$

The eigenvalue of the system is

$$\det(\mathbf{A} - \lambda\mathbf{I}) = \begin{vmatrix} 2-\lambda & 3 \\ 2 & 1-\lambda \end{vmatrix} = \lambda^2 - 3\lambda - 4 = (\lambda+1)(\lambda-4) = 0 \tag{5.163}$$

The eigenvalues are $\lambda = -1$ and 4. We have two distinct real roots for the eigenvalues, and this corresponds to case 1 discussed in Section 5.6.1. For $\lambda = -1$, the eigenvector can be calculated as

$$(\mathbf{A} - \lambda\mathbf{I})\mathbf{K_1} = \begin{bmatrix} 3 & 3 \\ 2 & 2 \end{bmatrix}\begin{bmatrix} k_1 \\ k_2 \end{bmatrix} = 0 \tag{5.164}$$

Both of these equations lead to

$$k_1 + k_2 = 0 \tag{5.165}$$

By choosing $k_1 = 1$, we have $k_2 = -1$ and the eigenvector becomes

$$\mathbf{K_1} = \begin{bmatrix} 1 \\ -1 \end{bmatrix} \tag{5.166}$$

Note that (5.166) is clearly not the only choice. In fact, any combination of k_1 and k_2 that satisfies (5.165) will give an appropriate eigenvector. The corresponding solution vector is

$$\mathbf{X_1} = \mathbf{K_1}e^{-t} = \begin{bmatrix} 1 \\ -1 \end{bmatrix} e^{-t} \tag{5.167}$$

For $\lambda = 4$, the eigenvector can be determined from the following system

$$(\mathbf{A}-\lambda\mathbf{I})\mathbf{K}_2 = \begin{bmatrix} -2 & 3 \\ 2 & -3 \end{bmatrix}\begin{bmatrix} k_1 \\ k_2 \end{bmatrix} = 0 \tag{5.168}$$

This provides a relation between the two components of the eigenvector:

$$2k_1 - 3k_2 = 0 \tag{5.169}$$

Choosing $k_1 = 3$, we have $k_2 = 2$ and obtain the eigenvector as

$$\mathbf{K}_2 = \begin{bmatrix} 3 \\ 2 \end{bmatrix} \tag{5.170}$$

The solution vector is

$$\mathbf{X}_2 = \mathbf{K}_2 e^{4t} = \begin{bmatrix} 3 \\ 2 \end{bmatrix} e^{4t} \tag{5.171}$$

Finally, adding these solutions with two unknown constants gives

$$\mathbf{X} = c_1\mathbf{X}_1 + c_2\mathbf{X}_2 = c_1\begin{bmatrix} 1 \\ -1 \end{bmatrix} e^{-t} + c_2\begin{bmatrix} 3 \\ 2 \end{bmatrix} e^{4t} \tag{5.172}$$

In a later section, we will discuss the behavior of this solution as $t \to \infty$.

Example 5.11 Find the general solution of the following system of ODEs

$$\begin{aligned} \frac{dx}{dt} &= -4x + y + z \\ \frac{dy}{dt} &= x + 5y - z \\ \frac{dz}{dt} &= \quad y - 3z \end{aligned} \tag{5.173}$$

Solution: Putting the system in matrix form yields

$$X' = AX = \begin{bmatrix} -4 & 1 & 1 \\ 1 & 5 & -1 \\ 0 & 1 & -3 \end{bmatrix} X, \quad \mathbf{X} = \begin{bmatrix} x \\ y \\ z \end{bmatrix} \tag{5.174}$$

The eigenvalue of the system can be determined from

$$\det(\mathbf{A}-\lambda\mathbf{I}) = -(\lambda+3)(\lambda+4)(\lambda-5) = 0 \tag{5.175}$$

The eigenvalues are $\lambda = -3, -4$, and 5. We have three distinct real roots for the eigenvalues, and this again corresponds to case 1 of Section 5.6.1. For $\lambda = -3$, the eigenvector can be calculated as

$$(\mathbf{A}-\lambda\mathbf{I})\mathbf{K}_1 = \begin{bmatrix} -1 & 1 & 1 \\ 1 & 8 & -1 \\ 0 & 1 & 0 \end{bmatrix}\begin{bmatrix} k_1 \\ k_2 \\ k_3 \end{bmatrix} = \begin{bmatrix} 0 \\ 0 \\ 0 \end{bmatrix} \tag{5.176}$$

The third row of the system gives $k_2 = 0$. The first row gives

$$-k_1 + k_3 = 0 \tag{5.177}$$

Therefore, we can set the eigenvector as

$$\mathbf{K}_1 = \begin{bmatrix} 1 \\ 0 \\ 1 \end{bmatrix} \tag{5.178}$$

For $\lambda = -4$, the eigenvector can be calculated as

$$(\mathbf{A} - \lambda\mathbf{I})\mathbf{K}_2 = \begin{bmatrix} 0 & 1 & 1 \\ 1 & 9 & -1 \\ 0 & 1 & 1 \end{bmatrix} \begin{bmatrix} k_1 \\ k_2 \\ k_3 \end{bmatrix} = \begin{bmatrix} 0 \\ 0 \\ 0 \end{bmatrix} \tag{5.179}$$

The first and third rows of the system give

$$k_2 + k_3 = 0 \tag{5.180}$$

The second gives

$$k_1 + 9k_2 - k_3 = 0 \tag{5.181}$$

Setting $k_3 = 1$, we have $k_2 = -1$. Substituting these values into (5.181) gives $k_1 = 10$ and the eigenvector becomes

$$\mathbf{K}_2 = \begin{bmatrix} 10 \\ -1 \\ 1 \end{bmatrix} \tag{5.182}$$

For $\lambda = 5$, the eigenvector can be calculated as

$$(\mathbf{A} - \lambda\mathbf{I})\mathbf{K}_3 = \begin{bmatrix} -9 & 1 & 1 \\ 1 & 0 & -1 \\ 0 & 1 & -8 \end{bmatrix} \begin{bmatrix} k_1 \\ k_2 \\ k_3 \end{bmatrix} = \begin{bmatrix} 0 \\ 0 \\ 0 \end{bmatrix} \tag{5.183}$$

The second row of this system gives

$$k_1 - k_3 = 0 \tag{5.184}$$

The third row of this system gives

$$k_2 - 8k_3 = 0 \tag{5.185}$$

Taking $k_1 = k_3 = 1$ into (5.185) gives $k_2 = 8$, and thus yields

$$\mathbf{K}_3 = \begin{bmatrix} 1 \\ 8 \\ 1 \end{bmatrix} \tag{5.186}$$

Combining all these eigenvectors, we finally obtain the general solution as

$$X = c_1 \begin{bmatrix} 1 \\ 0 \\ 1 \end{bmatrix} e^{-3t} + c_2 \begin{bmatrix} 10 \\ -1 \\ 1 \end{bmatrix} e^{-4t} + c_3 \begin{bmatrix} 1 \\ 8 \\ 1 \end{bmatrix} e^{5t} \tag{5.187}$$

Example 5.12 Find the general solution of the following system of ODEs

$$X' = AX = \begin{bmatrix} 1 & -2 & 2 \\ -2 & 1 & -2 \\ 2 & -2 & 1 \end{bmatrix} X \tag{5.188}$$

Solution: Solving the eigenvalue problem gives

$$\det(\mathbf{A} - \lambda \mathbf{I}) = -(\lambda + 1)^2 (\lambda - 5) \tag{5.189}$$

The eigenvalues are $\lambda = -1$, -1 and 5. Since A is Hermitian (or symmetric), this is Case 2.1 with a repeated eigenvalue. We should be able to obtain two independent eigenvectors for the repeated root for $\lambda = -1$.

For $\lambda = -1$, the eigenvector can be calculated as

$$(\mathbf{A} - \lambda \mathbf{I}) \begin{bmatrix} k_1 \\ k_2 \\ k_3 \end{bmatrix} = \begin{bmatrix} 2 & -2 & 2 \\ -2 & 2 & -2 \\ 2 & -2 & 2 \end{bmatrix} \begin{bmatrix} k_1 \\ k_2 \\ k_3 \end{bmatrix} = \begin{bmatrix} 0 \\ 0 \\ 0 \end{bmatrix} \tag{5.190}$$

This system implies

$$k_1 - k_2 + k_3 = 0 \tag{5.191}$$

There are three unknowns in (5.191) and two different solutions can be found from it. In particular, for the first solution, we set $k_1 = 0$ and $k_2 = 1$, then (5.191) gives $k_3 = 1$. Alternatively, we set $k_1 = 1$ and $k_2 = 0$, then (5.191) gives $k_3 = -1$. Thus, we have

$$\boldsymbol{K}_1 = \begin{bmatrix} 0 \\ 1 \\ 1 \end{bmatrix}, \quad \boldsymbol{K}_2 = \begin{bmatrix} 1 \\ 0 \\ -1 \end{bmatrix} \tag{5.192}$$

To check the independence of these two vectors, we require

$$c_1 \boldsymbol{K}_1 + c_2 \boldsymbol{K}_2 = \mathbf{0} \tag{5.193}$$

be satisfied only for $c_1 = c_2 = 0$. Substitution of (5.192) into (5.193) gives

$$c_1 \begin{bmatrix} 0 \\ 1 \\ 1 \end{bmatrix} + c_2 \begin{bmatrix} 1 \\ 0 \\ -1 \end{bmatrix} = \begin{bmatrix} 0 \\ 0 \\ 0 \end{bmatrix} \tag{5.194}$$

The first equation of it implies $c_2 = 0$, the second implies $c_1 = 0$, and the third requires $c_2 = c_1$. Thus, our random choice of picking two eigenvectors by (5.191) does arrive at the independent eigenvectors for the same repeated eigenvalue. In fact, setting $k_1 = 0$ and $k_2 = 1$ and vice versa to find the independent eigenvectors is one of the best choices. There are many different forms of these independent vectors $\boldsymbol{K}$ and the solutions may appear differently, depending on the choices for k_1 and k_2. However, one can show that when initial conditions are imposed, all of them will lead to the same solution (e.g., see Problem 5.15). The forms obtained in (5.192) are among the most efficient forms in satisfying the initial conditions.

For $\lambda = 5$, the eigenvector can be calculated as

$$(\mathbf{A} - \lambda \mathbf{I}) \begin{bmatrix} k_1 \\ k_2 \\ k_3 \end{bmatrix} = \begin{bmatrix} -4 & -2 & 2 \\ -2 & -4 & -2 \\ 2 & -2 & -4 \end{bmatrix} \begin{bmatrix} k_1 \\ k_2 \\ k_3 \end{bmatrix} = \begin{bmatrix} 0 \\ 0 \\ 0 \end{bmatrix} \tag{5.195}$$

The first row adding –2 multiplying the second row gives

$$k_2 + k_3 = 0 \tag{5.196}$$

The first row adding 2 multiplying the third row arrives at the same equation. We can set $k_3 = 1$ and $k_2 = -1$. The first row finally yields $k_1 = 1$ and, thus, the eigenvector becomes

$$\boldsymbol{K} = \begin{bmatrix} 1 \\ -1 \\ 1 \end{bmatrix} \tag{5.197}$$

The general solution becomes

$$\mathbf{X} = c_1 \begin{bmatrix} 0 \\ 1 \\ 1 \end{bmatrix} e^{-t} + c_2 \begin{bmatrix} 1 \\ 0 \\ -1 \end{bmatrix} e^{-t} + c_3 \begin{bmatrix} 1 \\ -1 \\ 1 \end{bmatrix} e^{5t} \tag{5.198}$$

Example 5.13 Find the general solution of the following system of ODEs

$$\boldsymbol{X}' = \boldsymbol{A}\boldsymbol{X} = \begin{bmatrix} 2 & 1 & 6 \\ 0 & 2 & 5 \\ 0 & 0 & 2 \end{bmatrix} \boldsymbol{X} \tag{5.199}$$

Solution: Solving the eigenvalue problem gives

$$\det(\mathbf{A} - \lambda \mathbf{I}) = (2-\lambda)^3 \tag{5.200}$$

The eigenvalues are $\lambda = 2$, 2, and 2. Since $\boldsymbol{A}$ is non-Hermitian (or unsymmetrical), this is Case 2.2 with an eigenvalue with multiplicity of m being 3. Let us illustrate how to find the eigenvectors.

With $\lambda = 2$, the matrix equation becomes

$$(\mathbf{A} - 2\mathbf{I}) \begin{bmatrix} k_1 \\ k_2 \\ k_3 \end{bmatrix} = \begin{bmatrix} 0 & 1 & 6 \\ 0 & 0 & 5 \\ 0 & 0 & 0 \end{bmatrix} \begin{bmatrix} k_1 \\ k_2 \\ k_3 \end{bmatrix} = \begin{bmatrix} 0 \\ 0 \\ 0 \end{bmatrix} \tag{5.201}$$

This system implies

$$5k_3 = 0, \quad k_2 + 6k_3 = 0 \tag{5.202}$$

This yields both $k_2 = k_3 = 0$, and subsequently we have

$$\boldsymbol{K}_{a,1} = \begin{bmatrix} 1 \\ 0 \\ 0 \end{bmatrix} \tag{5.203}$$

Recall from the second equation of (5.144) that

$$(\mathbf{A} - 2\mathbf{I})\mathbf{K}_{\mathbf{a,2}} = \mathbf{K}_{\mathbf{a,1}} \tag{5.204}$$

Substitution of (5.203) into (5.204) results in

$$\begin{bmatrix} 0 & 1 & 6 \\ 0 & 0 & 5 \\ 0 & 0 & 0 \end{bmatrix} \begin{bmatrix} k_1 \\ k_2 \\ k_3 \end{bmatrix} = \begin{bmatrix} 1 \\ 0 \\ 0 \end{bmatrix} \tag{5.205}$$

This system gives

$$5k_3 = 0, \quad k_2 + 6k_3 = 1, \quad k_1 = 0 \tag{5.206}$$

The corresponding vector is

$$\mathbf{K}_{\mathbf{a},2} = \begin{bmatrix} 0 \\ 1 \\ 0 \end{bmatrix} \tag{5.207}$$

Similar to (5.204) we can continue the calculation using

$$(\mathbf{A} - 2\mathbf{I})\mathbf{K}_{\mathbf{a},3} = \mathbf{K}_{\mathbf{a},2} \tag{5.208}$$

Using (5.207) in (5.208), we arrive at

$$\begin{bmatrix} 0 & 1 & 6 \\ 0 & 0 & 5 \\ 0 & 0 & 0 \end{bmatrix} \begin{bmatrix} k_1 \\ k_2 \\ k_3 \end{bmatrix} = \begin{bmatrix} 0 \\ 1 \\ 0 \end{bmatrix} \tag{5.209}$$

This is equivalent to

$$5k_3 = 1, \quad k_2 + 6k_3 = 0 \tag{5.210}$$

With these constraints, we can choose

$$\mathbf{K}_{\mathbf{a},3} = \begin{bmatrix} 0 \\ -6/5 \\ 1/5 \end{bmatrix} \tag{5.211}$$

Recall (5.142) that the general solution can be expressed as:

$$\mathbf{X} = c_1 \mathbf{K}_{\mathbf{a},1} e^{2t} + c_2 (\mathbf{K}_{\mathbf{a},1} t e^{2t} + \mathbf{K}_{\mathbf{a},2} e^{2t}) + c_3 (\mathbf{K}_{\mathbf{a},1} \frac{t^2}{2} e^{2t} + \mathbf{K}_{\mathbf{a},2} t e^{2t} + \mathbf{K}_{\mathbf{a},3} e^{2t}) \tag{5.212}$$

The general solution is therefore:

$$\mathbf{X} = c_1 \begin{bmatrix} 1 \\ 0 \\ 0 \end{bmatrix} e^{2t} + c_2 \left\{ \begin{bmatrix} 1 \\ 0 \\ 0 \end{bmatrix} t e^{2t} + \begin{bmatrix} 0 \\ 1 \\ 0 \end{bmatrix} e^{2t} \right\} + c_3 \left\{ \begin{bmatrix} 1 \\ 0 \\ 0 \end{bmatrix} \frac{t^2}{2} e^{2t} + \begin{bmatrix} 0 \\ 1 \\ 0 \end{bmatrix} t e^{2t} + \begin{bmatrix} 0 \\ -6/5 \\ 1/5 \end{bmatrix} e^{2t} \right\} \tag{5.213}$$

Note that the matrix A is unsymmetrical and thus we could not find 3 linearly independent eigenvectors.

Example 5.14 Find the general solution of the following system of ODEs

$$X' = AX = \begin{bmatrix} 2 & 8 \\ -1 & -2 \end{bmatrix} X \tag{5.214}$$

Solution: Solving the eigenvalue problem gives

$$\det(\mathbf{A}-\lambda\mathbf{I}) = \begin{vmatrix} 2-\lambda & 8 \\ -1 & -2-\lambda \end{vmatrix} = \lambda^2 + 4 = 0 \tag{5.215}$$

The eigenvalues are $\lambda = 2i$, and $-2i$. For $\lambda = 2i$, the eigenvector can be evaluated as:

$$(\mathbf{A}-\lambda\mathbf{I})\mathbf{K} = \begin{bmatrix} 2-2i & 8 \\ -1 & -2-2i \end{bmatrix} \begin{bmatrix} k_1 \\ k_2 \end{bmatrix} = \begin{bmatrix} 0 \\ 0 \end{bmatrix} \tag{5.216}$$

The second equation of (5.216) requires

$$-k_1 - 2(1+i)k_2 = 0 \tag{5.217}$$

Let $k_2 = -1$; we have $k_1 = 2+2i$. Thus, the corresponding eigenvector is

$$\mathbf{K}_1 = \begin{bmatrix} 2+2i \\ -1 \end{bmatrix} \tag{5.218}$$

Actually, it is straightforward to show that (5.217) is equivalent to the first equation of (5.216). To see this, we can multiply (5.217) by $-2\,(1-i)$ to get

$$2(1-i)k_1 + 4(1-i)(1+i)k_2 = (2-2i)k_1 + 8k_2 = 0 \tag{5.219}$$

Thus, we arrive at the first equation of (5.216). Recall that the eigenvector is defined as:

$$\boldsymbol{K}_1 = \boldsymbol{B}_1 + i\boldsymbol{B}_2 \tag{5.220}$$

Comparison of (5.218) and (5.220) gives

$$\mathbf{B}_1 = \begin{bmatrix} 2 \\ -1 \end{bmatrix}, \quad \mathbf{B}_2 = \begin{bmatrix} 2 \\ 0 \end{bmatrix} \tag{5.221}$$

Recalling (5.151) that

$$\mathbf{X}_\mathbf{a} = [\mathbf{B}_1 \cos\beta t - \mathbf{B}_2 \sin\beta t]e^{\alpha t}, \quad \mathbf{X}_\mathbf{b} = [\mathbf{B}_2 \cos\beta t + \mathbf{B}_1 \sin\beta t]e^{\alpha t} \tag{5.222}$$

Finally, the general solution is obtained by combining (5.221) and (5.222)

$$\mathbf{X} = c_1 \left\{ \begin{bmatrix} 2 \\ -1 \end{bmatrix} \cos 2t - \begin{bmatrix} 2 \\ 0 \end{bmatrix} \sin 2t \right\} + c_2 \left\{ \begin{bmatrix} 2 \\ 0 \end{bmatrix} \cos 2t + \begin{bmatrix} 2 \\ -1 \end{bmatrix} \sin 2t \right\} \tag{5.223}$$

5.7 NONHOMOGENEOUS SYSTEM WITH CONSTANT COEFFICIENT

So far, we have only considered the special case of the homogeneous ODE system. If $\boldsymbol{F}$ is nonzero, the general solution comprises two parts, just as the case of the n-th order ODE:

$$\begin{aligned} \mathbf{X} &= \mathbf{X_c} + \mathbf{X_p} \\ &= c_1\mathbf{X}_1 + c_2\mathbf{X}_2 + \cdots\cdots + c_n\mathbf{X_n} + \mathbf{X_p} \end{aligned} \tag{5.224}$$

where the complementary solution or homogeneous solution is denoted by $\boldsymbol{X}_c$ and the particular solution for the nonhomogeneous equation can be denoted by $\boldsymbol{X}_p$.

Written explicitly, the complementary solution can be expressed as

$$\mathbf{X}_c = c_1 \begin{bmatrix} x_{11}(t) \\ x_{21}(t) \\ \vdots \\ x_{n1}(t) \end{bmatrix} + c_2 \begin{bmatrix} x_{12}(t) \\ x_{22}(t) \\ \vdots \\ x_{n2}(t) \end{bmatrix} + \cdots\cdots + c_n \begin{bmatrix} x_{1n}(t) \\ x_{2n}(t) \\ \vdots \\ x_{nn}(t) \end{bmatrix} = \begin{Bmatrix} c_1 x_{11}(t) + c_2 x_{12}(t) + \ldots + c_n x_{1n}(t) \\ c_1 x_{21}(t) + c_2 x_{22}(t) + \ldots + c_n x_{2n}(t) \\ \vdots \\ c_1 x_{n1}(t) + c_2 x_{n2}(t) + \ldots + c_n x_{nn}(t) \end{Bmatrix} \tag{5.225}$$

Before we continue to consider the particular solution, we note that the complementary solution can be written in terms a fundamental matrix defined as:

$$\mathbf{\Phi}(t) = \begin{bmatrix} x_{11}(t) & x_{12}(t) & \cdots & x_{1n}(t) \\ x_{21}(t) & x_{22}(t) & \cdots & x_{2n}(t) \\ \vdots & \vdots & \ddots & \vdots \\ x_{n1}(t) & x_{n2}(t) & \cdots & x_{nn}(t) \end{bmatrix} \tag{5.226}$$

which is formed by combining the complementary solutions. The first column of $\boldsymbol{\Phi}$ is formed by the first vector solution and so on. More specifically, we can see that

$$\mathbf{X}_c = \mathbf{\Phi}(t)\mathbf{C} = \begin{bmatrix} x_{11}(t) & x_{12}(t) & \cdots & x_{1n}(t) \\ x_{21}(t) & x_{22}(t) & \cdots & x_{2n}(t) \\ \vdots & \vdots & \ddots & \vdots \\ x_{n1}(t) & x_{n2}(t) & \cdots & x_{nn}(t) \end{bmatrix} \begin{Bmatrix} c_1 \\ c_2 \\ \vdots \\ c_n \end{Bmatrix} = \begin{Bmatrix} c_1 x_{11}(t) + c_2 x_{12}(t) + \ldots + c_n x_{1n}(t) \\ c_1 x_{21}(t) + c_2 x_{22}(t) + \ldots + c_n x_{2n}(t) \\ \vdots \\ c_1 x_{n1}(t) + c_2 x_{n2}(t) + \ldots + c_n x_{nn}(t) \end{Bmatrix} \tag{5.227}$$

which is clearly equivalent to (5.225).

5.7.1 Undetermined Coefficients

The method of undetermined coefficients establishes a number of rules of how to assume the particular solution. Table 5.1 summarizes some simple rules for assuming a proper particular solution for various forms of nonhomogeneous terms. These terms involve a constant vector, polynomials of variables with a constant vector, an exponential function with a constant vector, a sine or cosine function with a constant vector, or finite terms or products of these functions. For other functions, there are no rules for guessing particular solutions, and we have to use the method of variation of parameters, which will be discussed in the next section. Note that the last four rows in Table 5.1 are for the cases of nonhomogeneous terms matching with the homogeneous solution.

Table 5.1. Table for the method of undetermined coefficients

root	F(t)	$X_p(t)$
	C	A
	Ct^n	$A_n t^n + ... + A_1 t + A_0$
	$C\cos bt, C\sin bt$	$A\cos bt + B\sin bt$
	$Ct^n e^{at}$	$e^{at}(A_n t^n + ... + A_1 t + A_0)$
	$Ce^{at}\cos bt, Ce^{at}\sin bt$	$e^{at}(A\cos bt + B\sin bt)$
	$Ct^n\cos bt, Ct^n\sin bt$	$\cos bt(A_n t^n + ... + A_1 t + A_0)$ $+\sin bt(B_n t^n + ... + B_1 t + B_0)$
$\pm i\lambda$	$C\cos\lambda t, C\sin\lambda t$	$(A_1 + tA_2)\cos\lambda t + (B_1 + tB_2)\sin\lambda t$
λ	$Ce^{\lambda t}$	$(At + B)e^{\lambda t}$
λ	$Ct^n e^{\lambda t}$	$e^{\lambda t}[(A_n + B_n t)t^n + ... + (A_1 + B_1 t)t + (A_0 + B_0 t)]$
$\alpha \pm i\beta$	$Ce^{\alpha t}\cos\beta t, Ce^{\alpha t}\sin\beta t$	$e^{\alpha t}[(A_1 + A_2 t)\cos\beta t + (B_1 + B_2 t)\sin\beta t]$

To illustrate this method, some examples are considered.

Example 5.15 Find the general solution of the following system of ODEs

$$\frac{dx}{dt} = 5x + 3y - 2e^{-t} + 1$$
$$\frac{dy}{dt} = -x + y + e^{-t} - 5t + 7 \tag{5.228}$$

Solution: First, we need to solve for the complimentary solution of the homogeneous system:

$$X_c' = AX_c \tag{5.229}$$

where

$$A = \begin{bmatrix} 5 & 3 \\ -1 & 1 \end{bmatrix} \tag{5.230}$$

The eigenvalues of the system are easily determined as

$$\det(\mathbf{A} - \lambda\mathbf{I}) = \begin{vmatrix} 5-\lambda & 3 \\ -1 & 1-\lambda \end{vmatrix} = \lambda^2 - 6\lambda + 8 = (\lambda - 2)(\lambda - 4) = 0 \tag{5.231}$$

Without going into detail, it is straightforward to obtain the following complimentary solution:

$$\mathbf{X_c} = c_1 \begin{bmatrix} 1 \\ -1 \end{bmatrix} e^{2t} + c_2 \begin{bmatrix} 3 \\ -1 \end{bmatrix} e^{4t} \tag{5.232}$$

The nonhomogeneous term can be written as:

$$\mathbf{F}(t) = \begin{bmatrix} -2e^{-t} + 1 \\ e^{-t} - 5t + 7 \end{bmatrix} = \begin{bmatrix} 1 \\ 7 \end{bmatrix} + \begin{bmatrix} 0 \\ -5 \end{bmatrix} t + \begin{bmatrix} -2 \\ 1 \end{bmatrix} e^{-t} \tag{5.233}$$

Clearly, there is no match of the homogeneous solution and according to the guidelines given in Table 5.1, we can assume the particular solution as

$$\mathbf{X_p}(t)=\begin{bmatrix} a_1 \\ b_1 \end{bmatrix}+\begin{bmatrix} a_2 \\ b_2 \end{bmatrix}t+\begin{bmatrix} a_3 \\ b_3 \end{bmatrix}e^{-t}=\begin{bmatrix} a_1+a_2t+a_3e^{-t} \\ b_1+b_2t+b_3e^{-t} \end{bmatrix} \tag{5.234}$$

Substitution of this particular solution into the original system gives

$$\begin{bmatrix} a_2-a_3e^{-t} \\ b_2-b_3e^{-t} \end{bmatrix}=\begin{bmatrix} 5a_1+3b_1+(5a_2+3b_2)t+(5a_3+3b_3)e^{-t} \\ -a_1+b_1+(-a_2+b_2)t+(-a_3+b_3)e^{-t} \end{bmatrix}+\begin{bmatrix} -2e^{-t}+1 \\ e^{-t}-5t+7 \end{bmatrix} \tag{5.235}$$

By matching the coefficients on both sides of the first equation of (5.235) gives

$$-5a_1+a_2-3b_1=1, \quad 5a_2+3b_2=0, \quad 6a_3+3b_3=2 \tag{5.236}$$

The second equation of (5.235) leads to

$$a_1-b_1+b_2=7, \quad -a_2+b_2=5, \quad a_3-2b_3=1 \tag{5.237}$$

The solution for this system of equations gives

$$a_1=\frac{35}{32}, \quad b_1=-\frac{89}{32}, \quad a_2=-\frac{15}{8}, \quad b_2=\frac{25}{8}, \quad a_3=\frac{7}{15}, \quad b_3=-\frac{4}{15} \tag{5.238}$$

Finally, the solution of the system becomes

$$\mathbf{X}=c_1\begin{bmatrix} 1 \\ -1 \end{bmatrix}e^{2t}+c_2\begin{bmatrix} 3 \\ -1 \end{bmatrix}e^{4t}+\begin{bmatrix} \frac{35}{32} \\ -\frac{89}{32} \end{bmatrix}+\begin{bmatrix} -\frac{15}{8} \\ \frac{25}{8} \end{bmatrix}t+\begin{bmatrix} \frac{7}{15} \\ -\frac{4}{15} \end{bmatrix}e^{-t} \tag{5.239}$$

Example 5.16 Find the general solution of the following system of ODEs

$$\mathbf{X}'=\mathbf{AX}+\mathbf{F}, \quad \mathbf{A}=\begin{bmatrix} 1 & 1 \\ 1 & 1 \end{bmatrix}, \quad \mathbf{F}=\begin{bmatrix} -8 \\ 3 \end{bmatrix} \tag{5.240}$$

Solution: The eigenvalues of the system are easily determined as

$$\det(\mathbf{A}-\lambda\mathbf{I})=\begin{vmatrix} 1-\lambda & 1 \\ 1 & 1-\lambda \end{vmatrix}=(1-\lambda)^2-1=\lambda(\lambda-2)=0 \tag{5.241}$$

The eigenvalues are $\lambda = 0, 2$ and the corresponding homogeneous solution is

$$\mathbf{X_c}=c_1\begin{bmatrix} 1 \\ -1 \end{bmatrix}+c_2\begin{bmatrix} 1 \\ 1 \end{bmatrix}e^{2t} \tag{5.242}$$

The nonhomogeneous vector $\boldsymbol{F}$ is a constant vector, which matches the solution corresponding to $\lambda = 0$ (see the first vector on the right hand side of (5.242)). Therefore, we seek the particular solution in the following form

$$\mathbf{X_p}(t)=\begin{bmatrix} a_1 \\ b_1 \end{bmatrix}+\begin{bmatrix} a_2 \\ b_2 \end{bmatrix}t \tag{5.243}$$

Substitution of (5.243) into (5.240) gives

$$\begin{bmatrix} a_2 \\ b_2 \end{bmatrix}=\begin{bmatrix} a_1+b_1 \\ a_1+b_1 \end{bmatrix}+t\begin{bmatrix} a_2+b_2 \\ a_2+b_2 \end{bmatrix}+\begin{bmatrix} -8 \\ 3 \end{bmatrix} \tag{5.244}$$

This equation leads to

$$a_2=a_1+b_1-8, \quad b_2=a_1+b_1+3, \quad a_2+b_2=0 \tag{5.245}$$

Subtraction of the second equation from the first equation gives

$$a_2 - b_2 = -11 \tag{5.246}$$

Solving this and the last equation of (5.245) simultaneously, we obtain

$$a_2 = -\frac{11}{2} \tag{5.247}$$

$$b_2 = \frac{11}{2} \tag{5.248}$$

Back substitution of (5.247) and (5.248) into (5.245) gives

$$a_1 + b_1 = \frac{5}{2} \tag{5.249}$$

Choosing $a_1 = 0$ (this is chosen for the sake of simplicity), we have

$$b_1 = \frac{5}{2} \tag{5.250}$$

In summary, we have the following constants

$$a_1 = 0, \quad b_1 = \frac{5}{2}, \quad a_2 = -\frac{11}{2}, \quad b_2 = \frac{11}{2} \tag{5.251}$$

Finally, the solution of the system becomes

$$\mathbf{X} = c_1 \begin{bmatrix} 1 \\ -1 \end{bmatrix} + c_2 \begin{bmatrix} 1 \\ 1 \end{bmatrix} e^{2t} + \begin{bmatrix} 0 \\ \frac{5}{2} \end{bmatrix} + \begin{bmatrix} -\frac{11}{2} \\ \frac{11}{2} \end{bmatrix} t \tag{5.252}$$

Note again that this solution form is not unique, depending on what we assume for a_1. However, there is a unique solution if initial conditions are prescribed.

If the given nonhomogeneous terms do not match those given in Table 5.1, we have no choice but to use the method of variation of parameters, which will be discussed next.

5.7.2 Variation of Parameters

The general solution has two parts, a complimentary solution with a constant vector $\boldsymbol{C}$ and a particular solution in terms of integration:

$$\mathbf{X} = \mathbf{\Phi}(t)\mathbf{C} + \mathbf{\Phi}(t)\int \mathbf{\Phi}^{-1}(t)\mathbf{F}(t)dt \tag{5.253}$$

where $\boldsymbol{\Phi}$ is the fundamental matrix defined in (5.226) and reported here again as

$$\mathbf{\Phi}(t) = \begin{bmatrix} x_{11}(t) & x_{12}(t) & \cdots & x_{1n}(t) \\ x_{21}(t) & x_{22}(t) & \cdots & x_{2n}(t) \\ \vdots & \vdots & \ddots & \vdots \\ x_{n1}(t) & x_{n2}(t) & \cdots & x_{nn}(t) \end{bmatrix}, \quad \mathbf{C} = \begin{bmatrix} c_1 \\ c_2 \\ \vdots \\ c_n \end{bmatrix} \tag{5.254}$$

For the case that initial values are given

$$\mathbf{X}(t_0) = \mathbf{X_0}, \tag{5.255}$$

the general solution becomes

$$\mathbf{X}(t) = \mathbf{\Phi}(t)\mathbf{\Phi}^{-1}(t_0)\mathbf{X}_0 + \mathbf{\Phi}(t)\int_{t_0}^{t} \mathbf{\Phi}^{-1}(\tau)\mathbf{F}(\tau)d\tau \tag{5.256}$$

To prove these results, we first recall from (5.227) that the homogeneous solution can be written as

$$\mathbf{X}_c(t) = \mathbf{\Phi}(t)C \tag{5.257}$$

This gives the first term in (5.253). The method of variation of parameters assumes that the particular solution of the system can be written as

$$\mathbf{X}_p(t) = \mathbf{\Phi}(t)\mathbf{U}(t) \tag{5.258}$$

where $\boldsymbol{U}$ is an unknown vector function. Differentiating (5.258) gives

$$\mathbf{X}_p' = \mathbf{\Phi}'(t)\mathbf{U}(t) + \mathbf{\Phi}(t)\mathbf{U}'(t) \tag{5.259}$$

Substitution of (5.259) into (5.121) leads to

$$\mathbf{\Phi}'(t)\mathbf{U}(t) + \mathbf{\Phi}(t)\mathbf{U}'(t) = \mathbf{A}\mathbf{\Phi}(t)\mathbf{U}(t) + \mathbf{F}(t) \tag{5.260}$$

Since the fundamental matrix $\boldsymbol{\Phi}$ is formed by the homogeneous solution, we have

$$\mathbf{\Phi}'(t) = \mathbf{A}\mathbf{\Phi}(t) \tag{5.261}$$

In view of (5.261), (5.260) becomes

$$\mathbf{\Phi}(t)\mathbf{U}'(t) = \mathbf{F}(t) \tag{5.262}$$

Multiplying both sides by the inverse of the fundamental matrix, we find

$$\mathbf{U}'(t) = \mathbf{\Phi}^{-1}(t)\mathbf{F}(t) \tag{5.263}$$

Thus, $\boldsymbol{U}$ can be found by integrating both sides with respect to t and this result leads to

$$\mathbf{X}_p(t) = \mathbf{\Phi}(t)\mathbf{U}(t) = \mathbf{\Phi}(t)\int \mathbf{\Phi}^{-1}(t)\mathbf{F}(t)dt \tag{5.264}$$

Summation of (5.257) and (5.264) gives the final required result.

For initial value problems with given condition

$$\mathbf{X}(t_0) = \mathbf{X}_0, \tag{5.265}$$

the integration of the general solution can be revised as

$$\mathbf{X}(t) = \mathbf{\Phi}(t)\mathbf{C} + \mathbf{\Phi}(t)\int_{t_0}^{t} \mathbf{\Phi}^{-1}(\tau)\mathbf{F}(\tau)d\tau \tag{5.266}$$

Substitution of the initial condition yields

$$\mathbf{X}(t_0) = \mathbf{\Phi}(t_0)\mathbf{C} = \mathbf{X}_0 \tag{5.267}$$

Multiplying the inverse of the fundamental matrix gives

$$\mathbf{C} = \mathbf{\Phi}^{-1}(t_0)\mathbf{X}_0 \tag{5.268}$$

Back substitution of this constant vector into the solution finally gives

$$\mathbf{X}(t) = \mathbf{\Phi}(t)\mathbf{\Phi}^{-1}(t_0)\mathbf{X}_0 + \mathbf{\Phi}(t)\int_{t_0}^{t} \mathbf{\Phi}^{-1}(\tau)\mathbf{F}(\tau)d\tau \tag{5.269}$$

This completes the proof of (5.256).

Example 5.17 Find the general solution of the following system of ODEs by variation of parameters

$$\mathbf{X}' = \mathbf{AX} + \mathbf{F}, \quad \mathbf{A} = \begin{bmatrix} -3 & 1 \\ 2 & -4 \end{bmatrix}, \quad \mathbf{F} = \begin{bmatrix} 3t \\ e^{-t} \end{bmatrix}, \quad X(0) = \begin{bmatrix} 1 \\ 0 \end{bmatrix} \tag{5.270}$$

Solution: The eigenvalues of the system are easily determined as

$$\det(\mathbf{A} - \lambda \mathbf{I}) = \begin{vmatrix} -3-\lambda & 1 \\ 2 & -4-\lambda \end{vmatrix} = \lambda^2 + 7\lambda + 10 = (\lambda+5)(\lambda+2) = 0 \tag{5.271}$$

The eigenvalues are $\lambda = -2, -5$ and the corresponding homogeneous solution is

$$\mathbf{X_c}(t) = c_1 \begin{bmatrix} 1 \\ 1 \end{bmatrix} e^{-2t} + c_2 \begin{bmatrix} 1 \\ -2 \end{bmatrix} e^{-5t} \tag{5.272}$$

The fundamental matrix can then be formulated as

$$\mathbf{\Phi}(t) = \begin{bmatrix} e^{-2t} & e^{-5t} \\ e^{-2t} & -2e^{-5t} \end{bmatrix} \tag{5.273}$$

The inverse of the fundamental matrix is

$$\mathbf{\Phi}^{-1}(t) = \frac{1}{\det(\mathbf{\Phi}(t))} \begin{bmatrix} -2e^{-5t} & -e^{-5t} \\ -e^{-2t} & e^{-2t} \end{bmatrix} = \begin{bmatrix} \frac{2}{3}e^{2t} & \frac{1}{3}e^{2t} \\ \frac{1}{3}e^{5t} & -\frac{1}{3}e^{5t} \end{bmatrix} \tag{5.274}$$

This inverse can be used to evaluate

$$\mathbf{\Phi}^{-1}(t)\mathbf{F}(t) = \begin{bmatrix} \frac{2}{3}e^{2t} & \frac{1}{3}e^{2t} \\ \frac{1}{3}e^{5t} & -\frac{1}{3}e^{5t} \end{bmatrix} \begin{bmatrix} 3t \\ e^{-t} \end{bmatrix} = \begin{bmatrix} 2te^{2t} + \frac{1}{3}e^{t} \\ te^{5t} - \frac{1}{3}e^{4t} \end{bmatrix} \tag{5.275}$$

Finally, the particular solution is obtained as

$$\mathbf{X}_p(t) = \mathbf{\Phi}(t) \int \mathbf{\Phi}^{-1}(t)\mathbf{F}(t)\,dt$$

$$= \begin{bmatrix} e^{-2t} & e^{-5t} \\ e^{-2t} & -2e^{-5t} \end{bmatrix} \begin{bmatrix} \int (2te^{2t} + \frac{1}{3}e^{t})dt \\ \int (te^{5t} - \frac{1}{3}e^{4t})dt \end{bmatrix} = \begin{bmatrix} \frac{6}{5}t - \frac{27}{50} + \frac{1}{4}e^{-t} \\ \frac{3}{5}t - \frac{21}{50} + \frac{1}{2}e^{-t} \end{bmatrix} \tag{5.276}$$

The final solution becomes

$$\mathbf{X}(t) = \mathbf{X}_c(t) + \mathbf{X}_p(t) = c_1 \begin{bmatrix} 1 \\ 1 \end{bmatrix} e^{-2t} + c_2 \begin{bmatrix} 1 \\ -2 \end{bmatrix} e^{-5t} + \begin{bmatrix} \frac{6}{5}t - \frac{27}{50} + \frac{1}{4}e^{-t} \\ \frac{3}{5}t - \frac{21}{50} + \frac{1}{2}e^{-t} \end{bmatrix} \tag{5.277}$$

Finally, we apply the initial condition to (5.277) to get

$$\begin{bmatrix} 1 \\ 0 \end{bmatrix} = c_1 \begin{bmatrix} 1 \\ 1 \end{bmatrix} + c_2 \begin{bmatrix} 1 \\ -2 \end{bmatrix} + \begin{bmatrix} -\frac{27}{50} + \frac{1}{4} \\ -\frac{21}{50} + \frac{1}{2} \end{bmatrix} = c_1 \begin{bmatrix} 1 \\ 1 \end{bmatrix} + c_2 \begin{bmatrix} 1 \\ -2 \end{bmatrix} + \begin{bmatrix} -\frac{29}{100} \\ \frac{2}{25} \end{bmatrix} \tag{5.278}$$

The solution for the constants is

$$c_1 = \frac{121}{200}, \quad c_2 = \frac{137}{200} \tag{5.279}$$

The final solution becomes

$$\mathbf{X}(t)=\frac{121}{200}\begin{bmatrix}1\\1\end{bmatrix}e^{-2t}+\frac{137}{200}\begin{bmatrix}1\\-2\end{bmatrix}e^{-5t}+\begin{bmatrix}\frac{6}{5}t-\frac{27}{50}+\frac{1}{4}e^{-t}\\ \frac{3}{5}t-\frac{21}{50}+\frac{1}{2}e^{-t}\end{bmatrix} \tag{5.280}$$

Note that there is no unknown constant in this solution because the initial conditions have been satisfied.

5.8 SYSTEM OF NONLINEAR ODE

So far, we have assumed that the system of first order ODEs is linear. Now we can extend our consideration to the more general case of a nonlinear system:

$$\frac{dx}{dt}=f_1(t,x,y),\quad \frac{dy}{dt}=f_2(t,x,y) \tag{5.281}$$

where f_1 and f_2 are generally a nonlinear function of the unknown functions x and y. It is straightforward to extend the system to equations with order higher than second. For equilibrium solutions not changing with time, we must have

$$\frac{dx}{dt}=0,\quad \frac{dy}{dt}=0 \tag{5.282}$$

Thus, the equilibrium solution can be found by solving:

$$f_1(t,x_e,y_e)=0,\quad f_2(t,x_e,y_e)=0 \tag{5.283}$$

Note, however, that (5.283) may not be easily solved to get the closed form equilibrium solution, depending on the given functions f_1 and f_2.

5.9 STABILITY OF AUTONOMOUS SYSTEM

The stability of an autonomous system was considered mainly by Poincare in 1881 and Lyapunov in 1892. If f_1 and f_2 in (5.283) are independent of t, the coupled system is called autonomous because its functions on the right hand side of (5.281) are independent of time (or are not affected by time variation and thus lead to the term autonomous). We can first rewrite the system of ODEs about the equilibrium solutions as:

$$\xi=x-x_e,\quad \eta=y-y_e \tag{5.284}$$

The magnitude of these variables indicates how far the current solution is from the equilibrium solution. Thus, as $t\to\infty$ if both ξ and $\eta\to 0$, the solution is clearly stable. The system of ODEs in the perturbation solution (in the sense that it is perturbed from the equilibrium solution) is

$$\frac{d\xi}{dt}=F_1(\xi,\eta),\quad \frac{d\eta}{dt}=F_2(\xi,\eta) \tag{5.285}$$

The stability of the solution of this system around an equilibrium solution can be examined by investigating the eigenvalue of the linearized system of (5.285). First, we can linearize (5.285) by expanding the functions on the right of (5.285) in polynomials of ξ and η. Dropping all higher order terms other than linear, we have

$$\frac{d\xi}{dt} = a_{11}\xi + a_{12}\eta, \quad \frac{d\eta}{dt} = a_{21}\xi + a_{22}\eta \tag{5.286}$$

where a_{ij} are constants. This is precisely the system of first order ODEs that we have discussed so far. Let us consider the eigenvalue of the system as:

$$\det(\mathbf{A} - \lambda\mathbf{I}) = \begin{vmatrix} a_{11} - \lambda & a_{12} \\ a_{21} & a_{22} - \lambda \end{vmatrix}$$
$$= \lambda^2 - (a_{11} + a_{22})\lambda + a_{11}a_{22} - a_{12}a_{21} \tag{5.287}$$
$$= \lambda^2 - p\lambda + q = 0$$

The eigenvalues are:

$$\lambda_{1,2} = \frac{p \pm \sqrt{p^2 - 4q}}{2} = \frac{p \pm \sqrt{\Delta}}{2} \tag{5.288}$$

As we have shown previously, the general solution of the perturbation around the equilibrium solution is

$$\mathbf{X} = \mathbf{K}_1 e^{\lambda_1 t} + \mathbf{K}_2 e^{\lambda_2 t} \tag{5.289}$$

Let us now consider the possible long-term behavior of the perturbation solution. If λ_1 and λ_2 are real, the solution either increases indefinitely with time or decays with time, depending on whether the eigenvalue is positive or negative. If λ_1 and λ_2 are a complex conjugate pair, the solution increases or decays depending totally on whether the real part of the eigenvalue is positive or negative. The imaginary part of the eigenvalue clearly will only lead to oscillating solutions of sine and cosine (recall Euler's formula).

A number of scenarios are possible. The following terminology of the stability of ODEs was introduced by Henri Poincare in 1881:

(i) If $q > 0$ and $p < 0$, the roots are either complex conjugates with negative real parts, or both real and negative. The solution is "asymptotically" stable, in the sense that the analysis of stability is valid if the domain being considered is so close to the equilibrium point that the effects of all nonlinear terms can be neglected. Therefore, such analysis is called linear stability analysis. If $\Delta = 0$, the eigenvalue is repeated and both of them are negative. Such an equilibrium point is called a *stable proper* or *stable improper node*, depending on whether the equal eigenvalue system has two or one independent eigenvectors (recalling from the fact that $\boldsymbol{A}$ is either a Hermitian matrix or non-Hermitian). If $\Delta > 0$, it is an *asymptotically stable node*. If $\Delta < 0$, the eigenvalues are complex conjugate pairs and it is an *asymptotically stable spiral point*. It is called spiral because of the oscillations accompanying with the decay.

(ii) If $q > 0$ and $p > 0$ with $\Delta > 0$, both the roots are positive. The equilibrium point is called an *unstable node*. If $\Delta = 0$, it is called an *unstable proper* or *unstable improper node*, depending on whether there are two or one independent eigenvectors (Hermitian or non-Hermitian $\boldsymbol{A}$). If $\Delta < 0$, the equilibrium point is an *unstable spiral point*.

(iii) If $q > 0$ and $p = 0$, then the roots are purely imaginary. The solution will neither increase nor decay. In fact, the solution is a sine or cosine function and it is a purely oscillating function. The equilibrium point is called a *stable centre*.

(iv) If $q < 0$, then the roots are real, and we must have $\Delta > 0$. If $p > 0$, both eigenvalues are real with one of them positive. If $p < 0$, both eigenvalues are real with one positive and one negative root. In either case, the behavior of the equilibrium point is the same and is called an *unstable saddle point*. It is called a saddle point in the sense that if the initial condition is given such that a solution is precisely in the direction of the eigenvector of the negative root, the solution will decay to zero. If any initial condition leads to a slight deviation from the direction of this eigenvector of the negative eigenvalue, then the solution is unstable. The situation is similar to walking on the ridge leading to a saddle of a mountain. Any step slightly deviated from the ridge will lead to falling down the cliff (unstable) and you will never travel safely to the saddle (stable).

These scenarios are summarized in Figure 5.2 in the p-q space.

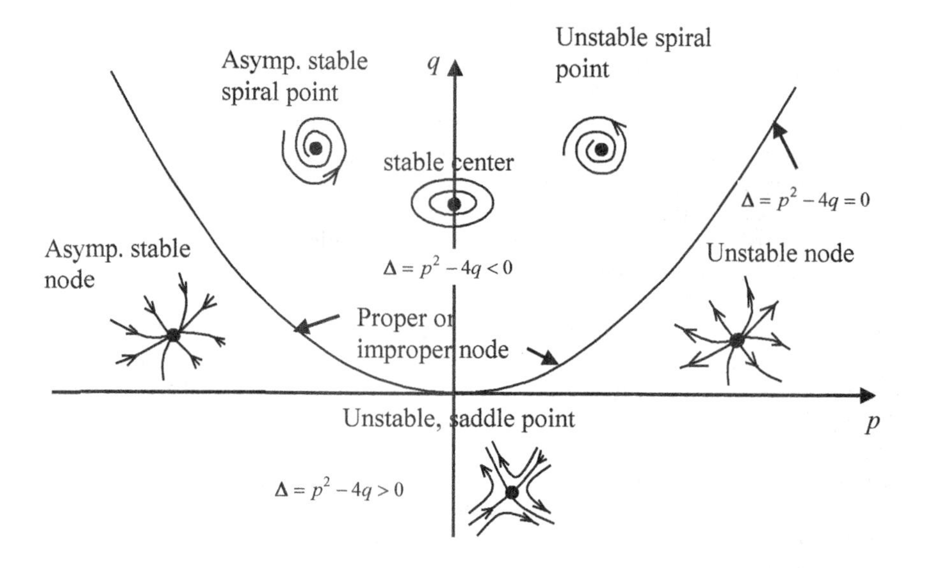

Figure 5.2 Stability classifications for 2-unknown-system

Table 5.2 summarizes the type of stability for an equilibrium point for the case of two unknowns. This system is proposed by Poincare in 1881. A system of three equations and three unknowns can be classified using a similar idea, but the classification is more complicated and difficult to present in graphical form like Figure 5.2. Reyn (1964) presented a systematic classification of the three differential equation systems. It has been successfully applied to analyzing landslides by Chau (1995, 1999b) using bifurcation theory on creeping slopes obeying a two-state variable friction law.

Table 5.2. Stability classification for 2-D autonomous system

q	p	Δ	Stability type of the equilibrium point
>0	<0	$=0$	Stable proper or improper node
		>0	Asymptotically stable node
		<0	Asymptotically stable spiral point
>0	>0	$=0$	Stable proper or improper node
		>0	Unstable proper or improper node
		<0	Unstable spiral point
>0	$=0$	<0	Stable center
<0		>0	Unstable saddle point

Example 5.18 Consider the stability of the following system

$$\frac{dx}{dt} = 2x + 3y$$
$$\frac{dy}{dt} = 2x + y \tag{5.290}$$

Solution: This solution is given in Example 5.10 as

$$\mathbf{X} = c_1\mathbf{X_1} + c_2\mathbf{X_2} = c_1\begin{bmatrix}1\\-1\end{bmatrix}e^{-t} + c_2\begin{bmatrix}3\\2\end{bmatrix}e^{4t} \tag{5.291}$$

For this case, we have

$$p = (a_{11} + a_{22}) = 3, \quad q = a_{11}a_{22} - a_{12}a_{21} = 2 - 6 = -4 \tag{5.292}$$

This is an unstable saddle point. The behavior of the solution is depicted in Figure 5.3.

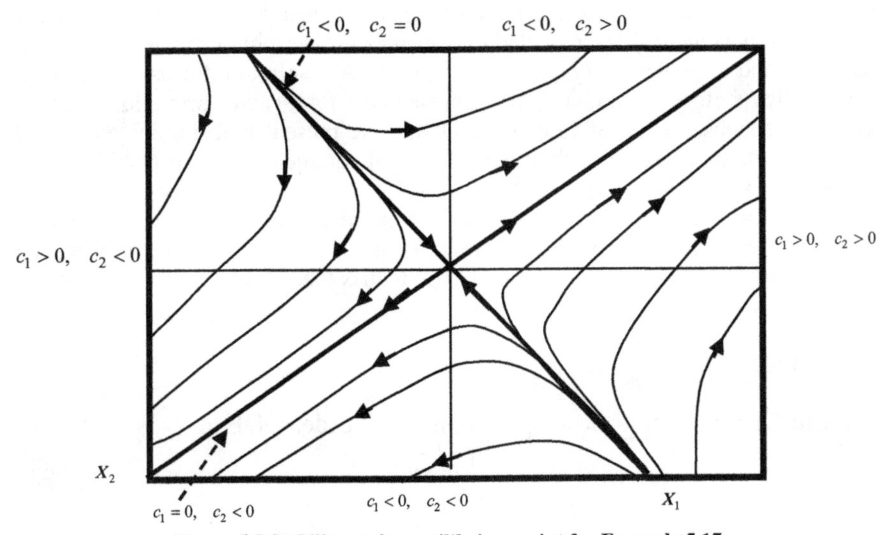

Figure 5.3 Stability at the equilibrium point for Example 5.17

If the initial condition of the system is such that it locates along the direction of the first eigenvector (1, −1), the solution is stable. As shown in Figure 5.3, the solution returns to the origin. However, if the initial condition does not locate exactly along this direction, the solution is unstable. Figure 5.3 shows that as $t \to \infty$ the solution goes to infinity along the direction of the second eigenvector of (3, 2). Note the exponential function for the first eigenvector is an exponentially decaying function whereas the exponential function for the second eigenvector is an exponentially increasing function. This is like a saddle at a mountain ridge. You will not fall down the cliff only if you walk along the edge of the saddle. Therefore, Poincare termed this an unstable saddle point.

5.10 SUMMARY AND FURTHER READING

This chapter discusses the solution technique for solving a system of first order ODE. It is important since all systems of ODEs, no matter the order and the number of coupled ODEs, can always be converted into a system of first order ODEs. For nonlinear systems, when we apply a numerical technique (such as the fourth order Runge-Kutta method discussed in Chapter 15) to solve the system of ODEs, it is always advisable to rewrite it as a system of first order ODE.

First, we demonstrate the technique of elimination for solving a system of ODEs analytically. The concept of a multiplier system for elimination of unknowns in a system of equations is introduced. The determinacy of a system of equations is also discussed, and using this technique, we can identify whether the solution is possible for a system of higher order ODEs. For a linear system of first order ODEs, we review the Hermitian matrix (which has a profound effect on the solution form of the system) and the eigenvalue problem. It is shown that a system of first order ODEs with constant coefficients can always be converted to solving an eigenvalue problem of a matrix. Four scenarios of the solutions are discussed. They are real distinct eigenvalues, repeated eigenvalues for Hermitian and non-Hermitian matrices, and complex conjugate eigenvalues. All solution forms are proved analytically before examples are presented for each scenario. For a nonhomogeneous system of first order ODEs, we present both the undetermined coefficient method and the method of variation of parameters. Linear stability of an autonomous system is briefly discussed.

For those readers who are interested in the historical development of matrix theory and its relation to solving differential equations, we highly recommend the review paper by Hawkins (1975a,b; 1977) on matrix theory.

5.11 PROBLEMS

Problem 5.1 Solve the following system of 1st order ODEs

$$\frac{d\mathbf{X}}{dt} = \mathbf{A}\mathbf{X} \tag{5.293}$$

where

$$\mathbf{A}=\begin{bmatrix}3 & -18\\ 2 & -9\end{bmatrix},\quad \mathbf{X}(0)=\begin{bmatrix}2\\ 1\end{bmatrix} \tag{5.294}$$

Ans:

$$\mathbf{X}=\begin{bmatrix}2\\ 1\end{bmatrix}e^{-3t}-2\begin{bmatrix}3\\ 1\end{bmatrix}te^{-3t} \tag{5.295}$$

Problem 5.2 Solve the following system of 1st order ODEs

$$\frac{d\mathbf{X}}{dt}=\mathbf{AX} \tag{5.296}$$

where

$$\mathbf{A}=\begin{bmatrix}1 & 3\\ 5 & 3\end{bmatrix} \tag{5.297}$$

Ans: Note the solution is not unique if we do not normalize the eigenvectors

$$\mathbf{X}=c_1\begin{bmatrix}1\\ -1\end{bmatrix}e^{-2t}+c_2\begin{bmatrix}3\\ 5\end{bmatrix}e^{6t} \tag{5.298}$$

Problem 5.3 Solve the following system of 1st order ODEs

$$\frac{d\mathbf{X}}{dt}=\mathbf{AX} \tag{5.299}$$

where

$$\mathbf{A}=\begin{bmatrix}-1 & 3\\ -3 & 5\end{bmatrix},\quad \mathbf{X}(0)=\begin{bmatrix}1\\ 2\end{bmatrix} \tag{5.300}$$

Ans:

$$\mathbf{X}=\begin{bmatrix}1\\ 2\end{bmatrix}e^{2t}+3\begin{bmatrix}1\\ 1\end{bmatrix}te^{2t} \tag{5.301}$$

Problem 5.4 Solve the following system of 1st order ODEs

$$\frac{d\mathbf{X}}{dt}=\mathbf{AX} \tag{5.302}$$

where

$$\mathbf{A}=\begin{bmatrix}4 & -1\\ 5 & 2\end{bmatrix} \tag{5.303}$$

Ans:

$$\mathbf{X}=c_1\left\{\begin{bmatrix}1\\ 1\end{bmatrix}\cos 2t-\begin{bmatrix}0\\ -2\end{bmatrix}\sin 2t\right\}e^{3t}+c_2\left\{\begin{bmatrix}0\\ -2\end{bmatrix}\cos 2t+\begin{bmatrix}1\\ 1\end{bmatrix}\sin 2t\right\}e^{3t} \tag{5.304}$$

Problem 5.5 Solve the following system of 1st order ODEs

$$2\frac{dx_1}{dt}-5\frac{dx_2}{dt}=4x_2-x_1$$
$$3\frac{dx_1}{dt}-4\frac{dx_2}{dt}=2x_1-x_2 \tag{5.305}$$

Ans:

$$\mathbf{X}=c_1\begin{bmatrix}3\\1\end{bmatrix}e^{t}+c_2\begin{bmatrix}1\\1\end{bmatrix}e^{-t} \tag{5.306}$$

Problem 5.6 Solve the following system of 1st order ODEs

$$\frac{d\mathbf{X}}{dt}=\mathbf{AX} \tag{5.307}$$

where

$$\mathbf{A}=\begin{bmatrix}-2 & 1\\1 & -2\end{bmatrix} \tag{5.308}$$

Ans:

$$\mathbf{X}=c_1\begin{bmatrix}1\\-1\end{bmatrix}e^{-3t}+c_2\begin{bmatrix}1\\1\end{bmatrix}e^{-t} \tag{5.309}$$

Problem 5.7 Solve the following system of 1st order ODEs by variation of parameters

$$\frac{d\mathbf{X}}{dt}=\mathbf{AX}+\boldsymbol{F} \tag{5.310}$$

where

$$\mathbf{A}=\begin{bmatrix}-2 & 1\\1 & -2\end{bmatrix},\quad \boldsymbol{F}=\begin{bmatrix}2e^{-t}\\3t\end{bmatrix} \tag{5.311}$$

Ans:

$$\mathbf{X}=c_1\begin{bmatrix}1\\-1\end{bmatrix}e^{-3t}+c_2\begin{bmatrix}1\\1\end{bmatrix}e^{-t}+\frac{1}{2}\begin{bmatrix}1\\-1\end{bmatrix}e^{-t}+\begin{bmatrix}1\\1\end{bmatrix}te^{-t}+\begin{bmatrix}1\\2\end{bmatrix}t-\frac{1}{3}\begin{bmatrix}4\\5\end{bmatrix} \tag{5.312}$$

Problem 5.8 Solve the following system of 1st order ODEs

$$\frac{d\mathbf{X}}{dt}=\mathbf{AX} \tag{5.313}$$

where

$$\mathbf{A}=\begin{bmatrix}3 & -1\\16 & -5\end{bmatrix},\quad \mathbf{X}(0)=\begin{bmatrix}1\\1\end{bmatrix} \tag{5.314}$$

Ans:

$$\mathbf{X}=\begin{bmatrix}1\\1\end{bmatrix}e^{-t}+3\begin{bmatrix}1\\4\end{bmatrix}te^{-t} \tag{5.315}$$

Problem 5.9 Solve the following system of 1st order ODEs

$$\frac{d\mathbf{X}}{dt} = \mathbf{AX} \tag{5.316}$$

where

$$\mathbf{A} = \begin{bmatrix} 1 & -2 \\ 3 & -4 \end{bmatrix}, \quad \mathbf{X}(0) = \begin{bmatrix} 1 \\ 0 \end{bmatrix} \tag{5.317}$$

Ans:

$$\mathbf{X} = 3\begin{bmatrix} 1 \\ 1 \end{bmatrix} e^{-t} - \begin{bmatrix} 2 \\ 3 \end{bmatrix} e^{-2t} \tag{5.318}$$

Problem 5.10 Solve the following system of 1st order ODEs

$$\frac{d\mathbf{X}}{dt} = \mathbf{AX} \tag{5.319}$$

where

$$\mathbf{A} = \begin{bmatrix} 3 & -18 \\ 2 & -9 \end{bmatrix}, \quad \mathbf{X}(0) = \begin{bmatrix} 1 \\ 2 \end{bmatrix} \tag{5.320}$$

Ans:

$$\mathbf{X} = \begin{bmatrix} 1 \\ 2 \end{bmatrix} e^{-3t} - 10\begin{bmatrix} 3 \\ 1 \end{bmatrix} te^{-3t} \tag{5.321}$$

Problem 5.11 Solve the following system of 1st order ODEs

$$\frac{d\mathbf{X}}{dt} = \mathbf{AX} \tag{5.322}$$

where

$$\mathbf{A} = \begin{bmatrix} -1 & 3 \\ -3 & 5 \end{bmatrix}, \quad \mathbf{X}(0) = \begin{bmatrix} 1 \\ 2 \end{bmatrix} \tag{5.323}$$

Ans:

$$\mathbf{X} = \begin{bmatrix} 1 \\ 2 \end{bmatrix} e^{2t} + 3\begin{bmatrix} 1 \\ 1 \end{bmatrix} te^{2t} \tag{5.324}$$

Problem 5.12 Solve the following system of 1st order ODEs

$$\frac{d\mathbf{X}}{dt} = \mathbf{AX} \tag{5.325}$$

where

$$\mathbf{A} = \begin{bmatrix} 3 & -18 \\ 2 & -9 \end{bmatrix}, \quad \mathbf{X}(0) = \begin{bmatrix} 2 \\ 1 \end{bmatrix} \tag{5.326}$$

Ans:

$$\mathbf{X} = \begin{bmatrix} 2 \\ 1 \end{bmatrix} e^{-3t} - 2\begin{bmatrix} 3 \\ 1 \end{bmatrix} te^{-3t} \tag{5.327}$$

Problem 5.13 Solve the following system of 1st order ODEs

$$\frac{d\mathbf{X}}{dt} = \mathbf{AX} \tag{5.328}$$

where

$$\mathbf{A} = \begin{bmatrix} 3 & -1 \\ 4 & -2 \end{bmatrix}, \quad \mathbf{X}(0) = \begin{bmatrix} 1 \\ 1 \end{bmatrix} \tag{5.329}$$

Ans:

$$\mathbf{X} = \begin{bmatrix} 1 \\ 1 \end{bmatrix} e^{2t} \tag{5.330}$$

Problem 5.14 For the weather system on Earth, air current may flow parallel to a contour of equal pressure (called isobars). This happens when the Coriolis force (force due to Earth's rotation) and pressure force balance each other. In geophysics, this is called geotrophic flow (see Figure 5.4). The following 2-D coupling system (a special form of Navier-Stokes equation) for velocity along North-South and East-West directions (v and u) can be used to model it:

$$\begin{aligned} \mu \frac{d^2 u}{dz^2} &= -\alpha v \\ \mu \frac{d^2 v}{dz^2} &= \alpha (u - u^*) \end{aligned} \tag{5.331}$$

where u and v are the flow speed in horizontal directions. The other parameters α, μ, and u^* are the Coriolis constant, kinematic viscosity of air, and uniform flow velocity of u far from earth's surface (all are assumed as constants).

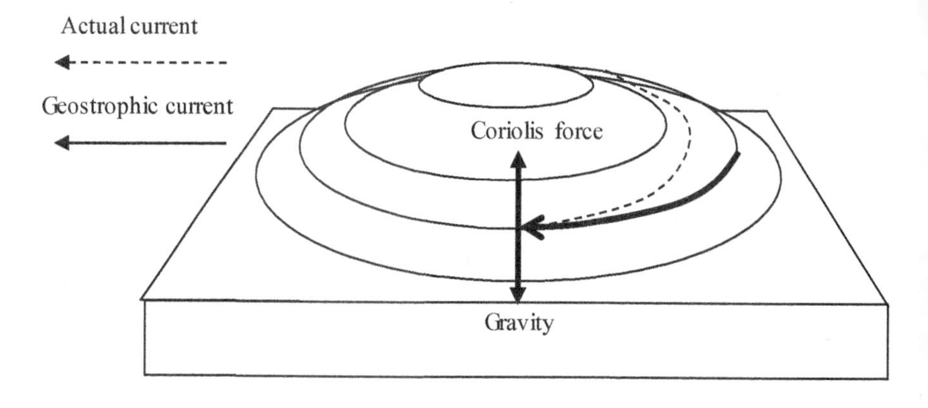

Figure 5.4 Illustration of geotrophic flow

(i) Assume the following solution form:

$$u = u^* + A\exp(\lambda z), \quad v = B\exp(\lambda z) \tag{5.332}$$

Find the characteristic equation of λ and solve for λ.

(ii) Redefine the variables as:

$$\begin{aligned} x_1 &= u - u^*, \\ x_2 &= du/dz, \\ x_3 &= v, \\ x_4 &= dv/dz \end{aligned} \tag{5.333}$$

(iii) Formulate the problem as a system of first order ODEs and find the matrix A

$$\mathbf{X}' = \mathbf{AX}, \quad \mathbf{X} = \begin{bmatrix} x_1 \\ x_2 \\ x_3 \\ x_4 \end{bmatrix} \tag{5.334}$$

(iv) Find the solution of the system.

Ans:

$$\mathbf{A} = \begin{bmatrix} 0 & 1 & 0 & 0 \\ 0 & 0 & -\dfrac{\alpha}{\mu} & 0 \\ 0 & 0 & 0 & 1 \\ \dfrac{\alpha}{\mu} & 0 & 0 & 0 \end{bmatrix} \tag{5.335}$$

$$\lambda_{1,2} = \pm\sqrt{\frac{\alpha}{2\mu}}(1+i), \quad \lambda_{3,4} = \pm\sqrt{\frac{\alpha}{2\mu}}(1-i) \tag{5.336}$$

$$\mathbf{X} = c_1\mathbf{X}_a + c_2\mathbf{X}_b + c_3\mathbf{X}_c + c_4\mathbf{X}_d \tag{5.337}$$

$$\mathbf{X_a} = \left[\mathbf{B_1}\cos\sqrt{\frac{\alpha}{2\mu}}t - \mathbf{B_2}\sin\sqrt{\frac{\alpha}{2\mu}}t\right]e^{\sqrt{\frac{\alpha}{2\mu}}t} \tag{5.338}$$

$$\mathbf{X}_b = \left[\mathbf{B}_2\cos\sqrt{\frac{\alpha}{2\mu}}t + \mathbf{B}_1\sin\sqrt{\frac{\alpha}{2\mu}}t\right]e^{\sqrt{\frac{\alpha}{2\mu}}t} \tag{5.339}$$

$$\mathbf{X}_c = \left[\boldsymbol{D}_1\cos\sqrt{\frac{\alpha}{2\mu}}t + \boldsymbol{D}_2\sin\sqrt{\frac{\alpha}{2\mu}}t\right]e^{-\sqrt{\frac{\alpha}{2\mu}}t} \tag{5.340}$$

$$\mathbf{X}_d = \left[\boldsymbol{D}_2\cos\sqrt{\frac{\alpha}{2\mu}}t - \boldsymbol{D}_1\sin\sqrt{\frac{\alpha}{2\mu}}t\right]e^{-\sqrt{\frac{\alpha}{2\mu}}t} \tag{5.341}$$

$$\boldsymbol{B}_1 = \begin{bmatrix} 1 \\ \sqrt{\alpha/(2\mu)} \\ 0 \\ \sqrt{\alpha/(2\mu)} \end{bmatrix}, \quad \boldsymbol{B}_2 = \begin{bmatrix} 0 \\ \sqrt{\alpha/(2\mu)} \\ -1 \\ -\sqrt{\alpha/(2\mu)} \end{bmatrix}, \quad \boldsymbol{D}_1 = \begin{bmatrix} 1 \\ -\sqrt{\alpha/(2\mu)} \\ 0 \\ -\sqrt{\alpha/(2\mu)} \end{bmatrix}, \quad \boldsymbol{D}_2 = \begin{bmatrix} 0 \\ -\sqrt{\alpha/(2\mu)} \\ -1 \\ \sqrt{\alpha/(2\mu)} \end{bmatrix} \tag{5.342}$$

Problem 5.15 Reconsider Example 5.16:

$$\mathbf{X}' = \mathbf{A}\mathbf{X} + \mathbf{F}, \quad \mathbf{A} = \begin{bmatrix} 1 & 1 \\ 1 & 1 \end{bmatrix}, \quad \mathbf{F} = \begin{bmatrix} -8 \\ 3 \end{bmatrix} \tag{5.343}$$

(i) By choosing $a_1 = a$ in (5.249) (where a is an arbitrary constant), show that the solution is

$$\mathbf{X} = c_1 \begin{bmatrix} 1 \\ -1 \end{bmatrix} + c_2 \begin{bmatrix} 1 \\ 1 \end{bmatrix} e^{2t} + \begin{bmatrix} a \\ \frac{5}{2} - a \end{bmatrix} + \begin{bmatrix} -\frac{11}{2} \\ \frac{11}{2} \end{bmatrix} t \tag{5.344}$$

(ii) Show that this result is indeed a particular solution.

(iii) Now consider the following initial condition, find the solution using either (5.252) or (5.344):

$$\mathbf{X}(0) = \begin{bmatrix} 1 \\ 2 \end{bmatrix} \tag{5.345}$$

(iv) Show that both solutions given in (5.252) and (5.344) give the same answer.

Ans:

$$\mathbf{X} = \frac{1}{4} \begin{bmatrix} 3 \\ 7 \end{bmatrix} + \frac{1}{4} \begin{bmatrix} 1 \\ 1 \end{bmatrix} e^{2t} + \frac{1}{2} \begin{bmatrix} -11 \\ 11 \end{bmatrix} t \tag{5.346}$$

CHAPTER SIX

First Order Partial Differential Equations (PDEs)

6.1 INTRODUCTION

Nearly all textbooks on partial differential equations will cover second order PDEs, like the wave equation, diffusion equation, and potential or Laplace equation. However, the discussion of first order PDEs is normally not covered in elementary textbooks on PDE. It is considered as a more advanced topic. The method of solution is quite different from that for second order and the general solution for first order PDEs will include an arbitrary function of some characteristics line.

Although we discuss in this chapter the first order PDE before discussing the second order PDE, historically the second order PDE was investigated before the first order. Bernoulli, Euler, D'Alembert and others considered the physical problem related to dynamics of rigid bodies, vibrations of membranes, and wave phenomena. These are second order PDE. The solution of first order PDEs was considered mainly by Lagrange, Clairaut, D'Alembert, Monge, Charpit, Jacobi, and others.

We will first discuss the geometric interpretation of the solution of linear first order PDEs. The Lagrange method for solving linear first order PDEs is then introduced. For the case of nonlinear first order PDEs, the Lagrange-Charpit method and Jacobi's method will be discussed, and the geometric interpretation of the solution of nonlinear first order PDEs in terms of Monge cones will also be covered. In the process, the idea of characteristics equations, a term introduced by Cauchy in 1819, and curves will be introduced. The concept of characteristics will be used again for second order PDEs of hyperbolic types (or wave types) in Chapter 7.

6.2 FIRST ORDER PDE

A nonlinear first order PDE can symbolically be written as

$$F(x, y, z, \frac{\partial z}{\partial x}, \frac{\partial z}{\partial y}) = 0 \tag{6.1}$$

The existence of the solution of a nonlinear PDE cannot be guaranteed. If the first order PDE is linear, we can express it explicitly as

$$P(x, y)\frac{\partial z}{\partial x} + Q(x, y)\frac{\partial z}{\partial y} = R(x, y) \tag{6.2}$$

where z is the unknown and x and y are the variables. A general feature of the solution of a first order PDE is that it normally involves arbitrariness in the solution. For example, consider the following first order PDE

$$\frac{\partial z}{\partial x} - a\frac{\partial z}{\partial y} = 0 \tag{6.3}$$

Now introduce a change of variable that

$$z = \phi(u), \quad u = ax + y \tag{6.4}$$

where ϕ is an arbitrary function. Differentiating (6.4) gives

$$\frac{\partial z}{\partial x} = \phi'(u)\frac{\partial u}{\partial x} = a\phi'(u), \quad a\frac{\partial z}{\partial y} = a\phi'(u)\frac{\partial u}{\partial y} = a\phi'(u) \tag{6.5}$$

Substitution of (6.5) into (6.3) shows that it is identically satisfied by (6.4). Thus, the general solution includes an arbitrary function ϕ. In fact, we will see in later sections that u is called the characteristics of the PDE.

6.3 HYPERBOLIC EQUATION

Physically, the first order PDE appears naturally in the modeling of the wave phenomenon. In the next chapter, we will see that the governing equation for a 1-D wave can be expressed as:

$$\frac{\partial^2 u}{\partial t^2} - c^2\frac{\partial^2 u}{\partial x^2} = 0 \tag{6.6}$$

Following the idea of factorization that we discussed in Chapter 3 for ODEs, we can factorize (6.6) as:

$$\frac{\partial^2 u}{\partial t^2} - c^2\frac{\partial^2 u}{\partial x^2} = (\frac{\partial}{\partial t} - c\frac{\partial}{\partial x})(\frac{\partial}{\partial t} + c\frac{\partial}{\partial x})u \tag{6.7}$$

Since c is a constant, the factorization in (6.7) is commutative. Therefore, the solution is the sum of the solutions of the following two first order PDE:

$$(\frac{\partial}{\partial t} - c\frac{\partial}{\partial x})u = 0, \quad (\frac{\partial}{\partial t} + c\frac{\partial}{\partial x})u = 0 \tag{6.8}$$

Referring to the discussion on (6.3) above, we can easily see that the solutions of (6.8) are

$$u = \phi(x+ct), \quad u = \psi(x-ct) \tag{6.9}$$

Thus, the solution of (6.6) can be formed by adding the solutions of (6.8) as:

$$u = \phi(x+ct) + \psi(x-ct) \tag{6.10}$$

The validity of this solution can be demonstrated by substituting (6.10) into (6.6). Let us consider the physical meaning of (6.8) more closely. For stationary waves, the wave undulation does not change with time or we have mathematically

$$\frac{\partial u}{\partial t} = 0 \tag{6.11}$$

If a wave is not stationary, the simplest type of wave equation is formed by adding an additional term as:

$$\frac{\partial u}{\partial t}+c\frac{\partial u}{\partial x}=0 \tag{6.12}$$

where c is the wave speed of the propagating disturbance. Note that (6.12) is the simplest type of transport equation and it is also known as the kinematic wave equation. We have seen that the solution of (6.12) appears as

$$u=\psi(x-ct) \tag{6.13}$$

Let us take a Galilean transform such that

$$u(t,x)=v(t,\xi)=v(t,x-ct) \tag{6.14}$$

Taking differentiation of (6.14) using the chain rule yields

$$\frac{\partial u}{\partial t}=\frac{\partial v}{\partial t}+\frac{\partial v}{\partial \xi}\frac{\partial \xi}{\partial t}=\frac{\partial v}{\partial t}-c\frac{\partial v}{\partial \xi},\quad \frac{\partial u}{\partial x}=\frac{\partial v}{\partial \xi} \tag{6.15}$$

Substitution of (6.15) into (6.12) yields

$$\frac{\partial u}{\partial t}+c\frac{\partial u}{\partial x}=\frac{\partial v}{\partial t}=0 \tag{6.16}$$

In other words, we must have $v = v(\xi)$. That is, if we travel along with the propagating wave at speed c, we see an unchanging shape of the waveform. Physically, the signals travel along the characteristics of $\xi = x-ct$. This forms the basis of the method of characteristics. Mathematically, all these wave equations are called hyperbolic, which will be explained in detail in Chapter 7.

6.3.1. Transport with Decay

For real waves, there must be damping. Mathematically, we can add an addition term as

$$\frac{\partial u}{\partial t}+c\frac{\partial u}{\partial x}+au=0 \tag{6.17}$$

Applying the change of variables given in (6.14), we have

$$\frac{\partial v}{\partial t}+av=0 \tag{6.18}$$

This is a separable PDE, and (6.18) can be integrated to get

$$\ln v=-at+C_0(\xi) \tag{6.19}$$

Thus, the solution is obtained as:

$$v(t,\xi)=f(\xi)e^{-at} \tag{6.20}$$

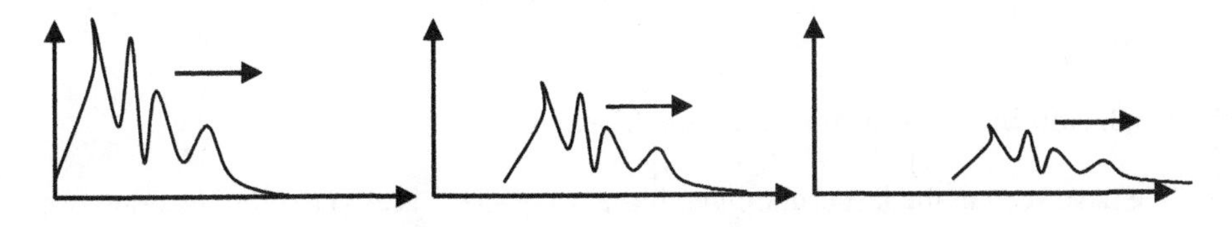

Figure 6.1 Decaying traveling waves of 1st order hyperbolic PDE

Expressed in terms of the original variables, we have a decay wave along the characteristics as

$$u(t,x) = f(x-ct)e^{-at} \tag{6.21}$$

This solution is illustrated in Figure 6.1.

6.3.2. Non-Uniform Transport

If the wave speed is not uniform, we can model the hyperbolic wave as

$$\frac{\partial u}{\partial t} + c(x)\frac{\partial u}{\partial x} = 0 \tag{6.22}$$

Let us consider a function h such that

$$h(t) = u(t, x(t)) \tag{6.23}$$

Taking the total differential with respect to t using the chain rule gives

$$\frac{dh(t)}{dt} = \frac{\partial u}{\partial t}(t, x(t)) + \frac{\partial u}{\partial x}(t, x(t))\frac{dx}{dt} \tag{6.24}$$

The speed of the propagating wave is defined

$$\frac{dx(t)}{dt} = c(x(t)) \tag{6.25}$$

Substitution of (6.25) into (6.24) and in view of (6.22), we have

$$\frac{dh(t)}{dt} = 0 \tag{6.26}$$

Therefore, a solution of (6.22) is given by

$$h(t) = \text{const.} \tag{6.27}$$

along a characteristics curve. We can also determine the characteristics as

$$\frac{dx}{c(x)} = dt \tag{6.28}$$

Integrating both sides gives

$$\alpha(x) = \int \frac{dx}{c(x)} = t + k \tag{6.29}$$

where k is a constant. The characteristics is thus

$$\xi = \alpha(x) - t = k \tag{6.30}$$

The final solution is then obtained as

$$u(t,x) = v(\alpha(x) - t) \tag{6.31}$$

6.3.3. Nonlinear Transport and Shock Waves

We have seen in the last section that if the wave speed is a function of position, we can express the solution in a characteristics curve. If the wave speed depends on the unknown magnitude u, the first order PDE becomes nonlinear. The simplest case is $c = u$ (Whitham, 1974):

$$\frac{\partial u}{\partial t}+u\frac{\partial u}{\partial x}=0 \tag{6.32}$$

This equation is called the Poisson-Riemann equation is found useful in modeling the shock wave phenomenon, including traffic flow and flood waves in rivers. For the case of traffic flow, u is the traffic density. Since the wave speed equals the wave magnitude, a larger wave travels faster than a smaller wave. Thus, a larger wave behind a smaller one will eventually catch up and pass the smaller wave.

Following a similar procedure in the last section, we seek a solution

$$h(t)=u(t,x(t)) \tag{6.33}$$

Taking the total differential with respect to t using the chain rule gives

$$\frac{dh(t)}{dt}=\frac{\partial u}{\partial t}(t,x(t))+\frac{\partial u}{\partial x}(t,x(t))\frac{dx}{dt}=\frac{\partial u}{\partial t}+u\frac{\partial u}{\partial x}=0 \tag{6.34}$$

Thus, we again have h = constant along a characteristics.

$$\frac{dx}{dt}=c=u \tag{6.35}$$

We can integrate (6.35) as

$$\xi=x-ut=k \tag{6.36}$$

Thus, the general solution of the Poisson-Riemann equation is

$$u=f(\xi)=f(x-ut) \tag{6.37}$$

We are going to see that this solution for shock waves will break down at a certain time. Let us consider that

$$\frac{\partial u}{\partial x}=\frac{\partial}{\partial x}f(\xi)=f'(\xi)\frac{\partial \xi}{\partial x}=f'(\xi)(1-t\frac{\partial u}{\partial x}) \tag{6.38}$$

Note that the derivative $\partial u/\partial x$ appears on both sides of (6.38). Solving for it gives

$$\frac{\partial u}{\partial x}=\frac{f'(\xi)}{1+tf'(\xi)} \tag{6.39}$$

We can see that

$$\frac{\partial u}{\partial x}\to\infty, \quad as \quad t\to t^*=-\frac{1}{f'(\xi)} \tag{6.40}$$

If the initial data of the wave profile is such that $f'(\xi) < 0$, we will have a breaking wave as shown in Figure 6.2. Physically, the wave breaks down (breaking wave) as u becomes a multi-valued solution of x for $t > t^*$. This is a major characteristics of a shock wave.

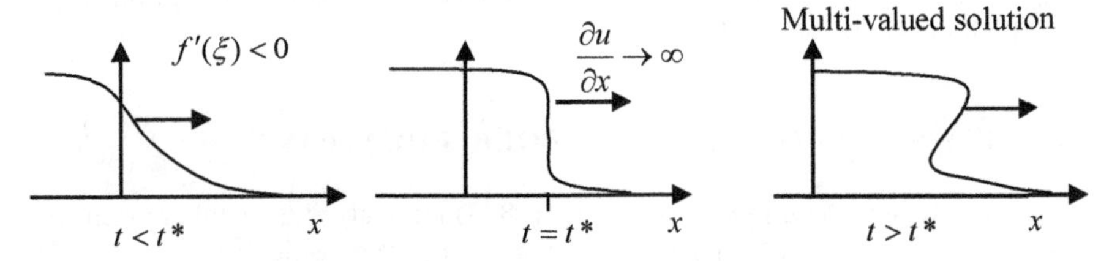

Figure 6.2 Evolution and breaking down of a shock wave

Actually, the Poisson-Riemann equation can be interpreted in terms of a generalized conservation law:

$$\frac{\partial T}{\partial t}+\frac{\partial X}{\partial x}=0 \tag{6.41}$$

where $T = T(t,u,x)$ is a conserved density and $X(t,u,x)$ is the flux through a control volume. We can see that if we take a particular case

$$T=u, \quad X=\frac{1}{2}u^2 \tag{6.42}$$

we have

$$\frac{\partial u}{\partial t}+\frac{\partial}{\partial x}\left(\frac{1}{2}u^2\right)=\frac{\partial u}{\partial t}+u\frac{\partial u}{\partial x}=0 \tag{6.43}$$

We recover the Poisson-Riemann equation. Thus, the shock dynamics can be interpreted as the mass conservation law in integral form:

$$\frac{d}{dt}\int_{x_1}^{x_2} T dx=\int_{x_1}^{x_2}\frac{\partial T}{\partial t}dx=-\int_{x_1}^{x_2}\frac{\partial X}{\partial x}dx=-X\Big|_{x_1}^{x_2} \tag{6.44}$$

Graphically, we can interpret this conservation law as shown in Figure 6.3.

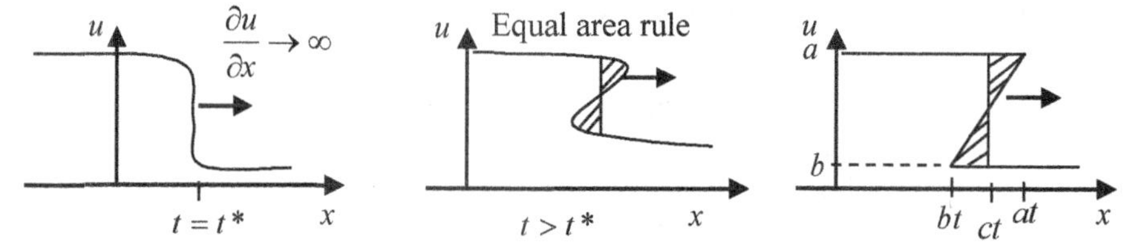

Figure 6.3 Physical law of conservation of mass

The value of u across the breaking wave can be estimated as

$$u(t,x)=\begin{cases} a & x<ct \\ b & x>ct \end{cases} \tag{6.45}$$

where

$$c=\frac{a+b}{2} \tag{6.46}$$

This is called the Rankine-Hugoniot condition.

We have seen from this section that a first order PDE relates directly to the applications of shock waves, which is important in traffic and flood wave modelling in engineering and mechanics.

6.4 AIRY'S METHOD FOR HYPERBOLIC EQUATION

In this section, a method presented in Airy (1873) for analyzing a nonhomogeneous hyperbolic equation will be discussed. In particular, we consider a first order PDE of the following form

$$\frac{\partial z}{\partial x} - a\frac{\partial z}{\partial y} = \alpha(x, y) \tag{6.47}$$

Airy proposed the following change of variables

$$u = ax + y, \quad v = ex + fy \tag{6.48}$$

in which e and f are unknown constants. If $af \neq e$, we can invert (6.48) as

$$x = \frac{fu - v}{af - e}, \quad y = \frac{-eu + av}{af - e} \tag{6.49}$$

Applying the chain rule we have

$$\frac{\partial z}{\partial x} = \frac{\partial z}{\partial u}\frac{\partial u}{\partial x} + \frac{\partial z}{\partial v}\frac{\partial v}{\partial x} = \frac{\partial z}{\partial u}a + e\frac{\partial z}{\partial v} \tag{6.50}$$

$$\frac{\partial z}{\partial y} = \frac{\partial z}{\partial u}\frac{\partial u}{\partial y} + \frac{\partial z}{\partial v}\frac{\partial v}{\partial y} = \frac{\partial z}{\partial u} + f\frac{\partial z}{\partial v} \tag{6.51}$$

Substitution of (6.50) and (6.51) into (6.47) gives

$$(e - af)\frac{\partial z}{\partial v} = \alpha(x, y) = \alpha(\frac{fu - v}{af - e}, \frac{-eu + av}{af - e}) \tag{6.52}$$

Integrating (6.52) with respect to v we have

$$z = \phi(u) + \frac{1}{e - af}\int_v \alpha(\frac{fu - v}{af - e}, \frac{-eu + av}{af - e})dv \tag{6.53}$$

where $u = ax+y$, which as we have seen in Section 6.2 is the characteristics. When the integration is conducted for v, we can treat u as constant.

Let us consider a more specific function α:

$$\alpha = x^p y^q \tag{6.54}$$

where p and q are constants. Substitution of (6.54) into (6.53) leads to

$$z = \phi(u) - \frac{1}{(af - e)^{p+q+1}}\int_v (fu - v)^p(-eu + av)^q dv \tag{6.55}$$

We can apply integration by parts to (6.55) as

$$\begin{aligned} z = \phi(u) &+ \frac{1}{(p+1)(e - af)^{p+q+1}}(fu - v)^{p+1}(-eu + av)^q \\ &- \frac{aq}{(p+1)(e - af)^{p+q+1}}\int_v (fu - v)^{p+1}(-eu + av)^{q-1}dv \end{aligned} \tag{6.56}$$

Repeat this process of integration by parts q times, and note the following result

$$\begin{aligned} \int_v (fu - v)^{p+q}dv &= -\int_v (fu - v)^{p+q}d(fu - v) \\ &= -\frac{(fu - v)^{p+q+1}}{p+q+1} = -\frac{(af - e)^{p+q+1}x^{p+q+1}}{p+q+1} \end{aligned} \tag{6.57}$$

we arrive at

$$z=\phi(u)+\frac{1}{p+1}x^{p+1}y^{q}+\frac{aq}{(p+1)(p+2)}x^{p+2}y^{q-1}$$
$$+\frac{a^2q(q-1)}{(p+1)(p+2)(p+3)}x^{p+3}y^{q-2}+...+\frac{a^q q(q-1)\cdots 2}{(p+1)\cdots(p+q)}x^{p+q} \tag{6.58}$$
$$+\frac{a^q q!p!}{(p+1+q)!}x^{p+q+1}$$

More importantly, we find that the solution is independent of e and f.

6.5 GEOMETRIC INTERPRETATION OF LINEAR PDE

The geometric interpretation was discovered by Monge in the 1770s and published much later in 1785. Geometrically, we first note that two different three-dimensional surfaces can be expressed in terms of two potential F and G as:

$$\nabla G=\frac{\partial G}{\partial x}\boldsymbol{e}_x+\frac{\partial G}{\partial y}\boldsymbol{e}_y+\frac{\partial G}{\partial z}\boldsymbol{e}_z \tag{6.59}$$

$$\nabla F=\frac{\partial F}{\partial x}\boldsymbol{e}_x+\frac{\partial F}{\partial y}\boldsymbol{e}_y+\frac{\partial F}{\partial z}\boldsymbol{e}_z \tag{6.60}$$

Figure 6.4 illustrates that ∇F is the normal of a spherical surface S_1 whereas ∇G is the normal to a conical surface S_2. The vector $\boldsymbol{E}$ can be interpreted as the tangent of the intersecting curve Γ,

$$\boldsymbol{E}=\nabla F\times\nabla G=(\frac{\partial F}{\partial y}\frac{\partial G}{\partial z}-\frac{\partial F}{\partial z}\frac{\partial G}{\partial y})\boldsymbol{e}_x$$
$$+(\frac{\partial G}{\partial x}\frac{\partial F}{\partial z}-\frac{\partial F}{\partial x}\frac{\partial G}{\partial z})\boldsymbol{e}_y+(\frac{\partial F}{\partial x}\frac{\partial G}{\partial y}-\frac{\partial F}{\partial y}\frac{\partial G}{\partial x})\boldsymbol{e}_z \tag{6.61}$$

The tangent of the intersecting curve is

$$d\xi=dx\boldsymbol{e}_x+dy\boldsymbol{e}_y+dz\boldsymbol{e}_z \tag{6.62}$$

Note that the coefficient of (6.61) can be expressed in terms of a Jacobian as

$$\frac{\partial(F,G)}{\partial(y,z)}=\frac{\partial F}{\partial y}\frac{\partial G}{\partial z}-\frac{\partial F}{\partial z}\frac{\partial G}{\partial y} \tag{6.63}$$

Since $\boldsymbol{E}$ and $d\xi$ are parallel, the components of these two vectors must be proportional:

$$\frac{dx}{\frac{\partial(F,G)}{\partial(y,z)}}=\frac{dy}{\frac{\partial(F,G)}{\partial(z,x)}}=\frac{dz}{\frac{\partial(F,G)}{\partial(x,y)}} \tag{6.64}$$

In a more compact form, we replace these Jacobians by P, Q, and R as

$$\frac{dx}{P(x,y,z)}=\frac{dy}{Q(x,y,z)}=\frac{dz}{R(x,y,z)} \tag{6.65}$$

We will discuss in more detail in the next section on solving these equations, which are also known as characteristics, and the final solution of its corresponding PDE.

We will now consider the relation between (6.65) and the solution of the first order PDE. Using the definition in (6.66), we have the vector $\boldsymbol{E}$

$$\boldsymbol{E} = P\boldsymbol{e}_x + Q\boldsymbol{e}_y + R\boldsymbol{e}_z \tag{6.66}$$

These components are the coefficients of the first order PDE given in (6.2). Note that the vector $\boldsymbol{E}$ will be parallel to a curve Γ which is an intersection of two surfaces denoting of their normal vectors ∇G and ∇F. The direction of the three-dimensional curve is shown in Figure 6.1 as $d\xi$. Therefore, the solution of any first order PDE can be interpreted as an intersection curve of two 3-D surfaces.

Clearly, (6.65) can be rearranged to give two characteristic equations

$$\frac{dy}{dx} = \frac{Q(x,y,z)}{P(x,y,z)}, \quad \frac{dz}{dx} = \frac{R(x,y,z)}{P(x,y,z)} \tag{6.67}$$

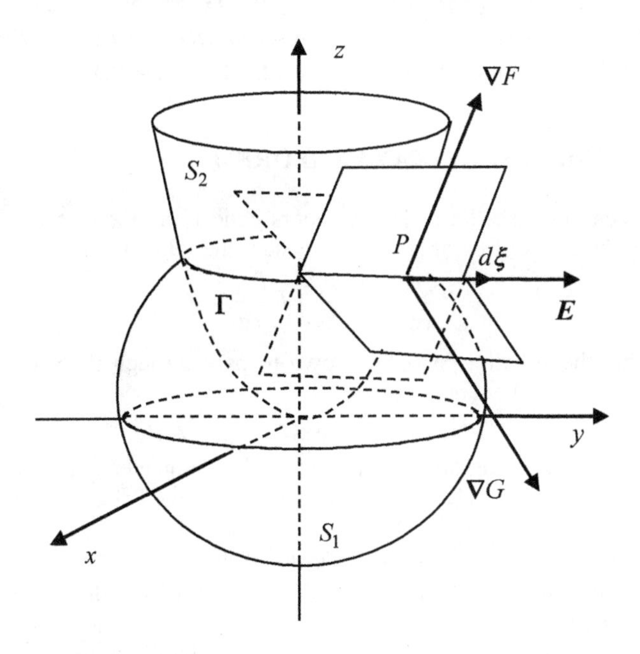

Figure 6.4 Geometric interpretation of the solution of 1st order PDE

This equation actually provides the idea of Cauchy's method of characteristics for wave equations that will be discussed in a later chapter. Solving these first order ODEs in (6.67), which are called characteristics equations, gives the following characteristic lines:

$$u_1(x,y,z) = c_1 \tag{6.68}$$

$$u_2(x,y,z) = c_2 \tag{6.69}$$

Equations (6.68) and (6.69) are the characteristics of the solution of the PDE. Physically, if the PDE reflects the wave phenomenon, these characteristics lines are the propagating directions of waves; in the theory of light, these are rays. The solution of the original PDE will be discussed in a later section. Taking the total differential of (6.68) gives

$$du_1 = \frac{\partial u_1}{\partial x} dx + \frac{\partial u_1}{\partial y} dy + \frac{\partial u_1}{\partial z} dz = 0 \tag{6.70}$$

If a first order PDE is given in the following form:

$$P(x, y, z)\frac{\partial u_1}{\partial x} + Q(x, y, z)\frac{\partial u_1}{\partial y} + R(x, y, z)\frac{\partial u_1}{\partial z} = 0 , \tag{6.71}$$

we see that the following subsidiary equation (parallelism between $\boldsymbol{E}$ and $d\xi$) of (6.71) is

$$\frac{dx}{P(x, y, z)} = \frac{dy}{Q(x, y, z)} = \frac{dz}{R(x, y, z)} = \frac{du_1}{0} \tag{6.72}$$

This is, of course, the same as (6.65). We will see in Section 6.7 that (6.72) is actually the subsidiary equation for the characteristics of the first order PDE.

6.6 CAUCHY-KOVALEVSKAYA THEOREM

In 1842, Cauchy published a series of papers concerning the existence of a PDE solution. For the case of first order PDE, consider the following PDE:

$$P\frac{\partial u_1}{\partial x} + Q\frac{\partial u_1}{\partial y} + R\frac{\partial u_1}{\partial z} = 0 \tag{6.73}$$

Assuming that the solution curve is known to pass through the following initial point

$$x_0 = x_0(s), \quad y_0 = y_0(s), \quad z_0 = z_0(s) \tag{6.74}$$

with all x_0, y_0, and z_0 being analytic functions of a parameter s (i.e., differentiable for real variables). Cauchy showed that the solution of the PDE with the initial data exists:

$$u_1(x, y, z, c_1, c_2) = 0 \tag{6.75}$$

The two constants in (6.75) indicate the fact that the final solution depends on both of the characteristics lines given in (6.68) and (6.69). The existence theorem was also investigated independently and improved by Sophie Kovalevskaya, who was a student of Weierstrass, in 1874. Darboux also considered the same problem in 1875. The work of Cauchy seems not to be known to Kovalevskaya and Darboux. In fact, in the paper by Kovalevskaya she cited both Briot and Bouquet but both of them refer to the work by Cauchy though without giving a precise citation of Cauchy. The proof by Cauchy and Kovalevskaya was later improved by Goursat in 1898 by using analytic functions. However, in the literature, the existence theorem is normally referred to as the Cauchy-Kovalevskaya existence theorem.

6.7 LAGRANGE'S METHOD FOR LINEAR FIRST ORDER PDE

The solution of first order PDEs was derived by Lagrange in 1772. Let us generalize the first order PDE to n-dimensional space (n variables). In particular, consider a first order PDE with n variables:

$$P_1 \frac{\partial z}{\partial x_1} + P_n \frac{\partial z}{\partial x_2} + ... + P_n \frac{\partial z}{\partial x_n} = R \tag{6.76}$$

Taking the total differential of z, we obtain:

$$\frac{\partial z}{\partial x_1} dx_1 + \frac{\partial z}{\partial x_2} dx_2 + ... + \frac{\partial z}{\partial x_n} dx_n = dz \tag{6.77}$$

If (6.76) and (6.77) are equivalent, we must have the following subsidiary equation (parallel between $\boldsymbol{E}$ and $d\xi$ discussed in Section 6.5) for characteristics

$$\frac{dx_1}{P_1} = \frac{dx_2}{P_2} = ... = \frac{dx_n}{P_n} = \frac{dz}{R} = \lambda \tag{6.78}$$

where λ is a constant. The term characteristics was introduced by Cauchy. The solution of the characteristics given in (6.78) is also a solution to (6.76). This equivalence was known to Euler.

The solutions of the characteristics defined in (6.78) can be expressed symbolically as

$$\begin{aligned} &u_1(x_1, x_2, ..., x_n, z) = a_1 \\ &u_2(x_1, x_2, ..., x_n, z) = a_2 \\ &\vdots \\ &u_n(x_1, x_2, ..., x_n, z) = a_n \end{aligned} \tag{6.79}$$

According to Lagrange, the general solution of (6.76) is given as

$$\phi(u_1, u_2, ..., u_n) = 0 \tag{6.80}$$

where ϕ is an arbitrary function. This result was obtained in 1785 by Lagrange. Let us consider the proof for the case of two independent variables here. The proof for more than two variables follows similarly. In particular, consider the differential equation and its subsidiary equation as

$$P_1 \frac{\partial z}{\partial x_1} + P_2 \frac{\partial z}{\partial x_2} = R, \quad \frac{dx_1}{P_1} = \frac{dx_2}{P_2} = \frac{dz}{R} = \lambda \tag{6.81}$$

The characteristics by solving the second of (6.81) are expressed symbolically as:

$$u_1(x_1, x_2, z) = c_1, \quad u_2(x_1, x_2, z) = c_2 \tag{6.82}$$

Taking the total differential of (6.82) gives

$$du_1 = \frac{\partial u_1}{\partial x_1} dx_1 + \frac{\partial u_1}{\partial x_2} dx_2 + \frac{\partial u_1}{\partial z} dz = 0 \tag{6.83}$$

$$du_2 = \frac{\partial u_2}{\partial x_1} dx_1 + \frac{\partial u_2}{\partial x_2} dx_2 + \frac{\partial u_2}{\partial z} dz = 0 \tag{6.84}$$

Using the results of Section 6.5, (6.83) and (6.84) correspond to two first order PDEs as

$$P_1 \frac{\partial u_1}{\partial x_1} + P_2 \frac{\partial u_1}{\partial x_2} + R \frac{\partial u_1}{\partial z} = 0 \tag{6.85}$$

$$P_1 \frac{\partial u_2}{\partial x_1} + P_2 \frac{\partial u_2}{\partial x_2} + R \frac{\partial u_2}{\partial z} = 0 \tag{6.86}$$

These two equations can be put in a matrix form as

$$\begin{bmatrix} \dfrac{\partial u_1}{\partial x_1} & \dfrac{\partial u_1}{\partial x_2} \\ \dfrac{\partial u_2}{\partial x_1} & \dfrac{\partial u_2}{\partial x_2} \end{bmatrix} \begin{Bmatrix} P_1 \\ P_2 \end{Bmatrix} = - \begin{Bmatrix} R \dfrac{\partial u_1}{\partial z} \\ R \dfrac{\partial u_2}{\partial z} \end{Bmatrix} \tag{6.87}$$

The solution of P_1 is

$$P_1 = \frac{-R \dfrac{\partial u_1}{\partial z} \dfrac{\partial u_2}{\partial x_2} + R \dfrac{\partial u_2}{\partial z} \dfrac{\partial u_1}{\partial x_2}}{\dfrac{\partial u_1}{\partial x_1} \dfrac{\partial u_2}{\partial x_2} - \dfrac{\partial u_1}{\partial x_2} \dfrac{\partial u_2}{\partial x_1}} = \frac{-R \dfrac{\partial (u_1, u_2)}{\partial (z, x_2)}}{\dfrac{\partial (u_1, u_2)}{\partial (x_1, x_2)}} \tag{6.88}$$

Rearranging (6.88), we obtain

$$\frac{P_1}{\dfrac{\partial u_1}{\partial z} \dfrac{\partial u_2}{\partial x_2} - \dfrac{\partial u_2}{\partial z} \dfrac{\partial u_1}{\partial x_2}} = \frac{-R}{\dfrac{\partial u_1}{\partial x_1} \dfrac{\partial u_2}{\partial x_2} - \dfrac{\partial u_1}{\partial x_2} \dfrac{\partial u_2}{\partial x_1}} \tag{6.89}$$

Similarly, the solution of P_2 leads to

$$\frac{P_2}{\dfrac{\partial (u_1, u_2)}{\partial (x_1, z)}} = \frac{-R}{\dfrac{\partial (u_1, u_2)}{\partial (x_1, x_2)}} \tag{6.90}$$

Characteristic equations relating R, P_1, and P_2 are

$$\frac{P_2}{\dfrac{\partial (u_1, u_2)}{\partial (x_1, z)}} = \frac{-R}{\dfrac{\partial (u_1, u_2)}{\partial (x_1, x_2)}} = \frac{P_1}{\dfrac{\partial (u_1, u_2)}{\partial (z, x_2)}} \tag{6.91}$$

We now return to the following characteristics

$$u_1(x_1, x_2, z) = c_2 \tag{6.92}$$

$$u_2(x_1, x_2, z) = c_2 \tag{6.93}$$

Consider an arbitrary function containing u_1 and u_2

$$\Phi(u_1, u_2) = 0 \tag{6.94}$$

Taking differentiation of (6.94) with respect to x_1 and x_2, we find

$$\frac{\partial \Phi}{\partial u_1} \frac{\partial u_1}{\partial x_1} + \frac{\partial \Phi}{\partial u_2} \frac{\partial u_2}{\partial x_1} = 0, \quad \frac{\partial \Phi}{\partial u_1} \frac{\partial u_1}{\partial x_2} + \frac{\partial \Phi}{\partial u_2} \frac{\partial u_2}{\partial x_2} = 0 \tag{6.95}$$

Differentiate (6.92) and (6.93) with respect to x_1

$$\frac{\partial u_1}{\partial x_1} = \frac{\partial u_1}{\partial x_1} + \frac{\partial u_1}{\partial z} \frac{\partial z}{\partial x_1}, \quad \frac{\partial u_2}{\partial x_1} = \frac{\partial u_2}{\partial x_1} + \frac{\partial u_2}{\partial z} \frac{\partial z}{\partial x_1} \tag{6.96}$$

Substitution of (6.96) into the first equation of (6.95) gives

$$\frac{\partial \Phi}{\partial u_1}\left(\frac{\partial u_1}{\partial x_1}+\frac{\partial u_1}{\partial z}\frac{\partial z}{\partial x_1}\right)+\frac{\partial \Phi}{\partial u_2}\left(\frac{\partial u_2}{\partial x_1}+\frac{\partial u_2}{\partial z}\frac{\partial z}{\partial x_1}\right)=0 \tag{6.97}$$

Similarly, the second equation of (6.95) can be rewritten as

$$\frac{\partial \Phi}{\partial u_1}\left(\frac{\partial u_1}{\partial x_2}+\frac{\partial u_1}{\partial z}\frac{\partial z}{\partial x_2}\right)+\frac{\partial \Phi}{\partial u_2}\left(\frac{\partial u_2}{\partial x_2}+\frac{\partial u_2}{\partial z}\frac{\partial z}{\partial x_2}\right)=0 \tag{6.98}$$

These two equations can be put in a matrix form as

$$\begin{bmatrix} \frac{\partial u_1}{\partial x_1}+\frac{\partial u_1}{\partial z}\frac{\partial z}{\partial x_1} & \frac{\partial u_2}{\partial x_1}+\frac{\partial u_2}{\partial z}\frac{\partial z}{\partial x_1} \\ \frac{\partial u_1}{\partial x_2}+\frac{\partial u_1}{\partial z}\frac{\partial z}{\partial x_2} & \frac{\partial u_2}{\partial x_2}+\frac{\partial u_2}{\partial z}\frac{\partial z}{\partial x_2} \end{bmatrix}\begin{Bmatrix} \frac{\partial \Phi}{\partial u_1} \\ \frac{\partial \Phi}{\partial u_2} \end{Bmatrix}=\begin{Bmatrix} 0 \\ 0 \end{Bmatrix} \tag{6.99}$$

For a nonzero solution for the derivatives of Φ, we require the determinant of the matrix being zero:

$$\left(\frac{\partial u_1}{\partial x_1}+\frac{\partial u_1}{\partial z}\frac{\partial z}{\partial x_1}\right)\left(\frac{\partial u_2}{\partial x_2}+\frac{\partial u_2}{\partial z}\frac{\partial z}{\partial x_2}\right)-\left(\frac{\partial u_2}{\partial x_1}+\frac{\partial u_2}{\partial z}\frac{\partial z}{\partial x_1}\right)\left(\frac{\partial u_1}{\partial x_2}+\frac{\partial u_1}{\partial z}\frac{\partial z}{\partial x_2}\right)=0 \tag{6.100}$$

Expanding the multiplication and simplifying the results gives

$$\begin{aligned} &\frac{\partial u_1}{\partial x_1}\frac{\partial u_2}{\partial x_2}-\frac{\partial u_1}{\partial x_2}\frac{\partial u_2}{\partial x_1}+\frac{\partial z}{\partial x_1}\left[\frac{\partial u_1}{\partial z}\frac{\partial u_2}{\partial x_2}-\frac{\partial u_2}{\partial z}\frac{\partial u_1}{\partial x_2}\right] \\ &+\frac{\partial z}{\partial x_2}\left[\frac{\partial u_2}{\partial z}\frac{\partial u_1}{\partial x_1}-\frac{\partial u_1}{\partial z}\frac{\partial u_2}{\partial x_1}\right]=0 \end{aligned} \tag{6.101}$$

Using the definition of the Jacobian, we can simplify (6.101) to

$$\frac{\partial(u_1,u_2)}{\partial(x_1,x_2)}+\frac{\partial z}{\partial x_1}\frac{\partial(u_1,u_2)}{\partial(z,x_2)}+\frac{\partial z}{\partial x_2}\frac{\partial(u_1,u_2)}{\partial(x_1,z)}=0 \tag{6.102}$$

Substitution of the results in (6.91) yields the following first order PDE

$$-R+P_1\frac{\partial z}{\partial x_1}+P_2\frac{\partial z}{\partial x_2}=0 \tag{6.103}$$

This is the PDE given in (6.81) and, thus, the solution of (6.103) is given by (6.94). The proof can be extended to the case of n variables to show the validity of (6.80).

6.8 LAGRANGE MULTIPLIER

In this section, we will solve the subsidiary equation given in (6.78) to get the characteristics. It is not always straightforward to solve these characteristics equations. Very often, the method of the Lagrange multiplier can be used to simplify the problem. In particular, for the case of n variables, we have the subsidiary equation as:

$$\frac{dx_1}{P_1}=\frac{dx_2}{P_2}=\ldots=\frac{dx_n}{P_n}=\frac{dz}{R}=\lambda=\frac{\mu_1 dx_1+\mu_2 dx_2+\ldots\mu_n dx_n+\mu_{n+1}dz}{\mu_1 P_1+\mu_2 P_2+\ldots\mu_n P_n+\mu_{n+1}R} \tag{6.104}$$

The last part of (6.104) is the result of the method of the Lagrange multiplier and $\mu_1, \mu_2, \ldots, \mu_{n+1}$ are the Lagrange multipliers and are arbitrary functions of the n variables.

To prove the last part of (6.104), we first notice that $dx_i = \lambda P_i$ or

$$\begin{aligned} &\mu_1 dx_1 = \mu_1 \lambda P_1, \quad \mu_2 dx_2 = \mu_2 \lambda P_2, \ldots, \\ &\mu_n dx_n = \mu_n \lambda P_n, \quad \mu_{n+1} dz = \mu_{n+1} \lambda R \end{aligned} \tag{6.105}$$

Substitution of (6.105) into the last of (6.104) gives

$$\begin{aligned} \frac{\mu_1 dx_1 + \mu_2 dx_2 + \ldots \mu_n dx_n + \mu_{n+1} dz}{\mu_1 P_1 + \mu_2 P_2 + \ldots \mu_n P_n + \mu_{n+1} R} &= \frac{\lambda \mu_1 P_1 + \lambda \mu_2 P_2 + \ldots + \lambda \mu_n P_n + \lambda \mu_{n+1} R}{\mu_1 P_1 + \mu_2 P_2 + \ldots \mu_n P_n + \mu_{n+1} R} \\ &= \frac{\lambda(\mu_1 P_1 + \mu_2 P_2 + \ldots + \mu_n P_n + \mu_{n+1} R)}{\mu_1 P_1 + \mu_2 P_2 + \ldots \mu_n P_n + \mu_{n+1} R} = \lambda \end{aligned} \tag{6.106}$$

This completes the proof. The role of $\mu_1, \mu_2, \ldots, \mu_{n+1}$ is like that of integrating factors. If we can find appropriate multipliers such that

$$\mu_1 P_1 + \mu_2 P_2 + \ldots \mu_n P_n + \mu_{n+1} R = 0, \tag{6.107}$$

from (6.106), it also implies

$$\mu_1 dx_1 + \mu_2 dx_2 + \ldots \mu_n dx_n + \mu_{n+1} dz = 0 \tag{6.108}$$

Thus, the solution can in principle be found

$$u(x_1, x_2, \ldots, x_n, z) = a \tag{6.109}$$

This method of the Lagrange multiplier will be illustrated in the following example.

Example 6.1 Find the solution for the following first order PDE

$$y \frac{\partial z}{\partial x} - x \frac{\partial z}{\partial y} = 0 \tag{6.110}$$

Solution: The subsidiary equation of (6.110) is

$$\frac{dx}{y} = -\frac{dy}{x} = \frac{dz}{0} \tag{6.111}$$

The last part of (6.111) implies

$$dz = 0 \tag{6.112}$$

Integration gives

$$u_1 = z = a_2 \tag{6.113}$$

The first two parts of (6.111) give

$$xdx + ydy = 0 \tag{6.114}$$

Integration of (6.114) results in

$$u_2 = x^2 + y^2 = a_1 \tag{6.115}$$

By (6.80), the solution can be expressed as

$$\phi(x^2 + y^2, z) = 0 \tag{6.116}$$

This solution can be rewritten as

$$z = f(x^2 + y^2) \tag{6.117}$$

As expected, the general solution involves an arbitrary function *f*.

Example 6.2 Find the solution for the following first order PDE

$$(x^2+y^2)\frac{\partial z}{\partial x}+2xy\frac{\partial z}{\partial y}+z^2=0 \tag{6.118}$$

Solution: The subsidiary equation of (6.118) is

$$\frac{dx}{x^2+y^2}=\frac{dy}{2xy}=-\frac{dz}{z^2} \tag{6.119}$$

Using the Lagrange multiplier method given in (6.104), we have

$$\frac{dx+dy}{x^2+2xy+y^2}=\frac{dx-dy}{x^2-2xy+y^2}=-\frac{dz}{z^2}=\frac{d(x+y)}{(x+y)^2}=\frac{d(x-y)}{(x-y)^2} \tag{6.120}$$

The last equation in (6.120) can be expressed as

$$d(\frac{1}{z})=-d(\frac{1}{x+y}) \tag{6.121}$$

$$d(\frac{1}{z})=-d(\frac{1}{x-y}) \tag{6.122}$$

Integration gives the following characteristics

$$u_1=\frac{1}{z}+\frac{1}{x+y}=a_1 \tag{6.123}$$

$$u_2=\frac{1}{z}+\frac{1}{x-y}=a_2 \tag{6.124}$$

Therefore, the general solution can be expressed as

$$\phi(\frac{1}{z}+\frac{1}{x+y},\frac{1}{z}+\frac{1}{x-y})=0 \tag{6.125}$$

where ϕ is an arbitrary function.

Example 6.3 Find the solution for the following first order PDE with boundary condition

$$\frac{\partial u}{\partial x}+\frac{\partial u}{\partial y}+2\frac{\partial u}{\partial z}=0, \tag{6.126}$$

$$u=yz, \quad x=1 \tag{6.127}$$

Solution: The subsidiary equation of (6.126) is

$$\frac{dx}{1}=\frac{dy}{1}=\frac{dz}{2}=\frac{du}{0} \tag{6.128}$$

This gives characteristics equations as:

$$dx = dy, \quad dx = \frac{1}{2}dz, \quad du = 0 \tag{6.129}$$

It is straightforward to see that the corresponding characteristics lines are:

$$u_1 = x - y = a_1, \quad u_2 = 2x - z = a_2, \quad u_3 = u = a_3 \tag{6.130}$$

The solution can be expressed in terms of these characteristics as

$$\bar{\phi}(u_1, u_2, u_3) = 0 \tag{6.131}$$

Alternatively, we can express u_3 in terms of the other characteristics as

$$u = \phi(x - y, 2x - z) \tag{6.132}$$

Substituting the boundary condition given in (6.127)

$$u(x = 1) = yz = \phi(1 - y, 2 - z) \tag{6.133}$$

Putting $x = 1$ into (6.130), we get

$$1 - y = a_1 \tag{6.134}$$

$$2 - z = a_2 \tag{6.135}$$

$$yz = a_3 \tag{6.136}$$

We can solve for y and z in terms of constants a_1 and a_2 as:

$$y = 1 - a_1 \tag{6.137}$$

$$z = 2 - a_2 \tag{6.138}$$

$$u = yz = (1 - a_1)(2 - a_2) = a_3 \tag{6.139}$$

We can now substitute the value of a_1 and a_2 from (6.130) into (6.131) to get the general solution

$$u = (1 - x + y)(2 - 2x + z) \tag{6.140}$$

As a final check, when $x = 1$, we have $u = yz$ from (6.140).

6.9 PFAFFIAN EQUATIONS

Recall from Chapter 3 that first order ODEs can be expressed in differential form as:

$$Mdx + Ndy = 0 \tag{6.141}$$

We note that first order PDEs can also be expressed in a similar fashion. This is called Pfaffian:

$$Pdx + Qdy + Rdz = 0 \tag{6.142}$$

which is named after Pfaff, who was the supervisor of the renowned mathematician Gauss.

6.10 EAXCT EQUATIONS

For the three-dimensional case, the first order PDE in Pfaffian form is

$$Pdx + Qdy + Rdz = 0 \tag{6.143}$$

This PDE can be interpreted as a dot product of the following two vectors:

$$\boldsymbol{E} = (P, Q, R), \ \boldsymbol{d\xi} = (dx, dy, dz) \tag{6.144}$$

Therefore, (6.143) is equivalent to

$$\boldsymbol{E} \cdot \boldsymbol{d\xi} = 0 \tag{6.145}$$

If the vector $\boldsymbol{E}$ can be expressed in terms of a potential function U

$$\boldsymbol{E} = \nabla U, \tag{6.146}$$

the PDE given in (6.145) can be rewritten as:

$$\nabla U \cdot \boldsymbol{d\xi} = 0 \tag{6.147}$$

Or equivalently, we can express it as

$$\frac{\partial U}{\partial x} dx + \frac{\partial U}{\partial y} dy + \frac{\partial U}{\partial z} dz = 0 \tag{6.148}$$

This is clearly a total differential of a function U given by

$$U(x, y, z) = c \tag{6.149}$$

where c is a constant. Thus, this is the solution the Pfaffian differential equation. Therefore, the existence of a potential U defined in (6.146) is the condition of the Pfaffian being exact. An equation of this type is first considered by Clairaut in 1739.

6.11 INTEGRABILITY OF PFAFFIAN

If the Pfaffian differential equation is not exact, we can assume the existence of an integrating factor μ such that

$$\mu P dx + \mu Q dy + \mu R dz = 0 \tag{6.150}$$

Note that this idea is exactly the same as that discussed for first order ODE. In vector form, we require that (compare with (6.146)):

$$\mu \boldsymbol{E} = \nabla U \tag{6.151}$$

The governing equations for this integrating factor are

$$\frac{\partial^2 U}{\partial y \partial x} = \frac{\partial(\mu P)}{\partial y} = \frac{\partial(\mu Q)}{\partial x} \tag{6.152}$$

$$\frac{\partial^2 U}{\partial y \partial z} = \frac{\partial(\mu Q)}{\partial z} = \frac{\partial(\mu R)}{\partial y} \tag{6.153}$$

$$\frac{\partial^2 U}{\partial z \partial x} = \frac{\partial(\mu R)}{\partial x} = \frac{\partial(\mu P)}{\partial z} \tag{6.154}$$

This is an extension of the condition for an integrating factor for first order ODEs discussed in Chapter 3. Expanding (6.152) to (6.154) gives

$$\mu(\frac{\partial P}{\partial y} - \frac{\partial Q}{\partial x}) = Q \frac{\partial \mu}{\partial x} - P \frac{\partial \mu}{\partial y} \tag{6.155}$$

$$\mu(\frac{\partial Q}{\partial z} - \frac{\partial R}{\partial y}) = R \frac{\partial \mu}{\partial y} - Q \frac{\partial \mu}{\partial z} \tag{6.156}$$

$$\mu(\frac{\partial R}{\partial x} - \frac{\partial P}{\partial z}) = P \frac{\partial \mu}{\partial z} - R \frac{\partial \mu}{\partial x} \tag{6.157}$$

We can add the results of (6.155) multiplied by R, (6.156) multiplied by P, and (6.157) multiplied by Q to yield

$$R(\frac{\partial P}{\partial y}-\frac{\partial Q}{\partial x})+P(\frac{\partial Q}{\partial z}-\frac{\partial R}{\partial y})+Q(\frac{\partial R}{\partial x}-\frac{\partial P}{\partial z})=0 \tag{6.158}$$

This condition, if expressed in terms of the vector $\boldsymbol{E}$, can be written as

$$\boldsymbol{E}\cdot(\nabla\times\boldsymbol{E})=0 \tag{6.159}$$

This is the condition of the Pfaffian differential equation being integrable. This condition (6.159) was derived by Clairaut in 1740 and by D'Alembert in 1744. Once (6.158) is satisfied, we can solve the problem using one of the following procedures. Note that the following procedure can only apply for the case where the Pfaffian is integrable.

Method 1:

Because the Pfaffian is integrable, we must have a function U such that

$$dU = Pdx + Qdy + Rdz \tag{6.160}$$

This implies that

$$P=\frac{\partial U}{\partial x},\quad Q=\frac{\partial U}{\partial y},\quad R=\frac{\partial U}{\partial z} \tag{6.161}$$

Integrating (6.160) we have

$$U(x,y,z)=C \tag{6.162}$$

where C is a constant. Therefore, the solution of the Pfaffian is a one-parameter curve.

Method 2:

First, (6.143) can be rearranged as:

$$dz=-\frac{P}{R}dx-\frac{Q}{R}dy=\frac{\partial z}{\partial x}dx+\frac{\partial z}{\partial y}dy \tag{6.163}$$

Comparison of the terms in (6.163) gives

$$\frac{\partial z}{\partial x}=-\frac{P}{R}=P_1(x,y,z) \tag{6.164}$$

$$\frac{\partial z}{\partial y}=-\frac{Q}{R}=Q_1(x,y,z) \tag{6.165}$$

Thus, we can integrate (6.164) with respect to x by holding y constant. We have

$$z=\varphi(x,y;C(y)) \tag{6.166}$$

where $C(y)$ is an arbitrary function that we add in the process of integration. Back substitution of (6.166) into (6.165) gives an ODE for $C(y)$. After its determination, (6.166) becomes the complete solution. Note that this procedure is similar to Section 3.2.4 for integrable ODEs.

Method 3:

Step 1: Hold z = constant or set $dz = 0$;

Step 2: integrate with respect to x and y and add an arbitrary function of $f(z)$;
Step 3: differentiate the result obtained in Step 2 with respect to all variables; and
Step 4: compare the result in Step 3 with the original PDE to identify $f(z)$ in Step 2.

Note that these methods are somewhat similar. These procedures are illustrated in the following example.

Example 6.4 Find the solution for the following PDE in Pfaffian form

$$3yzdx + 2xzdy + xydz = 0 \tag{6.167}$$

Solution: The functions P, Q, and R can be identified as

$$P = 3yz, \quad Q = 2xz, \quad R = xy \tag{6.168}$$

Taking partial differentiation of these functions gives

$$\frac{\partial P}{\partial y} = 3z, \frac{\partial Q}{\partial x} = 2z, \frac{\partial Q}{\partial z} = 2x, \frac{\partial R}{\partial y} = x, \frac{\partial R}{\partial x} = y, \frac{\partial P}{\partial z} = 3y \tag{6.169}$$

Substitution of (6.169) into the integrability condition (6.158) gives

$$\begin{aligned} &R(\frac{\partial P}{\partial y} - \frac{\partial Q}{\partial x}) + P(\frac{\partial Q}{\partial z} - \frac{\partial R}{\partial y}) + Q(\frac{\partial R}{\partial x} - \frac{\partial P}{\partial z}) \\ &= xy(3z - 2z) + 3yz(2x - x) + 2xz(y - 3y) = 0 \end{aligned} \tag{6.170}$$

Thus, (6.167) is integrable.

Method 1:

Dividing through (6.167) by xyz, we get

$$\frac{3}{x}dx + \frac{2}{y}dy + \frac{1}{z}dz = 0 . \tag{6.171}$$

Integrating (6.171), we arrive at

$$3\ln x + 2\ln y + \ln z = c \tag{6.172}$$

Combining these terms, we have

$$\ln(x^3 y^2 z) = c \tag{6.173}$$

Taking the exponential function on both sides, we find

$$x^3 y^2 z = C \tag{6.174}$$

This is the solution of the Pfaffian given in (6.167)

Method 2:

It is straightforward to identify that

$$\frac{\partial z}{\partial x} = -\frac{P}{R} = P_1(x, y, z) = -\frac{3z}{x} \tag{6.175}$$

$$\frac{\partial z}{\partial y} = -\frac{Q}{R} = Q_1(x, y, z) = -\frac{2z}{y} \tag{6.176}$$

Integrating (6.175), we obtain

$$\ln z + \ln x^3 = \bar{C}(y) \tag{6.177}$$

Thus, we have

$$zx^3 = C(y) \tag{6.178}$$

Substitution of (6.178) into (6.176) gives

$$\frac{dC}{dy} = -\frac{2C}{y} \tag{6.179}$$

Integrating both sides, we find C as

$$C(y) = \frac{C_1}{y^2} \tag{6.180}$$

Finally, substitution of (6.180) into (6.178) gives the solution as

$$x^3 y^2 z = C_1 \tag{6.181}$$

As expected, this is the same as (6.174).

Method 3:

Step 1: We can set

$$dz = 0, \text{ or } \ z = a_1 \tag{6.182}$$

By doing so, (6.167) is reduced to

$$3y dx + 2x dy = 0 \tag{6.183}$$

Step 2: Integrate (6.183) as

$$2\frac{dy}{y} = -3\frac{dx}{x} \tag{6.184}$$

This can be integrated exactly as

$$2\ln y = -3\ln x + \ln f(z) \tag{6.185}$$

We have put the arbitrary function of z in a compact form:

$$y^2 = \frac{1}{x^3} f(z) \tag{6.186}$$

Step 3: We can now differentiate (6.186) as

$$3x^2 y^2 dx + 2yx^3 dy = f' dz \tag{6.187}$$

Step 4: Comparison of (6.187) with the original PDE (6.167) yields f

$$f' = -\frac{x^3 y^2}{z} = -\frac{f}{z} \tag{6.188}$$

$$\ln f = -\ln z + C_0 \tag{6.189}$$

$$f = \frac{1}{z} C \tag{6.190}$$

Finally, we find

$$x^3 y^2 z = C \tag{6.191}$$

Again, all different procedures lead to the same answer.

Example 6.5 A complex function is defined by two functions $u(x,y)$ and $v(x,y)$ as:

$$w = u(x, y) + iv(x, y) = f(z) \tag{6.192}$$

where f is an arbitrary function and

$$z = x + iy \tag{6.193}$$

Show that

$$\frac{\partial u}{\partial y} = -\frac{\partial v}{\partial x}, \quad \frac{\partial u}{\partial x} = \frac{\partial v}{\partial y} \tag{6.194}$$

Solution: This problem actually provides a different proof of the Cauchy-Riemann equations, by solving first order PDEs. The main focus is to show that we must have $z = x+iy$, as the characteristics for this case. Consider the following derivatives:

$$\frac{\partial w}{\partial x} = \frac{dw}{dz}\frac{\partial z}{\partial x} = f'(z) = \frac{dw}{dz} \tag{6.195}$$

$$\frac{\partial w}{\partial y} = \frac{dw}{dz}\frac{\partial z}{\partial y} = if'(z) = i\frac{dw}{dz} \tag{6.196}$$

Comparison of (6.195) and (6.196) gives

$$f'(z) = \frac{\partial w}{\partial x} = \frac{1}{i}\frac{\partial w}{\partial y} \tag{6.197}$$

The last part of (6.197) actually provides a first order PDE:

$$\frac{\partial w}{\partial x} - \frac{1}{i}\frac{\partial w}{\partial y} = \frac{\partial w}{\partial x} + i\frac{\partial w}{\partial y} = 0 \tag{6.198}$$

Using the Lagrange method, we have the subsidiary equation as:

$$\frac{dx}{1} = \frac{dy}{i} = \frac{dw}{0} \tag{6.199}$$

The characteristics equation can be solved as

$$idx - dy = 0 \tag{6.200}$$

By multiplying i, we have

$$dx + idy = 0 \tag{6.201}$$

After integration we get

$$u_1 = x + iy = c_1 \tag{6.202}$$

The other characteristics are clearly

$$dw = 0, \quad or \quad u_2 = w = c_2 \tag{6.203}$$

$$\phi(w, x + iy) = 0 \tag{6.204}$$

Therefore, we can write it as

$$w = f(x + iy) \tag{6.205}$$

This completes the first proof.

Secondly, we can rearrange (6.198) as

$$i\frac{\partial w}{\partial x}=\frac{\partial w}{\partial y}=i\frac{\partial(u+iv)}{\partial x}=\frac{\partial(u+iv)}{\partial y} \tag{6.206}$$

Expanding (6.206), we obtain

$$i\frac{\partial u}{\partial x}-\frac{\partial v}{\partial x}=\frac{\partial u}{\partial y}+i\frac{\partial v}{\partial y} \tag{6.207}$$

Equating the real and imaginary parts of (6.207), we have

$$\frac{\partial u}{\partial y}=-\frac{\partial v}{\partial x}, \quad \frac{\partial u}{\partial x}=\frac{\partial v}{\partial y} \tag{6.208}$$

This proves the Cauchy-Riemann equations.

6.12 LAGRANGE-CHARPIT METHOD FOR NONLINEAR PDE

We are going to discuss a general technique called the Lagrange-Charpit method, Lagrange method, or Charpit method. There is a heated dispute on who discovered this technique. There is not much historical coverage of this method in any textbook. Some writers called this the Lagrange method as Charpit never published this method before he passed away and the original manuscript was reportedly "lost." The idea for this method was published in 1779 by Lagrange. It is clear that you should have heard about the name of Lagrange quite often (either from this book or from elsewhere) while Charpit is not well known and we don't even know when we was born. A vivid example of this view is given by Kline (1972):

> Lacroix said in 1798 that Charpit had submitted a paper in 1784 (which was not published) in which he reduced first order partial differential equations to system of ordinary differential equations. Jacobi found Lacroix's statement striking and expressed the wish that Charpit's work be published. But this was never done and we do not know whether Lacroix's statement is correct. Actualy Lagrange had done the full job and Charpit could have added nothing (p.535, Volume 2, Kline, 1972).

We do not agree with Kline on this, and Kline (1972) did not have a chance of reading the historical paper by Grattan-Guinness and Engelsman (1982), which will be summarized in a later paragraph.

Some authors called it the Charpit method (e.g., Sneddon, 1957) as there are records that Charpit presented this method in front of French Royal Academy of Sciences on June 30, 1784 shortly before Charpit passed away on December 28, 1784, although there was no published record of his work before Saltykow (1930, 1937) formally published it in 1930 for Charpit. However, the paper by Saltykow was not cited in textbooks, such as the one by Sneddon.

We follow the view that Lagrange laid down the fundamental ideas of this technique in 1779 and Charpit provided the actual details in 1784 and we prefer to call it the Lagrange-Charpit method (e.g., Delgado, 1997). Forsyth (1888) claimed that this method is partially due to Lagrange and partially due to Charpit, although

Forsyth called it Charpit's method. But again no detailed citation was given by Forsyth.

As remarked in the last paragraph, the history of this method is not covered in any textbook even though different names have been adopted by different authors. The best coverage was given by Grattan-Guinness and Engelsman (1982), and we reported its history briefly here. It was reported that Charpit had helped the course given by Monge, who also gave him private lectures in solid geometry. Charpit was clearly influenced by Lagrange. Paul Charpit was a "young" mathematician when he presented this method to the French Academy of Sciences on June 30 1874. Monge, Bossut, Condorcet, Cousin, Laplace, Vandermore, and de Borda were presented. At that time, Lagrange was still in Berlin. It was supposed that Laplace and Condorcet were nominated as reporters for Charpit's presentation, but such report was never found. Presumably Laplace kept the original manuscript of Charpit until he passed it to Lagrange in June 1793. In September 1793, Lagrange sent it to Arbogast, who made a copy for himself and sent the original to Lacroix, who made another copy of the manuscript. Lacroix mentioned Charpit's method in his book and publications in 1798, 1802, and 1814. When Jacobi asked in 1841 for the publication of Charpit's paper, it was reported lost for the first time. Apparently, Jacobi misunderstood Lacroix and claimed that it was lost. And since then, most authors simply quoted from Jacobi that Charpit's manuscript was lost. As remarked by Grattan-Guinness and Engelsman (1982), the extant manuscript of Charpit actually survived and has been archived at the Biblioteca Medicea-Laurenziana, Florence (Arbogast's copy) and at the Archives of the Academie des Sciences, Paris (both Arbogast's and Lacroix's copy). Thus, Jacobi's claim of a "lost" manuscript is not correct. In 1928, H. Villat found Lacroix's copy at the Archives of the Academie des Sciences and made a photocopy and sent it to N. Saltykow in Belgrade, who published it in Saltykow (1930, 1937). Although the paper of Saltykow (1930, 1937) was published in French, it is clear from the mathematical equations that the paper of Charpit did contain all the major steps of the present known Lagrange-Charpit method. Therefore, Kline's (1972) assertion is inaccurate.

Let us consider the following nonlinear first order PDE

$$F(x, y, z, \frac{\partial z}{\partial x}, \frac{\partial z}{\partial y}) = 0 \tag{6.209}$$

Next, we define

$$p = \frac{\partial z}{\partial x}, \quad q = \frac{\partial z}{\partial y} \tag{6.210}$$

The nonlinear PDE given in (6.209) can now be written as

$$F(x, y, z, p, q) = 0 \tag{6.211}$$

We seek a solution in the form:

$$U(x, y, z, p, q) = a \tag{6.212}$$

Mathematically, we can treat (6.211) and (6.212) as two equations for p and q:

$$p(x, y, z, a), \quad q(x, y, z, a) \tag{6.213}$$

This is the most fundamental idea of this method. Once p and q are solved, we can back substitute them into the total differential such that it is integrable:

$$dz = \frac{\partial z}{\partial x}dx + \frac{\partial z}{\partial y}dy = p(x,y,z,a)dx + q(x,y,z,a)dy \tag{6.214}$$

Then, we can integrate (6.214) to give a solution for the original PDE given in (6.209).

First, using the technique for linear PDEs, for (6.214) we can identify the vector $\boldsymbol{E}$ as

$$\boldsymbol{E} = (P,Q,R) = (p,q,-1) \tag{6.215}$$

Then, the integrability of the Pfaffian equation (6.158) becomes

$$\begin{aligned} &R(\frac{\partial P}{\partial y} - \frac{\partial Q}{\partial x}) + P(\frac{\partial Q}{\partial z} - \frac{\partial R}{\partial y}) + Q(\frac{\partial R}{\partial x} - \frac{\partial P}{\partial z}) \\ &= p\frac{\partial q}{\partial z} - q\frac{\partial p}{\partial z} - \frac{\partial p}{\partial y} + \frac{\partial q}{\partial x} = 0 \end{aligned} \tag{6.216}$$

Next, we will find the first derivatives involved in (6.216). To find them, we take the total differential of F and U with respect to x gives

$$\frac{dF}{dx} = \frac{\partial F}{\partial x} + \frac{\partial F}{\partial p}\frac{\partial p}{\partial x} + \frac{\partial F}{\partial q}\frac{\partial q}{\partial x} = 0 \tag{6.217}$$

$$\frac{dU}{dx} = \frac{\partial U}{\partial x} + \frac{\partial U}{\partial p}\frac{\partial p}{\partial x} + \frac{\partial U}{\partial q}\frac{\partial q}{\partial x} = 0 \tag{6.218}$$

These equations can be put in matrix form as

$$\begin{bmatrix} F_p & F_q \\ U_p & U_q \end{bmatrix} \begin{Bmatrix} p_x \\ q_x \end{Bmatrix} = -\begin{Bmatrix} F_x \\ U_x \end{Bmatrix} \tag{6.219}$$

or,

$$[A]\{p\} = -\{f\} \tag{6.220}$$

The inverse of the matrix A is

$$[A]^{-1} = \frac{1}{\det|A|}\begin{bmatrix} U_q & -F_q \\ -U_p & F_p \end{bmatrix} \tag{6.221}$$

The determinant of A is

$$\det|A| = \frac{\partial(F,U)}{\partial(p,q)} = F_p U_q - F_q U_p \neq 0 \tag{6.222}$$

Therefore, we have

$$\begin{Bmatrix} p_x \\ q_x \end{Bmatrix} = \frac{-1}{\dfrac{\partial(F,U)}{\partial(p,q)}} \begin{Bmatrix} U_q F_x - F_q U_x \\ -U_p F_x + F_p U_x \end{Bmatrix} \tag{6.223}$$

Thus, we have

$$p_x = -\frac{\partial(F,U)}{\partial(x,q)} / \frac{\partial(F,U)}{\partial(p,q)}, \quad q_x = \frac{\partial(F,U)}{\partial(x,p)} / \frac{\partial(F,U)}{\partial(p,q)} \tag{6.224}$$

Similarly, we can differentiate F and U with respect to y and z to obtain

$$p_y = -\frac{\partial(F,U)}{\partial(y,q)} / \frac{\partial(F,U)}{\partial(p,q)}, \quad p_z = -\frac{\partial(F,U)}{\partial(z,q)} / \frac{\partial(F,U)}{\partial(p,q)}, \quad q_z = \frac{\partial(F,U)}{\partial(z,p)} / \frac{\partial(F,U)}{\partial(p,q)} \tag{6.225}$$

Now, we can substitute these results into (6.216) to give

$$p\frac{\partial(F,U)}{\partial(z,p)} + q\frac{\partial(F,U)}{\partial(z,q)} + \frac{\partial(F,U)}{\partial(y,q)} + \frac{\partial(F,U)}{\partial(x,p)} = 0 \tag{6.226}$$

Equivalently, (6.226) can be expressed explicitly as

$$p(F_z U_p - F_p U_z) + q(F_z U_q - F_q U_z) + F_y U_q - F_q U_y + F_x U_p - F_p U_x = 0 \tag{6.227}$$

We can re-shuttle this equation by recalling that U actually is our unknown

$$F_p \frac{\partial U}{\partial x} + F_q \frac{\partial U}{\partial y} + (qF_q + pF_p)\frac{\partial U}{\partial z} - (F_x + pF_z)\frac{\partial U}{\partial p} - (F_y + qF_z)\frac{\partial U}{\partial q} = 0 \tag{6.228}$$

This is a first order PDE with five variables and the Lagrange method discussed in Section 6.7 can be used to write down the the subsidiary equation

$$\frac{dx}{F_p} = \frac{dy}{F_q} = \frac{dz}{qF_q + pF_p} = \frac{dp}{-(F_x + pF_z)} = \frac{dq}{-(F_y + qF_z)} \tag{6.229}$$

This is the main result of the Lagrange-Charpit method. From this equation, we can derive

$$U(x, y, z, p, q) = a \tag{6.230}$$

as long as

$$\frac{\partial(F,U)}{\partial(p,q)} \neq 0 \tag{6.231}$$

The characteristics equation can be set to dt where t is a parameter:

$$\frac{dx}{F_p} = \frac{dy}{F_q} = \frac{dz}{qF_q + pF_p} = \frac{dp}{-(F_x + pF_z)} = \frac{dq}{-(F_y + qF_z)} = dt \tag{6.232}$$

Therefore, the nonlinear PDE is transformed to a system of five coupled ODEs:

$$\frac{dx}{dt} = F_p, \quad \frac{dy}{dt} = F_q, \quad \frac{dz}{dt} = qF_q + pF_p \tag{6.233}$$

$$\frac{dp}{dt} = -(F_x + pF_z), \quad \frac{dq}{dt} = -(F_y + qF_z) \tag{6.234}$$

We now illustrate how to apply the Lagrange-Charpit method.

Example 6.6 Solve the following nonlinear PDE

$$(\frac{\partial z}{\partial x})^2 x + (\frac{\partial z}{\partial y})^2 y = z \tag{6.235}$$

Solution: This PDE can be recast as

$$F = p^2 x + q^2 y - z = 0 \tag{6.236}$$

Differentiation of (6.236) gives

$$F_p = 2px, \quad F_q = 2qy, \quad F_x = p^2, \quad F_y = q^2, \quad F_z = -1 \tag{6.237}$$

The subsidiary equation becomes

$$\frac{dx}{2px} = \frac{dy}{2qy} = \frac{dz}{2p^2x + 2q^2y} = \frac{dp}{p(1-p)} = \frac{dq}{q(1-q)} \tag{6.238}$$

Using the method of the Lagrange multiplier, we have

$$\frac{p^2dx + 2pxdp}{2p^3x + 2px(p-p^2)} = \frac{q^2dy + 2qydq}{2q^3y + 2qy(q-q^2)} \tag{6.239}$$

This can be simplified to

$$\frac{d(p^2x)}{p^2x} = \frac{d(q^2y)}{q^2y} \tag{6.240}$$

Integration of both sides gives

$$p^2x = aq^2y \tag{6.241}$$

On the other hand, we can rewrite (6.236)

$$p^2x = z - q^2y \tag{6.242}$$

Equating (6.241) and (6.242) gives

$$q = \left\{\frac{z}{y(1+a)}\right\}^{1/2} \tag{6.243}$$

Substitution of (6.243) into (6.241) gives

$$p = \left\{\frac{za}{x(1+a)}\right\}^{1/2} \tag{6.244}$$

Finally, from (6.214) we have

$$dz = \left\{\frac{za}{x(1+a)}\right\}^{1/2} dx + \left\{\frac{z}{y(1+a)}\right\}^{1/2} dy \tag{6.245}$$

Rearranging (6.245) leads to

$$\left\{\frac{(1+a)}{z}\right\}^{1/2} dz = \left\{\frac{a}{x}\right\}^{1/2} dx + \left\{\frac{1}{y}\right\}^{1/2} dy \tag{6.246}$$

Finally, integration of (6.246) gives the solution as

$$\{(1+a)z\}^{1/2} = (ax)^{1/2} + y^{1/2} + b \tag{6.247}$$

Example 6.6 shows that the analysis may not be straightforward if we do not recognize the appropriate multiplier for the subsidiary equation. Let us consider some special cases of nonlinear first order PDEs that can be solved easily by using the Lagrange-Charpit method.

6.12.1 Type I (only *p* and *q*)

Consider the case that, the PDE only involves p and q as

$$F(p,q)=0 \tag{6.248}$$

The subsidiary equation can be simplified to

$$\frac{dx}{F_p}=\frac{dy}{F_q}=\frac{dz}{qF_q+pF_p}=\frac{dp}{0}=\frac{dq}{0} \tag{6.249}$$

For such cases, we have

$$dp=0, \quad p=a \tag{6.250}$$

Substitution of (6.250) into (6.248) gives

$$f(a,q)=0 \tag{6.251}$$

If we can solve for q, we have

$$q=Q(a) \tag{6.252}$$

This is consistent with the last part of (6.249)

$$dq=0 \tag{6.253}$$

In addition, (6.248) is a first order PDE, and we expect only one unknown constant instead of two. This is precisely what we get in (6.252). Now we can integrate to get z as

$$p=\frac{\partial z}{\partial x}=a, \quad z=ax+f_1(y) \tag{6.254}$$

Similarly, we can integrate (6.252) to give

$$q=\frac{\partial z}{\partial y}=Q(a), \quad z=Q(a)y+f_2(x) \tag{6.255}$$

Comparison of (6.254) and (6.255) gives

$$f_2(x)=ax+C, \quad f_1(y)=Q(a)y+C \tag{6.256}$$

Finally, the function z is

$$z=ax+Q(a)y+C \tag{6.257}$$

Example 6.7 Solve the following nonlinear PDE

$$(\frac{\partial z}{\partial x})(\frac{\partial z}{\partial y})=1 \tag{6.258}$$

Solution: Note that this is Type I, and it can be rewritten in terms of p and q as:

$$pq=1 \tag{6.259}$$

Employing (6.249) of the Lagrange-Charpit method, from $dp = 0$ we have

$$p=a \tag{6.260}$$

Substitution of (6.260) into (6.259) gives

$$q=\frac{1}{a}=Q(a) \tag{6.261}$$

Integrating both (6.260) and (6.261), we have

$$z = ax + \frac{y}{a} + C \tag{6.262}$$

Alternatively, we can rewrite it as

$$a^2 x + y - az = C_1 \tag{6.263}$$

where C_1 is an arbitrary constant.

6.12.2 Type II (only z, p and q)

For type II, the functional form of the PDE is

$$F(z, p, q) = 0 \tag{6.264}$$

The subsidiary equation of the Lagrange-Charpit method is simplified to

$$\frac{dx}{F_p} = \frac{dy}{F_q} = \frac{dz}{qF_q + pF_p} = \frac{dp}{-pF_z} = \frac{dq}{-qF_z} \tag{6.265}$$

The last part of (6.265) gives

$$\frac{dp}{p} = \frac{dq}{q} \tag{6.266}$$

Integrating this, we obtain

$$p = aq \tag{6.267}$$

where a is a constant. Substitution of (6.267) into (6.264) gives

$$F(z, aq, q) = 0 \tag{6.268}$$

Suppose that we can solve for q such that

$$q = Q(z, a) \tag{6.269}$$

Finally, p becomes

$$p = aQ(z, a) \tag{6.270}$$

With known p and q, we can integrate (6.214).

Example 6.8 Solve the following nonlinear PDE

$$\left(\frac{\partial z}{\partial x}\right)^2 - \left(\frac{\partial z}{\partial y}\right)^2 = z \tag{6.271}$$

Solution: Note that this is Type II, and it can be rewritten as:

$$F(z, p, q) = p^2 - q^2 - z = 0 \tag{6.272}$$

Thus, we can use (6.267) as

$$p = aq \tag{6.273}$$

With (6.273), the PDE given in (6.271) becomes

$$a^2 q^2 - q^2 = z \tag{6.274}$$

Solving for q we have

$$q = \left(\frac{z}{a^2 - 1}\right)^{1/2} \tag{6.275}$$

Substitution of (6.275) into (6.273) gives

$$p = a(\frac{z}{a^2-1})^{1/2} \tag{6.276}$$

With these values of p and q, (6.214) becomes

$$dz = a(\frac{z}{a^2-1})^{1/2} dx + (\frac{z}{a^2-1})^{1/2} dy \tag{6.277}$$

Rewriting (6.277) gives

$$\frac{1}{z^{1/2}} dz = \frac{a}{(a^2-1)^{1/2}} dx + \frac{1}{(a^2-1)^{1/2}} dy \tag{6.278}$$

Integrating both sides gives

$$2z^{1/2} = \frac{ax}{(a^2-1)^{1/2}} + \frac{y}{(a^2-1)^{1/2}} + C \tag{6.279}$$

Finally, we have the final solution as

$$z = \frac{1}{4}\left\{\frac{ax}{(a^2-1)^{1/2}} + \frac{y}{(a^2-1)^{1/2}} + C\right\}^2 \tag{6.280}$$

6.12.3 Type III (Separable)

If the nonlinear PDE can be separated such that x and p only appear on the left whereas y and q only appear on the right, we have the Type III situation:

$$F(x,p) = G(y,q) \tag{6.281}$$

For such case, the subsidiary equation of the Lagrange-Charpit method becomes

$$\frac{dx}{F_p} = \frac{dy}{-G_q} = \frac{dz}{pF_p - qG_q} = \frac{dp}{-F_x} = \frac{dq}{G_y} \tag{6.282}$$

The first and fourth equations of (6.282) can be grouped to form an ODE as

$$\frac{dp}{dx} + \frac{F_x}{F_p} = 0 \tag{6.283}$$

which can be solved for p as a function of x because F is only a function of x and p (see (6.281)). In Pfaffian form, it is

$$F_p dp + F_x dx = 0 \tag{6.284}$$

This is a total differential of F = constant. Clearly, the solution is

$$F(x,p) = a \tag{6.285}$$

where a is a constant. Thus, (6.281) also leads to

$$G(y,q) = a \tag{6.286}$$

Thus, in principle, we can find p and q from (6.285) and (6.286) and eventually integrate (6.214).

Alternatively, we can also group the second and last equation of (6.282) to give

$$\frac{dq}{dy}+\frac{G_y}{G_q}=0 \tag{6.287}$$

Then, following the same logic we can find the solution for q since (6.287) is a total differential of G = constant. Thus, we can solve both p and q from the following equations

$$G(y,q)=a\,,\quad F(x,p)=a \tag{6.288}$$

The next example illustrates this case.

Example 6.9 Solve the following nonlinear PDE

$$(\frac{\partial z}{\partial x})^2 y(1+x^2)=(\frac{\partial z}{\partial y})x^2 \tag{6.289}$$

Solution: Note that this is Type III because it can be written as

$$\frac{p^2(1+x^2)}{x^2}=\frac{q}{y} \tag{6.290}$$

Because the given function G is much simpler than F, we can revise the procedure slightly as

$$G_q dq+G_y dy=0 \tag{6.291}$$

Using (6.290), we have

$$G_q=\frac{1}{y},\quad G_y=-\frac{q}{y^2} \tag{6.292}$$

Thus, we have

$$\frac{dq}{q}=\frac{dy}{y} \tag{6.293}$$

The solution is

$$q=ay \tag{6.294}$$

where a is a constant. Substitution of (6.294) into (6.290) gives

$$F(x,p)=p^2\frac{(1+x^2)}{x^2}=a \tag{6.295}$$

Solving for p we have

$$p=\frac{a^{1/2}x}{(1+x^2)^{1/2}} \tag{6.296}$$

Substitution of (6.296) and (6.294) into (6.214) gives

$$dz=pdx+qdy=\frac{a^{1/2}xdx}{(1+x^2)^{1/2}}+aydy \tag{6.297}$$

Integration gives

$$z=\sqrt{a}\sqrt{1+x^2}+\frac{a}{2}y^2+b \tag{6.298}$$

6.12.4 Type IV (Clairaut Type)

The type IV first order PDE can be considered as a generalization of the Clairaut equation of first order ODEs discussed in Chapter 3:

$$z = x\frac{\partial z}{\partial x} + y\frac{\partial z}{\partial y} + f(\frac{\partial z}{\partial x}, \frac{\partial z}{\partial y}) \tag{6.299}$$

Thus, the functional form F can be cast as

$$F = xp + yq + f(p,q) - z = 0 \tag{6.300}$$

Thus, we have

$$F_z = -1, \quad F_x = p, \quad F_y = q, \quad F_p = x + f_p, \quad F_q = y + f_q \tag{6.301}$$

The subsidiary equation of the Lagrange-Charpit method becomes

$$\frac{dx}{x+f_p} = \frac{dy}{y+f_q} = \frac{dz}{qy+px+qf_q+pf_p} = \frac{dp}{0} = \frac{dq}{0} \tag{6.302}$$

Clearly, similar to the Clairaut equation discussed in Chapter 3, we have

$$p = a, \quad q = b \tag{6.303}$$

Thus, the solution is (6.196) with the values of p and q given in (6.200) as:

$$z = ax + by + f(a,b) \tag{6.304}$$

Example 6.10 Solve the following nonlinear PDE

$$(\frac{\partial z}{\partial x} + \frac{\partial z}{\partial y})^2 \{z - x\frac{\partial z}{\partial x} - y\frac{\partial z}{\partial y}\} = 2 \tag{6.305}$$

Solution: Note that this is Type IV or Clairaut type

$$z - xp - yq = \frac{2}{(p+q)^2} \tag{6.306}$$

Thus, from the last two equations of (6.302) we have

$$p = a, \quad q = b \tag{6.307}$$

The general solution is

$$z = ax + by + \frac{2}{(a+b)^2} \tag{6.308}$$

6.12.5 Singular Solution

Similar to the discussion in Section 3.2.12, the singular solution can also be found accordingly. Consider again the following first order nonlinear PDE

$$F(x,y,z,p,q) = 0 \tag{6.309}$$

where p and q have been defined in (6.210). Let the general solution of it be

$$V(x,y,z,a,b) = 0 \tag{6.310}$$

Since the solution for z is a function of x and y, we can differentiate (6.310) as

$$\frac{\partial V}{\partial x}+\frac{\partial V}{\partial z}\frac{\partial z}{\partial x}=\frac{\partial V}{\partial x}+\frac{\partial V}{\partial z}p=0 \tag{6.311}$$

$$\frac{\partial V}{\partial y}+\frac{\partial V}{\partial z}\frac{\partial z}{\partial y}=\frac{\partial V}{\partial y}+\frac{\partial V}{\partial z}q=0 \tag{6.312}$$

Similar to the c-discriminate method discussed in Section 3.2.12, we allow a and b to be functions of x and y. Then, the differentiation of (6.310) leads to

$$\frac{\partial V}{\partial x}+\frac{\partial V}{\partial z}p+\frac{\partial V}{\partial a}\frac{\partial a}{\partial x}+\frac{\partial V}{\partial b}\frac{\partial b}{\partial x}=0 \tag{6.313}$$

$$\frac{\partial V}{\partial y}+\frac{\partial V}{\partial z}q+\frac{\partial V}{\partial a}\frac{\partial a}{\partial y}+\frac{\partial V}{\partial b}\frac{\partial b}{\partial y}=0 \tag{6.314}$$

Substitution of (6.311) and (6.312) into (6.313) and (6.314) gives

$$\frac{\partial V}{\partial a}\frac{\partial a}{\partial x}+\frac{\partial V}{\partial b}\frac{\partial b}{\partial x}=0 \tag{6.315}$$

$$\frac{\partial V}{\partial a}\frac{\partial a}{\partial y}+\frac{\partial V}{\partial b}\frac{\partial b}{\partial y}=0 \tag{6.316}$$

These two equations can be put in a matrix form as

$$\begin{bmatrix} \dfrac{\partial a}{\partial x} & \dfrac{\partial b}{\partial x} \\ \dfrac{\partial a}{\partial y} & \dfrac{\partial b}{\partial y} \end{bmatrix} \begin{Bmatrix} \dfrac{\partial V}{\partial a} \\ \dfrac{\partial V}{\partial b} \end{Bmatrix} = \begin{Bmatrix} 0 \\ 0 \end{Bmatrix} \tag{6.317}$$

In general, we can now use (6.310), (6.313), and (6.314) to eliminate a and b. There are three possible scenarios:

(i) For singular solutions, we require

$$\frac{\partial V}{\partial a}=\frac{\partial V}{\partial b}\equiv 0, \quad V=0\,. \tag{6.318}$$

Note that (6.313) and (6.314) are automatically satisfied by the first equation of (6.318).

(ii) To get back the general solution, we require

$$\frac{\partial a}{\partial x}=\frac{\partial b}{\partial x}=\frac{\partial a}{\partial y}=\frac{\partial b}{\partial y}=0 \tag{6.319}$$

(iii) For the third scenario, we have

$$\frac{\partial V}{\partial a}\neq 0, \quad \frac{\partial V}{\partial b}\neq 0 \tag{6.320}$$

From (6.317), we immediately see that the following Jacobian is zero:

$$J=\frac{\partial(a,b)}{\partial(x,y)}=0 \tag{6.321}$$

Thus, the variables a and b are not independent. We must have $a = a(b)$ or inversely we have

$$b = w(a) \tag{6.322}$$

where w is an arbitrary function. Differentiating (6.310) with respect to a and rewriting (6.310) in terms of (6.322), we have the following system of equations:

$$\frac{\partial V}{\partial a} + \frac{\partial V}{\partial b}\frac{\partial w}{\partial a} = 0 \tag{6.323}$$

$$V(x, y, z, a, w(a)) = 0 \tag{6.324}$$

Elimination of a from (6.323) and (6.324), we again obtain the general solution.

Example 6.11 Find the general solution as well as the singular solution of the following Clairaut equation:

$$z - x\frac{\partial z}{\partial x} - y\frac{\partial z}{\partial y} - \frac{\partial z}{\partial x}\frac{\partial z}{\partial y} = 0 \tag{6.325}$$

Solution: Expressing this in terms of p and q, we have

$$z = xp + yq + pq \tag{6.326}$$

The subsidiary equation of the Lagrange-Charpit method for the Type IV case is

$$\frac{dx}{x + f_p} = \frac{dy}{y + f_q} = \frac{dz}{qy + px + qf_q + pf_p} = \frac{dp}{0} = \frac{dq}{0} \tag{6.327}$$

From the last two parts of (6.327), we have

$$p = a, \quad q = b \tag{6.328}$$

The general solution becomes

$$V(x, y, z, a, b) = z - ax - by - ab = 0 \tag{6.329}$$

Applying (6.318) gives

$$\frac{\partial V}{\partial a} = -x - b = 0 \tag{6.330}$$

$$\frac{\partial V}{\partial b} = -y - a = 0 \tag{6.331}$$

Using (6.330) and (6.331) to eliminate a and b from (6.329), we have the singular solution as

$$z = -xy \tag{6.332}$$

6.13 GEOMETRIC INTERPRETATION OF NONLINEAR PDE

The geometric interpretation of a nonlinear PDE was given by Monge. Let us consider the linear form of PDE first

$$dz = \frac{\partial z}{\partial x}dx + \frac{\partial z}{\partial y}dy = pdx + qdy = 0 \tag{6.333}$$

Note that $d\xi$ is parallel to $\boldsymbol{E}$, and the normal vector to the increasing curve $\boldsymbol{E} = (P, Q, R)$ is the directional vector of the intersecting surfaces can be expressed:

$$N = (p, q, -1) \tag{6.334}$$

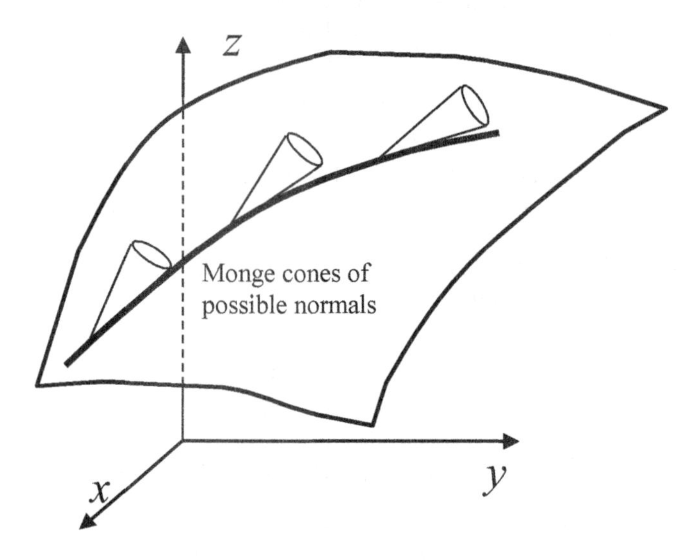

Figure 6.5 Geometric interpretation of the solution of nonlinear 1st order PDE

The normal direction of the vector $\boldsymbol{E}$ is

$$\boldsymbol{n} = \frac{\boldsymbol{N}}{|\boldsymbol{N}|} = \frac{1}{\sqrt{p^2 + q^2 + 1}}(p, q, -1) \tag{6.335}$$

More importantly, we see that the direction is an unique function of p and q. However, for a nonlinear PDE, there is no one-to-one correspondence between p and q. To illustrate this, we consider a special case that nonlinearity in p can be expressed as polynomials:

$$F(x, y, z, p, q) = a_n p^n + a_{n-1} p^{n-1} + \ldots + a_1 p + bq - c = 0 \tag{6.336}$$

Clearly, we have more than one pair of p and q that satisfies $F = 0$. In other words, we do not have a unique direction $\boldsymbol{n}$ for the intersecting curves. The direction of the curve is now non-unique and Monge interpreted the possible direction of the curve forming by a cone, called the Monge cone. The idea is showed in Figure 6.5.

6.14 JACOBI'S METHOD

We saw in the last section that the Lagrange-Charpit method is for two variables x and y. This technique, however, cannot be extended to first order PDEs of more than two independent variables. The technique discussed in this section was proposed by Jacobi in 1836. Jacobi's method can be applied to 3 or more variables. Therefore, it is more general than the Lagrange-Charpit method. However, for

Jacobi's method to be applicable, the unknown z cannot appear in the nonlinear PDE explicitly.

Let us consider the case of a first order PDE with three variables:

$$F(x_1, x_2, x_3, p_1, p_2, p_3) = 0 \tag{6.337}$$

where z is the unknown and its derivatives are defined by

$$p_1 = \frac{\partial z}{\partial x_1}, \quad p_2 = \frac{\partial z}{\partial x_2}, \quad p_3 = \frac{\partial z}{\partial x_3} \tag{6.338}$$

The main idea of Jacobi's method is to find two PDE involving two unknown constants:

$$F_1(x_1, x_2, x_3, p_1, p_2, p_3) = a_1 \tag{6.339}$$

$$F_2(x_1, x_2, x_3, p_1, p_2, p_3) = a_2 \tag{6.340}$$

In this way, we have now three equations to solve for p_1, p_2, and p_3. This idea is clearly a natural extension of the Lagrange-Charpit method.

The total differential of the unknown z is

$$dz = p_1 dx_1 + p_2 dx_2 + p_3 dx_3 \tag{6.341}$$

The integrable conditions of this equation are

$$\frac{\partial p_2}{\partial x_1} = \frac{\partial^2 z}{\partial x_1 \partial x_2} = \frac{\partial p_1}{\partial x_2} \tag{6.342}$$

$$\frac{\partial p_3}{\partial x_1} = \frac{\partial^2 z}{\partial x_1 \partial x_3} = \frac{\partial p_1}{\partial x_3} \tag{6.343}$$

$$\frac{\partial p_2}{\partial x_3} = \frac{\partial^2 z}{\partial x_3 \partial x_2} = \frac{\partial p_3}{\partial x_2} \tag{6.344}$$

Note that p_1, p_2, and p_3 are functions of the variables only. We can now take the total differential of F and F_1 with respective to x_1 :

$$\frac{\partial F}{\partial x_1} + \frac{\partial F}{\partial p_1}\frac{\partial p_1}{\partial x_1} + \frac{\partial F}{\partial p_2}\frac{\partial p_2}{\partial x_1} + \frac{\partial F}{\partial p_3}\frac{\partial p_3}{\partial x_1} = 0 \tag{6.345}$$

$$\frac{\partial F_1}{\partial x_1} + \frac{\partial F_1}{\partial p_1}\frac{\partial p_1}{\partial x_1} + \frac{\partial F_1}{\partial p_2}\frac{\partial p_2}{\partial x_1} + \frac{\partial F_1}{\partial p_3}\frac{\partial p_3}{\partial x_1} = 0 \tag{6.346}$$

The subtraction of the result of multiplying (6.346) by $\partial F/\partial p_1$ from the result of multiplying (6.345) by $\partial F_1/\partial p_1$ yields

$$\frac{\partial(F, F_1)}{\partial(x_1, p_1)} + \frac{\partial(F, F_1)}{\partial(p_2, p_1)}\frac{\partial p_2}{\partial x_1} + \frac{\partial(F, F_1)}{\partial(p_3, p_1)}\frac{\partial p_3}{\partial x_1} = 0 \tag{6.347}$$

where the Jacobian is

$$\frac{\partial(F, F_1)}{\partial(x_1, p_1)} = \frac{\partial F}{\partial x_1}\frac{\partial F_1}{\partial p_1} - \frac{\partial F}{\partial p_1}\frac{\partial F_1}{\partial x_1} \tag{6.348}$$

Similarly, we can differentiate F and F_1 with respect to x_2 and x_3 to get

$$\frac{\partial(F, F_1)}{\partial(x_2, p_2)} + \frac{\partial(F, F_1)}{\partial(p_1, p_2)}\frac{\partial p_1}{\partial x_2} + \frac{\partial(F, F_1)}{\partial(p_3, p_2)}\frac{\partial p_3}{\partial x_2} = 0 \tag{6.349}$$

$$\frac{\partial(F,F_1)}{\partial(x_3,p_3)}+\frac{\partial(F,F_1)}{\partial(p_1,p_3)}\frac{\partial p_1}{\partial x_3}+\frac{\partial(F,F_1)}{\partial(p_2,p_3)}\frac{\partial p_2}{\partial x_3}=0 \tag{6.350}$$

Adding (6.347), (6.349) and (6.350) leads to

$$\frac{\partial(F,F_1)}{\partial(x_1,p_1)}+\frac{\partial(F,F_1)}{\partial(x_2,p_2)}+\frac{\partial(F,F_1)}{\partial(x_3,p_3)}=0 \tag{6.351}$$

The result (6.351) is a consequence of applying the following identities:

$$\frac{\partial(F,F_1)}{\partial(p_2,p_1)}\frac{\partial p_2}{\partial x_1}+\frac{\partial(F,F_1)}{\partial(p_1,p_2)}\frac{\partial p_1}{\partial x_2}=\frac{\partial z}{\partial x_2\partial x_1}\{\frac{\partial(F,F_1)}{\partial(p_2,p_1)}+\frac{\partial(F,F_1)}{\partial(p_1,p_2)}\}=0 \tag{6.352}$$

Note that the last equation is the result of reversing the order of the two variables of a Jacobian leading to the negative value of the same Jacobian (see definition in (6.348)). Similarly, we also have

$$\frac{\partial(F,F_1)}{\partial(p_3,p_1)}\frac{\partial p_3}{\partial x_1}+\frac{\partial(F,F_1)}{\partial(p_1,p_3)}\frac{\partial p_1}{\partial x_3}=\frac{\partial z}{\partial x_3\partial x_1}\{\frac{\partial(F,F_1)}{\partial(p_3,p_1)}+\frac{\partial(F,F_1)}{\partial(p_1,p_3)}\}=0 \tag{6.353}$$

$$\frac{\partial(F,F_1)}{\partial(p_3,p_2)}\frac{\partial p_3}{\partial x_2}+\frac{\partial(F,F_1)}{\partial(p_2,p_3)}\frac{\partial p_2}{\partial x_3}=\frac{\partial z}{\partial x_3\partial x_2}\{\frac{\partial(F,F_1)}{\partial(p_3,p_2)}+\frac{\partial(F,F_1)}{\partial(p_2,p_3)}\}=0 \tag{6.354}$$

Rewriting (6.351) explicitly, we get

$$\frac{\partial F}{\partial x_1}\frac{\partial F_1}{\partial p_1}-\frac{\partial F}{\partial p_1}\frac{\partial F_1}{\partial x_1}+\frac{\partial F}{\partial x_2}\frac{\partial F_1}{\partial p_2}-\frac{\partial F}{\partial p_2}\frac{\partial F_1}{\partial x_2}+\frac{\partial F}{\partial x_3}\frac{\partial F_1}{\partial p_3}-\frac{\partial F}{\partial p_3}\frac{\partial F_1}{\partial x_3}=0 \tag{6.355}$$

This equation can be rewritten symbolically:

$$[F,F_1]=0 \tag{6.356}$$

This is actually the compatibility of the two PDEs $F = 0$ and $F_1 = 0$ (see Problem 4 of Section 2.9, Sneddon, 1956). Similarly, we can also obtain:

$$[F,F_2]=0 \tag{6.357}$$

$$[F_1,F_2]=0 \tag{6.358}$$

Note that $F = 0$ is given while $F_1 = 0$ is not. Thus, we can view (6.356) as a first order PDE for F_1. Its solution can be found readily using the Lagrange-Charpit method discussed in Section 6.12. The subsidiary equation of (6.355) using the Lagrange-Charpit method is

$$\frac{dp_1}{\frac{\partial F}{\partial x_1}}=\frac{dx_1}{-\frac{\partial F}{\partial p_1}}=\frac{dp_2}{\frac{\partial F}{\partial x_2}}=\frac{dx_2}{-\frac{\partial F}{\partial p_2}}=\frac{dp_3}{\frac{\partial F}{\partial x_3}}=\frac{dx_3}{-\frac{\partial F}{\partial p_3}} \tag{6.359}$$

From the characteristics equations of (6.359), we can find $F_1 = a_1$ and $F_2 = a_2$. However, after we obtain these solutions, we need to check whether they are compatible using

$$[F_1,F_2]\equiv\sum_{r=1}^{3}(\frac{\partial F_1}{\partial x_r}\frac{\partial F_2}{\partial p_r}-\frac{\partial F_1}{\partial p_r}\frac{\partial F_2}{\partial x_r}) \tag{6.360}$$

By now, we have three equations for p_1, p_2, and p_3:

$$F=0,\quad F_1=a_1,\quad F_2=a_2 \tag{6.361}$$

This completes Jacobi's method, which will be illustrated in the following example.

Example 6.12 Solve the following nonlinear PDE using Jacobi's method

$$2\frac{\partial z}{\partial x_1}x_1x_3 + 3\frac{\partial z}{\partial x_2}x_3^2 + (\frac{\partial z}{\partial x_2})^2\frac{\partial z}{\partial x_3} = 0 \tag{6.362}$$

Solution: Expressing this in terms of p_i ($i = 1,2,3$), we have

$$2p_1x_1x_3 + 3p_2x_3^2 + p_2^2p_3 = 0 \tag{6.363}$$

The subsidiary equation is

$$\frac{dx_1}{-2x_1x_3} = \frac{dp_1}{2p_1x_3} = \frac{dx_2}{-3x_3^2 - 2p_2p_3} = \frac{dp_2}{0} = \frac{dx_3}{-p_2^2} = \frac{dp_3}{2p_1x_1 + 6p_2x_3} \tag{6.364}$$

The first two parts of (6.364) give

$$\frac{dx_1}{-x_1} = \frac{dp_1}{p_1} \tag{6.365}$$

The solution of (6.365) is

$$F_1 = p_1x_1 = a_1 \tag{6.366}$$

The fourth part of the characteristics equation (6.364) gives

$$F_2 = p_2 = a_2 \tag{6.367}$$

Their compatibility can be checked as

$$[F_1, F_2] \equiv p_1 \cdot 0 - x_1 \cdot 0 + 0 \cdot 1 - 0 \cdot 0 + 0 \cdot 0 - 0 \cdot 0 = 0 \tag{6.368}$$

Thus, we can use (6.363), (6. 366), and (6.367) to find

$$p_1 = \frac{a_1}{x_1}, \quad p_2 = a_2, \quad p_3 = -\frac{2a_1x_3 + 3a_2x_3^2}{a_2^2} \tag{6.369}$$

The solution can now be integrated from the total differential

$$dz = \frac{a_1}{x_1}dx_1 + a_2dx_2 - (\frac{2a_1x_3 + 3a_2x_3^2}{a_2^2})dx_3 \tag{6.370}$$

Integration of (6.370) gives the final solution as

$$z = a_1 \ln x_1 + a_2x_2 - \frac{1}{a_2^2}(a_1x_3^2 + a_2x_3^3) + a_3 \tag{6.371}$$

Example 6.13 Solve the following nonlinear PDE using Jacobi's method

$$(x_2 + x_3)(\frac{\partial z}{\partial x_2} + \frac{\partial z}{\partial x_3})^2 + z\frac{\partial z}{\partial x_1} = 0 \tag{6.372}$$

Solution: We can rewrite (6.372) as

$$(x_2 + x_3)(p_2 + p_3)^2 + zp_1 = 0 \tag{6.373}$$

In (6.373), the unknown z appears explicitly, and we cannot apply Jacobi's method directly. Let us assume the following relation

$$u(x_1, x_2, x_3, z) = 0 \tag{6.374}$$

Note that (2.374) is clearly a solution of (6.373) because it relates the unknown z with the variables. We further assume that z is a new variable:

$$z = x_4, \quad p_1 = \frac{\partial z}{\partial x_1} = \frac{\partial x_4}{\partial x_1} = -\frac{\dfrac{\partial u}{\partial x_1}}{\dfrac{\partial u}{\partial x_4}} = -\frac{P_1}{P_4} \tag{6.375}$$

Similarly, we have

$$p_2 = -\frac{P_2}{P_4}, \quad p_3 = -\frac{P_3}{P_4} \tag{6.376}$$

The derivatives of the new unknown u are

$$P_1 = \frac{\partial u}{\partial x_1}, \quad P_2 = \frac{\partial u}{\partial x_2}, \quad P_3 = \frac{\partial u}{\partial x_3}, \quad P_4 = \frac{\partial u}{\partial x_4} \tag{6.377}$$

With the new unknown u and its derivatives defined in (6.377), we can rewrite the PDE symbolically as

$$F(x_1, x_2, x_3, x_4, P_1, P_2, P_3, P_4) = 0 \tag{6.378}$$

The original PDE now becomes

$$(x_2 + x_3)(P_2 + P_3)^2 - x_4 P_1 P_4 = 0 \tag{6.379}$$

Jacobi's method can now be applied to give a subsidiary equation:

$$\begin{aligned} &\frac{dx_1}{x_4 P_4} = \frac{dP_1}{0} = \frac{dx_2}{-2(x_2 + x_3)(P_2 + P_3)} = \frac{dP_2}{(P_2 + P_3)^2} = \frac{dx_3}{-2(x_2 + x_3)(P_2 + P_3)} \\ &= \frac{dP_3}{(P_2 + P_3)^2} = \frac{dx_4}{x_4 P_1} = \frac{dP_4}{-P_1 P_4} \end{aligned} \tag{6.380}$$

These characteristics equations can be solved to give

$$F_1 = P_1 = a_1, \quad F_2 = P_2 - P_3 = a_2, \quad F_3 = x_4 P_4 = a_3 \tag{6.381}$$

It is straightforward to show that

$$[F_1, F_2] = [F_1, F_3] = [F_3, F_2] = 0 \tag{6.382}$$

We can now find the derivatives from (6.379) and (6.381) to give

$$P_1 = a_1, \quad P_4 = \frac{a_3}{x_4}, \quad 2P_2 = a_2 \pm \sqrt{\frac{a_1 a_3}{x_2 + x_3}}, \quad P_3 = P_2 - a_2 \tag{6.383}$$

Thus, we can find u as:

$$du = a_1 dx_1 + \frac{a_3}{x_4} dx_4 + \frac{1}{2} a_2 (dx_2 - dx_3) \pm 2\sqrt{\frac{a_1 a_3}{x_2 + x_3}} (dx_2 + dx_3) \tag{6.384}$$

The solution can obtain by integration as

$$u = a_1 x_1 + a_3 \ln x_4 + \frac{1}{2} a_2 (x_2 - x_3) \pm \sqrt{a_1 a_3 (x_2 + x_3)} + a_4 \tag{6.385}$$

Finally, the solution of (6.373) can be found by making the following identifications:

$$u = 0, \quad x_4 \leftarrow z, \quad \frac{a_1}{a_3} \leftarrow A_1, \quad \frac{1}{2}\frac{a_2}{a_3} \leftarrow A_2, \quad \frac{a_4}{a_3} \leftarrow A_3 \tag{6.386}$$

The solution is

$$\ln z + A_1 x_1 + A_2 (x_2 - x_3) \pm \sqrt{A_1 (x_2 + x_3)} + A_3 = 0 \tag{6.387}$$

6.15 HODOGRAPH TRANSFORMATION

We will see in Chapter 7 that for fluid mechanics problems, Molenbroek in 1890 and Chaplygin in 1902 applied Legendre transform to rewrite the unknowns as x and y, and the variables as the velocity components. This technique is, in general, called a hodograph transformation. We will show in this section that such transformation can convert a system of nonlinear first order PDEs to a linear one if the variables do not appear explicitly in the differential equations.

In particular, we consider the following system of PDEs:

$$A_1(u,v)\frac{\partial u}{\partial t}+B_1(u,v)\frac{\partial u}{\partial x}+C_1(u,v)\frac{\partial v}{\partial t}+D_1(u,v)\frac{\partial v}{\partial x}=0 \tag{6.388}$$

$$A_2(u,v)\frac{\partial u}{\partial t}+B_2(u,v)\frac{\partial u}{\partial x}+C_2(u,v)\frac{\partial v}{\partial t}+D_2(u,v)\frac{\partial v}{\partial x}=0 \tag{6.389}$$

We can use the hodograph transformation as

$$x=x(u,v),\quad t=t(u,v) \tag{6.390}$$

where

$$\frac{\partial(u,v)}{\partial(x,y)}\neq 0 \tag{6.391}$$

We can use the transformation as

$$\frac{\partial u}{\partial t}=\frac{\dfrac{\partial x}{\partial v}}{\dfrac{\partial(x,t)}{\partial(u,v)}},\quad \frac{\partial u}{\partial x}=-\frac{\dfrac{\partial t}{\partial v}}{\dfrac{\partial(x,t)}{\partial(u,v)}},\quad \frac{\partial v}{\partial t}=-\frac{\dfrac{\partial x}{\partial u}}{\dfrac{\partial(x,t)}{\partial(u,v)}},\quad \frac{\partial v}{\partial x}=\frac{\dfrac{\partial t}{\partial u}}{\dfrac{\partial(x,t)}{\partial(u,v)}} \tag{6.392}$$

Substitution of (6.392) into (6.388) and (6.389) gives the following linear system of first order PDEs as

$$A_1(u,v)\frac{\partial x}{\partial v}-B_1(u,v)\frac{\partial t}{\partial v}-C_1(u,v)\frac{\partial x}{\partial u}+D_1(u,v)\frac{\partial t}{\partial u}=0 \tag{6.393}$$

$$A_2(u,v)\frac{\partial x}{\partial v}-B_2(u,v)\frac{\partial t}{\partial v}-C_2(u,v)\frac{\partial x}{\partial u}+D_2(u,v)\frac{\partial t}{\partial u}=0 \tag{6.394}$$

Since the unknowns are x and t, (6.393) and (6.394) are linear PDE.

6.16 SUMMARY AND FURTHER READING

Most textbooks on differential equations do not cover the solution technique for first order PDEs. However, we have illustrated that first order PDEs actually appears naturally as transport, kinematic wave, or shock wave equations in engineering and mechanics. Airy's method of solving nonhomogeneous hyperbolic equations is introduced. Geometric interpretation of the solution of linear first order PDE by Monge is discussed. For linear PDEs, we discuss the method of Lagrange, including the technique of the Lagrange multiplier, Pfaffian equations, exact equations, and integrability of Pfaffian. For nonlinear PDE, we cover Lagrange-Charpit method and Jacobi's method. Various special cases of the Lagrange-Charpit method are discussed and illustrated. Geometric interpretation of nonlinear PDEs is also discussed.

For further reading, we recommend the books by Lopez (2000) and Snedden (1957).

6.17 PROBLEMS

Problem 6.1 Solve the following PDE

$$\frac{\partial z}{\partial x} - a\frac{\partial z}{\partial y} = xy^2 \tag{6.395}$$

Ans: $z = \psi(ax+y) + \frac{1}{12a^2}\{6a^2x^2y^2 + 4a^3x^3y + a^4x^4\}$

Problem 6.2 Show also that the solution of Problem 6.1 can be expressed as

$$z = \phi(ax+y) + \frac{1}{12a^2}\{-4axy^3 - y^4\} \tag{6.396}$$

Problem 6.3 Solve the following PDE

$$\frac{\partial z}{\partial x} - a\frac{\partial z}{\partial y} = e^{\alpha x+\beta y} \tag{6.397}$$

Ans: $z = \psi(ax+y) + \frac{1}{(\alpha-\beta a)}e^{\alpha x+\beta y}$

Problem 6.4 Find the solution of

$$x\frac{\partial u}{\partial x} + y\frac{\partial u}{\partial y} + z\frac{\partial u}{\partial z} = 0 \tag{6.398}$$

Ans: $u = f(\frac{y}{z}, \frac{x}{y})$

Problem 6.5 Find the solution of

$$xz\frac{\partial z}{\partial x} + yz\frac{\partial z}{\partial y} = x \tag{6.399}$$

Ans: $z^2 = 2x + f(\frac{y}{x})$

Problem 6.6 Find the solution of

$$(x+3y)\frac{\partial z}{\partial x} - y\frac{\partial z}{\partial y} = 0 \tag{6.400}$$

Ans: $z = f(xy + \frac{3}{2}y^2)$

Problem 6.7 Find the solution of

$$y\frac{\partial z}{\partial x} - x\frac{\partial z}{\partial y} = 0 \tag{6.401}$$

Ans: $z = f(x^2 + y^2)$

Problem 6.8 Find the solution of

$$y\frac{\partial z}{\partial x} + x\frac{\partial z}{\partial y} = x - y \tag{6.402}$$

Ans: $z = y - x + f(x^2 - y^2)$

Problem 6.9 Find the solution of

$$(x^2 + y^2)\frac{\partial z}{\partial x} + 2xy\frac{\partial z}{\partial y} + z^2 = 0 \tag{6.403}$$

Ans: $\phi(\frac{1}{z} + \frac{1}{x+y}, \frac{1}{z} + \frac{1}{x-y}) = 0$

Problem 6.10 Find the solution of

$$x^2\frac{\partial z}{\partial x} - xy\frac{\partial z}{\partial y} + y^2 = 0 \tag{6.404}$$

Ans: $\phi(xy, 3xyz - y^3) = 0$

Problem 6.11 Find the solution of

$$x(x+y)\frac{\partial z}{\partial x} = \frac{\partial z}{\partial y}y(x+y) - (x-y)(2x+2y+z) \tag{6.405}$$

Ans: $\phi[xy, (x+y)(x+y+z)] = 0$

Problem 6.12 Find the solution of

$$y^2\frac{\partial z}{\partial x} - xy\frac{\partial z}{\partial y} = x(z - 2y) \tag{6.406}$$

Ans: $z = -\frac{x^2}{y} + \frac{1}{y}f(x^2 + y^2)$

Problem 6.13 Find the solution of

$$x(x^2+3y^2)\frac{\partial z}{\partial x}-y(3x^2+y^2)\frac{\partial z}{\partial y}=2z(y^2-x^2) \tag{6.407}$$

Ans: $\phi(\frac{xy}{z},\frac{x^2+y^2}{z})=0$

Problem 6.14 Find the solution of the following PDE

$$2y(z-3)\frac{\partial z}{\partial x}+(2x-z)\frac{\partial z}{\partial y}=y(2x-3) \tag{6.408}$$

passing through the curve

$$z=0, \quad x^2+y^2=2x \tag{6.409}$$

Ans: $x^2+y^2=2x+z^2-4z$

Problem 6.15 Find the solution of the following PDE

$$(2xy-1)\frac{\partial z}{\partial x}+(z-2x^2)\frac{\partial z}{\partial y}=2(x-yz) \tag{6.410}$$

passing through the curve

$$x=1, \quad y=0 \tag{6.411}$$

Ans: $x^2+y^2+z=1+xz+y$

Problem 6.16 Find the solution of

$$yz(\frac{\partial z}{\partial x})^2-(\frac{\partial z}{\partial y})=0 \tag{6.412}$$

Ans: $z^2=2ax+a^2y^2+b$

Problem 6.17 Find the solution of

$$\frac{\partial z}{\partial x}+\frac{\partial z}{\partial y}=(\frac{\partial z}{\partial x})(\frac{\partial z}{\partial y}) \tag{6.413}$$

Ans: $z=ax+\frac{ay}{a-1}+b$

Problem 6.18 Find the solution of

$$\frac{\partial z}{\partial x}+\frac{\partial z}{\partial y}=z(\frac{\partial z}{\partial x})(\frac{\partial z}{\partial y}) \tag{6.414}$$

Ans: $z^2=2(a+1)(x+\frac{y}{a})+b$

Problem 6.19 Find the solution of

$$(\frac{\partial z}{\partial x})^2(\frac{\partial z}{\partial y})^2 + x^2y^2 = x^2(\frac{\partial z}{\partial y})^2(x^2+y^2) \tag{6.415}$$

Ans: $z = \frac{1}{3}(x^2+a^2)^{3/2} + (y^2-a^2)^{1/2} + b$

Problem 6.20 Find the solution of

$$2xz - (\frac{\partial z}{\partial x})x^2 - 2xy(\frac{\partial z}{\partial y}) + (\frac{\partial z}{\partial x})(\frac{\partial z}{\partial y}) = 0 \tag{6.416}$$

Ans: $z = ay + b(x^2 - a)$

Problem 6.21 Two PDEs are given as:

$$F = (\frac{\partial z}{\partial x_1})^2 + (\frac{\partial z}{\partial x_2})(\frac{\partial z}{\partial x_3})x_2x_3^2 = 0 \tag{6.417}$$

$$F_1 = \frac{\partial z}{\partial x_1} + (\frac{\partial z}{\partial x_2})x_2 = 0 \tag{6.418}$$

(i) Check the compatibility of these two PDEs.
(ii) Use Jacobi's method to find $F_2 = a$.
(iii) Finally, find the solution for z.

Ans: (ii) $F_2 = p_1 = a$; (iii) $z = a(x_1 - \ln x_2 - \frac{1}{x_3}) + b$

Problem 6.22 Solve the following PDE by Jacobi's method

$$(\frac{\partial z}{\partial x_1})^2 x_1 + (\frac{\partial z}{\partial x_2})^2 x_2 = z \tag{6.419}$$

Ans: $c = 2(ax_1)^{1/2} + 2(bx_2)^{1/2} + 2\{(a+b)z\}^{1/2}$

Problem 6.23 Find the solution of the following PDE

$$(\frac{\partial z}{\partial x_1})^2 - (\frac{\partial z}{\partial x_2})^2 = 0 \tag{6.420}$$

Ans: $z = \pm ax + ay + b$

Problem 6.24 Find the solution of the following PDE

$$y^2(\frac{\partial z}{\partial x})^3 - x^3(\frac{\partial z}{\partial y})^2 = 0 \tag{6.421}$$

Ans: $z = \frac{1}{2}a^{1/3}x^2 \pm \frac{1}{2}a^{1/2}y^2 + b$

Problem 6.25 Find the solution of the following PDE

$$z-(\frac{\partial z}{\partial x})(\frac{\partial z}{\partial y})=0 \tag{6.422}$$

Ans: $z=\frac{1}{4}(a^{1/2}x+a^{-1/2}y+b)^2$

Problem 6.26 Find the solution of the following PDE

$$z(x\frac{\partial z}{\partial x}-y\frac{\partial z}{\partial y})=y^2-x^2 \tag{6.423}$$

Ans: $\phi(xy, z^2+x^2+y^2)=0$

Problem 6.27 Find the solution of the following PDE

$$x(y-z)\frac{\partial z}{\partial x}+y(z-x)\frac{\partial z}{\partial y}=z(x-y) \tag{6.424}$$

Ans: $\phi(x+y+z, xyz)=0$

Problem 6.28 Find the solution of the following PDE

$$z^2[(\frac{\partial z}{\partial x})^2+(\frac{\partial z}{\partial y})^2]=x^2+y^2 \tag{6.425}$$

Ans:

$$z^2=x\sqrt{x^2+a}+y\sqrt{y^2-a}+\frac{a}{2}\ln(\frac{x+\sqrt{x^2+a}}{y+\sqrt{y^2-a}})+b$$

Problem 6.29 Find the solution of the following PDE

$$(xz-y)\frac{\partial z}{\partial x}+(yz-x)\frac{\partial z}{\partial y}=1-z^2 \tag{6.426}$$

Ans:

$$\phi((x-y)(1-z),(x+y)(1+z))=0$$

CHAPTER SEVEN

Higher Order Partial Differential Equations (PDEs)

7.1 INTRODUCTION

We have considered the theory for first order PDEs in the last chapter. It seems logical to discuss first order PDEs before we discuss higher order in the present chapter. However, historically, investigation of partial differential equations starts with second order, as many physical problems, like wave propagation, heat diffusion, and incompressible and irrotational flow, have to be modelled by second order PDEs. They include the work of Bernoulli, Euler, D'Alembert, Laplace and many others. Because of its importance in physical applications, most of the available results for PDEs are developed for second order. A separation chapter (Chapter 9) will be devoted solely to the discussion of second order PDE, including wave, diffusion, and potential equations. We will consider the classification of second order PDEs in detail here, leading to the hyperbolic type, parabolic type, and elliptic type. We will show that the canonical forms of second order PDEs with constant coefficients can always convert to nonhomogeneous Klein-Gordon equations, nonhomogeneous diffusion equation and nonhomogeneous Helmholtz equations. The solution techniques for these equations are covered in some detail.

As reviewed by Selvadurai (2000a,b), nearly all existing textbooks on PDEs are restricted to the coverage of second order PDEs. For example, such PDE textbooks include some of the most popular textbooks on PDEs: Airy (1873), Bateman (1944), Carrier and Pearson (1976), Gustafson (1999), Evans et al. (2000), Gu (1989), John (1981), Heinbockel (2003), Myint-U (1987), Myint-U and Debnath (1987), Sneddon (1957), Petrovsky (1991), Sommerfeld (1949), Tricomi (1923), Drabek and Holubova (2007), Trim (1990), Farlow (1982), Zachmanoglou and Thoe (1986), Zill (1993), Zauderer (1989), and Zill and Cullen (2005). In reality, many phenomena need to be modeled by PDE of order higher than two, including the biharmonic equation, Onsager equation, Benjamin-Bonna-Mahony equation, Boussinesq equation, KdV equation, regularized long wave equation, and Hirota equation.

In view of the importance of the biharmonic equation in elasticity and in fluid flows, Selvadurai (2000b) considered the solution of the biharmonic equation in detail, whereas higher order PDEs with constant coefficients were considered in Chapters 31 and 32 of Ayres (1952).

Motivated by this shortcoming in most books, we present the biharmonic equation in detail, including a solution of the biharmonic equation expressed in the form of an integral equation similar to that of Poisson for potential theory. However, Green's function method for the biharmonic equation is deferred to Chapter 8 while the variational method for the biharmonic equation is covered in Chapter 14.

Another main difference of the present chapter from most of the pre-existing textbooks on PDEs is our discussion of the factorization technique for solving homogeneous higher order PDEs, and symbolic methods for solving nonhomogeneous PDEs of higher order. For second order PDEs with non-constant coefficients, we discuss Monge's method, and the Monge-Ampere method. These methods were not covered in most textbooks on PDEs, except Sneddon (1957) and Ayres (1952).

7.2 CLASSIFICATION OF SECOND ORDER PDE

There is something special about the second order PDE. For the case of two variables, all linear differential equations can be classified into three types of differential equations. Let us consider the following most general form of second order linear PDE:

$$A\frac{\partial^2 u}{\partial x^2}+B\frac{\partial^2 u}{\partial x\partial y}+C\frac{\partial^2 u}{\partial y^2}+D\frac{\partial u}{\partial x}+E\frac{\partial u}{\partial y}+Fu+G=0 \tag{7.1}$$

We apply a general change of variables as

$$u(x,y)\to u(\xi,\eta) \tag{7.2}$$

Inversely, the new variables can be expressed in x and y as

$$\xi=\xi(x,y),\quad \eta=\eta(x,y) \tag{7.3}$$

The mapping is arbitrary and the only requirement is having a nonzero Jacobian:

$$J=\begin{vmatrix}\xi_x & \xi_y\\ \eta_x & \eta_y\end{vmatrix}=\frac{\partial\xi}{\partial x}\frac{\partial\eta}{\partial y}-\frac{\partial\xi}{\partial y}\frac{\partial\eta}{\partial x}\neq 0 \tag{7.4}$$

The first derivatives of u with respect to x and y are

$$\frac{\partial u}{\partial x}=\frac{\partial u}{\partial\xi}\frac{\partial\xi}{\partial x}+\frac{\partial u}{\partial\eta}\frac{\partial\eta}{\partial x} \tag{7.5}$$

$$\frac{\partial u}{\partial y}=\frac{\partial u}{\partial\xi}\frac{\partial\xi}{\partial y}+\frac{\partial u}{\partial\eta}\frac{\partial\eta}{\partial y} \tag{7.6}$$

The second derivatives can be obtained by further differentiating (7.5) with respect to x as

$$\frac{\partial^2 u}{\partial x^2}=\frac{\partial}{\partial x}\left[\frac{\partial u}{\partial\xi}\frac{\partial\xi}{\partial x}+\frac{\partial u}{\partial\eta}\frac{\partial\eta}{\partial x}\right]=\frac{\partial}{\partial x}\left(\frac{\partial u}{\partial\xi}\right)\frac{\partial\xi}{\partial x}+\frac{\partial u}{\partial\xi}\frac{\partial^2\xi}{\partial x^2}+\frac{\partial}{\partial x}\left(\frac{\partial u}{\partial\eta}\right)\frac{\partial\eta}{\partial x}+\frac{\partial u}{\partial\eta}\frac{\partial^2\eta}{\partial x^2} \tag{7.7}$$

It is important to note that the differentiation of the bracket term on the right hand side of (7.7) is conducted using the chain rule similar to that for (7.5). Finally, we get

$$\frac{\partial^2 u}{\partial x^2}=\frac{\partial^2 u}{\partial\xi^2}\left(\frac{\partial\xi}{\partial x}\right)^2+2\frac{\partial^2 u}{\partial\xi\partial\eta}\frac{\partial\xi}{\partial x}\frac{\partial\eta}{\partial x}+\frac{\partial^2 u}{\partial\eta^2}\left(\frac{\partial\eta}{\partial x}\right)^2+\frac{\partial u}{\partial\xi}\frac{\partial^2\xi}{\partial x^2}+\frac{\partial u}{\partial\eta}\frac{\partial^2\eta}{\partial x^2} \tag{7.8}$$

Similarly, the other two second derivatives can be obtained as

$$\frac{\partial^2 u}{\partial x \partial y} = \frac{\partial^2 u}{\partial \xi^2}\frac{\partial \xi}{\partial x}\frac{\partial \xi}{\partial y} + \frac{\partial^2 u}{\partial \xi \partial \eta}(\frac{\partial \xi}{\partial x}\frac{\partial \eta}{\partial y} + \frac{\partial \eta}{\partial x}\frac{\partial \xi}{\partial y}) + \frac{\partial^2 u}{\partial \eta^2}\frac{\partial \eta}{\partial x}\frac{\partial \eta}{\partial y} + \frac{\partial u}{\partial \xi}\frac{\partial^2 \xi}{\partial x \partial y} + \frac{\partial u}{\partial \eta}\frac{\partial^2 \eta}{\partial x \partial y} \tag{7.9}$$

$$\frac{\partial^2 u}{\partial y^2} = \frac{\partial^2 u}{\partial \xi^2}(\frac{\partial \xi}{\partial y})^2 + 2\frac{\partial^2 u}{\partial \xi \partial \eta}\frac{\partial \xi}{\partial y}\frac{\partial \eta}{\partial y} + \frac{\partial^2 u}{\partial \eta^2}(\frac{\partial \eta}{\partial y})^2 + \frac{\partial u}{\partial \xi}\frac{\partial^2 \xi}{\partial y^2} + \frac{\partial u}{\partial \eta}\frac{\partial^2 \eta}{\partial y^2} \tag{7.10}$$

Finally, substitution of (7.5), (7.6), and (7.8) to (7.10) into (7.1) gives

$$\bar{A}\frac{\partial^2 u}{\partial \xi^2} + \bar{B}\frac{\partial^2 u}{\partial \xi \partial \eta} + \bar{C}\frac{\partial^2 u}{\partial \eta^2} + \bar{D}\frac{\partial u}{\partial \xi} + \bar{E}\frac{\partial u}{\partial \eta} + \bar{F}u = \bar{G} \tag{7.11}$$

$$\bar{A} = A(\frac{\partial \xi}{\partial x})^2 + B\frac{\partial \xi}{\partial x}\frac{\partial \xi}{\partial y} + C(\frac{\partial \xi}{\partial y})^2 \tag{7.12}$$

$$\bar{B} = 2A\frac{\partial \xi}{\partial x}\frac{\partial \eta}{\partial x} + B(\frac{\partial \xi}{\partial x}\frac{\partial \eta}{\partial y} + \frac{\partial \eta}{\partial x}\frac{\partial \xi}{\partial y}) + 2C\frac{\partial \xi}{\partial y}\frac{\partial \eta}{\partial y} \tag{7.13}$$

$$\bar{C} = A(\frac{\partial \eta}{\partial x})^2 + B\frac{\partial \eta}{\partial x}\frac{\partial \eta}{\partial y} + C(\frac{\partial \eta}{\partial y})^2 \tag{7.14}$$

$$\bar{D} = A\frac{\partial^2 \xi}{\partial x^2} + B\frac{\partial^2 \xi}{\partial x \partial y} + C\frac{\partial^2 \xi}{\partial y^2} + D\frac{\partial \xi}{\partial x} + E\frac{\partial \xi}{\partial y} \tag{7.15}$$

$$\bar{E} = A\frac{\partial^2 \eta}{\partial x^2} + B\frac{\partial^2 \eta}{\partial x \partial y} + C\frac{\partial^2 \eta}{\partial y^2} + D\frac{\partial \eta}{\partial x} + E\frac{\partial \eta}{\partial y} \tag{7.16}$$

$$\bar{F} = F(x(\xi,\eta), y(\xi,\eta)) \tag{7.17}$$

$$\bar{G} = -G(x(\xi,\eta), y(\xi,\eta)) \tag{7.18}$$

The mathematical form of (7.11) looks more complicated than (7.1). One may ask why we want to apply a more complicated change of variables to result in the system given in (7.11) to (7.18). However, the general mapping given in (7.11) allows us to search for a simpler form of second order PDE.

A major property of this transformation is that

$$B^2 - 4AC = \frac{\bar{B}^2 - 4\bar{A}\bar{C}}{J^2} \tag{7.19}$$

where J is the Jacobian of the mapping function defined in (7.4). The validity of (7.19) is proved as Problems 7.2 to 7.4 at the end of this chapter. That is, the sign of B^2-4AC in (7.18) will remain the same after any valid coordinate transformation with nonzero J. If it is positive, it is always positive. If it is negative, it is always negative. If it is zero, it is always zero. Its sign is thus an invariant. In fact, this property allows us to classify the PDE using the sign of (7.19). This idea is first explored by Laplace and later refined by Bois-Reymond. We will next consider the idea of characteristics and its relation to the sign of (7.19).

7.2.1 Physical Meaning of Characteristics

In particular, observing the mathematical similarity of (7.12) and (7.14), we look for the possibility of setting

$$\bar{A} = 0, \quad \bar{C} = 0 \tag{7.20}$$

The mathematical structure for both (7.12) and (7.14) is the same. Thus, they will be considered together here. In particular, we are looking for a solution that satisfies the following first order PDE:

$$A(\frac{\partial z}{\partial x})^2 + B\frac{\partial z}{\partial x}\frac{\partial z}{\partial y} + C(\frac{\partial z}{\partial y})^2 = 0 \tag{7.21}$$

There are two solutions for (7.21). Let us denote its solutions as

$$z = \xi(x, y) = C_1, \quad z = \eta(x, y) = C_2 \tag{7.22}$$

These curves are illustrated in Figure 7.1.

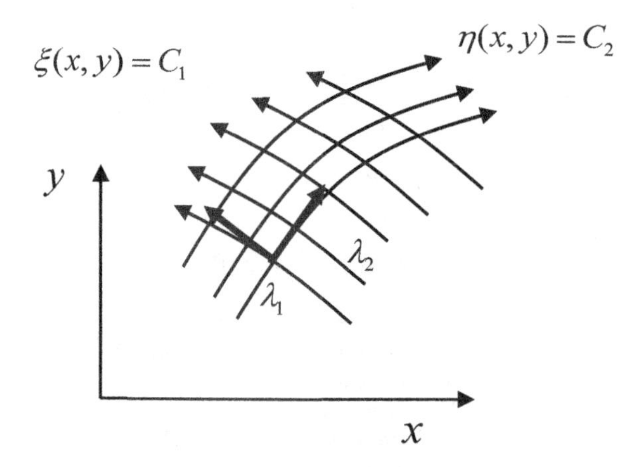

Figure 7.1 Solutions are propagating along two curves, which are called characteristics

Taking the total differential of (7.22) gives

$$d\xi = \frac{\partial \xi}{\partial x}dx + \frac{\partial \xi}{\partial y}dy = 0 \tag{7.23}$$

$$d\eta = \frac{\partial \eta}{\partial x}dx + \frac{\partial \eta}{\partial y}dy = 0 \tag{7.24}$$

Thus, the slopes of these can be found as

$$\lambda_1 = \frac{dy}{dx} = -\frac{\partial \xi / \partial x}{\partial \xi / \partial y} \tag{7.25}$$

$$\lambda_2 = \frac{dy}{dx} = -\frac{\partial \eta / \partial x}{\partial \eta / \partial y} \tag{7.26}$$

These slopes λ_1 and λ_2 are depicted in Figure 7.1. Along these curves, the original PDE becomes an ODE, and these curves are known as characteristics curves. If these curves exist, the solutions are propagating along these characteristics. Note, however, whether these characteristics exist or not, depends on the given coefficients A, B, and C given in the original PDE given in (7.1). On the other hand, we can see that (7.21) is actually a quadratic equation for the ratio of the partial derivative of ξ or η with respect to x to that with respect to y. In particular, we have

$$A(\frac{\partial \xi}{\partial x} / \frac{\partial \xi}{\partial y})^2 + B(\frac{\partial \xi}{\partial x} / \frac{\partial \xi}{\partial y}) + C = 0 \tag{7.27}$$

$$A(\frac{\partial \eta}{\partial x} / \frac{\partial \eta}{\partial y})^2 + B(\frac{\partial \eta}{\partial x} / \frac{\partial \eta}{\partial y}) + C = 0 \tag{7.28}$$

The two solutions of them are the same. Without loss of generality, we can pick the root of dy/dx for the two characteristics as:

$$\lambda_1 = \frac{dy}{dx} = -\frac{\partial \xi / \partial x}{\partial \xi / \partial y} = \frac{B + \sqrt{B^2 - 4AC}}{2A} \tag{7.29}$$

$$\lambda_2 = \frac{dy}{dx} = -\frac{\partial \eta / \partial x}{\partial \eta / \partial y} = \frac{B - \sqrt{B^2 - 4AC}}{2A} \tag{7.30}$$

Clearly, the slopes of these characteristics are functions of A, B, and C.

7.2.2 Sommerfeld's Interpretation of Characteristics

A closer look at the solutions given in (7.29) and (7.30) reveals a more in-depth physical meaning of the characteristics. The following interpretation was given by Monge in 1770 and summarized in the book by Sommerfeld (1949). Let us recast the second order PDE in (7.1) as

$$A\frac{\partial^2 u}{\partial x^2} + B\frac{\partial^2 u}{\partial x \partial y} + C\frac{\partial^2 u}{\partial y^2} = \Phi(u, \frac{\partial u}{\partial x}, \frac{\partial u}{\partial y}, x, y) \tag{7.31}$$

Next, we can rewrite (7.31) as:

$$Ar + Bs + Ct = \Phi \tag{7.32}$$

where

$$r = \frac{\partial^2 u}{\partial x^2}, s = \frac{\partial^2 u}{\partial x \partial y}, t = \frac{\partial^2 u}{\partial y^2} \tag{7.33}$$

By adopting the notation that we used for the first order ODE introduced in the last chapter, we have

$$p = \frac{\partial u}{\partial x}, \quad q = \frac{\partial u}{\partial y} \tag{7.34}$$

Taking the total differential of both p and q, we get

$$dp = d(\frac{\partial u}{\partial x}) = \frac{\partial^2 u}{\partial x^2}dx + \frac{\partial^2 u}{\partial x \partial y}dy = rdx + sdy \tag{7.35}$$

$$dq = d(\frac{\partial u}{\partial y}) = \frac{\partial^2 u}{\partial x \partial y}dx + \frac{\partial^2 u}{\partial y^2}dy = sdx + tdy \tag{7.36}$$

We now group (7.32), (7.35), and (7.36) in matrix form as

$$\begin{bmatrix} A & B & C \\ dx & dy & 0 \\ 0 & dx & dy \end{bmatrix} \begin{Bmatrix} r \\ s \\ t \end{Bmatrix} = \begin{Bmatrix} \Phi \\ dp \\ dq \end{Bmatrix} \tag{7.37}$$

We rewrite this symbolically as

$$[A]\{t\} = \{\Phi\} \tag{7.38}$$

For a certain problem, if Φ, dp, and dq are given on some curve, we can find r, s, and t if and only if the determinant of matrix $\boldsymbol{A}$ is nonzero. The determinant is obtained from (7.37) as:

$$\Delta = \det|A| = A(dy)^2 - Bdxdy + C(dx)^2 \tag{7.39}$$

Now, we observe that the slopes given in (7.29) and (7.30) satisfy the following equation

$$A(dy)^2 - Bdxdy + C(dx)^2 = 0 \tag{7.40}$$

which is precisely the determinant of the system given in (7.37). In other words, we cannot solve for r, s, and t if Φ, dp, and dq are given on the characteristics. This situation is illustrated in Figure 7.2. Given data cannot be prescribed on any line that parallels characteristics, otherwise the problem cannot be solved. The root of (7.40) is

$$\frac{dy}{dx} = \frac{B \mp \sqrt{B^2 - 4AC}}{2A} \tag{7.41}$$

which is precisely (7.29) and (7.30). Thus, $\Delta = 0$ implies characteristics exists.

When $\Delta \neq 0$, r, s, and t are nonzero, we can find their derivatives as:

$$r_x = \frac{\partial^3 u}{\partial x^3}, \quad s_x = r_y = \frac{\partial^3 u}{\partial x^2 \partial y}, \quad t_y = \frac{\partial^3 u}{\partial y^3} \tag{7.42}$$

Thus, differentiation of (7.32), (7.35), and (7.36) with respect to x gives

$$Ar_x + Bs_x + Ct_x = \Phi_x \tag{7.43}$$

$$dr = r_x dx + s_x dy \tag{7.44}$$

$$ds = s_x dx + t_x dy \tag{7.45}$$

Note that we have used the following identities in arriving at these results

$$r = \frac{\partial p}{\partial x}, \quad s = \frac{\partial q}{\partial x}, \quad s_y = t_x = \frac{\partial^3 u}{\partial x \partial y^2} \tag{7.46}$$

Again, we can put (7.43) to (7.45) into matrix form as

$$\begin{bmatrix} A & B & C \\ dx & dy & 0 \\ 0 & dx & dy \end{bmatrix} \begin{Bmatrix} r_x \\ s_x \\ t_x \end{Bmatrix} = \begin{Bmatrix} \Phi_x \\ dr \\ ds \end{Bmatrix} \tag{7.47}$$

Note that the matrix of coefficients is the same as that for (7.37). It means that we can solve for higher derivatives of u. This process of taking higher derivatives leads to the same coefficient matrix again and again. Therefore, if $\Delta \neq 0$, all higher derivatives of u can be found. In other words, u can be expanded in Taylor series expansion at any points. Thus, the function u must be analytic, smooth, and continuous. Therefore, there is no jump or no wave-like signal from the solution of u if $\Delta \neq 0$. On the contrary, if $\Delta = 0$ (and this results from assuming the validity of (7.20)), the solution is not continuous or representing a wave signal. Note that the arrival of the wave signal is an abrupt and discontinuous phenomenon. Consequently, we see that characteristics represent wave types of solutions. For differential equations with propagating solutions (or hyperbolic type), mathematicians use the term "characteristics" whereas physicists use the term "waves" for the same physical phenomenon. In fact, there is a Le Roux-Delassus Theorem in 1895 stating that any singular surface of a solution of a linear differential equation must be characteristics (Hadamard, 1923).

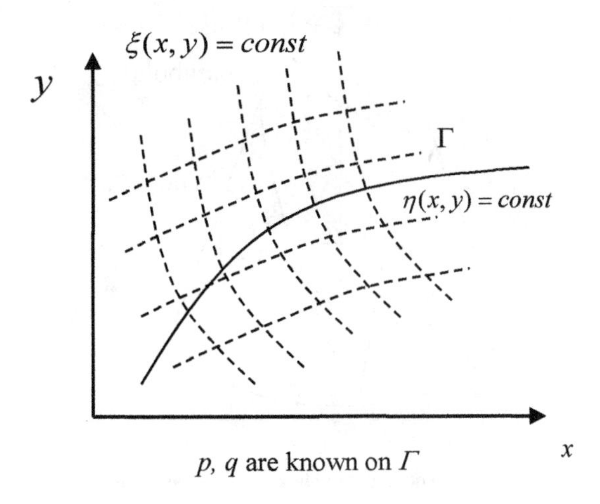

Figure 7.2 Discontinuous solutions or wave solution along characteristics. Curve Γ with initial derivatives data cannot be parallel to characteristics

Any second order PDE can now be classified according to how many roots exist for the characteristics, which only depends on the value of A, B, and C. Let us recall our terminology for second order hyperbola, parabolic, and hyperbolic curves. In particular, we have

$$Ax^2 + Bxy + Cy^2 + Dx + Ey + F = 0 \tag{7.48}$$

This equation for a curve is classified as a hyperbola, parabola, or ellipse according to:

$$B^2 - 4AC = \begin{cases} > 0 & \textit{hyperbola} \\ = 0 & \textit{parabola} \\ < 0 & \textit{ellipse} \end{cases} \tag{7.49}$$

These curves are also known as conic sections. As illustrated in Figure 7.3, they all result from cutting a section of a cone.

Recognizing the similarity of the mathematical structure between (7.1) and (7.48), Bois-Reymond in 1839 proposed the following classification of second order PDEs:

$$\text{Classification} \begin{cases} B^2 - 4AC < 0: \text{elliptic} \\ B^2 - 4AC = 0: \text{parabolic} \\ B^2 - 4AC > 0: \text{hyperbolic} \end{cases} \tag{7.50}$$

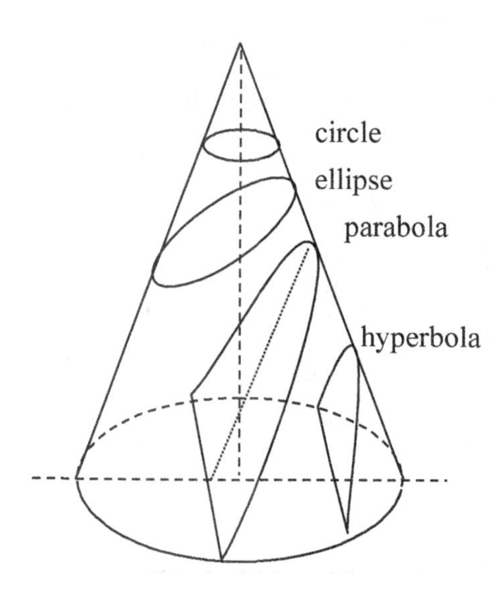

Figure 7.3 Hyperbola, parabola and ellipse as conic sections

Note that the hyperbolic type of second order PDE was first proposed by Laplace. It was subsequently extended to the classification summarized in (7.50) by Bois-Reymond. Note that the hyperbolic case corresponds to the existence of two characteristics or the existence of wave-type solutions. The parabolic case corresponds to the existence of one characteristics or physically corresponds to a diffusion type of phenomenon. Finally, the elliptic type corresponds to no characteristics or physically corresponds to an equilibrium type of phenomenon. More discussions on the physical meaning of this classification will be given later.

7.2.3 Hyperbolic PDE

Recalling that for the hyperbolic type, we have the validity of (7.20), thus (7.11) is reduced to

$$\bar{B}\frac{\partial^2 u}{\partial\xi\partial\eta}+\bar{D}\frac{\partial u}{\partial\xi}+\bar{E}\frac{\partial u}{\partial\eta}+\bar{F}u=\bar{G} \tag{7.51}$$

This can be rewritten as

$$\frac{\partial^2 u}{\partial\xi\partial\eta}+d\frac{\partial u}{\partial\xi}+\bar{e}\frac{\partial u}{\partial\eta}+fu=g \tag{7.52}$$

which is the first canonical form of the hyperbolic type of PDE, and where

$$d=\frac{\bar{D}}{\bar{B}},\quad \bar{e}=\frac{\bar{E}}{\bar{B}},\quad f=\frac{\bar{F}}{\bar{B}},\quad g=\frac{\bar{G}}{\bar{B}} \tag{7.53}$$

For the hyperbolic type of PDE, there exists another canonical form. To see this, we can introduce a change of variables as

$$\xi=s+t,\quad \eta=s-t \tag{7.54}$$

This can be inverted to give

$$s=\frac{1}{2}(\xi+\eta),\quad t=\frac{1}{2}(\xi-\eta) \tag{7.55}$$

The first derivative of u becomes

$$\frac{\partial u}{\partial\xi}=\frac{\partial u}{\partial s}\frac{\partial s}{\partial\xi}+\frac{\partial u}{\partial t}\frac{\partial t}{\partial\xi}=\frac{1}{2}(\frac{\partial u}{\partial s}+\frac{\partial u}{\partial t}) \tag{7.56}$$

$$\frac{\partial u}{\partial\eta}=\frac{\partial u}{\partial s}\frac{\partial s}{\partial\eta}+\frac{\partial u}{\partial t}\frac{\partial t}{\partial\eta}=\frac{1}{2}(\frac{\partial u}{\partial s}-\frac{\partial u}{\partial t}) \tag{7.57}$$

Subsequently, we have the second derivative as

$$\frac{\partial^2 u}{\partial\xi\partial\eta}=\frac{1}{2}\left[\frac{\partial}{\partial s}(\frac{\partial u}{\partial s}-\frac{\partial u}{\partial t})\frac{\partial s}{\partial\xi}+\frac{\partial}{\partial t}(\frac{\partial u}{\partial s}-\frac{\partial u}{\partial t})\frac{\partial t}{\partial\xi}\right]=\frac{1}{4}(\frac{\partial^2 u}{\partial s^2}-\frac{\partial^2 u}{\partial t^2}) \tag{7.58}$$

Substitution of (7.58) into (7.52) leads to

$$\frac{\partial^2 u}{\partial s^2}-\frac{\partial^2 u}{\partial t^2}+d_1\frac{\partial u}{\partial s}+e_1\frac{\partial u}{\partial t}+f_1 u=g_1 \tag{7.59}$$

where

$$d_1=2(d+\bar{e}),\quad e_1=2(d-\bar{e}),\quad f_1=4f,\quad g_1=4g \tag{7.60}$$

Equation (7.58) is called the second canonical form of hyperbolic type of PDE, which is mathematically equivalent to (7.52).

Let us recall the one-dimensional wave equation as:

$$\frac{\partial^2\varphi}{\partial t^2}-c^2\frac{\partial^2\varphi}{\partial x^2}=0 \tag{7.61}$$

Introduce the following change of variables

$$s=\frac{x}{c} \tag{7.62}$$

and we get

$$c^2 \frac{\partial^2 \varphi}{\partial x^2} = \frac{\partial^2 \varphi}{\partial s^2} \tag{7.63}$$

Thus, the second order derivative terms in (7.59) agree with those in the one-dimensional wave equation. In Section 7.3, we will discuss further simplification of the canonical form of the hyperbolic type PDE.

Note that for the cases of constant coefficients we can show that the characteristics are straight lines. In particular, for a special case of constant A, B, and C, we find the characteristics as

$$\xi = y - \frac{B - \sqrt{B^2 - 4AC}}{2A} x = C_1, \quad \eta = y - \frac{B + \sqrt{B^2 - 4AC}}{2A} x = C_2 \tag{7.64}$$

For such case, we can show that the coefficients in (7.52) can be found as

$$\begin{aligned} d &= \frac{2AE - D[B - \sqrt{B^2 - 4AC}]}{2[2AC(4A^2 + 1) - B^2]}, \\ \bar{e} &= \frac{2AE - D[B + \sqrt{B^2 - 4AC}]}{2[2AC(4A^2 + 1) - B^2]}, \\ f &= \frac{AF}{2AC(4A^2 + 1) - B^2}, \\ g &= \frac{AG}{2AC(4A^2 + 1) - B^2} \end{aligned} \tag{7.65}$$

Example 7.1 Classify the following second order PDE. If it is hyperbolic, find the two characteristics and the canonical form of the PDE.

$$u_{xx} - 4u_{yy} + u_x = 0 \tag{7.66}$$

Solution: Comparing (7.66) to the standard form (7.1), we have

$$A = 1, \quad B = 0, \quad C = -4, \quad D = 1, \quad E = 0, \quad F = 0 \tag{7.67}$$

Thus, we find

$$\sqrt{B^2 - 4AC} = 4 > 0 \tag{7.68}$$

Therefore, by the Bois-Reymond classification (7.66) is classified as hyperbolic type of PDE. The corresponding characteristics are

$$\lambda_1 = \frac{dy}{dx} = -\frac{\partial \xi / \partial x}{\partial \xi / \partial y} = \frac{B - \sqrt{B^2 - 4AC}}{2A} = -2 \tag{7.69}$$

$$\lambda_2 = \frac{dy}{dx} = -\frac{\partial \eta / \partial x}{\partial \eta / \partial y} = \frac{B + \sqrt{B^2 - 4AC}}{2A} = 2 \tag{7.70}$$

Integration of these two ODEs for y gives two equations

$$y = -2x + C_1 \tag{7.71}$$

$$y = 2x + C_2 \tag{7.72}$$

In view of (7.22), we find the two characteristics as

$$\xi = y + 2x = C_1 \tag{7.73}$$

$$\eta = y - 2x = C_2 \tag{7.74}$$

Thus, we have

$$\frac{\partial \xi}{\partial x} = 2, \quad \frac{\partial \xi}{\partial y} = 1, \quad \frac{\partial \eta}{\partial x} = -2, \quad \frac{\partial \eta}{\partial y} = 1 \tag{7.75}$$

Substitution of (7.75) into (7.13) to (7.18) yields

$$\bar{B} = -16, \quad \bar{D} = 2, \quad \bar{E} = -2, \quad \bar{A} = \bar{C} = \bar{F} = 0 \tag{7.76}$$

Finally, the canonical form becomes

$$\frac{\partial^2 u}{\partial \xi \partial \eta} = -\frac{1}{8}\left(\frac{\partial u}{\partial \eta} - \frac{\partial u}{\partial \xi}\right) \tag{7.77}$$

7.2.4 Parabolic PDE

For the parabolic type, we have only one characteristics. Let us set

$$\bar{A} = 0, \quad \bar{C} \neq 0 \tag{7.78}$$

Actually, for this case we must have

$$\bar{B} = 0 \tag{7.79}$$

To see this, we consider

$$\bar{A} = A\left(\frac{\partial \xi}{\partial x}\right)^2 + B\frac{\partial \xi}{\partial x}\frac{\partial \xi}{\partial y} + C\left(\frac{\partial \xi}{\partial y}\right)^2 = \left(\sqrt{A}\frac{\partial \xi}{\partial x} + \sqrt{C}\frac{\partial \xi}{\partial y}\right)^2 = 0 \tag{7.80}$$

In obtaining the last of (7.80), we have used the following identity

$$B = 2\sqrt{AC} \tag{7.81}$$

which is a natural consequence of the parabolic condition of

$$B^2 - 4AC = 0 \tag{7.82}$$

Note that (7.82) resulted from (7.29) and (7.30) if there is only one root for λ. Now let us consider

$$\begin{aligned} \bar{B} &= 2A\frac{\partial \xi}{\partial x}\frac{\partial \eta}{\partial x} + B\left(\frac{\partial \xi}{\partial x}\frac{\partial \eta}{\partial y} + \frac{\partial \eta}{\partial x}\frac{\partial \xi}{\partial y}\right) + 2C\frac{\partial \xi}{\partial y}\frac{\partial \eta}{\partial y} \\ &= 2\left(\sqrt{A}\frac{\partial \xi}{\partial x} + \sqrt{C}\frac{\partial \xi}{\partial y}\right)\left(\sqrt{A}\frac{\partial \eta}{\partial x} + \sqrt{C}\frac{\partial \eta}{\partial y}\right) = 0 \end{aligned} \tag{7.83}$$

This first bracket is zero from (7.80), and thus the second bracket must not be zero. Thus, we see

$$\bar{C} = A\left(\frac{\partial \eta}{\partial x}\right)^2 + B\frac{\partial \eta}{\partial x}\frac{\partial \eta}{\partial y} + C\left(\frac{\partial \eta}{\partial y}\right)^2 = \left(\sqrt{A}\frac{\partial \eta}{\partial x} + \sqrt{C}\frac{\partial \eta}{\partial y}\right)^2 \neq 0 \tag{7.84}$$

This is exactly the second bracket in (7.83) and therefore is not zero. This, of course, agrees with our expectation in (7.78) because there is only one characteristics. Recall from (7.80) that $\bar{A} = 0$ is equivalent to

$$\lambda_1 = \frac{dy}{dx} = -\frac{\partial \xi / \partial x}{\partial \xi / \partial y} = \frac{\sqrt{C}}{\sqrt{A}} = \frac{B}{2\sqrt{A}\sqrt{A}} = \frac{B}{2A} \tag{7.85}$$

Thus, for constant coefficients the characteristics is

$$\xi = y - \frac{B}{2A}x = C_1 \tag{7.86}$$

To find the canonical form, we need to have the other variable η. There is not much discussion on this in the literature. In fact, we can choose any η, which is not parallel to ξ. Because of this, the canonical form for the parabolic case is not unique.

Anyhow, the canonical form of the parabolic type is given symbolically as

$$\frac{\partial^2 u}{\partial \eta^2} + d_1 \frac{\partial u}{\partial \xi} + e_1 \frac{\partial u}{\partial \eta} + f_1 u = g_1 \tag{7.87}$$

where

$$d_1 = \frac{\bar{D}}{\bar{C}}, \quad e_1 = \frac{\bar{E}}{\bar{C}}, \quad f_1 = \frac{\bar{F}}{\bar{C}} \quad g_1 = \frac{\bar{G}}{\bar{C}} \tag{7.88}$$

If A, B, and C are constants and thus (7.86) is valid, we have

$$\xi = y - \frac{B}{2A}x = C_1, \quad \eta = x = C_2 \tag{7.89}$$

The second of (7.89) is chosen arbitrarily (any line not parallel to $\xi = C_1$) as long as the Jacobian of the mapping is not zero. For this case, we have $J = -1$. For this particular choice, we have

$$d_1 = \frac{2AE - DB}{2A^2}, \quad e_1 = \frac{D}{A}, \quad f_1 = \frac{F}{A}, \quad g_1 = -\frac{G}{A} \tag{7.90}$$

The most common parabolic type of second order PDE is a heat or diffusion equation:

$$\frac{\partial \varphi}{\partial t} = \alpha \frac{\partial^2 \varphi}{\partial x^2} \tag{7.91}$$

Again, the constant α can be easily absorbed into x to yield η (compare (7.62)). We will further simplify the canonical form in Section 7.3.

7.2.5 Elliptic PDE

For the elliptic case, we have

$$B^2 - 4AC < 0 : \text{elliptic} \tag{7.92}$$

This implies that the characteristics are complex (i.e., no real solution). Let us write the characteristics as

$$\zeta_1 = \xi + i\eta \tag{7.93}$$

$$\zeta_2 = \xi - i\eta \tag{7.94}$$

Using (7.93) as the characteristics, we have

$$A\left(\frac{\partial(\xi + i\eta)}{\partial x}\right)^2 + B\frac{\partial(\xi + i\eta)}{\partial x}\frac{\partial(\xi + i\eta)}{\partial y} + C\left(\frac{\partial(\xi + i\eta)}{\partial y}\right)^2 = 0 \tag{7.95}$$

Note that

$$(\frac{\partial(\xi+i\eta)}{\partial x})^2=(\frac{\partial\xi}{\partial x})^2+2i\frac{\partial\xi}{\partial x}\frac{\partial\eta}{\partial x}-(\frac{\partial\eta}{\partial x})^2 \tag{7.96}$$

$$(\frac{\partial(\xi+i\eta)}{\partial y})^2=(\frac{\partial\xi}{\partial y})^2+2i\frac{\partial\xi}{\partial y}\frac{\partial\eta}{\partial y}-(\frac{\partial\eta}{\partial y})^2 \tag{7.97}$$

$$\frac{\partial(\xi+i\eta)}{\partial x}\frac{\partial(\xi+i\eta)}{\partial y}=\frac{\partial\xi}{\partial x}\frac{\partial\xi}{\partial y}+i(\frac{\partial\eta}{\partial x}\frac{\partial\xi}{\partial y}+\frac{\partial\eta}{\partial y}\frac{\partial\xi}{\partial x})-\frac{\partial\eta}{\partial x}\frac{\partial\eta}{\partial y} \tag{7.98}$$

Substitution of (7.96) to (7.98) into (7.95) gives

$$(\bar{A}-\bar{C})+i\bar{B}=0 \tag{7.99}$$

This implies

$$\bar{A}=\bar{C}\neq 0, \quad \bar{B}=0 \tag{7.100}$$

Similarly, we can also use (7.94) into (7.95). It is straightforward to show that this also leads to (7.99) (see Problem 7.1).

The canonical form becomes

$$\frac{\partial^2 u}{\partial\xi^2}+\frac{\partial^2 u}{\partial\eta^2}+d\frac{\partial u}{\partial\xi}+\bar{e}\frac{\partial u}{\partial\eta}+fu=g \tag{7.101}$$

where

$$d=\frac{\bar{D}}{\bar{A}}, \quad \bar{e}=\frac{\bar{E}}{\bar{A}}, \quad f=\frac{\bar{F}}{\bar{A}}, \quad g=\frac{\bar{G}}{\bar{A}} \tag{7.102}$$

We will now work out the details of (7.12) to (7.18) for the case of complex characteristics to find the explicit forms for (7.102). In particular, we first let the characteristics given in (7.93) be

$$\zeta=\xi+i\eta \tag{7.103}$$

Then, substitution of (7.103) into the first equation of (7.100) gives

$$A(\xi_x^2-\eta_x^2)+B(\xi_x\xi_y-\eta_y\eta_x)+C(\xi_y^2-\eta_y^2)=0 \tag{7.104}$$

Substitution of (7.103) into the second equation of (7.100) gives

$$2A\xi_x\eta_x+B(\xi_x\eta_y+\xi_y\eta_x)+2C\xi_y\eta_y=0 \tag{7.105}$$

Differentiation of (7.103) gives

$$\zeta_x=\xi_x+i\eta_x, \quad \zeta_y=\xi_y+i\eta_y \tag{7.106}$$

By employing the results given in (7.106), we find

$$\zeta_x^2=\xi_x^2+2i\xi_x\eta_x-\eta_x^2 \tag{7.107}$$

$$\zeta_y^2=\xi_y^2+2i\xi_y\eta_y-\eta_y^2 \tag{7.108}$$

$$\zeta_y\zeta_x=(\xi_y+i\eta_y)(\xi_x+i\eta_x)=\xi_y\xi_x-\eta_y\eta_x+i(\eta_x\xi_y+\xi_x\eta_y) \tag{7.109}$$

Adding (7.104) and i times (7.105) and using (7.107) to (7.109), we obtain

$$A\zeta_x^2+B\zeta_x\zeta_y+C\zeta_y^2=0 \tag{7.110}$$

which is of course equal to (7.95). More importantly, the solution of (7.110) is simply the solution of a quadratic equation, and is given as

$$\frac{\zeta_x}{\zeta_y} = \frac{\xi_x + i\eta_x}{\xi_y + i\eta_y} = \frac{-B - i\sqrt{4AC - B^2}}{2A} \tag{7.111}$$

Rearranging (7.111) gives

$$\begin{aligned} \xi_x + i\eta_x &= (\frac{-B - i\sqrt{4AC - B^2}}{2A})(\xi_y + i\eta_y) \\ &= \left\{ \frac{-B}{2A}\xi_y + \frac{\eta_y}{2A}\sqrt{4AC - B^2} \right\} + i\left\{ \frac{-B}{2A}\eta_y - \frac{\xi_y}{2A}\sqrt{4AC - B^2} \right\} \end{aligned} \tag{7.112}$$

Comparing the real and imaginary parts of (7.112) results in

$$\xi_x = \left\{ \frac{-B}{2A}\xi_y + \frac{\eta_y}{2A}\sqrt{4AC - B^2} \right\}, \quad \eta_x = \left\{ \frac{-B}{2A}\eta_y - \frac{\xi_y}{2A}\sqrt{4AC - B^2} \right\} \tag{7.113}$$

The second equation of (7.113) can be used to solve for ξ_y and its result can be substituted into the first equation of (7.113) to give

$$\xi_y = -\frac{2A\eta_x + B\eta_y}{\sqrt{4AC - B^2}}, \quad \xi_x = \frac{2C\eta_y + B\eta_x}{\sqrt{4AC - B^2}} \tag{7.114}$$

These equations are called Beltrami equations. Clearly, we can eliminate ξ from (7.113) to get a second order PDE for η. However, the resulting equation is not easy to solve. If A, B, C, D, E, and F are constants, we can actually assume a linear dependence of ξ and η in terms of x and y (Kevorkian, 1990)

$$\xi = \alpha x + \beta y, \quad \eta = \gamma x + \delta y \tag{7.115}$$

Comparison of the mathematical form of (7.64) and (7.86) with (7.115) shows that these characteristics are all straight lines for the case of constant coefficients. Using (7.115), we find

$$\xi_y = \beta = -(2\tilde{A}\eta_x + \tilde{B}\eta_x) = -2\tilde{A}\gamma - \tilde{B}\delta \tag{7.116}$$

Comparing (7.116) and the first of (7.114), we obtain

$$\tilde{A} = \frac{A}{\sqrt{4AC - B^2}}, \quad \tilde{B} = \frac{B}{\sqrt{4AC - B^2}} \tag{7.117}$$

Alternatively, we can also find

$$\xi_x = \alpha = 2\tilde{A}\eta_y + \tilde{B}\eta_x = \tilde{B}\gamma + 2\tilde{A}\delta \tag{7.118}$$

For simplicity, we choose

$$\delta = 0, \quad \gamma = 1 \tag{7.119}$$

Thus, we have

$$\xi = \tilde{B}x - 2\tilde{A}y, \quad \eta = x \tag{7.120}$$

With this particular change of variables, we have

$$\xi_x = \tilde{B}, \quad \xi_y = -2\tilde{A}, \quad \eta_x = 1, \quad \eta_y = 0 \tag{7.121}$$

Consequently, we have

$$\overline{A} = A\xi_x^2 + B\xi_x\xi_y + C\xi_y^2 = \tilde{B}^2 A - 2B\tilde{B}\tilde{A} + 4\tilde{A}^2 C = A \tag{7.122}$$

$$\overline{C} = A\eta_x^2 + B\eta_x\eta_y + C\eta_y^2 = A \tag{7.123}$$

$$\bar{D} = A\xi_{xx} + B\xi_{xy} + C\xi_{yy} + D\xi_x + E\xi_y = D\tilde{B} - 2\tilde{A}E = \frac{DB - 2AE}{\sqrt{4AC - B^2}} \quad (7.124)$$

$$\bar{E} = A\eta_{xx} + B\eta_{xy} + C\eta_{yy} + D\eta_x + E\eta_y = D \quad (7.125)$$

$$\bar{F} = F \quad (7.126)$$

Finally, we obtain

$$u_{\xi\xi} + u_{\eta\eta} + \frac{DB - 2AE}{A\sqrt{4AC - B^2}} u_\xi + \frac{D}{A} u_\eta + \frac{F}{A} u = 0 \quad (7.127)$$

Therefore, comparison of (7.127) with (7.101) gives

$$d = \frac{DB - 2AE}{A\sqrt{4AC - B^2}}, \quad \bar{e} = \frac{D}{A}, \quad f = \frac{F}{A}, \quad g = \frac{G}{A} \quad (7.128)$$

A number of special cases of (7.101) will be considered. If the last four terms on the right hand side of (7.101) are zero, we have the Laplace equation. If d, e, and f are zeros and g is a function of ξ and η, we arrive at Poisson's equation. If only f is nonzero, we have the Helmholtz equation. Thus, the Laplace, Poisson, and Helmholtz equations are all elliptic type. Elliptic PDE are mainly for equilibrium type problems.

7.3 CANONICAL FORMS OF SECOND ORDER PDE

The canonical forms given in (7.51), (7.86), and (7.101) are the normal canonical form given in the literature. In this section, we will show that they can further be simplified. The following further reduction of canonical forms is adopted from the idea mentioned in a footnote on p.105 of Hadamard (1923).

7.3.1 Hyperbolic PDE

Recall the canonical form given in (7.59):

$$\frac{\partial^2 u}{\partial s^2} - \frac{\partial^2 u}{\partial t^2} + d_1 \frac{\partial u}{\partial s} + e_1 \frac{\partial u}{\partial t} + f_1 u = g_1 \quad (7.129)$$

Now introduce a change of variables:

$$u(s,t) = e^{\alpha s + \beta t} \psi(s,t) \quad (7.130)$$

Differentiation of (7.130) gives

$$u_s = e^{\alpha s + \beta t}(\alpha\psi + \psi_s) \quad (7.131)$$

$$u_t = e^{\alpha s + \beta t}(\beta\psi + \psi_t) \quad (7.132)$$

$$u_{ss} = e^{\alpha s + \beta t}(\psi_{ss} + 2\alpha\psi_s + \alpha^2\psi) \quad (7.133)$$

$$u_{tt} = e^{\alpha s + \beta t}(\psi_{tt} + 2\beta\psi_t + \beta^2\psi) \quad (7.134)$$

Substitution of (7.131) to (7.134) into (7.129) leads to

$$\psi_{ss} - \psi_{tt} + (2\alpha + d_1)\psi_s + (-2\beta + e_1)\psi_t + (\alpha^2 - \beta^2 + d_1\alpha + e_1\beta + f_1)\psi = g_1 e^{-(\alpha s + \beta t)} \quad (7.135)$$

Since α and β are arbitrary constants at our deposal, we can remove the two first order derivative terms by setting

$$\alpha = -d_1 / 2, \quad \beta = e_1 / 2 \tag{7.136}$$

Thus, the governing equation for ψ becomes

$$\psi_{ss} - \psi_{tt} + [\frac{(e_1^2 - d_1^2)}{4} + f_1]\psi = g_1 e^{(d_1 s - e_1 t)/2} \tag{7.137}$$

The idea of assuming the mathematical form of (7.130) to remove the first order derivative terms has been mentioned on page 105 of Hadamard (1923). Therefore, all hyperbolic second order PDEs with constant coefficients can be transformed to a nonhomogeneous Klein-Gordon equation, which is first derived by Klein in 1927 and Gordon in 1926 for the relativistic motion of charged particles in the electromagnetic field. This major and important conclusion has not been mentioned explicitly in any textbook on differential equations.

7.3.2 Parabolic PDE

Let us start with (7.87) that

$$\frac{\partial^2 u}{\partial \eta^2} + d_1 \frac{\partial u}{\partial \xi} + e_1 \frac{\partial u}{\partial \eta} + f_1 u = g_1 \tag{7.138}$$

Consider a change of variables as

$$u(\eta, \xi) = e^{-\alpha\eta + \beta\xi} \psi(\eta, \xi) \tag{7.139}$$

Differentiation of (7.139) gives

$$u_\eta = e^{-\alpha\eta + \beta\xi} \psi_\eta - \alpha \psi e^{-\alpha\eta + \beta\xi} \tag{7.140}$$

$$u_\xi = e^{-\alpha\eta + \beta\xi} \psi_\xi + \beta \psi e^{-\alpha\eta + \beta\xi} \tag{7.141}$$

$$u_{\eta\eta} = e^{-\alpha\eta + \beta\xi} \psi_{\eta\eta} - 2\alpha \psi e^{-\alpha\eta + \beta\xi} + \alpha^2 \psi e^{-\alpha\eta + \beta\xi} \tag{7.142}$$

Substitution of (7.140) to (7.142) into (7.138) gives

$$\psi_{\eta\eta} + (-2\alpha + e_1)\psi_\eta + d_1 \psi_\xi + (\alpha^2 + \beta d_1 - \alpha e_1 + f_1)\psi = g_1 e^{\alpha\eta - \beta\xi} \tag{7.143}$$

Since α and β are arbitrary constants at our deposal, we can remove the second and fourth terms on the left of (7.143):

$$\alpha = \frac{1}{2} e_1, \quad \beta = \frac{e_1^2 - 4 f_1}{4 d_1} \tag{7.144}$$

Thus, we can use the following transformation

$$u(\eta, \xi) = e^{-\frac{e_1}{2}\eta + \frac{(e_1^2 - 4f_1)}{4d_1}\xi} \psi(\eta, \xi) \tag{7.145}$$

such that the we have converted the mathematical problem into the solution of the following differential equation

$$\frac{\partial^2 \psi}{\partial \eta^2} + d_1 \frac{\partial \psi}{\partial \xi} = g_1 e^{\alpha\eta - \beta\xi} \tag{7.146}$$

This is a nonhomogeneous diffusion equation. Therefore, we can see that all parabolic type PDEs with constant coefficients can be converted to a diffusion equation with a nonhomogeneous term.

7.3.3 Elliptic PDE

In Section 7.2.5, the canonical form for the elliptic type PDE is obtained as:

$$\frac{\partial^2 u}{\partial \xi^2}+\frac{\partial^2 u}{\partial \eta^2}+d\frac{\partial u}{\partial \xi}+\overline{e}\frac{\partial u}{\partial \eta}+fu+g=0 \tag{7.147}$$

Adopting the same transformation used in the last two sections, we have

$$u(\eta,\xi)=e^{-(\alpha\eta+\beta\xi)/2}\psi(\eta,\xi) \tag{7.148}$$

Differentiation of (7.148) gives

$$u_\xi=e^{-(\alpha\eta+\beta\xi)/2}(\psi_\xi-\frac{1}{2}\beta\psi) \tag{7.149}$$

$$u_\eta=e^{-(\alpha\eta+\beta\xi)/2}(\psi_\eta-\frac{1}{2}\alpha\psi) \tag{7.150}$$

$$u_{\eta\eta}=e^{-(\alpha\eta+\beta\xi)/2}(\psi_{\eta\eta}-\alpha\psi_\eta+\frac{1}{4}\alpha^2\psi) \tag{7.151}$$

$$u_{\xi\xi}=e^{-(\alpha\eta+\beta\xi)/2}(\psi_{\xi\xi}-\beta\psi_\xi+\frac{1}{4}\beta^2\psi) \tag{7.152}$$

Substitution of (7.148) to (7.152) into (7.147) gives

$$\begin{aligned}&\psi_{\xi\xi}+\psi_{\eta\eta}+\psi_\xi(d-\beta)+\psi_\eta(\overline{e}-\alpha)\\&+[\frac{1}{4}(\alpha^2+\beta^2)-\frac{1}{2}(\beta d+\alpha\overline{e})+f]\psi=ge^{(\alpha\eta+\beta\xi)/2}\end{aligned} \tag{7.153}$$

We can remove the first derivative terms by setting

$$\alpha=\overline{e},\quad \beta=d \tag{7.154}$$

Consequently, we obtain

$$\psi_{\xi\xi}+\psi_{\eta\eta}+\left[f-\frac{(d^2+\overline{e}^2)}{4}\right]\psi=ge^{(\alpha\eta+\beta\xi)/2} \tag{7.155}$$

This is a nonhomogeneous Helmholtz equation.

We conclude here that, for linear second order PDEs of two independent variables with constant coefficients, hyperbolic, parabolic and elliptic type PDEs can be converted to solving the canonical forms of the nonhomogeneous Klein-Gordon equation, nonhomogeneous diffusion equation, and nonhomogeneous Helmholtz equation, respectively. To our best knowledge, this observation for canonical forms of hyperbolic, parabolic, and elliptic type PDEs has not been reported in any book on differential equations or mathematical physics.

7.4 SOLUTIONS OF CANONICAL FORMS OF SECOND ORDER PDE

As we have seen, there are three types of second order PDE. They are hyperbolic, parabolic, and elliptic. In the last section, we have demonstrated that all second order PDEs with constant coefficients eventually can be converted into three types of differential equations, namely the Klein-Gordon equation, diffusion equation and Helmholtz equation, all with nonhomogeneous terms. Clearly, the importance of these PDEs has not been noticed previously. We will briefly show here that all of these PDEs can be solved by using a technique called separation of variables. More discussion on the separation of variables will be given in Chapter 9.

7.4.1 Nonhomogeneous Klein-Gordon Equation

The Klein-Gordon equation is a dispersive wave equation, which was proposed independently by Oskar Klein in 1927 and Walter Gordon 1926 for modelling motion of a spinless charged particle in the electromagnetic field. In mechanics, the Klein-Gordon equation appears naturally in modelling the wave motion of a vibrating rope resting on a Winkler type foundation. Traditionally, the Klein-Gordon equation without a nonhomogeneous term can be expressed as:

$$\psi_{xx} = \frac{1}{c^2}\psi_{tt} + \alpha^2\psi \tag{7.156}$$

Let us assume the following separation of variables

$$\psi = X(x)T(t) \tag{7.157}$$

Substitution of (7.157) into (7.156) leads to

$$X''T = \frac{1}{c^2}X\ddot{T} + \alpha^2 XT \tag{7.158}$$

By dividing through by XT, this equation can be simplified to

$$\frac{X''}{X} = \frac{1}{c^2}\frac{\ddot{T}}{T} + \alpha^2 = -\lambda^2 \tag{7.159}$$

Since X is only a function of x whereas T is only a function of t, the only possibility is that the left hand side and the right hand side are both constant. Note that we have assumed a negative value for this constant in the last term of (7.159).

The two ODEs resulting from (7.159) are

$$X'' + \lambda^2 X = 0 \tag{7.160}$$

$$\ddot{T} + c^2(\alpha^2 + \lambda^2)T = 0 \tag{7.161}$$

The solutions of these equations are

$$X = A\sin\lambda x + B\cos\lambda x \tag{7.162}$$

$$T = C\sin\omega t + D\cos\omega t \tag{7.163}$$

where

$$\omega = c\sqrt{\alpha^2 + \lambda^2} \tag{7.164}$$

Thus, the general solution becomes

$$\psi = (A\sin\lambda x + B\cos\lambda x)(C\sin\omega t + D\cos\omega t) \tag{7.165}$$

The value of λ must be determined by the boundary condition. The following example illustrates the procedure in obtaining the eigenvalue λ.

Example 7.2 Solve the following Klein-Gordon equation with appropriate boundary conditions:

$$\psi_{xx} = \frac{1}{c^2}\psi_{tt} + \alpha^2\psi, \quad 0 < x < L, \quad t > 0 \tag{7.166}$$

$$\psi(0,t) = \psi(L,t) = 0, \quad t > 0 \tag{7.167}$$

$$\psi(x,0) = 0, \quad \psi_t(x,0) = v_0, \quad 0 < x < L \tag{7.168}$$

Solution: The boundary condition can be expressed as

$$\psi(0,t) = X(0)T(t) = 0, \quad t > 0 \tag{7.169}$$

$$\psi(L,t) = X(L)T(t) = 0, \quad t > 0 \tag{7.170}$$

whereas the first initial condition is

$$\psi(x,0) = X(x)T(0) = 0, \quad 0 < x < L \tag{7.171}$$

Thus, we must have

$$X(0) = X(L) = 0, \quad t > 0 \tag{7.172}$$

$$T(0) = 0, \quad 0 < x < L \tag{7.173}$$

Note that the boundary conditions are homogeneous and thus correspond to an eigenvalue problem of differential equations for X. Substitution of (7.162) into the first part of (7.172) leads to

$$X(0) = B = 0, \quad t > 0 \tag{7.174}$$

Substitution of (7.162) into the second part of (7.172) leads to

$$X(L) = A\sin\lambda L = 0, \quad t > 0 \tag{7.175}$$

Since A cannot be zero (otherwise the solution is identically zero), we must have the eigenvalue equation as

$$\lambda L = n\pi, \quad n = 1,2,3,... \tag{7.176}$$

Although there are infinite eigenvalues, the boundary condition cannot be satisfied by arbitrary values of the separation constant. This is precisely the property of the eigenvalue problem. Thus, the eigenvalues and their eigenfunctions are

$$\lambda_n = \frac{n\pi}{L}, \quad n = 1,2,3,... \tag{7.177}$$

$$X_n(x) = A\sin\lambda_n x = A\sin\frac{n\pi x}{L} \tag{7.178}$$

Note that if we choose a positive separation constant in (7.159), the fundamental solution for X is sinh and cosh. Consequently, the second part of (1.172) cannot be satisfied. Thus, hyperbolic sine and cosine cannot be used as the basis of eigenfunction expansion. This is the reason why we cannot take positive value in (7.159) for the separation constant λ^2.

For the initial condition, substitution of (7.163) into the first of (7.173) gives

$$T(0) = D = 0 \tag{7.179}$$

The fundamental solution becomes

$$\psi(x,t)=\sum_{n=1}^{\infty}c_n \sin\lambda_n x \sin\omega_n t \tag{7.180}$$

where

$$\omega_n = c\sqrt{\alpha^2+\lambda_n^2} \tag{7.181}$$

Differentiation of (7.180) gives

$$\psi_t(x,t)=\sum_{n=1}^{\infty}c_n\omega_n \sin\lambda_n x \cos\omega_n t \tag{7.182}$$

Substitution of (7.182) into (7.168) leads to

$$\psi_t(x,0)=v_0=\sum_{n=1}^{\infty}c_n\omega_n \sin\lambda_n x \tag{7.183}$$

Multiplying both sides by a sine function and integrating from 0 to L, we have

$$v_0\int_0^L \sin\frac{m\pi x}{L}dx=\sum_{n=1}^{\infty}c_n\omega_n\int_0^L \sin\frac{m\pi x}{L}\sin\frac{n\pi x}{L}dx \tag{7.184}$$

Note the following results, we have

$$\int_0^L \sin\frac{m\pi x}{L}dx=\frac{1}{m\pi}[(-1)^{m+1}+1] \tag{7.185}$$

$$\int_0^L \sin\frac{m\pi x}{L}\sin\frac{n\pi x}{L}dx=\begin{cases}=0, & m\neq n\\ =L/2, & m=n\end{cases} \tag{7.186}$$

Thus, substitution of (7.185) and (7.186) into (7.184) gives

$$c_n=\frac{2v_0}{\omega_n n\pi L}[(-1)^{n+1}+1] \tag{7.187}$$

Finally, we get the solution as

$$\psi(x,t)=\sum_{n=1}^{\infty}\frac{2v_0}{\omega_n n\pi L}[(-1)^{n+1}+1]\sin\lambda_n x\sin\omega_n t \tag{7.188}$$

where

$$\omega_n=c\sqrt{\alpha^2+\lambda_n^2}, \quad \lambda_n=\frac{n\pi}{L}, \quad n=1,2,3,... \tag{7.189}$$

Once the solution for the homogeneous Klein-Gordon equation is obtained, the nonhomogeneous case can be solved either by the method of undetermined coefficients or by the method of variation of parameters discussed in earlier chapters. Alternatively, for the nonhomogeneous Klein-Gordon equation, we can also solve the problem by Green's function method, which will be discussed in Chapter 8. Let us consider the following nonhomogeneous Klein-Gordon problem:

$$\psi_{xx}=\frac{1}{c^2}\psi_{tt}+\alpha^2\psi+f(x,t), \quad 0<x<L, \quad t>0 \tag{7.190}$$

$$\psi(0,t)=g_1(t), \quad \psi(L,t)=g_2(t), \quad t>0 \tag{7.191}$$

$$\psi(x,0)=f_0(x),\quad \psi_t(x,0)=f_1(x),\quad 0<x<L \tag{7.192}$$

We will not go through the details of analysis, but the solution can be found using Green's function method as (Section 4.1.3-4 of Polyanin, 2002)

$$\begin{aligned}\psi(x,t)=&\frac{\partial}{\partial t}\int_0^L f_0(\xi)G(x,\xi,t)d\xi+\int_0^L f_1(\xi)G(x,\xi,t)d\xi\\ &-\int_0^t\int_0^L f(\xi,\tau)G(x,\xi,t-\tau)d\xi d\tau\\ &+c^2\int_0^t g_1(\tau)[\frac{\partial G}{\partial \xi}(x,\xi,t-\tau)]_{\xi=0}d\tau-c^2\int_0^t g_2(\tau)[\frac{\partial G}{\partial \xi}(x,\xi,t-\tau)]_{\xi=L}d\tau\end{aligned} \tag{7.193}$$

where

$$G(x,\xi,t)=\frac{2}{L}\sum_{n=1}^{\infty}\sin(\lambda_n x)\sin(\lambda_n\xi)\frac{\sin[ct\sqrt{\lambda_n^2+\alpha^2}]}{c\sqrt{\lambda_n^2+\alpha^2}} \tag{7.194}$$

$$\lambda_n=\frac{n\pi}{L} \tag{7.195}$$

Green's function is denoted as $G(x,\xi,t)$ for the Helmholtz equation in a finite domain. It was derived by Sommerfeld in 1912. For another domain and another boundary or initial conditions, the reader can refer to the handbook by Polyanin (2001).

7.4.2 Nonhomogeneous Diffusion Equation

Let consider the following nonhomogeneous diffusion equation

$$u_t=\alpha^2 u_{xx}+f(x,t),\quad 0\le x\le L,\quad t>0 \tag{7.196}$$

subject to the following initial and boundary conditions:

$$u(0,t)=0,\quad u(L,t)=0,\quad t>0 \tag{7.197}$$

$$u(x,0)=\phi(x),\quad 0\le x\le L \tag{7.198}$$

First, let us consider the eigenvalue problem of the homogeneous form of (7.196) (i.e., $f=0$):

$$\alpha^2 u_{xx}=u_t,\quad 0\le x\le L,\quad t>0 \tag{7.199}$$

$$u(0,t)=0,\quad u(L,t)=0,\quad t>0 \tag{7.200}$$

$$u(x,0)=\phi(x),\quad 0\le x\le L \tag{7.201}$$

Assuming a separation of variables, we have:

$$u=X(x)T(t) \tag{7.202}$$

Substitution of (7.202) into (7.199) leads to

$$\alpha^2 X''T=XT' \tag{7.203}$$

By dividing through by XT, this equation can be simplified as

$$\frac{X''}{X}=\frac{1}{\alpha^2}\frac{T'}{T}=-\lambda^2 \tag{7.204}$$

Since X is only a function of x, whereas T is only a function of t, the only possibility is that the left hand side and the right hand side are both constant. Note again that we have assumed a negative value of this constant in the last term of (7.159).

In the following discussion, we will only focus on the discussion of X. In particular, we have from (7.204)

$$X'' + \lambda^2 X = 0 \tag{7.205}$$

The solutions of these equations are

$$X = A\sin\lambda x + B\cos\lambda x \tag{7.206}$$

The value of λ must be determined by the boundary condition. Substitution of (7.206) into (7.200) leads to

$$X(0) = X(L) = 0 \tag{7.207}$$

Substitution of (7.206) into the first part of (7.207) gives

$$B = 0 \tag{7.208}$$

The second part of (7.207) leads to

$$\sin\lambda L = 0 \tag{7.209}$$

That is, we require

$$\lambda L = n\pi \tag{7.210}$$

Thus, there are infinite discrete eigenvalues given by

$$\lambda_n = n\pi / L, \quad n = 1,2,3,\ldots \tag{7.211}$$

The eigenfunction that corresponds to this eigenvalue is

$$X_n(x) = \sin(\lambda_n x), \quad n = 1,2,3,\ldots \tag{7.212}$$

Thus, we have found eigenvalues and its eigenfunction expansion.

We now return to the nonhomogeneous problem given in (7.190) to (7.192). In particular, we assume the following eigenfunction expansions:

$$u(x,t) = \sum_{n=1}^{\infty} T_n(t)\sin(\lambda_n x) \tag{7.213}$$

$$f(x,t) = \sum_{n=1}^{\infty} f_n(t)\sin(\lambda_n x) \tag{7.214}$$

$$\phi(x) = \sum_{n=1}^{\infty} c_n \sin(\lambda_n x) \tag{7.215}$$

where T_n is to be determined and

$$f_n(t) = \frac{2}{L}\int_0^L f(x,t)\sin(n\pi x / L)\,dx \tag{7.216}$$

$$c_n = \frac{2}{L}\int_0^L \phi(x)\sin(n\pi x / L)\,dx \tag{7.217}$$

As we will discuss in Chapter 10, all functions can be expanded in an infinite series of the eigenfunctions (in this case sine function). More detailed discussions on eigenfunction expansion are given in Chapter 10.

Substitution of (7.213) and (7.14) into (7.196) gives

$$\sum_{n=1}^{\infty}[\frac{dT_n}{dt} + \alpha^2\lambda_n^2 T_n - f_n]\sin(n\pi x / L) = 0 \tag{7.218}$$

In view of the governing equation for X given in (7.205), we can rewrite (7.218) as

$$\frac{dT_n}{dt} + \alpha^2 k \lambda_n^2 T_n = f_n \tag{7.219}$$

Substitution of (7.213) and (7.215) into (7.198) gives

$$T_n(0) = c_n \tag{7.220}$$

Multiplying (7.219) by the integrating factor, we have

$$\frac{d}{dt}\{T_n e^{\alpha^2 k \lambda_n^2 t}\} = f_n e^{\alpha^2 k \lambda_n^2 t} \tag{7.221}$$

Integration of (7.221) gives

$$T_n e^{\alpha^2 k \lambda_n^2 t} = T_n(0) + \int_0^t f_n e^{\alpha^2 k \lambda_n^2 s} ds \tag{7.222}$$

Using the initial condition (7.220), we obtain

$$T_n = c_n e^{-\alpha^2 k \lambda_n^2 t} + \int_0^t f_n e^{\alpha^2 k \lambda_n^2 (s-t)} ds \tag{7.223}$$

Substitution of (7.223) into (7.213) gives the final solution for the nonhomogeneous diffusion equation given in (7.196) with boundary conditions (7.197) and (7.198) as

$$u(x,t) = \sum_{n=1}^{\infty} c_n e^{-\alpha^2 k \lambda_n^2 t} \sin(\frac{n\pi x}{L}) + \sum_{n=1}^{\infty}\left(\int_0^t f_n e^{\alpha^2 k \lambda_n^2 (s-t)} ds\right)\sin(\frac{n\pi x}{L}) \tag{7.224}$$

where λ_n is given in (7.210).

7.4.3 Nonhomogeneous Helmholtz Equation

The Helmholtz equation is also known as the reduced wave equation. To see why this is so, let us consider a two-dimensional wave equation as

$$\Psi_{xx} + \Psi_{yy} = \frac{1}{c^2}\Psi_{tt} \tag{7.225}$$

Physically, this PDE models the vibrations of a rectangular membrane with no bending. Let us now consider a harmonic wave type of solution in the form:

$$\Psi = \psi(x,y)e^{i\omega t} \tag{7.226}$$

where ω is the circular frequency of the wave. Substitution of (7.226) into (7.225) gives

$$\psi_{xx} + \psi_{yy} + \frac{\omega^2}{c^2}\psi = 0 \tag{7.227}$$

This can be rewritten as the standard form of the Helmholtz equation:

$$\psi_{xx} + \psi_{yy} + k^2\psi = 0 \tag{7.228}$$

This is the reason why it is referred to as the reduced wave equation. To solve (7.228), we can introduce the following separation of variables

$$\psi = X(x)Y(y) \tag{7.229}$$

Substitution of (7.229) into (7.228) yields

$$\frac{X''}{X}+\frac{Y''}{Y}+k^2=0 \tag{7.230}$$

Similar to the argument used in the previous sections, we can rewrite (7.230) as

$$\frac{X''}{X}=-\lambda_1^2,\quad \frac{Y''}{Y}=-\lambda_2^2,\quad \lambda_1^2+\lambda_2^2=k^2 \tag{7.231}$$

Thus, we have two ODEs as:

$$X''+\lambda_1^2 X=0 \tag{7.232}$$

$$Y''+\lambda_2^2 Y=0 \tag{7.233}$$

The general solutions of these equations are:

$$X=A\sin\lambda_1 x+B\cos\lambda_1 x \tag{7.234}$$

$$Y=A\sin\lambda_2 y+B\cos\lambda_2 y \tag{7.235}$$

Clearly, this set of solutions resembles vibrations of a membrane with a fixed boundary on the circumference of a rectangular membrane. However, if there is no physical meaning attached to a PDE, we may set the related differential equations as

$$\frac{X''}{X}=-\lambda_1^2,\quad \frac{Y''}{Y}=\lambda_2^2,\quad \lambda_1^2-\lambda_2^2=k^2 \tag{7.236}$$

In this case, we may have the solutions as

$$X=A\sin\lambda_1 x+B\cos\lambda_1 x \tag{7.237}$$

$$Y=A\sinh\lambda_2 y+B\cosh\lambda_2 y \tag{7.238}$$

This solution is valid provided that the third equation of (7.236) is satisfied. However, (7.238) cannot model vibration-type solutions for a membrane. Whether the solution set given in (7.235) or (7.238) is valid, it depends on the condition to be imposed on the boundary.

Example 7.3 Consider the waveguide problem of a rectangular hollow tube of a conductor. The problem can be mathematically prescribed as a three-dimensional Helmholtz equation with appropriate boundary conditions:

$$\psi_{xx}+\psi_{yy}+\psi_{zz}+k^2\psi=0,\quad 0<x<a,\quad 0<y<b \tag{7.239}$$

$$\psi(0,y)=0,\quad \psi(a,y)=0 \tag{7.240}$$

$$\psi(x,0)=0,\quad \psi(x,b)=0 \tag{7.241}$$

with k being positive. The boundary value problem is illustrated in Figure 7.4. The wave modes are normally referred as TM (transverse magnetic) or TE (transverse electric).

Solution: Using separation of variables, we can assume

$$\psi(x,y,z)=X(x)Y(y)Z(z) \tag{7.242}$$

Substitution of (7.242) into (7.239) leads to

$$\frac{X''}{X}=-\lambda_1^2,\quad \frac{Y''}{Y}=-\lambda_2^2,\quad \frac{Z''}{Z}=-\lambda_3^2 \tag{7.243}$$

Thus, the general solution for X is

$$X = A\sin\lambda_1 x + B\cos\lambda_1 x \tag{7.244}$$

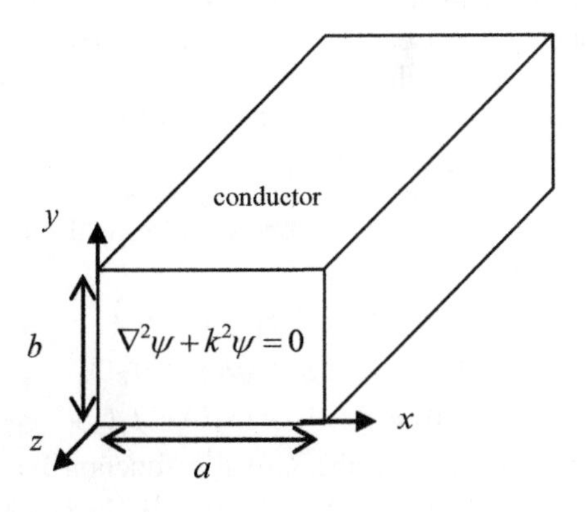

Figure 7.4 Problem of rectangular waveguide

The first condition of (7.240) leads to

$$B = 0, \quad \lambda_1 a = n\pi \tag{7.245}$$

Thus, the eigenvalue becomes

$$\lambda_1 = \frac{n\pi}{a} \tag{7.246}$$

Similarly, we can find the eigenvalue for λ_2 as:

$$\lambda_2 = \frac{m\pi}{b} \tag{7.247}$$

The solution for ψ becomes

$$\psi(x, y, z) = \sin(\frac{n\pi x}{a})\sin(\frac{m\pi y}{b})e^{i\lambda_3 z} \tag{7.248}$$

and

$$(\frac{n\pi}{a})^2 + (\frac{m\pi}{b})^2 + \lambda_3^2 = k^2 \tag{7.249}$$

We can rewrite (7.249) as

$$\lambda_3^2 = \beta_{mn} = \sqrt{k^2 - k_{cmn}^2} \tag{7.250}$$

where

$$k_{cmn} = \sqrt{(\frac{n\pi}{a})^2 + (\frac{m\pi}{b})^2} \tag{7.251}$$

This is called the cut-off frequency of the waveguide and it is clearly discrete. Physically, for the case of electromagnetic waves, the value of k is given by:

$$k^2 = \omega^2 \mu\varepsilon \tag{7.252}$$

where μ and ε are called the permeability and permittivity of the free space, and ω is the circular frequency of the electromagnetic waves. For small values of wave frequency, it clear from (7.250) that β_{mn} is purely imaginary. For this small frequency, (7.248) shows that the solution decays exponentially with z. That is, waves with low frequency attenuate quickly along the waveguides and therefore are cutoff. The phenomenon is caused by the fact that the rectangular tube is of finite size and only waves of certain discrete frequencies can propagate along it. More discussion of electromagnetic waves is presented in Volume 2 of this book series.

Let us now consider the following nonhomogeneous Helmholtz equation with prescribed boundary condition:

$$\psi_{xx} + \psi_{yy} + \lambda\psi = -f(x,y) \tag{7.253}$$

$$\psi(0,y) = f_1(y), \quad \psi(a,y) = f_2(y) \tag{7.254}$$

$$\psi(x,0) = f_3(x), \quad \psi(x,b) = f_4(x) \tag{7.255}$$

This problem can be solved by using Green's function method, which will be considered in Chapter 8. However, we will simply quote the following solution from Section 7.3.2-15 of Polyanin and Zaitsev (2002):

$$\begin{aligned}\psi(x,y) &= \int_0^a\int_0^b f(\xi,\eta)G(x,y,\xi,\eta)d\eta d\xi \\ &+\int_0^b f_1(\eta)[\frac{\partial G}{\partial \xi}(x,y,\xi,\eta)]_{\xi=0}d\eta - \int_0^b f_2(\eta)[\frac{\partial G}{\partial \xi}(x,y,\xi,\eta)]_{\xi=a}d\eta \\ &+\int_0^a f_3(\eta)[\frac{\partial G}{\partial \eta}(x,y,\xi,\eta)]_{\eta=0}d\xi - \int_0^a f_4(\eta)[\frac{\partial G}{\partial \eta}(x,y,\xi,\eta)]_{\eta=b}d\xi\end{aligned} \tag{7.256}$$

where

$$G(x,y,\xi,\eta) = \frac{4}{ab}\sum_{n=1}^{\infty}\sum_{m=1}^{\infty}\frac{\sin(p_n x)\sin(q_m y)\sin(p_n\xi)\sin(q_m\eta)}{p_n^2 + q_m^2 - \lambda} \tag{7.257}$$

$$p_n^2 = \frac{n\pi}{b}, \quad q_m^2 = \frac{m\pi}{b} \tag{7.258}$$

For other domain or boundary conditions, we refer to Polyanin and Zaitsev (2002).

We now consider the Helmholtz equation in polar form. In cylindrical coordinates, the Helmholtz equation can be expressed as

$$\frac{\partial^2 u}{\partial r^2} + \frac{1}{r}\frac{\partial u}{\partial r} + \frac{1}{r^2}\frac{\partial^2 u}{\partial \theta^2} + \frac{\partial^2 u}{\partial z^2} + k^2 u = 0 \tag{7.259}$$

Using the separation of variables, we have

$$u = R(r)\Theta(\theta)Z(z) \tag{7.260}$$

Substitution of (7.260) into (7.259) gives

$$\frac{1}{R}[\frac{d^2R}{dr^2} + \frac{1}{r}\frac{dR}{dr}] + \frac{1}{r^2}\frac{1}{\Theta}\frac{d^2\Theta}{d\theta^2} + \frac{1}{Z}\frac{d^2Z}{dz^2} + k^2 = 0 \tag{7.261}$$

This leads to the following ODEs as:

$$\frac{1}{\Theta}\frac{d^2\Theta}{d\theta^2} = -m^2 \tag{7.262}$$

$$\frac{1}{Z}\frac{d^2Z}{dz^2}=\alpha^2-k^2 \tag{7.263}$$

$$\frac{1}{R}[\frac{d^2R}{dr^2}+\frac{1}{r}\frac{dR}{dr}]-\frac{m^2}{r^2}+\alpha^2=0 \tag{7.264}$$

Note that (7.264) is a Bessel equation. The solution of Θ, Z, and R are

$$\Theta=A\cos m\theta+B\sin m\theta \tag{7.265}$$

$$Z=C\cosh[\sqrt{\alpha^2-k^2}\,z]+D\sinh[\sqrt{\alpha^2-k^2}\,z] \tag{7.266}$$

$$R=EJ_m(\alpha r)+FY_m(\alpha r) \tag{7.267}$$

In view of the physical requirement of θ, we must have the periodicity of Θ

$$\Theta(\theta+2\pi)=\Theta(\theta) \tag{7.268}$$

Consequently, m must be an integer.

For the special case that $\alpha=0$, we have u being independent of z, and (7.259) becomes

$$\frac{\partial^2u}{\partial r^2}+\frac{1}{r}\frac{\partial u}{\partial r}+\frac{1}{r^2}\frac{\partial^2u}{\partial\theta^2}+k^2u=0 \tag{7.269}$$

Introduction of a separation of variables gives

$$u=H(r)T(\theta) \tag{7.270}$$

Substitution of (7.270) into (7.269) gives

$$\frac{1}{H}[\frac{d^2H}{dr^2}+\frac{1}{r}\frac{dH}{dr}]+\frac{1}{T}\frac{1}{r^2}\frac{d^2T}{d\theta^2}+k^2=0 \tag{7.271}$$

This leads to

$$\frac{1}{T}\frac{d^2T}{d\theta^2}+m^2=0 \tag{7.272}$$

$$\frac{d^2H}{dr^2}+\frac{1}{r}\frac{dH}{dr}+(k^2-\frac{m^2}{r^2})H=0 \tag{7.273}$$

Again, we have the solution of T being

$$T=A\cos m\theta+B\sin m\theta \tag{7.274}$$

where m is an integer. Equation (7.273) is of Euler type and using the standard change of variable of Euler type yields

$$H=CJ_m(kr)+DY_m(kr) \tag{7.275}$$

In spherical coordinates, the Helmholtz equation can be expressed as

$$\frac{1}{r}\frac{\partial^2}{\partial r^2}(ru)+\frac{1}{r^2\sin\theta}[\frac{\partial}{\partial\theta}(\sin\theta\frac{\partial u}{\partial\theta})+\frac{1}{\sin\theta}\frac{\partial^2u}{\partial\varphi^2}]+k^2u=0 \tag{7.276}$$

Again using the separation of variables, we have

$$u=R(r)P(\theta)\Phi(\varphi) \tag{7.277}$$

Substitution of (7.277) into (7.276) gives

$$\frac{1}{R}\frac{1}{r}\frac{d^2}{dr^2}(rR)+\frac{1}{r^2}[\frac{1}{P}\frac{1}{\sin\theta}\frac{d}{d\theta}(\sin\theta\frac{dP}{d\theta})+\frac{1}{\sin^2\theta}\frac{1}{\Phi}\frac{d^2\Phi}{d\varphi^2}]+k^2=0 \tag{7.278}$$

The governing equations become

$$\frac{1}{\Phi}\frac{d^2\Phi}{d\varphi^2} = -m^2 \tag{7.279}$$

$$\frac{1}{\sin\theta}\frac{d}{d\theta}(\sin\theta\frac{dP}{d\theta}) + (\lambda - \frac{m^2}{\sin^2\theta})P = 0 \tag{7.280}$$

$$\frac{1}{R}[\frac{1}{r}\frac{d^2}{dr^2}(rR)] + k^2 - \frac{\lambda}{r^2} = 0 \tag{7.281}$$

where λ is a separation constant. The solution for Φ is

$$\Phi = A\cos m\varphi + B\sin m\varphi \tag{7.282}$$

The solution of P can be recognized by using the following change of variables:

$$x = \cos\theta \tag{7.283}$$

Using this change of variables, (7.280) can be transformed to

$$\frac{d}{dx}[(1-x^2)\frac{dP}{dx}] + (\lambda - \frac{m^2}{1-x^2})P = 0 \tag{7.284}$$

This can be shown as equivalent to

$$(1-x^2)\frac{d^2P}{dx^2} - 2x\frac{dP}{dx} + [l(l+1) - \frac{m^2}{1-x^2}]P = 0 \tag{7.285}$$

where we have set

$$\lambda = l(l+1) \tag{7.286}$$

Equation (7.285) is the associated Legendre equation and the solution is

$$P(\theta) = CP_l^m(\cos\theta) + DQ_l^m(\cos\theta) \tag{7.287}$$

Finally, to find the solution of R in (7.281) we first rewrite it as

$$\frac{d^2R}{d\rho^2} + \frac{2}{\rho}\frac{dR}{d\rho} + (1 - \frac{\lambda}{\rho^2})R = 0 \tag{7.288}$$

where

$$\rho = kr \tag{7.289}$$

We can introduce another change of variables as

$$R = \frac{1}{\sqrt{\rho}}S \tag{7.290}$$

Differentiation of (7.290) gives

$$\frac{dR}{d\rho} = -\frac{1}{2\sqrt{\rho}}\frac{S}{\rho} + \frac{1}{\sqrt{\rho}}\frac{dS}{d\rho} \tag{7.291}$$

$$\frac{d^2R}{d\rho^2} = \frac{3}{4}\frac{S}{\rho^{5/2}} - \frac{1}{\rho^{3/2}}\frac{dS}{d\rho} + \frac{1}{\sqrt{\rho}}\frac{d^2S}{d\rho^2} \tag{7.292}$$

Substitution of these results into (7.288) gives

$$\frac{d^2S}{d\rho^2} + \frac{1}{\rho}\frac{dS}{d\rho} + (1 - \frac{\lambda + 1/4}{\rho^2})S = 0 \tag{7.293}$$

This is precisely the Bessel equation and consequently the solution for R is

$$R(r)=\frac{1}{\sqrt{rk}}[EJ_{\sqrt{\lambda+1/4}}(rk)+FY_{\sqrt{\lambda+1/4}}(rk)] \tag{7.294}$$

Substitution of (7.282), (7.287) and (7.294) into (7.277) gives the final solution in spherical polar coordinates. More discussion of this spherical Helmholtz equation will be given in Chapter 9 (Section 9.4). In particular, a new function called the spherical Bessel function can be defined to simplify the solution form given in (7.294).

7.5 ADJOINT OF SECOND ORDER PDE

In an earlier chapter, we have discussed the adjoint differential equation of an ODE, and demonstrated that the adjoint equation is actually the governing equation for the integrating factor of the original ODE. In this section, we will extend this idea of the adjoint equation to second order PDEs. The result discussed in this section was first derived by De Bois-Reymond in 1889 and was also derived by Darboux in 1915. Recall from our earlier discussion that solving the adjoint problem of an ODE may be as difficult as solving the original ODE. Nevertheless, it is an important concept in solving higher order partial differential equations, but this is normally not covered in most textbooks on differential equations. We will, however, see the importance of adjoint PDEs in Green's function in Chapter 8. Let us consider a general form of linear second order PDE as

$$L(u)=A\frac{\partial^2 u}{\partial x^2}+B\frac{\partial^2 u}{\partial x\partial y}+C\frac{\partial^2 u}{\partial y^2}+D\frac{\partial u}{\partial x}+E\frac{\partial u}{\partial y}+Fu=0 \tag{7.295}$$

We now consider the product of another function v, which is the solution of the adjoint problem, with function $L(u)$ defined in (7.295) as:

$$vL(u)=Av\frac{\partial^2 u}{\partial x^2}+Bv\frac{\partial^2 u}{\partial x\partial y}+Cv\frac{\partial^2 u}{\partial y^2}+Dv\frac{\partial u}{\partial x}+Ev\frac{\partial u}{\partial y}+Fuv \tag{7.296}$$

We first note the following identities:

$$Av\frac{\partial^2 u}{\partial x^2}-u\frac{\partial^2 (Av)}{\partial x^2}=\frac{\partial}{\partial x}(Av\frac{\partial u}{\partial x}-u\frac{\partial Av}{\partial x}) \tag{7.297}$$

$$Bv\frac{\partial^2 u}{\partial x\partial y}-u\frac{\partial^2 (Bv)}{\partial x\partial y}=\frac{\partial}{\partial x}(Bv\frac{\partial u}{\partial y})-\frac{\partial}{\partial y}(u\frac{\partial Bv}{\partial x})=\frac{\partial}{\partial y}(Bv\frac{\partial u}{\partial x})-\frac{\partial}{\partial x}(u\frac{\partial Bv}{\partial y}) \tag{7.298}$$

$$Cv\frac{\partial^2 u}{\partial y^2}-u\frac{\partial^2 (Cv)}{\partial y^2}=\frac{\partial}{\partial y}(Cv\frac{\partial u}{\partial y}-u\frac{\partial Cv}{\partial y}) \tag{7.299}$$

$$Dv\frac{\partial u}{\partial x}-u\frac{\partial}{\partial x}(-Dv)=\frac{\partial}{\partial x}(Duv) \tag{7.300}$$

$$Ev\frac{\partial u}{\partial y}-u\frac{\partial}{\partial y}(-Ev)=\frac{\partial}{\partial y}(Euv) \tag{7.301}$$

Using the identity in (7.301), we note that (7.298) can be written as:

$$Bv\frac{\partial^2 u}{\partial x\partial y}-u\frac{\partial^2(Bv)}{\partial x\partial y}=\frac{1}{2}\left\{\frac{\partial}{\partial x}(Bv\frac{\partial u}{\partial y}-u\frac{\partial Bv}{\partial y})+\frac{\partial}{\partial y}(Bv\frac{\partial u}{\partial x}-u\frac{\partial Bv}{\partial x})\right\} \tag{7.302}$$

Substitution of (7.297) to (7.302) into (7.296), we get

$$vL(u)=\frac{\partial X}{\partial x}+\frac{\partial Y}{\partial y}+uM(v) \tag{7.303}$$

where

$$M(v)=\frac{\partial^2 Av}{\partial x^2}+\frac{\partial^2 Bv}{\partial x\partial y}+\frac{\partial^2 Cv}{\partial y^2}-\frac{\partial Dv}{\partial x}-\frac{\partial Ev}{\partial y}+Fv \tag{7.304}$$

$$X=Av\frac{\partial u}{\partial x}-u\frac{\partial Av}{\partial x}+\frac{1}{2}(vB\frac{\partial u}{\partial y}-u\frac{\partial vB}{\partial y})+Duv \tag{7.305}$$

$$Y=Cv\frac{\partial u}{\partial y}-u\frac{\partial Cv}{\partial y}+\frac{1}{2}(vB\frac{\partial u}{\partial x}-u\frac{\partial vB}{\partial x})+Euv \tag{7.306}$$

The functions X and Y given in (7.305) and (7.306) can further be rewritten as

$$X=A(v\frac{\partial u}{\partial x}-u\frac{\partial v}{\partial x})+\frac{1}{2}B(v\frac{\partial u}{\partial y}-u\frac{\partial v}{\partial y})+(D-\frac{\partial A}{\partial x}-\frac{1}{2}\frac{\partial B}{\partial y})uv \tag{7.307}$$

$$Y=\frac{1}{2}B(v\frac{\partial u}{\partial x}-u\frac{\partial v}{\partial x})+C(v\frac{\partial u}{\partial y}-u\frac{\partial v}{\partial y})+(E-\frac{\partial C}{\partial y}-\frac{1}{2}\frac{\partial B}{\partial x})uv \tag{7.308}$$

Equation (7.303) can be recast as

$$vL(u)-uM(v)=\frac{\partial X}{\partial x}+\frac{\partial Y}{\partial y} \tag{7.309}$$

This identity is sometimes referred to as the Lagrange identity. The adjoint of $L(u)$ is $M(v)$.

To find the condition of the self-adjoint PDE, we first expand $M(v)$ as:

$$\begin{aligned}M(v)&=A\frac{\partial^2 v}{\partial x^2}++B\frac{\partial^2 v}{\partial x\partial y}+C\frac{\partial^2 v}{\partial y^2}+D\frac{\partial v}{\partial x}+E\frac{\partial v}{\partial y}+Fv\\&+v\frac{\partial}{\partial x}(\frac{\partial A}{\partial x}+\frac{1}{2}\frac{\partial B}{\partial y}-D)+v\frac{\partial}{\partial y}(\frac{\partial C}{\partial y}+\frac{1}{2}\frac{\partial B}{\partial x}-E)\\&+2\frac{\partial v}{\partial x}(\frac{\partial A}{\partial x}+\frac{1}{2}\frac{\partial B}{\partial y}-D)+2\frac{\partial v}{\partial y}(\frac{\partial C}{\partial y}+\frac{1}{2}\frac{\partial B}{\partial x}-E)\end{aligned} \tag{7.310}$$

Therefore, the self-adjoint condition for linear second order PDEs are:

$$\frac{\partial A}{\partial x}+\frac{1}{2}\frac{\partial B}{\partial y}=D \tag{7.311}$$

$$\frac{\partial C}{\partial y}+\frac{1}{2}\frac{\partial B}{\partial x}=E \tag{7.312}$$

Many physical based second order PDEs are self-adjoint. Actually, the self-adjoint system occupies an essential role in the development of physical sciences and mechanics. The identity given in (7.309) can be used to form a generalized form of

Green's theorem of any linear second order PDE. However, such consideration will be deferred to the Chapter 8 on Green's function method.

This formulation can be easily generalized to the case of m variables as (Hadamard, 1923):

$$L(u)=\sum_{i,k=1}^{m}A_{ik}\frac{\partial^2 u}{\partial x_i\partial x_k}+\sum_{i=1}^{m}B_i\frac{\partial u}{\partial x}+Cu=0 \tag{7.313}$$

with

$$A_{ik}=A_{ki} \tag{7.314}$$

The associated adjoint PDE becomes

$$M(v)=\sum_{i,k=1}^{m}\frac{\partial^2}{\partial x_i\partial x_k}(A_{ik}v)-\sum_{i=1}^{m}\frac{\partial}{\partial x}(B_i v)+Cu=0 \tag{7.315}$$

The corresponding Lagrange identity is

$$vL(u)-uM(v)=\frac{\partial X_1}{\partial x_1}+\frac{\partial X_2}{\partial x_2}+\ldots+\frac{\partial X_m}{\partial x_m} \tag{7.316}$$

where

$$X_i(u,v)=\sum_{k=1}^{m}vA_{ik}\frac{\partial u}{\partial x_k}-\sum_{k=1}^{m}uA_{ik}\frac{\partial v}{\partial x_k}-uv(\sum_{k=1}^{m}\frac{\partial A_{ik}}{\partial x_k}-B_i) \tag{7.317}$$

To see the validity of these equations, we can set m = 2 to recover equations (7.303) to (7.308) as a special case. Readers are advised to check this by themselves.

7.6 SELF-ADJOINT CONDITION FOR SYSTEM OF SECOND ORDER PDE

In this section, we are going to extend the idea of finding adjoint PDEs to the case of a system of coupled second order PDEs. To the best of our knowledge, this topic has not been covered in any textbooks on differential equations. Let us consider a system of PDEs as:

$$L_{11}(u)+L_{12}(v)=0 \tag{7.318}$$

$$L_{21}(u)+L_{22}(v)=0 \tag{7.319}$$

where the differential operator L_{ij} can be defined as analogous to the second order PDE considered in the last section as:

$$L_{ij}(u)=A_{ij}\frac{\partial^2 u}{\partial x^2}+2B_{ij}\frac{\partial^2 u}{\partial x\partial y}+C_{ij}\frac{\partial^2 u}{\partial y^2}+D_{ij}\frac{\partial u}{\partial x}+E_{ij}\frac{\partial u}{\partial y}+F_{ij}u \tag{7.320}$$

with i, j = 1,2. From the result of the last section, we have the adjoint of (7.320) as

$$M_{ij}(v)=\frac{\partial^2}{\partial x^2}(A_{ij}v)+2\frac{\partial^2}{\partial x\partial y}(vB_{ij})+\frac{\partial^2}{\partial y^2}(vC_{ij})+\frac{\partial}{\partial x}(vD_{ij})+\frac{\partial}{\partial y}(vE_{ij})+F_{ij}v \tag{7.321}$$

Multiplying (7.318) by u and multiplying (7.319) by v, and adding the results of these, we obtain

$$uL_{11}(u)+uL_{12}(v)+vL_{21}(u)+vL_{22}(v)=0 \tag{7.322}$$

To consider the left hand side of (7.322), we first note the following identities, which can be obtained as special cases of (7.309) in the last section:

$$uL_{11}(u)=uM_{11}(u)+\frac{\partial X_1(u,u)}{\partial x}+\frac{\partial Y_1(u,u)}{\partial y} \tag{7.323}$$

$$uL_{12}(v)=vM_{12}(u)+\frac{\partial X_2(v,u)}{\partial x}+\frac{\partial Y_2(v,u)}{\partial y} \tag{7.324}$$

$$vL_{21}(u)=uM_{21}(v)+\frac{\partial X_3(u,v)}{\partial x}+\frac{\partial Y_3(u,v)}{\partial y} \tag{7.325}$$

$$vL_{22}(v)=vM_{22}(v)+\frac{\partial X_4(v,v)}{\partial x}+\frac{\partial Y_4(v,v)}{\partial y} \tag{7.326}$$

where

$$X_1(u,u)=(D_{11}-\frac{\partial A_{11}}{\partial x}-\frac{\partial B_{11}}{\partial y})u^2 \tag{7.327}$$

$$Y_1(u,u)=(E_{11}-\frac{\partial C_{11}}{\partial y}-\frac{\partial B_{11}}{\partial x})u^2 \tag{7.328}$$

$$X_2(u,v)=A_{12}(u\frac{\partial v}{\partial x}-v\frac{\partial u}{\partial x})+B_{12}(u\frac{\partial v}{\partial y}-v\frac{\partial u}{\partial y})+(D_{12}-\frac{\partial A_{12}}{\partial x}-\frac{\partial B_{12}}{\partial y})uv \tag{7.329}$$

$$Y_2(u,v)=B_{12}(u\frac{\partial v}{\partial x}-v\frac{\partial u}{\partial x})+C_{12}(u\frac{\partial v}{\partial y}-v\frac{\partial u}{\partial y})+(E_{12}-\frac{\partial C_{12}}{\partial y}-\frac{\partial B_{12}}{\partial x})uv \tag{7.330}$$

$$X_3(u,v)=A_{21}(v\frac{\partial u}{\partial x}-u\frac{\partial v}{\partial x})+B_{21}(v\frac{\partial u}{\partial y}-u\frac{\partial v}{\partial y})+(D_{21}-\frac{\partial A_{21}}{\partial x}-\frac{\partial B_{21}}{\partial y})uv \tag{7.331}$$

$$Y_3(u,v)=B_{21}(v\frac{\partial u}{\partial x}-u\frac{\partial v}{\partial x})+C_{21}(v\frac{\partial u}{\partial y}-u\frac{\partial v}{\partial y})+(E_{21}-\frac{\partial C_{21}}{\partial y}-\frac{\partial B_{21}}{\partial x})uv \tag{7.332}$$

$$X_4(v,v)=(D_{22}-\frac{\partial A_{22}}{\partial x}-\frac{\partial B_{22}}{\partial y})v^2 \tag{7.333}$$

$$Y_4(v,v)=(E_{22}-\frac{\partial C_{22}}{\partial y}-\frac{\partial B_{22}}{\partial x})v^2 \tag{7.334}$$

Adding (7.323) to (7.326) gives

$$\begin{aligned}&uL_{11}(u)-uM_{11}(u)+uL_{12}(v)-vM_{12}(u)+vL_{21}(u)-uM_{21}(v)\\&+vL_{22}(v)-vM_{22}(v)=\frac{\partial X_1(u,u)}{\partial x}+\frac{\partial Y_1(u,u)}{\partial y}+\frac{\partial X_2(u,v)}{\partial x}+\frac{\partial Y_2(u,v)}{\partial y}\\&+\frac{\partial X_3(u,v)}{\partial x}+\frac{\partial Y_3(u,v)}{\partial y}+\frac{\partial X_4(v,v)}{\partial x}+\frac{\partial Y_4(v,v)}{\partial y}\end{aligned} \tag{7.335}$$

Equation (7.335) can be rewritten as

$$u[L_{11}(u)-M_{11}(u)+L_{12}(v)-M_{21}(v)]$$
$$+v[L_{21}(u)-M_{12}(u)+L_{22}(v)-M_{22}(v)]=\frac{\partial P}{\partial x}+\frac{\partial Q}{\partial y} \tag{7.336}$$

where

$$P=X_1(u,u)+X_2(u,v)+X_3(u,v)+X_4(v,v) \tag{7.337}$$
$$Q=Y_1(u,u)+Y_2(u,v)+Y_3(u,v)+Y_4(v,v) \tag{7.338}$$

Equation (7.336) can be considered as a generalization of the Lagrange identity for the case of a coupled system of PDEs. The system of PDEs is called self-adjoint if

$$\frac{\partial P}{\partial x}+\frac{\partial Q}{\partial y}=0 \tag{7.339}$$

Therefore, if the left hand side of (7.336) is zero, the system of PDEs is self-adjoint. Thus, we will examine the condition that the left hand side of (7.336) is zero.

Similar to the procedure in arriving at the adjoint operator given in (7.310), we can easily verify that

$$M_{ij}(v)=L_{ij}(v)+v\frac{\partial}{\partial x}(\frac{\partial A_{ij}}{\partial x}+\frac{\partial B_{ij}}{\partial y}-D_{ij})+v\frac{\partial}{\partial y}(\frac{\partial C_{ij}}{\partial y}+\frac{\partial B_{ij}}{\partial x}-E_{ij})$$
$$+2\frac{\partial v}{\partial x}(\frac{\partial A_{ij}}{\partial x}+\frac{\partial B_{ij}}{\partial y}-D_{ij})+2\frac{\partial v}{\partial y}(\frac{\partial C_{ij}}{\partial y}+\frac{\partial B_{ij}}{\partial x}-E_{ij}) \tag{7.340}$$

Substitution of (7.340) into the left hand side of (7.336) yields

$$LHS=\Phi=-2u\frac{\partial u}{\partial x}(\frac{\partial A_{11}}{\partial x}+\frac{\partial B_{11}}{\partial y}-D_{11})-2u\frac{\partial u}{\partial y}(\frac{\partial B_{11}}{\partial x}+\frac{\partial C_{11}}{\partial y}-E_{11})$$
$$-u^2\frac{\partial}{\partial x}(\frac{\partial A_{11}}{\partial x}+\frac{\partial B_{11}}{\partial y}-D_{11})-u^2\frac{\partial}{\partial y}(\frac{\partial B_{11}}{\partial x}+\frac{\partial C_{11}}{\partial y}-E_{11})$$
$$+uL_{12}(v)-uL_{21}(v)-2u\frac{\partial v}{\partial x}(\frac{\partial A_{21}}{\partial x}+\frac{\partial B_{21}}{\partial y}-D_{21})-2u\frac{\partial v}{\partial y}(\frac{\partial B_{21}}{\partial x}+\frac{\partial C_{21}}{\partial y}-E_{21})$$
$$-vu\frac{\partial}{\partial x}(\frac{\partial A_{21}}{\partial x}+\frac{\partial B_{21}}{\partial y}-D_{21})-vu\frac{\partial}{\partial y}(\frac{\partial B_{21}}{\partial x}+\frac{\partial C_{21}}{\partial y}-E_{21})$$
$$+vL_{21}(u)-vL_{12}(u)-2v\frac{\partial u}{\partial x}(\frac{\partial A_{12}}{\partial x}+\frac{\partial B_{12}}{\partial y}-D_{12})-2v\frac{\partial u}{\partial y}(\frac{\partial B_{12}}{\partial x}+\frac{\partial C_{12}}{\partial y}-E_{12})$$
$$-vu\frac{\partial}{\partial x}(\frac{\partial A_{12}}{\partial x}+\frac{\partial B_{12}}{\partial y}-D_{12})-vu\frac{\partial}{\partial y}(\frac{\partial B_{12}}{\partial x}+\frac{\partial C_{12}}{\partial y}-E_{12})$$
$$-2v\frac{\partial v}{\partial x}(\frac{\partial A_{22}}{\partial x}+\frac{\partial B_{22}}{\partial y}-D_{22})-2v\frac{\partial v}{\partial y}(\frac{\partial B_{22}}{\partial x}+\frac{\partial C_{22}}{\partial y}-E_{22})$$
$$-v^2\frac{\partial}{\partial x}(\frac{\partial A_{22}}{\partial x}+\frac{\partial B_{22}}{\partial y}-D_{22})-v^2\frac{\partial}{\partial y}(\frac{\partial B_{22}}{\partial x}+\frac{\partial C_{22}}{\partial y}-E_{22})$$
$$\tag{7.341}$$

We will now consider the conditions that (7.341) becomes zero. First, we see that if

$$A_{12} = A_{21}, \quad B_{12} = B_{21}, \quad C_{12} = C_{21} \tag{7.342}$$

we have

$$uL_{12}(v) - uL_{21}(v) = u\frac{\partial v}{\partial x}(D_{12} - D_{21}) + u\frac{\partial v}{\partial y}(E_{12} - E_{21}) + uv(F_{12} - F_{21}) \tag{7.343}$$

$$vL_{21}(u) - vL_{11}(u) = -v\frac{\partial u}{\partial x}(D_{12} - D_{21}) - v\frac{\partial u}{\partial y}(E_{12} - E_{21}) - uv(F_{12} - F_{21}) \tag{7.344}$$

In addition, we find that many terms in (7.341) vanish if

$$\frac{\partial A_{11}}{\partial x} + \frac{\partial B_{11}}{\partial y} = D_{11}, \quad \frac{\partial B_{11}}{\partial x} + \frac{\partial C_{11}}{\partial y} = E_{11} \tag{7.345}$$

$$\frac{\partial A_{22}}{\partial x} + \frac{\partial B_{22}}{\partial y} = D_{22}, \quad \frac{\partial B_{22}}{\partial x} + \frac{\partial C_{22}}{\partial y} = E_{22} \tag{7.346}$$

Substitution of (7.342) to (7.346) into (7.341) gives

$$\begin{aligned}
\Phi = {} & u\frac{\partial v}{\partial x}(D_{12} - D_{21}) + u\frac{\partial v}{\partial y}(E_{12} - E_{21}) - v\frac{\partial u}{\partial x}(D_{12} - D_{21}) - v\frac{\partial u}{\partial y}(E_{12} - E_{21}) \\
& -2u\frac{\partial v}{\partial x}(\frac{\partial A_{21}}{\partial x} + \frac{\partial B_{21}}{\partial y} - D_{21}) - 2u\frac{\partial v}{\partial y}(\frac{\partial B_{21}}{\partial x} + \frac{\partial C_{21}}{\partial y} - E_{21}) \\
& -2v\frac{\partial u}{\partial x}(\frac{\partial A_{12}}{\partial x} + \frac{\partial B_{12}}{\partial y} - D_{12}) - 2v\frac{\partial u}{\partial y}(\frac{\partial B_{12}}{\partial x} + \frac{\partial C_{12}}{\partial y} - E_{12})
\end{aligned} \tag{7.347}$$

Finally, we observe that Φ in (7.347) is identically zero if

$$2(\frac{\partial B_{21}}{\partial x} + \frac{\partial C_{21}}{\partial y}) = E_{21} + E_{12} \tag{7.348}$$

$$2(\frac{\partial A_{12}}{\partial x} + \frac{\partial B_{12}}{\partial y}) = D_{12} + D_{21} \tag{7.349}$$

The conditions of (7.318) and (7.319) being self-adjoint can be summarized as

$$A_{ij} = A_{ji}, \quad B_{ij} = B_{ji}, \quad C_{ij} = C_{ji} \tag{7.350}$$

$$(\frac{\partial A_{ij}}{\partial x} + \frac{\partial B_{ij}}{\partial y}) = \frac{1}{2}(D_{ij} + D_{ji}) \tag{7.351}$$

$$(\frac{\partial B_{ij}}{\partial x} + \frac{\partial C_{ij}}{\partial y}) = \frac{1}{2}(E_{ij} + E_{ji}) \tag{7.352}$$

Note that self-adjoint conditions which are similar to (7.350) to (7.352) were obtained by Kuzmin (1963) when he considered the variational principle for electrodynamics fields. However, there is apparently a mistake in Kuzmin (1963), and there is no derivation given in Kuzmin (1963). Nevertheless, such analysis is related to Green's identity and the formulation of functionals.

7.7 MIXED TYPE PDE

In some situations, the behavior of the solution of a PDE may depend on the dependent variables of the problem. The most well-known mixed type PDE is probably the Tricomi equation.

$$\frac{\partial^2 u}{\partial x^2} - x\frac{\partial^2 u}{\partial y^2} = 0 \tag{7.353}$$

According to the procedure of classification given in Section 7.2.2, we have

$$B^2 - 4AC = 4x \tag{7.354}$$

Therefore, we have (7.353) being hyperbolic with $x > 0$, parabolic for $x = 0$, and elliptic for $x < 0$. This is called a mixed type of second order PDE. The origin of this equation comes from transonic flow in 2-D gas dynamics. This equation occupies an important place in the development of supersonic flows from transonic flow in the area of rocket science in the beginning of the twentieth century. We will sketch the origin of this equation briefly in the following section. We will first review two-dimensional steady gas flow in the next section.

7.7.1 Two-Dimensional Steady Gas Flows

For two-dimensional steady potential flows of a gas, the Euler equation of motion for a compressible fluid is

$$\frac{\partial \boldsymbol{v}}{\partial t} + (\boldsymbol{v}\cdot\nabla)\boldsymbol{v} = -\frac{1}{\rho}\nabla p + g \tag{7.355}$$

where g is the gravitational constant, p is the pressure, and $\boldsymbol{v}$ is the velocity field. This is a special case of Navier-Stokes equation for fluid mechanics. For steady flow with negligible gravity effect, we have

$$(\boldsymbol{v}\cdot\nabla)\boldsymbol{v} = -\frac{1}{\rho}\nabla p \tag{7.356}$$

To express (7.356) in incremental form, we have

$$v dv = -\frac{dp}{\rho} \tag{7.357}$$

On the other hand, sound wave speed in gas is defined as:

$$c = \sqrt{\frac{p}{\rho}} \tag{7.358}$$

Thus, density can be related to pressure as

$$\rho c^2 = p \tag{7.359}$$

We note that (7.359) can be expressed in incremental form as:

$$dp = c^2 d\rho \tag{7.360}$$

Rearranging (7.357) gives

$$\frac{dp}{dv} = -\rho v \tag{7.361}$$

Substitution of (7.360) into (7.361) results in

$$\frac{d\rho}{dv} = -\rho \frac{v}{c^2} \tag{7.362}$$

The total differential of the mass flux density along a streamline is

$$d(\rho v) = \rho dv + v d\rho \tag{7.363}$$

It is equivalent to

$$\frac{d(\rho v)}{dv} = \rho + v\frac{d\rho}{dv} = \rho - \rho\frac{v^2}{c^2} = \rho(1 - \frac{v^2}{c^2}) \tag{7.364}$$

Note that for supersonic flow, we have the flow speed faster than the sound speed c, we have $d(\rho v)/dv < 0$ or mass flux density as a decreasing function of velocity; whereas, for the subsonic flow the mass flux density is an increasing function of v.

From continuity, we have

$$\frac{\partial(\rho v_x)}{\partial x} + \frac{\partial(\rho v_y)}{\partial x} = 0 \tag{7.365}$$

By introducing a velocity potential ϕ, we have

$$d\phi = \frac{\partial \phi}{\partial x}dx + \frac{\partial \phi}{\partial y}dy = v_x dx + v_y dy \tag{7.366}$$

7.7.2 Hodograph Transformation

The formulation expressed in velocity is not easy to solve, Molenbroek in 1890 and Chaplygin in 1902 applied the Legendre transform to rewrite the unknowns of the gas flow equation (7.366) as x and y, and the variables as the velocity components. More discussion of the application of the Legendre transform can be found in Appendix C of Chau (2013). The velocity plane is known as a hodograph plane versus the physical plane of x-y. This approach is known as the hodograph method which has been discussed in Section 6.15 for the context of first order PDEs. In particular, we observe that (7.366) can be written as

$$d\phi = d(xv_x) - xdv_x + d(yv_y) - ydv_y \tag{7.367}$$

Thus, we can introduce a new function Φ such that

$$\Phi = -\phi + xv_x + yv_y \tag{7.368}$$

The total differential of (7.368) gives

$$d\Phi = -d\phi + xdv_x + ydv_y + v_x dx + v_y dy \tag{7.369}$$

Substitution of (7.366) into (7.369) gives

$$d\Phi = xdv_x + ydv_y \tag{7.370}$$

Therefore, we now have the unknown function being given in terms of the velocity as

$$\Phi = \Phi(v_x, v_y) \tag{7.371}$$

The total differential of (7.371) is

$$d\Phi = \frac{\partial \Phi}{\partial v_x}dv_x + \frac{\partial \Phi}{\partial v_y}dv_y = xdv_x + ydv_y \tag{7.372}$$

Consequently, we have

$$x = \frac{\partial \Phi}{\partial v_x}, \quad y = \frac{\partial \Phi}{\partial v_y} \tag{7.373}$$

The variables of v_x and v_y can be rewritten in polar form as:

$$v_x = v\cos\theta, \quad v_y = v\sin\theta \tag{7.374}$$

Clearly, we have

$$v^2 = v_x^2 + v_y^2, \quad \tan\theta = \frac{v_y}{v_x} \tag{7.375}$$

With these new variables, applying the chain rule to (7.373) we have

$$x = \frac{\partial \Phi}{\partial v_x} = \frac{\partial \Phi}{\partial v}\frac{\partial v}{\partial v_x} + \frac{\partial \Phi}{\partial \theta}\frac{\partial \theta}{\partial v_x} = \cos\theta\frac{\partial \Phi}{\partial v} - \frac{\sin\theta}{v}\frac{\partial \Phi}{\partial \theta} \tag{7.376}$$

$$y = \frac{\partial \Phi}{\partial v_y} = \frac{\partial \Phi}{\partial v}\frac{\partial v}{\partial v_y} + \frac{\partial \Phi}{\partial \theta}\frac{\partial \theta}{\partial v_y} = \sin\theta\frac{\partial \Phi}{\partial v} + \frac{\cos\theta}{v}\frac{\partial \Phi}{\partial \theta} \tag{7.377}$$

By virtue of (7.376) and (7.377), the last two terms in (7.368) can be found as

$$\begin{aligned} xv_x + yv_y &= (\cos\theta\frac{\partial \Phi}{\partial v} - \frac{\sin\theta}{v}\frac{\partial \Phi}{\partial \theta})v\cos\theta + (\sin\theta\frac{\partial \Phi}{\partial v} + \frac{\cos\theta}{v}\frac{\partial \Phi}{\partial \theta})v\sin\theta \\ &= v\frac{\partial \Phi}{\partial v} \end{aligned} \tag{7.378}$$

Substitution of (7.378) into (7.368) gives

$$\Phi = -\phi + v\frac{\partial \Phi}{\partial v} \tag{7.379}$$

7.7.3 Chaplygin's Equation

We now observe that the continuity equation can be expressed in terms of Jacobians as

$$\frac{\partial(\rho v_x, y)}{\partial(x, y)} - \frac{\partial(\rho v_y, x)}{\partial(x, y)} = \frac{\partial(\rho v_x)}{\partial x} + \frac{\partial(\rho v_y)}{\partial y} = 0 \tag{7.380}$$

Note that

$$\frac{\partial(\rho v_x, y)}{\partial(x, y)} = \begin{vmatrix} \frac{\partial \rho v_x}{\partial x} & \frac{\partial \rho v_x}{\partial y} \\ \frac{\partial y}{\partial x} & \frac{\partial y}{\partial y} \end{vmatrix} = \frac{\partial(\rho v_x)}{\partial x} \tag{7.381}$$

$$\frac{\partial(\rho v_y, x)}{\partial(x, y)} = \begin{vmatrix} \frac{\partial \rho v_y}{\partial x} & \frac{\partial \rho v_y}{\partial y} \\ \frac{\partial x}{\partial x} & \frac{\partial x}{\partial y} \end{vmatrix} = -\frac{\partial(\rho v_y)}{\partial y} \tag{7.382}$$

We can multiply (7.380) by $\partial(x,y)/\partial(v,\theta)$ and in view of (7.374) to get

$$\frac{\partial(\rho v_x, y)}{\partial(x,y)}\frac{\partial(x,y)}{\partial(v,\theta)} - \frac{\partial(\rho v_y, x)}{\partial(x,y)}\frac{\partial(x,y)}{\partial(v,\theta)} = \frac{\partial(\rho v\cos\theta, y)}{\partial(v,\theta)} - \frac{\partial(\rho v\sin\theta, x)}{(v,\theta)} = 0 \quad (7.383)$$

The first term on the right hand side of (7.383) can be evaluated as

$$\frac{\partial(\rho v\cos\theta, y)}{\partial(v,\theta)} = \begin{vmatrix} \dfrac{\partial \rho v\cos\theta}{\partial v} & \dfrac{\partial \rho v\cos\theta}{\partial \theta} \\ \dfrac{\partial y}{\partial v} & \dfrac{\partial y}{\partial \theta} \end{vmatrix} = \frac{\partial \rho v}{\partial v}\cos\theta\frac{\partial y}{\partial \theta} + \rho v\sin\theta\frac{\partial y}{\partial v} \quad (7.384)$$

Differentiation of (7.377) gives

$$\frac{\partial y}{\partial v} = \sin\theta\frac{\partial^2 \Phi}{\partial v^2} - \frac{\cos\theta}{v^2}\frac{\partial \Phi}{\partial \theta} + \frac{\cos\theta}{v}\frac{\partial^2 \Phi}{\partial \theta \partial v} \quad (7.385)$$

$$\frac{\partial y}{\partial \theta} = \cos\theta\frac{\partial \Phi}{\partial v} + \sin\theta\frac{\partial^2 \Phi}{\partial v \partial \theta} - \frac{\sin\theta}{v}\frac{\partial \Phi}{\partial \theta} + \frac{\cos\theta}{v}\frac{\partial^2 \Phi}{\partial \theta^2} \quad (7.386)$$

Similarly, the second term on the right of (7.383) can be evaluated as

$$\frac{\partial(\rho v\sin\theta, x)}{\partial(v,\theta)} = \begin{vmatrix} \dfrac{\partial \rho v\sin\theta}{\partial v} & \dfrac{\partial \rho v\sin\theta}{\partial \theta} \\ \dfrac{\partial x}{\partial v} & \dfrac{\partial x}{\partial \theta} \end{vmatrix} = \frac{\partial \rho v}{\partial v}\sin\theta\frac{\partial x}{\partial \theta} - \rho v\cos\theta\frac{\partial x}{\partial v} \quad (7.387)$$

Differentiation of (7.376) gives

$$\frac{\partial x}{\partial v} = \cos\theta\frac{\partial^2 \Phi}{\partial v^2} + \frac{\sin\theta}{v^2}\frac{\partial \Phi}{\partial \theta} - \frac{\sin\theta}{v}\frac{\partial^2 \Phi}{\partial \theta \partial v} \quad (7.388)$$

$$\frac{\partial x}{\partial \theta} = -\sin\theta\frac{\partial \Phi}{\partial v} + \cos\theta\frac{\partial^2 \Phi}{\partial v \partial \theta} - \frac{\cos\theta}{v}\frac{\partial \Phi}{\partial \theta} - \frac{\sin\theta}{v}\frac{\partial^2 \Phi}{\partial \theta^2} \quad (7.389)$$

Finally, combining (7.384) to (7.389) with (7.383) leads to

$$\frac{\partial(\rho v)}{\partial v}\left[\frac{\partial \Phi}{\partial v} + \frac{1}{v}\frac{\partial^2 \Phi}{\partial \theta^2}\right] + \rho v\frac{\partial^2 \Phi}{\partial v^2} = 0 \quad (7.390)$$

Substitution of (7.364) into (7.390) yields Chaplygin's equation:

$$\frac{\partial^2 \Phi}{\partial \theta^2} + \frac{v^2}{1 - v^2/c^2}\frac{\partial^2 \Phi}{\partial v^2} + v\frac{\partial \Phi}{\partial v} = 0 \quad (7.391)$$

This is a linear PDE, and is a good approximation of transonic flow (i.e., $0.8 \leq$ Mach number < 1.0). For this range, the speed v is about 965 km/hour to about 1236 km/hour. Thus, for steady state the nonlinear PDE of the Euler equation of motion and the continuity equation are now converted to a linear PDE through the use of the hodograph transformation. Once the solution of Φ is solved in terms of v and θ, (7.373) and (7.374) can be used to give the final solutions as:

$$x = x(v,\theta,\Phi), \quad y = y(v,\theta,\Phi) \quad (7.392)$$

Note that the validity of the Chaplygin equation given in (7.391) relies on the requirement of non-zero Jacobian of the following change of variables:

$$|J| = \frac{\partial(x,y)}{\partial(v,\theta)} = \frac{1}{v}\left[\left(\frac{\partial^2 \Phi}{\partial v \partial \theta} - v\frac{\partial \Phi}{\partial \theta}\right)^2 + \frac{v^2}{1 - v^2/c^2}\left(\frac{\partial^2 \Phi}{\partial v^2}\right)^2\right] \quad (7.393)$$

It is clear from (7.393) that the Jacobian can become zero if $v > c$ (or in the supersonic region). Once the Jacobian is zero, continuous flow throughout the region is impossible and shock waves must occur in the supersonic region. This is the reason that the supersonic airplane *Concord* is always accompanied by an unpleasant shock wave. In addition, the sound speed is, in general, a function of velocity or $c = c(v)$.

7.7.4 Tricomi's Equation

We now look at an approximation of the Chaplygin equation given in (7.391) when the flow is transonic (i.e., $v \to c$). In particular, the first derivative term is much smaller than the other two terms in (7.391):

$$v\frac{\partial \Phi}{\partial v} \ll \frac{v^2}{1-v^2/c^2}\left(\frac{\partial^2 \Phi}{\partial v^2}\right) \tag{7.394}$$

Thus, we have the approximation for (7.391) as

$$\frac{\partial^2 \Phi}{\partial \theta^2} + \frac{v^2}{1-v^2/c^2}\frac{\partial^2 \Phi}{\partial v^2} = 0 \tag{7.395}$$

At the transition from subsonic to supersonic, we passed through the transition or the so-called transonic flow. We have $v \approx c = c^*$, then

$$\frac{v^2}{1-v^2/c^2} \approx \frac{c_*^2}{(1+v/c)(1-v/c)} \approx \frac{c_*^2}{2(1-v/c)} = \frac{c_*^2}{2\alpha_*(1-v/c_*)} \tag{7.396}$$

The last part of (7.396) is the consequence of the following approximation near the transition from subsonic to supersonic:

$$\frac{v}{c} - 1 = \alpha_*\left(\frac{v}{c_*} - 1\right) \tag{7.397}$$

To prove (7.397) near the transition point $v \approx c = c^*$, the value of $c - c_*$ can first be expanded using the first term in Taylor series expansion as:

$$c - c_* = (v - c_*)\left(\frac{dc}{dv}\right)_{v=c*} \tag{7.398}$$

This expansion can be rewritten as:

$$c - v = (c_* - v)\left[1 - \left(\frac{dc}{dv}\right)_{v=c*}\right] \tag{7.399}$$

Since the sound speed is a function of density or $c = c(\rho)$, we can express it using the chain rule as:

$$\frac{dc}{dv} = \frac{dc}{d\rho}\frac{d\rho}{dv} = -\frac{\rho}{c}\frac{dc}{d\rho} \tag{7.400}$$

The last part of (7.400) is a result of the following special form of (7.362) near the transition point:

$$\frac{d\rho}{dv} = -\frac{\rho}{c} \tag{7.401}$$

Substitution of (7.400) into (7.399) gives

$$c-v=(c_*-v)[1+\frac{\rho}{c}\frac{dc}{d\rho}]=(\frac{c_*-v}{c})\frac{d(\rho c)}{d\rho}=\alpha_*(c_*-v) \tag{7.402}$$

where

$$\alpha_* = \frac{1}{c}\frac{d(\rho c)}{d\rho}=c\frac{d(\rho c)}{dp}=\frac{1}{2}\rho^3 c_*^4(\frac{\partial^2 V}{\partial p^2})_s \tag{7.403}$$

In obtaining the second part of (7.403), we have used (7.360). The last part of (7.403) can be found by considering the following relation for sound speed (see (7.358)):

$$\rho c=\rho\sqrt{\frac{\partial p}{\partial \rho}} \tag{7.404}$$

The definition of the sound wave can be written in terms of volume per unit mass V instead of density ρ. More specifically, its increment is related $d\rho$ to

$$d\rho=d(\frac{1}{V})=-\frac{dV}{V^2} \tag{7.405}$$

Substitution of (7.405) into (7.404) gives

$$\rho c=\frac{\rho V}{\sqrt{(-\frac{\partial V}{\partial p})}}=\frac{1}{\sqrt{(-\frac{\partial V}{\partial p})}} \tag{7.406}$$

Differentiation of (7.406) with respect to p gives

$$\frac{d(\rho c)}{d\rho}=c^2\frac{d(\rho c)}{dp}=\frac{1}{2}\frac{c^2}{(-\frac{\partial V}{\partial p})^{3/2}}(\frac{\partial^2 V}{\partial p^2})=\frac{1}{2}\rho^3 c^5(\frac{\partial^2 V}{\partial p^2}) \tag{7.407}$$

This particular form applies to the case of constant entropy. Using (7.407), the second part of (7.403) can be obtained without difficulty. Combining (7.396), (7.395), and (7.403) gives

$$\frac{\partial^2 \Phi}{\partial \theta^2}+\frac{c_*^2}{2\alpha_*(1-v/c_*)}\frac{\partial^2 \Phi}{\partial v^2}=0 \tag{7.408}$$

Introduction of the following change of variables

$$\eta=(2\alpha_*)^{1/3}(\frac{v-c_*}{c_*}) \tag{7.409}$$

gives

$$\frac{\partial^2 \Phi}{\partial v^2}=\frac{\partial^2 \Phi}{\partial \eta^2}\frac{(2\alpha_*)^{2/3}}{c_*^2} \tag{7.410}$$

Substitution of (7.410) into (7.408) gives the Tricomi equation:

$$\frac{\partial^2 \Phi}{\partial \eta^2}-\eta\frac{\partial^2 \Phi}{\partial \theta^2}=0 \tag{7.411}$$

Therefore, the Tricomi equation is an approximation of the Chaplygin equation when the speed is very close to the supersonic speed. The general solution of the Tricomi equation is considered next.

7.7.5 Solution of Tricomi's Equation

Using the classification for second order ODEs, we have for (7.411)

$$B^2 - 4AC = 4\eta > 0 \tag{7.412}$$

for supersonic flow (i.e., $v > c^*$). Therefore, the Tricomi equation is of hyperbolic type. The characteristics are

$$\frac{d\theta}{d\eta} = -\frac{\partial\xi / \partial\eta}{\partial\xi / \partial\theta} = \frac{B + \sqrt{B^2 - 4AC}}{2A} = \sqrt{\eta} \tag{7.413}$$

$$\frac{d\theta}{d\eta} = -\frac{\partial\zeta / \partial\theta}{\partial\zeta / \partial\eta} = \frac{B - \sqrt{B^2 - 4AC}}{2A} = -\sqrt{\eta} \tag{7.414}$$

Integrating both of these, we get the two characteristics as

$$\xi = \theta - \frac{2}{3}\eta^{3/2} = C_1 \tag{7.415}$$

$$\zeta = \theta + \frac{2}{3}\eta^{3/2} = C_2 \tag{7.416}$$

Instead of reducing it to the canonical form, we observe that the Tricomi equation remains unchanged if we make the following substitution:

$$\theta^2 \to a\theta^2, \quad \eta^3 \to a\eta^3 \tag{7.417}$$

This suggests that we can make the following change of variables:

$$\Phi = \theta^{2k} f(\xi), \quad \xi = 1 - \frac{4\eta^3}{9\theta^2} \tag{7.418}$$

Differentiation of (7.418) with respect to θ gives

$$\frac{\partial\Phi}{\partial\theta} = 2k\theta^{2k-1} f(\xi) + \theta^{2k} f'(\xi)\frac{\partial\xi}{\partial\theta} \tag{7.419}$$

Application of the chain rule to (7.418) gives

$$\frac{\partial\xi}{\partial\theta} = \frac{8\eta^3}{9\theta^3} = \frac{2}{\theta}(1-\xi) \tag{7.420}$$

Substitution of (7.420) into (7.419) gives

$$\frac{\partial\Phi}{\partial\theta} = 2\theta^{2k-1}[kf(\xi) + (1-\xi)f'(\xi)] \tag{7.421}$$

Differentiation of (7.421) one more time gives

$$\frac{\partial^2\Phi}{\partial\theta^2} = \theta^{2k-1}\left\{2k(2k-1)f(\xi) + (1-\xi)[8k-6]f'(\xi) + 4(1-\xi)^2 f''(\xi)\right\} \tag{7.422}$$

We now turn to the first term in (7.411). In particular, differentiation of (7.418) with respect to η gives

$$\frac{\partial\Phi}{\partial\eta} = \theta^{2k} f'(\xi)\frac{\partial\xi}{\partial\eta} = -\theta^{2k}\frac{4\eta^2}{3\theta^2} f'(\xi) \tag{7.423}$$

Application of the chain rule to (7.43) gives

$$\frac{\partial \xi}{\partial \eta} = -\frac{4\eta^2}{3\theta^2} = -\frac{3}{\eta}(1-\xi) \tag{7.424}$$

Differentiating (7.424) one more time and using (7.418), we find

$$\frac{\partial^2 \Phi}{\partial \eta^2} = \eta \theta^{2k-2}\{-\frac{8}{3}f'(\xi) + 4(1-\xi)f''(\xi)\} \tag{7.425}$$

Substitution of (7.422) and (7.425) into (7.411) gives

$$\xi(1-\xi)f''(\xi) + [\frac{5}{6} - 2k - \xi(\frac{3}{2} - 2k)]f'(\xi) - k(k - \frac{1}{2})f(\xi) = 0 \tag{7.426}$$

Recall the hypergeometric equation from (4.560) of Chapter 4 that

$$\xi(1-\xi)f''(\xi) + [\gamma - (\alpha + \beta + 1)\xi]f'(\xi) - \alpha\beta f(\xi) = 0 \tag{7.427}$$

Thus, comparison of (7.426) and (7.427) gives

$$\alpha = -k, \quad \beta = \frac{1}{2} - k, \quad \gamma = \frac{5}{6} - 2k \tag{7.428}$$

Therefore, according to Chapter 4 the solution of (7.426) can be expressed in terms of hypergeometric functions, and using (7.418), the final solution of the Tricomi equation given in (7.411) is

$$\begin{aligned} \Phi = \theta^{2k}\{ & AF(-k, \frac{1}{2} - k, \frac{5}{6} - 2k; 1 - \frac{4\eta^3}{9\theta^2}) \\ & + B(1 - \frac{4\eta^3}{9\theta^2})^{2k+\frac{1}{6}} F(k + \frac{1}{6}, k + \frac{2}{3}, 2k + \frac{7}{6}; 1 - \frac{4\eta^3}{9\theta^2})\} \end{aligned} \tag{7.429}$$

Note from Chapter 4 that we can use the bilinear transformation given in (4.610) to convert the solution given in (7.429) into five other different forms. In particular, we can use (4.599) to transform the solution to the form (see Table 4.4 second row for singular point near $\xi = 1$)

$$y = C_1(1-\xi)^{\mu_1} V_1 + C_2(1-\xi)^{\mu_2} V_2 \tag{7.430}$$

We further pick solution Numbers 1 and 5 in Table 4.7 to get

$$\begin{aligned} \Phi = \theta^{2k}\{ & AF(-k, -k + \frac{1}{2}, \frac{2}{3}; \frac{4\eta^3}{9\theta^2}) \\ & + B\frac{\eta}{\theta^{2/3}} F(-k + \frac{5}{6}, -k + \frac{1}{3}, \frac{4}{3}; \frac{4\eta^3}{9\theta^2})\} \end{aligned} \tag{7.431}$$

This result is obtained by noting that

$$(1-\xi)^{\gamma-\alpha-\beta} = (1-\xi)^{1/3} = (\frac{4}{9})^{1/3} \frac{\eta}{\theta^{2/3}} \tag{7.432}$$

Following Landau and Lifshitz (1987), we give one more different form of (7.429) by using the following transformation:

$$\xi = \frac{1}{1-u} \tag{7.433}$$

which is for the solution near the singular point of $\xi \to \infty$ (see Chapter 4). For this case, we can pick solution Numbers 3 and 7 from Table 4.8 as

$$\Phi = \eta^{3k}\{AF(-k,-k+\frac{1}{3},\frac{1}{2};\frac{9\theta^2}{4\eta^3})$$
$$+B\frac{\theta}{\eta^{2/3}}F(-k+\frac{1}{2},-k+\frac{5}{6},\frac{3}{2};\frac{9\theta^2}{4\eta^3})\} \quad (7.434)$$

This is obtained by noting the following:

$$(\xi-1)^{-\alpha} = -(\frac{9\theta^2}{4\eta^3})^{\alpha} = (-\frac{4}{9})^k \frac{\eta^{3k}}{\theta^{2k}} \quad (7.435)$$

$$(\xi-1)^{-\beta} = (-\frac{4\eta^3}{9\theta^2})^{k-1/2} = (-\frac{4}{9})^{k-1/2}\frac{\eta^{3k}}{\eta^{3/2}}(\frac{\theta}{\theta^{2k}}) \quad (7.436)$$

More importantly, we see from (7.431) that Φ can be expanded in integral powers of η near $\eta = 0$, and (7.434) shows that Φ can be expanded in integral powers of θ near $\theta = 0$. That is, the lines $\eta = 0$ and $\theta = 0$ are not singular lines because power expansion of infinite order exists. Equation (7.429) shows that the characteristics are singular lines if $2k+1/6$ is not an integer, and the factor $(9\theta^2-4\eta^3)^{2k+1/6}$ in (7.429) has branch points; whilst if $2k+1/6$ is an integer, the solution is degenerate for $2k+1/6 = 0$ and as shown in Chapter 4 that a second independent solution has logarithmic singularity (see cases 2 and 3 of Section 4.5). For more discussion of the solution near the lines $\eta = 0$ and $\theta = 0$, the readers are referred to Section 118 of Landau and Lifshitz (1987). Physically, this means that a continuous finite solution for the solution ceases to exist on the characteristics when the gas is moving supersonically. This singular solution suggests that a shock wave with jumps exists in a supersonic gas flow. Therefore, the study of the Tricomi equation is of fundamental importance in investigating supersonic flows around an object. Equivalently, to rephrase the same phenomenon in a reverse order, the Tricomi equation is of utmost importance in studying supersonic flight of aircraft (e.g., Mach number larger than 1). This is consistent with the so-called Le Roux-Delassus theorem that any singular surface of a solution of a linear differential equation must be characteristics. Thus, the jump character of the characteristics indicates shock waves.

Actually, the solution of the Tricomi equation can also be expressed in Fourier expansion as:

$$\Phi = u(\eta)e^{\pm i\nu\theta} \quad (7.437)$$

Substitution of (7.437) into (7.411) leads to

$$u'' + \eta\nu^2 u = 0 \quad (7.438)$$

This is an Airy equation and can be solved in terms of Bessel functions. By following the transformation proposed in Problem 4.52 in Chapter 4, we have

$$\Phi = \sqrt{\eta}e^{\pm i\nu\theta}\{AJ_{1/3}(\frac{2}{3}\nu\eta^{3/2}) + BY_{1/3}(\frac{2}{3}\nu\eta^{3/2})\} \quad (7.439)$$

As a final note, Landau and Lifshitz (1987) was first translated to English from Russian in 1958. Landau obtained a Nobel Prize in physics in 1962 for his work on liquid helium, whereas Lifshitz was a student of Landau and recipient of Landau Prize, USSR State Prize, Lomonosov Prize, and Lenin Prize. *Fluid Mechanics* by Landau and Lifshitz (1987) is volume six of the ten-volume book series *Course of*

Theoretical Physics by Landau and Lifshitz. This famous book series has been translated into six different languages (some individual volumes to more than 10 languages).

7.8 RIEMANN'S INTEGRAL FOR HYPERBOLIC PDE

For the hyperbolic equation, the following problem is known as the Goursat problem:

$$\frac{\partial^2 u}{\partial x \partial y} = F(x, y, u, \frac{\partial u}{\partial x}, \frac{\partial u}{\partial y}) \tag{7.440}$$

where $0 < x, y < 1$, and the boundary conditions are

$$u(0, y) = \phi(y), \quad u(x,1) = \psi(x), \quad \phi(1) = \psi(0) \tag{7.441}$$

This problem is also known as the Darboux problem. There is a related problem called the Cauchy problem that also involves the first derivative of u:

$$u(x(t), y(t)) = \alpha(t), \quad \frac{\partial u}{\partial x}\frac{dx}{dt} - \frac{\partial u}{\partial y}\frac{dy}{dt} = \beta(t) \tag{7.442}$$

Note that the boundary conditions are given in terms of a parameter t. We will not consider the so-called Cauchy problem here in detail. Note also that there are many different usages of the term "Cauchy problem" in the literature. In general, the term Cauchy problem is reserved for hyperbolic PDEs (or wave type solutions), instead of elliptic PDEs (or equilibrium type solutions). In particular, the so-called Cauchy problem is normally defined for an initial value problem not for a boundary value problem.

In this section, we discuss a method derived by Riemann in 1860 for two-dimensional space. This work by Riemann was not known until du Bois-Reymond noticed its importance in 1864 and was subsequently publicized by Darboux (this is the reason that the Goursat problem was sometimes referred as Darboux problem). We consider a special case of Goursat problem given in (7.440):

$$F(u) = \frac{\partial^2 u}{\partial x \partial y} + A\frac{\partial u}{\partial x} + B\frac{\partial u}{\partial y} + Cu = f(x, y) \tag{7.443}$$

This problem is clearly a linear form of the Goursat problem given in (7.440). The triangular domain of the problem is given in Figure 7.5. Boundary values of u and v and their first derivatives are given on the curve S, which is not a characteristics of the hyperbolic problem. The characteristics are assumed to be given as x = constant and y = constant. Riemann's method was originally proposed for gas dynamics and only a particular form of (7.443) was considered. The general form given in (7.443) was actually considered by Hadamard.

The associated adjoint problem is

$$G(v) = \frac{\partial^2 v}{\partial x \partial y} - \frac{\partial}{\partial x}(Av) - \frac{\partial}{\partial y}(Bv) + Cv = 0 \tag{7.444}$$

where v is not only a function of x, y but also a function of the position of a (i.e., x_0, y_0)

$$v = v(x, y; x_0, y_0) \tag{7.445}$$

Then, Riemann considered the function

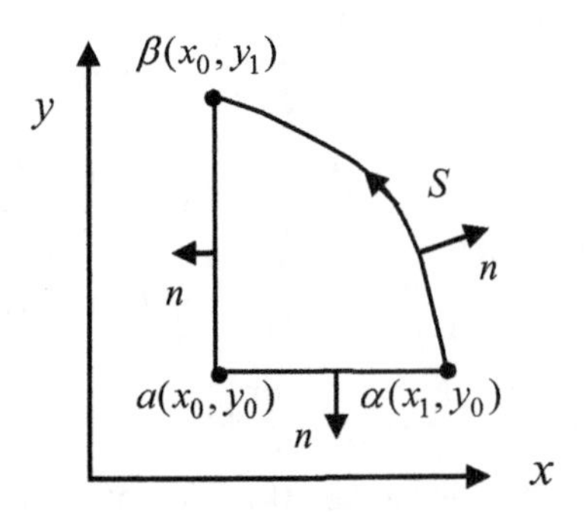

Figure 7.5 Domain of Riemann method

$$vF(u) = v\frac{\partial^2 u}{\partial x \partial y} + Av\frac{\partial u}{\partial x} + Bv\frac{\partial u}{\partial y} + Cvu \tag{7.446}$$

$$uG(v) = u\frac{\partial^2 v}{\partial x \partial y} - u\frac{\partial u}{\partial x}(Av) - u\frac{\partial u}{\partial y}(Bv) + Cuv = 0 \tag{7.447}$$

Subtracting (7.447) from (7.446), we find

$$vF(u) - uG(v) = v\frac{\partial^2 u}{\partial x \partial y} - u\frac{\partial^2 v}{\partial x \partial y} + \frac{\partial}{\partial x}(Auv) + \frac{\partial}{\partial y}(Buv) \tag{7.448}$$

The right hand side of (7.448) can be rearranged as

$$\begin{aligned}\frac{\partial P}{\partial x} + \frac{\partial Q}{\partial y} &= \frac{\partial}{\partial x}\{\frac{1}{2}(v\frac{\partial u}{\partial y} - u\frac{\partial v}{\partial y}) + Auv\} + \frac{\partial}{\partial y}\{\frac{1}{2}(v\frac{\partial u}{\partial x} - u\frac{\partial v}{\partial x}) + Buv\} \\ &= v\frac{\partial^2 u}{\partial x \partial y} - u\frac{\partial^2 v}{\partial x \partial y} + \frac{\partial}{\partial x}(Auv) + \frac{\partial}{\partial y}(Buv)\end{aligned} \tag{7.449}$$

where

$$P = \frac{1}{2}(v\frac{\partial u}{\partial y} - u\frac{\partial v}{\partial y}) + Auv \tag{7.450}$$

$$Q = \frac{1}{2}(v\frac{\partial u}{\partial x} - u\frac{\partial v}{\partial x}) + Buv \tag{7.451}$$

Equating (7.448) and (7.449) and integrating over the domain shown in Figure 7.5, we have

$$\iint [vF(u) - uG(v)]dxdy = \iint [\frac{\partial P}{\partial x} + \frac{\partial Q}{\partial y}]dxdy = \oint_S (Pdy - Qdx) \tag{7.452}$$

The line integral going anti-clockwise as shown in Figure 7.5 is taken as positive. Substituting (7.444) and (7.443) into (7.452), we get

$$\iint vfdxdy = \int_{\alpha\beta}(Pdy - Qdx) + \int_{\beta a}(Pdy - Qdx) + \int_{a\alpha}(Pdy - Qdx) \tag{7.453}$$

Since the boundary values of $\alpha\beta$ have been given, we only need to consider the line integrals along $a\alpha$ and βa as:

$$\int_{a\alpha}(Pdy-Qdx)=-\int_{a\alpha}Qdx=-\int_{a\alpha}[\frac{1}{2}\frac{\partial}{\partial x}(uv)+u(Bv-\frac{\partial v}{\partial x})]dx$$
$$=\frac{1}{2}(uv)_{\alpha}-\frac{1}{2}(uv)_{a}-\int_{a\alpha}u(Bv-\frac{\partial v}{\partial x})dx \tag{7.454}$$

$$\int_{\beta a}(Pdy-Qdx)=-\int_{\beta a}Pdy=-\int_{\beta a}[\frac{1}{2}\frac{\partial}{\partial y}(uv)+u(Av-\frac{\partial v}{\partial y})]dy$$
$$=\frac{1}{2}(uv)_{\beta}-\frac{1}{2}(uv)_{a}-\int_{\beta a}u(Av-\frac{\partial v}{\partial x})dx \tag{7.455}$$

where the subscripts α, β, a indicate the evaluation of the function at these points. Note that the direction cosines between the unit normal n and the characteristic lines $a\alpha$ and βa are both negative one. Now the boundary condition of the adjoint problem given in (7.444) can be chosen such that

$$G(v)=\frac{\partial^2 v}{\partial x\partial y}-\frac{\partial}{\partial x}(Av)-\frac{\partial}{\partial y}(Bv)+Cv=0 \tag{7.456}$$

$$Bv-\frac{\partial v}{\partial x}=0 \quad \text{on} \quad a\beta \tag{7.457}$$

$$Av-\frac{\partial v}{\partial y}=0 \quad \text{on} \quad \alpha a \tag{7.458}$$

Note that the last term in (7.454) and (7.455) becomes zero by such a choice for function v. The adjoint problem is introduced that (7.452) is valid. As we will see from the next chapter (7.452) is actually a generalized form of Green's theorem (the original Green's theorem is only for Laplace equation as in (8.37) given in Chapter 8). Equivalently, we have

$$G(v)=\frac{\partial^2 v}{\partial x\partial y}-\frac{\partial}{\partial x}(Av)-\frac{\partial}{\partial y}(Bv)+Cv=0 \tag{7.459}$$

$$v(x,y_0;x_0,y_0)=e^{\int_{x_0}^{x}B(\xi,y_0)d\xi} \quad \text{on} \quad y=y_0 \tag{7.460}$$

$$v(x_0,y;x_0,y_0)=e^{\int_{y_0}^{y}A(x_0,\zeta)d\zeta} \quad \text{on} \quad x=x_0 \tag{7.461}$$

$$v(a)=v(x_0,y_0;x_0,y_0)=1 \tag{7.462}$$

This function v is thus called Riemann's function. Substitution of (7.454) and (7.455) into (7.453) gives

$$u(a)=\frac{1}{2}(uv)_{\alpha}+\frac{1}{2}(uv)_{\beta}-\iint vfdxdy+\int_{\beta\alpha}(Pdy-Qdx) \tag{7.463}$$

This is the solution obtained by Riemann for the hyperbolic problem as long as the boundary data were not given on characteristics (i.e., S is not a characteristics). Note that Riemann's function is not symmetric with respect to x, y and x_0, y_0:

$$v(x,y;x_0,y_0)\neq v(x_0,y_0;x,y) \tag{7.464}$$

The equality in (7.464) only holds for the self-adjoint problem, but the original problem is not symmetric for general functions A and B.

This technique of using the Riemann function to solve the original hyperbolic PDE is called the Riemann method. The hyperbolic equation does not admit isolated singularities, and every singularity is continued along the characteristics. The key idea of Riemann's method is to reduce the BVP along the curved S to the BVP along two characteristics.

7.9 BIHARMONIC EQUATION

The biharmonic equation was found important in engineering applications. The biharmonic equation was obtained in the theory of thin plate bending, two-dimensional elastic stress analysis using Airy's stress function, two-dimensional highly viscous flow, and three-dimensional stress analysis of elastic solids. The analogy between plate bending and low Reynolds number viscous flow was noticed by Rayleigh in 1893, Sommerfeld in 1904, and Lamb in 1906. The development of theories for solving the biharmonic equation indeed closely relates to the analyses of Airy's stress function and plate bending. In 1892, Russian engineer Krylov visited Paris and posed the problem of the biharmonic equation to Hermite, and he related this problem to his son-in-law, Picard, another famous French mathematician. Consequently, the problem of solving the biharmonic equation in a rectangular domain was posed in Prix Vaillant in 1907 with prize money of 4000 francs. The initial judge panel included Poincare, Picard, and Painlevé and they wrote reports on the 12 submissions. The reports were then submitted to an authority commission including, Jordan, Appell, Humbert, Levy, Darboux, and Boussinesq. There were 4 winners who shared the prize money, and they are J. Hadamard, A. Korn, G Lauricella and T. Boggio. As a side note, W. Ritz (a student of David Hilbert) also submitted his paper but was "reported" missing and was not eventually awarded. It turns out that Ritz's analysis using a variational formulation related to plate bending was most influential for later development (see Chapter 14). It eventually becomes the focus of Krylov, who is the initiator of such analysis. Nevertheless, the biharmonic equation is clearly one of the most important PDEs higher than second order.

There are many famous mathematicians contributing to the development of theories for solving the biharmonic equation, including famous names like Airy, Maxwell, Clebsch, Ritz, Kirchhoff, Papkovich, Poisson, Sophie Germain, Love, Koialovich, Krylov, Navier, Boussinesq, Rayleigh, Sommerfeld, Galerkin, Dougall, Filon, Muskhelishvili, Levy, Michell, Pickett, and Timoshenko. Readers are referred to the excellent and comprehensive review article by Meleshko (2003) on the biharmonic equation.

The biharmonic equation can be expressed as a repeated application of the Laplacian operator as:

$$(\boldsymbol{\nabla}\cdot\boldsymbol{\nabla})(\boldsymbol{\nabla}\cdot\boldsymbol{\nabla})u = \nabla^2\nabla^2 u = 0 \tag{7.465}$$

In 3-D Cartesian coordinates, it can be expressed as

$$\nabla^4 u = (\frac{\partial^2}{\partial x^2}+\frac{\partial^2}{\partial y^2}+\frac{\partial^2}{\partial z^2})(\frac{\partial^2 u}{\partial x^2}+\frac{\partial^2 u}{\partial y^2}+\frac{\partial^2 u}{\partial z^2}) = 0 \tag{7.466}$$

More explicitly, it can be expressed as

$$\nabla^4 u = \frac{\partial^4 u}{\partial x^4} + \frac{\partial^4 u}{\partial y^4} + \frac{\partial^4 u}{\partial z^4} + 2\frac{\partial^2 u}{\partial x^2 \partial y^2} + 2\frac{\partial^2 u}{\partial x^2 \partial z^2} + 2\frac{\partial^2 u}{\partial z^2 \partial y^2} = 0 \tag{7.467}$$

In 2-D Cartesian coordinates, it is reduced to

$$\nabla_1^2 \nabla_1^2 u = \nabla_1^4 u = (\frac{\partial^2}{\partial x^2} + \frac{\partial^2}{\partial y^2})(\frac{\partial^2}{\partial x^2} + \frac{\partial^2}{\partial y^2})u = \frac{\partial^4 u}{\partial x^4} + 2\frac{\partial^4 u}{\partial x^2 \partial y^2} + \frac{\partial^4 u}{\partial y^4} = 0 \tag{7.468}$$

In cylindrical polar coordinates, we have

$$x = r\cos\theta, \quad y = r\sin\theta, \quad z = z \tag{7.469}$$

The corresponding biharmonic equation becomes

$$\nabla^4 u = (\frac{\partial^2}{\partial r^2} + \frac{1}{r}\frac{\partial}{\partial r} + \frac{1}{r^2}\frac{\partial^2}{\partial \theta^2} + \frac{\partial^2}{\partial z^2})(\frac{\partial^2}{\partial r^2} + \frac{1}{r}\frac{\partial}{\partial r} + \frac{1}{r^2}\frac{\partial^2}{\partial \theta^2} + \frac{\partial^2}{\partial z^2})u = 0 \tag{7.470}$$

where

$$r = \sqrt{x^2 + y^2} \tag{7.471}$$

In spherical polar coordinates, we have

$$x = R\sin\varphi\cos\theta, \quad y = R\sin\varphi\sin\theta, \quad z = R\cos\varphi \tag{7.472}$$

The biharmonic equation becomes

$$\begin{aligned} \nabla^4 u = &(\frac{\partial^2}{\partial R^2} + \frac{2}{R}\frac{\partial}{\partial R} + \frac{1}{R^2}\frac{\partial^2}{\partial \varphi^2} + \frac{\cot\varphi}{R^2}\frac{\partial}{\partial \varphi} + \frac{1}{R^2\sin^2\varphi}\frac{\partial^2}{\partial \theta^2}) \\ &\times(\frac{\partial^2}{\partial R^2} + \frac{2}{R}\frac{\partial}{\partial R} + \frac{1}{R^2}\frac{\partial^2}{\partial \varphi^2} + \frac{\cot\varphi}{R^2}\frac{\partial}{\partial \varphi} + \frac{1}{R^2\sin^2\varphi}\frac{\partial^2}{\partial \theta^2})u = 0 \end{aligned} \tag{7.473}$$

where

$$R = \sqrt{x^2 + y^2 + z^2} \tag{7.474}$$

7.9.1 Plane Elastic Stress Analysis for Solids

For two-dimensional solids under plane stress or plane strain conditions, the stress analysis can be formulated in Airy's stress function φ as:

$$\nabla^4 \varphi = -2(\frac{\kappa - 1}{\kappa + 1})\nabla^2 V \tag{7.475}$$

where V is the potential of body force and κ is defined as

$$\begin{aligned} \kappa &= \frac{3-\nu}{1+\nu}, \quad &\text{plane stress} \\ &= 3 - 4\nu, \quad &\text{plane strain} \end{aligned} \tag{7.476}$$

The proof of (7.475) can be found in Chau (2013) and will not be reported here. This is a nonhomogeneous biharmonic equation. In Cartesian coordinates, the stress components are defined in terms of φ

$$\sigma_{xx} = \frac{\partial^2 \varphi}{\partial y^2} + V, \quad \sigma_{yy} = \frac{\partial^2 \varphi}{\partial x^2} + V, \quad \sigma_{xy} = -\frac{\partial^2 \varphi}{\partial y \partial x}, \tag{7.477}$$

In polar coordinates, the stress components are defined in terms of φ

$$\sigma_{\phi\phi}=\frac{\partial^2\varphi}{\partial r^2}+V\,,\;\sigma_{r\phi}=-\frac{\partial}{\partial r}\Big(\frac{1}{r}\frac{\partial\varphi}{\partial\phi}\Big)\,,\quad \sigma_{rr}=\frac{1}{r}\frac{\partial\varphi}{\partial r}+\frac{1}{r^2}\frac{\partial^2\varphi}{\partial\phi^2}+V \qquad (7.478)$$

If body force is negligible, we recover the biharmonic equation

$$\nabla^4\varphi=0 \qquad (7.479)$$

The main difficulty encountered in solving Airy's stress function appears in satisfying the boundary condition.

7.9.2 Three-Dimensional Elastic Stress Analysis for Solids

For three-dimensional stress analysis in elastic solids, there are various formulations that can be used. They are classified into stress formulation and displacement formulation. The stress formulations include a Beltrami stress function, Maxwell stress function, and Morera stress function, while the displacement formulation includes Helmholtz decomposition, Lame strain potential for incompressible solids, the Galerkin vector, Love displacement potential for a cylindrical body, and Papkovitch-Neuber displacement potential (Chau, 2013). Among the displacement formulation, Helmholtz decomposition leads to one scalar biharmonic and one vector biharmonic equation as:

$$\nabla^4\phi=0\,,\;\;\nabla^4\psi=0 \qquad (7.480)$$

where the divergence of the vector ψ is zero (i.e., $\nabla\cdot\psi=0$). The displacement vector can be found as

$$\boldsymbol{u}=\nabla\phi+\nabla\times\psi \qquad (7.481)$$

Alternatively, displacement of elastic solids can also be expressed in terms of a single vector called the Galerkin vector $\boldsymbol{G}$. The governing equation for it is also a vector biharmonic equation

$$\nabla^4\boldsymbol{G}=0 \qquad (7.482)$$

For this case, the displacement vector is

$$\boldsymbol{u}=2(1-\nu)\nabla^2\boldsymbol{G}-\nabla\nabla\bullet\boldsymbol{G} \qquad (7.483)$$

For problems with axial symmetry, only one component of the Galerkin vector is needed, and it is known as Love's displacement potential. In particular, we have

$$\nabla^4 G_z=0 \qquad (7.484)$$

The displacement vector becomes

$$\boldsymbol{u}=\left[(1-2\nu)\nabla^2-\frac{\partial^2}{\partial z^2}\right]G_z\boldsymbol{e}_z-\frac{\partial^2 G_z}{\partial r\partial z}\boldsymbol{e}_r-\frac{1}{r}\frac{\partial^2 G_z}{\partial\theta\partial z}\boldsymbol{e}_\theta \qquad (7.485)$$

Thus, it can be seen that the biharmonic equation is closely related to three-dimensional stress analysis. The proofs for these formulations can be found in Chapter 4 of Chau (2013).

7.9.3 Bending of Thin Plates

The development of the theory for thin plate bending has a long history. Many great mathematicians were involved in its development, like Lagrange, Sophie Germain, Navier, Poisson, Kirchhoff, and Levy. The deflection of a thin plate is governed by the following nonhomogeneous biharmonic equation:

$$\nabla_1^2\nabla_1^2 w = \nabla_1^4 w = \frac{p(x,y)}{D} \tag{7.486}$$

where the 2-D Laplacian is defined in (7.468) and p is the distributed loading on the plate. The plate constant is defined as

$$D = \frac{Eh^3}{12(1-\nu^2)} \tag{7.487}$$

where h is the thickness of the plate, E is the Young's modulus, and ν is Poisson's ratio. In polar form, plate bending deflection is given by

$$\nabla_1^4 w = (\frac{\partial^2}{\partial r^2} + \frac{1}{r}\frac{\partial}{\partial r} + \frac{1}{r^2}\frac{\partial^2}{\partial \theta^2})(\frac{\partial^2}{\partial r^2} + \frac{1}{r}\frac{\partial}{\partial r} + \frac{1}{r^2}\frac{\partial^2}{\partial \theta^2})w = \frac{p}{D} \tag{7.488}$$

Bending of a thin plate finds applications in ship, aircraft, and building design. The proof of (7.488) can be found in the standard reference book by Timoshenko and Woinowsky-Krieger (1959).

7.9.4 Two-Dimensional Viscous Flow with Low Reynolds Number

When flow velocity is small in a highly viscous incompressible flow, Lamb (1932) showed that the stream function satisfies the biharmonic equation if the viscous effect is much larger than the inertia effect (see p. 607 of Lamb, 1932). This is also known as Stokes flow in the literature. In particular, the stream function ψ satisfies the biharmonic equation:

$$\nabla_1^4 \psi = 0 \tag{7.489}$$

where ψ is defined in terms of the velocity components u and v as

$$v = \frac{\partial \psi}{\partial x}, \quad u = -\frac{\partial \psi}{\partial y} \tag{7.490}$$

We sketch briefly the proof of (7.489) here. First, let us recall from (2.121) the material derivative as

$$\frac{D\boldsymbol{u}}{Dt} = \frac{\partial \boldsymbol{u}}{\partial t} + \nabla(\frac{|\boldsymbol{u}|^2}{2}) - \boldsymbol{u}\times\nabla\times\boldsymbol{u} \tag{7.491}$$

This can be rewritten as:

$$\frac{D\boldsymbol{u}}{Dt} = \frac{\partial \boldsymbol{u}}{\partial t} + \frac{1}{2}\nabla|\boldsymbol{u}|^2 - 2\boldsymbol{u}\times\boldsymbol{\omega} \tag{7.492}$$

where $\boldsymbol{\omega}$ is the vorticity of the flow and defined as:

$$\boldsymbol{\omega} = \frac{1}{2}\nabla\times\boldsymbol{u} \tag{7.493}$$

Physically, it is the angular velocity of the fluid element that would rotate if it were suddenly solidified. If the motion of the fluid is driven by an external force $\boldsymbol{F}$ and a pressure difference, the equation of motion becomes

$$\frac{D\boldsymbol{u}}{Dt}=\boldsymbol{F}-\frac{1}{\rho}\nabla p \tag{7.494}$$

where p and ρ are the fluid pressure and fluid density respectively. Combining (7.492) and (7.494) gives

$$\frac{\partial \boldsymbol{u}}{\partial t}+\frac{1}{2}\nabla|\boldsymbol{u}|^2-2\boldsymbol{u}\times\boldsymbol{\omega}-\boldsymbol{F}+\frac{1}{\rho}\nabla p=0 \tag{7.495}$$

Assuming the existence of an external force potential, we can rewrite $\boldsymbol{F}$ as

$$\boldsymbol{F}=-\nabla V \tag{7.496}$$

Next, we assume the flow is incompressible (ρ = constant), or

$$\frac{\partial \boldsymbol{u}}{\partial t}-2\boldsymbol{u}\times\boldsymbol{\omega}+\nabla(\frac{1}{2}|\boldsymbol{u}|^2+V+\frac{1}{\rho}p)=0 \tag{7.497}$$

If the viscosity in fluid is nonzero, the fluid pressure is no longer normal to the surface. Thus, we can add viscosity effect as

$$\frac{\partial \boldsymbol{u}}{\partial t}-2\boldsymbol{u}\times\boldsymbol{\omega}+\nabla(\frac{1}{2}|\boldsymbol{u}|^2+V+\frac{1}{\rho}p)+2\nu\nabla\times\boldsymbol{\omega}=0 \tag{7.498}$$

where ν is the kinematic viscosity and is defined as the coefficient of viscosity μ divided by the density ρ as

$$\nu=\frac{\mu}{\rho} \tag{7.499}$$

Next, we recall a vector identity that

$$\nabla^2\boldsymbol{u}=\nabla(\nabla\cdot\boldsymbol{u})-2\nabla\times\boldsymbol{\omega} \tag{7.500}$$

The proof of this identity will be left as an exercise for readers (see Problem 7.12). Incompressibility also implies the trace of velocity is zero, or the first term on the right of (7.500) is zero. Substitution of (7.500) into (7.498) gives

$$\frac{\partial \boldsymbol{u}}{\partial t}-2\boldsymbol{u}\times\boldsymbol{\omega}+\nabla(\frac{1}{2}|\boldsymbol{u}|^2+V+\frac{1}{\rho}p)-\nu\nabla^2\boldsymbol{u}=0 \tag{7.501}$$

Now consider a special case that

$$\boldsymbol{u}=(u,v,0),\quad \boldsymbol{\omega}=(0,0,\zeta) \tag{7.502}$$

Then, (7.501) gives two equations

$$\frac{\partial u}{\partial t}-2v\zeta+\frac{\partial\Omega}{\partial x}-\nu\nabla^2 u=0 \tag{7.503}$$

$$\frac{\partial v}{\partial t}+2u\zeta+\frac{\partial\Omega}{\partial y}-\nu\nabla^2 v=0 \tag{7.504}$$

where Ω can be identified easily from (7.498). It is not given here as we will eliminate it. Elimination of Ω from (7.503) and (7.504) gives

$$\frac{\partial\zeta}{\partial t}+u\frac{\partial\zeta}{\partial x}+v\frac{\partial\zeta}{\partial y}=\nu\nabla_1^2\zeta \tag{7.505}$$

In obtaining (7.505) we have used the definition of $\boldsymbol{\omega}$ given in (7.493) to get

$$\zeta = \frac{1}{2}(\frac{\partial v}{\partial x} - \frac{\partial u}{\partial y}) \tag{7.506}$$

Finally, we now introduce a stream function defined as

$$u = -\frac{\partial \psi}{\partial y}, \quad v = \frac{\partial \psi}{\partial x} \tag{7.507}$$

We note that this stream function automatically satisfies the incompressibility:

$$\nabla \cdot \boldsymbol{u} = \frac{\partial u}{\partial x} + \frac{\partial v}{\partial y} = -\frac{\partial^2 \psi}{\partial x \partial y} + \frac{\partial^2 \psi}{\partial y \partial x} = 0 \tag{7.508}$$

Substitution of (7.507) into (7.506) gives

$$\zeta = \frac{1}{2}\nabla_1^2 \psi \tag{7.509}$$

Finally, combining (7.505), (7.507), and (7.509) gives

$$\frac{\partial}{\partial t}\nabla_1^2 \psi + (\frac{\partial \psi}{\partial x}\frac{\partial}{\partial y} - \frac{\partial \psi}{\partial y}\frac{\partial}{\partial x})\nabla_1^2 \psi = \nu \nabla_1^4 \psi \tag{7.510}$$

This is a nonlinear PDE and not easy to solve. However, for slow moving fluid (such that ψ is small) with high viscosity ν, the terms on the left hand side of (7.510) are negligible compared to that on the right, and we finally obtain a biharmonic equation

$$\nabla_1^4 \psi = 0 \tag{7.511}$$

7.9.5 Uniqueness of the Solution of Biharmonic Equation

In this section, we will consider the uniqueness of the solution of the following two-dimensional biharmonic equation with boundary conditions

$$\nabla_1^4 u = \frac{\partial^4 u}{\partial x^4} + 2\frac{\partial^4 u}{\partial x^2 \partial y^2} + \frac{\partial^4 u}{\partial y^4} = 0 \tag{7.512}$$

$$u|_\Gamma = g(s), \quad \left.\frac{\partial u}{\partial n}\right|_\Gamma = h(s) \tag{7.513}$$

where s is the tangential coordinate along the boundary Γ of the two-dimensional domain. The following uniqueness proof was presented by Hua (2009, 2012). Suppose that u_1 and u_2 are two distinct solutions of the systems (7.512) and (7.513). We define a new function:

$$v = u_2 - u_1 \tag{7.514}$$

Taking the biharmonic operator to (7.514) gives

$$\nabla^4 v = \nabla^4 u_2 - \nabla^4 u_1 = 0 \tag{7.515}$$

The corresponding boundary conditions for v can be found as

$$v|_\Gamma = u_2|_\Gamma - u_1|_\Gamma = g(s) - g(s) = 0 \tag{7.516}$$

$$\left.\frac{\partial v}{\partial n}\right|_\Gamma = \left.\frac{\partial u_2}{\partial n}\right|_\Gamma - \left.\frac{\partial u_1}{\partial n}\right|_\Gamma = h(s) - h(s) = 0 \tag{7.517}$$

Now, we recall Green's theorem (or called Green's second identity) for the Laplacian operator:

$$\iiint_{\Omega}(\psi\nabla^2\varphi-\varphi\nabla^2\psi)d\Omega=\iint_{\Gamma}(\psi\frac{\partial\varphi}{\partial n}-\varphi\frac{\partial\psi}{\partial n})dS \tag{7.518}$$

The proof of this theorem will be given in the next chapter (see (8.37)). Let us make the following identification for ψ and φ:

$$\varphi=v,\quad \psi=\nabla^2 v \tag{7.519}$$

With this identification, (7.518) becomes

$$\iiint_{\Omega}(\nabla^2 v\nabla^2 v-v\nabla^2\nabla^2 v)d\Omega=\iint_{\Gamma}[\nabla^2 v\frac{\partial v}{\partial n}-v\frac{\partial}{\partial n}(\nabla^2 v)]dS \tag{7.520}$$

The second term on the left hand side is identically zero, and both terms on the right hand side are zeros in view of (7.516) and (7.517). Thus, we have

$$\iiint_{\Omega}(\nabla^2 v)^2 d\Omega=0 \tag{7.521}$$

Therefore, we must have

$$\nabla^2 v=0 \tag{7.522}$$

with

$$v\big|_{\Gamma}=0,\quad \frac{\partial v}{\partial n}\bigg|_{\Gamma}=0 \tag{7.523}$$

From the extremal principle to be derived in Section 9.7.7 of Chapter 9 for the Laplace equation, the maximum and minimum of v can only appear on the boundary. Thus, we must have v be identically zero. Then, (7.514) leads to

$$u_2=u_1 \tag{7.524}$$

This contradicts with our assumption that they are distinct solutions. Thus, the solution of biharmonic equation with boundary condition (7.497) must be unique.

Note the special case of $g = h = 0$, the same result was derived by Fugdele (1981) (see Footnote 1 on p. 450 of Fugdele, 1981).

7.9.6 Biharmonic Functions and Almansi Theorems

For the two-dimensional polar form, the general solution can be expressed as (Chau, 2013; Meleshko, 2003):

$$\begin{aligned}\varphi(r,\phi)=&A_0 r^2\ln r\phi+B_0\ln r\phi+C_0 r\ln r\phi\cos\phi+D_0 r\ln r\phi\sin\phi\\&+(A+Br^2)\ln(\frac{r}{R})+Cr^2+D+(E+Fr^2)\phi+r\phi(G\cos\phi+H\sin\phi)\\&+(A_1 r^3+B_1 r^{-1}+C_1 r\ln(\frac{r}{R}))(E_1\cos\phi+F_1\sin\phi)\\&+\sum_{n=2}^{\infty}[A_n r^{n+2}+B_n r^{-n}+C_n r^n+D_n r^{2-n}](E_n\cos n\phi+F_n\sin n\phi)\end{aligned} \tag{7.525}$$

For two-dimensional Cartesian coordinates, the general solution of the biharmonic equation can be expressed in two analytic functions ϕ and χ (Chau, 2013):

$$\varphi(x, y) = \mathrm{Re}[\bar{z}\phi(z) + \chi(z)] \tag{7.526}$$

where the complex variable z is defined as

$$z = x + iy \tag{7.527}$$

Alternatively, Chau (2013) mentioned two Almansi theorems in generating biharmonic functions in terms of harmonic functions. As discussed by Chau (2013) in Sections 4.5–4.7 in his book, there are infinite numbers of harmonic functions (i.e., solution of Laplace equation), and, thus, there are also infinite biharmonic functions. According to Meleshko (2003), Boussinesq actually derived the so-called Almansi theorem independently in 1885 before Almansi did in 1896. However, the Boussinesq name was not associated with these theorems. The proofs of Almansi theorems were given in Fung (1965). In this section, we will provide a simpler version of proofs given by Hua (2009, 2012).

The first Almansi theorem states that the following function is biharmonic

$$u = xu_1 + u_2 \tag{7.528}$$

where u_1 and u_2 are both harmonic. To prove this theorem, let us assume that u is biharmonic and u_2 is harmonic, and we want to show the following

$$\nabla^2 u_1 = 0 \tag{7.529}$$

Since u_2 is assumed as harmonic, from (7.528) we must have

$$\nabla^2 (u - xu_1) = 0 \tag{7.530}$$

Equivalently, it can be expressed as

$$\nabla^2 u = \nabla^2 (xu_1) = 2\frac{\partial u_1}{\partial x} + x(\frac{\partial^2 u_1}{\partial x^2} + \frac{\partial^2 u_1}{\partial y^2}) = 2\frac{\partial u_1}{\partial x} \tag{7.531}$$

In obtaining the last part of (7.531), we have employed the fact that u_1 is harmonic. First, we observe that a solution of (7.531) is

$$\overset{*}{u_1}(x, y) = \int_{x_0}^{x} \frac{1}{2} \nabla^2 u(\xi, y) d\xi \tag{7.532}$$

The validity of this solution can be proved by using Leibniz's rule of differentiation under the integral sign. Let us take the Laplacian of (7.531); we get

$$\frac{1}{2}\nabla^2\nabla^2 u = \nabla^2 \frac{\partial \overset{*}{u_1}}{\partial x} = \frac{\partial}{\partial x}\nabla^2 \overset{*}{u_1} = 0 \tag{7.533}$$

In getting the last part of (7.533), we observe that u is biharmonic. This leads to the fact that

$$\nabla^2 \overset{*}{u_1} = v(y) \tag{7.534}$$

Clearly, we can define another solution of u_1 as:

$$\nabla^2 \overset{**}{u_1} = \frac{\partial \overset{**}{u_1}}{\partial y^2} = -v(y) \tag{7.535}$$

Then, we can select u_1 as:

$$u_1 = \overset{**}{u_1} + \overset{*}{u_1} \tag{7.536}$$

Then, it is obvious that

$$\nabla^2 u_1 = \nabla^2 \overset{**}{u_1} + \nabla^2 \overset{*}{u_1} = -v(y) + v(y) = 0 \tag{7.537}$$

This completes the proof.

The second Almansi theorem states that the following function is biharmonic

$$u = (r^2 - r_0^2)u_1 + u_2 \tag{7.538}$$

where u_1 and u_2 are both harmonic. For u is biharmonic and u_2 is harmonic, we want to prove that u_1 is also harmonic. To prove this, we first recall the identity (Eq. (1.43) of Chau, 2013):

$$\begin{aligned}\nabla^2(\psi\varphi) &= \varphi\nabla^2\psi + 2(\nabla\psi)\bullet(\nabla\varphi) + \psi\nabla^2\varphi \\ &= \varphi\nabla^2\psi + \psi\nabla^2\varphi + 2\{\frac{\partial\varphi}{\partial x}\frac{\partial\psi}{\partial x} + \frac{\partial\varphi}{\partial y}\frac{\partial\psi}{\partial y}\}\end{aligned} \tag{7.539}$$

Taking the Laplacian of (7.538) and applying (7.539), we have

$$\begin{aligned}\nabla^2 u &= \nabla^2[(r^2 - r_0^2)u_1 + u_2] = \nabla^2[(r^2 - r_0^2)u_1] \\ &= (r^2 - r_0^2)\nabla^2 u_1 + u_1\nabla^2(r^2 - r_0^2) + 2\{\frac{\partial u_1}{\partial x}\frac{\partial r^2}{\partial x} + \frac{\partial u_1}{\partial y}\frac{\partial r^2}{\partial y}\} \\ &= 4u_1 + 4(x\frac{\partial u_1}{\partial x} + y\frac{\partial u_1}{\partial y})\end{aligned} \tag{7.540}$$

In obtaining (7.540), we have noted the following identity:

$$\nabla^2 r^2 = 4 \tag{7.541}$$

Taking the Laplacian of (7.540), we get

$$\nabla^2\nabla^2 u = 4\nabla^2 u_1 + 4[\nabla^2(x\frac{\partial u_1}{\partial x}) + \nabla^2(y\frac{\partial u_1}{\partial y})] = 0 \tag{7.542}$$

It is straightforward to show that

$$\nabla^2(x\frac{\partial u_1}{\partial x}) = 2\frac{\partial^2 u_1}{\partial x^2} + x\frac{\partial}{\partial x}(\nabla^2 u_1) \tag{7.543}$$

$$\nabla^2(y\frac{\partial u_1}{\partial y}) = 2\frac{\partial^2 u_1}{\partial y^2} + y\frac{\partial}{\partial y}(\nabla^2 u_1) \tag{7.544}$$

Substitution of (7.543) and (7.544) into (7.542)

$$12\nabla^2 u_1 + 4(x\frac{\partial}{\partial x} + y\frac{\partial}{\partial y})\nabla^2 u_1 = 0 \tag{7.545}$$

Thus, we finally get

$$\nabla^2 u_1 = 0 \tag{7.546}$$

This completes the proof.

7.9.7 Solution of Circular Domain

For the two-dimensional unit circular domain, the biharmonic equation becomes

$$\nabla^4 u = (\frac{\partial^2}{\partial r^2} + \frac{1}{r}\frac{\partial}{\partial r} + \frac{1}{r^2}\frac{\partial^2}{\partial\theta^2})(\frac{\partial^2}{\partial r^2} + \frac{1}{r}\frac{\partial}{\partial r} + \frac{1}{r^2}\frac{\partial^2}{\partial\theta^2})u = 0 \tag{7.547}$$

subjected to the following boundary conditions:

$$u\big|_{r=1} = g(\theta), \quad \frac{\partial u}{\partial r}\bigg|_{r=1} = h(\theta) \tag{7.548}$$

Recall from the second Almansi theorem that the biharmonic function can be expressed in terms of harmonic functions u_1 and u_2 as

$$u = (r^2 - 1)u_1 + u_2 \tag{7.549}$$

Applying boundary conditions (7.548) to (7.549) gives

$$u_2\big|_{r=1} = g(\theta), \quad \frac{\partial u}{\partial r}\bigg|_{r=1} = [2u_1 + \frac{\partial u_2}{\partial r}]_{r=1} = h(\theta) \tag{7.550}$$

For a unit circular domain, the harmonic function subject to the boundary condition given in the first equation of (7.550), there is a well-known Poisson integral formula for the Laplace equation (details of this formula will be discussed in Section 9.7.6 of Chapter 9):

$$u_2(r,\theta) = \frac{1}{2\pi}\int_{-\pi}^{\pi} \frac{(1-r^2)g(\psi)}{1+r^2 - 2r\cos(\theta-\psi)} d\psi \tag{7.551}$$

The most important step is to show that the following function is harmonic

$$\nabla^2(2u_1 + r\frac{\partial u_2}{\partial r}) = \nabla^2(r\frac{\partial u_2}{\partial r}) = 0 \tag{7.552}$$

To show the validity of (7.552), we have

$$\frac{\partial^2}{\partial r^2}(r\frac{\partial u_2}{\partial r}) = 2\frac{\partial^2 u_2}{\partial r^2} + r\frac{\partial^3 u_2}{\partial r^3} \tag{7.553}$$

$$\frac{1}{r}\frac{\partial}{\partial r}(r\frac{\partial u_2}{\partial r}) = \frac{1}{r}\frac{\partial u_2}{\partial r} + \frac{\partial^2 u_2}{\partial r^2} \tag{7.554}$$

$$\frac{1}{r^2}\frac{\partial^2}{\partial \theta^2}(r\frac{\partial u_2}{\partial r}) = \frac{1}{r}\frac{\partial}{\partial r}(\frac{\partial^2 u_2}{\partial \theta^2}) \tag{7.555}$$

Combining (7.553) to (7.555), we get the Laplacian of the second term in (7.552) as

$$\nabla^2(r\frac{\partial u_2}{\partial r}) = 3\frac{\partial^2 u_2}{\partial r^2} + r\frac{\partial^3 u_2}{\partial r^3} + \frac{1}{r}\frac{\partial u_2}{\partial r} + \frac{1}{r}\frac{\partial}{\partial r}(\frac{\partial^2 u_2}{\partial \theta^2}) \tag{7.556}$$

Next, we find that

$$r\frac{\partial}{\partial r}(\nabla^2 u_2) = r\frac{\partial^3 u_2}{\partial r^3} + \frac{1}{r}\frac{\partial}{\partial r}(\frac{\partial^2 u_2}{\partial \theta^2}) + \frac{\partial^2 u_2}{\partial r^2} - \frac{1}{r}\frac{\partial u_2}{\partial r} - \frac{2}{r}\frac{\partial^2 u_2}{\partial \theta^2} = 0 \tag{7.557}$$

It is because u_2 is a harmonic function. Therefore, we have

$$r\frac{\partial^3 u_2}{\partial r^3} + \frac{1}{r}\frac{\partial}{\partial r}(\frac{\partial^2 u_2}{\partial \theta^2}) = -\frac{\partial^2 u_2}{\partial r^2} + \frac{1}{r}\frac{\partial u_2}{\partial r} + \frac{2}{r}\frac{\partial^2 u_2}{\partial \theta^2} \tag{7.558}$$

Substitution of this result into (7.556) gives

$$\nabla^2(r\frac{\partial u_2}{\partial r}) = 2(\frac{\partial^2 u_2}{\partial r^2} + + \frac{1}{r}\frac{\partial u_2}{\partial r} + \frac{1}{r^2}\frac{\partial^2 u_2}{\partial \theta^2}) = 2\nabla^2 u_2 = 0 \tag{7.559}$$

This completes the proof of (7.552). Then, we can apply the Poisson integral formula for the Laplacian again:

$$2u_1 + r\frac{\partial u_2}{\partial r} = \frac{1}{2\pi}\int_{-\pi}^{\pi}\frac{(1-r^2)h(\psi)}{1+r^2-2r\cos(\theta-\psi)}d\psi \tag{7.560}$$

Differentiation of (7.551) with respect to r gives

$$\frac{\partial u_2}{\partial r} = \frac{1}{2\pi}\int_{-\pi}^{\pi} 2\frac{(1+r^2)\cos(\theta-\psi)-2r}{1+r^2-2r\cos(\theta-\psi)}g(\psi)d\psi \tag{7.561}$$

Substitution of (7.561) into (7.560) gives

$$u_1 = -\frac{r}{2\pi}\int_{-\pi}^{\pi}\frac{(1+r^2)\cos(\theta-\psi)-2r}{1+r^2-2r\cos(\theta-\psi)}g(\psi)d\psi + \frac{1}{4\pi}\int_{-\pi}^{\pi}\frac{(1-r^2)h(\psi)}{1+r^2-2r\cos(\theta-\psi)}d\psi \tag{7.562}$$

Finally, substitution of (7.551) and (7.562) into (7.549) gives

$$u = \frac{(r^2-1)^2}{2\pi}\{-\frac{1}{2}\int_{-\pi}^{\pi}\frac{h(\psi)}{1+r^2-2r\cos(\theta-\psi)}d\psi + \int_{-\pi}^{\pi}\frac{[1-r\cos(\theta-\psi)]g(\psi)}{[1+r^2-2r\cos(\theta-\psi)]^2}d\psi\} \tag{7.563}$$

Note that this formula was derived in Hua (2009). The kernel functions in Hua's integral are illustrated in Figure 7.6.

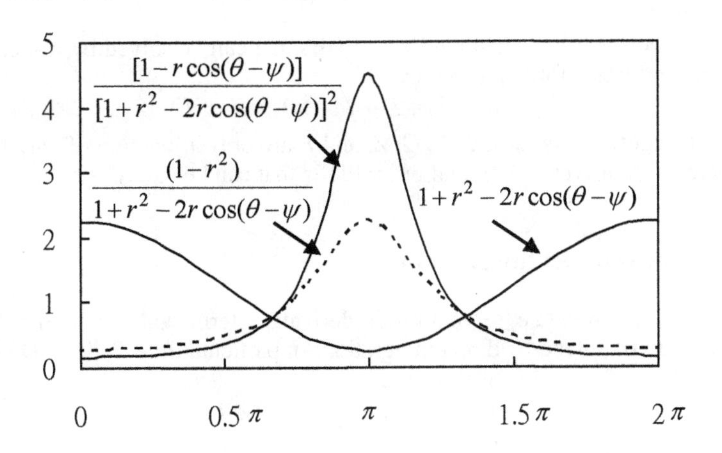

Figure 7.6 Kernel function for Hua's integral

This shows that the solution at any point inside the circular domain are functions of the weighted average of the boundary values $h(\psi)$ and $g(\psi)$. Actually, this four-volume series by Hua was originally published in 1962 (Volumes 1 and 2), 1978 (Volume 3) and 1981 (Volume 4). The 2009 edition was a reprint celebrating the 100th anniversary of Professor Hua Loo-Keng's birthday. This 2009 edition has been translated into English (Hua, 2012). The same result given in (7.563) has been re-derived by Dong et al. (2005) using the result of Zheng and Zheng (2000). Neither of them cited the book by Hua (earlier versions), and their derivations are more complicated than the analysis given here. It can be seen that Hua's recognition of the validity and usefulness of (7.538) simplifies the analysis tremendously. As a side note, we should mention that Hua Loo-Keng is the most influential mathematician of modern China. He did not have a formal education, and he is an excellent self-learner. A full biography is given at the end of this book.

7.10 SECOND ORDER PDE WITH NON-CONSTANT COEFFICIENTS

The main contributor for general nonlinear second order PDEs is Gaspard Monge. For PDEs of second and higher order, we normally use the following notations for the first and second derivatives:

$$r = \frac{\partial^2 z}{\partial x^2}, \quad s = \frac{\partial^2 z}{\partial x \partial y}, \quad t = \frac{\partial^2 z}{\partial y^2}, \quad p = \frac{\partial z}{\partial x}, \quad q = \frac{\partial z}{\partial y} \tag{7.564}$$

Some second order PDEs can be integrated immediately by inspection. Here are some examples. For the case of two variables, a general PDE of the second order can be expressed as:

$$F(x, y, z, r, s, t, p, q) = 0 \tag{7.565}$$

This PDE is in general nonlinear and very difficult to solve. In the following sections, we will consider some particular forms for which analytic methods can be used to find the solution.

First, we will look at some linear PDEs that can be solved by inspection. The most general linear form is given by

$$Rr + Ss + Tt + Pp + Qq + Zz = F \tag{7.566}$$

where the coefficients R, S, T, P, Q, Z, and F are only a function of variables x and y. There are four types of second order PDEs that can be solved easily.

7.10.1 Type I (Direct Integration)

In Type I, if only one second-order derivative term appears in our PDE, the problem can be solved by direct integration. In particular, the Type I PDE is

$$r = \frac{\partial^2 z}{\partial x^2} = \frac{F}{R} = F_1(x, y) \tag{7.567}$$

$$s = \frac{\partial^2 z}{\partial x \partial y} = \frac{F}{S} = F_2(x, y) \tag{7.568}$$

$$t = \frac{\partial^2 z}{\partial y^2} = \frac{F}{T} = F_3(x, y) \tag{7.569}$$

7.10.2 Type II (ODE of p or q)

If there is only one second order derivative term together with one first derivative term (either p or q) such that it becomes an ODE in either p or q. In particular, the Type II PDE appears as

$$Rr + Pp = R\frac{\partial p}{\partial x} + Pp = F(x, y) \tag{7.570}$$

$$Ss + Pp = S\frac{\partial p}{\partial y} + Pp = F(x, y) \tag{7.571}$$

$$Ss + Qq = S\frac{\partial q}{\partial y} + Qq = F(x, y) \tag{7.572}$$

$$Tt + Qq = T\frac{\partial q}{\partial y} + Qq = F(x, y) \tag{7.573}$$

Note that all of these become a first order ODE for either p or q.

7.10.3 Type III (First Order PDE)

Certain special cases of (7.566) can be expressed as a first order PDE for either p or q. They are

$$Rr + Ss + Pp = R\frac{\partial p}{\partial x} + S\frac{\partial p}{\partial y} + Pp = F(x, y) \tag{7.574}$$

$$Ss + Tt + Qq = S\frac{\partial q}{\partial x} + T\frac{\partial q}{\partial y} Pp + Qq = F(x, y) \tag{7.575}$$

Both of these are linear first order PDEs. The Lagrange method discussed in Section 6.7 of Chapter 6 can be used to solve them.

7.10.4 Type IV (Second Order ODE)

Certain special cases of (7.566) can be expressed as a second order PDE for either p or q. They are

$$Rr + Pp + Zz = R\frac{\partial^2 z}{\partial x^2} + P\frac{\partial z}{\partial x} + Zz = F(x, y) \tag{7.576}$$

$$Tt + Qq + Zz = T\frac{\partial^2 z}{\partial y^2} + Q\frac{\partial z}{\partial y} + Zz = F(x, y) \tag{7.577}$$

These are second order ODEs for z.

Example 7.4 Consider the solution of the following PDE:

$$\frac{\partial^2 z}{\partial x \partial y} = 2x + 2y \tag{7.578}$$

Solution: Integrating with respect to x, we get

$$\frac{\partial z}{\partial y} = x^2 + 2xy + \phi(y) \tag{7.579}$$

Integrating (7.579) with respect to y, we find

$$z = x^2 y + xy^2 + \int \phi(y) dy + \psi(x) \tag{7.580}$$

Since both ϕ and ψ are arbitrary functions, we can rewrite it as

$$z = x^2 y + xy^2 + f(x) + F(y) \tag{7.581}$$

Example 7.5 Consider the solution of the following PDE:

$$x\frac{\partial^2 z}{\partial x^2} + 2\frac{\partial z}{\partial x} = 0 \tag{7.582}$$

Solution: By using the definition of p given in (7.564), we have

$$x\frac{\partial p}{\partial x} + 2p = 0 \tag{7.583}$$

This can be rearranged as

$$\frac{\partial p}{p} = -\frac{2\partial x}{x} \tag{7.584}$$

Integration gives

$$p = \frac{f(y)}{x^2} \tag{7.585}$$

where f is an arbitrary function of y. Using the definition of p in (7.564), we get

$$\frac{\partial z}{\partial x} = p = \frac{1}{x^2} f(y) \tag{7.586}$$

Integrating (7.586) with respect to x, we find

$$z = -\frac{1}{x} f(y) + F(y) \tag{7.587}$$

7.11 MONGE AND MONGE-AMPERE METHODS

Recall from (7.565) that the following general second order PDE is considered:

$$F(x, y, z, r, s, t, p, q) = 0 \tag{7.588}$$

We will consider a general method in this section, called Monge's method for the case of quasi-linear. The extension of this method, called the Monge-Ampere method, will also be considered for the case of nonlinear.

7.11.1 Some Examples

Let us consider the following particular problem, and then we will generalize the observation to the so-called Monge method. In particular, a first order PDE is given as

$$2\frac{\partial z}{\partial x}x-\frac{\partial z}{\partial y}y=2px-qy=\phi(x^2y) \tag{7.589}$$

Differentiation of (7.589) with respect to x, we have

$$2\frac{\partial p}{\partial x}x+2p-\frac{\partial q}{\partial x}y=2xy\phi'(x^2y) \tag{7.590}$$

Using the definition given in (7.564), we can write (7.590) as

$$2rx+2p-sy=2xy\phi'(x^2y) \tag{7.591}$$

Differentiation of (7.589) with respect to y, we have

$$2sx-ty-q=x^2\phi'(x^2y) \tag{7.592}$$

Eliminating the arbitrary function from (7.591) and (7.592), we get

$$(2sx-ty-q)2y=x(2rx+2p-sy) \tag{7.593}$$

Rearranging (7.593), we obtain

$$2rx^2-5xys+2y^2t+2(px+qy)=0 \tag{7.594}$$

In explicit form, (7.574) is

$$2\frac{\partial^2 z}{\partial x^2}x^2-5xy\frac{\partial^2 z}{\partial x\partial y}+2y^2\frac{\partial^2 z}{\partial y^2}+2(\frac{\partial z}{\partial x}x+\frac{\partial z}{\partial y}y)=0 \tag{7.595}$$

Note that this is a linear second order PDE in r, s, and t. Inversely, we can consider (7.589) as a partial integral of PDE in (7.593). However, we can show that this is not the only partial integral of (7.595). In fact, applying the procedure in getting (7.595) to the following first order PDE, we find the same (7.595):

$$px-2qy=\psi(xy^2) \tag{7.596}$$

That is, two different first order PDEs lead to the same second order PDE. This is what Monge observed and this is the basis of formulating Monge's method.

For a second example, we consider a nonlinear first order PDE

$$p^2+q=\phi(2x+y) \tag{7.597}$$

Differentiation with respect to x and y respectively gives

$$2pr+s=2\phi'(2x+y) \tag{7.598}$$

$$2ps+t=\phi'(2x+y) \tag{7.599}$$

Eliminating the arbitrary function ϕ, we find

$$-2pr+(4p-1)s+2t=0 \tag{7.600}$$

This is a linear PDE in r, s, and t. In explicit form, it is

$$-2\frac{\partial z}{\partial x}\frac{\partial^2 z}{\partial x^2}+(4\frac{\partial z}{\partial x}-1)\frac{\partial^2 z}{\partial x\partial y}+2\frac{\partial^2 z}{\partial y^2}=0 \tag{7.601}$$

Finally, we consider the following example

$$y-p=\phi(x-q) \tag{7.602}$$

Differentiation with respect to x and y, respectively, gives

$$-r = (1-s)\phi'(x-q) \tag{7.603}$$

$$1-s = -t\phi'(x-q) \tag{7.604}$$

Eliminating the arbitrary function ϕ, we find

$$2s + (rt - s^2) = 1 \tag{7.605}$$

Note that the terms in the bracket are nonlinear in r, s, and t. This is equivalent to:

$$2\frac{\partial^2 z}{\partial x \partial y} + [\frac{\partial^2 z}{\partial x^2}\frac{\partial^2 z}{\partial y^2} - (\frac{\partial^2 z}{\partial x \partial y})^2] = 1 \tag{7.606}$$

7.11.2 Generalized Form

Guided by examples in the previous section, we will consider the more general form in this section. More specifically, two first integrals of the second order PDE given in (7.588) are assumed as:

$$u = u(x, y, z, p, q) \tag{7.607}$$

$$v = v(x, y, z, p, q) \tag{7.608}$$

In general, we can write intermediate integral of (7.588)

$$u = \phi(v) \tag{7.609}$$

That is, if both (7.607) and (7.608) correspond to the same second order PDE, u and v must relate to each other. Taking the total differential of (7.609), we find

$$\begin{aligned} &\frac{\partial u}{\partial x}dx + \frac{\partial u}{\partial y}dy + \frac{\partial u}{\partial z}dz + \frac{\partial u}{\partial p}dp + \frac{\partial u}{\partial q}dq \\ &= \phi'(v)\{\frac{\partial v}{\partial x}dx + \frac{\partial v}{\partial y}dy + \frac{\partial v}{\partial z}dz + \frac{\partial v}{\partial p}dp + \frac{\partial v}{\partial q}dq\} \end{aligned} \tag{7.610}$$

Dividing through dx and dy respectively, we find

$$\frac{\partial u}{\partial x} + \frac{\partial u}{\partial z}p + \frac{\partial u}{\partial p}r + \frac{\partial u}{\partial q}s = \phi'(v)\{\frac{\partial v}{\partial x} + \frac{\partial v}{\partial z}p + \frac{\partial v}{\partial p}r + \frac{\partial v}{\partial q}s\} \tag{7.611}$$

$$\frac{\partial u}{\partial y} + \frac{\partial u}{\partial z}q + \frac{\partial u}{\partial p}s + \frac{\partial u}{\partial q}t = \phi'(v)\{\frac{\partial v}{\partial y} + \frac{\partial v}{\partial z}q + \frac{\partial v}{\partial p}s + \frac{\partial v}{\partial q}t\} \tag{7.612}$$

Eliminating the unknown function ϕ, we get

$$Rr + Ss + Tt + U(rt - s^2) = V \tag{7.613}$$

where R, S, T, U, and V are all functions of x, y, z, p and q:

$$R = q\frac{\partial(u,v)}{\partial(p,z)} + \frac{\partial(u,v)}{\partial(p,y)}, \quad T = \frac{\partial(u,v)}{\partial(x,q)} + p\frac{\partial(u,v)}{\partial(z,q)} \tag{7.614}$$

$$S = \frac{\partial(u,v)}{\partial(x,p)} + p\frac{\partial(u,v)}{\partial(z,p)} + q\frac{\partial(u,v)}{\partial(q,z)} + \frac{\partial(u,v)}{\partial(q,y)} \tag{7.615}$$

$$U = \frac{\partial(u,v)}{\partial(p,q)}, \quad V = p\frac{\partial(u,v)}{\partial(y,z)} + q\frac{\partial(u,v)}{\partial(z,x)} + \frac{\partial(u,v)}{\partial(y,x)} \tag{7.616}$$

Note that (7.613) is a nonlinear second order PDE because of the bracket term. However, if u and v are not functions of p and q, U will be zero and in turn (7.613) is reduced to a linear PDE (or quasi-linear PDE). This special case is the focus of Monge's method.

7.11.3 Monge's Method

We first consider the case that U is zero, and thus the PDE becomes

$$Rr + Ss + Tt = V \tag{7.617}$$

The following method was proposed by Monge in 1784. Let us take the total differential of p and q as:

$$dp = \frac{\partial p}{\partial x} dx + \frac{\partial p}{\partial y} dy = rdx + sdy \tag{7.618}$$

$$dq = \frac{\partial q}{\partial x} dx + \frac{\partial q}{\partial y} dy = sdx + tdy \tag{7.619}$$

Substitution of (7.618) and (7.619) into (7.617) gives

$$R(\frac{dp - sdy}{dx}) + Ss + T(\frac{dq - sdx}{dy}) - V = 0 \tag{7.620}$$

This can be rewritten as

$$Rdpdy + Tdqdx - Vdxdy - s(Rdy^2 - Sdydx + Tdx^2) = 0 \tag{7.621}$$

Now we set the first three terms and the bracket term in (7.621) to zero separately. This can be rewritten as

$$Rdpdy + Tdqdx - Vdxdy = 0 \tag{7.622}$$

$$Rdy^2 - Sdydx + Tdx^2 = (A_1 dy + B_1 dx)(A_2 dy + B_2 dx) = 0 \tag{7.623}$$

These are the intermediate integrals of (7.617) and are known as Monge's equations. Note that we can always factorize (7.623) as shown in the second part of (7.623).

Then, this leads to two systems, namely

$$A_1 dy + B_1 dx = 0, \quad Rdpdy + Tdqdx - Vdxdy = 0 \tag{7.624}$$

$$A_2 dy + B_2 dx = 0, \quad Rdpdy + Tdqdx - Vdxdy = 0 \tag{7.625}$$

If these equations can be integrated, we have the intermediate integrals. We may have either one or two intermediate integrals, depending on whether both of them are integrable. The solution that satisfies (7.624) or (7.625) is also a solution of (7.617), but a solution of (7.617) may not satisfy (7.624) or (7.625).

Example 7.6 Consider the solution of the following PDE:

$$2x^2 \frac{\partial^2 z}{\partial x^2} - 5xy \frac{\partial^2 z}{\partial x \partial y} + 2y^2 \frac{\partial^2 z}{\partial y^2} + 2(\frac{\partial z}{\partial x} x + \frac{\partial z}{\partial y} y) = 0 \tag{7.626}$$

Solution: In simplified form, (7.626) can be written as

$$2x^2 r - 5xys + 2y^2 t + 2(px + qy) = 0 \tag{7.627}$$

Using the result of (7.622) and (7.623), we have

$$2x^2dy^2 + 5xydydx + 2y^2dx^2 = 0 \tag{7.628}$$

$$2x^2dpdy + 2y^2dqdx + 2(px + qy)dxdy = 0 \tag{7.629}$$

Factorizing (7.628), we have

$$(xdy + 2ydx)(2xdy + ydx) = 0 \tag{7.630}$$

Setting the first factor in (7.630) to zero, we get

$$xdy + 2ydx = 0 \tag{7.631}$$

Integration of (7.631) gives

$$u_1 = x^2 y = a \tag{7.632}$$

Dividing (7.629) by xdy and noting (7.631), we get

$$2xdp - ydq + 2pdx - qdy = 0 \tag{7.633}$$

This can be integrated as

$$u_2 = 2px - yq = b \tag{7.634}$$

These two characteristics give the solution as

$$2px - yq = \phi(x^2 y) \tag{7.635}$$

Similarly, setting the second factor of (7.630) to zero, we find

$$xy^2 = c \tag{7.636}$$

Following a similar procedure, we get

$$px - 2qy = \psi(xy^2) \tag{7.637}$$

Solving for p and q from (7.635) and (7.637), we get

$$p = \frac{1}{3x}\{2\phi(x^2 y) - \psi(xy^2)\} \tag{7.638}$$

$$q = \frac{1}{3y}\{\phi(x^2 y) - 2\psi(xy^2)\} \tag{7.639}$$

By definition, we can integrate z by its total differential

$$\begin{aligned} dz = pdx + qdy &= \frac{1}{3x}\{2\phi(x^2 y) - \psi(xy^2)\}dx + \frac{1}{3y}\{\phi(x^2 y) - 2\psi(xy^2)\}dy \\ &= \frac{1}{3}\phi(x^2 y)(\frac{2dx}{x} + \frac{dy}{y}) - \frac{1}{3}\psi(xy^2)(\frac{dx}{x} + \frac{2dy}{y}) \end{aligned} \tag{7.640}$$

Integrating both sides, we get

$$z = \frac{1}{3}\int \phi(x^2 y)d\ln(x^2 y) - \frac{1}{3}\int \psi(xy^2)d\ln(xy^2) \tag{7.641}$$

Therefore, a solution of (7.626) is

$$z = f(x^2 y) + F(xy^2) \tag{7.642}$$

However, this is only one of the possible solutions of (7.626), but not the most general solution.

Example 7.7 Consider the solution of the following PDE:

$$y^2\frac{\partial^2 z}{\partial x^2}-2y\frac{\partial^2 z}{\partial x\partial y}+\frac{\partial^2 z}{\partial y^2}=\frac{\partial z}{\partial x}+6y \tag{7.643}$$

Solution: In simplified form, (7.643) can be written as

$$y^2 r-2ys+t=p+6y \tag{7.644}$$

Using the result of (7.622) and (7.623), the two intermediate integrals are

$$y^2 dy^2+2ydydx+dx^2=0 \tag{7.645}$$

$$y^2 dpdy+dqdx-(p+6y)dxdy=0 \tag{7.646}$$

Factorizing (7.645), we obtain

$$(ydy+dx)^2=0 \tag{7.647}$$

Integration of (7.647) gives

$$2x+y^2=a \tag{7.648}$$

Dividing (7.646) by ydy and noting (7.647), we get

$$ydp-dq+(p+6y)dy=0 \tag{7.649}$$

Integration of (7.649) results in

$$py-q+3y^2=c \tag{7.650}$$

These two characteristics give the solution as

$$py-q+3y^2=\phi(2x+y^2) \tag{7.651}$$

In explicit form, we can express (7.651)

$$\frac{\partial z}{\partial x}y-\frac{\partial z}{\partial y}+3y^2=\phi(2x+y^2) \tag{7.652}$$

Applying the Lagrange method discussed in Section 6.7, we find the following subsidiary equation

$$\frac{dx}{y}=\frac{dy}{-1}=\frac{dz}{-3y^2+\phi(2x+y^2)} \tag{7.653}$$

The first two parts of (7.653) give the first characteristics as

$$u_1=2x+y^2=a \tag{7.654}$$

In view of (7.654), the last two of (7.653) gives the second characteristics as

$$dz+[-3y^2+\phi(a)]dy=0 \tag{7.655}$$

Integration of (7.655) gives the second characteristics and back substitution of the value of a from (7.654) leads to

$$u_2=z-y^3+y\phi(2x+y^2)=b \tag{7.656}$$

Thus, the solution for (7.643) is

$$\psi\{z-y^3+y\phi(2x+y^2),2x+y^2\}=0 \tag{7.657}$$

Finally, the solution in (7.657) can be given in explicit form

$$z=y^3-y\phi(2x+y^2)+f(2x+y^2) \tag{7.658}$$

7.11.4 Monge-Ampere Method

Monge's method, discussed in the last section, was extended to the nonlinear form given in (7.613) by Ampere in 1814. In particular, we consider the following PDE

$$Rr + Ss + Tt + U(rt - s^2) = V \tag{7.659}$$

Recall from (7.618) and (7.619) that r and t can be found as

$$r = \frac{dp - sdy}{dx} \tag{7.660}$$

$$t = \frac{dq - sdx}{dy} \tag{7.661}$$

Substitution of (7.660) and (7.661) into (7.659) gives

$$R\frac{dp - sdy}{dx} + Ss + T\frac{dq - sdx}{dy} + U\{(\frac{dp - sdy}{dx})(\frac{dq - sdx}{dy}) - s^2\} = V \tag{7.662}$$

Expanding (7.662), we get

$$\begin{aligned} &s[Rdy^2 - Sdxdy + Tdx^2 + U(dxdp + dydq)] \\ &-[Rdydp + Tdxdq + Udpdq - Vdxdy] = 0 \end{aligned} \tag{7.663}$$

Setting both the bracket terms to zero, we obtain

$$Rdy^2 - Sdxdy + Tdx^2 + U(dxdp + dydq) = 0 \tag{7.664}$$

$$Rdydp + Tdxdq + Udpdq - Vdxdy = 0 \tag{7.665}$$

We now assume the existence of a function $\lambda = \lambda(x,y,z,p,q)$ such that the following group can be factorized:

$$\begin{aligned} &\lambda[Rdy^2 - Sdxdy + Tdx^2 + U(dxdp + dydq)] + Rdydp + Tdxdq + Udpdq - Vdxdy \\ &= (ady + bdx + cdp)(\alpha dy + \beta dx + \gamma dq) \\ &= a\alpha dy^2 + (a\beta + b\alpha)dxdy + b\beta dx^2 + c\beta dxdp + a\gamma dydq + c\alpha dydp + b\gamma dxdq + c\gamma dqdp \\ &= 0 \end{aligned} \tag{7.666}$$

This is the most important step in the Monge-Ampere method. Equating coefficients of all differential terms on both sides of (7.666), we find

$$\begin{aligned} &\lambda R = a\alpha, \quad -S\lambda - V = a\beta + b\alpha, \quad \lambda T = b\beta, \quad c\beta = \lambda U = a\gamma, \\ &c\alpha = R, \quad b\gamma = T, \quad c\gamma = U \end{aligned} \tag{7.667}$$

We can choose the solution for the unknown coefficients as:

$$a = \lambda, \quad \alpha = R, \quad b = \frac{T}{U}, \quad \beta = \lambda U, \quad c = 1, \quad \gamma = U \tag{7.668}$$

Substitution of (7.668) into the second equation of (7.667) gives a governing equation for λ:

$$U\lambda^2 + \frac{TR}{U} = -S\lambda - V \tag{7.669}$$

Rewriting (7.669), we find a quadratic equation for λ

$$U^2\lambda^2 + SU\lambda + TR + UV = 0 \tag{7.670}$$

There are two distinct roots for λ:

$$\lambda = \lambda_1, \quad \lambda = \lambda_2 \tag{7.671}$$

Finally, we can factorize (7.666) as

$$(\lambda_1 Udy + Tdx + Udp)(Rdy + \lambda_1 Udx + Udq) = 0 \tag{7.672}$$

$$(\lambda_2 Udy + Tdx + Udp)(Rdy + \lambda_2 Udx + Udq) = 0 \tag{7.673}$$

By setting the bracket terms to zero in turn, we obtain four combinations. However, two of them give unacceptable results. In particular, the first and second systems are

$$(Rdy + \lambda_1 Udx + Udq) = 0, \quad (Rdy + \lambda_2 Udx + Udq) = 0 \tag{7.674}$$

$$(\lambda_1 Udy + Tdx + Udp) = 0, \quad (\lambda_2 Udy + Tdx + Udp) = 0 \tag{7.675}$$

Both of these lead to $\lambda_1 = \lambda_2$, which is not acceptable. Finally, we have the following two systems of equations:

$$(Rdy + \lambda_1 Udx + Udq) = 0, \quad (\lambda_2 Udy + Tdx + Udp) = 0 \tag{7.676}$$

$$(\lambda_1 Udy + Tdx + Udp) = 0, \quad (Rdy + \lambda_2 Udx + Udq) = 0 \tag{7.677}$$

If any of these two systems can be integrated, we have two equations for p and q. The solutions of them can be used to give the exact differential of z:

$$dz = pdx + qdy \tag{7.678}$$

We will illustrate this method in the following example. Thus, the failure or success of the Monge-Ampere method depends on whether we can solve for p and q from (7.676) and (7.677).

Example 7.8 Consider the solution of the following PDE:

$$2\frac{\partial^2 z}{\partial x \partial y} + [\frac{\partial^2 z}{\partial x^2}\frac{\partial^2 z}{\partial y^2} - (\frac{\partial^2 z}{\partial x \partial y})^2] = 1 \tag{7.679}$$

Solution: In simplified form, (7.679) is written as

$$2s + (rt - s^2) = 1 \tag{7.680}$$

This is the nonlinear PDE type given (7.659) considered by Ampere. Thus, we have

$$R = T = 0, \quad S = 2, \quad U = V = 1 \tag{7.681}$$

Therefore, the characteristic equation for λ given in (7.670) becomes

$$\lambda^2 + 2\lambda + 1 = 0 \tag{7.682}$$

Both roots are -1, and the corresponding system is

$$dy - dp = 0, \quad dx - dq = 0 \tag{7.683}$$

Integration gives the characteristics as

$$u_1 = y - p = a, \quad u_2 = x - q = b \tag{7.684}$$

Therefore, the intermediate integral is

$$y - p = f(x - q) \tag{7.685}$$

Note that an arbitrary function is involved in (7.685). The Monge-Ampere method does not lead to the general solution but instead yields a particular solution. Therefore, without loss of generality we can set

$$x - q = a \tag{7.686}$$

With (7.686), (7.685) gives

$$y - p = f(a) = b \tag{7.687}$$

By making this assumption, we can solve for p and q as:

$$p = y - b, \quad q = x - a \tag{7.688}$$

Substitution of (7.688) into (7.678) yields

$$dz = pdx + qdy = (y - b)dx + (x - a)dy \tag{7.689}$$

A particular solution is

$$z = xy - bx - ay + c \tag{7.690}$$

Another particular solution can be obtained by assuming a linear functional form for (7.685)

$$y - p = m(x - q) + n \tag{7.691}$$

Thus, we have

$$p - mq = \frac{\partial z}{\partial x} - m\frac{\partial z}{\partial y} = y - n - mx \tag{7.692}$$

Adopting the Lagrange method discussed in Section 6.7, the subsidiary equation is

$$\frac{dx}{1} = \frac{dy}{-m} = \frac{dz}{y - n - mx} \tag{7.693}$$

The first pair gives the following characteristics

$$u_1 = y + mx = a \tag{7.694}$$

The second pair of (7.693) gives

$$\frac{dx}{1} = \frac{dz}{y - n - mx} = \frac{dz}{a - n - 2mx} \tag{7.695}$$

Integrating both sides we get

$$u_2 = z - (a - n)x + mx^2 = b \tag{7.696}$$

Therefore, another particular solution is

$$z = -nx + xy + \phi(y + mx) \tag{7.697}$$

This solution is slightly more general, but the bottom line is that this is still a particular solution. Thus, the Monge-Ampere method will lead to a “particular” solution only.

Example 7.9 Consider the solution of the following PDE:

$$\frac{\partial^2 z}{\partial x^2} + 3\frac{\partial^2 z}{\partial x \partial y} + \frac{\partial^2 z}{\partial y^2} + [\frac{\partial^2 z}{\partial x^2}\frac{\partial^2 z}{\partial y^2} - (\frac{\partial^2 z}{\partial x \partial y})^2] = 1 \tag{7.698}$$

Solution: In simplified form, (7.698) is written as

$$r + 3s + t + (rt - s^2) = 1 \tag{7.699}$$

This is of the nonlinear PDE type given in (7.659) considered by Ampere. Thus, we have

$$R = T = 1, \quad S = 3, \quad U = V = 1 \tag{7.700}$$

Therefore, the characteristic equation for λ given in (7.670) becomes

$$2\lambda^2+3\lambda+1=0 \tag{7.701}$$

The two roots of λ are −1 and −1/2. From (7.676) and (7.677), we have two systems as:

$$-dy+dx+dp=0, \quad dy-\frac{1}{2}dx+dq=0 \tag{7.702}$$

$$-\frac{1}{2}dy+dx+dp=0, \quad dy-dx+dq=0 \tag{7.703}$$

Integration of (7.702) and (7.703) gives two intermediate integrals as

$$p+x-y=f(y-\frac{1}{2}x+q) \tag{7.704}$$

$$p+x-\frac{1}{2}y=g(y-x+q) \tag{7.705}$$

where f and g are arbitrary functions. Next, let us assume

$$q-\frac{1}{2}x+y=\alpha, \quad q-x+y=\beta \tag{7.706}$$

where α and β are not constants. Using (7.706), we can rewrite (7.704) and (7.705) as

$$p+x-y=f(\alpha), \quad p+x-\frac{1}{2}y=g(\beta) \tag{7.707}$$

Solving for x, y, p and q from (7.706) and (7.707), we get

$$x=2(\alpha-\beta), \quad y=2[g(\beta)-f(\alpha)] \tag{7.708}$$

$$p=y-x+f(\alpha), \quad q=x-y+\beta \tag{7.709}$$

Then, the total differential of z is expressed as

$$\begin{aligned} dz &= pdx+qdy=[y-x+f(\alpha)]dx+(x-y+\beta)dy \\ &= (y-x)(dx-dy)+f(\alpha)dx+\beta dy \end{aligned} \tag{7.710}$$

Note from (7.708) that

$$dx=2(d\alpha-d\beta), \quad dy=2[g'(\beta)d\beta-f'(\alpha)d\alpha] \tag{7.711}$$

Substitution of (7.711) into (7.710) gives

$$dz=-\frac{1}{2}d(x-y)^2-2f(\alpha)d\beta+2f(\alpha)d\alpha+2\beta g'(\beta)d\beta-2\beta f'(\alpha)d\alpha \tag{7.712}$$

Integration gives

$$\begin{aligned} z &= -\frac{1}{2}(x-y)^2+2\int f(\alpha)d\alpha+2\int \beta g'(\beta)d\beta-4\beta f(\alpha) \\ &= -\frac{1}{2}(x-y)^2+\phi(\alpha)+2\beta g(\beta)-2\int g(\beta)d\beta-2\beta\phi'(\alpha) \\ &= -\frac{1}{2}(x-y)^2+\phi(\alpha)+\beta\psi'(\beta)-\psi(\beta)-2\beta\phi'(\alpha) \end{aligned} \tag{7.713}$$

where the arbitrary functions ϕ and ψ are defined as

$$\phi(\alpha)=2\int f(\alpha)d\alpha, \quad \psi(\beta)=2\int g(\beta)d\beta \tag{7.714}$$

The final solution is summarized as

$$z = -\frac{1}{2}(x-y)^2 + \phi(\alpha) + \beta\psi'(\beta) - \psi(\beta) - 2\beta\phi'(\alpha) \tag{7.715}$$

$$x = 2(\alpha - \beta), \quad y = 2[g(\beta) - f(\alpha)] \tag{7.716}$$

In this solution, α and β are considered as parameters. Once they are given, x and y can be evaluated and, subsequently, z can be determined by (7.715). There are two arbitrary functions involved in this parametric solution, and this is the general solution.

7.12 FACTORIZATION OF PDE WITH CONSTANT COEFFICIENTS

The method considered here can be viewed as a generalization of Section 3.5.10 for ODEs. We now consider a special form of PDE with constant coefficients:

$$(D^n + a_1 D^{n-1}\bar{D} + a_2 D^{n-2}\bar{D}^2 + ... + a_n \bar{D}^n)z = f(x,y) \tag{7.717}$$

where the differentiation with respect to x and y are written symbolically

$$D = \frac{\partial}{\partial x}, \quad \bar{D} = \frac{\partial}{\partial y} \tag{7.718}$$

We first consider the homogeneous case

$$(D^n + a_1 D^{n-1}\bar{D} + a_2 D^{n-2}\bar{D}^2 + ... + a_n \bar{D}^n)z = 0 \tag{7.719}$$

Let us consider the simplest case

$$(D - m\bar{D})z = 0 \tag{7.720}$$

More explicitly, it can be written as

$$\frac{\partial z}{\partial x} - m\frac{\partial z}{\partial y} = 0 \tag{7.721}$$

Using the Lagrange method discussed in Section 6.7, the auxiliary equation of this first order PDE is

$$\frac{dx}{1} = \frac{dy}{-m} = \frac{dz}{0} \tag{7.722}$$

Integrating (7.722) gives two characteristics

$$u_1 = z = a \tag{7.723}$$

$$u_2 = mx + y = b \tag{7.724}$$

Therefore, the general solution is

$$z = F(mx + y) \tag{7.725}$$

This suggests that the solution of (7.719) can be expressed in the following form

$$z = F_1(y + m_1 x) + F_2(y + m_2 x) + ... + F_n(y + m_n x) \tag{7.726}$$

where m_i ($i = 1,2,3,...,n$) are the roots of

$$m^n + a_1 m^{n-1} + a_2 m^{n-2} + ... + a_n = 0 \tag{7.727}$$

For the case of equal roots, we can consider the problem as

$$(D - m\bar{D})^2 z = (D - m\bar{D})(D - m\bar{D})z = 0 \tag{7.728}$$

Thus, we can rewrite (7.728) as a system of equations as

$$(D - m\bar{D})z = u, \quad (D - m\bar{D})u = 0 \tag{7.729}$$

From the result in (7.725), we have

$$u = F(y + mx) \tag{7.730}$$

The first equation of (7.729) becomes

$$p - mq = F(y + mx) \tag{7.731}$$

The Lagrange method gives an auxiliary equation as

$$\frac{dx}{1} = \frac{dy}{-m} = \frac{dz}{F(y + mx)} \tag{7.732}$$

The first two parts of (7.732) give

$$u_1 = y + mx = a \tag{7.733}$$

The second and third parts of (7.732) result in

$$dz - F(a)dx = 0 \tag{7.734}$$

Integration gives

$$u_2 = z - xF(y + mx) = b \tag{7.735}$$

Thus, the general solution is

$$\phi\{z - xF(y + mx), y + mx\} = 0 \tag{7.736}$$

Therefore, equivalently the solution is

$$z = xF(y + mx) + F_1(y + mx) \tag{7.737}$$

This procedure can be generalized to give the solution of the following PDE

$$(D - m\bar{D})^n (D - m_1\bar{D}) \cdots (D - m_k\bar{D})z = 0 \tag{7.738}$$

The general solution is

$$\begin{aligned} z = x^{n-1}F(y + mx) + x^{n-2}F_1(y + mx) + \ldots + F_{n-1}(y + mx) \\ + H_1(y + m_1 x) + \ldots + H_k(y + m_k x) \end{aligned} \tag{7.739}$$

Note that in this technique the order of differentiation for each term must be of the same order.

Example 7.10 Consider the solution of the following PDE:

$$\frac{\partial^3 z}{\partial x^3} - 3\frac{\partial^3 z}{\partial x^2 \partial y} + 2\frac{\partial^3 z}{\partial x \partial y^2} = 0 \tag{7.740}$$

Solution: In factorized form, we have

$$(D^3 - 3D^2\bar{D} + 2D\bar{D}^2)z = 0 \tag{7.741}$$

The associated algebraic equation for m is

$$m^3 - 3m^2 + 2m = 0 \tag{7.742}$$

which can be factorized as

$$m(m - 2)(m - 1) = 0 \tag{7.743}$$

Using (7.739), we have the solution

$$z = F_1(y) + F_2(y + x) + F_3(y + 2x) \tag{7.744}$$

Example 7.11 Consider the solution of the following PDE:

$$2\frac{\partial^2 z}{\partial x^2}+5\frac{\partial^2 z}{\partial x\partial y}+2\frac{\partial^2 z}{\partial y^2}=0 \tag{7.745}$$

Solution: In factorized form, we have

$$(2D^2+5D\bar{D}+2\bar{D}^2)z=0 \tag{7.746}$$

The associated algebraic equation for m is

$$2m^2+5m+2=0 \tag{7.747}$$

which can be factorized as

$$(2m+1)(m+2)=0 \tag{7.748}$$

Using (7.739), we obtain the solution as

$$z=F_1(y-\frac{1}{2}x)+F_2(y-2x) \tag{7.749}$$

This solution can also be found by the classification discussed in an earlier section, and two characteristics can be found. The canonical form can be solved. Apparently, the present technique is more efficient.

7.13 PARTICULAR SOLUTION BY SYMBOLIC METHOD

The symbolic method for determining a particular solution of an ODE has been discussed in Section 3.5.11. In this section, this method is applied to consider a particular solution of a PDE. Consider a nonhomogeneous PDE of the form

$$F(D,\bar{D})z=f(x,y) \tag{7.750}$$

Using the symbolic method, we get the particular solution as:

$$z_p=\frac{1}{F(D,\bar{D})}f(x,y) \tag{7.751}$$

Following the procedure discussed in Section 3.5.11, the symbolic method involves factorization, finding partial fractions, and expanding functions in infinite series.

The following example illustrates this method.

Example 7.12 Consider the solution of the following PDE:

$$\frac{\partial^2 z}{\partial x^2}-6\frac{\partial^2 z}{\partial x\partial y}+9\frac{\partial^2 z}{\partial y^2}=12x^2+36xy \tag{7.752}$$

Solution: In factorized form, we have

$$(D^2-6D\bar{D}+9\bar{D}^2)z=12x^2+36xy \tag{7.753}$$

The homogeneous PDE becomes

$$(D^2-6D\bar{D}+9\bar{D}^2)z=0 \tag{7.754}$$

The associated characteristic equation is

$$m^2-6m+9=(m-3)^2=0 \tag{7.755}$$

The homogeneous solution becomes

$$z_h=F_1(y+3x)+xF_2(y+3x) \tag{7.756}$$

The particular solution can be expressed as

$$\begin{aligned} z_p &= \frac{1}{(D^2-6D\bar{D}+9\bar{D}^2)}(12x^2+36xy) \\ &= \frac{1}{D^2}\frac{1}{(1-\dfrac{3\bar{D}}{D})^2}(12x^2+36xy) \end{aligned} \tag{7.757}$$

Treating the differential operators as algebraic quantities, we can expand it using Taylor series as:

$$\frac{1}{(1-\dfrac{3\bar{D}}{D})^2}=1+6\frac{\bar{D}}{D}+27(\frac{\bar{D}}{D})^2+\dots \tag{7.758}$$

Substitution of (7.758) into (7.757) gives

$$\begin{aligned} z_p &= \frac{1}{D^2}(1+6\frac{\bar{D}}{D}+27(\frac{\bar{D}}{D})^2+..)(12x^2+36xy) \\ &= \frac{1}{D^2}(12x^2+36xy)+6\frac{1}{D^3}36x+\dots \\ &= x^4+6x^3y+9x^4=10x^4+6x^3y \end{aligned} \tag{7.759}$$

The general solution now becomes

$$z_h=F_1(y+3x)+xF_2(y+3x)+10x^4+6x^3y \tag{7.760}$$

Let us now consider a more general form of the symbolic method. Consider the following linear order PDE:

$$(D-m\bar{D})z=\frac{\partial z}{\partial x}-m\frac{\partial z}{\partial y}=p-mq=f(x,y) \tag{7.761}$$

Using the Lagrange method in Section 6.7, we have

$$\frac{dx}{1}=\frac{dy}{-m}=\frac{dz}{f(x,y)} \tag{7.762}$$

The first two parts of (7.762) give

$$-mdx=dy \tag{7.763}$$

The corresponding characteristics is

$$u_1=y+mx=c \tag{7.764}$$

Then, the first and third parts of (7.762) can be combined to give

$$dz=f(x,c-mx)dx \tag{7.765}$$

where we have used the first characteristics (7.764) to replace y by x. Integration gives

$$z = \int f(x, c - mx)dx \tag{7.766}$$

Therefore, symbolically we can write the particular solution as

$$z_p = \frac{1}{D - m\bar{D}} f(x, y) = \int f(x, c - mx)dx \tag{7.767}$$

Let us illustrate by example.

Example 7.13 Consider the solution of the following PDE:

$$(\frac{\partial}{\partial x} - 2\frac{\partial}{\partial y})(\frac{\partial}{\partial x} + \frac{\partial}{\partial y})z = (y-1)e^x \tag{7.768}$$

Solution: In factorized form, we have

$$(D - 2\bar{D})(D + \bar{D})z = (D^2 - D\bar{D} - 2\bar{D}^2)z = (y-1)e^x \tag{7.769}$$

The homogeneous PDE becomes

$$(D^2 - D\bar{D} - 2\bar{D}^2)z = 0 \tag{7.770}$$

The associated characteristic equation is

$$m^2 - m - 2 = (m-2)(m+1) = 0 \tag{7.771}$$

The homogeneous solution becomes

$$z_h = F_1(y + 2x) + F_2(y - x) \tag{7.772}$$

Define a function u in (7.769) such that it becomes

$$(D - 2\bar{D})u = (y-1)e^x \tag{7.773}$$

The first characteristics of (7.769) is

$$u_1 = y + 2x = c \tag{7.774}$$

The particular solution can be found by first considering

$$u = \int f(x, c-2x)dx = \int (c - 2x - 1)e^x dx = (c - 2x + 1)e^x = (y+1)e^x \tag{7.775}$$

By virtue of (7.767), we get

$$u = \frac{1}{(D - 2\bar{D})}(y-1)e^x = (y+1)e^x \tag{7.776}$$

Substitution of (7.776) into (7.769) gives

$$(D + \bar{D})z = u = (y+1)e^x$$

The first characteristics is

$$u_2 = y - x = c_1 \tag{7.777}$$

Thus, taking the inverse of the first fractional operator in (7.777)

$$z = \frac{1}{(D + \bar{D})}(y+1)e^x \tag{7.778}$$

Using (7.777), we have

$$z_p = \frac{1}{(D+\bar{D})}(y+1)e^x = \int (c_1 + x + 1)e^x dx = (c_1 + x)e^x = ye^x \tag{7.779}$$

The general solution now becomes

$$z_h = F_1(y+2x) + F_2(y-x) + ye^x \tag{7.780}$$

Let us now examine another form of nonhomogeneous PDE

$$(D - m\bar{D} - a)(D - n\bar{D} - b)z = f(x, y) \tag{7.781}$$

Let us consider the homogeneous case first. In particular, we first consider the simple case:

$$(D - m\bar{D} - a)z = p - mq - az = 0 \tag{7.782}$$

The Lagrange method gives an auxiliary condition of

$$\frac{dx}{1} = \frac{dy}{-m} = \frac{dz}{az} \tag{7.783}$$

The first two parts of (7.783) give

$$u_1 = y + mx = a \tag{7.784}$$

The first and third parts of (7.783) give

$$adx = \frac{dz}{z} \tag{7.785}$$

Integrating this on both sides, we find

$$u_2 = ze^{-ax} = b \tag{7.786}$$

Thus, the general solution is

$$\phi(ze^{-ax}, y + mx) = 0 \tag{7.787}$$

Equivalently, we can solve for z as

$$z = e^{ax}\psi(y + mx) \tag{7.788}$$

Extending this analysis to a more general form of homogeneous PDE, we find:

$$(D - m\bar{D} - a)(D - n\bar{D} - b)z = 0 \tag{7.789}$$

$$z = e^{ax} f(y + mx) + e^{bx} F(y + nx) \tag{7.790}$$

This is only true if a and b are constants (recall the factorization of ODEs discussed in Chapter 3). Let us illustrate the method with the following example.

Example 7.14 Consider the solution of the following PDE:

$$(\frac{\partial}{\partial x} + \frac{\partial}{\partial y} - 1)(\frac{\partial}{\partial x} + 2\frac{\partial}{\partial y} - 3)z = 4 + 3x + 6y \tag{7.791}$$

Solution: In factorized form, we have

$$(D + \bar{D} - 1)(D + 2\bar{D} - 3)z = 4 + 3x + 6y \tag{7.792}$$

According to (7.789) and (7.790), the homogeneous solution is

$$z_h = e^x f(y - x) + e^{3x} F(y - 2x) \tag{7.793}$$

The particular solution of (7.792) can be considered as

$$z_p = \frac{1}{(D+\bar{D}-1)(D+2\bar{D}-3)}(4+3x+6y) \tag{7.794}$$

The inverse operator can first be expanded in series form:

$$\begin{aligned}\frac{1}{(D+\bar{D}-1)(D+2\bar{D}-3)} &= \frac{1}{3}\{1-(D+\bar{D})\}^{-1}\{1-\frac{1}{3}(D+2\bar{D})\}^{-1} \\ &= \frac{1}{3}\{1+(D+\bar{D})+...\}\{1+\frac{1}{3}(D+2\bar{D})+...\} \\ &= \frac{1}{3}\{1+\frac{1}{3}(4D+5\bar{D})+...\}\end{aligned} \tag{7.795}$$

Substitution of (7.795) into (7.794) gives

$$\begin{aligned}z_p &= \frac{1}{3}\{1+\frac{1}{3}(4D+5\bar{D})+...\}(4+3x+6y) \\ &= \frac{1}{3}\{4+3x+6y+4+10\} \\ &= 6+x+2y\end{aligned} \tag{7.796}$$

Finally, combining the homogeneous solution and particular solution gives the general solution

$$z_h = e^x f(y-x)+e^{3x}F(y-2x)+ye^x+6+x+2y \tag{7.797}$$

7.14 SUMMARY AND FURTHER READING

In this chapter, we started with the classification of second order PDEs, leading to three different types of differential equations. They are hyperbolic, parabolic, and elliptic. The canonical forms of the three types are considered. It was shown that for any linear second order PDE with constant coefficients, we could always convert it to three types of second order PDE: they are the nonhomogeneous Klein-Gordon equation (for hyperbolic type), the nonhomogeneous diffusion equation (for parabolic type), and the nonhomogeneous Helmholtz equation (for elliptic type). The solutions of these PDEs are then investigated briefly. Adjoint and self-adjoint general second order PDEs are considered. The mixed type of PDE is discussed in the context of two-dimensional steady gas flows. By hodograph transform, ideal gas flows is converted to Chaplygin's equation and the Tricomi equation. The solution of the Tricomi equation is considered for the sake of completeness. Riemann's integral of hyperbolic PDEs is also discussed. Further discussions of the mixed type PDE are available in Smirnov (1978), Landau and Lifshitz (1987), and Tricomi (1923). In view of its importance in mechanics and elasticity, the biharmonic equation is considered in detail. Four mechanics problems leading to the biharmonic equation are defined and derived: they are plane elastic problems, three-dimensional elasticity, bending of thin elastic plates, and two-dimensional viscous flow with low Reynolds number. Uniqueness of solution of the biharmonic equation is demonstrated. The biharmonic functions are considered through the use of Almansi theorems. The biharmonic solution for the circular domain is considered in detail, using an integral formula that can be considered as an extension of the Poisson integral for potential theory. This solution was first obtained by Hua (2012). The solution technique for second order PDEs with non-constant coefficients is considered, including the Monge and Monge-Ampere methods. Finally, we discuss the factorization technique for higher order (higher than two) PDEs. Finally, we discuss the symbolic method for solving PDEs with constant coefficients.

The mathematical analysis of PDEs of second order has been covered in all textbooks on PDEs, but PDEs of higher than second order have been relatively untouched in most textbooks on PDE. The only exceptions are the coverage of biharmonic equations by Selvadurai (2000b) and by Ayres (1952). In this chapter, we cover the biharmonic equation in more detail and we also discuss the technique of factorization for solving higher order PDEs and the symbolic method for higher order PDEs of constant coefficients.

7.15 PROBLEMS

Problem 7.1 The validity of (7.99) can be established from the following equation:

$$A(\frac{\partial(\xi-i\eta)}{\partial x})^2+B\frac{\partial(\xi-i\eta)}{\partial x}\frac{\partial(\xi-i\eta)}{\partial y}+C(\frac{\partial(\xi-i\eta)}{\partial y})^2=0 \tag{7.798}$$

Show the details.

Problem 7.2 Show that

$$\begin{aligned}\overline{B}^2 &= 4A^2(\frac{\partial\xi}{\partial x})^2(\frac{\partial\eta}{\partial x})^2+4C^2(\frac{\partial\xi}{\partial y})^2(\frac{\partial\eta}{\partial y})^2+4AB\frac{\partial\xi}{\partial x}\frac{\partial\eta}{\partial x}[\frac{\partial\xi}{\partial y}\frac{\partial\eta}{\partial x}+\frac{\partial\xi}{\partial x}\frac{\partial\eta}{\partial y}]\\ &+B^2[(\frac{\partial\xi}{\partial x})^2(\frac{\partial\eta}{\partial y})^2+2(\frac{\partial\xi}{\partial x})(\frac{\partial\eta}{\partial y})(\frac{\partial\xi}{\partial y})(\frac{\partial\eta}{\partial x})+(\frac{\partial\eta}{\partial x})^2(\frac{\partial\xi}{\partial y})^2]\\ &+4BC(\frac{\partial\xi}{\partial y})(\frac{\partial\eta}{\partial y})[\frac{\partial\xi}{\partial y}\frac{\partial\eta}{\partial x}+\frac{\partial\xi}{\partial x}\frac{\partial\eta}{\partial y}]+8AC\frac{\partial\xi}{\partial y}\frac{\partial\eta}{\partial x}\frac{\partial\xi}{\partial x}\frac{\partial\eta}{\partial y}\end{aligned} \tag{7.799}$$

Problem 7.3 Show that

$$\begin{aligned}4\overline{A}\overline{C} &= 4A^2(\frac{\partial\xi}{\partial x})^2(\frac{\partial\eta}{\partial x})^2+4C^2(\frac{\partial\xi}{\partial y})^2(\frac{\partial\eta}{\partial y})^2+4B^2\frac{\partial\xi}{\partial y}\frac{\partial\eta}{\partial x}\frac{\partial\xi}{\partial x}\frac{\partial\eta}{\partial y}\\ &+4AB[(\frac{\partial\xi}{\partial x})^2\frac{\partial\eta}{\partial y}\frac{\partial\eta}{\partial x}+(\frac{\partial\eta}{\partial x})^2\frac{\partial\xi}{\partial x}\frac{\partial\xi}{\partial y}]+4AC[(\frac{\partial\eta}{\partial y})^2(\frac{\partial\xi}{\partial x})^2+(\frac{\partial\eta}{\partial x})^2(\frac{\partial\xi}{\partial y})^2]\\ &+4BC[(\frac{\partial\eta}{\partial y})^2\frac{\partial\xi}{\partial y}\frac{\partial\xi}{\partial x}+(\frac{\partial\xi}{\partial y})^2\frac{\partial\eta}{\partial y}\frac{\partial\eta}{\partial x}]\end{aligned} \tag{7.800}$$

Problem 7.4 From the results of Problems 7.2 and 7.3, show that

$$\bar{B}^2 - 4\bar{A}\bar{C} = B^2[(\frac{\partial \xi}{\partial x})^2(\frac{\partial \eta}{\partial y})^2 - 2(\frac{\partial \xi}{\partial x})(\frac{\partial \eta}{\partial y})(\frac{\partial \xi}{\partial y})(\frac{\partial \eta}{\partial x}) + (\frac{\partial \eta}{\partial x})^2(\frac{\partial \xi}{\partial y})^2]$$

$$-4AC[(\frac{\partial \xi}{\partial x})^2(\frac{\partial \eta}{\partial y})^2 - 2(\frac{\partial \xi}{\partial x})(\frac{\partial \eta}{\partial y})(\frac{\partial \xi}{\partial y})(\frac{\partial \eta}{\partial x}) + (\frac{\partial \eta}{\partial x})^2(\frac{\partial \xi}{\partial y})^2] \quad (7.801)$$

$$= (B^2 - 4AC)J^2$$

Thus, the validity of (7.19) is established.

Problem 7.5 Classify the following second order PDE:

$$x^2 \frac{\partial^2 u}{\partial x^2} + 2xy \frac{\partial^2 u}{\partial x \partial y} + y^2 \frac{\partial^2 u}{\partial y^2} = 0 \quad (7.802)$$

Also, find the canonical form of this differential equation.

Ans: Parabolic, $\frac{\partial^2 u}{\partial \eta^2} = 0$

Problem 7.6 Classify the following second order PDE:

$$x \frac{\partial^2 u}{\partial x^2} - y \frac{\partial^2 u}{\partial x \partial y} + \frac{\partial u}{\partial x} = 0 \quad (7.803)$$

Also, find the canonical form of this differential equation.

Ans: For $y > 0$, hyperbolic, $\frac{\partial^2 u}{\partial \eta \partial \xi} = 0$

Problem 7.7 Generalize the self-adjoint conditions given Section 7.6 to the three-dimensional case (i.e., three variables). In particular, the linear differential operator becomes

$$L_{ij}(u) = A_{ij} \frac{\partial^2 u}{\partial x^2} + B_{ij} \frac{\partial^2 u}{\partial y^2} + C_{ij} \frac{\partial^2 u}{\partial z^2} + 2D_{ij} \frac{\partial^2 u}{\partial x \partial y} + 2E_{ij} \frac{\partial^2 u}{\partial x \partial z}$$

$$+2F_{ij} \frac{\partial^2 u}{\partial y \partial z} + G_{ij} \frac{\partial u}{\partial x} + H_{ij} \frac{\partial u}{\partial y} + K_{ij} \frac{\partial u}{\partial z} + Q_{ij} u = 0 \quad (7.804)$$

Ans:

$$A_{ij} = A_{ji}, \quad B_{ij} = B_{ji}, \quad C_{ij} = C_{ji} \quad (7.805)$$

$$D_{ij} = D_{ji}, \quad E_{ij} = E_{ji}, \quad F_{ij} = F_{ji} \quad (7.806)$$

$$\frac{\partial A_{ij}}{\partial x} + \frac{\partial D_{ji}}{\partial y} + \frac{\partial E_{ji}}{\partial z} = \frac{1}{2}(G_{ij} + G_{ji}) \quad (7.807)$$

$$\frac{\partial D_{ij}}{\partial x} + \frac{\partial B_{ji}}{\partial y} + \frac{\partial F_{ji}}{\partial z} = \frac{1}{2}(H_{ij} + H_{ji}) \quad (7.808)$$

$$\frac{\partial E_{ij}}{\partial x}+\frac{\partial F_{ji}}{\partial y}+\frac{\partial C_{ji}}{\partial z}=\frac{1}{2}(K_{ij}+K_{ji}) \tag{7.809}$$

Problem 7.8 Find the solution of the following second order PDE:

$$\frac{\partial^2 W}{\partial \mu \partial \zeta}+\lambda W=0 \tag{7.810}$$

Ans:

$$W(\mu,\zeta)=AJ_0(2\sqrt{\lambda\mu\zeta})+BY_0(2\sqrt{\lambda\mu\zeta}) \tag{7.811}$$

Hint: Apply a change of variables of $\xi=\sqrt{\zeta\mu}$ and then $z=2\xi\sqrt{\lambda}$.

Problem 7.9 Find the Riemann function v for the following PDE

$$\frac{\partial^2 u}{\partial x \partial y}+\frac{C}{4}u=f(x,y) \tag{7.812}$$

Ans:

$$v(x,y)=J_0(\sqrt{C(x-x_0)(y-y_0)}) \tag{7.813}$$

Hint: See Problem 7.8.

Problem 7.10 Show that the Riemann function for the case of $A = B = C = 0$ in (7.443) is

$$v(x,y)=1 \tag{7.814}$$

Problem 7.11 Consider the following second order PDE, which is a gas dynamic problem considered by Riemann:

$$\frac{\partial^2 v}{\partial x \partial y}+\frac{a}{x+y}(\frac{\partial v}{\partial x}+\frac{\partial v}{\partial y})-\frac{2av}{(x+y)^2}=0 \tag{7.815}$$

(i) Consider the following change of variables

$$v=(\frac{\xi+\eta}{x+y})^a F(z), \quad z=-\frac{(x-\xi)(y-\eta)}{(x+y)(\xi+\eta)} \tag{7.816}$$

Show that

$$\frac{\partial v}{\partial x}=(\frac{\xi+\eta}{x+y})^a[\frac{-a}{x+y}F(z)+\frac{\partial z}{\partial x}F'(z)] \tag{7.817}$$

$$\frac{\partial v}{\partial y}=(\frac{\xi+\eta}{x+y})^a[\frac{-a}{x+y}F(z)+\frac{\partial z}{\partial y}F'(z)] \tag{7.818}$$

$$\frac{\partial^2 v}{\partial y \partial x}=(\frac{\xi+\eta}{x+y})^a[\frac{a(a+1)}{(x+y)^2}F(z)-\frac{a}{x+y}(\frac{\partial z}{\partial x}+\frac{\partial z}{\partial y})F'(z)$$
$$+\frac{\partial^2 z}{\partial y \partial x}F'(z)+\frac{\partial z}{\partial x}\frac{\partial z}{\partial y}F''(z)] \tag{7.819}$$

(ii) Show that (7.815) can be reduced to

$$\frac{\partial z}{\partial x}\frac{\partial z}{\partial y}F''(z)+\frac{\partial^2 z}{\partial y\partial x}F'(z)-\frac{a(a+1)}{(x+y)^2}F(z)=0 \tag{7.820}$$

(iii) Prove the following identities:

$$\frac{\partial z}{\partial x}\frac{\partial z}{\partial y}=\frac{1}{(x+y)^2}z(z-1) \tag{7.821}$$

$$\frac{\partial^2 z}{\partial y\partial x}=\frac{2z-1}{(x+y)^2} \tag{7.822}$$

(iv) Show that F satisfies the following equation

$$z(z-1)F''(z)+(1-2z)F'(z)+a(a+1)F(z)=0 \tag{7.823}$$

(v) Find the solution for v

Ans:

$$v=(\frac{\xi+\eta}{x+y})^a F[a+1,-a,1,-\frac{(x-\xi)(y-\eta)}{(x+y)(\xi+\eta)}] \tag{7.824}$$

Problem 7.12 Prove the following vector identity that we use in Section 7.9.4:

$$\nabla^2 \boldsymbol{u}=\nabla(\nabla\cdot\boldsymbol{u})-2\nabla\times\boldsymbol{\omega} \tag{7.825}$$

Problem 7.13 Prove that the biharmonic equation in Cartesian coordinates cannot be solved by using separation of variables.

Problem 7.14 It is given that

$$\nabla^4 w=0 \tag{7.826}$$

Consider a circular two-dimensional domain such that $w = w(r,\theta)$. Now define a new function w^* as:

$$w^*(r,\theta)=r^2 w(\frac{1}{r},\theta) \tag{7.827}$$

Prove that w^* also satisfies biharmonic equation (7.826).

Hint:
(i) Assume a change of variable of $r' = 1/r$ and $\theta' = \theta$.
(ii) Assume next that $w = w^*(r', \theta')/r^2$.
(iii) Use the second Almansi theorem to express w^*.
(iv) This result can be found in Eq. (13) of Duffy (1961).

Problem 7.15 It is given that u satisfies the following diffusion equation

$$\frac{\partial u}{\partial t}=\frac{\partial^2 u}{\partial x^2} \tag{7.828}$$

Consider a function v defined in terms of $u(x, t)$ as:

$$v(x,t)=\frac{1}{\sqrt{t}}e^{-\frac{x^2}{4t}}u(\frac{x}{t},\frac{-1}{t}) \tag{7.829}$$

Prove that v also satisfies the diffusion equation (7.828).

Hint: This is Problem 498 of Gelca and Andreescu (2007) as a training problem for the William Lowell Putnam Mathematical Competition for college students in North America.

Problem 7.16 Find a second order PDE that has a first or intermediate integral of the following form:

$$py-q+3y^2=\phi(2x+y^2) \tag{7.830}$$

Ans:

$$-2y^2\frac{\partial^2 z}{\partial x^2}+3y\frac{\partial^2 z}{\partial x\partial y}-\frac{\partial^2 z}{\partial y^2}+\frac{\partial z}{\partial x}+3y\frac{\partial z}{\partial y}=0 \tag{7.831}$$

Problem 7.17 Find the solution of the following second order PDE by direct integration

$$\frac{\partial^2 z}{\partial x\partial y}=x-y \tag{7.832}$$

Ans:

$$z=\frac{1}{2}x^2y-\frac{1}{2}y^2x+\phi_1(x)+\phi_2(y) \tag{7.833}$$

Problem 7.18 Find the solution of the following second order PDE by direct integration

$$\frac{\partial^2 z}{\partial x\partial y}=x-y \tag{7.834}$$

Ans:

$$z=\frac{1}{2}x^2y-\frac{1}{2}y^2x+\phi_1(x)+\phi_2(y) \tag{7.835}$$

Problem 7.19 Find the solution of the following nonhomogeneous diffusion equation

$$u_t=\alpha^2u_{xx}+f(x)\sin(t),\quad 0\le x\le 1,\quad t>0 \tag{7.836}$$

subject to the following initial and boundary conditions:

$$u(0,t)=0,\quad u(1,t)=0,\quad t>0 \tag{7.837}$$

$$u(x,0)=0,\quad 0\le x\le 1 \tag{7.838}$$

Ans:

$$u(x,t)=\sum_{n=1}^{\infty}\{\frac{\alpha^2\lambda_n^2\sin t+e^{-\alpha^2\lambda_n^2 t}-\cos t}{1+\alpha^4\lambda_n^4}\}f_n\sin(n\pi x) \tag{7.839}$$

$$f_n = 2\int_0^1 f(x)\sin(n\pi x)\,dx \tag{7.840}$$

$$\lambda_n = n\pi, \quad n = 1,2,3,\ldots \tag{7.841}$$

Problem 7.20 Extend Hua's formula derived in (7.563) to the case of a circular domain with radius r_0:

$$\begin{aligned} u = \frac{(r^2 - r_0^2)^2}{2\pi r_0}\{ & -\frac{1}{2}\int_{-\pi}^{\pi} \frac{h(\psi)}{r^2 + r_0^2 - 2rr_0\cos(\theta-\psi)}d\psi \\ & + \int_{-\pi}^{\pi} \frac{[r_0 - r\cos(\theta-\psi)]g(\psi)}{[r_0^2 + r^2 - 2rr_0\cos(\theta-\psi)]^2}d\psi\} \end{aligned} \tag{7.842}$$

Problem 7.21 Derive the following mean value theorem for biharmonic problems:

$$u(0,\theta) = \frac{1}{2\pi}\int_{-\pi}^{\pi} g(\psi)d\psi - \frac{r_0}{4\pi}\int_{-\pi}^{\pi} h(\psi)d\psi \tag{7.843}$$

Hint: Compare the mean value theorem given in Chapter 9 for harmonic problems.

CHAPTER EIGHT

Green's Function Method

8.1 INTRODUCTION

In 1828, self-taught genius George Green, at the age of thirty-five, published An Essay on the Application of Mathematical Analysis to the Theories of Electricity and Magnetism (Green, 1828). It is amazing that Green derived this result and published it as a book at his own expense before he received any formal education. The report was sent to 51 subscribers of the Nottingham Subscription Library. To be exact, he had only attended one year of primary school at the age of nine before this discovery. Green is probably the best self-learner in the history of mathematics and physics.

Green's work remained relatively unknown until Lord Kelvin in 1845, four years after George Green's death, rediscovered it. He recognized its importance and helped to publish Green's essay in *Crelle's Journal*. Poincare summarized our knowledge of Green's functions near the turn of the twentieth century. In 1946, P. M. Morse and H. Feshbach published their classnotes as *Methods of Theoretical Physics*. They laid out the four main properties that a Green's function must possess. Morse and Feshbach showed that "Green's function is the point source solution [to a boundary-value problem] satisfying appropriate boundary conditions." Thus, Green's function could be found by simply solving the differential equation subject to a Dirac delta function with homogeneous boundary conditions. With this understanding, the powerful techniques of eigenvalue expansions (e.g., Chapter 10) and integral transform methods (e.g., Chapter 11) could be used in a straightforward manner to find Green's functions. Green's function method is especially useful in solving nonhomogeneous differential equations. It always provides the basis of integral equations and consequently the boundary element method (Brebbia et al., 1983).

Shortly after the publication of Green's monograph, German mathematician Carl Gottfried Neumann (1832–1925) developed the concept of Green's function as it applies to the two-dimensional (in contrast to three-dimensional) potential equation. He defined the two-dimensional Green's function, showed that it possesses the property of reciprocity, and found that it behaves as $\ln(r)$ as $r \to \infty$. A. Harnack gave the Green's function for a circle and rectangle. All of these authors used eigenfunction expansions in obtaining the Green's function, which becomes one of the fundamental techniques in constructing a Green's function. Later on, John Dougall (1867–1960) derived three-dimensional Green's functions in cylindrical and spherical coordinates. Subsequently, Green's functions for many different differential equations were derived. For example, the Feynman diagram for elementary particle interactions actually describes the interaction between particles by Green's function. Nobel Prize laureate Julian Schwinger confessed that it was Green's function method led him to the Nobel Prize in physics. He shared the Nobel Prize with Tomonaga and Feynman in 1965 in quantum electrodynamics (Beiser, 2003).

8.2 POTENTIALS

Whenever a vector field can be determined by the vector derivative of a scalar function, we say that potentials exist. In nature, many phenomena can be modelled and expressed in terms of potentials, such as gravitational field, electric field, incompressible flow, etc.

Mathematically, a velocity field $\boldsymbol{v}$ is expressed as

$$\boldsymbol{v} = \nabla \phi \tag{8.1}$$

where ϕ is the scalar potential. From the divergence theorem of Gauss, the volume integral can be converted to the surface integral as

$$\iiint_{\Omega} \nabla \cdot \boldsymbol{v} d\Omega = \iint_{\Gamma} \boldsymbol{v} \cdot \boldsymbol{n} dS \tag{8.2}$$

Substitution of (8.1) into (8.2) gives

$$\iiint_{\Omega} \nabla^2 \phi d\Omega = \iint_{\Gamma} \nabla \phi \cdot \boldsymbol{n} dS \tag{8.3}$$

If the velocity field $\boldsymbol{v}$ represents fluid flow and it is incompressible (i.e., $\nabla \cdot \boldsymbol{v} = 0$), we have

$$\nabla^2 \phi = 0 \tag{8.4}$$

which is the Laplace equation. Therefore, for any problem that can be modelled by the Laplace equation, there must exist potentials.

8.3 GREEN'S FUNCTION FOR LAPLACE EQUATION

The Laplace equation is one of the most fundamental second order partial differential equations because of its repeated appearance in the modelling of physical problems, including electrostatics, incompressible fluid flow, gravitational theory, and membrane deflection. In Cartesian coordinates, it is given as

$$\nabla^2 u = \frac{\partial^2 u}{\partial x^2} + \frac{\partial^2 u}{\partial y^2} + \frac{\partial^2 u}{\partial z^2} = 0 \tag{8.5}$$

To consider the three-dimensional Green's function, it is more convenient to consider the Laplace equation in spherical problems

$$\nabla^2 u = \frac{1}{r^2} \frac{\partial}{\partial r}(r^2 \frac{\partial u}{\partial r}) + \frac{1}{r^2 \sin\theta} \frac{\partial}{\partial \theta}(\sin\theta \frac{\partial u}{\partial \theta}) + \frac{1}{r^2 \sin^2\theta} \frac{\partial^2 u}{\partial \varphi^2} = 0 \tag{8.6}$$

For the case of a point source or Green's function in an infinite domain, the solution must be symmetric with respect to the origin, and thus the solution is independent of θ and φ. For this case, we write the problem as

$$\nabla^2 G(\boldsymbol{r} - \boldsymbol{r}_0) = \delta(\boldsymbol{r} - \boldsymbol{r}_0) \tag{8.7}$$

where δ is the Dirac delta function, which is infinite at $\boldsymbol{r} = \boldsymbol{r}_0$ and otherwise zero:

$$\delta(\boldsymbol{r}-\boldsymbol{r}_0)=0 \quad \boldsymbol{r}\neq\boldsymbol{r}_0$$
$$=\infty \quad \boldsymbol{r}=\boldsymbol{r}_0 \tag{8.8}$$

In addition, we have that

$$\iiint_\Omega \delta(\boldsymbol{r}-\boldsymbol{r}_0)d\Omega=1 \tag{8.9}$$

where Ω is any volume integral embracing the source point $\boldsymbol{r}_0$. If the source is not included in Ω, the integral will be identically zero. This Dirac delta function is named in honor of electrical engineer and physicist, Paul Dirac, who received the Nobel Prize in physics for his major contribution to quantum mechanics. The idea of using the Dirac delta function actually occurred much earlier than its usage in quantum mechanics by Dirac. For example, it can be used to prescribe point force in the case of beam bending subject to a concentrated force. However, its successful application to quantum mechanics leads to the development of a completely new theory by mathematicians. To validate the mathematical analysis involving the Dirac delta function, the theory of distribution of Schwarz and Gelfand or the so-called generalized theory, was proposed. According to the generalized theory, the Dirac delta function is defined using the process of integration. An introduction on Dirac delta and the associated distribution theory will be given in a later section.

Note that strictly speaking, Green's function should be defined in (8.7) by using the adjoint differential operator of (8.5). But for the case of the Laplace equation, it is self-adjoint. Therefore, (8.5) and (8.7) are valid only for the case of the self-adjoint problem (Greenberg, 1971).

More generally, if a Green's function is formulated in a finite volume, we can express it into two parts:

$$G=G_0+G_1 \tag{8.10}$$

The first part, G_0, is called the fundamental solution with a singularity at the source point $\boldsymbol{r}=\boldsymbol{r}_0$ for an infinite domain (this is actually a particular solution of the PDE), whereas the second part is the homogeneous solution or satisfies the following homogeneous form with given a boundary condition (Greenberg, 1971):

$$\nabla^2 G_1=0 \tag{8.11}$$

This two-part Green function will be illustrated again in Section 8.6.

For functions independent of θ and φ, (8.6) is reduced to

$$\frac{1}{r^2}\frac{d}{dr}(r^2\frac{dG_0}{dr})=0 \tag{8.12}$$

except at point $\boldsymbol{r}=\boldsymbol{r}_0$. Integrating (8.12) once, we find

$$\frac{dG_0}{dr}=\frac{C_1}{r^2} \tag{8.13}$$

Integration of (8.13) gives

$$G_0=-\frac{C_1}{r}+C_2 \tag{8.14}$$

For the case of a point source at the origin, the solution G_0 should decay to zero as $r\to\infty$ (recalling that the fundamental solution is for the infinite domain), and this gives

$$C_2 = 0 \tag{8.15}$$

To find the first constant, we have to consider the volume integration embracing the source point $\boldsymbol{r}_0$ as shown in Figure 8.1. An arbitrary small spherical domain containing the source point is denoted by Ω_0 and the corresponding boundary is Σ_0. Then, we apply the Gauss theorem to get

$$\begin{aligned}\iiint_{\Omega_\varepsilon} \nabla^2 G_0 d\Omega &= \iint_{\Sigma_\varepsilon} \frac{\partial G_0}{\partial r} dS \\ &= \int_0^{2\pi}\int_0^{\pi} \frac{\partial}{\partial r}\left(-\frac{C_1}{r}\right) r^2 \sin\theta d\theta d\varphi \\ &= 4\pi C_1 = 1\end{aligned} \tag{8.16}$$

The last part of (8.16) results from (8.9). Therefore, we have

$$C_1 = \frac{1}{4\pi} \tag{8.17}$$

Finally, the fundamental solution is

$$G_0 = -\frac{1}{4\pi r} \tag{8.18}$$

However, in the literature related to electrostatic problems, a negative sign is used on the right hand side of (8.7). That is, (8.7) and its fundamental solution are normally given as

$$\nabla^2 G(\boldsymbol{r} - \boldsymbol{r}_0) = -\delta(\boldsymbol{r} - \boldsymbol{r}_0) \tag{8.19}$$

$$G_0 = \frac{1}{4\pi r} \tag{8.20}$$

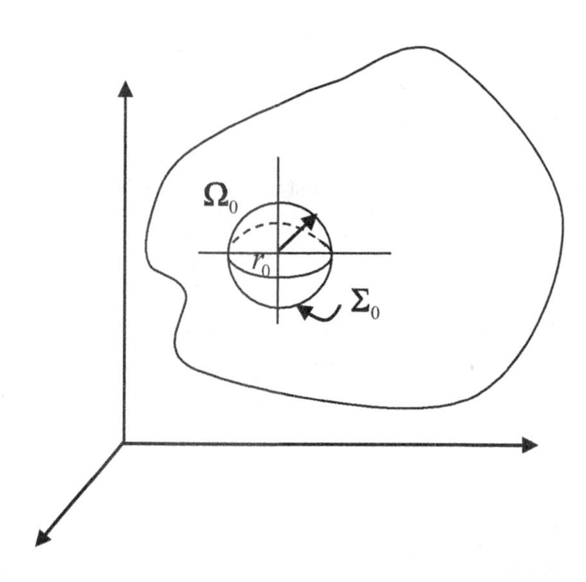

Figure 8.1 A 3-D body containing the singular point that a point source $\boldsymbol{r}_0$ is applied

The fundamental solution given in (8.18) or (8.20) is finite everywhere except at the origin (i.e., $r \neq 0$). This is the three-dimensional Green's function for the Laplace equation for an infinite domain. Actually, Chau (2013) showed that this harmonic solution (i.e., solution of Laplace equation) is the granddaddy of many other harmonic functions, and is of profound importance in solving elasticity problems using Papkovitch-Neuber displacement potentials (or the so-called P-N potentials).

For the two-dimensional Green's function, it is convenient to consider the cylindrical coordinates of the Laplace equation

$$\nabla^2 u = \frac{1}{r}\frac{\partial}{\partial r}(r\frac{\partial u}{\partial r}) + \frac{1}{r^2}\frac{\partial^2 u}{\partial \varphi^2} = \delta(\boldsymbol{r} - \boldsymbol{r}_0) \tag{8.21}$$

For this case, we write the solution as

$$u = g(r) \tag{8.22}$$

The governing equation of Green's function given in (8.21) becomes

$$\frac{d^2 g}{dr^2} + \frac{1}{r}\frac{dg}{dr} = \delta(r) \tag{8.23}$$

As discussed in an earlier chapter, we can introduce a change of variables to reduce the order of the ODE as

$$\frac{dg}{dr} = Z \tag{8.24}$$

Then, (8.23) becomes

$$\frac{dZ}{dr} + \frac{1}{r}Z = 0 \tag{8.25}$$

This is a separable first order ODE. Rearranging and integrating, we get

$$Z = \frac{C_1}{r} \tag{8.26}$$

Thus, combining (8.24) and (8.26), we obtain

$$g = C_1 \ln r + C_2 \tag{8.27}$$

Following the procedure for the 3-D Green's function for the infinite domain, we take an area integration of a circle containing the source point. That is,

$$\begin{aligned} \iint \nabla^2 g d\Omega &= \int \frac{\partial g}{\partial r} dS \\ &= \int_0^{2\pi} \frac{\partial}{\partial r}(C_1 \ln r + C_2) r d\theta \\ &= 2\pi C_1 = 1 \end{aligned} \tag{8.28}$$

We have set $C_2 = 0$ in (8.8) for simplicity. Thus, we have the fundamental solution as

$$g = \frac{1}{2\pi}\ln r \tag{8.29}$$

where $r \neq 0$. This is the two-dimensional Green's function for the Laplace equation. Again, in electrostatic literatures it is sometimes given as:

$$\nabla^2 u = \frac{1}{r}\frac{\partial}{\partial r}(r\frac{\partial u}{\partial r}) + \frac{1}{r^2}\frac{\partial^2 u}{\partial \varphi^2} = -\delta(\boldsymbol{r}-\boldsymbol{r}_0) \tag{8.30}$$

$$g = -\frac{1}{2\pi}\ln r = \frac{1}{2\pi}\ln(\frac{1}{r}) \tag{8.31}$$

For higher dimensions (i.e., $m > 2$), the infinite Green's function can be expressed as (Zachmanoglou and Thoe, 1986):

$$G_m = \frac{1}{2\pi(3-m)!}\frac{1}{r^{m-2}} \tag{8.32}$$

If we ignore the constant term, this agrees with the result given in Zachmanoglou-Thoe (1986). For $m = 3$, we recover the three-dimensional space Green's function given in (8.20).

8.4 GREEN'S IDENTITIES

Green's identities were derived in 1828 by George Green, in his attempt to provide a general mathematical theory for solving electricity and magnetism problems. It is amazing that Green derived this result and published it as a book at his own expense before he received any formal education. To be exact, he had only attended one year of primary school at the age of nine before this discovery. Green is probably the best self-learner in the history of mathematics and physics. Let us start with the divergence theorem of Gauss of the following function:

$$\iiint_\Omega \left(\frac{\partial P}{\partial x} + \frac{\partial Q}{\partial y} + \frac{\partial R}{\partial z}\right) d\Omega = \iint_\Gamma [P\frac{\partial x}{\partial n} + Q\frac{\partial y}{\partial n} + R\frac{\partial z}{\partial n}] dS \tag{8.33}$$

where P, Q, and R are some arbitrary functions of space within the body of Ω with boundary Γ as shown in Figure 8.2. The unit normal is denoted as $\boldsymbol{n}$.

We now assume particular forms of P, Q, and R, which are expressed in terms of two functions u and v:

$$P = u\frac{\partial v}{\partial x}, \quad Q = u\frac{\partial v}{\partial y}, \quad R = u\frac{\partial v}{\partial z} \tag{8.34}$$

Differentiation of (8.34) gives

$$\frac{\partial P}{\partial x} = \frac{\partial u}{\partial x}\frac{\partial v}{\partial x} + u\frac{\partial^2 v}{\partial x^2} \tag{8.35}$$

$$\frac{\partial Q}{\partial y} = \frac{\partial u}{\partial y}\frac{\partial v}{\partial y} + u\frac{\partial^2 v}{\partial y^2} \tag{8.36}$$

$$\frac{\partial R}{\partial z} = \frac{\partial u}{\partial z}\frac{\partial v}{\partial z} + u\frac{\partial^2 v}{\partial z^2} \tag{8.37}$$

Substitution of (8.35) to (8.37) into (8.33) gives

$$\iiint_\Omega u\nabla^2 v d\boldsymbol{\Omega} = -\iiint_\Omega (\frac{\partial u}{\partial x}\frac{\partial v}{\partial x} + \frac{\partial u}{\partial y}\frac{\partial v}{\partial y} + \frac{\partial u}{\partial z}\frac{\partial v}{\partial z}) d\boldsymbol{\Omega} + \iint_\Gamma u\frac{\partial v}{\partial n} dS \tag{8.38}$$

This is called Green's first identity.

Now, we can redefine P, Q, and R in (8.34) by reversing u and v:

$$P = v\frac{\partial u}{\partial x}, \quad Q = v\frac{\partial u}{\partial y}, \quad R = v\frac{\partial u}{\partial z} \tag{8.39}$$

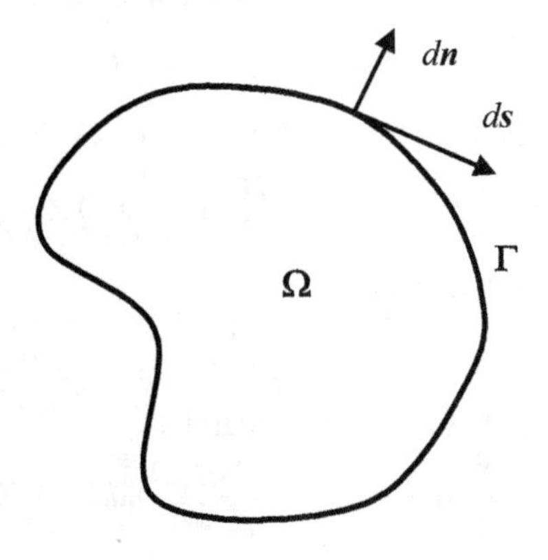

Figure 8.2 A body with domain Ω and boundary Γ showing the unit normal

Following the same procedure, we have

$$\iiint_{\Omega} v\nabla^2 u d\Omega = -\iiint_{\Omega}(\frac{\partial u}{\partial x}\frac{\partial v}{\partial x}+\frac{\partial u}{\partial y}\frac{\partial v}{\partial y}+\frac{\partial u}{\partial z}\frac{\partial v}{\partial z})d\Omega + \iint_{\Gamma} v\frac{\partial u}{\partial n}dS \tag{8.40}$$

Taking the difference of (8.38) and (8.40) gives

$$\iiint_{\Omega}(u\nabla^2 v - v\nabla^2 u)d\Omega = \iint_{\Gamma}(u\frac{\partial v}{\partial n} - v\frac{\partial u}{\partial n})dS \tag{8.41}$$

This is Green's second identity, also known as Green's theorem. This equation relates u, v, $\nabla^2 u$, and $\nabla^2 v$ inside the body to u, v, $\partial u/\partial n$, and $\partial v/\partial n$ on the surface of the body. This Green's identity is physically related to the uniqueness of the solution to the Laplace equation, the conservation of mass, and the Maxwell-Rayleigh reciprocity law. It lays the mathematical foundation for Green's function method and the boundary integral method. Therefore, it is a very important mathematical theorem. Note that the differential operators on the left hand side of (8.41) are Laplacian. In fact, a similar identity can also be formulated for a more general differential equation. This will be done in later sections.

We now consider a special case that

$$v = \frac{1}{r} = \frac{1}{\sqrt{(x-x_0)^2 + (y-y_0)^2 + (z-z_0)^2}} \tag{8.42}$$

where the observation point M_0 (x_0, y_0, z_0) and the distance r are shown in Figure 8.3. Note from the last section that we have actually set v as the three-dimensional Green's function for the Laplace equation. This is the idea behind Green's function method. Substitution of (8.42) into (8.41) gives

$$\iiint_{\Omega-\Omega_{\varepsilon}}[u\nabla^2(\frac{1}{r}) - \frac{1}{r}\nabla^2 u]d\Omega = \iint_{\Gamma+\Gamma_{\varepsilon}}[u\frac{\partial}{\partial n}(\frac{1}{r}) - \frac{1}{r}\frac{\partial u}{\partial n}]dS \tag{8.43}$$

A small sphere of radius ε around point M_0 has been excluded from the integration whereas an extra surface integral of the sphere is added to the surface integral. Note that once the point M_0 has been excluded in the integration, we have

$$\nabla^2(\frac{1}{r}) = 0, \quad \nabla^2 u = 0 \tag{8.44}$$

everywhere within the domain $\Omega - \Omega_\varepsilon$. Therefore, (8.43) is reduced to

$$\iint_\Gamma [u\frac{\partial}{\partial n}(\frac{1}{r}) - \frac{1}{r}\frac{\partial u}{\partial n}]dS + \iint_{\Gamma_\varepsilon} [u\frac{\partial}{\partial n}(\frac{1}{r}) - \frac{1}{r}\frac{\partial u}{\partial n}]dS = 0 \tag{8.45}$$

On the spherical surface Γ_ε, we have

$$\frac{\partial}{\partial n}(\frac{1}{r}) = -\frac{\partial}{\partial r}(\frac{1}{r}) = \frac{1}{r^2} = \frac{1}{\varepsilon^2} \tag{8.46}$$

Therefore, we can simplify the surface integral as

$$\iint_{\Gamma_\varepsilon} u\frac{\partial}{\partial n}(\frac{1}{r})dS = \frac{1}{\varepsilon^2}\iint_{\Gamma_\varepsilon} udS = 4\pi\bar{u} \tag{8.47}$$

where the superimposed bar indicates the average value of u on the spherical surface Γ_ε. On the other hand, on the spherical surface we have

$$\iint_{\Gamma_\varepsilon} \frac{1}{r}\frac{\partial u}{\partial n}dS = \frac{1}{\varepsilon}\iint_{\Gamma_\varepsilon} \frac{\partial u}{\partial n}dS = 4\pi\varepsilon\overline{\frac{\partial u}{\partial n}} \tag{8.48}$$

Back substitution of these results into (8.45) leads to

$$\iint_\Gamma [u\frac{\partial}{\partial n}(\frac{1}{r}) - \frac{1}{r}\frac{\partial u}{\partial n}]dS + 4\pi\bar{u} - 4\pi\varepsilon\overline{\frac{\partial u}{\partial n}} = 0 \tag{8.49}$$

We now take the limit $\varepsilon \to 0$:

$$\lim_{\varepsilon\to 0} 4\pi\bar{u} = 4\pi u(M_0) \tag{8.50}$$

$$\lim_{\varepsilon\to 0} 4\pi\varepsilon\overline{\frac{\partial u}{\partial n}} = 0 \tag{8.51}$$

In view of these results, (8.49) can be rewritten as:

$$u(M_0) = -\frac{1}{4\pi}\iint_\Gamma [u\frac{\partial}{\partial n}(\frac{1}{r}) - \frac{1}{r}\frac{\partial u}{\partial n}]dS \tag{8.52}$$

This is Green's third identity. In potential theory, the solution in (8.52) can be considered as the summation of two potentials on the boundary. In particular, the solution due to the so-called double layer potential is defined as:

$$u(M_0) = -\frac{1}{4\pi}\iint_\Gamma [\mu\frac{\partial}{\partial n}(\frac{1}{r})]dS \tag{8.53}$$

where μ is the double layer potential (given value of u on the surface). Alternatively, the solution due to the single layer potential is defined as:

$$u(M_0) = -\frac{1}{4\pi}\iint_\Gamma [\frac{1}{r}\sigma]dS \tag{8.54}$$

where σ is the single layer potential given on the boundary (as the normal derivative of u). The methods of single- and double-layer potentials are apparently due to Gustave Robin in his PhD thesis on potential theory. We will mention Robin again in Chapter 9 when we discuss the Robin problem in potential theory. For the case that u is not a harmonic function, (8.52) can be modified as

$$u(M_0) = -\frac{1}{4\pi}\iint_\Gamma [u\frac{\partial}{\partial n}(\frac{1}{r}) - \frac{1}{r}\frac{\partial u}{\partial n}]dS - \frac{1}{4\pi}\iiint_\Omega \frac{1}{r}\nabla^2 u d\Omega \tag{8.55}$$

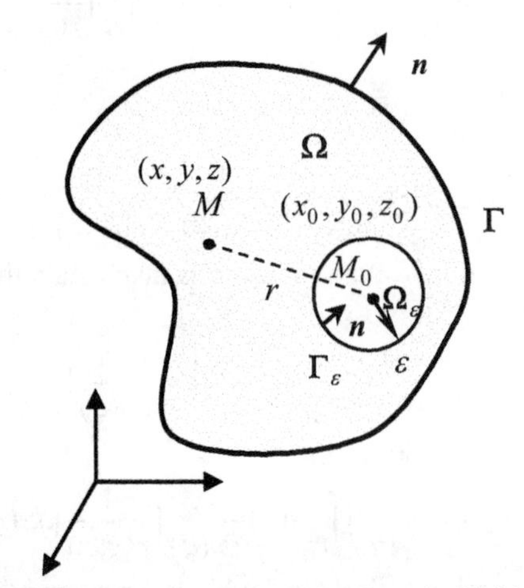

Figure 8.3 The integration domain for Green's third identity

From (8.55), if we know u and $\partial u/\partial n$ on the boundary, we can find the value of u anywhere within the body. However, in reality, we could not impose both of them on the same surface. This is similar to the fact that you cannot independently impose both moment and rotation at a hinged support of a beam at the same time. Once moment is applied, the beam will rotate according, depending on the bending stiffness of the beam, and vice versa.

The following example shows the reason why (8.53) and (8.54) are called double and single layer potentials (which was proposed by Robin) and what their relation is.

Example 8.1 It is given that a harmonic function for a single layer potential σ can be evaluated by the following surface integral

$$u(M_0) = -\frac{1}{4\pi}\iint_\Gamma [\frac{1}{r}\sigma]dS \tag{8.56}$$

where

$$\nabla^2 u = 0 \tag{8.57}$$

Consider two layers of single potentials distributed on two surfaces, which are separated by a distance h. One has strength σ whereas the other has strength $-\sigma$, such that the limit of $h \to 0$ and $\sigma \to \infty$ are taken such that $\sigma h \to \mu$, which is uniform on the whole surface. Show that the solution is given by

$$u(M_0) = -\frac{1}{4\pi}\iint_\Gamma [\mu\frac{\partial}{\partial n}(\frac{1}{r})]dS \tag{8.58}$$

Solution: The two single layers are the source and image as shown in Figure 8.4, and they can be related as:

$$\sigma(\xi)dS(\xi) = -\sigma(\xi')dS(\xi') \tag{8.59}$$

Thus, the solution for the two single layers can be superimposed as

$$\begin{aligned} u(M_0) &= -\frac{1}{4\pi}\iint_\Gamma [\frac{1}{r(\xi,x)}\sigma(\xi)]dS(\xi) - \frac{1}{4\pi}\iint_\Gamma [\frac{1}{r(\xi',x)}\sigma(\xi')]dS(\xi') \\ &= -\frac{1}{4\pi}\iint_\Gamma \sigma(\xi)[\frac{1}{r(\xi,x)} - \frac{1}{r(\xi',x)}]dS(\xi) \\ &= -\frac{1}{4\pi}\iint_\Gamma \sigma(\xi)h(\xi,\xi')\left\{\frac{1}{h(\xi,\xi')}[\frac{1}{r(\xi,x)} - \frac{1}{r(\xi',x)}]\right\}dS(\xi) \end{aligned} \tag{8.60}$$

We consider the limit that $h \to 0$ and $\sigma \to \infty$ is taken such that $\sigma h \to \mu$. Note that the bracket term can be replaced by

$$\left\{\frac{1}{h(\xi,\xi')}[\frac{1}{r(\xi,x)} - \frac{1}{r(\xi',x)}]\right\} = \frac{\partial}{\partial n(\xi)}[\frac{1}{r(\xi,x)}] \tag{8.61}$$

Substitution of (8.61) into (8.60) gives

$$u(M_0) = -\frac{1}{4\pi}\iint_\Gamma \mu(\xi)\frac{\partial}{\partial n(\xi)}[\frac{1}{r(\xi,x)}]dS(\xi) \tag{8.62}$$

This is exactly the integral equation for double layer potentials given in (8.58), and μ is known as the surface density or dipole of the source and its image.

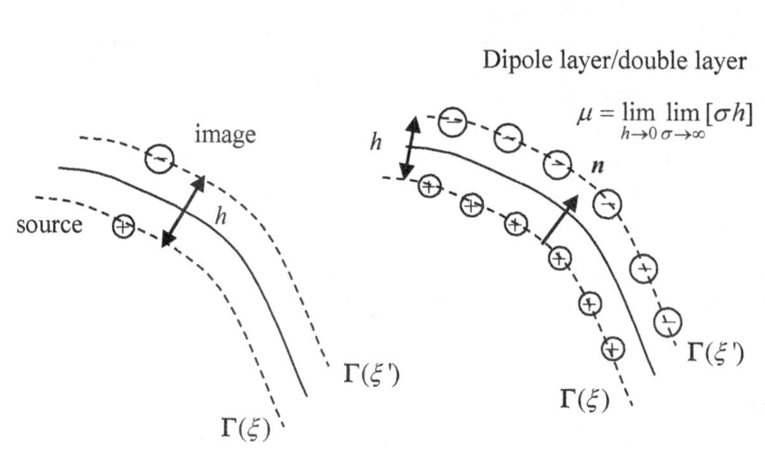

Figure 8.4 Double layer potential as a superposition of two single layer potentials

8.5 BOUNDARY INTEGRAL FOR HARMONIC FUNCTIONS

Green's third identity given in (8.52) is only valid when the observation point M_0 is not on the surface. If the observation point is on the boundary, special consideration needs to be made. Figure 8.5 illustrates the limiting process of an interior point M_0 moving in the direction of unit normal $\boldsymbol{n}$ to the boundary Γ. First, the surface is divided into two parts, one is the total surface excluding a circular disk containing the boundary singular point, and the other one is the circular disk containing the singular point. In view of this, the volume integral remains zero because and both u and $1/r$ are harmonic functions, and the singularity point is not inside the domain. Only the surface integral given in (8.52) needs to be evaluated and can be rewritten as

$$\lim_{\varepsilon\to 0}\{\iint_{\Gamma-\Gamma_\varepsilon}[u\frac{\partial}{\partial n}(\frac{1}{r})-\frac{1}{r}\frac{\partial u}{\partial n}]dS+\iint_{\Gamma_\varepsilon}[u\frac{\partial}{\partial n}(\frac{1}{r})-\frac{1}{r}\frac{\partial u}{\partial n}]dS\}=0 \quad (8.63)$$

Note that the circular disk is tangent to the surface and there is a unique tangent in the process of approaching the boundary. The first integral on the right of (8.63) over the domain $\Gamma-\Gamma_\varepsilon$ is analytic and regular, and thus

$$\lim_{\varepsilon\to 0}\iint_{\Gamma-\Gamma_\varepsilon}[u\frac{\partial}{\partial n}(\frac{1}{r})-\frac{1}{r}\frac{\partial u}{\partial n}]dS=\iint_{\Gamma}[u\frac{\partial}{\partial n}(\frac{1}{r})-\frac{1}{r}\frac{\partial u}{\partial n}]dS \quad (8.64)$$

The second integral on the left of (8.63) can be evaluated as

$$\lim_{\varepsilon\to 0}\iint_{\Gamma_\varepsilon}[u\frac{\partial}{\partial n}(\frac{1}{r})-\frac{1}{r}\frac{\partial u}{\partial n}]dS=\lim_{\varepsilon\to 0}\iint_{\Gamma_\varepsilon}u\frac{\partial}{\partial n}(\frac{1}{r})dS-\lim_{\varepsilon\to 0}\iint_{\Gamma_\varepsilon}\frac{1}{r}\frac{\partial u}{\partial n}dS \quad (8.65)$$

The second integral on the right of (8.65) can be determined as

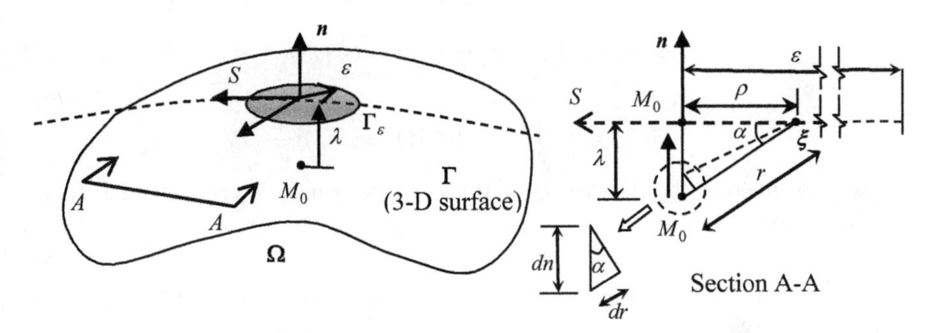

Figure 8.5 The observation point M_0 approaching the boundary Γ

$$\lim_{\varepsilon\to 0}\iint_{\Gamma_\varepsilon}\frac{1}{r}\frac{\partial u}{\partial n}dS=\lim_{\varepsilon\to 0}\frac{1}{\varepsilon}\iint_{\Gamma_\varepsilon}\frac{\partial u}{\partial n}dS=\lim_{\varepsilon\to 0}\frac{1}{\varepsilon}\pi\varepsilon^2\overline{\frac{\partial u}{\partial n}}=\lim_{\varepsilon\to 0}\pi\varepsilon\overline{\frac{\partial u}{\partial n}}=0 \quad (8.66)$$

The last part of (8.66) is zero provided that the double layer potentials prescribed on the surface are bounded.

The first integral on the right of (8.65) can be determined as

$$\lim_{\varepsilon\to 0}\iint_{\Gamma_\varepsilon}u\frac{\partial}{\partial n}(\frac{1}{r})dS=\lim_{\varepsilon\to 0}\{\iint_{\Gamma_\varepsilon}[u(M)-u(M_o)]\frac{\partial}{\partial n}(\frac{1}{r})dS+\iint_{\Gamma_\varepsilon}u(M_o)\frac{\partial}{\partial n}(\frac{1}{r})dS\} \quad (8.67)$$

The second integral on the right hand side of (8.67) can be evaluated as

$$I_2 = \lim_{\varepsilon\to 0}\iint_{\Gamma_\varepsilon} u(M_o)\frac{\partial}{\partial n}(\frac{1}{r})dS = -u(M_o)\lim_{\varepsilon\to 0}\int_0^{\varepsilon}\frac{1}{r^2}\frac{\partial r}{\partial n}2\pi\rho d\rho \tag{8.68}$$

Referring to Figure 8.5, we find that

$$\frac{\partial r}{\partial n} = \sin\alpha = \frac{\lambda}{r} \tag{8.69}$$

In addition, we apply the following change of variables from ρ to r:

$$\rho = \sqrt{r^2 - \lambda^2}\ , \text{ or } \ \rho d\rho = rdr \tag{8.70}$$

The integral in (8.68) becomes

$$I_2 = -2\pi u(M_o)\lim_{\varepsilon\to 0}\lambda\int_{\lambda}^{\sqrt{\lambda^2+\varepsilon^2}}\frac{1}{r^2}dr = 2\pi u(M_o)\lim_{\varepsilon\to 0}\lambda\{\frac{1}{\sqrt{\lambda^2+\varepsilon^2}} - \frac{1}{\lambda}\} \tag{8.71}$$

However, we can take the limit $\lambda \to 0$ (M_0 approaching the boundary Γ) much faster than $\varepsilon \to 0$, such that

$$\lim_{\varepsilon\to 0}\frac{\lambda}{\varepsilon} = 0 \tag{8.72}$$

Finally, we get

$$I_2 = -2\pi u(M_o) \tag{8.73}$$

By following a similar procedure, the first integral on the right hand side of (8.67) can be evaluated as

$$\begin{aligned} I_1 &= \lim_{\varepsilon\to 0}\iint_{\Gamma_\varepsilon}[u(M) - u(M_o)]\frac{\partial}{\partial n}(\frac{1}{r})dS \\ &= -2\pi\lim_{\varepsilon\to 0}\lambda\int_{\lambda}^{\sqrt{\lambda^2+\varepsilon^2}}[u(M) - u(M_o)]\frac{1}{r^2}dr \end{aligned} \tag{8.74}$$

Now, we impose the Holder condition on the single layer potential (e.g., Muskhelishvili, 1975):

$$|u(M) - u(M_o)| \le Ar^{\beta} \tag{8.75}$$

where A and $0 < \beta \le 1$ are positive constants. Substitution of the equality of (8.75) into (8.74) gives

$$\begin{aligned} I_1 &= -2\pi A\lim_{\varepsilon\to 0}\lambda\int_{\lambda}^{\sqrt{\lambda^2+\varepsilon^2}} r^{\beta-2}dr \\ &= -\frac{2\pi A}{\beta-1}\lim_{\varepsilon\to 0}\lambda\{(\varepsilon^2+\lambda^2)^{(\beta-1)/2} - \lambda^{\beta-1}\} \\ &= -\frac{2\pi A}{\beta-1}\lim_{\varepsilon\to 0}(\frac{\lambda}{\varepsilon})^{1-\beta}\lambda^{\beta}\{(1+\frac{\lambda^2}{\varepsilon^2})^{(\beta-1)/2} - (\frac{\lambda}{\varepsilon})^{\beta-1}\} \\ &= 0 \end{aligned} \tag{8.76}$$

as $\lambda \to 0$ (M_0 approaching the boundary) and $\varepsilon \to 0$ with $\lambda/\varepsilon \to 0$.

Substitution of these results into (8.63) gives the following boundary integral equation

$$u(M_o) = -\frac{1}{2\pi}\iint_{\Gamma}[u(M)\frac{\partial}{\partial n}(\frac{1}{r}) - \frac{1}{r}\frac{\partial u(M)}{\partial n}]dS \tag{8.77}$$

where M_0 is on the boundary and M is the source point for the surface integral. Note that (8.77) and (8.52) can be combined to give a single integral equation.

$$\alpha_b u(M_o) = -\frac{1}{4\pi}\iint_\Gamma [u\frac{\partial}{\partial n}(\frac{1}{r}) - \frac{1}{r}\frac{\partial u}{\partial n}]dS \tag{8.78}$$

where

$$\alpha_b = \begin{cases} 1 & M_0 \text{ in } \Omega \\ \frac{1}{2} & M_0 \text{ on } \Gamma \end{cases} \tag{8.79}$$

As discussed by Kellogg (1929), as long as the double layer potential is continuous and the single layer potential satisfies the Holder condition, the boundary value of u can be evaluated from a boundary integral.

The boundary integral shown in (8.78) also forms the basis of the boundary element method (Brebbia et al., 1983).

8.6 GREEN'S FUNCTION METHOD FOR LAPLACE EQUATION

In the Green theorem given in (8.41), if both u and v are harmonic functions, we have

$$\iint_\Gamma (u\frac{\partial v}{\partial n} - v\frac{\partial u}{\partial n})dS = 0 \tag{8.80}$$

Adding (8.80) to (8.52), we obtain

$$u(M_0) = \iint_\Gamma \left\{ u[\frac{\partial v}{\partial n} - \frac{\partial}{\partial n}(\frac{1}{4\pi r_{MM_0}})] + (\frac{1}{4\pi r_{MM_0}} - v)\frac{\partial u}{\partial n}] \right\} dS \tag{8.81}$$

If we impose the following boundary condition when we determine v:

$$v|_\Gamma = \left.\frac{1}{4\pi r_{MM_0}}\right|_\Gamma , \tag{8.82}$$

we can simplify (8.81) as

$$u(M_0) = -\iint_\Gamma u\frac{\partial}{\partial n}(\frac{1}{4\pi r_{MM_0}} - v)dS \tag{8.83}$$

Let us now define the function inside the bracket as

$$G(M, M_0) = \frac{1}{4\pi r_{MM_0}} - v \tag{8.84}$$

Then,

$$u(M_0) = -\iint_\Gamma u\frac{\partial G}{\partial n}dS \tag{8.85}$$

This function G is called Green's function.

On the other hand, if the boundary condition is given such that

$$\left.\frac{\partial v}{\partial n}\right|_{\Gamma}=\frac{\partial}{\partial n}\left(\frac{1}{4\pi r_{MM_0}}\right)\Bigg|_{\Gamma}, \tag{8.86}$$

for this boundary condition, (8.81) can be simplified to

$$u(M_0)=\iint_{\Gamma}\left\{G\frac{\partial u}{\partial n}\right\}dS \tag{8.87}$$

For this example, we can see that Green's function appears naturally and the structural form of Green's function depends only on the differential equation.

Note that this structural form of the two-part Green's function in (8.84) was mentioned in (8.10) in Section 8.3. Physically, these two parts of Green's function appears naturally. For example, for the case of the electrostatic problem for a finite domain V, the fundamental problem can be recast as:

$$\nabla^2 G=-\frac{1}{\varepsilon_0}\delta(\boldsymbol{r}-\boldsymbol{r}_0) \tag{8.88}$$

where ε_0 is the permittivity in a vacuum as discussed in (2.103) in Chapter 2. If this problem is recast into an infinite space:

$$\nabla^2 G=-\frac{1}{\varepsilon_0}[\delta(\boldsymbol{r}-\boldsymbol{r}_0)+\phi(\Gamma)\delta_{\Gamma}] \tag{8.89}$$

where δ_Γ is the delta function on the boundary of domain V (i.e., only nonzero on the boundary), the second term on the right of (8.89) is physically the felt electric potential on the boundary induced by the point source at location $\boldsymbol{r}=\boldsymbol{r}_0$ (i.e., induced by the first term on the right of (8.89) at the boundary of V). Thus, Green's function can be recast into two parts:

$$G=G_0+G_1 \tag{8.90}$$

where these two parts satisfy

$$\nabla^2 G_0=-\frac{1}{\varepsilon_0}\delta(\boldsymbol{r}-\boldsymbol{r}_0) \tag{8.91}$$

$$\nabla^2 G_1=-\frac{1}{\varepsilon_0}\phi(\Gamma)\delta_{\Gamma} \tag{8.92}$$

Physically, the first part G_0 is Green's function in an infinite space (or called the fundamental solution) and the second part of Green's function in this case is the electric potential induced by the felt potential on the finite boundary of domain V.

Thus, Green's function for the finite body is

$$G=-\frac{1}{2\pi\varepsilon_0}\ln r+G_1 \tag{8.93}$$

The exact form of G_1 depends on the geometry of the boundary as well as the boundary conditions.

8.7 GENERAL GREEN'S THEOREM FOR SECOND ORDER PDE

For the hyperbolic equation, the following generalized Green's theorem was derived by Riemann in 1860 in two-dimensional space. This work by Riemann was not known until du Bois-Reymond noticed it in 1864 and was subsequently publicized by Darboux. Actually, similar results were also obtained by Kirchhoff in 1895 and Volterra. In particular, the following hyperbolic differential operator is considered:

$$F(u) = \frac{\partial^2 u}{\partial x \partial y} + A\frac{\partial u}{\partial x} + B\frac{\partial u}{\partial y} + Cu \tag{8.94}$$

The adjoint differential operator of F is derived in Section 7.5 as

$$G(v) = \frac{\partial^2 v}{\partial x \partial y} - \frac{\partial}{\partial x}(Av) - \frac{\partial}{\partial y}(Bv) + Cv \tag{8.95}$$

From the result of Section 7.5, we have the following identity

$$vF(u) - uG(v) = \frac{\partial P}{\partial x} + \frac{\partial Q}{\partial y} \tag{8.96}$$

where

$$P = \frac{1}{2}(v\frac{\partial u}{\partial y} - u\frac{\partial v}{\partial y}) + Auv \tag{8.97}$$

$$Q = \frac{1}{2}(v\frac{\partial u}{\partial x} - u\frac{\partial v}{\partial x}) + Buv \tag{8.98}$$

Integrating (8.96) over the two-dimensional domain, we have

$$\iint [vF(u) - uG(v)]dxdy = \iint [\frac{\partial P}{\partial x} + \frac{\partial Q}{\partial y}]dxdy = \int (Pdy - Qdx) \tag{8.99}$$

This is generalization of Green's second identity for differential operators other than the Laplacian given in (8.41).

More generally, this Green's theorem has been extended to any linear general second order differential operator with m variables by Hadamard with the following differential operators:

$$F(u) = \sum_{i,k=1}^{m} A_{ik} \frac{\partial^2 u}{\partial x_i \partial x_k} + \sum_{i=1}^{m} B_i \frac{\partial u}{\partial x_i} + Cu \tag{8.100}$$

The corresponding adjoint operator of F is

$$G(v) = \sum_{i,k=1}^{m} \frac{\partial^2}{\partial x_i \partial x_k}(A_{ik}v) - \sum_{i=1}^{m} \frac{\partial}{\partial x_i}(B_i v) + Cv \tag{8.101}$$

The starting identity for Green's theorem becomes

$$vF(u) - uG(v) = \frac{\partial P_1}{\partial x_1} + ... + \frac{\partial P_m}{\partial x_m} \tag{8.102}$$

where

$$P_i = \sum_{k=1}^{m} vA_{ik} \frac{\partial u}{\partial x_k} - \sum_{k=1}^{m} uA_{ik} \frac{\partial v}{\partial x_k} - uv(\sum_{k=1}^{m} \frac{\partial A_{ik}}{\partial x_k} - B_i) \tag{8.103}$$

For this operator, the generalized Green's theorem becomes

$$\iiint [vF(u) - uG(v)]dV = -\iint (\pi_1 P_1 + \pi_1 P_1 + ... + \pi_n P_n)dS \tag{8.104}$$

where dV and the directional cosine are defined as:

$$dV = dx_1 dx_1 \cdots dx_m \tag{8.105}$$

$$\pi_i = \cos(n, x_i) \tag{8.106}$$

The hypersurface of the m-dimensional space is denoted by dS. In addition, π_i is the cosine of the angle between the normal to the surface and the x_i axis.

8.8 GREEN'S THEOREM FOR BIHARMONIC OPERATOR

More generally, Stakgold (1968) showed that the general Green's theorem can be formulated in terms of an n-th order differential equation:

$$L(u) = 0 \tag{8.107}$$

The associated adjoint problem is

$$L^*(u) = 0 \tag{8.108}$$

In fact, the adjoint problem is defined by the following generalized Lagrange identity:

$$vL(u) - uL^*(v) = \nabla \cdot \boldsymbol{J}(u, v) \tag{8.109}$$

where vector $\boldsymbol{J}$ is a bilinear function of u and v involving derivatives of up to order n–1. The formula given in (8.109) is known as the Lagrange identity. It was discussed in Chapter 3 for the case of ODEs. In this case, (8.109) is reduced to

$$vL(u) - uL^*(v) = \frac{d}{dx} J(u, v) \tag{8.110}$$

The general form of Green's theorem for (8.109) is

$$\iiint [vL(u) - uL^*(v)] dV = \iint \boldsymbol{n} \cdot \boldsymbol{J} dS \tag{8.111}$$

For the biharmonic operator, it can be shown that the adjoint operator is also a biharmonic operator (Stakgold, 1968). Equation (5.76) of Stakgold (1968) gives the following result for Green's theorem for the biharmonic operator as

$$\iiint [v\nabla^4 u - u\nabla^4 v] dV = -\iint (v\frac{\partial}{\partial n}\nabla^2 u - u\frac{\partial}{\partial n}\nabla^2 v + \nabla^2 v\frac{\partial u}{\partial n} - \nabla^2 u\frac{\partial v}{\partial n}) dS \tag{8.112}$$

This result was first obtained by Mathieu in 1869 and rederived by Koialovich in 1903.

Green's theorem in (8.112) can be proved easily by starting from Green's second identity given in (8.41). First, we make the following substitution:

$$v = \nabla^2 \phi, \quad u = \psi \tag{8.113}$$

Substitution of (8.109) into Green's second identity (8.41) gives

$$\iiint_\Omega (\psi\nabla^4\phi - \nabla^2\phi\nabla^2\psi) d\Omega = \iint_\Gamma (\psi\frac{\partial \nabla^2 \phi}{\partial n} - \nabla^2\phi\frac{\partial \psi}{\partial n}) dS \tag{8.114}$$

Secondly, we make the following substitution into (8.41):

$$v = \phi, \quad u = \nabla^2 \psi \tag{8.115}$$

$$\iiint_\Omega (\nabla^2\psi\nabla^2\phi - \phi\nabla^4\psi) d\Omega = \iint_\Gamma (\nabla^2\psi\frac{\partial \phi}{\partial n} - \phi\frac{\partial \nabla^2 \psi}{\partial n}) dS \tag{8.116}$$

Adding (8.114) and (8.116), we get

$$\iiint_{\Omega}(\psi\nabla^4\phi-\phi\nabla^4\psi)d\Omega=\iint_{\Gamma}(\psi\frac{\partial\nabla^2\phi}{\partial n}+\nabla^2\psi\frac{\partial\phi}{\partial n}-\phi\frac{\partial\nabla^2\psi}{\partial n}-\nabla^2\phi\frac{\partial\psi}{\partial n})dS \quad (8.117)$$

This is identical to (8.192) if we make the following identifications:

$$\psi\leftarrow v,\quad \phi\leftarrow u \quad (8.118)$$

This completes the proof of (8.112).

Now we can use (8.117) to formulate the solution of the biharmonic equation in terms of Green's function. Consider a nonhomogeneous biharmonic equation as:

$$\nabla^4 u=f(\boldsymbol{x}),\quad u=\frac{\partial u}{\partial n}=0,\quad \text{on}\,\boldsymbol{\Gamma} \quad (8.119)$$

To solve this problem, we consider the following Green's function defined by

$$\nabla^4 G=\delta(\boldsymbol{x}-\boldsymbol{x}_0),\quad G=\frac{\partial G}{\partial n}=0,\quad \text{on}\,\boldsymbol{\Gamma} \quad (8.120)$$

We can now make the following identifications for ϕ and ψ in (8.117)

$$\psi\leftarrow G,\quad \phi\leftarrow u \quad (8.121)$$

Then, we have

$$\iiint_{\Omega}(G\nabla^4 u-u\nabla^4 G)d\Omega=\iint_{\Gamma}(G\frac{\partial\nabla^2 u}{\partial n}+\nabla^2 G\frac{\partial u}{\partial n}-u\frac{\partial\nabla^2 G}{\partial n}-\nabla^2 u\frac{\partial G}{\partial n})dS \quad (8.122)$$

Substitution of the results from (8.119) and (8.120) into (8.122), we have all the boundary terms on the right of (8.122) zero and obtain

$$u(\boldsymbol{x})=\iiint_{\Omega}G(\xi,\boldsymbol{x})f(\xi)d\Omega \quad (8.123)$$

where $G(\xi,\boldsymbol{x})$ is the response at point ξ due to a force term applied at $\boldsymbol{x}$. This seems strange and confusing. For the present case of the biharmonic operator, it is self-adjoint and we have the reciprocity of

$$G(\xi,\boldsymbol{x})=G(\boldsymbol{x},\xi) \quad (8.124)$$

More discussion on this will be given in a later section.

If the boundary conditions are nonhomogeneous, the problem cannot be solved by using (8.123). This scenario is considered next using the boundary integral equation approach.

8.9 INTEGRAL EQUATION FOR BIHARMONIC PROBLEMS

The boundary integral equation for the two-dimensional biharmonic equation was derived by Christiansen and Hougaard in 1978 for the case of a circular domain. The formulation was extended to the general domain by Fuglede (1981). Weber et al. (2012) gave a slightly different derivation of Fuglede's result. In particular, Fuglede considered the following biharmonic Dirichlet problem with domain Ω and boundary S:

$$\nabla^4 u=0\quad \text{in}\,\Omega \quad (8.125)$$

$$u=f_0,\quad \frac{\partial u}{\partial n}=f_1\quad \text{on}\,\Gamma \quad (8.126)$$

Before we formulate the solution in terms of a pair of integral equations, the following fundamental functions for biharmonic and harmonic operators are defined:

$$\nabla^4 G = -\delta \quad \text{in} \, \Omega \tag{8.127}$$

$$\nabla^2 g = -\delta \quad \text{in} \, \Omega \tag{8.128}$$

For two-dimensional cases, we find

$$g = \frac{1}{2\pi} \ln(\frac{1}{r}) \tag{8.129}$$

$$G = \frac{1}{8\pi} r^2 \ln(\frac{1}{r}) \tag{8.130}$$

The proof of these expressions will be deferred to a later section. In addition, these two fundamental functions can be related by:

$$\nabla^2 G = g - \frac{1}{2\pi} \tag{8.131}$$

To show this, we can substitute (8.130) into the left hand side of (8.131) as

$$\begin{aligned} \nabla^2 G &= \frac{1}{r}\frac{\partial}{\partial r}\{r\frac{\partial}{\partial r}(\frac{1}{8\pi} r^2 \ln \frac{1}{r})\} \\ &= \frac{1}{8\pi r}\frac{\partial}{\partial r}(2r^2 \ln \frac{1}{r} - r^2) \\ &= \frac{1}{2\pi}\ln(\frac{1}{r}) - \frac{1}{2\pi} \end{aligned} \tag{8.132}$$

In view of (8.129), the validity of (8.131) is established.

Now, we recall Green's theorem from (8.41) with the following substitution $u = g$ to give

$$\iint_\Omega (g\nabla^2 v - v\nabla^2 g) d\Omega = \int_\Gamma (g\frac{\partial v}{\partial n} - v\frac{\partial g}{\partial n}) dS \tag{8.133}$$

Note that we have rewritten the integral in two-dimensional space. Substitution of (8.128) into (8.133) and the use of the integral property of the Dirac delta function gives

$$\alpha_b v(\boldsymbol{x}) + \iint_\Omega g\nabla^2 v d\Omega = \int_\Gamma (g\frac{\partial v}{\partial n} - v\frac{\partial g}{\partial n}) dS \tag{8.134}$$

where α_b has been defined in (8.79) and equals ½ when $\boldsymbol{x}$ is on the boundary or 1 when $\boldsymbol{x}$ is inside the domain Ω. In obtaining the first term, we have applied the following integral property of the Dirac delta function:

$$\iint_\Omega \delta(\boldsymbol{r} - \boldsymbol{r}_0) f(\boldsymbol{r}) d\Omega = f(\boldsymbol{r}_0) \tag{8.135}$$

More discussion of this integral property of the Dirac delta function will be discussed in a later section. Note that (8.134) is the first integral equation for the biharmonic operator.

Next, we make the following substitution into Green's theorem:

$$u \leftarrow G, \quad v \leftarrow \nabla^2 u \tag{8.136}$$

For the two-dimensional domain, we obtain

$$\iint_\Omega (G\nabla^4 u - \nabla^2 u \nabla^2 G) d\Omega = \int_\Gamma (G\frac{\partial \nabla^2 u}{\partial n} - \nabla^2 u \frac{\partial G}{\partial n}) dS \tag{8.137}$$

If u is a biharmonic function (i.e., satisfying (8.125)), (8.137) is reduced to

$$-\iint_\Omega \nabla^2 u \nabla^2 G d\Omega = \int_\Gamma (G\frac{\partial \nabla^2 u}{\partial n} - \nabla^2 u \frac{\partial G}{\partial n}) dS \tag{8.138}$$

Substitution of (8.131) into (8.138) gives

$$-\iint_\Omega g(\nabla^2 u) d\Omega + \frac{1}{2\pi}\iint_\Omega \nabla^2 u d\Omega = \int_\Gamma (G\frac{\partial \nabla^2 u}{\partial n} - \nabla^2 u \frac{\partial G}{\partial n}) dS \tag{8.139}$$

Using (8.134) to rewrite the first term on the left of (8.139) (i.e., use u instead of v in (8.134)), we get

$$\alpha_b u(\boldsymbol{x}) + \frac{1}{2\pi}\iint_\Omega \nabla^2 u d\Omega = \int_\Gamma (g\frac{\partial u}{\partial n} - u\frac{\partial g}{\partial n} + G\frac{\partial \nabla^2 u}{\partial n} - \nabla^2 u \frac{\partial G}{\partial n}) dS \tag{8.140}$$

The fundamental solution for harmonic operator g can be rewritten in terms of the biharmonic fundamental function G by using (8.131) as:

$$\alpha_b u(\boldsymbol{x}) + \frac{1}{2\pi}\iint_\Omega \nabla^2 u d\Omega = \int_\Gamma (\nabla^2 G\frac{\partial u}{\partial n} - u\frac{\partial \nabla^2 G}{\partial n} + G\frac{\partial \nabla^2 u}{\partial n} - \nabla^2 u \frac{\partial G}{\partial n}) dS + \frac{1}{2\pi}\int_\Gamma \frac{\partial u}{\partial n} dS \tag{8.141}$$

Finally, we can apply the following Gauss theorem to the second term on the left:

$$\iint_\Omega \nabla \cdot \boldsymbol{v} d\Omega = \int_\Gamma \boldsymbol{v} \cdot \boldsymbol{n} dS \tag{8.142}$$

Thus, we have

$$\iint_\Omega \nabla^2 u d\Omega = \iint_\Omega \nabla \cdot (\nabla u) d\Omega = \int_\Gamma (\nabla u) \cdot \boldsymbol{n} dS = \int_\Gamma \frac{\partial u}{\partial n} dS \tag{8.143}$$

With this result, (8.141) is simplified to

$$\alpha_b u(\boldsymbol{x}) = \int_\Gamma (\nabla^2 G\frac{\partial u}{\partial n} - u\frac{\partial \nabla^2 G}{\partial n} + G\frac{\partial \nabla^2 u}{\partial n} - \nabla^2 u \frac{\partial G}{\partial n}) dS \tag{8.144}$$

We can now specify (8.144) on the boundary and substitute the given boundary condition from (8.126) to get

$$\frac{1}{2} f_0(\boldsymbol{x}) = \int_\Gamma (f_1 \nabla^2 G - f_0 \frac{\partial \nabla^2 G}{\partial n} + G\frac{\partial \nabla^2 u}{\partial n} - \nabla^2 u \frac{\partial G}{\partial n}) dS \tag{8.145}$$

We now substitute the following identification for v in (8.134):

$$v \leftarrow \nabla^2 u \tag{8.146}$$

where u is a biharmonic function. Then, (8.134) becomes

$$\alpha_b \nabla^2 u + \iint_\Omega g \nabla^4 u d\Omega = \int_\Gamma (g\frac{\partial \nabla^2 u}{\partial n} - \nabla^2 u \frac{\partial g}{\partial n}) dS \tag{8.147}$$

The second term on the left of (8.147) vanishes identically as a consequence of the biharmonic function. Thus, on the boundary we get

$$\frac{1}{2}\nabla^2 u = \int_\Gamma (g\frac{\partial \nabla^2 u}{\partial n} - \nabla^2 u \frac{\partial g}{\partial n}) dS \tag{8.148}$$

Equations (8.145) and (8.148) provide a pair of integral equations on the boundary but the unknown is not u, but is instead

$$\nabla^2 u \quad \text{and} \quad \frac{\partial \nabla^2 u}{\partial n} \tag{8.149}$$

Making the following identifications, we have two unknowns defined as

$$\nabla^2 u \leftarrow v \quad \text{and} \quad \frac{\partial \nabla^2 u}{\partial n} \leftarrow w \tag{8.150}$$

The pair of coupled systems for v and w becomes

$$\frac{1}{2} f_0(\mathbf{x}) = \int_\Gamma (f_1 \nabla^2 G - f_0 \frac{\partial \nabla^2 G}{\partial n} + G w - v \frac{\partial G}{\partial n}) dS \tag{8.151}$$

$$\frac{1}{2} v = \int_\Gamma (g w - v \frac{\partial g}{\partial n}) dS \tag{8.152}$$

where f_0, f_1, are given on the boundary by (8.126), and g and G are the fundamental solutions for the Laplace and biharmonic equations respectively, and have been given in (8.129) and (8.130). Numerical evaluation of this pair of integrals is discussed in Weber et al. (2012) using the boundary element method in applications to smooth data in computer graphics.

It has been shown by Payne and Weinberger that the nonhomogeneous biharmonic Dirichlet problem can be transformed to the homogeneous one discussed in this section defined in (8.125) and (8.126). In particular, a nonhomogeneous Dirichlet problem can be formulated as:

$$\nabla^4 v = F \quad \text{in}\, \Omega \tag{8.153}$$

$$v = f, \quad \frac{\partial v}{\partial n} = g \quad \text{on}\, \Gamma \tag{8.154}$$

We can decompose the unknown v as

$$v = u_1 + u_2 \tag{8.155}$$

The formulation for u_1 and u_2 are Problems 1 and 2:

Problem 1:

$$\nabla^4 u_1 = F \quad \text{in}\, \Omega \tag{8.156}$$

$$u_1 = 0, \quad \frac{\partial u_1}{\partial n} = 0 \quad \text{on}\, \Gamma \tag{8.157}$$

Problem 2:

$$\nabla^4 u_2 = 0 \quad \text{in}\, \Omega \tag{8.158}$$

$$u_2 = f, \quad \frac{\partial u_2}{\partial n} = g \quad \text{on}\, \Gamma \tag{8.159}$$

Problem 1 can be solved by using Green's function method as

$$u_1 = \frac{1}{8\pi} \int_\Gamma F r^2 \ln r dS \tag{8.160}$$

Problem 2 is the homogeneous problem defined in (8.125) and (8.126). Thus, the solution u_2 can be obtained by solving the pair of integral equations given in

(8.151) and (8.152). Thus, substitution of the solution of u_2 and the solution in (8.160) into (8.155) gives the final solution for the nonhomogeneous problem.

Thus, the pair of integral equations approach discussed in this section can also be applied to the nonhomogeneous biharmonic problem.

8.10 DIRAC DELTA FUNCTION

For simplicity, we first recast the definition of a Dirac delta function defined in (8.8) and (8.9) for the one-dimensional case:

$$\begin{aligned} \delta(x) &= 0 \quad x \neq 0 \\ &= \infty \quad x = 0 \end{aligned} \tag{8.161}$$

$$\int_{-\infty}^{\infty} \delta(x) dx = 1 \tag{8.162}$$

This can be considered as a special case of the one given in (8.8) and (8.9). Another basic definition of the delta function is the shifting property of the Dirac delta function:

$$\int_{-\infty}^{\infty} f(x)\delta(x) dx = f(0) \tag{8.163}$$

This Dirac delta function is named in honor of electrical engineer and physicist, Paul Dirac, who received the Nobel Prize in physics for his major contribution to quantum mechanics. The idea of using the Dirac delta function was actually much earlier than its usage in quantum mechanics by Dirac. For example, it can be used to prescribe point force in the case of beam bending subject to a concentrated force. Physically, in the domain of mechanics, the delta function defined above is a point force or called a concentrated point.

According to our normal understanding in mathematics, there is no function that is nonzero only at one point and yet still has a finite integral as defined in (8.162). To justify the rigorous meaning and the use of the Dirac delta function, mathematicians developed a new theory for the delta function defined above. This theory is called the theory of distribution or the so-called generalized theory. The main contributors of this new mathematical theory include Bochner in 1932, Sobelev in 1936, Schwarz in 1940, Mikusinski in 1948, Temple in 1953, Lighthill in 1958, and Zemanian in 1965. It was Schwarz who put this theory on a firm foundation, and was awarded the inaugural Fields Medal in 1950 because of his contribution to distribution theory. He linked the distribution theory of the Dirac delta to the Fourier transform and such analysis was found very fruitful in solving partial differential equations. Schwartz's student Hormander was also awarded the Fields Medal in 1962 on related work. Some regard the Fields Medal as equivalent to the Noble Prize in physics (although their award criteria are quite different). In this sense, the importance of the Dirac delta function and the associated theory of distribution should not be overlooked in the development of mathematics.

8.11 DISTRIBUTION THEORY

The Dirac delta function is not a regular smooth continuous function that we are familiar with. However, its physical significance in physics and engineering leads to mathematicians attempting to rationalize its usage. It is a very powerful mathematical tool to model the point charge in electromagnetism or quantum mechanics, point mass in gravitational theory, point forces in mechanics, and point sources in heat conduction problems. A whole new branch of mathematics called distribution theory or theory of generalized function is introduced. The distribution theory is rather theoretical for most engineers and scientists. We will only discuss the basic concept of distribution theory in this section. The study of the Dirac function is also known as the study of generalized functions (in contrast to the regular functions that we have learned). A new set of rules on its manipulation are defined rigorously for this strangely behaving Dirac delta function. As mentioned in the last section, major contributors to its development include Bochner, Sobolev, Schwartz, Zemanian, Mikusinski, and Temple.

In the following subsections, the properties of the Dirac delta function in the sense of distribution will be summarized, and the concept of using generalized functions (called δ-sequence functions) to model the Dirac delta function is discussed.

8.11.1 Properties of Dirac Delta Function

According to the theory of distribution, the Dirac delta function is not treated as a pointwise function, but instead it is defined in terms of how it operates or integrates with other well-behaved functions. The integration given in (8.163) is first rewritten as:

$$\int_{-\infty}^{\infty} f(x)\delta(x)dx = < f(x), \delta(x) > = f(0) \tag{8.164}$$

where integration is replaced by $< , >$ and is called functional. The function $f(x)$ is now called a testing function. The function $f(x)$ needs to be infinitely differentiable and converge to zero at $\pm\infty$. We will replace the Dirac function by a distribution $\phi(x)$ such that

$$< f(x), \phi(x) > = \int_{-\infty}^{\infty} f(x)\phi(x)dx \tag{8.165}$$

According to the rules of integration, we have the following properties in the sense of distribution:

(i)

$$< \phi, \alpha f + \beta g > = \alpha < \phi, f > + \beta < \phi, g > \tag{8.166}$$

where α and β are constants. For a regular function $h(x)$, we have

(ii)

$$< h\phi, f > = < \phi, hf > \tag{8.167}$$

(iii)

$$< \phi(x-a), f(x) > = < \phi(x), f(x+a) > \tag{8.168}$$

(iv)

$$<\phi(ax), f(x)>=\frac{1}{|a|}<\phi(x), f(\frac{x}{a})> \tag{8.169}$$

(v)

$$<\phi'(x), f(x)>=-<\phi(x), f'(x)> \tag{8.170}$$

Properties (iii) and (iv) can be proved by change of variables. Note that Property (v) can be proved by integration by parts. The proofs of these identities will be left as an exercise for readers.

Some amazing properties of the delta function can be established by using these integral definitions of distribution theory by identifying the distribution ϕ as Dirac delta function δ.

Example 8.2 Find the value of the following function in the sense of distributions:

$$h(x)\delta(x) \tag{8.171}$$

where $h(x)$ is a regular function.

Solution: Let us consider the functional of this function with a testing function $f(x)$ as:

$$<h\delta, f>=<\delta, hf>=h(0)f(0) \tag{8.172}$$

The last result in (8.172) is a result of applying (8.167) and (8.164). On the other hand, we observe that

$$\begin{aligned} h(0)f(0) &= h(0)<\delta, f> \\ &= <h(0)\delta, f> \end{aligned} \tag{8.173}$$

By comparing (8.172) and (8.173), we obtain the following amazing result in the sense of distribution

$$h(x)\delta(x)=h(0)\delta(x) \tag{8.174}$$

Note that this result is true for all testing functions $f(x)$.

Let us look at the special case of $h(x) = x$; we have in the sense of distribution

$$x\delta(x)=0\delta(x)=0 \tag{8.175}$$

This is an amazing result that comes from distribution theory.

Using Property (iii) given in (8.168), we have

$$\int_{-\infty}^{\infty}\delta(x-a)f(x)dx=<\delta(x-a), f(x)>=<\delta(x), f(x+a)>=f(a) \tag{8.176}$$

This is the shifting property of the delta function that we reported in (8.163). Using Property (iv) given in (8.169)

$$\int_{-\infty}^{\infty} \delta(ax) f(x) dx = < \delta(ax), f(x) > = \frac{1}{|a|} < \delta(x), f(\frac{x}{a}) > = \frac{1}{|a|} f(0)$$

$$= \frac{1}{|a|} < \delta(x), f(x) > \tag{8.177}$$

$$= < \frac{1}{|a|} \delta(x), f(x) >$$

Thus, we have the scaling property of the Dirac delta function in the sense of distribution

$$\delta(ax) = \frac{1}{|a|} \delta(x) \tag{8.178}$$

Using Property (v) given in (8.170), we have in the sense of distribution

$$\int_{-\infty}^{\infty} \frac{d\delta(x)}{dx} f(x) dx = < \delta'(x), f(x) > = - < \delta(x), f'(x) > = - f'(0) \tag{8.179}$$

This expression can be generalized to a higher order derivative of the Dirac delta

$$\int_{-\infty}^{\infty} \frac{d^n \delta(x)}{dx^n} f(x) dx = (-1)^n f^{(n)}(0) \tag{8.180}$$

Let us define another function called the Heaviside step function as:

$$\begin{aligned} H(x) &= 1 \qquad x > 0 \\ &= 0 \qquad x \le 0 \end{aligned} \tag{8.181}$$

This step function is illustrated in Figure 8.6. Using distribution theory, we can establish a relation between this Heaviside delta function and the Dirac delta function.

We start with the functional of the Heaviside step function and a testing function

$$\int_{-\infty}^{\infty} H(x) f(x) dx = < H, f > = \int_{0}^{\infty} f(x) dx \tag{8.182}$$

The functional of the derivative of the Heaviside step function is

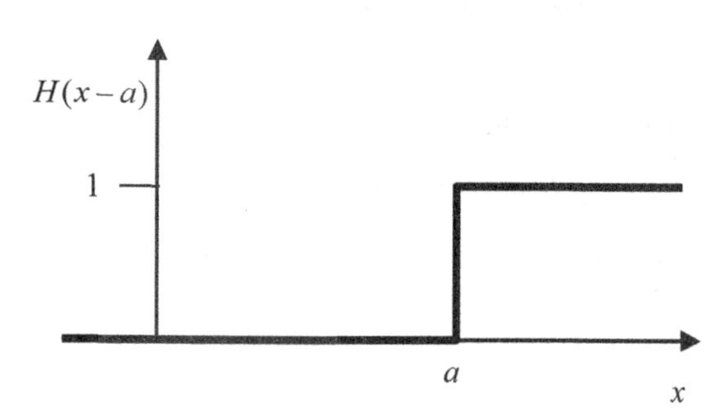

Figure 8.6 Heaviside step function

$$\int_{-\infty}^{\infty} \frac{dH(x)}{dx} f(x)dx =< H', f >= -\int_{0}^{\infty} f'(x)dx$$
$$= -f(x)\Big|_0^{\infty} = f(0) =< \delta, f > \tag{8.183}$$

Comparing the first and the last terms of (8.183), we find the following identity in the sense of distribution

$$\frac{dH(x)}{dx} = \delta(x) \tag{8.184}$$

Inversely, we can rewrite it as

$$H(x) = \int_{-\infty}^{x} \delta(\xi)d\xi \tag{8.185}$$

Using ordinary analysis, the derivative of a step function does not exist, but using the theory of distribution, its derivative is defined in (8.184). This property actually allows us to define the derivative of a discontinuous function by using Dirac delta function.

In particular, considering a piecewise continuous function containing a number of jumps, we can denote this discontinuous function $f(x)$ as depicted in Figure 8.7. We can define a continuous and piecewise differentiable function $g(x)$ in terms of $f(x)$ and a series of Heaviside step functions:

$$g(x) = f(x) - \sum_{j=1}^{k} \Delta f_j H(x - a_j) \tag{8.186}$$

where the jump is defined as

$$\Delta f_j = f(a_j^+) - f(a_j^-) \tag{8.187}$$

The function g is actually the dotted curve shown in Figure 8.7. We have assumed in (8.186) that there are k discontinuous points. By ignoring these jumps at a_1, a_2, ... a_n, we define the derivatives of function f as

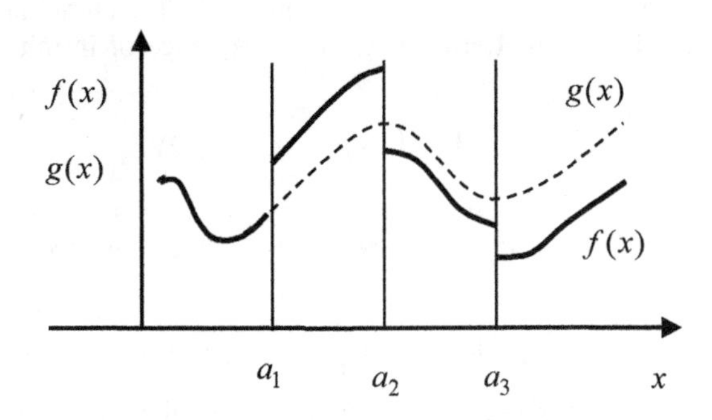

Figure 8.7 A discontinuous function in terms of Heaviside step function

$$[f'], \; [f''], \; [f'''], ..., [f^{(n)}] \tag{8.188}$$

The function $g(x)$ defined in (8.186) is a function whose derivative g' coincides with its distributional derivative. Note, however, that $[f']$ is in general different from the distributional derivative of function f. Let us now consider the differentiation of g in the sense of distributions as:

$$g'(x) = f'(x) - \sum_{j=1}^{k} \Delta f_j \delta(x - a_j) \tag{8.189}$$

But, since g is continuous, we must have $g' = [f']$ and thus from (8.189) we have

$$f'(x) = [f'(x)] + \sum_{j=1}^{k} \Delta f_j \delta(x - a_j) \tag{8.190}$$

Thus, each jump in the f multiplying delta function contributes to the distributional derivative of f as shown in (8.190).

We can also find a higher derivative of f with a similar procedure, and we have the distributional derivatives as

$$f'(x) = [f'(x)] + \sum_{j=1}^{k} \Delta f_j^{(0)} \delta(x) \tag{8.191}$$

$$f''(x) = [f''(x)] + \sum_{j=1}^{k} [\Delta f_j^{(1)} \delta(x) + \Delta f_j^{(0)} \delta'(x)] \tag{8.192}$$

$$f^{(m)}(x) = \left[f^{(m)}(x)\right] + \sum_{j=1}^{k}]\Delta f_j^{(m-1)} \delta(x) + \ldots + \Delta f_j^{(0)} \delta^{(m-1)}(x)] \tag{8.193}$$

Our discussion of the theory of distribution stops here.

8.11.2 Delta Function as Sequence of Functions

There is no function in a regular sense that satisfies the definition of (8.164) to (8.170). However, there are perfectly regular functions which get as close to the Dirac delta function as one like. Thus, we can define the Dirac delta function as a sequence of these functions. Let us consider a sequence of infinitely differentiable functions defined as f_n

$$\lim_{n\to\infty} \int_{-\infty}^{\infty} f_n(x) g(x) dx = g(0) \tag{8.194}$$

The Dirac delta function is considered as the limit of a sequence of functions. These functions are normally referred to as generalized functions. Using the sense of distribution, we define

$$\int_{-\infty}^{\infty} \phi(x) g(x) dx = \lim_{n\to\infty} \int_{-\infty}^{\infty} f_n(x) g(x) dx = g(0) \tag{8.195}$$

However, this definition is not equal to pointwise equivalence. In other words, we do not assume that the sequence function is equal to the delta function directly. That is,

$$\delta(x) \neq \lim_{n\to\infty} f_n(x) \tag{8.196}$$

Instead, we need the integration, or in a distributional sense, to enforce the closeness of this sequence function to the delta function (see (8.195)).

Here are some examples of sequence functions:

(i) First δ-sequence

$$f_n(x) = \frac{n}{\pi}\frac{1}{1+n^2x^2} \tag{8.197}$$

This function is illustrated in Figure 8.8, and, as expected, it approaches the shape of the Dirac delta function as n increases. Substitution of (8.197) into (8.195) gives

$$\begin{aligned}
\lim_{n\to\infty}\int_{-\infty}^{\infty} f_n(x)g(x)dx &= \lim_{n\to\infty}\frac{n}{\pi}\int_{-\infty}^{\infty}\frac{g(x)}{1+n^2x^2}dx \\
&= \lim_{n\to\infty}\frac{1}{\pi}\int_{-\infty}^{\infty}\frac{g(\zeta/n)}{1+\zeta^2}d\zeta \\
&= \frac{1}{\pi}\int_{-\infty}^{\infty}\frac{g(0)}{1+\zeta^2}d\zeta \\
&= \frac{g(0)}{\pi}[\tan^{-1}(\infty)-\tan^{-1}(-\infty)] \\
&= g(0)
\end{aligned} \tag{8.198}$$

where we have applied a change of variables of $\zeta = nx$. However, the calculation in (8.198) is not strictly valid as $\zeta \to 0$. This is because we do not have the behavior of a delta function, although it decays to zero as $\zeta \to \infty$. To fix the non-uniformity problem, we can integrate the left side of (8.198) as:

$$\frac{n}{\pi}\int_{-\infty}^{\infty}\frac{g(x)}{1+n^2x^2}dx = \frac{n}{\pi}\int_{-\infty}^{\infty}\frac{g(0)}{1+n^2x^2}dx + \frac{n}{\pi}\int_{-\infty}^{\infty}\frac{g(x)-g(0)}{1+n^2x^2}dx \tag{8.199}$$

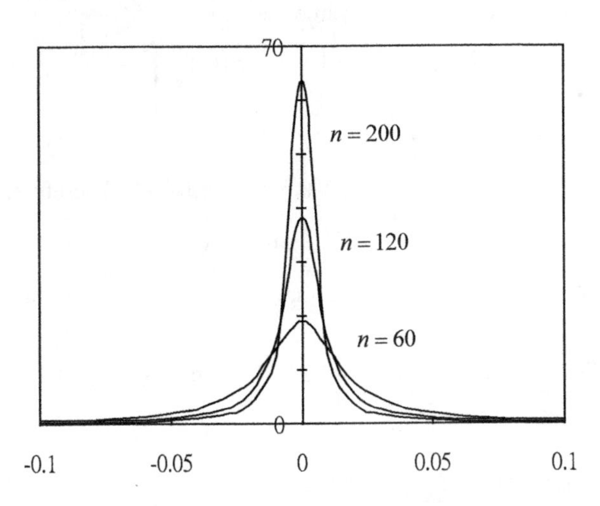

Figure 8.8 The first δ-sequence

The first term on the right of (8.199) has been evaluated in the second part of (8.198). Thus, we have

$$
\begin{aligned}
\frac{n}{\pi}\int_{-\infty}^{\infty}\frac{g(x)}{1+n^2x^2}dx &= g(0)+\frac{n}{\pi}\int_{-\infty}^{\infty}\frac{g(x)-g(0)}{1+n^2x^2}dx \\
&= g(0)+\frac{n}{\pi}\{\int_{-\infty}^{-\varepsilon}\frac{g(x)-g(0)}{1+n^2x^2}dx+\int_{-\varepsilon}^{\varepsilon}\frac{g(x)-g(0)}{1+n^2x^2}dx \\
&\quad +\int_{\varepsilon}^{\infty}\frac{g(x)-g(0)}{1+n^2x^2}dx\} \\
&= g(0)+I_{-\infty}+I_{\varepsilon}+I_{\infty}
\end{aligned}
\tag{8.200}
$$

The integrals for the tails that approach negative and positive infinity are denoted as $I_{-\infty}$ and I_{∞} respectively. If the value of $g(x)-g(0)$ is bounded such that

$$
|g(x)-g(0)|\le M, \tag{8.201}
$$

then, we can establish the inequality

$$
\begin{aligned}
I_{-\infty}=I_{\infty} &\le M\frac{n}{\pi}\int_{-\infty}^{\infty}\frac{1}{1+n^2x^2}dx \\
&\le \frac{2M}{\pi}[\frac{\pi}{2}-\tan^{-1}(n\varepsilon)]
\end{aligned}
\tag{8.202}
$$

Let us consider the limiting case that

$$
\varepsilon\to 0,\quad n\to\infty,\quad n\varepsilon\to\infty \tag{8.203}
$$

Note that this can be achieved easily by setting

$$
\varepsilon=\frac{1}{\sqrt{n}} \tag{8.204}
$$

This is clearly not the only choice for ε that satisfies (8.203). Thus, both $I_{-\infty}$ and I_{∞} approach zero as long as $g(x)$ is bounded near $x = 0$. Finally, the integral around the origin can be evaluated using the mean value theorem:

$$
\begin{aligned}
I_{\varepsilon} &= \frac{n}{\pi}\int_{-\varepsilon}^{\varepsilon}\frac{g(x)-g(0)}{1+n^2x^2}dx=[g(\xi)-g(0)]\frac{n}{\pi}\int_{-\varepsilon}^{\varepsilon}\frac{dx}{1+n^2x^2} \\
&= [g(\xi)-g(0)]\frac{2}{\pi}\tan^{-1}(n\varepsilon)
\end{aligned}
\tag{8.205}
$$

where $-\varepsilon<\xi<\varepsilon$. Since $n\varepsilon\to\infty$, and $\varepsilon\to 0$, we have $I_{\varepsilon}\to 0$. Therefore, we have

$$
\lim_{n\to\infty}\frac{n}{\pi}\int_{-\infty}^{\infty}\frac{g(x)}{1+n^2x^2}dx=g(0) \tag{8.206}
$$

In addition, if we set $g(x) = 1$, we see that the requirement of (8.162) is also fulfilled:

$$
\lim_{n\to\infty}\frac{n}{\pi}\int_{-\infty}^{\infty}\frac{1}{1+n^2x^2}dx=\frac{1}{\pi}[\tan^{-1}(\infty)-\tan^{-1}(-\infty)]=1 \tag{8.207}
$$

(ii) Second δ-sequence

$$
f_n(x)=\frac{n}{\sqrt{\pi}}e^{-n^2x^2} \tag{8.208}
$$

This is a bell-shaped function and is shown in Figure 8.9. We can see that this function approaches the Dirac delta function much faster than that given in (8.197). We will skip the details of proving (8.162) and (8.163).

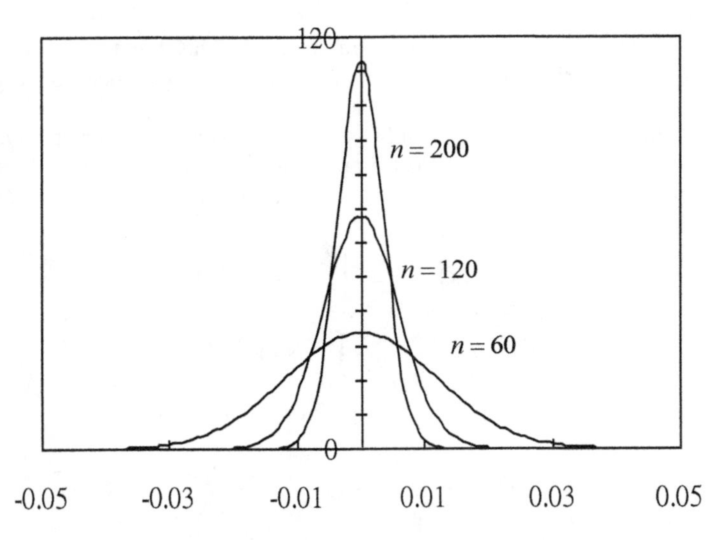

Figure 8.9 The second δ-sequence

(iii) Third δ-sequence

$$f_n(x) = \frac{1}{n\pi} \frac{\sin^2 nx}{x^2} \tag{8.209}$$

This δ-sequence is shown in Figure 8.9 for different values of n.

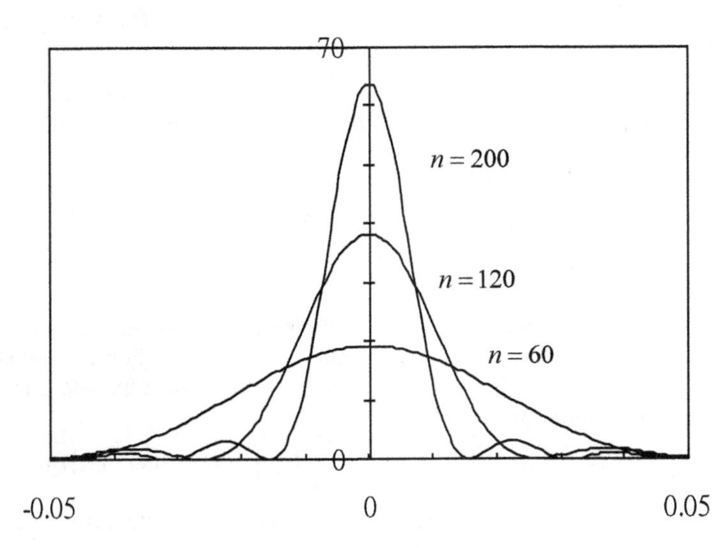

Figure 8.10 The third δ-sequence

(iv) Fourth δ-sequence

$$f_n(x) = \frac{\sin nx}{\pi x} \tag{8.210}$$

So far, all sequences discussed are non-negative. This function is illustrated in Figure 8.11 for various values of n, and, clearly, this sequence converges much slower than the first δ-sequence.

As shown in Figure 8.11, this is a weak δ-sequence for the Dirac delta function, but this sequence is very important since it is related to the Fourier transform of the Dirac delta function. The Fourier transform is defined as (see Section 11.2.2 in Chapter 11)

$$\hat{f}(k) = \frac{1}{\sqrt{2\pi}} \int_{-\infty}^{\infty} f(x) e^{-ikx} dx \tag{8.211}$$

$$f(x) = \frac{1}{\sqrt{2\pi}} \int_{-\infty}^{\infty} \hat{f}(k) e^{ikx} dk \tag{8.212}$$

We now formally substitute the Dirac delta function into f. That is,

$$f(x) = \delta(x - \xi) \tag{8.213}$$

By virtue of (8.211) and (8.212), we have

$$\hat{\delta}(k, \xi) = \frac{1}{\sqrt{2\pi}} \int_{-\infty}^{\infty} \delta(x - \xi) e^{-ikx} dx = \frac{1}{\sqrt{2\pi}} e^{-ik\xi} \tag{8.214}$$

Back substitution of this into (8.212), we have

$$\begin{aligned} \delta(x - \xi) &= \frac{1}{\sqrt{2\pi}} \int_{-\infty}^{\infty} \frac{1}{\sqrt{2\pi}} e^{-ik\xi} e^{ikx} dk \\ &= \frac{1}{2\pi} \int_{-\infty}^{\infty} e^{ik(x-\xi)} dk \end{aligned} \tag{8.215}$$

For the case of $\xi = 0$, we obtain an integral representation (or more precisely the Fourier transform) of the Dirac delta function

$$\delta(x) = \frac{1}{2\pi} \int_{-\infty}^{\infty} e^{ikx} dk \tag{8.216}$$

Finally, we can rewrite (8.216) as

$$\begin{aligned} \delta(x) &= \lim_{R\to\infty} \frac{1}{2\pi} \int_{-R}^{R} e^{ikx} dk = \lim_{R\to\infty} \frac{1}{2\pi} \int_{-R}^{R} (\cos kx + i \sin kx) dk \\ &= \lim_{R\to\infty} \frac{1}{2\pi x} \sin kx \Big|_{-R}^{R} = \lim_{R\to\infty} \frac{\sin Rx}{\pi x} \end{aligned} \tag{8.217}$$

If we take R as a large integer, we obtain the second δ-sequence given in (8.210). This δ-sequence actually gives the formal proof of the Fourier transform. In particular, if we multiply (8.213) by $f(\xi)$ and integrate with respect to ξ from minus infinity to plus infinity, we have

$$f(x) = \int_{-\infty}^{\infty} \delta(x - \xi) f(\xi) d\xi = \frac{1}{2\pi} \int_{-\infty}^{\infty} \int_{-\infty}^{\infty} e^{ik(x-\xi)} f(\xi) dk d\xi \tag{8.218}$$

We can reverse the order of integration and rewrite the last integral in (8.209) as

$$f(x)=\frac{1}{\sqrt{2\pi}}\int_{-\infty}^{\infty}e^{ikx}[\frac{1}{\sqrt{2\pi}}\int_{-\infty}^{\infty}e^{-ik\xi}f(\xi)d\xi]dk=\frac{1}{\sqrt{2\pi}}\int_{-\infty}^{\infty}e^{ikx}\hat{F}(k)dk \quad (8.219)$$

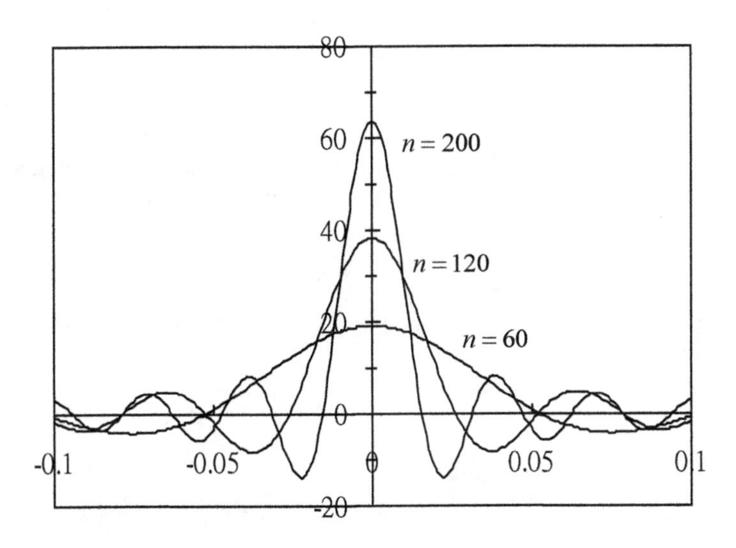

Figure 8.11 The fourth δ-sequence

This actually gives the formal representation of the Fourier transform. Thus, an appropriate form of the δ-sequence within the framework of the distribution theory can be used to derive the Fourier transform. In principle, other transforms can be recovered if proper δ-sequence representation is used.

In the next paragraph, we will see that Fourier series expansion can also be linked to the Dirac delta function if an appropriate form of δ-sequence is assumed.

(iv) Fifth δ-sequence

$$\delta(x)=\lim_{n\to\infty}\frac{\sin(n+\frac{1}{2})x}{2\pi\sin\frac{x}{2}}, \qquad |x|\le\pi$$
$$=0 \qquad |x|>\pi \quad (8.220)$$

The interval for x has been limited to from $-\pi$ to π such that (8.162) will be satisfied. This δ-sequence is essentially the same as that given in the fourth δ-sequence. To see this, we can rewrite (8.220) as

$$\delta(x)=\lim_{n\to\infty}[\frac{\sin(n+\frac{1}{2})x}{\pi x}][\frac{x}{2\sin\frac{x}{2}}] \quad (8.221)$$

We can see that for large n, the first square bracket term on the right of (8.221) is essentially the same as the fourth δ-sequence given in (8.210). The second square bracket term approaches 1 as $x \to 0$. Indeed, when we plot (8.221) using the same scale as in Figure 8.11, we essentially get the same plot (although the actual values are slightly different) and, thus, this plot will not be given.

Next, we use the following identity

$$\frac{\sin(n+\frac{1}{2})x}{2\pi \sin\frac{x}{2}} = \frac{1}{2\pi}\sum_{k=-n}^{n} e^{ikx} \tag{8.222}$$

The left hand side of (8.222) was known as the Dirichlet kernel having a period of 2π. To show the validity of (8.222), we first write the summation on the right as

$$\sum_{k=-n}^{n} e^{ikx} = -1 + \sum_{k=-n}^{0} e^{ikx} + \sum_{k=0}^{n} e^{ikx} \tag{8.223}$$

We now apply the following change of index $j = -k$ for the first sum on the right of (8.223). Then, we get

$$S = \sum_{k=-n}^{n} e^{ikx} = -1 + \sum_{k=0}^{n} e^{-ikx} + \sum_{k=0}^{n} e^{ikx} \tag{8.224}$$

More explicitly, we can express it as

$$\begin{aligned} S = -1 + \{1 + e^{ix} + e^{i2x} + \ldots + e^{inx} + \\ 1 + e^{-ix} + e^{-i2x} + \ldots + e^{-inx}\} \end{aligned} \tag{8.225}$$

It is clear that both of them are geometric series. That is, we can apply the following formula for the sum of geometric series

$$a + ar + ar^2 + \ldots + ar^{n-1} = \frac{a(1-r^n)}{1-r} \tag{8.226}$$

Application of (8.226) to both the series in (8.225) gives

$$\begin{aligned} S &= -1 + \frac{1-e^{i(n+1)x}}{1-e^{ix}} + \frac{1-e^{-i(n+1)x}}{1-e^{-ix}} \\ &= \frac{\cos(n+1)x - \cos nx}{1-\cos x} \end{aligned} \tag{8.227}$$

Note the following trigonometric identities

$$\cos A - \cos B = \frac{1}{2}\sin\frac{(A+B)}{2}\sin\frac{(B-A)}{2} \tag{8.228}$$

$$1 - \cos 2A = 2\sin^2 A \tag{8.229}$$

Substitution of (8.228) and (8.229) into (8.227) results in

$$S = \frac{\sin(n+\frac{1}{2})x}{\sin\frac{x}{2}} \tag{8.230}$$

Using (8.230) into (8.225), we finally get (8.222). This completes the proof.

Application of (8.222) to (8.220) gives another important form of the Dirac delta function:

$$\delta(x-\xi)=\frac{1}{2\pi}\sum_{k=-\infty}^{\infty}e^{ik(x-\xi)}, \qquad |x-\xi|\le\pi$$
$$=0 \qquad |x-\xi|>\pi \tag{8.231}$$

This indicates that there are infinite Dirac delta functions along the x-axis, and we must limit our consideration to 1 delta function only. We now multiply (8.231) by $f(\xi)$ and integrate from $-\pi$ to π to get

$$f(x)=\int_{-\pi}^{\pi}\delta(x-\xi)f(\xi)d\xi=\frac{1}{2\pi}\sum_{k=-\infty}^{\infty}e^{ikx}\int_{-\pi}^{\pi}f(\xi)e^{-ik\xi}d\xi, \qquad |x|\le\pi \tag{8.232}$$

This formula can be recast as:

$$f(x)=\frac{1}{2\pi}\sum_{k=-\infty}^{\infty}C_k e^{ikx}, \qquad |x|\le\pi \tag{8.233}$$

where

$$C_k=\int_{-\pi}^{\pi}f(\xi)e^{-ik\xi}d\xi \tag{8.234}$$

Therefore, we have shown that Fourier series expansion of an arbitrary function $f(x)$ can be proved using a special representation of the δ-sequence within the framework of the theory of distribution. More δ-sequences are given in Section 15.2.5 in Chapter 15.

In short, we have shown Schwartz's result that the theory of distribution of Dirac delta functions can be linked to both the Fourier series and Fourier transform. Without going into deep discussion, we mention that in solving PDEs, the Fourier series expansion technique is applicable to problems with a finite domain and PDEs with eigenvalues in the form of a discrete spectrum (like the buckling load of a column and their multiples for higher modes of buckling); whereas the Fourier transform technique is applicable to problems with an infinite domain and PDEs with eigenvalues in the form of a continuous spectrum (like wave propagation in a half-space). More discussions will be given in Chapter 10.

8.12 GREEN'S FUNCTION METHOD FOR PDE

Let us now consider Green's function method for a nonhomogeneous linear ODE of second order defined as:

$$L(u)=a_2\frac{d^2u}{dx^2}+a_1\frac{du}{dx}+a_0u=f(x) \tag{8.235}$$

where $a<x<b$ with boundary conditions:

$$B_1(u)=\alpha_{11}u(a)+\alpha_{12}u'(a)+\beta_{11}u(b)+\beta_{12}u'(b)=0 \tag{8.236}$$

$$B_2(u)=\alpha_{21}u(a)+\alpha_{22}u'(a)+\beta_{21}u(b)+\beta_{22}u'(b)=0 \tag{8.237}$$

We define Green's function of the operator L with homogeneous boundary conditions as the solution of

$$L(G)=\delta(x-\xi),\ \ B_1(G)=0,\ \ B_2(G)=0. \tag{8.238}$$

The adjoint Green's function of the adjoint operator L^* with homogeneous boundary conditions is the solution of

$$L^*(H) = \delta(x-\eta), \tag{8.239}$$

$$B_1^*(H) = 0, \quad B_2^*(H) = 0, \tag{8.240}$$

where

$$L^*(u) = \sum_{k=0}^{2} (-1)^k \frac{d^k(a_k u)}{dx^k} = a_2 \frac{d^2u}{dx^2} + (2a_2' - a_1)\frac{du}{dx} + (a_2'' - a_1' + a_1)u \tag{8.241}$$

$$B_1^*(u) = \bar{\alpha}_{11}u(a) + \bar{\alpha}_{12}u'(a) + \bar{\beta}_{11}u(b) + \bar{\beta}_{12}u'(b) = 0 \tag{8.242}$$

$$B_2^*(u) = \alpha_{21}u(a) + \alpha_{22}u'(a) + \beta_{21}u(b) + \beta_{22}u'(b) = 0 \tag{8.243}$$

This adjoint operator has been derived in an earlier chapter. We can multiply (8.238) by $H(x,\eta)$ and (8.241) by $G(x,\xi)$, subtract, and integrate from $x = a$ to $x = b$ to get

$$\begin{aligned}\int_a^b [HL(G) - GL^*(H)]\,dx &= \int_a^b H(x,\eta)\delta(x-\xi)dx - \int_a^b G(x,\xi)\delta(x-\eta)dx \\ &= H(\xi,\eta) - G(\eta,\xi) = J(G,H)\Big|_a^b\end{aligned} \tag{8.244}$$

where $a < \xi,\ \eta < b$ and

$$J(u,v)\Big|_a^b = \{a_2(vu' - uv') + (a_1 - a_2')uv\}_a^b \tag{8.245}$$

Thus, in explicit form we get

$$\begin{aligned}H(\xi,\eta) - G(\eta,\xi) = \{&a_2[H(x,\eta)G'(x,\xi) - G(x,\xi)H'(x,\eta)] \\ &+(a_1 - a_2')G(x,\xi)H(x,\eta)\}_a^b\end{aligned} \tag{8.246}$$

This is the relation between the two Green's functions of the operator L and adjoint operator L^*. In obtaining the above result, we have employed the well-known Green's formula given in (8.111). Depending on the boundary conditions of the original problem given in (8.236 and (8.237), we normally pick the boundary condition of the adjoint problem given in (8.242) and (8.243) such that (8.245) is identically zero. The following example demonstrates how to do it.

Example 8.3 It was given that the boundary conditions in (8.233) and (8.234) are

$$B_1(u) = u(a) = 0 \tag{8.247}$$

$$B_2(u) = u'(b) = 0 \tag{8.248}$$

Find the adjoint boundary conditions in (8.242) and (8.243) such that (8.245) is identically zero. What would be the relation between two Green's functions of the operator L and adjoint operator L^*?

Solution: With (8.247) and (8.248), (8.245) becomes

$$J(u,v)\Big|_a^b = \{[a_1(b) - a_2'(b)]v(b) - a_2(b)v'(b)\}u(b) - a_2(a)v(a)u'(a) \tag{8.249}$$

To set this to zero, we have

$$B_1^*(v) = v(a) = 0 \tag{8.250}$$

$$B_2^*(v) = v'(b) - [\frac{a_1(b) - a_2'(b)}{a_2(b)}]v(b) = 0 \tag{8.251}$$

Thus, we have

$$H(\xi,\eta) = G(\eta,\xi) \tag{8.252}$$

For the special case that the system is self-adjoint, we can further have

$$G(\xi,\eta) = G(\eta,\xi) \tag{8.253}$$

The reversibility of the source point and the observation point of Green's function is called the Maxwell-Rayleigh reciprocity law. In mechanics, the self-adjoint PDE implies existence of an energy function or energy conserves. The principle of virtual work only applies to the case that the corresponding PDE is self-adjoint. Most PDEs arising from classical mechanics and physics are for problems with the existence of an energy function and potential, and thus, the associated PDEs are mostly self-adjoint. For example, if we formulate problems with frictional forces, the resulting PDE is expected to be non-self-adjoint. Most of the existing books on Green's function only cover the case of self-adjoint, but without mentioning that their results are only valid for self-adjoint PDEs.

Now we return to seeking the solution of (8.235) using Green's function method. Let us consider a more general problem with nonhomogeneous boundary conditions

$$L(u) = f(x), \quad B_1(u) = \alpha, \quad B_2(u) = \beta \tag{8.254}$$

where α and β are constants. Now multiplying (8.254) by $H(x,\xi)$ and (8.241) by $u(x)$, subtracting and integrating from $x = a$ to $x = b$, we obtain

$$\int_a^b [HL(u) - uL^*(H)]\,dx = \int_a^b f(x)H(x,\xi)dx - \int_a^b u(x)\delta(x-\xi)dx \tag{8.255}$$

Applying integration by parts or Green's formula, we obtain

$$u(\xi) = \int_a^b f(x)H(x,\xi)dx - J(u,H)\big|_a^b \tag{8.256}$$

Let us rename the variables; we get

$$u(x) = \int_a^b f(\xi)H(\xi,x)d\xi - J(u(\xi),H(\xi,x))\big|_a^b \tag{8.257}$$

This is the formula for evaluating the solution of (8.254) using Green's function method.

Now we can consider the special case that $\alpha = \beta = 0$, and in view of the proper choice of adjoint boundary condition (see Example 8.3), we will have

$$J(u(x),H(\xi,x))\big|_a^b = 0 \tag{8.258}$$

For this particular case, we also have the validity of (8.252). Thus, for the homogeneous boundary condition, (8.257) can be simplified to

$$u(x) = \int_a^b f(\xi)G(x,\xi)d\xi \tag{8.259}$$

This explains the strange appearance of Green's function method that the Green's function in the integrand has the source point x and observation point ξ being reversed whilst on the left hand side, x is the observation point. Only for the special case of the self-adjoint PDE, we have in view of (8.253)

$$u(x)=\int_a^b f(\xi)G(\xi,x)d\xi \tag{8.260}$$

which is given in most textbooks without mentioning the importance of the adjoint operator.

Example 8.4 Consider the case of a simply supported beam subject to a non-uniform distributed load $q(x)$ as depicted in Figure 8.12. The problem is formulated as

$$\frac{d^2}{dx^2}(EI\frac{d^2u}{dx^2})=q(x) \tag{8.261}$$

where u is the deflection of the beam, and E and I are the Young's modulus and moment of inertia of the bean section. The boundary conditions are

$$u(a)=u(L)=0,\quad M(0)=EI\left.\frac{d^2u}{dx^2}\right|_{x=0}=0,\quad M(L)=EI\left.\frac{d^2u}{dx^2}\right|_{x=L}=0 \tag{8.262}$$

Find the solution in terms of Green's function method.

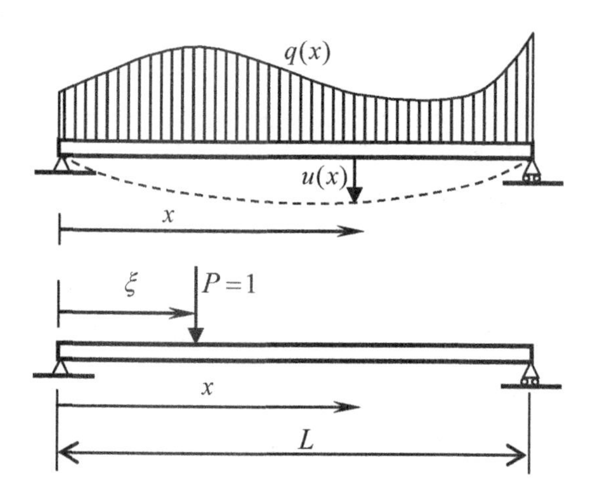

Figure 8.12 Green's function method for beam bending

Solution: For the case of a prismatic beam (i.e., beam with uniform cross-section), we can consider the beam deflection at point x subject to a unit point force applied at ξ. Mathematically, this is the formulation for Green's function

$$EI\frac{d^4u}{dx^4}=P\delta(x-\xi) \tag{8.263}$$

As shown in Figure 8.12, we have set the applied force at unity (i.e., P = 1). Balancing the unit on both sides of (8.263), we find that the dimension of the Dirac delta function is 1/length. Similarly, one can find that Dirac delta functions in 2-D and 3-D are of dimensions $1/\text{length}^2$ and $1/\text{length}^3$. By integrating (8.263), we have the following results

$$EI\frac{d^3u}{dx^3} = H(x-\xi) + C_1 \tag{8.264}$$

$$EI\frac{d^2u}{dx^2} = (x-\xi)H(x-\xi) + C_1x + C_2 \tag{8.265}$$

$$EI\frac{du}{dx} = \frac{1}{2}(x-\xi)^2 H(x-\xi) + C_1\frac{x^2}{2} + C_2x + C_3 \tag{8.266}$$

$$EIu = \frac{1}{6}(x-\xi)^3 H(x-\xi) + C_1\frac{x^3}{6} + C_2\frac{x^2}{2} + C_3x + C_4 \tag{8.267}$$

Applying the first two boundary conditions at $x = 0$ given in (8.262), we have

$$C_2 = C_4 = 0 \tag{8.268}$$

Using the boundary conditions at $x = L$ given in (8.262), we get

$$\frac{1}{6}(L-\xi)^3 + C_1\frac{L^3}{6} + C_3L = 0 \tag{8.269}$$

$$(L-\xi) + C_1L = 0 \tag{8.270}$$

The solutions of these equations are

$$C_3 = \frac{\xi}{6L}(L-\xi)(2L-\xi) \tag{8.271}$$

$$C_1 = -\frac{(L-\xi)}{L} \tag{8.272}$$

Therefore, Green's function for deflection is

$$G_u(x,\xi) = u(x,\xi) = \frac{1}{6EI}[(x-\xi)^3 H(x-\xi) + \frac{L-\xi}{L}(-x^2 + 2\xi L - \xi^2)x] \tag{8.273}$$

Similarly, we can get Green's function for moment as

$$G_M(x,\xi) = M(x,\xi) = (x-\xi)H(x-\xi) - \frac{(L-\xi)}{L}x \tag{8.274}$$

Different forms of G_u and G_M can be found in Problems 8.2 and 8.3. The beam deflection and moment can be expressed in terms of Green's functions as

$$u(x) = \int_0^L G_u(x,\xi)q(\xi)d\xi \tag{8.275}$$

$$M(x) = \int_0^L G_M(x,\xi)q(\xi)d\xi \tag{8.276}$$

In structural mechanics, Green's function for the moment given in (8.274) is also known as the influence line.

8.13 SOME RESULTS ON GREEN'S FUNCTIONS

In this chapter, we have discussed Green's function method mainly related to the Laplace equation and biharmonic equation. In this section, we will summarize some of the other commonly used Green's functions or fundamental solutions. The proofs of some of these Green's functions are out of the scope of the present chapter.

8.13.1 Helmholtz Equation

The Helmholtz equation is known as the reduced wave equation, and it results from the harmonic wave equation.

$$\nabla^2 G + k^2 G = -\delta(\boldsymbol{x}) \tag{8.277}$$

with $k^2 > 0$. The fundamental solution for 1-D, 2-D, and 3-D are respectively:

$$G_1(\boldsymbol{x}) = \frac{ie^{ik|x|}}{2k} \tag{8.278}$$

$$G_2(\boldsymbol{x}) = \frac{i}{4} H_0^{(1)}(k|\boldsymbol{x}|) \tag{8.279}$$

$$G_3(\boldsymbol{x}) = \frac{e^{ik|x|}}{4\pi|\boldsymbol{x}|} \tag{8.280}$$

where in (8.279) Green's function is expressed in terms of the Hankel function of the first kind of order zero. Note that the fundamental solution is of the wave type.

8.13.2 Diffusion Equation

For a diffusion equation subjected an initial pulse, the problem can be formulated as

$$\frac{\partial u}{\partial t} = a^2 \nabla^2 u, \quad u|_{t=0} = \delta(x-\xi) \tag{8.281}$$

The fundamental solution in 1-D is

$$\begin{aligned} u(x,t;\xi,\tau) &= \frac{1}{4\pi a^2 t} \exp\{-\frac{(x-\xi)^2}{4a^2(t-\tau)}\}, \quad t \ge \tau \\ &= 0 \qquad\qquad t < \tau \end{aligned} \tag{8.282}$$

Similarly, for the 2-D case we have

$$\begin{aligned} u(x,y,t;\xi,\eta,\tau) &= \frac{1}{4\pi a^2 t} \exp\{-\frac{(x-\xi)^2 + (y-\eta)^2}{4a^2(t-\tau)}\}, \quad t \ge \tau \\ &= 0 \qquad\qquad t < \tau \end{aligned} \tag{8.283}$$

Physically, the diffusion equation relates to problems of heat conduction, consolidation of soils, and diffusion of pollutants.

8.13.3 Wave Equation

The wave equation subject to an initial velocity pulse is formulated as

$$\frac{\partial u}{\partial t}=a^2\nabla^2 u,\quad u\big|_{t=0}=0,\quad \left.\frac{\partial u}{\partial t}\right|_{t=0}=\delta(\boldsymbol{x}-\boldsymbol{x}_0) \tag{8.284}$$

Physically, it is a velocity pulse. For a one-dimensional wave equation

$$\begin{aligned} u(x,t;\xi,\tau) &= \frac{1}{2a} && |x-\xi|\le a(t-\tau) \\ &= 0 && |x-\xi|> a(t-\tau) \end{aligned} \tag{8.285}$$

For a two-dimensional wave equation

$$\begin{aligned} u(x,t;\xi,\tau) &= \frac{1}{2\pi a\sqrt{a^2t^2-r^2}} && r\le at \\ &= 0 && r> at \end{aligned} \tag{8.286}$$

where

$$r=\sqrt{(\xi-x)^2+(\eta-y)^2} \tag{8.287}$$

For three-dimensional wave equation

$$u(x,y,z;\xi,\eta,\zeta,t)=\frac{\delta(r-at)}{4\pi ar} \tag{8.288}$$

where

$$r=\sqrt{(\xi-x)^2+(\eta-y)^2++(\zeta-z)^2} \tag{8.289}$$

8.13.4 Biharmonic Equation

For a two-dimensional biharmonic equation, Green's function can be obtained by solving the following PDE

$$\nabla^2\nabla^2 G=\delta(\boldsymbol{x}-\boldsymbol{x}_0) \tag{8.290}$$

The corresponding 2-D Green's solution for the biharmonic equation is

$$G=\frac{r^2}{8\pi}\{\ln r-1\}=\frac{(x^2+y^2)}{8\pi}\{\ln\sqrt{x^2+y^2}-1\} \tag{8.291}$$

This Green's function can be obtained by direct integration. Let us define

$$\nabla^2 G=\boldsymbol{\Phi} \tag{8.292}$$

Thus, (8.290) becomes

$$\nabla^2\boldsymbol{\Phi}=\delta(\boldsymbol{x}-\boldsymbol{x}_0) \tag{8.293}$$

This is, however, precisely the Green's function problem for the Laplace equation and its solution for the 2-D case is known

$$\boldsymbol{\Phi}=\frac{1}{2\pi}\ln r \tag{8.294}$$

Rewrite (8.292) in polar form (due to axisymmetry), and we get

$$\frac{1}{r}\frac{\partial}{\partial r}(r\frac{\partial}{\partial r}G)=\Phi \tag{8.295}$$

Substituting (8.294) into (8.295), rearranging and integrating, we find

$$\begin{aligned} r\frac{\partial G}{\partial r} &= \frac{1}{2\pi}\int r\ln r\,dr \\ &= \frac{1}{4\pi}[r^2\ln r-\frac{1}{2}r^2]+a \end{aligned} \tag{8.296}$$

The last part of (8.296) can easily be obtained by integration by parts. Thus, we have

$$\frac{\partial G}{\partial r}=\frac{1}{4\pi}[r\ln r-\frac{1}{2}r]+\frac{a}{r} \tag{8.297}$$

Integration of (8.297) gives

$$G=\frac{r^2}{8\pi}(\ln r-1)+a\ln r+b \tag{8.298}$$

The unknown constants can be found from the boundary conditions. If we set $a = b = 0$, we have (8.291). The fundamental solution is the singular part of (8.298), and thus the fundamental solution G_0 is normally given as:

$$G_0=\frac{r^2}{8\pi}\ln r \tag{8.299}$$

For the three-dimensional biharmonic equation, Green's function can be obtained by solving the following PDE

$$\nabla^2\nabla^2 G=\delta(\boldsymbol{x}-\boldsymbol{x}_0) \tag{8.300}$$

Using the same idea of the 2-D case, we can employ the fundamental solution of the 3-D Laplace equation as

$$\frac{1}{r^2}\frac{\partial}{\partial r}(r^2\frac{\partial G}{\partial r})=\Phi=\frac{1}{4\pi r} \tag{8.301}$$

This equation can be readily integrated to get

$$G=\frac{r}{8\pi}-\frac{a}{r}+b \tag{8.302}$$

Setting $a = b = 0$, we have the fundamental solution for the 3-D biharmonic equation as

$$G=\frac{r}{8\pi} \tag{8.303}$$

8.13.5 Multi-Harmonic Equation

For n-dimensional space, the fundamental function for the multi-harmonic equation is obtained from the following equation

$$(\nabla^2)^m G_m=-\delta(\boldsymbol{x}-\boldsymbol{x}_0) \tag{8.304}$$

$$\begin{aligned} G_m &= C_1 r^{2m-n}\ln r, \quad 2m\ge n \quad (n=\text{even}) \\ &= C_2 r^{2m-n} \end{aligned} \tag{8.305}$$

where m is an integer. Note that $m = 1$ corresponds to the harmonic fundamental solution whereas $m = 2$ corresponds to the biharmonic fundamental solution, and so on. For physical applications, we are mainly concerned with 2-D and 3-D cases. For such cases, the constants in (8.302) can be found explicitly and, more specifically, the fundamental function of an m-harmonic equation is

$$\begin{aligned} G_m &= \frac{1}{8\pi}\frac{1}{2^m[(m-1)!]^2} r^{2m-2}\ln r, \quad 2m \geq n \quad (n=2) \\ &= \frac{1}{4\pi}\frac{1}{(2m-2)!} r^{2m-3} \qquad (n=3) \end{aligned} \tag{8.306}$$

The validity of (8.302) can be proved by starting from the fundamental solution of the Laplace equation.

Two-dimensional case

For the 2-D case, the biharmonic Green's function is the solution of the following problem (i.e., $m = 2$)

$$\nabla^4 G_2 = \nabla^2 G_1 = -\delta(x - x_0) \tag{8.307}$$

where

$$G_1 = \nabla^2 G_2 \tag{8.308}$$

The second part of (8.307) is the Green's function problem for the Laplace equation, and has been solved previously. Thus, we have

$$G_1 = \frac{1}{2\pi}\ln r \tag{8.309}$$

Back substitution of (8.309) into (8.308) gives

$$\nabla^2 G_2 = \frac{1}{r}\frac{\partial}{\partial r}(r\frac{\partial G_2}{\partial r}) = \frac{1}{2\pi}\ln r \tag{8.310}$$

This was solved in the previous section (see Section (8.299)) and the fundamental solution is (i.e., only retaining the singular term)

$$G_2 = \frac{1}{8\pi} r^2 \ln r \tag{8.311}$$

Following a similar procedure, for $m = 3$ we have

$$\nabla^6 G_3 = \nabla^4 \nabla^2 G_3 = \nabla^4 G_2 = \delta(x - x_0) \tag{8.312}$$

The last two terms represent the biharmonic Green's function and the solution is given in (8.311). Thus, the tri-harmonic Green's function problem becomes

$$\nabla^2 G_3 = G_2 \tag{8.313}$$

More explicitly, it can be expressed as

$$\frac{1}{r}\frac{\partial}{\partial r}(r\frac{\partial G_3}{\partial r}) = \frac{1}{8\pi} r^2 \ln r \tag{8.314}$$

Integrating once, we obtain

$$\frac{\partial G_3}{\partial r} = \frac{1}{8\pi}\frac{r^3}{4}(\ln r - \frac{1}{4}) + \frac{a}{r} \tag{8.315}$$

Integrating one more time, we get

$$G_3 = \frac{1}{128\pi} r^4 (\ln r - \frac{1}{2}) + a \ln r + b \quad (8.316)$$

Thus, the fundamental solution is

$$G_3 = \frac{1}{128\pi} r^4 \ln r \quad (8.317)$$

By employing the same procedure, it is straightforward to show that the multi-harmonic Green's problem with $m = 4$ is reduced to solving the following equation:

$$\frac{1}{r}\frac{\partial}{\partial r}(r\frac{\partial G_4}{\partial r}) = \frac{1}{128\pi} r^4 \ln r \quad (8.318)$$

The fundamental solution of G_4 is found equal to

$$G_4 = \frac{1}{4608\pi} r^6 \ln r \quad (8.319)$$

By observation, we can generalize the coefficient in the multi-harmonic function as

$$G_m = \frac{1}{8\pi} \frac{1}{2^m [(m-1)!]^2} r^{2m-2} \ln r, \quad 2m \geq 2 \quad (8.320)$$

This is the formula given in (8.306).

Three-dimensional case

For the 3-D case, we can follow a similar procedure. In particular,

$$\nabla^4 G_2 = \nabla^2 G_1 = -\delta(\boldsymbol{x} - \boldsymbol{x}_0) \quad (8.321)$$

where

$$G_1 = \nabla^2 G_2 \quad (8.322)$$

The second part of (8.321) is the Green's function problem for the Laplace equation, and it has been solved previously for the three-dimensional case. Recall from (8.20) that

$$G_1 = \frac{1}{4\pi r} \quad (8.323)$$

Back substitution of (8.323) into (8.321) gives

$$\nabla^2 G_2 = \frac{1}{r^2}\frac{\partial}{\partial r}(r^2 \frac{\partial G_2}{\partial r}) = \frac{1}{4\pi r} \quad (8.324)$$

This was solved in the last section and the fundamental solution is given in (8.303) as

$$G_2 = \frac{r}{8\pi} \quad (8.325)$$

Following a similar procedure, for $m = 3$ we have

$$\nabla^6 G_3 = \nabla^4 \nabla^2 G_3 = \nabla^4 G_2 = \delta(\boldsymbol{x} - \boldsymbol{x}_0) \quad (8.326)$$

The last two terms represent the biharmonic Green's function and the solution is given in (8.325). Thus, the problem for the tri-harmonic Green's function problem becomes

$$\nabla^2 G_3 = G_2 \quad (8.327)$$

More explicitly, it can be expressed as

$$\frac{1}{r^2}\frac{\partial}{\partial r}(r^2\frac{\partial G_3}{\partial r})=\frac{r}{8\pi} \tag{8.328}$$

Integrating once, we obtain

$$\frac{\partial G_3}{\partial r}=\frac{r^2}{32\pi}+\frac{a}{r^2} \tag{8.329}$$

Integrating one more time, we get

$$G_3=\frac{r^3}{96\pi}-\frac{a}{r}+b \tag{8.330}$$

Thus, the fundamental solution is

$$G_3=\frac{r^3}{96\pi} \tag{8.331}$$

By employing the same procedure, it is straightforward to show that the multi-harmonic Green's problem with $m=4$ is reduced to solving the following equation:

$$\frac{1}{r^2}\frac{\partial}{\partial r}(r^2\frac{\partial G_4}{\partial r})=\frac{r^3}{96\pi} \tag{8.332}$$

Integration on both sides gives G_4 as

$$G_4=\frac{r^5}{2880\pi} \tag{8.333}$$

By inspection, we find that

$$G_m=\frac{1}{4\pi}\frac{1}{(2m-2)!}r^{2m-3} \tag{8.334}$$

This completes the proof of (8.306).

8.14 SUMMARY AND FURTHER READING

In this chapter, the powerful technique called Green's function method was discussed. The method was founded by George Green when he considered the solution of problems of electricity and magnetism. The method was founded on the validity of Green's first, second, and third identities, which were originally derived for the Laplace equation that governs the problems of electricity and magnetism. Green's second identity is also known as Green's theorem, whereas Green's third identity is actually the formula for Green's function method. We extend the discussion of Green's theorem to the biharmonic equation, in view of its importance in mechanics. The most general Green's theorem for second order linear PDEs was also derived. The evolution of the boundary integral equation from Green's function method is discussed in view of its importance in numerical analysis. The boundary integral equation for the biharmonic equation was also considered and it turns out that two coupled integral equations need to be solved. In view of the role of the Dirac delta function in the derivation of Green's function, the definition of the Dirac delta function and its justification using the theory of distribution is introduced. The Dirac delta function approximated by generalized functions called the δ-sequence is discussed. It was demonstrated that both Fourier series expansion and the Fourier transform were intimately linked to the theory of distribution if appropriate forms of the δ-sequence were chosen. Green's function method for general second order PDEs, which are not self-adjoint, is discussed. Finally, a number of Green's functions for commonly encountered PDEs were summarized. They include Helmholtz, diffusion, wave, biharmonic, and multi-harmonic equations.

The present chapter differs from most of the existing textbooks on differential equations in that detail of the Green's function method for biharmonic equations is discussed. The presentation on multi-harmonic equation is also original and cannot be found in any textbook on Green's function method (e.g., Duffy, 2001).

Green's function method is one of the most powerful mathematical techniques in solving PDEs and a short chapter like this can cover the main concepts and ideas of this method. An excellent and elementary introduction to Green's function method is given by Greenberg (1971). The book series by Stakgold (1967, 1968, 1979) also provided a comprehensive discussion of Green's function method and are also highly recommended. A handbook on Green's function is given by Butkovskiy (1982), whereas a compilation of exact solutions in terms of Green's function can be found in Polyanin and Zaitsev (2002). More colorful stories written on George Green can be found in Cannell and Lord (1993).

8.15 PROBLEMS

Problem 8.1 Show the following identity:

$$\cos x + \cos 2x + \ldots + \cos nx = \frac{\cos[(n+1)\frac{x}{2}]\sin\frac{nx}{2}}{\sin\frac{x}{2}} \tag{8.335}$$

Hint: Use the result related to the fifth δ-sequence.

Problem 8.2 Show that the Green's function of deflection given in (8.273) for a simply supported beam subject to unit point load can be recast into the following forms:

$$\begin{aligned} G_u(x,\xi) &= \frac{(L-\xi)x}{6EIL}[2L(L-x)-(L-\xi)^2-(L-x)^2] \quad (0 \le x \le \xi) \\ &= \frac{(L-x)\xi}{6EIL}[2L(L-x)-(L-x)^2-(L-\xi)^2] \quad (\xi \le x \le L) \end{aligned} \tag{8.336}$$

Problem 8.3 Show that the Green's function of moment given in (8.274) for a simply supported beam subject to unit point load can be recast into the following forms:

$$\begin{aligned} G_M(x,\xi) &= -\frac{(L-\xi)x}{L} \quad (0 \le x \le \xi) \\ &= -\frac{(L-x)\xi}{L} \quad (\xi \le x \le L) \end{aligned} \tag{8.337}$$

Problem 8.4 For the beam bending problem discussed in Example 8.4, show that Green's function of moment is continuous at $x = \xi$ and there is a unit discontinuity of the derivative of the G_M with respect to x at $x = \xi$. More specially, show the following:

$$[G_M(x,\xi)]_{\xi^-}^{\xi^+} = 0 \quad (8.338)$$

$$\left[\frac{\partial G_M(x,\xi)}{\partial x}\right]_{\xi^-}^{\xi^+} = 1 \quad (8.339)$$

Problem 8.5 Discuss the physical meaning of the results obtained in Problem 8.4.

Hint: What is the physical meaning of the derivative of moment? What changes suddenly within the beam across the unit point force?

Problem 8.6 Consider the case of a fixed end (or built-in support) supported beam subject to non-uniform distributed load $q(x)$ as depicted in Figure 8.13. The problem is formulated as

$$\frac{d^2}{dx^2}(EI\frac{d^2u}{dx^2}) = q(x) \quad (8.340)$$

where u is the deflection of the beam, and E and I are the Young's modulus and moment of inertia of the bean section. The boundary conditions for fixed end support are

$$u(a) = u(L) = 0, \quad \left.\frac{du}{dx}\right|_{x=0} = 0, \quad \left.\frac{du}{dx}\right|_{x=L} = 0 \quad (8.341)$$

Find the Green's function method of the problem.

Ans:

$$G_u(x,\xi) = \frac{1}{6EI}[(x-\xi)^3 H(x-\xi) + \frac{3(L-\xi)^2}{L^2}(-\frac{L+2\xi}{3L}x+\xi)x^2] \quad (8.342)$$

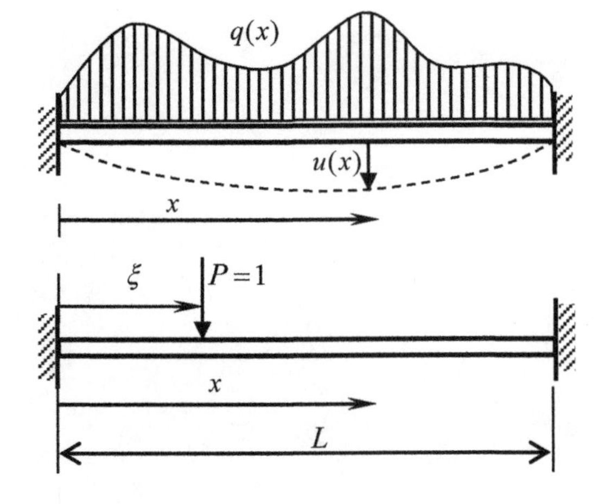

Figure 8.13 Green's function method for built-in beam bending

Problem 8.7 Define a possible form of Dirac delta function as:

$$\delta_\varepsilon(t) = \begin{cases} 0 & t < 0 \\ 1/\varepsilon & 0 < t < \varepsilon \\ 0 & t > \varepsilon \end{cases} \tag{8.343}$$

Show that

$$\int_{-\infty}^{\infty} \delta_\varepsilon(x) dx = 1 \tag{8.344}$$

$$\lim_{\varepsilon \to 0} \int_{-\infty}^{\infty} f(x)\delta_\varepsilon(x) dx = f(0) \tag{8.345}$$

Thus, it is a potential candidate for the Dirac delta function.

Problem 8.8 Derive the following result for G_4 for the 2-D case (multi-harmonic Green's function with $m = 4$):

$$G_4 = \frac{1}{4608\pi} r^6 \ln r - \frac{1}{13824\pi} r^6 + a \ln r + b \tag{8.346}$$

Thus, the validity of (8.319) can be demonstrated.

Problem 8.9 Show the following identity using the theory of distribution

$$x^n \delta(x) = 0 \tag{8.347}$$

CHAPTER NINE

Wave, Diffusion and Potential Equations

9.1 INTRODUCTION

Arguably, wave, diffusion and potential equations are the three most fundamental and important second order PDEs in engineering, science, and mathematical physics. They are the standard topics covered in nearly all textbooks on partial differential equations and mathematical physics. They are the PDEs that have been studied most extensively since the discovery of calculus. The main contributors in the studies of these equations include Bernoulli, Euler, D'Alembert, Huygens, Lagrange, Fourier, Laplace, Riemann, Hilbert, Poisson, Cauchy, Hadamard, and Sommerfeld, to name a few. In Chapter 7, we have demonstrated that there are three general types of second order PDEs, namely hyperbolic, parabolic, and elliptic types. The most commonly encountered type of hyperbolic, parabolic, and elliptic PDEs are wave, diffusion and Laplace equations.

The wave phenomenon has long been recognized as one of the most distinct and fundamental features observable in nature. When we throw a stone into a calm pond, a series of circular waves of water will radiate from the impact. This propagating wave seems to suggest that pockets of water particles are moving outward. In fact, energy is propagating along these waves, instead of water particles. Water molecules are simply oscillating around its equilibrium positions. This is a phenomenon of propagating energy. Wave phenomena influencing our daily lives include sound, light, other electromagnetic waves, earthquakes, and water waves. The direction of such wave signals is associated with characteristics. When a whip is lashed, a one-dimensional wave is generated. Similarly, vibrations of violin strings can be modelled as one-dimensional waves. Energy propagation in solids is considered as stress waves. As shown by Chau (2013), both shear stress waves and compressional stress waves result in classic three-dimensional elastodynamic equations. In Chapter 2, we have demonstrated that both Maxwell equations for electrodynamics and elastodynamics can be converted to nonhomogeneous wave equations. Therefore, the study of wave equations is of fundamental importance, especially nonhomogeneous wave equations.

Heat conduction has been considered since the time of Newton (Newton's cooling law) and Fourier (Fourier law of heat conduction). This phenomenon can be modelled as a diffusion equation. Subsidence of ground due to settlement is caused by consolidation of clay. When loading is applied to a fully saturated clay, the small value of permeability of clay does not allow the water to be squeezed out immediately. The pore water pressure will increase to balance this externally applied load. This excess pore water pressure (excess compared to the long term value) will, however, induce seepage through the clay. It is a very slow process of squeezing water out of the soil skeleton and at the same time the decreased pore water pressure will be converted to effective stress (portion of the total stress taken up by the soil skeleton interactions). This results in the diffusion of water and thus the phenomenon is governed by the diffusion equation.

Many physical problems are governed by the Laplace equation. These include the incompressible potential flow of fluid, electrostatic problems, gravitational attraction of bodies, steady state temperature distribution, twisting of elastic bars, and membrane deflection (such soap film). The solution of the Laplace equation is also known as the harmonic function. It is also known as potential theory as it is related to the problems of gravitational potential. For the two-dimensional case, it also closely relates to the analytic function in the complex variable technique.

9.2 WAVE EQUATIONS

Mathematically, the linear wave phenomenon can be modelled by the following wave equation

$$a^2\nabla^2\varphi = \frac{\partial^2\varphi}{\partial t^2} \tag{9.1}$$

where φ is a wave potential function, ∇^2 is the Laplacian operator, and a is the wave speed of the propagating energy. For the case of waves in solids, readers can refer to Section 9.4 of Chau (2013). For the special case that the function φ is independent of time, (9.1) reduces to the Laplace equation. In fact, the resulting Laplace equation governs the equilibrium problem of steady distribution of the potential subject to certain boundary conditions. Mathematically, as we have discussed in Chapter 7, (9.1) is a hyperbolic type of PDE. There exists a pair of curves called characteristics, along which there is a discontinuous solution propagating. The wave type phenomenon is the easiest to visualize in nature, compared to the equilibrium type of elliptic PDE and the diffusion type of parabolic PDE discussed in Chapter 7.

The next section will discuss the simplest case of 1-D waves.

9.2.1 D'Alembert Solution for 1-D Waves

The one-dimensional wave equation is one of the first differential equations ever formulated in modelling physical phenomena. For the case of a vibrating spring of a violin shown in Figure 9.1, the deflection of the spring is governed by

$$a^2\frac{\partial^2 u}{\partial x^2} = \frac{\partial^2 u}{\partial t^2} \tag{9.2}$$

where

$$a^2 = \frac{T}{\rho} \tag{9.3}$$

The tension in the vibrating spring is T and ρ is the mass per unit length of the spring.

Figure 9.2 shows the vertical and horizontal components of the spring tension T. Horizontal force equilibrium leads to

$$T(x+\Delta x,t)\cos(\theta+\Delta\theta) - T(x,t)\cos(\theta) = 0 \tag{9.4}$$

Similarly, vertical force equilibrium leads to the following equation of motion:

$$T(x+\Delta x,t)\sin(\theta+\Delta\theta) - T(x,t)\sin(\theta) = \rho\Delta x\, u_{tt}(\bar{x},t) \tag{9.5}$$

where $\bar{x}$ is the center of mass of the spring element shown in Figure 9.2. As shown in Figure 9.2, we can denote the vertical component of spring tension as V, and (9.5) can be simplified as:

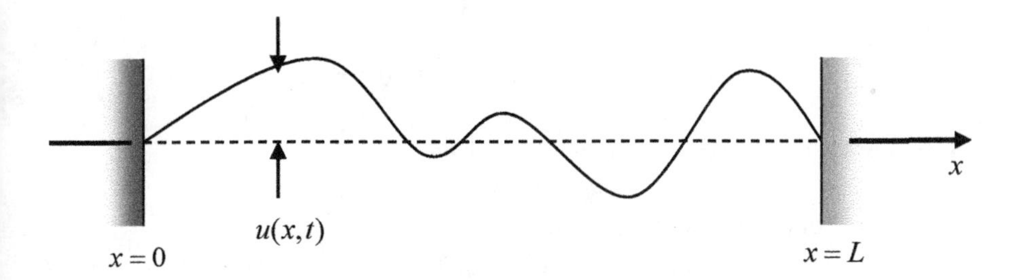

Figure 9.1 Vibrating violin string as 1-D wave phenomenon

$$\frac{V(x+\Delta x,t)-V(x,t)}{\Delta x}=\rho u_{tt}(\bar{x},t) \tag{9.6}$$

Using the fundamental definition of differentiation, we get

$$V_x(x,t)=\rho u_{tt}(x,t) \tag{9.7}$$

The vertical force component can be related to the horizontal force component by

$$V(x,t)=H(t)\tan\theta=H(t)u_x(x,t) \tag{9.8}$$

To obtain the last part of (9.8), we have set the slope of the spring equal to the derivative of u taken with respect to x.

Substitution of (9.8) into (9.7) gives

$$(Hu_x)_x=\rho u_{tt} \tag{9.9}$$

Finally, for small deflection (it is normally the case), we have $\cos\theta\to 1$ as $\theta\to 0$:

$$H(t)=T\cos\theta\approx T \tag{9.10}$$

Using (9.10), we can further simplify (9.9) as

$$a^2\frac{\partial^2 u}{\partial x^2}=\frac{\partial^2 u}{\partial t^2},\qquad a^2=T/\rho \tag{9.11}$$

This completes the proof for the one-dimensional wave equation given in (9.2).

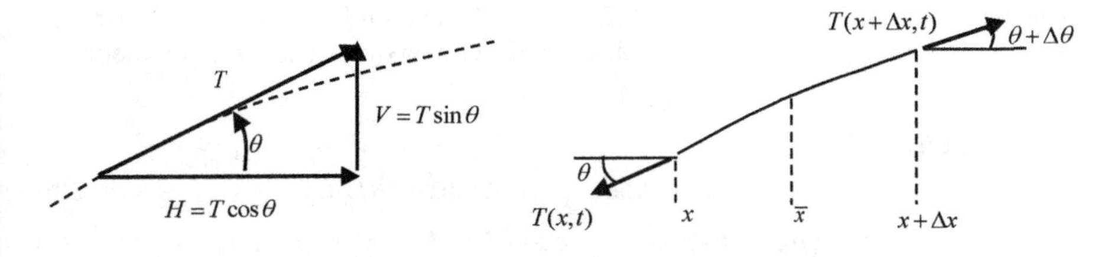

Figure 9.2 Force equilibrium in a vibrating spring of violin

Example 9.1 Consider the solution of the following finite spring of length L fixed at two supports and subject to an initial deflection $f(x)$:

$$a^2 u_{xx} = u_{tt}, \quad 0 < x < L, \quad t > 0$$
$$u(0,t) = 0, \quad u(L,t) = 0, \quad t \geq 0 \tag{9.12}$$
$$u(x,0) = f(x), \quad u_t(x,0) = 0, \quad 0 \leq x \leq L$$

Solution: Using separation of variables, we have

$$u(x,t) = X(x)T(t) \tag{9.13}$$

Substitution of (9.13) into (9.12) gives

$$a^2 X'' T = X T'' \tag{9.14}$$

This can be rearranged as

$$\frac{X''}{X} = \frac{1}{a^2}\frac{T''}{T} = -\lambda \tag{9.15}$$

This leads to two ODEs

$$X'' + \lambda X = 0 \tag{9.16}$$

$$T'' + a^2 \lambda T = 0 \tag{9.17}$$

Thus, the boundary and initial conditions become

$$u_t(x,0) = X(x)T'(0) = 0, \quad 0 \leq x \leq L \quad \Rightarrow \quad T'(0) = 0 \tag{9.18}$$

$$u(0,t) = X(0)T(t) = 0, \quad u(L,t) = X(L)T(t) = 0, \quad t \geq 0 \tag{9.19}$$

The governing equation for X can be summarized as

$$X'' + \lambda X = 0, \quad X(0) = X(L) = 0 \tag{9.20}$$

The general solution of (9.20) is

$$X = c_1 \cos\sqrt{\lambda} x + c_2 \sin\sqrt{\lambda} x \tag{9.21}$$

The boundary condition in (9.20) leads to

$$X(0) = 0 \Rightarrow c_1 = 0 \tag{9.22}$$

Substitution of (9.21) into the second boundary condition of (9.20) yields the eigenvalue equation

$$\sin\sqrt{\lambda} L = 0 \tag{9.23}$$

The eigenvalues are

$$\lambda_n = n^2\pi^2 / L^2, \quad n = 1,2,3,\ldots \tag{9.24}$$

The quantities $\lambda_n = n\pi a/L$, for $n = 1, 2, \ldots$, are the natural frequencies of the string, that is, the frequencies at which the string will freely vibrate. The vibration mode is

$$X_n(x) = \sin\left(n\pi x / L\right) \tag{9.25}$$

The solution of T is

$$T(t) = k_1 \cos\left(n\pi at / L\right) + k_2 \sin\left(n\pi at / L\right) \tag{9.26}$$

$$T'(t) = \left(n\pi a / L\right)\{-k_1 \sin\left(n\pi at / L\right) + k_2 \cos\left(n\pi at / L\right)\} \tag{9.27}$$

Boundary condition (9.18) will lead to

$$k_2 = 0 \tag{9.28}$$

The fundamental solution becomes

$$u_n(x,t) = \sin(n\pi x/L)\cos(n\pi at/L), \quad n = 1,2,3,..., \tag{9.29}$$

The general solution becomes

$$u(x,t) = \sum_{n=1}^{\infty} c_n u_n(x,t) = \sum_{n=1}^{\infty} c_n \sin(n\pi x/L)\cos(n\pi at/L) \tag{9.30}$$

$$u(x,0) = f(x) = \sum_{n=1}^{\infty} c_n \sin(n\pi x/L) \tag{9.31}$$

Multiplying both sides by the sine function and integrating from 0 to L, we have (in view of the orthogonal properties of the resulting sine integral given in (7.186)):

$$c_n = \frac{2}{L}\int_0^L f(x)\sin(n\pi x/L)dx \tag{9.32}$$

Thus, the final solution is

$$u(x,t) = \sum_{n=1}^{\infty} c_n \sin(n\pi x/L)\cos(n\pi at/L) \tag{9.33}$$

The constant c_n for the sine Fourier expansion of the initial deflected shape $f(x)$ can be carried out explicitly once the function is given. This completes the solution of the 1-D wave phenomenon of violin string vibrations.

We now introduce the classical solution proposed by D'Alembert for 1-D wave problems. This solution is beautiful and is a triumph in the history of solutions of the wave equation.

Let us rewrite the solution given in Example 9.1 by defining the following function:

$$h(x) = \sum_{n=1}^{\infty} c_n \sin(n\pi x/L) \tag{9.34}$$

Next, we consider the expansion of this function with two special arguments:

$$h(x-at) = \sum_{n=1}^{\infty} c_n \left[\sin(n\pi x/L)\cos(n\pi at/L) - \cos(n\pi x/L)\sin(n\pi at/L)\right] \tag{9.35}$$

$$h(x+at) = \sum_{n=1}^{\infty} c_n \left[\sin(n\pi x/L)\cos(n\pi at/L) + \cos(n\pi x/L)\sin(n\pi at/L)\right] \tag{9.36}$$

These expressions are the results of direct application of the following sum rule for the sine function:

$$\sin(A \pm B) = \sin A\cos B \pm \cos A\sin B \tag{9.37}$$

Adding (9.35) and (9.36) gives exactly twice (9.33), the solution that we obtained by solving the initial boundary value problem of Example 9.1. More precisely, we can write

$$
\begin{aligned}
u(x,t) &= \sum_{n=1}^{\infty} c_n \sin(n\pi x / L)\cos(n\pi at / L) \\
&= [h(x-at) + h(x+at)]/2
\end{aligned}
\tag{9.38}
$$

Physically the first h function is an outgoing wave or a right going wave and the second h is an incoming wave or a left going wave. The argument in the h function can be considered as the "characteristics" of the propagating solutions as we introduced in Chapter 7.

Naturally, one may ask whether all solutions of one-dimensional wave problems can be written as superposition of two waves. The answer is yes, and this general result is D'Alembert's seminal result. Let us consider the general 1-D wave equation as

$$
\frac{\partial^2 u}{\partial t^2} = a^2 \frac{\partial^2 u}{\partial x^2}
\tag{9.39}
$$

Introduce the following pair of change of variables:

$$
\xi = x + at, \quad \eta = x - at
\tag{9.40}
$$

Using the chain rule for partial differentiation, we get

$$
\frac{\partial u}{\partial x} = \frac{\partial u}{\partial \xi}\frac{\partial \xi}{\partial x} + \frac{\partial u}{\partial \eta}\frac{\partial \eta}{\partial x} = \frac{\partial u}{\partial \xi} + \frac{\partial u}{\partial \eta}
\tag{9.41}
$$

$$
\frac{\partial^2 u}{\partial x^2} = \frac{\partial}{\partial \xi}\left(\frac{\partial u}{\partial \xi} + \frac{\partial u}{\partial \eta}\right)\frac{\partial \xi}{\partial x} + \frac{\partial}{\partial \eta}\left(\frac{\partial u}{\partial \xi} + \frac{\partial u}{\partial \eta}\right)\frac{\partial \eta}{\partial x}
\tag{9.42}
$$

$$
\frac{\partial^2 u}{\partial x^2} = \frac{\partial^2 u}{\partial \xi^2} + 2\frac{\partial^2 u}{\partial \xi \partial \eta} + \frac{\partial^2 u}{\partial \eta^2}
\tag{9.43}
$$

Thus, we have

$$
a^2 \frac{\partial^2 u}{\partial x^2} = a^2 \frac{\partial^2 u}{\partial \xi^2} + 2a^2 \frac{\partial^2 u}{\partial \xi \partial \eta} + a^2 \frac{\partial^2 u}{\partial \eta^2}
\tag{9.44}
$$

$$
\frac{\partial u}{\partial t} = \frac{\partial u}{\partial \xi}\frac{\partial \xi}{\partial t} + \frac{\partial u}{\partial \eta}\frac{\partial \eta}{\partial t} = a\frac{\partial u}{\partial \xi} - a\frac{\partial u}{\partial \eta}
\tag{9.45}
$$

$$
\begin{aligned}
\frac{\partial^2 u}{\partial t^2} &= a\frac{\partial}{\partial \xi}\left(\frac{\partial u}{\partial \xi} - \frac{\partial u}{\partial \eta}\right)\frac{\partial \xi}{\partial x} + a\frac{\partial}{\partial \eta}\left(\frac{\partial u}{\partial \xi} - \frac{\partial u}{\partial \eta}\right)\frac{\partial \eta}{\partial x} \\
&= a^2 \left(\frac{\partial^2 u}{\partial \xi^2} - 2\frac{\partial^2 u}{\partial \xi \partial \eta} + \frac{\partial^2 u}{\partial \eta^2}\right)
\end{aligned}
\tag{9.46}
$$

Substitution of (9.44) and (9.46) into (9.39) leads to

$$
\frac{\partial^2 u}{\partial \xi \partial \eta} = 0
\tag{9.47}
$$

Integration once with respect to η gives

$$
\frac{\partial u}{\partial \xi} = f(\xi)
\tag{9.48}
$$

Integration one more time with respect to ξ gives

$$u(x,t) = \int f(\xi)\mathrm{d}\xi + f_2(\eta) \\ = f_1(x+at) + f_2(x-at) \tag{9.49}$$

In the derivation of (9.49), we have not specified any initial and boundary condition, and thus this result is general and can be applied to any initial boundary problem of a one-dimensional wave. More importantly, the solution in (9.49) is expressed in terms of an arbitrary function of the two arguments (or the two characteristics for this hyperbolic equation).

We will consider the case of an infinite spring subject to certain initial conditions using the solution in (9.49). More specifically, we consider an infinite spring subject to initial deflection and velocity:

$$u\,|_{t=0} = \varphi(x), -\infty < x < +\infty \tag{9.50}$$

$$\left.\frac{\partial u}{\partial t}\right|_{t=0} = \psi(x), -\infty < x < +\infty \tag{9.51}$$

Adopting the general solution form (9.49), we get

$$\frac{\partial u}{\partial t}(x,t) = f_1'(x+at)\frac{\partial x + at}{\partial t} + f_2'(x-at)\frac{\partial x - at}{\partial t} = af_1' - af_2' \tag{9.52}$$

Applying the initial conditions (9.50) and (9.51) (i.e., at $t = 0$) to (9.49) and (9.52), we obtain two equations:

$$f_1(x) + f_2(x) = \varphi(x) \tag{9.53}$$

$$af_1'(x) - af_2'(x) = \psi(x) \tag{9.54}$$

Integration of (9.54) gives

$$f_1(x) - f_2(x) = \frac{1}{a}\int_0^x \psi(\xi)\mathrm{d}\xi + C \tag{9.55}$$

Equations (9.53) and (9.55) provide a system of two equations for two unknown functions, and the solutions of it are:

$$f_1(x) = \frac{1}{2}\varphi(x) + \frac{1}{2a}\int_0^x \psi(\xi)\mathrm{d}\xi + \frac{C}{2} \tag{9.56}$$

$$f_2(x) = \frac{1}{2}\varphi(x) - \frac{1}{2a}\int_0^x \psi(\xi)\mathrm{d}\xi - \frac{C}{2} \tag{9.57}$$

Back substitution of these results into (9.49) leads to

$$u(x,t) = \frac{1}{2}[\varphi(x+at) + \varphi(x-at)] + \frac{1}{2a}\int_{x-at}^{x+at} \psi(\xi)\mathrm{d}\xi \tag{9.58}$$

This is the renowned solution by D'Alembert. For the case of zero initial velocity, Figure 9.3 illustrates the right going solution or outgoing solution. Because in the formulation of the wave equation we have not incorporated any damping term, the initial deflected shape will be preserved during the propagation. Similarly, the left going solution or incoming solution is illustrated in Figure 9.4.

More generally, once an infinitely long spring is initially deflected, waves will be generated to both left and right as illustrated in Figure 9.5 at three different times. In this illustration, the disturbance was imposed at $t = t_0$; as time increases, this initial disturbance generated a left and a right going wave. The wave is propagating at a speed of a.

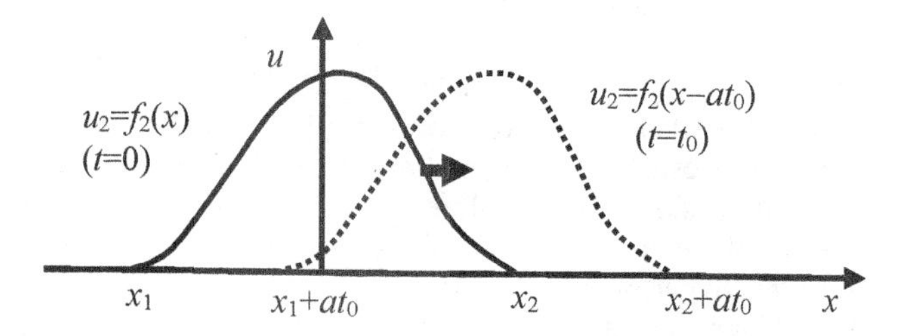

Figure 9.3 Right going solution of 1-D wave equation

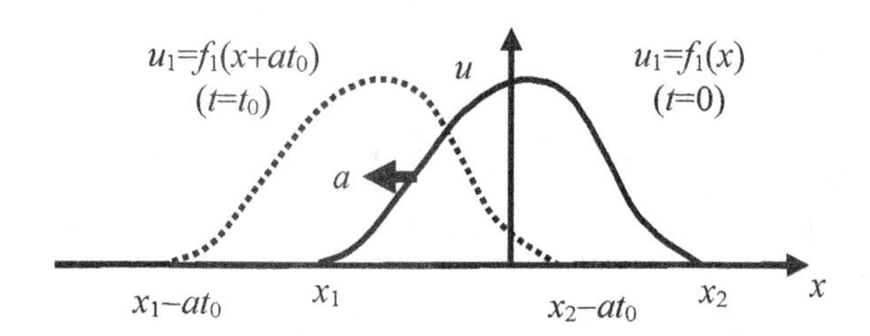

Figure 9.4 Left going solution of 1-D wave equation

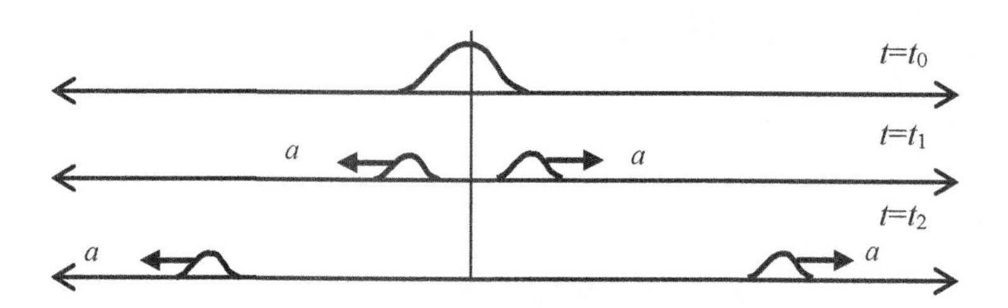

Figure 9.5 Both left and right going solutions of an infinitely long spring

9.2.2 Domain of Dependence and Influence Zone

Recall in the general solution of a 1-D wave that there are two characteristics and one corresponds to a left-going wave and the other the right-going wave. In the time-space, there are certain forbidden regions that the wave solutions cannot travel to. This is a consequence of causality.

Figure 9.6 illustrates the domain of dependence. Consider that at a point P in space-time, the wave signals that the point P received can only come from the influence of the space-time triangle of ABP. To see the mathematical details of this, we consider the following example of the nonhomogeneous wave problem.

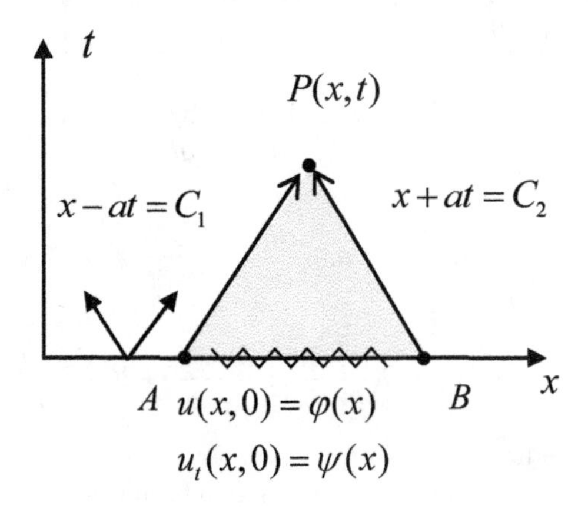

Figure 9.6 Domain of dependence

Example 9.2 Consider the solution of the following 1-D nonhomogeneous wave problem (with external forcing term given) with prescribed initial deflection and velocity:

$$
\begin{aligned}
&u_{tt}-a^2u_{xx}=f(x,t), \quad t>0 \\
&u(x,0)=\varphi(x), \ u_t(x,0)=\psi(x)
\end{aligned}
\tag{9.59}
$$

Solution: The problem can be solved by using superposition. First, we recognize that the original problem can be decomposed into two sub-problems:

Problem I

$$
\begin{aligned}
&u_{tt}-a^2u_{xx}=0, \quad t>0 \\
&u(x,0)=\varphi(x), \ u_t(x,0)=\psi(x)
\end{aligned}
\tag{9.60}
$$

Problem II

$$
\begin{aligned}
&u_{tt}-a^2u_{xx}=f(x,t), \quad t>0 \\
&u(x,0)=0, \ u_t(x,0)=0
\end{aligned}
\tag{9.61}
$$

Problem I has actually been solved by using the D'Alembert solution discussed in the last section as:

$$
u(x,t)=\frac{1}{2}[\varphi(x+at)+\varphi(x-at)]+\frac{1}{2a}\int_{x-at}^{x+at}\psi(\xi)\,\mathrm{d}\xi \tag{9.62}
$$

For Problem II, we can use the classification approach in Chapter 7 that

$$
A=1, \quad B=0, \quad C=-a^2 \tag{9.63}
$$

Thus, we have

$$
B^2-4AC=4a^2>0 \tag{9.64}
$$

This system is of course the hyperbolic or wave type. The two characteristics are

$$\frac{dx}{dt}=\frac{B+\sqrt{B^2-4AC}}{2A}=a,\quad \frac{dx}{dt}=\frac{B-\sqrt{B^2-4AC}}{2A}=-a \tag{9.65}$$

The corresponding characteristics can be obtained by integrating the ODE as

$$\xi=x+at=C_1,\quad \eta=x-at=C_2 \tag{9.66}$$

Then, we get

$$\frac{\partial\xi}{\partial x}=1,\quad \frac{\partial\xi}{\partial t}=a,\quad \frac{\partial\eta}{\partial x}=1,\quad \frac{\partial\eta}{\partial t}=-a \tag{9.67}$$

Thus,

$$\begin{aligned}\overline{B}&=2A\frac{\partial\xi}{\partial x}\frac{\partial\eta}{\partial x}+B(\frac{\partial\xi}{\partial x}\frac{\partial\eta}{\partial y}+\frac{\partial\eta}{\partial x}\frac{\partial\xi}{\partial y})+2C\frac{\partial\xi}{\partial y}\frac{\partial\eta}{\partial y}\\ &=-4a^2\end{aligned} \tag{9.68}$$

Therefore the canonical form of the wave equation becomes

$$\frac{\partial u}{\partial\xi\partial\eta}=-\frac{1}{4a^2}f(\xi,\eta) \tag{9.69}$$

Note that we have used the same symbols for u and f even though we have changed the variables. Integration of (9.69) with respect to η gives

$$\frac{\partial u}{\partial\xi}=-\frac{1}{4a^2}\int_\xi^\eta f(\xi,\overline{\eta})d\overline{\eta}+g(\xi) \tag{9.70}$$

The lower limit for the integral is obtained by virtue of the fact that at $t=0$, we have $\eta=\xi$ (see (9.66)). In addition, we note that

$$\frac{\partial u}{\partial\xi}=\frac{\partial u}{\partial t}\frac{\partial t}{\partial\xi}+\frac{\partial u}{\partial x}\frac{\partial x}{\partial\xi}=\frac{1}{a}\frac{\partial u}{\partial t}+\frac{\partial u}{\partial x} \tag{9.71}$$

At zero time, we have

$$\frac{\partial u}{\partial t}(x,0)=0,\quad \frac{\partial u}{\partial x}(0)=0 \tag{9.72}$$

The first is the second initial condition of Problem II given in (9.61) and the second is a direct consequence of the first initial condition in (9.61) (i.e., the initial data is identically zero for all x as is its derivative with respect to x). Thus, from (9.71) we must have

$$\frac{\partial u}{\partial\xi}(\xi,\xi)=0 \tag{9.73}$$

Using this information, we find that (9.70) becomes

$$g(\xi)=0 \tag{9.74}$$

$$\frac{\partial u}{\partial\xi}=\frac{1}{4a^2}\int_\eta^\xi f(\xi,\overline{\eta})d\overline{\eta} \tag{9.75}$$

Note that we absorb the negative by reversing the limits of integration. Integration of (9.70) one more time with respect to ξ yields

$$u=\frac{1}{4a^2}\int_\eta^\xi\left\{\int_\eta^{\overline{\xi}} f(\overline{\xi},\overline{\eta})d\overline{\eta}\right\}d\overline{\xi} \tag{9.76}$$

Note that the integration is for $\xi>\eta$, thus we have

$$\eta < \bar{\xi} < \xi, \quad \xi < \bar{\eta} < \bar{\xi} \tag{9.77}$$

The domain of integration can be illustrated in Figure 9.7. The domain of integration can be written as:

$$u = \frac{1}{4a^2} \iint_{\Omega_{\xi\eta}} f(\bar{\xi}, \bar{\eta}) d\bar{\eta} d\bar{\xi} \tag{9.78}$$

where $\Omega_{\xi\eta}$ is the triangular domain shown in Figure 9.7.

We map the variables back to the physical domain as:

$$\bar{\eta} = \bar{x} - c\bar{t}, \quad \bar{\xi} = \bar{x} + c\bar{t} \tag{9.79}$$

$$u = \frac{1}{4a^2} \iint_{\Omega_{xt}} f(\bar{x}, \bar{t}) |J| d\bar{x} d\bar{t} \tag{9.80}$$

where the domain of integration shown in Figure 9.8 can be mapped to another triangular domain shown in Figure 9.8.

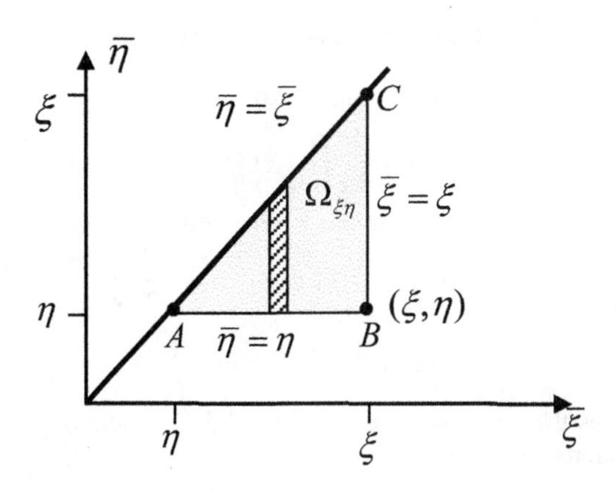

Figure 9.7 Domain of integration for (9.76)

The horizontal line *AB* in Figure 9.7 can be mapped as

$$\bar{\eta} = \eta, \quad \Rightarrow \quad \bar{x} = x - a(t - \bar{t}) \tag{9.81}$$

That is, in the physical domain, *AB* becomes an inclined line shown in Figure 9.8. Similarly, the vertical line *BC* can be mapped as an inclined line in the physical domain:

$$\bar{\xi} = \xi, \quad \Rightarrow \quad \bar{x} = x + a(t - \bar{t}) \tag{9.82}$$

Finally, the inclined line *AC* becomes a horizontal line in Figure 9.8 as:

$$\bar{\xi} = \bar{\eta}, \quad \Rightarrow \quad \bar{t} = 0 \tag{9.83}$$

In addition, the Jacobian for the mapping is

$$J = \begin{vmatrix} \dfrac{\partial \bar{\xi}}{\partial \bar{x}} & \dfrac{\partial \bar{\xi}}{\partial \bar{t}} \\ \dfrac{\partial \bar{\eta}}{\partial \bar{x}} & \dfrac{\partial \bar{\eta}}{\partial \bar{t}} \end{vmatrix} = \begin{vmatrix} 1 & a \\ 1 & -a \end{vmatrix} = -2a \tag{9.84}$$

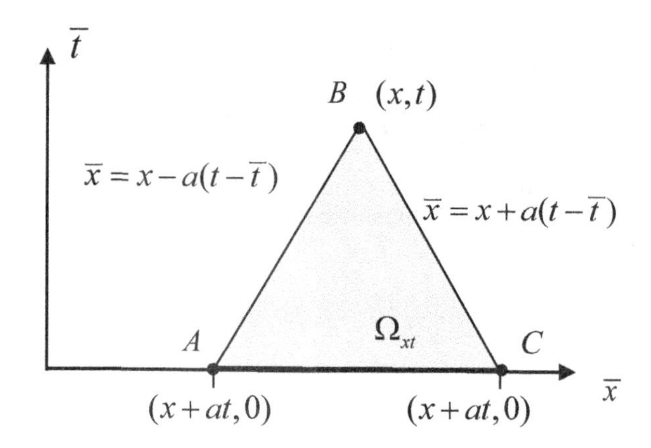

Figure 9.8 Domain of integration for (9.80)

Substitution of (9.81) and (9.84) into (9.80) yields

$$u = \frac{1}{2a} \iint_{\Omega_{xt}} f(\bar{x}, \bar{t}) d\bar{x} \, d\bar{t} \tag{9.85}$$

Finally, we can combine the solutions of Problems I and II as

$$u(x,t) = \frac{1}{2}[\varphi(x+at) + \varphi(x-at)] + \frac{1}{2a}\int_{x-at}^{x+at} \psi(\xi)\, d\xi + \frac{1}{2a} \iint_{\Omega_{xt}} f(\bar{x}, \bar{t}) d\bar{x} \, d\bar{t} \tag{9.86}$$

This completes the solution. The solution of 3-D nonhomogeneous wave equation will be considered in Section9.3.

The domain of integration found in Figure 9.8 is exactly the domain of dependence shown in Figure 9.6.

The region of influence is shown in Figure 9.9 for the case of initial deflection and velocity prescribed at $x = x_0$ and $t = 0$.

Figure 9.9 shows the space-time event horizon for the one-dimensional case. That is, we could not receive any signal from a point source as long as the chosen point in space-time is outside the wedging zone in Figure 9.9. The domain of dependence and region of influence can be considered as a consequence of causality. This is a distinct feature of the wave type solution. Figure 9.10 is the cone of the event horizon including both past and future cones.

In Chapter 15, we will see that this causality of the wave also provides a stability criterion on the numerical integration of wave equations (i.e., the Courant-Friedrichs-Lewy criterion or CFL criterion in (15.2.7)).

9.2.3 Two-Dimensional Waves

For the case of radially symmetric spherical waves (i.e., waves in unbounded solids), the wave equation is

$$\frac{1}{a^2}\frac{\partial^2 \varphi}{\partial t^2}=\frac{\partial^2 \varphi}{\partial r^2}+\frac{1}{r}\frac{\partial \varphi}{\partial r} \tag{9.87}$$

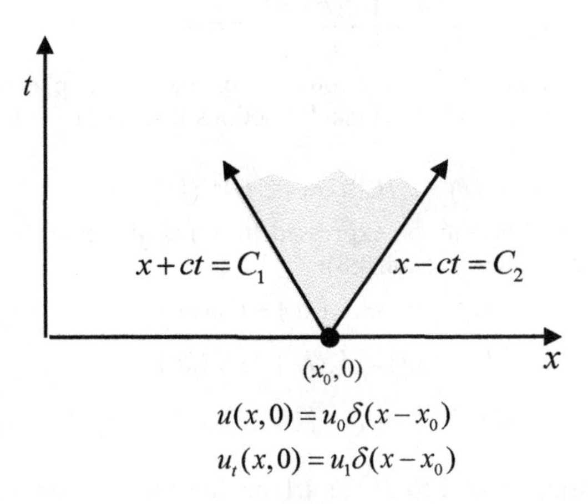

Figure 9.9 Domain of influence for a source point

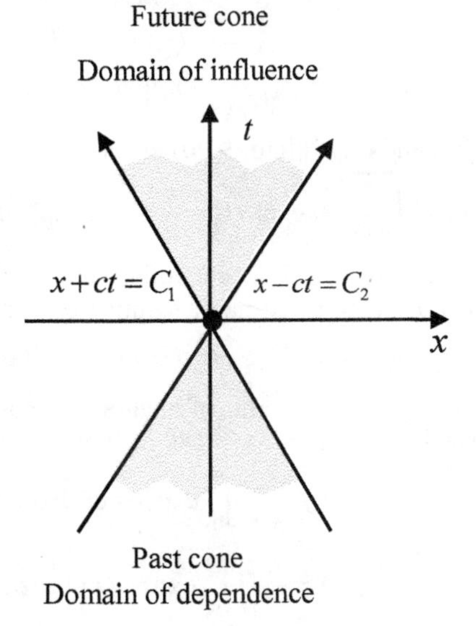

Figure 9.10 Cone of event horizon (past and future)

For the case of radially symmetric spherical waves (i.e., waves in unbounded solids), the wave equation is

$$\frac{1}{a^2}\frac{\partial^2 \varphi}{\partial t^2}=\frac{\partial^2 \varphi}{\partial r^2}+\frac{1}{r}\frac{\partial \varphi}{\partial r} \tag{9.88}$$

Consider a time harmonic wave as

$$\varphi=\psi(r)e^{-i\omega t} \tag{9.89}$$

$$\frac{d^2\psi}{dr^2}+\frac{1}{r}\frac{d\psi}{dr}+\frac{\omega^2}{a^2}\psi=0 \tag{9.90}$$

This is the Bessel equation of zero order. For the wave phenomenon, it is customary to express its solution in Hankel functions instead of the Bessel function as

$$\psi(r)=c_1 H_0^{(1)}(kr)+c_1 H_0^{(2)}(kr) \tag{9.91}$$

where the Hankel functions can be expressed in terms of Bessel functions of the first and second kinds as (see Section 4.8):

$$H_0^{(1)}(kr)=J_0(kr)+iY_0(kr) \tag{9.92}$$

$$H_0^{(2)}(kr)=J_0(kr)-iY_0(kr) \tag{9.93}$$

$$k=\frac{\omega}{a} \tag{9.94}$$

See also the discussion related to Table 4.1 in Chapter 4 on the role of Hankel function in wave phenomenon. Now, consider the asymptotic form of the Hankel functions for $r\to\infty$ as discussed in Chapter 4:

$$H_0^{(1)}(kr)\sim\sqrt{\frac{2}{\pi kr}}\exp[i(kr-\pi/4)] \tag{9.95}$$

$$H_0^{(2)}(kr)\sim\sqrt{\frac{2}{\pi kr}}\exp[-i(kr-\pi/4)] \tag{9.96}$$

Substitution of (9.94) and (9.95) into (9.90) gives

$$\psi(r)=\sqrt{\frac{2}{\pi kr}}\{c_1\exp[i(kr-\frac{\pi}{4})]+c_2\exp[-i(kr-\frac{\pi}{4})]\} \tag{9.97}$$

The first and second terms on the right reflect contracting wave and expanding waves respectively. For the time harmonic solution of (9.87), we have

$$\varphi(t,r)=e^{i\omega t}\left\{c_1 H_0^{(1)}(kr)+c_1 H_0^{(2)}(kr)\right\} \tag{9.98}$$

This solution can be expressed in terms of integrals by noting the following integral representation of Hankel functions as (p.180 of Watson, 1944):

$$H_0^{(1)}(kr)=\frac{2}{\pi i}\int_0^\infty \exp\{ikr\cosh\phi\}d\phi \tag{9.99}$$

$$H_0^{(2)}(kr)=-\frac{2}{\pi i}\int_0^\infty \exp\{-ikr\cosh\phi\}d\phi \tag{9.100}$$

Substitution of these results into (9.97) gives

$$\varphi(t,r) = c_1 \frac{2}{\pi i} \int_0^\infty \exp\{i\omega t + ikr\cosh\phi\} d\phi - c_2 \frac{2}{\pi i} \int_0^\infty \exp\{i\omega t - ikr\cosh\phi\} d\phi \quad (9.101)$$

The general solution of the arbitrary time function can be founded by summing different time harmonics as:

$$\varphi(t,r) = \int_{-\infty}^{\infty} \int_0^\infty \exp\{i\omega(t + \frac{r}{a}\cosh\phi)\} f(\omega) d\phi d\omega$$
$$+ \int_{-\infty}^{\infty} \int_0^\infty \exp\{i\omega(t - \frac{r}{a}\cosh\phi)\} g(\omega) d\phi d\omega \quad (9.102)$$

Therefore, this solution suggests that the general solution can be expressed as arbitrary functions of the form (Copson, 1975):

$$\varphi(t,r) = \int_0^\infty F(t - \frac{r}{a}\cosh\phi) d\phi + \int_0^\infty G(t + \frac{r}{a}\cosh\phi) d\phi \quad (9.103)$$

The validity of this solution can be verified by direct substitution of (9.102) into (9.87). The main feature of this solution is that for large time t the integral in (9.101) does not drop to zero. In other words, there are tails to the disturbance. The physical meaning of this will be discussed in a later section. Another solution for two-dimensional wave in a finite domain in integral terms will be given in a later section (or the so-called Poisson integral).

9.2.4 Three-Dimensional Waves

So far, we have considered the solution for both one-dimensional and two-dimensional wave equations. In this section, we will consider the solution of a three-dimensional wave problem defined as:

$$\frac{1}{a^2}\frac{\partial^2 u}{\partial t^2} = \frac{\partial^2 u}{\partial x^2} + \frac{\partial^2 u}{\partial y^2} + \frac{\partial^2 u}{\partial z^2} \quad (9.104)$$

$$u\big|_{t=0} = \varphi(x,y,z), \quad \frac{\partial u}{\partial t}\bigg|_{t=0} = \psi(x,y,z) \quad (9.105)$$

The solution is found expressed as:

$$u(t,x,y,z) = \frac{1}{4\pi a}\left[\frac{\partial}{\partial t}\iint_{S_{at}} \frac{\varphi}{at} dS + \iint_{S_{at}} \frac{\psi}{at} dS\right] \quad (9.106)$$

where S_{at} is the surface of a sphere with origin at (x, y, z) and radius at and dS denotes the surface integral over the sphere (see Figure 9.11). This formula is known as Poisson's mean value formula, and is also known as Kirchhoff's formula. The proof of (9.105) will be considered in Section 9.2.6. Physically, it means that the solution at the point (x, y, z) and time t only depends on the average value of the prescribed data on the surface of a sphere of radius at, and is independent of the value of the initial data within the sphere. As time t increases, once the spherical surface passes the initial disturbance zone, the spherical integral given in (9.105) will drop to zero. Thus, there is a sharp tail. This is referred to as the Huygens principle, which will be discussed further in Section 9.2.8. Actually, Poisson's formula can also be expressed as the mean value of the initial data on the sphere of integral as:

$$u(t,x,y,z)=\frac{1}{a}\left[\frac{\partial}{\partial t}[at\bar{\varphi}(at)]+at\bar{\psi}(at)\right] \tag{9.107}$$

where the superimposed bar denotes the average value taken over the spherical surface of radius *at*. In spherical coordinates, the mean value of φ over the sphere with center at (*x*, *y*, *z*) and radius of *at* can be evaluated as:

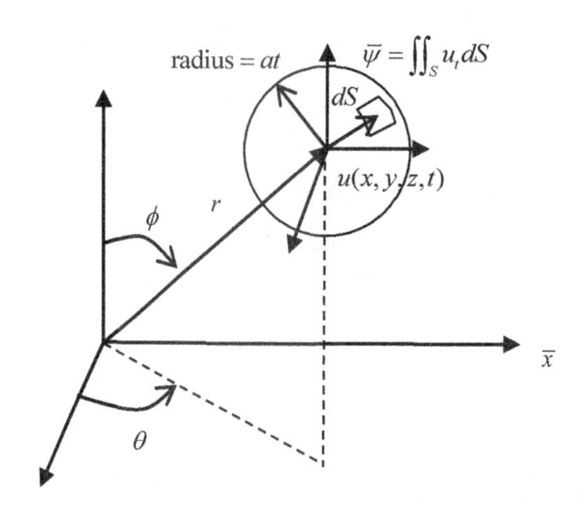

Figure 9.11 Poisson's mean value solution for 3-D wave

$$\bar{\varphi}(x,y,z;t)=\frac{1}{4\pi a^2t^2}\int_0^{\pi}\int_0^{2\pi}\varphi(x+at\sin\phi\cos\theta,y+at\sin\phi\sin\theta,\\ z+at\cos\phi)a^2t^2\sin\phi d\theta d\phi \tag{9.108}$$

The average of ψ can be defined similarly. The validity of this solution is shown in later sections.

9.2.5 Three-Dimensional Symmetric Waves

In this section, we will first consider the solution for spherically symmetric cases. More specifically, we have the prescribed values in (9.104) as:

$$\varphi=\varphi(r),\quad \psi=\psi(r) \tag{9.109}$$

Consequently, we must have the solution of u being:

$$u=u(r,t) \tag{9.110}$$

For this case, the wave equation becomes

$$\frac{1}{a^2}\frac{\partial^2 u}{\partial t^2}=\frac{\partial^2 u}{\partial r^2}+\frac{2}{r}\frac{\partial u}{\partial r} \tag{9.111}$$

It is straightforward to show that (9.110) can be written as:

$$\frac{\partial^2(ru)}{\partial t^2}=a^2\frac{\partial^2(ru)}{\partial r^2} \tag{9.112}$$

Accordingly, the initial conditions can be recast as

$$(ru)\big|_{t=0} = r\varphi(r), \quad \frac{\partial(ru)}{\partial t}\bigg|_{t=0} = r\psi(r) \tag{9.113}$$

In addition, we clearly have an additional condition that

$$(ru)\big|_{r=0} = 0 \tag{9.114}$$

This problem is mathematically equivalent to the one-dimensional wave problem for a vibrating string discussed earlier. Therefore, the D'Alembert solution discussed in the previous section can be applied directly as:

$$\begin{aligned} u(r,t) &= \frac{(r+at)\varphi(r+at)+(r-at)\varphi(r-at)}{2r} + \frac{1}{2ar}\int_{r-at}^{r+at} \xi\psi(\xi)d\xi, \quad r-at>0 \\ &= \frac{(r+at)\varphi(r+at)-(at-r)\varphi(at-r)}{2r} + \frac{1}{2ar}\int_{at-r}^{at+r} \xi\psi(\xi)d\xi, \quad r-at\le 0 \end{aligned} \tag{9.115}$$

This solution will be used again in the next section, when we discuss the general solution for a non-symmetric three-dimensional wave.

9.2.6 Poisson or Kirchhoff Formula for Three-Dimensional Waves

Let us define the surface average of the solution as

$$\bar{u}(r,t) = \frac{1}{4\pi r^2}\iint_S u(\xi,\eta,\zeta,t)dS \tag{9.116}$$

where the integral over dS is for variables (ξ, η, ζ). In spherical polar form defined in Fig. 1.27, we can rewrite it as:

$$\bar{u}(r,t) = \frac{1}{4\pi}\iint_{S_1} u(x+r\sin\theta\cos\phi, y+r\sin\theta\sin\phi, z+r\cos\theta, t)d\omega \tag{9.117}$$

where the surface integral is conducted over a unit sphere such that

$$dS = r^2 d\omega \tag{9.118}$$

Clearly, the value of u at the center of the sphere can be found by taking the following limit:

$$\lim_{r\to 0}\bar{u}(r,t) = u(x,y,z,t) \tag{9.119}$$

Differentiation of both sides of (9.116) gives

$$\begin{aligned} \frac{\partial \bar{u}}{\partial r} &= \frac{1}{4\pi}\iint_{S_1}\left(\frac{\partial u}{\partial \xi}\sin\theta\cos\phi + \frac{\partial u}{\partial \eta}\sin\theta\sin\phi + \frac{\partial u}{\partial \zeta}\cos\theta\right)d\omega \\ &= \frac{1}{4\pi r^2}\iint_{S_1}\left(\frac{\partial u}{\partial \xi}\sin\theta\cos\phi + \frac{\partial u}{\partial \eta}\sin\theta\sin\phi + \frac{\partial u}{\partial \zeta}\cos\theta\right)dS \end{aligned} \tag{9.120}$$

Applying Gauss's theorem to (9.119), we obtain

$$\begin{aligned} \frac{\partial \bar{u}}{\partial r} &= \frac{1}{4\pi r^2}\iiint_B \left(\frac{\partial^2 u}{\partial \xi^2} + \frac{\partial^2 u}{\partial \eta^2} + \frac{\partial^2 u}{\partial \zeta^2}\right)dV = \frac{1}{4\pi r^2}\iiint_B \nabla^2 u dV \\ &= \frac{1}{4\pi a^2 r^2}\iiint_B \frac{\partial^2 u}{\partial t^2}dV \end{aligned} \tag{9.121}$$

The last line of (9.120) results from the wave equation. The volume integral can be rewritten as

$$4\pi a^2 r^2 \frac{\partial \bar{u}}{\partial r} = \iiint_B \frac{\partial^2 u}{\partial t^2} dV = \frac{\partial^2}{\partial t^2} \int_0^r \iint_{S_{at}} udS\, d\rho \tag{9.122}$$

Differentiation of (9.121) gives

$$\frac{\partial}{\partial r}(4\pi a^2 r^2 \frac{\partial \bar{u}}{\partial r}) = \frac{\partial^2}{\partial t^2} \iint_{S_{at}} udS \tag{9.123}$$

This expression can be rearranged as

$$a^2 \frac{1}{r^2} \frac{\partial}{\partial r}(r^2 \frac{\partial \bar{u}}{\partial r}) = \frac{1}{4\pi r^2} \frac{\partial^2}{\partial t^2} \iint_{S_{at}} udS = \frac{\partial^2 \bar{u}}{\partial t^2} \tag{9.124}$$

The following identities can be proved easily

$$\frac{\partial^2 (r\bar{u})}{\partial r^2} = 2\frac{\partial \bar{u}}{\partial r} + r\frac{\partial^2 \bar{u}}{\partial r^2} \tag{9.125}$$

$$\frac{\partial}{\partial r}(r^2 \frac{\partial \bar{u}}{\partial r}) = 2r\frac{\partial \bar{u}}{\partial r} + r^2 \frac{\partial^2 \bar{u}}{\partial r^2} \tag{9.126}$$

By using (9.124) and (9.125), we finally obtain

$$\frac{\partial^2 (r\bar{u})}{\partial t^2} = a^2 \frac{\partial^2 (r\bar{u})}{\partial r^2} \tag{9.127}$$

The boundary conditions (9.104) can be rewritten accordingly

$$(r\bar{u})\big|_{t=0} = r\bar{\varphi}, \quad \frac{\partial (r\bar{u})}{\partial t}\bigg|_{t=0} = r\bar{\psi}, \quad (r\bar{u})\big|_{r=0} = 0 \tag{9.128}$$

where

$$\bar{\varphi}(r,t) = \frac{1}{4\pi r^2} \iint_{S_{at}} \varphi(\xi,\eta,\zeta,t) dS \tag{9.129}$$

$$\bar{\psi}(r,t) = \frac{1}{4\pi r^2} \iint_{S_{at}} \psi(\xi,\eta,\zeta,t) dS \tag{9.130}$$

The last boundary condition in (9.127) is added similarly to the argument in the last section.

$$\begin{aligned} \bar{u}(r,t) &= \frac{(r+at)\bar{\varphi}(r+at) + (r-at)\bar{\varphi}(r-at)}{2r} + \frac{1}{2ar}\int_{r-at}^{r+at} \xi\bar{\psi}(\xi)d\xi, \quad r-at>0 \\ &= \frac{(r+at)\bar{\varphi}(r+at) - (at-r)\bar{\varphi}(at-r)}{2r} + \frac{1}{2ar}\int_{at-r}^{at+r} \xi\bar{\psi}(\xi)d\xi, \quad r-at \le 0 \end{aligned} \tag{9.131}$$

Considering the limit of $r \to 0$, by virtue of (9.118) we have

$$\begin{aligned} u(x,y,z,t) = \lim_{r\to 0} \bar{u}(r,t) &= \lim_{r\to 0} \frac{(r+at)\bar{\varphi}(r+at) - (at-r)\bar{\varphi}(at-r)}{2r} \\ &+ \lim_{r\to 0} \frac{1}{2ar}\int_{at-r}^{at+r} \xi\bar{\psi}(\xi)d\xi \end{aligned} \tag{9.132}$$

It is clear that both the first and second terms on the right of (9.131) are of the form 0/0. Application of L'Hôpital's rule results in

$$\lim_{r\to 0}\frac{(r+at)\bar{\varphi}(r+at)-(at-r)\bar{\varphi}(at-r)}{2r}=\bar{\varphi}(at)+at\bar{\varphi}'(at)=\frac{1}{a}\frac{\partial}{\partial t}[at\bar{\varphi}(at)] \quad (9.133)$$

Applying L'Hôpital's rule and Leibniz's rule of differentiation on the integral given in (9.131), we find the second term as

$$\lim_{r\to 0}\frac{1}{2ar}\int_{at-r}^{at+r}\xi\bar{\psi}(\xi)d\xi=\frac{1}{a}[at\bar{\psi}(at)] \quad (9.134)$$

Substitution of (9.132) and (9.133) into (9.131) gives

$$u(x,y,z,t)=\frac{1}{a}\frac{\partial}{\partial t}[at\bar{\varphi}(at)]+\frac{1}{a}[at\bar{\psi}(at)] \quad (9.135)$$

Finally, in view of (9.128) and (9.129) we have

$$u(t,x,y,z)=\frac{1}{4\pi a}\left[\frac{\partial}{\partial t}\iint_{S_{at}}\frac{\varphi}{at}dS+\iint_{S_{at}}\frac{\psi}{at}dS\right] \quad (9.136)$$

This completes the proof of Poisson's formula or Kirchhoff's formula.

In spherical coordinates, Poisson's formula can be expressed as

$$\begin{aligned}u(r,\theta,\phi,t)=&\frac{\partial}{\partial t}\Big[\frac{t}{4\pi}\int_0^{\pi}\int_0^{2\pi}\varphi(x+atl,y+atm,z+atn)\sin\theta d\theta d\phi\Big]\\&+\frac{t}{4\pi}\int_0^{\pi}\int_0^{2\pi}\psi(x+atl,y+atm,z+atn)\sin\theta d\theta d\phi\end{aligned} \quad (9.137)$$

where

$$l=\sin\theta\cos\phi,\quad m=\sin\theta\sin\phi,\quad n=\cos\theta,\quad l^2+m^2+n^2=1 \quad (9.138)$$

9.2.7 Hadamard's Method of Descent for 2-D Wave

For two-dimensional wave problems, an integral solution similar to (9.135) can also be established by specifying it to the two-dimensional case. This method is normally referred as Hadamard's method of descent. First, the two-dimensional wave problem can be summarized as

$$\frac{1}{a^2}\frac{\partial^2 u}{\partial t^2}=\frac{\partial^2 u}{\partial x^2}+\frac{\partial^2 u}{\partial y^2} \quad (9.139)$$

$$u\big|_{t=0}=\varphi(x,y),\quad \frac{\partial u}{\partial t}\bigg|_{t=0}=\psi(x,y) \quad (9.140)$$

Since the surface integral is no longer a function of z, the incremental area dS can be projected to $d\Sigma$ as shown in Figure 9.12. The contribution from the upper hemisphere to the projected circular surface is the same as that from the lower hemisphere. Thus, we have

$$dS=2d\mathbf{\Sigma}=\frac{2}{\cos\theta}d\mathbf{\Sigma} \quad (9.141)$$

As shown in Figure 9.12, the cosine function can be determined as:

$$\cos\theta = \frac{\sqrt{a^2t^2 - (\xi - x)^2 - (\eta - y)^2}}{at} \tag{9.142}$$

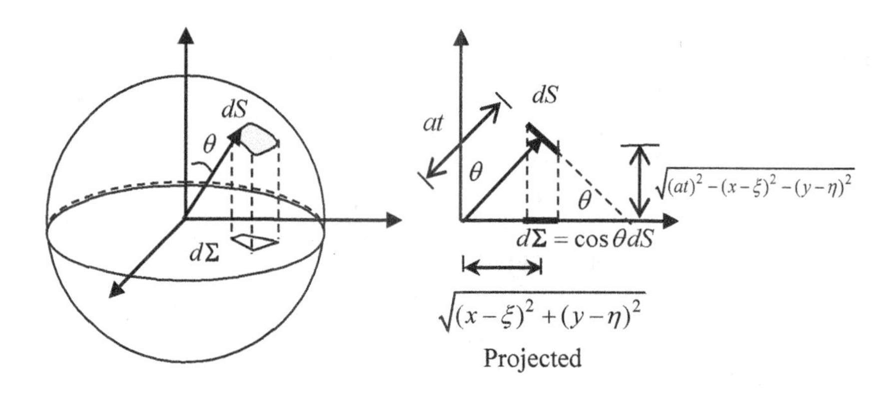

Figure 9.12 Spherical surface projected to circular

Using these results, we get from (9.135)

$$u(t,x,y,z) = \frac{1}{4\pi a}\left[\frac{\partial}{\partial t}\iint_{\Sigma_{at}} \frac{\varphi}{at}\frac{2}{\cos\theta} d\Sigma + \iint_{\Sigma_{at}} \frac{\psi}{at}\frac{2}{\cos\theta} d\Sigma\right] \tag{9.143}$$

Substitution of (9.141) into (9.142) gives

$$\begin{aligned} u(t,x,y,z) = \frac{1}{2\pi a}[\frac{\partial}{\partial t}\iint_{\Sigma_{at}} \frac{\varphi(\xi,\eta)}{\sqrt{a^2t^2 - (\xi - x)^2 - (\eta - y)^2}} d\xi d\eta \\ + \iint_{\Sigma_{at}} \frac{\psi(\xi,\eta)}{\sqrt{a^2t^2 - (\xi - x)^2 - (\eta - y)^2}} d\xi d\eta] \end{aligned} \tag{9.144}$$

This is the two-dimensional Poisson formula. In polar form, this formula can be written as

$$\begin{aligned} u(t,x,y,z) = \frac{1}{2\pi a}[\frac{\partial}{\partial t}\int_0^{at}\int_0^{2\pi} \frac{\varphi(x+\rho\cos\theta, y+\rho\sin\theta)}{\sqrt{a^2t^2 - \rho^2}} \rho d\theta d\rho] \\ + \frac{1}{2\pi a}\int_0^{at}\int_0^{2\pi} \frac{\psi(x+\rho\cos\theta, y+\rho\sin\theta)}{\sqrt{a^2t^2 - \rho^2}} \rho d\theta d\rho \end{aligned} \tag{9.145}$$

Note that the integration given in (9.143) and (9.144) are evaluated for the whole circular region instead of the spherical surface. This difference in the 2-D and 3-D Poisson formulas makes a big difference physically, which will be discussed next.

9.2.8 Huygen Principle

For the three-dimensional wave solution given by Poisson's formula, Figure 9.13 shows a particular situation that the initial disturbances φ and ψ are only given in domain *Ω*.

(i) Case I:

$$at < at_1 \tag{9.146}$$

For this case, the wave signal from Ω has not arrived the center of the sphere. Therefore the solution given by Poisson's formula gives a zero solution because the spherical surface has no initial values of φ and ψ.

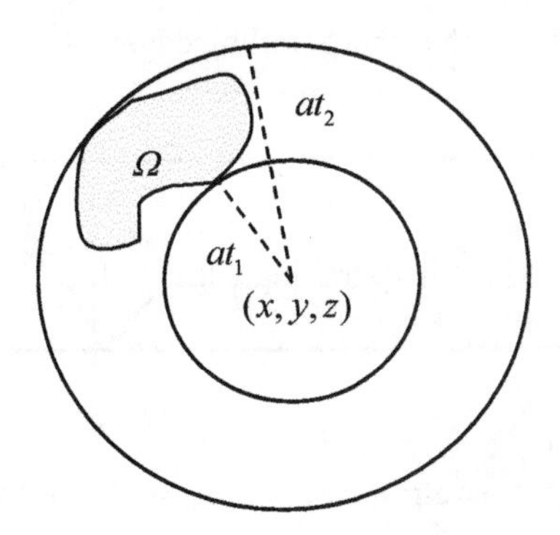

Figure 9.13 Interpretation of Poisson's formula

(ii) Case II:

$$at_2 \geq at \geq at_1 \tag{9.147}$$

Once $t = t_1$, the first wave signal arrives sharply. Within this time period, the intersection between the spherical surface of integration and the initial disturbance leads to a nonzero solution.

(iii) Case III:

$$at > at_2 \tag{9.148}$$

Once $t > t_2$, there is no intersection between the spherical surface of integration and the initial disturbance. Thus, the solution drops sharply to zero and there is no ripple once the wave energy has passed. That is, there is a sharp trailing edge of the wave solution. This is known as the Huygens principle. This Huygens principle actually does not apply to the two-dimensional wave solution.

For two-dimensional waves, the solution given in (9.144) shows that the solution is a circular area integration. Thus, we have only two scenarios.

(i) Case I:

$$at < at_1 \tag{9.149}$$

There is no wave signal overlapping the circular domain of integration, and thus the solution is zero.

(ii) Case II:

$$at \geq at_1 \tag{9.150}$$

The integration in (9.144) is conducted on the whole circular area instead of on the spherical surface in the three-dimensional case in (9.135). Therefore, there is always a tail of the solution in the two-dimensional case. Hence, the Huygens principle does not apply to the two-dimensional case.

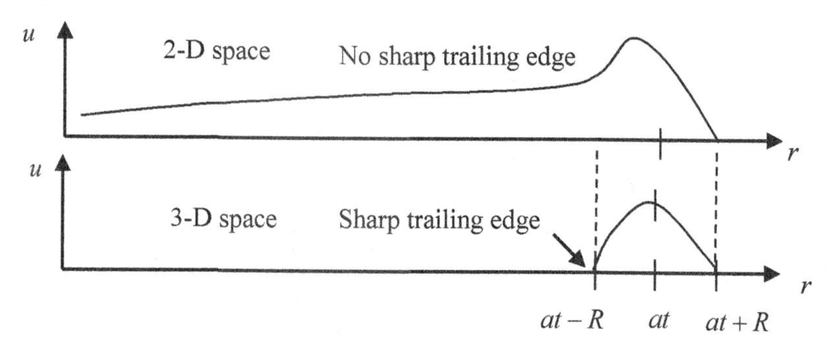

Figure 9.14 Wave solution in 2-D and 3-D space according to 2-D and 3-D Poisson's formula

The existence of tails in the two-dimensional solution can also be seen in (9.102). In particular, the solution expressed in terms of an infinite integration is never zero even though the region of disturbance is passed. Huygens principle has a significant impact in our daily lives. Luckily, we live in a three-dimensional world, in which we can hear a clear voice without infinite echoes and ripples when someone speaks to you. The sound wave in three-dimensional space dies off instantly after it simulates our ears. Imagine that an animal living in a two-dimensional world will never be able to hear a clear voice. There are always infinite echoes flying around in the two-dimensional flatland world. Sound sources initiated at different times will all mix up. The surface waves of the sea can be viewed as a two-dimensional domain, and you never see a perfectly calm sea surface.

More generally, the Huygens principle is true only for odd dimensions (e.g., 3,5,7,..., except for 1-D) and is false for all even dimensions (e.g., 2,4,6,...).

9.2.9 n-Dimensional Waves

In this section, we will extend the analysis of waves in n-dimensional space. In particular, the wave equation reads as

$$a^2(\varphi_{x_1x_1} + \varphi_{x_2x_2} + \ldots + \varphi_{x_nx_n}) = \varphi_{tt} \tag{9.151}$$

This distance in n-dimensional space can be defined as

$$r^2 = x_1^2 + x_2^2 + ... + x_n^2 = \sum_{i=1}^{n} x_i x_i = x_k x_k \tag{9.152}$$

Let us consider the case of symmetric radial waves such that φ = φ(*r*,*t*) (i.e., there is no angular dependence of the wave function φ.

Differentiation of *r* with respect to any arbitrary variable x_i gives

$$\frac{\partial r}{\partial x_i} = \frac{\partial}{\partial x_i}(x_1^2 + x_2^2 + ... + x_n^2)^{1/2} = \frac{1}{2}\frac{1}{(x_1^2 + x_2^2 + ... + x_n^2)^{1/2}} 2x_i = \frac{x_i}{r} \tag{9.153}$$

Using tensor notation and the chain rule, we have the following identities

$$\frac{\partial x_i}{\partial x_j} = \delta_{ij}, \quad \frac{\partial x_i}{\partial x_i} = \delta_{ii} = \delta_{11} + ... + \delta_{nn} = 1 + ... + 1 = n \tag{9.154}$$

$$\frac{\partial \varphi}{\partial x_i} = \frac{\partial \varphi}{\partial r}\frac{\partial r}{\partial x_i} = \frac{x_i}{r}\frac{\partial \varphi}{\partial r} \tag{9.155}$$

$$x_i \frac{\partial r}{\partial x_i} = \frac{x_i x_i}{r} = \frac{r^2}{r} = r \tag{9.156}$$

Using these identities, the Laplacian in *n*-dimensional becomes

$$\begin{aligned}\frac{\partial^2 \varphi}{\partial x_i \partial x_i} &= \frac{\partial}{\partial x_i}(\frac{x_i}{r}\frac{\partial \varphi}{\partial r}) = \frac{\partial x_i}{\partial x_i}(\frac{1}{r}\frac{\partial \varphi}{\partial r}) + x_i \frac{\partial}{\partial r}(\frac{1}{r}\frac{\partial \varphi}{\partial r})\frac{\partial r}{\partial x_i} \\ &= n(\frac{1}{r}\frac{\partial \varphi}{\partial r}) + \frac{\partial^2 \varphi}{\partial r^2} - \frac{1}{r}\frac{\partial \varphi}{\partial r} \\ &= \frac{\partial^2 \varphi}{\partial r^2} + \frac{n-1}{r}\frac{\partial \varphi}{\partial r}\end{aligned} \tag{9.157}$$

Consequently, the *n*-dimensional wave equation for radial waves becomes

$$\frac{1}{a^2}\frac{\partial^2 \varphi}{\partial t^2} = \frac{\partial^2 \varphi}{\partial r^2} + \frac{n-1}{r}\frac{\partial \varphi}{\partial r} = \frac{\partial}{\partial r}[r^{n-1}\frac{\partial \varphi}{\partial r}] \tag{9.158}$$

This *n*-dimensional wave is also known as the Euler-Poisson-Darboux equation. Now, we look at some special cases:

Special case: $n = 1$

$$\frac{1}{a^2}\frac{\partial^2 \varphi}{\partial t^2} = \frac{\partial^2 \varphi}{\partial r^2} \tag{9.159}$$

The solution obtained from the method of characteristics is

$$\varphi = f(r - at) + g(r + at) \tag{9.160}$$

Special case: $n = 3$

$$\frac{1}{a^2}\frac{\partial^2 \varphi}{\partial t^2} = \frac{\partial^2 \varphi}{\partial r^2} + \frac{2}{r}\frac{\partial \varphi}{\partial r} \tag{9.161}$$

It is straightforward to prove that it can be written as (compare Section 9.2.5)

$$\frac{1}{a^2}\frac{\partial^2 (r\varphi)}{\partial t^2} = \frac{\partial^2 (r\varphi)}{\partial r^2} \tag{9.162}$$

Again, the solution obtained from the method of characteristics is

$$\varphi = \frac{1}{r} f(r - at) + \frac{1}{r} g(r + at) \tag{9.163}$$

The first term on the right of (9.162) is an expanding wave whereas the second term on the right of (9.163) is a contracting wave. Therefore 3-D wave decay is

$$\varphi \propto \frac{1}{r} \tag{9.164}$$

Special case: $n = 2$

$$\frac{1}{a^2} \frac{\partial^2 \varphi}{\partial t^2} = \frac{\partial^2 \varphi}{\partial r^2} + \frac{1}{r} \frac{\partial \varphi}{\partial r} \tag{9.165}$$

This is the most difficult part, compared to the cases of 1-D and 3-D. The solution has been found in terms of Hankel functions in (9.90).

To give insight to the 2-D wave, we consider an approximate solution for the 2-D wave. In particular, we first consider the following identity

$$\begin{aligned}
\frac{1}{\sqrt{r}} \frac{\partial^2 (\sqrt{r}\varphi)}{\partial r^2} &= \frac{1}{\sqrt{r}} \frac{\partial}{\partial r} \left\{ \frac{1}{2\sqrt{r}} \varphi + \sqrt{r} \frac{\partial \varphi}{\partial r} \right\} \\
&= \frac{1}{\sqrt{r}} \left\{ -(\frac{1}{2}) \frac{1}{2r^{3/2}} \varphi + \frac{1}{2\sqrt{r}} \frac{\partial \varphi}{\partial r} + \frac{1}{2\sqrt{r}} \frac{\partial \varphi}{\partial r} + \sqrt{r} \frac{\partial^2 \varphi}{\partial r^2} \right\} \\
&= \frac{1}{\sqrt{r}} \left\{ -\frac{\varphi}{4r^{3/2}} + \frac{1}{\sqrt{r}} \frac{\partial \varphi}{\partial r} + \sqrt{r} \frac{\partial^2 \varphi}{\partial r^2} \right\} \\
&= \frac{1}{\sqrt{r}} \left\{ -\frac{\varphi}{4r^{3/2}} + \sqrt{r} (\frac{1}{r} \frac{\partial \varphi}{\partial r} + \frac{\partial^2 \varphi}{\partial r^2}) \right\} \\
&= \frac{1}{\sqrt{r}} \left\{ -\frac{\varphi}{4r^{3/2}} + \sqrt{r} (\frac{1}{a^2} \frac{\partial^2 \varphi}{\partial t^2}) \right\}
\end{aligned} \tag{9.166}$$

Using this identity, the 2-D wave can be expressed as

$$\frac{1}{a^2} \frac{\partial^2 (\sqrt{r}\varphi)}{\partial t^2} = \frac{\partial^2 (\sqrt{r}\varphi)}{\partial r^2} + \frac{\varphi}{4r^{3/2}} \tag{9.167}$$

This cannot be solved easily as in the 1-D or 3-D case. However, for $r \to \infty$, we have

$$\frac{1}{a^2} \frac{\partial^2 (\sqrt{r}\varphi)}{\partial t^2} \approx \frac{\partial^2 (\sqrt{r}\varphi)}{\partial r^2} \tag{9.168}$$

$$\varphi = \frac{1}{\sqrt{r}} f(r - at) + \frac{1}{\sqrt{r}} g(r + at) \tag{9.169}$$

Therefore, asymptotically the two-dimensional wave will behave as

$$\varphi \propto \frac{1}{\sqrt{r}} \tag{9.170}$$

This result of course agrees with the asymptotic solution for the two-dimensional waves given in Section 9.2.3 (see (9.96)).

9.2.10 Wavefront Condition

In this section, we will consider the wavefront condition. Recall a wave equation of the following form:

$$\frac{\partial^2 u}{\partial t^2} = a^2 \nabla^2 u \tag{9.171}$$

Consider the case of harmonic waves such that

$$u = \psi e^{-i\omega t} \tag{9.172}$$

Substitution of (9.171) into (9.170) leads to the following Helmholtz equation:

$$\nabla^2 \psi + k^2 \psi = 0 \tag{9.173}$$

where

$$k = \frac{\omega}{a} \tag{9.174}$$

Next, we look for a plane wave solution of the form:

$$\psi(\boldsymbol{x}) = e^{i\boldsymbol{k}\cdot\boldsymbol{x}} \tag{9.175}$$

Combining (9.171) and (9.174) gives the solution form as

$$u(\boldsymbol{x},t) = e^{i\boldsymbol{k}\cdot\boldsymbol{x} - i\omega t} \tag{9.176}$$

Since the wave type of the solution consists of a jump across the wavefront and at the wavefront the solution u is a constant, thus we have the following condition on the wavefront:

$$\boldsymbol{k}\cdot\boldsymbol{x} - \omega t = C_1 \tag{9.177}$$

where C_1 is a constant.

On the other hand, the wavefront can also be expressed generally as

$$F(\boldsymbol{x},t) = 0 \tag{9.178}$$

Taking the total differential of (9.177), we get

$$dF(\boldsymbol{x},t) = \nabla F \cdot d\boldsymbol{x} + \frac{\partial F}{\partial t} dt = 0 \tag{9.179}$$

Since there is no change of F along the wavefront, the kinetics compatibility condition gives

$$\nabla F \cdot \frac{d\boldsymbol{x}}{dt} + \frac{\partial F}{\partial t} = 0 \tag{9.180}$$

The gradient of the wavefront surface F is normal to the surface, or mathematically it requires

$$\nabla F = \left|\nabla F\right| \boldsymbol{n} \tag{9.181}$$

where $\boldsymbol{n}$ is the unit normal to the wavefront F as shown in Figure 9.15. The velocity of the wavefront is defined as

$$\boldsymbol{v} = \frac{d\boldsymbol{x}}{dt} \tag{9.182}$$

The normal propagating speed of the wavefront surface is

$$v_n = \boldsymbol{n}\cdot\boldsymbol{v} \tag{9.183}$$

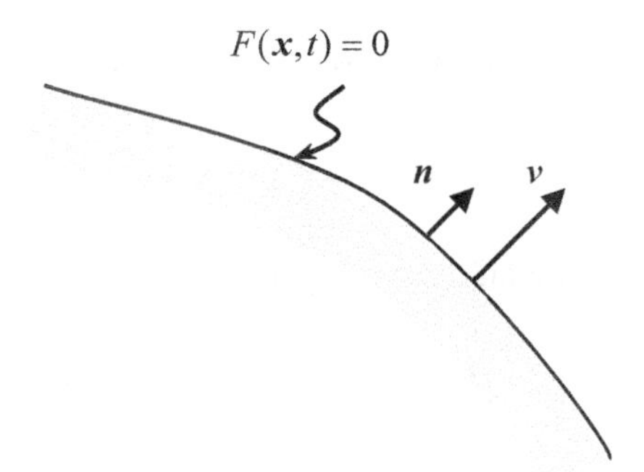

Figure 9.15 Moving wavefront with unit normal *n*

Substitution of (9.180), (9.181), and (9.182) into (9.179) gives

$$|\nabla F|v_n + \frac{\partial F}{\partial t} = 0 \tag{9.184}$$

Comparison of (9.176) and (9.178) gives the wave number and frequency as

$$|\nabla F| = |\boldsymbol{k}| = k\,, \quad \frac{\partial F}{\partial t} = -\omega \tag{9.185}$$

From (9.183) and (9.184), the normal component of the moving wavefront is then

$$v_n = -\frac{\partial F}{\partial t} / |\nabla F| = \frac{\omega}{k} = a \tag{9.186}$$

Therefore, the wavefront is moving at the wave speed of the differential equation.

9.3 NONHOMOGENEOUS WAVE EQUATION

We now consider the case of the nonhomogeneous wave equation. We have demonstrated in Chapter 2 that the nonhomogeneous wave equation appears naturally in the solution of the Maxwell equation as well as in the dynamic problem of elastic solids subject to body force.

The problem can be posed as:

$$\frac{\partial^2 u}{\partial t^2} = a^2 \nabla^2 u + f(x,y,z,t) \tag{9.187}$$

$$u\big|_{t=0} = \varphi(x,y,z), \quad \frac{\partial u}{\partial t}\bigg|_{t=0} = \psi(x,y,z) \tag{9.188}$$

By the principle of superposition, the problem can be decomposed into two associated problems.

Problem I:

$$\frac{\partial^2 u}{\partial t^2} = a^2 \nabla^2 u \tag{9.189}$$

$$u\big|_{t=0} = \varphi(x,y,z), \quad \frac{\partial u}{\partial t}\bigg|_{t=0} = \psi(x,y,z) \tag{9.190}$$

Problem II:

$$\frac{\partial^2 u}{\partial t^2} = a^2 \nabla^2 u + f(x,y,z,t) \tag{9.191}$$

$$u\big|_{t=0} = 0, \quad \frac{\partial u}{\partial t}\bigg|_{t=0} = 0 \tag{9.192}$$

The solution of Problem I has been given by Poisson's formula in Section 9.2.6.

The solution of Problem II can be found again by Poisson's formula if a proper change of variables is introduced. In particular, we use the Duhamel integral as:

$$u(x,y,z,t) = \int_0^t w(x,y,z,t-\tau;\tau)d\tau \tag{9.193}$$

Application of the Leibniz rule of differentiation gives

$$\frac{\partial u}{\partial t} = \int_0^t \frac{\partial w}{\partial t}(x,y,z,t-\tau;\tau)d\tau + w(x,y,z,0;t) \tag{9.194}$$

This can be simplified by setting an initial condition for w as:

$$w(x,y,z,0;t) = 0 \tag{9.195}$$

Differentiation of (9.193) again gives

$$\frac{\partial^2 u}{\partial t^2} = \int_0^t \frac{\partial^2 w}{\partial t^2}(x,y,z,t-\tau;\tau)d\tau + \frac{\partial w}{\partial t}(x,y,z,0;t) \tag{9.196}$$

Taking the Laplacian of (9.192), we get

$$\nabla^2 u(x,y,z,t) = \int_0^t \nabla^2 w d\tau \tag{9.197}$$

Substitution of (9.195) and (9.196) into (9.190) results in

$$\int_0^t \left(\frac{\partial^2 w}{\partial t^2} - a^2 \nabla^2 w\right)d\tau = 0, \quad \frac{\partial w}{\partial t}\bigg|_{t=0} = 0 \tag{9.198}$$

Therefore, the governing equation for w becomes

$$\frac{\partial^2 w}{\partial t^2} = a^2 \nabla^2 w \tag{9.199}$$

$$w\big|_{t=0} = 0, \quad \frac{\partial w}{\partial t}\bigg|_{t=0} = f(x,y,z,t) \tag{9.200}$$

The solution of w is clearly a special case of Poisson's formula that we considered in Section 9.2.6. Thus, the solution for w is readily obtained by using Poisson's formula as:

$$w(x,y,z,t;\tau) = \frac{1}{4\pi a^2} \iint_{S_{a(t-\tau)}} \frac{f(\xi,\eta,\zeta,\tau)}{t-\tau} dS \tag{9.201}$$

Substitution of (9.200) into (9.192) results in the solution of Problem II as:

$$u(x,y,z,t)=\frac{1}{4\pi a^2}\int_0^t d\tau\iint_{S_{a(t-\tau)}}\frac{f(\xi,\eta,\zeta,\tau)}{t-\tau}dS \tag{9.202}$$

Finally, the solution of the problem is obtained by combining (9.135) and (9.201) as

$$\begin{aligned}u(t,x,y,z)&=\frac{1}{4\pi a}\left[\frac{\partial}{\partial t}\iint_{S_{at}}\frac{\varphi(\xi,\eta,\zeta)}{at}dS+\iint_{S_{at}}\frac{\psi(\xi,\eta,\zeta)}{at}dS\right]\\&+\frac{1}{4\pi a^2}\int_0^t d\tau\iint_{S_{a(t-\tau)}}\frac{f(\xi,\eta,\zeta,\tau)}{t-\tau}dS\end{aligned} \tag{9.203}$$

The solution resulting from the nonhomogeneous term can be rewritten slightly by introducing the following change of variables:

$$R=a(t-\tau),\quad d\tau=-\frac{dR}{a} \tag{9.204}$$

The solution of Problem II becomes

$$\begin{aligned}u^{II}(x,y,z,t)&=\frac{1}{4\pi a^2}\int_{at}^{0}\iint_{S_R}\frac{f(\xi,\eta,\zeta,t-R/a)}{R/a}dS(-\frac{1}{a}dR)\\&=\frac{1}{4\pi a^2}\int_0^{at}\iint_{S_R}\frac{f(\xi,\eta,\zeta,t-R/a)}{R}dSdR\\&=\frac{1}{4\pi a^2}\iiint_{B_R}\frac{f(\xi,\eta,\zeta,t-R/a)}{R}dV\end{aligned} \tag{9.205}$$

This solution can be regarded a *retarded potential* (e.g., Jackson, 1999). To see this, we recall the solution of Poisson's equation in terms of the so-called Boussinesq potential. In mathematical terms, the Poisson equation is given as

$$\nabla^2 u(\boldsymbol{r})=-\rho(\boldsymbol{r}) \tag{9.206}$$

The solution of this equation can be expressed as

$$u(\boldsymbol{r})=\frac{1}{4\pi}\iiint\frac{\rho(\boldsymbol{r}')}{R}dV' \tag{9.207}$$

where $1/R$ is the Boussinesq potential or Newtonian potential. Comparison of (9.204) and (9.206) shows that the solution in the wave equation is indeed a solution resulting from a potential but at a reduced time $t-R/a$ instead of at time t. If there is a time dependent solution at a point $P(\boldsymbol{r})$ denoted by $u(\boldsymbol{r}, t)$ at time t, this solution at time t can only be dependent on a signal from an earlier time. The time required to arrive at point $P(\boldsymbol{r})$ is R/a (recall that a is the wave speed in (9.186)).

Finally, the solution given in (9.202) can be expressed in terms of the retarded potential as

$$\begin{aligned}u(t,x,y,z)&=\frac{1}{4\pi a}\left[\frac{\partial}{\partial t}\iint_{S_{at}}\frac{\varphi(\xi,\eta,\zeta)}{at}dS+\iint_{S_{at}}\frac{\psi(\xi,\eta,\zeta)}{at}dS\right]\\&+\frac{1}{4\pi a^2}\iiint_{B_R}\frac{f(\xi,\eta,\zeta,t-R/a)}{R}dV\end{aligned} \tag{9.208}$$

9.4 HELMHOLTZ EQUATION

In this section, we will consider the solution of the Helmholtz equation in spherical coordinates. According to classification discussed in Section 7.2.2, the Helmholtz equation is of the elliptic type. But as we discussed in an earlier chapter, Helmholtz is also known as the reduced wave equation. The Helmholtz equation is somewhat between the wave equation and the Laplace equation. In general, the Helmholtz equation can be expressed as

$$\nabla^2 u + k^2 u = 0 \tag{9.209}$$

More specifically, in spherical coordinates we have

$$\frac{1}{r^2}\frac{\partial}{\partial r}(r^2\frac{\partial u}{\partial r}) + \frac{1}{r^2 \sin\varphi}\frac{\partial}{\partial \varphi}(\sin\varphi\frac{\partial u}{\partial \varphi}) + \frac{1}{r^2 \sin^2\varphi}\frac{\partial^2 u}{\partial \theta^2} + k^2 u = 0 \tag{9.210}$$

Applying the following change of variables, we obtain

$$u = R(r)\Theta(\theta)\Phi(\varphi) \tag{9.211}$$

Substitution of (9.210) into (9.209) gives

$$\Theta\Phi\frac{1}{r^2}\frac{d}{dr}(r^2\frac{du}{dr}) + \frac{R\Theta}{r^2 \sin\varphi}\frac{d}{d\varphi}(\sin\varphi\frac{d\Phi}{d\varphi}) + \frac{R\Phi}{r^2 \sin^2\varphi}\frac{\partial^2 \Theta}{\partial \theta^2} + k^2 R\Theta\Phi = 0 \tag{9.212}$$

This equation can be rearranged as

$$[\frac{1}{R}\frac{d}{dr}(r^2\frac{dR}{dr}) + \frac{1}{\Phi \sin\varphi}\frac{d}{d\varphi}(\sin\varphi\frac{d\Phi}{d\varphi})]\sin^2\varphi + r^2 k^2 \sin^2\varphi = -\frac{1}{\Theta}\frac{d^2\Theta}{d\theta^2} = \mu^2 \tag{9.213}$$

where μ is a separation constant. The last part of (9.212) can be expressed as

$$\frac{d^2\Theta}{d\theta^2} + \mu^2\Theta = 0 \tag{9.214}$$

The solution of Θ is

$$\Theta = A\cos\mu\theta + B\sin\mu\theta \tag{9.215}$$

There must be periodicity in θ and this leads to $\mu = m$ ($m = 0,1,2,3,...$). The first equation of (9.212) can be rearranged as

$$[\frac{1}{R}\frac{d}{dr}(r^2\frac{dR}{dr}) + r^2 k^2 = \frac{\mu^2}{\sin^2\varphi} - \frac{1}{\Phi \sin\varphi}\frac{d}{d\varphi}(\sin\varphi\frac{d\Phi}{d\varphi}) = \nu(\nu+1) \tag{9.216}$$

where the separation constant is written as $\nu(\nu+1)$. Equation (9.215) leads to two ODEs:

$$\frac{d}{dr}(r^2\frac{dR}{dr}) + [r^2 k^2 - \nu(\nu+1)]R = 0 \tag{9.217}$$

$$\frac{1}{\sin\varphi}\frac{d}{d\varphi}(\sin\varphi\frac{d\Phi}{d\varphi}) + [\nu(\nu+1) - \frac{m^2}{\sin^2\varphi}]\Phi = 0 \tag{9.218}$$

For the differential equation of Θ, we can introduce a change of variables

$$x = \cos\varphi \tag{9.219}$$

Using the chain rule of differentiation, we get

$$\frac{d\Phi}{d\varphi} = \frac{d\Phi}{dx}\frac{dx}{d\varphi} = -\sqrt{1-x^2}\frac{d\Phi}{dx} \tag{9.220}$$

With this result, we can rewrite (9.217) as

$$(1+x^2)\frac{d^2\Phi}{dx^2}-2x\frac{d\Phi}{dx}+[\nu(\nu+1)-\frac{m^2}{1-x^2}]\Phi=0 \tag{9.221}$$

The general solution is the Associated Legendre functions:

$$\Phi = EP_\nu^m(\cos\varphi)+FQ_\nu^m(\cos\varphi) \tag{9.222}$$

For solid spheres, we must have boundedness at $\varphi = 0$ and π and this leads to $F = 0$ and $\nu = n$ (where $n = 0,1,2,3...$).

Equation (9.216) can be reduced to a standard ODE by using the following substitution:

$$R = r^{-1/2}V \tag{9.223}$$

$$r^2\frac{d^2V}{dr^2}+r\frac{dV}{dr}+[r^2k^2-n(n+1)-\frac{1}{4}]V=0 \tag{9.224}$$

This can further be rewritten as:

$$r^2\frac{d^2V}{dr^2}+r\frac{dV}{dr}+[r^2k^2-(n+1/2)^2]V=0 \tag{9.225}$$

This is the Bessel equation and the solution for v is

$$V = CJ_{n+1/2}(kr)+DY_{n+1/2}(kr) \tag{9.226}$$

Finally, combining all these results the general solution for Helmholtz equation in spherical coordinates is:

$$u = r^{-1/2}\{A\cos m\theta+B\sin m\theta\}\{CJ_{n+1/2}(kr)+DY_{n+1/2}(kr)\}P_n^m(\cos\varphi) \tag{9.227}$$

This can be shown to be the same as the one given in Section 7.4.3 of Chapter 7. Actually, the Bessel function together with $r^{-1/2}$ can be written as new functions called the spherical Bessel functions (Abramowitz and Stegun, 1964):

$$j_n(kr)=\sqrt{\frac{\pi}{2kr}}J_{n+1/2}(kr) \tag{9.228}$$

$$y_n(kr)=\sqrt{\frac{\pi}{2kr}}Y_{n+1/2}(kr) \tag{9.229}$$

which are called spherical Bessel functions of the first kind and second kind, respectively. In terms of these new functions, the solution of (9.209) can be expressed as:

$$u = \{A\cos m\theta+B\sin m\theta\}\{Cj_n(kr)+Dy_n(kr)\}P_n^m(\cos\varphi) \tag{9.230}$$

9.5 TELEGRAPH EQUATION

In this section, the telegraph equation is considered. First, starting from the Maxwell equations, we derive the telegraph equation. The hyperbolic equation is then transformed to the canonical form before a proper change of variables is applied. Eventually, it is shown that the solution can be expressed in terms of the Bessel function.

9.5.1 Formulation

Recall from Chapter 2 that the Maxwell equations can be written as:

$$\nabla \cdot \boldsymbol{E} = \frac{\rho}{\varepsilon}, \quad \nabla \cdot \boldsymbol{B} = 0,$$
$$\nabla \times \boldsymbol{E} = -\frac{\partial \boldsymbol{B}}{\partial t}, \quad \nabla \times \boldsymbol{B} = \mu(\boldsymbol{J} + \varepsilon \frac{\partial \boldsymbol{E}}{\partial t}) \tag{9.231}$$

where $\boldsymbol{B}$ and $\boldsymbol{E}$ are the magnetic field vector and electric field vector. In addition, ε is the permittivity of the material, μ is the permeability of the material, and ρ is the charge density. Let us introduce a constitutive relation for the conduction part of the electric current $\boldsymbol{J}$:

$$\boldsymbol{J} = \sigma \boldsymbol{E} \tag{9.232}$$

where σ is the electric conductivity. Note that σ = 0 for perfect dielectric. For the case of no electric charge ρ, we can rewrite the Maxwell equation as:

$$\nabla \cdot \boldsymbol{E} = 0, \quad \nabla \cdot \boldsymbol{B} = 0,$$
$$\nabla \times \boldsymbol{E} = -\frac{\partial \boldsymbol{B}}{\partial t}, \quad \nabla \times \boldsymbol{B} = \mu(\sigma \boldsymbol{E} + \varepsilon \frac{\partial \boldsymbol{E}}{\partial t}) \tag{9.233}$$

On the right hand side of the fourth equation of (9.232), the general current density comprises two terms, the first being the conduction current density and the second being the displacement current density. Maxwell was the first to realize this decomposition.

Taking the curl of the third equation of (9.230) and taking the time derivative of the fourth equation of (9.230), we obtain

$$\nabla \times \nabla \times \boldsymbol{E} = -\nabla \times \frac{\partial \boldsymbol{B}}{\partial t}, \quad \nabla \times \frac{\partial \boldsymbol{B}}{\partial t} = \mu(\sigma \frac{\partial \boldsymbol{E}}{\partial t} + \varepsilon \frac{\partial^2 \boldsymbol{E}}{\partial t^2}) \tag{9.234}$$

Recall from (1.362) of Chapter 1 the following vector identity:

$$\nabla \times (\nabla \times \boldsymbol{E}) = \nabla(\nabla \bullet \boldsymbol{E}) - \nabla^2 \boldsymbol{E} \tag{9.235}$$

Equating the two equations in (9.233) and applying the first equation of (9.232) and (9.234), we find

$$\mu\varepsilon \frac{\partial^2 \boldsymbol{E}}{\partial t^2} = \nabla^2 \boldsymbol{E} - \mu\sigma \frac{\partial \boldsymbol{E}}{\partial t} \tag{9.236}$$

This is Maxwell's equation for the electric intensity vector $\boldsymbol{E}$. Note that for a good conductor (ε ≈ 0), we have

$$\nabla^2 \boldsymbol{E} = \mu\sigma \frac{\partial \boldsymbol{E}}{\partial t} \tag{9.237}$$

The Maxwell equation becomes a three-dimensional heat equation. For perfect dielectric (σ ≈ 0), we have

$$\frac{\partial^2 \boldsymbol{E}}{\partial t^2} = c^2 \nabla^2 \boldsymbol{E} \tag{9.238}$$

where c is the light speed. Thus, the Maxwell equation reduces to a three-dimensional wave equation.

For the case of one-dimensional space, we have

$$\mu\varepsilon\frac{\partial^2 E}{\partial t^2}=\frac{\partial^2 E}{\partial x^2}-\mu\sigma\frac{\partial E}{\partial t} \tag{9.239}$$

where E becomes the scalar electric field for the 1-D case. This can be used to model the telegraph problem.

9.5.2 Solution

To solve (9.238), we first remove the first order derivative term by assuming

$$E(x,t)=W(x,t)e^{\kappa t} \tag{9.240}$$

Differentiation of (9.239) gives

$$\frac{\partial E}{\partial t}=\frac{\partial W}{\partial t}e^{\kappa t}+\kappa e^{\kappa t}W \tag{9.241}$$

$$\frac{\partial^2 E}{\partial t^2}=\frac{\partial^2 W}{\partial t^2}e^{\kappa t}+2\kappa e^{\kappa t}\frac{\partial W}{\partial t}+\kappa^2 e^{\kappa t}W \tag{9.242}$$

$$\frac{\partial^2 E}{\partial x^2}=\frac{\partial^2 W}{\partial x^2}e^{\kappa t} \tag{9.243}$$

Substitution of (9.240) to (9.242) into (9.238) leads to

$$\mu\varepsilon\frac{\partial^2 W}{\partial t^2}=\frac{\partial^2 W}{\partial x^2}-(\mu\sigma+2\kappa\mu\varepsilon)\frac{\partial W}{\partial t}-\kappa(\mu\sigma+\kappa\mu\varepsilon)W \tag{9.244}$$

To remove the first derivative term, we set

$$\kappa=-\frac{\sigma}{2\varepsilon} \tag{9.245}$$

Substitution of this value of κ into (9.243) gives

$$\frac{\partial^2 W}{\partial t^2}=\frac{1}{\mu\varepsilon}\frac{\partial^2 W}{\partial x^2}+(\frac{\sigma}{2\varepsilon})^2 W \tag{9.246}$$

To put this into the standard hyperbolic form, we can absorb $1/(\mu\varepsilon)$ of the first term on the right to give

$$\frac{\partial^2 W}{\partial t^2}=\frac{\partial^2 W}{\partial s^2}+(\frac{\sigma}{2\varepsilon})^2 W \tag{9.247}$$

where

$$s=\sqrt{\mu\varepsilon}x \tag{9.248}$$

Applying the standard change of variables for the hyperbolic equation, we introduce

$$\xi=s+t,\quad \eta=s-t \tag{9.249}$$

Or, equivalently, we can write

$$s=\frac{1}{2}(\xi+\eta),\quad t=\frac{1}{2}(\xi-\eta) \tag{9.250}$$

Using this change of variables, we get

$$\frac{\partial^2 W}{\partial t^2}=\frac{\partial^2 W}{\partial \xi^2}-2\frac{\partial^2 W}{\partial \eta \partial \xi}+\frac{\partial^2 W}{\partial \eta^2} \tag{9.251}$$

$$\frac{\partial^2 W}{\partial s^2}=\frac{\partial^2 W}{\partial \xi^2}+2\frac{\partial^2 W}{\partial \eta \partial \xi}+\frac{\partial^2 W}{\partial \eta^2} \tag{9.252}$$

Therefore, the canonical form (9.245) becomes

$$\frac{\partial^2 W}{\partial \xi \partial \eta}+(\frac{\sigma}{4\varepsilon})^2 W=0 \tag{9.253}$$

One major observation that we can make on (9.252) is that the differential equation is symmetric with respect to the two variables ξ and η. This observation suggests the following single variable

$$\zeta=(\xi-\xi_0)(\eta-\eta_0) \tag{9.254}$$

where

$$\xi_0=\sqrt{\mu\varepsilon}x_0+t_0, \quad \eta_0=\sqrt{\mu\varepsilon}x_0-t_0 \tag{9.255}$$

with x_0 and t_0 are the initial point and time that telegraph signals were sent. In terms of this new variable, we have

$$\zeta\frac{d^2 W}{d\zeta^2}+\frac{dW}{d\zeta}+cW=0 \tag{9.256}$$

where

$$c=\left(\frac{\sigma}{4\varepsilon}\right)^2 \tag{9.257}$$

This is a second order ODE with a non-constant coefficient. It can be transformed into a Bessel equation by introducing

$$\gamma=\sqrt{4c\zeta} \tag{9.258}$$

Differentiation of (9.257) gives

$$\frac{d\gamma}{d\zeta}=\sqrt{\frac{c}{\zeta}} \tag{9.259}$$

Using the chain rule, we get

$$\frac{dW}{d\zeta}=\sqrt{\frac{c}{\zeta}}\frac{dW}{d\gamma} \tag{9.260}$$

$$\frac{d^2 W}{d\zeta^2}=-\frac{1}{2\zeta}\sqrt{\frac{c}{\zeta}}\frac{dW}{d\gamma}+\frac{c}{\zeta}\frac{d^2 W}{d\gamma^2} \tag{9.261}$$

Substitution of (9.259) and (9.260) into (9.255) gives

$$\gamma^2\frac{d^2 W}{d\gamma^2}+\gamma\frac{dW}{d\gamma}+\gamma^2 W=0 \tag{9.262}$$

This is the Bessel equation of zero order. The electric field must be finite at the starting point. Thus, the solution becomes

$$E=AJ_0\{\frac{\sigma}{2\varepsilon}[\mu\varepsilon(x-x_0)^2-(t-t_0)^2]\}e^{-\frac{\sigma}{2\varepsilon}t} \tag{9.263}$$

This is the solution for the telegraph equation.

9.6 DIFFUSION EQUATION

As discussed in Chapter 7, second order PDEs can be classified into hyperbolic, elliptic and parabolic. It has been shown in Chapter 7 that all parabolic PDE types can be converted to nonhomogeneous diffusion equations. The solution for the nonhomogeneous diffusion equation has been considered in Section 7.4.2. In this section, we focus on the homogeneous diffusion equation.

9.6.1 Heat Conduction

Consider the heat conduction problem of a 1-D bar of length L modelled by the diffusion equation:

$$\alpha^2 \frac{\partial^2 u}{\partial x^2} = \frac{\partial u}{\partial t}, \quad 0 < x < L, \quad t > 0 \tag{9.264}$$

where u is the temperature field in the bar and α^2 is the coefficient of diffusion. In fact, this is one of the very first PDEs considered by scientists and mathematicians. We will derive this equation from the fundamental principle of heat conduction, or the so-called Fourier's law of heat conduction. According to the Fourier law of heat conduction, the heat flow at any cross section can be estimated by the temperature gradient at the section:

$$H(x,t) = -\lim_{\Delta x \to 0} \kappa A \frac{u(x+\Delta x,t) - u(x,t)}{\Delta x} = -\kappa A \frac{\partial u}{\partial x} \tag{9.265}$$

where κ is the coefficient of thermal conductivity and the negative sign indicates the decreasing nature of temperature along the direction of diffusion. The temperature change in this incremental element can be estimated as

$$\Delta u = \frac{1}{s} \frac{Q \Delta t}{\Delta m} = \frac{Q \Delta t}{s \rho A \Delta x} \tag{9.266}$$

where s is the specific heat of the material, ρ is the density of the bar, A is the cross-section area of the bar, Δm is the mass of this segment of cross-section, and Δt is the change in time. As illustrated in Figure 9.16, the net heat flow rate Q is related to H as:

$$Q = H(x,t) - H(x+\Delta x,t) = \kappa \left[A u_x(x+\Delta x,t) - A u_x(x,t) \right] = \kappa \frac{\partial}{\partial x}(A \frac{\partial u}{\partial x}) \Delta x \tag{9.267}$$

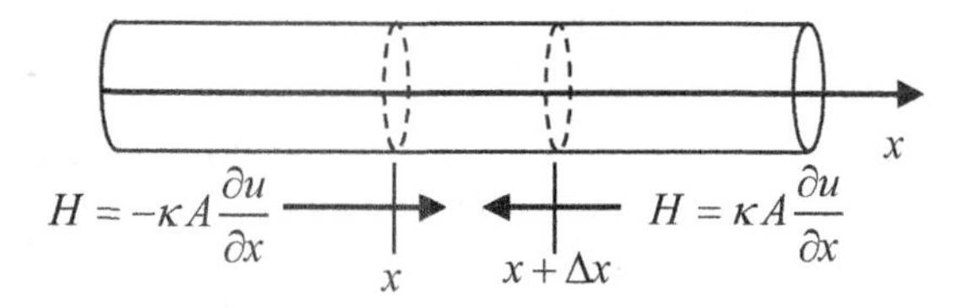

Figure 9.16 Fourier law of heat conduction

Substitution of (9.266) into (9.265) gives

$$s\rho A\frac{\partial u}{\partial t}=\kappa\frac{\partial}{\partial x}(A\frac{\partial u}{\partial x}) \tag{9.268}$$

In fact, the formulation of this relates to the well-known Sturm-Liouville problem of eigenfunction expansion (see discussion in the next chapter on eigenfunction expansion). If the bar is prismatic or uniform in cross-section, we find

$$\frac{\partial u}{\partial t}=\alpha^2\frac{\partial^2 u}{\partial x^2} \tag{9.269}$$

where the thermal diffusivity is defined as

$$\alpha^2=\frac{\kappa}{\rho s} \tag{9.270}$$

This completes the proof of the diffusion equation. In fact, diffusion of pollutants or chemicals in fluids can also be modelled by a similar equation:

$$\frac{\partial u}{\partial t}=\alpha^2\nabla^2 u \tag{9.271}$$

where the spatial derivative term is replaced by the Laplacian operator.

Let us assume the following separation of variables

$$u=X(x)T(t) \tag{9.272}$$

Substitution of (9.271) into (9.268) leads to

$$\alpha^2 X''T=XT' \tag{9.273}$$

By dividing through by XT, this equation can be simplified as

$$\frac{X''}{X}=\frac{1}{\alpha^2}\frac{T'}{T}=-\lambda^2 \tag{9.274}$$

Since X is only a function of x whereas T is only a function of t, the only possibility is that the left hand side and the right hand side are both constant. Note that we have assumed a negative value of this constant in the last part of (9.273). This choice is very important, and it is required by physical consideration of the problem. For initial boundary value problems with some initial non-uniform distribution of temperature along x and with conducting ends, we expect the temperature field is a function of time. More importantly, it must decay within time, and this fact leads to the choice of negative sign in (9.273).

The two ODEs that result from (9.268) are

$$X''+\lambda^2 X=0 \tag{9.275}$$

$$T'+\alpha^2\lambda^2 T=0 \tag{9.276}$$

The solutions of these equations are readily obtained as:

$$X=A\sin\lambda x+B\cos\lambda x \tag{9.277}$$

$$T=ce^{-\lambda^2\alpha^2 t} \tag{9.278}$$

It is now clear that the negative sign chosen in (9.273) indeed results in a decaying temperature field for the case of an imposing initial temperature field in a conducting bar. Thus, the general solution becomes

$$\psi=(A\sin\lambda x+B\cos\lambda x)e^{-\lambda^2\alpha^2 t} \tag{9.279}$$

The value of λ must be determined by boundary condition. The following example illustrates the procedure for obtaining the eigenvalue of λ. For more comprehensive coverage of heat conduction problems, we refer to the classic book by Carslaw and Jaeger (1959).

Example 9.3 Solve the following diffusion equation with prescribed boundary conditions:

$$\alpha^2 u_{xx} = u_t, \quad 0 < x < L, \quad t > 0 \tag{9.280}$$

$$u(0,t) = 0, \quad u(L,t) = 0, \quad t > 0 \tag{9.281}$$

$$u(x,0) = f(x), \quad 0 \le x \le L \tag{9.282}$$

Solution: Using the separation of variables given in (9.271), we have the formulation for X as

$$X'' + \lambda^2 X = 0 \tag{9.283}$$

$$X(0) = X(L) = 0, \quad t > 0 \tag{9.284}$$

Substitution of (9.276) into the first equation of (9.280) gives

$$B = 0 \tag{9.285}$$

The second equation of (9.280) leads to

$$\sin \sqrt{\lambda} L = 0 \tag{9.286}$$

That is, we require

$$\sqrt{\lambda} L = n\pi \tag{9.287}$$

Thus, there are infinite discrete eigenvalues

$$\lambda_n = n^2\pi^2 / L^2, \quad n = 1,2,3,... \tag{9.288}$$

The eigenfunction that corresponds to this eigenvalue is

$$X_n(x) = \sin(n\pi x / L), \quad n = 1,2,3,... \tag{9.289}$$

Substitution of this eigenvalue into (9.277) gives

$$T_n = k_n e^{-(n\pi\alpha/L)^2 t} \tag{9.290}$$

The fundamental solution is

$$u_n(x,t) = e^{-(n\pi\alpha/L)^2 t} \sin(n\pi x / L), \quad n = 1,2,3,..., \tag{9.291}$$

The general solution that can be used to fit any boundary condition can be expressed as:

$$u(x,t) = \sum_{n=1}^{\infty} c_n u_n(x,t) = \sum_{n=1}^{\infty} c_n e^{-(n\pi\alpha/L)^2 t} \sin(n\pi x / L) \tag{9.292}$$

Application of the initial condition given in (9.281) results in

$$u(x,0) = f(x) = \sum_{n=1}^{\infty} c_n \sin(n\pi x / L) \tag{9.293}$$

Multiplying both sides of (9.292) by the sine function of argument $m\pi x/L$ and integrating it from 0 to L gives:

$$\int_0^L f(x)\sin(m\pi x/L)dx = \sum_{n=1}^{\infty} c_n \int_0^L \sin(m\pi x/L)\sin(n\pi x/L)dx \quad (9.294)$$

Recall the following orthogonal property of circular functions (e.g., See Section 10.5 for proof):

$$\int_0^L \sin\frac{m\pi x}{L}\sin\frac{n\pi x}{L}dx = \begin{cases} = 0, & m \neq n \\ = L/2, & m = n \end{cases} \quad (9.295)$$

In view of (9.294), we have

$$c_n = \frac{2}{L}\int_0^L f(x)\sin(n\pi x/L)dx \quad (9.296)$$

Combining (9.295) and (9.291) gives the final solution of the problem.

9.6.2 Terzaghi 1-D Consolidation Theory

In the 1-D consolidation theory in soil mechanics, Terzaghi derived the following diffusion equation:

$$\frac{\partial u_e}{\partial t} = c_v \frac{\partial^2 u_e}{\partial z^2}, \quad 0 < z < 2d, \quad t > 0 \quad (9.297)$$

where u_e is the excess pore water pressure in the soil as a function of depth z and time t. The coefficient of consolidation c_v is defined as:

$$c_v = \frac{k}{\gamma_w m_v} \quad (9.298)$$

where k, m_v, and γ_w are the coefficient of permeability in Darcy's law (which has a similar physical meaning as the Fourier's law in heat conduction), coefficient of volume compressibility, and unit weight of water. The excess pore water pressure is defined as the difference between the pore water pressure in the soil and the long term steady state pore water pressure at the same point (typically hydrostatic pressure):

$$u_e = u(x,t) - u(x,\infty) \quad (9.299)$$

The consolidation theory describes the process of driving water from the clay due to the non-zero excess pore water pressure. We now look at its derivation.

The flow velocity through the soil element shown in Figure 9.17 is governed by Darcy's law as:

$$v_z = ki_z = -k\frac{\partial h}{\partial z} = -\frac{k}{\gamma_w}\frac{\partial u_e}{\partial z} \quad (9.300)$$

where k is the coefficient of permeability and is a function of soil type, i_z is the hydraulic gradient in the z direction and is defined as total head loss per flow distance, and h is the total head in the soil (total energy measured in length). This is because the total head change is due to the change in pore water only. The continuity condition is

$$\frac{\partial v_z}{\partial z}dxdydz = \frac{dV}{dt} \quad (9.301)$$

Combining (9.299) and (9.300) we obtain

$$-\frac{k}{\gamma_w}\frac{\partial^2 u_e}{\partial z^2}dxdydz = \frac{dV}{dt} = m_v \frac{\partial \sigma'}{\partial t}dxdydz \tag{9.302}$$

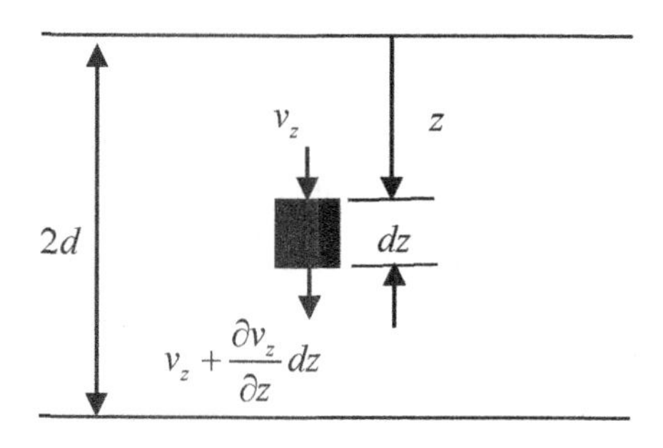

Figure 9.17 1-D consolidation theory of soil due to Terzaghi

where m_v is the coefficient of volume compressibility (a parameter indicating the compressibility of soil in the unit of inverse of stress). However, the increase of the effective stress σ' is due to the decrease in the excess pore pressure. In other words, the loading taken up temporarily by pore water pressure is transferred to the soil skeleton in terms of effective stress increment. Thus, we have

$$\frac{\partial \sigma'}{\partial t} = -\frac{\partial u_e}{\partial t} \tag{9.303}$$

Consequently, substitution of (9.302) into (9.301) gives

$$\frac{\partial u_e}{\partial t} = \frac{k}{\gamma_w m_v}\frac{\partial^2 u_e}{\partial z^2} \tag{9.304}$$

This is evidently equivalent to (9.296).

In an oedometer test in a soil laboratory, once a loading is suddenly applied to the clay of thickness $2d$, an excess pore water pressure u_i will build up at time zero. In this kind of test, porous stone is put at the bottom and the top of the clay so that drained conditions are created at the boundaries. Mathematically, the initial and boundary conditions are given as:

$$u_e(z,0) = u_i(z), \quad 0 \le z \le 2d \tag{9.305}$$

$$u_e(0,t) = 0, \quad u_e(2d,t) = 0, \quad t > 0 \tag{9.306}$$

We see that the solution for head conduction obtained in the last section equally applies here with the following identifications:

$$u(x,t) \leftarrow u_e(z,t),\ L \leftarrow 2d,\ \alpha^2 \leftarrow c_v,\ x \leftarrow z,\ f(x) \leftarrow u_i(z) \tag{9.307}$$

In addition, if we assume the initial excess pore water pressure is constant, we have

$$u_e = \frac{4u_i}{\pi}\sum_{m=0}^{\infty}\frac{1}{(2m+1)}\sin[\frac{(2m+1)\pi z}{2d}]e^{-(2m+1)^2\pi^2 T_v/4} \tag{9.308}$$

where the time factor T_v is defined as

$$T_v = \frac{c_v t}{d^2} \tag{9.309}$$

In obtaining (9.308), we have used the following identity:

$$\int_0^{2d} \sin\frac{n\pi z}{2d} dx = 1 - \cos n\pi = \begin{cases} 2, & n = \text{odd} \\ 0, & n = \text{even} \end{cases} \tag{2.310}$$

This solution is found very useful in devising the root time method as well as the logarithmic time method in estimating the coefficient of consolidation c_v defined in (9.297) in the laboratory.

9.6.3 Living Underground

One of the main reasons to live underground is that rock and soil can act as a thermal insulator for underground structures. Due to seasonal changes of temperature, the ground surface is subject to periodic heating and cooling. The problem can be formulated as heat diffusion for temperature field u as:

$$u_t = \gamma u_{zz} \tag{9.311}$$

$$u(t,0) = a\cos\omega t \tag{9.312}$$

The circular frequency for a yearly cycle is

$$\omega = \frac{2\pi}{365.5\times 24\times 60\times 60} = 2.0\times 10^{-7} s^{-1} \tag{9.313}$$

The temperature field must decay with the depth as

$$u(t,z) \to 0, \quad z \to \infty \tag{9.314}$$

To solve the problem efficiently, we can rewrite the boundary condition as

$$u(t,0) = ae^{i\omega t}, \quad \lim_{z\to\infty} u(t,z) = 0 \tag{9.315}$$

We seek a solution in the form:

$$u(t,z) = v(z)e^{i\omega t} \tag{9.316}$$

Substitution of (9.315) into (9.310) gives the following ODE for v

$$\gamma v'' - i\omega v = 0 \tag{9.317}$$

$$v(0) = a, \quad \lim_{z\to\infty} v(z) = 0\,. \tag{9.318}$$

The two independent solutions are

$$v(z) = C_1 e^{\sqrt{i\omega/\gamma z}} + C_1 e^{-\sqrt{i\omega/\gamma z}}\,. \tag{9.319}$$

We note by Euler's formula that

$$i = e^{\frac{\pi}{2}i}, \tag{9.320}$$

thus, we have

$$\sqrt{i} = e^{\frac{\pi}{4}i} = \cos\frac{\pi}{4} + i\sin\frac{\pi}{4} = \frac{1}{\sqrt{2}}(1+i). \tag{9.321}$$

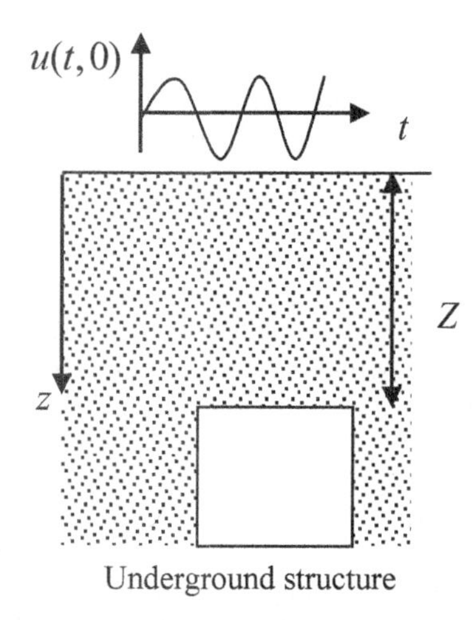

Figure 9.18 Temperature variations in underground structure due to seasonal temperature change on the ground surface

Substitution of (9.320) into (9.318) gives

$$v(z) = C_1 e^{\sqrt{\omega/(2\gamma)}(1+i)z} + C_1 e^{-\sqrt{\omega/(2\gamma)}(1+i)z} \tag{9.322}$$

Applying the boundary and decay conditions to (9.321) gives

$$v(z) = ae^{-\sqrt{\omega/(2\gamma)}(1+i)z} \tag{9.323}$$

Back substitution of (9.322) into (9.315) gives

$$u(t,z) = ae^{-\sqrt{\omega/(2\gamma)}z} e^{i(\omega t - \sqrt{\omega/(2\gamma)}z)} \tag{9.324}$$

For the boundary condition given in (9.311), we can take the real part of the solution given in (9.323) to give

$$u(t,z) = ae^{-z\sqrt{\omega/(2\gamma)}} \cos[\omega t - \sqrt{\omega/(2\gamma)}z] \tag{9.325}$$

where the phase lag is

$$\delta = \sqrt{\omega/(2\gamma)}z \tag{9.326}$$

For the case that the phase lag is an integer multiple of π, the temperature at the ground will be completely out of phase with that of the point of consideration. The first out-of-phase nodal point underneath the ground is at the depth

$$Z = \pi\sqrt{2\gamma/\omega} \tag{9.327}$$

For soil underneath the ground, a typical value of heat conduction is

$$\gamma = 10^{-6} m^2/s \tag{9.328}$$

Substitution of (9.327) and (9.312) into (9.326) gives

$$Z = 9.9m \tag{9.329}$$

and at this depth the surface value will decay by a factor of

$$e^{-\pi} = 0.043214 \tag{9.330}$$

In Hong Kong, the maximum temperature is about 34°C and the lowest temperature is about 7°C (or $a \approx 13.5$°C). This gives a maximum temperature variation of about 1.167°C at a depth of 9.9 m. In conclusion, we find that at a depth of about 10 m below the ground, the seasonal year-round temperature variation is from 19.9°C to 21.1°C. This makes the underground structure ideal for human usage in terms of the roughly constant temperature environment. It also provides an ideal condition for storage of goods and supplies. In fact, rock caverns and subsurface structures have been used for thousands of years in countries near the Arctic Circle, like Finland and Sweden. This provides a scientific judgment for using underground structures.

9.7 LAPLACE EQUATION

The Laplace equation is one of the very first second order PDEs studied extensively by mathematicians. Distribution of electrostatic potential, streamline and potential function of incompressible potential flow of fluid, deflection of membranes, and torsion of prismatic bars are phenomena governed by the Laplace equation. It was named after French mathematician Laplace. It is defined as

$$\nabla^2 u = 0 \tag{9.331}$$

where the Laplacian differential operator in 2-D and 3-D Cartesian coordinates are defined as:

$$\nabla^2 u = \frac{\partial^2 u}{\partial x^2} + \frac{\partial^2 u}{\partial y^2}, \quad \nabla^2 u = \frac{\partial^2 u}{\partial x^2} + \frac{\partial^2 u}{\partial y^2} + \frac{\partial^2 u}{\partial z^2} \tag{9.332}$$

Let us derive the Laplace equation from the seepage problem from a soil mechanics point of view here. In particular, consider an incompressible potential flow in 2-D space as shown in Figure 9.19. The net inflow to the element is

$$v_x dydz + v_z dxdy \tag{9.333}$$

The net outflow from the element is

$$(v_x + \frac{\partial v_x}{\partial x} dx)dydz + (v_z + \frac{\partial v_z}{\partial z} dz)dxdy \tag{9.334}$$

Subtracting the outflow from the inflow, we have

$$\frac{\partial v_x}{\partial x} + \frac{\partial v_z}{\partial z} = 0 \tag{9.335}$$

More generally, we can recast this continuity equation as

$$\nabla \cdot \boldsymbol{v} = 0 \tag{9.336}$$

Using Darcy's law, we have

$$v_x = k_x i_x = -k_x \frac{\partial h}{\partial x} = \frac{\partial \varphi}{\partial x} = \frac{\partial \psi}{\partial z} \tag{9.337}$$

$$v_z = k_z i_z = -k_z \frac{\partial h}{\partial z} = \frac{\partial \varphi}{\partial z} = -\frac{\partial \psi}{\partial x} \tag{9.338}$$

where the coefficient of permeability along the x and z directions are defined as k_x and k_z, and hydraulic gradients along the x and z directions are i_x and i_z. Darcy's law has been discussed in the last section when we discussed Terzaghi's theory of 1-D soil consolidation. There are two additional functions that we define in (9.336) and (9.337). The first one is the potential function φ, and the second one is the stream function ψ. If we assume isotropic flow (i.e., the coefficients of permeability along the x- and z-directions are the same), we can integrate both (9.336) and (9.337) to get

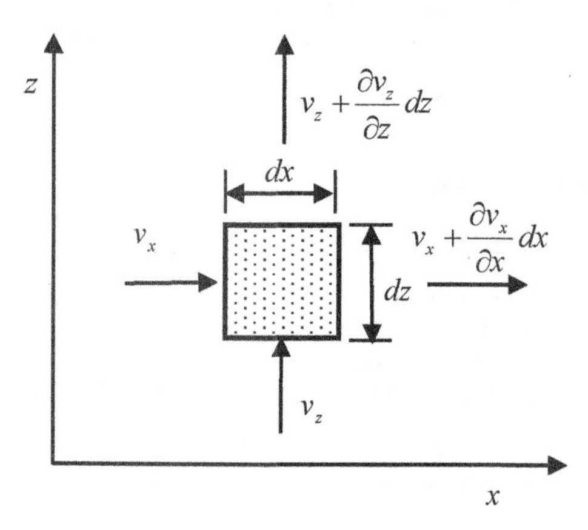

Figure 9.19 Continuity of fluid flow

$$\varphi(x,z) = -kh(x,z) + C \tag{9.339}$$

We can see that the contour plot of the potential function φ is similar to that of total head or total energy. In seepage theory, they are referred as the equi-potential lines of the flow. If we substitute the definition of φ given in (9.336) and (9.337) into (9.334), we obtain

$$\frac{\partial^2 \varphi}{\partial x^2} + \frac{\partial^2 \varphi}{\partial z^2} = 0 \tag{9.340}$$

Therefore, the flow potential function satisfies the Laplace equation. If the flow is irrotational, we have

$$\nabla \times v = 0 \tag{9.341}$$

This implies that there is no vorticity in the fluid and physically also implies there is no viscosity effect in the fluid. For the present 2-D case shown in Figure 9.19, we have the irrotational condition as:

$$\frac{\partial v_z}{\partial x} - \frac{\partial v_x}{\partial z} = 0 \tag{9.342}$$

If we substitute the definition of ψ given in (9.336) and (9.337) into (9.341), we obtain

$$\frac{\partial^2 \psi}{\partial x^2} + \frac{\partial^2 \psi}{\partial z^2} = 0 \tag{9.343}$$

This is again the Laplace equation. Thus, both the potential function and stream function satisfy the Laplace equation. Physically the stream function indicates the flow line in the fluid. To see this, we can take the total differential of the function equal to a constant value, i.e., $\psi = c$:

$$d\psi = \frac{\partial \psi}{\partial x} dx + \frac{\partial \psi}{\partial z} dz = -v_z dx + v_x dz = 0 \tag{9.344}$$

The direction of this stream function plot is

$$\left. \frac{dz}{dx} \right|_{\psi=\psi_1} = \frac{v_z}{v_x} \tag{9.345}$$

This shows that the slope of the plots of $\psi = c$ always equals the flow direction, regardless of the value of c. On the other hand, to study of the slope of plots of $\varphi = c$ we can consider the total differential of $\varphi = c$:

$$d\varphi = \frac{\partial \varphi}{\partial x} dx + \frac{\partial \varphi}{\partial z} dz = v_x dx + v_z dz = 0 \tag{9.346}$$

Rearranging this equation gives

$$\left. \frac{dz}{dx} \right|_{\varphi=\varphi_1} = -\frac{v_x}{v_z} \tag{9.347}$$

For two straight lines intercepting at a point, the angle of interception can be calculated as

$$\tan \vartheta = \frac{m_2 - m_1}{1 + m_1 m_2} \tag{9.348}$$

where m_1 and m_2 are the slopes of the two lines. If the two lines are perpendicular, we have $\vartheta = \pi/2$ or

$$m_1 = -\frac{1}{m_2} \tag{9.349}$$

Comparing the slope of stream function given in (9.344) and slope of the potential function given in (9.346), and in view of the result given in (9.348), we can conclude that plots of the potential function and stream function are always perpendicular.

Referring to Figure 9.20, we can also see that the value between different plots of stream function equal to the flow rate between these streamlines:

$$\Delta q = \int_{\psi_1}^{\psi_2} v_s dn = \int_{\psi_1}^{\psi_2} (-v_z dx + v_x dz) = \int_{\psi_1}^{\psi_2} (\frac{\partial \psi}{\partial x} dx + \frac{\partial \psi}{\partial z} dz) = \int_{\psi_1}^{\psi_2} d\psi = \psi_2 - \psi_1 \tag{9.350}$$

Considering the flow continuity between the two flowlines in element ABCD given in Figure 9.20, we have

$$\Delta q = \Delta \psi = v_s \Delta n = \frac{\partial \varphi}{\partial s} \Delta n \tag{9.351}$$

The last part of (9.350) is a consequence of the definition of φ given in (9.336) and (9.337). In particular, we can replace x or z in these equations by s and subsequently v_x or v_z by v_s. Replacing the incremental change in (9.350) by differentiation, we have

$$\frac{\partial \psi}{\partial n} = \frac{\partial \varphi}{\partial s} \tag{9.352}$$

This relation between ψ and φ gives the final condition for the potential function and stream function. Based upon them, the flownet technique can be developed. The details are, however, out of the scope of the present section.

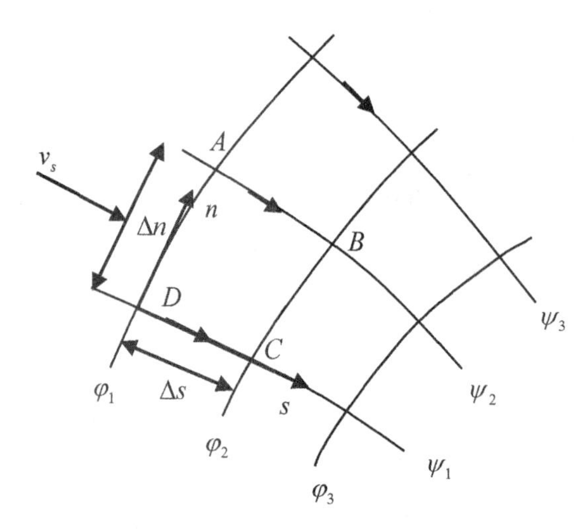

Figure 9.20 Flownet form by potential function and stream function in potential flow problems

9.7.1 Dirichlet Problem

This problem has been considered by many others (like Poisson, Green, and Gauss) long before German mathematician Dirichlet did. But because of Dirichlet's contribution to the analysis of the problem, it was named after him. Dirichlet was a German mathematician, who was born to a French family in Germany. Mathematically, it is formulated as

$$\nabla^2 u = 0 \tag{9.353}$$

$$u = u_0, \quad on\ S \tag{9.354}$$

where S denotes the surface of the domain of the problem. It is also referred to as the first boundary value problem in potential theory. For vibrations of a string, it gives a fixed end condition. For equilibrium problems of soap film, the Dirichlet problem is like a closed wire loop with prescribed deflection of the soap film. This problem has been studied by many well-known mathematicians, including Poincare, Lyapunov, Gauss, Lord Kelvin, Weierstrass, Neumann, Wiener, Lebesgue, and Kellogg regarding its uniqueness and existence of the solution for different domains. In finite element formulation using calculus of variations, the Dirichlet boundary condition is normally referred to as the essential boundary condition (see Chapters 13 and 14). We will see in Section 9.7.8 that uniqueness of the solution of the Dirichlet problem can be guaranteed.

9.7.2 Neumann Problem

Instead of imposing the unknown solution on the boundary as shown in (9.353), the Neumann problem prescribed the normal derivative of the unknown on the boundary. This formulation was named after German mathematician Carl Neumann (don't confuse with Nobel Prize winner von Neumann):

$$\nabla^2 u = 0 \tag{9.355}$$

$$\frac{\partial u}{\partial n} = g, \quad on\ S \tag{9.356}$$

It is also referred to as the second boundary value problem in potential theory. For example, for electrostatic problems with an insulating boundary, the Neumann problem is formulated. For acoustic problems, the Neumann boundary with $g = 0$ corresponds to a solid wall. For vibrations of string, the Neumann boundary condition corresponds to a "freely rotating end." For a nonhomogeneous Laplace equation or Poisson equation, we cannot arbitrarily impose the function g. There is a compatibility condition that g must satisfy. For the fluid flow problems, if there is an internal source (nonhomogeneous term in the PDE), then the outflow condition on the boundary (modelled by the Neumann boundary condition in (9.355)) must satisfy the continuity of the flow (i.e., fluid that comes in from the source must go out from the boundary). For the case of electrostatic problems, if there are internal charges (i.e., nonhomogeneous terms in the PDE), the net electric flux felt on the boundary as a whole must reflect the effect of the internal charges. For the case of heat flow, if there is an internal heat source, the total heat flux passing the boundary must equal that of the internal sources for conservation of energy. In finite element formulation using calculus of variations, the Neumann boundary condition corresponds to the natural boundary condition (see Chapters 13 and 14).

We will see in Section 9.7.8 that there is an integrability condition of the Neumann problem.

9.7.3 Robin Problem

Other than the Dirichlet or Neumann boundary condition, a more general type of boundary condition has been proposed. It is known as the Robin boundary condition or Robin problem. Many authors also simply refer to it as the third boundary condition (the first and second ones are referred to as Dirichlet and Neumann problems). Mathematically, it is formulated as

$$\nabla^2 u = 0 \tag{9.357}$$

$$au + b\frac{\partial u}{\partial n} = f, \quad on\ S \tag{9.358}$$

Note that the boundary condition involves both the unknown u and its normal derivative. For the special case $a = 0$, we recover the Neumann problem; and for the special case $b = 0$, we recover the Dirichlet problem. For a partially absorbing boundary, the Robin type of boundary condition can be used. For heat conduction problems, the Robin boundary condition corresponds to the heat conduction rate at the boundary being proportional to the temperature there. The problem is named after Gustave Robin, whose PhD advisor is the eminent mathematician Emil Picard and his thesis committees consisted of Hermite and Darboux. However, according to Gustafson and Abe (1998a,b), Robin never used this type of boundary condition himself. Nevertheless, they thought that it was related to his PhD thesis on potential theory. In potential theory, both the single- and double-layer potential methods for solving boundary value problems in electrostatics are attributed to Robin. Gustafson and Abe (1998a) speculated that it was Bergman in 1948 who called the third boundary condition Robin's boundary condition. But, it seems to be a mistake made by Bergman.

The solution form of the Laplace equation depends on the coordinate system that we employ. In general, we can use separation of variables to solve the Laplace equation. However, it is natural to ask whether the Laplace equation can always be solved by separation of variables. So far, it is known that the Laplace equation is separable in the following thirteen coordinate systems only: Cartesian, circular cylindrical, spherical, oblate spheroidal, prolate spheroidal, elliptic cylindrical, conical, paraboloidal, parabolic, parabolic cylindrical, ellipsoidal, bispherical, and toroidal. For most mechanics and engineering problems, Cartesian, circular cylindrical, and spherical coordinates are, however, found sufficient.

In particular, in Cartesian coordinate the general solutions are expressible in the product of circular functions and hyperbolic functions, in cylindrical coordinates the general solutions are expressible in the product of circular functions and Bessel functions, and in spherical coordinates the general solutions are expressible in the product of circular functions and Legendre polynomials. They are considered separately next.

9.7.4 Spherical Coordinate

In spherical coordinates, the Laplace equation can be expressed as (Chau, 2013):

$$\nabla^2 u = \frac{1}{r^2}\frac{\partial}{\partial r}(r^2\frac{\partial u}{\partial r}) + \frac{1}{r^2 \sin\varphi}\frac{\partial}{\partial \varphi}(\sin\varphi \frac{\partial u}{\partial \varphi}) + \frac{1}{r^2 \sin^2\varphi}\frac{\partial^2 u}{\partial \theta^2} = 0 \quad (9.359)$$

Applying the following separation of variables, we obtain

$$u = R(r)\Theta(\theta)\Phi(\varphi) \quad (9.360)$$

Similar to the discussion for the Helmholtz equation, this separation of variables leads to

$$[\frac{1}{R}\frac{d}{dr}(r^2\frac{dR}{dr}) + \frac{1}{\Phi \sin\varphi}\frac{d}{d\varphi}(\sin\varphi\frac{d\Phi}{d\varphi})]\sin^2\varphi = -\frac{1}{\Theta}\frac{d^2\Theta}{d\theta^2} = \mu^2 \quad (9.361)$$

where μ is a separation constant. The last part of (9.360) can be expressed as

$$\frac{d^2\Theta}{d\theta^2} + \mu^2\Theta = 0 \quad (9.362)$$

There must be periodicity in θ leading to $\mu = m$ ($m = 0,1,2,3,...$). The solution of Θ is

$$\Theta = A\cos m\theta + B\sin m\theta \quad (9.363)$$

The first part of (9.360) can be rearranged as

$$\frac{1}{R}\frac{d}{dr}(r^2\frac{dR}{dr})=\frac{\mu^2}{\sin^2\varphi}-\frac{1}{\Phi\sin\varphi}\frac{d}{d\varphi}(\sin\varphi\frac{d\Phi}{d\varphi})=n(n+1) \tag{9.364}$$

where the separation constant is written as $n\,(n+1)$ with n = 1,2,3... The reason for n being an integer has been given in Section 9.4 for the Helmholtz equation. Equation (9.214) leads to two ODEs, and the first one is:

$$\frac{1}{\sin\varphi}\frac{d}{d\varphi}(\sin\varphi\frac{d\Phi}{d\varphi})+[n(n+1)-\frac{m^2}{\sin^2\varphi}]\Phi=0 \tag{9.365}$$

For the differential equation of Θ, we can introduce a change of variables of $x = \cos\varphi$ and subsequently obtain the Associated Legendre equation:

$$(1+x^2)\frac{d^2\Phi}{dx^2}-2x\frac{d\Phi}{dx}+[n(n+1)-\frac{m^2}{1-x^2}]\Phi=0 \tag{9.366}$$

The general solution is the Associated Legendre functions:

$$\Phi=EP_n^m(\cos\varphi) \tag{9.367}$$

Note that we have only retained the Associated Legendre polynomials of the first kind in (9.366) due to the boundedness of u at $\varphi = 0$ and π (see Section 9.4). The second ODE resulting from (9.256) is

$$r^2\frac{d^2R}{dr^2}+2r\frac{dR}{dr}-n(n+1)R=0 \tag{9.368}$$

This ODE is of Euler type. Thus, we can use the standard technique for solving the Euler type equation and the solution is

$$R=Cr^n+Dr^{-n-1} \tag{9.369}$$

If the domain includes the origin of the spherical coordinate, we must have $D = 0$ due to boundedness. Finally, the general solution for solid spheres can be expressed as:

$$u=r^nP_n^m(\cos\varphi)\{A\cos m\theta+B\sin m\theta\} \tag{9.370}$$

For the case of rotational symmetry, the general solution can be reduced to:

$$R=A_nr^nP_n(\cos\varphi) \tag{9.371}$$

where P_n is Legendre polynomials of the first kind.

9.7.5 Cylindrical Coordinate

In cylindrical coordinate, the Laplace equation can be written as:

$$\nabla^2u=\frac{\partial^2u}{\partial x^2}+\frac{\partial^2u}{\partial y^2}+\frac{\partial^2u}{\partial z^2}=\frac{\partial^2u}{\partial r^2}+\frac{1}{r}\frac{\partial u}{\partial r}+\frac{1}{r^2}\frac{\partial^2u}{\partial\theta^2}+\frac{\partial^2u}{\partial z^2}=0 \tag{9.372}$$

To prove this, we can use the following change of variables

$$x=r\cos\theta,\ \ y=r\sin\theta, r^2=x^2+y^2,\quad \tan\theta=\frac{y}{x} \tag{9.373}$$

Applying the chain rule of partial differentiation we have

$$\frac{\partial u}{\partial x}=\frac{\partial u}{\partial r}\frac{\partial r}{\partial x}+\frac{\partial u}{\partial\theta}\frac{\partial\theta}{\partial x}=\cos\theta\frac{\partial u}{\partial r}-\frac{\sin\theta}{r}\frac{\partial u}{\partial\theta} \tag{9.374}$$

$$\frac{\partial u}{\partial y}=\frac{\partial u}{\partial r}\frac{\partial r}{\partial y}+\frac{\partial u}{\partial \theta}\frac{\partial \theta}{\partial y}=\sin\theta\frac{\partial u}{\partial r}+\frac{\cos\theta}{r}\frac{\partial u}{\partial \theta} \tag{9.375}$$

Reapplying the chain rule of partial differentiation again to (9.373) and (9.374), we obtain

$$\frac{\partial^2 u}{\partial x^2}=\cos^2\theta\frac{\partial^2 u}{\partial r^2}-\frac{2\sin\theta\cos\theta}{r}\frac{\partial^2 u}{\partial r\partial\theta}+\frac{\sin^2\theta}{r^2}\frac{\partial^2 u}{\partial \theta^2}+\frac{\sin^2\theta}{r}\frac{\partial u}{\partial r}+\frac{2\sin\theta\cos\theta}{r^2}\frac{\partial u}{\partial \theta} \tag{9.376}$$

$$\frac{\partial^2 u}{\partial y^2}=\sin^2\theta\frac{\partial^2 u}{\partial r^2}+\frac{2\sin\theta\cos\theta}{r}\frac{\partial^2 u}{\partial r\partial\theta}+\frac{\cos^2\theta}{r^2}\frac{\partial^2 u}{\partial \theta^2}+\frac{\cos^2\theta}{r}\frac{\partial u}{\partial r}-\frac{2\sin\theta\cos\theta}{r^2}\frac{\partial u}{\partial \theta} \tag{9.377}$$

Adding (9.375) and (9.376), we finally get

$$\nabla^2 u=\frac{\partial^2 u}{\partial r^2}+\frac{1}{r}\frac{\partial u}{\partial r}+\frac{1}{r^2}\frac{\partial^2 u}{\partial \theta^2}+\frac{\partial^2 u}{\partial z^2}=0 \tag{9.378}$$

By separation of variables, the solution can be assumed as:

$$u=R(r)Z(z)\Theta(\theta) \tag{9.379}$$

Substitution of (9.378) into (9.377) gives

$$\frac{1}{R}\left(\frac{d^2R}{dr^2}+\frac{1}{r}\frac{dR}{dr}\right)+\frac{1}{\Theta}\frac{1}{r^2}\frac{d^2\Theta}{d\theta^2}=-\frac{1}{Z}\frac{d^2Z}{dz^2}=-\lambda^2 \tag{9.380}$$

where λ is a constant of separation of variables. We introduce an additional constant as:

$$\frac{1}{\Theta}\frac{d^2\Theta}{d\theta^2}=-\mu^2 \tag{9.381}$$

where μ is also a constant of separation of variables. Consequently, we have the following ODEs:

$$\frac{d^2\Theta}{d\theta^2}+\mu^2\Theta=0 \tag{9.382}$$

$$\frac{d^2Z}{dz^2}-\lambda^2 Z=0 \tag{9.383}$$

$$\frac{d^2R}{dr^2}+\frac{1}{r}\frac{dR}{dr}+(\lambda^2-\frac{\mu^2}{r^2})R=0 \tag{9.384}$$

Note that the periodicity of Θ requires that $\mu=m$ (i.e., $m=1,2,3,...$).

The corresponding solutions for these ODEs are

$$\Theta=A\cos m\theta+B\sin m\theta \tag{9.385}$$

$$Z=C\cosh\lambda z+D\sinh\lambda z \tag{9.386}$$

$$R=EJ_m(\lambda r)+FY_m(\lambda r) \tag{9.387}$$

where μ is another constant of separation of variables, and J_m and Y_m are Bessel functions of the first and second kinds. Finally, the solution becomes

$$u=\{A\cos m\theta+B\sin m\theta\}\{C\cosh\lambda z+D\sinh\lambda z\}\{EJ_m(\lambda r)+FY_m(\lambda r)\} \tag{9.388}$$

In axisymmetric solid cylinder problems, we must set F to zero because $Y_m\to\infty$ as $r\to 0$. Thus, the solution becomes

$$u = J_0(\lambda r)\{C\cosh\lambda z + D\sinh\lambda z\} \tag{9.389}$$

The value of λ needs to be determined by boundary conditions. For the case of homogeneous boundary condition, the problem becomes eigenvalue problem and λ becomes eigenvalue, which will be discrete but infinite in number.

9.7.6 Poisson Integral

In this section, we consider the case of a two-dimensional unit circular disk. The two-dimensional Laplace equation is

$$\nabla^2 u = \frac{\partial^2 u}{\partial r^2} + \frac{1}{r}\frac{\partial u}{\partial r} + \frac{1}{r^2}\frac{\partial^2 u}{\partial \theta^2} = 0 \tag{9.390}$$

The boundary condition, condition of periodicity, and boundedness condition are respectively

$$u(1,\theta) = h(\theta) \tag{9.391}$$

$$u(r,\theta + 2\pi) = u(r,\theta) \tag{9.392}$$

$$\lim_{r\to 0} u(r,\theta) = \text{finite} \tag{9.393}$$

Again, application of separation of variables leads to

$$u = v(r)w(\theta) \tag{9.394}$$

Substitution of (9.393) into (9.389) results in

$$v''w + \frac{1}{r}v'w + \frac{1}{r^2}vw'' = 0 \tag{9.395}$$

Grouping functions of r and θ onto different sides of (9.394) gives

$$\frac{r^2 v'' + rv'}{v} = -\frac{w''}{w} = \lambda^2 \tag{9.396}$$

where λ is a constant of separation of variables. This gives two ODEs:

$$r^2\frac{d^2 v}{dr^2} + r\frac{dv}{dr} - \lambda^2 v = 0 \tag{9.397}$$

$$\frac{d^2 w}{d\theta^2} + \lambda^2 w = 0 \tag{9.398}$$

The solution of (9.397) is

$$w = A\cos\lambda\theta + B\sin\lambda\theta \tag{9.399}$$

The condition of periodicity requires that

$$\lambda = n, \quad n = 0,1,2,3,... \tag{9.400}$$

which is the eigenvalue of the problem. Using this value of λ, (9.399) can be written as

$$r^2\frac{d^2 v}{dr^2} + r\frac{dv}{dr} - n^2 v = 0 \tag{9.401}$$

This is again recognized as the Euler type of ODE, and thus the solution becomes

$$v = Cr^n + Dr^{-n} \tag{9.402}$$

However, at the center ($r \to 0$) v needs to be bounded as required by (9.392), and we have $D = 0$. For the special case that $n = 0$, we have

$$r^2 \frac{d^2 v}{dr^2} + r \frac{dv}{dr} = 0 \tag{9.403}$$

For this special case, we can let

$$t(r) = \frac{dv}{dr} \tag{9.404}$$

$$r \frac{dt}{dr} + t = 0 \tag{9.405}$$

This is a separable first order ODE and direct integration gives

$$t = C_1 \ln r + C_2 \tag{9.406}$$

However, due to boundedness we have to set $C_1 = 0$. Finally, the solution can be expressed as:

$$u(r,\theta) = \frac{a_0}{2} + \sum_{n=1}^{\infty} (a_n r^n \cos n\theta + b_n r^n \sin n\theta) \tag{9.407}$$

Application of (9.390) gives

$$u(1,\theta) = \frac{a_0}{2} + \sum_{n=1}^{\infty} (a_n \cos n\theta + b_n \sin n\theta) = h(\theta) \tag{9.408}$$

This is exactly the Fourier series expansion of $h(\theta)$ and thus we have

$$a_n = \frac{1}{\pi} \int_{-\pi}^{\pi} h(\theta) \cos n\theta d\theta \tag{9.409}$$

$$b_n = \frac{1}{\pi} \int_{-\pi}^{\pi} h(\theta) \sin n\theta d\theta \tag{9.410}$$

Thus, (9.406) can be written as

$$u(r,\theta) = \frac{1}{2\pi} \int_{-\pi}^{\pi} h(\phi) d\phi + \sum_{n=1}^{\infty} \Big(\frac{r^n \cos n\theta}{\pi} \int_{-\pi}^{\pi} h(\phi) \cos n\phi d\phi + \frac{r^n \sin n\theta}{\pi} \int_{-\pi}^{\pi} h(\phi) \sin n\phi d\phi\Big) \tag{9.411}$$

This can be further simplified as

$$u(r,\theta) = \frac{1}{\pi} \int_{-\pi}^{\pi} h(\phi) \{\frac{1}{2} + \sum_{n=1}^{\infty} r^n (\cos n\theta \cos n\phi + \sin n\theta \sin n\phi)\} d\phi \tag{9.412}$$

Using the sum rule of cosine functions, (9.304) is reduced to

$$u(r,\theta) = \frac{1}{\pi} \int_{-\pi}^{\pi} h(\phi) [\frac{1}{2} + \sum_{n=1}^{\infty} r^n \cos n(\theta - \phi)] d\phi \tag{9.413}$$

The bracket term in (9.412) can be summed exactly and this was done by Poisson. More specifically, the bracket term is

$$\frac{1}{2} + \sum_{n=1}^{\infty} r^n \cos n(\theta - \phi) = \mathrm{Re}(\frac{1}{2} + + \sum_{n=1}^{\infty} z^n) \tag{9.414}$$

where

$$z^n = \{re^{\theta-\phi}\}^n = r^n[\cos n(\theta-\phi) + i\sin n(\theta-\phi)] \tag{9.415}$$

Since we are considering a unit disk, we have

$$|z| < 1 \tag{9.416}$$

Applying Taylor series expansion, we find

$$\frac{z}{1-z} = z(1+z+z^2+z^3+...) = \sum_{n=1}^{\infty} z^n \tag{9.417}$$

With this summation formula, (9.413) is rewritten

$$\begin{aligned} \frac{1}{2}+\sum_{n=1}^{\infty} r^n \cos n(\theta-\phi) &= \mathrm{Re}(\frac{1}{2}+\frac{z}{1-z}) = \mathrm{Re}[\frac{1+z}{2(1-z)}] \\ &= \mathrm{Re}[\frac{(1+z)(1-\bar{z})}{2|1-z|^2}] = \frac{(1-|z|^2)}{2|1-z|^2} \end{aligned} \tag{9.418}$$

Note that

$$|1-z|^2 = [1-r\cos(\theta-\phi)]^2 + r^2\sin^2(\theta-\phi) = 1-2r\cos(\theta-\phi)+r^2 \tag{9.419}$$

$$|z|^2 = r^2 \tag{9.420}$$

Back substitution of these values into (9.413) and (9.412) gives

$$u(r,\theta) = \frac{1}{2\pi}\int_{-\pi}^{\pi} h(\theta)[\frac{1-r^2}{1+r^2-2r\cos(\theta-\phi)}]d\phi \tag{9.421}$$

This is called the Poisson integral formula, and is a solution of the Dirichlet type of boundary value problem. This formula shows that the solution at any interior point is a weighted average of its boundary potentials. The weighting function is given in the square bracket of (9.420). In fact, for $r \to 1-\delta$ (with $\delta \to 0$), the Poisson kernel behaves like a Dirac delta function (see Section 8.11) of $\delta(\theta-\phi)$ as $r \to 1^-$. Jesse Douglas's celebrated paper in proving the existence of the minimal surface of soap film in the Plateau problem was based on the Poisson integral given in (9.420) (he was awarded the Fields medal because of this theory).

The physical meaning of this weighting function can be understood better by referring to Figures 9.21 and 9.22.

The denominator and numerator of the weighting functions are shown in Figure 9.21. In particular, the square root of the denominator of the weighting function is actually the distance between the observation point (r,θ) and the boundary point at $(1,\phi)$. The square root of the numerator of the weighting function is also shown in the figure as the vertical height of the right angle triangle with the horizontal side being r and the hypotenuse being 1. If we consider the boundary potential at point B_1 with $\theta \approx \phi$, the denominator approaches its minimum. Consequently, the weighting function approaches its maximum. That is, the boundary point with the closest distance to (r, θ) has the biggest effect on the solution. Conversely, for boundary point B_2, the distance between the observation and boundary point is the farthest or the weighting function is the smallest. Figure 9.22 illustrates the angular variation of the denominator and the weighting function versus ϕ for $\theta = 3\pi/4$ and $r = 0.6$. It is clear that the minimum value of the denominator and the maximum weighting function occur at $\phi = 3\pi/4$.

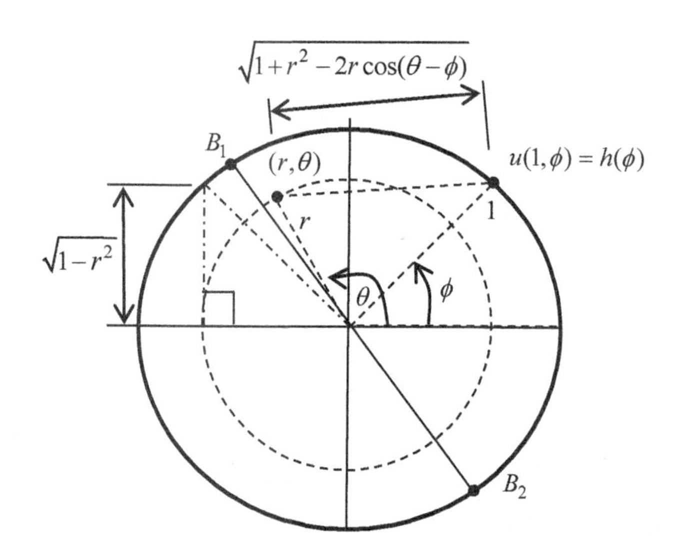

Figure 9.21 The weighting function for boundary potentials in Poisson integral

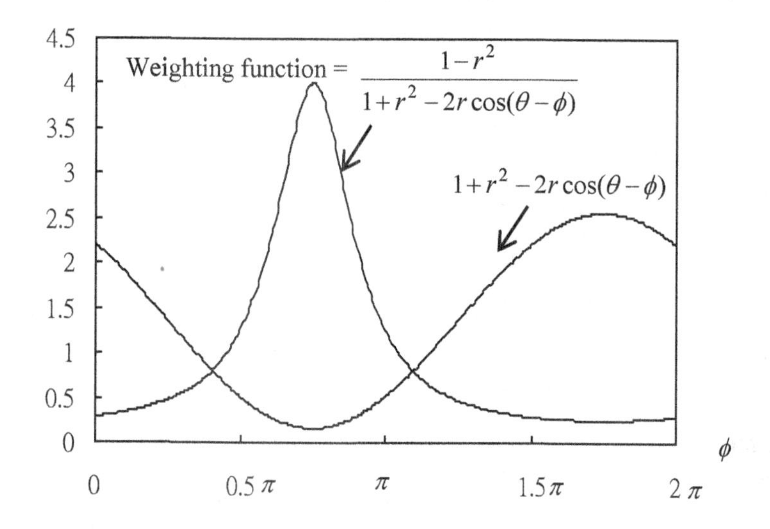

Figure 9.22 The denominator and weighting function versus ϕ for $\theta = 3\pi/4$ and r = 0.6

9.7.7 Extremal Principle

Physically, (9.420) has a very significant implication. Considering the solution at the center of the disk ($r = 0$) we have

$$u(0,\theta) = \frac{1}{2\pi}\int_{-\pi}^{\pi} h(\phi)d\phi \tag{9.422}$$

This shows that the value at the center of the disk is the average value of u on the boundary of the disk. For any non-constant function u, there cannot be a local maximum or a local minimum at any interior point of the unit circle. Physically, when a membrane is under equilibrium (Laplace equation is the governing equation for this kind of steady equilibrium type of problems), there cannot be any internal bump in the membrane. For problems of thermal equilibrium, the Poisson integral formula implies that a body can achieve its maximum and minimum temperature on the boundary of the domain only. Conversely, if a body contains a local maximum or minimum in its internal temperature, it could not be in thermal equilibrium. Although we interpret this extremal principle from the solution for a unit circular disk, it actually applies to equilibrium problems of domains other than circular shapes. Alternatively, for problems governed by the Laplace equation, conformal mapping can be applied to any non-circular domain (e.g., Section 3.13 of Chau, 2013) and to map it to a unit circle. In this way, we can also apply the extremal principle to other domains. In the next section, we will show a different approach in proving the extremal principle.

9.7.8 Properties of Harmonic Functions

In this section, we recall some of our observations for harmonic functions:

(i) Integrability of the Neumann problem:

Recall the definition of the Neumann problem given in Section 9.7.2. Note that for a harmonic u and setting $v = 1$ in Green's second identity (see (8.41)), we have

$$\iint_{\Gamma} \frac{\partial u}{\partial n} dS = 0 \tag{9.423}$$

Since the Laplace equation is for steady equilibrium problems, (9.422) implies that the fluid flow in and out of a control volume is the same in the case of potential flow, and the heat flow in and out of a volume is the same in the case of heat conduction. For a Neumann problem defined as:

$$\nabla^2 u = 0, \quad \left.\frac{\partial u}{\partial n}\right|_{\Gamma} = f(x, y, z), \tag{9.424}$$

substitution of (9.423) into (9.422) gives the integrability of the Neumann problem as

$$\iint f(x, y, z)dS = 0 \tag{9.425}$$

(ii) Mean Value Theorem

Let us recall the integral formula from (8.52) and specify on the surface Γ_a a small sphere of radius a embracing the point M_0 as:

$$u(M_0) = -\frac{1}{4\pi}\iint_{\Gamma_a}[u\frac{\partial}{\partial n}(\frac{1}{r}) - \frac{1}{r}\frac{\partial u}{\partial n}]dS \tag{9.426}$$

With the integrability given in (i), we have

$$\iint_{\Gamma_a}\frac{1}{r}\frac{\partial u}{\partial n}dS = \frac{1}{a}\iint_{\Gamma_a}\frac{\partial u}{\partial n}dS = 0 \tag{9.427}$$

where a is the radius of a small sphere around the point M_0. On the other hand, note that

$$\frac{\partial}{\partial n}(\frac{1}{r})\bigg|_{\Gamma_a} = \frac{\partial}{\partial r}(\frac{1}{r})\bigg|_{\Gamma_a} = -\frac{1}{a^2} \tag{9.428}$$

Substitution of (9.426) and (9.427) into (9.425) leads to

$$u(M_0) = \frac{1}{4\pi a^2}\iint_{\Gamma_a} udS \tag{9.429}$$

This result shows that the value of u at point M_0 equals the mean value of u on the surface of the sphere embracing the point. This is the mean value theorem for harmonic functions.

(iii) Extremal Principle

We can further use the mean value theorem to show the extreme principle. Suppose that u reaches a maximum value at a point M_1 inside the domain. Then, it must be larger than the average value on the surface of a sphere with an arbitrary radius R with point M_1 as the center, because there must be some point within less than the maximum value at M_1. That is,

$$\frac{1}{4\pi R^2}\iint_{S_R} udS < u(M_1) \tag{9.430}$$

However, the mean value theorem in (ii) shows that we must have

$$\frac{1}{4\pi R^2}\iint_{S_R} udS = u(M_1) \tag{9.431}$$

This is in contradiction with (9.429). Therefore, we cannot have a maximum inside the domain. In other words, we can only have the maximum value on the boundary of the domain. Following a similar procedure, we can also show that the minimum can only occur on the boundary.

(iv) Uniqueness of the Dirichlet problem

Recall the definition of the Dirichlet problem defined in Section 9.7.1. Assume both u_1 and u_2 are solutions and they are distinct for the following Dirichlet boundary value problem for the Laplace equation:

$$\nabla^2 u = 0, \quad u|_\Gamma = f(x, y, z) \tag{9.432}$$

Consider the function defined as

$$v = u_1 - u_2 \tag{9.433}$$

Since both u_1 and u_2 are solutions of the Laplace equation, substitution of (9.432) into (9.431) shows that v is also a harmonic function and v is identically zero on the boundary. The maximum principle in (iii) shows instantly that the only possibility is that v is zero. Thus, u_1 and u_2 must be equal or the solution of the Dirichlet problem must be unique.

9.8 PHYSICAL IMPLICATION OF LAPLACE EQUATION (SADDLE)

Recall the Laplace equation for two-dimensional space

$$\nabla^2 u = \frac{\partial^2 u}{\partial x^2} + \frac{\partial^2 u}{\partial y^2} = 0 \tag{9.434}$$

Since the sum of two second-derivative terms is zero, one must be negative while the other one must be positive. Figure 9.23 shows that a circular domain in which both of the second derivatives in (9.433) are positive and both are negative, respectively. It is clear that there is a local minimum if both derivatives are positive, and there is a local maximum if both derivatives are negative. Therefore, as equilibrium solution in the forms shown in Figure 9.23 cannot satisfy the Laplace equation.

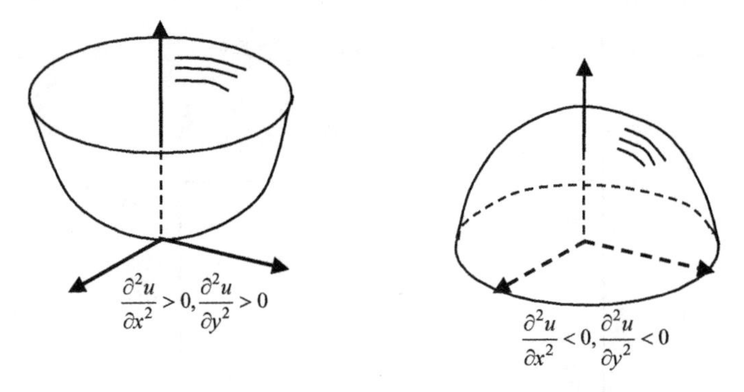

Figure 9.23 The distribution of *u* for both derivatives are positive and negative respectively

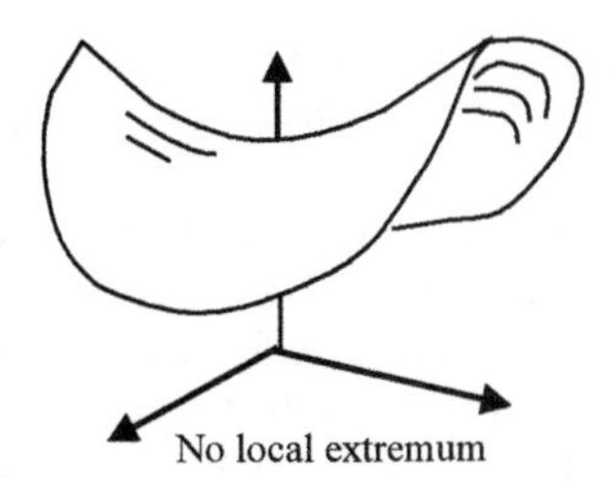

Figure 9.24 The illustration of a saddle plot for *u* satisfying the Laplace equation

Physically, for potential theory there is no maximum or minimum within the domain, or as expected from the extremal principle (see property (iii) in the last section) that maximum and minimum value can only appear on the boundary. Figure 9.24 illustrates a typical saddle plot of u satisfying the Laplace equation. The shape of the saddle also resembles a piece of potato chip. Note there is no maximum or minimum within the saddle in the circular domain.

9.9 LAPLACE EQUATION IN RECTANGULAR DOMAIN

9.9.1 Prescribed Function along y-axis

In this section, we consider the boundary value problem of a rectangular domain governed by the Laplace equation. In particular, consider the case that a non-zero boundary condition is imposed on the side at $x = a$ only (see Figure 9.25)

$$\nabla^2 u = \frac{\partial^2 u}{\partial x^2} + \frac{\partial^2 u}{\partial y^2} = 0, \quad 0 < x < a, \qquad 0 < y < b$$

$$u(x,0) = 0, \;\; u(x,b) = 0, \qquad 0 < x < a \tag{9.435}$$

$$u(0, y) = 0, \;\; u(a, y) = f(y), \;\; 0 \le y \le b$$

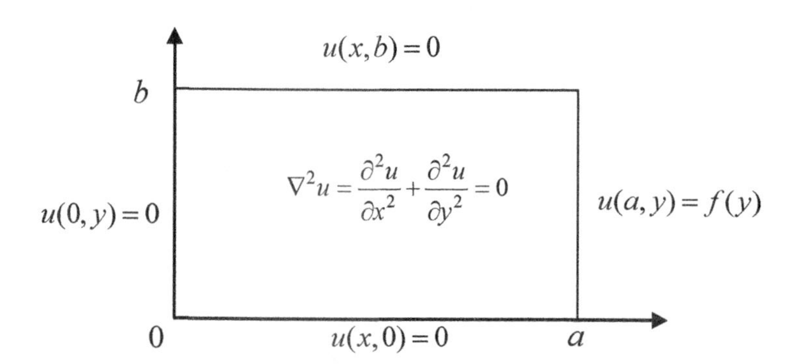

Figure 9.25 Laplace equation for a rectangular domain with prescribed function on vertical side $x = a$

This can be considered as a Dirichlet problem. Before we continue to solve this problem, it is instructive to note that for 2-D Cartesian coordinates our solution along the x- and y-axis will be either circular functions or hyperbolic functions (see below). It is well known from Fourier's work related to heat conduction that any arbitrary function can be expanded in a Fourier series expansion in terms of sine and cosine, but never be expandable in hyperbolic sine and hyperbolic cosine. For the problem given in (9.434), in order to satisfy the boundary condition we clearly need to expand the given function f in the y-direction in the sine or cosine. Thus, it is obvious that the fundamental solution in the y-axis must be sine and cosine, not hyperbolic functions.

Let us now consider the separation of variables:

$$u(x, y) = X(x)Y(y) \tag{9.436}$$

Substitution of (9.435) into the Laplace equation given in (9.434) gives

$$X''Y + XY'' = 0 \tag{9.437}$$

Dividing through (9.436) by XY, we get

$$\frac{X''}{X} = -\frac{Y''}{Y} = \lambda \tag{9.438}$$

Since we have assumed X is only a function, whereas Y is only a function of y, we must set both functions of x and y to a constant. We have picked a positive constant λ, and we will see that this leads to our desired functions of sine and cosine for Y. In particular, we have two ODEs resulting from (9.437):

$$X'' - \lambda X = 0 \tag{9.439}$$

$$Y'' + \lambda Y = 0 \tag{9.440}$$

The homogeneous boundary conditions given in (9.434) leads to the corresponding boundary conditions for functions X and Y as

$$\begin{aligned} u(0, y) &= X(0)Y(y) = 0, \quad 0 \le y \le b \Rightarrow X(0) = 0, \\ u(x, 0) &= X(x)Y(0) = 0, \quad 0 < x < a \Rightarrow Y(0) = 0, \\ u(x, b) &= X(x)Y(b) = 0, \quad 0 < x < a \Rightarrow Y(b) = 0. \end{aligned} \tag{9.441}$$

Note that we need not consider the nonhomogeneous boundary condition given in (9.434) at this moment. The governing equation and boundary conditions for function Y are

$$Y'' + \lambda Y = 0, \quad Y(0) = 0, \quad Y(b) = 0 \tag{9.442}$$

As expected, the solution for Y is

$$Y = c_1 \sin\sqrt{\lambda} y + c_2 \cos\sqrt{\lambda} y \tag{9.443}$$

The first boundary condition given in (9.441) leads to

$$Y(0) = c_2 = 0 \tag{9.444}$$

The second boundary condition given in (9.441) leads to

$$Y(b) = c_1 \sin\sqrt{\lambda} b = 0 \tag{9.445}$$

Since we cannot set c_1 to zero, we must have the sine function be zero and this leads to

$$\sqrt{\lambda} b = n\pi \tag{9.446}$$

This is the eigenvalue of the homogeneous boundary value problem. Thus, the eigenvalues and eigenfunctions are

$$\lambda_n = n^2\pi^2 / b^2, \quad Y_n(y) = \sin(n\pi y / b), \quad n = 1, 2, 3, \ldots \tag{9.447}$$

For the function X, we have

$$X'' - \lambda X = 0, \quad X(0) = 0 \tag{9.448}$$

The solution of (9.447) is expressible in terms of hyperbolic functions:

$$X(x) = k_1 \cosh(n\pi x / b) + k_2 \sinh(n\pi x / b) \tag{9.449}$$

Note that we have already substituted the eigenvalues of λ found in (9.446) into (9.448). The boundary condition given in (9.447) requires k_1 to be zero. Thus, we have

$$X(x) = k_2 \sinh(n\pi x / b) \tag{9.450}$$

We now combine solutions in (9.446) and (9.449) to get the fundamental solution

$$u_n(x, y) = \sinh(n\pi x / b)\sin(n\pi y / b), \quad n = 1, 2, 3, \ldots, \tag{9.451}$$

The general solution of the problem is the sum of all these eigenfunctions with unknown constants:

$$u(x, y) = \sum_{n=1}^{\infty} c_n u_n(x, y) = \sum_{n=1}^{\infty} c_n \sinh(n\pi x / b)\sin(n\pi y / b) \tag{9.452}$$

We are now ready to consider the nonhomogeneous boundary condition given in (9.434). More specifically, setting $x = a$ in (9.451) gives

$$u(a, y) = f(y) = \sum_{n=1}^{\infty} c_n \sinh(n\pi a / b)\sin(n\pi y / b) \tag{9.453}$$

Multiplying both sides by $\sin(m\pi y/b)$ and integrating from 0 to b, we get

$$\int_0^b f(y)\sin(m\pi y / b)dy = \sum_{n=1}^{\infty} c_n \sinh(n\pi a / b)\int_0^b \sin(m\pi y / b)\sin(n\pi y / b)dy \tag{9.454}$$

In view of the orthogonality for the sine given in (9.294), we have

$$\int_0^b f(y)\sin(n\pi y / b)dy = \sinh\frac{m\pi a}{b} c_n \int_0^b \sin^2(n\pi y / b)dy = \sinh\frac{m\pi a}{b} c_n \frac{b}{2} \tag{9.455}$$

Finally, we find the unknown constant c_n in terms of the given function f on the boundary as:

$$c_n = \frac{2}{b}\sinh\left(\frac{n\pi a}{b}\right)^{-1} \int_0^b f(y)\sin(n\pi y / b)dy \tag{9.456}$$

The final solution is

$$u(x, y) = \frac{2}{b}\sum_{n=1}^{\infty} \frac{\int_0^b f(\xi)\sin\left(\frac{n\pi \xi}{b}\right)d\xi}{\sinh\left(\frac{n\pi a}{b}\right)} \sinh\left(\frac{n\pi x}{b}\right)\sin\left(\frac{n\pi y}{b}\right) \tag{9.457}$$

We know that the hyperbolic sine goes to infinity as n approaches infinity. For large n, we have the following ratio:

$$\frac{\sinh(n\pi x / b)}{\sinh(n\pi a / b)} \cong \frac{e^{n\pi x/b}}{e^{n\pi a/b}} = e^{-n\pi(a-x)/b} \tag{9.458}$$

We can see that the sum converges very fast as long as $(a-x)$ is not too small.

9.9.2 Prescribed Function along x-axis

In this section, we consider the boundary value problem of a rectangular domain governed by the Laplace equation with the non-zero boundary condition being on side $y = b$ (see Figure 9.26):

$$\nabla^2 u = \frac{\partial^2 u}{\partial x^2} + \frac{\partial^2 u}{\partial y^2} = 0, \quad 0 < x < a, \qquad 0 < y < b$$

$$u(x,0) = 0, \;\; u(x,b) = g(x), \qquad 0 < x < a \tag{9.459}$$

$$u(0,y) = 0, \;\; u(a,y) = 0, \qquad 0 \le y \le b$$

Recall from our discussion in the last section that we need to expand the given function $g(x)$ in the Fourier series expansion of sine or cosine, we have to expand X as sine or cosine functions.

Following the analysis in the last section, we assume separation of variables as:

$$u(x,y) = X(x)Y(y) \tag{9.460}$$

Substitution of (9.459) into the Laplace equation given in (9.458) gives

$$X''Y + XY'' = 0 \tag{9.461}$$

Dividing through (9.460) by XY, we get

$$\frac{X''}{X} = -\frac{Y''}{Y} = -\lambda \tag{9.462}$$

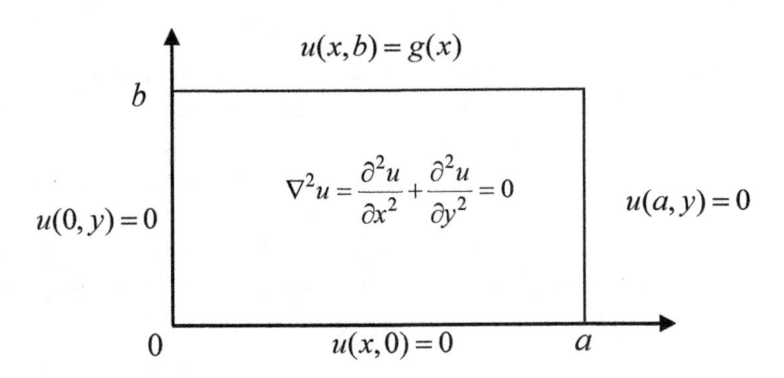

Figure 9.26 Laplace equation for a rectangular domain with a prescribed function on the horizontal side

Note that we have to use $-\lambda$ instead of $+\lambda$ as the constant of separation of variables for the present problem. Consequently, the PDE results in two ODEs:

$$X'' + \lambda X = 0 \tag{9.463}$$

$$Y'' - \lambda Y = 0 \tag{9.464}$$

The homogeneous boundary conditions given in (9.458) leads to the corresponding boundary conditions for functions X and Y as

$$u(0,y)=X(0)Y(y)=0,\quad 0\le y\le b \;\Rightarrow\; X(0)=0,$$
$$u(a,y)=X(a)Y(y)=0,\quad 0<x<a \;\Rightarrow\; X(a)=0, \tag{9.465}$$
$$u(x,0)=X(x)Y(0)=0,\quad 0<x<a \;\Rightarrow\; Y(0)=0.$$

The governing equation and boundary conditions for function Y are

$$X''+\lambda''X=0,\quad X(0)=0,\quad X(a)=0 \tag{9.466}$$

As expected, the solution for X is

$$X=c_1\sin\sqrt{\lambda}x+c_2\cos\sqrt{\lambda}x \tag{9.467}$$

The choice of taking the negative side (9.461) is important because it results in sine and cosine functions as the solutions of X. Only with these solutions, can we satisfy the nonhomogeneous boundary condition imposed on $y = b$. The first boundary condition given in (9.465) leads to

$$X(0)=c_2=0 \tag{9.468}$$

The second boundary condition given in (9.465) leads to

$$X(a)=c_1\sin\sqrt{\lambda}a=0 \tag{9.469}$$

Since we cannot set c_1 to zero, we must have the sine function be zero and this leads to

$$\sqrt{\lambda}a=n\pi \tag{9.470}$$

This is the eigenvalue of the homogeneous boundary value problem. Thus, the eigenvalues and eigenfunctions are

$$\lambda_n=n^2\pi^2/a^2,\quad X_n(x)=\sin\left(n\pi x/a\right),\quad n=1,2,3,\ldots \tag{9.471}$$

For the function Y, we have

$$Y''-\lambda''Y=0,\quad Y(0)=0 \tag{9.472}$$

The solution of (9.471) is expressible in terms of hyperbolic functions:

$$Y(y)=k_1\cosh\left(n\pi y/a\right)+k_2\sinh\left(n\pi y/a\right) \tag{9.473}$$

Note that we have already substituted the eigenvalues of λ found in (9.470) into (9.472). The boundary condition given in (9.471) requires k_1 to be zero. Thus, we have

$$Y(y)=k_2\sinh\left(n\pi y/a\right) \tag{9.474}$$

We now combine solutions in (9.470) and (9.473) to get the fundamental solution

$$u_n(x,y)=\sinh\left(n\pi y/a\right)\sin\left(n\pi x/a\right),\quad n=1,2,3,\ldots, \tag{9.475}$$

The general solution of the problem is the sum of all these eigenfunctions with unknown constants:

$$u(x,y)=\sum_{n=1}^{\infty}c_n u_n(x,y)=\sum_{n=1}^{\infty}c_n\sinh\left(n\pi y/a\right)\sin\left(n\pi x/a\right) \tag{9.476}$$

We are now ready to satisfy the nonhomogeneous boundary condition given in (9.458). More specifically, setting $x = 0$ in (9.475) gives

$$u(x,b)=g(x)=\sum_{n=1}^{\infty}c_n\sinh\left(n\pi b/a\right)\sin\left(n\pi x/a\right) \tag{9.477}$$

Multiplying both sides by $\sin(m\pi x/a)$ and integrating x from 0 to a, we get

$$\int_0^a g(x)\sin(m\pi x/a)dx = \sum_{n=1}^{\infty} c_n \sinh(n\pi b/a)\int_0^a \sin(m\pi x/a)\sin(n\pi x/a)dx \tag{9.478}$$

In view of the orthogonality for the sine given in (9.294), we have

$$\int_0^a g(x)\sin(n\pi x/a)dx = \sinh\frac{m\pi b}{a}c_n\int_0^a \sin^2(n\pi x/a)dx = \sinh\frac{m\pi b}{a}c_n\frac{a}{2} \tag{9.479}$$

Finally, we find the unknown constant in the solution in terms of the given function f on the boundary as:

$$c_n = \frac{2}{a}\sinh\left(\frac{n\pi b}{a}\right)^{-1}\int_0^a g(x)\sin(n\pi x/a)dx \tag{9.480}$$

The final solution is

$$u(x,y) = \frac{2}{a}\sum_{n=1}^{\infty}\frac{\int_0^a g(\xi)\sin\left(\frac{n\pi\xi}{a}\right)d\xi}{\sinh\left(\frac{n\pi b}{a}\right)}\sinh\left(\frac{n\pi y}{a}\right)\sin\left(\frac{n\pi x}{a}\right) \tag{9.481}$$

We note that the analyses in the last and the present sections are similar.

9.9.3 Prescribed Functions on All Four Sides

In this section, we consider the more general boundary value problem of a rectangular domain governed by the Laplace equation with non-zero boundary conditions on all four sides (see Figure 9.27):

$$\begin{gathered}\nabla^2 u = \frac{\partial^2 u}{\partial x^2} + \frac{\partial^2 u}{\partial y^2} = 0, \quad 0<x<a, \qquad 0<y<b \\ u(x,0) = g_1(x), \;\; u(x,b) = g_2(x), \qquad 0<x<a \\ u(0,y) = f_1(y), \;\; u(a,y) = f_2(y), \qquad 0<y<b\end{gathered} \tag{9.482}$$

Since the Laplace equation is a linear PDE, this problem can be solved by the method of superposition. As illustrated in Figure 9.28, we can break down the problem into four sub-problems, and each one of them has only one non-zero boundary condition. Mathematically, the four sub-problems are defined as:

Problem P_1:

$$\begin{gathered}\nabla^2 u_1 = \frac{\partial^2 u_1}{\partial x^2} + \frac{\partial^2 u_1}{\partial y^2} = 0, \quad 0<x<a, \qquad 0<y<b \\ u_1(x,0) = 0, \;\; u_1(x,b) = 0, \qquad 0<x<a \\ u_1(0,y) = f_1(y), \;\; u_1(a,y) = 0, \qquad 0<y<b\end{gathered} \tag{9.483}$$

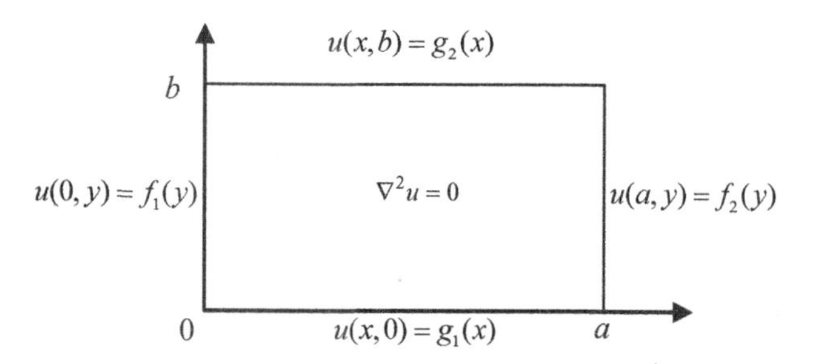

Figure 9.27 Laplace equation for a rectangular domain with nonzero conditions on all four sides

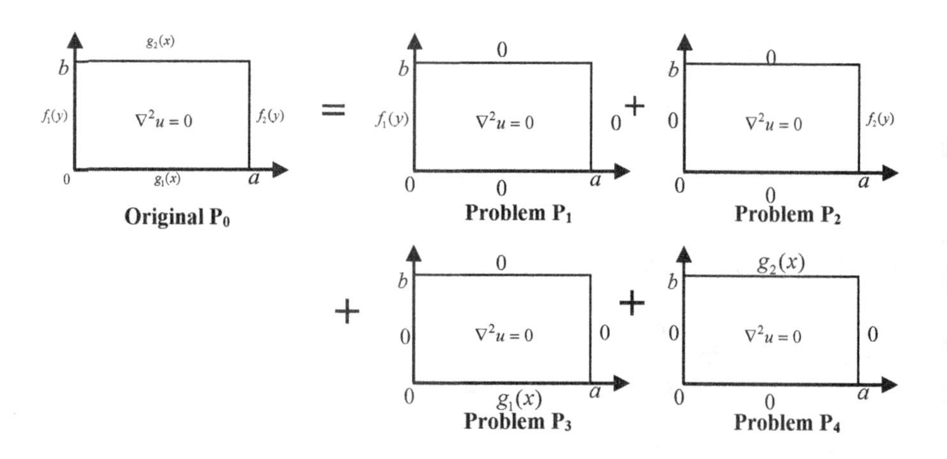

Figure 9.28 Method of superposition in solving Problem P_0 given in Figure 9.27

Problem P_2:

$$
\begin{aligned}
&\nabla^2 u_2 = \frac{\partial^2 u_2}{\partial x^2} + \frac{\partial^2 u_2}{\partial y^2} = 0, \quad 0 < x < a, \qquad 0 < y < b \\
&u_2(x,0) = 0, \;\; u_2(x,b) = 0, \qquad 0 < x < a \\
&u_2(0,y) = 0, \;\; u_2(a,y) = f_2(y), \qquad 0 < y < b
\end{aligned}
\tag{9.484}
$$

Problem P_3:

$$
\begin{aligned}
&\nabla^2 u_3 = \frac{\partial^2 u_3}{\partial x^2} + \frac{\partial^2 u_3}{\partial y^2} = 0, \quad 0 < x < a, \qquad 0 < y < b \\
&u_3(x,0) = g_1(x), \;\; u_3(x,b) = 0, \qquad 0 < x < a \\
&u_3(0,y) = 0, \;\; u_3(a,y) = 0, \qquad 0 < y < b
\end{aligned}
\tag{9.485}
$$

Problem P_4:

$$\nabla^2 u_4 = \frac{\partial^2 u_4}{\partial x^2} + \frac{\partial^2 u_4}{\partial y^2} = 0, \quad 0 < x < a, \qquad 0 < y < b$$
$$u_4(x,0) = 0, \;\; u_4(x,b) = g_2(x), \qquad 0 < x < a \tag{9.486}$$
$$u_4(0,y) = 0, \;\; u_4(a,y) = 0, \qquad 0 < y < b$$

The solution of the original problem is the superposition of each of the solutions of the four sub-problems:

$$u = u_1 + u_2 + u_3 + u_4 \tag{9.487}$$

Problem P_1 is solved in Problem 9.7 in Section 9.11, Problem P_2 is considered in Section 9.9.1, Problem P_3 is given in Problem 9.9 in Section 9.11, and Problem P_4 is solved in Section 9.9.2. Therefore, by the method of superposition, the solution is obtained in (9.486).

9.10 SUMMARY AND FURTHER READING

This chapter considers the most important types of second order PDEs. Nearly all textbooks on PDEs focus on the discussion of three second order PDEs. They are the wave, diffusion, and potential (Laplace) equations. The solution of these equations is the main focus of the present chapter. Inevitably, there is a slight overlap between the present chapter and Sections 7.2 to 7.4. The role of characteristics in wave type or hyperbolic PDEs has been discussed in detail in Chapter 7 and is not repeated in the present chapter.

For wave equations, we discuss the classic solution of D'Alembert for the 1-D wave, the consequence of characteristics in terms of the domain of dependence and of influence zone. We then continue to discuss the 2-D wave and the 3-D wave, and the 3-D symmetric wave. The classical formula of Kirchhoff or Poisson for 3-D wave is discussed and its degeneration to 2-D by Hadamard's method of descent is summarized. The consequence and implication of the Huygen principle is also discussed. The case of n-dimensional waves is discussed in detail. The jump condition at the wavefront is also presented. In view of its importance in solving the Maxwell equation and elastodynamics problems (see Chapter 2), the solution method for nonhomogeneous waves is discussed. Related to the wave phenomenon, we also discuss the Helmholtz equation and telegraph equation.

For the diffusion equation, we focus on the solution of the homogeneous diffusion equation in the present chapter, as the nonhomogeneous diffusion equation has been presented in Chapter 7. The one-dimensional heat conduction equation and 1-D consolidation equation are derived and solved. The heat conduction problems in underground structures subject to ground seasonal temperature variations are considered. The one-dimensional consolidation problem is also found expressible as a diffusion equation.

For the Laplace equation, the Dirichlet, Neumann, and Robin problems are defined and discussed. The Laplace equation is solved in both spherical and cylindrical coordinates, in terms of Legendre polynomials and Bessel functions, respectively. The classic result of the Poisson integral is considered together with the extremal principle. Some consequences of the extremal principle are discussed together with other properties of the harmonic functions, including the integrability of the Neumann problem, the uniqueness of the Dirichlet problem, and the mean value theorem. Graphical presentation of these consequences is demonstrated in Section 9.8. Finally, the boundary value problems of potential theory for the rectangular domain are considered using superpositions.

Nearly all textbooks on PDEs cover all wave, diffusion, and Laplace equations. The reader can refer any of these textbooks for further reading.

9.11 PROBLEMS

Problem 9.1. Derive the following Poisson integral formula for the problem of a circular disk of radius R governed by the Laplace equation. Mathematically, we have

$$\nabla^2 u = \frac{\partial^2 u}{\partial r^2} + \frac{1}{r}\frac{\partial u}{\partial r} + \frac{1}{r^2}\frac{\partial^2 u}{\partial \theta^2} = 0, \quad 0 < r \leq R \tag{9.488}$$

$$u(R,\theta) = h(\theta) \tag{9.489}$$

$$u(r,\theta+2\pi) = u(r,\theta) \tag{9.490}$$

$$\lim_{r\to 0} u(r,\theta) = \text{finite} \tag{9.491}$$

Ans:

$$u(r,\theta) = \frac{1}{2\pi}\int_{-\pi}^{\pi} h(\phi)[\frac{R^2 - r^2}{R^2 + r^2 - 2rR\cos(\theta-\phi)}]d\phi \tag{9.492}$$

Problem 9.2. Find the solution of the following problem of a Laplace equation of a unit disk:

$$\nabla^2 u = \frac{\partial^2 u}{\partial r^2} + \frac{1}{r}\frac{\partial u}{\partial r} + \frac{1}{r^2}\frac{\partial^2 u}{\partial \theta^2} = 0, \quad 0 < r \leq 1 \tag{9.493}$$

$$u(1,\theta) = A\cos\theta \tag{9.494}$$

Ans:

$$u(r,\theta) = Ar\cos\theta \tag{9.495}$$

Problem 9.3. Employing the procedure used in deriving the telegraph equation, derive the following equation for magnetic intensity $\boldsymbol{H}$:

$$\mu\varepsilon\frac{\partial^2 \boldsymbol{H}}{\partial t^2} = \nabla^2 \boldsymbol{H} - \mu\sigma\frac{\partial \boldsymbol{H}}{\partial t} \tag{9.496}$$

where $\boldsymbol{H}$ is defined as

$$\boldsymbol{B} = \mu\boldsymbol{H} \tag{9.497}$$

Problem 9.4. In the 2-D potential flow problem derived in Section 9.7 for the Laplace equation, derive the following relation between the potential function and stream function:

$$\frac{\partial \psi}{\partial s} = \frac{\partial \varphi}{\partial n} \tag{9.498}$$

Hint: Ask yourself what these values are physically!

Problem 9.5. Rederive (9.351) by the following steps:

(i) Prove the identity

$$\frac{\partial \varphi}{\partial s} = \frac{\partial \varphi}{\partial x}\frac{\partial x}{\partial s} + \frac{\partial \varphi}{\partial z}\frac{\partial z}{\partial s} = v_s \cos^2 \alpha + v_s \sin^2 \alpha = v_s \tag{9.499}$$

where α is the angle between the x-axis and the stream line ψ = constant.
(ii) Prove the identity

$$\frac{\partial \psi}{\partial n} = \frac{\partial \psi}{\partial x}\frac{\partial x}{\partial n} + \frac{\partial \psi}{\partial z}\frac{\partial z}{\partial n} = -v_s \sin\alpha(-\sin\alpha) + v_s \cos^2 \alpha = v_s \tag{9.500}$$

(iii) Use the results of parts (i) and (ii) to prove (9.351).

Problem 9.6. Use separation of variables to find the solution of the following diffusion problem:

$$\alpha^2 \frac{\partial^2 u}{\partial x^2} = \frac{\partial u}{\partial t} \tag{9.501}$$

$$u(0,t) = 0,\ u_x(L,t) + u(L,t) = 0, \quad u(x,0) = f(x),\ 0 \le x \le L \tag{9.502}$$

Ans:

$$u(x,t) = \sum_{n=1}^{\infty} c_n e^{-\lambda_n \alpha^2 t} \sin\left(\sqrt{\lambda_n}\, x\right) \tag{9.503}$$

$$c_n = \frac{2}{L}\int_0^L f(x) \sin\left(\sqrt{\lambda_n}\, x\right) dx \tag{9.504}$$

where λ_n satisfies the following eigenvalue equation:

$$\tan\sqrt{\lambda_n} L = -\sqrt{\lambda_n} \tag{9.505}$$

Problem 9.7. Use separation of variables to find the solution of the following Laplace equation for the rectangular domain shown in Figure 9.29:

$$\begin{aligned} &\frac{\partial^2 u}{\partial x^2} + \frac{\partial^2 u}{\partial y^2} = 0, \quad 0 < x < a, \qquad 0 < y < b \\ &u(x,0) = 0,\ u(x,b) = 0, \qquad 0 < x < a \\ &u(0,y) = f(y),\ u(a,y) = 0,\ 0 \le y \le b \end{aligned} \tag{9.506}$$

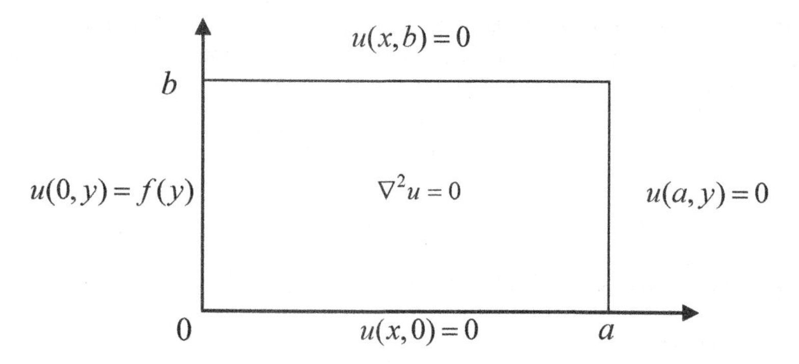

Figure 9.29 Laplace equation for a rectangular domain with nonzero boundary on $x = 0$

Ans:

$$u(x,y)=\sum_{n=1}^{\infty} c_n \sin(\frac{n\pi y}{b})[\sinh(\frac{n\pi x}{b})-\tanh(\frac{n\pi a}{b})\cosh(\frac{n\pi x}{b})] \tag{9.507}$$

$$c_n=-\frac{2}{b\tanh(\frac{n\pi a}{b})}\int_0^b f(y)\sin\left(\frac{n\pi y}{b}\right)dy \tag{9.508}$$

Problem 9.8. Use separation of variables to find the solution of the following Neumann problem of the Laplace equation for the rectangular domain shown in Figure 9.30:

$$\begin{aligned}
&\frac{\partial^2 u}{\partial x^2}+\frac{\partial^2 u}{\partial y^2}=0, \quad 0<x<a, \qquad 0<y<b \\
&u_y(x,0)=0, \ u_y(x,b)=0, \qquad 0<x<a \\
&u_x(0,y)=0, \ u_x(a,y)=f(y), \ 0\le y\le b
\end{aligned} \tag{9.509}$$

Show that

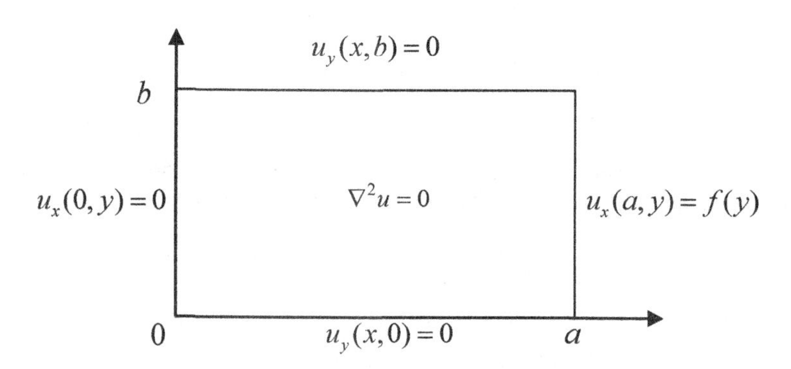

Figure 9.30 Neumann problem of the Laplace equation for a rectangular domain

(i) the solution is

$$u(x,y)=c_0+\sum_{n=1}^{\infty}c_n\cosh(\frac{n\pi x}{b})\cos(\frac{n\pi y}{b}) \tag{9.510}$$

where c_0 is an arbitrary constant and

$$c_n=\frac{2/(n\pi)}{\sinh(\frac{n\pi a}{b})}\int_0^b f(y)\cos\left(\frac{n\pi y}{b}\right)dy, \quad n=1,2,... \tag{9.511}$$

(ii) the necessary condition for the problem being solvable is

$$\int_0^b f(y)dy=0 \tag{9.512}$$

Hint: Physically, this condition corresponds to net influx equals net outflux.

Problem 9.9 Use separation of variables to find the solution of the following Laplace equation for rectangular domain shown in Figure 9.31:

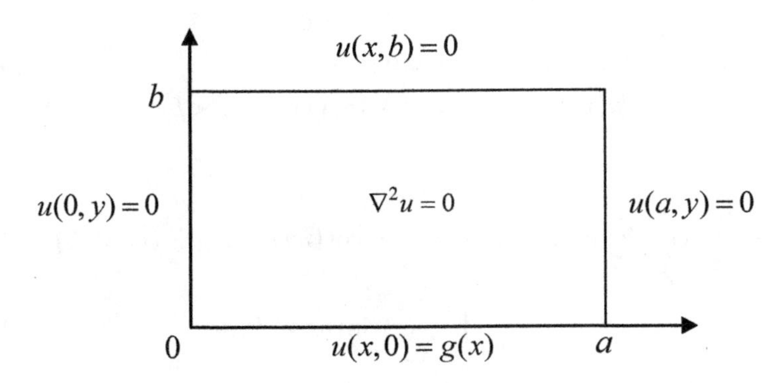

Figure 9.31 Laplace equation for a rectangular domain with prescribed function on $y = 0$

$$\begin{aligned}
&\frac{\partial^2 u}{\partial x^2}+\frac{\partial^2 u}{\partial y^2}=0, \quad 0<x<a, \qquad 0<y<b \\
&u(x,0)=g(x), \ u(x,b)=0, \qquad 0<x<a \\
&u(0,y)=0, \ u(a,y)=0, \ 0<y<b
\end{aligned} \tag{9.513}$$

Ans:

$$u(x,y)=\sum_{n=1}^{\infty}c_n\sin(\frac{n\pi x}{a})[\sinh(\frac{n\pi y}{a})-\tanh(\frac{n\pi b}{a})\cosh(\frac{n\pi y}{a})] \tag{9.514}$$

$$c_n=-\frac{2}{a\tanh(\frac{n\pi b}{a})}\int_0^a g(x)\sin\left(\frac{n\pi x}{a}\right)dx \tag{9.515}$$

Problem 9.10 Use separation of variables to find the solution of the following one-dimensional wave equation subject to initial velocity:

$$a^2 u_{xx} = u_{tt}, \quad 0 < x < L, \quad t > 0$$
$$u(0,t) = 0, \quad u(L,t) = 0, \quad t \geq 0 \tag{9.516}$$
$$u(x,0) = 0, \quad u_t(x,0) = g(x), \quad 0 < x < L$$

Ans:

$$u(x,t) = \sum_{n=1}^{\infty} c_n \sin(n\pi x / L) \sin(n\pi a t / L) \tag{9.517}$$

$$c_n = \frac{2}{n\pi a} \int_0^L g(x) \sin(n\pi x / L) dx \tag{9.518}$$

Problem 9.11 Use the method of superposition to solve the following one-dimensional wave equation subject to both initial deflection and initial velocity:

$$a^2 u_{xx} = u_{tt}, \quad 0 < x < L, \quad t > 0$$
$$u(0,t) = 0, \quad u(L,t) = 0, \quad t \geq 0 \tag{9.519}$$
$$u(x,0) = f(x), \quad u_t(x,0) = g(x), \quad 0 < x < L$$

Ans:

$$u(x,t) = \sum_{n=1}^{\infty} [c_n \sin(n\pi a t / L) + d_n \cos(n\pi a t / L)] \sin(n\pi x / L) \tag{9.520}$$

$$c_n = \frac{2}{n\pi a} \int_0^L g(x) \sin(n\pi x / L) dx \tag{9.521}$$

$$d_n = \frac{2}{L} \int_0^L f(x) \sin(n\pi x / L) dx \tag{9.522}$$

Problem 9.12 Use separation of variables to find the solution of the following Laplace equation for the rectangular domain shown in Figure 9.32:

$$\frac{\partial^2 u}{\partial x^2} + \frac{\partial^2 u}{\partial y^2} = 0, \quad 0 < x < a, \qquad 0 < y < b$$
$$u(x,0) =, \quad u(x,b) = Ax, \qquad 0 < x < a \tag{9.523}$$
$$u(0,y) = 0, \quad u(a,y) = By, \quad 0 \leq y \leq b$$

(i) Find the solution.

(ii) Find the compatibility condition in terms of A and B such that the boundary condition is continuous

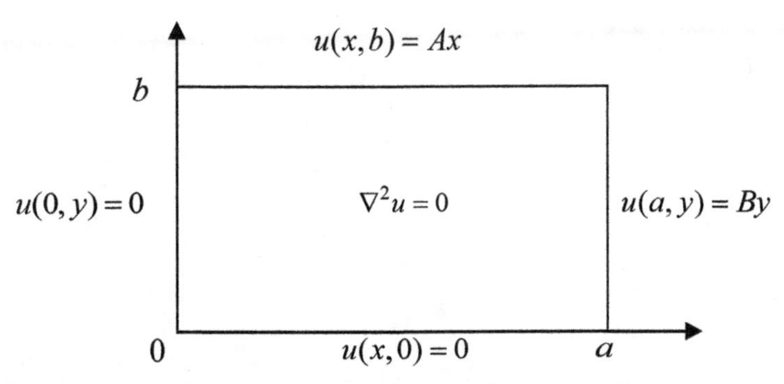

Figure 9.32 Laplace equation for a rectangular domain with prescribed functions on $y = b$ and $x = a$

Ans:

$$u(x,y) = \sum_{n=1}^{\infty} \frac{2A(-1)^{n+1}}{n\pi \sinh(\frac{n\pi b}{a})} \sin(\frac{n\pi x}{a}) \sinh(\frac{n\pi y}{a}) + \sum_{n=1}^{\infty} \frac{2B(-1)^{n+1}}{n\pi \sinh(\frac{n\pi a}{b})} \sin(\frac{n\pi y}{b}) \sinh(\frac{n\pi x}{b}) \tag{9.524}$$

$$Aa = Bb \tag{9.525}$$

CHAPTER TEN

Eigenfunction Expansions

10.1 INTRODUCTION

Eigenfunction expansion is a classical and a very powerful mathematical technique in solving certain types of boundary value problems of finite domain. When we talk about eigenvalues, we normally refer to the eigenvalue of a square matrix. However, in the context of the analysis of differential equations with certain homogeneous boundary conditions, we find that solutions exist only for certain values of the parameter embedded in the ODE. These certain values can be interpreted as eigenvalues and are determined from the particular set of boundary conditions, and play a similar role as the "normal" eigenvalues for a matrix. The corresponding solution form for each eigenvalue is called an eigenfunction, which is very much similar to the role of an eigenvector in the case of matrix analysis. In matrix analysis the number of eigenvalues depends on the rank of the matrix (e.g., a $j \times j$ matrix has j eigenvalues). Unlike the matrix analysis, the number of eigenvalues in homogeneous boundary value problems is typically infinite. In the case of beam vibrations, each eigenvalue is a natural vibration frequency of the beam and each vibration mode is an eigenfunction (e.g., with different number of nodes for the case of beam vibrations). When the same mathematical system is subject to external excitations (external in the sense that the excitations are independent of the response of the system), the general solution can be expressed as a summation of these eigenfunctions with a different contribution from each of these eigenfunctions. In the case of the dynamics of beam vibrations, a beam starts to vibrate when it is subject to continuous time-dependent forces (forcing or nonhomogeneous term). It turns out that the time-dependent vibrations can always be expressed in terms of a summation of the fundamental vibration modes, independent of the nature of the forcing terms. The contribution from each mode is determined from the eigenfunction expansion of the arbitrary forcing term in terms of the vibration modes.

This mathematical technique of eigenfunction expansion was normally associated with the name of Joseph Fourier (or Fourier series expansion). However, the first appearance of expanding a function in terms of an infinite sine series was given by Leonhard Euler in 1744 (twenty-four years before Fourier was born) when Euler wrote a letter to a friend. The formula that we are going to discuss in expressing an arbitrary function in terms of Fourier sine and cosine series was actually first proposed by Daniel Bernoulli and completed by Leonhard Euler as early as 1777. However, it is appropriate at this juncture to mention that there was actually a dispute on the correctness of such infinite expansion between Daniel Bernoulli and Leonhard Euler. This all starts with the investigation of the vibration of a violin string, which was posed by Johann Bernoulli in 1728, and solved by Jean D'Alembert and Leonhard Euler. But it was the solution posed by Daniel Bernoulli that led to the conclusion that any arbitrary function can be expressed as an infinite sum of sine functions. Both Euler and D'Alembert argued that it is impossible to do this, and the conclusion would not be right. However, to his credit, Bernoulli stood by his conclusion. Ironically, it was Euler who completed the correct analysis for Bernoulli in determining the coefficient of the expansion in 1777. In 1807 (twenty-five years after the death of Daniel Bernoulli), Fourier presented an astonishing paper at the Academy of Science in Paris asserting that an arbitrary function can be expressed in such an infinite series. Fourier specifically stated that any even function can be expressed as a sum of infinite terms of even functions of sine, and any odd function can also be expressed as a sum of infinite odd functions of cosine. Fourier analysis was based on his analysis of the heat equation, which he was most famous for (the heat diffusion law is also known as the Fourier law of heat conduction). Masters at the time, like Lagrange, simply thought that Fourier's claim was impossible. Resistance to his theory is so great that his paper was never published. His work on this eventually appeared in his book in 1822. The admirers of Fourier's work include Lord Kelvin, and his first paper at the age of fifteen was on Fourier series. The rest is history, and nowadays Fourier series expansion is accepted universally in the domains of mathematics, engineering and science. Therefore, do not feel embarrassed if you cannot understand this eigenfunction series expansion when you first learn it. Clearly, you are not alone and this topic is far from obvious.

Another well-known problem in expanding discontinuous functions in infinite series is that there are always wiggles around the point of sharp discontinuity. And these wiggles never go away no matter how many terms that we add in the infinite series. The following story is adopted from Nahin (2006). In 1898, a letter to the journal *Nature* was submitted by Albert Michelson, a recipient of the Nobel Prize in physics in 1907, and he disputed that the Fourier series expansion is not valid at the point of discontinuity because of this overshooting and undershooting. The well-known British geophysicist A.E.H. Love replied a week later, arguing that there was a mistake in Michelson's reasoning and the series that Michelson considered does not even converge. A few weeks later, Michelson replied in a short note stating that he was not convinced. In the same issue, J.W. Gibbs joined in by criticizing Michelson's reasoning. There was no reply from Michelson. A few months later, Gibbs published the most famous *Nature* letter that is cited in many textbooks. Without giving any details, Gibbs simply stated that the actual magnitude at the jump is overshot by 8.9% at one end estimated by the following formula:

$$\int_0^{\pi} \frac{\sin u}{u} du - \frac{\pi}{2} \doteq 0.089 \tag{10.1}$$

Remarkably, this overshooting percentage is a constant and does not improve by adding more terms. It was Maxime Bocher in his 1906 paper on the behavior of Fourier series at the discontinuity who coined this overshooting "Gibbs phenomenon." This term has been used ever since. The most interesting part of this story is that the result obtained by Gibbs in 1899 (given in (10.1)) was actually been published by a British mathematician Henry Wilbraham in 1848 (fifty-one years before Gibbs). The only difference is that Gibbs's paper contained no details while Wilbraham gave details of derivation.

We have covered enough history about eigenfunction expansions. In the next section, we start with the existence of the eigenvalue and eigenfunction for certain boundary value problems.

10.2 BOUNDARY VALUE PROBLEMS

In this section, we restrict our discussion to second order ODEs, but the same idea can be extended to the discussion of higher order PDEs. We have seen in an earlier chapter that a PDE can be converted to a number of ODEs through the use of separation of variables. Let us consider the following ODE subject to three different types of initial or boundary conditions:

$$y'' + p(t)y' + q(t)y = g(t), \quad y(t_0) = y_0,\ y'(t_0) = y_0' \tag{10.2}$$

$$y'' + p(x)y' + q(x)y = g(x), \quad y(\alpha) = y_0,\ y(\beta) = y_1 \tag{10.3}$$

$$y'' + p(x)y' + q(x)y = 0, \quad y(\alpha) = 0,\ y(\beta) = 0 \tag{10.4}$$

Although these boundary value or initial value problems have the same ODE structure, they are largely different problems. The mathematical techniques involved in solving them are vastly different. Note that for this second order ODE we need two conditions to fix the two unknown constants in the general solution, thus there two conditions to be satisfied in each of the problems defined in (10.2) to (10.4). Equation (10.2) is an initial value problem and the two initial conditions are given at the same point $t = t_0$. A typical example of this mathematical system arises from dynamic problems of mechanical oscillators, and $y(t)$ can be interpreted as the magnitude of the oscillations. The initial displacement and velocity were imposed at $t = t_0$. This type of problem can be solved readily by Laplace transform. Equation (10.3) is a boundary value problem with nonhomogeneous terms. The third type is the homogeneous boundary value problem, and is the main focus of the present chapter.

10.3 TWO-POINT BOUNDARY VALUE PROBLEMS

In general, the determination of the solution for (10.4) is not an easy business, depending on the exact forms of functions $p(x)$ and $q(x)$. For example, the general form (10.4) also embraces the Bessel equation as a special case, and thus finding the general solution may not be straightforward as shown in Chapter 4. In this section, we will consider the simplest case of (10.4) with $p = 0$ and $q = \lambda^2$. The solution form for this special case is easily obtained in terms of sine and cosine functions. However, the ideas discussed in the subsequent sections can also be applied to the more complicated case of (10.4).

We now illustrate the idea of the existence of eigenvalues and their associated eigenfunctions in the following examples.

Example 10.1 Consider the solution of the following two-point boundary value problem:

$$y'' + 2y = 0, \quad y(0) = 0,\ y(\pi) = 0 \tag{10.5}$$

Solution: The general solution of the ODE is

$$y = c_1 \cos\sqrt{2}x + c_2 \sin\sqrt{2}x \tag{10.6}$$

To find the unknown constants, we have

$$y(0) = c_1 = 0 \tag{10.7}$$

$$y(\pi) = c_2 \sin\sqrt{2}\pi = 0 \tag{10.8}$$

Since $\sin(\sqrt{2}\pi) \neq 0$, we must have both constants as zero. Thus, the only solution is the trivial solution $y = 0$.

Example 10.2 Consider the solution of the following two-point boundary value problem:

$$y'' + y = 0, \quad y(0) = 0,\ y(\pi) = 0 \tag{10.9}$$

Solution: The general solution of the ODE is

$$y = c_1 \cos x + c_2 \sin x \tag{10.10}$$

To find the unknown constants, we have

$$y(0) = c_1 = 0 \tag{10.11}$$

$$y(\pi) = -c_1 + c_2 0 = 0 \tag{10.12}$$

Both these end-point conditions lead to the same result that $c_1 = 0$. Thus, the solution is

$$y = c_2 \sin x \tag{10.13}$$

where c_2 is arbitrary. Therefore, there are infinitely many nontrivial solutions.

We see that the ODEs in Examples 10.1 and 10.2 are the same except for the coefficient in front of the second term on the left hand side. However, the solutions are very different.

We now consider a much more general form of ODE:

$$y'' + \lambda^2 y = 0, \quad y(0) = 0,\ y(\pi) = 0 \tag{10.14}$$

where $\lambda > 0$ is an arbitrary constant. We have seen that this equation has infinitely many solutions if $\lambda^2 = 1$ and has no non-trivial solution if $\lambda^2 = 2$. Thus, $\lambda^2 = 1$ can be interpreted as an eigenvalue of the ODE. To see whether there are other eigenvalues, we consider the general solution of (10.14)

$$y = c_1 \cos \lambda x + c_2 \sin \lambda x \tag{10.15}$$

The first boundary condition leads to $c_1 = 0$. The second boundary condition results in

$$\sin \lambda\pi = 0 \tag{10.16}$$

Thus, there are infinite eigenvalues: $\lambda = 1, 2, 3, 4,...$ and the corresponding eigenfunctions are

$$\sin \pi x, \quad \sin 2\pi x, \quad \sin 3\pi x, \quad \sin 4\pi x, \ldots \tag{10.17}$$

We have considered the simplest type of boundary conditions or the so-called Dirichlet type. In general, the boundary conditions can be specified as Neumann, Dirichlet, or Robin type, and they will be discussed in the next section.

10.4 NEUMANN, DIRICHLET AND ROBIN PROBLEMS

We have seen from the last section that homogeneous boundary conditions may lead to an eigenvalue problem, and the particular form of the eigenvalue is a function of the boundary conditions. For the present harmonic equation, the Dirichlet type boundary value problem is given as:

$$y'' + \lambda^2 y = 0, \quad 0 < x < L \tag{10.18}$$

$$y(0) = y(L) = 0 \tag{10.19}$$

The Neumann type boundary value problem is given as:

$$y'' + \lambda^2 y = 0, \quad 0 < x < L \tag{10.20}$$

$$y'(0) = y'(L) = 0 \tag{10.21}$$

The Robin type boundary value problem is defined as:

$$y'' + \lambda^2 y = 0, \quad 0 < x < L \tag{10.22}$$

$$-l_1 y' + h_1 y = 0, \qquad x = 0 \tag{10.23}$$

$$l_2 y' + h_2 y = 0, \qquad x = L \tag{10.24}$$

For $h_1 = h_2 = 1$ and $l_1 = l_2 = 0$ in the Robin type boundary value problem, we recover the Dirichlet type boundary value problem. For $h_1 = h_2 = 0$ and $l_1 = l_2 = 1$ in the Robin type boundary value problem, the Neumann type boundary is recovered. For each boundary, we can impose either of a Dirichlet, Neumann, or Robin type of boundary condition, and, thus, we have 3 by 3 or 9 combinations of boundary conditions. The general solution is

$$y = A\cos\lambda x + B\sin\lambda x \tag{10.25}$$

We now consider the eigenvalue and the corresponding eigenfunction for each of the nine cases. The boundary conditions for each case are given as:

Case 1:

$$y(0) = y(L) = 0 \tag{10.26}$$

Case 2:

$$y'(0) = y'(L) = 0 \tag{10.27}$$

Case 3:

$$y(0) = y'(L) = 0 \tag{10.28}$$

Case 4:

$$y'(0) = y(L) = 0 \tag{10.29}$$

Case 5:

$$-l_1 y'(0) + h_1 y(0) = y'(L) = 0 \tag{10.30}$$

Case 6:

$$-l_1 y'(0) + h_1 y(0) = y(L) = 0 \tag{10.31}$$

Case 7:

$$y'(0) = -l_2 y'(L) + h_2 y(L) = 0 \tag{10.32}$$

Case 8:

$$y(0) = -l_2 y'(L) + h_2 y(L) = 0 \tag{10.33}$$

Case 9:

$$-l_1 y'(0) + h_1 y(0) = 0, \; -l_2 y'(L) + h_2 y(L) = 0 \tag{10.34}$$

The results are summarized in Table 10.1. The eigenvalues for Cases 5–9 need to be solved numerically. Figures 10.1 to 10.3 show the evaluation of the eigenvalues for Cases 5, 6, and 9 respectively (open circles in the graphs).

Table 10.1 Eigenvalues and eigenfunctions for 9 different combinations of boundary conditions.

Case	BC: $x=0$	BC: $x=L$	Eigenvalue equation	Eigenfunction
1	$l_1=0, h_1=1$	$l_2=0, h_2=1$	$\sin\lambda L=0, \lambda_n=\dfrac{n\pi}{L}$	$\sin\lambda_n x$
2	$l_1=1, h_1=0$	$l_2=1, h_2=0$	$\sin\lambda L=0, \lambda_n=\dfrac{n\pi}{L}$	$\cos\lambda_n x$
3	$l_1=0, h_1=1$	$l_2=1, h_2=0$	$\cos\lambda L=0, \lambda_n=(\dfrac{2n-1}{2})\dfrac{\pi}{L}$	$\sin\lambda_n x$
4	$l_1=1, h_1=0$	$l_2=0, h_2=1$	$\cos\lambda L=0, \lambda_n=(\dfrac{2n-1}{2})\dfrac{\pi}{L}$	$\cos\lambda_n x$
5	$l_1, h_1 \neq 0$	$l_2=1, h_2=0$	$\tan\lambda L=\dfrac{h_1}{\lambda l_1}$	$\dfrac{\cos\lambda_n(L-x)}{\cos\lambda_n L}$
6	$l_1, h_1 \neq 0$	$l_2=0, h_2=1$	$\cot\lambda L=-\dfrac{h_1}{\lambda l_1}$	$\dfrac{\sin\lambda_n(L-x)}{\sin\lambda_n L}$
7	$l_1=1, h_1=0$	$l_2, h_2 \neq 0$	$\tan\lambda L=\dfrac{h_2}{\lambda l_2}$	$\cos\lambda_n x$
8	$l_1=0, h_1=1$	$l_2, h_2 \neq 0$	$\cot\lambda L=-\dfrac{h_2}{\lambda l_2}$	$\sin\lambda_n x$
9	$l_1, h_1 \neq 0$	$l_2, h_2 \neq 0$	$\tan\lambda L=\dfrac{\lambda(h_1/l_1+h_2/l_2)}{\lambda^2-h_1h_2/l_1l_2}$	$\cos\lambda_n x + \dfrac{h_1}{\lambda_n l_1}\sin\lambda_n x$

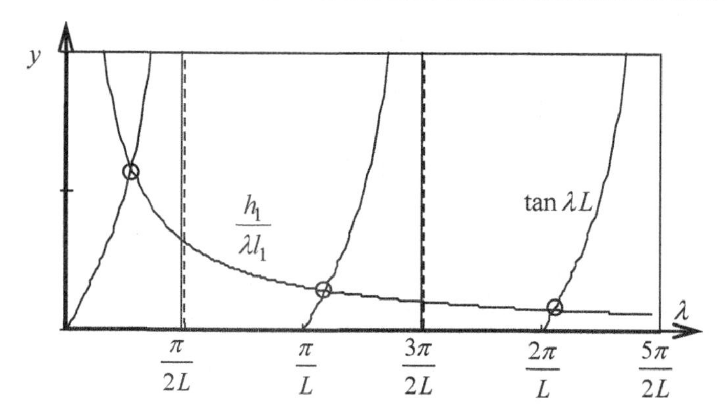

Figure 10.1 Evaluation of eigenvalues for Case 5

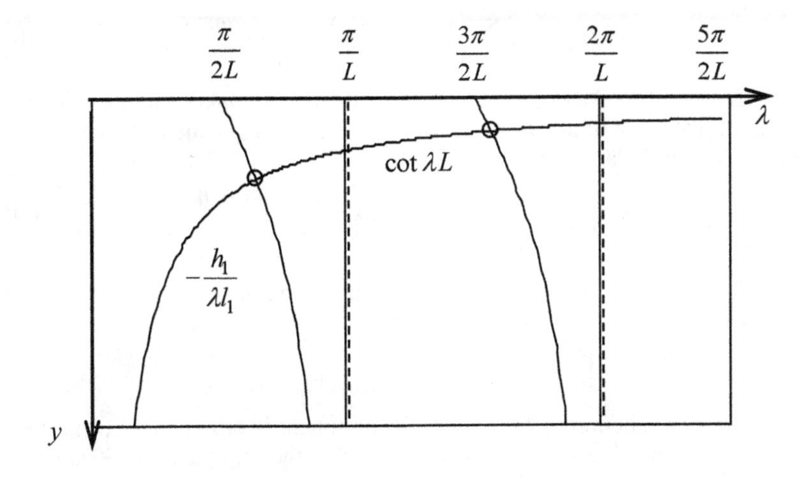

Figure 10.2 Evaluation of eigenvalues for Case 6

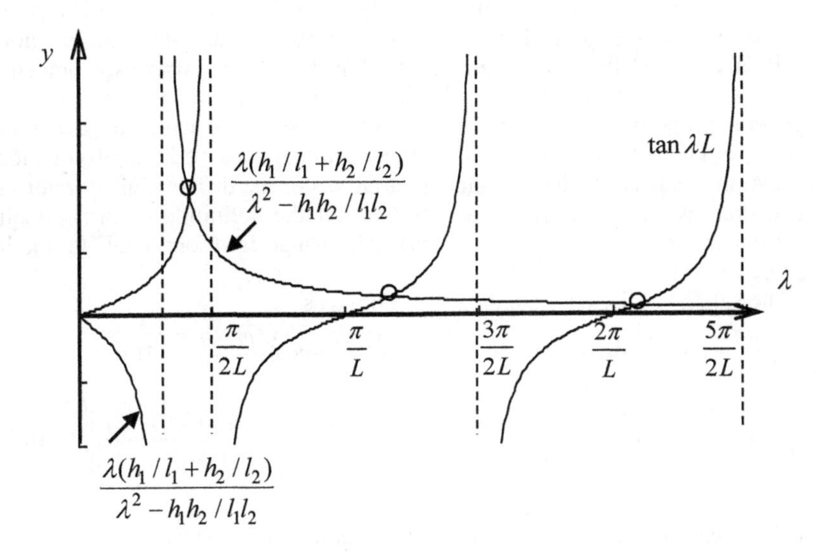

Figure 10.3 Evaluation of eigenvalues for Case 9

Cases 7 and 8 are essentially the same as those given in Figures 10.1 and 10.2 (only replacing h_1 by h_2).

10.5 FOURIER SERIES

In this section, we consider the Fourier series expansion of an arbitrary function $f(x)$. We assert that the following Fourier expansion exists:

$$f(x)=\frac{a_0}{2}+\sum_{m=1}^{\infty}\left(a_m\cos\frac{m\pi x}{L}+b_m\sin\frac{m\pi x}{L}\right) \tag{10.35}$$

where $-L < x < L$. The main objective is to find the unknown coefficients as a function of $f(x)$. First of all, we recall the so-called orthogonality of sine and cosine functions. In particular, we find that

$$\int_{-L}^{L}\cos\frac{m\pi x}{L}\cos\frac{n\pi x}{L}dx=\begin{cases}0, & m\neq n,\\ L, & m=n;\end{cases} \tag{10.36}$$

$$\int_{-L}^{L}\cos\frac{m\pi x}{L}\sin\frac{n\pi x}{L}dx=0,\ \text{all } m,n; \tag{10.37}$$

$$\int_{-L}^{L}\sin\frac{m\pi x}{L}\sin\frac{n\pi x}{L}dx=\begin{cases}0, & m\neq n,\\ L, & m=n.\end{cases} \tag{10.38}$$

That is, the sine function of order m in the argument is orthogonal (in the sense of integration) to the sine function with order n in the argument. The same is also observed for the cosine function. Sine and cosine are orthogonal regardless of the value of n and m.

Note that the definition of orthogonality given in (10.36) to (10.38) is also known as the inner product between two functions in the theory of functional analysis, also called linear operator theory. Functional analysis was a branch of rather "abstract" mathematics developed in the late nineteenth century that studies the general properties of the product of certain operators acting on a class of functions. Such linear operators may embrace the studies of integral equations, functional in the sense of the calculus of variations or the differential operator and its inversion. We will see in a later section that the definition of orthogonality between eigenfunctions (or inner product) will change from one ODE to another ODE (e.g., see (10.91)).

The validity of these formulas can be verified as:

$$\begin{aligned}\int_{-L}^{L}\cos\frac{m\pi x}{L}\cos\frac{n\pi x}{L}dx&=\frac{1}{2}\int_{-L}^{L}\{\cos[\frac{(m+n)\pi x}{L}]+\cos[\frac{(m-n)\pi x}{L}]\}dx\\&=\frac{1}{2}\frac{L}{\pi}\{\frac{\sin[(m+n)\pi x/L]}{(m+n)}+\frac{\sin[(m-n)\pi x/L]}{(m-n)}\}\bigg|_{-L}^{L}\\&=0\end{aligned} \tag{10.39}$$

In the first step, we have applied the following trigonometry identity:

$$\cos A\cos B=\frac{1}{2}[\cos(A+B)+\cos(A-B)] \tag{10.40}$$

When $n=m$, we have the special case

$$\begin{aligned}\int_{-L}^{L}(\cos\frac{m\pi x}{L})^2dx&=\frac{1}{2}\int_{-L}^{L}\{1+\cos(2m\pi x/L)\}dx=\frac{1}{2}\{x+\frac{\sin(2m\pi x/L)}{2m\pi x/L}\}\bigg|_{-L}^{L}\\&=L\end{aligned} \tag{10.41}$$

In obtaining the result in (10.41), we used the following double angle identity for the cosine:

$$\cos 2A=\cos^2 A-\sin^2 A=1-2\sin^2 A=2\cos^2 A-1 \tag{10.42}$$

Combining (10.39) and (10.41), we have the validity of (10.36). To prove (10.37), we have

$$\int_{-L}^{L} \sin\frac{m\pi x}{L}\cos\frac{n\pi x}{L}dx = \frac{1}{2}\int_{-L}^{L}\{\sin[\frac{(m+n)\pi x}{L}]+\sin[\frac{(m-n)\pi x}{L}]\}dx$$

$$= -\frac{1}{2}\frac{L}{\pi}\{\frac{\cos[(m+n)\pi x/L]}{(m+n)}+\frac{\cos[(m-n)\pi x/L]}{(m-n)}\}\Big|_{-L}^{L}$$

$$= -\frac{1}{2}\frac{L}{\pi}\{\frac{\cos[(m+n)\pi]-\cos[-(m+n)\pi]}{(m+n)}$$

$$+\frac{\cos[(m-n)\pi]-\cos[-(m-n)\pi]}{(m-n)}\}$$

$$= 0 \tag{10.43}$$

This demonstrates the validity of (10.37). Finally, orthogonality of the sine can be proved as:

$$\int_{-L}^{L} \sin\frac{m\pi x}{L}\sin\frac{n\pi x}{L}dx = \frac{1}{2}\int_{-L}^{L}\{\cos[\frac{(m-n)\pi x}{L}]-\cos[\frac{(m+n)\pi x}{L}]\}dx$$

$$= \frac{1}{2}\frac{L}{\pi}\{\frac{\sin[(m-n)\pi x/L]}{(m-n)}-\frac{\sin[(m+n)\pi x/L]}{(m-n)}\}\Big|_{-L}^{L} \tag{10.44}$$

$$= 0$$

In obtaining (10.44), we used the trigonometry identity

$$\cos(A-B)-\cos(A+B) = 2\sin A\sin B \tag{10.45}$$

Finally, for the case of $m = n$ and in view of (10.42), we have the special case

$$\int_{-L}^{L}(\sin\frac{m\pi x}{L})^2 dx = \frac{1}{2}\int_{-L}^{L}\{1-\cos(2m\pi x/L)\}dx = \frac{1}{2}\{x-\frac{\sin(2m\pi x/L)}{2m\pi x/L}\}\Big|_{-L}^{L} \tag{10.46}$$

$$= L$$

Therefore, (10.44) and (10.46) gives the identity (10.38). With these orthogonal properties for the sine and cosine, we can multiply both sides of (10.35) by $\cos(n\pi x/L)$ and $\sin(n\pi x/L)$ separately, and integrate with respect to x from $-L$ to L to find a_n and b_n respectively. In particular, the coefficients a_n ($n = 1, 2, \ldots$) can be found as follows:

$$\int_{-L}^{L} f(x)\cos\frac{n\pi x}{L}dx = \frac{a_0}{2}\int_{-L}^{L}\cos\frac{n\pi x}{L}dx + \sum_{m=1}^{\infty} a_m\int_{-L}^{L}\cos\frac{m\pi x}{L}\cos\frac{n\pi x}{L}dx$$

$$+\sum_{m=1}^{\infty} b_m\int_{-L}^{L}\sin\frac{m\pi x}{L}\cos\frac{n\pi x}{L}dx \tag{10.47}$$

In view of (10.36) to (10.38), only the second term on the left of (10.47) is nonzero. Therefore, we have

$$\int_{-L}^{L} f(x)\cos\frac{n\pi x}{L}dx = a_n\int_{-L}^{L}\cos^2\frac{n\pi x}{L}dx = La_n \tag{10.48}$$

Rearranging (10.48), we obtain the coefficient for the cosine term

$$a_n = \frac{1}{L}\int_{-L}^{L} f(x)\cos\frac{n\pi x}{L}dx, \quad n = 1,2,... \tag{10.49}$$

To find the constant coefficient a_0, we can integrate both sides of (10.35) to yield

$$\int_{-L}^{L} f(x)dx = \frac{a_0}{2}\int_{-L}^{L} dx + \sum_{m=1}^{\infty} a_m \int_{-L}^{L} \cos\frac{m\pi x}{L}dx + \sum_{m=1}^{\infty} b_m \int_{-L}^{L} \sin\frac{m\pi x}{L}dx = La_0 \tag{10.50}$$

Thus, combining (10.49) and (10.50) we find

$$a_n = \frac{1}{L}\int_{-L}^{L} f(x)\cos\frac{n\pi x}{L}dx, \quad n = 0,1,2,... \tag{10.51}$$

Similarly, we multiply both sides of (10.35) by $\sin(n\pi x/L)$, integrate with respect to x from $-L$ to L, and apply the orthogonal property to get

$$b_n = \frac{1}{L}\int_{-L}^{L} f(x)\sin\frac{n\pi x}{L}dx, \quad n = 1,2,... \tag{10.52}$$

The detailed steps of getting (10.52) are left for readers to fill in. The next example illustrates the procedure of Fourier series expansion.

Example 10.3 Find the Fourier series expansion for the following non-smooth periodic function:

$$f(x) = \begin{cases} -x, & -2 \le x < 0 \\ x, & 0 \le x < 2 \end{cases}, \qquad f(x+4) = f(x) \tag{10.53}$$

This function is shown in Figure 10.4.

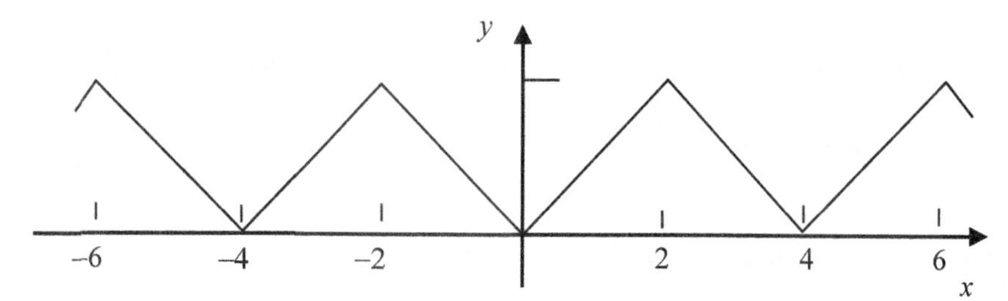

Figure 10.4 Periodic triangular function

Solution: The constant coefficient can be obtained by integrating from -2 to 2 as

$$a_0 = \frac{1}{2}\int_{-2}^{0}(-x)dx + \frac{1}{2}\int_{0}^{2} x\,dx = 1+1 = 2 \tag{10.54}$$

To find the unknown constants a_m, we can substitute (10.53) into (10.51) to get

$$a_m = \frac{1}{2}\int_{-2}^{0}(-x)\cos\frac{m\pi x}{2}dx + \frac{1}{2}\int_{0}^{2} x\cos\frac{m\pi x}{2}dx = \begin{cases} -8/(m\pi)^2, & m \text{ odd} \\ 0, & m \text{ even,} \end{cases} \tag{10.55}$$

Summarizing this result, we have

$$a_0 = 2, \; a_m = \begin{cases} -8/(m\pi)^2, & m \text{ odd} \\ 0, & m \text{ even.} \end{cases} \tag{10.56}$$

Since $f(x)$ is a symmetric function, we do not need to calculate b_n. Therefore, the periodic triangular function can be written as

$$\begin{aligned} f(x) &= \frac{a_0}{2} + \sum_{m=1}^{\infty}\left(a_m \cos\frac{m\pi x}{L} + b_m \sin\frac{m\pi x}{L}\right) \\ &= 1 - \frac{8}{\pi^2}\left(\cos\frac{\pi x}{2} + \frac{1}{3^2}\cos\frac{3\pi x}{2} + \frac{1}{5^2}\cos\frac{5\pi x}{2} + \dots\right) \\ &= 1 - \frac{8}{\pi^2}\sum_{m=1,3,5,\dots}^{\infty}\frac{\cos(m\pi x/2)}{m^2} \\ &= 1 - \frac{8}{\pi^2}\sum_{n=1}^{\infty}\frac{\cos(2n-1)\pi x/2}{(2n-1)^2} \end{aligned} \tag{10.57}$$

Figure 10.5 plots the 1-term, 3-term, and 5-term Fourier series expansions. The sharpness at the corner increases with more terms. The solutions for 3-term and 5-term expansions are indistinguishable in Figure 10.5.

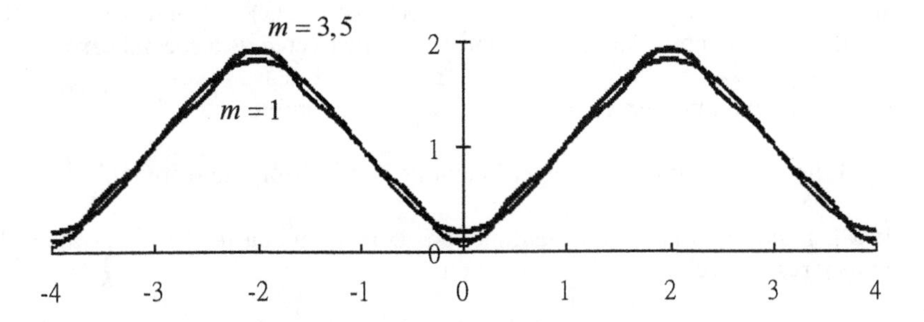

Figure 10.5 Fourier representation of triangular function

10.6 STURM-LIOUVILLE PROBLEM

In Sections 10.3 and 10.4, we have focused our discussion on the harmonic equation with the solution being the sine and cosine. The main observations discussed in previous sections also apply to a wider class of general differential equations. In 1836 and 1837, Sturm and Liouville in a series of papers considered the eigenvalue problems of a more general class of differential equation that includes many important differential equations in physics and mechanics as special cases. They showed that for such a system the eigenvalues are all real, the eigenfunctions of two distinct eigenvalues are orthogonal in the sense of integration or the inner product if a proper weighting function is included in the definition, and there are infinitely many eigenvalues. The Sturm-Liouville boundary value problem is self-adjoint and this property leads to real eigenvalues. It also bears a close resemblance between the so-called Hermitian matrix (or symmetric matrix for the case of a real matrix) and the Sturm-Liouville problem. In particular, both of them have real eigenvalues. The eigenvectors of the Hermitian matrix and eigenfunctions of the Sturm-Liouville boundary value problem are guaranteed orthogonal. The main difference is that the Hermitian matrix has a finite number of eigenvalues but the Sturm-Liouville boundary value problem has an infinite number of eigenvalues. Although most physical problems lead to self-adjoint PDEs or ODEs like the Sturm-Liouville problem, there are also non-conservative systems that result in non-self-adjoint PDEs or ODEs. That will be covered briefly in Section 10.10.

In particular, the Sturm-Liouville problem is formulated as:

$$\frac{d}{dx}[p(x)\frac{dy}{dx}]-q(x)y+\lambda r(x)y=-L[y]+\lambda r(x)y=0, \quad 0<x<1 \tag{10.58}$$

$$\alpha_1 y(0)+\alpha_2 y'(0)=0 \tag{10.59}$$

$$\beta_1 y(1)+\beta_2 y'(1)=0 \tag{10.60}$$

where $p(x) > 0$ and $r(x) > 0$ and p, dp/dx, q and r are continuous. All $p(x)$, $q(x)$, $r(x)$, α_1, α_2, β_1, and β_2 are real. Physically, $r(x)$ can be thought of as coming from separation of variables in the original associated PDE problems. The eigenvalue of the problem is denoted by λ. The linear differential operator L is defined in (10.58). The boundary conditions are separated or unmixed. That is, one boundary condition on one end point only. When we set $p(x) = r(x) = 1$ and $q(x) = 0$, the eigenvalue problem considered in Section 10.4 is recovered as a special case.

There are four main properties of this boundary value problem:

(i) All of the eigenvalues λ_n and eigenfunctions ϕ_n of this problem are real.

(ii) If $\phi_1(x)$ and $\phi_2(x)$ are two eigenfunctions corresponding to the eigenvalues λ_1 and λ_2, respectively, and if $\lambda_1 \neq \lambda_2$, then

$$\int_0^1 \phi_1(x)\phi_2(x)r(x)dx=0 \tag{10.61}$$

That is, $\phi_1(x)$ and $\phi_2(x)$ are orthogonal with respect to the weighting function $r(x)$. The integral defined in (10.61) actually provides a new definition for the inner product with $r(x)$ being the weighting function in the integrand.

(iii) All eigenvalues of the Sturm-Liouville problem are all simple and each eigenvalue has its own linearly independent eigenfunction. There are infinite eigenvalues:

$$\lambda_1<\lambda_2<\lambda_3<\cdots<\lambda_n<\cdots, \quad \lim_{n\to\infty}\lambda_n=\infty \tag{10.62}$$

(iv) The eigenfunctions are complete (i.e., they can be used for eigenfunction expansion of arbitrary functions).

The first two properties will be proved next. However, we will first consider the Lagrange identity in the next section.

10.6.1 Lagrange Identity

Because the Lagrange identity is central to the Sturm-Liouville problem, it is considered next. We start by considering an integral of operator L defined in (10.58) on a function u multiplying by another function v over the domain:

$$\int_0^1 L[u]v dx = \int_0^1 [-(pu')'v + quv] dx \tag{10.63}$$

Applying integration by parts to the first term on the right of (10.63), we get

$$-\int_0^1 (pu')'v dx = -[v(pu')]_0^1 + \int_0^1 (pu')v' dx \tag{10.64}$$

Next, reapplying integration by parts to the last term of (10.64), we find

$$\int_0^1 pv' du = [v'pu]_0^1 - \int_0^1 (pv')'u dx \tag{10.65}$$

Substituting these results into (10.63), we obtain

$$\begin{aligned} \int_0^1 L[u]v dx &= -p(x)u'(x)v(x)\Big|_0^1 + p(x)u(x)v'(x)\Big|_0^1 + \int_0^1 [-(pv')'u + quv] dx \\ &= -p(x)\left[u'(x)v(x) - u(x)v'(x)\right]\Big|_0^1 + \int_0^1 uL[v] dx \end{aligned} \tag{10.66}$$

The last term on the right can be moved to the other side to get

$$\int_0^1 \{L[u]v - uL[v]\} dx = -p(x)\left[u'(x)v(x) - u(x)v'(x)\right]\Big|_0^1 \tag{10.67}$$

This identity just obtained in (10.67) is known as the Lagrange identity. This identity closely resembles Green's theorem:

$$\int_V (u\nabla^2 v - v\nabla^2 u) dV = \int_S (u\frac{\partial v}{\partial n} - v\frac{\partial u}{\partial n}) dS \tag{10.68}$$

which has been discussed in Chapter 8. Now suppose that both u and v satisfy the boundary condition given in (10.59) and (10.60). That is, we have

$$\alpha_1 v(0) + \alpha_2 v'(0) = 0, \quad \beta_1 v(1) + \beta_2 v'(1) = 0 \tag{10.69}$$

$$\alpha_1 u(0) + \alpha_2 u'(0) = 0, \quad \beta_1 u(1) + \beta_2 u'(1) = 0 \tag{10.70}$$

Rearranging (10.69) and (10.70), we can rewrite the first derivative term on the two boundaries as

$$v'(1) = -\frac{\beta_1}{\beta_2} v(1), \quad u'(1) = -\frac{\beta_1}{\beta_2} u(1) \tag{10.71}$$

$$v'(0) = -\frac{\alpha_1}{\alpha_2} v(0), \quad u'(0) = -\frac{\alpha_1}{\alpha_2} u(0) \tag{10.72}$$

Substitution of (10.71) and (10.72) into the right hand side of (10.67) gives

$$\begin{aligned} &-p(x)\left[u'(x)v(x) - u(x)v'(x)\right]\Big|_0^1 \\ &= -p(1)\left[u'(1)v(1) - u(1)v'(1)\right] + p(0)\left[u'(0)v(0) - u(0)v'(0)\right] \\ &= p(1)\frac{\beta_1}{\beta_2}\left[u(1)v(1) - u(1)v(1)\right] - p(0)\frac{\alpha_1}{\alpha_2}\left[u(0)v(0) - u(0)v(0)\right] = 0 \end{aligned} \tag{10.73}$$

It is interesting to note that right hand side of (10.67) is identically zero regardless of the values of α_1, α_2, β_1, and β_2. Therefore, for the Sturm-Liouville problem defined in (10.58) to (10.60) the following form of the Lagrange identity applies:

$$\int_0^1 \{L[u]v - uL[v]\} dx = 0 \tag{10.74}$$

The Lagrange identity can be rewritten in another slightly different form:

$$(L[u], v) - (u, L[v]) = 0 \tag{10.75}$$

where the bracket is the inner product and defined as

$$(u, v) = \int_0^1 u\bar{v} dx \tag{10.76}$$

This definition of the inner space is the fundamental tool in functional analysis. The complex conjugate is denoted by a superimposed bar. However, for the present Sturm-Liouville problem both u and v are real.

10.6.2 Real Eigenvalues

We are going to prove the eigenvalues and eigenfunctions are real by contradiction (i.e., property (i) in Section 10.6). Suppose that the eigenvalue λ and its eigenfunction $\phi(x)$ are complex as:

$$\lambda = u + iv, \quad \phi(x) = U(x) + iV(x) \tag{10.77}$$

where u, v, U, and V are real. Substituting the eigenfunction $\phi(x)$ into the Lagrange's identity in (10.74), we have

$$\left(L[\phi], \phi\right) - \left(\phi, L[\phi]\right) = 0 \tag{10.78}$$

Using (10.58), we can write (10.78) as

$$\left(\lambda r\phi, \phi\right) - \left(\phi, \lambda r\phi\right) = 0 \tag{10.79}$$

$$\lambda \int_0^1 \phi(x)\bar{\phi}(x)r(x)dx = \bar{\lambda} \int_0^1 \phi(x)\bar{\phi}(x)\bar{r}(x)dx \tag{10.80}$$

Since $r(x)$ is real, we have

$$\left(\lambda - \bar{\lambda}\right) \int_0^1 \phi(x)\bar{\phi}(x)r(x)dx = 0 \tag{10.81}$$

$$\left(\lambda - \bar{\lambda}\right) \int_0^1 \left[U^2(x) + V^2(x)\right] r(x)dx = 0 \tag{10.82}$$

The integral in (10.82) is real positive, so we must have

$$\left(\lambda - \bar{\lambda}\right) = 2\,\mathrm{Im}(\lambda) = 0 \tag{10.83}$$

Therefore, λ must be real. Consequently, the corresponding eigenfunction is also real, but the details of the proof will not be discussed here.

10.6.3 Orthogonal Property of Eigenfunctions

Assume that there are two different eigenvalues λ_1 and λ_2 with corresponding eigenfunctions $\phi_1(x)$ and $\phi_2(x)$. Obviously, they satisfy

$$L[\phi_1] = \lambda_1 r \phi_1 \tag{10.84}$$

$$L[\phi_2] = \lambda_2 r \phi_2 \tag{10.85}$$

Let u and v be $\phi_1(x)$ and $\phi_2(x)$, and (10.74) becomes

$$(L[\phi_1], \phi_2) - (\phi_1, L[\phi_2]) = 0 \tag{10.86}$$

Using the definition in (10.75) and substituting (10.84) and (10.85) into (10.86), we get

$$\lambda_1 \int_0^1 \phi_1(x)\bar{\phi}_2(x) r(x) dx = \bar{\lambda}_2 \int_0^1 \phi_2(x)\bar{\phi}_1(x)\bar{r}(x) dx \tag{10.87}$$

Since both λ_1, λ_2, $r(x)$, $\phi_1(x)$ and $\phi_2(x)$ are real, we can simplify (10.87) as

$$(\lambda_1 - \lambda_2) \int_0^1 \phi_1(x)\phi_2(x) r(x) dx = 0 \tag{10.88}$$

As $\lambda_1 \neq \lambda_2$, then we must have the following orthogonal identity satisfied.

$$\int_0^1 \phi_1(x)\phi_2(x) r(x) dx = 0 \tag{10.89}$$

We are going to see that a lot of important differential equations can be recovered as special cases of the Sturm-Liouville problem. Thus, once we can identify $r(x)$ from the differential equation, we can find the orthogonal identity in the form of the integral given in (10.88). Thus, different ODEs require different definition of an inner product and a different orthogonal property. Problem 10.5 shows that the Bessel equation is a special case of the Sturm-Liouville problem with $r(x) = x$ and, thus, Problem 10.6 demonstrates that the orthogonal property of the Bessel function is given in (10.281) of Problem 10.6.

It is often convenient to multiply the eigenfunction by a constant such that the following normalization condition is satisfied:

$$\int_0^1 \phi_n^2(x) r(x) dx = 1, \quad n = 1, 2, \ldots \tag{10.90}$$

Thus, in view of (10.88) and (10.89), the orthogonality relation can be rewritten as

$$\int_0^1 \phi_m(x)\phi_n(x) r(x) dx = \delta_{mn} \tag{10.91}$$

This normalization process is also known as Gram-Schmidt normalization.

Example 10.4 Find the normalized orthogonal eigenfunctions of the following problem by Gram-Schmidt normalization

$$y'' + \lambda y = 0, \quad y(0) = 0, y(1) = 0 \tag{10.92}$$

Solution: We have considered this problem before. The weight function is $r(x) = 1$. The eigenvalues and eigenfunctions are

$$\lambda_n = n^2\pi^2, \quad y_n(x) = \sin n\pi x \tag{10.93}$$

To find the orthonormal eigenfunctions, we choose k_n so that

$$\int_0^1 (k_n \sin n\pi x)^2 dx = 1, \quad n = 1, 2, \ldots \tag{10.94}$$

This can be integrated to get

$$1 = k_n^2 \int_0^1 \sin^2 n\pi\, x dx = k_n^2 / 2 \int_0^1 (1 - \cos 2n\pi\, x) dx = k_n^2 / 2, \tag{10.95}$$

Therefore, the orthonormal eigenfunctions are

$$\phi_n(x) = \sqrt{2} \sin n\pi x, \quad n = 1, 2, \ldots, \tag{10.96}$$

10.6.4 Heat Conduction Problem

Although so far in this chapter we have been focused on the two-point boundary value problems for ODEs, actually the discussion here is motivated by its applications to problems governed by PDEs. In this section, we will demonstrate this by considering the heat conduction problem. Consider the one-dimensional heat conduction problem of a rod of finite length L subject an initial temperature distribution and to an isothermal boundary at the two end points, as shown in Figure 10.6.

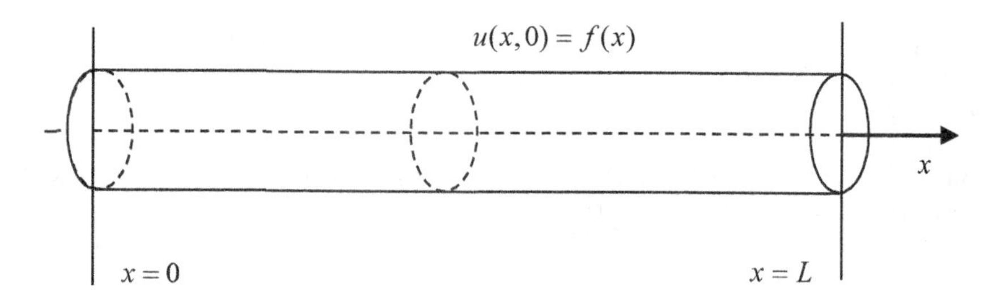

Figure 10.6 A one-dimensional conducting rod subject to initial nonhomogeneous temperature

The heat conduction problem along a rod is governed by a one-dimensional diffusion equation:

$$\alpha^2 u_{xx} = u_t, \quad 0 < x < L, \quad t > 0 \tag{10.97}$$

where α^2 is the thermal diffusivity and it depends on the material of the rod. The isothermal boundary condition is given as:

$$u(0,t) = 0, \quad u(L,t) = 0, \quad t > 0 \tag{10.98}$$

The initial condition of the temperature on the rod is prescribed as

$$u(x,0) = f(x), \quad 0 \le x \le L \tag{10.99}$$

Note that u should be interpreted as the temperature subtracting the constant temperature at the boundaries, such that the boundary condition becomes homogeneous. The following separation of variables is assumed:

$$u(x,t) = X(x)T(t) \tag{10.100}$$

Substitution of (10.100) into (10.97) gives

$$\alpha^2 X_{xx} T = X T_t \tag{10.101}$$

This equation can be rearranged as

$$\frac{X_{xx}}{X} = \frac{T_t}{\alpha^2 T} = -\lambda \tag{10.102}$$

Note that the minus sign chosen in the separation of variable constant is added to ensure the time dependence of the temperature to decay with time. This leads to two ODEs as

$$X_{xx} + \lambda X = 0, \quad T_t + \lambda \alpha^2 T = 0 \tag{10.103}$$

The boundary condition becomes

$$u(0,t) = X(0)T(t) = 0, \quad u(L,t) = X(L)T(t) = 0, \quad t > 0 \tag{10.104}$$

The initial condition needs to be applied after the fundamental solution for both x and t dependences are obtained. We can see that the x-dependent problem becomes

$$X_{xx} + \lambda X = 0, \quad X(0) = X(L) = 0 \tag{10.105}$$

This is precisely the problem that we consider in Section 10.4 and is a special case of the Sturm-Liouville problem.

We have only demonstrated that the special case of the harmonic oscillator equation of the Sturm-Liouville problem can result from the heat equation. Actually, other forms of Sturm-Liouville problems can also result from wave equation, Laplace equation, and other PDEs after the application of separation of variables.

10.6.5 Integrating Factors

In this section, we demonstrate that a general second order ODE can be converted to the Sturm-Liouville type problem by multiplying an integrating factor.

$$P(x)y'' + Q(x)y' + R(x)y = 0 \tag{10.106}$$

Suppose that there exists an integrating factor $\mu(x)$ such that after multiplying it by (10.106) we have the following structural form:

$$\frac{d}{dx}[\mu P \frac{dy}{dx}] + \mu R y = 0 \tag{10.107}$$

This can be expanded as

$$\mu P y'' + (\mu' P + \mu P')y' + R\mu y = 0 \tag{10.108}$$

Comparing (10.106) with (10.108), we obtain

$$\mu' P = \mu(Q - P') \tag{10.109}$$

This can be rearranged to give

$$\frac{\mu'}{\mu} = \frac{Q}{P} - \frac{P'}{P} \tag{10.110}$$

Integrating and rearranging, we get

$$\mu(x) = \frac{1}{P(x)} \exp[\int_{x_0}^{x} \frac{Q(s)}{P(s)} ds \tag{10.111}$$

where x_0 is the initial boundary ($x_0 = 0$ for the Sturm-Liouville problem). Now, we can use this result to convert the following ODE to a Sturm-Liouville problem:

$$P(x)y'' + Q(x)y' + [S(x) + \lambda R(x)]y = 0 \tag{10.112}$$

$$\frac{d}{dx}[\mu P\frac{dy}{dx}]-\mu Sy+\lambda\mu Ry=0 \tag{10.113}$$

Thus, we have

$$p=\mu P, \quad q=\mu S, \quad r=\mu R \tag{10.114}$$

where

$$\mu(x)=\frac{1}{P(x)}\exp[\int_{x_0}^{x}\frac{Q(s)}{P(s)}ds \tag{10.115}$$

A slightly different approach in obtaining this integrating factor is given in Problem 10.2. This illustrates a way to convert a second order operator to look like a Sturm-Liouville type problem. More precisely, we have made the operator formally self-adjoint but this change of variables does not help in changing the boundary conditions. In other words, the formally self-adjoint operators may not lead to self-adjoint problems.

10.6.6 Eigenfunction Expansion

Let ϕ_1, ϕ_2, ..., ϕ_n,... be the normalized eigenfunctions for the Sturm-Liouville problem, and let f and f' be piecewise continuous on $0 \le x \le 1$. The eigenfunction expansion becomes

$$f(x)=\sum_{n=1}^{\infty}c_n\phi_n(x) \tag{10.116}$$

$$c_n=\int_0^1 f(x)\phi_n(x)r(x)dx \tag{10.117}$$

To show this, we can multiply (10.116) by $r(x)\phi_m(x)$ and integrate over the domain; we have

$$\int_0^1 f(x)\phi_m(x)r(x)dx=\sum_{n=1}^{\infty}c_n\int_0^1 f(x)\phi_m(x)\phi_n(x)r(x)dx=\sum_{n=1}^{\infty}c_n\delta_{mn} \tag{10.118}$$

Hence, we have

$$c_m=\int_0^1 f(x)\phi_m(x)r(x)dx \tag{10.119}$$

This completes the proof of the eigenfunction expansion for the Sturm-Liouville problem.

10.7 EXAMPLES OF STURM-LIOUVILLE PROBLEMS

In this section, we list a number of important second order ODEs in physics and mechanics. All of these equations are special forms of the Sturm-Liouville problem. Thus, all the mathematical structures of eigenvalues and eigenfunctions listed in Section 10.6 apply equally to all of the following ODEs. The generality of this problem is the main reason why the Sturm-Liouville problem has received considerable attention in the past.

Legendre equation:

$$(1-x^2)\frac{d^2u}{dx^2}-2x\frac{du}{dx}+n(n+1)u=0 \tag{10.120}$$

Associated Legendre equation:

$$(1-x^2)\frac{d^2u}{dx^2}-2x\frac{du}{dx}+n(n+1)u-\frac{m^2}{1-x^2}u=0 \tag{10.121}$$

Bessel equation:

$$x^2\frac{d^2u}{dx^2}+x\frac{du}{dx}+(x^2-n^2)u=0 \tag{10.122}$$

Harmonic oscillator:

$$\frac{d^2u}{dx^2}+\omega^2u=0 \tag{10.123}$$

Schrodinger equation for the 1-D harmonic oscillator:

$$-\frac{d^2\psi}{dx^2}+x^2\psi=E\psi \tag{10.124}$$

Laguerre equation:

$$x\frac{d^2u}{dx^2}+(1-x)\frac{du}{dx}+\alpha u=0 \tag{10.125}$$

Associated Laguerre equation:

$$x\frac{d^2u}{dx^2}+(k+1-x)\frac{du}{dx}+(\alpha-k)u=0 \tag{10.126}$$

Hypergeometric equation:

$$x(1-x)\frac{d^2u}{dx^2}+[c-(a+b+1)x]\frac{du}{dx}-abu=0 \tag{10.127}$$

Chebyshev equation:

$$(1-x^2)\frac{d^2u}{dx^2}-x\frac{du}{dx}+n^2u=0 \tag{10.128}$$

Hermite equation:

$$\frac{d^2u}{dx^2}-2x\frac{du}{dx}+2\alpha u=0 \tag{10.129}$$

The proofs for some of them are set as problems at the end of the chapter.

10.8 NONHOMOGENEOUS BVP

10.8.1 Nonhomogeneous Differential Equation

Let us consider the following nonhomogeneous problem:

$$L[\phi]=-[p(x)\phi']'+q(x)\phi=\mu r(x)\phi(x)+f(x) \tag{10.130}$$

$$\alpha_1 y(0) + \alpha_2 y'(0) = 0 \tag{10.131}$$

$$\beta_1 y(1) + \beta_2 y'(1) = 0 \tag{10.132}$$

Note that we have considered this problem before in Section 10.6 if $f(x) = 0$. We first consider the eigenvalue problem of the corresponding homogeneous problem:

$$L[y] = \lambda r(x) y(x) \tag{10.133}$$

Note that λ is, in general, different from μ given in the corresponding nonhomogeneous equation. Let the eigenvalues and eigenfunctions for the following be found and they are

$$\lambda_n, \ \phi_n(x), \ n = 1, 2, ..., n, \quad n \to \infty \tag{10.134}$$

Let us expand the solution ϕ of the nonhomogeneous problem:

$$\phi(x) = \sum_{n=1}^{\infty} b_n \phi_n(x), \quad b_n = \int_0^1 \phi(x) \phi_n(x) r(x) dx \tag{10.135}$$

To find the coefficient b_n, we need to satisfy the governing equation given in (10.130). Thus, we have

$$L[\phi](x) = \sum_{n=1}^{\infty} b_n \lambda_n r(x) \phi_n(x) = \mu r(x) \phi(x) + f(x) \tag{10.136}$$

Since the eigenfunctions automatically satisfy the boundary conditions given in (10.131) and (10.132), we do not need to consider them here. Clearly, we want to expand the nonhomogeneous function $f(x)$ in terms of the eigenfunctions of the homogeneous problem as:

$$\frac{f(x)}{r(x)} = \sum_{n=1}^{\infty} c_n \phi_n(x) \tag{10.137}$$

Note that we are expanding $f(x)$ scaled with respect to $r(x)$. Using the formula derived in Section 10.6.6, we obtain

$$c_n = \int_0^1 \frac{f(x)}{r(x)} \phi_n(x) r(x) dx = \int_0^1 f(x) \phi_n(x) dx \tag{10.138}$$

Since we assume that the eigenvalue problem can be solved and $f(x)$ is given in the problem, thus (10.138) can be evaluated to give c_n. Substitution of (10.137) and (10.138) into (10.136) gives

$$\sum_{n=1}^{\infty} b_n \lambda_n r(x) \phi_n(x) = \mu r(x) \sum_{n=1}^{\infty} b_n \phi_n(x) + r(x) \sum_{n=1}^{\infty} c_n \phi_n(x) \tag{10.139}$$

Putting all terms on the same side, we get

$$r(x) \sum_{n=1}^{\infty} \left[(\lambda_n - \mu) b_n - c_n \right] \phi_n(x) = 0 \tag{10.140}$$

Since both $r(x)$ and $\phi_n(x)$ are not zero, we must have

$$b_n = \frac{c_n}{(\lambda_n - \mu)} \tag{10.141}$$

Now, we have the solution as

$$\phi(x) = \sum_{n=1}^{\infty} \frac{c_n}{\lambda_n - \mu} \phi_n(x) \tag{10.142}$$

This solution is valid only for $\lambda_n \neq \mu$.

If $\lambda_n = \mu$ and $c_n \neq 0$, there is no solution for the problem since there is no b_n that can satisfy (10.140). However, if $\lambda_n = \mu$ and $c_n = 0$, then b_n is arbitrary and there are infinite solutions with an arbitrary constant (non-unique solution). Note that $c_n = 0$ also implies from (10.138) that

$$\int_0^1 f(x)\phi_n(x)dx = 0 \tag{10.143}$$

Therefore, this will happen if and only if f is orthogonal to the eigenfunctions.

Example 10.5 Find the solution of the following nonhomogeneous problem with homogeneous boundary conditions using eigenfunction expansion:

$$-y'' = 3y + x, \quad y(0) = 0, y(1) + y'(1) = 0 \tag{10.144}$$

Solution: First, we consider the associated homogeneous problem:

$$y'' + \lambda y = 0, \quad y(0) = 0, y(1) + y'(1) = 0 \tag{10.145}$$

The Sturm-Liouville problem has been considered in Section 10.4 (Case 8) and the orthonormal eigenfunctions have been given in Problem 10.11:

$$\cot\sqrt{\lambda_n} = -\frac{1}{\sqrt{\lambda_n}} \tag{10.146}$$

$$\phi_n(x) = \frac{\sqrt{2}\sin\sqrt{\lambda_n}\, x}{(1+\cos^2\sqrt{\lambda_n})^{1/2}}, \quad n = 1, 2, \ldots \tag{10.147}$$

Using these results in (10.139) and (10.142), we have

$$y(x) = \sum_{n=1}^{\infty} \frac{c_n}{\lambda_n - \mu} \phi_n(x), \quad c_n = \int_0^1 x\phi_n(x)dx \tag{10.148}$$

Integrating the second equation of (10.148) by integration by parts, we find

$$c_n = \frac{2\sqrt{2}\sin\sqrt{\lambda_n}}{\lambda_n\sqrt{1+\cos^2\sqrt{\lambda_n}}}, \quad n = 1, 2, \ldots \tag{10.149}$$

Finally, the answer is written as

$$y(x) = 4\sum_{n=1}^{\infty} \frac{\sin\sqrt{\lambda_n}}{\lambda_n(\lambda_n - 3)\left(1+\cos^2\sqrt{\lambda_n}\right)} \sin\sqrt{\lambda_n}\, x \tag{10.150}$$

This completes the solution.

10.8.2 Nonhomogeneous Initial Value Problem

In the last section, we considered the case of a nonhomogeneous differential equation. Now we return to the initial value problem of heat conduction given in (10.97) to (10.99). We consider the general case of the Sturm-Liouville problem:

$$r(x)\frac{\partial u}{\partial t}=\frac{\partial}{\partial x}[p(x)\frac{\partial u}{\partial x}]-q(x)u, \quad 0<x<1, \quad t>0 \tag{10.151}$$

$$\alpha_1 u(0,t)+\alpha_2\frac{\partial u}{\partial x}(0,t)=0 \tag{10.152}$$

$$\beta_1 u(1,t)+\beta_2\frac{\partial u}{\partial x}(1,t)=0 \tag{10.153}$$

$$u(x,0)=f(x), \quad 0\le x\le 1 \tag{10.154}$$

This is a generalized heat conduction with variable material property $p(x)$ in the presence of heat source $q(x)$. Heat equation is recovered if $p(x) = \alpha^2$ and $q(x)=0$ and $r(x)=1$ (see (10.97) to (10.99)). Use the normally presumed separation of variables

$$u(x,t)=X(x)T(t) \tag{10.155}$$

Substitution of (10.155) into (10.151) gives

$$r(x)XT_t=\frac{d}{dx}[p(x)\frac{dX}{dx}]T-q(x)XT \tag{10.156}$$

This equation can be rearranged as

$$\frac{1}{r(x)X}\{\frac{d}{dx}[p(x)\frac{dX}{dx}]-q(x)X\}=\frac{T_t}{T}=-\lambda \tag{10.157}$$

where λ is the constant of separation of variables. Thus, we have the following ODEs

$$\frac{d}{dx}[p(x)\frac{dX}{dx}]-q(x)X=-\lambda r(x)X \tag{10.158}$$

$$T_t+\lambda T=0 \tag{10.159}$$

The boundary condition becomes

$$\alpha_1 X(0)+\alpha_2 X'(0)=0 \tag{10.160}$$

$$\beta_1 X(1)+\beta_2 X'(1)=0 \tag{10.161}$$

We have already shown that there are infinite eigenvalues and orthogonal eigenfunctions. For the initial value problem given in the second equation of (10.159), the solution is

$$T(t)=Ce^{-\lambda t} \tag{10.162}$$

The basic eigenfunction becomes

$$u(x,t)=\sum_{n=1}^{\infty}c_n\phi_n(x)e^{-\lambda_n t} \tag{10.163}$$

where the eigenvalues and eigenfunctions are

$$\lambda_n,\ \phi_n(x), \ n=1,2,...,n, \quad n\to\infty \tag{10.164}$$

The initial condition requires

$$u(x,0)=\sum_{n=1}^{\infty}c_n\phi_n(x)=f(x) \tag{10.165}$$

We now expand the initial temperature distribution by eigenfunction expansion

$$c_n=\int_0^1 r(x)\phi_n(x)f(x)dx \tag{10.166}$$

The final solution is

$$u(x,t)=\sum_{n=1}^{\infty}\phi_n(x)e^{-\lambda_n t}\int_0^1 r(\xi)\phi_n(\xi)f(\xi)d\xi \tag{10.167}$$

10.9 SINGULAR STURM-LIOUVILLE PROBLEM

So far, we have assumed that the problem is regular. We have also quoted in Section 10.7 that the Bessel equation is considered as a special case of the Sturm-Liouville problem. However, the Bessel equation is not regular at the origin (i.e., singular at the origin). We will demonstrate the singular Sturm-Liouville problem using the Bessel equation. Physically, this problem arises from the vibrations of an elastic membrane (e.g., Figure 10.7, which is generated by the software "*Mathematica*"). If we restrict vibrations to be axisymmetric, such a problem can be formulated as

$$a^2\left(\frac{\partial^2 u}{\partial r^2}+\frac{1}{r}\frac{\partial u}{\partial r}\right)=\frac{\partial^2 u}{\partial t^2},\quad 0<r<1,\quad t\geq 0 \tag{10.168}$$

$$u(1,t)=0,\quad t\geq 0 \tag{10.169}$$

$$u(r,0)=f(r),\quad 0\leq r\leq 1 \tag{10.170}$$

$$\frac{\partial u}{\partial t}(r,0)=0,\quad 0\leq r\leq 1 \tag{10.171}$$

At $r = 0$, the boundary condition needs not be $u(0,t) = 0$ (this will lead to a zero solution). For this singular Sturm-Liouville problem, we need to modify the boundary condition at the singular point (in this case, it is $r = 0$). The boundedness condition at the origin can be formulated as:

$$u, u' \text{ bounded},\quad \text{as } r\to 0 \tag{10.172}$$

That is, the membrane is supported at the circular edge and is subject to an initial deflection with zero initial velocity. Let us assume the following separation of variables:

$$u(r,t)=R(r)T(t) \tag{10.173}$$

Application of (10.172) into (10.168) gives

$$\frac{R''+(1/r)R'}{R}=\frac{T''}{a^2T}=-\lambda^2 \tag{10.174}$$

This results in two ODEs

$$r^2R''+rR'+\lambda^2r^2R=0 \tag{10.175}$$

$$T''+\lambda^2a^2T=0 \tag{10.176}$$

The general solutions for (10.176) are

$$T = k_1 \sin \lambda at + k_2 \cos \lambda at \tag{10.177}$$

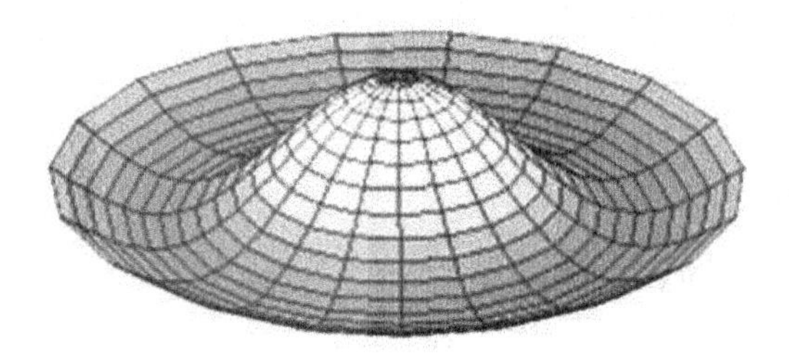

Figure 10.7 A typical vibration mode of circular membrane support at the circular edge

Introducing a change of variables $\xi = \lambda r$, we can convert (10.175) to

$$\xi^2 \frac{d^2 R}{d\xi^2} + \xi \frac{dR}{dx} + \xi^2 R = 0 \tag{10.178}$$

As mentioned in Section 10.7, this is a Sturm-Liouville problem. The solution of this Bessel function of zero order is

$$R(r) = c_1 J_0(\lambda r) + c_2 Y_0(\lambda r) \tag{10.179}$$

The solution must be bounded as $r \to 0$, and this leads to $c_2 = 0$. The boundary condition (10.169) gives

$$J_0(\lambda_n) = 0 \tag{10.180}$$

There are infinite roots for (10.180) and the general solution of the problem becomes

$$u(r,t) = \sum_{n=1}^{\infty} J_0(\lambda_n r)[k_n \sin \lambda_n at + c_n \cos \lambda_n at] \tag{10.181}$$

The initial condition given in (10.171) gives

$$k_n = 0 \tag{10.182}$$

The initial condition prescribed in (10.170) requires

$$u(r,0) = f(r) = \sum_{n=1}^{\infty} c_n J_0(\lambda_n r) \tag{10.183}$$

To find the unknown constant c_n, we multiply (10.183) by $rJ_0(\lambda_m r)$ and integrate from 0 to 1 to get

$$c_n = \frac{\int_0^1 rf(r)J_0(\lambda_n r)dr}{\int_0^1 r[J_0(\lambda_n r)]^2 dr}, \quad n = 1,2,... \tag{10.184}$$

Note that we have employed the orthogonal relation given in (10.281). Substitution of (10.184) into (10.181) with (10.182) gives the final solution as

$$u(r,t)=\sum_{n=1}^{\infty}\frac{\int_0^1 rf(r)J_0(\lambda_n r)dr}{\int_0^1 r[J_0(\lambda_n r)]^2\,dr}J_0(\lambda_n r)\cos\lambda_n at] \tag{10.185}$$

10.10 EIGENFUNCTION EXPANSION FOR NON-SELF-ADJOINT DE

So far in this chapter, we have implicitly assumed that the PDE or ODE is self-adjoint. In fact, the Sturm-Liouville problem is self-adjoint. In this section, we consider the eigenfunction expansion of the following non-self-adjoint ODE:

$$L[u]+\lambda u=0, \quad +B.C. \tag{10.186}$$

Let the eigenvalue and eigenfunction for this system be

$$\lambda_n, \quad u_n \tag{10.187}$$

Let the adjoint problem of (10.186) be

$$L^*[v]+\mu v=0, \quad +B.C. \tag{10.188}$$

The corresponding eigenvalue and eigenfunction for the adjoint problem are (Stakgold, 1979)

$$\mu_n=\bar{\lambda}_n, \quad v_n \tag{10.189}$$

Note that for differential operators, it can be shown that the eigenvalue of the original problem and the eigenvalue of the adjoint problem are related by the first equation of (10.189) (see p.357 of Stakgold, 1979)

For a non-self-adjoint operator, we have

$$L^*\neq L \tag{10.190}$$

Note, however, that $L^*=L$ is called formally self-adjoint, but it may not imply that the problem is self-adjoint. In particular, a formally self-adjoint problem may not be self-adjoint if the boundary conditions are not set properly.

Let u be the eigenfunction of the eigenvalue problem given in (10.187) and we apply the Lagrange identity as

$$([L+\lambda]u,v)=(u,[L^*+\bar{\lambda}]v) \tag{10.191}$$

Thus, in view of (10.186) we have

$$([L+\lambda]u,v)=(u,[L^*+\bar{\lambda}]v)=0 \tag{10.192}$$

For the solvability of any nonhomogeneous problem, for a particular value of μ we have either a solution for

$$[L^*+\mu]v=g(x) \tag{10.193}$$

or

$$[L^*+\mu]v=0 \tag{10.194}$$

has an eigenfunction and μ is the eigenvalue.

To prove (10.194), we now suppose that μ is not an eigenvalue so that

$$[L^*+\mu]v\neq 0 \tag{10.195}$$

for any v. Then, (10.193) always has a solution for any arbitrary g. This, in turn, requires the solvability condition for any function g being satisfied:

$$(u,g)=0 \tag{10.196}$$

This is clearly impossible unless u is identically zero. For example, g is any function and thus we can as well substitute $g = u$ into (10.196) to give

$$(u,u) = \|u\|^2 = 0 \tag{10.197}$$

Using the terminology of vector space, we have the double vertical bar being the norm of the function u. Thus, u is identically zero. This means that our assumption is false and thus $\mu = \bar{\lambda}$ is the eigenvalue and (10.194) must be satisfied.

If the boundary conditions of the adjoint problem L^* in (10.188) is chosen properly, the following Lagrange identity remains valid:

$$(v_n, Lu_m) - (L^* v_n, u_m) = 0 \tag{10.198}$$

Now if we subtract the scalar product of (10.188) with u_m from the scalar product of (10.186) with v_n, we find

$$(v_n, Lu_m) - (L^* v_n, u_m) = (\lambda_m - \lambda_n)(v_n, u_m) = 0 \tag{10.199}$$

Therefore, if

$$\lambda_m \neq \lambda_n \tag{10.200}$$

$$(v_n, u_m) = 0 \tag{10.201}$$

That is, we have established the orthogonal relation between the eigenfunction of L and the eigenfunction of L^*. This is referred to as the biorthogonal relation and this differs from the orthogonal relation between eigenfunctions of the same set of L or L^* (e.g., p.201 of Friedman, 1956). This is the main difference between self-adjoint ODEs and non-self-adjoint ODEs.

Now, we assume that a function $g(x)$ is Lebesgue square integrable:

$$\int_0^1 g(x)^2 dx < \infty \tag{10.202}$$

Then, we can expand g in the eigenfunctions of problem (10.186):

$$g(x) = \sum_{n=1}^{\infty} a_n u_n(x) \tag{10.203}$$

Multiplying both sides of (10.203) by $v_m(x)$ and integrating over the domain gives

$$(v_m, g) = \sum_{n=1}^{\infty} a_n (v_m, u_n) = a_m (v_m, u_m) \tag{10.204}$$

The last term of (10.203) is obtained in view of the biorthogonal relation derived in (10.201). Thus, we have

$$a_m = \frac{(v_m, g)}{(v_m, u_m)} \tag{10.205}$$

Finally, we get the eigenfunction for $g(x)$

$$g(x) = \sum_{n=1}^{\infty} \frac{(v_n, g)}{(v_n, u_n)} u_n(x) \tag{10.206}$$

For the special case of a self-adjoint problem, the eigenfunction expansion simplifies to

$$g(x) = \sum_{n=1}^{\infty} \frac{(u_n, g)}{(u_n, u_n)} u_n(x) \tag{10.207}$$

Example 10.6 Find the eigenfunction expansion for $f(x)$ in the following problem:

$$u''+\lambda u=f(x),\quad u(0)=0, u'(0)=u(1) \tag{10.208}$$

Solution: Note that the second boundary condition mixes the two end points. This mixed boundary condition will lead to a non-self-adjoint problem. The homogeneous problem is

$$u''+\lambda u=0,\quad u(0)=0,\ u'(0)=u(1) \tag{10.209}$$

The general solution is

$$u=c\sin\sqrt{\lambda}x+d\cos\sqrt{\lambda}x \tag{10.210}$$

The first boundary condition leads to $d = 0$. The second boundary condition leads to

$$\sqrt{\lambda_n}=\sin\sqrt{\lambda_n} \tag{10.211}$$

Apparently, there is only one root $\lambda_0 = 0$ with a corresponding eigenfunction of $u_0 = x$. However, we are going to see that this mixed boundary value problem is not self-adjoint, and thus there is no guarantee for the eigenvalue being real. Let us consider the adjoint problem:

$$\begin{aligned}(Lu,v)&=\int_0^1 u''\overline{v}dx=[u'\overline{v}-u\overline{v}']_0^1+\int_0^1 u\overline{v}''dx\\&=[u'\overline{v}-u\overline{v}']_0^1+(u,L\overline{v})\end{aligned} \tag{10.212}$$

Again the superimposed bar denotes the complex conjugate. Therefore, we find that $L^* = L$. Thus, the operator is formally self-adjoint. The boundary terms in (10.212) is

$$\begin{aligned}[u'\overline{v}-u\overline{v}']_0^1&=u'(1)\overline{v}(1)-u(1)\overline{v}'(1)-u'(0)\overline{v}(0)+u(0)\overline{v}'(0)\\&=u'(1)\overline{v}(1)-u(1)[\overline{v}(0)+\overline{v}'(1)]\end{aligned} \tag{10.213}$$

The last line of (10.213) is obtained by using the boundary conditions in (10.209). To set the result in (10.213) to zero, we require

$$v(1)=0,\quad v(0)+v'(1)=0 \tag{10.214}$$

Thus, the adjoint problem is

$$v''+\mu v=0,\quad v(1)=0,\quad v(0)+v'(1)=0 \tag{10.215}$$

Since the boundary conditions are not the same as the original problem, it is not self-adjoint although the operator is formally self-adjoint. In view of the second boundary conditions given in (20.215), we see that the solution of the first equation of (10.215) can be written as

$$v=c\sin\sqrt{\mu}(1-x) \tag{10.216}$$

Substitution of (10.216) into the second boundary condition in (20.215) results in the following eigenvalue equation

$$\sqrt{\mu_n}=\sin\sqrt{\mu_n} \tag{10.217}$$

which is identical to (10.211). Recall from (10.189) that the eigenvalues of the original operator of its adjoint operator are related by

$$\mu_n=\overline{\lambda}_n \tag{10.218}$$

For $\mu_0 = 0$, we can show that the corresponding eigenfunction is $v_0 = 1-x$. For $n > 0$, the eigenfunction of (10.215) becomes

$$v_n = c \sin \sqrt{\lambda_n}(1-x) \tag{10.219}$$

Since the problem is not self-adjoint, we shall also look for complex eigenvalues. To do that, let us assume

$$\sqrt{\lambda} = \alpha + i\beta \tag{10.220}$$

Substitution of (10.220) into (10.211) gives

$$\sqrt{\lambda} = \sin\sqrt{\lambda} = \sin(\alpha + i\beta) = \sin\alpha\cosh\beta + i\cos\alpha\sinh\beta \tag{10.221}$$

Thus, we have

$$\alpha = \sin\alpha\cosh\beta, \quad \beta = \cos\alpha\sinh\beta \tag{10.222}$$

Let us look for asymptotic behavior of the complex eigenvalues for large λ (since we want to see whether there are infinite roots). Suppose $\alpha \to \infty$; from the first of (10.222) we also have $\beta \to \infty$. The second equation of (10.222) gives

$$\frac{\beta}{\sinh\beta} \to 0, \quad \cos\alpha \to 0 \tag{10.223}$$

Therefore, we can set α as:

$$\alpha = (2m + \frac{1}{2})\pi + \varepsilon_{2m} \tag{10.224}$$

such that $\cos\alpha \to 0$, and ε_{2m} is a small number. Substitution of this result into the first equation of (10.222) gives

$$\begin{aligned} \alpha &= (2m+\frac{1}{2})\pi + \varepsilon_{2m} = \sin[(2m+\frac{1}{2})\pi + \varepsilon_{2m}]\cosh\beta \\ &= [\sin(2m+\frac{1}{2})\pi\cos\varepsilon_{2m} + \sin\varepsilon_{2m}\cos(2m+\frac{1}{2})\pi]\cosh\beta \\ &= \cosh\beta \end{aligned} \tag{10.225}$$

For large α and β, we have the approximation

$$\cosh\beta \approx (2m+\frac{1}{2})\pi \tag{10.226}$$

This can further be simplified as

$$\cosh\beta = \frac{1}{2}(e^{\beta} + e^{-\beta}) \approx \frac{1}{2}e^{\beta} \approx (2m+\frac{1}{2})\pi \tag{10.227}$$

The last part of (10.227) is from (10.226). This can be solved for β as:

$$\beta \approx \pm\ln[(4m+1)\pi] \tag{10.228}$$

The plus and minus signs result from the fact that the cosh is an even function. To find an approximation for α, we note that (10.220) can be rewritten in the form of Euler's formula as

$$\sqrt{\lambda} = \alpha + i\beta = \sqrt{\alpha^2 + \beta^2}\{\cos\theta + i\sin\theta\} \tag{10.229}$$

where

$$\theta = \tan^{-1}(\frac{\beta}{\alpha}) \tag{10.230}$$

The following approximation is made:

$$\sqrt{\alpha^2+\beta^2} = \{[(2m+\frac{1}{2})\pi+\varepsilon_{2m}]^2+\{\ln[(4m+1)\pi]\}^2\}^{1/2}$$
$$= (2m+\frac{1}{2})\pi\{1+\frac{1}{2}\delta^2+...+O(\varepsilon_{2m})\} \quad (10.231)$$

where

$$\delta = \frac{\ln[(4m+1)\pi]}{(2m+\frac{1}{2})\pi} \quad (10.232)$$

In arriving at (10.231), we have assumed that $\varepsilon_{2m} << \delta$ and it is obvious that $\delta < 1$. Similarly, we approximate $\cos\theta$ as

$$\cos\theta = \frac{\alpha}{\sqrt{\alpha^2+\beta^2}} = \frac{(2m+\frac{1}{2})\pi+\varepsilon_{2m}}{(2m+\frac{1}{2})\pi\{1+\frac{1}{2}\delta^2+...+O(\varepsilon_{2m})\}} \quad (10.233)$$
$$= 1-\frac{1}{2}\delta^2+...+O(\varepsilon_{2m})$$

We now back-substitute (10.231) and (10.233) into (10.229) to get the following approximation:

$$\alpha = \sqrt{\alpha^2+\beta^2}\cos\theta$$
$$\approx (2m+\frac{1}{2})\pi\{1+\frac{1}{2}\delta^2+...\}\{1-\frac{1}{2}\delta^2+...\} \quad (10.234)$$
$$= (2m+\frac{1}{2})\pi\{1-\frac{1}{4}\delta^4+...\}$$

Plugging in the definition of (10.232) into (10.234), we find

$$\alpha = (2m+\frac{1}{2})\pi - 2\frac{\{\ln[(4m+1)\pi]\}^4}{\{(4m+1)\pi\}^3}+... \quad (10.235)$$

Substitution of (10.235) into (10.229) gives

$$\sqrt{\lambda} \approx (2m+\frac{1}{2})\pi - 2\frac{\{\ln[(4m+1)\pi]\}^4}{\{(4m+1)\pi\}^3}+...\pm i\ln[(4m+1)\pi] \quad (10.236)$$

Note that this result does not agree with the result given on p. 204 of Friedman (1956). The powers 3 and 4 in the second term on the right are missing in Friedman (1956). Strictly speaking, this result is good for large values of *m*. Nevertheless, this approximation is illustrated in Figure 10.8.
For these eigenfunctions, we can check the bi-orthogonality as:

$$(u_n, v_m) = \int_0^1 \sin(\sqrt{\lambda_n}x)\overline{\sin[\sqrt{\overline{\lambda}_m}(1-x)]}dx$$
$$= \int_0^1 \sin(\sqrt{\lambda_n}x)\sin[\sqrt{\lambda_m}(1-x)]dx \quad (10.237)$$

Note the following trigonometry identity:

$$\sin A\sin B = \frac{1}{2}[\cos(A-B)-\cos(A+B)] \quad (10.238)$$

Using this (10.238), we get

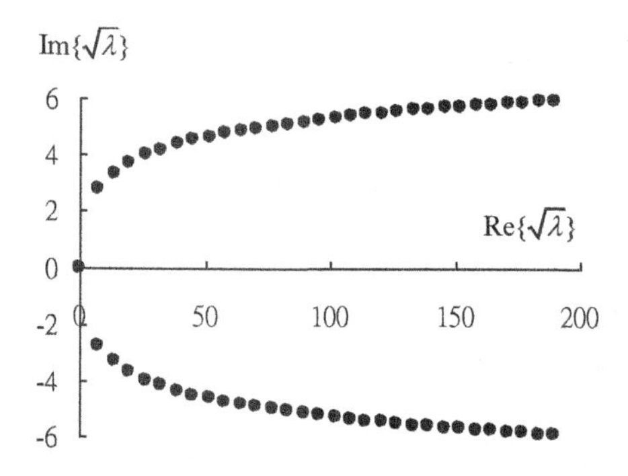

Figure 10.8 The complex roots of (10.236) for non-self-adjoint problem are plotted

$$
\begin{aligned}
(u_n, v_m) &= \frac{1}{2}\int_0^1 \{\cos[\sqrt{\lambda_n}x - \sqrt{\lambda_m}(1-x)] - \cos[\sqrt{\lambda_n}x + \sqrt{\lambda_m}(1-x)]\}dx \\
&= \frac{1}{2}\left\{\frac{\sin[\sqrt{\lambda_n}x + \sqrt{\lambda_m}x - \sqrt{\lambda_m}]}{\sqrt{\lambda_n} - \sqrt{\lambda_m}} - \frac{\sin[\sqrt{\lambda_n}x - \sqrt{\lambda_m}x + \sqrt{\lambda_m}]}{\sqrt{\lambda_n} + \sqrt{\lambda_m}}\right\}\Bigg|_0^1 \\
&= \frac{1}{2}\left\{\frac{\sin\sqrt{\lambda_n} + \sin\sqrt{\lambda_m}}{\sqrt{\lambda_n} + \sqrt{\lambda_m}} - \frac{\sin\sqrt{\lambda_n} - \sin\sqrt{\lambda_m}}{\sqrt{\lambda_n} - \sqrt{\lambda_m}}\right\}
\end{aligned}
\tag{10.239}
$$

In view of the eigenvalue equation (10.217), we can simplify (10.239) as

$$
(u_n, v_m) = \frac{1}{2}\left\{\frac{\sqrt{\lambda_n} + \sqrt{\lambda_m}}{\sqrt{\lambda_n} + \sqrt{\lambda_m}} - \frac{\sqrt{\lambda_n} - \sqrt{\lambda_m}}{\sqrt{\lambda_n} - \sqrt{\lambda_m}}\right\} = 0 \tag{10.240}
$$

When $n = m$, we can redo the integration to get

$$
(u_n, v_n) = \frac{1 - \cos\sqrt{\lambda_n}}{2} \tag{10.241}
$$

For the case $n = m = 0$, we have

$$
\int_0^1 x(1-x)dx = \frac{1}{6} \tag{10.242}
$$

Applying (10.206), we can write the expansion as

$$
f(x) = a_0 + \sum_{n=1}^{\infty} a_n \sin\sqrt{\lambda_n}x \tag{10.243}
$$

where

$$
a_0 = 6\int_0^1 f(x)(1-x)dx \tag{10.244}
$$

$$a_n = \frac{2}{1-\cos\sqrt{\lambda_n}} \int_0^1 f(x) \sin\sqrt{\lambda_n}(1-x) dx \tag{10.245}$$

As we have seen from this example, the most difficult part of the analysis is not getting the eigenfunction expansions for the function $f(x)$ for non-self-adjoint problems but finding the complex eigenvalues and corresponding eigenfunctions in the first place.

Another problem associated with non-self-adjoint eigenfunction expansion is that sometimes the eigenfunctions that we find may not be complete. As a consequence, we need an extra function (but not the actual eigenfunctions) to make it complete. Such eigenfunctions are known as "generalized eigenfunctions" (see Friedman, 1956). If such generalized functions exist, we must add them. Otherwise, the answer is not correct. Therefore, we have to be more careful in dealing with non-self-adjoint problems. That is, eigenvalues may be complex and generalized eigenfunctions may be necessary in ensuring completeness. In the next example, we will illustrate the existence of "generalized eigenfunctions."

Example 10.7 Find the eigenfunction expansion for $f(x)$ in the following problem:

$$u'' + \lambda u = f(x), \quad u(0) = 0, \ u'(0) = -u'(1) \tag{10.246}$$

Solution: The associated eigenvalue problem is

$$u'' + \lambda u = 0, \quad u(0) = 0, \ u'(0) = -u'(1) \tag{10.247}$$

The second boundary condition in (10.247) is mixed, including both boundary points in a single condition. The adjoint problem of (10.247) is found to be

$$v'' + \lambda v = 0, \quad v'(0) = 0, \ v(0) = -v(1) \tag{10.248}$$

Therefore, the problem is not self-adjoint. The solution for (10.247) is

$$u = A\cos\sqrt{\lambda}x + B\sin\sqrt{\lambda}x \tag{10.249}$$

The first boundary condition leads to $A = 0$, and the second boundary condition in (10.247) leads to the following eigenvalue equation:

$$\sqrt{\lambda} = -\sqrt{\lambda}\cos\sqrt{\lambda} \tag{10.250}$$

The only possible solution is

$$\cos\sqrt{\lambda} = -1 \tag{10.251}$$

Note that for this case λ is not zero. Therefore, we have

$$\sqrt{\lambda} = (2n+1)\pi \tag{10.252}$$

Thus, the eigenvalues and eigenfunctions become

$$\lambda_n = (2n+1)^2\pi^2, \quad n = 0,1,2,... \tag{10.253}$$

$$u_n = \sin(2n+1)\pi x, \quad n = 0,1,2,.. \tag{10.254}$$

Luckily, the eigenvalues are all real (unlike the last example of the non-self-adjoint problem). For the adjoint problem, we also have the eigenvalue being

$$\mu_n = \lambda_n = (2n+1)^2\pi^2, \quad n = 0,1,2,... \tag{10.255}$$

The corresponding eigenfunctions are

$$u_n = \cos[(2n+1)\pi(1-x)], \quad n = 0,1,2,.. \tag{10.256}$$

The biorthogonal condition for these two eigenfunctions is

$$(u_n, v_n) = \int_0^1 \sin[(2n+1)\pi x]\cos[(2n+1)\pi(1-x)]dx \tag{10.257}$$

Using the sum rule for the cosine function, we find

$$\begin{aligned} \cos[(2n+1)\pi(1-x)] &= \cos(2n+1)\pi\cos(2n+1)\pi x + \sin(2n+1)\pi\sin(2n+1)\pi x \\ &= -\cos(2n+1)\pi x \end{aligned} \tag{10.258}$$

Substitution of (10.258) into (10.257) gives

$$\begin{aligned} (u_n, v_n) &= -\int_0^1 \sin[(2n+1)\pi x]\cos[(2n+1)\pi x]dx \\ &= -\frac{1}{2}\int_0^1 \sin[(4n+2)\pi x]dx \\ &= -\frac{1}{2}\frac{\cos[(4n+2)\pi x]}{(4n+2)\pi}\bigg|_0^1 = 0 \end{aligned} \tag{10.259}$$

Thus, we have demonstrated that the bi-orthogonal relation is satisfied. Apparently, the eigenfunction expansion for $f(x)$ is

$$f(x) = \sum_{n=1}^{\infty} \frac{(v_n, f)}{(v_n, u_n)} \sin[(2n+1)\pi x] \tag{10.260}$$

But, this result is wrong because the eigenfunction is only for all the odd terms in the sine series and it is an incomplete expansion. The correct eigenfunction for this case should be:

$$f(x) = \sum_{n=1}^{\infty} [a_n \sin\sqrt{\lambda_n}x + b_n x\cos\sqrt{\lambda_n}(1-x)] \tag{10.261}$$

where

$$a_n = -4\int_0^1 f(x)(1-x)\sin\sqrt{\lambda_n}(1-x)dx \tag{10.262}$$

$$b_n = 4\int_0^1 f(x)\cos\sqrt{\lambda_n}(1-x)dx \tag{10.263}$$

$$\sqrt{\lambda_n} = (2n+1)\pi \tag{10.264}$$

Note that this is no longer a Fourier sine series expansion. Without going into detail in obtaining this, we should simply say that we need an additional function for completeness. The function in this case is

$$\tilde{u}_n = x\cos\sqrt{\lambda_n}x \tag{10.265}$$

This function satisfies the boundary conditions, but, however, does not satisfy the original eigenvalue problem. That is,

$$(L+\lambda_n)\tilde{u}_n = -2\sqrt{\lambda_n}\sin\sqrt{\lambda_n}x \neq 0 \tag{10.266}$$

Instead, it satisfies the following ODE

$$(L+\lambda_n)^2\tilde{u}_n = 0 \tag{10.267}$$

This is also equivalent to solving the following problem

$$(L+\lambda_n)\tilde{u}_n = u_n \tag{10.268}$$

Actually, (10.267) or (10.268) provides the definition of the generalized eigenfunctions. In a sense, this formulation is similar to Case 2.2 of Chapter 5 for a system of first order ODEs when the matrix is non-Hermitian. Recall the similarity between the properties of eigenvalues of a Hermitian matrix and the properties of eigenvalues of self-adjoint ODEs.

The main purpose of this example is to illustrate that generalized eigenfunctions may be needed for non-self-adjoint problems. Eigenfunction expansion for a non-self-adjoint problem is by no means straightforward and is very difficult. Their detailed discussion is, however, out of the scope of the present chapter.

10.11 SUMMARY AND FURTHER READING

In this chapter, we have summarized the essential ideas of eigenfunction expansions. We started with two point boundary value problems. Dirichlet, Neumann, and Robin problems are presented in the context of heat conduction. Fourier series expansion is introduced before we considered the general Sturm-Liouville problem, including derivation of the Lagrange identity, proof of the realness of eigenvalues, and the orthogonal property of eigenfunctions. We also summarized the list of classical equations that the eigenvalue problem can be studied under the framework of Sturm-Liouville problem. The use of eigenfunction expansion in solving the nonhomogeneous boundary value problem is demonstrated. A singular Sturm-Liouville problem is demonstrated by considering the vibration problems of circular membrane. Finally, the more advanced topic of eigenfunction expansion for non-self-adjoint ODEs is considered.

There are many good references for eigenfunction expansions, including Stakgold (1979) and Friedman (1956). The interested reader can consult these and other references.

10.12 PROBLEMS

Problem 10.1 A linear second order ODE is given by:

$$a_0(x)\frac{d^2u}{dx^2} + a_1(x)\frac{du}{dx} + [a_2(x) + \lambda a_3(x)]u = 0 \tag{10.269}$$

Show that it can be as a Sturm-Liouville problem if the following is satisfied:

$$a_1(x) = \frac{da_0(x)}{dx} \tag{10.270}$$

We can ignore the boundary condition for this problem.

Problem 10.2 A linear second order ODE is given as:

$$a_0(x)\frac{d^2u}{dx^2}+a_1(x)\frac{du}{dx}+[a_2(x)+\lambda a_3(x)]u=0 \tag{10.271}$$

(i) Compare this equation with the Sturm-Liouville problem and show that

$$\frac{p(x)}{a_0(x)}=\frac{p'(x)}{a_1(x)}=\frac{q(x)}{a_2(x)}=\frac{r(x)}{a_3(x)}=I(x) \tag{10.272}$$

(ii) Take the first two terms from the result obtained in (i) to show that

$$I(x)=\frac{1}{a_0(x)}\exp[\int\frac{a_1(x)}{a_0(x)}dx] \tag{10.273}$$

(iii) Show that p, q, and r in the Sturm-Liouville problem are

$$p(x)=\exp[\int\frac{a_1(x)}{a_0(x)}dx],\quad q(x)=I(x)a_2(x),\quad r(x)=a_3(x)I(x) \tag{10.274}$$

Therefore, (10.271) can be converted to the Sturm-Liouville problem of multiplying $I(x)$ obtained in Part (ii) if $a_1(x)\neq a'_0(x)$.

Problem 10.3 Use the result of Problem 10.2 to convert the following ODE to a Sturm-Liouville problem:

$$\frac{d^2y}{dx^2}-x\frac{dy}{dx}+\lambda y=0 \tag{10.275}$$

Ans:

$$\frac{d}{dx}(e^{-\frac{x^2}{2}}\frac{dy}{dx})+\lambda e^{-\frac{x^2}{2}}y=0 \tag{10.276}$$

Problem 10.4 Use the result of Problem 10.2 to convert the following ODE to a Sturm-Liouville problem:

$$\frac{d^2y}{dx^2}+\frac{dy}{dx}+\lambda y=0 \tag{10.277}$$

Ans:

$$\frac{d}{dx}(e^x\frac{dy}{dx})+\lambda e^x y=0 \tag{10.278}$$

Problem 10.5 Show that the Bessel equation given below can be converted to a Sturm-Liouville problem:

$$x^2\frac{d^2y}{dx^2}+x\frac{dy}{dx}+(\lambda x^2-\nu^2)y=0 \tag{10.279}$$

Ans:

$$\frac{d}{dx}(x\frac{dy}{dx})+(\lambda x-\frac{\nu^2}{x})y=0 \tag{10.280}$$

Problem 10.6 Show that the Bessel functions satisfy the following orthogonal relation:

$$\int_0^a x J_\nu(\frac{\alpha_n}{a}x) J_\nu(\frac{\alpha_m}{a}x) dx = 0, \quad n \neq m \tag{10.281}$$

where $J_\nu(\alpha_n x/a)$ satisfy (10.132).

Hint: Use the result of the Sturm-Liouville problem.

Problem 10.7 Show that the Laguerre equation can be converted to the Sturm-Liouville problem:

$$x\frac{d^2y}{dx^2} + (1-x)\frac{dy}{dx} + \lambda y = 0 \tag{10.282}$$

Ans:

$$\frac{d}{dx}(xe^{-x}\frac{dy}{dx}) + \lambda e^{-x} y = 0 \tag{10.283}$$

Problem 10.8 The Legendre equation is defined as

$$(1-x^2)\frac{d^2y}{dx^2} - 2x\frac{dy}{dx} + n(n+1)y = 0 \tag{10.284}$$

Recast the Legendre equation in the form of the Sturm-Liouville problem. Find the eigenvalue λ of the corresponding Sturm-Liouville problem.

Ans:

$$\frac{d}{dx}[(1-x^2)\frac{dy}{dx}) + n(n+1)y = 0 \tag{10.285}$$

Problem 10.9 Suppose that the boundary conditions given in (10.59) and (10.60) are replaced by the following periodic boundary conditions

$$p(0) = p(1), \quad y(0) = y(1), \quad y'(0) = y'(1) \tag{10.286}$$

Show that the following special form of the Lagrange identity remains valid

$$\int_0^1 \{L[u]v - uL[v]\} dx = 0 \tag{10.287}$$

Problem 10.10 Show that the following special form of the Lagrange identity remains valid

$$\int_0^1 \{L[u]v - uL[v]\} dx = 0 \tag{10.288}$$

for the following singular boundary conditions:

(i) $$p(0) = 0, \quad \alpha y(0) + \beta y'(1) = 0 \tag{10.289}$$

(ii) $$p(1) = 0, \quad \alpha y(0) + \beta y'(1) = 0 \tag{10.290}$$

(iii) $$p(0) = p(a) = 0 \tag{10.291}$$

Problem 10.11 Find the orthonormal eigenfunctions of the following problem

$$y'' + \lambda y = 0, \quad y(0) = 0, y(1) + y'(1) = 0 \tag{10.292}$$

Ans:

$$\phi_n(x) = \frac{\sqrt{2}\sin\sqrt{\lambda_n}\, x}{(1+\cos^2\sqrt{\lambda_n})^{1/2}}, \quad n = 1, 2, \ldots \tag{10.293}$$

CHAPTER ELEVEN

Integral and Integro-Differential Equations

11.1 INTRODUCTION

In the first part of this chapter, we will summarize the idea of integral transforms versus eigenfunction expansion that we discussed in the last chapter. Various types of integral transforms will be introduced, including the Fourier transform, Hankel transform, Mellin transform, Hilbert transform, and Laplace transform. When integral transforms are applied to ordinary differential equations, ODEs will become algebraic equations that can be solved readily most of the time. If integral transforms are applied to partial differential equations of two variables, the PDE will become an ODE. The resulting ODE is, of course, much easier to solve than the original PDE. For PDEs with n variables, each time an integral transform is applied, the resulting PDE will only involve $n-1$ variables. Therefore, when an integral transform is applied repeatedly, the PDE will eventually become an algebraic equation. However, due to space limitations we only cover the basics of integral transform and its introduction is primarily for setting the scene for the more difficult problems of integral equations and of integro-differential equations.

The second part deals with problems that need to be modeled by integral equations (governing equations involving integral of the unknown) and integro-differential equations (governing equations consisting of both differentiation and integrations of the unknown function). The word "integral equation" was coined by Du Bois-Reymond. Integral equations that we discuss include the Abel integral equation, Hilbert integral equation, Fredholm integral equation, and Volterra integral equation. Integro-differential equations will be considered using the Laplace transform.

Historically, the integral of the Laplace transform type and of the Mellin transform type was first considered by Laplace 1782. The Fourier transform was developed by Fourier in 1811 when the theory of heat conduction was considered. Formulating the mechanics problem of a sliding bead along a frictional wire, Abel formulated the Abel type of integral in 1823. In particular, Abel considered the tautochrone problem that for a given slide time what would be the shape of the wire, in contrast to the brachistochrone problem of the Bernoulli brothers. The Fredholm type of integral equation was first considered by Liouville in 1837 and 1838 based on the idea of successive substitutions of Neumann. However, the work of Neumann was not published until 1870. In 1890, French mathematician Picard formulated the successive approximation in a more general and a widely applicable form. In more recent literatures, the method of successive approximation has been mostly associated with Picard, instead of associated with Liouville and Neumann. This procedure was reconsidered by Volterra in 1896 and 1897 to solve the so-called Volterra integral equations. Volterra recognized that integral equations could be interpreted as a limiting form of a system of n linear algebraic equations for n unknowns as $n \to \infty$.

The eigenvalue problem of integral equations was considered by Poincare in 1896. In 1900 and 1903, Fredholm solved the more difficult problem of Fredholm integral equations and of asymmetric kernel functions in the integral. Following up the idea of Volterra, Fredholm again divided the interval into n equal parts and used linear algebraic equations to consider when the solution of a nonhomogeneous integral equation exists. He had been able to solve the Fredholm integral equation in terms of a determinant of the kernel K and the so-called first minor of the kernel K. This leads to a very general theorem called the Fredholm alternative theorem, which is also applicable to differential equations. Hilbert wrote a series of six papers from 1904 to 1910, and made major progress in the analysis and showed that the solution can be expressed in orthogonal eigenfunctions of the eigenvalue of kernel K. The analysis somewhat resemblances Fourier's eigenfunction expansion. This is called the Hilbert-Schmidt theorem of integral equations, as Schmidt in 1907 extended Hilbert's work to non-symmetric kernels. This eigenfunction expansion for an integral leads to the concept of the abstract functional analysis and functional space (like the Hilbert space). Hilbert was able to show that the Fredholm integral equation is equivalent to the solution of differential equation. The eigenvalue problem of ODEs is also equivalent to the eigenvalue problem of an integral equation considered by Poincare and Fredholm. The major concept of the spectrum theory of the eigenvalue of the differential equation was also formulated. In particular, when an eigenvalue of an ODE with associated boundary conditions is in the form of a discrete spectrum, the eigenfunction expansion should be used to obtain the solution. However, when there is a continuous spectrum of eigenvalues for the associated problem, typically for the infinite domain, the problem has to be solved by integral transform.

11.2 INTEGRAL TRANSFORMS

There exists different techniques of integral transforms, which are motivated by solving different types of linear differential equations. Eigenfunction expansion discussed in the last chapter is suitable for solving problems of finite domains, for which eigenvalues and eigenfunctions can be found, but they are useless in solving problems involving the infinite domain. For finite domains, the eigenvalues are discrete even though there are an infinite number of them; however, for the infinite domain, eigenvalues are continuous. We said that there is a continuous spectrum of eigenvalues. For problems with a spectrum of eigenvalues, we need to apply an integral transform instead of applying eigenfunction expansions.

In fact, the integral transform technique is developed for solving linear differential equations. We will discuss briefly the most commonly encountered integral transforms, including the Fourier transform, Hankel transform, Mellin transform, Hilbert transform, and Laplace transform. All of them can be cast into the following general transform and its inversion as:

$$F(s) = \int_{\alpha}^{\beta} K(s,t) f(t) dt, \quad -\infty \le \alpha < \beta \le \infty \tag{11.1}$$

$$f(t) = \int_{\alpha_1}^{\beta_1} H(s,t) F(s) ds, \quad -\infty \le \alpha < \beta \le \infty \tag{11.2}$$

Different kernel functions $K(s,t)$ lead to different types of integral transforms. The limits of integration also vary depending on the type of transform.

11.2.1 Laplace Transform

One of the most popular integral transforms is the Laplace transform, which is applicable to problems in the semi-infinite domain. It can be shown that the Fourier transform to be presented next is equivalent to the Laplace transform for the infinite domain (Sneddon, 1951). The Laplace transform was originally proposed by Euler and developed to the present form by Laplace. It had been known as operational calculus and its application in engineering is mainly publicized by Oliver Heaviside in the 1980s. The kernel of the Laplace transform is

$$K(s,t) = e^{-st} \tag{11.3}$$

The resulting integral transform is

$$F(s) = L[f(t)] = \int_0^\infty e^{-st} f(t)dt, \tag{11.4}$$

$$f(x) = L^{-1}[F(s)] = \frac{1}{2\pi i}\int_{c-i\infty}^{c+i\infty} e^{st} F(s)ds \tag{11.5}$$

Its application has been mainly for the time variable in differential equations. A brief summary of the Laplace transform can be found in the appendix of Chau (2013) and in Spiegel (1965). Since the solutions of linear differential equations with constant coefficients are based on the exponential function, the Laplace transform is particularly useful for such differential equations.

In view of its importance in application to differential equations, we will discuss the Laplace transform in more detail. Formula (11.5) is called Bromwich's contour integral for the inversion of the Laplace transform. This formula is now the standard form of the inverse of the Laplace transform. Bromwich learned Heaviside's operational calculus (i.e., the Laplace transform of today) and derived this important result. Bromwich died at the age of 44 and his result was unnoticed until J.R. Carson and H. Jeffrey made it well known.

Suppose that we want to evaluate the following integral (i.e., definition of Laplace transform):

$$F(s) = L[f(t)] = \int_0^\infty e^{-st} f(t)dt, \tag{11.6}$$

To solve for $f(t)$ in (11.6) or to derive (11.5), we can consider the following complex Fourier transform:

$$e^{-cx} f(x) = \frac{1}{2\pi}\int_{-\infty}^{\infty} e^{i\omega x}[\int_{-\infty}^{\infty} e^{-i\omega t} e^{-ct} f(t)dt]d\omega \tag{11.7}$$

The transform will be discussed more in the next section. If we impose the condition that

$$f(x) = 0, \quad \text{for} \quad x < 0, \tag{11.8}$$

using (11.8), we have (11.23) becoming

$$e^{-cx}f(x)=\frac{1}{2\pi}\int_{-\infty}^{\infty}e^{i\omega x}[\int_{0}^{\infty}e^{-i\omega t}e^{-ct}f(t)dt]d\omega \tag{11.9}$$

Now, c is an arbitrary constant and we want to choose its value such that c will give the limit of convergence of the solution or c will be larger than the real part of all singularities of the function $F(s)$. Mathematically, we want to choose c such that the following integral exists:

$$\int_{0}^{\infty}e^{-ct}\left|f(t)\right|dt \tag{11.10}$$

Multiplying (11.9) by e^{cx}, we get

$$f(x)=\frac{1}{2\pi}\int_{-\infty}^{\infty}e^{(c+i\omega)x}[\int_{0}^{\infty}e^{-(c+i\omega)t}f(t)dt]d\omega \tag{11.11}$$

Now we can apply the following change of variables:

$$s=c+i\omega\,,\ \text{or}\quad ds=id\omega \tag{11.12}$$

The limits of the outer integral will be shifted by

$$\omega\to\infty,\quad s=c+i\infty;\quad \omega\to-\infty,\quad s=c-i\infty \tag{11.13}$$

With this change of variables, (11.11) can be rewritten as

$$f(x)=\frac{1}{2\pi i}\int_{c-i\infty}^{c+i\infty}e^{sx}[\int_{0}^{\infty}e^{-st}f(t)dt]ds=\frac{1}{2\pi i}\int_{c-i\infty}^{c+i\infty}e^{sx}F(s)ds \tag{11.14}$$

This is the Bromwich contour integral, and the proof is completed.

We will look at a number of special cases.

Example 11.1 Find the Laplace transform of a constant C.

Solution: The Laplace transform of a constant is

$$L\{C\}=\int_{0}^{\infty}Ce^{-st}dt \tag{11.15}$$

This can be solved easily by direct integration

$$\begin{aligned}L\{C\}&=C\int_{0}^{\infty}e^{-st}dt\\&=-\frac{Ce^{-st}}{s}\bigg|_{0}^{\infty}=\frac{C}{s},\quad s>0\end{aligned} \tag{11.16}$$

Example 11.2 Find the Laplace transform of

$$L\left\{e^{at}\right\}=\int_{0}^{\infty}e^{-st}e^{at}dt \tag{11.17}$$

Solution: This can be solved directly by integration:

$$L\{e^{at}\} = \int_0^{\infty} e^{-st} e^{at} dt = -\frac{e^{-(s-a)t}}{s-a}\Bigg|_0^{\infty} \\ = \frac{1}{s-a}, \quad s > a \tag{11.18}$$

Thus, the formula can be obtained in a straightforward manner.

Example 11.3 Find the Laplace transform of the cosine

$$L\{\cos(at)\} = \int_0^{\infty} e^{-st} \cos(at) dt \tag{11.19}$$

Solution: We have already seen from Example 11.2 that an exponential function can be integrated readily. We can first express the cosine function by Euler's formula as

$$L[\cos(at)] = \int_0^{\infty} \frac{(e^{iat} + e^{-iat})}{2} e^{-st} dt = \frac{1}{2}\left[\frac{1}{s-ia} - \frac{1}{s+ia}\right] \\ = \frac{s}{s^2 + a^2} \tag{11.20}$$

Note that the order of s is higher in the denominator than in the numerator, and this observation also applies to the former two examples.

Example 11.4 Find the Laplace transform of the sine

$$F(s) = L\{\sin(at)\} = \int_0^{\infty} e^{-st} \sin atdt \tag{11.21}$$

Solution: Although we have shown in Example 11.3 that the Laplace transform of the cosine can be evaluated using an exponential function via Euler's formula. In general, we can also use this technique for (11.21), but we will proceed following integration by parts here

$$F(s) = \left[-(e^{-st} \cos at)/a\Big|_0^{\infty} - \frac{s}{a}\int_0^{\infty} e^{-st} \cos at \right] \\ = \frac{1}{a} - \frac{s}{a}\left[\int_0^{\infty} e^{-st} \cos at\right] \tag{11.22}$$

Reapplying integration by parts again, we have

$$F(s) = \frac{1}{a} - \frac{s}{a}\left[(e^{-st} \sin at)/a\Big|_0^{\infty} + \frac{s}{a}\int_0^{\infty} e^{-st} \sin at \right] \\ = \frac{1}{a} - \frac{s^2}{a^2} F(s) \tag{11.23}$$

We observe that the unknown appears on the right hand side as well. We can put all unknowns on the left hand side to obtain

$$F(s) = \frac{a}{s^2 + a^2}, \quad s > 0 \tag{11.24}$$

This is the required formula.

Example 11.5 Find the Laplace transform of the Heaviside step function:

$$H(t-c) = \begin{cases} 0, & t < c \\ 1, & t \geq c \end{cases} \tag{11.25}$$

This is the Heaviside step function.

Solution: Substitution of (11.25) into (11.6) gives

$$L\{H(t-c)\} = \int_0^\infty e^{-st} H(t-c) dt = \int_c^\infty e^{-st} dt \tag{11.26}$$

This can be integrated directly as

$$L\{H(t-c)\} = \int_c^\infty e^{-st} dt = -\frac{1}{s} e^{-st} \bigg|_c^\infty = \frac{e^{-cs}}{s} \tag{11.27}$$

One of the main applications of the Laplace transform is to solve ODEs. Similar to the Fourier transform, the Laplace transform can convert differentiation to an algebraic equation. In particular, we can find the following transform of differentiation of a function:

$$\begin{aligned} L\left[\frac{df}{dt}\right] &= f(t)e^{-st}\Big|_0^\infty - \int_0^\infty f(t)\left[-se^{-st}\right] dt \\ &= 0 - f(0) + s\int_0^\infty f(t)e^{-st} dt \end{aligned} \tag{11.28}$$

Therefore, we get

$$L\left[\frac{df(t)}{dt}\right] = sF(s) - f(0) \tag{11.29}$$

The process of integration by parts can be repeated to get

$$L\left[\frac{d^2 f(t)}{dt^2}\right] = s^2 F(s) - sf(0) - f'(0) \tag{11.30}$$

$$L\left[\frac{d^3 f(t)}{dt^3}\right] = s^3 F(s) - s^2 f(0) - sf'(0) - f''(0) \tag{11.31}$$

$$L\left[\frac{d^n f(t)}{dt^n}\right] = s^n F(s) - s^{n-1} f(0) - s^{n-2} f'(0) - \ldots - f^{(n-1)}(0) \tag{11.32}$$

Using these formulas, we can easily convert linear ODEs to algebraic equations. Without going into detail, we also record the following formulas:

$$L\left[t^n f(t)\right]=(-1)^n \frac{d^n F(s)}{ds^n}=(-1)^n F^{(n)}(s) \tag{11.33}$$

$$L\left[\frac{f(t)}{t}\right]=\int_s^{\infty} F(u)du \tag{11.34}$$

These formulas can convert ODEs with non-constant coefficients to ODEs with constant coefficients.

Before we apply to these formulas to solve ODEs, we note that

$$\begin{aligned} L\{c_1 f(t)+c_2 g(t)\} &= c_1\int_0^{\infty} e^{-st} f(t)dt + c_2\int_0^{\infty} e^{-st} g(t)dt \\ &= c_1 L\{f(t)\} + c_2 L\{g(t)\} \end{aligned} \tag{11.35}$$

This is the property of a linear operator. Similarly, we can also show that the inverse Laplace transform is also a linear operator. If we define $F(s)$ as a sum of a number of transformed functions:

$$F(s)=F_1(s)+F_2(s)+\cdots+F_n(s), \tag{11.36}$$

we have

$$f(t)=L^{-1}\{F(s)\}=L^{-1}\{F_1(s)\}+\cdots+L^{-1}\{F_n(s)\}. \tag{11.37}$$

There is also a shifting property of the Laplace transform:

$$L\left\{e^{ct} f(t)\right\}=\int_0^{\infty} e^{-st}e^{ct} f(t)dt=\int_0^{\infty} e^{-(s-c)t} f(t)dt=F(s-c). \tag{11.38}$$

where

$$L\{f(t)\}=F(s). \tag{11.39}$$

We now consider a very powerful theorem called the convolution theorem. Consider the following inversion of the Laplace transform:

$$\begin{aligned} \frac{1}{2\pi i}\int_{\gamma-i\infty}^{\gamma+i\infty} e^{st}F(s)G(s)ds &= \frac{1}{2\pi i}\int_{\gamma-i\infty}^{\gamma+i\infty} e^{st}F(s)\int_0^{\infty} g(\tau)e^{-s\tau}d\tau ds \\ &= \int_0^{\infty} g(\tau)\frac{1}{2\pi i}\int_{\gamma-i\infty}^{\gamma+i\infty} F(s)e^{s(t-\tau)}dsd\tau \\ &= \int_0^{\infty} g(\tau)f(t-\tau)d\tau \end{aligned} \tag{11.40}$$

In obtaining (11.40), we have reversed the order of integration. Therefore, we have

$$L^{-1}\{F(s)G(s)\}=\int_0^{\infty} g(\tau)f(t-\tau)d\tau \tag{11.41}$$

This is the convolution theorem and its application is demonstrated below.

Example 11.6 Find the inverse Laplace transform of the following function in the transformed space:

$$H(s)=\frac{a}{s^2(s^2+a^2)} \tag{11.42}$$

Solution: We can interpret F and G as

$$F(s)=\frac{1}{s^2},\quad G(s)=\frac{a}{s^2+a^2} \tag{11.43}$$

To find the inversion, we first note that

$$L\{t^n\}=\int_0^\infty t^n e^{-st}dt=-\frac{1}{s}[t^n e^{-st}\Big|_0^\infty-\int_0^\infty nt^{n-1}e^{-st}dt] \tag{11.44}$$

By applying L'Hôpital's rule repeatedly, the boundary term for $t\to\infty$ is found to be identically zero:

$$\lim_{t\to\infty}[\frac{t^n}{e^{st}}]=\lim_{t\to\infty}\frac{nt^{n-1}}{se^{st}}=\cdots=\lim_{t\to\infty}\frac{n!}{s^n e^{st}}=0 \tag{11.45}$$

Thus, we have

$$L\{t^n\}=\int_0^\infty t^n e^{-st}dt=\frac{n}{s}\int_0^\infty t^{n-1}e^{-st}dt \tag{11.46}$$

Repeating the integration by parts n times, we obtain

$$L\{t^n\}=\int_0^\infty t^n e^{-st}dt=\frac{n!}{s^n}\int_0^\infty e^{-st}dt=\frac{n!}{s^{n+1}} \tag{11.47}$$

Applying of formula (11.47), we have

$$L^{-1}\{\frac{1!}{s^2}\}=t \tag{11.48}$$

In view of this result and the result of Example 11.4, we can apply the convolution theorem to get

$$L^{-1}\{H(s)\}=L^{-1}\{(\frac{1}{s^2})\frac{a}{s^2(s^2+a^2)}\}=L^{-1}\{F(s)G(s)\}=\int_0^t(t-\tau)\sin(a\tau)d\tau \tag{11.49}$$

This is the required inverse of $H(s)$ in terms of integral.

Example 11.7 Solve the following ODE with prescribed initial conditions:

$$y''+y=\sin 2t,\quad y(0)=2,\ y'(0)=1 \tag{11.50}$$

Solution: Applying the Laplace transform to both sides of the ODE, we have

$$L\{\frac{d^2y}{dt^2}\}+L\{y\}=L\{\sin 2t\} \tag{11.51}$$

Using the results from (11.24) and (11.32), we get

$$\left[s^2L\{y\}-sy(0)-y'(0)\right]+L\{y\}=L\{\sin 2t\}=2/(s^2+4) \tag{11.52}$$

Letting $L\{y\}=Y(s)$ and grouping terms, we get

$$\left(s^2+1\right)Y(s)-sy(0)-y'(0)=2/(s^2+4) \tag{11.53}$$

Substitution of the boundary conditions gives

$$\left(s^2+1\right)Y(s)-2s-1=2/(s^2+4) \tag{11.54}$$

Solving for Y we find

$$Y(s)=\frac{2s^3+s^2+8s+6}{(s^2+1)(s^2+4)} \tag{11.55}$$

Applying a partial fraction, we can divide the right hand side of (11.55) as

$$Y(s)=\frac{2s^3+s^2+8s+6}{(s^2+1)(s^2+4)}=\frac{As+B}{s^2+1}+\frac{Cs+D}{s^2+4} \tag{11.56}$$

Balancing terms of the numerators on both sides of (11.56), we have

$$\begin{aligned} 2s^3+s^2+8s+6 &= (As+B)(s^2+4)+(Cs+D)(s^2+1) \\ &= (A+C)s^3+(B+D)s^2+(4A+C)s+(4B+D) \end{aligned} \tag{11.57}$$

Solving for the constants, we get

$$A=2, \quad B=5/3, \quad C=0, \quad D=-2/3 \tag{11.58}$$

Thus, we have

$$Y(s)=\frac{2s}{s^2+1}+\frac{5/3}{s^2+1}-\frac{2/3}{s^2+4} \tag{11.59}$$

Taking the inverse of the Laplace transform, we find

$$L^{-1}\{Y(s)\}=L^{-1}\{\frac{2s}{s^2+1}\}+L^{-1}\{\frac{5/3}{s^2+1}\}-L^{-1}\{\frac{2/3}{s^2+4}\} \tag{11.60}$$

Applying the results from Examples 11.3 and 11.4, we obtain the final result

$$y(t)=2\cos t+\frac{5}{3}\sin t-\frac{1}{3}\sin 2t \tag{11.61}$$

11.2.2 Fourier Transform

The Fourier transform is one of the most important integral transform techniques. It can be used to solve linear differential equations in Cartesian coordinates, and has been widely used in many engineering problems. For example, FFT (Fast Fourier Transform) is a standard technique in analyzing vibration data. There is a close relation between Fourier series expansion and Fourier transform. Mathematically, for differential equations defined for finite domain eigenvalues and eigenfunctions can be found and used to solve boundary value problems for such finite domain. These eigenvalues are discrete and not continuous. Fourier series expansion is the most fundamental series expansion (see Chapter 10). For the infinite domain, the spectrum of eigenvalues is continuous, and integral transform needs to be used. Fourier transform is the most important transform technique for Cartesian coordinates.

The kernel for the Fourier transform is

$$K(s,t)=e^{-ist} \tag{11.62}$$

For this kernel, the pair of Fourier and inverse Fourier transforms are

$$F(s)=\Im[f(t)]=\frac{1}{\sqrt{2\pi}}\int_{-\infty}^{\infty} e^{ist} f(t)dt, \tag{11.63}$$

$$f(t) = \Im^{-1}[F(s)] = \frac{1}{\sqrt{2\pi}} \int_{-\infty}^{\infty} e^{-ist} F(s) ds, \tag{11.64}$$

Be cautious that there are many slightly different definitions adopted in the literature, mainly regarding the constant $(2\pi)^{1/2}$. Here are the most popular definitions of Fourier transform:

$$F(s) = \Im[f(t)] = \frac{1}{\alpha} \int_{-\infty}^{\infty} e^{i\beta st} f(t) dt, \tag{11.65}$$

where α and β are

$$\alpha = \sqrt{2\pi}, \quad \beta = \pm 1 \tag{11.66}$$

$$\alpha = 1, \quad \beta = \pm 2\pi \tag{11.67}$$

$$\alpha = 1, \quad \beta = \pm 1 \tag{11.68}$$

There are six popular choices, and we are using $\alpha = (2\pi)^{1/2}$ and $\beta = -1$. Look at the definition of the Fourier transform carefully before using any table of transforms prepared in textbooks or derived by others. For real functions, we can take the real part and imaginary part of the above transform to give the following Fourier cosine and sine transforms:

$$F(\xi) = \Im_c[f(x)] = \sqrt{\frac{2}{\pi}} \int_0^{\infty} \cos(\xi x) f(x) dx, \tag{11.69}$$

$$f(x) = \Im_c^{-1}[F(\xi)] = \sqrt{\frac{2}{\pi}} \int_0^{\infty} \cos(\xi x) F(\xi) d\xi \tag{11.70}$$

$$F(\xi) = \Im_s[f(x)] = \sqrt{\frac{2}{\pi}} \int_0^{\infty} \sin(\xi x) f(x) dx, \tag{11.71}$$

$$f(x) = \Im_s^{-1}[F(\xi)] = \sqrt{\frac{2}{\pi}} \int_0^{\infty} \sin(\xi x) F(\xi) d\xi \tag{11.72}$$

They are useful in solving one-dimensional wave problems in a semi-infinite line subject to different boundary conditions. More detailed discussions on these transforms are found in Sneddon (1951), who reported their applications to solve vibration problems, heat conduction problems, hydrodynamic problems, nuclear physics, and elastic stress of 2-D or axisymmetric solids. For a table of Fourier transforms, readers should consult Erdelyi (1953) (i.e., Bateman Manuscript Project).

Here we will only demonstrate the technique in solving a simple PDE problem. Figure 11.1 shows a half-plane with a prescribed function f(x) on the surface on $y = 0$.

Mathematically, the problem is defined as:

$$\nabla^2 u(x,y) = \frac{\partial^2 u}{\partial x^2} + \frac{\partial^2 u}{\partial y^2} = 0, \quad 0 < y < \infty, -\infty < x < \infty \tag{11.73}$$

with boundary conditions

$$u(x,0) = f(x), \quad u \to 0, \quad \text{as } r = \sqrt{x^2 + y^2} \to \infty \tag{11.74}$$

The surface function $f(x)$ and its derivative $f'(x)$ must approach zero as $r \to \infty$. Multiplying the Laplace equation given in (11.73) by e^{isx} and integrating with respect to x from $-\infty$ to ∞ (or equivalently taking the Fourier transform), we get

$$\Im[\frac{\partial^2 u}{\partial x^2}] + \Im[\frac{\partial^2 u}{\partial y^2}] = \Im[\frac{\partial^2 u}{\partial x^2}] + \frac{d^2}{dy^2}\Im[u] = \Im[\frac{\partial^2 u}{\partial x^2}] + \frac{d^2 U(s,y)}{dy^2} = 0 \quad (11.75)$$

The first term can be found by integration by parts as:

$$\begin{aligned}\Im[\frac{\partial^2 u}{\partial x^2}] &= \int_{-\infty}^{\infty} \frac{\partial^2 u}{\partial x^2} e^{isx} dx = [\frac{\partial u}{\partial x} e^{isx}]_{-\infty}^{\infty} - is\int_{-\infty}^{\infty} \frac{\partial u}{\partial x} e^{isx} dx \\ &= -is[ue^{isx}]_{-\infty}^{\infty} + (-is)^2 \int_{-\infty}^{\infty} ue^{isx} dx = (-is)^2 U\end{aligned} \quad (11.76)$$

All boundary terms vanish in view of the boundedness of the solution and its derivative at infinity given in (11.74). Thus, the PDE becomes an ODE as:

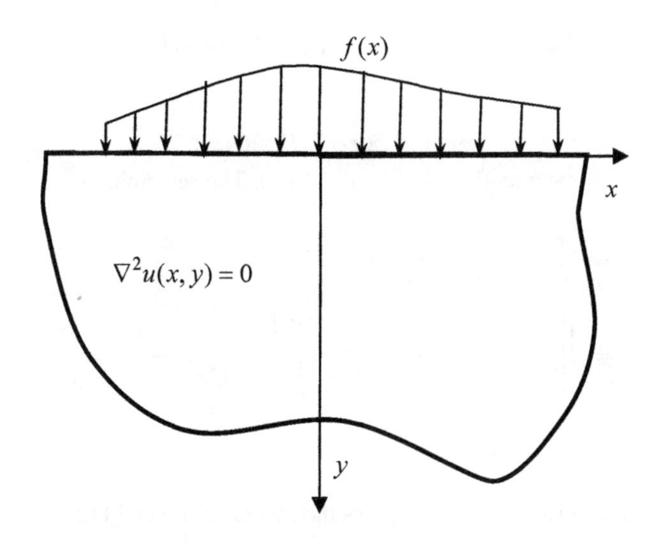

Figure 11.1 A half-plane satisfying the Laplace equation with a prescribed function f(x) on the surface

$$-s^2 U + \frac{d^2 U(s,y)}{dy^2} = 0 \quad (11.77)$$

Applying the Fourier transform to the boundary condition gives

$$\Im\{u(x,0)\} = U(s,0) = \Im\{f(x)\} = F(s) \quad (11.78)$$

The solution of (11.77) is

$$U = Ae^{|s|y} + Be^{-|s|y} \quad (11.79)$$

The boundary condition requires $A = 0$, and thus we have

$$U = Be^{-|s|y} \quad (11.80)$$

Enforcing the boundary condition given in (11.78), we obtain

$$U = F(s)e^{-|s|y} \tag{11.81}$$

To find the solution, we are going to see that the Fourier transform of an integral can result in a very powerful theorem called the convolution theorem or Faltung theorem (such theorems also exist for other transforms discussed in this chapter):

$$\Im\{f * g(x)\} = \Im\{\int_{-\infty}^{\infty} f(x-u)g(u)du\} = \frac{1}{\sqrt{2\pi}}\int_{-\infty}^{\infty}\int_{-\infty}^{\infty} f(x-u)g(u)du\, e^{-isx}dx \tag{11.82}$$

Here the asterisk represents the integral involving the two functions. We then apply the following change of variables:

$$v = x - w, \quad w = u \tag{11.83}$$

It is straightforward to see that the Jacobian is -1, thus we have

$$\begin{aligned}\Im\{f * g(x)\} &= \frac{1}{\sqrt{2\pi}}\int_{-\infty}^{\infty}\int_{-\infty}^{\infty} f(v)g(w)e^{-is(v+w)}du\,dx \\ &= \frac{1}{\sqrt{2\pi}}\int_{-\infty}^{\infty} f(v)e^{-isv}dv\int_{-\infty}^{\infty} g(w)e^{-isw}\,dw = \sqrt{2\pi}F(s)G(s)\end{aligned} \tag{11.84}$$

Therefore, we get

$$f * g(x) = \sqrt{2\pi}\,\Im^{-1}\{F(s)G(s)\} \tag{11.85}$$

Using this idea, we can identify $G(s)$ from (11.81). The remaining job is to find the following inversion:

$$\begin{aligned}\Im^{-1}\{e^{-|s|y}\} &= \frac{1}{\sqrt{2\pi}}\{\int_{-\infty}^{0} e^{-isx}e^{sy}ds + \int_{0}^{\infty} e^{-isx}e^{-sy}ds\} \\ &= \frac{1}{\sqrt{2\pi}}\left\{\left[\frac{e^{s(y-ix)}}{y-ix}\right]_{-\infty}^{0} - \left[\frac{e^{-s(y+ix)}}{y+ix}\right]_{0}^{\infty}\right\} = \frac{1}{\sqrt{2\pi}}\left\{\frac{1}{y-ix}+\frac{1}{y+ix}\right\} \\ &= \sqrt{\frac{2}{\pi}}(\frac{y}{y^2+x^2})\end{aligned} \tag{11.86}$$

Using (11.85) and (11.86), we can express the inversion of (11.81) as

$$\begin{aligned}u(x,y) = \Im^{-1}\{F(s)e^{-|s|y}\} &= \frac{1}{\sqrt{2\pi}}\int_{-\infty}^{\infty} f(x-\zeta)\sqrt{\frac{2}{\pi}}(\frac{y}{y^2+\zeta^2})d\zeta \\ &= \frac{y}{\pi}\int_{-\infty}^{\infty}\frac{f(x-\zeta)}{y^2+\zeta^2}d\zeta\end{aligned} \tag{11.87}$$

Therefore, the problem is solved in terms of the given function f. In conclusion, we also find the following formulas as by-products:

$$\Im[\frac{d^n u}{dx^n}] = (-is)^n U(s) \tag{11.88}$$

$$\Im^{-1}[F(s)G(s)] = \frac{1}{\sqrt{2\pi}}\int_{-\infty}^{\infty} f(x-u)g(u)du \tag{11.89}$$

$$\Im^{-1}\{e^{-|s|y}\} = \sqrt{\frac{2}{\pi}}(\frac{y}{y^2+x^2}) \quad (11.90)$$

The first one is the main reason, and it is the reason why we can convert differentiation to algebraic manipulation and thus solve PDEs. The second one is called the convolution theorem and is a very powerful technique. For the given example, we have yet provided the exact form of the given function f, but we have found a general formula for the result in terms of its integral. If the given function f is so complicated that we cannot integrate analytically, we can always integrate the resulting integral numerically. We can also generalize (without proof) here that the Fourier transform is applicable to PDEs formulated in the Cartesian coordinate system, and the variable of the problem is of infinite extent (see the range of x: $-\infty < x < \infty$). Therefore, the Fourier transform is suitable for solving linear PDEs with a domain of half-plane, half-space, full-plane, or half-space and formulated in Cartesian coordinates.

11.2.3 Hankel Transform

For potential problems that can be formulated in cylindrical coordinates, the Hankel transform can be applied to convey the radial dependency to algebraic equations. The kernel in the Hankel transform can be expressed as:

$$K(s,t) = J_\nu(st)t \quad (11.91)$$

With this kernel, the resulting Hankel transform and its inversion are

$$F(\xi) = H[f(x)] = \int_0^\infty xf(x)J_\nu(\xi x)dx, \quad (11.92)$$

$$F(x) = H^{-1}[F(\xi)] = \int_0^\infty \xi F(\xi)J_\nu(\xi x)d\xi \quad (11.93)$$

Many problems in cylindrical coordinates can be solved by Hankel transform. It can be shown that the Hankel transform of order zero ($\nu = 0$) is equivalent to a double Fourier transform for axisymmetric problems in Cartesian coordinates.

Let us consider the application of Hankel transform to the following PDE (e.g., Section 4.9.2 of Chau, 2013):

$$(\frac{\partial^2}{\partial r^2} + \frac{1}{r}\frac{\partial}{\partial r} - \frac{m^2}{r^2} + \frac{\partial^2}{\partial z^2})\psi_m = 0, \quad 0 < r < \infty \quad (11.94)$$

where m is an integer.

This is the Laplace equation in cylindrical coordinates for m-harmonics in circumferential coordinates (i.e., θ) (see Figure 11.2). In particular, the Hankel transform pair of ψ_m are defined as:

$$\Psi_m(\xi,z) = \int_0^\infty r\psi_m(r,z)J_m(\xi r)dr \quad (11.95)$$

$$\psi_m(r,z) = \int_0^\infty \xi\Psi_m(\xi,z)J_m(\xi r)d\xi \quad (11.96)$$

Substituting (11.96) into (11.94), that is, we have

$$(\frac{\partial^2}{\partial r^2}+\frac{1}{r}\frac{\partial}{\partial r}-\frac{m^2}{r^2}+\frac{\partial^2}{\partial z^2})\int_0^\infty \xi \Psi_m(\xi,z)J_m(\xi r)d\xi=0 \tag{11.97}$$

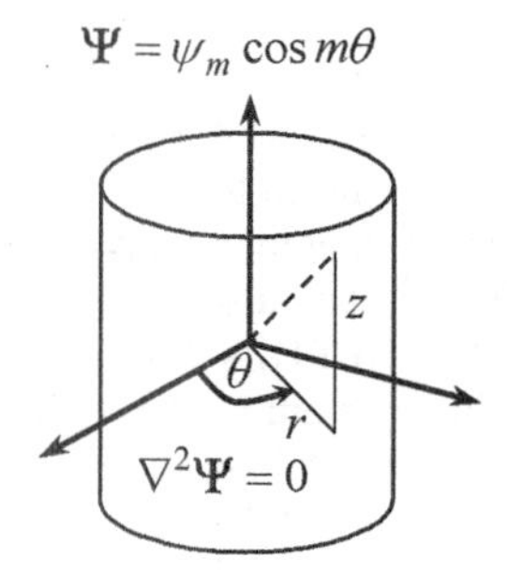

Figure 11.2 Cylindrical coordinates

Reversing the order of differentiation and integration, we obtain

$$\int_0^\infty \xi(\frac{d^2 J_m(\xi r)}{dr^2}+\frac{1}{r}\frac{dJ_m(\xi r)}{dr}-\frac{m^2 J_m(\xi r)}{r^2})\Psi_m(\xi,z)d\xi$$
$$+\int_0^\infty \xi\frac{d^2\Psi_m(\xi,z)}{dz^2}J_m(\xi r)d\xi=0 \tag{11.98}$$

Recall that the governing equation of $J_m(r\xi)$ is

$$\frac{d^2 J_m(\xi r)}{dr^2}+\frac{1}{r}\frac{dJ_m(\xi r)}{dr}+(\xi^2-\frac{m^2}{r^2})J_m(\xi r)=0 \tag{11.99}$$

Substitution of (11.99) into (11.98) yields

$$\int_0^\infty \xi[\frac{d^2\Psi_m(\xi,z)}{dz^2}-\xi^2\Psi_m(\xi,z)]J_m(\xi r)d\xi=0 \tag{11.100}$$

Therefore, the bracket within the integral must be zero

$$\frac{d^2\Psi_m(\xi,z)}{dz^2}-\xi^2\Psi_m(\xi,z)=0 \tag{11.101}$$

The solution of this ODE is easily obtained as:

$$\Psi_m(\xi,z)=A_m e^{\xi z}+B_m e^{-\xi z} \tag{11.102}$$

Finally, we have the solution of the PDE in (11.90) is

$$\psi_m(r,z)=\int_0^\infty \xi[A_m e^{\xi z}+B_m e^{-\xi z}]J_m(\xi r)d\xi \tag{11.103}$$

The constants, of course, need to be fixed by boundary conditions. In this example, we are content to get the general solution for (11.94). We observe that the Hankel transform is applied to the Laplace equation in cylindrical coordinates of infinite extent. In fact, one of the main differences between eigenfunction expansion discussed in Chapter 10 and integral transform is that eigenfunction expansion is used for the finite domain whereas integral transform is applicable to the infinite domain. In Muki's formulism for elastic solids in cylindrical coordinates is based on the Hankel transform (e.g., Chau, 2013). In fact, many physical problems in cylindrical coordinates are expressible in Bessel functions via Hankel transform.

11.2.4 Mellin Transform

For the two-dimensional elastic and potential problem in the shape of a wedge formulated in cylindrical coordinates, the so-called Mellin transform is found useful. It also finds applications in finding the sum of infinite series, the asymptotic value of an integral involving a large parameter, signal analysis, and imaging technique. Much of the results in the book of Bleistein and Handelsman on asymptotic analysis of integrals is in fact based on the Mellin transform. More detailed discussions on asymptotic analysis of integrals will be discussed in Chapter 12, but we will avoid the use of the Mellin transform in Chapter 12. In probability theory, the Mellin transform is an important tool in studying the distributions of products of two random variables. In particular, the Mellin transform of the product of two independent random variables equals the product of the Mellin transforms of the two variables. The Mellin transform is closely related to the two-sided Laplace transform.

The so-called Mellin transform has been considered by Laplace and used by Riemann in his study of the zeta function. It was, however, Mellin who provided a systematic formation of the transform and its application to solve ODEs and to estimate the value of integrals. Thus, the Mellin transform is named after Finnish mathematician Hjalmar Mellin, who was a student of Mittag-Leffler and Weierstrass. The kernel for the Mellin transform is

$$K(s,t)=t^{s-1} \tag{11.104}$$

The Mellin transform and its inversion are defined as:

$$F(s)=M[f(x)]=\int_0^{\infty} x^{s-1} f(x)dx, \tag{11.105}$$

$$f(x)=M^{-1}[F(s)]=\frac{1}{2\pi i}\int_{c-i\infty}^{c+i\infty} x^{-s}F(s)ds \tag{11.106}$$

where c is a constant that it lies on the right of all singularities of the kernel function. With a proper change of variables, the Mellin transform can be converted to a two-sided Laplace transform. In particular, a two-sided Laplace transform can be written as:

$$L[g(t)]=\int_{-\infty}^{\infty} g(t)e^{-st}dt \tag{11.107}$$

Let us consider the following change of variables:

$$t=-\ln x, \quad dt=-\frac{1}{x}dx \tag{11.108}$$

Applying (11.108) to (11.107), we get

$$L[g(t)]=-\int_{\infty}^{0} f(x)x^{s-1}dx=\int_0^{\infty} f(x)x^{s-1}dx=M\{f(x)\} \tag{11.109}$$

where we have defined

$$g(-\ln x)=f(x) \tag{11.110}$$

Thus, we have the following identity:

$$L\{g(-\ln x)\} = M\{f(x)\} = F(s) \tag{11.111}$$

Let us recall the inverse Laplace transform as

$$g(t) = \frac{1}{2\pi i}\int_{c-i\infty}^{c+i\infty} L\{g(t)\}e^{st}\,ds \tag{11.112}$$

Applying (11.108) in (11.112), we find

$$g(-\ln x) = \frac{1}{2\pi i}\int_{c-i\infty}^{c+i\infty} L\{g(-\ln x)\}e^{-s\ln x}\,ds \tag{11.113}$$

Substitution of (11.110) and (11.111) into (11.113) gives

$$f(x) = \frac{1}{2\pi i}\int_{c-i\infty}^{c+i\infty} M\{f(x)\}x^{-s}\,ds \tag{11.114}$$

This gives the inversion of the Mellin transform in (11.106).

There is a Parseval formula for the Mellin transform. To see this, let us assume the existence of two Mellin transforms of two functions as:

$$M[f(x)] = F(s), \quad M[g(x)] = G(s) \tag{11.115}$$

We now consider the following Mellin transform:

$$\begin{aligned} M[f(x)g(x)] &= \int_0^\infty f(x)g(x)x^{s-1}dx \\ &= \frac{1}{2\pi i}\int_0^\infty x^{s-1}g(x)\int_{c-i\infty}^{c+i\infty} F(z)x^{-z}\,dz\,dx \\ &= \frac{1}{2\pi i}\int_{c-i\infty}^{c+i\infty} F(z)dz\int_0^\infty g(x)x^{s-z-1}dx \\ &= \frac{1}{2\pi i}\int_{c-i\infty}^{c+i\infty} F(z)G(s-z)dz \end{aligned} \tag{11.116}$$

Substituting $s = 1$, we get Parseval's formula:

$$\int_0^\infty f(x)g(x)dx = \frac{1}{2\pi i}\int_{c-i\infty}^{c+i\infty} F(z)G(1-z)dz \tag{11.117}$$

Example 11.8 Consider the Mellin transform of the function

$$f(t) = e^{-pt} \tag{11.118}$$

Solution: The Mellin transform of $f(t)$ is

$$M[f(t)] = \int_0^\infty e^{-pt}t^{s-1}dt \tag{11.119}$$

This can be converted to Euler's gamma function by using the following change of variables

$$\zeta = pt, \quad d\zeta = pdt \tag{11.120}$$

Using this new variable we have

$$M[f(t)] = \frac{1}{p^s}\int_0^\infty e^{-\zeta}\zeta^{s-1}d\zeta = \frac{\Gamma(s)}{p^s} \tag{11.121}$$

One major application of the Mellin transform is of course to solve differential equations. Thus, we consider the following formula for taking the Mellin transform of the derivative of function *f*:

$$M[t^k \frac{d^k f(t)}{dt^k}] = (-1)^k (s)_k F(s) \tag{11.122}$$

where $(s)_k$ is the Pochhammer's symbol defined in Chapter 4 as:

$$(s)_k = s(s+1)\cdots(s+k-1) \tag{11.123}$$

To derive this formula, we can apply integration by parts to the following

$$\begin{aligned} M[t^k \frac{d^k f(t)}{dt^k}] &= \int_0^\infty t^k \frac{d^k f(t)}{dt^k} t^{s-1} dt \\ &= \{t^{k+s-1} f^{(k-1)}(t)\Big|_0^\infty - (k+s-1)\int_0^\infty t^{k+s-2} f^{(k-1)}(t)\, dt\} \end{aligned} \tag{11.124}$$

We can drop the boundary terms if $f \sim 1/t^\beta$ and $0 < \text{Re}(s) < \beta$. Repeating the integration by parts *k* times, we obtain

$$\begin{aligned} M[t^k \frac{d^k f(t)}{dt^k}] &= (-1)^k (k+s-1)(k+s)\cdots s \int_0^\infty t^{s-1} f(t)\, dt \\ &= (-1)^k (s)_k F(s) \end{aligned} \tag{11.125}$$

This completes the proof.

Example 11.9 Find the solution of the following problem of a wedge governed by the potential theory (see Figure 11.3):

$$\nabla^2 u = \frac{\partial^2 u}{\partial r^2} + \frac{1}{r}\frac{\partial u}{\partial r} + \frac{1}{r^2}\frac{\partial^2 u}{\partial \theta^2} = 0 \tag{11.126}$$

$$u(r, \pm\alpha) = f(r) \tag{11.127}$$

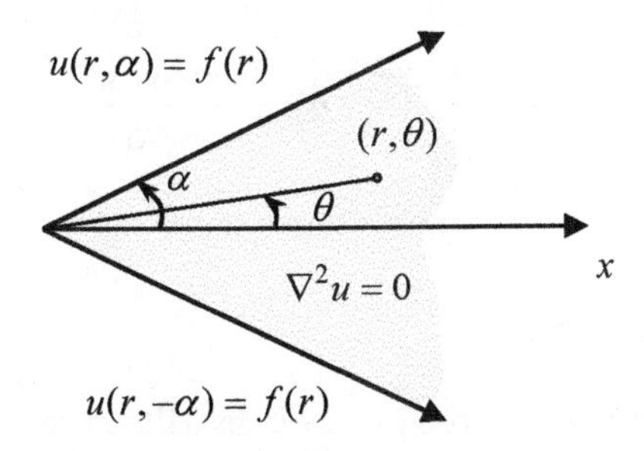

Figure 11.3 Wedge domain of potential theory

Solution: We are expecting

$$r \to 0, \quad u(r,\theta) = \text{bounded} \tag{11.128}$$

$$r \to \infty, \quad u(r,\theta) \sim \frac{1}{r^{\beta}}, \quad \beta > 0 \tag{11.129}$$

We can first rewrite the Laplacian as

$$r^2 \frac{\partial^2 u}{\partial r^2} + r\frac{\partial u}{\partial r} + \frac{\partial^2 u}{\partial \theta^2} = 0 \tag{11.130}$$

Applying the Mellin transform to (11.130), we get

$$\int_0^{\infty} r^{s-1}(r^2 \frac{\partial^2 u}{\partial r^2} + r\frac{\partial u}{\partial r} + \frac{\partial^2 u}{\partial \theta^2})dr = 0 \tag{11.131}$$

Substitution of (11.122) into (11.131) gives

$$s(s+1)U - sU + \frac{d^2 U}{d\theta^2} = 0 \tag{11.132}$$

Simplification of (11.132) gives

$$\frac{d^2 U}{d\theta^2} + s^2 U = 0 \tag{11.133}$$

The general solution is

$$U(s,\theta) = A(s)\cos(s\theta) + B(s)\sin(s\theta) \tag{11.134}$$

Boundary condition (11.127) leads to

$$U(s,\pm\alpha) = F(s) \tag{11.135}$$

Substitution of (11.134) into (11.135) leads to

$$U(s,\alpha) = A(s)\cos(s\alpha) + B(s)\sin(s\alpha) = F(s) \tag{11.136}$$

$$U(s,-\alpha) = A(s)\cos(s\alpha) - B(s)\sin(s\alpha) = F(s) \tag{11.137}$$

Solving A and B we get

$$A(s) = \frac{F(s)}{\cos(s\alpha)}, \quad B(s) = 0 \tag{11.138}$$

Therefore, we have

$$U(s,\theta) = \frac{F(s)\cos(s\theta)}{\cos(s\alpha)} \tag{11.139}$$

Finally, we get

$$u(r,\theta) = \frac{1}{2\pi i}\int_{c-i\infty}^{c+i\infty} r^{-s} \frac{F(s)\cos(s\theta)}{\cos(s\alpha)} ds \tag{11.140}$$

11.2.5 Hilbert Transform

The Hilbert transform was proposed in 1905 by Hilbert in considering a problem posed by Riemann. Subsequently, it was developed by Weyl, Schur, and Riesz. The kernel for the Hilbert transform is

$$K(s,t) = \frac{1}{\pi}\frac{1}{s-t} \tag{11.141}$$

$$F(\xi) = \mathscr{H}[f(x)] = \int_{-\infty}^{\infty} \frac{f(x)}{\pi(x-\xi)} dx, \tag{11.142}$$

$$f(x) = \mathscr{H}^{-1}[F(\xi)] = -\int_{-\infty}^{\infty} \frac{F(\xi)}{\pi(\xi - x)} d\xi \tag{11.143}$$

where the integral is singular at $x = \xi$ and in the above formulas we have taken the Cauchy principal value as:

$$\int_{-\infty}^{\infty} \frac{f(x)}{\pi(x-\xi)} dx = \lim_{\varepsilon \to 0} [\int_{-\infty}^{\xi-\varepsilon} \frac{f(x)}{\pi(x-\xi)} dx + \int_{\xi+\varepsilon}^{\infty} \frac{f(x)}{\pi(x-\xi)} dx] \tag{11.144}$$

The term Hilbert transform was coined by Hardy in 1909. As shown in Mura (1987), Weertman (1996) and Chau (2013), the two-dimensional dislocation pile-up in elastic bodies can be formulated as Hilbert transform. As demonstrated in Broberg (1999) and Chau (2013), the two-dimensional crack problem can also be solved using the Hilbert transform in terms of dislocation pile-up. Its application is also found in Muskhelishvili's (1975) formalist for solving two-dimensional elastic problems using complex variables. If the limits of the integral are finite numbers, we have the finite Hilbert transform (Tricomi, 1957):

$$F(\xi) = \mathscr{F}[f(y)] = \int_{-1}^{1} \frac{f(x)}{\pi(x-\xi)} dx, \tag{11.145}$$

$$f(x) = \mathscr{F}^{-1}[F(\xi)] = -\frac{1}{\pi}\int_{-1}^{1} \sqrt{\frac{1-\xi^2}{1-x^2}} \frac{F(\xi)}{(\xi - x)} d\xi + \frac{C}{\sqrt{1-x^2}} \tag{11.146}$$

The proof of this result is found in Tricomi (1957). This transform arises from the analysis of an airfoil using aerodynamics (Tricomi, 1957), analysis of electronics (Nahin, 2006), and in the analysis of reflected SV waves (Ben-Menahem and Singh, 2000).

11.2.6 Other Transforms

The more popular types of integral transforms have been summarized in earlier sections. In order to provide a more comprehensive summary of the integral transform, a summary of eleven other integral transforms is given below.

(i) Hartley transform

$$F(\xi) = h[f(x)] = \frac{1}{\sqrt{2\pi}} \int_{-\infty}^{\infty} (\cos \xi x + \sin \xi x) f(x) dx \tag{11.147}$$

$$f(x) = h^{-1}[F(\xi)] = \frac{1}{\sqrt{2\pi}} \int_{-\infty}^{\infty} (\cos \xi x + \sin \xi x) F(\xi) d\xi \tag{11.148}$$

This was proposed by Hartley in 1942 as an alternative to the Fourier transform. It deals with real functions exclusively, compared to the traditional Fourier transform using complex functions.

(ii) K-transform

$$F(\xi)=K[f(x)]=\int_0^{\infty}K_\nu(x\xi)\sqrt{x\xi}\,f(x)dx \tag{11.149}$$

$$f(x)=K^{-1}[F(\xi)]=\frac{1}{\pi i}\int_{c-i\infty}^{c+i\infty}I_\nu(x\xi)\sqrt{x\xi}\,F(\xi)d\xi \tag{11.150}$$

where K_ν and I_ν are the modified Bessel functions. Table of K-transform can be found in Erdelyi (1954).

(iii) Kontorovich-Lebedev transform

$$F(\xi)=K_L[f(x)]=\int_0^{\infty}\frac{K_{i\xi}(x)}{x}f(x)dx \tag{11.151}$$

$$f(x)=K_L^{-1}[F(\xi)]=\frac{2}{\pi^2}\int_0^{\infty}\xi\sin(\pi\xi)K_{i\xi}(x)F(\xi)d\xi \tag{11.152}$$

where $K_{i\nu}$ is the modified Bessel function of the second kind of imaginary order. This was proposed by Kontorovich and Lebedev in 1938 for solving diffraction problems, and further detail can be found in Lebedev et al. (1965).

(iv) Mehler-Fock transform

$$F(\xi)=M_F[f(x)]=\int_0^{\infty}\sinh x P_{i\xi-1/2}^{m}(\cosh x)f(x)dx \tag{11.153}$$

$$f(x)=M_F^{-1}[F(\xi)]=\int_0^{\infty}\xi\tanh(\pi\xi)P_{i\xi-1/2}^{m}(\cosh x)F(\xi)d\xi \tag{11.154}$$

where the kernel is the associated Legendre polynomial of the first kind of imaginary order. This was proposed by Mehler in 1881 and its basic theorems were proved by Fock in 1943. Detail can be found in Nasim (1984) and Sneddon (1972).

(v) Weber-Orr transform

$$F(\xi)=W_O[f(x)]=\int_a^{\infty}\sqrt{x}[J_\nu(x\xi)Y_\nu(a\xi)-Y_\nu(x\xi)J_\nu(a\xi)]f(x)dx \tag{11.155}$$

$$f(x)=W_O^{-1}[F(\xi)]=\sqrt{x}\int_0^{\infty}\frac{J_\nu(x\xi)Y_\nu(a\xi)-Y_\nu(x\xi)J_\nu(a\xi)}{J_\nu^2(a\xi)+Y_\nu^2(a\xi)}F(\xi)d\xi \tag{11.156}$$

This was proposed by Weber in 1873 for considering problems of infinite regions outside a circular cylindrical hole. A similar result was re-discovered by Orr in 1909, and thus it is also referred to as the Weber-Orr integral transform (e.g., Olesiak, 1990). Rigorous justification for the method was done by Watson in 1944 and Titchmarsh in 1922.

(vi) Associated Weber transform

$$F(\xi)=W_{\mu\nu}[f(x)]=\int_0^{\infty}[J_\mu(x\xi)Y_\nu(a\xi)-Y_\mu(x\xi)J_\nu(a\xi)]xf(x)dx \tag{11.157}$$

$$f(x)=W_{\mu\nu}^{-1}[F(\xi)]=\int_0^{\infty}\frac{J_\mu(x\xi)Y_\nu(a\xi)-Y_\mu(x\xi)J_\nu(a\xi)}{J_\nu^2(a\xi)+Y_\nu^2(a\xi)}\xi F(\xi)d\xi \tag{11.158}$$

The term of associated Weber transform was apparently coined by Krajewski and Olesiak in 1982 (Nasim, 1989). It has been found useful in thermoelastic problems of an infinite domain containing a circular cylindrical hole (Olesiak, 1990).

(vii) Weierstrass transform

$$F(\xi) = W[f(x)] = \frac{1}{\sqrt{4\pi}} \int_{-\infty}^{\infty} e^{(\xi - x)^2/4} f(x) dx \tag{11.159}$$

$$f(x) = W^{-1}[F(\xi)] = \frac{1}{\sqrt{4\pi}} \int_{-\infty}^{\infty} e^{(x - i\xi)^2/4} F(i\xi) d\xi \tag{11.160}$$

This transform was also known as Gauss transform, Gauss-Weierstrass transform, or Hille transform. It is related to heat or diffusion problem.

(viii) Y-transform

$$F(\xi) = y_\nu[f(x)] = \int_0^{\infty} (x\xi)^{1/2} f(x) Y_\nu(x\xi) dx \tag{11.161}$$

$$f(x) = y_\nu^{-1}[F(\xi)] = \int_0^{\infty} (x\xi)^{1/2} F(\xi) \boldsymbol{H}_\nu(x\xi) d\xi \tag{11.162}$$

where $\boldsymbol{H}_\nu$ is the Struve function, which is a particular solution of a nonhomogeneous Bessel equation. It is useful for solving problems with singular behavior at the axis of symmetry. This is closely related to Hankel transform.

(ix) **H**-transform

$$F(\xi) = \mathscr{H}_\nu[f(x)] = \int_0^{\infty} (x\xi)^{1/2} f(x) \boldsymbol{H}_\nu(x\xi) dx \tag{11.163}$$

$$f(x) = \mathscr{H}_\nu^{-1}[F(\xi)] = \int_0^{\infty} (x\xi)^{1/2} F(\xi) Y_\nu(x\xi) d\xi \tag{11.164}$$

The Y- and **H**-transforms are complementary pairs of integral transforms.

(x) R-transform

$$F(\xi) = \mathscr{R}[f(x)] = \sqrt{\frac{2}{\pi}} \int_0^{\infty} f(x)\{\xi \cos(x\xi) + h \sin(x\xi)\} dx \tag{11.165}$$

$$f(x) = \mathscr{R}^{-1}[F(\xi)] = \sqrt{\frac{2}{\pi}} \int_0^{\infty} F(\xi)\{\frac{\xi \cos(x\xi) + h \sin(x\xi)}{\xi^2 + h^2}\} d\xi \tag{11.166}$$

This is for heat conduction problems with a boundary as

$$\frac{\partial f}{\partial x}(0, y, z) - hf(0, y, z) = \Psi(y, z) \tag{11.167}$$

where h is the parameter prescribed in the boundary condition.

(xi) Abel transform

$$F(\xi) = A[f(x)] = 2\int_{\xi}^{\infty} \frac{f(x)x}{\sqrt{x^2 - \xi^2}} dx \tag{11.168}$$

$$f(x) = A^{-1}[F(\xi)] = -\frac{1}{\pi}\int_x^{\infty} \frac{1}{\sqrt{\xi^2 - x^2}} \frac{dF}{d\xi} d\xi \tag{11.169}$$

The Abel integral transform is useful for axially or spherically symmetry problems.

11.2.7 Governing Equation of Kernel Functions

In this section, we will illustrate how to derive the proper integral transform for a differential equation. In particular, consider the following second order PDE for a semi-finite quarter space and boundary conditions as:

$$a(x)\frac{\partial^2 u}{\partial x^2} + b(x)\frac{\partial u}{\partial x} + c(x)u + \frac{\partial^2 u}{\partial y^2} = 0, \quad 0 \le x, y < \infty \tag{11.170}$$

$$u(0, y) = 0, \quad u(x, 0) = f(x) \tag{11.171}$$

$$R = \sqrt{x^2 + y^2} \to \infty, \quad u(x, y) \to 0 \tag{11.172}$$

The problem is illustrated in Figure 11.4. Multiplying (11.170) by a kernel function $K(x, \xi)$ and integrating the result from 0 to ∞ with respect to x, we have

$$\int_0^{\infty} \left\{ K(x,\xi)a(x)\frac{\partial^2 u}{\partial x^2} + K(x,\xi)b(x)\frac{\partial u}{\partial x} + K(x,\xi)c(x)u + K(x,\xi)\frac{\partial^2 u}{\partial y^2} \right\} dx = 0 \tag{11.173}$$

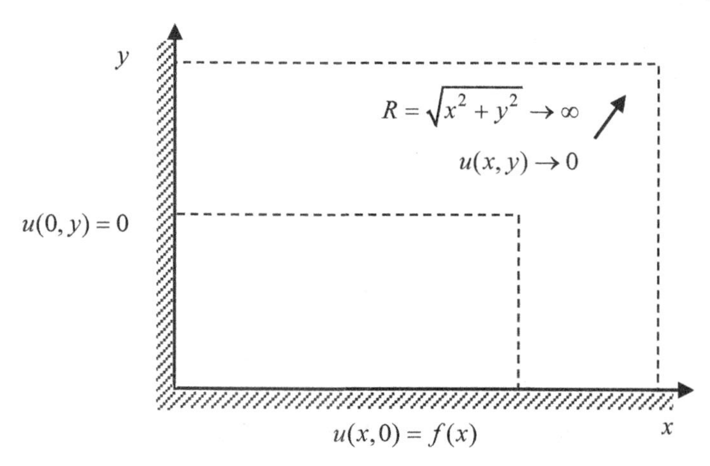

Figure 11.4 Domain of the PDE defined in (11.45)

At this moment, the kernel function is the unknown of the problem. We are going to find what condition needs to be satisfied by $K(x,\xi)$. Using integration by parts, the first term (11.173) can be integrated as

$$
\begin{aligned}
&\int_0^\infty K(x,\xi)a(x)\frac{\partial^2 u}{\partial x^2}dx \\
&= K(x,\xi)a(x)\frac{\partial u}{\partial x}\bigg|_0^\infty - \int_0^\infty \frac{\partial u}{\partial x}\frac{\partial}{\partial x}[K(x,\xi)a(x)]dx \\
&= K(x,\xi)a(x)\frac{\partial u}{\partial x}\bigg|_0^\infty - \left\{u\frac{\partial}{\partial x}[K(x,\xi)a(x)]\right\}_0^\infty + \int_0^\infty u\frac{\partial^2}{\partial x^2}[K(x,\xi)a(x)]dx \\
&= K(x,\xi)a(x)\frac{\partial u}{\partial x}\bigg|_0^\infty + \int_0^\infty u\frac{\partial^2}{\partial x^2}[K(x,\xi)a(x)]dx
\end{aligned}
\tag{11.174}
$$

We can impose an additional constraint on the boundary conditions for K as:

$$K(0,\xi)a(0)=0, \quad \frac{\partial u}{\partial x}(\infty,\xi)=0 \tag{11.175}$$

With these constraints, we have (11.174) becoming

$$\int_0^\infty K(x,\xi)a(x)\frac{\partial^2 u}{\partial x^2}dx = \int_0^\infty u\frac{\partial^2}{\partial x^2}[K(x,\xi)a(x)]dx \tag{11.176}$$

Similarly, we have the other terms being

$$
\begin{aligned}
\int_0^\infty K(x,\xi)b(x)\frac{\partial u}{\partial x}dx &= K(x,\xi)b(x)u\big|_0^\infty - \int_0^\infty u\frac{\partial}{\partial x}[K(x,\xi)b(x)]dx \\
&= -\int_0^\infty u\frac{\partial}{\partial x}[K(x,\xi)b(x)]dx
\end{aligned}
\tag{11.177}
$$

$$\int_0^\infty K(x,\xi)\frac{\partial^2 u}{\partial y^2}dx = \frac{\partial^2}{\partial y^2}\int_0^\infty K(x,\xi)u dx \tag{11.178}$$

Now we define the integral transform as

$$U(\xi,y) = \int_0^\infty K(x,\xi)u(x,y)dx \tag{11.179}$$

In view of the decay condition as $y \to \infty$, we can assume U has the following form:

$$U(\xi,y) = F(\xi)e^{-|\xi|y} \tag{11.180}$$

Substitution of (11.180) into the boundary condition given in the second equation of (11.171) leads to the following identity

$$F(\xi) = \int_0^\infty K(x,\xi)f(x)dx \tag{11.181}$$

The differential equation given in (11.170) becomes

$$\int_0^\infty u\left\{\frac{\partial^2 u}{\partial x^2}[K(x,\xi)a(x)] - \frac{\partial}{\partial x}[K(x,\xi)b(x)] + K(x,\xi)[c(x)+\xi^2]\right\}dx = 0 \tag{11.182}$$

Since u cannot be identically zero for all u, we must have

$$\frac{\partial^2}{\partial x^2}[K(x,\xi)a(x)] - \frac{\partial}{\partial x}[K(x,\xi)b(x)] + [c(x)+\xi^2]K(x,\xi) = 0 \tag{11.183}$$

Therefore, we have arrived at a differential equation for the kernel function $K(x,\xi)$. Depending on the exact mathematical form of functions $a(x)$, $b(x)$ and $c(x)$, this non-constant coefficient PDE is in general not easy to solve.

Example 11.10 For the special case of $a(x) = 1$, and $b(x) = c(x) = 0$, find the kernel function $K(x, \xi)$.

Solution: For this special case, we have the governing equation for $K(x, \xi)$ as:

$$\frac{\partial^2}{\partial x^2} K(x,\xi) + \xi^2 K(x,\xi) = 0 \tag{11.184}$$

The solution is clearly

$$K(x,\xi) = A_1 \sin(\xi x) + A_2 \cos(\xi x) \tag{11.185}$$

Substitution of (11.185) into (11.175) yields

$$K(x,\xi) = A_1 \sin(\xi x) \tag{11.186}$$

The proper integral transform for this problem is therefore

$$U(\xi, y) = \int_0^\infty \sin(x\xi) u(x,y) dx \tag{11.187}$$

Therefore, the Fourier sine transform can be used to solve the following Laplace equation:

$$\frac{\partial^2 u}{\partial x^2} + \frac{\partial^2 u}{\partial y^2} = 0, \quad 0 \le x, y < \infty \tag{11.188}$$

11.3 ABEL INTEGRAL EQUATION

The Abel integral equation is formulated as

$$\int_0^\rho \frac{d\varphi(t)}{dt} \frac{1}{\sqrt{\rho^2 - t^2}} dt = f(\rho) \tag{11.189}$$

where φ is the unknown to be determined. This integral is known as the Abel integral. The solution of it is well known and is given as

$$\varphi'(s) = \frac{2}{\pi} \frac{d}{ds} \int_0^s \frac{\rho f(\rho) d\rho}{\sqrt{s^2 - \rho^2}} \tag{11.190}$$

This equation appears naturally in crack problems (e.g., Chapter 5 of Mura, 1987). This integral was first considered by Abel in 1923 when he studied a mass sliding along a frictionless curved vertical plane under the influence of gravity such that the arrival time at the lowest point is independent of its starting point on this curve. The solution is called a tautochrone curve, as in Greek *tauto* has the same meaning as iso- or equal and *chrone* is time. The tautochrone curve is also the brachistochrone (shortest time in Greek) curve. That is also the curve that takes the shortest time to travel to the lowest point. It is this problem that leads to the latter development of calculus of variations by Euler and Lagrange.

In particular, Figure 11.5 shows the tautochrone problem of a sliding bead along a frictionless wire.

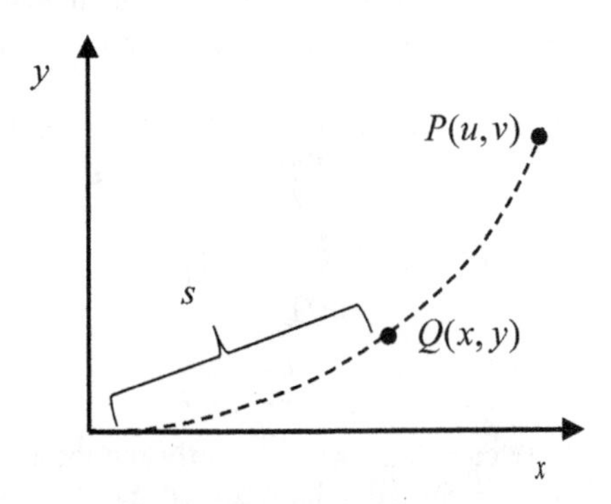

Figure 11.5 Tautochrone problem of a frictionless sliding mass along a curve with the same arrival time independent of the starting point

Using conservation of energy, we have

$$mgv + 0 = mgy + \frac{1}{2}m(\frac{ds}{dt})^2 \tag{11.191}$$

Rearranging (11.191), we get

$$\frac{ds}{dt} = -\sqrt{2g(v-y)} \tag{11.192}$$

In obtaining (11.192), we have taken the negative sign in front of the square root in view of the fact that s shown in Figure 11.5 is a decreasing function of t. Integrating both sides, we have

$$T = \int_0^T dt = \int_v^0 \frac{-ds}{\sqrt{2g(v-y)}} = \int_0^v \frac{ds}{\sqrt{2g(v-y)}} \tag{11.193}$$

Clearly, the sliding distance s is an unknown function of the shape of the unknown curve, and we assume

$$ds = F(y)dy \tag{11.194}$$

where $F(y)$ is an unknown function to be determined such that the arrival time T is a constant. Thus, we have

$$T = \frac{1}{\sqrt{2g}} \int_0^v \frac{F(y)dy}{\sqrt{(v-y)}} \tag{11.195}$$

This is the Abel type of integral equation. Niels Henrik Abel was a Norwegian mathematician and was born in 1802 and passed away at the age of 26. Abel had shown that it is impossible to solve the quintic equation (5th order algebraic equation) in radical forms. He also discovered elliptic function, which was subsequently improved by Jacobi and is now called the Jacobi elliptic function (see Chapter 1 and Appendix D). Independent of Galois, he also developed the group theory. He lived in poverty for his whole life, and passed away just two days short of receiving the good news from his friend Crelle (editor of Crelle's journal) that he was appointed professor at University of Berlin. The renowned French mathematician Charles Hermite (the discoverer of Hermite polynomials) remarked on Abel's six years of works as "Abel left mathematicians enough to keep them busy for five hundred years." Another renowned French mathematician Adrien-Marie Legendre (the discoverer of Legendre polynomials) commented, "What a head the young Norwegian has!"

We now go to back to the technical side of (11.189) and (11.190). To prove the validity of (11.190), we consider the following generalized Abel integral equation:

$$\int_a^x \frac{g(t)dt}{[h(x)-h(t)]^{\alpha}} = f(x) \tag{11.196}$$

where

$$0<\alpha<1, \quad a<x<b \tag{11.197}$$

In addition, we consider the case that $h(t)$ is strictly monotonically increasing and its first derivative is not zero for all t between a and b. Note that $h(t) = t^2$ in (11.189).

To start the proof, we consider the following related integral:

$$I(x) = \int_a^x \frac{h'(u)f(u)du}{[h(x)-h(u)]^{1-\alpha}} \tag{11.198}$$

Substitution of (11.196) into (11.198) results in

$$\begin{aligned} I(x) &= \int_a^x \frac{h'(u)}{[h(x)-h(u)]^{1-\alpha}} \int_a^u \frac{g(t)dt}{[h(u)-h(t)]^{\alpha}} du \\ &= \int_a^x g(t) \int_t^x \frac{h'(u)du}{[h(x)-h(u)]^{1-\alpha}[h(u)-h(t)]^{\alpha}} dt \end{aligned} \tag{11.199}$$

The last line of (11.199) is the result of reversing the order of integration. In addition, the upper limit for variable t has been set to x (instead of using u).

Applying the following change of variables, we get

$$h(u) = \xi, \quad \rho_2 = h(x), \quad \rho_1 = h(t) \tag{11.200}$$

$$\begin{aligned} \int_t^x \frac{h'(u)}{[h(x)-h(u)]^{1-\alpha}[h(u)-h(t)]^{\alpha}} du &= \int_{h(t)}^{h(x)} \frac{d\xi}{(\rho_2-\xi)^{1-\alpha}(\xi-\rho_1)^{\alpha}} \\ &= \int_{\rho}^{\rho_2} \frac{d\xi}{(\rho_2-\xi)^{1-\alpha}(\xi-\rho_1)^{\alpha}} \end{aligned} \tag{11.201}$$

This integral cannot be found in most handbooks on tables of integrations, including Gradshteyn and Ryzhik (1980). However, the integration given in (11.201) can be evaluated through an integral covered in Chapter 1 and it is also reported on p. 118 of Whittaker and Watson (1927), In particular, we recall the following result from Chapter 1:

$$I_1 = \int_0^{\infty} \frac{x^{\alpha-1}dx}{1+x} = \frac{\pi}{\sin(\alpha\pi)} \tag{11.202}$$

Let us apply a change of variables with the new variable being y as:

$$x = \frac{z-y}{y-\xi} \tag{11.203}$$

Then, the differential of x is

$$dx = \frac{\xi - z}{(y-\xi)^2} dy \tag{11.204}$$

Noting that $x = 0$ gives $y = z$, and $x = \infty$ gives $y = \xi$, we can convert (11.202) to

$$I_1 = \int_{\xi}^{z} \frac{dy}{(z-y)^{1-\alpha}(y-\xi)^{\alpha}} = \frac{\pi}{\sin(\alpha\pi)} \tag{11.205}$$

This is precisely the integral we got in (11.201). Therefore, substitution of this result into (11.201) and (11.199) gives

$$I(x) = \int_a^x g(t)dt \frac{\pi}{\sin(\pi\alpha)} = \int_a^x \frac{h'(u)f(u)du}{[h(x)-h(u)]^{1-\alpha}} \tag{11.206}$$

Differentiating (11.206) with respect to x and using the Leibniz rule of differentiation on integral, we get

$$g(x) = \frac{\sin(\pi\alpha)}{\pi} \frac{d}{dx} \int_a^x \frac{h'(u)f(u)du}{[h(x)-h(u)]^{1-\alpha}} \tag{11.207}$$

This is the solution of (11.196) and in addition we now make the following identifications:

$$h(u) = u^2, \quad g(\rho) = \varphi'(\rho), \quad \alpha = 1/2, \quad x = \rho, \quad a = 0 \tag{11.208}$$

We finally obtain

$$\varphi'(\rho) = \frac{2}{\pi} \frac{d}{d\rho} \int_0^{\rho} \frac{\xi f(\xi) d\xi}{\sqrt{\rho^2 - \xi^2}} \tag{11.209}$$

Integrating on both sides gives

$$\varphi(\rho) = \frac{2}{\pi} \int_0^{\rho} \frac{\xi f(\xi) d\xi}{\sqrt{\rho^2 - \xi^2}} \tag{11.210}$$

This completes the proof of (11.190).

11.4 FREDHOLM INTEGRAL EQUATION

The linear Fredholm integral equation of the first kind is defined as

$$f(x) = \int_a^b K(x,s)\phi(s)ds \tag{11.211}$$

where ϕ is the unknown in this equation. The linear Fredholm integral equation of the second kind is defined as

$$\phi(x) = f(x) + \lambda \int_a^b K(x,s)\phi(s)ds \tag{11.212}$$

where f is a given continuous function, ϕ is the unknown in this equation, and λ is a parameter of the equation. More generally, if the unknown ϕ does not appear in the integrand linearly, we have the nonlinear Fredholm integral equation of the second kind as

$$\phi(x) = f(x) + \lambda \int_a^b K[x, s, \phi(s)] ds \tag{11.213}$$

11.4.1 Solution of the Fredholm Integral Equation of the Second Kind

The following Fredholm formula gives the solution of (11.212)

$$\phi(x) \equiv f(x) + \frac{\lambda}{D(\lambda)} \int D(x, y; \lambda) f(y) dy \tag{11.214}$$

where the $D(\lambda)$ and $D(x, y; \lambda)$ are defined as

$$D(\lambda) \equiv 1 + \sum_{m=1}^{\infty} \frac{(-\lambda)^m}{m!} \iint \dots \int K \begin{pmatrix} \xi_1, \xi_2, \dots, \xi_m \\ \xi_1, \xi_2, \dots, \xi_m \end{pmatrix} d\xi_1 d\xi_2 \dots d\xi_m \tag{11.215}$$

$$D(x, y; \lambda) \equiv \sum_{m=0}^{\infty} \frac{(-\lambda)^m}{m!} \iint \dots \int K \begin{pmatrix} x, \xi_1, \xi_2, \dots, \xi_m \\ y, \xi_1, \xi_2, \dots, \xi_m \end{pmatrix} d\xi_1 d\xi_2 \dots d\xi_m \tag{11.216}$$

The integrands in (11.215) and (11.216) are defined in the form of determinants

$$K \begin{pmatrix} x_1, x_2, \dots, x_m \\ y_1, y_2, \dots, y_m \end{pmatrix} = \begin{vmatrix} K(x_1, y_1) & K(x_1, y_2) & \dots & K(x_1, y_m) \\ K(x_2, y_1) & K(x_2, y_2) & \dots & K(x_2, y_m) \\ \dots & \dots & \dots & \dots \\ K(x_m, y_1) & K(x_m, y_2) & \dots & K(x_m, y_m) \end{vmatrix} \tag{11.217}$$

$$K \begin{pmatrix} x, \xi_1, \xi_2, \dots, \xi_m \\ y, \xi_1, \xi_2, \dots, \xi_m \end{pmatrix} = \begin{vmatrix} K(x, y) & K(x, \xi_1) & \dots & K(x, \xi_m) \\ K(\xi_1, y) & K(\xi_1, \xi_1) & \dots & K(\xi_1, \xi_m) \\ \dots & \dots & \dots & \dots \\ K(\xi_m, y) & K(\xi_m, \xi_1) & \dots & K(\xi_m, \xi_m) \end{vmatrix} \tag{11.218}$$

Although the formula given in (11.214) is exact, the calculation of the determinant, evaluation of the integration, and the sum of the infinite series are very tedious procedures.

To derive Fredholm's formula, we first divide the limit of the integral as n equal intervals:

$$a, \quad x_1 = a + \delta, \quad x_2 = a + 2\delta, \dots, \quad x_n = a + n\delta = b \tag{11.219}$$

Then, we use the fundamental principle of integration by expressing (11.213) in the form of an infinite sum:

$$\phi_n(x) = f(x) + \lambda \sum_{j=1}^{n} K(x, x_j) \phi_n(x_j) \delta \tag{11.220}$$

This solution form is supposed to be valid for all x within the interval from a to b. Thus, we can substitute each x_i defined in (11.219) into (11.220) to obtain a system of n equations:

$$-\lambda\sum_{j=1}^{n}K(x_i,x_j)\phi_n(x_j)\delta+\phi_n(x_i)=f(x_i),\quad i=1,2,\ldots,,n \tag{11.221}$$

This is a system of n nonhomogeneous equations for n unknowns $\phi_n(x_i)$, where i = 1,2,...,n. The solution of this system is unique if the determinant of coefficients in (11.221) is not zero. We denote the determinant of (11.221) as:

$$D_n(\lambda)=\begin{vmatrix}1-\lambda\delta K(x_1,x_1) & -\lambda\delta K(x_1,x_2) & \ldots & -\lambda\delta K(x_1,x_n)\\ -\lambda\delta K(x_2,x_1) & 1-\lambda\delta K(x_2,x_2) & \ldots & -\lambda\delta K(x_2,x_n)\\ \ldots & \ldots & \ldots & \ldots\\ -\lambda\delta K(x_n,x_1) & -\lambda\delta K(x_n,x_2) & \ldots & 1-\lambda\delta K(x_n,x_n)\end{vmatrix} \tag{11.222}$$

Next, we observe that from the theory of determinant expansion, the determinant of the following matrix

$$[S_n]=\begin{bmatrix}1+a_{11} & a_{12} & \ldots & a_{1n}\\ a_{21} & 1+a_{22} & \ldots & a_{2n}\\ \ldots & \ldots & \ldots & \ldots\\ a_{n1} & a_{n2} & \ldots & 1+a_{nn}\end{bmatrix} \tag{11.223}$$

can be expanded as (see p.88 of Aitken, 1944)

$$\det|S_n|=1+\frac{1}{1!}\sum_{r_1}^{n}a_{r_rr_1}+\frac{1}{2!}\sum_{r_1,r_2}^{n}\begin{vmatrix}a_{r_1r_1} & a_{r_1r_2}\\ a_{r_2r_1} & a_{r_2r_2}\end{vmatrix}+\ldots+\frac{1}{n!}\sum_{r_1,r_2,\ldots,r_n}^{n}\begin{vmatrix}a_{r_1r_1} & a_{r_1r_2} & \ldots & a_{r_1r_n}\\ a_{r_2r_1} & a_{r_2r_2} & \ldots & a_{r_2r_n}\\ \ldots & \ldots & \ldots & \ldots\\ a_{r_nr_1} & a_{r_nr_2} & \ldots & a_{r_nr_n}\end{vmatrix} \tag{11.224}$$

We should mention a story of Alexander Aitken here. Alexander Aitken was a New Zealand–born mathematician, who possessed extraordinary memory. He was a PhD student of E.T. Whittaker at Edinburgh University, but was awarded a D.Sc. instead because of his outstanding thesis on data smoothing. He was known as one of the greatest mental calculators, and he could multiply two nine-digit numbers in his head in 30 seconds, and render fractions to 26 decimal places in under five seconds. Aitken made significant contributions to statistics and econometrics.

The formal proof of (11.224) will not be discussed here, however, for cases of n = 2 and 3 the author is advised to check its validity (see Problems 11.8 and 11.9 set at the end of the chapter). Thus, we can write the determinant in (11.222) as

$$D_n(\lambda)=1-\lambda\sum_{r_1}^{n}\delta K(x_{r_1},x_{r_1})+\frac{\lambda^2}{2!}\sum_{r_1,r_2,}^{n}\delta^2\begin{vmatrix}K(x_{r_1},x_{r_1}) & K(x_{r_1},x_{r_2})\\ K(x_{r_2},x_{r_1}) & K(x_{r_2},x_{r_2})\end{vmatrix}$$
$$+\dots+\frac{(-\lambda)^n}{n!}\sum_{r_1,r_2,\dots,r_n}^{n}\delta^n\begin{vmatrix}K(x_{r_1},x_{r_1}) & K(x_{r_1},x_{r_2}) & \dots & K(x_{r_1},x_{r_n})\\ K(x_{r_2},x_{r_1}) & K(x_{r_2},x_{r_2}) & \dots & K(x_{r_2},x_{r_n})\\ \dots & \dots & \dots & \dots\\ K(x_{r_n},x_{r_1}) & K(x_{r_n},x_{r_2}) & \dots & K(x_{r_n},x_{r_n})\end{vmatrix}$$

(11.225)

Now, we take the limit of $\delta \to 0$ and $n \to \infty$, and replace the summation by integration. This turns (11.225) into

$$D(\lambda)=1-\lambda\int_a^b K(\xi_1,\xi_1)d\xi_1+\frac{\lambda^2}{2!}\int_a^b\int_a^b\begin{vmatrix}K(\xi_1,\xi_1) & K(\xi_1,\xi_2)\\ K(\xi_2,\xi_1) & K(\xi_2,\xi_2)\end{vmatrix}d\xi_1 d\xi_2$$
$$+\dots+\frac{(-\lambda)^n}{n!}\int_a^b\dots\int_a^b\begin{vmatrix}K(\xi_1,\xi_1) & K(\xi_1,\xi_2) & \dots & K(\xi_1,\xi_n)\\ K(\xi_2,\xi_1) & K(\xi_2,\xi_2) & \dots & K(\xi_2,\xi_n)\\ \dots & \dots & \dots & \dots\\ K(\xi_n,\xi_1) & K(\xi_n,\xi_2) & \dots & K(\xi_n,\xi_n)\end{vmatrix}d\xi_1\dots d\xi_n+\dots$$

(11.226)

where $D(\lambda)$ is also known as the Fredholm determinant of the kernel K. Using Cramer's rule, we can solve for (11.221) as:

$$\phi(x_\mu)=\frac{f(x_1)D_n(x_\mu,x_1)+f(x_2)D_n(x_\mu,x_2)+\dots+f(x_n)D_n(x_\mu,x_n)}{D_n(\lambda)} \tag{11.227}$$

where $D_n(x_\mu, x_\nu)$ is the cofactor of the determinant $D_n(\lambda)$ for the term $K(x_\nu, x_\mu)$. It is defined as:

$$D_n(x_\mu,x_\nu)=\lambda\delta\{K(x_\mu,x_\nu)-\lambda\delta\sum_{r_1}^{n}\begin{vmatrix}K(x_\mu,x_\nu) & K(x_\mu,x_{r_1})\\ K(x_{r_1},x_\nu) & K(x_{r_1},x_{r_1})\end{vmatrix}$$
$$+\dots+\frac{(-\lambda\delta)^{n-1}}{(n-1)!}\sum_{r_1,r_2,\dots,r_n}^{n}\begin{vmatrix}K(x_\mu,x_\nu) & K(x_\mu,x_{r_2}) & \dots & K(x_\mu,x_{r_n})\\ K(x_{r_2},x_\nu) & K(x_{r_2},x_{r_2}) & \dots & K(x_{r_2},x_{r_n})\\ \dots & \dots & \dots & \dots\\ K(x_{r_n},x_\nu) & K(x_{r_n},x_{r_2}) & \dots & K(x_{r_n},x_{r_n})\end{vmatrix}\}$$

(11.228)

Take the limiting form of $\delta \to 0$ and $n \to \infty$ such that $\delta^{-1}D(x_\mu, x_\nu)$ is redefined as:

$$D(x_\mu, x_\nu; \lambda) = \lambda K(x_\mu, x_\nu) - \lambda^2 \int_a^b \begin{vmatrix} K(x_\mu, x_\nu) & K(x_\mu, \xi_1) \\ K(\xi_1, x_\nu) & K(\xi_1, \xi_1) \end{vmatrix} d\xi_1 +$$

$$+\frac{\lambda^3}{2!}\int_a^b\int_a^b \begin{vmatrix} K(x_\mu, x_\nu) & K(x_\mu, \xi_1) & K(x_\mu, \xi_2) \\ K(\xi_1, x_\nu) & K(\xi_1, \xi_1) & K(\xi_1, \xi_2) \\ K(\xi_2, x_\nu) & K(\xi_2, \xi_1) & K(\xi_2, \xi_2) \end{vmatrix} d\xi_1 d\xi_2 - ... \tag{11.229}$$

For $D(\lambda) \neq 0$, Fredholm inferred that the solution of (11.212) is

$$\phi(x) = f(x) + \frac{1}{D(\lambda)} \int_a^b D(x, \xi; \lambda) f(\xi) d\xi \tag{11.230}$$

This is Fredholm's celebrated formula, and it is a breakthrough in the analysis of the Fredholm type of integral equation. This is an elegant formula but the actual calculation of the determinant and integration may be tedious. Although Fredholm's approach is kind of intuitive, its formal justification was considered by Hilbert in 1904 and subsequently led to the development of functional analysis, which will be discussed further in a later section.

Example 11.11 Solve the following Fredholm integral of the second kind:

$$\phi(x) = x + \lambda \int_0^1 xy\phi(y)dy \tag{11.231}$$

Solution: For this special case, we have

$$D(\lambda) = 1 - \lambda \int_0^1 \xi^2 d\xi = 1 - \frac{1}{3}\lambda, \quad D(x, y; \lambda) = \lambda xy \tag{11.232}$$

Note for this case that only the first term in (11.229) is nonzero. The solution is clearly

$$\phi(x) \equiv x + \frac{3}{3-\lambda} \int_0^1 \lambda x y^2 dy = \frac{3x}{3-\lambda} \tag{11.233}$$

Example 11.12 Solve the following Fredholm integral of the second kind.

$$\phi(x) = x + \lambda \int_0^1 (xy + y^2)\phi(y)dy \tag{11.234}$$

Solution: For this special case, the determinants are:

$$D(\lambda) = 1 - \frac{2}{3}\lambda - \frac{1}{72}\lambda^2 \tag{11.235}$$

$$D(x, y; \lambda) = \lambda(xy + y^2) + \lambda^2(\frac{1}{2}xy^2 - \frac{1}{3}xy - \frac{1}{3}y^2 + \frac{1}{4}y) \tag{11.236}$$

The solution is clearly

$$\phi(x) \equiv x + \lambda^2(\frac{\lambda x + 18 + 24x}{72 - 48\lambda - \lambda^2}) \tag{11.237}$$

11.4.2 Solution of the Fredholm Integral Equation of the First Kind

Note from (11.214) that we must have $D(\lambda) \neq 0$. For the case $D(\lambda) = 0$, suppose that its roots are called eigenvalues and denoted as:

$$\lambda = \lambda_1, \lambda_2, ..., \lambda_n, \quad n \to \infty \tag{11.238}$$

It corresponds to the following eigenvalue problem:

$$\psi_j(x) = \lambda_j \int_a^b K(x,\xi)\psi_j(\xi)d\xi \tag{11.239}$$

where ψ_j ($j = 1,2,...,n$) is the eigenfunction. This so-called eigenvalue problem defined in (11.239) can be considered as a generalization of the traditional eigenvalue problem for a matrix. This eigenfunction has been normalized as:

$$\int_a^b \{\psi_j(\xi)\}^2 d\xi = 1 \tag{11.240}$$

It can be shown that these eigenfunctions are orthogonal in the sense that

$$\int_a^b \psi_i(\xi)\psi_j(x)d\xi = \delta_{ij} \tag{11.241}$$

Note that (11.241) is actually the homogeneous form of the original Fredholm equation given in (11.212). We will see in a later section that (11.241) is actually the inner product defined for the Hilbert space.

With this eigenfunction, we are now ready to establish the solution of the following Fredholm integral equation of the first kind:

$$f(x) = \int_a^b K(x,s)\phi(s)ds \tag{11.242}$$

The solution can be established using the so-called Hilbert-Schmidt theorem. In essence, this theorem asserts that both $f(x)$ and $\phi(x)$ can be expanded in a Fourier series expansion in the eigenfunction ψ_j:

$$f(x) = \sum_{n=1}^{\infty} a_n \psi_n(x) \tag{11.243}$$

$$\phi(x) = \sum_{n=1}^{\infty} b_n \psi_n(x) \tag{11.244}$$

Multiplying both sides of (11.243) by $\psi_m(x)$ and integrating from a to b gives

$$\int_a^b f(x)\psi_m(x)dx = \sum_{n=1}^{\infty} a_n \int_a^b \psi_m(x)\psi_n(x)dx = a_m \tag{11.245}$$

The last part of (11.245) is a consequence of the orthogonal property of the eigenfunctions given in (11.241). Thus, we have

$$a_m = \int_a^b f(x)\psi_m(x)dx \tag{11.246}$$

Substitution of (11.244) and (11.243) into (11.242) gives

$$\sum_{n=1}^{\infty} a_n \psi_n(x) = \sum_{n=1}^{\infty} b_n \int_a^b K(x,s)\psi_n(s)ds \tag{11.247}$$

In view of (11.239), we have

$$a_n\lambda_n \int_a^b K(x,\xi)\psi_n(\xi)d\xi = b_n \int_a^b K(x,s)\psi_n(s)ds \tag{11.248}$$

This gives

$$b_n = a_n\lambda_n \tag{11.249}$$

Therefore, the solution of the Fredholm integral equation of the first kind is

$$\phi(x) = \sum_{n=1}^{\infty} a_n\lambda_n\psi_n(x) \tag{11.250}$$

This solution is analogous to the eigenfunction expansion discussed in Chapter 10.

11.5 FREDHOLM ALTERNATIVE THEOREM

There is a major by-product of Fredholm's formula discussed in Section 11.4.1. We recall that the solution is valid only if $D(\lambda) \neq 0$. For the case $D(\lambda) = 0$, it is an eigenvalue problem of the following problem:

$$u(x) = \lambda \int_a^b K(x,\xi)u(\xi)d\xi \tag{11.251}$$

This is actually the homogeneous equation of (11.212). If the kernel function is a square integrable function:

$$\int_a^b \int_a^b |K(x,\xi)|^2 d\xi dx < \infty , \tag{11.252}$$

the following theorem can be established.

In particular, for $D(\lambda) \neq 0$ there will be no solution for the eigenvalue (11.251). For $D(\lambda) = 0$, we have an infinite number of roots called eigenvalues:

$$\lambda = \lambda_1, \lambda_2, ..., \lambda_n, \quad n \to \infty \tag{11.253}$$

This λ is also called the characteristics of kernel $K(x,\xi)$ and the set of roots of $D(\lambda) = 0$ is called the spectrum of kernel $K(x,\xi)$. The solutions of (11.251) corresponding to the eigenvalues λ_i are called eigenfunctions or characteristic functions:

$$u = u_1, u_2, ..., u_n, \quad n \to \infty \tag{11.254}$$

The general solution of (11.251) is

$$u(x) = c_1u_1(x) + c_2u_2(x) + ... + c_nu_n(x), \quad n \to \infty \tag{11.255}$$

Fredholm defined an associated integral equation as:

$$u(x) = \lambda \int_a^b K(\xi,x)u(\xi)d\xi \tag{11.256}$$

Note that this is also the adjoint problem of the original integral (11.251). Corresponding to the eigenvalues given in (11.230), the characteristic functions of the adjoint equation in (11.256) are

$$\psi_1(x), \psi_2(x), ..., \psi_n(x) \tag{11.257}$$

Then, the condition of having the solution of the nonhomogeneous form of Fredholm integral equation given in (11.242) is:

$$\int_a^b \psi_i(x)f(x)dx = 0 \tag{11.258}$$

This is called the Fredholm alternative theorem, and if it is satisfied, the Fredholm equation in (11.242) is solvable. This is related to the existence of a solution of a nonhomogeneous equation, and thus it is a very important and powerful theorem in the theory of integral equations. In fact, the Fredholm alternative theorem can also be applied to investigate the existence of nonhomogeneous ODEs (e.g., Boyce and DiPrima, 2010).

11.6 ADJOINT INTEGRAL EQUATION

Let us formally define the adjoint integral equation. Let us consider an integral equation defined symbolically as:

$$g(x) = I[f(x)] \tag{11.259}$$

where I is an integral operator. For example, the linear integral operator can be defined:

$$g(x) = \int_a^b K(x,\xi) f(\xi) d\xi \tag{11.260}$$

The corresponding adjoint problem is defined as:

$$\bar{I}[f(x)] = \int_a^b K(\xi,x) f(\xi) d\xi \tag{11.261}$$

The adjoint problem has to be studied by employing the following definition of the inner product defined by Hilbert as:

$$[f_1, f_2] = \int_a^b f_1(\xi) f_2(\xi) d\xi \tag{11.262}$$

Note that the concept of the inner product is an important ingredient of the functional space concept. Of course, the inner product defined by Hilbert is for the so-called Hilbert space (or Hilbert functional space). In fact, the functional analysis or sometimes referred to as abstract space analysis in the last hundred years was initiated by Hilbert's studies related to the Fredholm integral equation. More discussions on this are presented in a later section.

Then, the adjoint integral operator of an integral operator I is defined by

$$[I(f_1), f_2] = [f_1, \bar{I}(f_2)] \tag{11.263}$$

where the superimposed bar indicates the adjoint integral operator. For a self-adjoint integral equation, we require:

$$[I(f_1), f_2] = [f_1, I(f_2)] \tag{11.264}$$

with zero boundary terms. This leads to

$$\int_a^b \int_a^b K(x,\xi) f_1(\xi) f_2(x) d\xi dx = \int_a^b \int_a^b K(\xi,x) f_1(x) f_2(\xi) d\xi dx \tag{11.265}$$

Consequently, the conditions of the self-adjoint integral equation leads to

$$K(x,\xi) = K(\xi,x) \tag{11.266}$$

Similar to the self-adjoint differential equation, the corresponding eigenvalue of the self-adjoint integral equation has real eigenvalue λ_i and eigenfunctions ψ_i. In addition, the eigenfunctions are orthogonal.

11.7 VOLTERRA INTEGRAL EQUATION

11.7.1 Volterra Integral Equations of the First and Second Kinds

In this section, we will consider a special case of the Fredholm equation. If the upper limit is not the constant b, instead the upper limit becomes x. We have the linear Volterra integral equation of the first kind

$$f(x)=\int_a^x K(x,s)y(s)ds \tag{11.267}$$

Similarly, the linear Volterra integral equation of the second kind is defined as

$$y(x)=f(x)+\lambda\int_a^x K(x,s)y(s)ds \tag{11.268}$$

More generally, the nonlinear Volterra integral equation of the second kind can be defined as

$$y(x)=f(x)+\lambda\int_a^x K[x,s,y(s)]ds \tag{11.269}$$

We observe that the only difference between Fredholm and Volterra integral equations is the upper limit. In particular, if the kernel function in (11.245) or (11.212) is defined as

$$\begin{aligned} K(x,s)&=0, \quad s>x \\ &\neq 0, \quad x>s, \end{aligned} \tag{11.270}$$

when (11.270) is substituted into (11.211) and (11.212), (11.267) and (11.268) are recovered as a special case. It is illustrated in Figure 11.5.

Despite the similarity between the Fredholm integral equation and the Volterra integral equation, there is a major difference between them. In particular, if f and K in (11.267) and (11.268) are continuous, there is a unique solution for the linear nonhomogeneous Volterra integral equation of the second kind. However, for a linear nonhomogeneous Fredholm integral equation of the second kind to have a unique solution, the Fredholm alternative theorem must be satisfied.

11.7.2 Liouville-Neumann or Picard Successive Approximation

Historically, Volterra's solution of the linear Volterra-type integral equation of the second kind was obtained in 1896 and 1897, and, thus, preceded Fredholm's solution. Let us consider the solution scheme by Volterra, which was obtained by the idea of successive iteration (p. 1057 of Kline, 1972). In particular, Volterra assumed that the solution can be expressed as an infinite sum as:

$$y(x)=f(x)+\sum_{p=1}^{\infty} f_p(x) \tag{11.271}$$

where every term in the infinite series is an integral defined as:

$$
\begin{aligned}
f_1(x) &= \int_a^x K(x,y) f(y) dy \\
f_2(x) &= \int_a^x K(x,y) f_1(y) dy \\
&\dots \\
f_n(x) &= \int_a^x K(x,y) f_{n-1}(y) dy
\end{aligned}
\tag{11.272}
$$

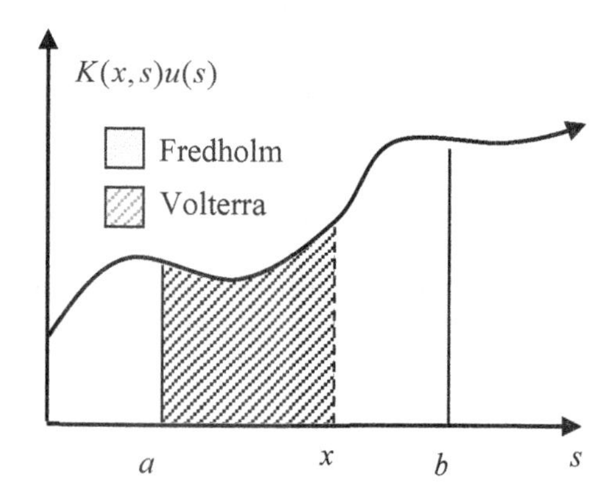

Figure 11.6 Domain of integrand of Fredholm and Volterra type integrals

The first integral depends on the given function f, whereas the second integral depends on the result of the first integral, and so on. Stakgold (1967) called this infinite series given in (11.271) the Neumann series, which was used by Neumann for many years before it was published in 1877. The first publication of using this kind of series was by Liouville in 1832 and 1837. Therefore, some researchers called it the Liouville-Neumann series. The same idea was published again in 1890 by Picard and he established the formulation in a general and widely applicable form (Tricomi, 1957). Therefore, it was also called the Picard process of successive approximation. Apparently, the most popular adopted name is Picard's method of successive approximation.

Now, substitution of (11.272) into (11.271) and then the result into the right hand side of (11.268) gives

$$
y(x) = f(x) + \int_a^x R(x,s) f(s) ds \tag{11.273}
$$

where $R(x,s)$ is called the resolvent kernel (versus the kernel $K(x,s)$) and is defined as

$$
R(x,s) = K(x,s) + \int_a^b K(x,r) K(r,s) dr + \int_a^b \int_a^b K(x,r) K(r,w) K(w,s) dr dw + \dots \tag{11.274}
$$

This result can also be recast into a slightly different form:

$$y(x) = f(x) + \int_a^x \sum_{j=1}^{\infty} K_j(x,s) f(s) ds \tag{11.275}$$

where

$$K_1(x,s) = K(x,s) \tag{11.276}$$

$$K_j(x,s) = \int_a^x K(x,v) K_{j-1}(v,s) dv \tag{11.277}$$

In order for this solution to be valid, we must require that the kernel function $K(x,s)$ be bounded.

If the kernel function is of convolution form (i.e., $K(x,s) = K(x-s)$), the linear Volterra integral equation can be solved by using the Laplace transform. This will be discussed in Section 11.7.5 and illustrated in Example 11.16.

Example 11.13 Solve the following integral equation by Picard's iterative method:

$$y(t) = 1 - \int_0^t (t-x) y(s) ds, \quad y(0) = 1 \tag{11.278}$$

Solution: We can start with the boundary condition as the first approximation

$$y_0 = y(0) = 1 \tag{11.279}$$

Then, the second approximation is

$$y_1 = 1 - \int_0^t (t-s) ds = 1 - \frac{1}{2} t^2 \tag{11.280}$$

The third approximation is

$$\begin{aligned} y_2 &= 1 - \int_0^t (t-s) y_1(s) ds = 1 - \int_0^t (t-s)(1 - \frac{1}{2} s^2) ds \\ &= 1 - \int_0^t t(1 - \frac{1}{2} s^2) ds + \int_0^t (s - \frac{1}{2} s^3) ds \end{aligned} \tag{11.281}$$

Carrying out the integration, we find

$$y_2 = 1 - \frac{1}{2} t^2 + \frac{1}{24} t^4 \tag{11.282}$$

If we continue the process, we find

$$y_\infty = 1 - \frac{1}{2!} t^2 + \frac{1}{4!} t^3 + \ldots = \cos t \tag{11.283}$$

This is also the exact solution.

11.7.3 Solution of Volterra Integral Equation of the First Kind

For the solution of the linear Volterra integral equation of the first kind, we find that we can actually convert it to the Volterra integral of the second kind such that the above iterative solution scheme proposed by Volterra applies. In particular, rewrite the linear Volterra integral equation of the first kind as

$$g(x) = \int_a^x H(x,s)y(s)ds \tag{11.284}$$

Differentiation of both sides with respect to x and application of the Leibniz rule of differentiation on integral gives

$$g'(x) = H(x,x)y(x) + \int_a^x \frac{\partial H(x,s)}{\partial x} y(s)ds \tag{11.285}$$

If the kernel $H(x,x)$ is not zero, we can rearrange the equation as

$$y(x) = \frac{g'(x)}{H(x,x)} + \int_a^x \frac{-1}{H(x,x)} \frac{\partial H(x,s)}{\partial x} y(s)ds \tag{11.286}$$

This is, in fact, the Volterra integral equation of the second kind if we make the following identifications:

$$f(x) = \frac{g'(x)}{H(x,x)}, \quad K(x,s) = \frac{-1}{H(x,x)} \frac{\partial H(x,s)}{\partial x} \tag{11.287}$$

Thus, we know how to solve it.

Example 11.14 Solve the following Fredholm integral equation

$$\phi(x) - \lambda \int_0^x e^{x-y} \phi(y)dy = f(x) \tag{11.288}$$

Solution: Using (11.275), for this special case we have the first kernel $K_1(x, \xi)$ as:

$$K(x,y) = e^{x-y} = e^x e^{-y} \tag{11.289}$$

The second kernel given in (11.277) is

$$K_2(x,y) = \int_y^x e^{x-z} e^{z-y} dz = e^{x-y} \int_y^x dz = e^{x-y}(x-y) \tag{11.290}$$

Substitution of (11.290) into (11.277) yields

$$K_3(x,y) = \int_y^x e^{x-z} e^{z-y} (z-y)dz = e^{x-y} \int_y^x (z-y)dz = e^{x-y} \frac{(x-y)^2}{2} \tag{11.291}$$

Substitution of (11.291) into (11.277) again gives

$$K_4(x,y) = \frac{1}{2}\int_y^x e^{x-z} e^{z-y} (z-y)^2 dz = \frac{1}{2} e^{x-y} \int_y^x (z-y)^2 dz = e^{x-y} \frac{(x-y)^3}{3!} \tag{11.292}$$

Obviously, we can generalize the observation from (11.289) to (11.292) to

$$K_{n+1}(x,y) = e^{x-y} \frac{(x-y)^n}{n!} \quad (n = 0,1,2,...) \tag{11.293}$$

Substitution of (11.293) into (11.274) gives

$$R(x,y;\lambda) = -\sum_{n=0}^{\infty} \lambda^n K_{n+1}(x,y) = -e^{x-y} \sum_{n=0}^{\infty} \frac{\lambda^n (x-y)^n}{n!} = -e^{(1+\lambda)(x-y)} \tag{11.294}$$

Using this result, we finally get

$$\phi(x) = f(x) + \lambda \int_0^x e^{(1+\lambda)(x-y)} f(y)dy \tag{11.295}$$

Two particular forms of kernels are of particular interest because of their physical significance and because of the easiness of solving these particular cases. They are the separable kernel (or called Pincherle-Goursat kernel) and the convolution type kernel.

11.7.4 Volterra Integral Equation with Separable Kernel

The kernel of the Volterra integral equation is called separable if the following equation is satisfied:

$$K(t,s) = p(t)q(s) \tag{11.296}$$

Substitution of (11.296) into the Volterra integral equation of the second kind given in (11.268) leads to

$$y(x) = f(x) + \lambda \int_a^x p(x)q(s)y(s)ds \tag{11.297}$$

This can be rewritten as

$$y(x) = f(x) + \lambda p(x)Y(x) \tag{11.298}$$

where

$$Y(x) = \int_a^x q(s)y(s)ds \tag{11.299}$$

Now, we can differentiate (11.299) with respect to x to give

$$\frac{dY(x)}{dx} = q(x)y(x) \tag{11.300}$$

Substitution of (11.298) into (11.300) leads to the following first order ODE for $Y(x)$:

$$\frac{dY(x)}{dx} = q(x)f(x) + \lambda q(x)p(x)Y(x) \tag{11.301}$$

This is the general linear form of the first order ODE and can be solved readily. We will illustrate this solution method in the following example.

Example 11.15 Solve the following Volterra integral equation of the second kind with separable kernel:

$$y(t) = \sqrt{t} - \int_0^t 2\sqrt{ts}\, y(s)ds \tag{11.302}$$

Solution: Let us define:

$$Y(t) = \int_0^t \sqrt{s}\, y(s)ds \tag{11.303}$$

Substitution of (11.303) into (11.302) gives

$$y(t) = \sqrt{t} - 2\sqrt{t}\,Y(t) \tag{11.304}$$

Differentiation of (11.304) with respect to t gives

$$Y'(t) = \sqrt{t}\,y = t - 2tY(t) \tag{11.305}$$

The solution of the corresponding homogeneous ODE of (11.305) is

$$Y_h(t) = Ce^{-t^2} \tag{11.306}$$

Using the procedure of variation of parameters, we seek

$$Y(t) = C(x)e^{-t^2} \tag{11.307}$$

Substitution of (11.307) into (11.305) gives an equation for C and the solution of (11.305) is

$$Y(t) = \frac{1}{2}(1 - e^{-t^2}) + C_1 e^{-t^2} \tag{11.308}$$

From the definition of (11.302), we have

$$Y(0) = 0 \tag{11.309}$$

Using this boundary condition for Y, we obtain

$$Y(t) = \frac{1}{2}(1 - e^{-t^2}) \tag{11.310}$$

Finally, back substitution of this Y into (11.303) gives the solution of (11.302):

$$y(t) = \sqrt{t}e^{-t^2} \tag{11.311}$$

More generally, Tricomi (1957) called the following separable kernel as the Pincherle-Goursat kernel

$$K(x, y) = \sum_{k=1}^{n} X_k(x) Y_k(y) \tag{11.312}$$

It was shown that for this case the solution of the Volterra integral equation can be converted to solving a system of linear algebraic equations. The details are found in Tricomi (1957).

11.7.5 Volterra Integral Equation of Convolution Type

The second special form of the Volterra integral equation of the second kind that can be solved easily is the convolution type kernel. That is,

$$K(t, s) = K(t - s) \tag{11.313}$$

The argument in the kernel only appears as the difference between t and s. For such cases, the Laplace transform can be used to solve the integral equation. In particular, we have

$$y(x) = f(x) + \lambda \int_a^x K(x - s) u(s) ds \tag{11.314}$$

If the Laplace transforms of two functions f_1 and f_2 exist, we have

$$F_1(s) = L\{f_1(x)\}, \quad F_2(s) = L\{f_2(x)\} \tag{11.315}$$

The convolution product is defined as:

$$(f_1 * f_2)(x) = (f_2 * f_1)(x) = \int_0^x f_1(x - t) f_2(t) dt \tag{11.316}$$

By the convolution theorem for the Laplace transform, we have

$$L\{(f_1 * f_2)(x)\} = L\{\int_0^x f_1(x-t)f_2(t)dt\} = F_1(s)F_2(s) \tag{11.317}$$

It is clear that the integral in (11.314) is of the convolution type, thus (11.317) can be applied to (11.314). We will illustrate this method by the following example.

Example 11.16 Solve the following integral equation of convolution type by using the Laplace transform:

$$y(t) = 1 + 2\int_0^t \cos(t-u)y(u)du \tag{11.318}$$

Solution: Applying the Laplace transform and convolution theorem, we get

$$L\{y(t)\} = L\{1\} + 2L\{\int_0^t \cos(t-u)y(u)du\} = L\{1\} + 2L\{\cos t\}L\{y(t)\} \tag{11.319}$$

Using the table of Laplace transforms, we obtain

$$Y(s) = \frac{1}{s} + 2(\frac{s}{s^2+1})Y(s) \tag{11.320}$$

Solving for $Y(s)$, we obtain

$$Y(s) = \frac{s^2+1}{s(s-1)^2} \tag{11.321}$$

Application of partial fraction gives

$$Y(s) = \frac{1}{s} + \frac{2}{(s-1)^2} \tag{11.322}$$

Taking the inverse Laplace transform, we obtain

$$\begin{aligned} y(t) &= L^{-1}\{Y(s)\} = L^{-1}\{\frac{1}{s}\} + L^{-1}\{\frac{2}{(s-1)^2}\} \\ &= 1 + 2te^t \end{aligned} \tag{11.323}$$

Example 11.17 Solve the following integral equation of convolution type using the Laplace transform:

$$\int_0^t \frac{y(u)}{\sqrt{t-u}} du = \sqrt{t} \tag{11.324}$$

Solution: Applying the Laplace transform and convolution theorem, we get

$$L\{\int_0^t \frac{y(u)}{\sqrt{t-u}} du\} = L\{\sqrt{t}\} \tag{11.325}$$

Noting that

$$L\{\sqrt{t}\} = \frac{\Gamma(3/2)}{s^{3/2}} = \frac{\sqrt{\pi}}{2s^{3/2}}, \quad L\{\frac{1}{\sqrt{t}}\} = \frac{\sqrt{\pi}}{s^{1/2}}, \tag{11.326}$$

substitution of (11.326) into (11.325) gives

$$L\{y(t)\}L\{\frac{1}{\sqrt{t}}\} = Y(s)\frac{\sqrt{\pi}}{s^{1/2}} = \frac{\sqrt{\pi}}{2s^{3/2}} \tag{11.327}$$

Thus, we have

$$Y(s) = \frac{1}{2s} \tag{11.328}$$

Note that the following inversion formulas for the Laplace transform

$$L^{-1}\{\frac{1}{s}\} = \frac{1}{\Gamma(1)} = 1, \quad y(t) = L^{-1}\{Y(s)\} \tag{11.329}$$

Taking the inverse Laplace transform of (11.328), we obtain

$$y(t) = \frac{1}{2} \tag{11.330}$$

11.8 FUNCTIONAL ANALYSIS AND VECTOR SPACE

In this section, we continue to discuss the work of Volterra, Fredholm, and Hilbert and its subsequent development into functional analysis. The term functional analysis was coined by Levy. We recall from the previous section that in our discussion on the Fredholm alternative theorem we needed to introduce the concept of the inner product (in terms of integration), adjoint integral equation, the orthogonality between the nonhomogeneous terms and the eigenfunctions of the adjoint problem. In fact, the inner product and orthogonal properties are the main assumptions in the analysis of Hilbert, and functions that satisfy such conditions (plus some more standard operational requirements of the inner product, definition of norm, and convergence requirements) are said to form the Hilbert space. Such functional analysis mainly deals with the existence of the solution (similar to the role of the Fredholm alternative theorem) and is also referred to as abstract space analysis. Some of the major contributors to such development include Volterra, Hilbert, Riesz, Schmidt, Gelfand, Wiener, Fischer, Frechet, and Banach. There are many functional spaces being proposed, and the more notable ones include Hilbert space, Banach space, Schwartz space, Sobolev space, and Holder space. For example, the so-called Banach space does not have the orthogonal requirement between functions and is less strict compared with Hilbert space. In mechanics, it was discovered that Hilbert spaces are applicable to the analysis of quantum mechanics. In this context, the eigenvalues are the quantum energy levels. The work of Hilbert is also related to the concept of the discrete and continuous spectrum of the eigenvalues. It turns out that the Hilbert integral equation is a singular type of integral equation, and for such case the eigenvalue becomes continuous (at least within some interval).

11.9 INTEGRAL EQUATION VERSUS DIFFERENTIAL EQUATION

In 1904 and 1905, Hilbert showed that the eigenvalues and eigenfunctions of the following Sturm-Liouville problem

$$\frac{d}{dx}[p(x)\frac{du}{dx}]+q(x)u+\lambda u=0 \tag{11.331}$$

subject to the boundary conditions

$$u(a)=0, \quad u(b)=0 \tag{11.332}$$

are also the eigenvalues and eigenfunctions of the following integral equation:

$$\phi(x)-\lambda\int_0^x G(x,\xi)\phi(\xi)d\xi=0 \tag{11.333}$$

The kernel function $G(x,\xi)$ is the Green's function of the following ODE:

$$\frac{d}{dx}[p(x)\frac{du}{dx}]+q(x)u=0 \tag{11.334}$$

The derivative of this Green's function $\partial G(x,\xi)/\partial x$ has a jump of $-1/p(\xi)$. In other words, an integral equation is a way of solving ordinary or partial differential equations.

To further investigate the relation between an ODE and an integral equation, we consider the following first order initial value problem:

$$\frac{du}{dt}=a(t)u(t)+b(t) \tag{11.335}$$

subject to initial condition

$$u(0)=u_0 \tag{11.336}$$

We note that (11.335) can be obtained by differentiating the following Volterra integral equation of the second kind:

$$u(t)=g(t)+\int_0^t K(t,s)u(s)ds \tag{11.337}$$

where

$$g(t)=u_0+\int_0^t b(s)ds, \quad K(t,s)=a(s) \tag{11.338}$$

For this particular case, the kernel K is independent of t.

Secondly, we consider the following initial value problem model by second order ODE:

$$\frac{d^2u}{dt^2}=a(t)u(t)+b(t) \tag{11.339}$$

subject to initial conditions:

$$u(0)=u_0, \quad u'(0)=v_0 \tag{11.340}$$

This equation can be obtained by differentiating the following Volterra integral of the second kind twice:

$$u(t)=g(t)+\int_0^t K(t,s)u(s)ds \tag{11.341}$$

where

$$g(t)=u_0+v_0t+\int_0^t (t-s)b(s)ds, \quad K(t,s)=(t-s)a(s) \tag{11.342}$$

Therefore, the second order ODE with non-constant coefficients given in (11.339) can be solved by considering the Volterra integral equation of the second kind in (11.341).

However, if we start by differentiating a general form of Volterra integral equation of the second kind given in (11.341), we have

$$\frac{du(t)}{dt} = \frac{dg(t)}{dt} + K(t,t)u(t) + \int_0^t \frac{\partial K(t,s)}{\partial t} u(s) ds \tag{11.343}$$

This is not an ODE, but instead it is a Volterra integro-differential equation (which will be discussed in more detail in a later section). Therefore, an ODE can be converted to a Volterra integral equation, whereas a Volterra integral equation of the second kind with arbitrary kernel K may not correspond to any ODE.

Consider the following n-th order ODE with initial conditions

$$\frac{d^n u}{dx^n} + a_1(x)\frac{d^{n-1}u}{dx^{n-1}} + ... + a_n(x)u = F(x) \tag{11.344}$$

$$u(0) = c_0,\ u'(0) = c_1, ...,\ u^{(n-1)}(0) = c_{n-1} \tag{11.345}$$

This is equivalent to the following linear Volterra integral equations of second order:

$$\phi(x) = f(x) - \int_0^x K(x,y)\phi(y) dy \tag{11.346}$$

where

$$f(x) = F(x) - c_{n-1}a_1(x) - (c_{n-1}x + c_{n-2})a_2(x) - ... - [c_{n-1}\frac{x^{n-1}}{(n-1)!} + ... + c_1 x + c_0]a_n(x) \tag{11.347}$$

$$K(x,y) = \sum_{m=1}^{n} a_m(x)\frac{(x-y)^{m-1}}{(m-1)!} \tag{11.348}$$

Consider the following boundary value problem modelled by an ODE:

$$\frac{d^2 w}{dx^2} + C(x)\frac{dw}{dx} + D(x)w = F(x) \tag{11.349}$$

$$w(a) = \gamma, \quad w(b) = \delta \tag{11.350}$$

This is equivalent to the following Fredholm integral equation of the second kind (Zwillinger, 1997):

$$w(x) = h(x) + \int_a^b H(x,\zeta)w(\zeta)d\zeta \tag{11.351}$$

$$h(x) = \gamma + \int_a^x (x-\zeta)F(\zeta)d\zeta + \frac{x-a}{b-a}[\delta - \gamma - \int_a^b (b-\zeta)F(\zeta)d\zeta]$$

$$H(x,\zeta) = \begin{cases} \dfrac{x-b}{b-a}[C(\zeta) - (a-\zeta)(C'(\zeta) - D(\zeta))], & x > \zeta \\ \dfrac{x-a}{b-a}[C(\zeta) - (b-\zeta)(C'(\zeta) - D(\zeta))], & x < \zeta \end{cases} \tag{11.352}$$

Therefore, the boundary value problem can be expressed as a Fredholm integral equation of the second kind.

For multi-variable function u, we can also have the following combined or mixed Volterra-Fredholm integral equation of the second kind (Zwillinger, 1997):

$$u(t,x) = g(t,x) + \int_0^t \int_a^b K(t,s,x,\xi)u(s,\xi)d\xi ds \quad (11.353)$$

In contrast to the Fredholm integral equation, this mixed Volterra-Fredholm integral equation has a unique solution, and it can be expressed in a resolvent kernel as

$$u(t,x) = g(t,x) + \int_0^t \int_a^b R(t,s,x,\xi)g(s,\xi)d\xi ds \quad (11.354)$$

Example 11.18 Find an equivalent differential equation for the following integral equation:

$$y(t) = 5\cos t + \int_0^t (t-u)y(u)du \quad (11.355)$$

Solution: Differentiation of (11.355) with respect to t once and twice gives

$$y'(t) = -5\sin t + \int_0^t y(u)du \quad (11.356)$$

$$y''(t) = -5\cos t + y(t) \quad (11.357)$$

Rearrangement of (11.357) gives

$$y''(t) - y(t) = -5\cos t \quad (11.358)$$

We can substitute $t = 0$ into (11.355) and (11.356), and we find the following initial conditions for the ODE found in (11.340):

$$y(0) = 5, \quad y'(0) = 0 \quad (11.359)$$

11.10 INTEGRO-DIFFERENTIAL EQUATION

Many physical problems need to be modelled by integro-differential equation, such as airfoil studies, temperature variation in melted glass under heat conduction and heat radiation, Boltzman's model of the distribution of particles of an ideal gas in an enclosure, probability of a customer's waiting time in a queue, probability of brightness of a star being reduced by clouds of interstellar dust, and the theory of atomic scattering. Literally, it means that the unknown of the problem appears in derivative form as well as inside an integral in the same equation. For the case of one variable, a rather general form of integro-differential equation is written symbolically as

$$y^{(n)}(x) = f(x, y(x), y', ...y^{(n-1)}, \lambda \int_a^b K[x, s, y(s), ..., y^{(m)}(s)]ds) \quad (11.360)$$

For $n \geq m$, (11.360) has a unique solution. The boundary conditions for (11.360) are

$$y^{(k)}(a) = y_k, \quad k = 0, 1, ..., n-1 \quad (11.361)$$

A number of more notable integro-differential equations are reported here. Integro-differential equations may also appear in a pair of coupled systems. For example, the growth of two conflicting populations can be modelled by the following Volterra nonlinear integro-differential equation

$$\frac{dx}{dt} = ax - bxy - x\int_0^t K_1(t-s)y(s)ds \tag{11.362}$$

$$\frac{dy}{dt} = -\alpha x + \beta xy + y\int_0^t K_2(t-s)x(s)ds \tag{11.363}$$

This is a prey-predator model proposed by Volterra, taking into account heredity factors. The integro-differential equation of Abel is given as

$$\int_0^x \frac{F[x, y, \varphi(y), \varphi'(y)]}{(x-y)^\alpha} dy = f(x) \tag{11.364}$$

The Darboux-Picard integro-differential equation is

$$z(x, y) = \sigma(x) + \tau(y) - z_0 + \int_0^x \int_0^y f(u, v, z, z_x, z_y)dudv \tag{11.365}$$

Relating to airfoil analysis, Prandtl's circulation equation is given by

$$cy(x) + \frac{1}{\pi}\int_{-1}^1 (\frac{1}{\xi - x})\frac{dy}{d\xi} d\xi = f(x) \tag{11.366}$$

where the integral is taken as a Cauchy principal value. For radioactive transfer, we have the following partial integro-differential equation:

$$\mu\frac{\partial\varphi(r,\mu)}{\partial r} + \frac{1-\mu^2}{r}\frac{\partial\varphi(r,\mu)}{\partial\mu} + \varphi(r,\mu) - \frac{1}{2}\int_{-1}^1 \varphi(r,\xi)d\xi = 0 \tag{11.367}$$

For modeling multi-electron atoms, the following Thomas-Fermi equation has been proposed:

$$\frac{du}{dx} = B + \int_0^x u^{3/2}(t)t^{-1/2}dt \tag{11.368}$$

We now quote the two most popular integro-differential equations, namely the Volterra type and Fredholm type of integro-differential equations. Mathematically, they are written as

$$u^{(n)}(x) = f(x) + \lambda\int_a^x K(x,t)u(t)dt \tag{11.369}$$

$$u^{(n)}(x) = f(x) + \lambda\int_a^b K(x,t)u(t)dt \tag{11.370}$$

Sometimes, the Volterra and Fredholm integral equations may appear simultaneously in a single equation as the linear Volterra-Fredholm integro-differential equation:

$$u^{(n)}(x) = f(x) + \lambda_1\int_a^x K_1(x,t)u(t)dt + \lambda_2\int_0^b K_2(x,t)u(t)dt \tag{11.371}$$

They may also appear as a combined form in a mixed Volterra-Fredholm integro-differential equation as:

$$u^{(n)}(x) = f(x) + \lambda\int_0^x \int_a^b K(r,t)u(t)dtdr \tag{11.372}$$

There are various techniques that we can use to solve for their solutions, such as the variational iterative method, Laplace transform technique, wavelet method, Taylor series method, Adomian decomposition method, modified Adomian decomposition method, and direct computation method. For details, we refer readers to the excellent coverage by Wazwaz (2011) and more recent publications on this topic. Since the linear integro-differential equation is a special case of the nonlinear integro-differential equations, in the next section we will demonstrate two techniques via nonlinear Volterra and Fredholm integro-differential equations. More specifically, they are the Adomian decomposition technique for the nonlinear Volterra integro-differential equation and the direct computation technique for the nonlinear Fredholm integro-differential equation.

11.10.1 Nonlinear Volterra Integro-Differential Equation

If the unknown function does not appear linearly inside the integral of (11.369), it becomes a nonlinear Volterra integro-differential equation and is expressed as:

$$u^{(n)}(x) = f(x) + \lambda \int_a^x K(x,t)F(u(t))dt \tag{11.373}$$

Although there are various techniques that can give approximations to (11.373), we will only cover two of them, namely the Adomian decomposition method and the modified Adomian decomposition method. This method was discovered by George Adomian who was a professor at the University of Georgia and had published a number of books on this decomposition method.

11.10.1.1 Adomian Decomposition Method

We now introduce a method called the Adomian decomposition method. We can, in general, integrate both sides n times with respect to x to get

$$u = L^{-1}\{f(x)\} + \lambda L^{-1}\{\int_a^x K(x,t)F(u(t))dt\} \tag{11.374}$$

If $n = 1$ and $a = 0$ (this choice is only for the sake of simplicity), we have

$$u = g(x) + \lambda L^{-1}\{\int_0^x K(x,t)F(u(t))dt\} \tag{11.375}$$

where

$$g(x) = \int_0^x f(t)dt \tag{11.376}$$

Next, we assume that the solution u can be decomposed in an infinite series

$$u(x) = \sum_{n=0}^{\infty} u_n(x) \tag{11.377}$$

In addition, the nonlinear integrand can be decomposed into an infinite series of Adomian polynomials as

$$G(u) = K(x,t)F(u(t)) = \sum_{k=0}^{\infty} A_k \tag{11.378}$$

where A_k are the Adomian polynomials and are defined as

$$A_k = \frac{1}{k!}\left\{\frac{d^k}{d\lambda^k}[G(\sum_{i=0}^{k}\lambda^i u_i)]\right\}_{\lambda=0} \tag{11.379}$$

Substitution of (11.378) and (11.379) into (11.375) gives

$$\sum_{n=1}^{\infty} u_n(x) = g(x) + \lambda L^{-1}\{\int_0^x \sum_{k=0}^{\infty} A_k(s)ds\} \tag{11.380}$$

Comparing each term from both sides starting with the zero order term with the first term on the right hand side:

$$u_0(x) = g(x) \tag{11.381}$$

$$u_{n+1}(x) = L^{-1}\left\{\int_0^x A_n(t)dt\right\}, \quad n \geq 0 \tag{11.382}$$

The explicit forms of these Adomian polynomials are

$$\begin{aligned}
A_0 &= G(u_0) \\
A_1 &= u_1 G'(u_0) \\
A_2 &= u_2 G'(u_0) + \frac{1}{2} u_1^2 G''(u_0) \\
A_3 &= u_3 G'(u_0) + u_1 u_2 G''(u_0) + \frac{1}{3!} u_1^3 G'''(u_0) \\
A_4 &= u_4 G'(u_0) + (\frac{1}{2!} u_2^2 + u_1 u_3) G''(u_0) + \frac{1}{3!} u_1^2 u_2 G'''(u_0) + \frac{1}{4!} u_1^4 G^{(iv)}(u_0)
\end{aligned} \tag{11.383}$$

To see the physical meaning of these Adomian polynomials, we can substitute (11.383) into (11.378) to give

$$\begin{aligned}
G(u) &= A_0 + A_1 + A_2 + A_3 + \ldots \\
&= G(u_0) + (u_1 + u_2 + u_3 + \ldots) G'(u_0) + \ldots \\
&+ \frac{1}{2!}(u_1^2 + 2u_1 u_2 + 2u_1 u_3 + u_2^2 + \ldots) G''(u_0) + \ldots \\
&+ \frac{1}{3!}(u_1^3 + 3u_1^2 u_2 + 3u_1^2 u_3 + 6u_1 u_2 u_3 + \ldots) G'''(u_0) + \ldots \\
&= G(u_0) + (u - u_0) G'(u_0) + \frac{1}{2!}(u - u_0)^2 G''(u_0) + \ldots
\end{aligned} \tag{11.384}$$

We see that Adomian polynomials are a Taylor series expansion about a function u_0 instead of the traditional Taylor series expansion about a point.

Before we consider some examples, it is instructive to consider some nonlinear functional forms of $G(u)$:

Case 1: $G(u) = u^2$

$$\begin{aligned}
A_0 &= G(u_0) = u_0^2 \\
A_1 &= u_1 G'(u_0) = 2u_0 u_1 \\
A_2 &= u_2 G'(u_0) + \frac{1}{2} u_1^2 G''(u_0) = 2u_0 u_1 + u_1^2 \\
A_3 &= u_3 G'(u_0) + u_1 u_2 G''(u_0) + \frac{1}{3!} u_1^3 G'''(u_0) = 2u_0 u_3 + 2u_1 u_2
\end{aligned} \tag{11.385}$$

Case 2: $G(u) = u^3$

$$\begin{aligned} A_0 &= u_0^3 \\ A_1 &= 3u_0^2 u_1 \\ A_2 &= 3u_0^2 u_2 + 3u_0 u_1^2 \\ A_3 &= 3u_0^2 u_3 + 6u_0 u_1 u_2 + u_1^3 \end{aligned} \tag{11.386}$$

Case 3: $G(u) = u^4$

$$\begin{aligned} A_0 &= u_0^4 \\ A_1 &= 4u_0^3 u_1 \\ A_2 &= 4u_0^3 u_2 + 6u_0^2 u_1^2 \\ A_3 &= 4u_0^3 u_3 + 4u_1^3 u_0 + 12u_0^2 u_1 u_2 \end{aligned} \tag{11.387}$$

Case 4: $G(u) = \sin u$

$$\begin{aligned} A_0 &= \sin u_0 \\ A_1 &= u_1 \cos u_0 \\ A_2 &= u_2 \cos u_0 - \frac{1}{2!} u_1^2 \sin u_0 \\ A_3 &= u_3 \cos u_0 - u_1 u_2 \sin u_0 - \frac{1}{3!} u_1^3 \cos u_0 \end{aligned} \tag{11.388}$$

Case 5: $G(u) = \cos u$

$$\begin{aligned} A_0 &= \cos u_0 \\ A_1 &= -u_1 \sin u_0 \\ A_2 &= -u_2 \sin u_0 - \frac{1}{2!} u_1^2 \cos u_0 \\ A_3 &= -u_3 \sin u_0 - u_1 u_2 \cos u_0 + \frac{1}{3!} u_1^3 \sin u_0 \end{aligned} \tag{11.389}$$

Case 6: $G(u) = e^{u_0}$

$$\begin{aligned} A_0 &= e^{u_0} \\ A_1 &= u_1 e^{u_0} \\ A_2 &= (u_2 + \frac{1}{2!} u_1^2) e^{u_0} \\ A_3 &= (u_3 + u_1 u_2 + \frac{1}{3!} u_1^3) e^{u_0} \end{aligned} \tag{11.390}$$

Example 11.19 Solve the following nonlinear integro-differential equation by Adomian decomposition method

$$\frac{du}{dx} = -1 + \int_0^x u^2(t)dt, \quad u(0) = 0 \tag{11.391}$$

Solution: Integration on both sides and in view of (11.378) gives

$$u = -x + L^{-1}\{\int_0^x \sum_{n=0}^{\infty} A_n(t)dt\} \tag{11.392}$$

Substitution of (11.377) into (11.392) gives

$$\sum_{n=1}^{\infty} u_n(x) = -x + L^{-1}\{\int_0^x \sum_{n=0}^{\infty} A_n(t)dt\} \tag{11.393}$$

Thus, we have

$$u_0(x) = g(x) = -x \tag{11.394}$$

Application of (11.385) and (11.394) for the first term

$$u_1(x) = \int_0^x \int_0^s A_0(t)dtds = \int_0^x \int_0^s u_0^2(t)dtds = \int_0^x \int_0^s t^2 dtds = \frac{x^4}{12} \tag{11.395}$$

Similarly, for higher order terms we have

$$u_2(x) = \int_0^x \int_0^s A_1(t)dtds = -\frac{1}{6}\int_0^x \int_0^s t^5 dtds = -\frac{x^7}{252} \tag{11.396}$$

$$u_3(x) = \int_0^x \int_0^s A_2(t)dtds = \frac{5}{336}\int_0^x \int_0^s t^8 dtds = \frac{x^{10}}{6048} \tag{11.397}$$

$$u_4(x) = \int_0^x \int_0^s A_3(t)dtds = -\frac{1}{1008}\int_0^x \int_0^s t^{11} dtds = -\frac{x^{13}}{157248} \tag{11.398}$$

Thus, we have an approximation as

$$u = -x + \frac{x^4}{12} - \frac{x^7}{252} + \frac{x^{10}}{6048} - \frac{x^{13}}{157248} + \ldots \tag{11.399}$$

Clearly, the Adomian decomposition technique can also be used to solve nonlinear Volterra integral equations. This is illustrated in the next example.

Example 11.20 Solve the following nonlinear Volterra integral equation by Adomian decomposition method

$$u = x + \int_0^x u^2(t)dt, \quad u(0) = 0 \tag{11.400}$$

Solution: Using the Adomian decomposition method to (11.400), we obtain

$$\sum_{n=1}^{\infty} u_n(x) = x + \int_0^x \sum_{n=0}^{\infty} A_n(t)dt \tag{11.401}$$

Comparison term by term on both sides gives

$$u_0(x) = x \tag{11.402}$$

$$u_{k+1}(x) = \int_0^x A_k(t)dt, \quad k \geq 0 \tag{11.403}$$

Direct integration of (11.403) term by term gives

$$u_1(x) = \int_0^x u_0^2(t)dt = \frac{1}{3}x^3 \tag{11.404}$$

$$u_2(x) = \int_0^x 2u_0u_1 dt = \frac{2}{15}x^5 \tag{11.405}$$

$$u_3(x) = \int_0^x [2u_0u_2 + u_1^2]dt = \frac{17}{315}x^7 \tag{11.406}$$

Therefore, the Adomian decomposition method gives the following approximation:

$$u = x + \frac{x^3}{3} + \frac{2}{15}x^5 + \frac{17}{315}x^7 + ... \tag{11.407}$$

In fact, the exact solution for this case can be shown to be

$$u(x) = \tan x = x + \frac{x^3}{3} + \frac{2}{15}x^5 + \frac{17}{315}x^7 + ... \tag{11.408}$$

Thus, the Adomian composition does converge to the exact solution.

11.10.1.2 Modified Adomian Decomposition Method

For certain nonlinear integro-differential equations with the nonhomogeneous term consisting of a series of terms, a modified Adomian decomposition method was proposed by Wazwaz (2011). Under certain restrictions, we may arrive at an exact solution due to a so-called noise term phenomenon.

In particular, we will illustrate this using the following nonlinear integral equation:

$$u(x) = f_1(x) + f_2(x) + \lambda \int_0^x K(x,t)F(u(t))dt \tag{11.409}$$

Using the Adomian decomposition method, we have

$$\sum_{n=0}^{\infty} u_n(x) = f_1(x) + f_2(x) + \lambda \int_0^x K(x,t) \sum_{n=0}^{\infty} A_n(t)dt \tag{11.410}$$

Thus, we have

$$u_0(x) = f_1(x) \tag{11.411}$$

$$u_1(x) = f_2(x) + \lambda \int_0^x K(x,t)A_0(t)dt \tag{11.412}$$

$$u_{k+1}(x) = \lambda \int_0^x K(x,t)A_k(t)dt, \quad k \geq 1 \tag{11.413}$$

We will illustrate this in the next example.

Example 11.21 Solve the following nonlinear integral equation by modified Adomian decomposition method

$$u(x) = 1 + 3x - \frac{1}{2}x^2 - x^3 - \frac{3}{4}x^4 + \int_0^x (x-t)u^2(t)dt \tag{11.414}$$

Solution: Using the modified Adomian decomposition method by Wazwaz (2011), we can split the nonhomogeneous term into

$$f_1(x) = 1 + 3x - \frac{1}{2}x^2, \quad f_2(x) = -x^3 - \frac{3}{4}x^4 \tag{11.415}$$

Thus, we have

$$u_0(x) = f_1(x) = 1 + 3x - \frac{1}{2}x^2 \tag{11.416}$$

$$\begin{aligned} u_1(x) &= -x^3 - \frac{3}{4}x^4 + \int_0^x (x-t)A_0(t)dt \\ &= -x^3 - \frac{3}{4}x^4 + \int_0^x (x-t)u_0^2(t)dt \end{aligned} \tag{11.417}$$

Substitution of the result in (11.416) into (11.417) gives

$$\begin{aligned} u_1(x) &= -x^3 - \frac{3}{4}x^4 + x\int_0^x (1 + 6t + 8t^2 - 3t^3 + \frac{1}{4}t^4)dt \\ &- \int_0^x (t + 6t^2 + 8t^3 - 3t^4 + \frac{1}{4}t^5)dt \\ &= \frac{1}{2}x^2 - \frac{1}{12}x^4 - \frac{3}{20}x^5 + \frac{1}{120}x^6 \end{aligned} \tag{11.418}$$

Adding the first two terms, we have

$$u(x) \approx u_0(x) + u_1(x) = 1 + 3x - \frac{1}{2}x^2 + \frac{1}{2}x^2 - \frac{1}{12}x^4 - \frac{3}{20}x^5 + \frac{1}{120}x^6 \tag{11.419}$$

We see that the last term of u_0 cancels the first term in u_1, and these are called noise terms. Wazwaz (2011) discovered that when these noise terms appear, the exact solution may be the first two terms of u_0. However, we must check this by substituting it into the original integral equation.

In particular, we assume that u is $1+3x$ and, thus, we have

$$u^2(x) \approx (1+3x)^2 = 1 + 6x + 9x^2 \tag{11.420}$$

$$\int_0^x (x-t)(1 + 6t + 9t^2)dt = \frac{x^2}{2} + x^3 + \frac{3}{4}x^4 \tag{11.421}$$

Using these results, we find that (11.414) is satisfied exactly. This is called the noise term phenomenon.

We want to emphasize that only certain types of integral equations will contain noise terms and will lead to the exact solution. We see that the exact solution must be contained in the first nonhomogeneous term $f_1(x)$. Even if there are noise terms in the first two terms in the solution series, it does not automatically lead to the exact solution. It is necessary to show that the non-canceled terms of $u_0(x)$ indeed satisfied the original equation. If it is not the exact solution, we have to continue our normal calculation for the remaining Adomian polynomials. In general, it has been found that the Adomian decomposition method converges faster than the Picard iterative method. It can be used to solve either linear or nonlinear, and either an integro-differential equation or integral equation. So, it is highly recommended for solving Volterra type integral equations or Volterra integro-differential equations.

11.10.2 Nonlinear Fredholm Integro-Differential Equation

In this section, we consider another commonly encountered nonlinear Fredholm integro-differential equation:

$$u^{(n)}(x) = f(x) + \int_a^b K(x,t)F(u(t))dt, \quad u^{(k)}(0) = b_k, \quad 0 \le k \le n-1 \qquad (11.422)$$

We will only consider the case that the kernel $K(x,t)$ is separable or a P-G (Pincherle-Goursat) kernel. That is, the kernel is written as:

$$K(x,t) = \sum_{k=1}^{n} g_k(x)h_k(t) \qquad (11.423)$$

Substitution of (11.423) into (11.422) gives

$$\begin{aligned} u^{(n)}(x) &= f(x) + g_1(x)\int_a^b h_1(t)F(u(t))dt + g_2(x)\int_a^b h_2(t)F(u(t))dt + \dots \\ &\quad + g_n(x)\int_a^b h_n(t)F(u(t))dt \\ &= f(x) + \alpha_1 g_1(x) + \alpha_2 g_2(x) + \dots + \alpha_n g_n(x) \end{aligned} \qquad (11.424)$$

where

$$\alpha_k = \int_a^b h_k(t)F(u(t))dt, \quad 1 \le k \le n \qquad (11.425)$$

We can integrate both sides of (11.424) n times to get

$$\begin{aligned} u(x) &= u(0) + xu'(0) + \frac{1}{2!}x^2 u''(0) + \dots + \frac{1}{(n-1)!}x^{n-1}u^{(n-1)}(0) \\ &\quad + L^{-1}\{f(x) + \alpha_1 g_1(x) + \alpha_2 g_2(x) + \dots + \alpha_n g_n(x)\} \end{aligned} \qquad (11.426)$$

where L^{-1} is the n-fold integral operator. When (11.426) is substituted into (11.425), we will have n equations for all the constants α_k ($k = 1,2,\dots,n$).

This procedure is called the direct computation method by Wazwaz (2011), and it is illustrated in the next example.

Example 11.22 Solve the following nonlinear Fredholm integro-differential equation by the direct computation method:

$$\frac{du}{dx} = \cos x - \frac{\pi^2}{4}x + \int_0^{\pi} xtu^2(t)dt, \quad u(0) = 0 \qquad (11.427)$$

Solution: Clearly, the kernel K is separable and thus we have

$$\frac{du}{dx} = \cos x + (\alpha - \frac{\pi^2}{4})x \tag{11.428}$$

where

$$\alpha = \int_0^{\pi} tu^2(t)dt \tag{11.429}$$

Integrating (11.428) with respect to x and applying the boundary condition given in (11.427) leads to

$$u = \sin x + (\frac{\alpha}{2} - \frac{\pi^2}{8})x^2 \tag{11.430}$$

Back substitution of (11.430) into (11.429) gives an equation for α. The two solutions for α are:

$$\alpha = \frac{\pi^2}{4}, \frac{\pi^8 - 96\pi^3 + 576\pi + 96}{4\pi^6} \tag{11.431}$$

There two corresponding solutions for $u(x)$, and they are

$$u(x) = \sin x,\ \sin x - (\frac{12}{\pi^3} + \frac{72}{\pi^5} + \frac{12}{\pi^6})x^2 \tag{11.432}$$

Since the integro-differential is nonlinear, we can have more than one solution.

11.11 SUMMARY AND FURTHER READING

In this chapter, we review the concept of the integral transform, including the Laplace transform, Fourier transform, Hankel transform, Mellin transform, and Hilbert transform. It was shown that for different equations we need a different integral transform technique. The Abel integral equation is discussed with its connection to the tautochrone problem in mechanics. Fredholm integral equations of the first and second kinds are presented, and Fredholm's solution is summarized. We also discuss the adjoint integral equation and its role in the Fredholm alternative theorem. The application of the Fredholm alternative theorem to investigate the existence of a particular solution for a nonhomogeneous integral equation is discussed. Volterra integral equations of the first and second kinds are summarized and the classical solution using the Liouville-Neumann series or Picard successive approximation method is presented. Two special kinds of Volterra integral equations are considered: separable type (that can be solved by direct computation) and convolution type (that can be solved by using the Laplace transform). The relation between the existence of a solution of the integral equation and functional analysis is discussed. We also discuss how to convert a differential equation into an integral equation, or vice versa. Finally, both linear and nonlinear integro-differential equations are discussed. Nonlinear Volterra integro-differential equations are solved using the Adomian decomposition technique, whereas nonlinear Fredholm integro-differential equations are solved by using the direct computation method for the case of separable kernels.

The best textbook on integral transform is Sneddon (1972). An introduction to integral equations can be found in Davis (1962) and Tricomi (1957). For compilation of the solutions of integral equations, we refer to the comprehensive handbook by Polyanin and Manzhirov (2008). For more comprehensive coverage of integro-differential equations, we highly recommend the book by Wazwaz (2011). There are other types of solution techniques for solving integro-differential equations that are not covered in the current chapter. They include the variational iterative method, wavelet-Galerkin method, Taylor series method, and least square method. In the current chapter, for the case of the nonlinear Volterra integro-differential, we cover the Adomian decomposition method and its modified version proposed by Wazwaz (2011), and for the case of nonlinear Fredholm integro-differential we present the direct computation method if the kernel is separable.

11.12 PROBLEMS

Problem 11.1 Consider the following double Fourier transform for three-dimensional space:

$$u(x,y,z)=\frac{1}{2\pi}\int_{-\infty}^{\infty}\int_{-\infty}^{\infty}U(\xi,\eta,z)e^{-i(\xi x+\eta y)}d\xi d\eta \tag{11.433}$$

$$U(\xi,\eta,z)=\frac{1}{2\pi}\int_{-\infty}^{\infty}\int_{-\infty}^{\infty}u(x,y,z)e^{i(\xi x+\eta y)}dxdy \tag{11.434}$$

Applying this double Fourier transform to a three-dimensional PDE, we can convert simultaneously the differentiation with respect to x and y to algebraic forms in ξ and η respectively. Prove the Faltung theorem or convolution theorem for this double Fourier transform:

$$\begin{aligned}&\frac{1}{2\pi}\int_{-\infty}^{\infty}\int_{-\infty}^{\infty}F(\xi,\eta,z)G(\xi,\eta,z)e^{-i(\xi x+\eta y)}d\xi d\eta\\ &=\frac{1}{2\pi}\int_{-\infty}^{\infty}\int_{-\infty}^{\infty}f(u,v,z)\,g(x-u,y-v,z)dudv\end{aligned} \tag{11.435}$$

where F is the double Fourier transform of f:

$$F(\xi,\eta,z)=\frac{1}{2\pi}\int_{-\infty}^{\infty}\int_{-\infty}^{\infty}f(x,y,z)e^{i(\xi x+\eta y)}dxdy \tag{11.436}$$

The function G relates to g in a similar manner.

Problem 11.2 Solve the following half-space problem satisfying the Laplace equation by using the Fourier transform for both x and y axes (see Figure 11.7)

$$\nabla^2 u(x,y,z)=\frac{\partial^2 u}{\partial x^2}+\frac{\partial^2 u}{\partial y^2}+\frac{\partial^2 u}{\partial z^2}=0,\quad 0<z<\infty,-\infty<x,y<\infty \tag{11.437}$$

with boundary conditions

$$u(x,y,0)=f(x,y),\quad u\to 0,\quad \text{as } r=\sqrt{x^2+y^2+z^2}\to\infty \tag{11.438}$$

Ans:

$$u(x,y,z)=\frac{1}{2\pi}\int_{-\infty}^{\infty}\int_{-\infty}^{\infty}F(\xi,\eta)e^{-\sqrt{\xi^2+\eta^2}z}e^{-i(\xi x+\eta y)}d\xi d\eta \tag{11.439}$$

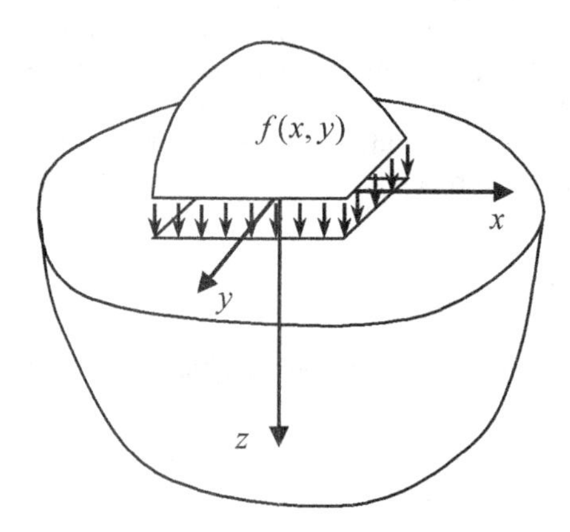

Figure 11.7 Half-space subject to boundary condition

Problem 11.3 Show that the following double Fourier transform can be converted to the Hankel transform of zero order:

$$F(\xi_1,\xi_2)=\frac{1}{2\pi}\int_{-\infty}^{\infty}\int_{-\infty}^{\infty}f(\sqrt{x_1^2+x_2^2})e^{i(\xi_1x_1+\xi_2x_2)}dx_1dx_2 \tag{11.440}$$

That is,

$$F(\rho)=\frac{1}{2\pi}\int_{0}^{\infty}rf(r)J_0(\rho r)dr \tag{11.441}$$

where

$$\rho=\sqrt{\xi_1^2+\xi_2^2},\quad r=\sqrt{x_1^2+x_2^2} \tag{11.442}$$

Hints:

(i) Use the following change of variables

$$\xi_1=\rho\cos\varphi,\quad \xi_2=\rho\sin\varphi\quad x_1=r\cos\theta,\quad x_2=r\sin\theta \tag{11.443}$$

(ii) Show that the area integral can be converted as (use Jacobian):

$$dx_1dx_2=rdrd\theta \tag{11.444}$$

(iii) Show that the following equality

$$\xi_1x_1+\xi_2x_2=r\rho\cos(\theta-\varphi) \tag{11.445}$$

(iv) Show the following identity:

$$F(\xi_1,\xi_2)=\frac{1}{2\pi}\int_{0}^{\infty}rf(r)\int_{0}^{2\pi}e^{i\rho r\cos(\theta-\varphi)}\,d\theta dr \tag{11.446}$$

(v) Because of the periodicity of 2π for θ, we must have:

$$\int_0^{2\pi} e^{i\rho r\cos(\theta-\varphi)}d\theta = \int_0^{2\pi} e^{i\rho r\cos\theta}d\theta \tag{11.447}$$

(vi) Note that the integral representation of the Bessel function of zero order is:

$$J_0(\rho r) = \frac{1}{2\pi}\int_0^{2\pi} e^{i\rho r\cos\theta}d\theta \tag{11.448}$$

Problem 11.4 Consider the Fourier transform of the following function

$$f(t) = \begin{cases} e^{-xt}\Phi(t), & t > 0 \\ 0, & t < 0 \end{cases} \tag{11.449}$$

Show that the result is

$$\Im[f(t)] = \int_0^{\infty} e^{-pt}\Phi(t)dt = L\{\Phi(t)\} \tag{11.450}$$

where $p = x+is$. Note that this is a Laplace transform of $\Phi(t)$. Thus, this problem relates the Fourier transform to the Laplace transform.

Problem 11.5 Find the Fourier transform of the following rectangular function

$$\Pi(t) = \begin{cases} 1, & |t| < 1/2 \\ 0, & |t| \ge 1/2 \end{cases} \tag{11.451}$$

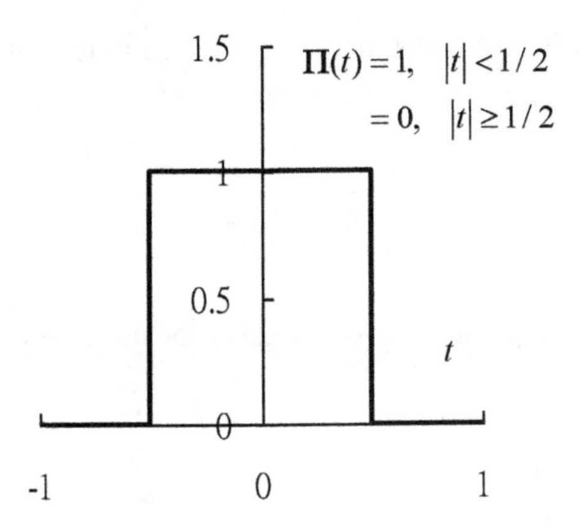

Figure 11.8 Rectangular function

Ans:

$$F(s) = \frac{1}{\sqrt{2\pi}}\frac{\sin(s/2)}{s/2} \tag{11.452}$$

Problem 11.6 Consider the Mellin transform of the following function

$$f(t) = \frac{1}{1+t} \tag{11.453}$$

Hint:

(i) Use a change of variables of

$$1+t=\frac{1}{1-x} \tag{11.454}$$

to show that

$$M[f(t)]=\int_0^1 (1-x)^{-s}x^{s-1}dt \tag{11.455}$$

(ii) Note that the beta function is defined as

$$B(x,y)=\int_0^1 (1-t)^{y-1}t^{x-1}dt \tag{11.456}$$

and can be expressed in terms gamma function as (Lebedev, 1972):

$$B(x,y)=\frac{\Gamma(x)\Gamma(y)}{\Gamma(x+y)} \tag{11.457}$$

(iii) Recall from Chapter 4 that

$$\Gamma(s)\Gamma(1-s)=\frac{\pi}{\sin \pi s} \tag{11.458}$$

Ans:

$$M[\frac{1}{1+t}]=\frac{\pi}{\sin \pi s} \tag{11.459}$$

Problem 11.7 Consider the Mellin transform of the following function

$$f(t)=H(t-t_0)t^z \tag{11.460}$$

Ans:

$$M[f(t)]=-\frac{t_0^{z+s}}{z+s} \tag{11.461}$$

Problem 11.8 Show the validity of (11.224) for the case of n = 2. In particular, show the following

(i)
$$\begin{vmatrix} 1+a_{11} & a_{12} \\ a_{21} & 1+a_{22} \end{vmatrix} = 1+a_{11}+a_{22}+(a_{11}a_{22}-a_{21}a_{12}) \tag{11.462}$$

(ii)
$$\begin{vmatrix} a_{11} & a_{12} \\ a_{21} & a_{22} \end{vmatrix} = \begin{vmatrix} a_{22} & a_{21} \\ a_{12} & a_{11} \end{vmatrix} \tag{11.463}$$

(iii) Use the above results to show that

$$\begin{vmatrix} 1+a_{11} & a_{12} \\ a_{21} & 1+a_{22} \end{vmatrix} = 1+\sum_{r_1}^{2} a_{r_1 r_1} + \frac{1}{2!}\sum_{r_1,r_2} \begin{vmatrix} a_{r_1 r_1} & a_{r_1 r_2} \\ a_{r_2 r_1} & a_{r_2 r_2} \end{vmatrix} \tag{11.464}$$

Problem 11.9 Show the validity of (11.224) for the case of n = 3. In particular, show the following

(i)
$$\begin{vmatrix} 1+a_{11} & a_{12} & a_{13} \\ a_{21} & 1+a_{22} & a_{23} \\ a_{31} & a_{32} & 1+a_{33} \end{vmatrix} = 1+a_{11}+a_{22}+a_{33}+(a_{11}a_{22}-a_{21}a_{12}) \\ +(a_{11}a_{33}-a_{13}a_{31})+(a_{22}a_{33}-a_{23}a_{32}) \\ +a_{11}a_{22}a_{33}+a_{12}a_{23}a_{31}+a_{13}a_{21}a_{32} \\ -a_{13}a_{31}a_{22}-a_{33}a_{21}a_{12}-a_{11}a_{23}a_{32} \quad (11.465)$$

(ii)
$$\begin{vmatrix} a_{11} & a_{12} & a_{13} \\ a_{21} & a_{22} & a_{23} \\ a_{31} & a_{32} & a_{33} \end{vmatrix} = \begin{vmatrix} a_{11} & a_{13} & a_{12} \\ a_{31} & a_{33} & a_{32} \\ a_{21} & a_{23} & a_{22} \end{vmatrix} = \begin{vmatrix} a_{22} & a_{21} & a_{23} \\ a_{12} & a_{11} & a_{13} \\ a_{32} & a_{31} & a_{33} \end{vmatrix} \\ = \begin{vmatrix} a_{22} & a_{23} & a_{21} \\ a_{32} & a_{33} & a_{31} \\ a_{12} & a_{13} & a_{11} \end{vmatrix} = \begin{vmatrix} a_{33} & a_{31} & a_{32} \\ a_{13} & a_{11} & a_{12} \\ a_{23} & a_{21} & a_{22} \end{vmatrix} = \begin{vmatrix} a_{33} & a_{32} & a_{31} \\ a_{23} & a_{22} & a_{21} \\ a_{13} & a_{12} & a_{11} \end{vmatrix} \quad (11.466)$$

(iii) Use the above results to show that

$$\begin{vmatrix} 1+a_{11} & a_{12} & a_{13} \\ a_{21} & 1+a_{22} & a_{23} \\ a_{31} & a_{32} & 1+a_{33} \end{vmatrix} = 1+\sum_{r_1}^{3} a_{r_1r_1} + \frac{1}{2!}\sum_{r_1,r_2}^{3} \begin{vmatrix} a_{r_1r_1} & a_{r_1r_2} \\ a_{r_2r_1} & a_{r_2r_2} \end{vmatrix} + \frac{1}{3!}\sum_{r_1,r_2,r_3}^{3} \begin{vmatrix} a_{r_1r_1} & a_{r_1r_2} & a_{r_1r_3} \\ a_{r_2r_1} & a_{r_2r_2} & a_{r_2r_3} \\ a_{r_3r_1} & a_{r_3r_2} & a_{r_3r_3} \end{vmatrix} \quad (11.467)$$

Problem 11.10 Solve the following integral equation by Laplace transform:

$$y(t) = t + 2\int_0^t \cos(t-u)y(u)du \quad (11.468)$$

Ans:

$$y(t) = t + 2 + 2(t-1)e^t \quad (11.469)$$

Problem 11.11 Solve the following Abel integral equation by Laplace transform:

$$\int_0^t \frac{y(u)}{\sqrt{t-u}} du = 1 + t + t^2 \quad (11.470)$$

Ans:

$$y(t) = \frac{t^{-1/2}}{3\pi}\{3 + 6t + 8t^2\} \quad (11.471)$$

Problem 11.12 Solve the following Abel integral equation by Laplace transform:

$$\int_0^t \frac{y(u)}{(t-u)^{1/3}} du = t + t^2 \tag{11.472}$$

Hint: Using the following identity

$$\Gamma(p)\Gamma(1-p) = \frac{\pi}{\sin p\pi}, \tag{11.473}$$

show that

$$\Gamma(\frac{2}{3}) = \frac{2\pi}{\Gamma(1/3)\sqrt{3}} \tag{11.474}$$

Ans:

$$y(t) = \frac{3\sqrt{3}t^{1/3}}{4\pi}\{3t+2\} \tag{11.475}$$

Problem 11.13 Find the differential equation that is equivalent to the following integral equation

$$y(t) = t^2 - 3t + 4 - 3\int_0^t (t-u)^2 y(u) du \tag{11.476}$$

Ans:

$$y'''(t) + 6y(t) = 0, \quad y(0) = 4, \quad y'(0) = -3, \quad y''(0) = 2 \tag{11.477}$$

Problem 11.14 Find the differential equation that is equivalent to the following integral equation

$$y(t) + \int_0^t (t^2 + 4t - ut - u - 2) y(u) du = 0 \tag{11.478}$$

Ans:

$$y''' + (3t-2)y'' + (t+10)y' + 3y = 0, \quad y(0) = 0, \quad y'(0) = 0, \quad y''(0) = 0 \tag{11.479}$$

Problem 11.15 Solve the following integro-differential equation by Laplace transform:

$$\frac{dy}{dx} = \int_0^t \cos(t-u) y(u) du, \quad y(0) = 1 \tag{11.480}$$

Ans:

$$y(t) = 1 + \frac{1}{2}t^2 \tag{11.481}$$

Problem 11.16 Solve the following integro-differential equation by Laplace transform:

$$\int_0^t y'(u) y(t-u) du = 24t^3, \quad y(0) = 0 \tag{11.482}$$

Ans:

$$y(t) = \pm\frac{16}{\sqrt{\pi}}t^{3/2} \tag{11.483}$$

Problem 11.17 Solve the following nonlinear integro-differential equation by the Adomian decomposition method and collect seven terms in the series

$$\frac{du}{dx} = -1 + \int_0^x u^2(t)dt, \quad u(0) = 0 \tag{11.484}$$

Ans:

$$u = -x + \frac{x^4}{12} - \frac{x^7}{252} + \frac{x^{10}}{6048} - \frac{x^{13}}{157248} + \frac{79x^{16}}{264176640} - \frac{1177x^{19}}{135522616320} + \dots \tag{11.485}$$

Problem 11.18 Solve the following nonlinear integro-differential equation by the direct computation method

$$\frac{du}{dx} = e^x + \frac{1-e^2}{2}x + \int_0^1 xu^2(t)dt, \quad u(0) = 1 \tag{11.486}$$

Ans:

$$u(x) = e^x, e^x + (30-10e)x^2 \tag{11.487}$$

Problem 11.19 Prove the validity of (11.385) to (7.390).

Problem 11.20 Solve the following ODE with non-constant coefficient by the Laplace transform:

$$t\frac{d^2y}{dt^2} + \frac{dy}{dt} + 9ty = 0, \quad y(0) = 3, \quad y'(0) = 0 \tag{11.488}$$

Hint: Refer to (11.33) and note the following result

$$J_0(at) = L^{-1}\{\frac{1}{\sqrt{s^2+a^2}}\} \tag{11.489}$$

where J_0 is the Bessel function of the first kind of zero order defined in Chapter 4.

Ans:

$$y(t) = 3J_0(3t) \tag{11.490}$$

Problem 11.21 Solve the following ODE by the Laplace transform:

$$y'' - 3y' + 2y = 4e^{2t}, \quad y(0) = -3, \quad y'(0) = 5 \tag{11.491}$$

Ans:

$$y(t) = -7e^t + 4e^{2t} + 4te^{2t} \tag{11.492}$$

Problem 11.22 Apply Mellin transform to solve the following wedge problem:

$$\nabla^2 u = \frac{\partial^2 u}{\partial r^2} + \frac{1}{r}\frac{\partial u}{\partial r} + \frac{1}{r^2}\frac{\partial^2 u}{\partial \theta^2} = 0 \tag{11.493}$$

$$u(r, \pm\alpha) = H(R - r) \tag{11.494}$$

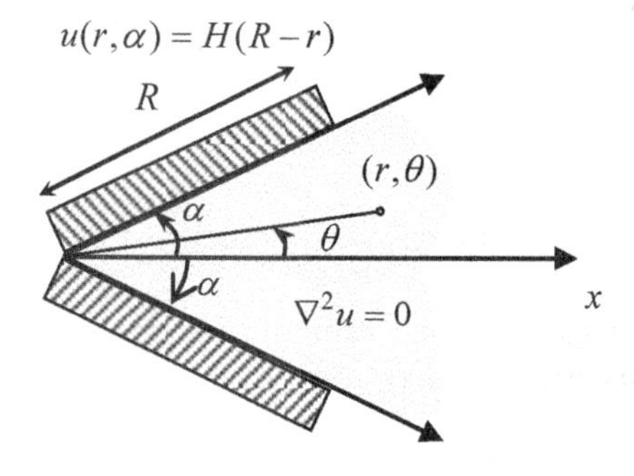

Figure 11.9 Wedge domain of potential theory subject to finite uniform boundary conditions

Ans:

$$u(r,\theta) = \frac{1}{2\pi i}\int_{c-i\infty}^{c+i\infty} r^{-s} \frac{R^s \cos(s\theta)}{s\cos(s\alpha)} ds \tag{11.495}$$

CHAPTER TWELVE

Asymptotic Expansion and Perturbation

12.1 INTRODUCTION

In this chapter, we will discuss asymptotic expansion and perturbation analysis. Although these two methods can be studied independently, they are related. In fact, asymptotic series lays the fundamentals for perturbation analysis. The method of asymptotic expansion has been studied since the time of Euler, and, in fact, it is one of major weapons of Euler. However, asymptotic series are non-convergent, and thus caution must be taken in using them. Nevertheless, many famous results of Euler were, in fact, obtained by manipulating non-convergent asymptotic series. Although Euler never found any problems in his analysis with non-convergent series, Pringsheim did find counter-examples that non-convergent series led to erroneous conclusions.

The works of Abel and Cauchy on non-convergent series led to the banishing of the non-convergent series by mathematicians for more than 25 years. It is the works by Poincaré in 1886 and Stieltjes in 1886 that refreshed the interest in divergent series and founded the development of asymptotic series analysis. It turns out that solutions expressed in terms of asymptotic series are accurate and useful if a finite number of terms are used, but its accuracy deteriorates as "too many" terms are used (because of its eventual non-convergent nature). The term "asymptotic series" was coined by Poincaré whereas it was called "semi-convergent series" by Stieltjes and "convergently beginning series" by Emde. Many mathematicians had studied asymptotic series, including Cesaro, Borel, Le Roy, Mellin, Mittag-Leffler, Van Vleck, Barnes, Hardy and Littlewood. A number of studies considered transforming asymptotic expansions into convergent series, including Airey, van der Corput, Miller, van Wijngaarden and Watson. Note that Poincaré's theory of asymptotic series considered both summability of divergent series and the asymptotic solution of differential equations. These asymptotic solutions are useful in calculating planetary positions. In fact, perturbation methods using asymptotic series were motivated by its application in celestial mechanics. The perturbation theory is related to the three-body problem, which considered the perturbations of the motion of two bodies being influenced by the existence of a much smaller third body. An exact solution could not be found and series solutions of finite terms were formulated to obtain an approximation. The main contributors to the three-body problem were Euler, Clairaut, Lagrange and Laplace.

In considering Prandtl's boundary layer in viscous fluids, Friedrichs in 1941 matched the inner and outer solutions of the boundary layer. Friedrichs was a student of Courant and a post-doc of von Karman. This "matched asymptotic expansion" was popularized by the books of van Dyke in 1964 and Cole in 1968. The method of multiple scales was introduced by Swedish astronomer Lindstedt in 1883 and by Poincaré in 1886 for solving problems with more than one intrinsic time scale. The method of perturbation is particularly useful in obtaining accurate approximate solutions for nonlinear differential equations. Such approximate solutions provide insight into the physics and mechanics of the problem, which computer generated simulations cannot produce. Sometimes, it also happens that a different approximate solution expressed in convergent series actually converges to the true value much slower than a divergent asymptotic solution (as long as we are not too greedy on the number of terms in the divergent series) (e.g., Bleistein and Handelsman, 1986).

12.2 ASYMPTOTIC EXPANSION

12.2.1 Order Symbols

We first introduce the so-called order symbols or Bachmann-Landau symbols, which were introduced by Bachmann in 1894 and popularized by Landau in 1909. In particular, the "big oh" and "small oh" are defined as

$$f = O(\phi), \quad g = o(\phi) \tag{12.1}$$

where f and g satisfy the following equations respectively

$$\lim_{\varepsilon \to \varepsilon_0} \frac{f(\varepsilon)}{\phi(\varepsilon)} = \text{finite}, \quad \lim_{\varepsilon \to \varepsilon_0} \frac{g(\varepsilon)}{\phi(\varepsilon)} = 0 \tag{12.2}$$

where ε_0 is typically a small parameter larger than zero. For example, if $\phi = \varepsilon^2$ and $\varepsilon_0 = 0$, we have

$$f = O(\varepsilon^2), \text{ then } \lim_{\varepsilon \to 0} \frac{f}{\varepsilon^2} = k \tag{12.3}$$

where k is a constant, and we have

$$g = o(\varepsilon^2), \text{ then } \lim_{\varepsilon \to 0} \frac{g}{\varepsilon^2} = 0 \tag{12.4}$$

In other words, f contains a leading order term of ε^2 and g contains a leading order term of ε^3 or higher.

12.2.2 Asymptotic Series

Among all divergent series, a particular type is known as the asymptotic series. Despite the fact that this series diverges, the value of the functions can be calculated with a high degree of accuracy if we take the sum of a suitable number of terms of such series. The asymptotic series have many properties that are analogous to those of the convergent series. The asymptotic expansion of a function is denoted as

$$f(z) \sim \sum_{n=0}^{\infty} A_n z^{-n} \tag{12.5}$$

for $z \to \infty$. Let us consider a particular function, which closely resembles exponential integral (Abramowitz and Stegun, 1964):

$$f(z) = ze^z \int_z^\infty \frac{e^{-t}}{t} dt \tag{12.6}$$

We will now derive an asymptotic series for (12.6). Applying integration by parts, we get

$$\begin{aligned} f(z) &= -ze^z \int_z^\infty \frac{de^{-t}}{t} = -ze^z \{ \frac{e^{-t}}{t} \Big|_z^\infty - (-1) \int_z^\infty \frac{e^{-t}}{t^2} dt \} \\ &= 1 - ze^z \int_z^\infty \frac{e^{-t}}{t^2} dt \end{aligned} \tag{12.7}$$

We can apply integration by parts another time to (12.7) to get

$$f(z) = 1 - \frac{1!}{z} + (-1)^2 ze^z 2! \int_z^\infty \frac{e^{-t}}{t^3} dt \tag{12.8}$$

Integration by parts again yields

$$f(z) = 1 - \frac{1!}{z} + \frac{2!}{z^2} + (-1)^3 ze^z 3! \int_z^\infty \frac{e^{-t}}{t^4} dt \tag{12.9}$$

Finally, repeating the process of integration by parts gives

$$f(z) = 1 - \frac{1!}{z} + \frac{2!}{z^2} - \frac{3!}{z^3} + \ldots + \frac{(-1)^{n-1}(n-1)!}{z^{n-1}} + (-1)^n n! ze^z \int_z^\infty \frac{e^{-t}}{t^{n+1}} dt \tag{12.10}$$

The last term is the error of the expansion for the first n terms. This series diverges for all values of z, and appears useless. However, we are going to show that it does give an accuracy solution if a finite number of terms are used. It was given that an exact convergent series of $f(z)$ is

$$f(z) = ze^z(-\gamma - \ln z + z - \frac{z^2}{2 \cdot 2!} + \frac{z^2}{3 \cdot 3!} - \frac{z^2}{4 \cdot 4!} + \ldots) \tag{12.11}$$

where γ is Euler's constant, which is discussed in Appendix E. Numerical results of the sum of these series for $z = 10$ are compiled in Table 12.1. We see that the sum of the first 9 terms for $f(10)$ in the asymptotic series given in (12.10) is 0.9158192 and for the first 10 terms is 0.91545632. In fact, the exact value of $f(10)$ is between these values. The divergent series actually converges for the first 10 terms before it diverges (see the result of 40 terms or more in Table 12.1). We need only 6 terms to achieve 3-decimal accuracy. A more accurate calculation shows that the sum is about 0.915633339264773. We also evaluate the sum of the exact convergent series given in (12.11), and these results are also compiled in Table 12.1 for comparison. If we take the first 9 terms, the result is totally out of order. We need 40 terms to achieve 3-decimal accuracy. This series does converge but it converges very slowly compared to the asymptotic series. Therefore, we have illustrated that for a large value of z (in this case we use $z = 10$) the asymptotic series is practically much better than the exact series. This is the reason why the asymptotic series, even though it diverges, has been found very useful in getting approximate solutions. This observation is also true for approximate solution of differential equations in terms of asymptotic series.

In general, the following properties of asymptotic series should be noted:

1. Operations of addition, substitution, multiplication and rising to a power can be performed on asymptotic series just as on absolutely convergent series.

2. One asymptotic series can be divided by another asymptotic series provided that A_0 in (12.5) is not zero. The series obtained as a result of division will also be asymptotic.
3. An asymptotic series can be integrated term by term, and the resultant series will also be asymptotic. In contrast, differentiation of an asymptotic series is, in general, not permissible.
4. A single asymptotic expansion can represent more than one function. On the other hand, a given function can be expanded in an asymptotic series in only one manner.

Table 12.1 Comparison of convergence asymptotic divergent series and convergent series of $f(10)$

	$f(10)$	
Term	Asymptotic divergent series	Exact convergent series
1	1.000000000	1568328.251
2	0.900000000	−3938288.198
3	0.920000000	8298637.243
4	0.914000000	−14645597.96
5	0.916400000	22065178.37
6	0.915200000	−28922010.97
7	0.915920000	33511282.09
8	0.915416000	−34775132.2
9	0.915819200	32668239.94
10	0.915456320	−28030794.99
11	0.915819200	22133696.69
12	0.915420032	−16186401.12
40	−16252520.109560400	0.915620319
41	65339008.215229400	0.915636379
42	−269186257.916409000	0.915632646
43	1135819859.836470000	0.915633494
44	−4905706446.500910000	0.915633306
45	21677009301.383600000	0.915633346
46	−97945211564.096600000	0.915633338
47	452317004417.112000000	0.915633339

When two functions are asymptotically equivalent,

$$f(\varepsilon) \sim g(\varepsilon) \tag{12.12}$$

we mean that as $\varepsilon \to 0$ we have

$$f(\varepsilon) = g(\varepsilon)\{1 + o(1)\} \tag{12.13}$$

For example, we have

$$x + 2 \sim x, \quad x \to \infty \tag{12.14}$$

$$\varepsilon + \varepsilon^3 \sim \varepsilon, \quad \varepsilon \to 0 \tag{12.15}$$

$$2\sinh x \sim e^x, \quad x \to \infty \tag{12.16}$$

We normally express an asymptotic sequence as

$$f(x) \sim f_0(x) + \varepsilon f_1(x) + \varepsilon^2 f_2(x) + ..., \quad \varepsilon \to 0 \tag{12.17}$$

Whenever an asymptotic series in the form of (12.17) exists, the perturbation method for regular expansion can be used. In general, there are other forms of asymptotic sequences ϕ_n for the perturbation of a function. Here are some examples of asymptotic sequences:

$$\{\phi_n(x)\} = (x - x_0)^n, \quad x \to x_0 \tag{12.18}$$

$$\{\phi_n(x)\} = \frac{1}{x^n}, \quad x \to \infty \tag{12.19}$$

$$\{\phi_n(x)\} = \frac{1}{x^{\lambda_n}}, \quad x \to \infty, \quad \lambda_{n+1} > \lambda_n \tag{12.20}$$

12.3 REGULAR PERTURBATION METHOD FOR ODE

In this section, we consider the perturbation method for ODE. In general, three main steps are involved in the perturbation method:

1. Identify a small parameter ε in the original problem.
2. Assume an expression for the solution in the form of a perturbation series and find the differential equation for each order of approximation of the perturbation series with appropriate boundary conditions.
3. The solution of the differential equation for each order is solved accordingly.
4. The final answer to the original ODE can be found by substituting the solution of each order into the assumed perturbation series.

To illustrate the idea, we consider the following ODE

$$y'' + 2\varepsilon y' - y = 0 \tag{12.21}$$

where ε is a small parameter of the ODE. We seek the following regular perturbation series

$$y \sim y_0(x) + \varepsilon y_1(x) + \varepsilon^2 y_2(x) + ... \tag{12.22}$$

Substitution of (12.22) into (12.21) yields:

$$\begin{aligned} &[y_0''(x) + \varepsilon y_1''(x) + \varepsilon^2 y_2''(x) + ...] + 2\varepsilon[y_0'(x) + \varepsilon y_1'(x) + \varepsilon^2 y_2'(x) + ...] \\ &-[y_0(x) + \varepsilon y_1(x) + \varepsilon^2 y_2(x) + ...] \sim 0 \end{aligned} \tag{12.23}$$

By collecting coefficients of different orders of ε, we have

$$O(1): \quad y_0''(x) - y_0(x) = 0 \tag{12.24}$$

$$O(\varepsilon): \quad y_1''(x) - y_1(x) = -2y_0'(x) \tag{12.25}$$

$$O(\varepsilon^2): \quad y_2''(x) - y_2(x) = -2y_1'(x) \tag{12.26}$$

$$\cdots$$

$$O(\varepsilon^n): \quad y_n''(x) - y_n(x) = -2y_{n-1}'(x) \tag{12.27}$$

This set of ODEs for each level of approximation illustrates that the ODE is the same except that the first order does not have the nonhomogeneous term whereas all other nonhomogeneous terms depend on the solution of the previous solution (e.g., the ODE for y_n depends on y_{n-1} for $n \geq 1$). Thus, we have to solve the solution order by order.

The ODE given in (12.21) is of course linear. In general, for nonlinear ODEs the situation is much more difficult. For differential equations arising from practical problems, the solution normally behaves well at most points of the domain, except at isolated singular points (such as the singularity at the origin in the Bessel equation). However, solutions of nonlinear differential equations possess a richer spectrum of singular behaviors. For linear ODEs, the singularities are fixed by the coefficient functions of the ODE and are independent of the choice of the initial or boundary conditions. They are called fixed singularities. For nonlinear ODEs, in addition to fixed singularities, there are also singularities that move around as the initial or boundary conditions vary. These are called movable singularities (compare Section 4.14). This makes asymptotic analysis very difficult for nonlinear ODEs.

The perturbation method for nonlinear ODEs is illustrated by the projectile problem considered in the next section.

12.3.1 Projectile Problem

In this section, we illustrate the perturbation method by considering the nonlinear problem of a projectile. Assume that a projectile is shooting up from the surface of the Earth, as shown in Figure 12.1. Using Newton's second law, we obtain the equation of motion of the projectile as

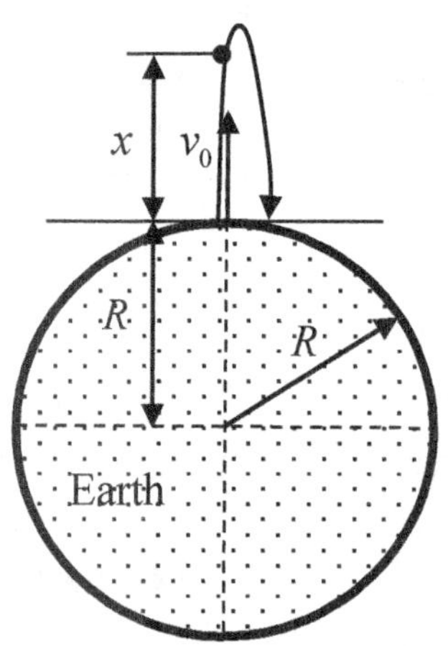

Figure 12.1 Projectile problem on surface of the Earth

$$m\frac{d^2x}{dt^2} = -\frac{GmM}{(R+x)^2} \tag{12.28}$$

for $t > 0$, where G is the universal gravitational constant, m is the mass of the projectile, R and M are the radius and mass of the Earth, and x is the position of the projectile measured from the surface of the Earth. The initial conditions are

$$x(0) = 0, \quad \frac{dx(0)}{dt} = v_0 \tag{12.29}$$

We note that at the ground surface the weight of the projectile can be expressed as:

$$mg = \frac{GmM}{R^2} \tag{12.30}$$

Therefore, the mass of the earth M and universal gravitational constant G can be replaced by

$$GM = gR^2 \tag{12.31}$$

Substitution of (12.31) into (12.28) leads to the following differential equation

$$\frac{d^2x}{dt^2} = -\frac{gR^2}{(R+x)^2} \tag{12.32}$$

For most applications, we have $x << R$, and (12.32) leads to the usual approximation of

$$\frac{d^2x_0}{dt^2} = -g \tag{12.33}$$

where x_0 is used to denote the fact that the solution of (12.33) only leads to the first approximation of x. We recover the gravitational constant of 9.81 m/s^2 on the surface of the Earth, but for an intercontinental missile or rocket projectile, the vertical distance x becomes not negligible compared to Earth's radius R. The initial conditions (12.29) can be used as

$$x_0(0) = 0, \quad \left.\frac{dx_0}{dt}\right|_{t=0} = v_0 \tag{12.34}$$

Integrating (12.33) once and twice, we get

$$\frac{dx_0}{dt} = -gt + a \tag{12.35}$$

$$x_0 = -\frac{1}{2}gt^2 + at + b \tag{12.36}$$

By using the boundary condition given in (12.34), the unknown constants can be found as $a = v_0$ and $b = 0$ and the final solution is

$$x_0 = -\frac{1}{2}gt^2 + v_0 t \tag{12.37}$$

The maximum height that the projectile reaches can be found by setting the upward velocity to zero in (12.35) and substituting the result in (12.37) as

$$x_0 = \frac{v_0^2}{2g} \tag{12.38}$$

We now return to the nonlinear ODE given in (12.32). We first normalize the ODE by using the following normalized parameters:

$$\tau = \frac{t}{t_c} = \frac{gt}{v_0}, \quad y(\tau) = \frac{x(t)}{x_c} = \frac{x(t)g}{v_0^2} \tag{12.39}$$

where t_c and x_c are chosen such that the initial condition can be simplified. Substitution of (12.39) into (12.32) gives

$$\frac{d^2y}{d\tau^2} = -\frac{1}{(1+\varepsilon y)^2} \tag{12.40}$$

where the small parameter ε is defined as

$$\varepsilon = \frac{x_c}{R} = \frac{v_0^2}{Rg} \tag{12.41}$$

For the radius of the Earth being 6300 km, we have

$$\varepsilon = \frac{v_0^2}{Rg} = 1.618 \times 10^{-8} v_0^2 (s^2/m^2) \tag{12.42}$$

If the initial velocity $v_0 \approx 7.86$ km/s, we have $\varepsilon \approx 1$. This is extremely difficult to achieve technically, so we can assume ε is a small parameter. Thus, we seek the following asymptotic series

$$y \sim y_0(\tau) + \varepsilon^\beta y_1(\tau) + \ldots \tag{12.43}$$

where $\beta > 0$ is a parameter to be determined. Substitution of (12.43) into (12.40) gives

$$\begin{aligned} y_0''(\tau) + \varepsilon^\beta y_1''(\tau) + \ldots &= -\frac{1}{\{1+\varepsilon[y_0(\tau) + \varepsilon^\beta y_1(\tau) + ..]\}^2} \\ &= -1 + 2\varepsilon y_0(\tau) + \ldots \end{aligned} \tag{12.44}$$

To balance the term of ε on both sides, we must have $\beta = 1$. The initial conditions given in (12.34) lead to

$$y(0) = \frac{x(0)}{x_c} = 0, \quad y'(0) = \frac{t_c}{x_c} x'(0) = \frac{t_c}{x_c} v_0 = 1 \tag{12.45}$$

For the t_c and x_c defined in (12.39), we arrive at the simple form of the second condition given in (12.45). Collecting the constant and linear order term of ε, we get

$$\begin{aligned} O(1):\quad & y_0''(\tau) = -1, \\ & y_0(0) = 0, \quad y_0'(0) = 1 \end{aligned} \tag{12.46}$$

$$\begin{aligned} O(\varepsilon):\quad & y_1''(\tau) = 2y_0(\tau), \\ & y_1(0) = 0, \quad y_1'(0) = 0 \end{aligned} \tag{12.47}$$

The first order solution can be integrated directly from (12.46) as:

$$y_0(\tau) = -\frac{1}{2}\tau^2 + a\tau + b \tag{12.48}$$

Using the initial conditions, we have

$$y_0(\tau) = -\frac{1}{2}\tau^2 + \tau = \tau(1 - \frac{\tau}{2}) \tag{12.49}$$

Substitution of (12.49) into (12.47) and integration gives

$$y_1(\tau) = -\frac{1}{12}\tau^4 + \frac{1}{3}\tau^3 + a\tau + b \tag{12.50}$$

Using the initial conditions given in (12.47), we get

$$y_1(\tau) = \frac{1}{3}\tau^3(1 - \frac{1}{4}\tau) \tag{12.51}$$

Finally, we obtain a two-term asymptotic solution as

$$y(\tau) \sim \tau(1 - \frac{1}{2}\tau) + \frac{1}{3}\varepsilon\tau^3(1 - \frac{1}{4}\tau) + ... \tag{12.52}$$

The first term in (12.52) is the approximation for the case of a uniform gravitational field. This solution is only valid for

$$0 \le \tau \le \tau_h \tag{12.53}$$

where τ_h is the time for the projectile to come back to the surface of the Earth.

12.3.2 Projectile Problem with Air Resistance

In this section, we extend the projectile problem to include the effect of air resistance approximately. The equation of motion of the projectile becomes:

$$\frac{d^2x}{dt^2} = -\frac{gR^2}{(R+x)^2} - \frac{k}{(R+x)}\frac{dx}{dt} \tag{12.54}$$

Using the same normalized time and distance defined in (12.39), we have

$$\frac{d^2y}{d\tau^2} = -\frac{1}{(1+\varepsilon y)^2} - \frac{\alpha}{(1+\varepsilon y)}\frac{dy}{d\tau} \tag{12.55}$$

where

$$\alpha = \frac{kv_0}{gR} \tag{12.56}$$

The definition of ε given in (12.42) suggests that α is a function of ε. However, to simplify our analysis, we assume that α is a constant. The following asymptotic series is assumed

$$y \sim y_0(\tau) + \varepsilon^\beta y_1(\tau) + ... \tag{12.57}$$

Substitution of (12.57) into (12.55) gives

$$\begin{aligned} y_0''(\tau) + \varepsilon^\beta y_1''(\tau) + ... &= -\frac{1}{\{1 + \varepsilon[y_0(\tau) + \varepsilon^\beta y_1(\tau) + ...]\}^2} \\ &\quad - \frac{\alpha}{1 + \varepsilon[y_0(\tau) + \varepsilon^\beta y_1(\tau) + ...]}[y_0'(\tau) + \varepsilon^\beta y_1'(\tau) + ...] \end{aligned} \tag{12.58}$$

Using Taylor series expansion, we have

$$y_0''(\tau) + \varepsilon^\beta y_1''(\tau) + ... = -(1 + \alpha y_0') + (2y_0 + \alpha y_0 y_0')\varepsilon - \varepsilon^\beta \alpha y_1' + ... \tag{12.59}$$

The corresponding initial conditions become

$$y(0) = y_0(0) + \varepsilon y_1(0) + ... = 0 \tag{12.60}$$

$$y'(0) = y_0'(0) + \varepsilon y_1'(0) + ... = 1 \tag{12.61}$$

Collecting the constant and linear order terms of ε, we get

$$O(1): \quad y_0'' + \alpha y_0' = -1, \qquad y_0(0) = 0, \quad y_0'(0) = 1 \tag{12.62}$$

$$O(\varepsilon): \quad y_1'' + \alpha y_1' = y_0(2 + \alpha y_0'), \qquad y_1(0) = 0, \quad y_1'(0) = 0 \tag{12.63}$$

For the first order solution, we assume an exponential solution form for the homogeneous form as

$$y_{0h}(\tau) = e^{r\tau} \tag{12.64}$$

The corresponding characteristics equation of the homogeneous form of (12.62) becomes

$$r(r + \alpha) = 0 \tag{12.65}$$

The homogeneous solution becomes

$$y_{0h} = A + Be^{-\alpha\tau} \tag{12.66}$$

Since the constant is a homogeneous solution (i.e., matching a single characteristic root), we assume the following particular solution form (see Section 3.3.3)

$$y_{0p}(\tau) = C\tau \tag{12.67}$$

Substitution of (12.67) into the original ODE in (12.62) gives the particular solution and combining the homogeneous solution with the particular solution gives

$$y_0 = A + Be^{-\alpha\tau} - \frac{1}{\alpha}\tau \tag{12.68}$$

Substitution of (12.68) into the initial conditions given in (12.62) leads to

$$A = -B = \frac{1}{\alpha}(1 + \frac{1}{\alpha}) \tag{12.69}$$

Finally, we have the first order of approximation as

$$y_0 = \frac{1}{\alpha}(1 + \frac{1}{\alpha})\left\{1 - e^{-\alpha\tau}\right\} - \frac{1}{\alpha}\tau \tag{12.70}$$

Substitution of (12.70) into the right hand side of (12.63) gives

$$y_1'' + \alpha y_1' = \frac{1}{\alpha}(1 + \frac{1}{\alpha}) + \frac{(1+\alpha)}{\alpha}e^{-\alpha\tau} - \frac{(1+\alpha)^2}{\alpha^2}e^{-2\alpha\tau} - \frac{\tau}{\alpha} - \frac{(1+\alpha)}{\alpha}\tau e^{-\alpha\tau} \tag{12.71}$$

The homogeneous solution is again

$$y_{1h} = C + De^{-\alpha\tau} \tag{12.72}$$

Using the method of undetermined coefficient, we assume the particular solution as

$$y_{1p} = E\tau + F\tau e^{-\alpha\tau} + Ge^{-2\alpha\tau} + H\tau^2 + I\tau^2 e^{-\alpha\tau} \tag{12.73}$$

Differentiation of (12.73) gives

$$y_{1p}' = E + Fe^{-\alpha\tau} + (2I - F\alpha)\tau e^{-\alpha\tau} - 2\alpha Ge^{-2\alpha\tau} + 2H\tau - \alpha I\tau^2 e^{-\alpha\tau} \tag{12.74}$$

The second derivative of (12.73) gives

$$y_{1p}'' = 2(I - F\alpha)e^{-\alpha\tau} + (\alpha^2 F - 4\alpha I)\tau e^{-\alpha\tau} + 4\alpha^2 G e^{-2\alpha\tau} + 2H + \alpha^2 I \tau^2 e^{-\alpha\tau} \tag{12.75}$$

Combining (12.74) and (12.75) gives

$$y_{1p}'' + \alpha y_{1p}' = (2I - F\alpha)e^{-\alpha\tau} - 2\alpha I \tau e^{-\alpha\tau} + 2H + \alpha E + 2\alpha^2 G e^{-2\alpha\tau} + 2H\alpha\tau \tag{12.76}$$

Equating (12.71) and (12.76) gives

$$\begin{gathered} 2H + \alpha E = \frac{1}{\alpha}(1 + \frac{1}{\alpha}); \quad 2I - F\alpha = \frac{(1+\alpha)}{\alpha}; \\ 2\alpha I = \frac{(1+\alpha)}{\alpha}, \quad 2\alpha^2 G = -\frac{(1+\alpha)^2}{\alpha^2}; \quad 2H\alpha = -\frac{1}{\alpha} \end{gathered} \tag{12.77}$$

This provides a system of 5 equations for 5 unknowns, and the solution of the system gives

$$E = \frac{2+\alpha}{\alpha^3} \tag{12.78}$$

$$F = \frac{1-\alpha^2}{\alpha^3} \tag{12.79}$$

$$G = -\frac{(1+\alpha)^2}{2\alpha^4} \tag{12.80}$$

$$H = -\frac{1}{2\alpha^2} \tag{12.81}$$

$$I = \frac{1+\alpha}{2\alpha^2} \tag{12.82}$$

Finally, the general solution is

$$y_1 = C + De^{-\alpha\tau} + \frac{2+\alpha}{\alpha^3}\tau + \frac{1-\alpha^2}{\alpha^3}\tau e^{-\alpha\tau} - \frac{(1+\alpha)^2}{2\alpha^4}e^{-2\alpha\tau} - \frac{1}{2\alpha^2}\tau^2 + \frac{1+\alpha}{2\alpha^2}\tau^2 e^{-\alpha\tau} \tag{12.83}$$

Substitution of these into the initial conditions leads to

$$C + D = \frac{(1+\alpha)^2}{2\alpha^4} \tag{12.84}$$

$$E + F = \alpha D + 2\alpha G \tag{12.85}$$

Solving for C and D, we have

$$C = \frac{\alpha^2 - 7 - 4\alpha}{2\alpha^4} \tag{12.86}$$

$$D = \frac{4 + 3\alpha}{\alpha^4} \tag{12.87}$$

Finally, we obtain the following second order solution

$$y_1=\frac{\alpha^2-7-4\alpha}{2\alpha^4}+\frac{4+3\alpha}{\alpha^4}e^{-\alpha\tau}+\frac{2+\alpha}{\alpha^3}\tau+\frac{1-\alpha^2}{\alpha^3}\tau e^{-\alpha\tau}-\frac{(1+\alpha)^2}{2\alpha^4}e^{-2\alpha\tau}$$
$$-\frac{1}{2\alpha^2}\tau^2+\frac{1+\alpha}{2\alpha^2}\tau^2e^{-\alpha\tau} \tag{12.88}$$

Combining (12.70) and (12.88), we obtain the approximation

$$y\sim\{\frac{1}{\alpha}(1+\frac{1}{\alpha})[1-e^{-\alpha\tau}]-\frac{1}{\alpha}\tau\}+\varepsilon\{\frac{\alpha^2-7-4\alpha}{2\alpha^4}+\frac{4+3\alpha}{\alpha^4}e^{-\alpha\tau}+\frac{2+\alpha}{\alpha^3}\tau$$
$$+\frac{1-\alpha^2}{\alpha^3}\tau e^{-\alpha\tau}-\frac{(1+\alpha)^2}{2\alpha^4}e^{-2\alpha\tau}-\frac{1}{2\alpha^2}\tau^2+\frac{1+\alpha}{2\alpha^2}\tau^2e^{-\alpha\tau}\}+O(\varepsilon^2) \tag{12.89}$$

This solution appears quite different from the case of no air resistance obtained in (12.52). It is not obvious that for $\alpha\to 0$ we can recover (12.52). To show this, for the case that the air resistance is small (i.e., $\alpha\to 0$), we can expand the exponential functions as a Taylor series expansion as:

$$e^{-\alpha\tau}=1-\alpha\tau+\frac{(\alpha\tau)^2}{2}-\frac{(\alpha\tau)^3}{6}+\frac{(\alpha\tau)^4}{24}+... \tag{12.90}$$

$$e^{-2\alpha\tau}=1-2\alpha\tau+\frac{4(\alpha\tau)^2}{2}-\frac{8(\alpha\tau)^3}{6}+\frac{16(\alpha\tau)^4}{24}+... \tag{12.91}$$

Substitution of (12.90) into (12.70), we get

$$y_0=\tau-\frac{\tau^2}{2}(1+\alpha)+O(\alpha^2\tau^2) \tag{12.92}$$

On the other hand, substituting (12.90) and (12.91) into (12.88) and collecting terms of the same order, we find that coefficients for each order are:

$$O(1):\quad \frac{\alpha^2-7-4\alpha}{2\alpha^4}+\frac{4+3\alpha}{\alpha^4}-\frac{(1+\alpha)^2}{2\alpha^4}=0 \tag{12.93}$$

$$O(\tau):\quad -\frac{4+3\alpha}{\alpha^3}+\frac{2+\alpha}{\alpha^3}+\frac{1-\alpha^2}{\alpha^3}+\frac{(1+\alpha)^2}{\alpha^3}=0 \tag{12.94}$$

$$O(\tau^2):\quad \frac{4+3\alpha}{2\alpha^2}-\frac{1-\alpha^2}{\alpha^2}-\frac{(1+\alpha)^2}{\alpha^2}-\frac{1}{2\alpha^2}+\frac{1+\alpha}{2\alpha^2}=0 \tag{12.95}$$

$$O(\tau^3):\quad -\frac{4+3\alpha}{6\alpha}+\frac{1-\alpha^2}{2\alpha}+\frac{2(1+\alpha)^2}{3\alpha}-\frac{1+\alpha}{2\alpha}=\frac{1}{6}(2+\alpha) \tag{12.96}$$

$$O(\tau^4):\quad \frac{4+3\alpha}{24}-\frac{1-\alpha^2}{6}-\frac{(1+\alpha)^2}{3}+\frac{1+\alpha}{4}=-\frac{1}{24}(2+7\alpha+4\alpha^2) \tag{12.97}$$

Using these results, (12.88) finally becomes

$$y_1=\frac{1}{6}(2+\alpha)\tau^3-\frac{1}{24}(2+7\alpha+4\alpha^2)\tau^4+O(\alpha^5\tau^5) \tag{12.98}$$

Finally, for $\varepsilon\to 0$ and $\alpha\to 0$, we have the following asymptotic series solution:

$$y\sim\tau-\frac{\tau^2}{2}(1+\alpha)+\frac{\varepsilon}{6}\{(2+\alpha)\tau^3-\frac{1}{4}(2+7\alpha+4\alpha^2)\tau^4\}+... \tag{12.99}$$

In this asymptotic form, we can set $\alpha = 0$ to recover the solution obtained in the last section for no air resistance.

To illustrate the effect of a non-uniform gravitational field and the effect of air resistance, Figure 12.2 plots the projectile flying height versus normalized time for y_0, $y_0+\varepsilon y_1$, and $y_0(\alpha)+\varepsilon y_1(\alpha)$ (with $\varepsilon = 0.1$ and $\alpha = 0.1$). We can see that the nonlinear effect due to ε increases the maximum reached height and flight time (dotted line), whereas the effect due to air resistance reduces the flying height as well as the flight time (dashed line). This example precisely illustrates that asymptotic solutions can provide qualitative behavior of certain parameters in the problem that numerical simulations cannot provide.

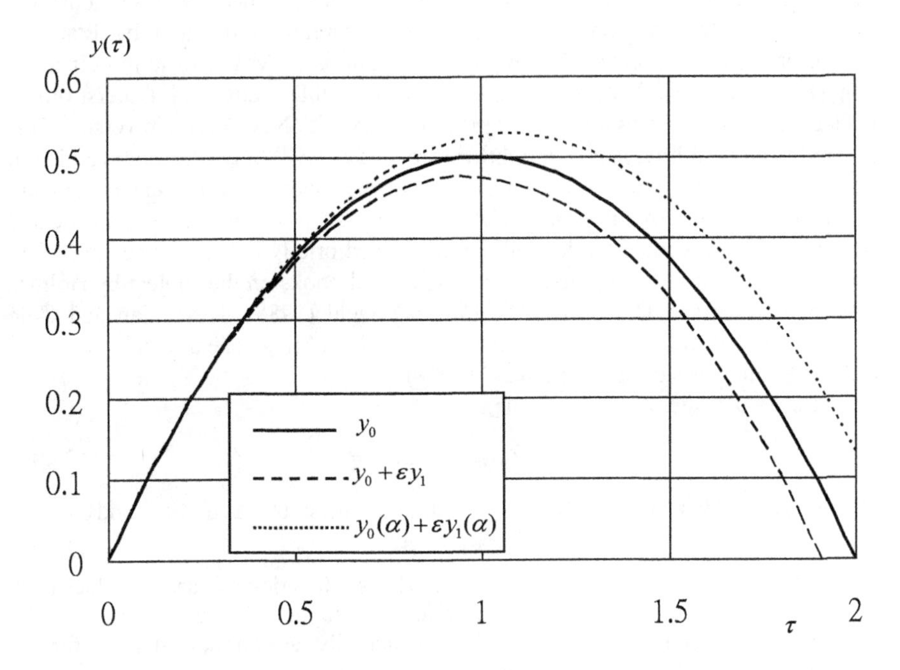

Figure 12.2 Effects of non-uniform gravitational field and air resistance on projectile maximum reached height and flight time for $\alpha = 0.1$ and $\varepsilon = 0.1$

12.4 METHOD OF MATCHED ASYMPTOTIC EXPANSIONS (MMAE)

In the last section, we have introduced the idea for regular perturbation analysis. However, it was discovered that for some problems the regular perturbation method breaks down. For such problems, there is typically a rapid change of the solution at one of the boundaries. This development is strongly associated with fluid mechanics. Prandtl in 1905 developed the theory of boundary layer for fluid mechanics. It was discovered that no matter how small the viscosity, there exists a narrow layer of fluid near the surface of a solid. The fluid within the boundary layer is different from the flow around the body, such that there is a rapid change of the fluid field from the outer field to the inner boundary layer. The idea of a boundary layer for fluid mechanics was so influential that Prandtl was nominated for a Nobel Prize in physics (see Appendix B). It was Friedrichs in 1941 who fully developed the boundary layer problem systematically. In this technique, an asymptotic expansion is developed for the inner layer whereas another asymptotic expansion is developed for the outer region. The two different expansions are matched at the boundary region where both asymptotic expansions are valid. If a single expansion is used, the solution is found to be singular at the boundary point. Therefore, different names have been used for this perturbation technique, including the singular perturbation method (a term coined by Friedrichs and Wasow in 1946), boundary layer analysis (apparently motivated by Prandtl's work), or method of matched asymptotic expansions (MMAE) (a term coined by van Dyke in 1964). Note that van Dyke was a PhD student of Lagerstrom at Caltech, and Wasow was a PhD student of Friedrichs at New York University. This method is powerful but is not straightforward, and we will only illustrate the idea of MMAE in this section. The complication that the inner and outer expansions are expressed in terms of different variables indeed suggests the more sophisticated multiple scale procedure (which will be introduced briefly in the next section). Full discussions of MMAE are found in more technical books on the subject by Holmes (1995), Bender and Orszag (1978), Lin and Segal (1988), Kevorkian and Cole (1981), and O'Malley (2014). Application of MMAE to shear crack in elastic diffusive solids was done by Rudnicki (1991).

Let us consider a simple problem of

$$\varepsilon \frac{du}{dx} + u = 0, \quad u(0) = 1 \tag{12.100}$$

where $\varepsilon \to 0$. This is a linear first order ODE, and the solution can be found as

$$u(x,\varepsilon) = e^{-x/\varepsilon} \tag{12.101}$$

Figure 12.3 plots the solution given in (12.101) as a function of various values of ε. Note that a small parameter ε multiplies the higher derivative term in the differential equation in (12.100). This is actually a common feature for all problems that contain a boundary layer.

For a fixed $x \neq 0$, we have the following limit as $\varepsilon \to 0$

$$\lim_{\varepsilon \to 0} u(x,\varepsilon) = 0 \tag{12.102}$$

For a fixed $\varepsilon \neq 0$, we have the following limit as $x \to 0$

$$\lim_{x \to 0} u(x,\varepsilon) = 1 \tag{12.103}$$

As demonstrated in Figure 12.3, for small ε the solution of $u \approx 0$ suddenly changes to $u = 1$ as $x \to 0$. Thus, we find that

$$\lim_{x \to 0} \lim_{\varepsilon \to 0} u(x,\varepsilon) \neq \lim_{\varepsilon \to 0} \lim_{x \to 0} u(x,\varepsilon) \tag{12.104}$$

Therefore, $x = 0$ is considered a singular point of u for small ε. There is a very narrow boundary of solution near the singular point at $x = 0$. This suggests the use of the term "singular perturbation method," but we will stick to the name of MMAE in this section. The existence of a narrow boundary suggests the use of "boundary layer analysis."

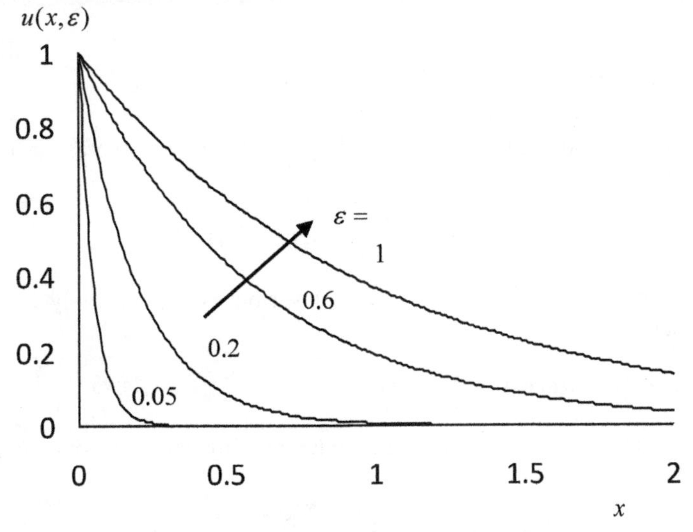

Figure 12.3 Illustration of boundary layer type solution in (12.101) for ε = 0.05, 0.2, 0.6 and 1

We will summarize the procedure of MMAE as follows:

(i) Outer expansion

We assume that an asymptotic expansion outside the boundary layer can be established as

$$u(x,\varepsilon)=\sum_{n=0}^{\infty}u_n(x)f_n(x,\varepsilon) \tag{12.105}$$

as $\varepsilon \to 0$. This solution is not uniformly valid in x as $\varepsilon \to 0$, and it is definitely not valid at $x = 0$.

(ii) Stretching transformation

We want to stretch out the neighborhood of the singular point. In particular, we assume a new variable for the boundary layer near the singular point:

$$\xi=\frac{x}{\phi(\varepsilon)} \tag{12.106}$$

where the scaling parameter ϕ is subject to the following condition as $\varepsilon \to 0$:

$$\phi(0)=0 \tag{12.107}$$

In the problem above, we have $\phi = \varepsilon$ or $\xi = x/\varepsilon$. We can see that the solution given in (12.101) is uniformly valid in ξ as $\varepsilon \to 0$. There are two main properties for this stretching coordinate ξ:

$$x\neq 0, \quad \xi\to\infty, \quad \varepsilon\to 0 \tag{12.108}$$

$$x=0, \quad \xi=0, \quad \text{for all } \varepsilon \tag{12.109}$$

In other words, we want to stretch the boundary layer to infinity.

(iii) Inner expansion

The inner expansion for the solution is expressed as

$$u(x,\varepsilon) = u(\xi\phi(\varepsilon),\varepsilon) = U(\xi,\varepsilon) \tag{12.110}$$

We assume the solution in the boundary layer can be expressed in another asymptotic sequence as:

$$U(\xi,\varepsilon) = \sum_{n=0}^{\infty} U_n(\xi) g_n(\xi,\varepsilon) \tag{12.111}$$

This solution is valid within the boundary layer but is not uniformly valid outside the boundary layer.

(iv) The overlap interval

Next, we assume that there is an overlap interval of the inner and outer expansions. That is, the overlap interval is defined as

$$x_O(\varepsilon) \le x \le x_I(\varepsilon) \tag{12.112}$$

In particular, we have assumed that

(i) $$x_O(\varepsilon) \le x_I(\varepsilon), \quad \varepsilon \to 0 \tag{12.113}$$

(ii) $$x_O(\varepsilon), x_I(\varepsilon) \to 0, \quad \varepsilon \to 0 \tag{12.114}$$

(iii) $$\frac{x_O(\varepsilon)}{\phi(\varepsilon)}, \frac{x_I(\varepsilon)}{\phi(\varepsilon)} \to \infty, \quad \varepsilon \to 0 \tag{12.115}$$

The existence of such an overlap interval cannot be proved analytically. Its existence depends on the problem itself. The presumed existence of the overlap interval is also known as Kaplun's hypothesis. Kaplun was a polish-born PhD student of Lagerstrom at Caltech in 1950s.

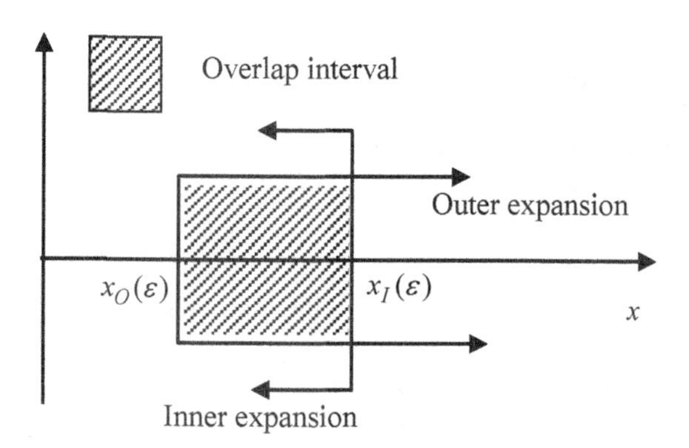

Figure 12.4 Overlap interval for the inner and outer expansions

(v) The matching condition

To match the two expansions, we expand the outer expansion in (12.105) in terms of the inner variable and the inner asymptotic sequences:

$$u(\xi,\varepsilon)=\sum_{n=0}^{\infty}u_n(\xi\phi(\varepsilon))f_n(\xi\phi(\varepsilon),\varepsilon) \tag{12.116}$$

More precisely, we expand it as

$$u(\xi,\varepsilon)=\sum_{n=0}^{\infty}\tilde{u}_n(\xi)g_n(\xi,\varepsilon) \tag{12.117}$$

Within the overlap interval, we expect that the difference of the inner and outer expansions approaches zero as $\varepsilon \to 0$ and $\xi \to \infty$. This leads to the following matching conditions:

$$\lim_{\xi\to\infty}[\tilde{u}_n(\xi)-U_n(\xi)]=0, \quad n=0,1,2,... \tag{12.118}$$

This condition should not depend on the choice of the scaling parameter $\phi(\varepsilon)$.

(vi) The composite expansion

To find a composite expansion valid for both inner and outer layers, we can add (12.105) and (12.111) and subtract (12.117):

$$u(x,\varepsilon)=\sum_{n=0}^{\infty}\left\{u_n(x)f_n(x,\varepsilon)+[U_n(\frac{x}{\phi(\varepsilon)})-\tilde{u}_n(\frac{x}{\phi(\varepsilon)})]g_n(\frac{x}{\phi(\varepsilon)},\varepsilon)\right\} \tag{12.119}$$

The result in (12.119) is the sum of the two expansions subtracting the overlapping term within the common interval. We will illustrate this procedure by the following example.

Example 12.1 Use MMAE to find an asymptotic composite expansion which is uniformly valid throughout the interval as $\varepsilon \to 0$:

$$\varepsilon y''+2y'+y=0, \quad 0<x<1 \tag{12.120}$$

$$y(0)=0, \quad y(1)=1, \quad 0<\varepsilon \ll 1 \tag{12.121}$$

where $\varepsilon \to 0$. Assume that a boundary layer of the solution locates at $x = 0$.

Solution: Again, the highest derivative term is multiplied by a small parameter ε. Therefore, we expect a boundary layer, and we assume the boundary layer is at $x = 0$. To find the outer expansion, we assume that

$$\varepsilon y'' \approx 0 \tag{12.122}$$

Thus, the outer solution can be determined by solving the following ODE

$$2y_O'+y_O=0, \quad y_O(1)=1 \tag{12.123}$$

Since this is the outer solution, we should not expect the boundary condition to be satisfied at $x = 0$. Instead, we only need to enforce the boundary condition at $x = 1$. The solution for the first order ODE given in (12.123) can be found easily as:

$$y_O(x)=e^{\frac{1}{2}(1-x)} \tag{12.124}$$

A stretching transformation is assumed as:

$$\xi = \frac{x}{\phi(\varepsilon)} \tag{12.125}$$

Substituting this change of variables into (12.120), we get

$$\frac{\varepsilon}{\phi^2}\frac{d^2Y}{d\xi^2} + \frac{2}{\phi}\frac{dY}{d\xi} + Y = 0 \tag{12.126}$$

If the stretching is appropriate, we expect that the coefficient for all three terms are comparable. In particular, we want to balance any pair of the following coefficients:

$$\frac{\varepsilon}{\phi^2}, \quad \frac{2}{\phi}, \quad 1 \tag{12.127}$$

The first coefficient must be used in the pair balance since the highest derivative always controls the main behavior of the solution. Pairing the first and third terms of (12.127), we get $\phi = \varepsilon^{1/2}$. As $\varepsilon \to 0$, the second coefficient goes to infinity. Therefore, this choice is not correct. Pairing the first and second terms of (12.127), we get $\phi = \varepsilon$, and the third coefficient is 1. Thus, we must have $\phi = \varepsilon$. Using this information, (12.126) becomes

$$\frac{d^2Y}{d\xi^2} + 2\frac{dY}{d\xi} + \varepsilon Y = 0 \tag{12.128}$$

Thus, the inner expansion can be found by

$$\frac{d^2y_I}{d\xi^2} + 2\frac{dy_I}{d\xi} = 0, \quad y_I(0) = 0 \tag{12.129}$$

This ODE can be solved easily to get

$$y_I(\xi) = C(1 - e^{-2\xi}) \tag{12.130}$$

To do the matching, we assume the intermediate region as

$$x \approx O[\Theta(\varepsilon)] \tag{12.131}$$

where Θ satisfies the following conditions:

$$\lim_{\varepsilon\to 0}\frac{\Theta}{\phi} = \infty, \quad \lim_{\varepsilon\to 0}\Theta = 0 \tag{12.132}$$

We introduce a new variable η in the intermediate region such that:

$$\eta = \frac{x}{\Theta}, \quad \xi = \frac{\eta\Theta}{\phi} \tag{12.133}$$

As $\varepsilon \to 0$, the matching condition for a fixed η is

$$y_O(x)\big|_{x=\eta\Theta} = y_I(\xi)\big|_{\xi=\eta\Theta/\phi} \tag{12.134}$$

Substitution of (12.124) and (12.130) into (12.134) gives

$$e^{\frac{1}{2}(1-\eta\Theta)} = C(1 - e^{-\frac{2\eta\Theta}{\phi}}) = C, \quad \Theta \to 0, \xi \to \infty \tag{12.135}$$

Therefore, we get

$$C = e^{1/2} \tag{12.136}$$

With the constant found in (12.136), we have the inner expansion in (12.130) becoming:

$$y_I(\xi) = e^{1/2}(1 - e^{-2\xi}) \tag{12.137}$$

Finally, the composite expansion can be found as

$$\begin{aligned} y_C(x) &\approx y_O(x) + y_I(\frac{x}{\phi}) - \lim_{\varepsilon \to 0} y_O(\eta\Theta) \\ &= e^{\frac{1}{2}(1-x)} + e^{1/2}(1 - e^{-\frac{2x}{\varepsilon}}) - e^{1/2} \end{aligned} \tag{12.138}$$

After simplification, we get the final expansion as

$$y_C(x) \approx e^{1/2}(e^{-x/2} - e^{-2x/\varepsilon}) \tag{12.139}$$

Actually, the exact solution for this case can be obtained as

$$y(x) = \frac{e^{m_1 x} - e^{m_2 x}}{e^{m_1} - e^{m_2}} \tag{12.140}$$

where

$$m_{1,2} = \frac{-1 \pm \sqrt{1-\varepsilon}}{\varepsilon} \tag{12.141}$$

Figure 12.5 compares the results of MMAE with the exact solution for the case of ε = 0.05. The inner solution is accurate near the boundary at $x = 0$, whereas the outer solution is good outside the boundary layer. The composite expansion and the exact solution are basically indistinguishable from one and other. This example illustrates that when the coefficient of the highest derivative term is multiplied by a small parameter (in this case ε), there exists a boundary layer and MMAE can be used to provide a composite expansion which is accurate for the whole domain (in the case $0 < x < 1$).

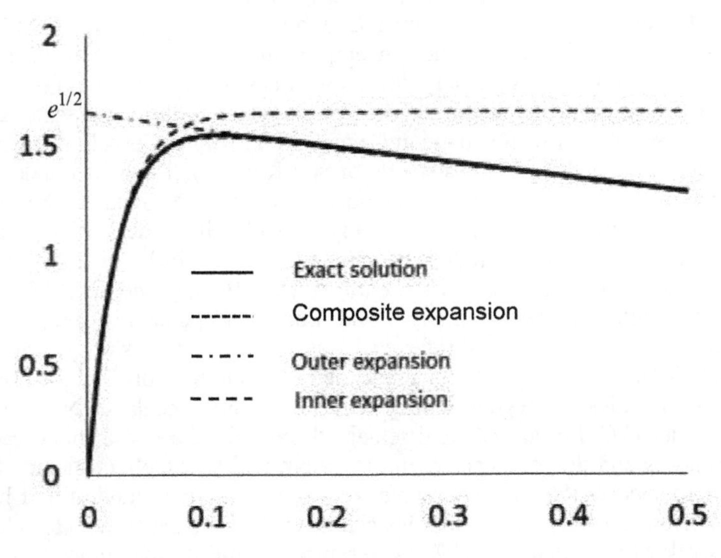

Figure 12.5 Comparison of MMAE solution with exact solution for ε = 0.05

12.5 MULTIPLE SCALE PERTURBATION

In this section, we will consider the cases that a single small scale in asymptotic perturbation analysis is not enough to give an accurate approximation. In such cases, multiple scales are needed in the perturbation analysis. This situation arises in singular perturbation problems discussed in the last section. The multiple-scale perturbation analysis constructs uniformly valid approximations to the solutions of perturbation problems, both for small as well as large values of the independent variables. This is done by introducing fast-scale and slow-scale variables (can be multiple scales) and treating them as independent. In fact, both MMAE and multi-scale perturbation can be used to solve singular perturbation problems.

The two-time scales analysis was originated from the Poincaré-Lindstedt method of solving the nonlinear Duffing equation related to three-body problem in astronomy. The idea of multi-scale analysis was originated by Lindstedt in 1883 and studied in detail by Poincaré in 1889. In 1889, Poincaré won the King Oscar II Prize, which was set up to celebrate the king's 60th birthday, for his work on the three-body problem. There was an error in his prize submission and the corrected version was published in 1890. The judging panels consisted of Mittag-Leffler, Hermite, and Weierstrass. According to O'Malley (2014), this multi-scale method has been repeatedly rediscovered by many mathematicians and scientists in different disciplines (including physics, engineering and applied mathematics), using different names. Full details are found in O'Malley (2014) who compiled a lot of results in a reader-friendly manner, and is highly recommended. Half-jokingly, Nayfeh (1973) claimed in his celebrated book on perturbation analysis that "the method of multiple scales is so popular that it is being rediscovered just about every 6 months." In view of the importance of such analysis, Mittag-Leffler tried to get a Nobel Prize for Poincaré from their initiation in 1901, but he was never successful. The father of Chinese rocket science, H.S. Tsien, called the multi-scale perturbation analysis the PLK-method (in honoring the contributions from Poincaré, Lighthill, and Kuo). Sir James Lighthill used the technique to study gas dynamics in 1949 while Kuo studied incompressible air flow over a thin plate in 1953 by combining both boundary layer and multi-scale analysis. Tsien was a student of von Karman at Caltech, but he got interested in the multi-scale analysis through his interactions with Erdelyi at Caltech, H.Y. Kuo at Cornell, and C.C. Lin at MIT (both Kuo and Lin were students of von Karman). Tsien was under house arrest for 5 years in California in 1950 before he returned to China and became the father of rocket science in China. Nayfeh (1973) called it the derivative expansion method. Nayfeh was a PhD student of van Dyke at Stanford University. In their studies of nonlinear interactions of wave trains, D. J. Benney and his MIT students (Newel, Ablowitz, Haberman, Lange, and Luke) independently developed the multi-scale methods and it was referred to as the method of slow variations by Haberman. Many of his students became experts in soliton theory and professors in applied mathematics. Professor Benney was born in New Zealand (N.Z.) and got his PhD under C.C. Lin at MIT and remained there for the rest of his career. He was a hilarious and devoted teacher at classroom and an excellent mentor for his PhD student supervision. He was recognized as an extremely modest and humble person. There is a lovely story about his character. When Benney was a student, he used to work as a gardener at N.Z. Government House during summer holidays.

The Governor General, Sir Freyberg, sometimes would walk around in the garden and talked to Benney about tomatoes, etc. In the meantime, Benney was nominated for a prestigious scholarship and Freyberg was one of the judge panel members. On the day of the interview, Benney worked at the Government House garden in the morning as usual before he went for the interview in the afternoon. When the interview started, Freyberg asked: "Haven't I met you? You look familiar!" Benney however never revealed in the interview that he was his gardener!

In this section, we will consider a case that the exact solution can be found and we will demonstrate that the regular perturbation of a single scale fails. Then, we will extend the perturbation analysis to multiple scale analysis. Let us consider multiple scale perturbation through the following example:

$$u''+2\varepsilon u'+u=0, \quad u(0)=0, \quad u'(0)=1 \tag{12.142}$$

Mathematically, this is the same as the dynamic equation of a lump mass oscillator subject to initial velocity, but the damping term is very small as $\varepsilon \to 0$.

By regular perturbation, we assume the following asymptotic expansion for u:

$$u(t,\varepsilon)=u_0(t)+u_1(t)\varepsilon+\frac{1}{2}u_2(t)\varepsilon^2+... \tag{12.143}$$

To get the first order approximation, we can substitute (12.143) into (12.142) and set $\varepsilon = 0$ to form the ODE for the constant order:

$$u_0''+u_0=0, \quad u_0(0)=0, \quad u_0'(0)=1 \tag{12.144}$$

Mathematically, this is the same as the dynamic equation of a lump mass oscillator subject to initial velocity, but the damping term is very small as $\varepsilon \to 0$. The exact solution can be found easily as:

$$u_0(t)=\sin t \tag{12.145}$$

For the first order of ε, we substitute (12.143) into (12.142), and differentiate the result with respect to ε to get

$$\dot{u}''+2u'+2\varepsilon\dot{u}'+\dot{u}=0, \quad u_1(0)=0, \quad u_1'(0)=0 \tag{12.146}$$

where

$$\dot{u}=\frac{\partial u}{\partial \varepsilon} \tag{12.147}$$

More explicitly, we get

$$[u_1''+u_2''\varepsilon+...]+2[u_0'+u_1'\varepsilon+...]+2\varepsilon[u_1'+u_2'\varepsilon+...]+[u_1+u_2\varepsilon+...]=0 \tag{12.148}$$

Setting $\varepsilon = 0$, we obtain

$$u_1''+u_1=-2u_0' \tag{12.149}$$

The solution can be found as

$$u_1(t)\approx -t\sin t \tag{12.150}$$

Adding the first two terms, we finally get the asymptotic expansion as

$$u(t)\approx \sin t-(t\sin t)\varepsilon \tag{12.151}$$

We see that for $t \to \infty$, (12.151) gives $u \to \infty$. This is not only unphysical but also incorrect. The exact solution of (12.142) can be found easily as:

$$u(t,\varepsilon)=\frac{e^{-\varepsilon t}\sin\sqrt{1-\varepsilon^2}t}{\sqrt{1-\varepsilon^2}} \tag{12.152}$$

This exact solution predicts $u \to 0$ for $t \to \infty$. Therefore, the result of the regular perturbation method is not uniformly valid as $t \to \infty$. The exponential function contains a time scale of

$$\tau = \varepsilon t \tag{12.153}$$

The time scale in the sine function is

$$\sqrt{1-\varepsilon^2 t} \approx t \tag{12.154}$$

Strictly speaking, we can expand the sine function as

$$\sin\sqrt{1-\varepsilon^2 t} = \sqrt{1-\varepsilon^2 t} + (1-\varepsilon^2)^{3/2} t^3 + ... \tag{12.155}$$

More generally, we can have infinite time scales. For the time being, we are content with the two time scales given in (12.153) and (12.154). The slow time scale is given by τ whereas the fast time scale is given by t. The asymptotic expansion becomes:

$$u(t,\varepsilon) = u^0(t,\tau) + u^1(t,\tau)\varepsilon + \frac{1}{2}u^2(t,\tau)\varepsilon^2 + ... \tag{12.156}$$

The asymptotic series now includes functions of both time scales t and τ. Note that the superscript of u^i (where i = 1,2,3,...) denotes the number in the asymptotic sequence. In contrast to a single time scale, the subscript is reserved for denoting differentiation. By the chain rule, we have

$$\frac{du}{dt} = \frac{\partial u}{\partial t} + \frac{\partial \tau}{\partial t}\frac{\partial u}{\partial \tau} = \frac{\partial u}{\partial t} + \varepsilon\frac{\partial u}{\partial \tau} \tag{12.157}$$

Using this chain rule, we get

$$u'(t,\varepsilon) = \frac{du}{dt} = u_t^0 + \varepsilon u_\tau^0 + [u_t^1 + \varepsilon u_\tau^1]\varepsilon + [u_t^2 + \varepsilon u_\tau^2]\frac{\varepsilon^2}{2} + ... \tag{12.158}$$

$$u''(t,\varepsilon) = u_{tt}^0 + \varepsilon u_{t\tau}^0 + [u_{tt}^1 + u_{t\tau}^1\varepsilon + u_{\tau t}^0 + u_{\tau\tau}^0\varepsilon]\varepsilon + [u_{tt}^2 + u_{\tau t}^2\varepsilon + 2u_{t\tau}^1 + 2u_{\tau\tau}^1\varepsilon]\frac{\varepsilon^2}{2} + ... \tag{12.159}$$

Substituting (12.156) and (12.159) into (12.142) and setting $\varepsilon = 0$, we find

$$u_{tt}^0 + u^0 = 0, \quad u^0(0,0) = 0, \quad u_t^0(0,0) = 1 \tag{12.160}$$

The solution for (12.160) is

$$u^0 = A(\tau)\cos t + B(\tau)\sin t \tag{12.161}$$

Note that A and B are not constants but functions of the slow time scale τ. The boundary conditions in (12.160) become

$$u^0(0,0) = A(0) = 0, \quad u_t^0(0,0) = B(0) = 1 \tag{12.162}$$

To find the functions A and B, we have to go to a higher order to find conditions that they satisfy. Substituting (12.156) and (12.159) into (12.142), differentiating the result with respect to ε, and setting $\varepsilon = 0$, we get

$$u_{tt}^1 + u^1 = -2[u_{t\tau}^0 + u_t^0] \tag{12.163}$$

$$\dot{u}(0) = u^1(0,0) = 0, \quad \dot{u}'(0) = u_t^1(0,0) + u_\tau^0(0,0) = 0 \tag{12.164}$$

Substitution of (12.161) into (12.163) gives

$$u_{tt}^1 + u^1 = 2[A_\tau + A]\sin t - 2[B_\tau + B]\cos t \tag{12.165}$$

The two bracket terms are known as "wanted" secular terms and we need to remove them. We learned from the undetermined coefficient method for solving nonhomogeneous ODE that the particular solution will contain the following terms:

$$u^1 \propto t\sin t, t\cos t \tag{12.166}$$

No matter how small is ε, the solution eventually becomes unbounded as $t \to \infty$. Lin and Segal (1988) called them resonant terms. These secular terms appear in the analysis at each power of ε. In fact, this boundedness requirement from the next order of problem forms the heart of multiple scale methods.

Thus, we get two additional conditions for solving A and B:

$$A_\tau + A = 0, \quad B_\tau + B = 0 \tag{12.167}$$

The general solutions for (12.167) are

$$A(\tau) = \alpha_0 e^{-\tau}, \quad B(\tau) = \beta_0 e^{-\tau} \tag{12.168}$$

Imposing boundary conditions in (12.162), we obtain

$$A(\tau) = 0, \quad B(\tau) = e^{-\tau} \tag{12.169}$$

Finally, we get the first order expansion as

$$u \sim u^0 + O(\varepsilon) = e^{-\varepsilon t}\sin t \tag{12.170}$$

We now see that this asymptotic expansion indeed decays to zero as $t \to \infty$. Figure 12.6 plots the results of single and multiple scale perturbations given in (12.151) and (12.170) versus the analytical solution given in (12.152). We have used a relatively large value of $\varepsilon = 0.2$ in Figure 12.6 because for small values of ε the multiple-scale plot and exact solution essentially overlap. The single scale expansion oscillates with a much larger magnitude and extends beyond the range of our plotting area for large x.

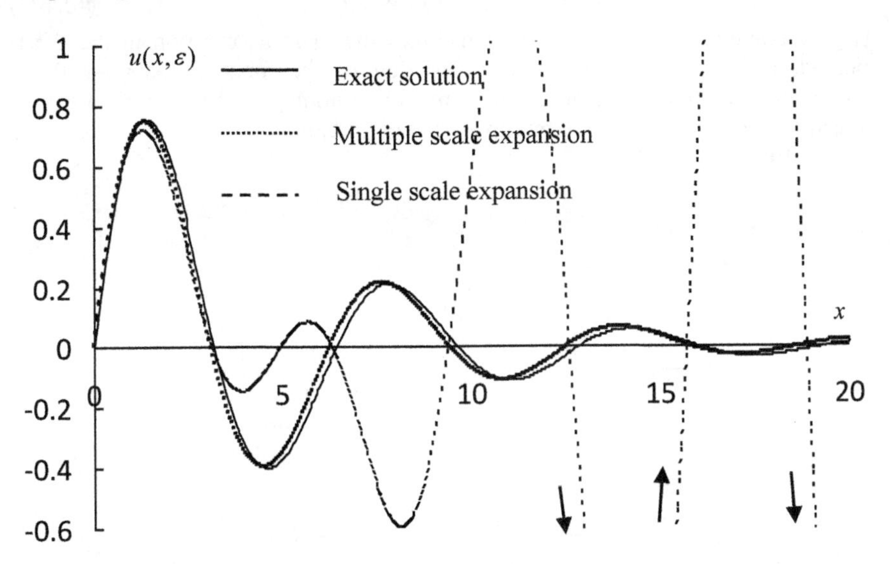

Figure 12.6 Single-scale and multiple-scale expansions with exact solution for ε = 0.2

12.6 LIOUVILLE-GREEN OR WKB EXPANSION

In quantum mechanics, the WKB method was developed in 1926 and was named after Wentzel, Kramers, and Brillouin, who studied the solution of the Schrodinger equation. Some authors also referred to it as the WKBJ method (J stands for Jefferys). However, in mathematics the method and the idea can trace back to the celestial analysis by Carlini in 1817, and was also referred to as the method of Liouville and Green, both of whom published the method in 1837. Therefore, it is known as the L-G approximation. The WKB method can be applied to a singular perturbation problem (i.e., a small parameter multiplying the highest derivative term), but the problem must be linear. Review of the WKB method in solid mechanics is given by Steele (1976). In this sense, although WKB is easier than both MMAE and a multiple scales method, it is less powerful than either one of them.

Let us consider the following linear ODE

$$\varepsilon^2 y'' - q(x)y = 0 \tag{12.171}$$

In order to get an approximation for the problem, we see that if q is a constant, we can solve the problem instantly as:

$$y(x) = Ae^{-\frac{x}{\varepsilon}\sqrt{q}} + Be^{\frac{x}{\varepsilon}\sqrt{q}} \tag{12.172}$$

This solution is the main idea behind the WKB method. It was assumed that an exponential function provides a good approximation to equation (12.171). To allow a more flexible form of exponential functions that can handle the non-constant function $q(x)$ in (12.171), the following form is assumed in the WKB method:

$$y(x) \sim e^{\theta(x)/\varepsilon^\alpha}\{y_0(x) + \varepsilon^\alpha y_1(x) + ...\} \tag{12.173}$$

The exponential form in (12.173) constitutes the main assumption in the WKB method, and it focuses on the fast variation of the function (as $\varepsilon \to 0$ the exponential function is the dominant term). In addition, the WKB method can be considered as a special case of multiple scale perturbation.

Differentiation of (12.173) gives

$$\begin{aligned} y' &= \frac{\theta'}{\varepsilon^\alpha} e^{\theta/\varepsilon^\alpha}\{y_0 + \varepsilon^\alpha y_1 + ...\} + e^{\theta/\varepsilon^\alpha}\{y_0' + \varepsilon^\alpha y_1' + ...\} \\ &= e^{\theta/\varepsilon^\alpha}\left\{\frac{\theta'}{\varepsilon^\alpha} y_0 + y_0' + \theta' y_1 + ...\right\} \end{aligned} \tag{12.174}$$

$$\begin{aligned} y'' &= \frac{\theta'}{\varepsilon^\alpha} e^{\theta/\varepsilon^\alpha}\left\{\frac{\theta'}{\varepsilon^\alpha} y_0 + y_0' + \theta' y_1 + ...\right\} \\ &+ e^{\theta/\varepsilon^\alpha}\left\{\frac{\theta''}{\varepsilon^\alpha} y_0 + \frac{\theta'}{\varepsilon^\alpha} y_0' + y_0'' + \theta' y_1' + \theta'' y_1 + ...\right\} \\ &= e^{\theta/\varepsilon^\alpha}\left\{\frac{(\theta')^2}{\varepsilon^{2\alpha}} y_0 + \frac{1}{\varepsilon^\alpha}\left[\theta'' y_0 + 2\theta' y_0' + (\theta')^2 y_1\right] + ...\right\} \end{aligned} \tag{12.175}$$

Substitution of (12.174) and (12.175) into (12.171) gives

$$\varepsilon^2 \left\{ \frac{(\theta')^2}{\varepsilon^{2\alpha}} y_0 + \frac{1}{\varepsilon^{\alpha}} \left[\theta'' y_0 + 2\theta' y_0' + (\theta')^2 y_1 \right] + ... \right\} \tag{12.176}$$
$$- q(x)\{y_0 + \varepsilon^{\alpha} y_1 + ...\} = 0$$

In order to balance the order of the equations, we must have α = 1. Collecting the coefficients of different order and setting all of them to zero (in order to satisfy the ODE), we get the constant order term

O(1):
$$(\theta')^2 = q(x) \tag{12.177}$$

This is called the eikonal equation, and it is related to the studies of nonlinear waves and optics. The solution can be expressed as

$$\theta(x) = \pm \int^x \sqrt{q(s)} ds \tag{12.178}$$

The linear order term of ε is

O(ε):
$$\theta'' y_0 + 2\theta' y_0' + (\theta')^2 y_1 = q y_1 \tag{12.179}$$

This is called the transport equation. Using the result of (12.177), we have

$$\theta'' y_0 + 2\theta' y_0' = 0 \tag{12.180}$$

This can be rearranged as

$$\frac{dy_0}{y_0} = -\frac{\theta''}{2\theta'} dx = -\frac{d\theta'}{2\theta'} \tag{12.181}$$

Integrating both sides we find the solution as

$$y_0 = \frac{C}{\sqrt{\theta'}} \tag{12.182}$$

where C is a constant. Substitution of (12.182) and (12.178) into (12.173) gives

$$y(x) \sim e^{\theta/\varepsilon} \{ \frac{C}{\sqrt{\theta'}} + \varepsilon y_1 + ... \} \tag{12.183}$$

Since there are two values of θ from (12.178), we obtain the following WKB approximation

$$y(x) \sim q(x)^{-1/4} \{ A e^{-\frac{1}{\varepsilon}\int^x \sqrt{q(s)} ds} + B e^{\frac{1}{\varepsilon}\int^x \sqrt{q(s)} ds} \} \tag{12.184}$$

Example 12.2 Use the WKB method to find the approximation of the following ODE with non-constant coefficient

$$\varepsilon^2 y'' + e^{2x} y = 0 \tag{12.185}$$
$$y(0) = a, \quad y(1) = b \tag{12.186}$$

Solution: We see that

$$q(x) = -e^{2x} \tag{12.187}$$

Thus, we find

$$\sqrt{q(x)} = i e^x \tag{12.188}$$
$$q(x)^{-1/4} = (-1)^{-1/4} e^{-x/2} = e^{-i\pi/4} e^{-x/2} \tag{12.189}$$

Substitution of (12.187) and (12.188) into (12.184) gives

$$y(x) \sim e^{-i\pi/4} e^{-x/2} \{A e^{-i\lambda e^x} + B e^{i\lambda e^x}\} \tag{12.190}$$

where $\lambda = 1/\varepsilon$. Note, however, that we start with an ODE with real variables and real parameters, thus our approximation must be real. Thus, A and B must be complex constant and conjugate to each other. Using this information, we can recast the approximation in (12.190) as

$$y(x) \sim e^{-x/2} \{C \cos(\lambda e^x) + D \sin(\lambda e^x)\} \tag{12.191}$$

where C and D are now real constants. Substitution of (12.191) into the boundary conditions given in (12.186) gives

$$y(0) = C \cos \lambda + D \sin \lambda = a \tag{12.192}$$

$$y(1) = e^{-1/2} \{C \cos(\lambda e) + D \sin(\lambda e)\} = b \tag{12.193}$$

This provides two equations for two unknowns. The solutions of the system can be found as

$$C = \frac{a \sin(\lambda e) - b e^{1/2} \sin \lambda}{\sin \lambda (e-1)} \tag{12.194}$$

$$D = \frac{b e^{1/2} \cos \lambda - a \cos(\lambda e)}{\sin \lambda (e-1)} \tag{12.195}$$

With these constants, the WKB approximation can be expressed as

$$y(x) \sim e^{-x/2} \{\frac{b e^{1/2} \sin \lambda (e^x - 1) - a \sin \lambda (e^x - e)}{\sin \lambda (e-1)}\} \tag{12.196}$$

12.7 LAPLACE METHOD

One of the main applications of asymptotic analysis is to estimate the value of an integral for certain particular limit of the parameters (i.e., one of the parameters approaches infinity). Since many problems of differential equations can be solved by integral transforms, as introduced in the last chapter, the final solutions of differential equations are often expressed in terms of some definite integral because an inverse transform (often expressed in the form of integration) cannot be obtained analytically in most cases. Very often numerical evaluations of these integrals are needed. Asymptotic analysis provides the dominant behavior when one of the parameters in the integrand is large.

12.7.1 Erdelyi's Derivation

We first consider the following integral, known as the Laplace integral:

$$f(x) = \int_{\alpha}^{\beta} g(t) e^{xh(t)} dt \tag{12.197}$$

where x approaches infinity and $h(t)$ is real. Note for the special case that $\alpha = 0$, $\beta \to \infty$, and $h(t) = -t$, (12.197) is in fact the definition of the Laplace transform. If $h(t)$ has a number of maxima, we can always break up the integral in a finite number of integrals in a way that $h(t)$ reaches its maximum at one of the end-points and at no other points. For the case that the maximum is at $t = \alpha$ and $x \to \infty$, it can be shown by the Laplace method that the function on the left of (12.197) can be expressed as:

$$f(x) \sim \left[\frac{-\pi}{2xh''(\alpha)}\right]^{1/2} g(\alpha)e^{xh(\alpha)} \tag{12.198}$$

where α is the maximum point of the function $h(t)$ such that

$$h'(\alpha) = 0, \quad h''(\alpha) < 0 \tag{12.199}$$

Equation (12.199) illustrates that the main contribution from the integrand to the integration is from the maximum point of function $h(t)$.

To prove (12.198), Laplace introduced the following change of variables

$$h(\alpha) - h(t) = u^2 \tag{12.200}$$

Thus, we have from (12.200) that

$$dt = -\frac{2udu}{h'(t)} \tag{12.201}$$

The lower limit for u becomes

$$t = \alpha, \quad \Rightarrow u = 0 \tag{12.202}$$

Note that the main contribution of the integrand is from the neighborhood of the maximum point $t = \alpha$ because we have $h'(t) < 0$ for the range

$$\alpha < t < \alpha + \eta, \quad \eta \to 0 \tag{12.203}$$

Therefore, the integral can be approximated by replacing the upper limit β by $\alpha+\eta$

$$t = \alpha + \eta, \quad \Rightarrow u = [h(\alpha) - h(\alpha+\eta)]^{1/2} = U \tag{12.204}$$

That is, we have

$$f(x) \sim \int_{\alpha}^{\alpha+\eta} g(t)e^{xh(t)}dt = -\int_0^U 2u\frac{g(t)}{h'(t)}\{\exp x[h(\alpha) - u^2]\}du \tag{12.205}$$

For $x \to \infty$, the following term changes very fast with u:

$$e^{x[h(\alpha)-u^2]} \tag{12.206}$$

whereas, as $\eta \to 0$, $g(t)$ does not change rapidly. Thus, we can set

$$g(t) \approx g(\alpha) \tag{12.207}$$

In addition, the following limit needs to be considered

$$\lim_{t\to\alpha}\frac{u}{h'(t)} = \lim_{t\to\alpha}\frac{-\frac{h'(t)}{2u}}{h''(t)} = -\lim_{t\to\alpha}\frac{h'(t)}{2uh''(t)} \tag{12.208}$$

At the limit $t \to \alpha$, we have the form 0/0. Applying L'Hôpital's rule, we obtain the second part of (12.208). Amazingly, note that the left hand side appears on the right of (12.208). The whole success of this method comes from this beautiful result. We can rearrange (12.208) as

$$\lim_{t\to\alpha}[\frac{u}{h'(t)}]^2 = \lim_{t\to\alpha}[-\frac{1}{2h''(t)}] \tag{12.209}$$

Thus, we find the following limit

$$\lim_{t\to\alpha}[\frac{u}{h'(t)}] = [-\frac{1}{2h''(\alpha)}]^{1/2} \tag{12.210}$$

Substitution of (12.207) and (12.210) into (12.205) gives

$$\begin{aligned} f(x) &\approx [-\frac{2}{h''(\alpha)}]^{1/2} g(\alpha)\int_0^U \exp[-xu^2 + xh(\alpha)]du \\ &\approx [-\frac{2}{h''(\alpha)}]^{1/2} g(\alpha)e^{xh(\alpha)}\int_0^U \exp(-xu^2)du \end{aligned} \tag{12.211}$$

If the only maximum is at α, we can well extend U to infinity and note the following Laplace integral:

$$\int_0^\infty \exp(-xu^2)du = \frac{1}{2}\sqrt{\frac{\pi}{x}} \tag{12.212}$$

With this result, (12.211) becomes

$$f(x) \sim \left[\frac{-\pi}{2xh''(\alpha)}\right]^{1/2} g(\alpha)\exp\{xh(\alpha)\} \tag{12.213}$$

This is the required result.

12.7.2 Bleistein-Handelsman Derivation

A slightly different proof of the Laplace method was given by Bleistein and Handelsman (1986). In particular, the following form of integral is considered (Bleistein and Handelsman, 1986)

$$I(\lambda) = \int_a^b f(t)e^{-\lambda\phi(t)}dt \tag{12.214}$$

where λ is real and approaches infinity and again both $\phi(t)$ and $f(t)$ are real. Note that (12.214) is essentially the same as (12.197) except that a minus sign has been included explicitly in defining the index of the exponential function. Thus, we search for the minimum $\phi(t)$, instead of the maximum as we did in the last section. Suppose that a minimum can be found such that

$$\phi'(t_0) = 0, \quad \phi''(t_0) > 0 \tag{12.215}$$

In addition, we assume that the function $f(t)$ exists at the minimum point $t = t_0$. Similar to the argument used in the last section, we expect that the main contribution from the integrand comes from the minimum point t_0. Thus, we define the following integral around this point as

$$e^{\lambda\phi(t_0)}I_0(\lambda) = \int_{t_0-\varepsilon}^{t_0+\varepsilon} f(t)e^{-\lambda[\phi(t)-\phi(t_0)]}dt \tag{12.216}$$

where ε is a small number. We expect the following limit holds

$$\lim_{\lambda \to \infty} \frac{I_0(\lambda)}{I(\lambda)} = 1 \tag{12.217}$$

That is, when $\lambda \to \infty$, the integral defined in (12.216) approaches the original integral (12.214). To justify this assertion, Figure 12.4 shows that the area under the exponential function in (12.216) mainly comes from the neighborhood of t_0 for large λ. This justifies the limit given in (12.217).

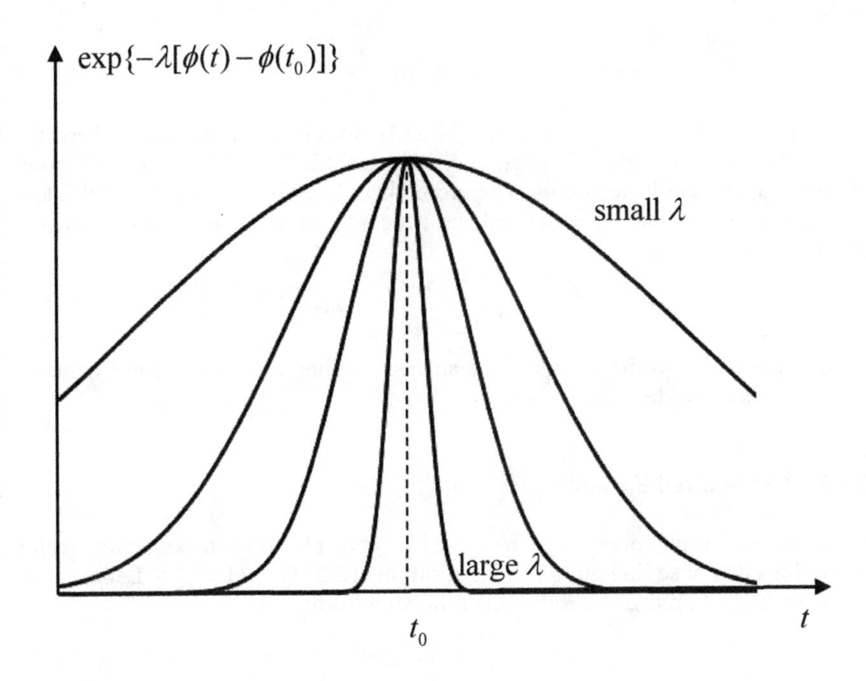

Figure 12.7 The behavior of the exponential function in (12.11) around the critical point t_0

Using Taylor series expansion and assuming that $f(t)$ is not a highly oscillating function near t_0, we have

$$e^{\lambda\phi(t_0)} I_0(\lambda) \approx f(t_0) \int_{t_0-\varepsilon}^{t_0+\varepsilon} e^{-\frac{\lambda}{2}\phi''(t_0)(t-t_0)^2} dt + \dots \tag{12.218}$$

Now, we can apply a change of variables

$$\tau = \sqrt{\frac{\lambda}{2}\phi''(t_0)}(t-t_0), \quad d\tau = \sqrt{\frac{\lambda}{2}\phi''(t_0)}dt \tag{12.219}$$

Applying (12.219) and (12.217), we find the following approximation:

$$I(\lambda) \approx e^{-\lambda\phi(t_0)} \sqrt{\frac{2}{\lambda\phi''(t_0)}} f(t_0) \int_{-\sqrt{\frac{\lambda}{2}\phi''(t_0)}\varepsilon}^{\sqrt{\frac{\lambda}{2}\phi''(t_0)}\varepsilon} \exp\{-\tau^2\} d\tau + \dots \tag{12.220}$$

As seen in Figure 12.7, we can extend this integration from $-\infty$ to $+\infty$ without changing the integration (as long as $\lambda \to \infty$):

$$I(\lambda) \approx e^{-\lambda\phi(t_0)} \sqrt{\frac{2}{\lambda\phi''(t_0)}} f(t_0) \int_{-\infty}^{\infty} \exp\{-\tau^2\} d\tau + ... \tag{12.221}$$

The integration in (12.221) can be carried out exactly by (see Chapter 1):

$$\int_{-\infty}^{\infty} \exp\{-\tau^2\} d\tau = \sqrt{\pi} \tag{12.222}$$

Finally, we get

$$I(\lambda) \approx \exp\{-\lambda\phi(t_0)\} f(t_0) \sqrt{\frac{2\pi}{\lambda\phi''(t_0)}} \tag{12.223}$$

This result is essentially the same as (12.213), which is for the case where the critical point is on the limit of integration, whereas (12.223) is for the critical point within the upper and lower limits. Therefore, they differ by a factor of 2. Bleistein and Handelsman (1986) also derived the accuracy, or more precisely found the order of error of (12.223):

$$I(\lambda) \approx \exp\{-\lambda\phi(t_0)\} f(t_0) \sqrt{\frac{2\pi}{\lambda\phi''(t_0)}} + O(\frac{\exp\{-\lambda\phi(t_0)\}}{\lambda^{3/2}}) \tag{12.224}$$

We refer the reader to the details of the analysis leading to this result in Chapter 5 of Bleistein and Handelsman (1986).

12.7.3 Generalized Formula

The previous result presented in (12.213) and (12.223) makes two major assumptions that pose limitations on the result in (12.213) or (12.223). Let us recap here. Consider a Laplace integral of the following form:

$$I(\lambda) = \int_a^b g(t) e^{\lambda h(t)} dt \tag{12.225}$$

for $\lambda \to \infty$. In the last section, the two major assumptions are made:

$$g(a) \neq 0, \quad h'(a) = 0 \tag{12.226}$$

where a is the maximum boundary point. We will generalize this result to the cases that

$$g(a) = g'(a) = ... = g^{(m-1)}(a) = 0, \quad g^{(m-1)}(a) \neq 0 \tag{12.227}$$

$$h(a) = h'(a) = ... = h^{(n-1)}(a) = 0, \quad h^{(n-1)}(a) \neq 0 \tag{12.228}$$

where m and n are integers. To do this, let us assume a much more general form of functions g and h, and then specialize them to (12.227) and (12.228). In particular, we assume for $\lambda \to \infty$, the asymptotic forms of $g(t)$ and $h(t)$ as

$$g(t) \sim A(t-a)^{\alpha}, \quad \alpha > -1 \tag{12.229}$$

$$h(t) \sim h(a) - B(t-a)^{\beta} \tag{12.230}$$

Substitution of (12.229) and (12.230) into (12.225) gives

$$I(\lambda) \approx A e^{\lambda h(a)} \int_a^{\infty} (t-a)^{\alpha} e^{-\lambda B(t-a)^{\beta}} dt \tag{12.231}$$

Using previous arguments, we can extend the upper limit to infinity. Applying a change of variables of $s = t-a$, we obtain

$$I(\lambda) = Ae^{\lambda h(a)} \int_0^{\infty} s^{\alpha} e^{-\lambda B s^{\beta}} ds \tag{12.232}$$

Let a new variable u such that

$$u = \lambda B s^{\beta} \tag{12.233}$$

Then, we have

$$du = \beta \lambda B s^{\beta-1} ds, \quad s^{\alpha} = (\frac{u}{\lambda B})^{\frac{\alpha}{\beta}} \tag{12.234}$$

Using (12.234), we can rewrite the integral in (12.232)

$$I(\lambda) = Ae^{\lambda h(a)} \int_0^{\infty} (\frac{u}{\lambda B})^{\frac{\alpha}{\beta}} e^{-u} \frac{du}{\beta \lambda B s^{\beta-1}} \tag{12.235}$$

Using (12.234) and rearranging (12.235) we have

$$\begin{aligned} I(\lambda) &= \frac{Ae^{\lambda h(a)}}{\beta (\lambda B)^{(\alpha+1)/\beta}} \int_0^{\infty} u^{(\alpha+1)/\beta-1} e^{-u} du \\ &= \frac{Ae^{\lambda h(a)}}{\beta (\lambda B)^{(\alpha+1)/\beta}} \Gamma(\frac{\alpha+1}{\beta}) \end{aligned} \tag{12.236}$$

where Γ is Euler's gamma function. If α and β are integers and the coefficients A and B are those from Taylor series expansions, we have

$$\alpha = m, \quad \beta = n \tag{12.237}$$

$$A = \frac{g^{(m)}(a)}{m!}, \quad B = -\frac{h^{(n)}(a)}{n!} \tag{12.238}$$

This special case agrees with the conditions imposed in (12.227) and (12.228). Thus, we have the asymptotic value for the integral (12.225) subject to the condition of (12.227) and (12.228):

$$I(\lambda) \approx [-\frac{n!}{h^{(n)}(a)}]^{(m+1)/n} \frac{g^{(m)}(a)}{m!} \frac{e^{\lambda h(a)}}{n \lambda^{(m+1)/n}} \Gamma(\frac{m+1}{n}) \tag{12.239}$$

If $m = 0$ (i.e., function $g(t)$ is regular at $t = a$), we have the special case:

$$I(\lambda) \approx [-\frac{n!}{h^{(n)}(a)}]^{1/n} \frac{g(a) e^{\lambda h(a)}}{n \lambda^{1/n}} \Gamma(\frac{1}{n}) \tag{12.240}$$

If $m = 0$ and $n = 1$, we have the special case:

$$I(\lambda) \approx -\frac{g(a) e^{\lambda h(a)}}{\lambda h'(a)} \tag{12.241}$$

If $m = 0$ and $n = 2$, we have the special case:

$$I(\lambda) \approx \sqrt{\frac{\pi}{-2\lambda h''(a)}} g(a) \exp\{\lambda h(a)\} \tag{12.242}$$

Of course, (12.242) is the same as (12.198) that we obtained in the last section. Therefore, (12.239) is a very general result.

Example 12.3 Find the asymptotic form of Legendre polynomials for large order, which is defined in integral form as follow:

$$P_n(z) = \frac{1}{\pi}\int_0^{\pi} (z + \sqrt{z^2 - 1}\cos\theta)^n d\theta \tag{12.243}$$

for $z > 1$ and $n \to \infty$.

Solution: We look for the asymptotic behavior of the Legendre function for a fixed z (> 1) and for $n \to \infty$. First, we note the following formula:

$$t^{\alpha} = e^{\alpha \ln t} \tag{12.244}$$

Using this idea, we can recast (12.243) in the required form:

$$P_n(z) = \frac{1}{\pi}\int_0^{\pi} e^{n\ln(z+\sqrt{z^2-1}\cos\theta)} d\theta \tag{12.245}$$

Comparing (12.245) and (12.225), we have

$$g(\theta) = 1, \quad h(\theta) = \ln q(\theta), \quad q(\theta) = z + \sqrt{z^2 - 1}\cos\theta \tag{12.246}$$

Clearly, the maximum occurs at $\theta = 0$ because of the cosine function. The differentiation of q gives

$$q'(\theta) = -\sqrt{z^2 - 1}\sin\theta \tag{12.247}$$

For $0 < \theta < \pi$, we have $q'(\theta) < 0$ and, thus, we do have a maximum and the maximum is at the lower limit of integration. We now expand the cosine function about $\theta = 0$:

$$\cos\theta \approx 1 - \frac{\theta^2}{2} + ... \tag{12.248}$$

Substitution of (12.248) into (12.245) leads to

$$P_n(z) = \frac{1}{\pi}\int_0^{\pi} e^{n\ln[z+\sqrt{z^2-1}(1-\theta^2/2)]} d\theta \tag{12.249}$$

To proceed further, we employ the following mathematical trick

$$P_n(z) = \frac{1}{\pi}\int_0^{\pi} \exp\{n\ln[(z+\sqrt{z^2-1})(1-\frac{\sqrt{z^2-1}\theta^2}{2(z+\sqrt{z^2-1})})]\}d\theta \tag{12.250}$$

Then, we have

$$\begin{aligned} P_n(z) &= \frac{1}{\pi}\int_0^{\pi} \exp\left\{ n[\ln(z+\sqrt{z^2-1}) + \ln(1-\frac{\sqrt{z^2-1}\theta^2}{2(z+\sqrt{z^2-1})})] \right\} d\theta \\ &= \frac{\exp\{n\ln(z+\sqrt{z^2-1})\}}{\pi}\int_0^{\pi} \exp\left\{ n\ln(1-\frac{\sqrt{z^2-1}\theta^2}{2(z+\sqrt{z^2-1})}) \right\} d\theta \end{aligned} \tag{12.251}$$

Note the following series expansion

$$\ln(1-x) \sim -x + ... \tag{12.252}$$

Then, (12.251) becomes

$$P_n(z) = \frac{\exp\{n\ln(z+\sqrt{z^2-1})\}}{\pi}\int_0^{\pi}\exp\{-\frac{n\sqrt{z^2-1}\theta^2}{2(z+\sqrt{z^2-1})}\}d\theta \tag{12.253}$$

Then, we can apply the following change of variables

$$u^2 = \frac{n\sqrt{z^2-1}}{2(z+\sqrt{z^2-1})}\theta^2 \tag{12.254}$$

After change of variables, we get

$$P_n(z) = \frac{(z+\sqrt{z^2-1})^n}{\pi}\frac{(z+\sqrt{z^2-1})^{1/2}}{(z^2-1)^{1/4}}\sqrt{\frac{2}{n}}\int_0^{\infty}\exp\{-u^2\}du \tag{12.255}$$

Note that we have shifted the upper limit to infinity as the main contribution is from the lower limit. From (1.102) of Chapter 1, we get

$$P_n(z) \approx \frac{1}{\sqrt{2\pi n}}\frac{(z+\sqrt{z^2-1})^{n+1/2}}{(z^2-1)^{1/4}} \tag{12.256}$$

Example 12.4 Find the asymptotic form of the following integral

$$I(x) = \int_0^{\infty} e^{-t-\frac{x}{t^2}}dt \tag{12.257}$$

for $x \to \infty$.

Solution: For this case, we let

$$h(t) = (t+\frac{x}{t^2}) \tag{12.258}$$

We look for the minimum of $h(t)$ which will have maximum contribution to the integral. Then, we can set

$$h'(t) = 1-\frac{2x}{t^3} = 0 \tag{12.259}$$

Thus, we have the minimum point at

$$t = (2x)^{1/3} \tag{12.260}$$

Guided by this observation, we let

$$s = \frac{t}{(2x)^{1/3}} \tag{12.261}$$

Substitution of (12.261) into (12.257) leads to

$$I(x) = (2x)^{1/3}\int_0^{\infty} e^{-(2x)^{1/3}[s+1/(2s^2)]}ds \tag{12.262}$$

This change of variables allows us to convert the integral in (12.257) to the standard Laplace type integral as $x \to \infty$. Comparing (12.262) with (12.214), we identify that

$$\phi(s) = s+\frac{1}{2s^2} \tag{12.263}$$

We can easily find that

$$\phi'(s) = 1 - \frac{1}{s^3}, \quad \phi'(1) = 0, \quad h(1) = \frac{3}{2}, \quad \phi''(s) = \frac{3}{s^4}, \quad \phi''(1) = 3 \qquad (12.264)$$

The minimum is at $s = 1$. Expanding ϕ in Taylor series expansion about $s = 1$, we have

$$\phi(s) = \frac{3}{2} + \frac{3(s-1)^2}{2} + ... \qquad (12.265)$$

Using the result in (12.224), we get

$$I(x) \sim 2(2x)^{1/3} \exp\{-\frac{3}{2}(2x)^{1/3}\}\sqrt{\frac{\pi}{6(2x)^{1/3}}} \qquad (12.266)$$

Note that we have added an extra 2 because the minimum point ($s = 1$) is not at the lower limit of integration of $s = 0$. Simplifying (12.266), we get

$$I(x) \sim \sqrt{\frac{2\pi}{3}}(2x)^{1/6} \exp\{-\frac{3}{2}(2x)^{1/3}\} \qquad (12.267)$$

12.7.4 Laplace Type Integrals in Higher Dimensions

So far, we have restricted our discussion to integrals with one variable. In this section, we will consider a multiple integral of functions depending on multi-variables. In general, consider the case of n variables:

$$I(\beta) = \int ... \int \varphi(x_1, x_2, ..., x_n) e^{-\beta f(x_1, x_2, ..., x_n)} dx_1 \cdots dx_n \qquad (12.268)$$

To find the critical point or minimum point for the above integral, we set

$$\nabla \cdot f(\boldsymbol{a}) = 0 \qquad (12.269)$$

More explicitly, we have

$$\frac{\partial f}{\partial x_1}(\boldsymbol{a}) = \cdots = \frac{\partial f}{\partial x_n}(\boldsymbol{a}) = 0 \qquad (12.270)$$

where

$$\boldsymbol{a} = (a_1, ..., a_n) \qquad (12.271)$$

The critical point or stationary point $\boldsymbol{a}$ is said to be non-degenerate if the Hessian matrix is nonzero and, in particular, larger than zero for the present case of the minimum point:

$$\det\left|\boldsymbol{H}(\boldsymbol{a})\right| = \det\left|\frac{\partial^2 f}{\partial x_i \partial x_j}\right| > 0 \qquad (12.272)$$

Assuming that φ is regular at point $\boldsymbol{a}$ and applying Taylor series expansion for the vector form, we get

$$I(\beta) \approx \varphi(a_1, a_2, ..., a_n) \int ... \int e^{-\beta[f(a_1,...,a_n) + \frac{1}{2}(\boldsymbol{x}-\boldsymbol{a})^T \boldsymbol{H}(\boldsymbol{x}-\boldsymbol{a})]} d\boldsymbol{x} \qquad (12.273)$$

Introducing a new vector $\boldsymbol{y} = \boldsymbol{x} - \boldsymbol{a}$, we obtain

$$I(\beta) \approx \varphi(\boldsymbol{a}) e^{-\beta f(\boldsymbol{a})} \int \ldots \int e^{-\frac{\beta}{2} \boldsymbol{y}^T \boldsymbol{H} \boldsymbol{y}} d\boldsymbol{y} \tag{12.274}$$

According to a well-known theorem in matrix algebra, since the Hessian is real and symmetric (see definition in (12.272)) we can also decompose it as (see p. 288 of Lipschutz, 1987)

$$\boldsymbol{P}^T \boldsymbol{D} \boldsymbol{P} = \boldsymbol{H} \tag{12.275}$$

where $\boldsymbol{P}$ is the orthogonal matrix and its column is composed of eigenvectors of the Hessian matrix. Equation (12.275) is known as the spectral decomposition of Hessian matrix $\boldsymbol{H}$. Because of the orthogonal properties of eigenvectors, $\boldsymbol{D}$ is a diagonal matrix with eigenvalues of $\boldsymbol{H}$ as the diagonal terms:

$$\boldsymbol{D} = \begin{bmatrix} \lambda_1 & 0 & 0 \\ \vdots & \ddots & \vdots \\ 0 & 0 & \lambda_n \end{bmatrix} \tag{12.276}$$

Thus, (12.274) can be expressed as

$$I(\beta) \approx \varphi(\boldsymbol{a}) e^{-\beta f(\boldsymbol{a})} \int \ldots \int e^{-\frac{\beta}{2} \boldsymbol{z}^T \boldsymbol{D} \boldsymbol{z}} d\boldsymbol{z} \tag{12.277}$$

where

$$\boldsymbol{z} = \boldsymbol{P} \boldsymbol{y} \tag{12.278}$$

Equation (12.278) is also known as the shearing transformation. In obtaining (12.180), we note that since the matrix $\boldsymbol{P}$ is composed of unit eigenvectors, we must have the determinant of $\boldsymbol{P}$ be unity or:

$$\det |\boldsymbol{P}| = \det \left| \frac{\partial y_i}{\partial z_j} \right| = J = 1 \tag{12.279}$$

Note also that

$$\boldsymbol{z}^T \boldsymbol{D} \boldsymbol{z} = \lambda_1 z_1^2 + \ldots + \lambda_n z_n^2 \tag{12.280}$$

Thus, we have

$$I(\beta) \approx \varphi(\boldsymbol{a}) e^{-\beta f(\boldsymbol{a})} \int e^{-\frac{\beta}{2} \lambda_1 z_1^2} dz_1 \ldots \int e^{-\frac{\beta}{2} \lambda_n z_n^2} dz_n \tag{12.281}$$

For each integral in (12.281), we can apply the result from the last section and obtain

$$I(\beta) \approx \varphi(\boldsymbol{a}) e^{-\beta f(\boldsymbol{a})} \left(\frac{2\pi}{\beta}\right)^{n/2} \frac{1}{\sqrt{\lambda_1 \cdots \lambda_n}} \tag{12.282}$$

Note that the determinant of the Hessian equals the product of the eigenvalues or (see definition in (12.276)):

$$\prod_{i=1}^{n} \lambda_i = \det \boldsymbol{D} = \det \boldsymbol{H} \tag{12.283}$$

Substitution of (12.283) into (12.282) results in

$$I(\beta) \approx \varphi(\boldsymbol{a}) e^{-\beta f(\boldsymbol{a})} \left(\frac{2\pi}{\beta}\right)^{n/2} \frac{1}{\sqrt{\det \boldsymbol{H}}} \tag{12.284}$$

This result agrees with Equation (5.15) on p. 498 of Wong (2001). However, the present proof is somewhat simpler and does not need to introduce the Morse lemma in bifurcation theory.

12.8 STEEPEST DESCENT OF RIEMANN

The Laplace method that we discussed in the last section is for real functions h and g. When both of these functions are complex and analytic, the asymptotic analysis is called the method of steepest descent and was originated by Riemann and further developed by Debye. This original work by Riemann was never formally published. It was only published posthumously in 1876, and it is among the unpublished notes found in his house (or *nachlass*) that was subsequently included in his collected works. The method was first published by Debye in 1902 and he applied it to find an integral representation of Bessel functions for a large argument or order. In Debye's (1909) paper, he did mention its appearance in Riemann's Collected Works. Riemann is one of the most influential German mathematicians and he studied number theory, differential geometry, and complex analysis. His work on the complex zeta function related to prime numbers led to the unsolved Riemann hypothesis (see Problem 4.56 of Chapter 4 and Sabbagh, 2003), and his Riemann geometry is the basis for Einstein's theory of relativity. The Riemann hypothesis is probably the most well-known unsolved mathematics problem after the celebrated proof of Fermat's Last Theorem by Andrew Wiles (Singh, 1997). Debye is a Dutch physicist and was a student of Sommerfeld, and he studied the specific heat of solids (Debye model), ionic solutions (Debye potential and Debye-Hückel method), and he received the Nobel Prize in chemistry in 1936. In fact, Debye's undergraduate degree was in electrical engineering and his PhD was in physics under the supervision of Sommerfeld. Debye had been nominated for a Nobel Prize in physics as well. Friedrich Hund, his colleague at the University of Leipzig, described him as "clever but lazy" as he could often spot Debye watering roses in the institute garden when he should have been working. Debye responded that he had a certain tendency to take things easy. It was speculated that many great ideas behind his over 200 publications (one led to the Nobel Prize) were germinated in this outdoor garden. The steepest descent method is also known as the saddle point method. The method of steepest descent belongs to a special theory of contour integration. This idea of steepest descent has also been used in optimization problems with success, but such consideration is out of the scope of the present book.

12.8.1 Direction of Steepest Descent

When $h(z)$ and $g(z)$ are complex functions of a complex variable z, the integral in (12.225) is of Fourier type (compare the integral in the Fourier transform given in Section 11.2.2):

$$I(\lambda) = \int_C g(z) e^{\lambda w(z)} dz \tag{12.285}$$

where $z = x+iy$ is a complex variable and integration is carried out along a certain contour C. The saddle point of the exponential function is found by setting

$$w'(z)\big|_{z=z_0} = 0 \tag{12.286}$$

If the first $(n-1)$-th derivatives of w are also zeros, we call the saddle point is of order n. Mathematically, it is defined by:

$$w'(z)\big|_{z=z_0} = w''(z)\big|_{z=z_0} = ... = w^{(n-1)}(z)\big|_{z=z_0} = 0, \quad w^{(n)}(z)\big|_{z=z_0} \neq 0 \tag{12.287}$$

If $n = 2$, the saddle is called a simple saddle point. If we plot the exponential function in (12.285) on the complex plane (with axes being $x = \text{Re}[z]$ and $y = \text{Im}[z]$), the function will appear as mountains with valleys, cols, and saddles between them. The exponential function in (12.285) can be written as:

$$e^{\lambda w(z)} = e^{\lambda \text{Re}[w(z)]} e^{i\lambda \text{Im}[w(z)]} = e^{\lambda u(x,y)} e^{i\lambda v(x,y)} \tag{12.288}$$

The contour lines and constant phase lines of such a plot are represented respectively by:

$$u(x,y) = \text{Re}[w(z)] = C_1, \quad v(x,y) = \text{Im}[w(z)] = C_2 \tag{12.289}$$

where C_1 and C_2 are constants (see Fig 12.8). Note that the imaginary part in (12.288) will lead to oscillation in the functions but does not contribute to the integral, while the real part contributes mainly to the integration. The constant phase lines also represent the steepest descent or ascend paths. Physically, it is because the changes in the real part of the function is fastest (or the steepest descent or ascend along the contour) if the imaginary part is a constant. Mathematically, the change in w can be expressed as

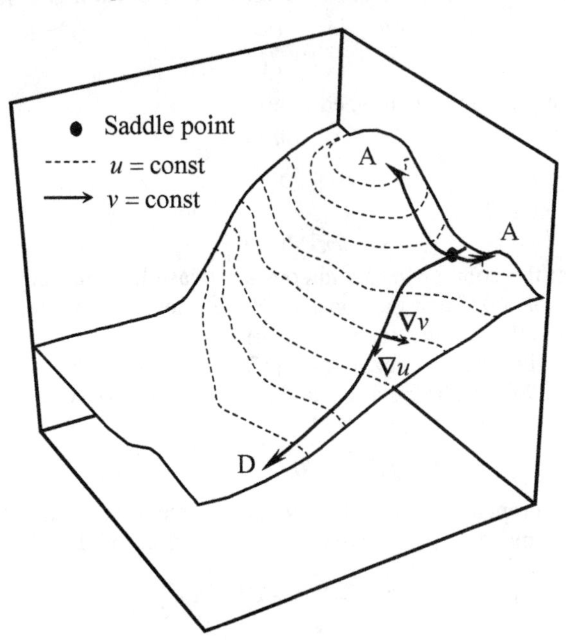

Figure 12.8 Contour lines and constant phase lines

$$|\delta w|^2 = |\delta u|^2 + |\delta v|^2 \tag{12.290}$$

Along the lines of constant phase, we have $\delta v = 0$ or

$$\mathrm{Im}[w(z)] = \mathrm{Im}[w(z_0)] \tag{12.291}$$

Then, the change of δu is maximal when (12.291) is satisfied. The method of steepest descent aims to deform the contour path of integration passing through the saddle point z_0 such that it coincides with the steepest path as far as possible (i.e., a path that (12.291) is satisfied). If (12.291) is satisfied, the integrand would not oscillate rapidly on the steepest descent path. Then, an approximate value of the integral will be determined from the integrand in the neighborhood of the saddle point.

Alternatively, the equivalence between the constant phase lines and steepest descent path can be proved by using directional or vector differentiation. The gradients of the functions u and v are

$$\nabla u = \frac{\partial u}{\partial x}\boldsymbol{e}_x + \frac{\partial u}{\partial y}\boldsymbol{e}_y, \quad \nabla v = \frac{\partial v}{\partial x}\boldsymbol{e}_x + \frac{\partial v}{\partial y}\boldsymbol{e}_y \tag{12.292}$$

The gradient ∇u is perpendicular to the contour lines of $u =$ const. (dashed lines in Figure 12.8) and indicates the direction of maximum changes of u whereas ∇v is perpendicular to the constant phase lines of $v =$ const. (solid arrow lines). Emerging from the saddle point, the deepest descent line is denoted by D and the steepest ascent line is denoted by A. Since w is an analytic function, u and v must satisfy the Cauchy-Riemann equations as (recall from Section 1.7.1):

$$\frac{\partial u}{\partial x} = \frac{\partial v}{\partial y}, \quad \frac{\partial u}{\partial y} = -\frac{\partial v}{\partial x} \tag{12.293}$$

These two equations can be combined to give

$$\frac{\partial u}{\partial x}\frac{\partial v}{\partial x} + \frac{\partial u}{\partial y}\frac{\partial v}{\partial y} = 0 \tag{12.294}$$

In tensor notation, (12.294) is

$$\nabla u \cdot \nabla v = 0 \tag{12.295}$$

This shows that the contour lines and constant phase lines are perpendicular. This also gives another proof that the constant phase lines are the steepest descent path (corresponding to the maximum change of u).

Differentiating the first equation of (12.293) with respect to x and the second equation of (12.293) with respect to y and adding these two, we get

$$\frac{\partial^2 u}{\partial x^2} + \frac{\partial^2 u}{\partial y^2} = \nabla^2 u = 0 \tag{12.296}$$

Similarly, we can repeat this procedure with a reverse of the order of differentiation with respect to y and x for the first and second equations of (12.293) to yield

$$\frac{\partial^2 v}{\partial x^2} + \frac{\partial^2 v}{\partial y^2} = \nabla^2 v = 0 \tag{12.297}$$

Thus, as expected, both u and v are harmonic functions. From the maximum principle of potential theory from Chapter 9 (e.g., Sections 9.7.7 and 9.8), there is no maximum or minimum of u within the domain, except on the boundary. The same conclusion can also be made in view of the maximum modulus principle of complex variable theory (e.g., p.146 of Silverman, 1974).

We now consider the direction of the steepest descent path at the saddle point. Using the information on the derivative of w, we can express the change of function w at the saddle point as:

$$\delta w = w(z) - w(z_0) = \frac{1}{n!}\frac{d^n w}{dz^n}|z - z_0|^n \{1 + O(|z - z_0|)\} \tag{12.298}$$

Using Euler's formula to write the terms in (12.298) in polar form, we get

$$\frac{d^n w}{dz^n} = ae^{i\alpha}, \quad z - z_0 = \rho e^{i\theta} \tag{12.299}$$

Clearly, we are looking for θ along the steepest descent path. Substitution of (12.299) into (12.298) yields

$$\delta w = \frac{ae^{i\alpha}}{n!}\rho^n e^{in\theta}\{1 + O(\rho)\} \tag{12.300}$$

This can be rearranged as

$$\frac{\delta w}{\rho^n} = \frac{ae^{i(\alpha+n\theta)}}{n!}\{1 + O(\rho)\} \approx \frac{a}{n!}[\cos(\alpha + n\theta) + i\sin(\alpha + n\theta)] \tag{12.301}$$

Along the steepest descent path, we must have (12.291) satisfied or (12.301) must be real and, in addition, $\delta w < 0$. Thus, we have to set

$$\alpha + n\theta = (2p+1)\pi, \quad p = 0,1,...,n-1 \tag{12.302}$$

Therefore, the directions for the steepest descent path are

$$\theta = -\frac{\alpha}{n} + (2p+1)\frac{\pi}{n}, \quad p = 0,1,...,n-1 \tag{12.303}$$

Along the steepest ascent path, we must have $\delta w > 0$ instead, and this leads to

$$\alpha + n\theta = 2p\pi, \quad p = 0,1,...,n-1 \tag{12.304}$$

The corresponding θ are

$$\theta = -\frac{\alpha}{n} + \frac{2p\pi}{n}, \quad p = 0,1,...,n-1 \tag{12.305}$$

The direction for constant u (the divide between hills and valleys) is given by

$$\alpha + n\theta = (p + \frac{1}{2})\pi, \quad p = 0,1,...,2n-1 \tag{12.306}$$

The directions θ for the separation lines between hills and valleys are

$$\theta = -\frac{\alpha}{n} + (p + \frac{1}{2})\frac{\pi}{n}, \quad p = 0,1,...,2n-1 \tag{12.307}$$

The most important type of saddle point is for $n = 2$. The values of θ for special cases are:

$$\text{Descent: } \theta = -\frac{\alpha}{2} + \frac{\pi}{2}, \quad -\frac{\alpha}{2} + \frac{3\pi}{2} \tag{12.308}$$

$$\text{Ascent: } \theta = -\frac{\alpha}{2}, \quad -\frac{\alpha}{2} + \pi \tag{12.309}$$

$$\text{Constant } u\text{:}\quad \theta = -\frac{\alpha}{2}+\frac{\pi}{4},\quad -\frac{\alpha}{2}+\frac{3\pi}{4},\quad -\frac{\alpha}{2}+\frac{5\pi}{4},\quad -\frac{\alpha}{2}+\frac{7\pi}{4} \tag{12.310}$$

The simple saddle for $n = 2$ is illustrated in Figure 12.9.

A more detailed view of the directions of steepest descent and ascent are depicted in Figure 12.10. The steepest descent path is labeled as D and those for steepest ascent are labeled as A. The projections of hills, valleys and the horizontal plane cutting the saddle point are defined as:

$$\text{hills:}\quad u(x,y) > u(x_0,y_0) \tag{12.311}$$

$$\text{valleys:}\quad u(x,y) < u(x_0,y_0) \tag{12.312}$$

$$\text{plane:}\quad u(x,y) = u(x_0,y_0) \tag{12.313}$$

Physically, along the steepest ascent (A in Figure 12.10) the real part of $w(z)$ will diverge to infinity as the contour passes the saddle point (except for very special case). It means that the integral will diverge. Therefore, we must consider the steepest descent path indicated by D in Figure 12.10.

12.8.2 Steepest Descent of Regular Saddle Point

We now return to the asymptotic expansion of (12.285). We first rewrite it as

$$I(\lambda) = \exp\{\lambda w(z_0)\}\int_C g(z)\exp\{\lambda w(z) - \lambda w(z_0)\}dz \tag{12.314}$$

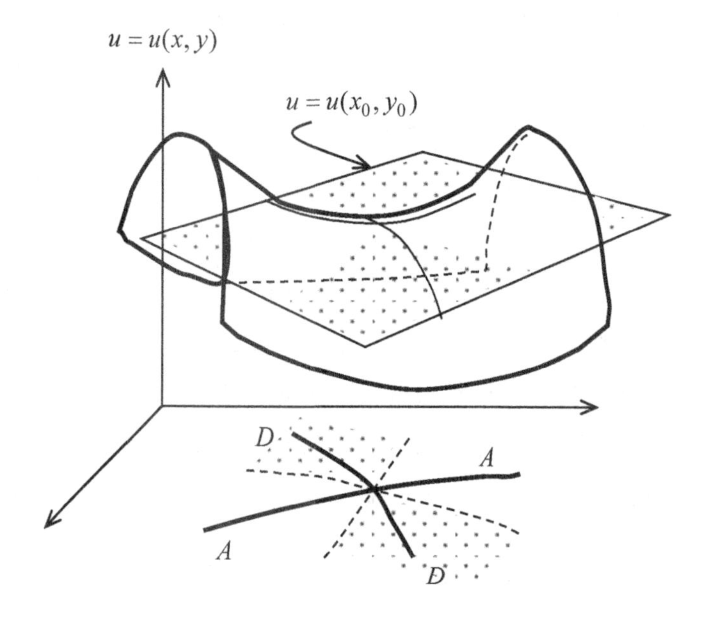

Figure 12.9 A simple saddle with $n = 2$

The exponential function in the integrand is exactly the change of the function $w(z)$ and thus is related to the steepest descent. Since the contour C is supposed to pass through the saddle point such that we have

$$w'(z_0) = 0, \tag{12.315}$$

hence, the Taylor series expansion for $w(z)$ about the saddle point is

$$w(z) - w(z_0) = \frac{1}{2!} \frac{d^2 w}{dz^2}\bigg|_{z=z_0} (z - z_0)^2 + \ldots \tag{12.316}$$

Substitution of (12.316) into (12.314) gives

$$I(\lambda) = \exp\{\lambda w(z_0)\} \int_C g(z) \exp\{\frac{\lambda}{2} w''(z_0)(z - z_0)^2\} dz \tag{12.317}$$

We now follow the Euler form given in (12.299), then (12.317) can be expressed as

$$I(\lambda) = \exp\{\lambda w(z_0)\} \int_C g(z) \exp\{\frac{\lambda a}{2} \rho^2 [\cos(\alpha + 2\theta) + i \sin(\alpha + 2\theta)\} dz \tag{12.318}$$

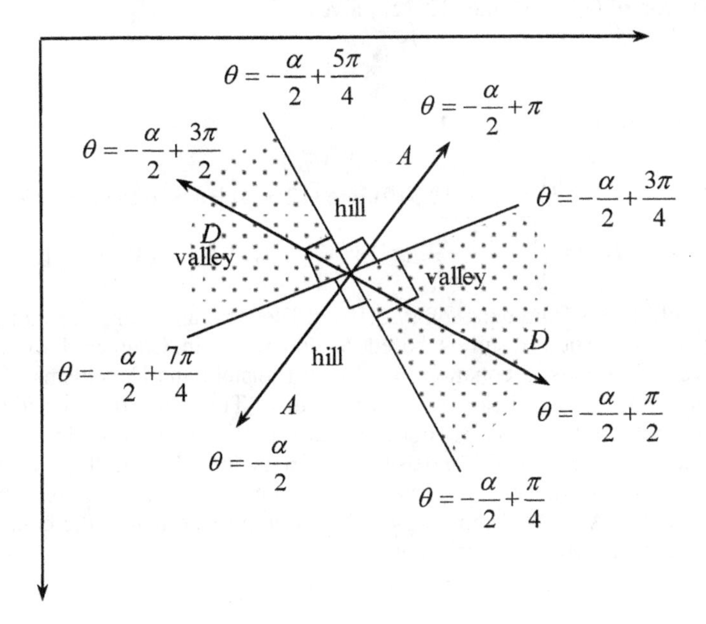

Figure 12.10 Orientations of steepest descent and ascent for saddle point with $n = 2$

Applying the idea of steepest descent for contour C, we deform it to from $-\infty$ to ∞ or from 0 to ∞. The choice on the limits of integration depends on the nature of $w(z)$. More discussion on this choice of contour to real integration will be given later in this section. As discussed earlier in the section, we can set

$$\alpha + n\theta = (2p + 1)\pi, \quad p = 0, 1, \ldots, n-1 \tag{12.319}$$

such that the imaginary term in (12.318) will be zero (i.e., constant phase lines). Thus, we have for the case of $n = 2$

$$I(\lambda)=\exp\{\lambda w(z_0)\}\int_{-\infty}^{\infty} g(z)\exp\{-\frac{\lambda a}{2}\rho^2\}dz \tag{12.320}$$

From the second equation of (12.299), we have

$$dz=e^{i\theta}d\rho \tag{12.321}$$

Substitution of (12.321) into (12.320) gives

$$I(\lambda)=\exp\{\lambda w(z_0)\}g(z_0)e^{i\theta}\int_{-\infty}^{\infty}\exp\{-\frac{\lambda a}{2}\rho^2\}d\rho \tag{12.322}$$

In obtaining (12.322), we have assumed $g(z)$ is a well-behaved function near the saddle point. Note that integration of the complex function is now converted to integration of the real variable. The integral is known as the Gauss or Laplace integral and can be evaluated as (see (1.102) of Chapter 1):

$$\int_{-\infty}^{\infty}\exp\{-\frac{\lambda a}{2}\rho^2\}d\rho=\sqrt{\frac{2\pi}{\lambda a}} \tag{12.323}$$

Substitution of (12.323) into (12.322) gives

$$I(\lambda)=\sqrt{\frac{2\pi}{\lambda a}}\exp\{\lambda w(z_0)\}g(z_0)e^{i\theta} \tag{12.324}$$

From the first of (12.299), we find

$$a=|w''(z_0)| \tag{12.325}$$

Substitution of (12.325) and (12.319) into (12.324) leads to the final result:

$$I(\lambda)=\sqrt{\frac{2\pi}{\lambda|w''(z_0)|}}g(z_0)\exp\{\lambda w(z_0)+i[(2p+1)\frac{\pi}{n}-\frac{\alpha}{n}]\} \tag{12.326}$$

This result differs from equation (7.2.10) of Bleistein and Handelsman (1986) by a factor of two. Their result is based on the result in Chapter 4 of their book. However, if we look at equation (4.4.24) of Bleistein and Handelsman (1986), the formulation is for integration limits from 0 to ∞. Therefore, if we divide our result by 2, we recover the result of Bleistein and Handelsman (1986). The derivation of the results of Bleistein and Handelsman (1986) is done through the consideration of the Mellin transform and involves a triple summation. The current approach is much simpler. As a final remark, we should emphasize that if the contour can be deformed to the following real limits

$$\int_C ...dz \to \int_0^{\infty} ...d\rho, \tag{12.327}$$

we should use half of (12.326). If the contour can be deformed to the following real limits

$$\int_C ...dz \to \int_{-\infty}^{\infty} ...d\rho, \tag{12.328}$$

we should use (12.326).

In summary, the method of steepest descent can be divided into 5 steps:

(i) Identify all critical points, including saddle points.

(ii) Determine the path of the steepest descent from these critical points.

(iii) Justify the deformation of the original contour C onto one or more of the paths of steepest descent found in (ii).

(iv) Determine the asymptotic expansion of the integrals on the deformed contour such that the complex integral becomes a real Laplace type integral.
(v) Sum the asymptotic expansions to give an approximation for the integral.

Among these, Step (iii) is the most difficult, and the process of deforming the contour is a consequence of the Cauchy integral theorem.

12.8.3 Steepest Descent of Higher Order Saddle Point

The previous result presented in (12.326) makes two major assumptions that pose a limitation on the result in (12.326). Let us recap here. Recasting the Laplace integral slightly, we have:

$$I(\lambda) = \int_a^b g(t)\exp\{\lambda w(t)\}dt \tag{12.329}$$

In the last section, the two major assumptions are:

$$g(t_0) = \text{regular}, \quad w'(z_0) = 0 \tag{12.330}$$

First we generalize the second condition in (12.330), and suppose now that the saddle point is not a simple saddle and the function w satisfies:

$$w'(z_0) = w''(z_0) = ... = w^{(n-1)}(z_0) = 0, \quad w^{(n)}(z_0) \neq 0 \tag{12.331}$$

where n is an integer (can be both even and odd). The analysis given in Section 12.8.2 can easily be modified. In particular, we recall from (12.299) that:

$$w^{(n)}(z_0) = ae^{i\alpha}, \quad z - z_0 = \rho e^{i\theta} \tag{12.332}$$

In addition, the integral (12.329) can be rewritten as

$$I(\lambda) = \exp\{\lambda w(z_0)\}\int_a^b g(z)\exp\{\lambda[w(z) - w(z_0)]\}dz \tag{12.333}$$

Taylor series expansion gives

$$\begin{aligned} w(z) - w(z_0) &= \frac{w^{(n)}(z_0)}{n!}(z - z_0)^n + ... = \frac{a\rho^n}{n!}e^{i(n\theta+\alpha)} \\ &= \frac{a\rho^n}{n!}[\cos(n\theta+\alpha) + i\sin(n\theta+\alpha)] \end{aligned} \tag{12.334}$$

By requiring that the imaginary part of (12.334) is a constant, we have

$$\theta = (2k+1)\frac{\pi}{n} - \frac{\alpha}{n}, \quad k = 0,1,...,n-1 \tag{12.335}$$

In view of (12.335), we have

$$w(z) - w(z_0) = -\frac{a\rho^n}{n!} \tag{12.336}$$

Substitution of (12.336) into (12.333) gives

$$I(\lambda) = \exp\{\lambda w(z_0)\}\exp(i\theta)g(z_0)\int_0^\infty \exp\{-\frac{\lambda a\rho^n}{n!}\}d\rho \tag{12.337}$$

where $g(z)$ is an analytic function at $z = z_0$. If n is not an integer, the integral is not an even function. Thus, we cannot in general extend the limits of the integral to $-\infty$ to ∞. Now consider a change of variables:

$$(\frac{\lambda a}{n!})^{1/n}\rho=\zeta \tag{12.338}$$

With this change of variables, (12.337) can be written as

$$I(\lambda)=\exp\{\lambda w(z_0)+i\theta\}g(z_0)(\frac{n!}{\lambda a})^{1/n}\int_0^{\infty}\exp\{-\zeta^n\}d\rho \tag{12.339}$$

Finally, the integral in (12.339) can be evaluated in terms of the gamma function through the following change of variables:

$$s=\zeta^n,\quad ds=n\zeta^{n-1}d\zeta \tag{12.340}$$

Using this change of variables, the integral in (12.339) becomes

$$I_G=\int_0^{\infty}\exp\{-\zeta^n\}d\rho=\frac{1}{n}\int_0^{\infty}e^{-s}\rho^{\frac{1}{n}-1}d\rho=\frac{1}{n}\Gamma(\frac{1}{n}) \tag{12.341}$$

The last result is obtained by noting the definition of the gamma function given in Chapter 4. Back substitution of (12.341) into (12.339) gives

$$I(\lambda)\approx\frac{g(z_0)}{n}(\frac{n!}{\lambda\left|w^{(n)}(z_0)\right|})^{1/n}\Gamma(\frac{1}{n})\exp\{\lambda w(z_0)+i[(2k+1)\frac{\pi}{n}-\frac{\alpha}{n}]\} \tag{12.342}$$

This result agrees with the result given by Bleistein and Handelsman (1986).

Similarly, we can also generalize the first condition in (12.330) that g is not a regular function at $z=z_0$:

$$g(z_0)=g'(z_0)=...=g^{(m-1)}(z_0)=0,\quad g^{(m)}(z_0)\neq 0, \tag{12.343}$$

For this case, we expand $g(z)$ in a Taylor series expansion as (since all lower derivatives vanish at $z=z_0$)

$$g(z)-g(z_0)=\frac{g^{(m)}(z_0)}{m!}(z-z_0)^m+...\approx[\frac{g^{(m)}(z_0)}{m!}]\rho^m e^{im\theta}, \tag{12.344}$$

Substituting (12.344) into (12.333) and in view of (12.343) and (12.344), we can express (12.333) as

$$I(\lambda)=[\frac{g^{(m)}(z_0)}{m!}]e^{im\theta}\exp\{\lambda w(z_0)\}\int_0^{\infty}\rho^m\exp\{-\frac{\lambda a\rho^n}{n!}\}d\rho \tag{12.345}$$

To determine the integral, we can apply the change of variables proposed in (12.338) to simplify the integral in (12.345) to

$$\int_0^{\infty}\rho^m\exp\{-\frac{\lambda a\rho^n}{n!}\}d\rho=(\frac{n!}{\lambda a})^{(m+1)/n}\int_0^{\infty}\zeta^m\exp\{-\zeta^n\}d\zeta \tag{12.346}$$

Introducing another round of change of variables $\zeta=s^{1/n}$, we get

$$\int_0^{\infty}\rho^m\exp\{-\frac{\lambda a\rho^n}{n!}\}d\rho=(\frac{n!}{\lambda a})^{(m+1)/n}\frac{1}{n}\int_0^{\infty}s^{(m+1)/n-1}\exp\{-s\}ds \tag{12.347}$$

The last integral in (12.347) is again recognized as the gamma function. Thus, we finally get

$$I(\lambda)\approx\frac{1}{n}[\frac{g^{(m)}(z_0)}{m!}](\frac{n!}{\lambda\left|w^{(n)}(z_0)\right|})^{\frac{m+1}{n}}\Gamma(\frac{m+1}{n})\exp\{\lambda w(z_0)+i(m+1)[(2k+1)\frac{\pi}{n}-\frac{\alpha}{n}]\} \tag{12.348}$$

This result again agrees with that given by Bleistein and Handelsman (1986), but the present proof is much simpler that theirs.

12.9 STATIONARY PHASE OF STOKES-KELVIN

According to the discussion by Watson (1944), the method of stationary phase was first published explicitly by Kelvin in 1887 when he considered the asymptotic value of the following integral with a large argument, which was related to two-dimensional waves in water (Kelvin, 1887):

$$u = \frac{1}{2\pi}\int_0^\infty \cos[m\{x - tf(m)\}]dm \tag{12.349}$$

However, Watson (1944) also discovered that the same idea had been used by Cauchy in 1815, Stokes in 1856, and Riemann in 1876. Nevertheless, in most literatures this method was attributed to either Stokes or Kelvin.

In particular, the following integral was considered:

$$I(x) = \int_\alpha^\beta g(t)\exp\{ixh(t)\}dt \tag{12.350}$$

where $x \to \infty$. Note that the Laplace method deals with a real function in the exponential function, the method of steepest descent considers a complex function in the exponential function, whereas in the present case, the method of stationary phase treats a purely imaginary function in the exponential function. Actually, the method of analysis closely resembles that used in the method of steepest descent. It is well known from Euler's formula that

$$\exp\{ixh(t)\} = \cos[xh(t)] + i\sin[xh(t)] \tag{12.351}$$

When $x \to \infty$, this is a highly oscillating function. In mechanics and physics, such functions are mostly related to wave phenomena, and $xh(t)$ is actually the phase of the wave motion. The integral is essentially zero as the integrand is rapidly oscillatory such that the contributions from adjacent sub-intervals nearly cancel one another, except at the point where the function is stationary (i.e., $h'(t) = 0$). Kelvin asserted that the main contribution is from the end points and stationary points, but the contribution from the stationary point is more important than those from the end points. In other words, we are looking at the point where the phase of the wave motion is stationary. This suggests the name "method of stationary phase."

12.9.1 Erdelyi's Derivation

Assume that the stationary point is at $t = \tau$ such that

$$h'(\tau) = 0, \quad h''(\tau) \neq 0, \quad \alpha < \tau < \beta \tag{12.352}$$

Since the contribution is from the neighborhood of the stationary point, we have

$$I(x) \approx \int_{\tau-\varepsilon}^{\tau+\varepsilon} g(t)\exp\{ixh(t)\}dt \tag{12.353}$$

where ε is a small positive number. Following the procedure discussed in the Laplace method, we can apply the following change of variables:

$$h(t)-h(\tau)=u^2, \quad dt=\frac{2udu}{h'(t)} \tag{12.354}$$

Thus, the integral in (12.353) becomes

$$I(x)\approx\int_{-[h(\tau-\varepsilon)-h(\tau)]^{1/2}}^{[h(\tau+\varepsilon)-h(\tau)]^{1/2}} g(t)\frac{2u}{h'(t)}\exp\{ix[h(\tau)+u^2]\}du \tag{12.355}$$

We now consider the following limit as $t\to\tau$

$$\lim_{t\to\tau}\frac{2u(t)}{h'(t)}=\frac{0}{0} \tag{12.356}$$

Since the limit is of the form of 0/0, we can apply L'Hôpital's rule to get

$$\lim_{t\to\tau}\frac{2u(t)}{h'(t)}=\lim_{t\to\tau}\frac{2u'(t)}{h''(t)}=\lim_{t\to\tau}\frac{2}{h''(t)}\frac{h'(t)}{2u} \quad . \tag{12.357}$$

We note that the limit that we want to find on the left hand side appears as an inverse on the right hand side. Thus, we can rearrange the result as

$$\lim_{t\to\tau}[\frac{2u(t)}{h'(t)}]^{1/2}=\frac{2}{h''(\tau)} \tag{12.358}$$

Using this result, we can rewrite (12.355) as

$$I(x)\approx\left[\frac{2}{h''(\tau)}\right]^{1/2} g(\tau)\exp\{ixh(\tau)\}\int_{-\infty}^{\infty}\exp\{ixu^2\}du \tag{12.359}$$

To evaluate the integral in (12.359), we make the following change of variables

$$u=e^{\pm i\pi/4}\left[\frac{\zeta}{x}\right]^{1/2} \tag{12.360}$$

where the positive sign is for $h''(\tau)>0$ and the negative sign is for $h''(\tau)<0$. Effectively, we rotate the angle or we convert the oscillating exponential function to become a decaying function as

$$I(x)\approx\left[\frac{2}{\pm h''(\tau)}\right]^{1/2} g(\tau)\exp\{ixh(\tau)\}\frac{e^{\pm i\pi/4}}{x^{1/2}}\int_0^{\infty}\zeta^{1/2-1}\exp\{-\zeta\}d\zeta \tag{12.361}$$

The last integral becomes a gamma function and thus we have

$$I(x)\approx\left[\frac{2}{\pm xh''(\tau)}\right]^{1/2} g(\tau)\exp\{ixh(\tau)\pm\frac{i\pi}{4}\}\Gamma(\frac{1}{2}) \tag{12.362}$$

Finally, we can use the well-known result for the gamma function that (see Example 4.2 in Chapter 4)

$$\Gamma(\frac{1}{2})=\sqrt{\pi} \tag{12.363}$$

Consequently, we have the asymptotic result as

$$I(x)\approx\left[\frac{2\pi}{\pm xh''(\tau)}\right]^{1/2} g(\tau)\exp\{ixh(\tau)\pm\frac{i\pi}{4}\} \tag{12.364}$$

This result of course agrees with that of Erdelyi (1956), who only considered the positive sign. Note however that there is a typo in equation (6.1.5) on p. 220 of Bleistein and Handelsman (1986).

Example 12.5 Consider the leading order term of the following integral form of the Bessel function defined by:

$$J_n(\lambda) = \frac{1}{\pi}\int_0^{\pi} \cos(nt - \lambda \sin t) dt \tag{12.365}$$

where $\lambda \to \infty$.

Solution: We first note from Euler's formula that

$$\cos\theta = \frac{1}{2}(e^{i\theta} + e^{-i\theta}) \tag{12.366}$$

Using (12.366), we have

$$\begin{aligned} J_n(\lambda) &= \frac{1}{2\pi}\left\{\int_0^{\pi} \exp(int)\exp(-i\lambda \sin t)dt + \int_0^{\pi} \exp(-int)\exp(i\lambda \sin t)dt\right\} \\ &= \frac{1}{2\pi}\{J_{n1}(\lambda) + J_{n2}(\lambda)\} \end{aligned} \tag{12.367}$$

We recognize that both integrals are of the type of (12.350). For the first integral, we find

$$h(t) = -\sin t, \quad h'(t) = -\cos t, \quad h''(t) = \sin t \tag{12.368}$$

The stationary point is at $t = \pi/2$, and thus

$$h(\pi/2) = -1, \quad h'(\pi/2) = 0, \quad h''(\pi/2) = 1 \tag{12.369}$$

Employing the formula derived in (12.364), we get

$$J_{n1}(\lambda) \approx \exp\{-i\lambda\}\exp(i\frac{n\pi}{2})\sqrt{\frac{2\pi}{\lambda}}\exp(\frac{i\pi}{4}) \tag{12.370}$$

Similarly, for the second integral we can follow the same procedure in deriving

$$J_{n2}(\lambda) \approx \exp\{i\lambda\}\exp(-i\frac{n\pi}{2})\sqrt{\frac{2\pi}{\lambda}}\exp(-\frac{i\pi}{4}) \tag{12.371}$$

Substitution of (12.370) and (12.371) into (12.367), we find

$$J_n(\lambda) \approx \sqrt{\frac{2\pi}{\lambda}}\cos(\lambda - \frac{n\pi}{2} - \frac{\pi}{4}) \tag{12.372}$$

This can be found in 9.2.1 of Abramowitz and Stegun (1964).

12.9.2 Stationary Phase of Higher Order

This section follows closely the presentation by Bender and Orszag (1978). Let us recast the problem as

$$I(x) = \int_a^b f(t)\exp\{ix\psi(t)\}dt \tag{12.373}$$

as $x \to \infty$. Without loss of generality, we can assume the higher order saddle point is at the boundary point $t = a$ defined by:

$$\psi'(a) = ... = \psi^{(p-1)}(a) = 0 \quad \psi^{(p)}(a) \neq 0 \tag{12.374}$$

First, we can decompose (12.373) into two integrals:

$$I(x) = \int_a^{a+\varepsilon} f(t)\exp\{ix\psi(t)\}dt + \int_{a+\varepsilon}^{b} f(t)\exp\{ix\psi(t)\}dt = I_1(x) + I_2(x) \quad (12.375)$$

We are going to show that the order of magnitude of these two integrals is not the same as $x \to \infty$. The leading order term is from the first integral as the second one decays much faster than the first for large x. Let us apply integration by parts to the second part of (12.375) as:

$$\begin{aligned} I_2(x) &= \int_{a+\varepsilon}^{b} f(t)\exp\{ix\psi(t)\}dt = \frac{1}{ix\psi'(t)}\int_{a+\varepsilon}^{b} f(t)d\{\exp[ix\psi(t)]\} \\ &= \frac{1}{ix\psi'(t)} f(t)\exp[ix\psi(t)]\Big|_{a+\varepsilon}^{b} - \frac{1}{ix}\int_{a+\varepsilon}^{b} \frac{\exp[ix\psi(t)]}{\psi'(t)}\frac{df(t)}{dt}dt \end{aligned} \quad (12.376)$$

If there is no stationary point within the limit of integrations, the integrals in (12.376) decays as $1/x$, and so does the second integral in (12.375).

To evaluate the first integral in (12.376), we use Taylor series expansion for $\psi(t)$ as

$$\psi(t) = \psi(a) + \psi^{(p)}(a)\frac{(t-a)^p}{p!} + \dots \quad (12.377)$$

Thus, if $f(t)$ is a regular function, we get

$$I_1(x) \sim f(a)\exp\{ix\psi(a)\}\int_a^{a+\varepsilon} \exp\{\frac{ix}{p!}\psi^{(p)}(a)(t-a)^p\}dt \quad (12.378)$$

The following change of variables is adopted

$$s = t - a \quad (12.379)$$

Then, with $\varepsilon \to \infty$ (12.379) is simplified to

$$I_1(x) \sim f(a)\exp\{ix\psi(a)\}\int_0^{\infty} \exp\{\frac{ix}{p!}\psi^{(p)}(a)s^p\}ds \quad (12.380)$$

To evaluate the integral, we apply the following change of variables

$$s = e^{\pm\frac{i\pi}{2p}}\left[\frac{p!u}{x\left|\psi^{(p)}(a)\right|}\right]^{1/p} \quad (12.381)$$

where the plus sign is used if $\psi^{(p)}(a) > 0$ and the minus sign is used if $\psi^{(p)}(a) < 0$. Employing this change of variables, (12.380) becomes

$$I_1(x) \sim \left[\frac{p!}{x\left|\psi^{(p)}(a)\right|}\right]^{1/p} f(a)\exp\{i[x\psi(a) + \frac{\pi}{2p}]\}\frac{1}{p}\int_0^{\infty} u^{1/p-1}\exp\{-u\}du \quad (12.382)$$

In obtaining this, we set upper limit $\varepsilon \to \infty$. The integral is again the gamma function, and thus we finally arrive at the following result:

$$I_1(x) \sim \left[\frac{p!}{x\left|\psi^{(p)}(a)\right|}\right]^{1/p} f(a)\exp\{i[x\psi(a) + \frac{\pi}{2p}]\}\Gamma(\frac{1+p}{p}) \quad (12.383)$$

Since p is an integer larger than 1, this integral decays as $1/x^{1/p}$, and it is much slower than $1/x$ compared to the second integral in (12.375). Therefore, the result is

$$I(x) \sim f(a)\exp\{i[x\psi(a)+\frac{\pi}{2p}]\}[\frac{p!}{x\left|\psi^{(p)}(a)\right|}]^{1/p}\Gamma(\frac{1+p}{p}) \tag{12.384}$$

This result agrees with Equation (6.5.12) on p.279 of Bender and Orszag (1978).

Example 12.6 Consider the leading order term of the following Bessel function with both the argument and order approaching infinity:

$$J_n(n) = \frac{1}{\pi}\int_0^\pi \cos(nt - n\sin t)dt \tag{12.385}$$

where $n \to \infty$.

Solution: For this case, the phase function is identified as

$$\psi(t) = \sin t - t \tag{12.386}$$

Therefore, we have

$$\psi'(t) = \cos t - 1, \quad \psi''(t) = -\sin t, \quad \psi'''(t) = -\cos t \tag{12.387}$$

The stationary point is at $t = 0$. Substituting the stationary point into (12.387) we get

$$\psi'(0) = 0, \quad \psi''(t) = 0, \quad \psi'''(t) = -1 \tag{12.388}$$

Therefore, we have $p = 3$ for this case. Employing the formula derived in (12.384), we obtain

$$J_n(n) \sim \frac{1}{\pi}\mathrm{Re}\{\frac{1}{3}e^{-i\pi/6}(\frac{6}{n})^{1/3}\Gamma(\frac{1}{3})\} \tag{12.389}$$

Finally, taking the real part we get

$$J_n(n) \sim \frac{1}{\pi}2^{-2/3}3^{-1/6}n^{-1/3}\Gamma(\frac{1}{3}) \tag{12.390}$$

This result was first obtained by Cauchy in 1854, and this was also obtained by Nicholson in 1909, Rayleigh in 1910, and Watson in 1918 (Watson, 1918).

12.9.3 Stationary Phase of Higher Dimensions

So far, our results are for the one-dimensional case. We can extend our analysis to multi-dimensional integral as:

$$I(\lambda) = \int \ldots \int \varphi(x_1,\ldots,x_n)\exp\{i\lambda f(x_1,\ldots,x_n)\}dx_1\cdots dx_n \tag{12.391}$$

The derivation of the result is similar to that employed in Section 12.7.4. In particular, the stationary point can be determined as

$$\nabla f(\boldsymbol{a}) = 0 \tag{12.392}$$

This is equivalent to

$$\frac{\partial f}{\partial x_1}(\boldsymbol{a}) = \cdots = \frac{\partial f}{\partial x_n}(\boldsymbol{a}) = 0 \tag{12.393}$$

where

$$\boldsymbol{a} = (a_1, ..., a_n) \tag{12.394}$$

As discussed before, the determinant of the Hessian matrix is larger than zero:

$$\det\left|\boldsymbol{H}(\boldsymbol{a})\right| = \det\left|\frac{\partial^2 f}{\partial x_i \partial x_j}\right| > 0 \tag{12.395}$$

Expanding $f(x)$ above $\boldsymbol{a}$, we get

$$I(\lambda) \approx \varphi(a_1, a_2, ..., a_n) \int ... \int e^{i\lambda[f(a_1,...,a_n) + \frac{1}{2}(x-a)^T H(x-a)]} dx \tag{12.396}$$

Denoting $\boldsymbol{y} = \boldsymbol{x} - \boldsymbol{a}$, we obtain

$$I(\lambda) \approx \varphi(\boldsymbol{a}) e^{i\lambda f(\boldsymbol{a})} \int ... \int e^{\frac{i\lambda}{2} \boldsymbol{y}^T \boldsymbol{H} \boldsymbol{y}} d\boldsymbol{y} \tag{12.397}$$

We can decompose the Hessian matrix as (see p. 288 of Lipschutz, 1987)

$$\boldsymbol{P}^T \boldsymbol{D} \boldsymbol{P} = \boldsymbol{H} \tag{12.398}$$

where $\boldsymbol{P}$ is the orthogonal matrix and its column is composed of eigenvectors of the Hessian matrix. Consequently, $\boldsymbol{D}$ is a diagonal matrix with eigenvalues of $\boldsymbol{H}$ as the diagonal terms:

$$\boldsymbol{D} = \begin{bmatrix} \lambda_1 & 0 & 0 \\ \vdots & \ddots & \vdots \\ 0 & 0 & \lambda_n \end{bmatrix} \tag{12.399}$$

Thus, (12.177) can be expressed as

$$I(\lambda) \approx \varphi(\boldsymbol{a}) e^{i\lambda f(\boldsymbol{a})} \int ... \int e^{\frac{i\lambda}{2} \boldsymbol{z}^T \boldsymbol{D} \boldsymbol{z}} d\boldsymbol{z} \tag{12.400}$$

where

$$\boldsymbol{z} = \boldsymbol{P} \boldsymbol{y} \tag{12.401}$$

The matrix $\boldsymbol{P}$ is composed of unit eigenvectors as:

$$\det\left|\boldsymbol{P}\right| = \det\left|\frac{\partial y_i}{\partial z_j}\right| = J = 1 \tag{12.402}$$

It is straightforward to see that

$$\boldsymbol{z}^T \boldsymbol{D} \boldsymbol{z} = \lambda_1 z_1^2 + ... + \lambda_n z_n^2 \tag{12.403}$$

Substitution of (12.403) into (12.400) gives

$$I(\lambda) \approx \varphi(\boldsymbol{a}) e^{i\lambda f(\boldsymbol{a})} \int e^{\frac{i\lambda}{2} \lambda_1 z_1^2} dz_1 ... \int e^{\frac{i\lambda}{2} \lambda_n z_n^2} dz_n \tag{12.404}$$

We now recall the previous result obtained in Section 12.9.1 that

$$\int_0^\infty \exp\{\pm i \frac{\lambda}{2} \lambda_i u^2\} du = (\frac{1}{2\lambda})^{1/2} e^{\pm \frac{i\pi}{4}} \frac{1}{\sqrt{\lambda_1}} \Gamma(\frac{1}{2}) = (\frac{\pi}{2\lambda})^{1/2} e^{\pm \frac{i\pi}{4}} \frac{1}{\sqrt{\lambda_1}} \tag{12.405}$$

The plus or minus sign depends on whether the eigenvalue λ_i defined as the diagonal terms in (12.399) is positive or negative. In view of the symmetric property of the integral, (12.405) can be rewritten as

$$\int_{-\infty}^{\infty} \exp\{\pm i\frac{\lambda}{2}\lambda_i u^2\}du = (\frac{2\pi}{\lambda})^{1/2} e^{\pm\frac{i\pi}{4}} \frac{1}{\sqrt{\lambda_1}} \tag{12.406}$$

where $i = 1,2,...,n$. Substitution of this result into (12.404) gives

$$I(\lambda) \approx \varphi(\boldsymbol{a}) e^{i\lambda f(\boldsymbol{a})} e^{\frac{i\pi}{4}\sum_{j=1}^{n}\operatorname{sgn}\lambda_j} (\frac{2\pi}{\lambda})^{n/2} \frac{1}{\sqrt{\lambda_1 \cdots \lambda_n}} \tag{12.407}$$

Recall that the determinant of the Hessian equals the product of the eigenvalues:

$$\prod_{i=1}^{n} \lambda_i = \det \boldsymbol{D} = \det \boldsymbol{H} \tag{12.408}$$

Substitution of (12.408) into (12.407) gives

$$I(\lambda) \approx \varphi(\boldsymbol{a})(\frac{2\pi}{\lambda})^{n/2} \frac{1}{\sqrt{\det \boldsymbol{H}}} \exp\{i\lambda f(\boldsymbol{a}) + \frac{i\pi}{4}\sum_{j=1}^{n}\operatorname{sgn}\lambda_j\} \tag{12.409}$$

If the stationary point is at the boundary, the value of the integral will be half

$$I(\lambda) \approx \frac{1}{2}\varphi(\boldsymbol{a})(\frac{2\pi}{\lambda})^{n/2} \frac{1}{\sqrt{\det \boldsymbol{H}}} \exp\{i\lambda f(\boldsymbol{a}) + \frac{i\pi}{4}\sum_{j=1}^{n}\operatorname{sgn}\lambda_j\} \tag{12.410}$$

This result agrees with Equation (2.32) on p. 487 of Wong (2001), but the present derivation is much simpler than that of Wong (2001).

12.10 NAVIER-STOKES EQUATIONS FOR SURFACE WAVES

As derived in Chapter 2, the most general form of incompressible fluid flow can be modelled by the following Navier-Stokes equation

$$\frac{\partial \boldsymbol{u}}{\partial t} + \boldsymbol{u}\cdot\nabla\boldsymbol{u} = -\frac{1}{\rho}\nabla p + \nu\nabla^2\boldsymbol{u} \tag{12.411}$$

Continuity of the fluid for a 2-D, incompressible, inviscid flow with no capillary is

$$\frac{\partial u}{\partial x} + \frac{\partial w}{\partial z} = 0 \tag{12.412}$$

The equation of motion along the x-axis is

$$\frac{\partial u}{\partial t} + u\frac{\partial u}{\partial x} + w\frac{\partial u}{\partial z} = -\frac{1}{\rho}\frac{\partial p}{\partial x} \tag{12.413}$$

The equation of motion along the z-axis is

$$\frac{\partial w}{\partial t} + u\frac{\partial w}{\partial x} + w\frac{\partial w}{\partial z} = -g - \frac{1}{\rho}\frac{\partial p}{\partial z} \tag{12.414}$$

As shown in Figure 12.11, there are two sets of boundary conditions, one at the free surface and one at the bottom. The dynamic condition on the free water surface is

$$p = 0, \quad z = \eta \tag{12.415}$$

Another surface condition is the kinematic condition on the water surface

$$\frac{\partial \eta}{\partial t}+u\frac{\partial \eta}{\partial x}=w, \quad z=\eta \tag{12.416}$$

whilst the kinematic condition at the sea bottom is

$$w=0, \quad z=-h \tag{12.417}$$

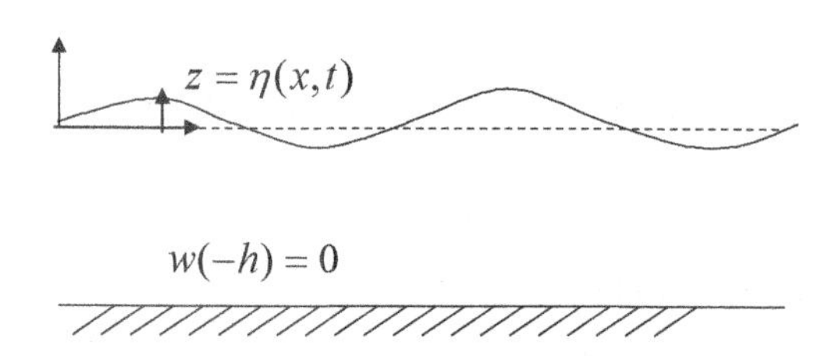

Figure 12.11 Boundary condition for shallow water waves

Because of the incompressibility given in (12.412), we can introduce a velocity potential ϕ

$$u=-\frac{\partial \phi}{\partial x}, \quad w=-\frac{\partial \phi}{\partial z} \tag{12.418}$$

In terms of the velocity potential, the continuity condition in (12.412) can be expressed as a Laplace equation

$$\frac{\partial^2 \phi}{\partial x^2}+\frac{\partial^2 \phi}{\partial z^2}=0, \quad \nabla^2\phi=0 \tag{12.419}$$

Equations of motion along the x- and z- axes given in (12.413) and (12.414) can be rewritten in terms of velocity potential as:

$$-\frac{\partial}{\partial x}(\frac{\partial \phi}{\partial t})+\frac{1}{2}\frac{\partial}{\partial x}[(\frac{\partial \phi}{\partial x})^2+(\frac{\partial \phi}{\partial z})^2]=-\frac{1}{\rho}\frac{\partial p}{\partial x} \tag{12.420}$$

$$-\frac{\partial}{\partial z}(\frac{\partial \phi}{\partial t})+\frac{1}{2}\frac{\partial}{\partial z}[(\frac{\partial \phi}{\partial x})^2+(\frac{\partial \phi}{\partial z})^2]=-g-\frac{1}{\rho}\frac{\partial p}{\partial z} \tag{12.421}$$

The boundary conditions become

$$p=0, \quad z=\eta \tag{12.422}$$

$$\frac{\partial \eta}{\partial t}-\frac{\partial \phi}{\partial x}\frac{\partial \eta}{\partial x}=-\frac{\partial \phi}{\partial z}, \quad z=\eta(x,t) \tag{12.423}$$

And the kinematic condition at the sea bottom is

$$\frac{\partial \phi}{\partial z}=0, \quad z=-h \tag{12.424}$$

Integrating (12.420) once, we obtain

$$-(\frac{\partial \phi}{\partial t})+\frac{1}{2}[(\frac{\partial \phi}{\partial x})^2+(\frac{\partial \phi}{\partial z})^2]+\frac{p}{\rho}+gz=0 \tag{12.425}$$

We now specify (12.425) on the free surface ($z=\eta$) to get

$$-(\frac{\partial \phi}{\partial t})+\frac{1}{2}[(\frac{\partial \phi}{\partial x})^2+(\frac{\partial \phi}{\partial z})^2]+g\eta=0, \quad z=\eta(x,t) \tag{12.426}$$

Finally, the equations to be solved are (12.419), (12.423), (12.424) and (12.426).

12.11 PERTURBATION FOR NAVIER-STOKES EQUATION

Note that the governing equation for the velocity potential is nonlinear and, in addition, the shape of the free surface, which is one of the boundaries, is unknown. This nonlinear problem can be handled by the perturbation method discussed in this chapter. In particular, we assume that motion in the water is small. Without wave motion, the surface of the water is flat, and with wave motions, the water surface becomes uneven. Thus, the small motion also means that the vertical displacement of the free surface is small. This assumption normally breaks down when a wave motion is plunging onto a sloping bottom along the coastline. It can be shown that momentum change due to surface undulation with wavelength L is proportional to the wave steepness defined by $\varepsilon = H/L$ where H is twice the amplitude of the wave function on the free surface, as illustrated in Figure 12.12. For small amplitude wave motions, we have ε being small.

12.11.1 Shallow Water Waves

For small amplitude waves with small ε (say <1/7), we assume that the wave motion is small and thus we have

$$\phi = \varepsilon\phi_1 + \varepsilon^2\phi_2 + \varepsilon^3\phi_3 + ... \tag{12.427}$$

$$\eta = \varepsilon\eta_1 + \varepsilon^2\eta_2 + \varepsilon^3\eta_3 + ... \tag{12.428}$$

$$p = p_0 + \varepsilon p_1 + \varepsilon^2 p_2 + \varepsilon^3 p_3 + ... \tag{12.429}$$

where $\varepsilon = H/L$ is the wave steepness (the nonlinear parameter of the problem). Note that we have finite water pressure p even for the case of no wave motions. Therefore, only the first order terms of the surface undulation η and wave function ϕ is proportional to ε, not p.

(i) Equation of continuity

Substitution of (12.427) into the equation of continuity (12.419) gives

$$\varepsilon\frac{\partial^2\phi_1}{\partial x^2} + \varepsilon^2\frac{\partial^2\phi_2}{\partial x^2} + ... + \varepsilon\frac{\partial^2\phi_1}{\partial z^2} + \varepsilon^2\frac{\partial^2\phi_2}{\partial z^2} + ... = 0 \tag{12.430}$$

Thus, we have the following governing equations for the first few approximations as

$$O(\varepsilon): \quad \frac{\partial^2\phi_1}{\partial x^2} + \frac{\partial^2\phi_1}{\partial z^2} = 0 \tag{12.431}$$

$$O(\varepsilon^2): \quad \frac{\partial^2\phi_2}{\partial x^2} + \frac{\partial^2\phi_2}{\partial z^2} = 0 \tag{12.432}$$

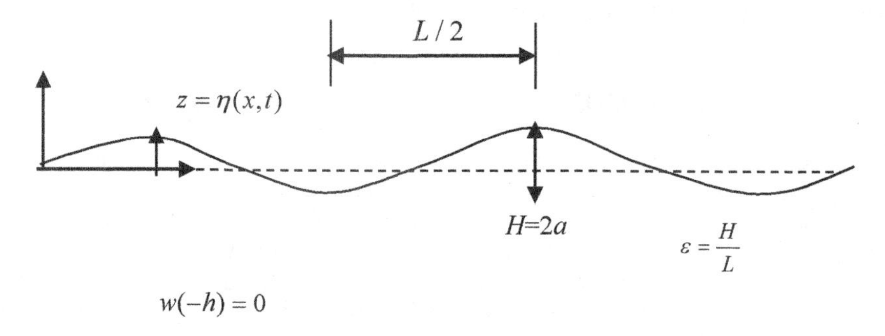

Figure 12.12 Boundary condition for shallow water waves

$$O(\varepsilon^3): \quad \frac{\partial^2 \phi_3}{\partial x^2} + \frac{\partial^2 \phi_3}{\partial z^2} = 0 \tag{12.433}$$

(ii) Dynamic condition on free surface

The free surface condition given in (12.426) is expanded around $z = 0$ with respect to η

$$\begin{aligned} &-(\frac{\partial \phi}{\partial t})_{z=0} - \eta(\frac{\partial^2 \phi}{\partial t \partial z})_{z=0} - \frac{\eta^2}{2!}(\frac{\partial^3 \phi}{\partial t \partial z^2})_{z=0} + ... + \frac{1}{2}[(\frac{\partial \phi}{\partial x})^2 + (\frac{\partial \phi}{\partial z})^2]_{z=0} \\ &+ \eta \frac{1}{2}\left\{\frac{\partial}{\partial z}[(\frac{\partial \phi}{\partial x})^2 + (\frac{\partial \phi}{\partial z})^2]\right\}_{z=0} + \frac{\eta^2}{2!}\frac{1}{2}\left\{\frac{\partial^2}{\partial z^2}[(\frac{\partial \phi}{\partial x})^2 + (\frac{\partial \phi}{\partial z})^2]\right\}_{z=0} \\ &+ ... + g\eta = 0 \end{aligned} \tag{12.434}$$

All these terms can be expanded in a series of ε as:

$$\frac{\partial \phi}{\partial t} = \varepsilon \frac{\partial \phi_1}{\partial t} + \varepsilon^2 \frac{\partial \phi_2}{\partial t} + \varepsilon^3 \frac{\partial \phi_3}{\partial t} + ... \tag{12.435}$$

$$\begin{aligned} \eta(\frac{\partial^2 \phi}{\partial t \partial z}) &= (\varepsilon \eta_1 + \varepsilon^2 \eta_2 + ...)[\varepsilon \frac{\partial^2 \phi_1}{\partial t \partial z} + \varepsilon^2 \frac{\partial^2 \phi_2}{\partial t \partial z} + ...] \\ &= \varepsilon^2 \eta_1 \frac{\partial^2 \phi_1}{\partial t \partial z} + \varepsilon^3 \eta_2 \frac{\partial^2 \phi_1}{\partial t \partial z} + \varepsilon^3 \eta_1 \frac{\partial^2 \phi_2}{\partial t \partial z} + O(\varepsilon^4) \end{aligned} \tag{12.436}$$

$$\frac{\eta^2}{2}(\frac{\partial^3 \phi}{\partial t \partial z^2}) = \frac{\varepsilon^3}{2} \eta_1^2 (\frac{\partial^3 \phi_1}{\partial t \partial z^2}) + O\left(\varepsilon^4\right) \tag{12.437}$$

$$\frac{1}{2}[(\frac{\partial\phi}{\partial x})^2+(\frac{\partial\phi}{\partial z})^2]=\frac{1}{2}[(\varepsilon\frac{\partial\phi_1}{\partial x}+\varepsilon^2\frac{\partial\phi_2}{\partial x}+...)^2+(\varepsilon\frac{\partial\phi_1}{\partial z}+\varepsilon^2\frac{\partial\phi_2}{\partial z}+...)^2]$$
$$=\frac{\varepsilon^2}{2}[(\frac{\partial\phi_1}{\partial x})^2+(\frac{\partial\phi_1}{\partial z})^2]+\varepsilon^3[(\frac{\partial\phi_1}{\partial x})(\frac{\partial\phi_2}{\partial x})+(\frac{\partial\phi_1}{\partial z})(\frac{\partial\phi_2}{\partial z})]+O(\varepsilon^4)$$
(12.438)

$$\eta\frac{1}{2}\left\{\frac{\partial}{\partial z}[(\frac{\partial\phi}{\partial x})^2+(\frac{\partial\phi}{\partial z})^2]\right\}=\eta\left\{\frac{\partial\phi}{\partial x}\frac{\partial^2\phi}{\partial z\partial x}+\frac{\partial\phi}{\partial z}\frac{\partial^2\phi}{\partial z^2}\right\}$$
$$=(\varepsilon\eta_1+\varepsilon^2\eta_2+...)\{(\varepsilon\frac{\partial\phi_1}{\partial x}+...)(\varepsilon\frac{\partial^2\phi_1}{\partial z\partial x}+...)+(\varepsilon\frac{\partial\phi_1}{\partial z}+...)(\varepsilon\frac{\partial^2\phi_1}{\partial z^2}+...)\} \tag{12.439}$$
$$=\varepsilon^3\eta_1[\frac{\partial\phi_1}{\partial x}\frac{\partial^2\phi_1}{\partial z\partial x}+\frac{\partial\phi_1}{\partial z}\frac{\partial^2\phi_1}{\partial z^2}]+O(\varepsilon^4)$$

Substituting of these results into (12.434) and collecting the first and second order terms, we have

$$O(\varepsilon):\quad -\frac{\partial\phi_1}{\partial t}+g\eta_1=0 \tag{12.440}$$

$$O(\varepsilon^2):\quad -\frac{\partial\phi_2}{\partial t}+g\eta_2=\eta_1\frac{\partial^2\phi_1}{\partial t\partial z}-\frac{1}{2}[(\frac{\partial\phi_1}{\partial x})^2+(\frac{\partial\phi_1}{\partial z})^2] \tag{12.441}$$

Note from (12.441) that the second approximation depends on the solution of the first approximation.

(iii) Kinetic condition on water surface

We expand the kinetic condition given (12.423) around $z = 0$ to get

$$\frac{\partial\eta}{\partial t}-[\frac{\partial\phi}{\partial x}+\eta\frac{\partial^2\phi}{\partial x\partial z}+\frac{\eta^2}{2!}\frac{\partial^3\phi}{\partial x\partial z^2}+...]\frac{\partial\eta}{\partial x}=-\frac{\partial\phi}{\partial z}-\eta\frac{\partial^2\phi}{\partial z^2}-\frac{\eta^2}{2!}\frac{\partial^3\phi}{\partial z^3}+...$$
(12.442)

$$\varepsilon\frac{\partial\eta_1}{\partial t}+\varepsilon^2\frac{\partial\eta_2}{\partial t}+...-[\varepsilon\frac{\partial\phi_1}{\partial x}+\varepsilon^2(\frac{\partial\phi_2}{\partial x}+\eta_1\frac{\partial^2\phi_1}{\partial x\partial z})+...]\left\{\varepsilon\frac{\partial\eta_1}{\partial x}+\varepsilon^2\frac{\partial\eta_2}{\partial x}+...\right\}$$
$$=-[\varepsilon\frac{\partial\phi_1}{\partial z}+\varepsilon^2\frac{\partial\phi_2}{\partial z}+...]-\varepsilon^2\eta_1\frac{\partial^2\phi_1}{\partial z^2}+...$$
(12.443)

The first and second order approximations are:

$$O(\varepsilon):\quad \frac{\partial\eta_1}{\partial t}+\frac{\partial\phi_1}{\partial z}=0 \tag{12.444}$$

$$O(\varepsilon^2):\quad \frac{\partial\eta_2}{\partial t}+\frac{\partial\phi_2}{\partial z}=\frac{\partial\phi_1}{\partial x}\frac{\partial\eta_1}{\partial x}-\eta_1\frac{\partial^2\phi_1}{\partial z^2} \tag{12.445}$$

(iv) Kinetic condition on sea bottom

The boundary conditions given in (12.424) for the first two approximations are

$$O(\varepsilon): \quad \frac{\partial \phi_1}{\partial z} = 0 \tag{12.446}$$

$$O(\varepsilon^2): \quad \frac{\partial \phi_2}{\partial z} = 0 \tag{12.447}$$

We now summarize the formulation for the first order approximation

$$\frac{\partial^2 \phi_1}{\partial x^2} + \frac{\partial^2 \phi_1}{\partial z^2} = 0 \tag{12.448}$$

$$-\frac{\partial \phi_1}{\partial t} + g\eta_1 = 0, \quad \frac{\partial \eta_1}{\partial t} + \frac{\partial \phi_1}{\partial z} = 0, \quad z = 0 \tag{12.449}$$

$$\frac{\partial \phi_1}{\partial z} = 0, \quad z = -h \tag{12.450}$$

The governing equation for the first approximation is a Laplace equation. The two conditions given in (12.449) can be combined by elimination. Thus, the surface boundary condition becomes

$$\frac{\partial^2 \phi_1}{\partial t^2} = -g\frac{\partial \phi_1}{\partial z}, \quad z = 0 \tag{12.451}$$

The problem of the second order approximation is summarized as

$$\frac{\partial^2 \phi_2}{\partial x^2} + \frac{\partial^2 \phi_2}{\partial z^2} = 0 \tag{12.452}$$

$$-\frac{\partial \phi_2}{\partial t} + g\eta_2 = \eta_1 \frac{\partial^2 \phi_1}{\partial t \partial z} - \frac{1}{2}[(\frac{\partial \phi_1}{\partial x})^2 + (\frac{\partial \phi_1}{\partial z})^2], \quad z = 0 \tag{12.453}$$

$$\frac{\partial \eta_2}{\partial t} + \frac{\partial \phi_2}{\partial z} = \frac{\partial \phi_1}{\partial x}\frac{\partial \eta_1}{\partial x} - \eta_1 \frac{\partial^2 \phi_1}{\partial z^2}, \quad z = 0 \tag{12.454}$$

$$\frac{\partial \phi_2}{\partial z} = 0, \quad z = -h \tag{12.455}$$

Again, nonhomogeneous terms on the right of (12.453) and (12.454) depend on the solution of the first approximation.

12.11.2 Airy Surface Waves

The first order of approximation for a surface water wave is called an Airy surface wave. We adopt a separation of variables as

$$\phi(x,z,t) = X(x)Z(z)T(t) \tag{12.456}$$

Substitution of (12.456) into (12.448) gives

$$\frac{d^2 X}{dx^2} ZT = -XT \frac{d^2 Z}{dz^2} \tag{12.457}$$

This can be rearranged as

$$\frac{1}{X}\frac{d^2X}{dx^2}=-\frac{1}{Z}\frac{d^2Z}{dz^2}=-k^2 \tag{12.458}$$

where k is the constant of separation of variables. This leads to two ODEs, one for X and one for Z:

$$\frac{d^2X}{dx^2}+k^2X=0,\quad \frac{d^2Z}{dz^2}-k^2Z=0 \tag{12.459}$$

The solutions for the first and second equations of (12.459) are respectively

$$X=A_1\sin kx+A_2\cos kx \tag{12.460}$$

$$Z=B_1\sinh kz+B_2\cosh kz \tag{12.461}$$

There is no governing equation for $T(t)$, but we look for a periodic solution with respect to time

$$T=C_1\sin\omega t+C_2\cos\omega t \tag{12.462}$$

Combining the time and space variables, we obtain

$$\phi(x,z,t)=[B_1\sinh kz+B_2\cosh kz]\{C_1\sin(kx-\omega t)+C_2\cos(kx-\omega t)\} \tag{12.463}$$

which is a forward moving wave. Note for simplicity that we have dropped the subscript "1" in the wave potential. Similarly, the backward moving wave is

$$\phi(x,z,t)=[B_1\sinh kz+B_2\cosh kz]\{C_1\sin(kx+\omega t)+C_2\cos(kx+\omega t)\} \tag{12.464}$$

Without loss of generality, we look for a particular solution form

$$\phi(x,z,t)=[D_1\cosh kz+D_2\sinh kz]\cos(kx-\omega t) \tag{12.465}$$

The bottom boundary condition requires:

$$\left(\frac{\partial\phi}{\partial z}\right)_{z=-h}=k[-D_1\sinh kh+D_2\cosh kh]\cos(kx-\omega t)=0 \tag{12.466}$$

This leads to

$$D_2=D_1\tanh kh \tag{12.467}$$

Using this result in (12.465) and differentiating with respect to time and z, we get

$$\frac{\partial^2\phi}{\partial t^2}=-\omega^2D_1[\cosh kz+\tanh kh\sinh kz]\cos(kx-\omega t) \tag{12.468}$$

$$-g\frac{\partial\phi}{\partial z}=-gkD_1[\sinh kz+\tanh kh\cosh kz]\cos(kx-\omega t) \tag{12.469}$$

Substitution of (12.468) and (12.469) into the following surface condition

$$\frac{\partial^2\phi_1}{\partial t^2}=-g\frac{\partial\phi_1}{\partial z} \tag{12.470}$$

gives the dispersion equation for frequency ω

$$\omega^2=gk\tanh kh \tag{12.471}$$

This dispersive relation relates the wave frequency ω with the wave number k. Whenever ω is a function of k, the wave is called a dispersive wave. The main characteristic of a dispersive wave is that the wave amplitude decreases with propagation. Now, we can let

$$A=\frac{D_1}{\cosh kh} \tag{12.472}$$

In addition, we note the following sum formula for the hyperbolic cosine

$$\cosh(x+y) = \cosh x \cosh y + \sinh x \sinh y \tag{12.473}$$

Then, (12.465) can be rewritten in a compact form as:

$$\phi = A\cosh[k(z+h)]\cos(kx-\omega t) \tag{12.474}$$

Differentiation with respect to time gives

$$\frac{\partial \phi}{\partial t} = \omega A\cosh[k(z+h)]\sin(kx-\omega t) \tag{12.475}$$

The unknown constant A must depend on the wave amplitude a. Let us assume the surface undulation as

$$\eta = a\sin(kx-\omega t) \tag{12.476}$$

To find A in terms of a, we can substitute (12.476) and (12.474) into the (12.440) or

$$-\frac{\partial \phi}{\partial t} + g\eta = 0 \tag{12.477}$$

Thus, we have

$$\omega A\cosh kh \sin(kx-\omega t) = ga\sin(kx-\omega t) \tag{12.478}$$

Solving for A, we get

$$A = \frac{ga}{\omega}\frac{1}{\cosh kh} = \frac{a\omega}{k}\frac{1}{\sinh kh} \tag{12.479}$$

The second part of (12.479) results from the dispersive relation derived in (12.471). Thus, we obtain the flow potential as

$$\begin{aligned}\phi &= \frac{a\omega}{k}\frac{\cosh[k(z+h)]}{\sinh kh}\cos(kx-\omega t) \\ &= \frac{ga}{\omega}\frac{\cosh[k(z+h)]}{\cosh kh}\cos(kx-\omega t)\end{aligned} \tag{12.480}$$

The time dependent function can be expressed as

$$\cos(kx-\omega t) = \cos[k(x-\frac{\omega}{k}t)] = \cos[k(x-ct)] \tag{12.481}$$

where the phase speed c is defined as

$$c = \frac{\omega}{k} \tag{12.482}$$

Finally, the velocity components can be obtained as

$$u = -\frac{\partial \phi}{\partial x} = a\omega\frac{\cosh[k(z+h)]}{\sinh kh}\sin(kx-\omega t) \tag{12.483}$$

$$w = -\frac{\partial \phi}{\partial z} = -a\omega\frac{\sinh[k(z+h)]}{\sinh kh}\cos(kx-\omega t) \tag{12.484}$$

The trajectories of water particles can be evaluated by using a Lagrangian formulation. Assume that the initial position of the water particle is at (x_0, z_0) and the position of the same point at time t is $(x = x_0+\delta x, z = z_0+\delta z)$. The change of the position can be approximated as

$$\frac{d(x-x_0)}{dt} = u_L(x_0,z_0,t) = u_E(x_0+\delta x, z_0+\delta z, t)$$

$$= u_E(x_0,z_0,t) + \delta x \left.\frac{\partial u_E}{\partial x}\right|_{x_0,z_0,t} + \delta z \left.\frac{\partial u_E}{\partial z}\right|_{x_0,z_0,t} + \dots \qquad (12.485)$$

$$= u_E(x_0,z_0,t) + (\int_0^t u_E dt) \left.\frac{\partial u_E}{\partial x}\right|_{x_0,z_0,t} + (\int_0^t w_E dt) \left.\frac{\partial u_E}{\partial z}\right|_{x_0,z_0,t} + \dots$$

$$\frac{d(z-z_0)}{dt} = w_L(x_0,z_0,t) = w_E(x_0+\delta x, z_0+\delta z, t)$$

$$= w_E(x_0,z_0,t) + \delta x \left.\frac{\partial w_E}{\partial x}\right|_{x_0,z_0,t} + \delta z \left.\frac{\partial w_E}{\partial z}\right|_{x_0,z_0,t} + \dots \qquad (12.486)$$

where the subscripts "E" and "L" denote Eulerian and Lagrangian velocities. Using the (12.485) and (12.486) as the first terms in (12.483) and (12.484), we get

$$\frac{d(x-x_0)}{dt} = a\omega \frac{\cosh[k(z_0+h)]}{\sinh kh} \sin(kx_0 - \omega t) \qquad (12.487)$$

$$\frac{d(z-z_0)}{dt} = -a\omega \frac{\sinh[k(z_0+h)]}{\sinh kh} \cos(kx_0 - \omega t) \qquad (12.488)$$

Integrating with respect to time, we have

$$x - x_0 = a \frac{\cosh[k(z_0+h)]}{\sinh kh} \cos(kx_0 - \omega t) \qquad (12.489)$$

$$z - z_0 = a \frac{\sinh[k(z_0+h)]}{\sinh kh} \sin(kx_0 - \omega t) \qquad (12.490)$$

These two components can be combined to get the following elliptic trajectory of particle motions:

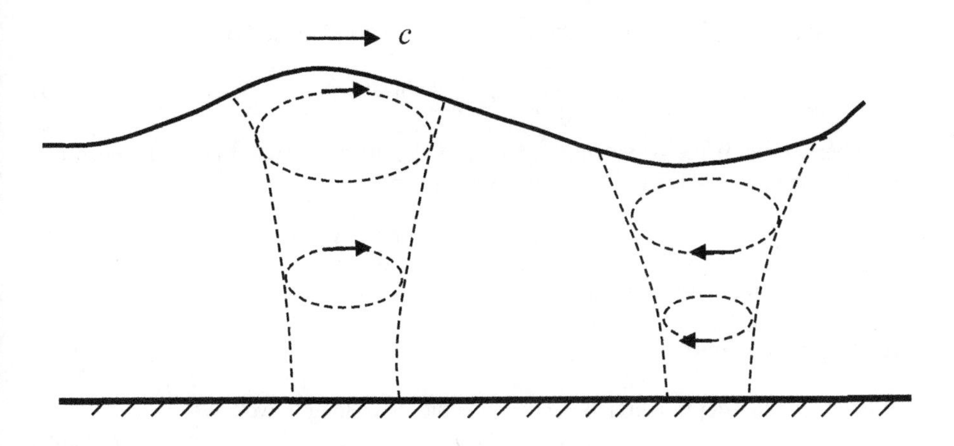

Figure 12.13 Elliptic trajectories motion for the first approximation

$$\left\{\frac{x-x_0}{a\dfrac{\cosh[k(z_0+h)]}{\sinh kh}}\right\}^2+\left\{\frac{z-z_0}{a\dfrac{\sinh[k(z_0+h)]}{\sinh kh}}\right\}^2=1 \tag{12.491}$$

This elliptic trajectory is illustrated in Figure 12.13. This is the Airy wave. Note however that if the nonlinear effect is included in the wave analysis, the motion of the water particle is no longer a closed circuit. That is, the water particle is being drifted forward in each cycle.

12.11.3 Shallow and Deep Water Limits

We will now consider the limiting cases of Airy's wave that were first considered by George Green (recall Green's theorem in Chapter 8). We first recall some relations for wave characteristics. The wave speed is defined as

$$c=\frac{\omega}{k}=\frac{L}{T} \tag{12.492}$$

where T is the period of the surface wave. The wave frequency relates to the wave period as

$$\omega=\frac{2\pi}{T} \tag{12.493}$$

The wave number is defined by

$$k=\frac{2\pi}{L} \tag{12.494}$$

Using these and (12.471), we write the phase speed in terms of the wave number, gravitational constant g, and depth of water h as

$$\begin{aligned} c^2&=\frac{\omega^2}{k^2}=\frac{g}{k}\tanh kh \\ &=\frac{gL}{2\pi}\tanh\left(\frac{2\pi h}{L}\right)=\frac{gcT}{2\pi}\tanh\left(\frac{2\pi h}{L}\right) \end{aligned} \tag{12.495}$$

The phase speed given in (12.495) can be put in a normalized form as:

$$\frac{c}{\sqrt{gh}}=\sqrt{\frac{1}{kh}\tanh(kh)} \tag{12.496}$$

In other words, if the wavelength and water depth are known, the wave speed can be estimated. The wavelength can be expressed as

$$L=cT=\frac{gT^2}{2\pi}\tanh\left(\frac{2\pi h}{L}\right) \tag{12.497}$$

We now consider two particular limits.

(i) Deep water limit

When the water depth is much deeper than the wave length, we have

$$h/L\to\infty \tag{12.498}$$

For this deep water limit, we have

$$\tanh kh \to 1 \tag{12.499}$$

By virtue of (12.492) and (12.496), we have the following relation for a deep water wave:

$$L = \frac{gT^2}{2\pi} = 1.56T^2 \tag{12.500}$$

where the unit of wavelength is in meters while the period is in seconds. The wave relates to the period as

$$c = \frac{gT}{2\pi} = 1.56T \tag{12.501}$$

where the wave speed is in m/s and the period is in seconds. The normalized speed can be specified from (12.496) as

$$\frac{c}{\sqrt{gh}} = \frac{1}{\sqrt{kh}} \tag{12.502}$$

(ii) Shallow water limit:

When the wavelength is much larger than the depth, we have

$$h / L \to 0 \tag{12.503}$$

Using the first term of Taylor series expansion for hyperbolic tangent, we find

$$\tanh(\frac{2\pi h}{L}) \approx \frac{2\pi h}{L} = \frac{2\pi h}{cT} \tag{12.504}$$

Substitution of (12.504) into (12.495) gives

$$c = \frac{gT}{2\pi}\frac{2\pi h}{cT} = \frac{gh}{c} \tag{12.505}$$

Rearranging (12.505) we find the classical result of water speed for a shallow water wave:

$$c = \sqrt{gh} \tag{12.506}$$

The water length L depends on the period T as

$$L = \sqrt{gh}T \tag{12.507}$$

We now consider a scenario for a large submarine earthquake and apply the shallow wave approximation to calculate the wave speed of a tsunami wave. When a large shallow submarine earthquake of magnitude 7.5 or higher occurs in deep sea (on the order of 8000 m), large sea bottom movements occurs. Water will be displaced upward or downward, and a so-called tsunami (literally means harbor wave in Japanese) will be generated. Because the size of the rupture surface for a large destructive earthquake is on the order of 200 km or more, the wavelength of such an initial disturbance is also on the order of 200 km. Using these data, (12.507) shows that the period of such an initial wave form is in the order of 12 minutes. In addition, we have $h/L \approx 0.04$ such that the shallow water wave assumption is valid.

The wave speed of such a tsunami wave can be estimated by (12.506) as

$$c = \sqrt{9.81 \times 8000} = 280 m/s$$
$$= 280 \times \frac{60 \times 60}{10^3} km/hr = 1008 km/hr \tag{12.508}$$

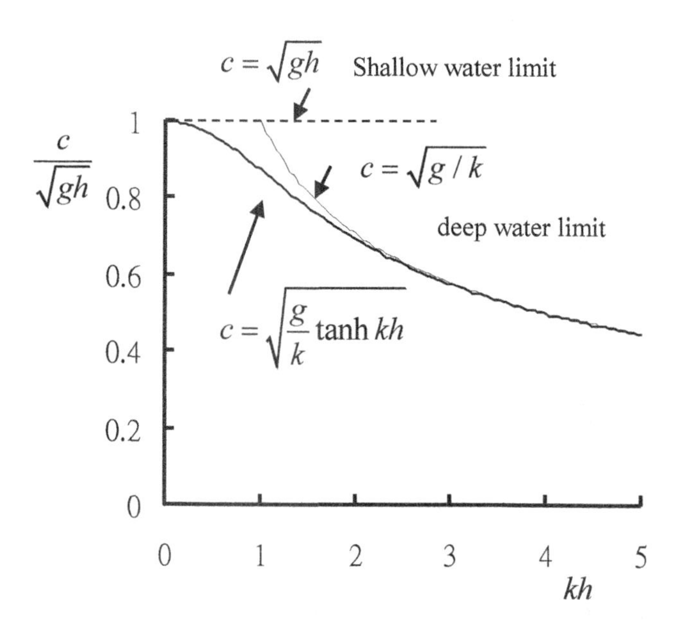

Figure 12.14 Phase speed for small amplitude wave, shallow water wave, and deep water wave

The tsunami wave travels in deep sea as fast as a jet airplane. However, the wave speed decreases rapidly when a tsunami wave comes close to the continental shelf, and at the same time the amplitude also increases. When such a tsunami wave comes ashore, the amplitude of the wave will increase rapidly and our small amplitude wave assumption will break down. Therefore, many scientists and geophysicists claimed after the 2004 South Asian tsunami that a tsunami propagating to shore at the speed of jet airplane is totally incorrect and is misleading.

For shallow water waves, h decreases and the wave speed decreases as $\sqrt{(gh)}$. As illustrated in Figure 12.15, the wave crest line always turns roughly parallel to the coastline. This is the reason why we seem to see waves are coming toward the coastline no matter where you go along a beach.

12.12 SUMMARY AND FURTHER READING

Using asymptotic and perturbation methods, especially singular perturbation, in solving ODEs or PDEs is not a straightforward business. This method is normally covered in graduate courses in the area of applied mathematics. It is a topic still undergoing fundamental developments. The perturbation method was developed to solve nonlinear ODEs containing a physically small parameter, and the technique turns *one nonlinear* ODE into an *infinite system of linear* ODEs by expanding the unknown function in asymptotic series in the small parameter of the problem. Normally one or two term expansions will give very accurate and meaningful results. In view of the asymptotic nature of the technique, more terms do not necessarily give a more accurate result. Very often, even powerful numerical methods will also break down when it is applied to solve highly nonlinear problems, unless you have an idea of how the solution is going to behave and under what circumstances the solution may change rapidly. For example, solutions of nonlinear systems are often unpredictable near singular points or bifurcation points.

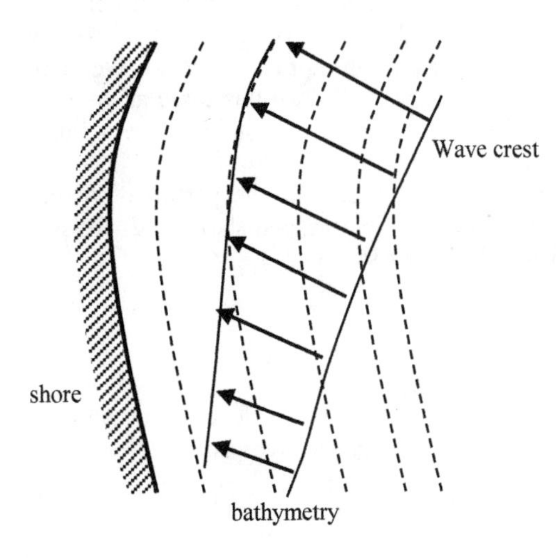

Figure 12.15 Turning of wave toward the coastline as predicted by (12.506)

In a chapter of this size, we can only introduce the basic ideas of this advanced technique that can provide approximate but insightful results for highly nonlinear problems, which are otherwise unsolvable. Our discussion only targets elementary and introductory levels. In particular, asymptotic expansions and their application in the regular perturbation method in solving ODEs are discussed. Singular perturbation methods are discussed briefly, and they include the method of matched asymptotic expansion (or boundary layer analysis), multiple scale analysis, and WKB approximation. The application of asymptotic expansion in evaluating integrals with a large parameter is discussed in light of the Laplace method, Riemann (or Debye) method of steepest descent (or saddle point method), and the method of stationary phase of Kelvin and Stokes. We conclude the chapter by looking at how to apply the perturbation method to convert the Navier-Stokes equation of fluid flows into a linear solvable problem. The Airy water wave and shallow water wave are considered as an example.

There are a lot of good and advanced books in asymptotic and perturbation methods. They include Kevorkian and Cole (1981), Nayfeh (1973), van Dyke (1975), Bender and Orszag (1978), Lagerstrom (1988), Lin and Segel (1988), Holmes (1995), and O'Malley (2014). Holmes (1995) and O'Malley (2014) are particularly useful references in terms of the historical development of the method, and both of them contain a lot of examples. The scope and coverage of Bender and Orszag (1978) are very comprehensive and provide detailed discussion on many issues that you may likely encounter in applying the perturbation method. The book by van Dyke (1975) aims at solving fluid mechanics problems. Nayfeh (1973) probably provides the most comprehensive references of the applications of perturbation methods in engineering, science, and applied mathematics. Most of these books are not targeted at the undergraduate level and readers may find it difficult to read. At the introductory level, Lin and Segel (1988) provided the best insights and is very easy to read for beginners.

For asymptotic analysis of integrals, we refer to Erdelyi (1956), Olver (1997), Bleistein and Handelsman (1986), and Wong (2001). These books target graduate students and researchers and thus are not easy to read. We also highly recommend the book by Bender and Orszag (1978) on this topic as well. Their presentation is probably more appealing to undergraduates and engineers, but at the same time are very comprehensive.

In the area of structural instability, the perturbation method has also been applied to the buckling of beams, plates, and shells (e.g., Reiss, 1969, 1977, 1980a-b, 1982, 1984, Reiss and Matkowsky, 1971).

12.13 PROBLEMS

Problem 12.1 Repeat the analysis in Example 12.1 but with boundary layer at x = 1. Redefine the variable in the boundary layer as:

$$\bar{\xi} = \frac{1-x}{\phi(\varepsilon)} \tag{12.509}$$

Find the inner expansion, outer expansion, and the composite expansion.

Ans:

$$y_O(x) = 0, \quad y_I(\bar{\xi}) = C(e^{2\bar{\xi}} - 1) + 1, \quad y_C \sim e^{2(1-x)/\varepsilon} \tag{12.510}$$

Problem 12.2 Prove the following Stirling asymptotic formula for Euler's gamma function:

$$\Gamma(k+1) \approx \sqrt{2\pi k} k^k e^{-k} \tag{12.511}$$

for $k \to \infty$, where the gamma function is defined by

$$\Gamma(k+1) = \int_0^\infty e^{-t} t^k dt \tag{12.512}$$

Answer the following questions one by one:
(i) Show that

$$\Gamma(k+1) = \int_0^\infty e^{-t+k\ln t} dt \tag{12.513}$$

(ii) Use a change of variable $t = ks$ to show that

$$\Gamma(k+1) = kk^k \int_0^\infty e^{k[-s+\ln s]} ds \tag{12.514}$$

(iii) Use the Laplace method to prove that for $k \to \infty$

$$\Gamma(k+1) \approx \sqrt{2\pi k} k^k e^{-k} \tag{12.515}$$

Problem 12.3 Consider the asymptotic form of the following double integral

$$I(k) = \iint e^{-k(x^2-xy+y^2)} \varphi(x,y) dx dy \tag{12.516}$$

for $k \to \infty$ and φ is a regular function for all values of x and y.

(i) Find the stationary point of the function

$$f(x,y) = x^2 - xy + y^2 \tag{12.517}$$

(ii) Apply the following change of variables to the integral

$$u = x - \frac{y}{2}, \quad v = y \tag{12.518}$$

(iii) Show that the Jacobian of the change of variables in (ii) is 1.
(iv) Show that the asymptotic form of the integral given in (12.516) is

$$I(k) = \sqrt{\frac{4}{3}\frac{\pi}{k}} \varphi(0,0) \tag{12.519}$$

Hint: Use the definition of the gamma function.

Problem 12.4 Reconsider the asymptotic form of Problem 12.2 using the formula given in (12.284). Show that the answer is the same as that given in (12.519).

Problem 12.5 Find the asymptotic form of the following integral by the Laplace method:

$$I(x) = \int_0^\infty e^{-x\sinh^2 t} dt \tag{12.520}$$

Ans:

$$I(x) \sim \frac{1}{2}\sqrt{\frac{\pi}{x}} \tag{12.521}$$

Problem 12.6 Find the asymptotic form of the following integral representation of modified Bessel function:

$$K_\nu(x) = \int_0^\infty e^{-x\cosh t} \cosh(\nu t) dt \tag{12.522}$$

Ans:

$$K_\nu(x) \sim \sqrt{\frac{\pi}{2x}} e^{-x} \tag{12.523}$$

Problem 12.7 Extend the result for the Laplace method to include the fourth order derivative term of ϕ and the second order derivative term of f for the following integral

$$I(x)=\int_{-\infty}^{\infty} f(t)e^{x\phi(t)}dt \tag{12.524}$$

where

$$\phi'(c)=0, \quad \phi''(c)<0, \quad f(c)\neq 0 \tag{12.525}$$

Hint: See Section 6.4 of Bender and Orszag (1978) for the details of derivation of the result.

Ans:

$$I(x)\sim\sqrt{\frac{2\pi}{-x\phi''(c)}}e^{x\phi(c)}\left\{f(c)+\frac{1}{x}\left[-\frac{f''(c)}{2\phi''(c)}+\frac{f(c)\phi^{(4)}(c)}{8[\phi''(c)]^2}\right.\right.$$
$$\left.\left.+\frac{f'(c)\phi'''(c)}{2[\phi''(c)]^2}-\frac{5f(c)[\phi'''(c)]^2}{24[\phi''(c)]^3}\right]\right\} \tag{12.526}$$

Problem 12.8 Repeat the analysis in Problem 12.5 with the result derived in Problem 12.7. Find a two-term asymptotic expansion in the series.

Ans:

$$I(x)\sim\frac{1}{2}\sqrt{\frac{\pi}{x}}(1+\frac{1}{4x}) \tag{12.527}$$

Problem 12.9 Assume the following two-scale perturbation expansions for a function u:

$$u(t,\varepsilon)=u^0(t,\tau)+u^1(t,\tau)\varepsilon+\frac{1}{2}u^2(t,\tau)\varepsilon^2+... \tag{12.528}$$

where

$$\tau=\varepsilon t \tag{12.529}$$

Show the validity of the following differentiation formulas:

$$u'(t,\varepsilon)=\frac{du}{dt}=u_t^0+[u_\tau^0+u_t^1]\varepsilon+[u_\tau^1+\frac{1}{2}u_t^2]\varepsilon^2+... \tag{12.530}$$

$$u''(t,\varepsilon)=u_{tt}^0+[u_{tt}^1+2u_{\tau t}^0]\varepsilon+[2u_{t\tau}^1+u_{\tau\tau}^0+\frac{1}{2}u_{tt}^2]\varepsilon^2+... \tag{12.531}$$

$$u'''(t,\varepsilon)=u_{ttt}^0+[3u_{tt\tau}^0+u_{ttt}^1]\varepsilon+[3u_{tt\tau}^1+3u_{t\tau\tau}^0+\frac{1}{2}u_{ttt}^2]\varepsilon^2+... \tag{12.532}$$

$$u^{(iv)}(t,\varepsilon)=u_{tttt}^0+[4u_{ttt\tau}^0+u_{tttt}^1]\varepsilon+[6u_{tt\tau\tau}^0+4u_{ttt\tau}^1+\frac{1}{2}u_{tttt}^2]\varepsilon^2+... \tag{12.533}$$

CHAPTER THIRTEEN

Calculus of Variations

13.1 INTRODUCTION

One of the main applications of the calculus of variations is related to the variational formulation in mechanics problems, which is the topic to be covered in the next chapter (i.e., Chapter 14). In traditional mechanics, the governing equation of a mechanics problem can be formulated via two independent paths: (i) Newtonian mechanics formulation by considering the equation of motion or force equilibrium for an small free body cut out from the original body; and (ii) variational formulation that requires the minimization of energy or some functional (in the form of integral). The calculus of variations provides the backbone for the second approach. The origin of the calculus of variations can be traced back to the time of Bernoulli and Euler. Its formal development was, however, mainly done by Euler and Lagrange. Seeking a functional (a function of admissible functions in integral form) that is stationary, Euler in 1736 and Lagrange in 1755 derived independently that a second order PDE called the Euler-Lagrange equation of the functional. The resulting governing equation of the problem is known as the Euler-Lagrange equation. A special case of it is known as the Beltrami identity if the integrand function inside the integral of the functional is independent of the independent variable of the problem. The term calculus of variations was introduced by Euler.

The application of the calculus of variations is mainly used in searching an optimum solution of problems. For example, these problems include what is the shortest distance between two points in space, what is the shape of the strongest column (proposed by Lagrange in 1773), what is the shape of the column strongest against torsion (St. Venant problem solved by George Polya in 1948), what is the shape of a drum of minimized tone for a given area (Rayleigh conjecture solved by Courant, Faber, and Krahn in 1920s), what is the shortest curve between two points on a curved surface (geodesics problem first considered Euler in 1755), what is the shape of a simply connected electric capacitor that maximizes capacity (solved by Poincare and Szego), what is the shape of a soap film form between two metal circular rings (catenoid problem), what is the profile of a wire for a frictionless sliding bead giving the shortest travel time (brachistochrone problem posed by Johann Bernoulli), what is the least-perimeter of a soap bubble enclosing a given volume of air (solved by Schwarz in 1884 as sphere bubble), what solid with 3-D shape minimizes heat loss (Polya's cat curling problem), what is the shape of a rocket nose that minimizes the air resistance in supersonic flight (Newton's minimal resistance problem), and what is the shape of a closed curve of fixed length giving the greatest enclosed area on a surface (isoperimetric problem or Dido problem).

The geodesic problem is of great importance in surveying, navigation on the surface of the earth, signals traveling on earth's surface, and defining maritime boundaries. The term "geodesic line" was coined by Laplace in 1799. If the earth's surface is approximated by a sphere, the shortest distance between two points is an arc of a great circle (Euler in 1755). For the more realistic ellipsoidal earth, the problem is more complicated and has been considered by many famous mathematicians and scientists, including Newton in 1687, Clairaut in 1735, Legendre in 1806, Oriani in 1806, Bessel in 1825, Gauss in 1828, and Poincare in 1905.

In the 19th century, Weierstrass realized that there could be subtle problems involved as the Euler-Lagrange equation is the necessary but not sufficient condition, and thus some of the solution is not the minimum or maximum. The study of the existence of an extremal value of functionals results in the so-called direct method of the calculus of variation. The main contributors are Weierstrass, Schwarz, Poincare, and Hilbert and their works are based on functional analysis and topology. This topic is, however, out of the scope of the present study.

In this chapter, we will discuss the calculus of variations and the associated Euler-Lagrange equation for the case of a single variable and the case of multi-independent and dependent variables. The idea of the Lagrange multiplier will also be discussed.

For example, the brachistochrone problem, catenoid problem, Dido's problem of isoperimeter, and geodesics will be considered as illustrations in the present chapter. Among these problems, we should mention the problem of isoperimetry (i.e., solving a problem of isoperimeter like the Dido problem) in particular. In the 19th century, Belgian physicist J. Plateau experimented with soap film and conjectured that every nice closed wire loop bounds a soap film or minimal surface. This is referred to as the Plateau problem in the literature. This conjecture was subsequently proved by J. Douglas in 1931, who was awarded the Fields Medal on this achievement (Douglas, 1931). Another Fields Medal recipient, Enrico Bombieri, received the medal because of work on higher dimensional minimal surfaces. Thus, calculus is an important topic in mathematics.

13.2 FUNCTIONAL

A functional I is defined in terms of a function F, which is in turn a function of another function y and its derivatives:

$$I(y) = \int_a^b F[x, y(x), y'(x)]dx \tag{13.1}$$

where x is the variable of the function y. The objective of the calculus of variation is to find what admissible functions $y(x)$ will lead to a maximum or a minimum value of the functional I or so-called stationary. Physically, I can be an arc-length that we want to minimize, an energy function of a problem, the shortest time of travel, the maximum or minimum area, etc. More generally, the functional may also involve higher derivative terms, or involve more dependent variables. Such situations will be considered in later sections.

13.3 ANALOGOUS TO CALCULUS

There is close resemblance between differential calculus and the calculus of variations. In particular, differential calculus always involves the determination of the point x_0 of a single variable function $f(x)$ within a bounded domain such that the function f will achieve a maximum or minimum value. Mathematically, to find the maximum or minimum, we are seeking the solution of

$$\left.\frac{df}{dx}\right|_{x=x_0} = 0 \tag{13.2}$$

To ensure that the point x_0 with vanishing derivative is a maximum, we need to impose an additional condition of

$$\left.\frac{d^2 f}{dx^2}\right|_{x=x_0} < 0 \tag{13.3}$$

Similarly, the condition for a minimum is

$$\left.\frac{d^2 f}{dx^2}\right|_{x=x_0} > 0 \tag{13.4}$$

In the calculus of variations, we are looking for a permissible function $y(x)$ such that the following integral I that we called "functional" attains an extremal value (either a maximum or a minimum)

$$I(\varepsilon) = \int_a^b F[x, y(x), y'(x)]dx \tag{13.5}$$

where

$$y(x) = \overline{y}(x) + \varepsilon\eta(x) \tag{13.6}$$

$$y'(x) = \overline{y}'(x) + \varepsilon\eta'(x) \tag{13.7}$$

In a sense, the functional can be considered as the function of a function. The admissible function is $\eta(x)$ which will vanish at the end points as shown in Figure 13.1. The necessary condition for the extremal to occur is that

$$\left.\frac{dI(\varepsilon)}{d\varepsilon}\right|_{\varepsilon=0} = 0 \tag{13.8}$$

Note that there is a close resemblance between (13.2) and (13.8). Figure 13.1 illustrates the similarity between calculus and the calculus of variations.

13.4 EULER-LAGRANGE EQUATION

If (13.8) is satisfied, we also said that the functional I is stationary. Using the chain rule and noting (13.6) and (13.7), we get

$$\frac{dI}{d\varepsilon} = \int_a^b \left[\frac{\partial F}{\partial y}\frac{\partial y}{\partial \varepsilon} + \frac{\partial F}{\partial y'}\frac{\partial y'}{\partial \varepsilon}\right]dx \tag{13.9}$$

Differentiation of (13.5) and (13.6) with respect to ε gives

$$\frac{\partial y}{\partial \varepsilon} = \eta, \quad \frac{\partial y'}{\partial \varepsilon} = \eta' \tag{13.10}$$

Substitution of (13.10) into (13.9) results in

$$\frac{dI}{d\varepsilon} = \int_a^b \left[\frac{\partial F}{\partial y}\eta + \frac{\partial F}{\partial y'}\eta' \right] dx \tag{13.11}$$

Applying the condition (13.8) gives

$$\left.\frac{dI}{d\varepsilon}\right|_{\varepsilon=0} = \int_a^b \left[\frac{\partial F}{\partial \bar{y}}\eta + \frac{\partial F}{\partial \bar{y}'}\eta' \right] dx = 0 \tag{13.12}$$

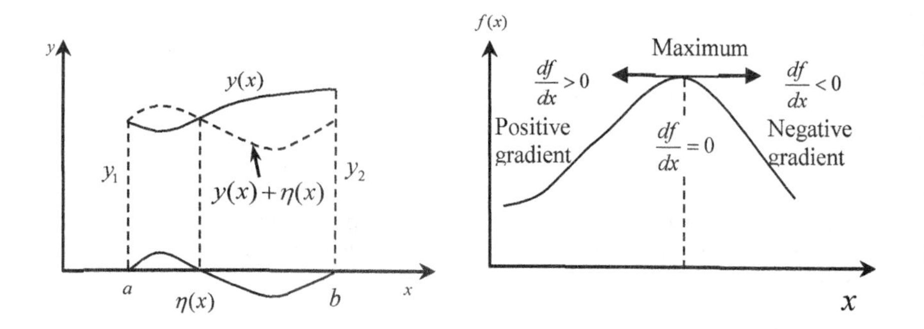

Figure 13.1 Differential calculus versus the calculus of variations

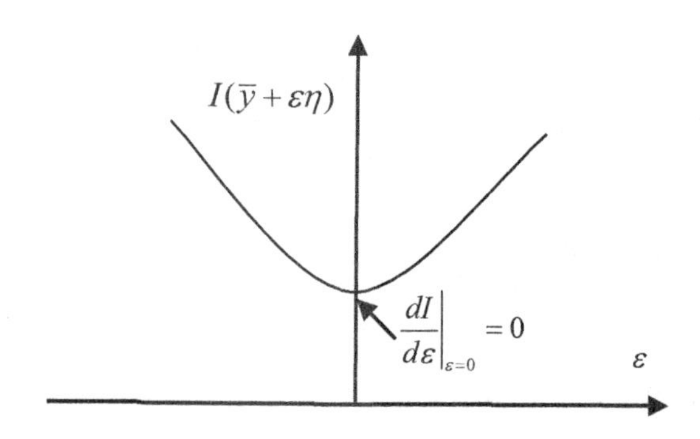

Figure 13.2 Variations of a functional

Applying integration by parts to the second term on the right hand side of (13.12), we find

$$\left.\frac{dI}{d\varepsilon}\right|_{\varepsilon=0} = \int_a^b \left[\frac{\partial F}{\partial \bar{y}} - \frac{d}{dx}[\frac{\partial F}{\partial \bar{y}'}] \right] \eta dx + [\frac{\partial F}{\partial \bar{y}'}\eta]_a^b = 0 \tag{13.13}$$

However, we have imposed that the admissible function vanishes at the end points, and this yields

$$\eta(a) = \eta(b) = 0 \tag{13.14}$$

This gives a differential equation for F as

$$\frac{\partial F}{\partial \bar{y}} - \frac{d}{dx}[\frac{\partial F}{\partial \bar{y}'}] = 0 \tag{13.15}$$

For simplicity, we will drop the bar in all y in (13.15) in subsequent presentation. This equation is called the Euler-Lagrange equation, Euler equation, or Lagrange equation. The more popular choice is the Euler-Lagrange equation. This equation can be recast in a more explicit form by noting

$$\frac{df}{dx} = \frac{\partial f}{\partial x} + \frac{\partial f}{\partial y}\frac{dy}{dx} + \frac{\partial f}{\partial y'}\frac{dy'}{dx} = \frac{\partial f}{\partial x} + \frac{\partial f}{\partial y}y' + \frac{\partial f}{\partial y'}y'' \tag{13.16}$$

Substituting $f = \partial F/\partial y$ in (13.15), we find another form for the Euler-Lagrange equation as

$$y''\frac{\partial^2 F}{\partial y'^2} + y'\frac{\partial^2 F}{\partial y \partial y'} + \frac{\partial^2 F}{\partial x \partial y'} - \frac{\partial F}{\partial y} = 0 \tag{13.17}$$

This is an alternate form for the Euler-Lagrange equation given in (13.15), which may not be easy to solve. We should note that the Euler-Lagrange equation is only the necessary condition for the functional being stationary, but it is not the sufficient condition. Therefore, if we get more than one solution, we have to check whether it indeed gives the required stationary state and whether it is a maximum or a minimum, or even a saddle point.

It is also possible to rewrite the Euler-Lagrange equation in yet another form, in addition to (13.15) and (13.17). In particular, we note that

$$\frac{d}{dx}(y'\frac{\partial F}{\partial y'}) = y'\frac{d}{dx}(\frac{\partial F}{\partial y'}) + \frac{\partial F}{\partial y'}y'' \tag{13.18}$$

Next, we can subtract (13.18) from (13.16) (with f being replaced by F) to get

$$\frac{dF}{dx} - \frac{d}{dx}(y'\frac{\partial F}{\partial y'}) = \frac{\partial F}{\partial x} + \frac{\partial F}{\partial y}y' - y'\frac{d}{dx}(\frac{\partial F}{\partial y'}) \tag{13.19}$$

This can be rearranged as

$$\frac{d}{dx}\{F - y'\frac{\partial F}{\partial y'}\} - \frac{\partial F}{\partial x} = y'[\frac{\partial F}{\partial y} - \frac{d}{dx}(\frac{\partial F}{\partial y'})] \tag{13.20}$$

However, the bracket term on the right of (13.20) is exactly zero in view of the Euler-Lagrange equation obtained in (13.15). Therefore, we end up with the following form of Euler-Lagrange equation:

$$\frac{d}{dx}\{F - y'\frac{\partial F}{\partial y'}\} - \frac{\partial F}{\partial x} = 0 \tag{13.21}$$

This is equivalent to (13.15) and (13.17).

A special form of the Euler-Lagrange equation is called the Beltrami identity, which is equivalent to the Euler-Lagrange equation for the case that F is not an explicit function of x. In particular, we have

$$F = F(y, y') \tag{13.22}$$

Consider the second term on the right hand side of (13.15) by using the chain rule, keeping in mind that F does not depend on x

$$\frac{d}{dx}[\frac{\partial F}{\partial y'}]=\frac{\partial^2 F}{\partial y \partial y'}y'+\frac{\partial^2 F}{\partial y'^2}y'' \tag{13.23}$$

Substitution of (13.23) into (13.15) gives the special form of the Euler-Lagrange equation

$$\frac{\partial F}{\partial y}=\frac{\partial^2 F}{\partial y \partial y'}y'+\frac{\partial^2 F}{\partial y'^2}y'' \tag{13.24}$$

Note that this is exactly the same as (13.17) if we drop the third term in terms of the x derivative. Finally, to prove Beltrami's identity, we consider the following function:

$$H=\frac{\partial F}{\partial y'}y'-F \tag{13.25}$$

Differentiation of (13.25) with respect to x using the chain rule gives

$$\begin{aligned}\frac{dH}{dx}&=(\frac{\partial F}{\partial y'}y''+\frac{\partial^2 F}{\partial y'\partial y}y'^2+\frac{\partial^2 F}{\partial y'^2}y'y'')-(\frac{\partial F}{\partial y}y'+\frac{\partial F}{\partial y'}y'')\\&=y'(\frac{\partial^2 F}{\partial y \partial y'}y'+\frac{\partial^2 F}{\partial y'^2}y''-\frac{\partial F}{\partial y})=0\end{aligned} \tag{13.26}$$

The last identity of (13.26) is a consequence of (13.24). Thus, we must have

$$\frac{\partial F}{\partial y'}y'-F=C \tag{13.27}$$

where C is an arbitrary constant. This is called the Beltrami identity. Note that the Beltrami identity can also be recovered by setting the last term in the alternative form of the Euler-Lagrange equation given in (13.21) to zero [i.e., $F \neq F(x)$ or $F = F(y, y')$ given in (13.22)].

13.5 DEGENERATE CASES OF EULER-LAGRANGE EQUATIONS

In this section, we consider some special degenerate cases of the Euler-Lagrange equation.

Case (i) $F=F(y,y')$

This is exactly the case of the Beltrami identity and we have

$$\frac{\partial F}{\partial y'}y'-F=C \tag{13.28}$$

where C is an arbitrary constant.

Case (ii) $F=F(x,y')$

From (13.15), we have the first term zero being zero and the Euler-Lagrange equation is reduced to

$$\frac{d}{dx}[\frac{\partial F}{\partial y'}] = 0 \tag{13.29}$$

Integrating (13.29), we get the following form

$$\frac{\partial F}{\partial y'} = C \tag{13.30}$$

where C is a constant. Integrating one more time, we obtain

$$F = Cy' + f(x) \tag{13.31}$$

Case (iii) $F = F(y')$

For this case, (13.17) gives

$$y''\frac{\partial^2 F}{\partial y'^2} = 0 \tag{13.32}$$

In general, the second derivative of F with respect to y' is not zero, and thus we have

$$\frac{d^2 y}{dx^2} = 0 \tag{13.33}$$

Or, we have y being a straight line as

$$y = C_1 x + C_2 \tag{13.34}$$

where C_1 and C_2 are constants.

Case (iv) $F = F(x, y)$

For this case, the second term on the left of (13.15) is zero, and thus we obtain

$$\frac{\partial F}{\partial y} = 0 \tag{13.35}$$

Finally, we must have the special case of

$$F = F(x) \tag{13.36}$$

13.6 FUNCTIONAL OF SEVERAL VARIABLES

When there is more than one dependent variable, the functional may be formulated as

$$I = \int_a^b F(x, y, z\ ; y', z')dx \tag{13.37}$$

The corresponding Euler-Lagrange equations are

$$\frac{\partial F}{\partial y} - \frac{d}{dx}(\frac{\partial F}{\partial y'}) = 0 \tag{13.38}$$

$$\frac{\partial F}{\partial z}-\frac{d}{dx}(\frac{\partial F}{\partial z'})=0 \tag{13.39}$$

When there is more than one independent variable, the functional may be formulated as

$$I=\iint_R F(x,y,u,u_x,u_y)dxdy \tag{13.40}$$

The corresponding Euler-Lagrange equations are

$$\frac{\partial F}{\partial u}-\frac{\partial}{\partial x}(\frac{\partial F}{\partial u_x})-\frac{\partial}{\partial y}(\frac{\partial F}{\partial u_y})=0 \tag{13.41}$$

For a functional containing derivative of higher than first order, we have

$$I=\int_{x_0}^{x_1} F(x,y,y',y'')dx \tag{13.42}$$

The corresponding Euler-Lagrange equations are

$$\frac{\partial F}{\partial y}-\frac{d}{dx}(\frac{\partial F}{\partial y'})+\frac{d^2}{dx^2}(\frac{\partial F}{\partial y''})=0 \tag{13.43}$$

For the cases of several dependent and independent variables, the functional may be formulated as

$$I=\iint_R F(x,y,u,u_x,u_y,v,v_x,v_y)dxdy \tag{13.44}$$

The corresponding Euler-Lagrange equations are

$$\frac{\partial F}{\partial u}-\frac{\partial}{\partial x}(\frac{\partial F}{\partial u_x})-\frac{\partial}{\partial y}(\frac{\partial F}{\partial u_y})=0 \tag{13.45}$$

$$\frac{\partial F}{\partial v}-\frac{\partial}{\partial x}(\frac{\partial F}{\partial v_x})-\frac{\partial}{\partial y}(\frac{\partial F}{\partial v_y})=0 \tag{13.46}$$

It is not difficult to extend these Euler-Lagrange equations to more variables and higher derivatives.

A number of examples will be considered next.

13.7 CATENOID

When a soap film is formed between two circular metal rings, the optimum shape of the soap film can be expressed in terms of hyperbolic cosine functions. This particular problem is called catenoid. It was first considered by Euler. The actual shape of the catenoid is shown in Figure 13.3.

By symmetry, we put the origin of the coordinate system at the mid-section of the soap film, as shown in Figure 13.3. The surface area of the ring (shown as dotted lines in Figure 13.3) at a distance x from the origin can be formulated as:

$$dA=2\pi y ds \tag{13.47}$$

Along this unknown curve, the length increment of the curve ds between the two rings can be evaluated as:

$$(ds)^2=(dx)^2+(dy)^2 \tag{13.48}$$

This formula can be recast as:

$$\frac{ds}{dx} = \sqrt{1 + (\frac{dy}{dx})^2} \tag{13.49}$$

Substitution of (13.49) into (13.47) and integration of dx from 0 to a gives

$$A = \int_0^a 2\pi y \sqrt{1 + y'^2}\, dx \tag{13.50}$$

where $2a$ is the distance between the two steel rings. In the calculus of variations, we are searching for an optimum function that gives a stationary functional. In the problem of a catenoid, we are looking for a function of the shape of the soap film such that the area of the soap film formed between the rings is a minimum. Thus, mathematically we expect the functional to be

$$I = \int_0^a y \sqrt{1 + y'^2}\, dx \tag{13.51}$$

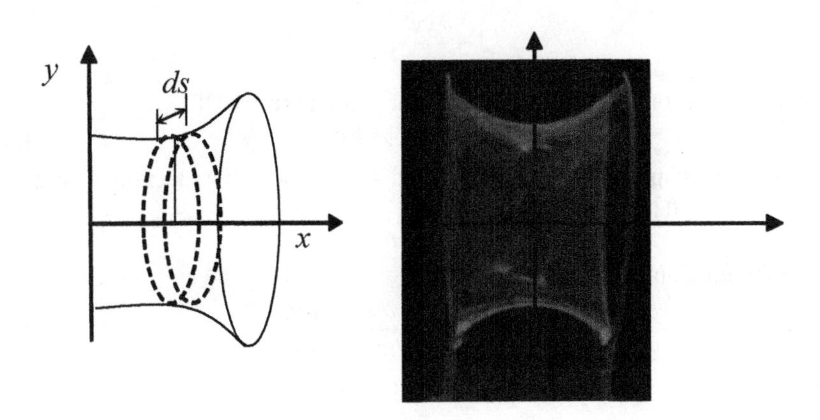

Figure 13.3 Mathematical formulation of the catenoid (photo on the right is reproduced from soapbubble.dk with permission)

Comparing to (13.5), we obtain

$$F(x, y, y') = y\sqrt{1 + y'^2} \tag{13.52}$$

Differentiation of (13.52) respect to both y and y' gives:

$$\frac{\partial F}{\partial y'} = \frac{yy'}{\sqrt{1 + y'^2}} \tag{13.53}$$

$$\frac{\partial F}{\partial y} = \sqrt{1 + y'^2} \tag{13.54}$$

Then, using (13.15) the Euler-Lagrange equation becomes

$$\sqrt{1 + y'^2} - \frac{d}{dx}[\frac{yy'}{\sqrt{1 + y'^2}}] = 0 \tag{13.55}$$

Recall from (13.53) that

$$\frac{\partial F}{\partial y'} = f(y, y') \tag{13.56}$$

Using the chain rule of differentiation, we get

$$\frac{d}{dx}(\frac{\partial F}{\partial y'}) = \frac{\partial}{\partial y}(\frac{\partial F}{\partial y'})\frac{dy}{dx} + \frac{\partial}{\partial y'}(\frac{\partial F}{\partial y'})\frac{dy'}{dx} \tag{13.57}$$

Substitution of (13.53) into (13.57) gives

$$\frac{\partial}{\partial y}(\frac{\partial F}{\partial y'})\frac{dy}{dx} = \frac{y'^2}{\sqrt{1+y'^2}} \tag{13.58}$$

$$\frac{\partial}{\partial y'}(\frac{\partial F}{\partial y'})\frac{dy'}{dx} = \frac{yy''}{(1+y'^2)^{3/2}} \tag{13.59}$$

Substitution of (13.58), (13.59), and (13.53) into (13.55) yields

$$1 - \frac{yy''}{1+y'^2} = 0 \tag{13.60}$$

This is equivalent to the following nonlinear second order ODE:

$$1 + y'^2 - yy'' = 0 \tag{13.61}$$

In an earlier chapter, we have introduced the method of reduction of order of differentiation. In particular, we introduced

$$y' = p \tag{13.62}$$

The differentiation of (13.62) gives

$$y'' = \frac{dp}{dx} = \frac{dp}{dy}\frac{dy}{dx} = p\frac{dp}{dy} \tag{13.63}$$

Thus, (13.61) can be reduced to a first order ODE as

$$1 + p^2 - yp\frac{dp}{dy} = 0 \tag{13.64}$$

This first order ODE is clearly separable, and can be rearranged as:

$$\frac{pdp}{1+p^2} = \frac{dy}{y} \tag{13.65}$$

It is straightforward to see that

$$\frac{1}{2}\frac{d(1+p^2)}{1+p^2} = \frac{dy}{y} \tag{13.66}$$

This can be integrated immediately to give

$$c_2\sqrt{1+p^2} = y \tag{13.67}$$

Recalling the definition of p from (13.62), we get

$$c_2\sqrt{1+(\frac{dy}{dx})^2} = y \tag{13.68}$$

Rearranging (13.68) gives

$$\frac{dy}{dx} = \sqrt{(\frac{y}{c_2})^2 - 1} \tag{13.69}$$

This can be integrated as

$$\int \frac{dy}{\sqrt{(\frac{y}{c_2})^2 - 1}} = \int dx + c_1 \tag{13.70}$$

It is clear that we can introduce the following change of variables:

$$y = c_2 \cosh\theta, \quad dy = c_2 \sinh\theta d\theta \tag{13.71}$$

This change of variables is an obvious choice by noting the following identity of hyperbolic functions:

$$\cosh^2\theta - 1 = \sinh^2\theta \tag{13.72}$$

Substitution of (13.71) and (13.72) into (13.70) gives

$$\theta = \frac{x}{c_2} + \frac{c_1}{c_2} \tag{13.73}$$

Combining (13.71) and (13.73) gives

$$y = c_2 \cosh(\frac{x}{c_2} + \frac{c_1}{c_2}) \tag{13.74}$$

As expected, we have two unknown constants for second order ODEs given (13.61). Let the boundary conditions for the soap film be given as

$$y(0) = b, \quad \frac{dy(0)}{dx} = 0 \tag{13.75}$$

Substitution of (13.74) into the boundary conditions given in (13.75) gives

$$c_2 = b, \quad c_1 = 0 \tag{13.76}$$

Finally, we have

$$y = b\cosh(\frac{x}{b}) \tag{13.77}$$

We have just obtained the sectional profile of a catenoid of the soap film shown in Figure 13.3.

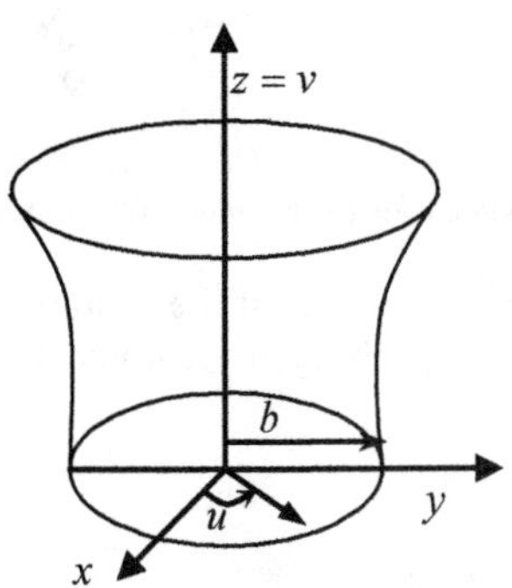

Figure 13.4 Coordinate system for a 3-D surface of the catenoid given in (13.78)–(13.80)

For a three-dimensional shape, we can recast the solution as:

$$x = b\cosh\frac{v}{b}\cos u \tag{13.78}$$

$$y = b\cosh\frac{v}{b}\sin u \tag{13.79}$$

$$z = v \tag{13.80}$$

where x, y, z, u and v are defined in Figure 13.4. For this case, intuition does not work: the minimum soap film is in the shape of revolution of hyperbolic cosine, instead of a cylindrical surface.

13.8 BRACHISTOCHRONE

The brachistochrone was first formulated by Galileo in 1638, but he was unable to solve the problem. This problem was considered one of the founding problems in the calculus of variations. This problem was originally posed as a challenge to other mathematicians by Johann Bernoulli in 1696. The word "brachistochrone" is from Greek, and literally means "the least time." The problem was solved independently by the Bernoulli brothers, Leibniz, and Newton. This problem investigates the optimum shape of a frictionless wire along which a bead will slide down with the shortest time, as shown in Figure 13.5. This problem was studied by Galileo Galilei in 1638 but he mistakenly got the answer as a quarter of a circle. As a side note, his last name is actually Galilei although most people just called him Galileo because it is how he referred to himself.

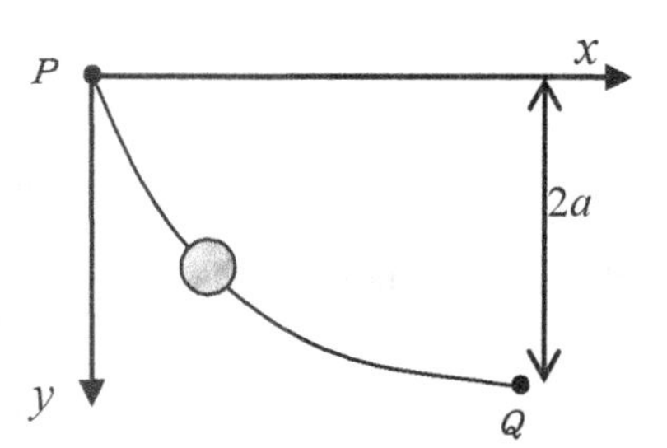

Figure 13.5 Brachistochrone problem of a frictionless sliding bead

Since the sliding is frictionless, the energy of the bead must conserve. In particular, the kinetic energy gained in the sliding must come from the drop of potential energy from points P to Q:

$$\frac{1}{2}mv^2 = mgy \tag{13.81}$$

Thus, the velocity can be evaluated as

$$v = \sqrt{2gy} \tag{13.82}$$

From (13.49) in the last section, we have

$$v = \frac{ds}{dt} = \sqrt{1+y'^2}\,\frac{dx}{dt} \tag{13.83}$$

Combining (13.82) and (13.83), we get

$$dt = \frac{\sqrt{1+y'^2}}{\sqrt{2gy}}dx \tag{13.84}$$

The sliding time for the bead can be evaluated by integrating (13.84) as

$$T = \int \frac{\sqrt{1+y'^2}}{\sqrt{2gy}}dx \tag{13.85}$$

Now we are searching for an optimum profile $y(x)$ such that the functional T (in this case is the time of travel) being the minimum. Clearly, this is another problem that can be solved by using the calculus of variations. In particular, the integrand function F can be identified as:

$$F(y,y') = \frac{\sqrt{1+y'^2}}{\sqrt{2gy}} \tag{13.86}$$

Recalling the Euler-Lagrange equation, we have

$$\frac{\partial F}{\partial y} - \frac{d}{dx}[\frac{\partial F}{\partial y'}] = 0 \tag{13.87}$$

In particular, using (13.86) we have

$$\frac{\partial F}{\partial y} = -\frac{1}{\sqrt{2gy}}\frac{\sqrt{1+y'^2}}{2y} \tag{13.88}$$

$$\begin{aligned}\frac{d}{dx}(\frac{\partial F}{\partial y'}) &= \frac{\partial}{\partial y}(\frac{\partial F}{\partial y'})\frac{dy}{dx} + \frac{\partial}{\partial y'}(\frac{\partial F}{\partial y'})\frac{dy'}{dx} \\ &= -\frac{1}{\sqrt{2gy}}\frac{y'^2}{2y\sqrt{1+y'^2}} + \frac{y''}{\sqrt{2gy}}\left\{\frac{1}{\sqrt{1+y'^2}} - \frac{y'^2}{(1+y'^2)^{3/2}}\right\}\end{aligned} \tag{13.89}$$

Using (13.88) and (13.89) in (13.87), we have

$$-2y(1+y'^2) + y'^2 - \frac{2yy''}{1+y'^2} = 0 \tag{13.90}$$

Equation (13.90) can be simplified as

$$2yy'' + 1 + y'^2 = 0 \tag{13.91}$$

Similar to the catenoid problem, we can reduce the order of the ODE by using

$$y' = p \tag{13.92}$$

In view of (13.92), we can reduce (13.91) to a first order ODE as

$$2yp\frac{dp}{dy} + 1 + p^2 = 0 \tag{13.93}$$

This is a separable ODE and can be integrated readily to give

$$\ln y = -\ln(1+p^2) + \ln c \tag{13.94}$$

Taking the exponential function, we find

$$y[1+y'^2]=c_1 \tag{13.95}$$

This can be rearranged as

$$\frac{dy}{dx}=\sqrt{\frac{c_1-y}{y}} \tag{13.96}$$

To integrate (13.96), we introduce

$$y=c_1\sin^2\phi, \quad dy=2c_1\sin\phi\cos\phi d\phi \tag{13.97}$$

Finally, (13.96) is reduced to

$$\frac{dy}{dx}=\sqrt{\frac{c_1\cos^2\phi}{c_1\sin^2\phi}}=\cot\phi \tag{13.98}$$

Combining (13.97) and (13.98), we have

$$2c_1\sin^2\phi d\phi=dx \tag{13.99}$$

This can be integrated readily as

$$x=\frac{c_1}{2}(2\phi-\sin 2\phi)+c_2 \tag{13.100}$$

At point P, the initial condition of the sliding bead can be formulated as:

$$x=0, t=0, y=0, \phi=0 \tag{13.101}$$

Thus, we have $c_2 = 0$. To simplify the presentation, we define

$$\theta=2\phi \tag{13.102}$$

Then, the solutions of x and y in terms of θ are

$$x=\frac{c_1}{2}(\theta-\sin\theta) \tag{13.103}$$

$$y=\frac{c_1}{2}(1-\cos\theta) \tag{13.104}$$

Finally, at Point Q shown in Fig. 13.5, we have the condition

$$\theta=\pi, y=2a \tag{13.105}$$

The final profile of the brachistochrone can be expressed in a single parameter θ as

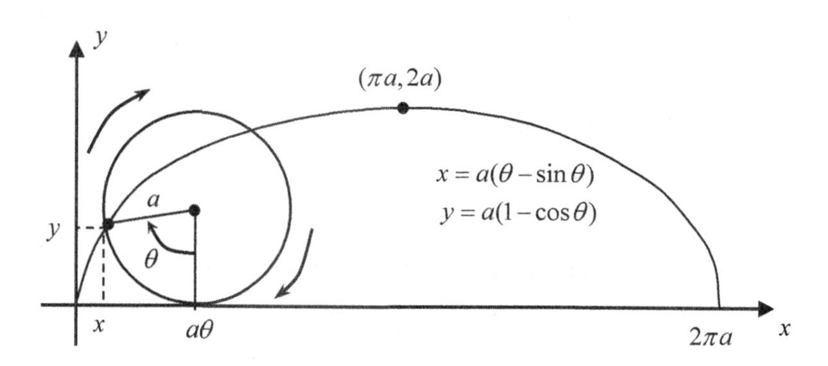

Figure 13.6 The locus of a cycloid

$$x = a(\theta - \sin\theta) \tag{13.106}$$

$$y = a(1 - \cos\theta) \tag{13.107}$$

The profile of this parametric curve is also known as a cycloid. Physically, it is also the locus of a point on a rolling wheel, as illustrated in Figure 13.6. Note however from Figure 13.5 that the y-axis of the brachistochrone is going downward. Thus, the solution of the brachistochrone is actually an inverted cycloid.

13.9 TAUTOCHRONE

Although the tautochrone problem is not a problem that needs to be solved by using the calculus of variations, it should be considered together with the brachistochrone problem. In particular, Huygens in 1673 discovered that no matter where we start to slide the bead along the cycloid, it always takes the same amount of time for the bead to slide from the starting point to the bottom. This is indeed an amazing result. To show this, we differentiate both (13.103) and (13.104) to give

$$\frac{dx}{d\theta} = a(1 - \cos\theta) \tag{13.108}$$

$$\frac{dy}{d\theta} = a\sin\theta \tag{13.109}$$

Using these results, we find the increment of the curve of the cycloid as

$$\begin{aligned}(\frac{dx}{d\theta})^2 + (\frac{dy}{d\theta})^2 &= a^2[(1 - 2\cos\theta + \cos^2\theta) + \sin^2\theta] \\ &= 2a^2(1 - \cos\theta)\end{aligned} \tag{13.110}$$

Recalling from (13.83), we have

$$v = \frac{ds}{dt} = \sqrt{2gy} \tag{13.111}$$

The travel time can then be written

$$dt = \frac{ds}{\sqrt{2gy}} = \frac{\sqrt{dx^2 + dy^2}}{\sqrt{2gy}} = \frac{a\sqrt{2(1-\cos\theta)}d\theta}{\sqrt{2ga(1-\cos\theta)}} = \sqrt{\frac{a}{g}}d\theta \tag{13.112}$$

Since the vertical drop from the top of the cycloid to the bottom is $2a$, as shown in Figure 13.5, we have the initial and final conditions from (13.94) as

$$t = 0, \quad y = 0, \quad \theta = 0 \tag{13.113}$$

$$t = T_1, \quad y = 2a, \quad \theta = \pi \tag{13.114}$$

Thus, (13.112) leads to the travel time from the top of the cycloid to the bottom

$$T_1 = \int_0^{\pi} \sqrt{\frac{a}{g}} d\theta = \sqrt{\frac{a}{g}}\pi \tag{13.115}$$

Next, we are considering the drop of the bead from any initial height, which is measured y_0 from the top of the cycloid as shown in Figure 13.7. The coordinate of the sliding motion is given by

$$y_0 = a(1 - \cos\theta_0) \tag{13.116}$$

The velocity is accordingly revised as

$$v = \frac{ds}{dt} = \sqrt{2g(y - y_0)} \tag{13.117}$$

Note from (13.117) that we must have $y > y_0$ and at the starting point we have $y = y_0$ or $v = 0$. Combining (13.116) and (13.117) yields

$$T_2 = \int_{\theta_0}^{\pi} \frac{a\sqrt{2(1-\cos\theta)}}{\sqrt{2ga(\cos\theta_0 - \cos\theta)}} d\theta = \sqrt{\frac{a}{g}} \int_{\theta_0}^{\pi} \sqrt{\frac{1-\cos\theta}{\cos\theta_0 - \cos\theta}} d\theta \tag{13.118}$$

Using the following identities for cosθ

$$\cos\theta_0 - \cos\theta = 2[\cos^2(\frac{\theta_0}{2}) - \cos^2(\frac{\theta}{2})], \quad \sin\frac{\theta}{2} = \sqrt{\frac{1-\cos\theta}{2}} \tag{13.119}$$

we can convert (13.118) to

$$T_2 = \sqrt{\frac{a}{g}} \int_{\theta_0}^{\pi} \frac{\sin(\theta/2)}{\sqrt{\cos^2(\theta_0/2) - \cos^2(\theta/2)}} d\theta \tag{13.120}$$

To evaluate this integration, we introduce the following change of variables

$$u = \frac{\cos(\theta/2)}{\cos(\theta_0/2)}, \quad du = -\frac{\sin(\theta/2)}{2\cos(\theta_0/2)} d\theta \tag{13.121}$$

The travel time is reduced to

$$T_2 = 2\sqrt{\frac{a}{g}} \int_0^1 \frac{1}{\sqrt{1-u^2}} du \tag{13.122}$$

Finally, we can introduce another change of variables as

$$u = \sin\alpha \tag{13.123}$$

Equation (13.112) is reduced to

$$T_2 = 2\sqrt{\frac{a}{g}} \int_0^{\pi/2} d\alpha = \sqrt{\frac{a}{g}}\pi \tag{13.124}$$

This is precisely equal to (13.115). Therefore, we have established the validity of the tautochrone, which means "equal time" in Greek, with "tauto" for same and "chrone" for time. Figure 13.7 illustrates that although different beads are starting from various y, they all arrive the bottom at the same time.

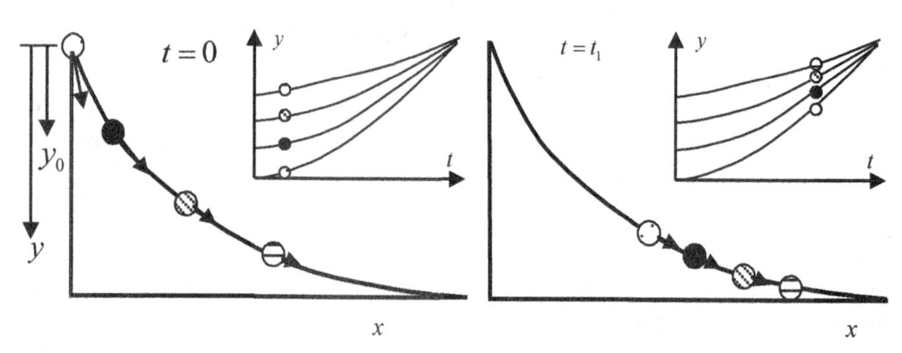

Figure 13.7 Illustration of the tautochrone problem at $t = 0$ and $t = t_1$

13.10 LAGRANGE MULTIPLIER

Sometimes we have to minimize a functional subject to certain constraints. This problem can also be considered by the calculus of variations by using the Lagrange multiplier. This technique was proposed by Lagrange in 1788. In particular, we want to consider the stationary value of the following functional:

$$I = \int_a^b F(t, z, \dot{z})dt \tag{13.125}$$

which is subjected to a constraint that

$$\int_a^b G(t, z, \dot{z})dt = C \tag{13.126}$$

where C is a constant. The superimposed "dot" implies a derivative taken with respect to t. We can formulate the functional in terms of Λ which is defined as:

$$I_1 = \int_a^b [F(t, z, \dot{z}) + \lambda G(t, z, \dot{z})]dt = \int_a^b \Lambda(t, z, \dot{z})dt \tag{13.127}$$

where λ is called the Lagrange multiplier and Λ is now the new Lagrangian. Following the same procedure of deriving the Euler-Lagrange equation shown in Section 13.4, we have

$$\frac{\partial \Lambda}{\partial z} - \frac{d}{dt}[\frac{\partial \Lambda}{\partial \dot{z}}] = 0 \tag{13.128}$$

We will apply the Lagrange multiplier technique to the isoperimetric problem in the next section.

13.11 DIDO'S PROBLEM (ISOPERIMETER)

According to legend, Dido arrived in Tunisia in 814 BC with her entourage after a power struggle with her brother at Tyre in Lebanon. She requested a piece of land and founded the city of Carthage. The land given to her could only be enclosed by a bull's hide, so she smartly cut the hide into long thin strips and used it to embrace a circular piece of land. Dido eventually became the first queen of Carthage. Therefore, the optimum shape of a closed curve that can enclose the greatest area is a circle.

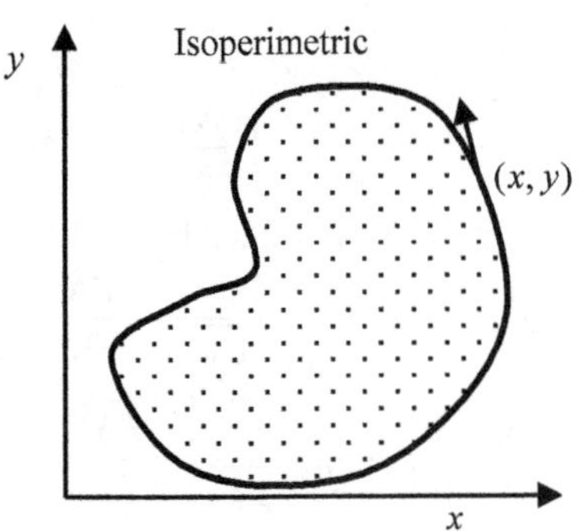

Figure 13.8 Dido's problem of greatest covered area

Therefore, the problem of Dido is to find the greatest area being enclosed by a closed curve of a fixed length. Recall from (13.48) that the length of the curve can be written as:

$$(ds)^2 = (dx)^2 + (dy)^2 \tag{13.129}$$

Writing the increment along the curve ds in terms of a time parameter t:

$$ds = \sqrt{\dot{x}^2 + \dot{y}^2}\, dt \tag{13.130}$$

where the superimposed dot implies differentiation with respect to t. Thus, the length of the perimeter can be expressed as

$$L = \oint \sqrt{\dot{x}^2 + \dot{y}^2}\, dt = p \tag{13.131}$$

where p is a fixed constant. To express the area covered by the closed curve, we recall Green's theorem that

$$\int_{\partial D} (f dx + g dy) = \int_{D} (\frac{\partial g}{\partial x} - \frac{\partial f}{\partial y}) dx dy \tag{13.132}$$

Taking the following values of f and g

$$f = -\frac{y}{2}, \quad g = \frac{x}{2} \tag{13.133}$$

we have (13.132) becoming

$$\frac{1}{2} \int_{\partial D} (x dy - y dx) = \int_{D} dx dy \tag{13.134}$$

Making a closed curve, we arrive at the following enclosed area

$$A = \frac{1}{2} \oint (x\dot{y} - y\dot{x}) dt \tag{13.135}$$

The functional in (13.135) and the constraint (13.131) can be combined using the Lagrange multiplier as

$$I = \oint \Lambda(t, x, \dot{x}, y, \dot{y}) dt = \oint \{\frac{1}{2}(x\dot{y} - y\dot{x}) + \lambda \sqrt{\dot{x}^2 + \dot{y}^2}\} dt \tag{13.136}$$

Therefore, for the present case of two dependent variables, we have the Euler-Lagrange equations as

$$\frac{\partial \Lambda}{\partial x} - \frac{d}{dt}[\frac{\partial \Lambda}{\partial \dot{x}}] = 0 \tag{13.137}$$

$$\frac{\partial \Lambda}{\partial y} - \frac{d}{dt}[\frac{\partial \Lambda}{\partial \dot{y}}] = 0 \tag{13.138}$$

Using the definition of Lagrangian Λ defined in (13.136), we find

$$\frac{\partial \Lambda}{\partial \dot{x}} = -\frac{1}{2} y + \frac{\lambda \dot{x}}{\sqrt{\dot{x}^2 + \dot{y}^2}}, \quad \frac{\partial \Lambda}{\partial x} = \frac{1}{2} \dot{y} \tag{13.139}$$

$$\frac{\partial \Lambda}{\partial \dot{y}} = \frac{1}{2} x + \frac{\lambda \dot{y}}{\sqrt{\dot{x}^2 + \dot{y}^2}}, \quad \frac{\partial \Lambda}{\partial x} = -\frac{1}{2} \dot{x} \tag{13.140}$$

Substitution of (13.139) and (13.140) into (13.137) and (13.138) leads to the following Euler-Lagrange equations

$$\frac{d}{dt}[-\frac{1}{2}y+\frac{\lambda\dot{x}}{\sqrt{\dot{x}^2+\dot{y}^2}}]-\frac{1}{2}\dot{y}=0 \tag{13.141}$$

$$\frac{d}{dt}[\frac{1}{2}x+\frac{\lambda\dot{y}}{\sqrt{\dot{x}^2+\dot{y}^2}}]+\frac{1}{2}\dot{x}=0 \tag{13.142}$$

Integrating with respect to t, we immediately obtain

$$y-\frac{\lambda\dot{x}}{\sqrt{\dot{x}^2+\dot{y}^2}}=b \tag{13.143}$$

$$x+\frac{\lambda\dot{y}}{\sqrt{\dot{x}^2+\dot{y}^2}}=a \tag{13.144}$$

These two equations can be combined to give

$$(x-a)^2+(y-b)^2=\lambda^2 \tag{13.145}$$

This is an equation of a circle with radius λ at center (a,b) and thus physically the Lagrange multiplier is the radius. Using the constraint in (13.131), we have the Lagrange multiplier and the enclosed area being

$$\lambda=\frac{p}{2\pi},\quad A=\frac{p^2}{4\pi} \tag{13.146}$$

As expected by Dido, the optimum shape is indeed a circle.

13.12 GEODESICS

In this section, we will present the formulation for finding the geodesics, or the shortest curve between two given points on a given surface. Mathematically, the equation of the three-dimensional surface can be expressed as:

$$G(x,y,z)=0 \tag{13.147}$$

Introducing a time parameter t, the arc length between any two points a and b can be evaluated as

$$I=\int_a^b f(\dot{x},\dot{y},\dot{z})\,dt=\int_a^b \sqrt{\dot{x}^2+\dot{y}^2+\dot{z}^2}\,dt \tag{13.148}$$

where

$$\dot{x}=\frac{dx}{dt},\quad \dot{y}=\frac{dy}{dt},\quad \dot{z}=\frac{dz}{dt} \tag{13.149}$$

The arc must be on the surface $G = 0$, thus we can impose a relation between the velocities of the variables. First, we can rewrite

$$z=g(x(t),y(t)) \tag{13.150}$$

Differentiation of z in (13.150) gives

$$\dot{z}=g_x\dot{x}+g_y\dot{y} \tag{13.151}$$

Taking a second derivative of (13.151) with respect to x gives

$$\frac{\partial \dot{z}}{\partial x} = \frac{\partial}{\partial x}(g_x \dot{x} + g_y \dot{y}) = g_{xx}\dot{x} + g_{xy}\dot{y} \tag{13.152}$$

On the other hand, we observe that from the chain rule

$$\frac{d}{dt}(g_x) = g_{xx}\dot{x} + g_{xy}\dot{y} \tag{13.153}$$

Comparison of (13.152) and (13.153) gives the following identity

$$\frac{\partial \dot{z}}{\partial x} = \frac{d}{dt}(g_x) \tag{13.154}$$

If we follow the same procedure in obtaining (13.152) to (13.154) but taking differentiation with respect to y, we get the following identity

$$\frac{\partial \dot{z}}{\partial y} = \frac{d}{dt}(g_y) \tag{13.155}$$

Let us now rewrite our functional by eliminating the time derivative of z in (13.148)

$$I = \int_a^b F(x, \dot{x}, y, \dot{y})dt = \int_a^b \sqrt{\dot{x}^2 + \dot{y}^2 + (g_x \dot{x} + g_y \dot{y})^2}\, dt \tag{13.156}$$

Referring to Section 13.6, the corresponding Euler-Lagrange equations for a two-variable system are

$$\frac{\partial F}{\partial x} - \frac{d}{dt}(\frac{\partial F}{\partial \dot{x}}) = 0 \tag{13.157}$$

$$\frac{\partial F}{\partial y} - \frac{d}{dt}(\frac{\partial F}{\partial \dot{y}}) = 0 \tag{13.158}$$

The first term on the left of (13.157) can be evaluated as

$$\frac{\partial F}{\partial x} = \frac{\partial f}{\partial \dot{z}}\frac{\partial}{\partial x}(g_x \dot{x} + g_y \dot{y}) = \frac{\partial f}{\partial \dot{z}}(g_{xx}\dot{x} + g_{xy}\dot{y}) \tag{13.159}$$

In view of (13.151), (13.159) can be simplified to

$$\frac{\partial F}{\partial x} = \frac{\partial f}{\partial \dot{z}}\frac{\partial}{\partial x}(g_x \dot{x} + g_y \dot{y}) = \frac{\partial f}{\partial \dot{z}}\frac{\partial \dot{z}}{\partial x} \tag{13.160}$$

In parallel to this development, we can repeat the procedure for the first term of (13.158), and we eventually obtain

$$\frac{\partial F}{\partial y} = \frac{\partial f}{\partial \dot{z}}\frac{\partial}{\partial y}(g_x \dot{x} + g_y \dot{y}) = \frac{\partial f}{\partial \dot{z}}\frac{\partial \dot{z}}{\partial y} \tag{13.161}$$

For the second term of (13.157), we have

$$\frac{\partial F}{\partial \dot{x}} = \frac{\partial f}{\partial \dot{x}} + \frac{\partial f}{\partial \dot{z}}\frac{\partial \dot{z}}{\partial \dot{x}} = \frac{\partial f}{\partial \dot{x}} + \frac{\partial f}{\partial \dot{z}} g_x \tag{13.162}$$

The last part of (13.162) is obtained in view of (13.151). Taking the time derivative of (13.162), we arrive at

$$\begin{aligned} \frac{d}{dt}(\frac{\partial F}{\partial \dot{x}}) &= \frac{d}{dt}(\frac{\partial f}{\partial \dot{x}}) + \frac{d}{dt}(\frac{\partial f}{\partial \dot{z}} g_x) = \frac{d}{dt}(\frac{\partial f}{\partial \dot{x}}) + \frac{d}{dt}(\frac{\partial f}{\partial \dot{z}})g_x + \frac{\partial f}{\partial \dot{z}}\frac{d}{dt}(g_x) \\ &= \frac{d}{dt}(\frac{\partial f}{\partial \dot{x}}) + \frac{d}{dt}(\frac{\partial f}{\partial \dot{z}})g_x + \frac{\partial f}{\partial \dot{z}}\frac{\partial \dot{z}}{\partial x} \end{aligned} \tag{13.163}$$

The last line of (13.163) results from the substitution of (13.154). Finally, (13.160) and (13.163) can be substituted into the first Euler-Lagrange equation given in (13.157) to give

$$\frac{d}{dt}(\frac{\partial f}{\partial \dot{x}})+g_x\frac{d}{dt}(\frac{\partial f}{\partial \dot{z}})=0 \tag{13.164}$$

On the other hand, we can deal with the second term of (13.158) in a similar manner; we get

$$\frac{d}{dt}(\frac{\partial F}{\partial \dot{y}})=\frac{d}{dt}(\frac{\partial f}{\partial \dot{y}})+\frac{d}{dt}(\frac{\partial f}{\partial \dot{z}})g_y+\frac{\partial f}{\partial \dot{z}}\frac{\partial \dot{z}}{\partial y} \tag{13.165}$$

Finally, combining (13.161) and (13.165) into the second Euler-Lagrange equation given in (13.158), we obtain

$$\frac{d}{dt}(\frac{\partial f}{\partial \dot{y}})+g_y\frac{d}{dt}(\frac{\partial f}{\partial \dot{z}})=0 \tag{13.166}$$

Now, we apply the most crucial step in our analysis in which we set

$$\frac{d}{dt}(\frac{\partial f}{\partial \dot{z}})=\lambda(t)G_z \tag{13.167}$$

where $G = 0$ has been defined as our surface of consideration in (13.147) and $\lambda(t)$ plays the role of the Lagrange multiplier (we will show this in a short while). We note from definition (13.147) that

$$G_x dx+G_y dy+G_z dz=0 \tag{13.168}$$

which can be rewritten as

$$G_x+G_y\frac{dy}{dx}+G_z\frac{dz}{dx}=G_x+G_z g_x=0 \tag{13.169}$$

Note that $dy/dx = 0$ since x and y are two independent variables. Or equivalently, we have

$$g_x=-\frac{G_x}{G_z} \tag{13.170}$$

In similar fashion, we can also have

$$g_y=-\frac{G_y}{G_z} \tag{13.171}$$

Employing (13.170), (13.171) and (13.167) in the special form of Euler-Lagrange equations (13.164) and (13.166), we finally get

$$\frac{d}{dt}(\frac{\partial f}{\partial \dot{x}})=\lambda(t)G_x \tag{13.172}$$

$$\frac{d}{dt}(\frac{\partial f}{\partial \dot{y}})=\lambda(t)G_y \tag{13.173}$$

By eliminating λ from (13.167), (13.172) and (13.173), we have

$$\lambda(t)=\frac{\frac{d}{dt}(\frac{\partial f}{\partial \dot{x}})}{G_x}=\frac{\frac{d}{dt}(\frac{\partial f}{\partial \dot{y}})}{G_y}=\frac{\frac{d}{dt}(\frac{\partial f}{\partial \dot{z}})}{G_z} \tag{13.174}$$

Finally, recalling our arc length problem defined in (13.148), we can simplify (13.172) to

$$\frac{f\ddot{x}-\dot{x}\dot{f}}{G_x f^2}=\frac{f\ddot{y}-\dot{y}\dot{f}}{G_y f^2}=\frac{f\ddot{z}-\dot{z}\dot{f}}{G_z f^2} \tag{13.175}$$

This is the final governing equation for our geodesic problem.

Example 13.1 Re-derive (13.174) by combining the Lagrange multiplier method discussed in Section 13.10 in conjunction with the functional of several variables discussed in Section 13.6.

Solution: The functional with constraint can be formulated by combining (13.147) and (13.148) as:

$$\begin{aligned} I &= \int_a^b \{f(\dot{x},\dot{y},\dot{z})+\lambda(t)G(x(t),y(t),z(t))\}dt \\ &= \int_a^b \Lambda(x,\dot{x},y,\dot{y},z,\dot{z})dt \end{aligned} \tag{13.176}$$

where $\lambda(t)$ is the Lagrange multiplier.

Applying the Euler-Lagrange equation for three variables, we have

$$\frac{\partial\Lambda}{\partial x}-\frac{d}{dt}[\frac{\partial\Lambda}{\partial\dot{x}}]=0 \tag{13.177}$$

$$\frac{\partial\Lambda}{\partial y}-\frac{d}{dt}[\frac{\partial\Lambda}{\partial\dot{y}}]=0 \tag{13.178}$$

$$\frac{\partial\Lambda}{\partial z}-\frac{d}{dt}[\frac{\partial\Lambda}{\partial\dot{z}}]=0 \tag{13.179}$$

Substitution of Λ defined in (13.176) into (13.177) to (13.179) yields

$$\lambda\frac{\partial G}{\partial x}-\frac{d}{dt}[\frac{\partial f}{\partial\dot{x}}]=0 \tag{13.180}$$

$$\lambda\frac{\partial G}{\partial x}-\frac{d}{dt}[\frac{\partial f}{\partial\dot{y}}]=0 \tag{13.181}$$

$$\lambda\frac{\partial G}{\partial x}-\frac{d}{dt}[\frac{\partial f}{\partial\dot{z}}]=0 \tag{13.182}$$

Thus, we have

$$\lambda(t)=\frac{\frac{d}{dt}(\frac{\partial f}{\partial\dot{x}})}{G_x}=\frac{\frac{d}{dt}(\frac{\partial f}{\partial\dot{y}})}{G_y}=\frac{\frac{d}{dt}(\frac{\partial f}{\partial\dot{z}})}{G_z} \tag{13.183}$$

This completes the proof. Therefore, we see that the physical meaning of $\lambda(t)$ introduced in (13.167) actually plays the role of Lagrange multiplier. It links the original functional with the constraint of the problem to form the new functional for minimization. Note thar the Lagrange multiplier method is much simpler than the intuitive step employed in (13.167).

Example 13.2 Find the shortest curve between two points A and B on a plane. The arc length is defined as

$$I = \int_{x_1}^{x_2} F(y')dx = \int_{x_1}^{x_2} \sqrt{1+y'^2}\,dx \tag{13.184}$$

The problem is shown in Figure 13.9.

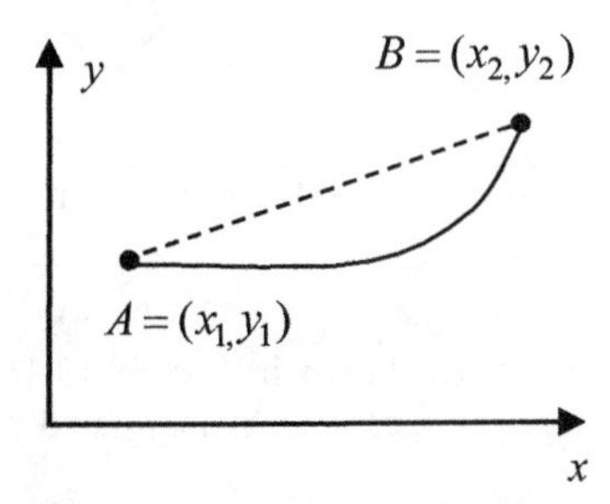

Figure 13.9 Shortest curve between points A and B on a plane

Solution: For this problem, the Lagrangian is identified as

$$F(y') = \sqrt{1+y'^2} \tag{13.185}$$

Thus, we have

$$\frac{\partial F}{\partial y} = 0, \quad \frac{\partial F}{\partial y'} = \frac{y'}{\sqrt{1+y'^2}} \tag{13.186}$$

The Euler-Lagrange equation requires

$$\frac{d}{dx}\left[\frac{y'}{\sqrt{1+y'^2}}\right] = 0 \tag{13.187}$$

Integration of (13.187) gives

$$\frac{y'}{\sqrt{1+y'^2}} = c \tag{13.188}$$

Rearranging (13.188), we obtain

$$y' = c\sqrt{1+y'^2} \tag{13.189}$$

Squaring both sides and solving for y', we get

$$y' = \sqrt{\frac{c^2}{1-c^2}} \tag{13.190}$$

This can be integrated immediately to give

$$y = \sqrt{\frac{c^2}{1-c^2}}x + c_2 = mx + c_2 \tag{13.191}$$

This is an equation of a straight line. Therefore, the shortest distance between two points is a straight line joining them, as expected.

However, if we want to minimize the surface area formed by the revolution of a curve between two points A and B shown in Figure 13.9, one is tempted to speculate that it must be a revolution of the straight line obtained here. Sometimes, we have to set aside our intuition and to rely on mathematics. In fact, we find that the answer is not a straight line but a hyperbolic cosine. This is the catenoid problem that we discussed in Section 13.7, and the minimal area of revolution is not formed by a straight line.

Example 13.3 Consider the shortest curve between two points A and B on a sphere defined by:

$$G(x,y,z) = x^2 + y^2 + z^2 - r^2 = 0 \tag{13.192}$$

where r is the radius of the sphere shown in Figure 13.10. This problem is also known as the Columbus problem because it is related to the shortest path in navigation on a voyage.

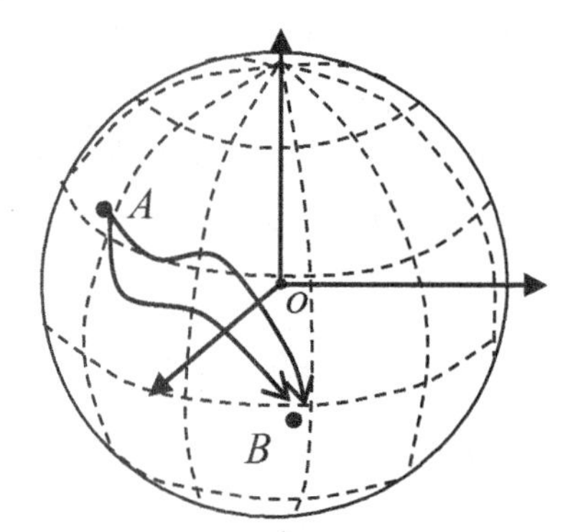

Figure 13.10 Shortest curve between points A and B on a sphere

Solution: Differentiating (13.192), we obtain

$$G_x = 2x, \quad G_y = 2y, \quad G_z = 2z \tag{13.193}$$

Subsequently, (13.175) is reduced to

$$\frac{f\ddot{x} - \dot{x}\dot{f}}{2xf^2} = \frac{f\ddot{y} - \dot{y}\dot{f}}{2yf^2} = \frac{f\ddot{z} - \dot{z}\dot{f}}{2zf^2} \tag{13.194}$$

Rearranging terms in (13.194), we get

$$\frac{\dot{f}}{f} = \frac{y\ddot{x} - x\ddot{y}}{y\dot{x} - x\dot{y}} = \frac{y\ddot{z} - z\ddot{y}}{y\dot{z} - z\dot{y}} \tag{13.195}$$

Note that

$$\frac{d(y\dot{x} - x\dot{y})}{dt} = \dot{y}\dot{x} + y\ddot{x} - \dot{y}\dot{x} - x\ddot{y} = y\ddot{x} - x\ddot{y} \tag{13.196}$$

$$\frac{d(y\dot{z}-z\dot{y})}{dt}=y\ddot{z}-z\ddot{y} \tag{13.197}$$

In view of these identities, (13.195) can be integrated as

$$\ln(y\dot{x}-x\dot{y})=\ln(y\dot{z}-z\dot{y})+C \tag{13.198}$$

Taking an exponential function on both sides, we get

$$y\dot{x}-x\dot{y}=C_1(y\dot{z}-z\dot{y}) \tag{13.199}$$

This can be further rearranged as

$$\frac{\dot{x}-C_1\dot{z}}{x-C_1 z}=\frac{\dot{y}}{y} \tag{13.200}$$

Or equivalently it can be written as

$$\frac{d(x-C_1 z)}{x-C_1 z}=\frac{dy}{y} \tag{13.201}$$

We can integrate both sides one more time to get

$$\ln(x-C_1 z)=\ln y+C^* \tag{13.202}$$

Thus, the geodesics must be on the following plane

$$x-C_1 z=C_2 y \tag{13.203}$$

This is an equation for a plane passing through the origin. Thus, the geodesics must be on the intersection between the sphere and the plane through the origin as shown in Figure 13.11. This result was first obtained by Euler in 1755. This is the reason why the shortest path of flight from Hong Kong to Los Angeles is not flying over the Pacific Ocean via Hawaii but instead flying over Alaska.

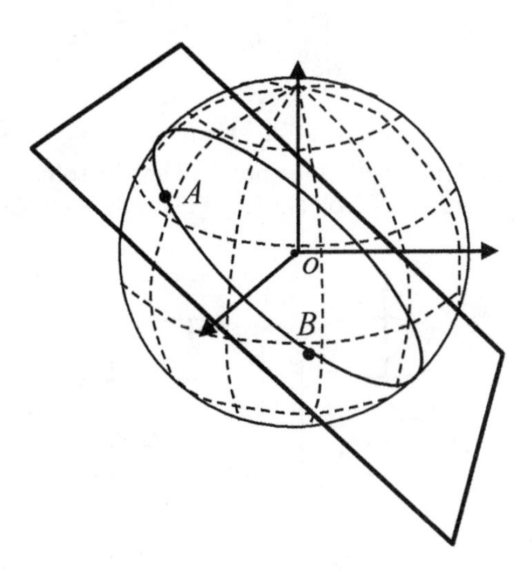

Figure 13.11 Great circle as the shortest curve between points A and B on a sphere

---

Example 13.4 Find the shortest curve between two points A and B on the surface of a circular cylinder of radius a. The cylindrical coordinate is $(r,\theta z)$.

Solution: The curve length on the surface of a circular cylinder is

$$(ds)^2 = (dr)^2 + (rd\theta)^2 + (dz)^2 \tag{13.204}$$

On the surface of the cylinder, we have $r = a$. Thus, we have $dr = 0$, and (13.204) becomes

$$(\frac{ds}{d\theta})^2 = a^2 + (\frac{dz}{d\theta})^2 \tag{13.205}$$

Using (13.205), we have the curve path

$$ds = \sqrt{a^2 + (\frac{dz}{d\theta})^2}\, d\theta \tag{13.206}$$

The curve length between two points A (a, θ_1, z_1) and B (a, θ_2, z_2) is

$$s = \int_{\theta_1}^{\theta_2} \sqrt{a^2 + (\frac{dz}{d\theta})^2}\, d\theta \tag{13.207}$$

Clearly, we want to minimize the functional s for the shortest curve. Thus, we identify that

$$F = \sqrt{a^2 + (\frac{dz}{d\theta})^2} = F(z') \tag{13.208}$$

Note that F is independent of z and θ, and thus we have case (iii) in Section 13.5 or the Euler-Lagrange equation becomes

$$\frac{d^2 z}{d\theta^2} = 0 \tag{13.209}$$

Or, the shortest curve is

$$z = C_1\theta + C_2, \quad r = a \tag{13.210}$$

This is the equation of a circular helix. The problem and its solution in terms of the helix are illustrated in Figure 3.12. Problem 13.14 gives the formulas for z and s.

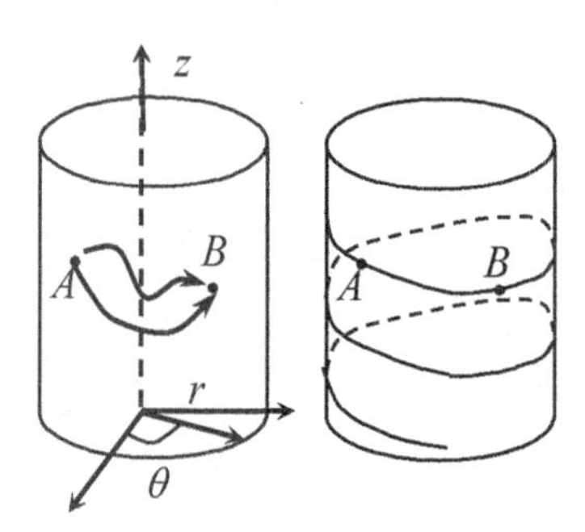

Figure 13.12 Shortest curve between points A and B on a circular cylinder

---

13.13 CATENARY AS A SAGGING ROPE

When a rope or chain is hanging between two supports, nature chooses the shape of the hanging rope by allowing its center of gravity at its lowest possible position. The shape is called catenary. The functional I in this case is the y-coordinate of the center of gravity together with the constraint G of the length of the hanging rope:

$$I[y(x)] = \frac{\int_{x_1}^{x_2} y\sqrt{1+y'^2}\,dx}{\int_{x_1}^{x_2} \sqrt{1+y'^2}\,dx} \tag{13.211}$$

$$G[y(x)] = \int_{x_1}^{x_2} \sqrt{1+y'^2}\,dx = L \tag{13.212}$$

Note that the numerator is essentially the same as that of (13.52) of the catenoid problem. The denominator becomes a constant by virtue of the constraint given in (13.212). Thus, we are minimizing the numerator of (13.211) and the constraint in (13.212) by the Lagrange multiplier, and we have

$$I = \int_a^b \{y\sqrt{1+y'^2} + \lambda\sqrt{1+y'^2}\}dx \tag{13.213}$$

In other words, we have

$$\Lambda = y\sqrt{1+y'^2} + \lambda\sqrt{1+y'^2} = (y+\lambda)\sqrt{1+y'^2} \tag{13.214}$$

which is not a function x. This corresponds to Case (i) in Section 13.5 and thus the Euler-Lagrange equation given in (13.28) becomes

$$\Lambda - y'\frac{\partial \Lambda}{\partial y'} = k_1 \tag{13.215}$$

Substitution of (13.214) into (13.215) gives

$$(y+\lambda)\sqrt{1+y'^2} - y'(y+\lambda)\frac{y'}{\sqrt{1+y'^2}} = k_1 \tag{13.216}$$

Rearranging (13.216) gives

$$(y+\lambda) = k_1\sqrt{1+y'^2} \tag{13.217}$$

The following change of variables is introduced:

$$y' = \frac{dy}{dx} = \sinh t \tag{13.218}$$

Using (13.218), (13.217) becomes

$$(y+\lambda) = k_1 \cosh t \tag{13.219}$$

Combining (13.218) and (13.219), we have

$$dx = \frac{dy}{y'} = \frac{k_1 \sinh t dt}{\sinh t} = k_1 dt \tag{13.220}$$

Integration of (13.220) gives

$$x = k_1 t + k_2 \tag{13.221}$$

Therefore, back substitution of (13.221) into (13.219) yields the hyperbolic cosine function as the catenary:

$$(y+\lambda) = k_1 \cosh(\frac{x-k_2}{k_1}) \tag{13.222}$$

For the values of k_1, k_2, and λ, we have to satisfy the boundary condition of the hanging rope.

Alternatively, (13.222) can be obtained using a mechanics approach. In particular, Figure 13.13 shows a particular case of a hanging rope with two supports at different elevations and the mass per length of the rope is assumed as μ (kg/m) and the horizontal component of the tension in the rope is T_0. For this case, k_1, k_2, and λ in (13.222) can also be determined with the boundary conditions given in Figure 13.13.

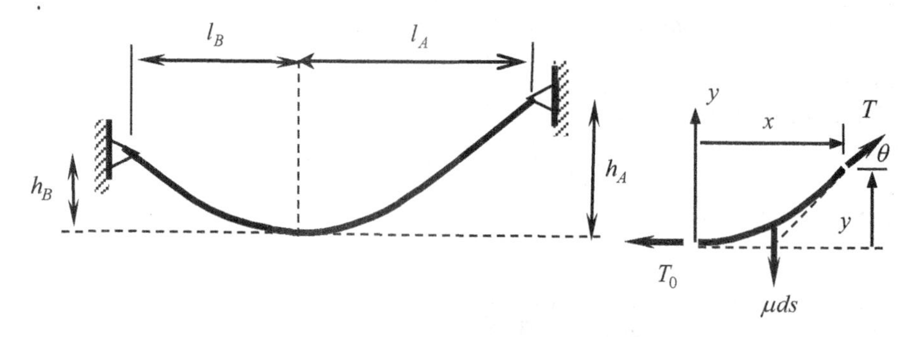

Figure 13.13 Boundary conditions of a hanging rope

A small rope of length ds can be cut out as a free body and the forces applied on this segment are shown in Figure 13.14. The vertical and horizontal force equilibriums give

$$(T+dT)\sin(\theta+d\theta) = T\sin\theta + \mu ds \tag{13.223}$$

$$(T+dT)\cos(\theta+d\theta) = T\cos\theta \tag{13.224}$$

By neglecting the higher order terms, we have

$$d(T\sin\theta) = \mu ds, \quad d(T\cos\theta) = 0 \tag{13.225}$$

The second equation of (13.225) gives

$$T_0 = T\cos\theta \tag{13.226}$$

Physically, (13.226) shows that the horizontal component of the tension in the rope is a constant because there is no net horizontal force applied on this rope element. Then, substitution of (13.226) into the first of (13.225) gives

$$\frac{d}{ds}(T_0 \tan\theta) = \mu \tag{13.227}$$

Note that the slope is defined as

$$\tan\theta = \frac{dy}{dx} \tag{13.228}$$

Thus, we can rewrite (13.227) as

$$\frac{d}{ds}(\frac{dy}{dx}) = \frac{\mu}{T_0} \tag{13.229}$$

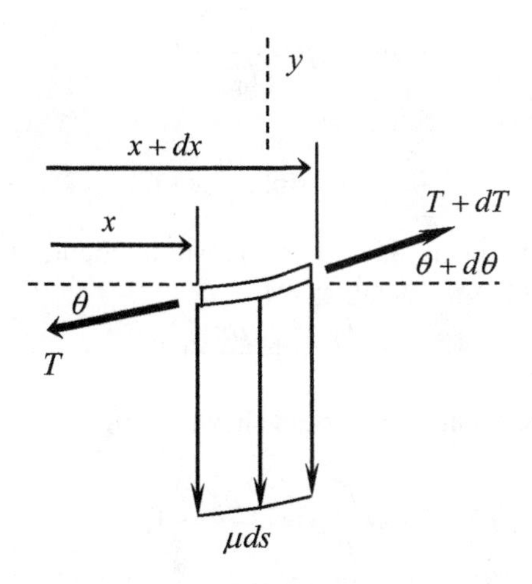

Figure 13.14 Shortest curve between points A and B on a circular cylinder

This can be integrated with respect to s leading to

$$\frac{dy}{dx} = \frac{\mu}{T_0} s + C \tag{13.230}$$

On the other hand, we can apply the chain rule to get

$$\frac{d}{ds}(\frac{dy}{dx}) = \frac{d}{dx}[(\frac{dy}{dx})]\frac{dx}{ds} = \frac{\mu}{T_0} \tag{13.231}$$

This can be rearranged to give

$$\frac{d^2 y}{dx^2} = \frac{\mu}{T_0}\frac{ds}{dx} \tag{13.232}$$

In view of the following identity,

$$\frac{ds}{dx} = \sqrt{1+(\frac{dy}{dx})^2} \tag{13.233}$$

we can rewrite (13.232) as

$$\frac{d^2 y}{dx^2} = \frac{\mu}{T_0}\sqrt{1+(\frac{dy}{dx})^2} \tag{13.234}$$

Since the differential equation does not depend on y and x, we can apply the standard rule of reduction of order as:

$$p = \frac{dy}{dx} \tag{13.235}$$

Substitution of (13.235) into (13.234) results in

$$\frac{dp}{\sqrt{1+p^2}} = \frac{\mu}{T_0} dx \tag{13.236}$$

Integration on both sides gives

$$\sinh^{-1} p = \frac{\mu}{T_0} x + C \tag{13.237}$$

Alternatively, this can be rewritten as

$$\frac{dy}{dx} = p = \sinh(\frac{\mu}{T_0} x + C) \tag{13.238}$$

The zero slope condition at the origin shown in Figure 13.13 gives C = 0. Integration with respect to x instantly gives

$$y = \frac{T_0}{\mu} \cosh(\frac{\mu x}{T_0}) + K \tag{13.239}$$

The zero slope displacement at the origin shown in Figure 13.13 gives $K = -T_0/\mu$. Finally, we get

$$y = \frac{T_0}{\mu} [\cosh(\frac{\mu x}{T_0}) - 1] \tag{13.240}$$

Comparison of (13.240) with (13.222) gives

$$\lambda = k_1 = \frac{T_0}{\mu}, \quad k_2 = 0 \tag{13.241}$$

in (13.222). We can see that the Lagrange multiplier plays the role of normalized length in terms of rope tension divided by mass per length.

The catenary appears naturally in nature in the form of a hanging spider web, of hanging rope, and of hanging chain. We have seen that it also forms the catenoid of soap film spreading over two rings. Many man-made structures were inspired by the catenary. The most vivid example is the inverted catenary of the Gateway Arch built in St. Louis, Missouri, USA. It stands 192 m above ground level. The vault at the Casa Mila, Barcelona, Spain and vault at Ctesiphon, Iraq are also inverted catenaries. The vault of Ctesiphon stands 37 m above ground level and is believed to be the tallest vault in a structure in the world. As a side note, the trademark of McDonald's is also made of two inverted catenaries.

13.14 NEWTON'S PROBLEM OF LEAST RESISTANCE

13.14.1 Introduction

In 1685, Newton included in his celebrated *Principia* the problem of minimum resistance on a solid of revolution moving in a rare gas modeled as non-viscous flow, regarding the optimum shape of the solid of revolution (Newton, 1685). Tacitly, Newton assumed that the section of the solid must be of convex shape and the body must be axisymmetric. The gas flow is assumed so rare and dispersed that resistance of the gas flow on the solid can be considered as particle impact. Thus, gas flow is assumed as corpuscular flow. There are also no interactions between these globally evenly distributed-particles. The assumption of convexity of the solid surface ensures that there is only a single impact of these particles on the body. Newton claimed that this problem might be useful in ship design. One main problem is that there is no proof of Newton's optimum shape reported in *Principia*. This problem was considered by Huygens and David Gregory. The rigorous proof of Newton's solution was given in 1902 by Kneser. Nevertheless, it is commonly believed that Newton's problem is one of the first problems of its type that prompted the development of the calculus of variations (isoperimetric problems are other examples of such problems).

It turns out that, instead of its use in ship design, as originally proposed by Newton, Newton's formulation was found applicable to bodies traveling at high supersonic speed in air, such as missiles. In particular, the so-called Newton's cosine-square law for resistance (which will be discussed later in this section) was coincidentally obtained for the pressure coefficient when Riemann's shock wave conditions are taken into consideration. Thus, Newton's model was found applicable to supersonic flow.

13.14.2 Newton's Sine-Square or Cosine-Square Law of Resistance

Newton formulated the impact-induced resistance on a solid with a spherical surface in terms of the resistance of the normal impact on a flat surface. It turns out that the formulation is general for any curved surface of revolution (Goldstine, 1980). In particular, Figure 13.15 shows the impact force on a flat cylindrical surface as well as on a curved surface of revolution. The resistance on normal impact is assumed as f. The projection of the force f along the inclination of the line drawn from the center of the projectile is $f\cos\theta$. A second projection of this inclined force on the curved surface of the projectile is

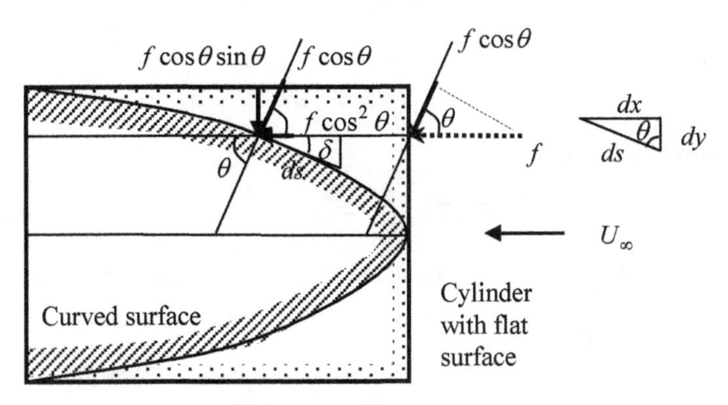

Figure 13.15 Illustration of the Newton's cosine-square law

$$f_{||} = f\cos^2\theta = f\sin^2\delta \tag{13.242}$$

Physically, this is the force of impact on the curved surface along the flow direction. This is normally referred to as the sine-square law or cosine-square resistance law of Newton. Note that the force component normal to the flow direction is

$$f_{\perp} = f\cos\theta\sin\theta \tag{13.243}$$

However, by symmetry this force perpendicular to the flow will be cancelled out. Therefore, the net resistance force on a segment of the curved surface is

$$dF_{\parallel} = f\cos^2\theta ds \tag{13.244}$$

Therefore, the total resistance force on the curved surface is

$$R = 2\pi f\int_{y_1}^{y_2} y\cos^2\theta dy = 2\pi f\int_{s_1}^{s_2} y\cos^3\theta ds \tag{13.245}$$

From Figure 13.15, we see that

$$\cos\theta = \frac{dy}{ds} \tag{13.246}$$

Consequently, (13.245) can be expressed as

$$R = 2\pi f\int_{y_1}^{y_2} y(\frac{dy}{ds})^2 dy \tag{13.247}$$

Note that the curved segment is given by

$$(ds)^2 = (dx)^2 + (dy)^2 \tag{13.248}$$

This can be rewritten as

$$(\frac{dy}{ds})^2 = \frac{1}{1+(\frac{dx}{dy})^2} \tag{13.249}$$

Substitution of (13.249) into (13.247) gives

$$R = 2\pi f\int_{y_1}^{y_2} \frac{y}{1+(\frac{dx}{dy})^2} dy \tag{13.250}$$

By noting

$$dy = \frac{dy}{dx}dx = y'dx \tag{13.251}$$

we can recast (13.250) as

$$R = 2\pi f\int_{x_1}^{x_2} \frac{yy'^3}{1+y'^2} dx \tag{13.252}$$

Another form of (13.250) can be formulated by assuming a parameter t such that

$$x = x(t), \quad y = y(t) \tag{13.253}$$

With this parametric form, we have

$$dx = \frac{dx}{dt}dt = \dot{x}dt, \quad dy = \frac{dy}{dt}dt = \dot{y}dt, \quad \frac{dx}{dy} = \frac{\dot{x}}{\dot{y}} \tag{13.254}$$

Substitution of (13.254) into (13.252) gives

$$R = 2\pi f\int_{t_1}^{t_2} \frac{y\dot{y}^3}{\dot{x}^2+\dot{y}^2} dt \tag{13.255}$$

Note that all (13.250), (13.252) and (13.255) are equivalent and all of them have been adopted in the literature. The next question is to minimize the resistance R, and we are going to see that finding its solution is not straightforward.

13.14.3 Newton's Resistance Law for Supersonic Flow on Solid

Although Newton was thinking about ship design when he proposed the impact theory discussed in the last section, it was subsequently found applicable to consider the minimal drag on bodies of revolution at high supersonic speed when a shock wave was formed on the surface of the bodies as shown in Figure 13.16.

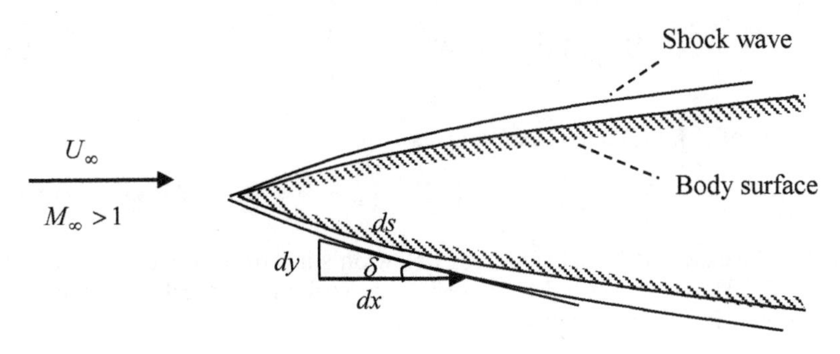

Figure 13.16 Resistance on supersonic flow

According to Eggers et al. (1953), the pressure coefficient is defined in terms of δ as:

$$C_p = \frac{p - p_\infty}{q_\infty} = 2\sin^2\delta \tag{13.256}$$

where p and q are the static and dynamic pressure and subscript "∞" implies values at infinity or far from the body of revolution. The angle δ is the angle in the meridian plane between the free stream and the tangent to the body surface. When the curvature of the body is small in the stream direction, the hypersonic layer will be thin. Subsequently, (13.256) can also be used to estimate the pressure coefficient, and thus the pressure drag, on the surface of the body. This formula is generally acceptable for the case that the hypersonic similarity parameter K is greater than one:

$$K = M_\infty \frac{d}{l} > 1 \tag{13.257}$$

where d and l are the diameter and length of the body, and M_∞ is the Mach number at far field. When the pressure coefficient over the body is known, by neglecting the base drag at the far end, the pressure drag of a body can be integrated as:

$$D = \frac{C_D q_\infty \pi d^2}{4} = 2\pi q_\infty \int_0^{d/2} C_p y dy = 2\pi q_\infty \int_0^{l} C_p y y' dx \tag{13.258}$$

To simplify the presentation, we can also define a drag parameter as

$$I_D = \frac{D}{2\pi q_\infty} = \int_0^l C_p yy' dx \tag{13.259}$$

Substitution of (13.256) into (13.259) gives

$$I_D = 2\int_0^l \sin^2 \delta yy' dx \tag{13.260}$$

To find $\sin\delta$, we note from Figure 13.16 that

$$(ds)^2 = (dx)^2 + (dy)^2, \quad \sin\delta = \frac{dy}{ds} \tag{13.261}$$

Combining these results, we get

$$\sin^2 \delta = \frac{(\frac{dy}{dx})^2}{1+(\frac{dy}{dx})^2} = \frac{y'^2}{1+y'^2} \tag{13.262}$$

Substitution of (13.262) into (13.260) gives

$$I_D = \int_0^l \frac{2yy'^3}{1+y'^2} dx \tag{13.263}$$

This is of the exact mathematical form of Newton's law of resistance that we found in (13.252). Thus, Newton's law of resistance is found applicable to hypersonic flow, and thus it is useful for missile shape design.

13.14.4 Eggers et al. (1953) Parameter Solution

Eggers et al. (1953) proposed that the drag parameter be modified to allow for any finite region of flat nose of radius y_1 and of infinite slope at the front. Thus, (13.263) can be modified as

$$I_D = y_1^2 + \int_0^l \frac{2yy'^3}{1+y'^2} dx \tag{13.264}$$

Eggers et al. (1953) proposed to consider three different cases of constraint conditions. We will, however, restrict our discussion to the case of a given length and base diameter (case (a) in their report). In particular, we have

$$F = \frac{2yy'^3}{1+y'^2} \tag{13.265}$$

Since the Lagrangian is not a function of variable x, the Euler-Lagrange equation is reduced to the Beltrami identity as:

$$y'\frac{\partial F}{\partial y'} - F = C_1 \tag{13.266}$$

Differentiation of (13.265) gives

$$\frac{\partial F}{\partial y'} = \frac{6yy'^3}{1+y'^2} - \frac{4yy'^4}{(1+y'^2)^2} = \frac{2}{(1+y'^2)^2}\{3yy'^2 + yy'^4\}$$
$$= \frac{2yy'^2}{(1+y'^2)^2}\{3+y'^2\} \tag{13.267}$$

Substitution of (13.267) into (13.266) leads to

$$\frac{4yy'^3}{(1+y'^2)^2} = C_1 \tag{13.268}$$

Note that we have allowed a non-pointed tip at the nose (such that $y(0) = y_1$), and the end points y_1 of the minimizing curve are not fixed yet. Thus, we have to impose another condition at the terminal points. Following from Courant and Hilbert, Eggers at al. (1953) imposed the following condition at $y = y_1$:

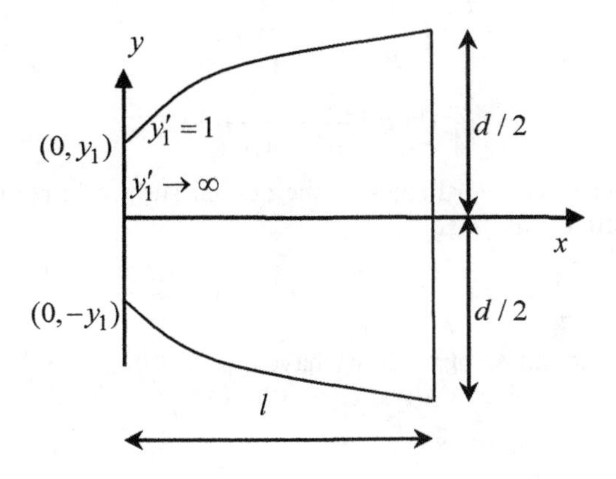

Figure 13.17 The enlargement of the nose condition of the body of revolution

$$\left(\frac{\partial F}{\partial y'} - \frac{d}{dy}y^2\right)_{y=y_1} = 2y\frac{y'^2 - 1}{(1+y'^2)^2}\bigg|_{y=y_1} = 0 \tag{13.269}$$

Thus, we find

$$y'(0) = y_1' = 1 \tag{13.270}$$

The system of (13.268) and (13.270) can be solved in parametric form. In particular, we can rewrite (12.268) as

$$y = \frac{C_1}{4}\frac{(1+y'^2)^2}{y'^3} \tag{13.271}$$

Let us rewrite the first derivative of y as p

$$y = \frac{C_1}{4}\frac{(1+p^2)^2}{p^3}, \quad p = \frac{dy}{dx} = y' \tag{13.272}$$

Taking the total differential on both sides, we find

$$dy = \frac{C_1}{4}\frac{(1+p^2)}{p^2}\{4 - \frac{3(1+p^2)}{p^2}\}dp$$
$$= \frac{C_1}{4}\frac{(1+p^2)(p^2-3)}{p^4}dp \tag{13.273}$$

Substitution of (13.273) into the second equation of (13.272) gives

$$dx = \frac{1}{p}dy = \frac{C_1}{4}\frac{(1+p^2)(p^2-3)}{p^5}dp \tag{13.274}$$

Integration of both sides leads to

$$x = \frac{C_1}{4}\int \frac{(1+p^2)(p^2-3)}{p^5}dp + C_2$$
$$= \frac{C_1}{4}\int (\frac{1}{p} - \frac{2}{p^3} - \frac{3}{p^5})dp + C_2 \tag{13.275}$$
$$= \frac{C_1}{4}(\ln p + \frac{1}{p^2} + \frac{3}{4p^4}) + C_2$$

In summary, we have the coordinates of the curved surface in parametric form in terms of the derivative p as:

$$x = \frac{C_1}{4}(\ln p + \frac{1}{p^2} + \frac{3}{4p^4}) + C_2, \quad y = \frac{C_1}{4}\frac{(1+p^2)^2}{p^3} \tag{13.276}$$

From the boundary condition at $x = 0$, we have

$$y(0) = y_1, \quad p(0) = 1 \tag{13.277}$$

Using these boundary conditions, we have

$$C_1 = y_1, \quad C_2 = -\frac{y_1}{4}(\frac{7}{4}) \tag{13.278}$$

Substitution of (13.278) into (13.276) finally yields

$$x = \frac{y_1}{4}(\ln p + \frac{1}{p^2} + \frac{3}{4p^4} - \frac{7}{4}), \quad y = \frac{y_1}{4}\frac{(1+p^2)^2}{p^3} \tag{13.279}$$

By referring to (13.270), we can start with $p = 1$ for $x = 0$. Figure 13.17 suggests that we should decrease p (less than 1) as both x and y increase.

In order to give a plot of the curved surface on the meridian plane, we plot on Figure 13.18 the profile that gives the minimal resistance based on Newton's impact theory for least resistance for the case of $l/d = 3$. In addition, in Figure 13.18, we also plot the following profile of a 3/4 power law:

$$y = \frac{d}{2}(\frac{x}{l})^{3/4} \tag{13.280}$$

Numerical results show that the solution given by the least resistance of (13.279) (solid line) is virtually the same as that of (13.280) (dashed line). We will demonstrate in the next section that (13.280) is indeed an approximate solution of the minimal resistance problem.

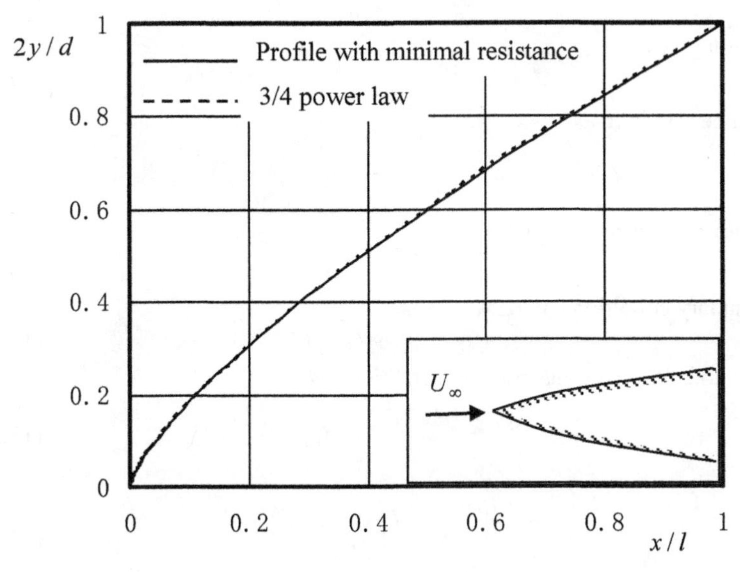

Figure 13.18 Profile of the surface with minimal resistance and of 3/4 power law for $l/d = 3$

13.14.5 Approximation by 3/4 Power Law

Let us consider the following approximation of (13.263) that y' is small compared to unity (i.e., $y' \ll 1$)

$$I_D = \int_0^l \frac{2yy'^3}{1+y'^2} dx \approx 2\int_0^l yy'^3 dx \tag{13.281}$$

That is, we will only minimize the numerator of (13.263). For such approximation, we have

$$F \approx yy'^3 \tag{13.282}$$

Using this F, we get

$$\frac{\partial F}{\partial y'} = 3yy'^2, \quad \frac{\partial F}{\partial y} = y'^3 \tag{13.283}$$

Differentiation of the first term with respect to x gives

$$\frac{d}{dx}\frac{\partial F}{\partial y'} = \frac{d}{dx} 3yy'^2 = 3y' + 6yy'y'' \tag{13.284}$$

Substitution of (13.283) and (13.284) into the Euler-Lagrange equation given in (13.15) gives

$$\frac{\partial F}{\partial y} - \frac{d}{dx}\frac{\partial F}{\partial y'} = -2(y'^3 + 3yy'y'') = 0 \tag{13.285}$$

To further integrate (13.285), we multiply it by y' to get

$$y'^4 + 3yy'^2 y'' = \frac{d}{dx}(yy'^3) = 0 \tag{13.286}$$

Integration gives

$$yy'^3 = C_1^3 \tag{13.287}$$

This can be rewritten as

$$\frac{dy}{dx} = \frac{1}{y^{1/3}} C_1 \tag{13.288}$$

Integration one more time yields

$$y = (C_3 x + C_4)^{3/4} \tag{13.289}$$

The boundary condition can be written as

$$y(0) = 0, \quad y(l) = d/2 \tag{13.290}$$

These boundary conditions give the conditions as:

$$C_3 = \frac{(d/2)^{4/3}}{l}, \quad C_4 = 0 \tag{13.291}$$

Finally, we get

$$y = \frac{d}{2}(\frac{x}{l})^{3/4} \tag{13.292}$$

This gives the approximation given in (13.280) and this gives an excellent approximation for the optimum shape with minimal resistance, as shown in Figure 13.18. As reported by Eggers et al. (1953), experiments conducted in the supersonic wind tunnel at Ames Aeronautical Laboratory for objects satisfying the power law with $n = 1$ (cone), 3/4, 1/2, 1/4, and an Ogive (an object with a roundly tapered end as shown in Figure 13.19) show that the shape with $n = 3/4$ indeed gives the minimal resistance. The experiments were conducted for Mach numbers ranging from 2.73 to 6.28. Figure 13.19 shows the experimental results obtained by Eggers et al. (1953). Thus, Newton's impact theory for minimal resistance is verified for a body of revolution in supersonic flow.

13.15 SUMMARY AND FURTHER READING

The calculus of variations is a classical mathematical topic in applied mathematics and engineering. The historical development of the calculus of variations has been reported in excellent detail by Bliss (1925, 1930) and by Goldstine (1980). The book by Goldstine (1980) covered many technical details of its development. Readers are referred to them for more discussion of the calculus of variations. There are many good books on the calculus of variations, including Forsyth (1960), Goldstine (1980), and Weinstock (1974), just to name a few. Simmons and Krantz (2007) provided a very good elementary introduction to the calculus of variations. The history of isoperimetric problems was covered by Fraser (1992).

In the 1990s, Newton's minimal resistance problems were revisited by many researchers for the cases in which the body of revolution is non-symmetric and concave, which allow multiple impacts of the particles. However, this new analysis depends heavily on differential geometry which is out of the scope of the present chapter.

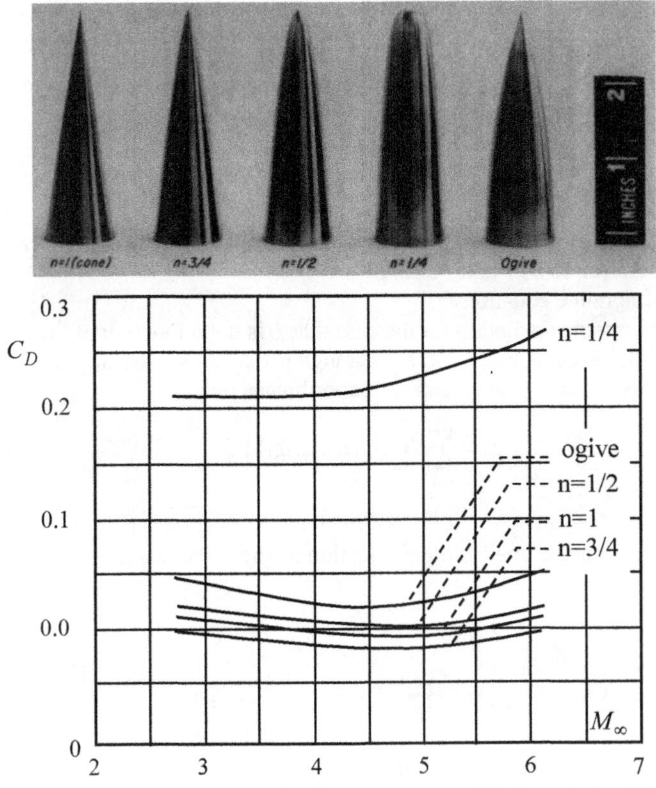

Figure 13.19 Pressure foredrag C_D for various shapes of solid of revolution (adopted from Eggers et al., 1953).

13.16 PROBLEMS

Problem 13.1 Find the function $y(x)$ that minimizes the following functional:

$$I(y(x)) = \int_a^b (y'^2 - y^2)dx \tag{13.293}$$

Ans:

$$y(x) = C_1 \sin x + C_1 \cos x \tag{13.294}$$

Problem 13.2 Find the function $y(x)$ that minimizes the following functional:

$$I(y(x)) = \int_a^b x(y'^2 - y^2)dx \tag{13.295}$$

Ans:

$$y(x) = C_1 J_0(x) + C_1 Y_0(x) \tag{13.296}$$

Problem 13.3 Hamilton's principle is the most general fundamental principle for analyzing rigid-body mechanics. It can be formulated using the calculus of variations. More specifically, there are n generalized coordinates in the Lagrangian defined in the functional:

$$I(y(t)) = \int_a^b L(t, q_1(t), ..., q_n(t), \dot{q}_1(t), ..., \dot{q}_n(t))dt \tag{13.297}$$

where

$$L(t, q_1(t), ..., q_n(t), \dot{q}_1(t), ..., \dot{q}_n(t)) = T - V \tag{13.298}$$

The kinetic energy and potential energy of the system of n generalized coordinates are normally referred to as T and V. Find

(i) The Euler-Lagrange equation

(ii) The Euler-Lagrange equation for the case that L is not a function of time

(iii) Consider the special case of the Lagrangian that T is quadratic in generalized velocity and V is a function of generalized coordinates only

$$T = \sum_{i=1}^{n}\sum_{j=1}^{n} a_{ij}(q_1, ..., q_n)\dot{q}_i\dot{q}_j \tag{13.299}$$

$$V = V(q_1, ..., q_n) \tag{13.300}$$

Show that the constant in the Beltrami equation in (ii) is the negative sign of the total energy.

Ans:

(i)
$$\frac{\partial L}{\partial q_i} - \frac{d}{dt}[\frac{\partial L}{\partial \dot{q}_i}] = 0, \quad i = 1, 2, ..., n \tag{13.301}$$

(ii)
$$L - \sum_{i=1}^{n} \dot{q}_i \frac{\partial L}{\partial \dot{q}_i} = C \tag{13.302}$$

(iii)
$$C = -(T + V) \tag{13.303}$$

Problem 13.4 Find the Euler-Lagrange equation for a particle in a conservative force field (e.g., gravitational field) by considering the minimum of the functional

$$I(y(x)) = \int_a^b \{\frac{1}{2}(m\dot{x}_1^2 + m\dot{x}_2^2 + m\dot{x}_3^2) - V(x_1, x_2, x_3)\}dt \tag{13.304}$$

Ans:

$$\frac{d}{dt}(m\dot{\boldsymbol{r}}) = -\nabla V \text{ where } \boldsymbol{r} = x_i \boldsymbol{e}_i \tag{13.305}$$

Problem 13.5 Find the Euler-Lagrange equation for the following functional

$$I(y(x)) = \iint \frac{1}{2}(u_x^2 + u_y^2)dxdy \tag{13.306}$$

Ans:

$$\nabla^2 u = 0 \tag{13.307}$$

Problem 13.6 Reconsider the shortest curve on the sphere by using polar coordinates:

$$x = R\sin\theta\cos\phi, \quad y = R\sin\theta\sin\phi, \quad z = R\cos\theta \tag{13.308}$$

(i) Show that the corresponding functional in polar form is

$$I(y(x)) = R\int [1+\sin^2\theta(\frac{d\phi}{d\theta})^2]^{1/2} d\theta \tag{13.309}$$

(ii) Show that the shortest curve is on a great circle.

Problem 13.7 Find the Euler-Lagrange equation for the following functional for beam bending

$$I(w(x)) = \int_0^L [\frac{EI}{2}(\frac{d^2w}{dx^2})^2 - qw]dx \tag{13.310}$$

Ans:

$$EI\frac{d^4w}{dx^4} = q(x) \tag{13.311}$$

Problem 13.8 Find the Euler-Lagrange equation for the following functional for a circular plate bending under axisymmetric loading $q(r)$ and fixed at the edges:

$$I(w(x)) = D\pi\int_0^a [r(\frac{d^2w}{dr^2})^2 + \frac{1}{r}(\frac{dw}{dr})^2 + 2\nu\frac{dw}{dr}\frac{d^2w}{dr^2} - \frac{2q}{D}rw]dr \tag{13.312}$$

Ans:

$$r\frac{d^4w}{dr^4} + 2\frac{d^3w}{dr^3} - \frac{1}{r}\frac{d^2w}{dr^2} + \frac{1}{r^2}\frac{dw}{dr} = \frac{qr}{D} \tag{13.313}$$

Problem 13.9 Find the geodesics on a right circular cone of semi-vertical angle α (see Figure 13.20). It is given that the differential of an arc *ds* on a right circular cone is given by

$$(ds)^2 = (dr)^2 + (rd\theta)^2 + (r\sin\alpha d\phi)^2 \tag{13.314}$$

If the vertex of the cone is at the origin and the *z*-axis is the axis of the cone, the polar equation of the cone is

$$\theta = \alpha \tag{13.315}$$

The functional is the arc length defined by

$$s = \int_{\phi_1}^{\phi_2} \sqrt{(\frac{dr}{d\phi})^2 + r^2\sin^2\alpha}\, d\phi \tag{13.316}$$

Ans:

$$\frac{C_1}{r\sin\alpha} = \cos(\phi\sin\alpha + C_2) \tag{13.317}$$

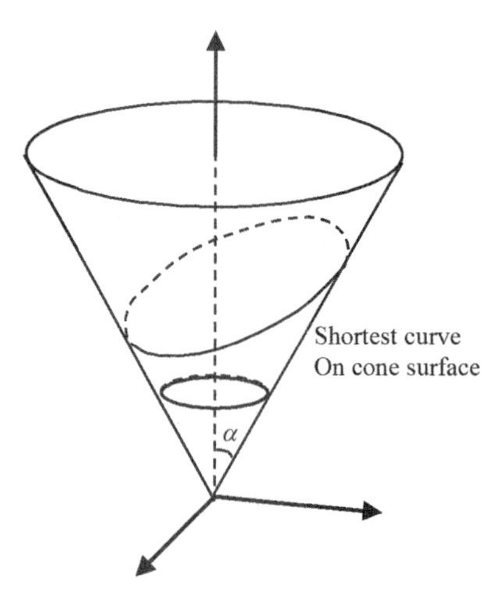

Figure 13.20 Shortest curve on cone between two points

Problem 13.10 In the text, we have considered the catenoid problem, a case of minimal surface for soap film. The catenoid problem is actually a one-dimensional problem because of the surface of revolution (see (13.47)). In this problem, we will consider the problem of minimal surface of soap film formed in a closed wire loop defined by $z = u(x,y)$. In particular, the functional of minimal surface is

$$I = \iint_D \sqrt{1+u_x^2+u_y^2}\, dxdy \tag{13.318}$$

Find the following:

(i) Referring to (13.41) for a two-variable case, show that the Euler-Lagrange equation can be written as

$$\frac{\partial}{\partial x}\left(\frac{u_x}{F}\right)+\frac{\partial}{\partial x}\left(\frac{u_y}{F}\right)=0 \tag{13.319}$$

where F is defined as

$$F = \sqrt{1+u_x^2+u_y^2} \tag{13.320}$$

(ii) Show that if the first derivative of u is small, the Euler-Lagrange equation becomes the Laplace equation (i.e., membrane equilibrium of small deflection is governed by the Laplace equation).

(iii) Show that the Euler-Lagrange equation for the minimal surface can also be written as:

$$(1+u_y^2)u_{xx}+(1+u_x^2)u_{yy}-2u_x u_y u_{xy}=0 \tag{13.321}$$

(Note that this PDE is extremely difficult to solve.)

Problem 13.11 Continued from Problem 13.10, if the domain D is given by a rectangle of size a and b. Show that a helicoid defined by

$$u(x, y) = \tan^{-1}(\frac{y}{x}) \tag{13.322}$$

is a solution for the minimal surface in a rectangular domain.

Problem 13.12 Fermat's principle of least time is a powerful tool in studying geometric optics. The least time can be formulated as functional and the path of light can be solved by using the resulting Euler-Lagrange equation. Figure 13.21 shows the formation of a mirage in desert areas. The travel time by light path is

$$T = \int dt = \int \frac{dl}{v} = \frac{1}{c}\int ndl = \frac{1}{c}\int f(y)dl \tag{13.323}$$

where v is the velocity of light in air and c is the speed of light in a vacuum and n is the index of refraction. In desert areas, the ground surface is much hotter than the air above the ground. The refraction index approximately increases with height, as shown in Figure 13.21. This is the mirage phenomenon that occurs in desert areas. The functional in terms of the path of travel can be set as

$$I = \int f(y)\sqrt{1+x'^2}\,dy \tag{13.324}$$

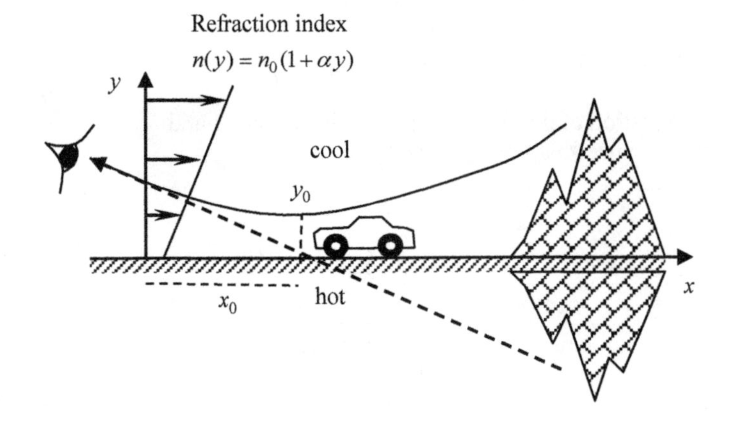

Figure 13.21 Mirage formation and the path of the light with least travel time

(i) Show that Euler-Lagrange equation leads to

$$\frac{dx}{dy} = \frac{C}{\sqrt{f^2(y) - C^2}} \tag{13.325}$$

(ii) Show the solution for y is

$$y = -\frac{1}{\alpha} + \frac{C}{\alpha n_0}\cosh[\frac{\alpha n_0}{C}(x - x_0)] \tag{13.326}$$

where x_0 is, as shown in Figure 13.17, the center of symmetry.

(iii) Find the unknown constant in terms of y_0 and show that the final solution is

$$y = -\frac{1}{\alpha} + \frac{(\alpha y_0 + 1)}{\alpha} \cosh[\frac{\alpha(x - x_0)}{\alpha y_0 + 1}] \tag{13.327}$$

(iv) Show that for small α we can approximate the hyperbolic solution given in (13.327) as a parabola:

$$y = y_0 + \frac{\alpha(x - x_0)^2}{2(\alpha y_0 + 1)} \tag{13.328}$$

Problem 13.13 Find the Euler-Lagrange equations for the following functional:

$$\begin{aligned} I_1 &= \int_a^b [F(t, x, \dot{x}, y, \dot{y}) + \lambda G(t, x, \dot{x}, y, \dot{y})] dt \\ &= \int_a^b \Lambda(t, x, \dot{x}, y, \dot{y}) dt \end{aligned} \tag{13.329}$$

Ans:

$$\frac{\partial \Lambda}{\partial x} - \frac{d}{dt}(\frac{\partial \Lambda}{\partial \dot{x}}) = 0 \tag{13.330}$$

$$\frac{\partial \Lambda}{\partial y} - \frac{d}{dt}(\frac{\partial \Lambda}{\partial \dot{y}}) = 0 \tag{13.331}$$

Problem 13.14 Continue the calculation of Example 13.4.

(i) Find the equation of the shortest path joining points *A* and *B*;
(ii) Find the distance *s* between points *A* and *B*.

Ans:

$$z = (\frac{z_2 - z_1}{\theta_2 - \theta_1})\theta + \frac{\theta_2 z_1 - \theta_1 z_2}{\theta_2 - \theta_1} \tag{13.332}$$

$$s = \sqrt{a^2(\theta_2 - \theta_1)^2 + (z_2 - z_1)^2} \tag{13.333}$$

CHAPTER FOURTEEN

Variational and Related Methods

14.1 INTRODUCTION

The idea of using the variational principle (or the minimum energy hypothesis) can be traced back to Archimedes, Aristotle, and Galileo. Variational methods have been developed since the time of the development of the calculus of variations. The calculus of variations developed mainly by Euler and Lagrange aims to maximize or minimize some functionals, such as Dido's problem of maximizing area, brachistochrone (path of quickest descent), tautochrone (optimum curve for sliding bead on wire), geodesics (shortest path on surface), and catenoid problems (minimum area of revolution of soap films). The reason why nature chooses to minimize or maximize certain quantities has puzzled the greatest philosophers, scientists, and mathematicians. In view of this, Fermat formulated his principle of least time for optics, Maupertuis formulated the principle of least action, and Hamilton formulated his principle (the path of any motion minimizes the integral of the difference between kinetic and potential energies over interval of time). In 1760 and 1761, Lagrange developed the principle of virtual work and the Lagrange multiplier in the context of variational mechanics. The principle of least action had been discussed by Euler in 1744 for the case of column buckling, and by Leibniz in 1705, although credit is normally given to Maupertuis. All these principles can be considered as some kind of variational principle. Hamilton's principle is considered the most general of all, linking all phenomena in mechanics, optics, gravitation, electricity and magnetism, and quantum mechanics by a single integral (Kline, 1959). The development of the functional includes Legendre's work on distinguishing maxima and minima, and Jacobi's and Weierstrass's work on existence of extrema in a functional.

Our discussion in this chapter focuses on minimization or maximization of functionals and their relation and application to solving problems of differential equations. In the last century, variational methods have been associated with the formulation of differential equations for physical phenomena, and associated with approximation methods such as the finite element method. In solid mechanics, the Veubeke-Hu-Washizu (VHW) principle and Hellinger-Reissner (HR) principle are of fundamental importance in finite element formulation. Both of these are important variational principles in solid mechanics. The VHW variational principle was apparently formulated independently by Veubeke in 1951, by Hu in 1955, and by Washizu in 1955. Although both Veubeke and Washizu were visitors at MIT (Massachusetts Institute of Technology) in 1952, they never cited one another. The HR variational principle was studied by Hellinger in 1914 and by Reissner in 1950. The HR variational principle can be considered as a special case of the VHW principle. Interestingly, Reissner was a professor at MIT when both Veubeke and Washizu were visiting. The starting functional normally possesses the physical meaning of some kind of energies. Alternatively, numerical methods can also be formulated in terms of integrals with the kernel being the product of the differential equation and some weighting functions, which is somewhat arbitrary as long as they are admissible kinetically. This method is known as the weighted residue method, which is closely related to the variational principle. The most notable weighted residue methods is the Galerkin method, whereas the most notable variational methods is the Rayleigh-Ritz method. Incidentally, both Galerkin and Ritz aimed to provide approximate solutions of problems of plate bending in 1915 and 1908 respectively. Both beam and plate bending problems will be considered as examples.

This chapter introduces the fundamental concepts and ideas of the variational principle and its associated methods. More references will be given in the summary section for more serious readers.

14.2 VIRTUAL WORK PRINCIPLE

The idea of virtual work can date back to the time of Bernoulli, and it is based on the conservation of energy. Implicitly, work done by frictional forces or other irreversible processes are neglected. Let us consider the equilibrium of a body V and the corresponding prescribed boundary conditions ($S = S_\sigma + S_u$):

$$\nabla \cdot \sigma + f = 0, \quad \text{in} \quad V \tag{14.1}$$

$$\sigma \cdot \boldsymbol{n} = \overline{\boldsymbol{T}}, \quad \text{on } S_\sigma, \quad \boldsymbol{u} = \overline{\boldsymbol{u}}, \quad \text{on } S_u \tag{14.2}$$

where $\boldsymbol{\sigma}$, f, $\boldsymbol{T}$ and $\boldsymbol{n}$ are stress tensor, body force vector, traction vector, and unit normal vector to the surface of the body S. The superimposed bar are used to denote those given traction and displacement on the corresponding boundaries. The traction and displacement boundaries are denoted by S_σ and S_u respectively.

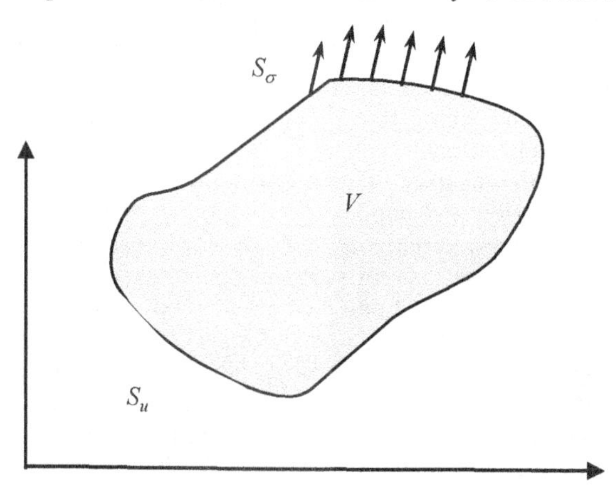

Figure 14.1 A deformation body with volume V and with boundaries S_σ and S_u

For two unrelated states of a deformable body, one is an arbitrary admissible stress $\boldsymbol{\sigma}$ that satisfies equilibrium in V and traction boundary conditions on S_σ, and the other is an arbitrary admissible displacement $\boldsymbol{u}^*$ that satisfies a compatibility condition in V and displacement boundary conditions on S_u. Note that the asterisk for the displacement field is used to emphasize that the stress (without asterisk) and displacement fields (with asterisk) are unrelated. These two states are unrelated and not necessary real (this is the reason why we call them virtual in the first place). We said that virtual work is done, if the admissible stress state has undergone an unrelated admissible displacement state. In this sense the work done is not necessarily real (or virtual). The external virtual work done by an external force (body force $\boldsymbol{f}$ and applied traction and displacement) on the body undergoing an admissible displacement field $\boldsymbol{u}$ is

$$\Pi_e = \int_V \boldsymbol{f} \cdot \boldsymbol{u}^* dV + \int_{S_\sigma} \overline{\boldsymbol{T}} \cdot \boldsymbol{u}^* dS + \int_{S_u} \boldsymbol{T} \cdot \overline{\boldsymbol{u}} dS \tag{14.3}$$

The internal virtual work is the work done by the admissible stress, which is in equilibrium with the external applied force and traction, on the unrelated associated strain (i.e., resulting from $\boldsymbol{u}^*$):

$$\Pi_i = \int_V \sigma : \varepsilon^* dV \tag{14.4}$$

The principle of virtual work says that the external virtual work must be equal to the internal virtual work when the equilibrated forces and stresses undergo unrelated but consistent displacement and strain:

$$\int_V \sigma : \varepsilon^* dV = \int_V \boldsymbol{f} \cdot \boldsymbol{u}^* dV + \int_{S_\sigma} \overline{\boldsymbol{T}} \cdot \boldsymbol{u}^* dS + \int_{S_u} \boldsymbol{T} \cdot \overline{\boldsymbol{u}} dS \tag{14.5}$$

If the body is rigid, the internal virtual work will be zero. In the process of analysis, we never impose any constitutive response of the deformable body, and thus, the principle is valid regardless of the material behavior. The principle can also be applied to consider the case of large deformation as long as the appropriate stresses are used in the energy calculation in (14.5). The principle can also be extended to dynamic problems, if the inertia force is interpreted as body forces in (14.5). For the case of a dynamic system of rigid bodies, it is equivalent to the Lagrange-D'Alembert principle for dynamics.

In applying (14.5), we may impose equilibrium of the "real" stresses and forces that are in equilibrium on a consistent virtual displacement and strain in the formulation. This will lead to a special case called the *principle of virtual displacement*, which will be discussed next. On the other hand, we may impose a consistent real displacement and strain on self-equilibrated virtual stresses and forces in the virtual work equation. This will lead to a special case called the *principle of virtual traction.* The principle of virtual work occupies a major place in the area of structural analysis, solid mechanics, and finite element analysis.

14.3 VIRTUAL DISPLACEMENT PRINCIPLE

If we impose self-equilibrated "real" stresses and forces on consistent virtual displacement and strain in (14.5), we have

$$\boldsymbol{u}^* = \delta \boldsymbol{u}, \quad \boldsymbol{\varepsilon}^* = \delta \boldsymbol{\varepsilon} \tag{14.6}$$

This virtual displacement must be zero on the displacement boundary S_u. Applying this notation and constraint on the virtual displacement, we get

$$\int_V \sigma : \delta \boldsymbol{\varepsilon} dV = \int_V \boldsymbol{f} \cdot \delta \boldsymbol{u} dV + \int_{S_\sigma} \overline{\boldsymbol{T}} \cdot \delta \boldsymbol{u} dS \tag{14.7}$$

This is called the *principle of virtual displacement*. When the body is elastic, (14.7) is related to Castigliano's first theorem. In other words, if a body is in equilibrium, the total virtual work done is zero. In structural mechanics, the principle of virtual displacement is normally used to find the real force or stress of a body.

14.4 VIRTUAL TRACTION PRINCIPLE

If we impose consistent "real" displacement and strain on equilibrated virtual stresses and forces in the virtual work principle, we have

$$\boldsymbol{\sigma} = \delta \boldsymbol{\sigma}, \quad \boldsymbol{T} = \delta \boldsymbol{T}, \quad \boldsymbol{f} = \delta \boldsymbol{f} \tag{14.8}$$

The virtual force is zero on the traction boundary S_σ.

$$\int_V \delta \boldsymbol{\sigma} : \varepsilon^* dV = \int_V \delta \boldsymbol{f} \cdot \boldsymbol{u}^* dV + \int_{S_u} \delta \boldsymbol{T} \cdot \overline{\boldsymbol{u}} dS \tag{14.9}$$

This will lead to a special case called the *principle of virtual traction.* This is also called the *principle of complementary virtual work.* When the body is elastic, (14.9) is related to Castigliano's second theorem. The principle of virtual force is normally used to find the real displacement of a body.

The principle of virtual work will be illustrated by example in a later section.

14.5 HAMILTON'S PRINCIPLE

As we remarked in the introduction, the Hamilton principle is very general in the sense that it unifies optics, mechanics, electricity, and magnetism by a single minimum principle. The following integral is defined in terms of kinetic energy T and potential energy V as:

$$I = \int_{t_0}^{t_1} (T - V) dt \tag{14.10}$$

The principle states that I is at minimum for the path traversed by an object during its motion (including the path of light or other electromagnetic waves!) from time t_0 to time t_1. For deformable solid, it can be rewritten as:

$$H = \int_{t_0}^{t_1} (U(\varepsilon_{ij}) + T - W) dt \tag{14.11}$$

where W is the work done by external forces and U is the strain energy density. For static problems, the total potential energy for a deformable body can be defined as

$$\Pi_p = \int_V \{U(\varepsilon_{ij}) - \bar{f}_i u_i\} dV - \int_{S_\sigma} \bar{T}_i u_i dS \tag{14.12}$$

On the other hand, the total complementary energy for a deforming body is defined as

$$\Pi_c = \int_V \{U_c(\sigma_{ij}) dV - \int_{S_u} T_i \bar{u}_i dS \tag{14.13}$$

For solving equilibrium problems, we need to minimize the total potential energy or maximize the total complementary energy. These two functionals in terms of energies given in (14.12) and (1.13) are the basis for the variational principle in formulating differential equations of physical or mechanics problems. Hamilton's principle will lead to an equation of motions whereas either total potential energy or total complementary energy will lead to equations of equilibrium. As demonstrated in Figure 2.10 of Chau (2013), these two energy functions are complementary to one another in the following sense:

$$\Pi_c + \Pi_p = \sigma_{ij}\varepsilon_{ij} \tag{14.14}$$

When (14.12) or (14.13) is used in numerical analyses, such as the Rayleigh-Ritz method, Galerkin method, or finite element method, it is well known that both of them have their limitations. In particular, methods based on minimum total potential energy use displacement as the unknown and are more accurate for displacement prediction but less accurate for stress prediction (as they were calculated based on the numerically obtained displacement); whereas those based on maximum total complementary energy use stress as the unknown and are accurate for stress prediction but less accurate for displacement prediction. Trefftz in 1926 demonstrated that the upper bound of torsional rigidity could be found by using minimum potential energy, whereas the lower bound of torsional rigidity can be found by using maximum complementary energy. A similar idea was proposed by Prager and Synge in 1947 for determining the upper and lower bounds of elastic modulus by using the functional space concept. This method has been used successfully for finding the Young's modulus for both cylindrical and rectangular specimens of arbitrary shape under compression with end constraints (Chau, 1997, 1999a). In general, the minimum total potential energy formulation will lead to a lower bound for the displacement prediction (i.e., the numerical model appears to be stiffer than the actual system), whereas the maximum total complementary energy formulation will lead to an upper bound for the displacement prediction (i.e., the numerical model appears to be softer than the actual system). This general observation is illustrated in Figure 14.2 using the beam problem as an example.

In finite element methods, the first one is normally referred as the displacement method whereas the second one is normally referred as the force method. In view of these limitations for both formulations, more general variational principles have been proposed. They are normally referred to as mixed variational principles because both displacements and stresses are unknowns of the problems. In the next two sections, we will consider two generalized (or mixed) variational principles; they are the Veubeke-Hu-Washizu (VHW) principle and the Hellinger-Reissner (HR) principle, which are more powerful in numerical analysis. These principles are the basis for such mixed or hybrid finite element methods.

However, the finite element method is a topic that cannot be covered in the present or later chapter. We will only briefly summarize the idea of the variational principle and give a simple example of the case of incompressible flow in Chapter 15.

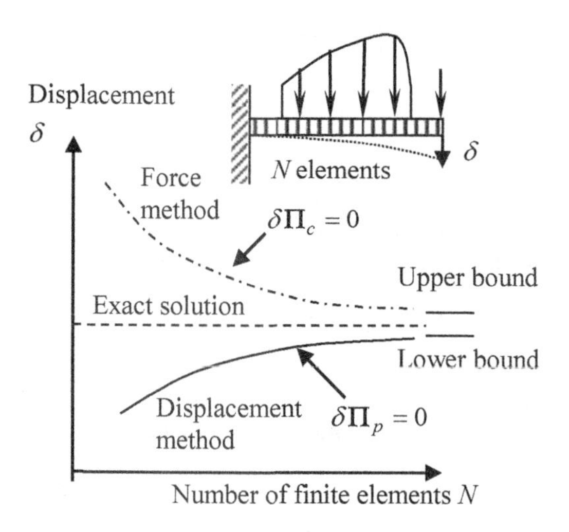

Figure 14.2 Illustration of the upper and lower displacements by force and displacement methods

14.6 VEUBEKE-HU-WASHIZU PRINCIPLE

In this section, we will consider a general variational principle called the Veubeke-Hu-Washizu principle. It is formulated for elastic bodies and aims for numerical methods, such as the finite element method. Let us consider the mathematical formulation for an elastic body V as:

$$\sigma_{ij,j} + \bar{f}_i = 0, \quad \text{in } V \tag{14.15}$$

$$\varepsilon_{ij} = \frac{1}{2}(u_{i,j} + u_{j,i}), \quad \text{in } V \tag{14.16}$$

$$\sigma_{ij} = a_{ijkl}\varepsilon_{kl}, \quad \text{or} \quad \varepsilon_{ij} = b_{ijkl}\sigma_{kl}, \quad \text{in } V \tag{14.17}$$

The first of these is an equilibrium equation, the second is a strain-displacement relation, and the third one is a constitutive law (or Hooke's law). The boundary conditions are:

$$u_i = \bar{u}_i, \quad \text{on } S_u \tag{14.18}$$

$$T_i = \sigma_{ij} n_j = \bar{T}_i, \quad \text{on } S_\sigma \tag{14.19}$$

The first of these is called an essential boundary condition (i.e., displacement boundary condition), and the second is called a natural boundary condition (i.e., traction boundary condition). The Veubeke-Hu-Washizu principle can be expressed as

$$\Pi_{VHW} = \int_V \{U(\varepsilon_{ij}) - \bar{f}_i u_i - \sigma_{ij}[\varepsilon_{ij} - \frac{1}{2}(u_{i,j} + u_{j,i})]\} dV - \int_{S_\sigma} \bar{T}_i u_i dS - \int_{S_u} T_i(u_i - \bar{u}_i) dS \tag{14.20}$$

where independent functions subject to variations are σ_{ij}, ε_{ij} and u_i and, that is, there are 15 unknowns (note that both stress and strain are symmetric). In terms of the finite element method, stress, strain, and displacement are also independent. That is, we have relaxed the relations given in (14.15) to (14.18). Since both stress and displacement are unknowns, they are called the mixed variational principle.

Comparing the VHW variational principle in (14.20) with the total potential energy in (14.12), we observe that there are two extra terms appearing in (14.20). The principle can be interpreted in the following form

$$\Pi_{VHW} = \int_V \{U(\varepsilon_{ij}) - \bar{f}_i u_i - \lambda_{ij}[\varepsilon_{ij} - \frac{1}{2}(u_{i,j} + u_{j,i})]\} dV - \int_{S_\sigma} \bar{T}_i u_i dS - \int_{S_u} p_i(u_i - \bar{u}_i) dS \tag{14.21}$$

where both λ_{ij} and p_i are Lagrange multipliers and they are

$$\lambda_{ij} = \sigma_{ij}, \quad p_i = T_i \tag{14.22}$$

We can see that by adding these two constraints we are actually relaxing the strain-displacement relation and essential boundary condition. We allow the flexibility that the approximation is sought such that all (14.15) to (14.19) are satisfied approximately in a global sense (integrating over the whole body and whole boundary). By applying the Gauss theorem, we note that

$$\begin{aligned}\int_V \{\frac{1}{2}\sigma_{ij}(u_{i,j} + u_{j,i})\} dV &= \int_V \sigma_{ij} u_{i,j} dV = -\int_V \sigma_{ij,j} u_i dV + \int_{S_\sigma + S_u} \sigma_{ij} n_j u_i dS \\ &= -\int_V \sigma_{ij,j} u_i dV + \int_{S_\sigma + S_u} T_i u_i dS\end{aligned} \tag{14.23}$$

Substitution of (14.23) into (14.20) gives

$$\Pi_{VHW} = \int_V \{U(\varepsilon_{ij}) - \sigma_{ij}\varepsilon_{ij} - (\sigma_{ij,j} + \bar{f}_i)u_i\} dV - \int_{S_\sigma} (T_i - \bar{T}_i) u_i dS + \int_{S_u} T_i \bar{u}_i dS \tag{14.24}$$

From (14.24), we see that the equilibrium equation will not be satisfied automatically by our approximate solution.

The variational principle requires the functional of the VHW formulation to be stationary or

$$\delta \Pi_{VHW} = 0 \tag{14.25}$$

where δ denotes the variation. Taking variations of (14.20), we get

$$\delta\Pi_{VHW} = \int_V \{\frac{\partial U(\varepsilon_{ij})}{\partial \varepsilon_{ij}}\delta\varepsilon_{ij} - \bar{f}_i\delta u_i - [\varepsilon_{ij} - \frac{1}{2}(u_{i,j} + u_{j,i})]\delta\sigma_{ij}\}dV$$

$$+\int_V \{\frac{1}{2}\sigma_{ij}(\delta u_{i,j} + \delta u_{j,i}) - \sigma_{ij}\delta\varepsilon_{ij}\}dV - \int_{S_\sigma} \bar{T}_i\delta u_i dS \tag{14.26}$$

$$-\int_{S_u} [(u_i - \bar{u}_i)n_j\delta\sigma_{ij} + \sigma_{ij}n_j\delta u_i]dS$$

By applying the Gauss theorem, we have

$$\int_V \sigma_{ij}\delta u_{i,j}dV = -\int_V \sigma_{ij,j}\delta u_i dV + \int_{S_\sigma + S_u} \sigma_{ij}n_j\delta u_i dS \tag{14.27}$$

Using this result, we can rewrite (14.26) as

$$\delta\Pi_{VHW} = \int_V \{(\frac{\partial U}{\partial \varepsilon_{ij}} - \sigma_{ij})\delta\varepsilon_{ij} - (\sigma_{ij,j} + \bar{f}_i)\delta u_i + [\frac{1}{2}(u_{i,j} + u_{j,i}) - \varepsilon_{ij}]\delta\sigma_{ij}\}dV$$

$$-\int_{S_\sigma} (\bar{T}_i - T_i)\delta u_i dS - \int_{S_u} (u_i - \bar{u}_i)n_j\delta\sigma_{ij}dS \tag{14.28}$$

In view of (14.25), we have the following Euler equations and boundary conditions:

$$\sigma_{ij} = \frac{\partial U}{\partial \varepsilon_{ij}}, \quad \text{in } V \tag{14.29}$$

$$\sigma_{ij,j} + \bar{f}_i = 0, \quad \text{in } V \tag{14.30}$$

$$\varepsilon_{ij} = \frac{1}{2}(u_{i,j} + u_{j,i}), \quad \text{in } V \tag{14.31}$$

$$T_i = \sigma_{ij}n_j = \bar{T}_i, \quad \text{on } S_\sigma \tag{14.32}$$

$$u_i = \bar{u}_i, \quad \text{on } S_u \tag{14.33}$$

Therefore, we do not assume any of these five equations are satisfied in the VHW variational principle.

In the next section, we will consider another mixed variational principle.

14.7 HELLINGER-REISSNER PRINCIPLE

Another variational principle in close relation to the VHW variational principle is called the Hellinger-Reissner (HR) principle. The functional of the HR principle is

$$\Pi_{HR} = \int_V \{U_c(\sigma_{ij}) - \frac{1}{2}\sigma_{ij}(u_{i,j} + u_{j,i}) + \bar{f}_i u_i\}dV - \int_{S_\sigma} \bar{T}_i u_i dS - \int_{S_u} T_i(u_i - \bar{u}_i)dS \tag{14.34}$$

where the independent functions subject to variations are σ_{ij}, u_i and T_i. Again the equation can be interpreted as total complementary energy by adding two constraints as:

$$\Pi_{HR} = \int_V \{U_c(\sigma_{ij}) - \frac{1}{2}\lambda_{ij}(u_{i,j}+u_{j,i}) + \bar{f}_i u_i\}dV - \int_{S_\sigma} \bar{T}_i u_i dS - \int_{S_u} p_i(u_i - \bar{u}_i)dS \quad (14.35)$$

where the Lagrange multipliers are defined as

$$\lambda_{ij} = \sigma_{ij}, \quad p_i = T_i \quad (14.36)$$

If we set $\lambda_{ij} = 0$ and $p_i = 0$, we recover the total complementary energy given in (14.13). Using the Gauss theorem, we obtain another form of the HR functional:

$$\Pi_{HR} = \int_V \{U_c(\sigma_{ij}) + (\sigma_{ij,j} + \bar{f}_i)u_i\}dV - \int_{S_\sigma} (T_i - \bar{T}_i)u_i dS - \int_{S_u} T_i \bar{u}_i dS \quad (14.37)$$

Let us consider the corresponding Euler equation by taking the variations of (14.37):

$$\delta\Pi_{HR} = 0 \quad (14.38)$$

The variation of (14.38) is

$$\begin{aligned} \delta\Pi_{HR} = &\int_V \{\frac{\partial U_c}{\partial \sigma_{ij}}\delta\sigma_{ij} + u_i\delta\sigma_{ij,j} + (\sigma_{ij,j} + \bar{f}_i)\delta u_i\}dV \\ &- \int_{S_\sigma} [u_i n_j \delta\sigma_{ij} + (\sigma_{ij}n_j - \bar{T}_i)\delta u_i dS - \int_{S_u} \bar{u}_i \delta\sigma_{ij} n_j dS \end{aligned} \quad (14.39)$$

By applying the Gauss theorem, we have

$$\int_V u_i \delta\sigma_{ij,j} dV = -\int_V u_{i,j}\delta\sigma_{ij} dV + \int_{S_\sigma + S_u} u_i \delta\sigma_{ij} n_j dS \quad (14.40)$$

Substitution of (14.40) into (14.39) gives

$$\begin{aligned} \delta\Pi_{HR} = &\int_V \{(\frac{\partial U_c}{\partial \sigma_{ij}} - u_{i,j})\delta\sigma_{ij} + (\sigma_{ij,j} + \bar{f}_i)\delta u_i\}dV \\ &- \int_{S_\sigma} (\sigma_{ij}n_j - \bar{T}_i)\delta u_i dS - \int_{S_u} (\bar{u}_i - u_i)n_j \delta\sigma_{ij} dS \end{aligned} \quad (14.41)$$

Since the variations can be arbitrary, we require the following equation be satisfied:

$$\varepsilon_{ij} = \frac{\partial U}{\partial \sigma_{ij}}, \quad \text{in } V \quad (14.42)$$

$$\sigma_{ij,j} + \bar{f}_i = 0, \quad \text{in } V \quad (14.43)$$

$$T_i = \sigma_{ij}n_j = \bar{T}_i, \quad \text{on } S_\sigma \quad (14.44)$$

$$u_i = \bar{u}_i, \quad \text{on } S_u \quad (14.45)$$

We see that there is one less equation compared to the case of the VHW variational principle discussed in the previous section. In this sense, we can see that VHW is more general.

We can substitute (14.14) into (14.34) and get the following identity:

$$\Pi_{HR} = -H_{VHW} \quad (14.46)$$

Therefore, the HR variational principle is consistent with the VHW variational principle. However, they are not the same since we need to impose one more condition (14.14) to link them.

14.8 RAYLEIGH-RITZ METHOD

In this section, we introduce a method developed by Rayleigh in 1877 and by Ritz in 1908. In this approach, a finite number of approximations are assumed:

$$u = u_0 + \sum_{r=1}^{n} a_r u_r \tag{14.47}$$

$$v = v_0 + \sum_{r=1}^{n} b_r v_r \tag{14.48}$$

$$w = w_0 + \sum_{r=1}^{n} c_r w_r \tag{14.49}$$

Note that we have used the physical components u, v, and w to represent tensor notation u_1, u_2 and u_3. However, these approximations must be selected in such a way that the essential boundary condition is satisfied identically by choosing

$$u_0 = \bar{u}, \quad v_0 = \bar{v}, \quad w_0 = \bar{w}, \quad \text{on } S_u \tag{14.50}$$

$$u_r = 0, \quad v_r = 0, \quad w_r = 0, \quad (r = 1,2,...,n) \text{ on } S_u \tag{14.51}$$

Note, however, we do not need to satisfy the traction boundary condition imposed by (14.19). Substitution of this approximation into (14.12) and let us express the total potential energy in terms of the unknown constants a_r, b_r, and c_r. Next we can set the variation to zero as

$$\delta \Pi_p = 0 \tag{14.52}$$

We use this condition to seek for the optimum values of the constants a_r, b_r, and c_r ($r = 1,2,...,n$). Therefore, the variables in the formulation are those constants given in (14.47) to (14.49). Thus, (14.52) implies

$$\frac{\partial \Pi_p}{\partial a_r} = 0, \quad \frac{\partial \Pi_p}{\partial b_r} = 0, \quad \frac{\partial \Pi_p}{\partial c_r} = 0 \tag{14.53}$$

We can see that we have exactly $3n$ equations from (14.53) for the $3n$ unknowns. Solving for theses constants, we find the approximations assumed in (14.47) to (14.49). This procedure is called the Rayleigh-Ritz method. Note that the Rayleigh-Ritz approach is only an approximate method. Alternatively, we can also substitute (14.47) to (14.49) into the virtual work principle with virtual displacements as

$$\delta u = \sum_{r=1}^{n} \delta a_r u_r, \quad \delta v = \sum_{r=1}^{n} \delta b_r v_r, \quad \delta w = \sum_{r=1}^{n} \delta c_r w_r \tag{14.54}$$

Equation (14.7) leads to the following $3n$ equations:

$$L_r = -\int_V \left(\frac{\partial \sigma_{xx}}{\partial x} + \frac{\partial \tau_{xy}}{\partial y} + \frac{\partial \tau_{zx}}{\partial z} + \bar{f}_x\right) u_r dV + \int_S (T_x - \bar{T}_x) u_r dS = 0 \tag{14.55}$$

$$M_r = -\int_V \left(\frac{\partial \tau_{xy}}{\partial x} + \frac{\partial \sigma_{yy}}{\partial y} + \frac{\partial \tau_{yz}}{\partial z} + \bar{f}_y\right) v_r dV + \int_S (T_y - \bar{T}_y) v_r dS = 0 \tag{14.56}$$

$$N_r = -\int_V \left(\frac{\partial \tau_{zx}}{\partial x} + \frac{\partial \tau_{yz}}{\partial y} + \frac{\partial \sigma_{zz}}{\partial z} + \bar{f}_z\right) w_r dV + \int_S (T_z - \bar{T}_z) w_r dS = 0 \tag{14.57}$$

where the stresses are

$$\sigma_{xx} = 2G\{\frac{\partial u_0}{\partial x} + \sum_{r=1}^{n} a_r \frac{\partial u_r}{\partial x} + \frac{\nu \varepsilon_v}{1-2\nu}\} \tag{14.58}$$

$$\sigma_{yy} = 2G\{\frac{\partial v_0}{\partial y} + \sum_{r=1}^{n} b_r \frac{\partial v_r}{\partial y} + \frac{\nu \varepsilon_v}{1-2\nu}\} \tag{14.59}$$

$$\sigma_{zz} = 2G\{\frac{\partial w_0}{\partial z} + \sum_{r=1}^{n} c_r \frac{\partial w_r}{\partial z} + \frac{\nu \varepsilon_v}{1-2\nu}\} \tag{14.60}$$

$$\varepsilon_v = \frac{\partial u_0}{\partial x} + \frac{\partial v_0}{\partial y} + \frac{\partial w_0}{\partial z} + \sum_{r=1}^{n} \{a_r \frac{\partial u_r}{\partial x} + b_r \frac{\partial v_r}{\partial y} + c_r \frac{\partial w_r}{\partial z}\} \tag{14.61}$$

$$\tau_{xy} = G\{\frac{\partial u_0}{\partial y} + \frac{\partial v_0}{\partial x} + \sum_{r=1}^{n} (a_r \frac{\partial u_r}{\partial y} + b_r \frac{\partial v_r}{\partial x})\} \tag{14.62}$$

$$\tau_{yz} = G\{\frac{\partial v_0}{\partial z} + \frac{\partial w_0}{\partial y} + \sum_{r=1}^{n} (b_r \frac{\partial v_r}{\partial z} + c_r \frac{\partial w_r}{\partial y})\} \tag{14.63}$$

$$\tau_{xz} = G\{\frac{\partial w_0}{\partial x} + \frac{\partial u_0}{\partial z} + \sum_{r=1}^{n} (c_r \frac{\partial w_r}{\partial x} + a_r \frac{\partial u_r}{\partial z})\} \tag{14.64}$$

The $3n$ equations from (14.53) are found to be identical to (14.55) to (14.57). Thus, the Rayleigh-Ritz method is equivalent to the virtual work formulation.

Example 14.1 Find the system of equations for the coefficients a_r for the following Rayleigh-Ritz approximation

$$y_n = \frac{x}{l} y_1 + \frac{l-x}{l} y_0 + \sum_{k=1}^{n} a_k \sin \frac{k\pi x}{l} \tag{14.65}$$

for the functional:

$$I[y] = \int_0^l [p(y')^2 + qy^2 + 2fy] dx \tag{14.66}$$

with boundary conditions:

$$y(0) = y_0, \quad y(l) = y_1 \tag{14.67}$$

Solution: Note that the assumed y_n given in (14.65) automatically satisfies the essential boundary condition given in (14.67). Substitution of (14.65) into the functional (14.66) gives

$$I[y_n]=\sum_{k=1}^{n}\sum_{h=1}^{n}a_k a_h\int_0^l[p\frac{kh\pi^2}{l^2}\cos\frac{k\pi x}{l}\cos\frac{h\pi x}{l}+q\sin\frac{k\pi x}{l}\sin\frac{h\pi x}{l}]dx$$
$$+2\sum_{k=1}^{n}a_k\int_0^l[p(\frac{y_1 k\pi}{l^2}-\frac{y_0 k\pi}{l^2})\cos\frac{k\pi x}{l}+q(\frac{xy_1}{l}+\frac{l-x}{l}y_0)\sin\frac{k\pi x}{l}$$
$$+f\sin\frac{k\pi x}{l}]dx+\int_0^l[p(\frac{y_1^2}{l^2}+\frac{y_0^2}{l^2}-2\frac{y_0 y_1}{l^2})+q(\frac{x^2 y_1^2}{l^2}+\frac{(l-x)^2}{l^2}y_0^2 \tag{14.68}$$
$$+2\frac{x(l-x)}{l^2}y_0 y_1)+2f(\frac{xy_1}{l}+\frac{l-x}{l}y_0)]dx$$

The unknown coefficients can be found by the following conditions:

$$\frac{\partial I}{\partial a_1}=0,\quad \frac{\partial I}{\partial a_2}=0,\cdots,\quad \frac{\partial I}{\partial a_n}=0 \tag{14.69}$$

Thus, we have

$$0=\sum_{h=1}^{n}a_h\int_0^l[p\frac{kh\pi^2}{l^2}\cos\frac{k\pi x}{l}\cos\frac{h\pi x}{l}+q\sin\frac{k\pi x}{l}\sin\frac{h\pi x}{l}]dx$$
$$+\sum_{h=1}^{n}\int_0^l[p(\frac{y_1 h\pi}{l^2}-\frac{y_0 h\pi}{l^2})\cos\frac{h\pi x}{l}+q(\frac{xy_1}{l}+\frac{l-x}{l}y_0)\sin\frac{h\pi x}{l} \tag{14.70}$$
$$+f\sin\frac{h\pi x}{l}]dx$$

where $k = 1,2,3,...,n$. Once the functions p and q are given, integrations can be conducted and the unknown constants can be determined analytically.

14.9 WEIGHTED RESIDUE METHOD

Consider a differential equation given in a symbolic form as:

$$L(u)=f \tag{14.71}$$

We normally cannot satisfy the equation pointwise (if we can, we actually have the analytic solution). The weighted residue method is an approximate technique that requires the differential equation be satisfied in a global sense. That is, the integral of this differential equation multiplying by an arbitrary weighting function is zero. Mathematically, it is

$$\int_V[L(\bar{u})-f]\psi_i dV=0 \tag{14.72}$$

where ψ_i is the weighting function and the approximate function is assumed in series form similar to that in the Rayleigh-Ritz method

$$\bar{u}=\phi_0+\sum_{r=1}^{n}a_r\phi_r \tag{14.73}$$

We are searching for some approximate functions that satisfy (14.72). More discussions of the weighted residue method in the context of finite difference will be given in Chapter 15. For examples, the following cases are of particular interest.

14.9.1 Least Square Method

The least square method of Gauss is recovered if the weighting function equals the error function

$$\psi_i = L[\bar{u}(\phi_i)] - f \tag{14.74}$$

where the approximation of the unknown has been defined in (14.73). This least square of the error concept has been found very powerful in fitting data to a straight line (or so-called linear regression).

14.9.2 Point Collocation Method

The point collocation method is recovered if the weighting function equals the Dirac delta function

$$\psi_i = \delta(\boldsymbol{x} - \boldsymbol{x}_i) \tag{14.75}$$

The point collocation method is a standard procedure in solving boundary integral equations (e.g., Ho and Chau, 1999).

14.9.3 Petrov-Galerkin Method

The weighted residue method is also known as the Petrov-Galerkin method if the weighting function does not equal the fundamental function ϕ_i defined in (14.73)

$$\psi_i \neq \phi_i \tag{14.76}$$

The method is more general than the Galerkin method.

14.9.4 Galerkin Method

The Galerkin method is recovered if the weighting function equals the fundamental function used in (14.73):

$$\psi_i = \phi_i \tag{14.77}$$

The Galerkin method is a powerful method because it can be shown that it originates from the principle of virtual work. It will be further discussed in the next section.

14.10 GALERKIN METHOD

In this section, we will discuss the Galerkin method in more details. Let us consider the two-dimensional cases such that an approximation for u can be expressed as

$$\bar{u}(x,y)=\sum_{i=1}^{n}c_i\varphi_i(x,y) \tag{14.78}$$

where the unknown constants c_i are to be determined. Note that φ_i needs to satisfy both essential and natural boundary conditions (or displacement and traction boundary conditions in the case of solid mechanics). For the case of $f = 0$, substitution of (14.78) into (14.72) gives

$$\int_V L(\bar{u}(x,y))\psi_i dV=\int_V L(\sum_{i=1}^{n}c_i\varphi_i(x,y))\psi_i(x,y)dxdy=0 \tag{14.79}$$

If we pick the approximate function defined in (14.78) as the weighting function, we have

$$\int_V L(\sum_{i=1}^{n}c_i\varphi_i(x,y))\varphi_i(x,y)dxdy=0 \tag{14.80}$$

This is called the Galerkin method and it can be shown that it is the same as the principle of virtual displacements. For second order PDEs, it can be applied to elliptic, hyperbolic, and parabolic types. In the Rayleigh-Ritz method, we need the functional of the problem, however we do not need that in the Galerkin method. All we need is the differential equation of the problem. Strictly speaking, the Galerkin method is not a variational method since we do not need a functional. For physical problems in which the energy function exists, Galerkin is similar to the Rayleigh-Ritz method except for the choice of the approximate functions in (14.78). The trying functions in the Rayleigh-Ritz method need only to satisfy the essential (or displacement) boundary condition, but not the natural (or traction) boundary condition; whereas the trying functions in the Galerkin method needs to satisfy both essential and natural boundary conditions. This is the main limitation of the Galerkin method.

Example 14.2 Reconsider the minimization problem given in Example 14.1 and derive the Galerkin method for it:

$$I[y]=\int_0^l [p(y')^2+qy^2+2fy]dx \tag{14.81}$$

with boundary conditions:

$$y(0)=y_0,\quad y(l)=y_1 \tag{14.82}$$

Use the following approximation:

$$y=\frac{x}{l}y_1+\frac{l-x}{l}y_0+a_1w_1+a_2w_2+...+a_nw_n+... \tag{14.83}$$

Solution: Substitution of (14.83) into (14.81) and differentiation the functional with respect to the unknown constants a_i (i = 1,2,...) gives

$$\frac{\partial I}{\partial a_n}=2\int_0^l [(py')w_n'+qyw_n+fw_n]dx=0 \tag{14.84}$$

Note that this is the same as the procedure of the Rayleigh-Ritz method. In obtaining (14.84), we have used the following identities:

$$\frac{\partial y'}{\partial a_n} = w_n', \quad \frac{\partial y}{\partial a_n} = w_n \tag{14.85}$$

The first term in (14.84) can be evaluated by integration by parts as:

$$\int_0^l (py')w_n' dx = \int_0^l (py')d(w_n) = py'w_n\Big|_0^l - \int_0^l (py')'w_n dx \tag{14.86}$$

To remove the boundary terms, we must impose

$$w_n(0) = w_n(l) = 0 \tag{14.87}$$

Thus, (14.84) becomes

$$\int_0^l [-(py')' + qy + f]w_n dx = 0 \tag{14.88}$$

This is precisely the Galerkin method for the following differential equation:

$$-(py')' + qy + fy = 0 \tag{14.89}$$

Through this example, we show that the Rayleigh-Ritz method and the Galerkin method is equivalent if the boundary conditions given in (14.82) are identically satisfied.

We have shown that for the particular functional given in (14.81), the Raleigh-Ritz method and Galerkin method are equivalent if the boundary conditions given in (14.82) are identically satisfied. In fact, we can formulate this observation more generally. Putting in a general term, we consider the following boundary value problem of a differential equation:

$$L[u] = 0 \tag{14.90}$$

subject to the following boundary conditions:

$$u(a) = \psi \tag{14.91}$$

$$B[u] = \phi \tag{14.92}$$

where L and B are given differential operators. The Galerkin method can be cast as:

$$u = u_0 + a_1 w_1 + a_2 w_2 + \ldots \tag{14.93}$$

such that

$$u_0(a) = \psi \tag{14.94}$$

$$B[u_0] = \phi \tag{14.95}$$

$$w_n(a) = 0 \tag{14.96}$$

$$B[w_n] = 0 \tag{14.97}$$

The coefficients for a_i, $i = 1,2,3,\ldots$ can be determined from:

$$\int_V L[u]w_n dV = 0 \tag{14.98}$$

where V is the domain of the differential equation being defined.

The Galerkin method can also be shown as equivalent to the principle of virtual work. In particular, for the more general functional for the two-dimensional case the functional can be cast as:

$$I[u] = \int_V F[x, y, u, u_x, u_{xx}, \ldots, u_y, u_{yy}, \ldots, u_{xy}, \ldots]dxdy \tag{14.99}$$

Using the calculus of variations, we can obtain the first variation symbolically as

$$\delta I = \int_V L[u]\delta u \, dxdy = 0 \tag{14.100}$$

Note that (14.98) is equivalent to the principle of virtual work. We see that if δu is replaced by the fundamental function w_n, the Galerkin method is therefore consistent with the principle of virtual work. It is a direct method for finding the stationary value of the functional (14.99). However, we do not need a functional of the problem in the Galerkin formulation and we only need the Euler differential equation of the functional. Therefore, it is more general than the Rayleigh-Ritz method and is applicable to problems in which a functional does not exist. We will further illustrate this idea using the following example.

Example 14.3 Solve the following cantilever beam problem shown in Figure 14.3 using the principle of virtual displacement together with the Galerkin method.

The boundary conditions at $x = 0$ are

$$w(0) = 0, \quad w'(0) = 0, \tag{14.101}$$

You can assume the Euler-Bernoulli beam theory and the following two term trial functions:

$$w = a_1 x^2 + a_2 x^3 \tag{14.102}$$

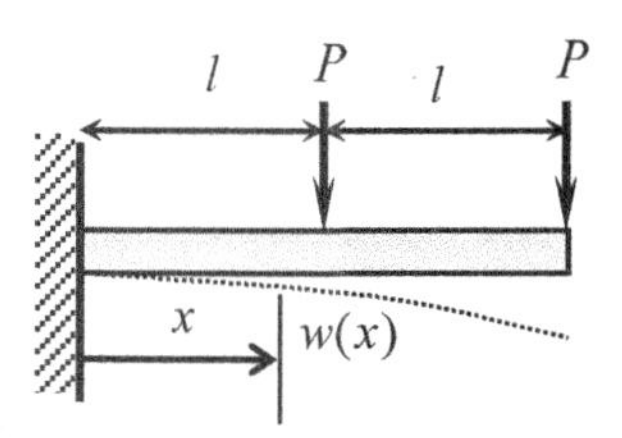

Figure 14.3 A cantilever beam subject to two concentrated forces

Solution: Note that by choosing (14.102), both of the boundary conditions given (14.101) are satisfied. In a sense, both essential and natural boundary conditions are satisfied, and thus it is a feasible choice for the trying functions of the Galerkin method. It is also the reason why we need a trying function of a power cube. In addition, we cannot add a linear term of x in (14.102), if we do, the second boundary condition of (14.101) will not be satisfied.

The bending moment corresponding to the deflection given in (14.102) is (Timoshenko, 1956)

$$M = -EIw'' = -EI(2a_1 + 6a_2 x) \tag{14.103}$$

The bending stress induced by this moment is (Timoshenko, 1956)

$$\sigma_x = \frac{M(x)z}{I} \tag{14.104}$$

The nonzero internal virtual work is

$$W_I = \int_V \sigma_x \delta\varepsilon_x dxdydz \tag{14.105}$$

Using Euler-Bernoulli beam theory, we have the bending strain proportional to z measured from the neutral surface (mid-surface) of the beam:

$$\varepsilon_x = -z\eta(x) \tag{14.106}$$

where $\eta(x)$ is an unknown function to be determined. Substitution of (14.106) and (14.104) into (14.105) gives:

$$W_I = \int_V \frac{Mz}{I}(-z\delta\eta)dxdydz = -\int_0^{2l}\frac{1}{I}(\int_A z^2 dydz)M\delta\eta dx = -\int_0^{2l} M\delta\eta dx \tag{14.107}$$

Figure 14.4 shows the more general case of external virtual work due to distributed load, concentrated load, and concentrated moment. The external virtual work due to virtual deflection δw is

$$W_E = \int_0^{2l} q(x)\delta w dx + \bar{P}_z \delta w(2l) + [-\bar{M}\delta w'(2l)] \tag{14.108}$$

where the minus sign in front of the moment and rotation product term indicate that the convention of moment is opposite to the positive slope.

For the present case shown in Figure 14.3, we have

$$W_E = -P\delta w(l) - P\delta w(2l) \tag{14.109}$$

Using the principle of virtual displacement, we balance the internal and external virtual work to give

$$-\int_0^{2l} M\delta\eta dx = -P\delta w(l) - P\delta w(2l) \tag{14.110}$$

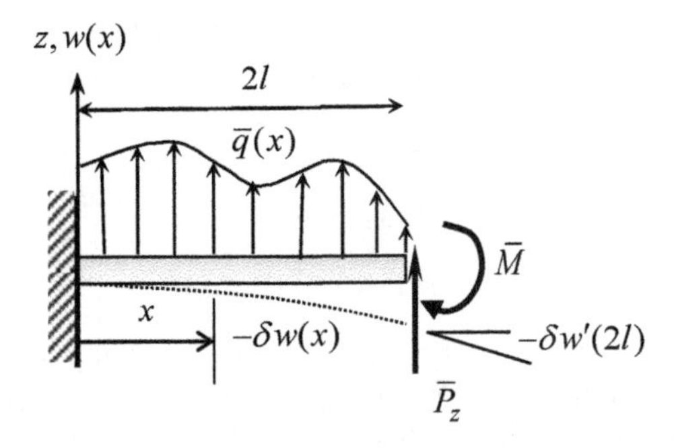

Figure 14.4 The general case of a beam subject to distributed load, point load, and concentrated moment

Taking the virtual displacement of (14.102), we obtain

$$\delta w = \delta a_1 x^2 + \delta a_2 x^3 \tag{14.111}$$

Using Hooke's law to relate (14.104) and (14.106), we get

$$M = -EI\frac{d^2 w}{dx^2} = -EI\eta \tag{14.112}$$

Thus, we have

$$\delta\eta = \frac{d^2\delta w}{dx^2} \tag{14.113}$$

Substitution of (14.111) into (14.113) gives

$$\delta\eta = \frac{d^2\delta w}{dx^2} = 2\delta a_1 + 6\delta a_2 x \tag{14.114}$$

Substituting (14.111) and (14.114) into (14.110) gives

$$(-2\int_0^{2l} Mdx + 5Pl^2)\delta a_1 + (-6\int_0^{2l} Mxdx + 9Pl^3)\delta a_2 = 0 \tag{14.115}$$

This implies the following two equations to be satisfied:

$$-2\int_0^{2l} Mdx + 5Pl^2 = 0 \tag{14.116}$$

$$-6\int_0^{2l} Mxdx + 9Pl^3 = 0 \tag{14.117}$$

Substitution of (14.103) into (14.116) and (14.117) and integration of the resulting equations gives

$$8a_1 + 24a_2 l = -5\frac{Pl}{EI} \tag{14.118}$$

$$24a_1 + 96a_2 l = -9\frac{Pl}{EI} \tag{14.119}$$

Solving for the unknown constants, we find

$$a_1 = -\frac{11}{8}\frac{Pl}{EI}, \quad a_2 = \frac{1}{4}\frac{Pl}{EI} \tag{14.120}$$

From (14.102), we have the approximation as:

$$\tilde{w}(x) = -\frac{11}{8}\frac{Pl}{EI}x^2 + \frac{1}{4}\frac{Pl}{EI}x^3 \tag{14.121}$$

Substitution of (14.121) into (14.103) gives

$$\tilde{M}(x) = \frac{1}{4}(11l - 6x)P \tag{14.122}$$

For the present problem, it can be solved exactly by taking the moment by cutting a free body of the beam, and the resulting moment can be plugged into (14.103) to solve for the deflection. The final exact solutions for the deflection and bending moment are found as:

$$\begin{aligned} M(x) &= (3l - 2x)P, \quad 0 < x < l \\ &= (2l - x)P, \quad l < x < 2l \end{aligned} \tag{14.123}$$

$$\begin{aligned} w(x) &= -\frac{3}{2}\frac{Pl}{EI}x^2 + \frac{1}{3}\frac{Pl}{EI}x^3, \quad 0 < x < l \\ &= -\frac{Pl}{EI}(lx^2 - \frac{x^3}{6} + \frac{l^2 x}{2} - \frac{l^3}{6}), \quad l < x < 2l \end{aligned} \tag{14.124}$$

The details of this evaluation will be left as an exercise for readers. The comparison of the results of the Galerkin method and the exact solution is given in Figure 14.5.

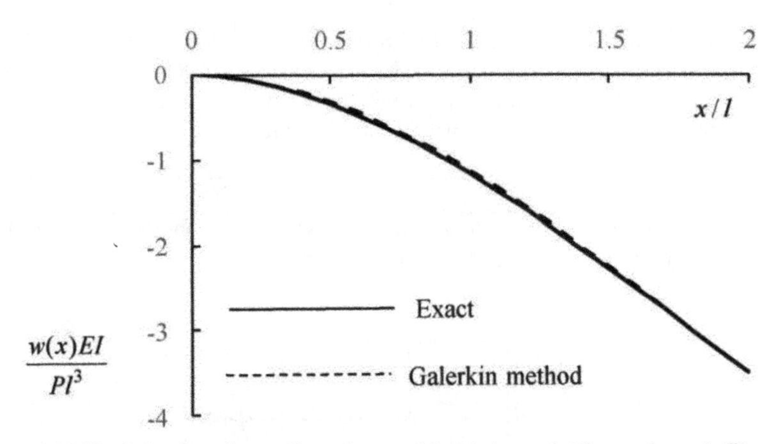

Figure 14.5 The deflection of a cantilever beam subject to two point forces shown in Figure 14.3. The solid and dotted lines are the exact solution and the Galerkin approximation

The comparison of the exact solution given in (14.123) with the Galerkin approximation given in (14.122) is plotted in Figure 14.6. It can be seen that the approximation for deflection shown in Figure 14.5 is much better than that of the bending moment as shown in Figure 14.6. Actually, this illustrates a very general observation that the virtual displacement method (the present case) gives a more accurate approximation for the deflection in (14.121), whereas the virtual force method will give a better approximation for moment. In the present case, the bending moment is evaluated as the second derivative of deflection given in (14.121), and this indirect evaluation always gives a less accurate prediction.

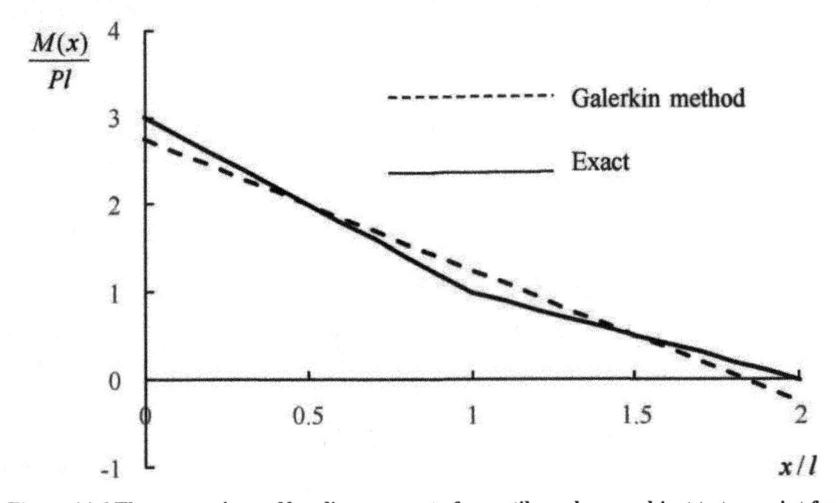

Figure 14.6 The comparison of bending moment of a cantilever beam subject to two point forces shown in Figure 14.3

In the present example, we consider a fixed end cantilever beam subject to concentrated loads that can be solved quite easily because we want to compare the approximation of the Galerkin method with the exact solution. More generally, for more complicated mechanics problems where an exact solution is not possible, an approximation like the Galerkin method will be found very useful to provide the first order approximation and gives insight into the solution of the problem.

Example 14.4 Consider the two-dimensional stress analysis shown in Figure 14.7. Find the stress state in the plate by the Galerkin method.

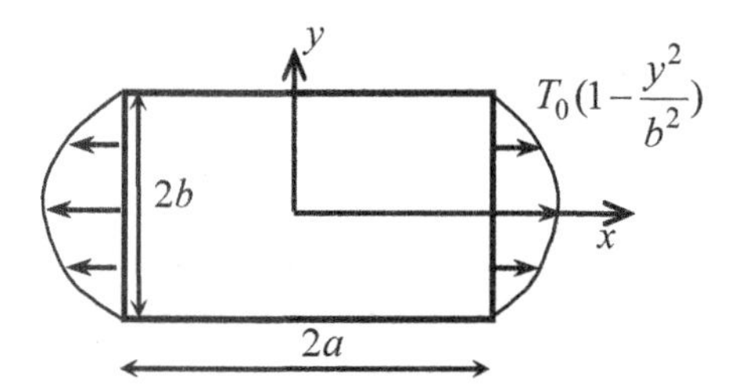

Figure 14.7 Rectangular plate subject to parabolic pulling from two ends

Solution: As shown in Chau (2013), this kind of plane stress problem can be solved by using Airy's stress function:

$$\sigma_{xx} = \frac{\partial^2 \phi}{\partial y^2}, \quad \sigma_{yy} = \frac{\partial^2 \phi}{\partial x^2}, \quad \sigma_{xy} = -\frac{\partial^2 \phi}{\partial y \partial x} \tag{14.125}$$

The stress function satisfies the biharmonic equation

$$\frac{\partial^4 \phi}{\partial y^4} + 2\frac{\partial^4 \phi}{\partial y^2 \partial x^2} + \frac{\partial^4 \phi}{\partial x^4} = 0 \tag{14.126}$$

The boundary conditions can be prescribed as:

$$\frac{\partial^2 \phi}{\partial y^2} = T_0(1-\frac{y^2}{b^2}), \quad \frac{\partial^2 \phi}{\partial y \partial x} = 0, \quad x = \pm a \tag{14.127}$$

$$\frac{\partial^2 \phi}{\partial x^2} = 0, \quad \frac{\partial^2 \phi}{\partial y \partial x} = 0, \quad y = \pm b \tag{14.128}$$

We choose the following fundamental function:

$$\phi = \frac{1}{2}T_0 y^2 (1-\frac{1}{6}\frac{y^2}{b^2}) + (x^2-a^2)^2(y^2-b^2)^2(a_1 + a_2 x^2 + a_3 y^2 + \dots) \tag{14.129}$$

Differentiating (14.129) with respect to x and y, we have

$$\begin{aligned}\phi_{yy} &= T_0 y^2 (1-\frac{y^2}{b^2}) + 4(x^2-a^2)^2(3y^2-b^2)(a_1 + a_2 x^2 + a_3 y^2 + \dots) \\ &+2a_3(x^2-a^2)^2(y^2-b^2)^2 + 16a_3 y^2 (y^2-b^2)(x^2-a^2)^2\end{aligned} \tag{14.130}$$

$$\begin{aligned}\phi_{xy} &= 16xy(x^2-a^2)(y^2-b^2)(a_1 + a_2 x^2 + a_3 y^2 + \dots) \\ &+8xya_2(x^2-a^2)^2(y^2-b^2) + 8xya_3(x^2-a^2)(y^2-b^2)^2\end{aligned} \tag{14.131}$$

$$\phi_{xx} = 4(3x^2 - a^2)(y^2 - b^2)^2(a_1 + a_2x^2 + a_3y^2 + ...) \\ +16x^2a_2(x^2 - a^2)(y^2 - b^2)^2 + 2a_2(x^2 - a^2)^2(y^2 - b^2)^2 \tag{14.132}$$

It is obvious that all boundary conditions given in (14.127) and (14.128) are satisfied identically. It is a matter of art, however, to identify the feasible form of the fundamental solution given in (12.129). Using (14.100), we have

$$\int_{-a}^{a}\int_{-b}^{b}(\nabla^4\phi)\delta\phi\, dxdy = 0 \tag{14.133}$$

The variation of the fundamental function is

$$\delta\phi = (x^2 - a^2)^2(y^2 - b^2)^2(\delta a_1 + \delta a_2 x^2 + \delta a_3 y^2 + ...) \tag{14.134}$$

To further simplify the problem, we only retain a_1. Substitution of (14.130) to (14.132) into (14.133) gives

$$\int_{-a}^{a}\int_{-b}^{b}\{24(y^2 - b^2)^2 a_1 + 32(3x^2 - a^2)(3y^2 - b^2)a_1 + 24(x^2 - a^2)a_1 - \frac{2T_0}{b^2}\} \\ \times(x^2 - a^2)^2(y^2 - b^2)^2\, dxdy = 0 \tag{14.135}$$

Therefore, conducting integration, we have

$$a_1 = T_0 / \{b^2(\frac{64}{7}b^4 + \frac{256}{49}a^2b^2 + \frac{64}{7}a^4)\} \tag{14.136}$$

This gives the approximate stress field:

$$\sigma_{xx} = T_0(1 - \frac{y^2}{b^2}) + 4(x^2 - a^2)^2(3y^2 - b^2)a_1 \tag{14.137}$$

$$\sigma_{yy} = 4(y^2 - b^2)^2(3x^2 - a^2)a_1 \tag{14.138}$$

$$\sigma_{xy} = -16(x^2 - a^2)(y^2 - b^2)xya_1 \tag{14.139}$$

For the special case of a square plate $a = b$, we have

$$a_1 = 0.04253\frac{T_0}{b^6} \tag{14.140}$$

14.11 KANTOROVICH'S METHOD

In this section, we introduce a method closely related to the Galerkin method by considering the more complicated problems of plate bending. Consider the case of a rectangular plate subject to two-dimensional bending, as shown in Figure 14.8. The plate has fixed supports on $y = \pm b$, and is free on $x = \pm a$. The governing equation of the problem is found to be a nonhomogeneous biharmonic equation (Timoshenko and Woinowsky-Krieger, 1959):

$$\nabla^2\nabla^2 u = \nabla^4 u = \frac{\partial^4 u}{\partial x^4} + 2\frac{\partial^4 u}{\partial y^2 \partial x^2} + \frac{\partial^4 u}{\partial y^4} = \frac{q(x,y)}{D} \tag{14.141}$$

where D is

$$D = \frac{Eh^3}{12(1-\nu^2)} \tag{14.142}$$

The boundary conditions are

$$u = 0, \quad \frac{\partial u}{\partial y} = 0, \quad \text{on } y = \pm b \tag{14.143}$$

We assume the following fundamental function:

$$u = \varphi_1(y) f_1(x) + \varphi_2(y) f_2(x) + \ldots + \varphi_n(y) f_n(x) \tag{14.144}$$

For the special case that the loading is symmetric with respect to y, we have u being an even function of y. One such function is

$$\varphi_k(y) = (y^2 - b^2)^2 y^{2k-2} \tag{14.145}$$

As a first approximation, we consider the following term:

$$\tilde{u}(x, y) = (y^2 - b^2)^2 f(x) \tag{14.146}$$

where f is an unknown function to be determined. Note that this method is not exactly the Galerkin method since we have not specified the boundary conditions on $x = \pm a$. The function f will be determined analytically. Thus, the current method is a semi-Galerkin method. This method was first proposed in Kantorovich and Krylov (1964) and was called Kantorovich's method by Reiss (1965).

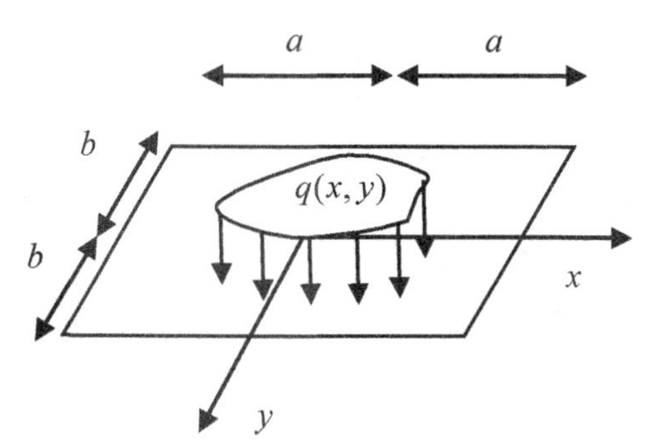

Figure 14.8 Rectangular plate with two free edges and two fixed support edges

To proceed with the first approximation, we have

$$\varphi_1(y) = (y^2 - b^2)^2 \tag{14.147}$$

Using the Galerkin method, we have

$$\int_{-b}^{b} (\nabla^4 \tilde{u} - \frac{q}{D}) \varphi_1(y) dy = 0 \tag{14.148}$$

Substitution of (14.147) and (14.146) into (14.148) gives

$$\int_{-b}^{b} [24 f + 2(12 y^2 - 4b^2) f'' + (y^2 - b^2)^2 f^{(IV)} - \frac{q}{D}](y^2 - b^2)^2 dy = 0 \tag{14.149}$$

Integration gives the following results:

$$\int_{-b}^{b} 24 (y^2 - b^2)^2 dy = \frac{128}{5} b^5 \tag{14.150}$$

$$\int_{-b}^{b} 8(3y^2 - b^2)(y^2 - b^2)^2 dy = -\frac{512}{105} b^7 \tag{14.151}$$

$$\int_{-b}^{b} (y^2 - b^2)^4 dy = \frac{256}{315} b^9 \tag{14.152}$$

Substitution of (14.150) to (14.152) into (14.149) gives

$$\frac{256}{315} b^9 f^{(IV)} - \frac{512}{105} b^7 f'' + \frac{128}{5} b^5 f = p_1(x) \tag{14.153}$$

where

$$q_1(x) = \int_{-b}^{b} \frac{q(x,y)}{D} (y^2 - b^2)^2 dy \tag{14.154}$$

For the homogeneous case, we have the exponential form for f as:

$$f = e^{\lambda x} \tag{14.155}$$

The characteristic equation of the homogeneous form of (14.153) is

$$\frac{256}{315} b^4 \lambda^4 - \frac{512}{105} b^2 \lambda^2 + \frac{128}{5} = 0 \tag{14.156}$$

The roots of λ in (14.156) are:

$$\lambda_{1,2,3,4} = \frac{1}{b}(\pm\alpha \pm i\beta), \quad \alpha = 2.075, \quad \beta = 1.143 \tag{14.157}$$

The general solution becomes

$$\begin{aligned} f(x) &= A_1 \cosh(\alpha \frac{x}{b}) \cos(\beta \frac{x}{b}) + A_2 \cosh(\alpha \frac{x}{b}) \sin(\beta \frac{x}{b}) \\ &+ B_1 \sinh(\alpha \frac{x}{b}) \sin(\beta \frac{x}{b}) + B_2 \sinh(\alpha \frac{x}{b}) \cos(\beta \frac{x}{b}) + f_p(x) \end{aligned} \tag{14.158}$$

For the special case of uniform distributed load, we have

$$q_1(x) = \int_{-b}^{b} \frac{q}{D} (y^2 - b^2)^2 dy = \frac{16}{15} \frac{q}{D} \tag{14.159}$$

The particular solution becomes

$$\frac{128}{5} b^5 f_p = \frac{16}{15} \frac{q}{D} b^5 \tag{14.160}$$

Thus, we have

$$f_p = \frac{1}{24} \frac{q}{D} \tag{14.161}$$

For the case of a symmetric boundary condition on $x = +a$ and $x = -a$, we must have $u(x,y)$ and $f(x)$ an even function of x. Thus, (14.158) is reduced to

$$f(x) = A \cosh(\alpha \frac{x}{b}) \cos(\beta \frac{x}{b}) + B \sinh(\alpha \frac{x}{b}) \sin(\beta \frac{x}{b}) + \frac{1}{24} \frac{q}{D} \tag{14.162}$$

We further suppose that for the case of fixed end supports on $x = \pm a$, we have

$$f(\pm a) = f'(\pm a) = 0 \tag{14.163}$$

Substitution of (14.162) into (14.163) gives

$$A \cosh(\alpha \frac{a}{b}) \cos(\beta \frac{a}{b}) + B \sinh(\alpha \frac{a}{b}) \sin(\beta \frac{a}{b}) + \frac{1}{24} \frac{q}{D} = 0 \tag{14.164}$$

$$\begin{aligned}&A[\frac{\alpha}{b}\sinh(\alpha\frac{a}{b})\cos(\beta\frac{a}{b})-\frac{\beta}{b}\cosh(\alpha\frac{a}{b})\sin(\beta\frac{a}{b})]\\&+B[\frac{\alpha}{b}\cosh(\alpha\frac{a}{b})\sin(\beta\frac{a}{b})+\frac{\beta}{b}\sinh(\alpha\frac{a}{b})\cos(\beta\frac{a}{b})]=0\end{aligned} \tag{14.165}$$

This provides two equations for two unknowns A and B and the solutions are

$$A=\frac{\gamma_1}{\gamma_0}\frac{1}{24}\frac{q}{D},\quad B=\frac{\gamma_2}{\gamma_0}\frac{1}{24}\frac{q}{D} \tag{14.166}$$

where

$$\gamma_0=\beta\sinh\alpha\mu\cosh\alpha\mu+\alpha\sin\beta\mu\cos\beta\mu \tag{14.167}$$

$$\gamma_1=-(\alpha\cosh\alpha\mu\sin\beta\mu+\beta\sinh\alpha\mu\cos\beta\mu) \tag{14.168}$$

$$\gamma_2=\alpha\sinh\alpha\mu\cos\beta\mu-\beta\cosh\alpha\mu\sin\beta\mu \tag{14.169}$$

$$\mu=\frac{a}{b} \tag{14.170}$$

For an infinitely long plate, we have $a\to\infty$ and $\mu\to\infty$, we have

$$\tilde{u}(x,y)=\frac{1}{24}\frac{q}{D}(y^2-b^2)^2 \tag{14.171}$$

This equals the exact solution. For a square plate $\mu=1$ and $\nu=0.3$, we have the solution at the center as

$$\tilde{u}(0,0)=0.479\frac{1}{24}\frac{q}{D}b^4=0.01362\frac{q(2b)^4}{Eh^3} \tag{14.172}$$

The exact solution for this case of a square plate is

$$u_{\max}=0.0138\frac{q(2b)^4}{Eh^3} \tag{14.173}$$

Thus, the approximation differs only by 1.3% from the exact solution. Referring to Timoshenko and Woinowsky-Krieger (1959), we can also find the maximum bending stress in the plate accordingly (p. 42 of Timoshenko and Woinowsky-Krieger, 1959)

$$(\sigma_x)_{\max}=\frac{6M_x}{h^2},\quad (\sigma_y)_{\max}=\frac{6M_y}{h^2} \tag{14.174}$$

The bending moment can be found from deflection as

$$M_x=-D(\frac{\partial^2 u}{\partial x^2}+\nu\frac{\partial^2 u}{\partial y^2}),\quad M_y=-D(\frac{\partial^2 u}{\partial y^2}+\nu\frac{\partial^2 u}{\partial x^2}) \tag{14.175}$$

Using these expressions, we find

$$(\sigma_x)_{\max}=0.140\frac{q(2b)^2}{h^2},\quad (\sigma_y)_{\max}=0.138\frac{q(2b)^2}{h^2} \tag{14.176}$$

The exact solution for the bending moment is

$$(\sigma_x)_{\max}=(\sigma_y)_{\max}=0.137\frac{q(2b)^2}{h^2} \tag{14.177}$$

Thus, the error of bending moment by the semi-Galerkin method (or Kantorovich method) is about 1~2%.

14.12 FUNCTIONAL FOR BIHARMONIC EQUATION

As remarked in the introduction, both the Rayleigh-Ritz and Galerkin methods were originally motivated by plate bending problems. In this section, we will consider the following functional form for the plate bending problem:

$$\begin{aligned} U = &\iint_A [(\nabla^2 u)^2 - 2(1-\nu)\{\frac{\partial^2 u}{\partial x^2}\frac{\partial^2 u}{\partial x^2} - (\frac{\partial^2 u}{\partial x \partial y})^2\} - 2fu]dxdy \\ &-2\int_\Gamma p(s)uds + 2\int_\Gamma m(s)\frac{\partial u}{\partial n}ds \end{aligned} \tag{14.178}$$

where distributed load q divided by D is denoted by $f\,(= q/D)$, point force on the boundary by p, and concentrated moment on the boundary by m. Physically, the functional is the total energy minus the external work done as:

$$U = \iint_A (dV - dU_1) - U_2 \tag{14.179}$$

where

$$dV = -\frac{1}{2}(M_x \frac{\partial^2 u}{\partial x^2} + M_y \frac{\partial^2 u}{\partial y^2} + 2M_{xy}\frac{\partial^2 u}{\partial y \partial x})dxdy \tag{14.180}$$

$$dU_1 = u\,q dxdy \tag{14.181}$$

$$U_2 = -2\int_\Gamma p(s)uds + 2\int_\Gamma m(s)\frac{\partial u}{\partial n}ds \tag{14.182}$$

Substitution of the following definitions of bending and twisting moments into (14.180) and the result into (14.179) gives (14.178):

$$M_x = -D(\frac{\partial^2 u}{\partial x^2} + \nu\frac{\partial^2 u}{\partial y^2}), \quad M_y = -D(\frac{\partial^2 u}{\partial y^2} + \nu\frac{\partial^2 u}{\partial x^2}) \tag{14.183}$$

$$M_{xy} = -D(1-\nu)\frac{\partial^2 u}{\partial x \partial y} \tag{14.184}$$

The calculus of variations requires

$$\left.\frac{dU(u+\alpha\eta)}{d\alpha}\right|_{\alpha=0} = 0 \tag{14.185}$$

That is, we have to substitute

$$u \leftarrow u + \alpha\eta \tag{14.186}$$

into (14.178) and differentiate it with respect to α. More specifically, we find

$$\begin{aligned} \frac{d}{d\alpha}(\nabla^2(u+\alpha\eta))^2 &= \frac{d}{d\alpha}[(\nabla^2 u)^2 + 2\alpha\nabla^2 u\nabla^2\eta + \alpha^2(\nabla^2\eta)^2] \\ &= 2\nabla^2 u\nabla^2\eta + 2\alpha(\nabla^2\eta)^2 \end{aligned} \tag{14.187}$$

$$\begin{aligned} \frac{d}{d\alpha}\{\frac{\partial^2(u+\alpha\eta)}{\partial x^2}\frac{\partial^2(u+\alpha\eta)}{\partial x^2}\} &= \frac{d}{d\alpha}\{(\frac{\partial^2 u}{\partial x^2} + \alpha\frac{\partial^2\eta}{\partial x^2})(\frac{\partial^2 u}{\partial y^2} + \alpha\frac{\partial^2\eta}{\partial y^2})\} \\ &= (\frac{\partial^2 u}{\partial y^2}\frac{\partial^2\eta}{\partial x^2} + \frac{\partial^2 u}{\partial x^2}\frac{\partial^2\eta}{\partial y^2}) + 2\alpha\frac{\partial^2\eta}{\partial x^2}\frac{\partial^2\eta}{\partial y^2} \end{aligned} \tag{14.188}$$

$$\frac{d}{d\alpha}[\frac{\partial^2(u+\alpha\eta)}{\partial x\partial y}]^2 = \frac{d}{d\alpha}\{(\frac{\partial^2 u}{\partial x\partial y})^2 + 2\alpha(\frac{\partial^2 u}{\partial x\partial y})(\frac{\partial^2 \eta}{\partial x\partial y}) + \alpha^2(\frac{\partial^2 \eta}{\partial x\partial y})^2\}$$
$$= 2(\frac{\partial^2 u}{\partial x\partial y})(\frac{\partial^2 \eta}{\partial x\partial y}) + 2\alpha(\frac{\partial^2 \eta}{\partial x\partial y})^2 \tag{14.189}$$

Setting α = 0 into (14.187) to (14.189) and substituting the results into (14.178) gives

$$\iint_A [2\nabla^2 u\nabla^2\eta - 2(1-\nu)(\frac{\partial^2 u}{\partial x^2}\frac{\partial^2 \eta}{\partial y^2} + \frac{\partial^2 u}{\partial y^2}\frac{\partial^2 \eta}{\partial x^2} - 2\frac{\partial^2 u}{\partial x\partial y}\frac{\partial^2 \eta}{\partial x\partial y}) - 2f\eta]dxdy$$
$$-2\int_\Gamma p(s)\eta ds + 2\int_\Gamma m(s)\frac{\partial\eta}{\partial n}ds = 0 \tag{14.190}$$

This equation can be further be simplified using the following identity

$$\nabla^2 u\nabla^2\eta = \nabla^2 u(\frac{\partial^2\eta}{\partial y^2} + \frac{\partial^2\eta}{\partial x^2}) = \frac{\partial\nabla^2 u}{\partial x}\frac{\partial\eta}{\partial x} + \nabla^2 u\frac{\partial^2\eta}{\partial x^2} + \frac{\partial\nabla^2 u}{\partial y}\frac{\partial\eta}{\partial y}$$
$$+\nabla^2 u\frac{\partial^2\eta}{\partial y^2} - \frac{\partial\eta}{\partial x}\frac{\partial\nabla^2 u}{\partial x} - \eta\frac{\partial^2\nabla^2 u}{\partial x^2} - \frac{\partial\eta}{\partial y}\frac{\partial\nabla^2 u}{\partial y} - \eta\frac{\partial^2\nabla^2 u}{\partial y^2}$$
$$+\eta\frac{\partial^2\nabla^2 u}{\partial x^2} + \eta\frac{\partial^2\nabla^2 u}{\partial y^2} = [\frac{\partial}{\partial x}(\nabla^2 u\frac{\partial\eta}{\partial x}) + \frac{\partial}{\partial y}(\nabla^2 u\frac{\partial\eta}{\partial y})] \tag{14.191}$$
$$-[\frac{\partial}{\partial x}(\eta\frac{\partial\nabla^2 u}{\partial x}) + \frac{\partial}{\partial y}(\eta\frac{\partial\nabla^2 u}{\partial y})] + \eta[\frac{\partial^2\nabla^2 u}{\partial x^2} + \frac{\partial^2\nabla^2 u}{\partial y^2}]$$

Applying (14.191), we have

$$\iint_A \nabla^2 u\nabla^2\eta\, dxdy = \iint_A [\frac{\partial}{\partial x}(\nabla^2 u\frac{\partial\eta}{\partial x}) + \frac{\partial}{\partial y}(\nabla^2 u\frac{\partial\eta}{\partial y})]dxdy$$
$$-\iint_A [\frac{\partial}{\partial x}(\eta\frac{\partial\nabla^2 u}{\partial x}) + \frac{\partial}{\partial y}(\eta\frac{\partial\nabla^2 u}{\partial y})]dxdy + \iint_A \eta[\frac{\partial^2\nabla^2 u}{\partial x^2} + \frac{\partial^2\nabla^2 u}{\partial y^2}]dxdy \tag{14.192}$$

The first and second terms on the right hand side of (14.192) can be reduced to

$$\iint_A [\frac{\partial}{\partial x}(\nabla^2 u\frac{\partial\eta}{\partial x}) + \frac{\partial}{\partial y}(\nabla^2 u\frac{\partial\eta}{\partial y})]dxdy = \int_\Gamma \nabla^2 u(\frac{\partial\eta}{\partial x}dy - \frac{\partial\eta}{\partial y}dx) \tag{14.193}$$

$$\iint_A [\frac{\partial}{\partial x}(\eta\frac{\partial\nabla^2 u}{\partial x}) + \frac{\partial}{\partial y}(\eta\frac{\partial\nabla^2 u}{\partial y})]dxdy = \int_\Gamma \eta(\frac{\partial\nabla^2 u}{\partial x}dy - \frac{\partial\nabla^2 u}{\partial y}dx) \tag{14.194}$$

Using these results, we finally get

$$\iint_A \nabla^2 u\nabla^2\eta\, dxdy = \iint_A \eta\nabla^4 u dxdy + \int_\Gamma \nabla^2 u\frac{\partial\eta}{\partial n}ds - \int_\Gamma \eta\frac{\partial\nabla^2 u}{\partial n}ds \tag{14.195}$$

In addition, it is straightforward to show that

$$I_1 = \iint_A \left(\frac{\partial^2 u}{\partial x^2}\frac{\partial^2 \eta}{\partial y^2} + \frac{\partial^2 u}{\partial y^2}\frac{\partial^2 \eta}{\partial x^2} - 2\frac{\partial^2 u}{\partial x \partial y}\frac{\partial^2 \eta}{\partial x \partial y}\right) dxdy$$
$$= \iint_A \left[\frac{\partial}{\partial x}\left(\frac{\partial^2 u}{\partial y^2}\frac{\partial \eta}{\partial x} - \frac{\partial^2 u}{\partial y \partial x}\frac{\partial \eta}{\partial y}\right) + \frac{\partial}{\partial y}\left(\frac{\partial^2 u}{\partial x^2}\frac{\partial \eta}{\partial y} - \frac{\partial^2 u}{\partial y \partial x}\frac{\partial \eta}{\partial x}\right)\right] dxdy \tag{14.196}$$

where (14.196) is the second term in the first integral of (14.190). It can further be reduced to

$$I_1 = \int_\Gamma \left[\left(\frac{\partial^2 u}{\partial y^2}\frac{\partial \eta}{\partial x} - \frac{\partial^2 u}{\partial y \partial x}\frac{\partial \eta}{\partial y}\right)x_n + \left(\frac{\partial^2 u}{\partial x^2}\frac{\partial \eta}{\partial y} - \frac{\partial^2 u}{\partial y \partial x}\frac{\partial \eta}{\partial x}\right)y_n\right] ds \tag{14.197}$$

Note also that by the chain rule we have

$$\frac{\partial \eta}{\partial x} = \frac{\partial \eta}{\partial s}\frac{\partial s}{\partial x} + \frac{\partial \eta}{\partial n}\frac{\partial n}{\partial x} = -\frac{\partial \eta}{\partial s}x_s + \frac{\partial \eta}{\partial n}x_n \tag{14.198}$$

$$\frac{\partial \eta}{\partial y} = \frac{\partial \eta}{\partial s}\frac{\partial s}{\partial y} + \frac{\partial \eta}{\partial n}\frac{\partial n}{\partial y} = \frac{\partial \eta}{\partial s}y_s + \frac{\partial \eta}{\partial n}y_n \tag{14.199}$$

The directional cosines are given as:

$$x_s = \cos(s, x), \quad y_s = \cos(s, y) \tag{14.200}$$

$$x_n = \cos(n, x), \quad y_n = \cos(n, y) \tag{14.201}$$

Substitution of (14.198) and (14.199) into (14.197) gives

$$I_1 = \int_\Gamma \left(\frac{\partial^2 u}{\partial y^2}x_n^2 + \frac{\partial^2 u}{\partial x^2}y_n^2 - 2x_n y_n \frac{\partial^2 u}{\partial y \partial x}\right)\frac{\partial \eta}{\partial n} ds$$
$$+ \int_\Gamma \left[\frac{\partial^2 u}{\partial x^2}y_s y_n - \frac{\partial^2 u}{\partial y^2}x_s x_n + (x_s y_n - y_s x_n)\frac{\partial^2 u}{\partial y \partial x}\right]\frac{\partial \eta}{\partial s} ds \tag{14.202}$$

Applying integration by parts to the last term, we have

$$\int_\Gamma \left[\frac{\partial^2 u}{\partial x^2}y_s y_n - \frac{\partial^2 u}{\partial y^2}x_s x_n + (x_s y_n - y_s x_n)\frac{\partial^2 u}{\partial y \partial x}\right]\frac{\partial \eta}{\partial s} ds =$$
$$-\int_\Gamma \frac{\partial}{\partial s}\left[\frac{\partial^2 u}{\partial x^2}y_s y_n - \frac{\partial^2 u}{\partial y^2}x_s x_n + (x_s y_n - y_s x_n)\frac{\partial^2 u}{\partial y \partial x}\right]\eta ds \tag{14.203}$$

Using (14.195), (14.202), and (14.203), we obtain the following formula for (14.190):

$$\iint_A \left[\nabla^4 u - \frac{q}{D}\right]\eta dxdy + \int_\Gamma [M(u) + m(s)]\frac{\partial \eta}{\partial n} ds - \int_\Gamma [P(u) + p(s)]\eta ds \tag{14.204}$$

where

$$M(u) = \nabla^2 u - (1-\nu)\left(\frac{\partial^2 u}{\partial y^2}x_n^2 + \frac{\partial^2 u}{\partial x^2}y_n^2 - 2x_n y_n \frac{\partial^2 u}{\partial y \partial x}\right) \tag{14.205}$$

$$P(u) = \frac{\partial}{\partial n}\nabla^2 u + (1-\nu)\frac{\partial}{\partial s}\left[\frac{\partial^2 u}{\partial x^2}y_s y_n - \frac{\partial^2 u}{\partial y^2}x_s x_n + (x_s y_n - y_s x_n)\frac{\partial^2 u}{\partial y \partial x}\right] \tag{14.206}$$

Therefore, the Euler equation of the problem is

$$\nabla^4 u = \frac{q}{D} \tag{14.207}$$

and the boundary condition on Γ

$$M(u) + m(s) = 0 \tag{14.208}$$

$$P(u) + p(s) = 0 \tag{14.209}$$

We can see that the variational method provides a systematic approach to derive the governing equation of plate bending given in (14.207). For the case of a built-in or fixed support, we have boundary conditions as

$$u = 0, \quad \frac{\partial u}{\partial n} = 0, \quad \text{on } \Gamma \tag{14.210}$$

Employing the Galerkin method, we have

$$\bar{u} = \sum_{k=1}^{n} a_k \varphi_k \tag{14.211}$$

In the context of numerical analysis, the so-called strong form requires

$$\left| u(x,y) - \sum_{k=1}^{n} a_k \varphi_k(x,y) \right| < \varepsilon \tag{14.212}$$

where ε is the error control. This is a rather strong point-wise requirement. The weak form, however, relaxes this requirement to

$$\iint_A [u(x,y) - \sum_{k=1}^{n} a_k \varphi_k(x,y)] dxdy < \varepsilon \tag{14.213}$$

The required accuracy is satisfied in a global sense in the weak form. According to the previous section, the Galerkin method requires:

$$\iint_A [\nabla^4 u_n - \frac{q}{D}] \varphi_s dxdy = 0 \tag{14.214}$$

where $s = 1,2,..., n$ and

$$u_n(x,y) = \sum_{k=1}^{n} a_k \varphi_k(x,y) \tag{14.215}$$

14.13 VIBRATIONS OF CIRCULAR PLATES

For simplicity, let us consider the axisymmetric natural vibrations of circular plates with built-in edges (see Figure 14.9). In this case, we have

$$\nabla^4 u - \lambda u = (\frac{d^2}{dr^2} + \frac{1}{r}\frac{d}{dr})(\frac{d^2 u}{dr^2} + \frac{1}{r}\frac{du}{dr}) - \lambda u = 0 \tag{14.216}$$

where λ is the square of the normalized vibration frequency:

$$\lambda = \frac{12(1-\nu^2)\rho\omega^2}{Eh^2} \tag{14.217}$$

where ω is the circular frequency of vibrations and ρ is the density of the plate.

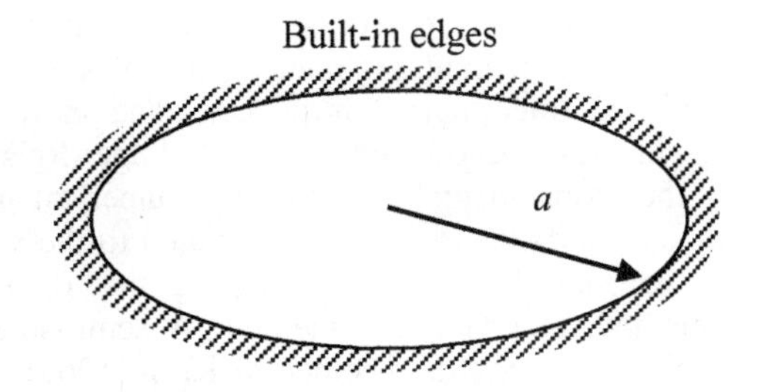

Figure 14.9 Axisymmetric vibrations of circular plates of radius *a* with built-in edges

The built-in edge requires

$$u\big|_{r=a} = \frac{\partial u}{\partial r}\bigg|_{r=a} = 0 \tag{14.218}$$

The following fundamental solutions can be shown to satisfy the built-in conditions:

$$\begin{aligned} u_n &= a_1\varphi_1(r) + a_2\varphi_2(r) + \dots + a_n\varphi_n(r) \\ &= a_1(1-\frac{r^2}{a^2})^2 + a_2(1-\frac{r^2}{a^2})^3 + \dots + a_n(1-\frac{r^2}{a^2})^{n+1} \end{aligned} \tag{14.219}$$

Substitution of (14.219) into the following Galerkin method gives:

$$\iint_A (\nabla^4 u_n - \lambda u_n)\varphi_s dr = 0 \tag{14.220}$$

Taking the first two terms in (14.219) we have

$$a_1(\frac{192}{9} - \frac{\lambda a^4}{5}) + a_2(\frac{144}{9} - \frac{\lambda a^4}{6}) = 0 \tag{14.221}$$

$$a_1(\frac{144}{9} - \frac{\lambda a^4}{6}) + a_2(\frac{96}{5} - \frac{\lambda a^4}{7}) = 0 \tag{14.222}$$

The characteristic equation becomes

$$(\lambda a^4)^2 - \frac{9792}{5}\lambda a^4 + 193536 = 0 \tag{14.223}$$

The smallest root of (14.223) is

$$\lambda = \frac{104.4}{a^4} \tag{14.224}$$

With this value of λ, we find

$$0.455a_1 - 1.4a_2 = 0 \tag{14.225}$$

The approximation of the deflection in (14.219) is

$$u_2 = a_1[(1-\frac{r^2}{a^2})^2 + 0.325(1-\frac{r^2}{a^2})^3] \tag{14.226}$$

14.14 SUMMARY

In this chapter, we have presented an overview of variational methods. Because of its relation with the Galerkin method, the related principles of virtual work, virtual displacement, and virtual traction are introduced. The more general Hamilton principle, Veubeke-Hu-Washizu principle, and Hellinger-Reissner principle are reviewed in view of their fundamental importance in numerical analysis, such as the finite element method. Approximate techniques related to the variational principle are introduced in Sections 14.8–14.10, including the Rayleigh-Ritz method, weighted residue method, and Galerkin method. A semi-Galerkin type method known as Kantorovich's method, a term coined by Reiss (1965), is introduced using plate bending problems as an example in Section 14.11. In essence, for problems governed by partial differential equations, the Galerkin method is used to approximate one of the variables whereas the other one is solved analytically. Therefore, it is a semi-Galerkin type or semi-analytic technique. Functional formulation for plate bending is considered in Section 14.12 before the vibrations of circular plates are considered.

There are a number of good books on the variational method, including Mura and Koya (1992), Washizu (1982), Kantorovich and Krylov (1964), Reiss (1965), and Reddy (2002). There are also more specialized methods similar to the Galerkin method, such as the Trefftz method. In this method, the approximate solution is selected such that the governing equation is exactly satisfied and the boundary conditions are satisfied approximately in a variational sense. However, it is not easy to find the approximation that satisfies the governing equation, and thus the Trefftz method is not discussed in the present chapter.

14.15 PROBLEMS

Problem 14.1 Consider the two-dimensional stress analysis shown in Figure 14.7. Find the stress state in the plate by the Galerkin method using the following approximation of Airy's stress function:

$$\phi = \frac{1}{2}T_0 y^2 (1 - \frac{1}{6}\frac{y^2}{b^2}) + (x^2 - a^2)^2 (y^2 - b^2)^2 (a_1 + a_2 x^2 + a_3 y^2 + ...) \quad (14.227)$$

(i) Show that the constants a_1, a_2, and a_3 are governed by

$$a_1(\frac{64}{7} + \frac{256}{49}\frac{b^2}{a^2} + \frac{64}{7}\frac{b^4}{a^4}) + a_2 a^2 (\frac{64}{77} + \frac{64}{49}\frac{b^4}{a^4}) + a_3 a^2 (\frac{64}{49}\frac{b^2}{a^2} + \frac{64}{77}\frac{b^6}{a^6}) = \frac{T_0}{a^4 b^2} \quad (14.228)$$

$$a_1(\frac{64}{11} + \frac{64}{7}\frac{b^4}{a^4}) + a_2 a^2 (\frac{192}{143} + \frac{256}{77}\frac{b^2}{a^2} + \frac{192}{7}\frac{b^4}{a^4}) + a_3 a^2 (\frac{64}{77}\frac{b^2}{a^2} + \frac{64}{77}\frac{b^6}{a^6}) = \frac{T_0}{a^4 b^2} \quad (14.229)$$

$$a_1(\frac{64}{7} + \frac{64}{11}\frac{b^4}{a^4}) + a_2 a^2 (\frac{64}{77} + \frac{64}{77}\frac{b^4}{a^4}) + a_3 a^2 (\frac{192}{7}\frac{b^2}{a^2} + \frac{256}{77}\frac{b^4}{a^4} + \frac{192}{143}\frac{b^6}{a^6}) = \frac{T_0}{a^4 b^2} \quad (14.230)$$

(ii) For the case of square plates, show that

$$a_1 = 0.04040\frac{T_0}{a^6}, \quad a_2 = a_3 = 0.01174\frac{T_0}{a^8} \tag{14.231}$$

Problem 14.2 Find the deflection of the cantilever beam problem shown in Figure 14.10 using the principle of virtual displacement together with the Galerkin method.

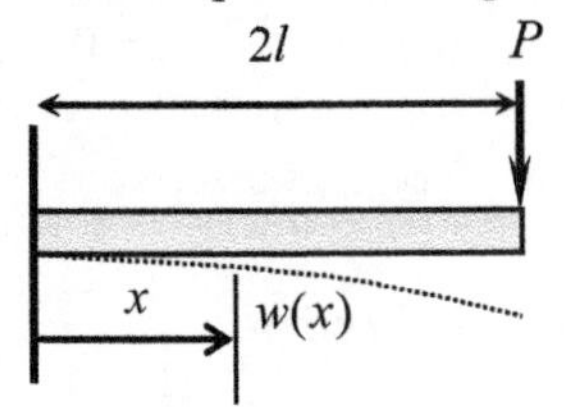

Figure 14.10 A cantilever beam subject to a concentrated force at the free end

Ans:

$$w(x) = -\frac{Pl}{EI}x^2 + \frac{1}{6}\frac{P}{EI}x^3 \tag{14.232}$$

Problem 14.3 Find the deflection of the cantilever beam problem shown in Figure 14.11 using the principle of virtual displacement together with the Galerkin method.

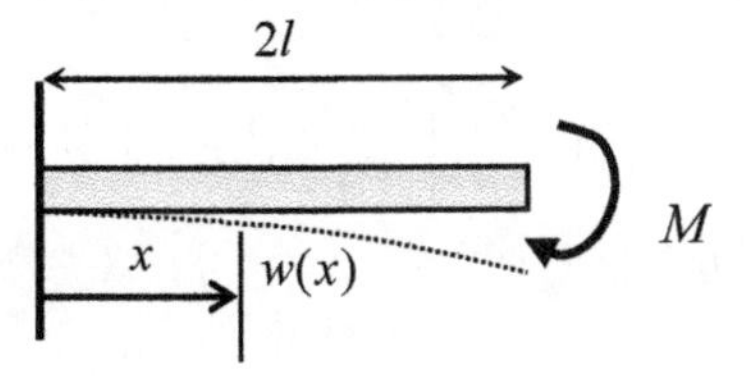

Figure 14.11 A cantilever beam subject to a concentrated moment

Ans:

$$w(x) = -\frac{M}{2EI}x^2 \tag{14.233}$$

Problem 14.4 Find the deflection of the cantilever beam problem shown in Figure 14.12 using the principle of virtual displacement together with Galerkin method.

Ans:

$$w(x) = -\frac{1}{8EI}(19Pl + 4M)x^2 + \frac{5}{12}\frac{P}{EI}x^3 \tag{14.234}$$

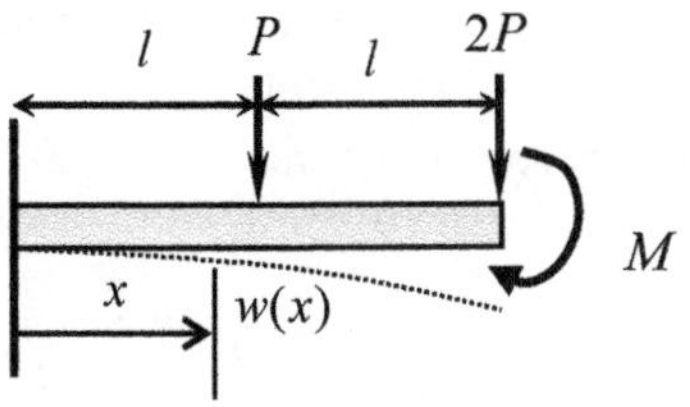

Figure 14.12 A cantilever beam subject to two concentrated forces and a concentrated moment

Problem 14.5 Show that the functional of the following Sturm-Liouville problem:

$$\frac{d}{dx}[p(x)\frac{dy}{dx}]+q(x)y+\lambda r(x)y=0 \tag{14.235}$$

$$y'(a)=0, \quad y'(b)=0 \tag{14.236}$$

is

$$J=a\int_a^b [p(\frac{dy}{dx})^2-q(x)y^2-\lambda r(x)y^2]dx \tag{14.237}$$

Hint: Find Euler's equation of this function.

Problem 14.6 Show that the functional of the Laplace equation:

$$\nabla^2\psi=0 \tag{14.238}$$

is

$$J=\frac{1}{2}\iint[(\frac{\partial\psi}{\partial x})^2+(\frac{\partial\psi}{\partial y})^2]dxdy \tag{14.239}$$

Hint: Find Euler's equation of this function.

Problem 14.7 Following the procedure used in Section 14.12 for plate bending, show that for functional:

$$I[y]=\frac{EI}{2}\int_0^l (\frac{d^2y}{dx^2})^2 dx-\int_0^l q\, ydx \tag{14.240}$$

we have, by substituting $y = u+\alpha\eta$,

$$\left.\frac{dI(u+\alpha\eta)}{d\alpha}\right|_{\alpha=0}=0 \tag{14.241}$$

The stationary value of the functional requires

$$EI\int_0^l (\frac{d^2u}{dx^2})(\frac{d^2\eta}{dx^2})dx-\int_0^l q\,\eta dx=0 \tag{14.242}$$

Problem 14.8 Show that Euler's equation for the function given in (14.243) is the Euler-Bernoulli beam theory:

$$EI\frac{d^4y}{dx^4}-q(x)=0 \tag{14.243}$$

Problem 14.9 Consider the case of a simply supported beam subject to a uniform distributed load shown in Figure 14.13. The essential boundary conditions are

$$u(0)=u(l)=0 \tag{14.244}$$

whereas the natural boundary conditions are

$$u''(0)=u''(l)=0 \tag{14.245}$$

Use the Rayleigh-Ritz method by adopting the following approximation:

$$u(x) = c_1\phi_1(x) + c_2\phi_2(x) \tag{14.246}$$

where

$$\phi_1(x) = x(l-x), \quad \phi_2(x) = x^2(l-x). \tag{14.247}$$

Note that the approximation given in (14.247) only satisfies the essential boundary condition. Find the Rayleigh-Ritz approximation.

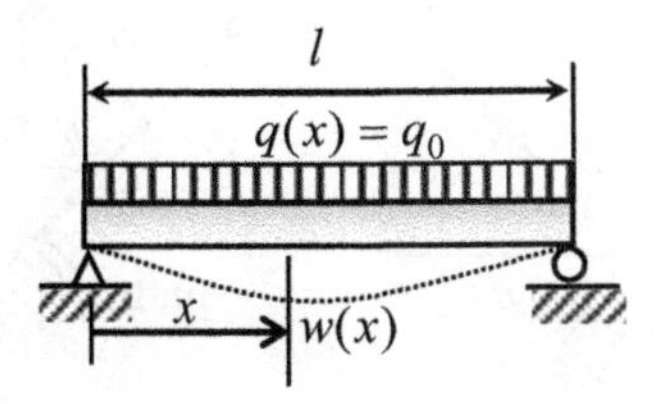

Figure 14.13 A simply-supported beam subject to uniform distributed load

Ans:

$$u(x) = \frac{q_0 l^4}{24EI}\left(\frac{x}{l} - \frac{x^2}{l^2}\right) \tag{14.248}$$

Problem 14.10 Find Euler's equation of the following functional for membrane deflection:

$$I[u] = \iint_A \left[\frac{1}{2}T\left\{\left(\frac{\partial u}{\partial x}\right)^2 + \left(\frac{\partial u}{\partial y}\right)^2\right\} - f\,u\right]dxdy \tag{14.249}$$

where T is the tension in the membrane and f is the applied load on the membrane. The boundary condition is

$$u = 0, \quad \text{on } \Gamma \tag{14.250}$$

Ans:

$$T\left(\frac{\partial^2 u}{\partial x^2} + \frac{\partial^2 u}{\partial y^2}\right) + f = 0 \tag{14.251}$$

Problem 14.11 Further simplify (14.205) and (14.206) by referring to the following diagram:

(i) Referring to Figure 14.14, show that

$$x_n = y_s = n_x = \cos(n, x), \quad y_n = x_s = n_y = \cos(n, y) \tag{14.252}$$

(ii) Show that (14.205) and (14.206) can further be simplified to

$$M(u) = \left(\frac{\partial^2 u}{\partial x^2} + \nu\frac{\partial^2 u}{\partial y^2}\right)n_x^2 + \left(\frac{\partial^2 u}{\partial y^2} + \nu\frac{\partial^2 u}{\partial x^2}\right)n_y^2 + 2(1-\nu)n_x n_y \frac{\partial^2 u}{\partial y \partial x} \tag{14.253}$$

$$P(u)=\frac{\partial}{\partial n}\nabla^2 u+\frac{\partial}{\partial s}[(\frac{\partial^2 u}{\partial x^2}+\nu\frac{\partial^2 u}{\partial y^2})n_x n_y-(\frac{\partial^2 u}{\partial y^2}+\nu\frac{\partial^2 u}{\partial x^2})n_x n_y$$
$$+(1-\nu)(n_y^2-n_x^2)\frac{\partial^2 u}{\partial y\partial x}] \tag{14.254}$$

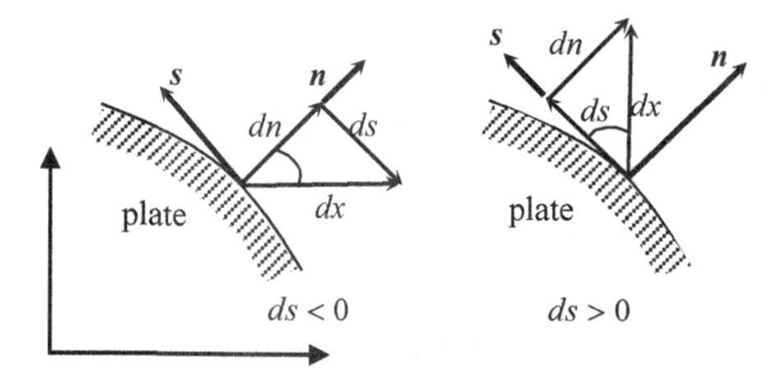

Figure 14.14 The relation between directional cosines between normal and tangent

Problem 14.12 Show that along a straight edge along $x = a$ the boundary conditions given in (14.208) and (14.209) become

$$\frac{\partial^2 u}{\partial x^2}+\nu\frac{\partial^2 u}{\partial y^2}+m=0,\quad \frac{\partial^3 u}{\partial x^3}+(2-\nu)\frac{\partial^3 u}{\partial x\partial y^2}+p=0 \tag{14.255}$$

CHAPTER FIFTEEN

Finite Difference and Numerical Methods

15.1 INTRODUCTION

The origin of the finite difference method (FDM) probably traces back to the time of Leibniz and Euler (e.g., Euler's method in 1768), and subsequently evolved into different techniques (e.g., Runge-Kutta method). The FDM became more established after 1928 after the Courant-Friedrichs-Lewy (CFL) stability condition was derived for hyperbolic type partial differential equations (Courant et al. 1928). Its day-to-day application, of course, starts with the popularization of computers, especially personal computers. Although for solid mechanics and structural analysis the emergence of the finite element method in 1960 took over the role of finite difference in numerical analysis, in the area of fluid mechanics, the finite difference method remains a popular choice.

There are many different kinds of finite difference schemes. In general, it can be classified into explicit and implicit finite difference schemes. For differential equations in time variables, the explicit methods naturally suggest that the unknown function of the next time step can be expressed explicitly in terms of the numerical results of the previous time steps, whereas in the implicit methods the unknown function of the next time step cannot be expressed explicitly in terms of the numerical results of the previous time steps. In structural dynamics, the most popular methods of finite difference for second order differential equations are the Wilson θ method and the Newmark β method (Bathe, 1982). These methods will be discussed in Section 15.3. Multi-step methods, such as the Adams-Bashforth method and Adams-Moulton method, and the predictor-corrector method, use more than just the time step before the current step, but the solutions from the last few time steps. Polynomials of different orders are used to fit the first derivative in the differential equations. The predictor-corrector method combines the explicit together with the implicit method.

A major topic in numerical integration for time dependent nonlinear systems is the Newton-Raphson method. For searching roots in nonlinear equations, the Newton-Raphson method has been proposed. For structural dynamics problems, the stiffness matrix of the numerical model needed to be evaluation at each iteration step within a time step of integration. Therefore, it is computationally very demanding, especially for a large system (i.e., stiffness matrix of very large size). There are various modified versions of the Newton-Raphson method. If we use the initial stiffness of the model in all time steps as well as in iteration steps, this method is called the initial stress method. Clearly, this scheme will not converge very fast or may not converge at all. The modified Newton-Raphson method takes the stiffness matrix from the last time step during the whole process of iterations within a time step. Yet, the convergence may not converge fast enough for some highly nonlinear problems. Therefore, it has been proposed that the stiffness matrix should be updated once the convergence is not up to expectation within the iterative steps within a time step. This leads to the so-called quasi-Newton and BFGS methods. If the stiffness matrix is formulated between the starting point of each iteration process and the current iterative point, the method is called fixed-point iteration or the secant method. For mechanics problems, the control algorithms for nonlinear problems are very important. They are displacement, load, and arc-length controls, and pros and cons of these algorithms will be discussed.

In the final sections, we will illustrate the application of the finite difference method as well as the finite element method to the incompressible potential flow problems.

15.2 FINITE DIFFERENCE FOR FIRST ORDER ODE

Let us consider the following first order ODE in this section:

$$\frac{dy}{dt} = f(t, y) \tag{15.1}$$

More generally, we can extend this first ODE to a system of n coupled first order ODEs as:

$$\boldsymbol{C}\frac{d\boldsymbol{y}}{dt} + \boldsymbol{k}(\boldsymbol{y}) = \boldsymbol{f}(t) \tag{15.2}$$

where $\boldsymbol{C}$ is a matrix of size $n \times n$ and $\boldsymbol{y}$ and $\boldsymbol{f}$ are $n \times 1$ vectors. The linearized form of (15.2) can be written as

$$\boldsymbol{C}\frac{d\boldsymbol{y}}{dt} + \boldsymbol{K}\boldsymbol{y} = \boldsymbol{f}(t) \tag{15.3}$$

All kinds of finite difference schemes can be formulated by starting with Taylor series expansion:

$$y(t_{n+1}) = y(t_n + \Delta t) = y(t_n) + \Delta t \left.\frac{dy}{dt}\right|_{t=t_n} + \frac{\Delta t^2}{2!}\left.\frac{d^2 y}{dt^2}\right|_{t=t_n} + \frac{\Delta t^3}{3!}\left.\frac{d^3 y}{dt^3}\right|_{t=t_n} + \ldots \tag{15.4}$$

In the following subsections, we will apply the general mean value theorem (see Article 150 of Hardy, 1944) to (15.4) in arriving at different types of finite difference schemes.

15.2.1 Forward Difference (Euler) Method

If we evaluate the second derivative at an appropriate point, say $t_n+\theta_1\Delta t$, all higher derivative terms in the Taylor series expansion can be dropped

$$y(t_{n+1}) = y(t_n + \Delta t) = y(t_n) + \Delta t \left.\frac{dy}{dt}\right|_{t=t_n} + \frac{\Delta t^2}{2}\left.\frac{d^2 y}{dt^2}\right|_{t=t_n+\theta_1\Delta t} \tag{15.5}$$

where $0 \le \theta_1 \le 1$. This is called the general mean value theorem or Taylor theorem (e.g., Article 150 of Hardy, 1944). To simplify the following presentations, we should adopt the following notation:

$$y(t_{n+1}) = y_{n+1} \tag{15.6}$$

With this notation, (15.5) can be rearranged as:

$$\left.\frac{dy}{dt}\right|_{t=t_n} = \frac{y_{n+1} - y_n}{\Delta t} - \frac{\Delta t}{2}\left.\frac{d^2 y}{dt^2}\right|_{t=t_n+\theta_1 \Delta t} \tag{15.7}$$

If we drop the last term, we get the following forward difference approximation or the Euler's finite difference scheme as:

$$\left.\frac{dy}{dt}\right|_{t=t_n} \approx \frac{y_{n+1} - y_n}{\Delta t} \tag{15.8}$$

Comparing (15.7) and (15.8), it is clear that the error is proportional to Δt. We say that this error is of the order Δt or known as the order of the error $O(\Delta t)$ (e.g., Erdelyi, 1956; Bleistein and Handelsman, 1986).

Substitution of (15.8) into (15.1) yields

$$y_{n+1} = y_n + (t_{n+1} - t_n) f(t_n, y_n) \tag{15.9}$$

Therefore, once we know y_n and t_n, we can find y_{n+1} from (15.9). Since the unknown is on the left hand side only, it is an explicit finite difference scheme. Similarly, we can substitute (15.8) into (15.3) to get

$$\frac{\boldsymbol{C}}{\Delta t}\boldsymbol{y}_{n+1} + (-\frac{\boldsymbol{C}}{\Delta t} + \boldsymbol{K})\boldsymbol{y}_n = \boldsymbol{f}_n \tag{15.10}$$

When $\boldsymbol{C}$ is a diagonal matrix (i.e., all off-diagonal terms are zeros), the unknown vector $\boldsymbol{y}_{n+1}$ can be expressed explicitly as:

$$\boldsymbol{y}_{n+1} = \Delta t \boldsymbol{C}^{-1}\{(\boldsymbol{K} - \frac{\boldsymbol{C}}{\Delta t})\boldsymbol{y}_n + \boldsymbol{f}_n\} \tag{15.11}$$

However, when $\boldsymbol{C}$ is non-diagonal, (15.10) is not explicit but implicit. That is, we need to solve the matrix equation to find $\boldsymbol{y}_{n+1}$.

15.2.2 Backward Difference (Euler) Method

Alternatively, we can evaluate the function at a previous step ϕ_{n-1} or substitute $-\Delta t_n$ for Δt_n in (15.5) to get

$$y(t_{n-1}) = y(t_n - \Delta t) = y(t_n) - \Delta t \left.\frac{dy}{dt}\right|_{t=t_n} + \frac{\Delta t^2}{2}\left.\frac{d^2 y}{dt^2}\right|_{t=t_n-\theta_2 \Delta t} \tag{15.12}$$

where $0 \le \theta_2 \le 1$. By rearranging (15.12), we obtain the first derivative term as

$$\left.\frac{dy}{dt}\right|_{t=t_n} = \frac{y_n - y_{n-1}}{\Delta t} + \frac{\Delta t}{2}\left.\frac{d^2 y}{dt^2}\right|_{t=t_n-\theta_2 \Delta t} \tag{15.13}$$

Therefore, by dropping the second term on the right of (15.13), we have the backward difference scheme as

$$\left.\frac{dy}{dt}\right|_{t=t_n} \approx \frac{y_n - y_{n-1}}{\Delta t} \tag{15.14}$$

It is clear from (15.13) and (15.14) that the error is again of $O(\Delta t)$ which is the same as that of Euler's forward difference scheme.

Substitution of (15.14) into (15.1) yields

$$y_{n+1} = y_n + (t_{n+1} - t_n) f(t_{n+1}, y_{n+1}) \tag{15.15}$$

Therefore, once we know y_n and t_n, we can find y_{n+1} from (15.15). Unless the function f is given as a simple function such that y_{n+1} can be solved explicitly from (15.15), the backward difference scheme is an implicit method. Similarly, we can substitute (15.14) into (15.3) to get

$$(\frac{\boldsymbol{C}}{\Delta t} + \boldsymbol{K}) \boldsymbol{y}_{n+1} - \frac{\boldsymbol{C}}{\Delta t} \boldsymbol{y}_n = \boldsymbol{f}_{n+1} \tag{15.16}$$

Even when $\boldsymbol{C}$ is a diagonal matrix (i.e., all off-diagonal terms are zeros), it is unlikely that $\boldsymbol{K}$ will be diagonal. If this is the case, the original problem (15.3) is totally uncoupled. That is, each component of vector $\boldsymbol{y}$ can be determined separately. Therefore, the backward difference scheme is implicit.

15.2.3 Central Difference (Crank-Nicholson) Method

The idea of applying the general mean value theorem can be extended to consider higher order derivative terms in (15.4). In particular, we can apply the general mean value theorem to the third derivative as:

$$y_{n+1} = y_n + \Delta t \frac{dy}{dt}\bigg|_{t=t_n} + \frac{\Delta t^2}{2} \frac{d^2 y}{dt^2}\bigg|_{t=t_n} + \frac{\Delta t^3}{6} \frac{d^3 y}{dt^3}\bigg|_{t=t_n+\theta_3 \Delta t} \tag{15.17}$$

$$y_{n-1} = y_n - \Delta t \frac{dy}{dt}\bigg|_{t=t_n} + \frac{\Delta t^2}{2} \frac{d^2 y}{dt^2}\bigg|_{t=t_n} - \frac{\Delta t^3}{6} \frac{d^3 y}{dt^3}\bigg|_{t=t_n-\theta_4 \Delta t} \tag{15.18}$$

for some $0 \le \theta_3 \le 1$ and $0 \le \theta_4 \le 1$. Subtracting (15.18) from (15.17) we have

$$y_{n+1} - y_{n-1} = 2\Delta t \frac{dy}{dt}\bigg|_{t=t_n} + \frac{\Delta t^3}{6} \left[\frac{d^3 y}{dt^3}\bigg|_{t=t_n+\theta_4 \Delta t} + \frac{d^3 y}{dt^3}\bigg|_{t=t_n-\theta_4 \Delta t} \right] \tag{15.19}$$

Dropping the higher order term, we obtain the central difference scheme for the first derivative as

$$\frac{dy}{dt}\bigg|_{t=t_n} \approx \frac{y_{n+1} - y_{n-1}}{2\Delta t} \tag{15.20}$$

When we apply this central difference scheme to the heat conduction problem, it is also known as the Crank-Nicholson method (Zienkiewicz, 1977). The first derivative in (15.20) is expressed in terms of the solution at time $t_n+\Delta t$ and $t_n-\Delta t$, and thus it is natural to approximate the right hand side of (15.1) as

$$f(t, y) = \frac{f(t_{n+1}, y_{n+1}) + f(t_{n-1}, y_{n-1})}{2} \tag{15.21}$$

By using (15.21), the finite difference will remain a two-level scheme.

By applying (15.20) and (15.21) to (15.1), we have

$$y_{n+1} = y_{n-1} + \frac{1}{2}(t_{n+1} - t_{n-1}) \left[f(t_{n+1}, y_{n+1}) + f(t_{n-1}, y_{n-1}) \right] \tag{15.22}$$

Since y_{n+1} appears implicitly on the right of (15.22), it is an implicit method.

Following a similar two-level scheme for y and f, we have

$$y=\frac{1}{2}(y_{n-1}+y_{n+1}),\quad f=\frac{1}{2}(f_{n-1}+f_{n+1}) \tag{15.23}$$

Thus, substitution of (15.23) and (15.20) into (15.3) results in

$$(\frac{C}{2\Delta t}+\frac{K}{2})y_{n+1}+(-\frac{C}{2\Delta t}+\frac{K}{2})y_{n-1}=\frac{1}{2}(f_{n-1}+f_{n+1}) \tag{15.24}$$

Following the remarks made earlier, (15.24) is clearly an implicit finite difference scheme.

The central difference scheme can sometimes appear in a slightly different form. For example, we can expand the Taylor series expansion using half time step $\Delta t/2$ as:

$$y_{n+1/2}=y_n+\frac{\Delta t}{2}\frac{dy}{dt}\bigg|_{t=t_n}+\frac{\Delta t^2}{8}\frac{d^2y}{dt^2}\bigg|_{t=t_n}+\frac{\Delta t^3}{48}\frac{d^3y}{dt^3}\bigg|_{t=t_n+\theta_3\Delta t/2} \tag{15.25}$$

$$y_{n-1/2}=y_n-\frac{\Delta t}{2}\frac{dy}{dt}\bigg|_{t=t_n}+\frac{\Delta t^2}{8}\frac{d^2y}{dt^2}\bigg|_{t=t_n}-\frac{\Delta t^3}{48}\frac{d^3y}{dt^3}\bigg|_{t=t_n-\theta_4\Delta t/2} \tag{15.26}$$

The difference between (15.25) and (15.26) is

$$y_{n+1/2}-y_{n-1/2}=\Delta t\frac{dy}{dt}\bigg|_{t=t_n}+\frac{\Delta t^3}{48}\left[\frac{d^3y}{dt^3}\bigg|_{t=t_n+\theta_4\Delta t/2}+\frac{d^3y}{dt^3}\bigg|_{t=t_n-\theta_4\Delta t/2}\right] \tag{15.27}$$

Now, the first derivative of y becomes

$$\frac{dy}{dt}\bigg|_{t=t_n}\approx\frac{y_{n+1/2}-y_{n-1/2}}{\Delta t} \tag{15.28}$$

Using a two-level scheme similar to (15.21), we find that (15.1) can be evaluated in a slightly different form

$$y_{n+1/2}=y_{n-1/2}+\frac{1}{2}(t_{n+1/2}-t_{n-1/2})\left[f(t_{n+1/2},y_{n+1/2})+f(t_{n-1/2},y_{n-1/2})\right] \tag{15.29}$$

Similarly, the central difference scheme for (15.3) becomes

$$(\frac{C}{\Delta t}+\frac{K}{2})y_{n+1/2}+(-\frac{C}{\Delta t}+\frac{K}{2})y_{n-1/2}=\frac{1}{2}(f_{n-1/2}+f_{n+1/2}) \tag{15.30}$$

Alternatively, we can also shift half of the time step to rewrite (15.30) as

$$(\frac{C}{\Delta t}+\frac{K}{2})y_{n+1}+(-\frac{C}{\Delta t}+\frac{K}{2})y_n=\frac{1}{2}(f_n+f_{n+1}) \tag{15.31}$$

This is the Crank-Nicholson (or central difference) scheme given in Zienkiewicz and Morgan (1983).

The interpretation of these three finite difference schemes can be seen in Fig. 15.1. The actual slope BD at time t_n can be represented by slope AB in the backward finite difference scheme, by slope BC in the forward finite difference scheme, and by AC in the central finite difference scheme.

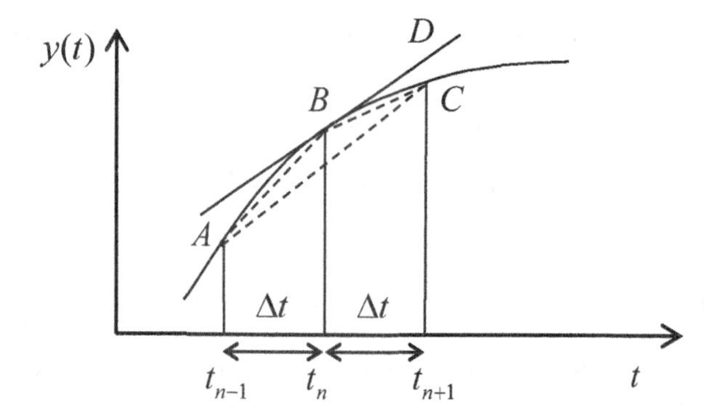

Figure 15.1 Interpretation of the first derivative using the forward, central, and backward finite difference schemes

15.2.4 Weighted Residue Approach for Finite Difference Scheme

As shown by Zienkiewicz and Morgan (1983), all of the forward, central, and backward finite difference schemes can be recovered as a special case of a more general formulation using the weighted residue approach, which has been introduced in Section 14.9.

In particular, we can adopt a time discretization using linear shape functions as shown in Fig. 15.2. In particular, the unknown function y can be approximated by

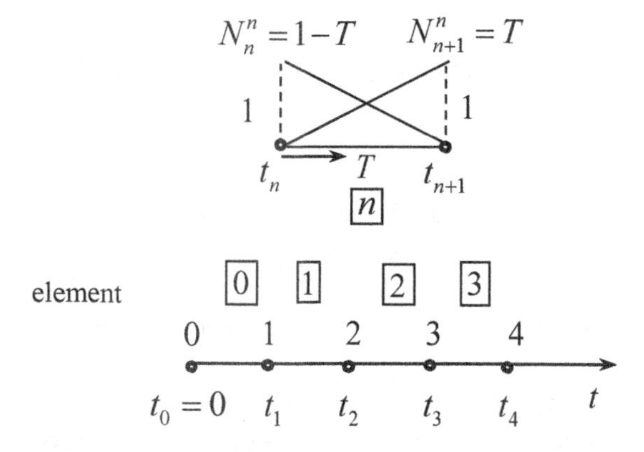

Figure 15.2 Time discretization using nodal values with interpolation function

$$y \approx \tilde{y} = \sum_{m=1}^{\infty} y_m N_m \tag{15.32}$$

If we define the time element n between times t_n and t_{n+1}, the approximation for the unknown function y can be rewritten as

$$y \approx \tilde{y} = y_n N_n^n + y_{n+1} N_{n+1}^n \tag{15.33}$$

where

$$N_n^n = 1 - T\,,\; N_{n+1}^n = T \tag{15.34}$$

$$T = \frac{t - t_n}{\Delta t_n}\,,\;\; \Delta t_n = t_{n+1} - t_n \tag{15.35}$$

By formulating the weighted residue of the approximation of (15.3) when (15.32) is substituted into it, we have

$$\int_0^\infty \left[\boldsymbol{C}\frac{d\tilde{\boldsymbol{y}}}{dt} + \boldsymbol{K}\tilde{\boldsymbol{y}} - \boldsymbol{f}(t) \right] W_n dt = 0\,, \tag{15.36}$$

for $n = 0,1,2,...\infty$, and W_n is called the weighting function. Applying different forms of the weighting function will result in different kinds of numerical methods. For the case that W_n is only nonzero within the n element, (15.36) can be simplified to

$$\int_{a_n}^{a_{n+1}} \left[\boldsymbol{C}\frac{d\tilde{\boldsymbol{y}}}{dt} + \boldsymbol{K}\tilde{\boldsymbol{y}} - \boldsymbol{f}(t) \right] W_n dt = 0\,. \tag{15.37}$$

In view of (15.34), the time derivative of (15.33) becomes

$$\frac{d\tilde{\boldsymbol{y}}}{dt} = -\frac{\boldsymbol{y}_n}{\Delta t_n} + \frac{\boldsymbol{y}_{n+1}}{\Delta t_n}\,. \tag{15.38}$$

Substitution of (15.33) and (15.38) into (15.37) and change of variables from t to T leads to

$$\left\{ \frac{\boldsymbol{C}}{\Delta t_n}\int_0^1 W_n dT + \boldsymbol{K}\int_0^1 W_n T dT \right\} \boldsymbol{y}_{n+1} + \left\{ -\frac{\boldsymbol{C}}{\Delta t_n}\int_0^1 W_n dT + \boldsymbol{K}\int_0^1 W_n (1-T) dT \right\} \boldsymbol{y}_n$$
$$= \int_0^1 \boldsymbol{f}(t_n + T\Delta t_n) W_n dT \tag{15.39}$$

Equation (15.39) can be simplified as

$$\left\{ \frac{\boldsymbol{C}}{\Delta t_n} + \gamma_n \boldsymbol{K} \right\} \boldsymbol{y}_{n+1} + \left\{ -\frac{\boldsymbol{C}}{\Delta t_n} + (1-\gamma_n)\boldsymbol{K} \right\} \boldsymbol{y}_n = \tilde{\boldsymbol{f}}_n \tag{15.40}$$

where

$$\gamma_n = \int_0^1 W_n T dT \Big/ \int_0^1 W_n dT\,,\;\; \tilde{\boldsymbol{f}}_n = \int_0^1 \boldsymbol{f}(t_n + T\Delta t_n) W_n dT \Big/ \int_0^1 W_n dT \tag{15.41}$$

If the applied force $\boldsymbol{f}$ is smooth, we can also approximate the forcing term by node values by using the same interpolation or shape function given in (15.33) as

$$\tilde{\boldsymbol{f}}(t_n + T\Delta t_n) \approx \boldsymbol{f}_n N_n^n(T) + \boldsymbol{f}_{n+1} N_{n+1}^n(T) \tag{15.42}$$

Thus, we can approximate (15.3) as

$$\left[\frac{\boldsymbol{C}}{\Delta t_n} + \gamma_n \boldsymbol{K} \right] \boldsymbol{y}_{n+1} + \left[-\frac{\boldsymbol{C}}{\Delta t_n} + (1-\gamma_n)\boldsymbol{K} \right] \boldsymbol{y}_n = (1-\gamma_n)\tilde{\boldsymbol{f}}_n + \gamma_n \tilde{\boldsymbol{f}}_{n+1} \tag{15.43}$$

15.2.5 Dirac Delta Function and Point Collocation Method

The simplest type of weighted residue formulation is obtained by using the Dirac delta function for the weighting function. This method has been mentioned in Section 14.9.2. of Chapter 14. This type of numerical technique is also known as the point collocation method (e.g., Section 2.2.1 of Zienkiewicz and Morgan, 1983; Section 1.4.1 of Brebbia et al., 1983). For example, the point collocation method has been found useful in investigating stress concentration due to rivet loading in a finite strip (see Ho and Chau, 1999).

That is, we can assume

$$W_n = \delta(T-\theta) \tag{15.44}$$

where the Dirac delta function δ can be defined as

$$\delta(T-\theta) = \begin{cases} 0 & T \neq \theta \\ \lim\limits_{\Delta T \to 0} \dfrac{1}{\Delta T} & T = \theta \end{cases} \tag{15.45}$$

We will summarize some properties of the Dirac delta function here without discussing the distribution theory as we did in Section 8.11. This function tends to infinity at $T = \theta$, but however its integral over the entire domain is finite and defined by

$$\int_0^\infty \delta(T-\theta)dt = 1 \tag{15.46}$$

The following sifting property of the Dirac delta function is most useful for our discussion in this chapter:

$$\int_0^\infty h(T)\delta(T-\theta)dT = h(\theta) \tag{15.47}$$

where $\theta > 0$ (named by van der Pol). Before we continue to consider (15.43) by using the Dirac function given in (15.44), an informal discussion of Dirac delta will be informative.

This delta function was motivated by its application in quantum mechanics and was proposed by Dirac to deal with jump properties of physical quantities (Dirac, 1947). Dirac shared the 1933 Nobel Prize in physics with Schrödinger. In civil engineering and mechanics, they include impulsive force, concentrated force (in contrast to distributed force), concentrated moment in beams and structures, heat source and dipole in heat conduction, fluid sources (such as point sink or point source as in Chau, 1996 and Kanok-Nukulchai and Chau, 1990 in fluid-infiltrated solids) or fluid dipole (see Rudnicki, 1986; Chau, 2013), point charges, dipoles, and surface layers in electrostatics. However, if we restrict to our basic assumptions on the existence of continuous, smooth, differentiable functions, the Dirac delta function is certainly not qualified to even be called a "function" in the normal mathematical sense. Dirac was aware of its limitation and called it an "improper function," and he recommended its use in mathematical analysis provided that no inconsistency would follow from its usage. That is, once the solution of a physical problem is solved by adopting the Dirac delta function, it should be subject to classical mathematical analysis to show rigorously that it does satisfy all the conditions posed in the original formulation of the problem. Strictly speaking, all procedures or formulas derived for differentiable functions cannot be applied to deal with the Dirac delta function unless we can generalize the concept of function to include "strange" functions like the Dirac function. Because of this reason, a whole new area of mathematics appeared and it is now called "generalized functions" or "theory of distribution" (Chapter 9 of Sneddon, 1972; Chapter 2 of Stakgold, 1979). This is a classic example of how mathematical development is motivated by physical problems.

The subject of generalized functions or theory of distribution was first considered by Bochner in 1932 and Sobolev in 1936, but it was the work of Schwartz in the 1940s that put the generalized functions on a firm foundation. Later contributors include Gelfand and Shilov (1964) and Zemanian (1965).

The idea of the theory of distribution has been covered in Section 8.11 and is briefly summarized here. First, a so-called "support" (or range) around the Dirac function was defined. Then, some very smooth testing functions with rapid descent were defined such that they are differentiable within the support. At the same time, these testing functions are required to be absolutely integrable over the domain of the variable. Outside the support, these testing functions would vanish identically. The functional space of these admissible testing functions is called the Schwartz space. Rules of mathematical operations are formally defined in this Schwartz space. In general, many distributed forms of the Dirac delta functions can be found and each mathematical form also consists of infinite sequences of such testing functions (Sneddon, 1972). Although these admissible testing functions in Schwartz space are very smooth, their functional (i.e., testing function that satisfies the finite integration requirement) can describe the "wild" nature of the Dirac delta function.

Here we list some admissible choices for the Dirac delta functions using Schwartz's theory of distribution (Stakgold, 1979):

$$\delta(x) = \lim_{t \to 0^+} \frac{t}{\pi(t^2 + x^2)} \tag{15.48}$$

$$\delta(x) = \lim_{t \to 0^+} \frac{e^{-x^2/4t}}{\sqrt{4\pi t}} \tag{15.49}$$

$$\delta(x) = \lim_{t \to 0^+} H(x) \frac{t e^{-t^2/4x}}{\sqrt{4\pi} x^{3/2}} \tag{15.50}$$

$$\delta(x) = \lim_{R \to \infty} \frac{\sin^2 Rx}{\pi R x^2} \tag{15.51}$$

$$\delta(x) = \lim_{t \to 0^+} \frac{e^{-|x|^2/4t}}{(4\pi t)^{n/2}} \tag{15.52}$$

Note that some of these delta sequences have been introduced in Chapter 8 in a slightly different manner (see (8.197), (8.208) and (8.209)). If t is closer to 0^+ or R is closer to ∞, the steeper these functions are and the more they resemble the actual delta function. Stakgold (1979) showed that these functions can be put into a general form as a theorem:

$$f_\alpha(x) = \frac{1}{\alpha^n} f(\frac{x}{\alpha}) \tag{15.53}$$

where $\alpha > 0$ and $\{f_\alpha(x)\}$ is a delta family as $\alpha \to 0$ provided that f is a nonnegative locally integrable function satisfying

$$\int_{-\infty}^{\infty} f(x)dx = 1 \tag{15.54}$$

Or equivalently, by setting $k = 1/\alpha$ we have

$$S_k(x) = k^n f(kx) \tag{15.55}$$

where $\{S_k(x)\}$ is a delta sequence $k \to \infty$. Thus, we have

$$\delta(x) = \lim_{\alpha \to 0} f_\alpha(x), \quad \delta(x) = \lim_{k \to \infty} S_k(x) \tag{15.56}$$

Another type of Dirac delta sequence is (Stakgold, 1979)

$$\delta(x) = \lim_{R \to \infty} \frac{\sin Rx}{\pi x} \tag{15.57}$$

$$\delta(x) = \begin{cases} \lim_{k \to \infty} \dfrac{(1-x^2)^k}{\int_{-1}^{1} (1-x^2)^k dx}, & |x| < 1 \\ 0 & |x| \geq 1 \end{cases}, \tag{15.58}$$

$$\delta(\theta) = \begin{cases} \lim_{r \to 1^-} \dfrac{1}{2\pi} (\dfrac{1-r^2}{1+r^2-2r\cos\theta}), & |\theta| \leq \pi \\ 0 & |\theta| > \pi \end{cases} \tag{15.59}$$

$$\delta(x) = \begin{cases} \lim_{k \to \infty} \sum_{m=-k}^{k} \dfrac{1}{2\pi} e^{imx} = \dfrac{\sin(k+1/2)x}{2\pi \sin(\frac{1}{2}x)}, & |x| \leq \pi \\ 0 & |x| > \pi \end{cases} \tag{15.60}$$

Note that (15.57) and (15.60) have been covered in Section 8.11.2; whereas, (15.59) appears in the Poisson integral in potential theory (see Section 9.7.6).

Example 15.1 Construct a delta function sequence for the following function:

$$g(x) = \frac{y}{y^2 + x^2} \tag{15.61}$$

where x is the variable and y is a parameter.

Solutions:

Let us make the following observation of this function. We note that when $x = 0$,

$$g(x) = \frac{1}{y} \tag{15.62}$$

We note that when $y \to 0$,

$$g(x) \sim \frac{y}{x^2} \tag{15.63}$$

Figure 15.3 plots $g(x)$ versus x for three different values of y. As shown, the smaller is y the steeper is the function. As expected, the Dirac delta function type is expected as $y \to 0$.

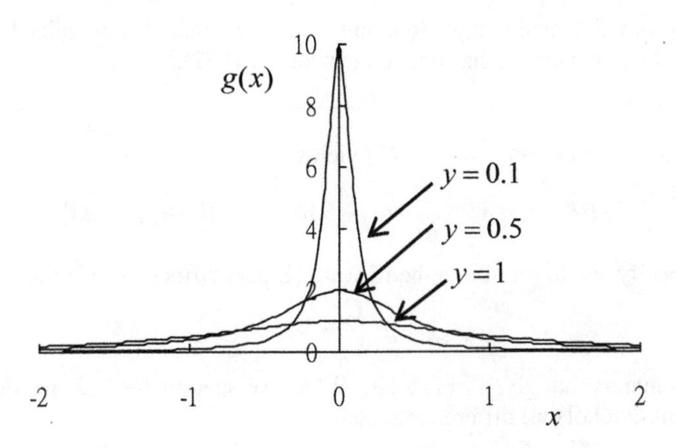

Figure 15.3 Dirac delta function like testing functions

Note that

$$\int \frac{dx}{y^2 + x^2} = \frac{1}{y}\tan^{-1}\left(\frac{x}{y}\right) \tag{15.64}$$

$$\int_{-\infty}^{\infty} \frac{dx}{y^2 + x^2} = \frac{1}{y}\tan^{-1}\left(\frac{x}{y}\right)_{-\infty}^{\infty} = \begin{cases} \frac{1}{y}[\tan^{-1}(\infty) - \tan^{-1}(-\infty)] = \frac{\pi}{y} & y > 0 \\ \frac{1}{y}[\tan^{-1}(-\infty) - \tan^{-1}(\infty)] = -\frac{\pi}{y} & y < 0 \end{cases} \tag{15.65}$$

$$\int_{-\infty}^{\infty} \frac{dx}{y^2 + x^2} = \frac{\pi}{|y|} \tag{15.66}$$

Thus, we can define a testing function as

$$f(x) = \frac{|y|}{\pi(y^2 + x^2)} = \frac{1}{|y|}\frac{|y|^2}{\pi(y^2 + x^2)} = \frac{1}{|y|}\frac{1}{\pi[1 + (x/|y|)^2]} \tag{15.67}$$

such that (15.54) is satisfied identically. The last part of (15.67) is clearly in the structural form of (15.53) and (15.56), as remarked by Stakgold (1979). Therefore, we have the delta function as

$$\delta(x) = \lim_{|y| \to 0} f(x) = \lim_{|y| \to 0} \frac{1}{|y|}\frac{1}{\pi[1 + (x/|y|)^2]} \tag{15.68}$$

Thus, we also have

$$\lim_{y\to 0}\frac{y}{x^2+y^2}=\lim_{y\to 0}g(x)=\pi\frac{y}{|y|}\delta(x)=\pi\,\mathrm{sgn}(y)\delta(x) \tag{15.69}$$

where sgn (x) is defined as the sign of x

$$\mathrm{sgn}(x)=\begin{cases}1 & x>0\\ -1 & x<0\end{cases} \tag{15.70}$$

It should be noted that this sign function is also a kind of generalized function similar to that of the Dirac delta function (Sneddon, 1972).

Substitution of (15.44) into (15.43) yields

$$\left[\frac{\boldsymbol{C}}{\Delta t_n}+\theta\boldsymbol{K}\right]\boldsymbol{y}_{n+1}+\left[-\frac{\boldsymbol{C}}{\Delta t_n}+(1-\theta)\boldsymbol{K}\right]\boldsymbol{y}_n=(1-\theta)\tilde{\boldsymbol{f}}_n+\theta\tilde{\boldsymbol{f}}_{n+1} \tag{15.71}$$

When we specify $\theta=0$, we obtain the forward (Euler) difference scheme

$$\frac{\boldsymbol{C}}{\Delta t_n}\boldsymbol{y}_{n+1}+(-\frac{\boldsymbol{C}}{\Delta t_n}+\boldsymbol{K})\boldsymbol{y}_n=\boldsymbol{f}_n \tag{15.72}$$

This is the same as that given in (15.10). When we specify $\theta=1/2$, we obtain the central (Crank-Nicholson) difference scheme

$$(\frac{\boldsymbol{C}}{\Delta t_n}+\frac{\boldsymbol{K}}{2})\boldsymbol{y}_{n+1}+(-\frac{\boldsymbol{C}}{\Delta t_n}+\frac{\boldsymbol{K}}{2})\boldsymbol{y}_n=\frac{1}{2}(\boldsymbol{f}_n+\boldsymbol{f}_{n+1}) \tag{15.73}$$

This is the same as that given in (15.31). When we specify $\theta=1$, we obtain the backward (Euler) difference scheme

$$(\frac{\boldsymbol{C}}{\Delta t_n}+\boldsymbol{K})\boldsymbol{y}_{n+1}-\frac{\boldsymbol{C}}{\Delta t_n}\boldsymbol{y}_n=\boldsymbol{f}_{n+1} \tag{15.74}$$

This is the same as that given in (15.6). Thus, all finite difference schemes can be interpreted as the weighted residue method with the appropriate delta function as the weighting function.

15.2.6 Stability Condition of 2-Level Scheme

Let us consider the case without the forcing function (i.e., $\boldsymbol{f}=0$) in (15.71).

$$\left[\frac{\boldsymbol{C}}{\Delta t_n}+\theta\boldsymbol{K}\right]\boldsymbol{y}_{n+1}+\left[-\frac{\boldsymbol{C}}{\Delta t_n}+(1-\theta)\boldsymbol{K}\right]\boldsymbol{y}_n=\boldsymbol{0} \tag{15.75}$$

To consider the stability of the 2-level finite difference scheme, we first consider the eigenvalue of the original ODE given in (15.3) by seeking a solution of the form

$$y=\boldsymbol{\alpha}e^{\lambda t} \tag{15.76}$$

Thus, we have

$$(\lambda\boldsymbol{C}+\boldsymbol{K})\boldsymbol{\alpha}=\boldsymbol{0} \tag{15.77}$$

For physical problems, the matrices $\boldsymbol{C}$ and $\boldsymbol{K}$ are normally positive definite. In other words, we have for any nonzero vector $\boldsymbol{x}$

$$x^T Cx \geq 0\,, \quad x^T Kx \geq 0 \tag{15.78}$$

For such cases, there are infinite distinct eigenvalues and all of them are negative and real. The eigenvalue problem becomes

$$\det(\lambda C + K) = 0 \tag{15.79}$$

Let the eigenvalues and eigenvectors of (15.79) be λ_m and $\boldsymbol{\alpha}_m (m = 1,2,...,M)$. Using the concept of modal decomposition, we can express the nodal unknown in terms of the eigenvectors as:

$$y_n = \sum_{m=1}^{M} y_m^n \boldsymbol{\alpha}_m\,, \quad y_{n+1} = \sum_{m=1}^{M} y_m^{n+1} \boldsymbol{\alpha}_m \tag{15.80}$$

where M is the dimension of the unknown vector $\boldsymbol{y}$. This model analysis closely resembles the eigenfunction expansion discussed in Chapter 10.

Substitution of (15.80) into (15.75) gives

$$\left[\frac{\boldsymbol{I}}{\Delta t_n} + \theta \boldsymbol{C}^{-1}\boldsymbol{K}\right] y_m^{n+1} \boldsymbol{\alpha}_m + \left[-\frac{\boldsymbol{I}}{\Delta t_n} + (1-\theta)\boldsymbol{C}^{-1}\boldsymbol{K}\right] y_m^n \boldsymbol{\alpha}_m = 0 \tag{15.81}$$

Note, however, that the eigenvector $\boldsymbol{\alpha}_m$ satisfies

$$\lambda_m \boldsymbol{C}\boldsymbol{\alpha}_m + \boldsymbol{K}\boldsymbol{\alpha}_m = \boldsymbol{0} \tag{15.82}$$

Rewriting (15.82) in another form, we obtain

$$\boldsymbol{C}^{-1}\boldsymbol{K}\boldsymbol{\alpha}_m = -\lambda_m \boldsymbol{\alpha}_m \tag{15.83}$$

Substitution of (15.83) into (15.81) results in

$$\left(\frac{1}{\Delta t_n} - \theta\lambda_m\right)\boldsymbol{\alpha}_m y_m^{n+1} = \left[\frac{1}{\Delta t_n} + (1-\theta)\lambda_m\right]\boldsymbol{\alpha}_m y_m^n \tag{15.84}$$

Therefore, the coefficient of the n+1 mode can be expressed in terms of that of the n mode as:

$$y_m^{n+1} = \left[\frac{\frac{1}{\Delta t_n} + (1-\theta)\lambda_m}{\left(\frac{1}{\Delta t_n} - \theta\lambda_m\right)}\right] y_m^n \tag{15.85}$$

To ensure a converged solution for the finite difference solution given in (15.80), we must have

$$\left|y_m^{n+1}\right| < \left|y_m^n\right| \tag{15.86}$$

for $m = 1,2, ..., M$. Therefore, we require that

$$-1 < \left[\frac{1/\Delta t_n + (1-\theta)\lambda_m}{1/\Delta t_n - \theta\lambda_m}\right] < 1 \tag{15.87}$$

as the condition of stability. However, if we want to converge monotonically to the true solution, the modal participation factor has the same sign at each time level n and (15.85) yields

$$\frac{y_m^{n+1}}{y_m^n} = \left[\frac{1/\Delta t_n + (1-\theta)\lambda_m}{1/\Delta t_n - \theta\lambda_m}\right] > 0 \tag{15.88}$$

If (15.88) is not satisfied, the solution will be oscillating around the true solution. Therefore, the 2-level finite difference scheme will be stable and free of oscillation if

$$0<[\frac{1/\Delta t_n+(1-\theta)\lambda_m}{1/\Delta t_n-\theta\lambda_m}]<1 \tag{15.89}$$

Recall that we use different θ for different finite difference schemes. Thus, for difference methods with a fixed θ, (15.89) imposes the corresponding maximum time step that can be used such that numerical instability can be avoided.

Example 15.2 Use the finite difference method to consider the special case of a single ODE with initial condition of $y = 0$ at $t = 0$, and determine time step to ensure numerical stability:

$$k\frac{dy}{dt}+\bar{\lambda}y=0 \tag{15.90}$$

Solutions:
The eigenvalue equation of (15.79) becomes

$$\det(k\lambda+\bar{\lambda})=0 \tag{15.91}$$

That is, the eigenvalue is $\lambda= -\bar{\lambda}/k$. Thus, (15.84) is simplified to

$$(\frac{1}{\Delta t_n}+\theta\frac{\bar{\lambda}}{k})y_{n+1}+[-\frac{1}{\Delta t_n}+(1-\theta)\frac{\bar{\lambda}}{k}]y_n=0 \tag{15.92}$$

The stability condition (15.87) becomes

$$-1<\frac{1/\Delta t_n-(1-\theta)\bar{\lambda}/k}{1/\Delta t_n+\theta\bar{\lambda}/k}<1 \tag{15.93}$$

Since $\bar{\lambda}/k > 0$ and $0 \le \theta \le 1$, the second inequality of (15.93) is automatically satisfied. The first inequality of (15.93) can be simplified to

$$(1-2\theta)\bar{\lambda}\Delta t_n<2k \tag{15.94}$$

Therefore, the finite difference method is unconditionally stable if

$$\theta\ge 1/2 \tag{15.95}$$

Since $\theta = 1$ for the backward Euler difference method and $\theta = 1/2$ for the central (Crank-Nicholson) method, these schemes are unconditionally stable. If $1/2 > \theta \ge 0$, we have conditional stability if the time step satisfies the following condition

$$\bar{\lambda}\Delta t_n<\frac{2k}{(1-2\theta)} \tag{15.96}$$

The numerical solution will be stable and free from oscillation if

$$\frac{1/\Delta t_n-(1-\theta)\bar{\lambda}/k}{1/\Delta t_n+\theta\bar{\lambda}/k}>0 \tag{15.97}$$

This condition can be simplified to

$$(1-\theta)\bar{\lambda}\Delta t_n/k<1 \tag{15.98}$$

Therefore, the forward (Euler) difference scheme may not converge if a "big" time step is used. The regions of stability are illustrated in Figure 15.4.

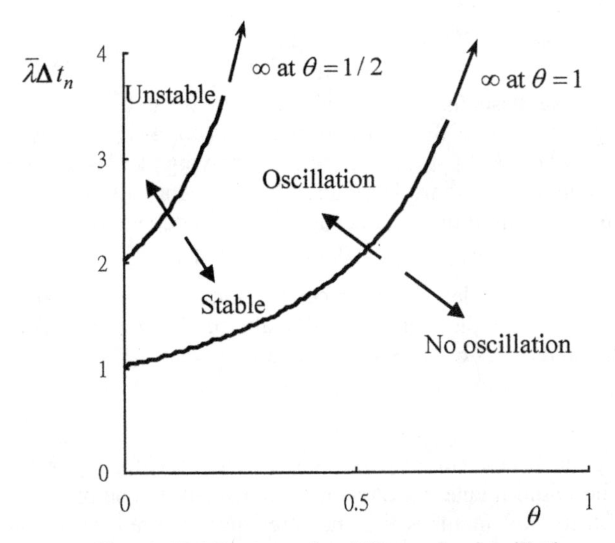

Figure 15.4 Regions of stability and no oscillations

15.2.7 Courant-Friedrichs-Lewy Condition

The propagation of mechanical disturbances in solids is of profound importance in many disciplines including physical sciences and engineering. In these kinds of problems, the loading or disturbance is applied at such a "fast" rate that the effect of inertia cannot be ignored. These loadings are normally described as suddenly applied.

Numerical integration of the wave signal cannot violate the causality of the arrival of wave signals. In particular, in Chapter 9 we saw that the method of characteristics leads to the concept of the domain of dependence. This concept leads to the so-called CFL condition (or Courant-Friedrichs-Lewy condition) discussed in this section.

Let us consider the simplest wave equation:

$$u_{tt} = c^2 u_{xx} \tag{15.99}$$

with initial conditions:

$$u(x,0) = f(x), \quad u_t(x,0) = g(x) \tag{15.100}$$

Introduce a new notation for the solution at time $t = t_n$ and $x = x_j$ as:

$$v_{n,j} = u(t_n, x_j) \tag{15.101}$$

where

$$t_n = n\Delta t, \quad x_j = j\Delta x \tag{15.102}$$

Using the second order centered difference scheme, the wave equation can be approximated as

$$\frac{u_{n+1,j} - 2u_{n,j} + u_{n-1,j}}{(\Delta t)^2} = c^2 \frac{u_{n,j+1} - 2u_{n,j} + u_{n,j-1}}{(\Delta x)^2} \tag{15.103}$$

We can rearrange this to give

$$u_{n+1,j}=2[1-(\frac{c\Delta t}{\Delta x})^2]u_{n,j}+(\frac{c\Delta t}{\Delta x})^2(u_{n,j+1}+u_{n,j-1})-u_{n-1,j} \tag{15.104}$$

Therefore, (15.104) suggests that $u_{n+1,j}$ depends on $u_{n,j+k}$ where $k = 0, \pm 1$ and $u_{n-1,j}$. Then, reapplying (15.104) to the solution at time step $n-1$, we have that $u_{n+1,j}$ depends on $u_{n-1,j+k}$ where $k = 0,\pm 1, \pm 2$. Further reapplying (15.104) again, we have that $u_{n+1,j}$ depends on $u_{n-1,j+k}$ where $k = 0,\pm 1, \pm 2, \pm 3$. Eventually, we can trace back all the way down to the initial time $n = 0$, such that $u_{n+1,j}$ depends on

$$\{u_{0,j+k},\quad k=0,\pm1,\pm2,...,\pm n\} \tag{15.105}$$

In other words, the solution depends on a certain domain of the initial data. These initial data are known as the numerical domain of dependence. As shown in Chapter 9, the characteristics of (15.99) are the two curves:

$$C_-:\quad x-ct=x_i \tag{15.106}$$

$$C_+:\quad x+ct=x_i \tag{15.107}$$

where x_i is any point on the initial curve defined in (15.105). Figure 15.5 shows two choices of the initial numerical domain of dependence. The numerical domain of dependence on the left includes the characteristics (dashed lines), whereas the numerical domain of dependence on the right does not include the characteristics. Thus, the left one may converge to the actual solution whereas the one on the right can never converge to the exact solution because parts of the initial data have not been employed in the numerical integration. This can be summarized as:

$$Stable:\quad c\frac{\Delta t}{\Delta x}<1 \tag{15.108}$$

$$Unstable:\quad c\frac{\Delta t}{\Delta x}>1 \tag{15.109}$$

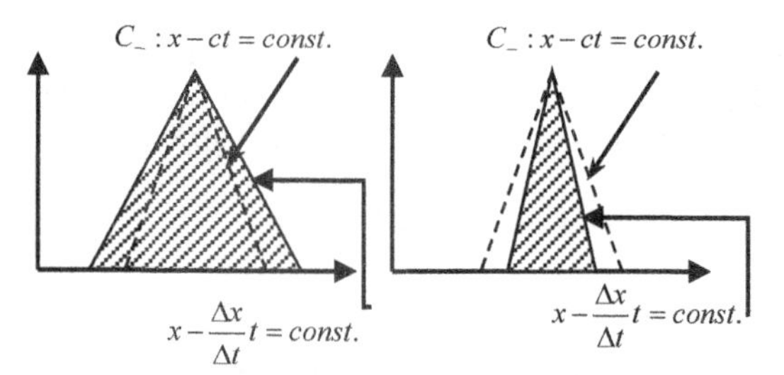

Figure 15.5 The importance of including the domain of dependence

This is called the Courant-Friedrichs-Lewy criterion for the stability of numerical integration for the hyperbolic type of PDE. Therefore, this criterion only applies to wave type propagation of information. Once the spatial discretization is selected, the time step must satisfy the following inequality:

$$Stable:\quad \Delta t<\frac{\Delta x}{c} \tag{15.110}$$

15.2.8 von Neumann Test

The von Neumann test can be used to determine whether a finite difference scheme is stable (i.e., converges to the exact solution) or unstable (i.e., does not converge to the exact solution).

Let us consider the following diffusion equation:

$$u_t = c_v u_{xx} \tag{15.111}$$

Applying the following finite difference to the derivative terms, we have

$$u_t = \frac{1}{k}[u(x,t+k) - u(x,t)] \tag{15.112}$$

$$u_{xx} = \frac{1}{h^2}[u(x+h,t) - 2u(x,t) + u(x-h,t)] \tag{15.113}$$

To simplify the notation, we introduce

$$u_{m,n} = u(m\Delta x, n\Delta t) \tag{15.114}$$

Then, (15.112) and (15.113) becomes

$$u_t = \frac{1}{\Delta t}[u_{m,n+1} - u_{m,n}] \tag{15.115}$$

$$u_{xx} = \frac{1}{(\Delta x)^2}[u_{m+1,n} - 2u_{m,n} + u_{m-1,n}] \tag{15.116}$$

Thus, we can substitute these results into the diffusion equation given in (15.111) and rearrange this result as

$$u_{m,n+1} = u_{m,n} + \frac{c_v \Delta t}{(\Delta x)^2}(u_{m+1,n} - 2u_{m,n} + u_{m-1,n}) \tag{15.117}$$

We now seek an exponential solution as

$$u_{m,n} = e^{im\theta} e^{in\lambda} \tag{15.118}$$

Substitution of (15.118) into (15.117) gives

$$e^{im\theta} e^{i(n+1)\lambda} = e^{im\theta} e^{in\lambda} + \frac{c_v \Delta t}{(\Delta x)^2}(e^{i(m+1)\theta} e^{in\lambda} - 2e^{im\theta} e^{in\lambda} + e^{i(m-1)\theta} e^{in\lambda}) \tag{15.119}$$

Thus, θ and λ resemble spatial and temporal variables. This can be simplified to

$$e^{i\lambda} = 1 + \frac{c_v \Delta t}{(\Delta x)^2}[e^{i\theta} - 2 + e^{-i\theta}] = 1 - \frac{c_v \Delta t}{(\Delta x)^2} 2(1 - \cos\theta) \tag{15.120}$$

This can further be rewritten as

$$e^{i\lambda} = 1 - \frac{4c_v \Delta t}{(\Delta x)^2}\sin^2(\frac{\theta}{2}) \tag{15.121}$$

Note further that

$$e^{i[\mathrm{Re}(\lambda) + i\,\mathrm{Im}(\lambda)]} = e^{i\mathrm{Re}(\lambda)} e^{-\mathrm{Im}(\lambda)} \tag{15.122}$$

For a large time variable λ, we have

$$\mathrm{Im}(\lambda) > 0 \Rightarrow e^{-\mathrm{Im}(\lambda)} \to 0, \quad \text{as}\, \lambda \to \infty \tag{15.123}$$

Equivalently, we also have

$$\left|e^{i\lambda}\right| \le 1 \Rightarrow \text{stable} \tag{15.124}$$

This is because the solution will converge to zero as suggested by (15.118). Imposing condition (15.124) onto (15.121) we obtain

$$\frac{4c_v\Delta t}{(\Delta x)^2}\sin^2(\frac{\theta}{2}) \le \frac{4c_v\Delta t}{(\Delta x)^2} \le 2 \tag{15.125}$$

Thus, the von Neumann test gives the condition

$$\frac{c_v\Delta t}{(\Delta x)^2} \le \frac{1}{2} \tag{15.126}$$

Consequently, the von Neumann test leads to the stability condition of

$$\Delta t \le \frac{(\Delta x)^2}{2c_v} \tag{15.127}$$

If the time step is not set according to (15.127), the numerical scheme will not converge to the exact solution.

Note, however, that a numerical scheme being stable does not necessarily imply that the answer must be accurate. If the time step is taken to be too small, the round-off error may accumulate in the large number of operations. It is because all computers can only retain a finite number of digits in each numerical calculation. If the number of steps becomes unnecessary large, the error from round-off accumulation may eventually degrade a "good" numerical scheme. Similar to the asymptotic series expansion, we should not be too greedy on the accuracy. For this reason, modern numerical codes normally allow the time step to vary depending on the error control scheme such that a larger time step is used whenever possible and a very small step size only where necessary.

15.3 FINITE DIFFERENCE FOR SECOND ORDER ODE

We now consider the second order ODE, which repeatedly appears in dynamics formulation using force equilibrium of mechanical systems. In particular, consider the following linear system of ODEs:

$$\boldsymbol{M}\frac{d^2\boldsymbol{a}}{dt^2} + \boldsymbol{C}\frac{d\boldsymbol{a}}{dt} + \boldsymbol{K}\boldsymbol{a} = \boldsymbol{f}(t) \tag{15.128}$$

For the time variable, we can discretize using a trial function in time as:

$$\boldsymbol{a} \approx \hat{\boldsymbol{a}} = \sum_{m=1}^{\infty} \boldsymbol{a}^m N_m(t) \tag{15.129}$$

where N_m has to be at least of degree two. Figure 15.6 shows a three-node quadratic element in time.

As illustrated in Figure 15.6, an approximation for $\boldsymbol{a}$ depends on the value of three time nodes as:

$$\hat{\boldsymbol{a}} = \boldsymbol{a}^{2n} N_{2n}^n + \boldsymbol{a}^{2n+1} N_{2n+1}^n + \boldsymbol{a}^{2n+2} N_{2n+2}^n \tag{15.130}$$

where the approximation functions are

$$N_{2n}^n = -\frac{T(1-T)}{2}, \quad N_{2n+1}^n = 1-T^2, \quad N_{2n+2}^n = \frac{T(1+T)}{2} \tag{15.131}$$

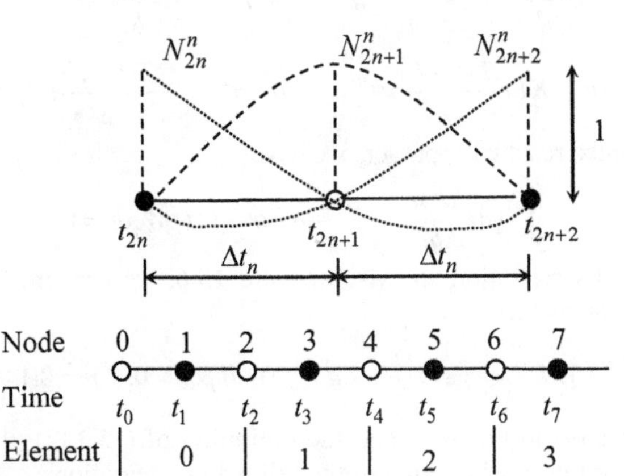

Figure 15.6 Discretization of time in nodes and elements

with

$$T = \frac{t - t_{2n+1}}{\Delta t_n} \tag{15.132}$$

As shown in Figure 15.6, we have

$$\Delta t_n = t_{2n+2} - t_{2n+1} = t_{2n+1} - t_{2n} \tag{15.133}$$

It is straightforward to see that at time t_{2n}

$$T = -1, \quad N_{2n}^n = 1, \quad N_{2n+1}^n = 0, \quad N_{2n+2}^n = 0\,; \tag{15.134}$$

at time t_{2n+1}

$$T = 0, \quad N_{2n}^n = 0, \quad N_{2n+1}^n = 1, \quad N_{2n+2}^n = 0\,; \tag{15.135}$$

and at time t_{2n+2}

$$T = 1, \quad N_{2n}^n = 0, \quad N_{2n+1}^n = 0, \quad N_{2n+2}^n = 1\,. \tag{15.136}$$

Note that the differentiation of the trial functions given in (15.131) leads to

$$\frac{dN_{2n}^n}{dt} = \frac{-\frac{1}{2} + T}{\Delta t_n}, \quad \frac{dN_{2n+1}^n}{dt} = -\frac{2T}{\Delta t_n}, \quad \frac{dN_{2n+2}^n}{dt} = \frac{\frac{1}{2} + T}{\Delta t_n} \tag{15.137}$$

$$\frac{d^2 N_{2n}^n}{dt^2} = \frac{1}{\Delta t_n^2}, \quad \frac{d^2 N_{2n+1}^n}{dt^2} = -\frac{2}{\Delta t_n^2}, \quad \frac{d^2 N_{2n+2}^n}{dt^2} = \frac{1}{\Delta t_n^2} \tag{15.138}$$

Using these approximations, we have

$$\boldsymbol{M}\frac{d^2\boldsymbol{a}}{dt^2} = \boldsymbol{M}\left(\frac{1}{\Delta t_n^2}\boldsymbol{a}^{2n} - \frac{2}{\Delta t_n^2}\boldsymbol{a}^{2n+1} + \frac{1}{\Delta t_n^2}\boldsymbol{a}^{2n+2}\right) \tag{15.139}$$

$$\boldsymbol{C}\frac{d\boldsymbol{a}}{dt}=\boldsymbol{C}(\frac{-\frac{1}{2}+T}{\Delta t_n}\boldsymbol{a}^{2n}-\frac{2T}{\Delta t_n}\boldsymbol{a}^{2n+1}+\frac{\frac{1}{2}+T}{\Delta t_n}\boldsymbol{a}^{2n+2}) \tag{15.140}$$

$$\boldsymbol{K}\boldsymbol{a}=\boldsymbol{K}(-\frac{T(1-T)}{2}\boldsymbol{a}^{2n}+(1-T^2)\boldsymbol{a}^{2n+1}+\frac{T(1+T)}{2}\boldsymbol{a}^{2n+2}) \tag{15.141}$$

Using the weighted residual approach, we have

$$\int_0^{\infty}[\boldsymbol{M}\frac{d^2\boldsymbol{a}}{dt^2}+\boldsymbol{C}\frac{d\boldsymbol{a}}{dt}+\boldsymbol{K}\boldsymbol{a}-\boldsymbol{f}(t)]W_n dt=0 \tag{15.142}$$

With the time discretization shown in Figure 15.6, we can break down the time integration into segments:

$$\int_{t_{2n}}^{t_{2n+2}}[\boldsymbol{M}\frac{d^2\boldsymbol{a}}{dt^2}+\boldsymbol{C}\frac{d\boldsymbol{a}}{dt}+\boldsymbol{K}\boldsymbol{a}-\boldsymbol{f}(t)]W_n dt=0,\quad n=0,1,2,... \tag{15.143}$$

In view of (15.139) to (15.141) and the definition of (15.132), the integration in (15.143) leads to the consideration of the following integrations:

$$\int_{t_{2n}}^{t_{2n+2}}(\frac{1}{2}+T)W_n dt=\Delta t_n\gamma\int_{-1}^{1}W_n dT \tag{15.144}$$

$$\int_{t_{2n}}^{t_{2n+2}}\frac{T(1+T)}{2}W_n dt=\Delta t_n\beta\int_{-1}^{1}W_n dT \tag{15.145}$$

$$\begin{aligned}\int_{t_{2n}}^{t_{2n+2}}(1-T^2)W_n dt&=\int_{t_{2n}}^{t_{2n+2}}\{\frac{1}{2}+(\frac{1}{2}+T)-2[\frac{1}{2}T(1+T)]\}W_n dt\\&=\Delta t_n\{\frac{1}{2}+\gamma-2\beta\}\int_{-1}^{1}W_n dT\end{aligned} \tag{15.146}$$

$$\begin{aligned}\int_{t_{2n}}^{t_{2n+2}}(-\frac{1}{2}+T)W_n dt&=\int_{t_{2n}}^{t_{2n+2}}(-1+\frac{1}{2}+T)W_n dt\\&=\Delta t_n\{-1+\gamma\}\int_{-1}^{1}W_n dT\end{aligned} \tag{15.147}$$

$$\begin{aligned}\int_{t_{2n}}^{t_{2n+2}}(-\frac{T(1-T)}{2})W_n dt&=\int_{t_{2n}}^{t_{2n+2}}[\frac{1}{2}-(\frac{1}{2}+T)+\frac{1}{2}T(1+T)]W_n dt\\&=\Delta t_n\{\frac{1}{2}-\gamma+\beta\}\int_{-1}^{1}W_n dT\end{aligned} \tag{15.148}$$

$$\begin{aligned}\int_{t_{2n}}^{t_{2n+2}}(-2T)W_n dt&=\int_{t_{2n}}^{t_{2n+2}}[1-2(\frac{1}{2}+T)]W_n dt\\&=\Delta t_n\{1-2\gamma\}\int_{-1}^{1}W_n dT\end{aligned} \tag{15.149}$$

where

$$\gamma=\frac{\int_{-1}^{1}(\frac{1}{2}+T)W_n dT}{\int_{-1}^{1}W_n dT} \tag{15.150}$$

$$\beta = \frac{\int_{-1}^{1} \frac{T(1+T)}{2} W_n dT}{\int_{-1}^{1} W_n dT} \tag{15.151}$$

Substitution of these results into (15.143) gives

$$\begin{aligned} &\boldsymbol{a}^{2n+2}(\boldsymbol{M} + \gamma\Delta t_n \boldsymbol{C} + \beta \Delta t_n^2 \boldsymbol{K}) \\ &+\boldsymbol{a}^{2n+1}[-2\boldsymbol{M} + (1-2\gamma)\Delta t_n \boldsymbol{C} + (\frac{1}{2} - 2\beta + \gamma)\Delta t_n^2 \boldsymbol{K}] \\ &+\boldsymbol{a}^{2n}[\boldsymbol{M} - (1-\gamma)\Delta t_n \boldsymbol{C} + (\frac{1}{2} + \beta - \gamma)\Delta t_n^2 \boldsymbol{K}] = \boldsymbol{f}^n \Delta t_n^2 \end{aligned} \tag{15.152}$$

where

$$\boldsymbol{f}^n = \frac{\int_{-1}^{1} \boldsymbol{f}(t_{2n+1} + T\Delta t_n) W_n dT}{\int_{-1}^{1} W_n dT} \tag{15.153}$$

Note that the actual value of β and γ depend on the weighting function that we used in (15.142) and (15.143). The formulation given in (15.152) is related to the Newmark method, which will be introduced in a later section. If we used the same interpolation function for the forcing term, we would have

$$\boldsymbol{f}^n = \boldsymbol{f}^{2n} N_{2n}^n + \boldsymbol{f}^{2n+1} N_{2n+1}^n + \boldsymbol{f}^{2n+2} N_{2n+2}^n \tag{15.154}$$

It is straightforward to see that

$$\boldsymbol{f}^n = \beta \boldsymbol{f}^{2n+2} + (\frac{1}{2} - 2\beta + \gamma)\boldsymbol{f}^{2n+1} + (\frac{1}{2} + \beta - \gamma)\boldsymbol{f}^{2n} \tag{15.155}$$

In the next two sections, we will introduce the two most popular methods in structural dynamics, namely the Wilson θ method and Newmark β method.

15.3.1 Wilson θ Method

The Wilson θ method is an extension of the linear acceleration method. According to Figure 15.7, the acceleration at a time $t+\tau$ can be expressed by linear extrapolation as:

$$\ddot{\boldsymbol{U}}^{t+\tau} = \ddot{\boldsymbol{U}}^t + \frac{\tau}{\theta\Delta t}(\ddot{\boldsymbol{U}}^{t+\theta\Delta t} - \ddot{\boldsymbol{U}}^t) \tag{15.156}$$

where $\theta \geq 1$. The Wilson θ method is an implicit method because the stiffness matrix $\boldsymbol{K}$ has to be evaluated at the unknown displacement $\boldsymbol{U}^{t+\theta\Delta t}$. This method is unconditionally stable for $\theta \geq 1.37$, but in actual numerical calculation, $\theta = 1.4$ is normally used. The idea of the Wilson θ method is to allow a bigger time step but at the same time the numerical scheme remains stable. The acceleration is assumed to be linear from t to $t+\theta\Delta t$. When $\theta = 1$, it reduces to the linear acceleration scheme.

Integration of (15.156) gives the velocity as well as the displacement as:

$$\dot{U}^{t+\tau} = \dot{U}^t + \ddot{U}^t \tau + \frac{\tau^2}{2\theta\Delta t}(\ddot{U}^{t+\theta\Delta t} - \ddot{U}^t) \tag{15.157}$$

$$U^{t+\tau} = U^t + \dot{U}^t \tau + \frac{1}{2}\ddot{U}^t \tau^2 + \frac{\tau^3}{6\theta\Delta t}(\ddot{U}^{t+\theta\Delta t} - \ddot{U}^t) \tag{15.158}$$

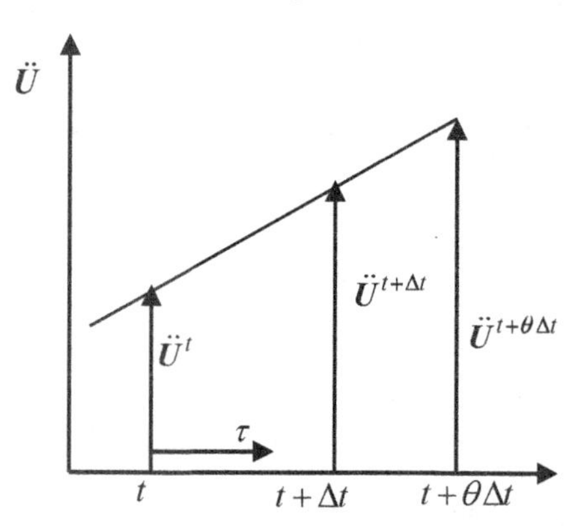

Figure 15.7 Sketch showing the Wilson θ method

Substitution of $\tau = \theta\Delta t$ into (15.157) and (15.158) gives

$$\dot{U}^{t+\theta\Delta t} = \dot{U}^t + \ddot{U}^t \tau + \frac{\theta\Delta t}{2}(\ddot{U}^{t+\theta\Delta t} - \ddot{U}^t) \tag{15.159}$$

$$U^{t+\theta\Delta t} = U^t + \dot{U}^t \theta\Delta t + \frac{(\theta\Delta t)^2}{6}(\ddot{U}^{t+\theta\Delta t} + 2\ddot{U}^t) \tag{15.160}$$

Solving for the velocity and acceleration in terms of the displacement at $t = t+\theta\Delta t$, we obtain

$$\ddot{U}^{t+\theta\Delta t} = \frac{6}{\theta^2\Delta t^2}(U^{t+\theta\Delta t} - U^t) - \frac{6}{\theta\Delta t}\dot{U}^t - 2\ddot{U}^t \tag{15.161}$$

$$\dot{U}^{t+\theta\Delta t} = \frac{3}{\theta\Delta t}(U^{t+\theta\Delta t} - U^t) - 2\dot{U}^t - \frac{\theta\Delta t}{2}\ddot{U}^t \tag{15.162}$$

The equilibrium is considered at time $t+\theta\Delta t$ as:

$$M\ddot{U}^{t+\theta\Delta t} + C\dot{U}^{t+\theta\Delta t} + KU^{t+\theta\Delta t} = f^{t+\theta\Delta t} \tag{15.163}$$

Since the acceleration is assumed to be linear, the force vector must also be linear as

$$f^{t+\theta\Delta t} = f^t + \theta(f^{t+\Delta t} - f^t) \tag{15.164}$$

Substitution of (15.161) and (15.162) into (15.163) gives an equation for displacement $U^{t+\theta\Delta t}$. This solution can then be substituted into (15.161) and (15.162) to obtain the acceleration and velocity.

Before we consider the Newmark β method in the next section, we first set $\theta = 1$ in (15.160) and obtain the special case of linear acceleration:

$$U^{t+\Delta t} = U^t + \dot{U}^t \Delta t + (\frac{1}{6}\ddot{U}^{t+\Delta t} + \frac{1}{3}\ddot{U}^t)(\Delta t)^2 \tag{15.165}$$

15.3.2 Newmark β Method

The Newmark β method can be formulated by starting from Taylor series expansion:

$$U^{t+\Delta t} = U^t + \frac{\Delta t}{1!}\dot{U}^t + \frac{(\Delta t)^2}{2!}\ddot{U}^t + \frac{(\Delta t)^3}{3!}\dddot{U}^t + \dots \tag{15.166}$$

For small Δt, we can estimate the rate of acceleration as

$$\dddot{U}^t = \frac{\ddot{U}^{t+\Delta t} - \ddot{U}^t}{\Delta t} \tag{15.167}$$

Substitution of (15.167) into (15.166) gives

$$U^{t+\Delta t} = U^t + \Delta t\dot{U}^t + (\Delta t)^2\{(\frac{1}{2} - \frac{1}{6})\ddot{U}^t + \frac{1}{6}\ddot{U}^{t+\Delta t}\} + \dots \tag{15.168}$$

Note that we have only retained the third order term in the Taylor series expansion. The idea of the Newmark β method is to modify the factor 1/6 in (15.168) to β such that all higher order terms in Taylor series expansion can be neglected. In particular, we have the Newmark β method as

$$U^{t+\Delta t} = U^t + \Delta t\dot{U}^t + (\Delta t)^2\{(\frac{1}{2} - \beta)\ddot{U}^t + \beta\ddot{U}^{t+\Delta t}\} \tag{15.169}$$

Similarly, we can also use Taylor series expansion for the velocity as

$$\dot{U}^{t+\Delta t} = \dot{U}^t + \Delta t\ddot{U}^t + \frac{(\Delta t)^2}{2!}\dddot{U}^t + \dots \tag{15.170}$$

Again for small Δt, we can substitute the rate of acceleration given in (15.167) into (15.170) as

$$\dot{U}^{t+\Delta t} = \dot{U}^t + \Delta t\ddot{U}^t + \frac{(\Delta t)}{2!}[\ddot{U}^{t+\Delta t} - \ddot{U}^t]\dots \tag{15.171}$$

This can be rewritten as

$$\dot{U}^{t+\Delta t} = \dot{U}^t + \Delta t\{(1 - \frac{1}{2})\ddot{U}^t + \frac{1}{2}\ddot{U}^{t+\Delta t}\} + \dots \tag{15.172}$$

Using the same idea, we can replace 1/2 by γ such that all higher order terms can be dropped:

$$\dot{U}^{t+\Delta t} = \dot{U}^t + \Delta t\{(1-\gamma)\ddot{U}^t + \gamma\ddot{U}^{t+\Delta t}\} \tag{15.173}$$

Newmark originally proposed to use $\beta = 1/4$ and $\gamma = 1/2$. Consider the special case that

$$\beta = 1/6, \quad \gamma = 1/2. \tag{15.174}$$

Substitution of these values into (15.166) gives

$$U^{t+\Delta t} = U^t + \Delta t\dot{U}^t + (\Delta t)^2\{\frac{1}{3}\ddot{U}^t + \frac{1}{6}\ddot{U}^{t+\Delta t}\} \tag{15.175}$$

Note that (15.175) is exactly the same as (15.165) obtained in the last section, which is for linear acceleration. Thus, the special case given in (15.174) for the Newmark β method gives the linear acceleration method. Note also that for β = 0, it can be seen from (15.169) that the Newmark β method becomes an explicit scheme.

Thus, in conclusion, the Wilson θ method with θ = 1 is equivalent to the Newmark β method with β = 1/6, and γ = 1/2. In addition, both the Wilson and Newmark methods are exact if the acceleration in the problem is constant (this is clearly not true for the case of seismic loads). We should note that the Wilson θ method and the Newmark β method are the most popular finite difference methods in time integration in the area of structural dynamics. All dynamic finite element programs include the options of using either of these methods.

Prof. Wilson was a professor at University of California at Berkeley and made significant contributions to nonlinear analyses of structures under seismic loadings, whereas Prof. Newmark was a professor at University of Illinois at Urbana Champaign and made significant contributions to seismic analysis of structures. In geotechnical engineering, Newmark chart was developed to estimate the vertical stress increment under arbitrary surface loading and Newmark sliding block model was developed to investigate seismic slope stability.

15.3.3 Stability Condition of 3-Level Scheme

We now return to study the stability condition of the 3-level scheme discussed in Section 15.3. In particular, a special form of (15.128) with no damping and no forcing term is considered

$$\boldsymbol{M}\frac{d^2\boldsymbol{a}}{dt^2}+\boldsymbol{K}\boldsymbol{a}=\boldsymbol{0} \tag{15.176}$$

The corresponding numerical scheme given in (15.152) is simplified to

$$\begin{aligned}&(\boldsymbol{I}+\beta\Delta t_n^2\boldsymbol{M}^{-1}\boldsymbol{K})\boldsymbol{a}^{2n+2}+[-2\boldsymbol{I}+(\frac{1}{2}-2\beta+\gamma)\Delta t_n^2\boldsymbol{M}^{-1}\boldsymbol{K}]\boldsymbol{a}^{2n+1}\\&+[\boldsymbol{I}+(\frac{1}{2}+\beta-\gamma)\Delta t_n^2\boldsymbol{M}^{-1}\boldsymbol{K}]\boldsymbol{a}^{2n}=0\end{aligned} \tag{15.177}$$

We first return to the original system and consider the natural frequency of vibrations of the system. That is, we seek for a solution of the form:

$$\boldsymbol{a}=\boldsymbol{\alpha}\cos(\omega t-\delta) \tag{15.178}$$

Substitution of (15.178) into (15.176) leads to

$$(-\omega^2\boldsymbol{M}+\boldsymbol{K})\boldsymbol{\alpha}=\boldsymbol{0} \tag{15.179}$$

It is clear that if the mass matrix $\boldsymbol{M}$ is invertible, we can rewrite (15.179) as

$$(\boldsymbol{M}^{-1}\boldsymbol{K}-\omega^2\boldsymbol{I})\boldsymbol{\alpha}=\boldsymbol{0} \tag{15.180}$$

This is the well-known eigenvalue problem. For nonzero vector $\boldsymbol{\alpha}$, we require

$$\det(\boldsymbol{M}^{-1}\boldsymbol{K}-\omega^2\boldsymbol{I})=0 \tag{15.181}$$

If the size of the matrices $\boldsymbol{M}$ and $\boldsymbol{K}$ is $M\times M$, the solution can be expressed as:

$$\boldsymbol{a} = \sum_{m=1}^{M} \boldsymbol{\alpha}_m \cos(\omega_m t - \delta_m) \tag{15.182}$$

where the eigenvector is normalized as

$$\boldsymbol{\alpha}_m^T \boldsymbol{M} \boldsymbol{\alpha}_m = 1 \tag{15.183}$$

$$\boldsymbol{\alpha}_l^T \boldsymbol{M} \boldsymbol{\alpha}_m = 0, \quad m \neq l \tag{15.184}$$

$$\boldsymbol{\alpha}_l^T \boldsymbol{K} \boldsymbol{\alpha}_m = 0, \quad m \neq l \tag{15.185}$$

We can express the numerical approximation vector $\boldsymbol{a}^{2n}$ as:

$$\boldsymbol{a}^{2n} = \sum_{m=1}^{M} y_m^{2n} \boldsymbol{\alpha}_m \tag{15.186}$$

Substitution of (15.186) into (15.177) gives

$$\begin{aligned} &(1+\beta \Delta t_n^2 \omega_m^2) y_m^{2n+2} + [-2 + (\frac{1}{2} - 2\beta + \gamma)\Delta t_n^2 \omega_m^2] y_m^{2n+1} \\ &+ [1 + (\frac{1}{2} + \beta - \gamma)\Delta t_n^2 \omega_m^2] y_n^{2n} = 0 \end{aligned} \tag{15.187}$$

For numerical calculations, we can further define:

$$y_m^{2n} = A\mu^{2n} \tag{15.188}$$

Using this particular form, (15.187) is reduced to

$$(1+\beta \Delta t_n^2 \omega_m^2)\mu^2 + [-2 + (\frac{1}{2} - 2\beta + \gamma)\Delta t_n^2 \omega_m^2]\mu + [1 + (\frac{1}{2} + \beta - \gamma)\Delta t_n^2 \omega_m^2] = 0 \tag{15.189}$$

There are two roots for μ, namely μ_1 and μ_2. We can see that μ is in general complex. If μ_1 and μ_2 are complex and the modulus is less than 1, we have stability criteria as

Stable: $$|\mu_i| \leq 1, \quad i = 1,2 \tag{15.190}$$

Stable and undamped: $$|\mu_i| = 1, \quad i = 1,2 \tag{15.191}$$

Stable and artificially damped: $$|\mu_i| < 1, \quad i = 1,2 \tag{15.192}$$

The roots of μ can be expressed as:

$$\mu = \frac{(2-g) \pm \sqrt{(2-g)^2 - 4(1+h)}}{2} \tag{15.193}$$

where

$$g = \frac{(\frac{1}{2}+\gamma)\omega_m^2 \Delta t_n^2}{(1+\beta\omega_m^2 \Delta t_n^2)}, \quad h = \frac{(\frac{1}{2}-\gamma)\omega_m^2 \Delta t_n^2}{(1+\beta\omega_m^2 \Delta t_n^2)} \tag{15.194}$$

Using these definitions, we find

$$2-g = \frac{2-(\frac{1}{2}-2\beta+\gamma)\omega_m^2 \Delta t_n^2}{(1+\beta\omega_m^2 \Delta t_n^2)}, \quad 1+h = \frac{1+(\frac{1}{2}-\gamma+\beta)\omega_m^2 \Delta t_n^2}{(1+\beta\omega_m^2 \Delta t_n^2)} \tag{15.195}$$

It is straightforward to see that

$$|\mu| = \sqrt{\mu\bar{\mu}} = \sqrt{1+h} \tag{15.196}$$

Substitution of (15.196) into (15.190) to (15.192) leads to

$$-1 < h \leq 0 \tag{15.197}$$

This leads to

$$\gamma \geq \frac{1}{2}, \quad \frac{1}{2} + \beta - \gamma > 0 \tag{15.198}$$

It is clear that Newmark's original proposal of $\beta = 1/4$ and $\gamma = 1/2$ will lead to an unconditional stable solution. The linear acceleration method is also unconditional stable.

15.4 MULTI-STEP METHOD

So far, our discussion has been focused on the numerical scheme that the current time step depends only on the solution of the last time step. This is known as the one-step method. It is natural to ask whether we can estimate the solution at the current time step as a function of previous time steps (i.e., more than just the last time step). This technique is referred as the multi-step method. Consider a first order ODE as

$$\frac{dy}{dt} = f(t, y), \quad y(t_0) = y_0 \tag{15.199}$$

Assume that an approximation of y is given as:

$$y(t) = \phi(t) \tag{15.200}$$

The numerical solution can be written as:

$$\phi(t_{n+1}) - \phi(t_n) = \int_{t_n}^{t_{n+1}} \phi'(t)dt \tag{15.201}$$

15.4.1 Adams-Bashforth Method

The Adams-Bashforth method assumes a polynomial form of ϕ' in (15.201) and carries the explicit integration on the right hand side of (15.201). For example, if a second order function is assumed for ϕ', we have

$$P_1(t) = \phi'(t) = At + B \tag{15.202}$$

First, we can evaluate the constants A and B by considering the equations at two points (t_n, y_n) and (t_{n-1}, y_{n-1}):

$$At_n + B = f(t_n, y_n) = f_n \tag{15.203}$$

$$At_{n-1} + B = f(t_{n-1}, y_{n-1}) = f_{n-1} \tag{15.204}$$

This provides a system of two equations for two unknowns and the solutions are

$$A = \frac{f_n - f_{n-1}}{t_n - t_{n-1}} = \frac{f_n - f_{n-1}}{h} \tag{15.205}$$

$$B = \frac{f_{n-1}t_n - f_n t_{n-1}}{h} \tag{15.206}$$

where h is the time step. Thus, substitution of (15.205) and (15.206) into (15.202) gives

$$\begin{aligned}\int_{t_n}^{t_{n+1}} \phi'(t)dt &= \int_{t_n}^{t_{n+1}} (At_n + B)dt = \frac{A}{2}(t_{n+1}^2 - t_n^2) + B(t_{n+1} - t_n) \\ &= \frac{f_n - f_{n-1}}{2h}(t_{n+1} - t_n)(t_{n+1} + t_n) + \frac{f_{n-1}t_n - f_n t_{n-1}}{h}(t_{n+1} - t_n) \\ &= f_n\left(\frac{t_{n+1} + t_n - 2t_{n-1}}{2}\right) + f_{n-1}\left(\frac{2t_n - t_{n+1} - t_n}{2}\right) \\ &= \frac{3h}{2} f_n - \frac{1}{2} h f_{n-1}\end{aligned} \quad (15.207)$$

Substitution of this result into (15.201) gives the following second order Adams-Bashforth finite difference scheme:

$$y_{n+1} = y_n + \frac{3h}{2} f_n - \frac{1}{2} h f_{n-1} \quad (15.208)$$

This is an explicit scheme as the unknown only appears on the left hand side and no iteration is needed. The truncation error is proportional to h^3. If we use only the constant term in (15.202) and follow the same procedure, we will arrive at Euler's forward difference method that we discussed in Section 15.2.1. Following a similar idea, we can extend the analysis to higher orders. For example, if we employ the results of previous steps to fit an approximation of third order,

$$P_3(t) = At^3 + Bt^2 + Ct + D \quad (15.209)$$

we have the following fourth order Adams-Bashforth formula

$$y_{n+1} = y_n + \frac{h}{24}(55 f_n - 59 f_{n-1} + 37 f_{n-2} - 9 f_{n-3}) \quad (15.210)$$

Again, this is an explicit numerical scheme.

15.4.2 Adams-Moulton Method

There is another variation of such derivation and it leads to the Adams-Moulton scheme. The only difference is that instead of using the previous solution in fitting the power law for the differential form, one may use the unknown solution in fitting the power law. More specifically, for the second order case we have

$$P_1(t) = \phi'(t) = \alpha t + \beta \quad (15.211)$$

First, we can evaluate the constants by considering the equations at the current point (t_n, y_n) and the unknown point (t_{n+1}, y_{n+1}):

$$\alpha t_n + \beta = f(t_n, y_n) = f_n \quad (15.212)$$

$$\alpha t_{n+1} + \beta = f(t_{n+1}, y_{n+1}) = f_{n+1} \quad (15.213)$$

Thus, solving we have

$$\alpha = \frac{f_{n+1} - f_n}{h} \quad (15.214)$$

$$\beta = \frac{f_n t_{n+1} - f_{n+1} t_n}{h} \quad (15.215)$$

Following exactly the same procedure, we obtain the second order Adams-Moulton formula:

$$y_{n+1} = y_n + \frac{1}{2} hf_n + \frac{1}{2} hf_{n+1}(t_{n+1}, y_{n+1}) \tag{15.216}$$

This formulation is an implicit finite difference scheme as the unknown also appears on the right hand side of (15.216). If we take simply the constant term in (15.211), we arrive at Euler's backward method. This idea can easily be extended to the case of higher orders (such as the one given in (15.209)), and we can obtain the fourth order Adams-Moulton formula

$$y_{n+1} = y_n + \frac{h}{24}(9f_{n+1} + 19f_n - 5f_{n-1} + f_{n-2}) \tag{15.217}$$

Again, this is an implicit numerical scheme. Moulton was an American astronomer who derived this formula during World War I when he worked on the ballistics trajectories for the US Army. Moulton also published a number of books related to astronomy, including celestial mechanics (Moulton, 1914).

Since the implicit method is more complicated (and iterations is needed), one may ask why it is being proposed and used in the first place. It turns out that some differentiation equations are stiff, in a way that a much smaller step size is needed for stability than for the accuracy requirement. For such problems, the backward or implicit scheme is found stable independent of the time step whereas the time step in the explicit method must be constrained to arrive at a stable solution (e.g., compare Example 15.2).

15.4.3 Predictor-Corrector Method

A popular method is to combine the explicit method and the implicit method. In particular, we approximate ϕ' by polynomials passing through several previous points and possibly also passing through the current point t_{n+1}. The evaluation of the integral in (15.201), in general, leads to the following form:

$$y_{n+1} = y_n + h(\alpha_0 f_{n+1} + \alpha_1 f_n + \alpha_2 f_{n-1} + \alpha_3 f_{n-2} + ...) \tag{15.218}$$

If α_0 is zero, we have the explicit method; otherwise, we have the implicit method. To solve (15.218), we can use either functional iteration or Newton's method. For the iteration, we have to make an initial guess for y_{n+1} and substitute it into the right hand side of (15.218) to get an updated value of y_{n+1}. If the change of the updated value is too large, we can repeat the iteration process. To get an initial guess, we can use the explicit method, such as the Adams-Bashforth method. This is the predictor step. It is essentially an extrapolation of previous data points. Once we have the predicted value, we can use it to interpolate the derivative term to get a corrected result. This is the corrector step. The comparison of the predicted and corrected values provides information on the local truncation error and this can lead to error control and to adjusting step size.

More specifically, a popular predictor-corrector method is the Adams-Bashforth-Moulton scheme. For example, we can use the third order Adams-Bashforth as the predictor equation:

$$y_{n+1} = y_n + \frac{h}{12}(23f_n - 16f_{n-1} + 5f_{n-2}) \tag{15.219}$$

Once the prediction for y_{n+1} is made, we can use the following third order Adams-Moulton method as the corrector equation:

$$y_{n+1} = y_n + \frac{h}{12}(5f_{n+1} + 8f_n - f_{n-1}) \tag{15.220}$$

Note that with the help of the predictor equation in (15.219) the formula given in (15.220) is actually an explicit scheme. This is the beauty of this method. That is, we don't have to solve the nonlinear implicit equation. If the predicted and corrected solutions differ too much, we can repeat the iteration process using (15.220). If after a number of iterations the accuracy is still not up to our expectation, we may consider reducing the step size instead of continuing our iteration process. Note, however, that this predictor-corrector scheme is essentially an explicit scheme and, thus, the strong stability property of the implicit methods is lost and should be used for stiff ODEs.

15.4.4 Backward Differentiation Formula

An alternative to the Adams type of numerical scheme is to assume a polynomial for the unknown y instead as for y' in the Adams approach. In particular, we assume in the backward differentiation formula:

$$P_1(t) = \phi(t) = \alpha t + \beta \tag{15.221}$$

First, we can evaluate the constants by considering the equations at the current point (t_n, y_n) and the unknown point (t_{n+1}, y_{n+1}):

$$\alpha t_n + \beta = y_n \tag{15.222}$$

$$\alpha t_{n+1} + \beta = y_{n+1} \tag{15.223}$$

Thus, solving we have

$$\alpha = \frac{y_{n+1} - y_n}{h} \tag{15.224}$$

However, since we have

$$P_1'(t) = \alpha = f(t_{n+1}, y_{n+1}), \tag{15.225}$$

combining (15.225) and (12.224), we obtain

$$y_{n+1} = y_n + hf(t_{n+1}, y_{n+1}) = y_n + hf_{n+1} \tag{15.226}$$

This is an implicit method, and equals (15.15) or the Euler backward difference method. This method can be extended easily to higher order polynomials. In particular, the fourth order backward differentiation formula is

$$y_{n+1} = \frac{1}{25}\left[48y_n - 36y_{n-1} + 16y_{n-2} - 3y_{n-3} + 12hf(t_{n+1}, y_{n+1})\right] \tag{15.227}$$

This method has been found useful in solving stiff ODEs.

Example 15.3 Consider the solution of the following "stiff" ODE by using Euler's forward and backward schemes, and the fourth order backward differentiation formula:

$$\frac{dy}{dt} = -cy \tag{15.228}$$

where c is a very large number.

Solutions: Using Euler's forward difference scheme, we have

$$y_{n+1} = y_n + hf_n = y_n - chy_n = y_n(1-ch) \tag{15.229}$$

If the absolute value of $(1-ch)$ is larger than 1, the finite difference will not be unstable in the sense that $y \to \infty$ as $n \to \infty$. In other words, for a stable solution we need to control the time step h as

$$h < \frac{2}{c} \tag{15.230}$$

For example, if $c = 10000$, we need to have a time step of less than 0.0002 just for the stability of the solution.

If we apply (15.226) or the Euler backward difference scheme, we get

$$y_{n+1} = y_n - chy_{n+1} \tag{15.231}$$

Solving for y_{n+1}, we obtain

$$y_{n+1} = \frac{y_n}{1+ch} \tag{15.232}$$

We can see that no matter how large c is, the implicit scheme is stable for all time steps of h.

Using the fourth order backward differentiation scheme, we get

$$y_{n+1} = \frac{1}{25(1+12ch)}\left[48y_n - 36y_{n-1} + 16y_{n-2} - 3y_{n-3}\right] \tag{15.233}$$

Again, we see that (15.233) is unconditional stable. Therefore, the implicit backward differentiation formula is a good choice for solving stiff ODEs.

15.5 RUNGE-KUTTA METHOD

People often save the best for the last. We now discuss one of the most popular and successful finite difference schemes, which is known as the Runge-Kutta method. It was originally proposed by Runge in 1895 and was extended to solve systems of equations in 1901 by Kutta. Runge was well known for his work on spectroscopy and Kutta was famous for his airfoil theory.

The idea behind the Runge-Kutta method is to increase the order of accuracy without reducing the time step. In particular, we consider the following first order ODE:

$$\frac{dy}{dx} = f(t, y(t)) \tag{15.234}$$

The Taylor series expansion of the solution at time step $n+1$ about that at time step n is:

$$y_{n+1} = y_n + \frac{h}{1!}f_n + \frac{h^2}{2!}(\frac{df}{dt})_n + \frac{h^3}{3!}(\frac{d^2 f}{dt^2})_n + \frac{h^4}{4!}(\frac{d^3 f}{dt^3})_n + \dots \tag{15.235}$$

where the time step is

$$h = t_{n+1} - t_n \tag{15.236}$$

The idea of the second order Runge-Kutta method is to look for the constants a and b in the following formula:

$$y_{n+1} = y_n + h[af(t_n, y_n) + bf(t_{n+1}, k_{n1})] + O(h^3) \tag{15.237}$$

where

$$k_{n1} = y_n + hf(t_n, y_n) \tag{15.238}$$

We want to find the values of a and b such that the accuracy of the finite difference scheme is of accuracy $O(h^3)$ instead of just $O(h^2)$. In a sense, we are using the slope of y' at both initial point (t_n, y_n) and final point (t_{n+1}, y_{n+1}) in (15.237); and we want to adjust their corresponding weightings such that the accuracy would be one order higher. Note that for $a = 0$ and $b = 1$ we recover the Euler backward difference scheme and for $a = 1$ and $b = 0$ we recover the Euler forward difference scheme. Physically, k_{n1} is the first estimation of the solution of y_{n+1} using the Euler forward difference scheme. In essence, (15.237) is an explicit scheme.

Note by definition that the slope at the end point is

$$f(t_{n+1}, k_{n1}) = f(t_n + h, y_n + hf_n) = f_n + h(\frac{df}{dt})_n \tag{15.239}$$

Substitution of (15.239) into (15.237) gives

$$\begin{aligned} y_{n+1} &= y_n + h[af(t_n, y_n) + bf_n + bh(\frac{df}{dt})_n] + O(h^3) \\ &= y_n + h(a+b)f_n + bh^2(\frac{df}{dt})_n + O(h^3) \end{aligned} \tag{15.240}$$

Comparison of (15.240) with the Taylor series expansion (15.235) leads to

$$a = b = \frac{1}{2} \tag{15.241}$$

Finally, we obtain the second order Runge-Kutta method

$$y_{n+1} = y_n + h[\frac{1}{2}f(t_n, y_n) + \frac{1}{2}f(t_{n+1}, k_{n1})] + O(h^3) \tag{15.242}$$

This is also known as the trapezoidal rule. We note that we can raise the order accuracy of the finite difference scheme by an order of magnitude, if we are willing to evaluate the slope (i.e., the right hand side of (15.234)) twice.

This idea can be extended to higher orders; fourth order Runge-Kutta method is by far most popular and defined by the following approximation:

$$y_{n+1} = y_n + h[af(t_n, y_n) + bf(t_{n+1/2}, k_{n1}) + cf(t_{n+1/2}, k_{n2}) + df(t_{n+1}, k_{n3})] + O(h^5) \tag{15.243}$$

where

$$k_{n1} = y_n + hf(t_n, y_n) \tag{15.244}$$

$$k_{n2} = y_n + hf(t_{n+1/2}, k_{n1}) \tag{15.245}$$

$$k_{n3} = y_n + hf(t_{n+1/2}, k_{n2}) \tag{15.246}$$

In general, for m-th order Runge-Kutta method, we have

$$\begin{aligned} k_{n1} &= y_n + hf(t_n, y_n) \\ k_{n(p)} &= y_n + hf(t_{n+1/2}, k_{n(p-1)}), \quad 2 \le p < m \\ k_{n(m)} &= y_n + hf(t_{n+1}, k_{n(m-1)}) \end{aligned} \tag{15.247}$$

Following the same idea, we can expand the slopes in (15.243) as

$$f(t_{n+1/2},k_{n1})=f(t_n+\frac{h}{2},y_n+\frac{h}{2}f_n)=f_n+\frac{h}{2}(\frac{df}{dt})_n \tag{15.248}$$

$$\begin{aligned} f(t_{n+1/2},k_{n2}) &= f\{t_n+\frac{h}{2},y_n+\frac{h}{2}f(t_{n+1/2},k_{n1})\} \\ &= f_n+\frac{h}{2}\frac{d}{dt}[f_n+\frac{h}{2}(\frac{df}{dt})]_n \end{aligned} \tag{15.249}$$

$$\begin{aligned} f(t_{n+1},k_{n3}) &= f(t_n+h,y_n+hf(t_{n+1/2},k_{n2})) \\ &= f_n+h(\frac{d}{dt}\{f_n+\frac{h}{2}\frac{d}{dt}[f_n+\frac{h}{2}(\frac{df}{dt})]\})_n \end{aligned} \tag{15.250}$$

Substitution of (15.248) to (15.250) into (15.243) leads to

$$\begin{aligned} y_{n+1} = y_n + h[af_n &+ b\{f_n+\frac{h}{2}(\frac{df}{dt})_n\}+c\{f_n+\frac{h}{2}\frac{d}{dt}[f_n+\frac{h}{2}(\frac{df}{dt})]_n\} \\ &+d\{f_n+h(\frac{d}{dt}\{f_n+\frac{h}{2}\frac{d}{dt}[f_n+\frac{h}{2}(\frac{df}{dt})]\})_n\}]+O(h^5) \end{aligned} \tag{15.251}$$

This can be rewritten as

$$\begin{aligned} y_{n+1} &= y_n+h\{a+b+c+d\}f_n+h^2(\frac{df}{dt})_n\{\frac{b}{2}+\frac{c}{2}+d\} \\ &+h^3(\frac{d^2f}{dt^2})_n\{\frac{c}{4}+\frac{d}{2}\}+h^4(\frac{d^3f}{dt^3})_n\{\frac{d}{4}\}+O(h^5) \end{aligned} \tag{15.252}$$

Comparison of (15.251) with the Taylor series expansion in (15.235) gives four equations for four unknowns:

$$a+b+c+d=1 \tag{15.253}$$

$$\frac{b}{2}+\frac{c}{2}+d=\frac{1}{2!} \tag{15.254}$$

$$\frac{c}{4}+\frac{d}{2}=\frac{1}{3!} \tag{15.255}$$

$$\frac{d}{4}=\frac{1}{4!} \tag{15.256}$$

This system of equations can be solved easily and the solutions are

$$a=d=\frac{1}{6}, \quad b=c=\frac{1}{3} \tag{15.257}$$

Finally, we obtain the popular fourth order Runge-Kutta method:

$$y_{n+1}=y_n+\frac{h}{6}[f(t_n,y_n)+2f(t_{n+1/2},k_{n1})+2f(t_{n+1/2},k_{n2})+f(t_{n+1},k_{n3})]+O(h^5) \tag{15.258}$$

This is called the fourth order in the sense that the order of error is $O(h^5)$. Physically, the first slope in the bracket term on the right of (15.258) is the slope evaluated at the initial point, the second term is the first approximation of the slope evaluated at the mid-point, the third term is the second approximation of the slope evaluated at the mid-point, and finally the fourth term is the slope evaluated at the end point. There is also a clear symmetry in the coefficients of a, b, c and d. Thus, we keep the time step as h, but the order of error becomes $O(h^5)$ instead of $O(h^2)$. The price to pay is that we have to evaluate the right hand side of (15.234) four times.

Without going into the details, we quote the following Runge-Kutta formulas for different orders. The third order Runge-Kutta formula is

$$y_{n+1} = y_n + \frac{h}{3}[f(t_n, y_n) + f(t_{n+1/2}, k_{n1}) + f(t_{n+1}, k_{n2})] + O(h^4) \tag{15.259}$$

The fifth order Runge-Kutta formula is

$$\begin{aligned} y_{n+1} = y_n + \frac{h}{15}[f(t_n, y_n) + 5f(t_{n+1/2}, k_{n1}) + 5f(t_{n+1/2}, k_{n2}) \\ +3f(t_{n+1/2}, k_{n3}) + f(t_{n+1}, k_{n4})] + O(h^6) \end{aligned} \tag{15.260}$$

The sixth order Runge-Kutta formula is

$$\begin{aligned} y_{n+1} = y_n + \frac{h}{45}[f(t_n, y_n) + 15f(t_{n+1/2}, k_{n1}) + 15f(t_{n+1/2}, k_{n2}) \\ +9f(t_{n+1/2}, k_{n3}) + 4f(t_{n+1/2}, k_{n4}) + f(t_{n+1}, k_{n5})] + O(h^7) \end{aligned} \tag{15.261}$$

For the proof of these formulas, we refer to the problems at the back of the chapter. We note that there is no more symmetry in the coefficients for scheme beyond the fourth order.

Let us consider the general k-th order Runge-Kutta method as:

$$\begin{aligned} y_{n+1} = y_n + h[a_1 f(t_n, y_n) + a_2 f(t_{n+1/2}, k_{n1}) + a_3 f(t_{n+1/2}, k_{n2}) + \ldots \\ +a_{n-1} f(t_{n+1/2}, k_{n(n-2)}) + a_n f(t_{n+1}, k_{n(n-1)})] + O(h^{k+1}) \end{aligned} \tag{15.262}$$

Following the procedure in obtaining (15.253) to (15.256), it can be shown that the coefficients in (15.262) satisfy the following system of k equations for the k unknowns:

$$\begin{aligned}
a_1 + a_2 + a_3 + a_4 + \ldots\ldots\ldots\ldots + a_{n-1} + a_n &= 1 \\
\frac{a_2}{2} + \frac{a_3}{2} + \frac{a_4}{2} + \ldots\ldots\ldots\ldots + \frac{a_{n-1}}{2} + a_n &= \frac{1}{2!} \\
\frac{a_3}{2^2} + \frac{a_4}{2^2} + \ldots\ldots\ldots\ldots + \frac{a_{n-1}}{2^2} + \frac{a_n}{2} &= \frac{1}{3!} \\
\ldots\ldots\ldots\ldots\ldots\ldots\ldots\ldots\ldots\ldots & \\
\frac{a_k}{2^{k-1}} + \frac{a_{k+1}}{2^{k-1}} + \ldots\ldots + \frac{a_n}{2^{k-2}} &= \frac{1}{k!} \\
\ldots\ldots\ldots\ldots\ldots\ldots\ldots\ldots & \\
\frac{a_{n-1}}{2^{n-2}} + \frac{a_n}{2^{n-3}} &= \frac{1}{(n-1)!} \\
\frac{a_n}{2^{n-2}} &= \frac{1}{n!}
\end{aligned} \tag{15.263}$$

The last equation gives the constant a_n instantly. The second to last equation gives a_{n-1}, and so on. The solution of these equations can be compiled as:

$$a_n = \frac{2^{n-2}}{n!} \tag{15.264}$$

$$a_{n-1} = \frac{2^{n-2}}{(n-1)!} - 2a_n \tag{15.265}$$

$$a_k = \frac{2^{k-1}}{k!} - 2a_n - a_{n-1} - ... - a_{k+1}, \quad 2 \le k \le n-2, \quad n \ge 3 \tag{15.266}$$

$$a_1 = 1 - a_n - a_{n-1} - ... - a_2 \tag{15.267}$$

Using (15.264) and (15.265), we can first find the last two coefficients, then use (15.266) to find the other coefficients. The last coefficient must be found by (15.267) as suggested by the first equation of (15.263). Although we give the solution scheme for generating the higher order Runge-Kutta method in (15.264) to (15.267), the Runge-Kutta higher than 4th order is seldom used because a higher order scheme does not necessarily give a more accurate result (Press et al., 1992).

The Runge-Kutta method (especially the fourth order method) has been used extensively by researchers. Although it may not be the most efficient numerical scheme, it is very stable and reliable especially when those with an adaptive step size algorithm are incorporated in the numerical code. For example, for the same accuracy the predictor-corrector technique may be more efficient. The fifth order Runge-Kutta formula can be embedded in the algorithm to estimate the plausible error at a certain step for the fourth order Runge-Kutta method. The step size can then be lengthened or shortened according to the estimated error. Section 16.2 of Press et al. (1992) gives a more detailed discussion of this error control algorithm. In short, for fourth order Runge-Kutta method the error estimate Δ can be determined by comparing the fourth order prediction (say y_{n+1} in (15.260)) with the fifth order Runge-Kutta method (say y^*_{n+1}) as:

$$\Delta \equiv y_{n+1} - y^*_{n+1} \propto h^5 \tag{15.268}$$

Since we are using the fourth order method, we are expecting the error is proportional to h^5. Suppose that the error estimate at the current step with step size h_1 is Δ_1 and the required error is Δ_0 with a corresponding projected step size h_0. Then, we must have

$$(\frac{\Delta_0}{\Delta_1}) = (\frac{h_0}{h_1})^5 \tag{15.269}$$

If the current estimated error is smaller than the requirement, we can have $\Delta_1/\Delta_0 < 1$. Thus, we can extrapolate to increase the step size to:

$$h_0 = h_1 (\frac{\Delta_0}{\Delta_1})^{1/5} \tag{15.270}$$

If the estimated error is larger than the requirement, we have $\Delta_1/\Delta_0 > 1$. We can use the same formula given in (15.270) to reduce the step size. However, due to the inherit uncertainty in the error estimations, Press et al. (1992) proposed a more conservative approach. In particular, it was proposed that a larger exponent index is used (say 1/4 instead of 1/5) for the case of reducing the step size. In summary, we have (Press et al., 1992):

$$h_0 = h_1 (\frac{\Delta_0}{\Delta_1})^{1/5}, \quad \Delta_0 \geq \Delta_1$$
$$= h_1 (\frac{\Delta_0}{\Delta_1})^{1/4}, \quad \Delta_0 < \Delta_1 \tag{15.271}$$

This approach is also equivalent to assuming that the error is actually proportional to h^4 (more conservative thinking).

In solving the highly nonlinear landslide model based on state- and velocity-dependent friction law, Chau (1995, 1996b) found that the fourth order Runge-Kutta method with adaptive step size algorithm recommended by Press et al. (1992) was very reliable.

15.6 NEWTON-RAPHSON AND RELATED METHODS

In this section, we consider the Newton-Raphson iteration method for the nonlinear problem. This method was developed by Isaac Newton in 1671 but it was not published until 1736, while essentially the same method was published by Joseph Raphson in 1690. The currently adopted version of the Newton-Raphson method is actually due to Raphson, which is simpler than Newton's version. The method was originally developed for searching the roots of an equation. In addition, some related methods are also introduced, namely the initial stress method, modified Newton-Raphson method, and quasi-Newton or BFGS method.

15.6.1 Newton-Raphson Method

Let us illustrate the idea by considering the following second order ODE, which is obtained for nonlinear undamped structural dynamic problems:

$$\boldsymbol{M}\frac{d^2\boldsymbol{U}}{dt^2} + \boldsymbol{f}(\boldsymbol{U}) - \boldsymbol{R}(t) = \boldsymbol{0} \tag{15.272}$$

where $\boldsymbol{U}$ is the displacement vector for the dynamic system, $\boldsymbol{M}$ is the mass matrix, $\boldsymbol{f}$ is the restoring force vector, and $\boldsymbol{R}$ is the external applied force vector on the system. Clearly, the nonlinearity in (15.272) comes from the nonlinear restoring force $\boldsymbol{f}$. For simplicity, we focus on the static problem and assume that we have the following solution at time t

$$\boldsymbol{F}(\boldsymbol{U}^t) = \boldsymbol{f}(\boldsymbol{U}^t) - \boldsymbol{R}^t(t) = \boldsymbol{0} \tag{15.273}$$

where $\boldsymbol{U}^t$ denotes the actual solution at any time. Considering the Taylor series expansion of $\boldsymbol{F}$ at $\boldsymbol{U}^{t+\Delta t}$ about an approximation after $i-1$ iteration, we have

$$\boldsymbol{F}(\boldsymbol{U}^{t+\Delta t}) = \boldsymbol{F}(\boldsymbol{U}_{i-1}^t) + \left[\frac{\partial \boldsymbol{F}(\boldsymbol{U}^t)}{\partial \boldsymbol{U}}\right]_{\boldsymbol{U}_{i-1}^{t+\Delta t}} [\boldsymbol{U}^{t+\Delta t} - \boldsymbol{U}_{i-1}^{t+\Delta t}] + ... \tag{15.274}$$

Assuming the external force $\boldsymbol{R}$ is not a function of displacement $\boldsymbol{U}$, we have

$$\left[\frac{\partial \boldsymbol{F}}{\partial \boldsymbol{U}}\right]_{\boldsymbol{U}_{(i-1)}^{t+\Delta t}}[\boldsymbol{U}^{t+\Delta t}-\boldsymbol{U}_{(i-1)}^{t+\Delta t}] = \left[\frac{\partial \boldsymbol{f}}{\partial \boldsymbol{U}}\right]_{\boldsymbol{U}_{(i-1)}^{t+\Delta t}}\Delta \boldsymbol{U}_{(i)} = \boldsymbol{f}_{(i-1)}^{t+\Delta t}-\boldsymbol{R}^{t+\Delta t} \tag{15.275}$$

Note that the superscript indicates the current time whereas the subscript denotes the iteration number within the time step. The right hand side of (15.275) is the unbalanced force at the current iteration (in general not zero) and this term must be calculated exactly. The tangent stiffness matrix is denoted by $[\partial f/\partial U]$, which is evaluated based on the displacement at the new time step and from the last iteration step $i-1$. If we use $\boldsymbol{K}$ as the tangent stiffness matrix, we can write (15.275) as

$$\left[\frac{\partial \boldsymbol{f}}{\partial \boldsymbol{U}}\right]_{\boldsymbol{U}_{(i-1)}^{t+\Delta t}}\Delta \boldsymbol{U}_{(i)} = \boldsymbol{K}_{(i-1)}^{t+\Delta t}\Delta \boldsymbol{U}_{(i)} = \boldsymbol{f}_{(i-1)}^{t+\Delta t}-\boldsymbol{R}^{t+\Delta t} \tag{15.276}$$

This system is used to solve for the displacement increment, and with this the updated displacement after the i-th iteration is calculated as

$$\boldsymbol{U}_{(i)}^{t+\Delta t} = \boldsymbol{U}_{(i-1)}^{t+\Delta t} + \Delta \boldsymbol{U}_{(i)} \tag{15.277}$$

This iteration process starts with the stiffness matrix, restoring force vector, and displacement vector of the last time step as follows:

$$\boldsymbol{K}_{(0)}^{t+\Delta t} = \boldsymbol{K}^{t} \tag{15.278}$$

$$\boldsymbol{f}_{(0)}^{t+\Delta t} - = \boldsymbol{f}^{t} \tag{15.279}$$

$$\boldsymbol{U}_{(0)}^{t+\Delta t} = \boldsymbol{U}^{t} \tag{15.280}$$

Equations (15.276) and (15.277) constitute the Newton-Raphson method, and it is illustrated in Figure 15.8. We can see from (15.276) that the tangent stiffness matrix needs to be updated and solved (or inverted) at each iteration. For large systems and highly nonlinear problems, it is the most time consuming process for the method.

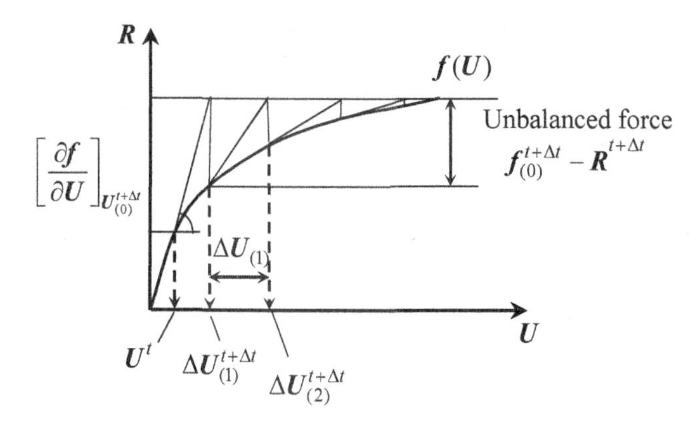

Figure 15.8 Illustration of the Newton-Raphson method

15.6.2 Initial Stress Method

In view of the computationally demanding effort in the Newton-Raphson method, several variations of the Newton-Raphson method have been proposed. The simplest and fastest one is to use the stiffness matrix of the system before the loading is applied. That is, we replace (15.276) by

$$\boldsymbol{K}^0 \Delta \boldsymbol{U}_{(i)} = \boldsymbol{f}_{(i-1)}^{t+\Delta t} - \boldsymbol{R}^{t+\Delta t} \tag{15.281}$$

This is known as the initial stress method. If the system of equations results from the finite element method (FEM), it is the linearized response about the initial configuration of the FEM model. For highly nonlinear problems, the convergence of the method is slow and sometimes the solution may even diverge.

15.6.3 Modified Newton-Raphson Method

The modified Newton-Raphson method lies somewhere between the initial stress method and the full iteration method of Newton-Raphson and is called modified Newton-Raphson. It uses the tangent stiffness matrix from the last time step throughout the iteration process:

$$\boldsymbol{K}^t \Delta \boldsymbol{U}_{(i)} = \boldsymbol{f}_{(i-1)}^{t+\Delta t} - \boldsymbol{R}^{t+\Delta t} = \Delta R_{(i-1)} \tag{15.282}$$

The modified Newton-Raphson method is less computationally demanding but then converges slower than the Newton-Raphson method. Physically, (15.282) gives the displacement vector increment at the i-th iteration to balance the difference between the internal restoring force vector and the external applied force vector at the $(i-1)$-th iteration. The modified Newton-Raphson method is illustrated in Figure 15.9. As expected, the modified Newton-Raphson method converges much slower than those in Figure 15.8, but, however, we do not need to solve the tangent stiffness at every iteration step. Thus, the modified Newton-Raphson is, in general, faster than the full Newton-Raphson method.

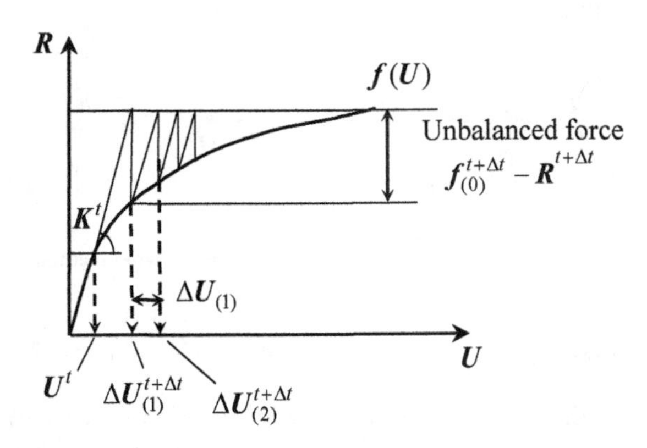

Figure 15.9 Illustration of the modified Newton-Raphson method

15.6.4 Quasi-Newton and BFGS Method

The shortcoming of the modified Newton-Raphson method leads further to the development that we should update the tangent stiffness as required by the accuracy as well as the degree of nonlinearity of the problems. This variation is called the quasi-Newton method. In this method, the tangent stiffness is updated according to whether a certain error control condition is satisfied. For example, the following criterion is used for the so-called BFGS method:

$$c_{(i)} = [\frac{\boldsymbol{\delta}_{(i)}^T \boldsymbol{\beta}_{(i)}}{\boldsymbol{\delta}_{(i)}^T \boldsymbol{K}_{(i-1)}^{t+\Delta t} \boldsymbol{\beta}_{(i)}}] \leq \varepsilon_C \tag{15.283}$$

where

$$\boldsymbol{\delta}_{(i)}^T = \boldsymbol{U}_{(i)}^{t+\Delta t} - \boldsymbol{U}_{(i-1)}^{t+\Delta t} \tag{15.284}$$

$$\boldsymbol{\beta}_{(i)} = \Delta \boldsymbol{R}_{(i-1)} - \Delta \boldsymbol{R}_{(i)} \tag{15.285}$$

Whenever the condition number defined in (15.283) is larger than the prescribed number ε_c (or (15.283) is violated), an update of the tangent stiffness is performed. This BFGS method is illustrated in Figure 15.10.

Therefore, in a sense this is an error-driven scheme for updating the tangent stiffness. This particular version of the quasi-Newton method is known as the BFGS (Broyden-Fletcher-Goldfarb-Shanno) method, which was developed independently by Broyden, Fletcher, Goldfarb, and Shanno all in 1970 (see Press et al., 1992). It is believed that the quasi-Newton method is most efficient, and is also widely used in hill-climbing types of unconstrained optimization problems.

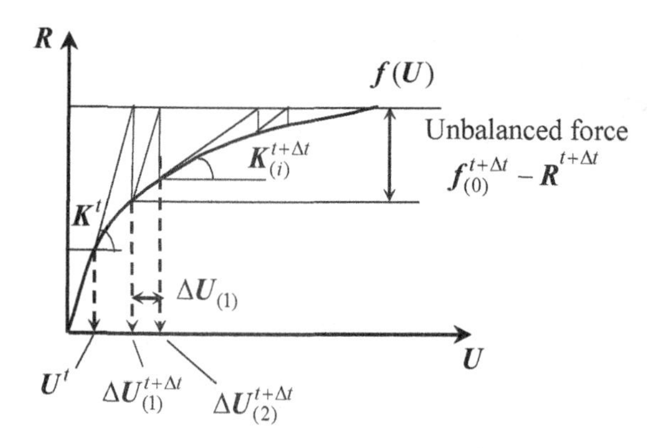

Figure 15.10 Illustration of the BFGS or quasi-Newton method

15.6.5 Secant or Fixed Point Iteration Method

There is another variation the called fixed point iteration or secant method. The stiffness is formed between the fixed initial point and the current point, and thus is a secant stiffness. The unbalanced force vector can be formulated:

$$\left[\frac{\partial \boldsymbol{F}}{\partial \boldsymbol{U}}\right]_{\boldsymbol{U}_{(i-1)}^{t+\Delta t}-\boldsymbol{U}_{(0)}^{t+\Delta t}}[\boldsymbol{U}^{t+\Delta t}-\boldsymbol{U}_{(0)}^{t+\Delta t}] \\ =\left[\frac{\partial \boldsymbol{f}}{\partial \boldsymbol{U}}\right]_{\boldsymbol{U}_{(i-1)}^{t+\Delta t}-\boldsymbol{U}_{(0)}^{t+\Delta t}}\Delta \boldsymbol{U}_{(i)}=\boldsymbol{f}_{(i-1)}^{t+\Delta t}-\boldsymbol{R}^{t+\Delta t} \tag{15.286}$$

This secant method is illustrated in Figure 15.11.

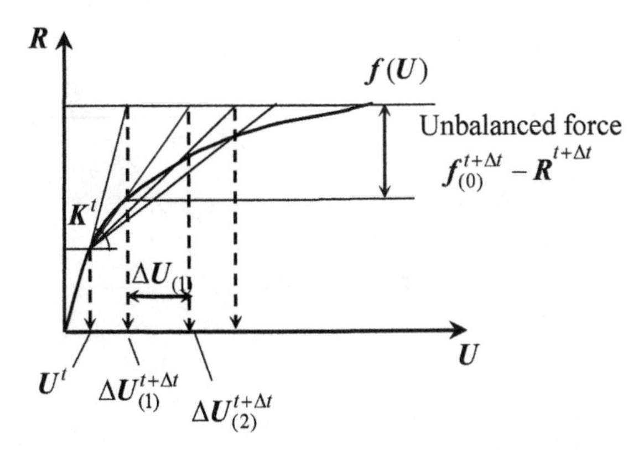

Figure 15.11 Illustration of the secant or fixed point method

All Newton-Raphson type methods may or may not converge to the actual solution, depending on whether the function has a quadratic Taylor expansion near the solution point and on how far the initial point of iteration is from the true solution. In general, there is no single nonlinear algorithm that can guarantee convergence for every nonlinear problem. In particular, Figure 15.12 shows some situations in which at least one of the above nonlinear iterative methods breaks down.

15.6.6 Convergence Criteria

Regarding the convergence criteria for the iterations of the Newton-Raphson type methods, we can have at least three different approaches.

(i) Displacement error control

The increment of displacement after the current iteration step can be compared to the current value of the displacement. If the contribution in the displacement increment is less than a prescribed error, iteration can be stopped. The current value of displacement is regarded as the solution for the current time step. For example, we have

$$\frac{\left\|\Delta \boldsymbol{U}_{(i)}\right\|_2}{\left\|\Delta \boldsymbol{U}_{(i)}^{t+\Delta t}\right\|_2}\leq \varepsilon_D \tag{15.287}$$

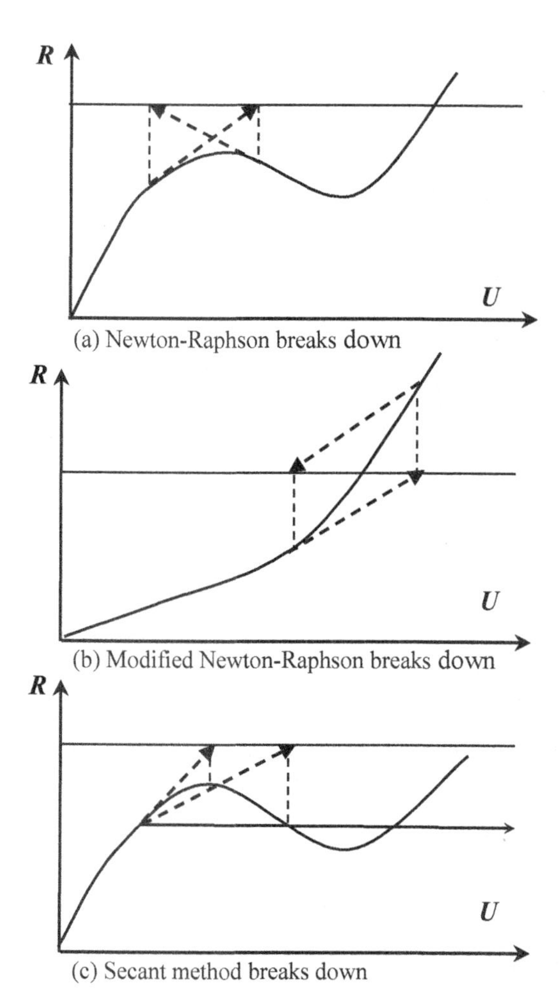

Figure 15.12 Pitfall of some softening behavior of a nonlinear system

where the norm of the displacement vector is defined as

$$\|U\|_2 = \left[\sum_{i=1}^{n} |U_i|^2\right]^{1/2} \tag{15.288}$$

This definition is also known as the Euclidean norm. The dimension of the vector is denoted by n in (15.288). However, we should not be too greedy about this choice and its value should also depend on the current time step that we are using. As illustrated in Figure 15.13, it shows the case of a stiffening system and for such stiffening, the response displacement convergence criterion will fail. In view of this limitation, load error control is introduced.

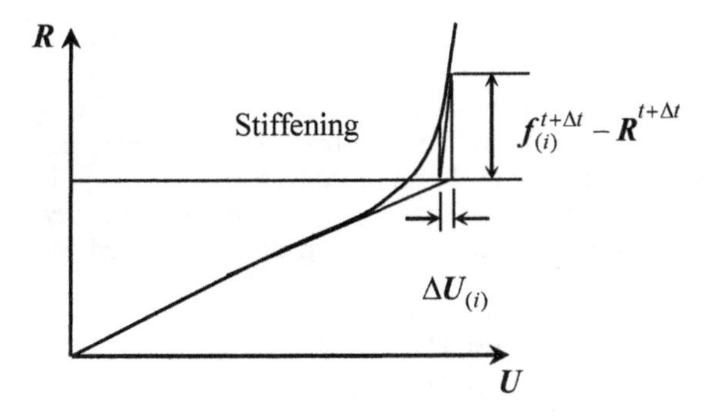

Figure 15.13 Stiffening behavior of a nonlinear system

(ii) Load error control

When a nonlinear system becomes excessively stiff with the increase of the displacement, displacement increment will be small but the magnitude or norm of the unbalanced force vector can remain large. It is sometimes more meaningful to control the error in the unbalanced force:

$$\frac{\left\| \boldsymbol{f}_{(i)}^{t+\Delta t} - \boldsymbol{R}^{t+\Delta t} \right\|_2}{\left\| \boldsymbol{f}^{t} - \boldsymbol{R}^{t+\Delta t} \right\|_2} \leq \varepsilon_L \tag{15.289}$$

This load convergence criterion is not effective for the case in which a very soft response of the nonlinear system becomes apparent with the progressive displacement. A typical value of $\varepsilon_L = 0.0001$ can normally be prescribed; for example, a truss undergoing a large deformation due to elastic-plastic behavior of the material. In this case, we should use the displacement control convergence criterion. Figure 15.14 shows such a case of a softening system and for which the load convergence criterion will fail.

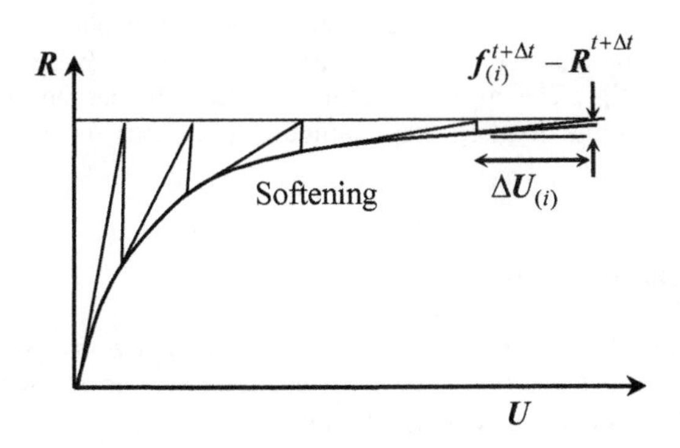

Figure 15.14 Softening behavior of a nonlinear system

(iii) Energy error control

For highly nonlinear and complicated models of composite structures, it is sometimes difficult to pre-determine whether the structure is getting softer and stiffer under loading. Therefore, the third convergence criterion based on energy change in the system due to the unbalanced force at each iteration can be used as an error control criterion. It is because energy change is involved both displacement increment and unbalanced force:

$$\frac{\Delta \boldsymbol{U}_{(i)}^{t+\Delta t}(\boldsymbol{f}_{(i)}^{t+\Delta t} - \boldsymbol{R}^{t+\Delta t})}{\Delta \boldsymbol{U}_{(1)}^{t+\Delta t}(\boldsymbol{f}^{t} - \boldsymbol{R}^{t+\Delta t})} \leq \varepsilon_E \tag{15.290}$$

where ε_E is a prescribed error.

15.7 CONTROL ALGORITHMS FOR NONLINEAR ODE

The static nonlinear system considered in (15.273) is first rewritten in the following form:

$$\boldsymbol{f}(\boldsymbol{U}) - \boldsymbol{R} = \boldsymbol{f}(\boldsymbol{U}) - \lambda \overline{\boldsymbol{R}} = 0 \tag{15.291}$$

where λ is the load parameter. Tacitly, we have assumed that the loading is of the proportional type. That is, the ratio between different loading components of $\boldsymbol{R}$ for each degree of freedom of the displacement vector $\boldsymbol{U}$ is constant, and this is the reason why we can extract a common loading parameter λ in (15.291). We can also rewrite this vector equation in scalar form as:

$$\frac{\overline{\boldsymbol{R}}^T \boldsymbol{f}(\boldsymbol{U})}{\overline{\boldsymbol{R}}^T \overline{\boldsymbol{R}}} - \lambda = 0, \quad or\ f(\boldsymbol{U}) - \lambda = 0 \tag{15.292}$$

The load control algorithm can be unified with the displacement control algorithm in the following constraint:

$$\sum_{k=1}^{n} \beta_k (U_k^i - U_k^0)^2 + \beta_{n+1}(\lambda^i - \lambda^0)^2 = c^2 \tag{15.293}$$

where β_i is a controlling parameter for each component of displacement or loading, and c is the size of the loading control. Clearly the unit of β_i (where i = 1,2,..., n) differs from that of β_{n+1}. The superscript for U denotes the degree of freedom and the subscript denotes the loading step number. A number of special cases are considered next.

15.7.1 Displacement Control

For the special case $\beta_j = 1$ and all other $\beta_i = 0$ ($i \neq j$), we may select the most dominant displacement U_j as the controlling parameter (typically at the degree of freedom subjected to the largest external load):

$$U_j^i - U_j^0 = c \tag{15.294}$$

Displacement control may fail for the case of a very stiff system, and this is illustrated in Figure 15.15.

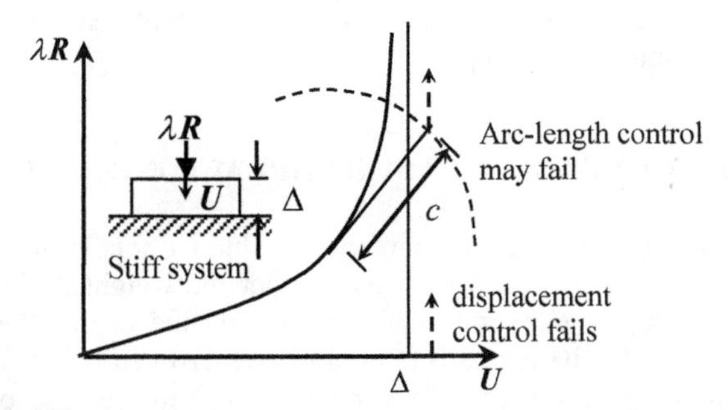

Figure 15.15 Stiff system in which displacement control fails and arc-length control may also fail

15.7.2 Load Control

For the special case $\beta_{n+1} = 1$ and all other $\beta_i = 0$ ($i \neq n+1$), we select the loading step as the nonlinear parameter in searching for the next equilibrium solution. Mathematically, it can be expressed as:

$$\lambda^i - \lambda^0 = c \tag{15.295}$$

This type of loading control parameter is the most commonly employed approach in numerical analysis. Figure 15.16 illustrates the case of snap through buckling of cylindrical shells in which load control fails.

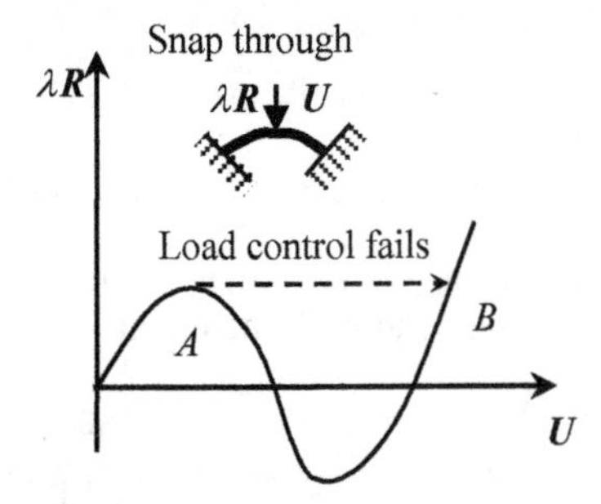

Figure 15.16 Snap through buckling problem in which load control fails

15.7.3 Arc-Length Control

If we set all $\beta_i = 1$ (where $i = 1,2,...,n+1$), we have the arc-length control algorithm. We select the loading step as the nonlinear parameter in searching for the next equilibrium solution. Mathematically, it can be expressed as:

$$\sum_{k=1}^{n}(U_k^i - U_k^0)^2 + (\lambda^i - \lambda^0)^2 = c^2 \tag{15.296}$$

Figure 15.15 shows that if c is too large, the arc-length control may also fail. In short, there is no single algorithm that works for all nonlinear problems, if the loading step is not controlled properly.

5.8 FINITE ELEMENT METHOD FOR LAPLACE EQUATION

The finite element method (FEM) is a numerical technique for finding approximate solutions to boundary value problems. It uses either the weighted residue method (e.g., Galerkin method) or the variational methods via the use of the calculus of variations (e.g., Rayleigh-Ritz method) to minimize an error function (in the case of the weighted residue approach) or a functional (energy in the case of variational approach) in a global sense and produces a stable solution. FEM divides the origin domain into many small finite sub-domains, named finite elements, to approximate a larger domain. The origin of FEM is not so straightforward. As summarized by Oden (1990), it probably dates back to the time of Hrennikoff in 1941 and Courant in 1943. However, this idea was not further pursued then since computers were still largely unavailable in 1940s. Decades later, the term "finite element method" was coined by the renowned structural and earthquake engineer Ray Clough in 1960 when he was involved in the design of the wings of the Boeing 747 airplane (Clough, 1980). The first finite element book was *The Finite Element Method in Structural and Continuum Mechanics* by O.C. Zienkiewicz in 1967.

In this section, we will illustrate the main concepts of FEM using the Laplace equation. In particular, for potential flow around a circular cylinder shown in Figure 15.17, we can formulate the problem as:

$$\nabla^2 \psi = 0 \tag{15.297}$$

$$\psi = \psi_i, \quad \text{on } S_i \quad (i = 1,2,3) \tag{15.298}$$

$$\frac{\partial \psi}{\partial n} = -q_n, \quad \text{on } S_4; \quad \frac{\partial \psi}{\partial n} = q_n, \quad \text{on } S_5 \tag{15.299}$$

where ψ is the stream function. As shown in Figure 15.17, the surface of the circular cylinder is impermeable, and both the upper and lower boundaries are also impermeable.

In view of symmetry with respect to both vertical and horizontal axes, we can consider a quarter of the domain. This quarter domain is further discretized into elements (in this case triangular elements). Within each element, we assume there is a fixed and assumed variations of the streamline function ψ between the nodal values of the element:

$$\psi(x, y) = \sum_{i=1}^{3} N_i \psi_i = N_1(x, y)\psi_1 + N_2(x, y)\psi_2 + N_3(x, y)\psi_3 \tag{15.300}$$

where N_i (i = 1,2,3) are called shape functions. The nodal values of ψ_i are the unknowns. At node 1, we have $N_1 = 1$ whereas the others' shape functions are zero, similar to other shape functions. Without going into detail, it was found that

$$N_1(x, y) = \frac{1}{2\Delta}[(x_2 y_3 - x_3 y_2) + (y_2 - y_3)x + (x_3 - x_2)y] \tag{15.301}$$

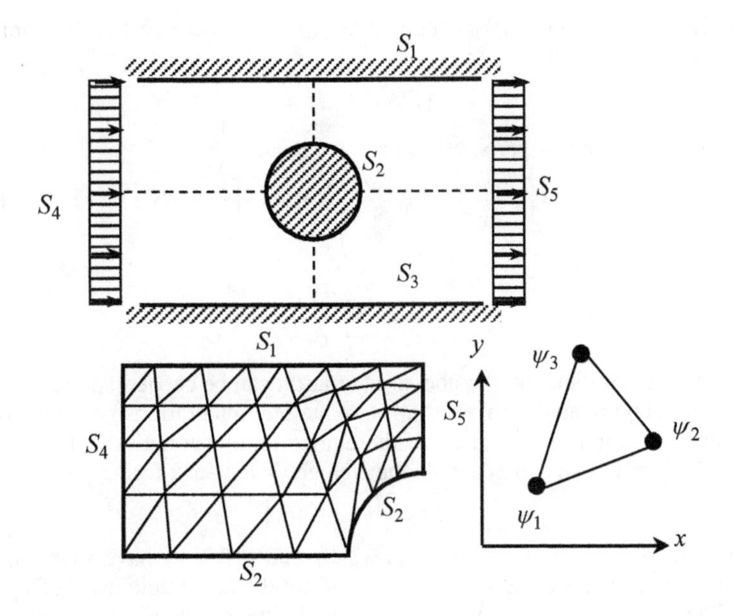

Figure 15.17 Finite element model for potential flow around a cylinder in a channel

$$N_2(x,y) = \frac{1}{2\Delta}[(x_3 y_1 - x_1 y_3) + (y_3 - y_1)x + (x_1 - x_3)y] \tag{15.302}$$

$$N_3(x,y) = \frac{1}{2\Delta}[(x_1 y_2 - x_2 y_1) + (y_1 - y_2)x + (x_2 - x_1)y] \tag{15.303}$$

where Δ is the area of the triangular element and (x_i, y_i) are the coordinates of the i node in the element. Physically, this shape function can be interpreted as:

$$N_1(x,y) = \frac{\Delta_{023}}{\Delta_{123}}, \quad N_2(x,y) = \frac{\Delta_{013}}{\Delta_{123}}, \quad N_3(x,y) = \frac{\Delta_{012}}{\Delta_{123}} \tag{15.304}$$

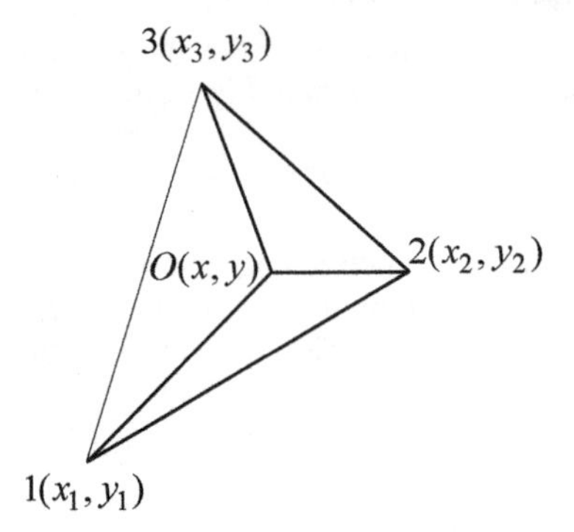

Figure 15.18 Interpretation of shape functions in a triangular element

where the area of triangle ijk is denoted as Δ_{ijk} (see Figure 15.18). By adopting the Galerkin method, we have

$$\iint \nabla^2 \psi N_i dA = 0 \tag{15.305}$$

Applying integration by parts, we can rewrite it as:

$$\iint \{\frac{\partial \psi}{\partial x}\frac{\partial N_i}{\partial x} + \frac{\partial \psi}{\partial y}\frac{\partial N_i}{\partial y}\} dA = \int q_n N_i dS \tag{15.306}$$

where

$$\frac{\partial \psi}{\partial x} = \sum_{j=1}^{3} \frac{\partial N_j}{\partial x}\psi_j \tag{15.307}$$

For the triangular element, the above integration can be carried analytically but for other more complicated element shapes, numerical integration should be used (like the Gauss quadrature integration). If there are a total of n nodes and m elements in the whole domain, we can assemble the element in a matrix form:

$$\left[K_{ij}\right]\{\psi_j\} = \{q_i\} \tag{15.308}$$

The boundary conditions given in (15.298) and (15.299) have to be applied to (15.308). The matrix equation can be solved for the nodal unknowns of ψ_j.

The total error associated with FEM modelling can be expressed as:

$$Error = E_c + E_b + E_p + E_I \tag{15.309}$$

where E_c is the continuity error, E_b is boundary error, E_p is the interpolation error, and E_I is the integration error. The first three errors are caused by discretization. The continuity error depends on the highest derivative in the formulation and can be removed if the shape function used in the element is raised. The boundary errors can be reduced if the number of elements is increased and if the size of the elements is reduced. The interpolation error is reduced if the element size is reduced and the order of the interpolation function is increased.

5.9 FINITE DIFFERENCE METHOD FOR LAPLACE EQUATION

In this section, we will illustrate how to apply FDM to PDEs, and again the Laplace equation is selected for the sake of simplicity. In particular, for potential flow around a circular cylinder shown in Figure 15.17, we can formulate the problem as:

$$\nabla^2 \psi = \frac{\partial^2 \psi}{\partial x^2} + \frac{\partial^2 \psi}{\partial y^2} = 0 \tag{15.310}$$

where ψ is the stream function. As shown in Figure 15.19, the domain of the quarter problem of the potential flow around a circular cylinder can be expressed in a finite difference scheme. The following notation is adopted to express the streamline function at different grid points:

$$\psi_{i,j} = \psi(x_0 + i\Delta x, y_0 + j\Delta y) \tag{15.311}$$

where Δx and Δy are the grid sizes along the x-direction and y-direction respectively.

Using the finite difference scheme, differentiation of the streamline with respect to x can be approximated as

$$\frac{\partial \psi}{\partial x} \approx \frac{\psi(x+\Delta x, y) - \psi(x, y)}{\Delta x} = \frac{1}{\Delta x}(\psi_{i+1,j} - \psi_{i,j}) \tag{15.312}$$

Similarly, differentiation of the streamline with respect to y can be expressed as

$$\frac{\partial \psi}{\partial y} \approx \frac{\psi(x, y+\Delta y) - \psi(x, y)}{\Delta y} = \frac{1}{\Delta y}(\psi_{i,j+1} - \psi_{i,j}) \tag{15.313}$$

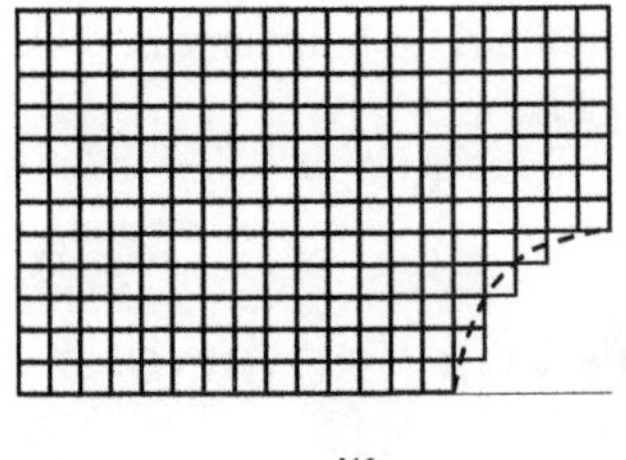

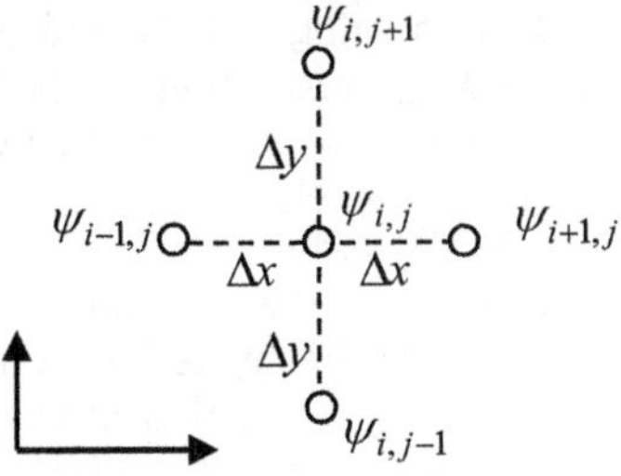

Figure 15.19 Grid network used in the finite difference scheme

Repeat the same procedure once more, and the second derivative can be approximated as:

$$\frac{\partial^2 \psi}{\partial x^2} \approx \frac{1}{\Delta x}\left[\frac{\psi(x+\Delta x, y) - \psi(x, y)}{\Delta x} - \frac{\psi(x, y) - \psi(x-\Delta x, y)}{\Delta x}\right]$$
$$= \frac{1}{\Delta x^2}(\psi_{i+1,j} - 2\psi_{i,j} + \psi_{i-1,j}) \tag{15.314}$$

Similarly, the second derivative taken with respect to y can be approximated as:

$$\frac{\partial^2 \psi}{\partial y^2} \approx \frac{1}{\Delta y^2}(\psi_{i,j+1} - 2\psi_{i,j} + \psi_{i,j-1}) \tag{15.315}$$

For further simplification, we can set the grid size in both directions as equal:

$$\Delta y = \Delta x \tag{15.316}$$

Substitution of (15.314) and (15.315) into the Laplace equation and utilization of (15.316) leads to

$$\psi_{i,j+1} - 2\psi_{i,j} + \psi_{i,j-1} + \psi_{i+1,j} - 2\psi_{i,j} + \psi_{i-1,j} = 0 \tag{15.317}$$

Solving for $\psi_{i,j}$ we get

$$\psi_{i,j} = \frac{1}{4}(\psi_{i,j+1} + \psi_{i,j-1} + \psi_{i+1,j} + \psi_{i-1,j}) \tag{15.318}$$

As illustrated in Figure 15.19, the value of $\psi_{i,j}$ depends on the neighboring points. Once the boundary values of ψ are known on the essential boundaries, we can use (15.318) to generate the rest. For the natural boundary conditions, we can use either forward difference or backward difference schemes, depending on the location of the natural boundaries. For example, for problems shown in Figure 15.17 and 15.19, the inflow boundary on the left boundary is

$$\frac{\partial \psi}{\partial n} = \frac{\partial \psi}{\partial x} = \frac{1}{\Delta x}(\psi_{i+1,j} - \psi_{i,j}) = -q_x \tag{15.319}$$

whereas the inflow boundary on the right boundary is

$$\frac{\partial \psi}{\partial n} = \frac{\partial \psi}{\partial x} = \frac{1}{\Delta x}(\psi_{i,j} - \psi_{i-1,j}) = q_x \tag{15.320}$$

15.10 SUMMARY AND FURTHER READING

Numerical methods employed in solving ODEs and PDEs are under rapid expansion in the last few decades. Many restrictions of the finite difference method and finite element method have been removed in these new methods. A number of them are summarized briefly here, namely, the finite volume method (FVM), smooth particle hydrodynamics (SPH), and the material point method (MPM). Full discussions of them are out of the scope of the present chapter.

A numerical technique called the material point method (MPM) is particularly useful in modelling large deformation problems, such as landslides, runouts, or dynamic fragmentations. This formulation uses a dual description of the media by using Lagrangian material points and a Eulerian numerical mesh. The MPM is an extension of the particle-in-cell method (a method developed in Los Alamos National Laboratory in 1957) in computational fluid dynamics to computational solid dynamics, and is a finite element method (FEM)-based particle method. It is primarily used for multiphase simulations, because of the ease of detecting contact without inter-penetration. It can also be used as an alternative to dynamic FEM methods in simulating large material deformations, because there is no re-meshing required by the MPM. It was originally proposed by Sulsky et al. (1995).

The smoothed particle hydrodynamics (SPH) method belongs to mesh-free techniques that have been widely adopted in many areas of mechanics. Smoothed particle hydrodynamics (SPH) is a computational method used mainly for simulating fluid flows. It was developed by Gingold and Monaghan (1977) and Lucy (1977) initially for astrophysical problems. It is a mesh-free Lagrangian method in which the coordinates move with the fluid, and the resolution of the method can easily be adjusted with respect to variables such as density. This technique can handle very large deformations and is more suitable for post-failure analysis in the case of solid mechanics.

In recent years, the finite volume method (FVM) has become a popular numerical method used in fluid mechanics. "Finite volume" refers to the small volume surrounding each node point on a mesh, resulting from discretization of the body. In the finite volume method, a partial differential equation is converted to surface integrals. These integrals are then evaluated as fluxes at the surfaces of each finite volume. Because the flux entering a given volume is identical to that leaving the adjacent volume, these methods are conservative. FVM is best for solving conservation law in integral form and can solve for discontinuous solutions. The most fundamental hyperbolic wave problem with a jump discontinuity is called the Riemann problem (Le Veque, 2002). In fact, most of the current finite volume methods make use of the Riemann problem as the building block, and therefore FVM literally uses the Riemann solver. Most of the FVM solution schemes used nowadays are of the Godunov type (Le Veque, 2002). Another advantage of the finite volume method is that it is easily formulated to allow for unstructured meshes.

15.11 PROBLEMS

Problem 15.1 Consider the following third order Runge-Kutta method:

$$y_{n+1} = y_n + h[af(t_n, y_n) + bf(t_{n+1/2}, k_{n1}) + cf(t_{n+1}, k_{n2})] + O(h^4) \quad (15.321)$$

where k_{n1} and k_{n2} have been defined in (15.244) and (15.245).

(i) Show that the coefficients satisfy the following system of equations:

$$a + b + c = 1 \quad (15.322)$$

$$\frac{b}{2} + \frac{c}{1} = \frac{1}{2!} \quad (15.323)$$

$$\frac{c}{2} = \frac{1}{3!} \quad (15.324)$$

(ii) Show the validity of (15.259).

Problem 15.2 Consider the following fifth order Runge-Kutta method:

$$y_{n+1} = y_n + h[af(t_n, y_n) + bf(t_{n+1/2}, k_{n1}) + cf(t_{n+1/2}, k_{n2}) + df(t_{n+1/2}, k_{n3}) + ef(t_{n+1}, k_{n4})] + O(h^6) \quad (15.325)$$

(i) Show that the coefficients satisfy the following system of equations:

$$a + b + c + d + e = 1 \quad (15.326)$$

$$\frac{b}{2} + \frac{c}{2} + \frac{d}{2} + e = \frac{1}{2!} \quad (15.327)$$

$$\frac{c}{2^2} + \frac{d}{2^2} + \frac{e}{2^1} = \frac{1}{3!} \quad (15.328)$$

$$\frac{d}{2^3} + \frac{e}{2^2} = \frac{1}{4!} \quad (15.329)$$

$$\frac{e}{2^3} = \frac{1}{5!} \quad (15.330)$$

(ii) Show the validity of (15.260).

Problem 15.3 Consider the following sixth order Runge-Kutta method:

$$y_{n+1} = y_n + h[af(t_n, y_n) + bf(t_{n+1/2}, k_{n1}) + cf(t_{n+1/2}, k_{n2}) + df(t_{n+1/2}, k_{n3}) + ef(t_{n+1/2}, k_{n4}) + gf(t_{n+1}, k_{n5})]] + O(h^7) \tag{15.331}$$

(i) Show that the coefficients satisfy the following system of equations:

$$a + b + c + d + e + g = 1 \tag{15.332}$$

$$\frac{b}{2} + \frac{c}{2} + \frac{d}{2} + \frac{e}{2} + g = \frac{1}{2!} \tag{15.333}$$

$$\frac{c}{2^2} + \frac{d}{2^2} + \frac{e}{2^2} + \frac{g}{2^1} = \frac{1}{3!} \tag{15.334}$$

$$\frac{d}{2^3} + \frac{e}{2^3} + \frac{g}{2^2} = \frac{1}{4!} \tag{15.335}$$

$$\frac{e}{2^4} + \frac{g}{2^3} = \frac{1}{5!} \tag{15.336}$$

$$\frac{g}{2^4} = \frac{1}{6!} \tag{15.337}$$

(ii) Show the validity of (15.261).

Problem 15.4 Derive the formulas governing the coefficients in the k-th Runge-Kutta method given in (15.263).

Problem 15.5 In view of (15.263), show the validity of formulas given in (15.264) to (15.267).

Problem 15.6 Prove the following seventh order Runge-Kutta method:

$$y_{n+1} = y_n + \frac{h}{315}[2f(t_n, y_n) + 77f(t_{n+1/2}, k_{n1}) + 133f(t_{n+1/2}, k_{n2}) + 63f(t_{n+1/2}, k_{n3}) + 28f(t_{n+1/2}, k_{n4}) + 10f(t_{n+1/2}, k_{n5}) + 2f(t_{n+1}, k_{n6})] + O(h^8) \tag{15.338}$$

Problem 15.7 Prove the following fifth order Adams-Bashforth formula:

$$y_{n+1} = y_n + \frac{h}{720}[1901f_n - 2774f_{n-1} + 2616f_{n-2} - 1274f_{n-3} + 251f_{n-4}] \tag{15.339}$$

Problem 15.8 Prove the following fifth order Adams-Moulton formula:

$$y_{n+1} = y_n + \frac{h}{720}[251f_{n+1} + 646f_n - 264f_{n-1} + 106f_{n-2} - 19f_{n-3}] \tag{15.340}$$

Appendix A: Greek Letters

For mathematical and engineering analysis, Greek alphabets are normally adopted as mathematical symbols or scientific symbols. Engineering and science students who do not know how to pronounce Greek letters will have trouble communicating with others. It is of paramount importance to recognize and to pronounce Greek letters. Here we compile a list for easy reference by readers. There are a total of 24 letters in Greek and they are summarized below:

Table A.1 Table of Greek letters

Number	Capital	lower	Pronoun.	Number	Capital	lower	Pronoun.
1	Α	α	Alpha	13	Ν	ν	Nu
2	Β	β	Beta	14	Ξ	ξ	Xi
3	Γ	γ	Gamma	15	Ο	ο	Omicron
4	Δ	δ	Delta	16	Π	π	Pi
5	Ε	ε	Epsilon	17	Ρ	ρ	Rho
6	Ζ	ζ	Zeta	18	Σ	σ	Sigma
7	Η	η	Eta	19	Τ	τ	Tau
8	Θ	θ	Theta	20	Υ	υ	Upsilon
9	Ι	ι	Iota	21	Φ	φ	Phi
10	Κ	κ	Kappa	22	Χ	χ	Chi
11	Λ	λ	Lambda	23	Ψ	ψ	Psi
12	Μ	μ	Mu	24	Ω	ω	Omega

Note that the partial differentiation sign ∂ (pronounced as “round”) is not a Greek letter.

Appendix B: Nobel Prize and Mathematicians

Alfred Bernhard Nobel (1833–1896) was a Swedish chemist and engineer. He invented dynamite in 1867, gelignite in 1875, and ballistite (a kind of smokeless gunpowder) in 1887. When Nobel passed away in 1896, he left all of his considerable fortune to a foundation that funds five Nobel Prizes annually in the areas of physics, physiology or medicine, chemistry, literature, and contributions to peace. The first prizes were awarded in 1901, five years after the death of Nobel. The award also comes with cash of about 1 million US dollars for each prize. It is, however, not the cash prize but the instant fame that makes the Nobel winners a household name and an iconic figure of his or her generation. Even today, it is safe to claim that it remains the highest recognition and reward that can be earned by a "scientist." In view of the important role of mathematics in the development of scientific and technological breakthroughs, many people, mathematicians and scientists alike, question why there is no "mathematics" category of the Nobel Prize. The renowned Fields Medal in mathematics is awarded to a young mathematician once every 4 years, but is no match for the Nobel Prize. A more recent establishment of the Shaw Prize, named after the Hong Kong-based movie tycoon Mr. Shaw, does include a category of "mathematical science." Clearly, this reflects the views of scientists and mathematicians on this issue.

However, there is an unconfirmed rumor about why Nobel did not choose to award his prize in "mathematics." When Alfred Nobel set up his Nobel Prize, mathematics was one of the potential subject areas to be awarded. However, when he knew that Mittag-Leffler was a potential candidate for the prize in mathematics, Nobel crossed out mathematics and no such prize has ever been awarded. This negative impact by Mittag-Leffler on mathematics is definitely more far-reaching than his mathematical contributions.

Nevertheless, some mathematicians have been nominated to receive Nobel Prizes and they also have played an important role in promoting the success of Nobel Prizes. Ironically, it was S. Arrhenius and G. Mittag-Leffler (the Swedish mathematician that Nobel disliked) who helped to mobilize large numbers of scientists from various nations participating in the nomination process. Arrhenius received the Nobel Prize in chemistry in 1903, and, of course, Mittag-Leffler never received the prize as a mathematician. In 1974, the Nobel foundation and associated institutions agreed to open their archives for historical research on materials at least 50 years old. Crawford (1985, 1998, 2001) is among the forerunner on the analysis of the selection process of Nobel Prizes.

According to the released data, the renowned French engineer, mathematician, and physicist Henri Poincare (pioneer in stability analysis of differential equations) was nominated 34 times in 1910. In fact, Poincare received a total of 51 nominations in 1904–1912 before he passed away in 1913. Poincare himself had made 10 nominations and four nominations were successful (including Lorentz in Physics 1902, Becquerel in Physics 1903, Lippmann in Physics 1908, and Curie in Chemistry 1911). Mittag-Leffler made 6 nominations with one successful case (Lorentz in Physics 1902). The famous German mathematician

David Hilbert was nominated 6 times between 1929 to 1933. He also nominated the successful winner Peter Debye (Chemistry in 1936), and, incidentally, Debye received 47 nominations in both Physics and Chemistry. More importantly, being a student of Sommerfeld, Debye made a major contribution in 1912 to the mathematical development of the integral representation of the Bessel function of the first kind of large order using the saddle point method (see Watson, 1944) when he considered the vibrations of large spheres.

We should note the story of Sommerfeld (supervisor of Debye) before continuing our discussion of the other mathematicians and their role in the Nobel Prize. Sommerfeld was nominated a record 81 times for the Nobel Prize within a period of 34 years before he was run down by a car in 1951, and he never received it. Otto Stern also shared the honor with the highest number of nominations of 81, but Stern was awarded the Nobel Prize in 1943 for demonstrating the wavelike properties of elementary particles. Ironically, he himself had made two successful nominations, one to Albert Einstein (Physics in 1921) and one to Max Planck (Physics in 1918). Even ironically, four of his PhD students did receive Nobel Prizes (Heisenberg in 1932 (PhD in 1923); Bethe in 1967 (PhD in 1928); Pauli in 1945 (PhD in 1921); and Debye in 1936 (PhD in 1908)). Three of his other postgraduate and post-doctoral students also received Nobel Prizes (including L. Pauling in Chemistry in 1954 and in Peace in 1962, I.I. Rabi in 1944, and M.T. von Laue in 1914). Sommerfeld had published a six-volume series of classic textbooks, and one is *Partial Differential Equations in Physics*. This is a very good book on PDEs even by today's standard. One of his mathematical contributions is Sommerfeld's wave radiation condition in wave equation analysis. Sommerfeld was also a pioneer in the mathematical theory of diffraction of waves.

Another famous French mathematician Jacques Hadamard made 32 nominations and his successful nominees include Einstein, Perrin, Richardson, Yukawa, and Blackett. French mathematician Emile Borel (a pioneer in game theory and a main contributor on the theory of real variables) also made 6 nominations and Nobel laureate Perrin is among them (Physics in 1926). Recall in Chapter 11 on Volterra's integral and integro-differential equations, the renowned Italian mathematician V. Volterra also made 9 nominations (1 in chemistry and 8 in physics).

There is also a long list of renowned mathematicians and mechanicians, most of their names are mentioned in this book, who made nominations or had been nominated. They include G. Darboux (Darboux's formula in series expansion in Chapter 1), O. Heaviside (Heaviside step function throughout the text), Lord Kelvin (method of stationary phase in Chapter 12), I. Fredholm (Fredholm integral equation in Chapter 11), O. Backlund (Backlund transform in soliton), P. Painlevé (differential equation with movable poles), T. Levi-Civita (permutation tensor in Chapter 1), P. Levy (solution on plate bending), A.E.H. Love (Love's wave in half-space), G.I. Taylor (main contributor in stability of fluid), P.M. Morse (author of the classic book on *Method of Theoretical Physics*), van der Pol (van der Pol nonlinear oscillator), L. Fuchs (Fuchsian singular differential equation in Chapter 4), H. Bateman (Bateman's manuscript project), L. Prandtl (boundary layer in fluids in Chapter 12), J. Boussinesq (shallow wave equation), and L. Boltzmann (kinetic gas theory). Among them, Prandtl received 4 nominations in 1928 and 1937 because of his boundary layer theory in fluids, and G.I. Taylor was nominated in 1937 by H. Jeffreys (geophysicist and author of the book *The Earth*) and in 1945 by Sir N.F. Mott (Nobel Prize in physics 1977). A master among mechanicians, Lord Kelvin was actually one of the eleven people being nominated to receive the Nobel Prize in physics in 1901 (the first year Nobel Prizes were awarded). The first recipient in physics in 1901 was Wilhelm Conrad Rontgen (for his discovery of X-ray) and in that year Rontgen actually nominated Lord Kelvin to get the Nobel Prize! Recall from Chapter 12 that Kelvin was the main inventor of the method of stationary phase and from Chapter 1 that Kelvin was the founder of the Kelvin-Stokes theorem. We can see that many Nobel Prize nominees and winners could be considered as applied mathematicians themselves and made significant contributions to the development of various branches of applied mathematics, especially related to the analysis of differential equations.

In Crawford's mind, there are certainly many well-deserved winners who ended up as losers. Among others, they include Arnold Sommerfeld, Henri Poincare, Oliver Heaviside, Ludwig Boltzmann, and Vilhelm Bjerknes (father of modern meteorology and proposed the theory of polar front or jet stream in weather systems).

Appendix C: Proof of Ramanujan's Master and Integral Theorems

C.1 RAMANUJAN'S MASTER THEOREM

To prove Ramanujan's Master Theorem, let us first recall the following definition of gamma function $\Gamma(x)$ proposed by Euler

$$\Gamma(x) = \int_0^\infty e^{-t} t^{x-1} dt \tag{C.1}$$

Next, consider the Laplace transform of a power series:

$$\mathcal{L}\{t^n\} = \int_0^\infty e^{-st} t^n dt \tag{C.2}$$

Applying the change of variables of $st = z$, we obtain

$$\mathcal{L}\{t^n\} = \int_0^\infty e^{-z} (\frac{z}{s})^n \frac{dz}{s} = \frac{1}{s^{n+1}} \int_0^\infty e^{-z} z^n dz = \frac{\Gamma(n+1)}{s^{n+1}} = \int_0^\infty e^{-st} t^n dt \tag{C.3}$$

The last equation in (C.3) can be rewritten as

$$\frac{\Gamma(n)}{s^n} = \int_0^\infty e^{-st} t^{n-1} dt \tag{C.4}$$

Applying another change of variable of $s = r^k$ gives:

$$\frac{\Gamma(n)}{r^{kn}} = \int_0^\infty e^{-r^k t} t^{n-1} dt \tag{C.5}$$

Multiplying both sides by the k-th derivative of a function f(x) evaluated at zero divided by k factorial as:

$$\frac{f^{(k)}(0)}{k!} \int_0^\infty e^{-r^k t} t^{n-1} dt = \frac{f^{(k)}(0)}{k!} \frac{\Gamma(n)}{r^{kn}} \tag{C.6}$$

Consider an infinite sum on both sides with respect to the index k

$$\sum_{k=0}^\infty \frac{f^{(k)}(0)}{k!} \int_0^\infty e^{-r^k t} t^{n-1} dt = \sum_{k=0}^\infty \frac{f^{(k)}(0)}{k!} \frac{\Gamma(n)}{r^{kn}} \tag{C.7}$$

The exponential function inside the integral is now expanded in Taylor series as

$$\exp(-r^k x) = \sum_{m=0}^\infty \frac{(-r^k x)^m}{m!} = 1 - \frac{r^k x}{1!} + \frac{r^{2k} x^2}{2!} - \frac{r^{3k} x^3}{3!} + \ldots \tag{C.8}$$

Substitute (C.8) into (C.7) and reverse the order of integration and summation to yield

$$\int_0^\infty \sum_{k=0}^\infty \frac{f^{(k)}(0)}{k!} \sum_{m=0}^\infty \frac{(-r^k x)^m}{m!} x^{n-1} dx = \sum_{k=0}^\infty \Gamma(n) \frac{f^{(k)}(0) r^{-kn}}{k!} \tag{C.9}$$

We now further reverse the order of summation within the integral on the left hand side of (C.7)

$$LHS = \int_0^\infty \sum_{m=0}^\infty \frac{(-x)^m}{m!} \sum_{k=0}^\infty \frac{f^{(k)}(0)}{k!} r^{km} x^{n-1} dx = \int_0^\infty x^{n-1} \sum_{m=0}^\infty \frac{(-x)^m}{m!} f(r^m) dx \quad \text{(C.10)}$$

The last part of (C.10) follows from the Maclaurin series expansion the function $f(x)$ as:

$$f(x) = \sum_{k=0}^\infty \frac{f^{(k)}(0) x^k}{k!} \quad \text{(C.11)}$$

Equation (C.11) can also be applied to the right hand side of (C.9) to get

$$\int_0^\infty x^{n-1} \sum_{k=0}^\infty \frac{f(r^k)(-x)^k}{k!} dx = \Gamma(n) f(r^{-n}) \quad \text{(C.12)}$$

Finally, we can rewrite using the following identifications:

$$f(r^k) \leftarrow \varphi(k), \quad \text{(C.13)}$$

we have

$$\int_0^\infty x^{n-1} \sum_{k=0}^\infty \frac{\varphi(k)(-x)^k}{k!} dx = \Gamma(n) \varphi(-n) \quad \text{(C.14)}$$

This completes the proof of Ramanujan's Master Theorem. This is a clever and original technique employed by Ramanujan. The main success of the proof rests on the expansion of the exponential function, the reverse of the order of integral and summation, and the reverse of the order of different summations.

C.2 RAMANUJAN'S INTEGRAL THEOREM

In this appendix, we consider Ramanujan's integral theorem, which is a generalization of the Frullani-Cauchy integral theorem reported in Section 1.8. This integral theorem is expressed as:

$$\int_0^\infty \frac{\{f(ax) - g(bx)\}}{x} dx = \{f(0) - f(\infty)\} \left[\ln(\frac{b}{a}) + \frac{d}{ds} \left(\ln[\frac{v(s)}{u(s)}] \right)_{s=0} \right] \quad \text{(C.15)}$$

provided that $f(0) = g(0)$ and $f(\infty) = g(\infty)$ with a, $b > 0$. In addition, functions u and v are defined as the coefficients of the Borel theorem of functions f and g as:

$$f(x) - f(\infty) = \sum_{k=0}^\infty \frac{u(k)(-x)^k}{k!}, \quad g(x) - g(\infty) = \sum_{k=0}^\infty \frac{v(k)(-x)^k}{k!} \quad \text{(C.16)}$$

For the special case that $f = g$, we have the last term in (C.15) vanishing and giving

$$\int_0^\infty \frac{\{f(ax) - f(bx)\}}{x} dx = \{f(0) - f(\infty)\} \ln(\frac{b}{a}) \quad \text{(C.17)}$$

which is the Frullani-Cauchy integral given in (1.202). Thus, the Frullani-Cauchy integral is recovered as a special case of Ramanujan's integral theorem.

To prove (C.15), we consider the following integral

$$I_n = \int_0^\infty x^{n-1}\left(\{f(ax)-f(\infty)\}-\{g(bx)-g(\infty)\}\right)dx$$
$$= \Gamma(n)\left[\varphi_1(-n)-\varphi_2(-n)\right] \tag{C.18}$$

where we have applied (C.16) to get

$$f(ax)-f(\infty)=\sum_{k=0}^{\infty}\frac{u(k)(-ax)^k}{k!}=\sum_{k=0}^{\infty}\frac{u(k)(-x)^k a^k}{k!}=\sum_{k=0}^{\infty}\frac{\varphi_1(k)(-x)^k}{k!}, \tag{C.19}$$

$$g(bx)-g(\infty)=\sum_{k=0}^{\infty}\frac{v(k)(-bx)^k}{k!}=\sum_{k=0}^{\infty}\frac{v(k)(-x)^k b^k}{k!}=\sum_{k=0}^{\infty}\frac{\varphi_2(k)(-x)^k}{k!}, \tag{C.20}$$

In other words, we have

$$\varphi_1(k)=u(k)a^k, \quad \varphi_2(k)=v(k)b^k \tag{C.21}$$

Clearly, substitution of (C.19) and (C.20) into the first line of (C.18) and application of Ramanujan's Master Theorem given in (C.14) results in the second line of (C.18). Using of the definition of φ_1 and φ_2 given in (C.19) and (C.20) gives

$$I_n=\Gamma(n)\left[a^{-n}u(-n)-b^{-n}v(-n)\right]=\Gamma(n+1)\{\frac{a^{-n}u(-n)-b^{-n}v(-n)}{n}\} \tag{C.22}$$

Now, we consider the limit of $n \to 0$ for I_n given in (C.21) as:

$$\lim_{n\to 0} I_n = \lim_{n\to 0}\frac{a^{-n}u(-n)-b^{-n}v(-n)}{n}$$
$$=\lim_{n\to 0}\frac{b^n u(-n)-a^n v(-n)}{a^n b^n n}=\lim_{n\to 0}\frac{b^n u(-n)-a^n v(-n)}{n} \tag{C.23}$$

In obtaining (C.23), we have taken the following limits

$$\lim_{n\to 0}\Gamma(n+1)=\Gamma(1)=0!=1, \quad \lim_{n\to 0} a^n b^n = a^0 b^0 = 1 \tag{C.24}$$

Note that $u(0) = v(0)$ because $f(0) = g(0)$. Application of L'Hôpital's rule to the last part of (C.23) leads to

$$\lim_{n\to 0} I_n = \lim_{n\to 0}\{b^n u(-n)\ln b - b^n u'(-n) - a^n v(-n)\ln a + a^n v'(-n)\}$$
$$= u(0)\ln b - v(0)\ln a + v'(0) - u'(0)$$
$$= u(0)\ln(\frac{b}{a}) + v'(0) - u'(0)$$
$$= \{f(0)-f(\infty)\}\ln(\frac{b}{a}) + v'(0) - u'(0) \tag{C.25}$$

Finally, Ramanujan found a more compact form for the last two terms in the last part of (C.25) by noting

$$\frac{d}{ds}\left(\ln[\frac{v(s)}{u(s)}]\right)=\frac{u(s)}{v(s)}(\frac{v'}{u}-\frac{vu'}{u^2})=\frac{v'}{v}-\frac{u'}{u} \tag{C.26}$$

Taking the limit $s \to 0$, we have

$$\frac{d}{ds}\left(\ln[\frac{v(s)}{u(s)}]\right)_{s=0}=\frac{v'(0)}{v(0)}-\frac{u'(0)}{u(0)}=\frac{1}{u(0)}[v'(0)-u'(0)] \tag{C.27}$$

To visualize the last part of (C.27), we have used the following identity

$$f(0)-f(\infty)=u(0), \quad g(0)-g(\infty)=v(0) \tag{C.28}$$

Since, we require $f(0) = g(0)$ and $f(\infty) = g(\infty)$ thus we must have

$$u(0)=v(0) \tag{C.29}$$

Combining (C.27) and (C.28) gives

$$\begin{aligned} v'(0)-u'(0) &= u(0)\frac{d}{ds}\left(\ln[\frac{v(s)}{u(s)}]\right)_{s=0} \\ &= \{f(0)-f(\infty)\}\frac{d}{ds}\left(\ln[\frac{v(s)}{u(s)}]\right)_{s=0} \end{aligned} \tag{C.30}$$

Finally, substitution of (C.30) into the last line of (C.25), we have Ramanujan's integral theorem.

Appendix D: Jacobi Elliptic Functions

D.1 JACOBI ELLIPTIC FUNCTIONS

Jacobi elliptic functions are related to the solution of the following integral when u and k are given:

$$u = \int_0^{\phi} \frac{d\theta}{\sqrt{1-k^2 \sin^2 \theta}} = F(\phi, k) \tag{D.1}$$

where $F(\phi,\ k)$ is the elliptic integral of the first kind. If the evaluation of the amplitude ϕ of the integral can be done, we write its solution as:

$$\mathrm{sn}(u,k) = \sin\phi, \quad \mathrm{cn}(u,k) = \cos\phi, \quad \mathrm{dn}(u,k) = \sqrt{1-k^2\sin^2\phi} = \Delta\phi \tag{D.2}$$

$$\mathrm{tn}(u,k) = \frac{\mathrm{sn}(u,k)}{\mathrm{cn}(u,k)} = \tan\phi, \quad \mathrm{am}(u,k) = \phi, \tag{D.3}$$

It can be seen that sn, cn, and tn closely resemble the circular functions sine, cosine and tangent. The amplitude function of the integral is denoted by $\mathrm{am}(u,k)$ or ϕ. For a fixed value of k, the Jacobi elliptic sine, cosine, and tangent functions can be simplified as:

$$\mathrm{sn}(u,k) = \mathrm{sn}(u), \quad \mathrm{cn}(u,k) = \mathrm{cn}(u), \quad \mathrm{tn}(u,k) = \mathrm{tn}(u), \quad \mathrm{dn}(u,k) = \mathrm{dn}(u) \tag{D.4}$$

Abramowitz and Stegun (1964) also defined additional types of Jacobi elliptic functions, namely,

$$\begin{aligned} &\mathrm{cd}(u) = \frac{\mathrm{cn}(u)}{\mathrm{dn}(u)}, \quad \mathrm{dc}(u) = \frac{\mathrm{dn}(u)}{\mathrm{cn}(u)}, \quad \mathrm{ns}(u) = \frac{1}{\mathrm{sn}(u)}, \\ &\mathrm{sd}(u) = \frac{\mathrm{sn}(u)}{\mathrm{dn}(u)}, \quad \mathrm{nc}(u) = \frac{1}{\mathrm{cn}(u)}, \quad \mathrm{ds}(u) = \frac{\mathrm{dn}(u)}{\mathrm{sn}(u)}, \\ &\mathrm{nd}(u) = \frac{1}{\mathrm{dn}(u)}, \quad \mathrm{sc}(u) = \frac{\mathrm{sn}(u)}{\mathrm{cn}(u)}, \quad \mathrm{cs}(u) = \frac{\mathrm{cn}(u)}{\mathrm{sn}(u)}. \end{aligned} \tag{D.5}$$

The values of the integral u can be evaluated by using inverse functions that closely resemble arcsine, arccosine, etc.

$$\begin{aligned} &u = \mathrm{sn}^{-1}(\sin\phi, k) \quad u = \mathrm{cn}^{-1}(\cos\phi, k), \quad u = \mathrm{tn}^{-1}(\tan\phi, k), \\ &u = \mathrm{dn}^{-1}(\Delta\phi, k), \quad u = \mathrm{am}^{-1}(\phi, k), \end{aligned} \tag{D.6}$$

For example, the inverse elliptic function can be evaluated as

$$\mathrm{sn}^{-1}(\frac{1}{2}, \frac{\sqrt{3}}{2}) = F(30^\circ \setminus 60^\circ) = 0.54222911 \tag{D.7}$$

where $F(\phi,\alpha)$ is the elliptic integral of the first kind expressed in terms of the parameter α instead of k as

$$F(\phi \setminus \alpha) = \int_0^{\phi} \frac{d\theta}{\sqrt{1-\sin^2\alpha \sin^2\theta}} \tag{D.8}$$

The numerical value in (D.7) was looked up from Table 17.5 of Abramowitz and Stegun (1964).

D.2 IDENTITIES OF JACOBI ELLIPTIC FUNCTIONS

Some special values of Jacobi elliptic functions are:

$$\mathrm{sn}(0)=0, \quad \mathrm{cn}(0)=1, \quad \mathrm{dn}(0)=1, \quad \mathrm{am}(0)=0 \tag{D.9}$$

$$\begin{aligned} &\mathrm{sn}(u,0)=\sin u, \quad \mathrm{cn}(u,0)=\cos u, \quad \mathrm{dn}(u,0)=1, \\ &\mathrm{sn}(u,1)=\tanh u, \quad \mathrm{cn}(u,1)=\mathrm{dn}(u,1)=\mathrm{sec\,h}u, \end{aligned} \tag{D.10}$$

A number of identities can be proved easily

$$\mathrm{sn}^2 u+\mathrm{cn}^2 u=\sin^2\phi+\cos^2\phi=1 \tag{D.11}$$

$$\mathrm{dn}^2 u-k^2\mathrm{cn}^2 u=1-k^2=k'^2 \tag{D.12}$$

$$k^2\mathrm{sn}^2 u+\mathrm{dn}^2 u=1 \tag{D.13}$$

$$\mathrm{sn}(-u)=-\mathrm{sn}\, u \tag{D.14}$$

$$\mathrm{cn}(-u)=\mathrm{cn}\, u \tag{D.15}$$

$$\mathrm{dn}(-u)=\mathrm{dn}\, u \tag{D.16}$$

$$\mathrm{am}(-u)=-\mathrm{am}\, u \tag{D.17}$$

Jacobi elliptic functions are doubly periodic functions

$$\mathrm{sn}(u+2mK+2nK'i)=(-1)^m\,\mathrm{sn}\, u \tag{D.18}$$

$$\mathrm{cn}(u+2mK+2nK'i)=(-1)^{m+n}\,\mathrm{cn}\, u \tag{D.19}$$

$$\mathrm{dn}(u+2mK+2nK'i)=(-1)^n\,\mathrm{dn}\, u \tag{D.20}$$

where

$$K=K(m)=\int_0^{\pi/2}\frac{d\theta}{\sqrt{1-m\sin^2\theta}} \tag{D.21}$$

$$K'=K'(m)=\int_0^{\pi/2}\frac{d\theta}{\sqrt{1-m_1\sin^2\theta}}, \qquad m_1+m=1 \tag{D.22}$$

Thus, K and iK' are called real and imaginary quarter-periods.

We will illustrate a simple case of periodicity by considering

$$\begin{aligned} v&=\int_0^{\phi+\pi}\frac{d\theta}{\sqrt{1-k^2\sin^2\theta}}=\int_0^{\pi}\frac{d\theta}{\sqrt{1-k^2\sin^2\theta}}+\int_{\pi}^{\phi+\pi}\frac{d\theta}{\sqrt{1-k^2\sin^2\theta}} \\ &=\int_0^{\pi/2}\frac{d\theta}{\sqrt{1-k^2\sin^2\theta}}+\int_{\pi/2}^{\pi}\frac{d\theta}{\sqrt{1-k^2\sin^2\theta}}+\int_{\pi}^{\phi+\pi}\frac{d\theta}{\sqrt{1-k^2\sin^2\theta}} \\ &=K+\int_{\pi/2}^{\pi}\frac{d\theta}{\sqrt{1-k^2\sin^2\theta}}+\int_{\pi}^{\phi+\pi}\frac{d\theta}{\sqrt{1-k^2\sin^2\theta}} \end{aligned} \tag{D.23}$$

For the second integral on the right of (D.23), we can apply the following change of variables

$$\theta=\pi-\varphi \tag{D.24}$$

Thus, we have

$$\int_{\pi/2}^{\pi}\frac{d\theta}{\sqrt{1-k^2\sin^2\theta}}=\int_{\pi/2}^{0}\frac{-d\varphi}{\sqrt{1-k^2\sin^2\varphi}}=\int_0^{\pi/2}\frac{d\varphi}{\sqrt{1-k^2\sin^2\varphi}}=K \tag{D.25}$$

For the last integral on the right of (D.23), the following change of variables is applied

$$\theta = \pi + \varphi \tag{D.26}$$

Thus, the final integral becomes

$$\int_{\pi}^{\phi+\pi} \frac{d\theta}{\sqrt{1-k^2\sin^2\theta}} = \int_0^{\phi} \frac{d\varphi}{\sqrt{1-k^2\sin^2\varphi}} = u \tag{D.27}$$

Substitution of (D.25) and (D.27) into (D.23) gives

$$v = 2K + u \tag{D.28}$$

By applying the Jacobi elliptic function to solve for the amplitude of the first integral of (D.23) gives

$$\mathrm{sn}(v) = \mathrm{sn}(2K+u) = \sin(\phi+\pi) = -\sin\phi = -\mathrm{sn}(u) \tag{D.29}$$

Therefore, comparing the second and the fifth terms gives the required periodicity

$$\mathrm{sn}(2K+u) = -\mathrm{sn}(u) \tag{D.30}$$

D.3 DIFFERENTIATION OF JACOBI ELLIPTIC FUNCTIONS

Some formulas of differentiation for Jacobi elliptic functions are

$$\frac{d}{du}\mathrm{sn}\,u = \mathrm{cn}\,u\,\mathrm{dn}\,u \tag{D.31}$$

$$\frac{d}{du}\mathrm{cn}\,u = -\mathrm{sn}\,u\,\mathrm{dn}\,u \tag{D.32}$$

$$\frac{d}{du}\mathrm{dn}\,u = -k^2\mathrm{sn}\,u\,\mathrm{cn}\,u \tag{D.33}$$

$$\frac{d^2}{du^2}\mathrm{sn}\,u = 2k^2\mathrm{sn}^3 u - (1+k^2)\mathrm{sn}\,u \tag{D.34}$$

$$\frac{d^2}{du^2}\mathrm{cn}\,u = (2k^2-1)\mathrm{cn}\,u - 2k^2\mathrm{cn}^3 u \tag{D.35}$$

$$\frac{d^2}{du^2}\mathrm{dn}\,u = (2-k^2)\mathrm{dn}\,u - 2\mathrm{dn}^3 u \tag{D.36}$$

$$\frac{d}{du}\mathrm{am}\,u = \mathrm{dn}\,u \tag{D.37}$$

To see the validity of these formulas, we first consider the differentiation of u with respect to ϕ

$$\frac{du}{d\phi} = \frac{d}{d\phi}\int_0^{\phi} \frac{d\theta}{\sqrt{1-k^2\sin^2\theta}} = \frac{1}{\sqrt{1-k^2\sin^2\phi}} = \frac{1}{\mathrm{dn}\,u} \tag{D.38}$$

In obtaining the above result, we have applied the Leibniz rule for differentiating integrals. Recalling that $\mathrm{sn}\,u = \sin\phi$, we have

$$\frac{d}{du}\mathrm{sn}\,u = \frac{d}{du}\sin\phi = \cos\phi\frac{d\phi}{du} = \mathrm{cn}\,u\;\mathrm{dn}\,u \tag{D.39}$$

We have used (D.38) to obtain the last part of (D.39). Thus, the identity given in (D.31) is established. Similarly, we note that cn u = cosϕ and thus

$$\frac{d}{du}\operatorname{cn}u = \frac{d}{du}\cos\phi = -\sin\phi\frac{d\phi}{du} = -\operatorname{sn}u\operatorname{dn}u \tag{D.40}$$

This is identical to (D.32).

Note from (D.2) that dn u = $(1-k^2\sin^2\phi)^{1/2}$, and we have

$$\begin{aligned}\frac{d}{du}\operatorname{dn}u &= \frac{d}{du}\sqrt{1-k^2\sin^2\phi} = \frac{1}{2\sqrt{1-k^2\sin^2\phi}}(-2k^2\sin\phi\cos\phi\frac{d\phi}{du}) \\ &= -k^2\sin\phi\cos\phi = -k^2\operatorname{sn}u\operatorname{cn}u\end{aligned} \tag{D.41}$$

This proved (D.33). For the second derivative, we differentiate (D.31) one more time with respect to u to get

$$\frac{d^2}{du^2}\operatorname{sn}u = \operatorname{dn}u\frac{d}{du}\operatorname{cn}u + \operatorname{cn}u\frac{d}{du}\operatorname{dn}u \tag{D.42}$$

Substitution of (D.32) and (D.33) into (D.42) results in

$$\frac{d^2}{du^2}\operatorname{sn}u = -\operatorname{sn}u\operatorname{dn}^2u - k^2\operatorname{sn}u\operatorname{cn}^2u = -\operatorname{sn}u(\operatorname{dn}^2u + k^2\operatorname{cn}^2u) \tag{D.43}$$

By virtue of (D.11) and (D.12), all Jacobi elliptic functions in the right hand side of (D.43) can be expressed in terms of a Jacobi elliptic sine function as:

$$\begin{aligned}\frac{d^2}{du^2}\operatorname{sn}u &= -\operatorname{sn}u[1-k^2\operatorname{sn}^2u + k^2(1-\operatorname{sn}^2u)] \\ &= -\operatorname{sn}u[1+k^2-2k^2\operatorname{sn}^2u] = 2k^2\operatorname{sn}^3u - (1+k^2)\operatorname{sn}u\end{aligned} \tag{D.44}$$

To prove (D.35), we differentiate (D.32) one more time with respect to u to get

$$\frac{d^2}{du^2}\operatorname{cn}u = -\operatorname{sn}u\frac{d}{du}\operatorname{dn}u - \operatorname{dn}u\frac{d}{du}\operatorname{sn}u \tag{D.45}$$

Substitution of (D.31) and (D.33) into (D.45) results in

$$\frac{d^2}{du^2}\operatorname{cn}u = k^2\operatorname{sn}^2u\operatorname{cn}u - \operatorname{dn}^2u\operatorname{cn}u \tag{D.46}$$

Utilizing (D.11) and (D.12), (D.46) becomes

$$\begin{aligned}\frac{d^2}{du^2}\operatorname{cn}u &= k^2(1-\operatorname{cn}^2u)\operatorname{cn}u - (1-k^2+k^2\operatorname{cn}^2u)\operatorname{cn}u \\ &= (2k^2-1)\operatorname{cn}u - 2k^2\operatorname{cn}^3u\end{aligned} \tag{D.47}$$

This establishes the formula given in (D.35). Differentiation of (D.33) with respect to u gives

$$\begin{aligned}\frac{d^2}{du^2}\operatorname{dn}u &= -k^2(\operatorname{sn}u\frac{d}{du}\operatorname{cn}u + \operatorname{cn}u\frac{d}{du}\operatorname{sn}u) \\ &= -k^2(-\operatorname{sn}^2u\operatorname{dn}u + \operatorname{cn}^2u\operatorname{dn}u) \\ &= -k^2\operatorname{dn}u(\operatorname{cn}^2u - \operatorname{sn}^2u) \\ &= (2-k^2)\operatorname{dn}u - 2\operatorname{dn}^3u\end{aligned} \tag{D.48}$$

In proving this formula, we have employed (D.12), (D.13), (D.31) and (D.32) in (D.48). Finally, the proof of (D.37) is more straightforward by noting that am $u = \phi$

$$\frac{d}{du}\operatorname{am} u = \frac{d\phi}{du} = \operatorname{dn} u \tag{D.49}$$

D.4 INTEGRATION OF JACOBI ELLIPTIC FUNCTIONS

Integration of Jacobi elliptic functions is less straightforward. Here are some formulas of integration:

$$\int \operatorname{sn} u\, du = \frac{1}{k}\ln(\operatorname{dn} u - k \operatorname{cn} u) \tag{D.50}$$

$$\int \operatorname{cn} u\, du = \frac{1}{k}\cos^{-1}(\operatorname{dn} u) \tag{D.51}$$

$$\int \operatorname{dn} u\, du = \sin^{-1}(\operatorname{sn} u) \tag{D.52}$$

$$\int \frac{1}{\operatorname{sn} u} du = \ln(\frac{\operatorname{dn} u - \operatorname{cn} u}{\operatorname{sn} u}) \tag{D.53}$$

$$\int \frac{1}{\operatorname{cn} u} du = \frac{1}{k'}\ln(\frac{\operatorname{dn} u + k'\operatorname{sn} u}{\operatorname{cn} u}) \tag{D.54}$$

$$\int \frac{1}{\operatorname{dn} u} du = \frac{1}{k'}\cos^{-1}(\frac{\operatorname{cn} u}{\operatorname{dn} u}) \tag{D.55}$$

$$\int \frac{\operatorname{cn} u}{\operatorname{sn} u} du = \ln(\frac{1 - \operatorname{dn} u}{\operatorname{sn} u}) \tag{D.56}$$

$$\int \frac{\operatorname{sn} u}{\operatorname{cn} u} du = \frac{1}{k'}\ln(\frac{\operatorname{dn} u + k'}{\operatorname{cn} u}) \tag{D.57}$$

$$\int \frac{\operatorname{dn} u}{\operatorname{cn} u} du = \ln(\frac{1 + \operatorname{sn} u}{\operatorname{cn} u}) \tag{D.58}$$

where $k' = (1-k^2)^{1/2}$. These formulas can be proved easily by directly differentiating both sides with respect to u. For example, (D.50) can be proved as

$$\begin{aligned}\frac{d}{du}\int \operatorname{sn} u\, du = \operatorname{sn} u &= \frac{1}{k}(\frac{1}{\operatorname{dn} u - k\operatorname{cn} u})[\frac{d}{du}\operatorname{dn} u + k\frac{d}{du}\operatorname{cn} u] \\ &= \frac{1}{k}(\frac{1}{\operatorname{dn} u - k\operatorname{cn} u})[-k^2 \operatorname{sn} u \operatorname{cn} u + k \operatorname{sn} u \operatorname{dn} u] \\ &= \operatorname{sn} u\end{aligned} \tag{D.59}$$

Similarly, all other formulas can be checked.

D.5 OTHER PROPERTIES OF JACOBI ELLIPTIC FUNCTIONS

Maclaurin series expansion can be used to expand the Jacobi elliptic functions in a power series as

$$\mathrm{sn}(u,k)=u-\frac{1}{3!}(1+k^2)u^3+\frac{1}{5!}(1+14k^2+k^4)u^5-... \tag{D.60}$$

$$\mathrm{cn}(u,k)=1-\frac{1}{2!}u^2+\frac{1}{4!}(1+4k^2)u^4-\frac{1}{6!}(1+44k^2+16k^4)u^6+... \tag{D.61}$$

$$\mathrm{dn}(u,k)=1-\frac{1}{2!}k^2u^2+\frac{1}{4!}k^2(4+k^2)u^4-\frac{1}{6!}k^2(16+44k^2+k^4)u^6+... \tag{D.62}$$

The sum rules for Jacobi elliptic functions are also somewhat similar to those for circular functions

$$\mathrm{sn}(u\pm v)=\frac{\mathrm{sn}\,u\,\mathrm{cn}\,v\,\mathrm{dn}\,v\pm\mathrm{cn}\,u\,\mathrm{sn}\,v\,\mathrm{dn}\,u}{1-k^2\mathrm{sn}^2u\,\mathrm{sn}^2\,v} \tag{D.63}$$

$$\mathrm{cn}(u\pm v)=\frac{\mathrm{cn}\,u\,\mathrm{cn}\,v\mp\mathrm{sn}\,u\,\mathrm{sn}\,v\,\mathrm{dn}\,u\,\mathrm{dn}\,v}{1-k^2\mathrm{sn}^2u\,\mathrm{sn}^2\,v} \tag{D.64}$$

$$\mathrm{dn}(u\pm v)=\frac{\mathrm{dn}\,u\,\mathrm{dn}\,v\mp k^2\mathrm{sn}\,u\,\mathrm{sn}\,v\,\mathrm{cn}\,u\,\mathrm{cn}\,v}{1-k^2\mathrm{sn}^2u\,\mathrm{sn}^2\,v} \tag{D.65}$$

$$\mathrm{tn}(u\pm v)=\frac{\mathrm{tn}\,u\,\mathrm{dn}\,v\pm\mathrm{tn}\,v\,\mathrm{dn}\,u}{1\mp\mathrm{tn}\,u\,\mathrm{tn}\,v\,\mathrm{dn}\,u\,\mathrm{dn}\,v} \tag{D.66}$$

Davis (1962) provided a sizable introduction to elliptic integrals and Jacobi elliptic functions and their applications in nonlinear differential equations. One can refer to Abramowitz and Stegun (1964) and Olver et al. (2010) for comprehensive coverage of the properties of Jacobi elliptic functions.

Appendix E: Euler's Constant

The birth of Euler's constant is related to the infinite series of

$$\lim_{m\to\infty}\sum_{k=1}^{m}\frac{1}{k} \tag{E.1}$$

It was known to Euler and other mathematicians that this infinite series does not converge. For example, we find numerically that

$$\sum_{i=1}^{20}\frac{1}{k}\approx 3.60,\quad \sum_{i=1}^{220}\frac{1}{k}\approx 5.98,\quad \sum_{i=1}^{20220}\frac{1}{k}\approx 10.49 \tag{E.2}$$

Although the initial convergence looks promising, but it never converges

$$\sum_{i=1}^{\infty}\frac{1}{k}\to\infty \tag{E.3}$$

Thus, people were not interested in it, except Euler. It turns out that the difference between this infinite series and ln m when $m\to\infty$ is finite

$$\gamma=\lim_{m\to\infty}\left(\sum_{k=1}^{m}\frac{1}{k}-\ln m\right)\cong 0.5772157 \tag{E.4}$$

This finite value constant appears naturally in many mathematical analyses and thus becomes a very important mathematical constant. To start with, let us consider the area under a hyperbolic curve shown in Figure E.1:

$$\ln(1+x)=\int_1^x\frac{dt}{1+t}=\int_1^x\left(1-t+t^2-t^3+...\right)dt=x-\frac{x^2}{2}+\frac{x^3}{3}-\frac{x^4}{4}+\frac{x^5}{5}+... \tag{E.5}$$

Euler was an expert in playing around with infinite series (although at his time the convergence test for infinite series did not exist). He substituted $x = 1/n$ and expanded the left hand side using Taylor series expansion as

$$\ln(1+\frac{1}{n})=\frac{1}{n}-\frac{1}{2n^2}+\frac{1}{3n^3}-\frac{1}{4n^4}+... \tag{E.6}$$

This can be rearranged as

$$\frac{1}{n}=\ln(\frac{n+1}{n})+\frac{1}{2n^2}-\frac{1}{3n^3}+\frac{1}{4n^4}+... \tag{E.7}$$

Taking different values of n, we have

$$1=\ln(2)+\frac{1}{2}-\frac{1}{3}+\frac{1}{4}+... \tag{E.8}$$

$$\frac{1}{2}=\ln(\frac{3}{2})+\frac{1}{8}-\frac{1}{24}+\frac{1}{64}+... \tag{E.9}$$

$$\frac{1}{3}=\ln(\frac{4}{3})+\frac{1}{18}-\frac{1}{81}+\frac{1}{324}+... \tag{E.10}$$

Summing all the left hand sides of (E.8) to (E.10), we have

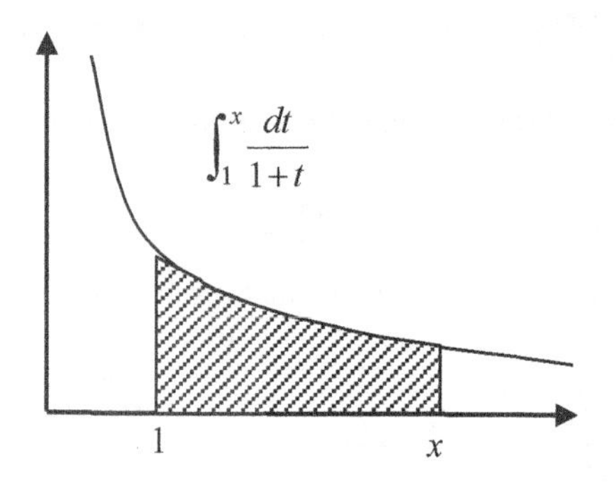

Figure E.1 Area under a hyperbola

$$\sum_{i=1}^{\infty}\frac{1}{k}=[\ln 2+\ln\frac{3}{2}+\ln\frac{4}{3}+...+\ln(\frac{n+1}{n})]+\frac{1}{2}[1+\frac{1}{4}+\frac{1}{9}+...+\frac{1}{n^2}]$$
$$-\frac{1}{3}[1+\frac{1}{8}+\frac{1}{27}+...+\frac{1}{n^3}]+\frac{1}{4}[1+\frac{1}{16}+\frac{1}{81}+...+\frac{1}{n^4}]-... \quad (E.11)$$

The first sum in the square bracket on the right of (E.11) can be simplified as

$$[\ln 2+\ln\frac{3}{2}+\ln\frac{4}{3}+...+\ln(\frac{n+1}{n})]$$
$$=\ln 2+\ln 3-\ln 2+\ln 4-\ln 3+...+\ln(n+1)-\ln n \quad (E.12)$$
$$=\ln(n+1)$$

All the other terms on the right of (E.12) do converge and can be summed to approximately 0.5772157. Thus, we have the following approximation

$$\sum_{i=1}^{\infty}\frac{1}{k}\approx\ln(n+1)+0.5772157 \quad (E.13)$$

For large n, we have the approximation

$$\lim_{n\to\infty}\ln(n+1)=\lim_{n\to\infty}\ln n \quad (E.14)$$

Using this condition, we finally arrive at Euler's constant

$$\gamma=\lim_{m\to\infty}\left(\sum_{k=1}^{m}\frac{1}{k}-\ln m\right)\cong 0.57721566490153286060651 2... \quad (E.15)$$

There are many amazing properties of γ, and here are some of them:

$$\gamma=-\int_0^{\infty}e^{-x}\ln x dx=-\int_0^1\ln(\ln x)dx=-4\int_0^{\infty}e^{-x^2}x\ln x dx \quad (E.16)$$

$$\gamma=\int_0^{\infty}(\frac{e^{-x}}{1-e^{-x}}-\frac{e^{-x}}{x})dx=\int_0^{\infty}\frac{1}{x}(\frac{1}{1+x}-e^{-x})dx=\int_0^1(\frac{1-e^{-x}-e^{-1/x}}{x})dx \quad (E.17)$$

$$\gamma=1+\int_0^{\infty}\frac{1}{x}(\frac{e^{-x}-1}{x}+\frac{1}{1+x})dx=\int_0^1(\frac{1}{\ln x}+\frac{1}{1-x})dx \quad (E.18)$$

$$\gamma = -\int_0^\infty \frac{2}{x}(e^{-x^2} - \frac{1}{1+x^2})dx \tag{E.19}$$

$$\gamma = \int_0^\infty \frac{4}{x}(e^{-x^4} - e^{-x^2})dx \tag{E.20}$$

$$\gamma = \lim_{x \to 1^+} \sum_{i=1}^{n} (\frac{1}{n^x} - \frac{1}{x^n}) \tag{E.21}$$

$$\gamma = [\frac{1}{2 \cdot 2!} - \frac{1}{4 \cdot 4!} + \frac{1}{6 \cdot 6!} - ...] - \int_1^\infty \frac{\cos x}{x} dx \tag{E.22}$$

$$\gamma = \int_0^\infty \frac{2}{x}(e^{-x^2} - e^{-x})dx \tag{E.23}$$

$$\gamma = \int_0^\infty \frac{4}{3x}(e^{-x^4} - e^{-x})dx \tag{E.24}$$

$$\gamma = \sum_{m=2}^{\infty} (-1)^m \frac{\zeta(m)}{m}, \quad \zeta(m) = \sum_{n=1}^{\infty} n^{-m} = \frac{1}{1^m} + \frac{1}{2^m} + ... \tag{E.25}$$

Ramanujan derived some amazing formulas for γ

$$\begin{aligned} \gamma &= \ln 2 - \sum_{n=2}^{\infty} 2n \sum_{k=\frac{3^{n-1}+1}{2}}^{\frac{3^n+1}{2}} \frac{1}{(3k)^3 - 3k} \\ &= \ln 2 - \frac{2}{3^3 - 3} - 4\left(\frac{1}{6^3 - 6} + \frac{1}{9^3 - 9} + \frac{1}{12^3 - 12}\right) \\ &\quad -6\left(\frac{1}{15^3 - 15} + \frac{1}{18^3 - 18} + ... + \frac{1}{39^3 - 39}\right) - ... \end{aligned} \tag{E.26}$$

$$\gamma = 1 - \int_0^1 \frac{1 + \frac{1}{2}\sqrt{x}}{(1+\sqrt{x})(1+\sqrt{x}+x)} \sum_{k=1}^{\infty} x^{(3/2)^k} dx \tag{E.27}$$

Appendix F: π

It has been known for thousands of year that the circumference of a circle depends on the diameter of the same circle. The ratio between them is known to be a constant and is normally denoted by the Greek alphabet π, which was originally proposed by Euler.

The symbol π is defined as:

$$\pi = \frac{circumference}{diameter} \tag{D.1}$$

Mathematically, π can be evaluated as the area of a unit circle

$$\pi = 4\int_0^1 y dx = 4\int_0^1 \sqrt{1-x^2}\, dx \tag{F.2}$$

where x and y are defined in Figure F.1.

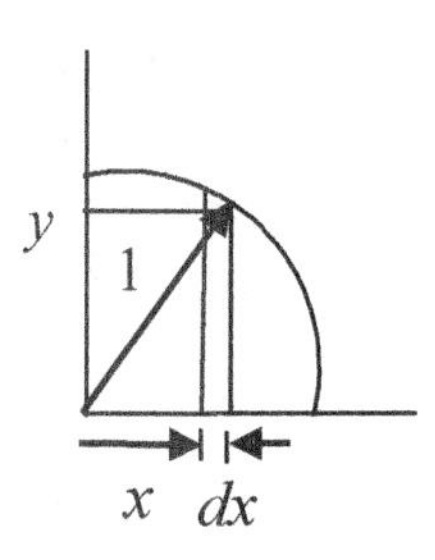

Figure F.1 Area of a quarter of a circle

Alternatively, we can also define π as

$$\pi = 4\int_0^1 \frac{1}{\sqrt{1-x^2}} dx = 4\tan^{-1} x \Big|_0^1 \tag{F.3}$$

By employing Euler's formula, Fagnano defined π as:

$$\pi = 4\ln\left[\frac{1-i}{1+i}\right]^{i/2} \tag{F.4}$$

This can be shown easily by observing that

$$1-i = \sqrt{2}e^{-i\pi/4}, \quad 1+i = \sqrt{2}e^{i\pi/4} \tag{F.5}$$

It turns out that it is transcendental, or it is not the solution of any algebraic equation (proved by Lindemann in 1882). It is also an irrational number, or it could not be written as the ratio of two integers (proved by Lambert and Legendre in 1700s). In other words, there are infinite decimal digits for π. This is one of the most fundamental constant of mathematics.

Mathematicians had struggled in the last few thousand years to give the most "accurate" result for this π. In 2009, π had been computed to 2.7 trillion decimal digits. In fact, the speed of a supercomputer is normally gauged by measuring the time that it takes to calculate the trillion digits of π (this apparently was advocated by John von Neumann).

For practical purposes, these most "accurate" π's normally do not have much significance. For example, the approximation by Zu proposed in 462 AD:

$$\pi \approx \frac{355}{113} \tag{F.6}$$

This gives 3.14159292... compared to the more accurate result 3.14159265... If you draw a circle 10 km in diameter, this approximation gives a circumference of the circle of less than 4 mm. Thus, a 6-digit accuracy appears to be good enough for most engineering applications. It is, however, of fundamental importance and of theoretical interest to summarize some commonly used formulas and some less commonly known formulas of π.

It has been known for thousands of years that a first approximation in rational form is

$$\pi \approx \frac{22}{7} \tag{F.7}$$

This number and 3.14 are the π values that all primary school students are asked to remember. This approximation gives only 0.04% error. If you draw a circle of 10 km in diameter, this approximation gives a circumference of the circle of about 3 m. In 825, Archimedes gave the result as

$$\frac{223}{71} < \pi < \frac{22}{7} \tag{F.8}$$

In fact, the number 355/113 is the most accurate fraction approximating π for a denominator of less than 16,586. To see this, if there is fraction more accurate than 355/113, we must have

$$\left| \frac{355}{113} - \frac{q}{p} \right| < 2 \times 0.00000026677 \tag{F.9}$$

This can be rewritten as

$$\frac{1}{113p} |355p - 113q| < 2 \times 0.00000026677 \tag{F.10}$$

Since $355p-113q$ must be an integer larger than 1, we must have a less restrictive inequality:

$$p > \frac{1}{(113 \times 2 \times 0.00000026677)} > 16586 \tag{F.11}$$

Thus, this completes the proof. Other commonly adopted fractions for π include:

$$\begin{gathered} \pi \approx \frac{333}{106} = 3.141509..., \pi \approx \frac{754}{240} = 3.14166..., \pi \approx \frac{730}{232} = 3.14655..., \\ \pi \approx \frac{195,882}{62,351} = 3.141602..., \pi \approx \frac{211,869}{67,440} = 3.14159253... \end{gathered} \tag{F.12}$$

Historically, the number of significant digits was improved by cutting a circle geometrically into more segments. One such geometrically obtained formula is by Viète in 1615 (published posthumously):

$$\frac{2}{\pi}=\frac{\sqrt{2}}{2}\frac{\sqrt{2+\sqrt{2}}}{2}\frac{\sqrt{2+\sqrt{2+\sqrt{2}}}}{2}\cdots \tag{F.13}$$

This formula is of theoretical significance, since it is the first formula that extends to infinity. We outline the proof of the Viète's formula briefly.

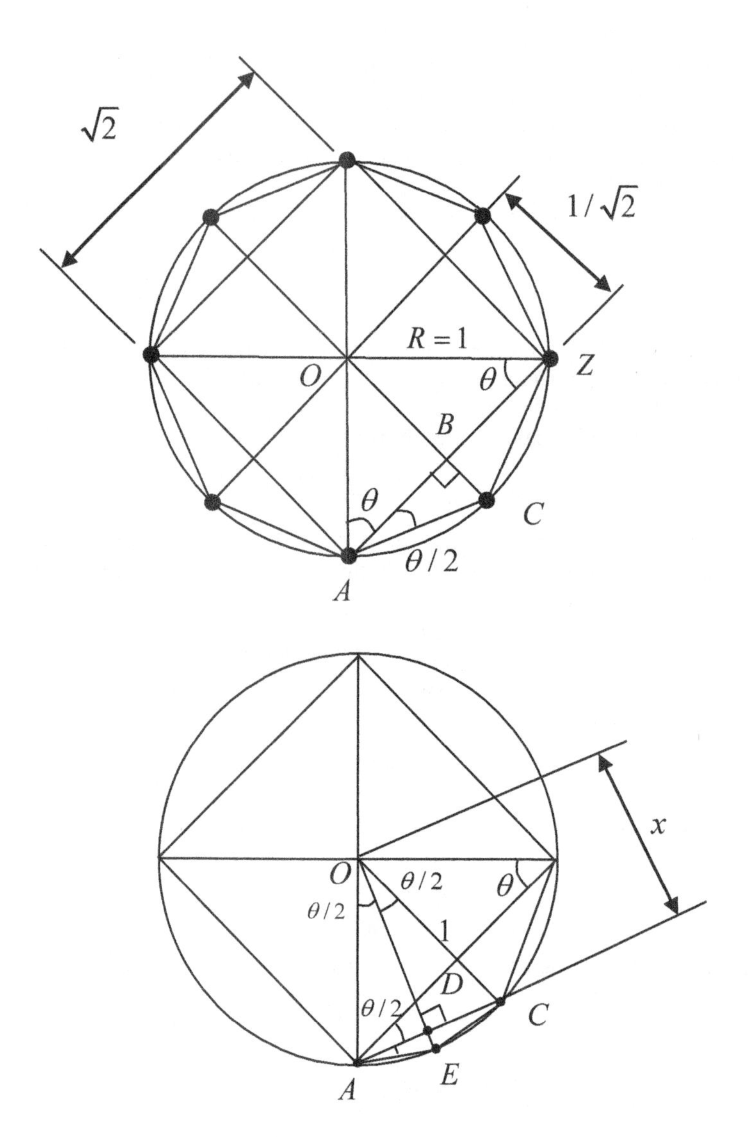

Figure F.2 Viète scheme for calculating the circumference of a circle

Consider the case that $\theta=\pi/4$ in Figure F.2; we have

$$\cos(\frac{\pi}{4}) = \frac{1}{\sqrt{2}} = \frac{\sqrt{2}}{2} \tag{F.14}$$

Next, we can use the sum rule for the cosine function as

$$\cos(\frac{\pi}{4}) = \cos(\frac{\pi}{8} + \frac{\pi}{8}) = 2\cos^2(\frac{\pi}{8}) - 1 \tag{F.15}$$

Solving (F.15) and noting the result of (F.14) gives

$$\cos^2(\frac{\pi}{8}) = \frac{1+\cos(\frac{\pi}{4})}{2} = \frac{2+\sqrt{2}}{4} \tag{F.16}$$

Therefore, we have

$$\cos(\frac{\pi}{8}) = \frac{\sqrt{2+\sqrt{2}}}{2} \tag{F.17}$$

The process of taking half of the angle again leads to

$$\cos(\frac{\pi}{8}) = \cos(\frac{\pi}{16} + \frac{\pi}{16}) = 2\cos^2(\frac{\pi}{16}) - 1 \tag{F.18}$$

Thus, the solution of it and in view of (F.17) gives

$$\cos(\frac{\pi}{16}) = \frac{\sqrt{2+\sqrt{2+\sqrt{2}}}}{2} \tag{F.19}$$

Clearly, we can repeat the process of halving the angle to infinitely small angles. This is the basis for Viète's formula going to infinity. For example, we can easily see that

$$\cos(\frac{\pi}{128}) = \frac{\sqrt{2+\sqrt{2+\sqrt{2+\sqrt{2+\sqrt{2+\sqrt{2}}}}}}}{2} \tag{F.20}$$

Or more generally, we can express the general form as:

$$\cos(\frac{\pi}{4k}) = \frac{\sqrt{2+\sqrt{2+\sqrt{2+\sqrt{2+\sqrt{2+...+\sqrt{2}}}}}}}{2} \tag{F.21}$$

In (F.21), there is a k square root sign on the right hand side. When $k \to \infty$, we will have an infinitely long square root term on the right.

Our main task is now to link the above identity of the cosine function to the circular arcs AC and AE shown in Figure F.2. First, we recall that

$$\sec(\frac{\pi}{4}) = \sqrt{2} = \sec\theta \tag{F.22}$$

From right angle triangle OAB, the length of AB is

$$AB = R\cos\theta = \frac{1}{\sec\theta} \tag{F.23}$$

From right angle triangle OAZ, we see that AZ equals

$$AZ = \sqrt{1^2+1^2} = \sqrt{2} = \sec\theta \tag{F.24}$$

Thus, AB equals

$$AB = \frac{AZ}{2} = \frac{\sqrt{2}}{2} \tag{F.25}$$

Considering triangle ABC and assuming that ∠BAC = θ/2 (this will be verified as true), we have

$$AC = AB\sec\frac{\theta}{2} = \frac{1}{2}\sec\theta\sec\frac{\theta}{2} \tag{F.26}$$

Noting (F.17) and (F.22) that

$$\sec\theta = \sqrt{2}, \quad \sec\frac{\theta}{2} = \frac{2}{\sqrt{2+\sqrt{2}}} \tag{F.27}$$

Substitution of (F.27) into (F.26) gives

$$AC = \frac{2}{\sqrt{2}}\frac{1}{\sqrt{2+\sqrt{2}}} = \frac{2}{\sqrt{2}}\frac{\sqrt{2-\sqrt{2}}}{\sqrt{2+\sqrt{2}}}\frac{1}{\sqrt{2-\sqrt{2}}} = \sqrt{2-\sqrt{2}} \tag{F.28}$$

Let us now use the Pythagoras theorem (independent of the assumption that angle ∠BAC is θ/2) as

$$AC = \sqrt{AB^2 + BC^2} = \sqrt{\frac{2}{4} + \frac{(\sqrt{2}-1)^2}{2}} = \sqrt{\frac{4-2\sqrt{2}}{2}} = \sqrt{2-\sqrt{2}} \tag{F.29}$$

This result is consistent with (F.26) which is a result of the assumption that ∠BAC = θ/2. Thus, we have proved that ∠BAC is indeed θ/2. This also proves the validity of (F.26).

However, taking the first approximation that arc AC equals straight line AC, we have the circumference as

$$2\pi \approx 8AC = 8\sqrt{2-\sqrt{2}} \tag{F.30}$$

Inverting this formula, we have

$$\frac{2}{\pi} \approx \frac{1}{2\sqrt{2-\sqrt{2}}} = \frac{\sqrt{2+\sqrt{2}}}{2\sqrt{2-\sqrt{2}}\sqrt{2+\sqrt{2}}} = \frac{1}{2}\left(\frac{\sqrt{2+\sqrt{2}}}{\sqrt{2}}\right) = \frac{\sqrt{2}}{2}\left(\frac{\sqrt{2+\sqrt{2}}}{2}\right) \tag{F.31}$$

This is clearly the first two terms in the Viète's formula given in (F.13).

Now, let us take the next approximation shown in the lower figure in Figure F.2. Now, let us first apply the Pythagoras theorem that

$$\begin{aligned} AE &= \sqrt{AD^2 + DE^2} = \sqrt{\sin^2(\frac{\theta}{2}) + (1-x)^2} = \sqrt{\sin^2(\frac{\theta}{2}) + [1-\cos(\frac{\theta}{2})]^2} \\ &= \sqrt{2-2\cos(\frac{\theta}{2})} = \sqrt{2-\sqrt{2+\sqrt{2}}} \end{aligned} \tag{F.32}$$

In obtaining the last result of (F.32), we have employed the result of (F.17). Finally, we want to show that it can be expressed as:

$$AE = \sqrt{2-\sqrt{2+\sqrt{2}}} = \frac{1}{4}\sec\theta\sec(\frac{\theta}{2})\sec(\frac{\theta}{4}) \tag{F.33}$$

In view of (F.19). the right hand side of (F.33) is

$$RHS = \frac{1}{4}\sec\theta\sec(\frac{\theta}{2})\sec(\frac{\theta}{4}) = \frac{1}{4}\frac{2}{\sqrt{2}}\frac{2}{\sqrt{2+\sqrt{2}}}\frac{2}{\sqrt{2+\sqrt{2+\sqrt{2}}}}$$

$$= \frac{2}{\sqrt{2}}\frac{1}{\sqrt{2+\sqrt{2}}}\frac{1}{\sqrt{2+\sqrt{2+\sqrt{2}}}}\frac{\sqrt{2-\sqrt{2+\sqrt{2}}}}{\sqrt{2-\sqrt{2+\sqrt{2}}}} \tag{F.34}$$

$$= \frac{2}{\sqrt{2}}\frac{\sqrt{2-\sqrt{2+\sqrt{2}}}}{\sqrt{2+\sqrt{2}}\sqrt{2-\sqrt{2}}}$$

$$= \sqrt{2-\sqrt{2+\sqrt{2}}}$$

Therefore, it equals the left hand side of (F.33). One half of the circumference is approximately given as:

$$\frac{\pi}{2} \approx 4AC = \sec\theta\sec(\frac{\theta}{2})\sec(\frac{\theta}{4}) \tag{F.35}$$

In view of (F.34), inversion of (F.35) gives

$$\cos\theta\cos(\frac{\theta}{2})\cos(\frac{\theta}{4}) = \frac{\sqrt{2}}{2}\frac{\sqrt{2+\sqrt{2}}}{2}\frac{\sqrt{2+\sqrt{2+\sqrt{2}}}}{2} \tag{F.36}$$

When we repeat the halving procedure infinitely, we have Viète's formula:

$$\frac{2}{\pi} = \frac{\sqrt{2}}{2}\frac{\sqrt{2+\sqrt{2}}}{2}\frac{\sqrt{2+\sqrt{2+\sqrt{2}}}}{2}\cdots \tag{F.37}$$

Viète's formula can actually be obtained by setting $A = \pi/2$ in the following formula for the infinite product of cosines (discovered by Euler):

$$A = \frac{\sin A}{\cos\frac{A}{2}\cos\frac{A}{4}\cos\frac{A}{8}\cdots} \tag{F.38}$$

But such technique is tedious and the best that can be done was only up to 40 digits in 1630 by Grienberger. Beyond this, calculus and infinite series are needed.

A particularly useful formula in calculating π involves the arctangent. In particular, using calculus and Taylor series expansion it was found that

$$\tan^{-1}(x) = x - \frac{x^3}{3} + \frac{x^5}{5} - \frac{x^7}{7} + \ldots \tag{F.39}$$

Setting $x = 1$ in (F.39), Leibniz obtained the following result in 1673

$$\frac{\pi}{4} = \tan^{-1}(1) = 1 - \frac{1}{3} + \frac{1}{5} - \frac{1}{7} + \ldots \tag{F.40}$$

With this value of the arctangent, John Machin in 1706 proposed a more refined formula that involves two arctangent functions:

$$\frac{\pi}{4} = 4\tan^{-1}(\frac{1}{5}) - \tan^{-1}(\frac{1}{239}) \tag{F.41}$$

Using it, Machin obtained 101 digits of π. Euler in 1755 followed the same line of analysis and obtained 205 digits of π. Gauss proposed another form as:

$$\frac{\pi}{4} = 3\tan^{-1}(\frac{1}{4}) + \tan^{-1}(\frac{1}{20}) + \tan^{-1}(\frac{1}{1985}) \tag{F.42}$$

Although these kinds of formulas are more complicated, they converge faster. The human record on the digits of π was obtained by Levi Smith in 1949 with 1121 digits. This is the end of human computation of digits of π, but it also marks the emergence of computer calculation of π. Actually, there are an infinite number of formulas similar to (F.41) and (F.42). To see how to generate them, we follow Casper Wessel's (a Danish surveyor) approach of using complex numbers. Let us start with a simple example as:

$$(2+i)(3+i) = 5 + i5 \tag{F.43}$$

Rewriting it in terms of the polar form we have

$$r_1 e^{i\theta_1} r_2 e^{i\theta_2} = r_1 r_2 e^{i(\theta_1+\theta_2)} = re^{i\theta} \tag{F.44}$$

However, from the original complex numbers in (F.43), we have

$$\theta_1 = \tan^{-1}(\frac{1}{2}), \quad \theta_1 = \tan^{-1}(\frac{1}{3}), \quad \theta = \tan^{-1}(1) = \frac{\pi}{4} \tag{F.45}$$

Equation (F.44) instantly gives another simple formula of the type of (F.41) and (F.42) as:

$$\frac{\pi}{4} = \tan^{-1}(\frac{1}{2}) + \tan^{-1}(\frac{1}{3}) \tag{F.46}$$

This analysis can be extended to more general cases

$$(n+1+in)(2n+1+i) = 2n^2 + 2n + 1 + i(2n^2 + 2n + 1) \tag{F.47}$$

Using the same line of argument, we get

$$\frac{\pi}{4} = \tan^{-1}(\frac{n}{n+1}) + \tan^{-1}(\frac{1}{2n+1}) \tag{F.48}$$

Alternatively, we find

$$\begin{aligned} &(n+2k+1+in)[2n+2k+1+i(2k+1)] \\ &= 2n^2 + 2n(2k+1) + (2k+1)^2 + i[2n^2 + 2n(2k+1) + (2k+1)^2] \end{aligned} \tag{F.49}$$

Thus, we have

$$\frac{\pi}{4} = \tan^{-1}(\frac{n}{n+2k+1}) + \tan^{-1}(\frac{2k+1}{2n+2k+1}) \tag{F.50}$$

When $k = 0$ and $n = 1$, (F.46) is recovered. Clearly, (F.50) is true for any integer of n and k, therefore we have an infinite number of formulas of this type.

Other similar forms are

$$[4\cdot 10^n + 1 + i(4\cdot 10^n - 1)][4\cdot 10^n + i] = (4\cdot 10^n)^2 + 1 + i[(4\cdot 10^n)^2 + 1] \tag{F.51}$$

The corresponding arctangent formula is

$$\frac{\pi}{4} = \tan^{-1}(\frac{4\cdot 10^n - 1}{4\cdot 10^n + 1}) + \tan^{-1}(\frac{1}{4\cdot 10^n}) \tag{F.52}$$

A more elaborate formula involves the power of a complex number, for example

$$(5+i)^4(-239+i) = -114244 - i114244 \tag{F.53}$$

The author is encouraged to show the validity of the above formula. This gives

$$\frac{\pi}{4} = 4\tan^{-1}(\frac{1}{5}) - \tan^{-1}(\frac{1}{239}) \tag{F.54}$$

This is the Machin formula that we noted in (F.41). This kind of formula may not be easily obtained but there are an infinite number of them as illustrated in (F.50) and (F.52). The convergence of each formula varies from one to another. We report here some of the more well-known formulas from Euler, Gauss, and others:

$$\frac{\pi}{4} = 2\tan^{-1}(\frac{1}{3}) - \tan^{-1}(\frac{1}{7}) \tag{F.55}$$

$$\frac{\pi}{4} = 5\tan^{-1}(\frac{1}{3}) + 2\tan^{-1}(\frac{3}{79}) \tag{F.56}$$

$$\frac{\pi}{4} = \tan^{-1}(\frac{1}{2}) + \tan^{-1}(\frac{1}{5}) + \tan^{-1}(\frac{1}{8}) \tag{F.57}$$

$$\frac{\pi}{4} = 4\tan^{-1}(\frac{1}{5}) - \tan^{-1}(\frac{1}{70}) + \tan^{-1}(\frac{1}{99}) \tag{F.58}$$

$$\frac{\pi}{4} = 4\tan^{-1}(\frac{1}{5}) - 2\tan^{-1}(\frac{1}{408}) + \tan^{-1}(\frac{1}{1393}) \tag{F.59}$$

$$\frac{\pi}{4} = 6\tan^{-1}(\frac{1}{8}) + 2\tan^{-1}(\frac{1}{57}) + \tan^{-1}(\frac{1}{239}) \tag{F.60}$$

$$\frac{\pi}{4} = 12\tan^{-1}(\frac{1}{18}) + 8\tan^{-1}(\frac{1}{57}) - 5\tan^{-1}(\frac{1}{239}) \tag{F.61}$$

$$\frac{\pi}{4} = 4\tan^{-1}(\frac{1}{5}) - \tan^{-1}(\frac{1}{240}) - \tan^{-1}(\frac{1}{57361}) \tag{F.62}$$

$$\frac{\pi}{4} = \tan^{-1}(\frac{1}{4}) + \tan^{-1}(\frac{1}{5}) + \tan^{-1}(\frac{1}{12}) + \tan^{-1}(\frac{1}{13}) + \tan^{-1}(\frac{5}{27}) \tag{F.63}$$

The steps to prove the identities involving three arctangent functions are somewhat similar, although it is more tedious. For example, to prove (F.62) we can consider

$$\begin{aligned}(5+i)^4(240-i)(57361-i) &= (476+480i)(240-i)(57361-i) \\ &= 6580568644(1+i)\end{aligned} \tag{F.64}$$

Thus, the identity of (F.62) is established by using the polar form of the complex numbers on both sides.

Other formulas for the evaluation of π include:

John Wallis:

$$\frac{\pi}{4} = \frac{2\cdot4\cdot4\cdot6\cdots}{3\cdot3\cdot5\cdot5\cdots} \tag{F.65}$$

Lord Brouncker:

$$\frac{2}{\pi}=\cfrac{1}{1+\cfrac{9}{2+\cfrac{25}{2+\cfrac{49}{2+\ldots}}}} \tag{F.66}$$

Al-Kashi:

$$2\pi=6+\frac{16}{60}+\frac{59}{60^2}+\frac{28}{60^3}+\frac{1}{60^4}+\frac{34}{60^5}+\frac{51}{60^6}+\frac{46}{60^7}+\frac{14}{60^8}+\frac{50}{60^9} \tag{F.67}$$

Takano:

$$\pi=48\tan^{-1}(\frac{1}{49})+128\tan^{-1}(\frac{1}{57})-20\tan^{-1}(\frac{1}{239})+48\tan^{-1}(\frac{1}{110443}) \tag{F.68}$$

Gregory-Leibniz:

$$\frac{\pi}{4}=1-\frac{1}{3}+\frac{1}{5}-\frac{1}{7}+\frac{1}{9}-\frac{1}{11}+\ldots \tag{F.69}$$

Euler:

$$\frac{\pi}{4}=\frac{1}{2}-\frac{1}{3\cdot 2^3}+\frac{1}{5\cdot 2^5}-\frac{1}{7\cdot 2^7}+\ldots+\frac{1}{3}-\frac{1}{3\cdot 3^3}+\frac{1}{5\cdot 3^5}-\frac{1}{7\cdot 3^7}+\ldots \tag{F.70}$$

Simon Plouffe:

$$\pi=\sum_{n=0}^{\infty}\frac{1}{16^n}(\frac{4}{8n+1}-\frac{2}{8n+4}-\frac{1}{8n+5}-\frac{1}{8n+6}) \tag{F.71}$$

It would be incomplete in any discussion of π without mentioning the amazing formulas by the Indian mathematics genius Ramanujan. Some of them are the results from the analyses of the elliptic integral and modular equation. Here is one of his amazing formulas:

$$\frac{1}{\pi}=\frac{2\sqrt{2}}{9801}\sum_{n=0}^{\infty}\frac{(4n)!(1103+26390n)}{(n!)^4 396^{4n}} \tag{F.72}$$

The first term in this series already gives 7 digits of π

$$\pi\approx\frac{9801}{2206\sqrt{2}}=3.14159273.. \text{ (7 digits)} \tag{F.73}$$

We can add eight correct digits with each additional term in the summation. This convergent rate is amazing. He also gave other very simple but peculiar forms of approximations of π, which are largely different from other approximations. Here are some of them (Ramanujan, 1910):

$$\pi\approx\frac{3}{\sqrt{163}}\ln(640320)=3.141592654\ldots \text{ (15 digits)} \tag{F.74}$$

$$\pi\approx\frac{3}{\sqrt{67}}\ln(5280)=3.141592653\ldots \text{ (9 digits)} \tag{F.75}$$

$$\frac{1}{\pi}=\sum_{n=0}^{\infty}(\frac{C_n^{2n}}{16^n})^3\frac{42n+5}{16} \tag{F.76}$$

$$\pi = \sqrt[4]{9^2 + \frac{19^2}{22}} = 3.141592652.. \text{ (9 digits)} \tag{F.77}$$

$$\pi = \sqrt[4]{\frac{2143}{22}} = 3.141592653.. \text{ (9 digits)} \tag{F.78}$$

$$\pi = \frac{12}{\sqrt{130}} \ln\{\frac{(2+\sqrt{5})(3+\sqrt{13})}{\sqrt{2}}\} \tag{F.79}$$

$$\pi = \frac{24}{\sqrt{142}} \ln\{\sqrt{\frac{10+11\sqrt{2}}{4}} + \sqrt{\frac{10+7\sqrt{2}}{4}}\} \tag{F.80}$$

$$\pi = \frac{12}{\sqrt{190}} \ln\{(2\sqrt{2}+\sqrt{10})(3+\sqrt{10})\} \tag{F.81}$$

$$\pi \approx (97.5 - \frac{1}{11})^{1/4} = 3.14159273.. \text{ (7 digits)} \tag{F.82}$$

$$\pi \approx \frac{63}{25}(\frac{17+15\sqrt{5}}{7+15\sqrt{5}}) = 3.14159265380... \text{ (11 digits)} \tag{F.83}$$

$$\pi \approx \frac{355}{113}(1 - \frac{0.0003}{3533}) = 3.141592744... \text{ (7 digits)} \tag{F.84}$$

$$\pi \approx \frac{1}{2\sqrt{2}}(\frac{1}{0.112539539}) \text{ (7 digits)} \tag{F.85}$$

$$\frac{1}{2\pi\sqrt{2}} \approx \frac{1103}{99^2} + \frac{27493}{99^6}(\frac{1}{2})(\frac{1\cdot 3}{4^2}) + \frac{53883}{99^{10}}(\frac{1\cdot 3}{2\cdot 4})(\frac{1\cdot 3\cdot 5\cdot 7}{4^2\cdot 8^2}) + \ldots \tag{F.86}$$

$$\pi \approx \frac{12}{\sqrt{310}} \ln\{\frac{1}{4}(3+\sqrt{5})(2+\sqrt{2})[(5+2\sqrt{10}) + \sqrt{61+20\sqrt{10}}]\} \text{ (22 digits)} \tag{F.87}$$

$$\pi \approx \frac{4}{\sqrt{522}} \ln\{(\frac{5+\sqrt{29}}{\sqrt{2}})^3 (5\sqrt{29}+11\sqrt{6})[\sqrt{(\frac{9+3\sqrt{6}}{4})} + \sqrt{(\frac{5+3\sqrt{6}}{4})}]^6\} \text{ (31 digits)} \tag{F.88}$$

An improved version of Ramanujan's formula given in (F.76) is known as the David-Gregory-Chudnovsky formula

$$\frac{1}{\pi} = 12 \sum_{n=0}^{\infty} \frac{(-1)^n (6n)!(13591409 + 545140134n)}{(3n!)(n!)^3 640320^{3n+3/2}} \tag{F.89}$$

This improved result will lead to 14 correct digits just by the first term. In 1998, this formula was used to calculate 1 billion decimals.

There are also some very peculiar observations about π. Here are the first 38 digits of π

$$\pi \approx 3.1415926535897932384626433832795028841... \tag{F.90}$$

Amazingly, this number appears to have intimate relation with prime (i.e., an integer that cannot be written in terms of multiples of integers of other than 1 and itself). We see that the first digit is 3, which is a prime. The first two digits are 31, which is also a prime. The first six digits are 314159 which is also a prime. More amazingly, the first 38 digits of π is also prime. That is, the following are primes:

$$\begin{aligned} &3 \\ &31 \\ &314159 \\ &31415926535897932384626433832795028841 \end{aligned} \tag{F.91}$$

More interestingly, it was discovered that the reverse of the first three primes are also primes:

$$\begin{aligned} &3 \\ &13 \\ &951413 \end{aligned} \tag{F.92}$$

But, it has been checked up to the first 432 digits of π and no more primes were found. It is still not known whether there is another prime for more digits of π. In addition, 314159 is a very peculiar prime. The complement number of this prime (each digit of the number is replaced by the difference of it with 10) is 796951, which is also a prime. If this prime is chopped into 3 two-digit numbers as 31, 41, and 59, we obtain another three primes. They are, by the way, twin primes. That is, 29 and 31 are twin primes, 41 and 43 are twin primes, and finally 59 and 61 are also twin primes. The sum of them, i.e., 31+41+59 = 131, is also a prime, and the sum of the cube of them, i.e., $31^3+41^3+59^3$ = 304091, is also a prime. In conclusion, 314159 is a very peculiar prime and is the first 6 digits of π.

There are too many coincidences between prime and π. There may exist a more in-depth relation between them, and it remains to be discovered.

Problem F.1 Prove Euler's formula given in (F.38)

$$A = \frac{\sin A}{\cos\frac{A}{2}\cos\frac{A}{4}\cos\frac{A}{8}\cdots} \tag{F.93}$$

Hints: The following formulas are useful

$$\sin A = 2\sin\frac{A}{2}\cos\frac{A}{2} \tag{F.94}$$

$$\lim_{n\to\infty}\frac{\sin\frac{A}{2^n}}{\frac{A}{2^n}} = 1 \tag{F.95}$$

Problem F.2 Use the result of Problem F.1 to show the validity (F.13):

$$\frac{2}{\pi} = \frac{\sqrt{2}}{2}\frac{\sqrt{2+\sqrt{2}}}{2}\frac{\sqrt{2+\sqrt{2+\sqrt{2}}}}{2}\cdots \tag{F.96}$$

Problem F.3 Prove that

$$\tan^{-1}(a) + \tan^{-1}(b) = \tan^{-1}\left[\frac{a+b}{1-ab}\right] \tag{F.97}$$

Problem F.4 Use the result of Problem F.3 to show that

$$\tan^{-1}(\frac{1}{p+q})+\tan^{-1}(\frac{q}{p^2+pq+1})=\tan^{-1}(\frac{1}{p}) \tag{F.98}$$

Problem F.5 Use the result of Problem F.4 to show that

$$\frac{\pi}{4}=\tan^{-1}(\frac{1}{2})+\tan^{-1}(\frac{1}{3}) \tag{F.99}$$

Problem F.6 Use the following formula

$$\tan^{-1}(a)-\tan^{-1}(b)=\tan^{-1}\left[\frac{a-b}{1+ab}\right] \tag{F.100}$$

to show that

$$\frac{\pi}{4}=\tan^{-1}(\frac{1}{100})+\tan^{-1}(\frac{99}{101}) \tag{F.101}$$

Problem F.7 Prove that

$$\tan^{-1}(\frac{1}{b})=\tan^{-1}(\frac{1}{a+b})+\tan^{-1}\left[\frac{1}{b+(b^2+1)/a}\right] \tag{F.102}$$

and subsequently show that

$$\frac{\pi}{4}=\tan^{-1}(\frac{1}{99})+\tan^{-1}(\frac{49}{50}) \tag{F.103}$$

Problem F.8 Use Wessel's complex number approach to show that

$$\frac{\pi}{4}=\tan^{-1}(\frac{n}{n+2k})+\tan^{-1}(\frac{k}{n+k}) \tag{F.104}$$

and subsequently show that

$$\frac{\pi}{4}=\tan^{-1}(\frac{1}{4})+\tan^{-1}(\frac{3}{5}) \ . \tag{F.105}$$

SELECTED BIOGRAPHIES

Knowing the biographies of mathematicians and scientists is essential for readers to appreciate the significance of these mathematical theories. Many of these mathematical techniques in solving the associated differential equations have great influence on the technological development of our daily lives. Biographies of a number of mathematicians, scientists, and engineers whose works are covered or mentioned in this book are included here. It is sometimes difficult to distinguish applied mathematicians from scientists and engineers, especially theoretical scientists. The main references for this section are Jenkins-Jones (1996), Struik (1987), James (2002), Kline (1972), and Millar et al. (2002). This section will hopefully form a mini-Who's Who in differential equations, and their applications to mechanics and engineering.

Abel, N.H. (1802–1829) was a Norwegian mathematician who made major contributions to elliptic functions, integral equations, infinite series, binomial theorem, and group theory. He provided the first stringent proof of the binominal theorem. He revolutionized elliptic integrals. He discovered Abelian functions. He proved that no algebraic solution of the general fifth degree (quintic) equation exists, but ironically Gauss threw his proof away unread when Abel sent it to him. Abel was extremely poor throughout his life, and was unrecognized. In 1825, he visited France and Germany but was largely ignored by mathematicians like Gauss and Cauchy, except Leopold Crelle. He died at age 26 due to tuberculosis. Two days after his death, a letter arrived from Crelle offering him a professorship at Berlin.

Adams, J.C. (1819–1892) was a British mathematician and astronomer. His most famous achievement was his prediction of the existence of Neptune using "perturbation theory" while he was still an undergraduate student, but his prediction was not followed up by G.B. Airy. He was the Lowndean Professor at University of Cambridge and a recipient of the Gold Medal of the Royal Astronomical Society in 1866. He was also a foreign honorary member of the American Academy of Arts and Sciences. He was supposed to be knighted by Queen Victoria but declined. The Adams-Moulton formula in the finite difference method in Chapter 15 is named after him.

Airy, G.B. (1801–1892) was a British mathematician, astronomer, and geophysicist. A special function called the Airy function was the result of a kind of water wave. In 2-D elasticity, the Airy stress function has been of great importance to the analysis of 2-D elasticity problems. He considered the bending of beams and published in 1862 the use of the stress function on a rectangular beam, but failed to consider the compatibility condition. He was also involved in laying the transatlantic telegraph cable, and the construction of the clock of Big Ben. Airy was, however, better known for serving as the Astronomer Royal for 46 years and for measuring Greenwich mean time by stars crossing the meridian observed through his telescope. Airy was arrogant and perhaps best known for his failure to exploit Adam's prediction of a new planet, Neptune. While still an undergraduate, Adam sent his prediction to Airy, but Airy was skeptical. He asked Adams for clarification but did not receive a complete reply from Adams. Nine months later Le Verrier of France made the same prediction, which led to the discovery of Neptune.

Bateman, H. (1882–1946) was an English mathematician and moved to USA in 1910 and took up his permanent position at the California Institute of Technology (Caltech) in 1917. He published his textbook *Differential Equations* in 1915, and *Partial Differential Equations in Mathematical Physics* in 1932. He was elected to the National Academy of Sciences in 1930 and to the Royal Society of London in 1928. When he passed away in 1946, a huge number of drafts of manuscripts on special functions and mathematical tables for integral transforms were left behind at Caltech, and Erdelyi was recruited to form a team to finish his classical works of "Bateman Manuscript Project." These are significant textbooks on differential equations and engineering mathematics.

Beltrami, E. (1835–1900) was an Italian mathematician who made notable contributions to differential geometry, non-Euclidean geometry, and mathematical physics. He developed the singular decomposition theory for matrices. We have mentioned Beltrami equations in Chapter 7 when we talked about the canonical form of elliptic PDEs.

Bernoulli, Daniel (1700–1782) was a Swiss mathematician and physicist. He was one of the main pioneers of differential equations through interactions with Euler. It was Bernoulli who found Euler a job at St. Petersburg. The Bernoulli equation in fluid mechanics was named after him and is the main principle that the airplane wing was based upon.

Bernoulli, Jacob (1655–1705) was a Swiss mathematician and a brother of Johann Bernoulli. He studied theology first and then shifted to mathematics and astronomy. He was one of the main contributors to calculus, differential equations, and the calculus of variations. The Bernoulli equation in first order ODEs was named after him. He was initially a tutor of his younger brother Johann (who was major in medicine), but soon their rivalry turned sour. When he considered the isochrone problem (originally posed by Galileo), or the curve of constant descent along which a particle will descend under gravity from any point to the bottom in exactly the same time, he used the term *integral* for the first time. He also discovered the Bernoulli numbers in probability and logarithmic spirals (such a spiral even appeared on his gravestone).

Bernoulli, Johann (1667–1748) was a Swiss mathematician and was one of the many prominent mathematicians in the Bernoulli family. He is known as John Bernoulli. He studied medicine at Basel University. He was one of the main contributors to differential equations and educated the youth Euler. He also solved the isochrone problem independent of Jacob. He and his brother Jacob are among the first mathematicians using calculus to solve real problems. He was also the discoverer of the L'Hôpital's rule.

Bessel, F.W. (1784–1846) was a German astronomer and mathematician. He was the first to measure a star's distance by parallax. He studied the perturbation of planetary and stellar motions and he developed a mathematical function, now called the Bessel function. This function had wide applications in many other areas of mechanics. The Hankel transform for cylindrical coordinates is based on the Bessel function. Bessel made fundamental contributions to positional astronomy, geodesy, and calculating the sizes of stars, galaxies, and clusters of galaxies. Based on the irregularities of Uranus's orbit, he predicted the existence of Neptune in 1840, but died a few months before its discovery.

Cauchy, A.L. (1789–1857) was a French civil engineer, mathematician, and mechanician who founded complex analysis and contour integration (on which the inverse Laplace transform is based). He also made major contributions in continuum mechanics and elasticity. He published 7 books and over 700 papers, on such topics as calculus, definite integrals, limits, probability, convergence of infinite series, mechanics, astronomy, geometry, wave modulation, and complex functions. There are 16 concepts and theorems named after him, the most of any mathematician. The story is told that when Cauchy presented his theory of convergence of series, Laplace rushed home and checked those series that he used in his books on celestial mechanics (luckily they all converged). He was a devoted teacher, was the most careful in citing other people's works, and candidly admitted errors in his publications. He is also a founder of the matrix method for systems of first order ODEs.

Cayley, A. (1821–1895) was a British mathematician who knew many languages, including Greek, French, German, and Italian. He worked as a lawyer for 14 years. He developed matrix algebra, algebraic invariants, and n-dimensional geometry. He was a recipient of the Royal medal and Copley medal of the Royal Society. He finished his undergraduate course by winning the place of Senior Wrangler and the first Smith's Prize. He is among the most prolific researchers. A total of 967 papers were assembled in his 14 volumes of collected works published by Cambridge University Press. He worked in nearly all areas of mathematics. His hobbies include novel reading, painting, architecture, traveling, and hiking. He wrote many papers in French and was totally at home with German and Italian. He was popular among continental mathematicians. For example, Hermite compared him to Cauchy and others even compared him to Euler. There are also criticisms of him. Tait remarked "Is it not a pity that such an outstanding man puts his abilities to such entirely useless questions?" and when G.H. Hardy was asked whether he thought Cayley was a great mathematician, he just glared. Indeed, it is not too difficult to spot careless typos in his collected works with careful reading. Quality and quantity are not always compatible. The factorization of the ODEs discussed in Section 3.5.10 was introduced by Cayley.

Charpit, P. (??–1784) was a "young" French mathematician who died in 1784. His birth year is not known. The so-called Lagrange-Charpit method for nonlinear

1st order PDEs (also referred to as the "Lagrange method" or "Charpit method" in the literature) was named after him. Paul Charpit was a nephew of Laplace, and had assisted in a course of Monge. There is a dispute in recent literatures as to whether Charpit contributed anything to the so-called Lagrange-Charpit method (e.g., Kline, 1972, p 535). Charpit submitted a paper to the French Academy of Sciences in 1784, the year of his death. The work was never published. His work was first reported and publicized by Sylvestre-Francois Lacroix in 1814 (Johnson, 2010). Kline (1972) reported that when Jacobi learned about the method from Lacroix, he expressed the wish that Charpit's work be published. Kline (1972) further claimed that it was never done. Kline (1972) further questioned the reliability of Lacroix's claim in his book, and he inferred that "Lagrange had done the full job and Charpit would have added nothing." In fact, the assertion by Kline is not accurate. Grattan-Guinness and Engelsman (1982) did find the original manuscript of Charpit that Lacroix got at the Archives of the Academie des Sciences, Paris. A copy owned by Charpit's friend Arbogast was also found at the Biblioteca Medicea-Lauenziana, Florence. According to Grattan-Guinness and Engelsman (1982), Charpit presented his paper on June 30, 1784, but died on December 28, 1784. When Charpit read his paper to the Academie des Sciences, Monge, Bossut, Condorcet, Cousin, Laplace, Vandermonde, and de Borda were present. The original paper was in Laplace's hands for 9 years before he passed it to Lagrange on June 13, 1793. In September 1973, Lagrange sent the text to Arbogast, who made a copy (which ended up at Florence's library mentioned above). Then, Arbogast sent the original paper to Lacroix (which ended up at Paris's Archives of the Academie des Sciences mentioned above). When H. Villat sorted documents in the Archives of the Academie des Sciences in 1928, he unearthed Lacroix's copy of Charpit's paper. He made a photocopy and sent it to N. Saltykow in Belgrade, who published the main content of Charpit's paper in French in two separate papers (Saltykow, 1930, 1937). The content of these papers did contain the main equations that we today call the Lagrange-Charpit method. Therefore, we disagree with Kline's (1972) verdict that Charpit would have added nothing to the method.

Clairaut, A.C. (1713–1765) was a French mathematician, astronomer, and geophysicist. He was a prodigy, published his first paper at the age of twelve, read in front of Academie of France at fourteen, and published his first book at eighteen in 1731. He found the exact condition for 1st order ODEs. The Clairaut equation was named after him, and in the process he found his singular solution of differential equation. His second book in 1743 was on the equilibrium of fluids and the attraction of ellipsoids of revolution. In the book, Clairaut's theorem was applied to find Earth's ellipticity based on surface measurement of gravity. His third book in 1752 was on the three-body problem in astronomy. He calculated the path of Halley's comet as affected by the perturbation of the planets. He was elected to fellow of the Royal Society at the age of twenty-four.

Courant, R. (1888–1972) was one of the most influential applied mathematicians of our time, and was a student of Hilbert and Klein. He helped Hilbert to search literature and write lectures. As a result, he co-authored with Hilbert the book *Methods of Mathematical Physics*. It is a classic for both applied mathematicians and physicists (in fact Hilbert's contribution to the writing of this book is limited).

His charisma made him an educational leader. The CFL stability criterion of the finite difference method for solving PDEs was named after him, Friedrichs, and Lewy. According to J.T. Oden (1987), Courant was also one of the founders of the present-day finite element method. He left Germany and joined New York University and pretty much single-handedly founded the Courant Institute of Mathematical Sciences in 1936. He recruited K.O. Friedrichs and J.J. Stoker, making "Courant Institute" the home of the next generation of applied mathematicians in the United States. He was a member of the National Academy of Sciences of USA.

D'Alembert, J.L.R. (1717–1783) was a French mathematician, mechanician, physicist, philosopher, and music theorist. He studied law and medicine before he shifted to mathematics and mechanics. He worked on dynamics, celestial mechanics, and partial differential equation. The D'Alembert principle in dynamics was named after him. He was one of the founders of the three-body problem, together with Clairaut, Euler, Lagrange, and Laplace. The D'Alembert solution for 1-D wave equations was named after him. He was the illegitimate son of writer Tencin and artillery officer Destouches, and was left on the steps of the Saint-Jean-le-Rond de Paris Church. His first name Jean-le-Rond came from the name of the church. He was put in an orphanage and was later adopted by the wife of a glazier. D'Alembert's education was secretly supported by his father Destouches, who left a modest sum to support him when D'Alembert was only nine. He was elected to the Academie des Sciences in 1741, and he was the co-editor (with Denis Diderot) of 17 volumes of an encyclopedia. In his later years, he helped launch the careers of Lagrange and Laplace. He was the first man proposing the idea of 4-dimensional time-space.

Darboux, G. (1842–1917) was a French mathematician. He made several important contributions to geometry and linear differential equation. His results on differential geometry of surfaces were compiled in four volumes of collected works from 1887 to 1896. In Darboux's hand, differential geometry became connected with ordinary differential equations, partial differential equations, and mechanics. His influence in France was compared to that of Klein in Germany. He was a biographer of Henri Poincaré and edited the *Selected Works of Joseph Fourier*. In 1884, he was elected to Academie des Sciences, elected as a fellow of the Royal Society in 1902, and received the Sylvester Medal in 1916.

Debye, P.J.W. (1884–1966) was a Dutch-American physicist, physical chemist, and applied mathematician, and Nobel laureate in Chemistry (1936). His degree was in electrical engineering and he got his PhD following theoretical physicist Arnold Sommerfeld, who later claimed that his most important discovery was Peter Debye. In 1909, Debye obtained a new integral representation of the Bessel function of large order using the method of steepest descent of Riemann. He developed the theory for dipole moment to charge distribution in asymmetric molecules in 1912 (the units of molecular dipole moments are termed "Debyes" in honor of him). He derived the Planck radiation formula using a method which Plank agreed was simpler than his own. In 1913, he extended Bohr's atomic structure, introducing elliptical orbits. In 1914–1915, he and Paul Scherrer calculated the effect of temperature on X-ray diffraction of crystalline solids (the Debye-Waller factor). He and his assistant developed the Debye-Huckel equation to model the conductivity of an electrolyte solution in 1923. He also developed a theory to explain the Compton effect, the shifting of the frequency of X-rays when they interact with electrons. After serving as professor at the University of Zurich (succeeding Einstein in 1911), Utrecht, Gottingen, ETH Zurich, Leipzig, and finally Berlin (succeeding Einstein in 1934), Debye moved to Cornell University in 1940 for good just before the Nazi invasion of the Netherlands (we will come back to this later). He was awarded the Rumford medal in 1930, the Lorentz medal in 1935, the Nobel Prize in 1936, the Franklin medal in 1937, the Max Plank medal in 1950, the Priestley medal in 1963, and the National Medal of Science in 1965. Actually, Debye was nominated 47 times to receive the Nobel Prize, both in Physics and Chemistry. Two of these nominations were made by the great mathematician David Hilbert. Debye's name was never mentioned in Struik (1987) and Kline (1972), Jenkins-Jones (1996) listed him as physicist, and Millar et al. (2002) listed him as chemical-physicist. In Chapter 13, we mentioned the saddle point method by Debye. He is also known for the Debye potential in solving Maxwell equations for spherical waves. Therefore, Debye should at least be identified as an applied mathematician, as we did here. According to Beiser (2003), Heisenberg, a colleague of Debye for a time, thought him lazy ("I frequently see him walking around his garden and watering roses even during duty hours of the Institute"), but Debye published nearly 250 papers and received the Nobel Prize in chemistry in 1936.

Forty years after death, a Dutch book written by Rispens accused Einstein of actively trying to prevent Debye from being appointed in the United States and further accused Debye of being a Nazi activist, and stirred an international debate (leading to the 2007 NIOD report and the 2008 Terlow report). This initially resulted in the "Debye Institute" at University of Utrecht being renamed in 2006 but the name was reinstated in 2008 after the 2008 Terlow report, but otherwise the claims by Rispens were mainly dismissed. In 2010, a publication by Reiding asserted that Debye may have been an MI6 spy because of Debye's close friend Rosbaud being a well-documented spy, and because of Debye's timely departure to the United States on January 16, 1940, coinciding with the planned German invasion of the Netherlands a day later.

Dirichlet, J.P.G.L. (1805–1859) was a German mathematician who made contributions to partial differential equations and number theory. He proved Fermat's last theorem for the case of $n = 5$, an exceptional feat for a 20-year-old without a degree. He read the result at the French Academy of Sciences, and that brought him immediate fame. This put him in close contact with Fourier and Poisson. They later introduced him to German explorer and scientist Alexander von Humboldt who secured a recommendation from Gauss for Dirichlet and helped to secure a teaching position for Dirichlet at the University of Breslau. However, he failed to pass the doctoral dissertation submitted to the University of Bonn because of his poor Latin. But because of his work on the last Fermat theorem, the university bypassed the problem by awarding him an honorary doctorate in 1827. He became the youngest member of the Prussian Academy of Sciences at age 27 in 1832. His Dirichlet theorem proves the existence of an infinite number of primes.

He was a friend of Gauss and Jacobi. He succeeded Gauss's chair at Gottingen in 1855. His students at Gottingen included R. Dedekind, B. Riemann, and M. Cantor. He found the sufficient conditions for Fourier series to converge in 1829. The Dirichlet boundary value problem in PDEs was named after him. His PhD students included L. Kronecker and R. Lipschitz. His collected work was edited by Kronecker and L. Fuchs, initiated by the Academy of Berlin. He was a foreign member of Royal Society and French Academy of Sciences.

Du Bois-Reymond, P. (1831–1889) was a German mathematician who worked on mathematical physics, including the Sturm-Liouville theory, integral equations, variational calculus, and Fourier series. He coined the terms elliptic, parabolic, and hyperbolic for the classification of the 2nd order PDEs.

Erdelyi, A. (1908–1977) was a Hungarian-born British mathematician who was a leading expert on special functions, especially hypergeometric functions. He got a degree in electrical engineering from Czechoslovakia. He got a DSc in 1940 and joined the University of Edinburgh with the help of E.T. Whittaker. In 1946, Whittaker recommended Erdelyi to the California Institute of Technology to take up the task of publishing Harry Bateman's manuscripts: the Bateman manuscript project. He was elected fellow of the Royal Society of Edinburgh in 1945, and fellow of the Royal Society in 1975.

Euler, L. (1707–1783) was a Swiss mathematician, physicist, and astronomer. He is recognized as the greatest mathematician genius of all time. He wrote almost 900 papers, memoirs, books, and other works, and is one of the most prolific mathematicians ever. He made seminal contributions to differential equations. In terms of mechanics, he contributed to the principle of superposition, the principle of virtual work, the free-body and section principle, tidal theory, and the Laplace equation in potential flow. His investigation of the seven-bridge problem of Konigsberg marked the beginning of graph theory. Euler made major contributions to all areas in mathematics, engineering, and science, including calculus, differential equations, analytic and differential geometry of curves and surfaces, number theory, infinite series (such as Euler's constant in infinite series), calculus of variations, optics, acoustics, light, and hydrodynamics. It was estimated that three-quarters of analytical mechanics consists of Euler's contributions. He also contributed to the design of telescopes, microscopes, and ships. His solution of the three-body problem of Earth, Moon, and Sun improved navigational tables. He developed much of classical perturbation theory. In geometry, the beautiful Euler formula for polyhedron relates numbers of vertices, edges, and faces. Euler's formula of $e^{\pi i}+1 = 0$ is considered by many to be the most famous and beautiful formula in mathematics. In structural mechanics, Euler's buckling formula for columns remains a classical result today. He investigated the base of the natural logarithm e (Euler's number). The Eulerian formulation for large deformations is named in honor of him. Most of our modern mathematical notations are those of Euler. After Euler lost one of his eyes in Russia, he said "now I have less distraction and can focus more." Euler had a prodigious memory and could perform complex calculations in his head when he became blind in his old age.

Forsyth, A.R. (1858–1942) was a Scottish mathematician who worked on theory of functions and differential equations. He was elected a fellow of the Royal Society in 1886 and awarded the Royal medal in 1897. He was forced to resign his chair at the University of Liverpool as a result of his adultery with the wife of physicist Boys. His only student was E.T. Whittaker, whose biography is also covered in this section. His most important contribution to differential equations was his 6 volumes of *Theory of Differential Equations*.

Fourier, J.B.J. (1768–1830) was a French mathematician and physicist. The Fourier series, Fourier transform, and Fourier law of heat conduction were named in his honor. He accompanied Napoleon on his Egyptian expedition. His name is inscribed on the Eiffel Tower. He also contributed to dimensional analysis, and the heat equation or diffusion equation in 2nd order PDEs. He was the first in discovering the greenhouse effect of the Earth's atmosphere. He attended Ecole Normale in Paris and was taught by Lagrange, Laplace, and Monge. His advisor was J.L. Lagrange and his students included G. Dirichlet. He was elected to the Academie des Sciences in 1817.

Fredholm, E.I. (1866–1927) was a Swedish mathematician who contributed to integral equations and the theory of Hilbert space. His teacher was Mittag-Leffler. The Fredholm integrals of the first and second kinds are named in honor of him. His Fredholm Alternative Theorem considered the existence of a solution to the nonhomogeneous differential as well as integral equations, and is a very important theorem in differential equations.

Friedrichs, K.O. (1901–1982) was a German American mathematician. When he taught at Technische Hochschule, he fell in love with a young Jewish student, Nellie Bruell. With the anti-Semitic rules under Hitler's regime, they managed to emigrate separately to New York City where they married. In New York, he joined his former teacher Courant, and became the co-founder of the Courant Institute at New York University. He received the National Medal of Science in 1977. He worked on partial differential equations, the finite difference method, differential operators in Hilbert space, and nonlinear buckling of plates. The CFL criteria for numerical stability in time step in finite difference is named after him, Courant, and Levy. He formalized the boundary layer analysis proposed by L. Prandtl to become the method of matched asymptotic expansion.

Frobenius, F.G. (1848–1917) was a German mathematician who made contributions to differential equations, elliptic functions, number theory, and group theory. He gave the first proof of the Cayley-Hamilton theorem. His teachers include Kronecker, Kummer, and Weierstrass. He succeeded Kronecker at Berlin in 1891 upon the recommendation of Weierstrass. He was elected to the Prussian Academy of Sciences.

Frullani, G. (1795–1834) was an Italian mathematician. He was a professor at the University of Pisa. He worked on definite integrals and trigonometric functions in series and in integrals. The Frullani integral discussed in Section 1.8 was named after him.

Fuchs, L. (1833–1902) was a German mathematician who worked on differential equations. He was a student of Weierstrass and a contemporary of Riemann. His works resulted in a Fuchsian theory of linear ODEs. He mainly considered solutions of linear differential equations with singular points, which include the hypergeometric equation as a special case. His work was a great influence on Henri Poincaré. The term *adjoint of differential equation* was coined by Fuchs in 1873.

Galerkin, B.G. (1871–1945) was a Russian/Soviet structural engineer, mathematician, elastician, and engineer who made significant contributions to numerical methods for solving differential equations and to the theory of three-dimensional elasticity by extending Love's potential to 3-D cases. He grew up in a poor family and went to work in the Russian Court as a calligrapher at the age of 12. In his college years, he had to work as a private tutor and draftsman to support himself. His involvement in political activities when he worked as a railway engineer resulted in a 1.5-year jail sentence. It was the turning point in his life. He lost interest in politics and devoted himself to science and engineering. He wrote his first paper (130 pages) while in prison. In 1915, Galerkin published a paper on the approximate solution of differential equations applied to plate bending problems. This method is now known as the Galerkin method. This method forms the basis of the finite element method. He was a member of the Academy of Sciences (USSR).

Gauss, C.F. (1777–1855) was a German mathematician considered by many to be one of the greatest of all mathematicians. He contributed to all areas of mathematics, especially number theory, statistics, and topology. In statistics, normal distribution is called Gaussian distribution. Gauss also originated the method of least squares for best fit curves among data points. In science, Gauss made contributions in geodesy, electric telegraph, crystallography, optics, mechanics, electricity, magnetism, and capillarity. His book on arithmetic is the basis of modern number theory. The Gauss theorem, introduced in Chapter 1, is of great importance in mechanics.

Goursat, E. (1850–1936) was a French mathematician who contributed to complex analysis, differential equations, and hypergeometric series. The proof of the Cauchy theorem was extended by Goursat to the general situation that no Riemann-Cauchy relation is needed (Spiegel, 1964). It was therefore also known as the Cauchy-Goursat theorem (see Chapter 1). His doctoral thesis advisor was Gaston Darboux.

Green, G. (1793–1841) was a British mathematical physicist who introduced Green's theorem and Green's function method for partial differential equations. The entire Chapter 8 discussed Green's function method. These methods had huge impacts in applied mathematics and mechanics. His work on potential theory ran parallel to that of Gauss. Green's story is remarkable in that he was almost entirely self-taught; he only had one year of formal education at the age of eight. The son of a baker, he worked his childhood years in a bakery, except for one year of formal schooling at Robert Goodacre Academy. He published his famous Green's theorem in "An Essay on the Application of Mathematical Analysis to the Theories of Electricity and Magnetism" at his own expense at the age of 35 in 1828. This work was considered by some to be one of the most significant mathematical works of all time. The way that he acquired his mathematical skill remains a mystery. He was encouraged by Sir Bromhead to enroll as an undergraduate at Cambridge University at the age of 40. He died before his work were discovered and publicized by Lord Kelvin (see biography of Lord Kelvin). His works were further developed by James Maxwell to formulate the electromagnetic theory. To commemorate the 200th anniversary of his birth in 1993, a plaque bearing Green's name was placed in Westminster Abbey near Isaac Newton's grave. Similar honors have been given to Michael Faraday, William Thomson (Lord Kelvin), and James Clerk Maxwell.

Hadamard, J. (1865–1963) was a French mathematician who founded the area functional analysis. Hadamard was one of the most influential mathematicians of his time. He published over 300 papers containing novel and highly creative works. He made contributions to logic, complex analytic functions, number theory, geodesics, and hydrodynamics. He proved the prime number theorem (proposed by Gauss and Riemann) independently with Poussin that the number of prime numbers less than x approach $x/\ln x$ as $x \to \infty$. This remains perhaps the most important result in number theory. He published a book on psychology of mathematical minds and initiated the concept of "well posed" in differential equations. He was an acclaimed and inspiring lecturer. Hadamard's method of descent for 2-D waves is discussed in Section 9.2.7.

Hamilton, Sir W.R. (1805–1865) was an Irish physicist and mathematician who made major contributions to optics, mechanics and quaternions (an extension of the complex number to higher dimensions). His Hamilton principle is covered in Chapter 14. In classical mechanics, the Hamiltonian is named after him. He was a fellow of the Royal Society of Edinburgh.

Hankel, H. (1839–1873) was a German mathematician who made significant contributions to complex and hypercomplex numbers, and the theory of function. The Hankel functions provided a solution to the Bessel equation. The Hankel transform used in Chapter 11 bears his name. He originated the "measure" theory of point sets which are useful in probability, cybernetics, and electronic.

Helmholtz, H. von (1821–1894) was a German physicist, mathematician, and physiologist. He discovered the law of conservation of energy, developed a theory on the nature of harmony and musical sound (he was a skillful musician), and invented the ophthalmoscope for viewing the human retina. Boltzmann was one of his students. Helmholtz was considered the most versatile scientist of his century. He has been called the last scholar whose work covered science, physiology, and the arts. Helmholtz believed that his diversified interests helped him adopt novel ideas in research. Together with Kirchhoff, he was one of the main contributors to mathematical physics in Germany in the 19th century. His work on Riemann's quadratic measures led to the Lie–Helmholtz space problem which is important to Einstein's relativity, group theory, and physiology (Struik, 1987). Helmholtz's equation (a reduced wave equation) is discussed in Chapters 7 and 9.

Hilbert, D. (1862–1943) was a German mathematician who worked on geometry, logic, and functional analysis for differential and integral equations (see Chapter 11). Hilbert space was named after him, and is important in the spectral theory of self-adjoint linear differential operators (like in quantum mechanics). The Einstein-Hilbert action for the field theory of gravitation was named after him. He also coauthored the famous textbook on mathematical physics with Courant. At Gottingen, one of his students was H. Weyl and one of his assistants was J. von Neumann. He listed 23 unsolved problems at the International Congress of Mathematics in Paris in 1900. He was a foreign member of the Royal Society.

Hua, Loo-Keng, (1910–1985) was a Chinese mathematician who made important contributions to number theory. After three years of middle school, Hua attended the Chinese Vocational College in Shanghai but could not graduate due to the lack of funding. In 1927, he returned home in Jintan (Jiangsu Province) to help his father's store and studied mathematics by himself. In 1929, Hua suffered from typhoid fever, resulting a partially paralyzed left leg. Hua did not receive a formal university education, although he was awarded several honorary doctorates (from the University of Nancy, the Chinese University of Hong Kong, and the University of Illinois). After reading his early paper, Prof. Xiong Qinglai invited Hua to study mathematics at Tsinghua University. Despite the lack of formal qualification, Hua was exceptionally hired by Tsinghua, initially at Library and eventually rose to the rank of lecturer because of his research papers. In 1935–36, Hua attended classes by visiting French mathematician Jacques Hadamard and American mathematician Norbert Wiener. It was reported that he was the only one who could follow through to the end of the lecture series. Wiener was impressed and mentioned Hua to G.I. Hardy at Cambridge. Hardy invited Hua to Cambridge to visit (probably envisioned another Ramanujan from China) for 2 years. Hua quickly established his name in the area of number theory. In 1938, in view of the full outbreak of the Sino-Japan war, Hua decided to return to Tsinghua, where Hua was hired as a full professor despite not having any degree. Due to Japanese occupation, Hua followed Tsinghua's retreat to Yunnan. Despite the hardship of poverty, enemy bombing, and academic isolation, Hua continued to produce important mathematical papers. After the war, Hua visited Ivan Vinogradov in the Soviet Union for three months and then the Institute of Advanced Study at Princeton University in the United States. In the spring of 1948, Hua accepted the appointment of full professor at the University of Illinois at Urban-Champaign. In October 1949, he gave up his comfortable life in the United States and returned China with his wife and kids. Gradually, Hua moved from pure mathematics to applied mathematics, including linear programming, operation research, and multidimensional numerical integration. He was involved in solving all practical problems that new China faced using mathematics. His Chinese book *An Introduction to Higher Mathematics* was translated into English in 2009, with an excellent introduction to differential equations. He was elected a foreign associate of the National Academy of Sciences in 1982. Chapter 7 reported his solution for a circular domain governed by the biharmonic equation.

Huygens, C. (1629–1695) was a prominent Dutch mathematician and scientist. He made notable contributions to astronomy, physics, and probability. His father was a diplomat and was a friend of G. Galilei, M. Mersenne, and R. Descartes. Huygens was educated at home until the age of 16. He studied the rings of Saturn and discovered its moon, Titan. He invented the pendulum clock. The Huygens or Huygens-Fresnel principle in wave propagation was named after him. In 1646, he demonstrated that a catenary is not a parabola. Huygens was the first to derive the period of an ideal pendulum. In 1659, he derived the centripetal force of circular motions. He tutored the young diplomat Gottfried Leibniz in mathematics from 1673. In 1673, Huygens found the curve down which a mass will slide under the influence of gravity in the same amount of time, regardless of its starting point; he solved this so-called tautochrone problem by geometric method. In 1675, Huygens patented a pocket watch. He was elected to the Royal Society in 1663. The Huygens principle for a 3-D wave is discussed in Chapter 9.

Ince, E.L. (1891–1941) was a British mathematician who worked on differential equations with periodic coefficients, such as the Mathieu equation and Lame equation. He also introduced the Ince equation, which is a generalization of the Mathieu equation. He received the Smith Prize in 1918 and the Makdougall Brisbane Prize in 1938–1940. His famous book *Ordinary Differential Equations* in 1956 remains a classic today.

Jacobi, C.G.J. (1804–1851) was a German mathematician who made fundamental contributions to elliptic functions, dynamics, differential equations, and number theory. Jacobi elliptic functions formed the solution of the pendulum problem (Chapter 2 and Appendix D). In mapping, we check the existence of a nonzero Jacobian, which was named in his honor (see Chapters 1 and 6). For first order ODEs, we have discussed the Jacobi method in Chapter 3. For 1st order PDEs, there is the Jacobi method (see Chapter 6). In celestial mechanics, Jacobi's integral gave the first integral of the three-body problem (Moulton, 1914).

Kelvin, Lord (Thomson, William) (1824–1907) was an Irish mathematician, physicist, and mechanician. Kelvin is probably best known for his introduction of the absolute temperature scale Kelvin. As a young man, he discovered Green's work, then little known, and publicized it. Since then, Green's method has become a powerful tool in mathematical physics. His work on the conservation of energy led to the second law of thermodynamics. He was an unusual scientist with unparalleled enthusiasm, energy, and talent. He invented the tide gauge, an improved compass, and a simpler method for fixing a ship's position at sea. He investigated many different areas of science. He published 661 papers and many books and was the author of several patents. He coined the term "turbulence" in fluid mechanics. The Kelvin solution in elasticity remains one of the most fundamental contributions to applied mechanics. For his role in the Kelvin-Stokes theorem, see the biography of G.G. Stokes (see also Chapter 1). His stationary phase method with Stokes is discussed in Chapter 12. He directed the first successful project for a transatlantic cable telegraph, which became operational in 1866, and brought him considerable wealth. The *Cambridge Dictionary of*

Scientists says he was "probably the first scientist to become wealthy through science" (Millar et al., 2002).

Kirchhoff, G.R. (1824–1887) was a German physicist and a pioneer in spectroscopy. He also made major contributions to plate theory and elasticity. An early accident made him a wheelchair user but did not alter his cheerful character or hinder his scientific curiosity. He formulated Kirchhoff's law for electrical networks. Kirchhoff and his lifelong friend and colleague Bunsen established spectroscopy as an analytical technique in chemical analysis. Using spectroscopy, they discovered the elements cesium and rubidium, and were able to analyze the chemical element present in the Sun's atmosphere (see, however, the biography of Stokes for his role in the development of spectroscopy). The spectrometer, telescope, and microscope are the most dominant scientific instruments of our time. Kirchhoff's formula for 3-D waves is discussed in Chapter 9.

Kline, M. (1908–1992) was a physicist and applied mathematician who studied the history of mathematics and philosophy on physical sciences. He graduated from New York University (NYU) and continued to teach at NYU as an instructor. In 1946–1966, he was the director of the division for electromagnetic research at the Courant Institute of Mathematical Sciences. He made significant contributions to mathematics teaching and was an advocator of changes. He published more than 12 books and mathematics and sciences. His 3-volume series, *Mathematical Thought from Ancient to Modern Times*, is probably the best book on the historical development of differential and integral equations.

Kovalevskaya, S. (1850–1891) was probably one of the most influential female mathematicians in the 19th century. She was the mathematical protégé of Weierstrass and Mittag-Leffler. She was also a novelist. But she died in her prime at the age of 40. She was known for her ability and originality. The Cauchy-Kovalevskaya theorem mentioned in Chapter 6 was published at the age of 25. She obtained the prestigious Bordin prize on her mathematical theory on the dynamics of top (now known as Kovalevskaya top). She was promoted to full professor at Hogskola because of her originally, charm, and mathematical knowledge. She was particularly popular among her students.

Lagrange, J.L. (1736–1813) was an Italian mathematician. He is one of the most influential mathematicians on differential equations, in addition to his major contributions in number theory and algebra. Lagrange almost single-handedly developed the major part of the methods for solving first order PDEs (see Chapter 6). The calculus of variations presented in Chapter 13 was also mainly formulated by Lagrange, including the use of the Lagrange multiplier method. We also discussed the Lagrange identity in Chapters 7 and 10 when we discussed the adjoint ODE problem. He also made major contributions to celestial mechanics, winning him prizes of the French Academy of Sciences many times. In classical mechanics, the Lagrangian formulation is the standard (although Hamiltonian may be more powerful for non-classical mechanics, like quantum mechanics). He was elected a fellow of the Royal Society of London, Royal Society of Edinburgh, foreign member of the Swedish Academy of Sciences, and member of the Berlin Academy.

Lambert, J.H. (1728–1777) was a Swiss mathematician, physicist, philosopher, and astronomer. He is best known for proving the irrationality of π. After he left school, he had been working as an assistant to his father (a tailor), a clerk, a private tutor, and a secretary to an editor. In 1758, he published first book on optics and cosmology, and this allowed him to start an academic career. After a few posts, he was invited to a position at the Prussian Academy of Sciences in Berlin. Lambert was the first to introduce hyperbolic functions. Lambert also derived a theorem on conic sections to make the calculations of the orbit of comets much simpler. He also derived hyperbolic triangles for a concave surface. Lambert was the first mathematician to consider the general properties of map projection. For analyzing 3-D planes on a plane surface, Lambert invented the azimuthal equal-area projection and the equal-area stereonet, which finds application in rock slope stability analysis.

Laplace, P.-S. (1749–1827) was a French mathematician, astronomer, and mathematical physicist. The story has often been told of how D'Alembert gave him difficult mathematical problems to test his ability, and found that Laplace was able to solve them overnight. Much impressed, D'Alembert helped secure Laplace a teaching job at the École Militaire in Paris. He is one of the founders of probability, and he made his name in celestial mechanics by publishing a five-volume survey of celestial mechanics. He theorized that the solar system originated from a cloud of gas (called the nebular hypothesis). Laplace developed the concept of potential and the study of the Laplace equation. He was from a poor family, but he was appointed minister and later senator by Napoleon. The Laplace transform that bears his name is of fundamental importance for solving differential equations. Many considered Laplace as the most illustrious scientist of France's golden age, and one of the most influential scientists of all time. Our current unit of length, the meter, was proposed by Laplace in 1790. Laplace is considered only second to Newton in scientific talent. He was known for his arrogance, and he frequently neglected to acknowledge the sources of his results. He was notorious for overusing the term "it is obvious" in mathematical derivations when it was far from obvious (James, 2002).

Legendre, A.M. (1752–1833) was a French mathematician who contributed to number theory, celestial mechanics, and elliptical functions. In celestial mechanics, he derived the Legendre equations and the Legendre polynomials (see Chapter 4). As shown in Chapter 9, it is closely related to the Laplace equation in spherical coordinates.

Leibniz, G.W. (1646–1716) was a German mathematician, philosopher, and diplomat. His Bachelor and Master degrees are both in philosophy, and he had a doctoral degree in law. He could write articles in Latin, French, and German. He assisted redrafting of the legal code for electorate for the Elector of Mainz, and was involved in politics in proposing an Egyptian plan for German-speaking Europe to diverge France's attention from them to Egypt. In Paris, he met Dutch physicist and mathematician Christiaan Huygens and Irish chemist Robert Boyle, and began a program of self-study with Huygens as his mentor. Eventually, Leibniz developed differential and integral calculus independently of Newton, and his notations in calculus have been adopted today. He also contributed to the development of mechanical calculators and refined the binary number system for later development of the digital computer. When he visited the Royal Society, he demonstrated his calculating machine carrying all four arithmetical operations and finding square roots, and the Society quickly made him a foreign member. Leibniz was the first to see that the coefficients of a system of linear equations could be arranged into an array or matrix. His integral sign $\int$ represents an elongated S, from the Latin word summa and *d* representing differentiate or Latin differentia. Product and quotient rules of differentiation are due to Leibniz. The Leibniz rule for differentiation of integrals is named after him. In dynamics, he favored the conservation of energy instead of the conservation of momentum used by Newton. He also envisioned a universal language for all humans.

L'Hôpital, G.F.A. (1661–1704) was a French mathematician. The first calculus textbook was written by L'Hôpital in 1696. The so-called L'Hôpital's rule in taking the limit of the indeterminate form of 0/0 and ∞/∞ appeared in his book, but actually was discovered by Johann Bernoulli. He was elected to the French Academy of Sciences in 1693.

Liouville, J. (1809–1882) was a French mathematician who made contributions to number theory, complex analysis, differential geometry and topography, linear differential equations, mathematical physics, and astronomy. His teachers at Ecole Polytechnique included A.M. Ampere. Liouville theorem in complex analysis is in honor of him. In mathematical physics, Sturm-Liouville problem resulted from collaboration with C.F. Sturm; it studied the eigenfunction expansions of self-adjoint type boundary value problems and found important application in physics. In Hamiltonian dynamics, the Liouville-Arnold theorem was named in honor of him. He proved the existence of transcendental numbers. According to Mittag-Leffler, Liouville's greatest grief was meeting Abel, but failed to realize his mathematical talent. He was the first to recognize the importance of the unpublished works of Galois.

Mittag-Leffler, G. (1846–1927) is regarded as the father of Swedish mathematics; he spent altogether 3 years in France and Germany. He became friends of Hermite, Poincaré, and Weierstrass. He considered himself as a disciple of Weierstrass, particularly following his power-series approach to function theory. The Mittag-Leffler expansion of trigonometrical functions was reported in Chapter 1. He also found a job for the extremely talented Sonya Kovalevskaya at Hogskola (see biography of Kovalevskaya). He was a member of King Oscar II's circle (the king himself was a mathematician) and was an international celebrity, although he was not so popular among outsiders. When Alfred Nobel set up his Nobel Prize, mathematics was one of the potential subject areas to be awarded. However, when he knew that Mittag-Leffler was a potential candidate for the prize in mathematics, Nobel crossed out mathematics and no such prize has ever been awarded. This negative impact by Mittag-Leffler on mathematics is definitely more far-reaching than his mathematical contributions.

Monge, G. (1746–1818) was a French mathematician and the father of descriptive geometry (the mathematical basis for technical drawing) and of differential geometry. He was involved in the reform of the French education system during the French Revolution, helping to found Ecole Polytechnique. He devised a graphical method to optimize the defensive arrangement of fortification. He joined Napoleon's expedition to Egypt. His name was inscribed on the base of the Eiffel Tower. His method in solving second order partial differential equations was reported in Chapter 7.

Moulton, F.R. (1872–1952) was an American astronomer. He was a professor at the University of Chicago. He proposed that small satellites in the orbit around Jupiter were actually gravitationally captured planetesimals (small bodies that formed planets). This theory has become well accepted among astronomers. The Adams-Moulton method discussed in Chapter 15 for solving differential equation was named after him.

Painlevé, P. (1863–1933) was a French mathematician and politician. Painlevé is best known for his studies of those nonlinear ODEs that can be transformed to linear ODEs or the so-called Painlevé transcendents. He was a student of Flex Klein. He served as French Prime Minister twice.

Pfaff, J.F. (1765–1825) was a German mathematician who worked on series, integral calculus, and partial differential equations. The Pfaffian of 1st order PDEs was named after him. He was the PhD supervisor of C.F. Gauss, and also a teacher of A. Mobius.

Picard, C.E. (1856–1941) was a French mathematician who made contributions to analytic functions and linear differential equations, including the Picard successive approximation for solving first order ODEs, and Painlevé transcendents for ODEs. He also worked on theories of telegraphy and elasticity. Charles Hermite was his father in law.

Poincaré, H. (1854–1912) was a renowned French mathematician, mining engineer and theoretical physicist. He was a mining engineer by training, but got his PhD in mathematics under Charles Hermit. His thesis is on differential equations. Henri Poincaré, a pioneer in stability analysis of differential equations, was nominated to receive the Nobel Prize 34 times in 1910. In fact, Poincaré received a total of 51 nominations in 1904-1912 before he passed away in 1913, but he never received it. His Poincaré conjecture was one of the ten millennium problems of the Clay Mathematics Society. It was eventually "solved" by Perelman, but it was bigger news when he declined both the Fields medal and the Millennium Prize from the Clay Mathematics Institute. The *New Yorker*'s article stirs the controversy of Perelman with former Fields medalist S.T. Yau.

Poisson, S.D. (1781–1840) was a French mathematician and physicist who made contributions to probability theory, elasticity, electricity, magnetism, heat, and sound. In probability, we have the Poisson distribution (the basis for modern hazard analysis) and in elasticity we have the Poisson ratio. In complex analysis, he was the first to carry out path or contour integration of complex functions (called contour integration). The solutions of 2-D and 3-D wave equations are expressible in Poisson's integral (Chapter 9). The nonhomogeneous Laplace equation is also called Poisson's equation.

Ramanujan, S. (1887–1920) was an Indian mathematician who made significant contributions to mathematical analysis, number theory, infinite series, and continued fractions. He received no formal training in mathematics but rediscovered known theorems and produced new ones. He was working as a clerk when he conducted his mathematical research. He was a prodigy discovered by G.H. Hardy, and Hardy compared Ramanujan with Jacobi or Euler. After Ramanujan sent some of his theorems to G.H. Hardy, Ramanujan was invited to work with him at Cambridge. He compiled a total of 3900 identities and equations, and many of them are highly original and unconventional, including the Ramanujan prime, and Ramanujan theta function. He became a fellow of the Royal Society and died at the age of 32. His 87 pages of lost notebook was rediscovered by George Andrews in 1976. The number 1729 was called the Ramanujan-Hardy number: Once Hardy took a taxi to visit Ramanujan in hospital, and found the cab number of 1729 rather dull and believed it was an omen. But when he told Ramanujan about this, Ramanujan told him that it is a very interesting number which is the smallest number that can be expressed as sum of two cubes in two different ways (i.e. $1729=1^3+12^3 = 9^3+10^3$) (Kanigel, 1991). One of Ramanujan's formulas has been adopted to generate a huge amount of digits of π, and it became one way to test the speed of supercomputers. Ramanujan's Master Theorem was introduced in Chapter 1 and Appendix C. Some of his amazing formulas on π were reported in Appendix F. The 2015 British biographical film *The Man Who Knew Infinity* was based on his biography written by Kanigel (1991).

Rayleigh, Lord (Strutt, J.W.) (1842–1919) was a British mathematician and physicist and Nobel Prize winner for his work on gas density and on argon. Rayleigh made major contributions to sound, light, surface waves, and electricity. He wrote his classic book *Theory of Sound* partly on a boathouse on the Nile. He inherited the title Lord Rayleigh from his father, and succeeded Maxwell at Cambridge. Rayleigh explained the blue color of sky from the scattering of light by dust particles in the air. His enthusiasm on precise measurement led him to the standardization of electrical units in 1884: the ohm, ampere, and volt. The inconsistency of the Rayleigh-Jeans equation (published by Rayleigh in 1900), which describes the distribution of wavelengths in black-body radiation, led Planck to the formulation of quantum theory. In numerical analysis, the Rayleigh-Ritz method discussed in Chapter 14 is a powerful approximate method that bears his name.

Riemann, G.F.B. (1826–1866) was a German mathematician, who originated Riemann geometry, which was used by Einstein in the theory of general relativity. He also made breakthroughs in conceptual understanding of the theory of functions, vector analysis, differential geometry, and topology. He was a student of Gauss. In complex variable theory, he developed the concept of the Riemann surface which separates multi-connected surfaces by branch cuts. The differentiable condition for complex variables is now known as the Cauchy-Riemann relation. He defined the Riemann zeta function and formulated a Riemann hypothesis of this function. It remains one of the most important unsolved problems of number theory and analysis. The Clay Mathematical Institute of Cambridge offered US $1 million for its proof (Sabbagh, 2003). Riemann died at the age of 39 because of tuberculosis.

Robin, G. (1855–1897) was a French mathematician whose method on single- and double-layer potentials boundary value problems of electrostatics is an important technique in potential theory (see Section 8.4 of Chapter 8). In short, the Dirichlet and Neumann boundary value problems were reduced to integral equations for electric densities and to power series expansions of potentials. The single-layer potentials were also known as Robin-Steklov potentials (Steklov was a student of Russian mathematician A. Lyapunov and Robin was a student of the well-known French mathematician Picard). The criterion for uniform convergence of his series is called the Robin principle. The Robin constant is associated with the logarithmic electric capacitor. His name has been associated with the third type of boundary value problem of potential theory (as the Robin boundary condition or Robin Problem), but Gustafson and Abe (1998a,b) concluded that Robin had actually never used or studied such a boundary condition (see Chapter 9). He was a recipient of the Francoeur Prize and Poncelet Prize.

Sommerfeld, A.J.W. (1868–1951) was a German theoretical physicist who was a pioneer of quantum theory. He was nominated a record of 81 times for the Nobel Prize within a period of 34 years before his death, but he never received it. Very likely, this record will not be broken easily in the near future. Ironically, he made two nominations, one of Albert Einstein (Nobel laureate in 1921) and one of Max Planck (Nobel laureate in 1918), and thus, in a sense, his nominations had a 100% success rate. In addition, he served as PhD supervisor for more Nobel Prize winners in physics than any other supervisor before or since. His PhD students getting Nobel Prizes include W. Heisenberg in 1932 (PhD in 1923), H. Bethe in 1967 (PhD in 1928), W. Pauli in 1945 (PhD in 1921), and P. Debye in 1936 (PhD in 1908). His other postgraduate and post-doctoral students getting Nobel Prizes include L. Pauling in Chemistry in 1954 and in Peace in 1962, I.I. Rabi in 1944, and M.T. von Laue in 1914. Many of his other students became famous in their own right. His professors include German mathematicians Lindemann, Hurwitz, and Hilbert, and after graduation Sommerfeld was an assistant of Felix Klein and wrote up his lecture notes for the reading room. This resulted in a six-volume series of classical textbooks of Lectures on Theoretical Physics, including *Mechanics*, *Mechanics for Deformable Bodies*, *Electrodynamics*, *Optics*, *Mathematical Theory of Diffraction*, and *Partial Differential Equations in Physics*. They are all influential and all have been translated into English. Albert Einstein also admired Sommerfeld for nurturing a large number of young talents. Sommerfeld was one of the pioneers of quantum theory. He was elected to the Royal Society of London, US National Academy of Sciences, Indian Academy of Sciences, and USSR Academy of Sciences. Example 2.7 in Chapter 2 presents Sommerfeld's solution for the Thomas-Fermi equation.

Stokes, G.G. (1819–1903) was an Irish mathematician and physicist who made fundamental contributions to fluid dynamics. The most general governing equations for fluid dynamics are called the Navier-Stokes equations. Stokes described the phenomenon of fluorescence in 1852. His Stokes law for a sphere settling in a fluid also bears his name. He was the first to explain the fundamentals of spectroscopy. However, the Stokes theorem discussed in Chapter 1 was in fact discovered by Lord Kelvin and communicated to Stokes in 1850, and Stokes set the theorem as a question for the 1854 Smith's prize examination, which led to the theorem bearing his name. Therefore, some mathematicians called it the Kelvin-Stokes theorem. Stokes served as president of the Royal Society.

Sturm, J.C.F. (1803–1855) was a French mathematician and best known for the Sturm-Liouville problem discussed in Chapter 10. He pointed out the relevance of eigenvalue analysis for systems of first order ODEs to Cauchy in 1828 (see Chapter 5). Related to his works on stability of ODEs, Sturm's theorem is a basic result of finding and counting real roots of polynomials. He was a foreign member of the Royal Society of London.

Taylor, B. (1685–1731) was an English mathematician who is known for his Taylor series expansion. He was elected fellow of the Royal Society in 1712. However, he failed to express his ideas fully and clearly. His Taylor series expansion was found important by Lagrange only in 1772.

Titchmarsh, E.T. (1899–1963) was an English mathematician. He was a student of Hardy and succeeded his position at Oxford in 1931. He contributed on the entire function of complex variables, integral equations, Riemann zeta function, eigenfunction expansions of differential equations, and Fourier series. He was elected fellow of the Royal Society in 1931 and received the Sylvester Medal in 1955. He also received the De Morgan Medal and the Berwick Prize from London Mathematical Society. The Titchmarsh contour integral is discussed in Chapter 1.

Tricomi, F.G. (1897–1978) was an Italian mathematician famous for his studies of mixed type partial differential equations. His book on the mixed type second order partial differential equations in 1923 was translated to Russian, then from Russian to Chinese. He also authored *Integral Equations*. He spent two years at Caltech to work on the Bateman Manuscript Project with Erdelyi, Magnus, and Oberhettinger. He was the president of the Turin Academy of Sciences. The Tricomi equation is discussed in Section 7.7.

Volterra, V. (1860–1940) was an Italian mathematician who contributed to functional theory, nonlinear integro-differential equations, biological and population growth, and dislocation theory. Volterra integral equations were named after him. His contribution to dislocation was introduced in Chapter 2. During World War I, he was involved in designing armaments, and he was also the first to propose the use of helium to replace hydrogen in airships.

Watson, G.N. (1886–1965) was an English mathematician who was an expert in the application of complex variables to the theory of special functions, especially

Bessel functions. He co-authored, with his supervisor E.T. Whittaker, the 2nd edition of *A Course of Modern Analysis* (1915). His book *Treatise on the Theory of Bessel Functions* in 1922 is a classic even using today's standard. In 1918, he proved Watson's lemma, which has applications on the asymptotic behavior of exponential integrals. Convergence of certain infinite series can be improved by his Watson's transform. Ramanujan's lost notebook was in his hands before it was rediscovered by George E. Andrews in 1976 from Watons's left-behind boxes at Wren Library at Cambridge. He was elected fellow of the Royal Society and received the Sylvester Medal from the Society.

Weierstrass, K.T.W. (1815–1897) was a German mathematician who is often cited as the father of modern analysis. He studied law, economics, and finance at University of Bonn but dropped out because of its conflict with his hope to study mathematics. He then moved to the University of Munster to study mathematics. He formally laid down the rigorous foundation of calculus and clarified the concepts of uniform convergence, derivative, and continuity. He also paved the way for the modern study of the calculus of variations. The Weierstrass elliptic functions were named after him. This contribution to the gamma function was reported in Chapter 4.

Whittaker, E.T. (1873–1956) was an English mathematician who contributed to applied mathematics, mathematical physics, and the theory of special functions. The Whittaker function in the theory of confluent hypergeometric functions (or Kummer functions) was named after him. He also derived Bessel functions in terms of integrals of Legendre functions. He married the daughter of the minister of the Presbyterian Church. His co-authored book *A Course of Modern Analysis* with his student G.N. Watson (Whittaker and Watson, 1927) is a classic text on transcendental complex functions. He also worked on the history of physics and celestial mechanics. As a historian of science, he wrote *A History of the Theories of Aether and Electricity* in 1910, which was revised and expanded to be two books in 1951 and 1953. In this book, he attributed the discovery of special relativity more to Henri Poincaré and Lorentz, and less to Einstein. He also attributed the formula $E=mc^2$ to Poincaré. Whittaker's view on this appears not in the main stream. In Bodanis's (2000) biography on $E=mc^2$, he did mention that Henri Poincaré made a presentation at the St. Louis World Fair entitled "theory of relativity" and came close to giving the famous formula which Einstein got a year later. Whittaker wrote *The Calculus of Observations: A Treatise on Numerical Mathematics* in 1924 and *Treatise on the Analytical Dynamics of Particles and Rigid Bodies: With an Introduction to the Problem of Three Bodies* in 1937. He received the Copley Medal and Sylvester Medal from the Society. The Whittaker equation is discussed in (3.584) in Chapter 3.

Wronski, J.M.H. (1776–1853) was a Polish mathematician. The Wronskian repeatedly used in Chapter 3 was named after him by Thomas Muir in 1882. He was also a metaphysician and proposed to build a machine that could predict the future.

REFERENCES

Ablowitz, M.J. and Clarkson, P.A., 1991, *Solitons, Nonlinear Evolution Equations and Inverse Scattering,* London Mathematical Society Lecture Note Series 149 (Cambridge: Cambridge University Press).

Abramowitz, M. and Stegun, I.A., 1964, *Handbook of Mathematical Functions* (New York: Dover).

Airy, G.B., 1873, *An Elementary Treatise on Partial Differential Equations,* 2nd ed. (London: MacMillan).

Aitken, A.C., 1944, *Determinants and Matrices* (Edinburg: Oliver and Boyd).

Alzer, H., 2009, A superadditive property of Hadamard's gamma function. Abhandlungen aus dem Mathematischen Seminar der Universität Hamburg, **79**, 11–23.

Assur, A., 1959, *Criteria for Landing Bomber and Fighter Aircraft on Floating Ice Sheets* (U), US Army Snow Ice and Permafrost Research Establishment, Corps of Engineers, Wilmette, Illinois.

Ayres, Jr. F., 1952, *Theory and Problem of Differential Equations* (New York: Schaum Publishing Co.).

Bateman, H., 1918, *Differential Equations* (London: Longmans).

Bateman, H., 1944, *Partial Differential Equations of Mathematical Physics* (New York: Dover).

Bathe, K.J., 1982, *Finite Element Procedures in Engineering Analysis* (Englewood Cliffs: Prentice-Hall).

Beiser, A., 2003, *Concepts of Modern Physics,* 6th ed. (New York: McGraw-Hill).

Ben-Menahem, A. and Singh, S.J., 2000, *Seismic Waves and Sources* (New York: Dover).

Bender, C.M. and Orszag, S.A., 1978, *Advanced Mathematical Methods for Scientists and Engineers* (New York: McGraw Hill).

Bell, W.W., 1968, *Special Functions for Scientists and Engineers* (Mineola: Dover).

Berndt, B.C., 1985, Ramanujan's Lost Notebooks, Part I (New York: Springer-Verlag).

Berndt, B.C., 1989, Ramanujan's Lost Notebooks, Part II (New York: Springer-Verlag).

Bleistein, N. and Handelsman, R.A., 1986, *Asymptotic Expansions of Integrals* (New York: Dover).

Bliss, G.A., 1925, *Calculus of Variations,* Carus Mathematical Monograph No. 1 (Chicago: American Mathematical Society).

Bliss, G.A., 1930, The problem of Lagrange in the calculus of variations. *American Journal of Mathematics*, **52** (4), pp. 673–744.

Bodanis, D., 2000 *$E=mc^2$: A Biography of the World's Most Famous Equation* (New York: Walker).

Boisvert, R.F. and Lozier, D.W., 2001, "Handbook of Mathematical Functions". In Lide, D.R.A *Century of Excellence in Measurements Standards and Technology: A Chronicle of Selected NBS/NIST Publications 1901–2000*. Washington, D.C., USA: U.S. Department of Commerce, National Institute of Standards and Technology (NIST) / CRC Press. pp. 135–139.

Boyce, W.E. and DiPrima, R.C., 2010, *Elementary Differential Equations and Boundary Value Problems*, 9th ed. (New York: Wiley).

Brebbia, C.A., Telles, J.C.F. and Wrobel, L.C., 1983, *Boundary Element Techniques: Theory and Applications in Engineering* (Berlin: Springer-Verlag).

Broberg, K.B., 1999, *Cracks and Fracture* (San Diego: Academic Press).

Butkovskiy, A.G., 1982, *Green's Functions and Transfer Functions Handbook* (Chichester: Ellis Horwood Ltd).

Cajori, F., 1993, *A History of Mathematical Notations* (New York: Dover).

Cannell, D.M. and Lord, N.J., 1993, George Green, Mathematician and Physicist 1793–1841. *The Mathematical Gazette*, **77**(478), pp. 26–51.

Carrier, G.F. and Pearson, C.E., 1976, *Partial Differential Equations: Theory and Technique* (New York: Academic Press).

Carslaw, H.S. and Jaeger, J.C., 1959. *Conduction of Heat in Solids*, 2nd ed. (Oxford: Oxford University Press).

Chau, K.T., 1995, Landslides modeled as bifurcations of creeping slopes with nonlinear friction law. *International Journal of Solids and Structures,* **32**(23), pp. 3451–3464.

Chau, K.T., 1996, Fluid point source and point forces in linear elastic diffusive half-spaces. *Mechanics of Materials*, **23**, pp. 241–253.

Chau, K.T., 1997, Young's modulus interpreted from compression tests with end friction. *Journal of Engineering Mechanics ASCE*, **123**, pp. 1–7.

Chau, K.T., 1999a, Young's modulus interpreted from plane compressions of geomaterials between rough end blocks. *International Journal of Solids and Structures,* **36**, pp. 4963–4974.

Chau, K.T., 1999b, Onset of natural terrain landslides modeled by linear stability analysis of creeping slopes with a two state variable friction law. *International Journal of Numerical and Analytical Methods in Geomechanics,* **23**, pp. 1835–1855.

Chau, K.T., 2013, *Analytic Methods in Geomechanics* (Boca Raton: CRC Press).

Chau, K.T. and Choi, S.K., 1998, Bifurcations of thick-walled hollow cylinders of geomaterials under axisymmetric compression. *International Journal of Numerical and Analytical Methods in Geomechanics*, **22,** pp. 903–919.

Clough, R. W., 1980, The Finite Element Method after twenty-five years: A personal view. *Computer and Structures*, **12**, pp. 361–370.

Copson, E.T., 1935, *An Introduction to the Theory of Functions of a Complex Variable* (Oxford: Clarendon Press).

Copson, E.T., 1975, *Partial Differential Equations* (Cambridge: Cambridge University Press).

Craig, T., 1889, *A Treatise on Linear Differential Equations* (New York: John Wiley and Sons).

Crawford, E., 1985, *The Beginnings of the Nobel Institutions.* The Science Prizes, 1901–1915 (New York: Cambridge University Press).

Crawford, E., 1998, Nobel: Always the Winners, Never the Losers. *Science* (by AAAS) **262**, 5392, pp. 1256–1257.

Crawford, E., 2001, Nobel population 1901–1950: Anatomy of a scientific elite. *Physics World*, **14** (11), pp. 31–35.

Crowe, M.J., 1993, *A History of Vector Analysis: The Evolution of the Idea of a Vectorial System* (New York: Dover).

Davis, H.T., 1962, *Introduction to Nonlinear Differential and Integral Equations* (New York: Dover).

Davis, P.J., 1959, Leonhard Euler's integral: A historical profile of the gamma function: In memoriam: Milton Abramowitz. *The American Mathematical Monthly,* **66**(10), pp. 849–869.

Dirac, P.A.M., 1947, *The Principle of Quantum Mechanics* (Oxford: Clarendon Press).

Douglas, J., 1931, Solutions of the problem of Plateau. *Transaction of American Mathematical Society,* **33**, pp. 263–321.

Drabek, P., and Holubova, G., 2007, *Elements of Partial Differential Equations* (Berlin: Walter de Gruyter).

Duffy, D.G., 2001, *Green's Functions with Applications* (Boca Raton: Chapman & Hall).

Duffy, R.J., 1961, The maximum principle and biharmonic functions. *Journal of Mathematical Analysis and Applications,* **3**, pp. 399–405.

Dunham, W., 1999, *Euler: Master of Us All* (Washington DC: The Mathematical Society of America).

Edwards, C.H. and Penney, D.E., 2005, *Elementary Differential Equations* (Englewood Cliffs, N.J.: Prentice Hall).

Eggers, A.J. Jr., Resnikoff, M.M. and Dennis, D.H., 1953, *Bodies of Revolution having Minimum Drag at High Supersonic Airspeeds,* NACA Technical Note 3666 (Washington: NACA).

Erdelyi, A., 1953, *Tables of Integral Transforms*, Bateman Manuscript Project, Vols. 1–2 (New York: McGraw-Hill).

Erdelyi, A. 1956, *Asymptotic Expansions* (New York: Dover).

Evans, G., Blackledge, J. and Yardley, P., 2000, *Analytic Methods for Partial Differential Equations* (London: Springer).

Farlow, S.J., 1982, *Partial Differential Equations for Scientists and Engineers* (New York: John Wiley & Sons).

Forsyth, A.R., 1890, *Theory of Differential Equations: Part I Exact Equations and Pfaff's Problem* (Cambridge: Cambridge University Press).

Forsyth, A.R., 1893, *Theory of Functions of a Complex Variable* (Cambridge: Cambridge University Press).

Forsyth, A.R., 1900, *Theory of Differential Equations: Part II Ordinary Equations, Not Linear* (Cambridge: Cambridge University Press).

Forsyth, A.R., 1902, *Theory of Differential Equations: Part III Ordinary Linear Equations* (Cambridge: Cambridge University Press).

Forsyth, A.R., 1906, *Theory of Differential Equations: Part IV* (Cambridge: Cambridge University Press).

Forsyth, A.R., 1918, *Solutions of the Examples in a Treatise on Differential Equations* (London: MacMillan).

Forsyth, A.R., 1956, *A Treatise on Differential Equations*, 6th ed. (London: MacMillan).

Forsyth, A.R., 1960, *Calculus of Variations* (New York: Dover).

Fraser, C.G., 1992, Isoperimetric problems in the variational calculus of Euler and Lagrange, *Historia Mathematica,* 19, 4–23.

French, A.P., 1971, *Newtonian Mechanics*, MIT Introductory Physics Series (New York: W.W. Norton & Company).

Friedman, B., 1956, *Principles and Techniques of Applied Mathematics* (New York: Dover).

Fuglede, B., 1981, On a direct method of integral equations for solving the biharmonic Dirichlet problem. *ZAMM (Zeitschrift für Angewandte Mathematik und Mechanik)*, **61**, pp. 449–459.

Fung, Y.C., 1965, *Foundations of Solid Mechanics* (Englewood Cliffs, N.J.: Prentice-Hall).

Gelca R. and Andreescu T., 2007, *Putnam and Beyond.* (New York: Springer).

Gelfand, I.M. and Shilov, G.E., 1964, *Generalized Functions*, Vol. 1 (New York: Academic Press).

Gingold, R.A. and Monaghan, J.J., 1977, Smoothed particle hydrodynamics: Theory and application to non-spherical stars. *Monthly Notices of the Royal Astronomical Society*, **181**, pp. 375–89.

Gjertsen, D., 1986, *The Newton Handbook* (New York: Routledge & Kegan Paul).

Gleick, J., 1988, *Chaos: Making a New Science* (New York: Penguin Books).

Goldstine, H.H., 1980, *A History of the Calculus of Variations from 17th through the 19th Century* (New York: Springer-Verlag).

Golomb, M. and Shanks, M., 1965, *Elements of Ordinary Differential Equations*, 2nd ed. (New York: McGraw-Hill).

Gradshteyn, I.S. and Ryzhik, I.M., 1980, *Table of Integrals, Series, and Products* (New York: Academic Press).

Grattan-Guinness, I. and Engelsman, S., 1982, The manuscripts of Paul Charpit. In *Sources*, HM9, pp. 65–75 (New York: Academic Press).

Green, G., 1828, An Essay on the Application of Mathematical Analysis to the Theories of Electricity and Magnetism. Originally published in Nottingham (1828).

Greenberg, M.D., 1971, *Application of Green's Functions in Science and Engineering* (Englewood Cliffs, N.J.: Prentice-Hall).

Gu, C.H., 1989, *Partial Differential Equations* (Beijing: Science and Technical Press) (in Chinese).

Gustafson, K., 1999, *Introduction to Partial Differential Equations and Hilbert Space Methods,* 3rd ed. (New York: Dover).

Gustafson, K. and Abe, T., 1998a, The third boundary condition: Was it Robin's? *Mathematical Intelligencer,* **20**(1), pp. 63–71.

Gustafson, K. and Abe, T., 1998b, (Victor) Gustave Robin: 1855–1897. *Mathematical Intelligencer,* **20**(2), pp. 47–53.

Hadamard, J., 1923, *Lectures on Cauchy's Problem in Linear Partial Differentiation Equations* (New York: Yale University Press).

Hardy, G.H., 1944, *A Course of Pure Mathematics*, 10th ed. (Cambridge: Cambridge University Press).

Havil, J., 2003, *Gamma: Exploring Euler's Constant* (Princeton: Princeton University Press).

Hawking, S.W., 1988, *A Brief History of Time* (Toronto: Bantam Books).

Hawkins, T., 1975a, Cauchy and the spectral theory of matrices. *Historia Mathematica*, **2,** pp. 1–29.

Hawkins, T., 1975b, The theory of matrices in the 19th century. *Proceedings of the International Congress of Mathematicians,* Vancouver, Canada, pp. 561–570.

Hawkins, T., 1977, Weierstrass and the theory of matrices. *Archive for History of Exact Sciences*, **17**(2), pp. 119–163.

Heinbockel, J.H., 2003, *Mathematical Methods for Partial Differential Equations* (Victoria: Trafford).

Hill, J.M., Berndt, B.C. and Huber, T., 2007, Solving Ramanujan's differential equations for Eisenstein series via a first order Riccati equation. *Acta Arithmetica*, **128**(3), pp. 281–294.

Ho, K.C. and Chau, K.T., 1999, A finite strip loaded by a bonded-rivet of a different material. *Computers and Structures*, **70**, pp. 203–218.

Holmes, M.H., 1995, *Introduction to Perturbation Methods,* Texts in Applied Mathematics 20 (New York: Springer).

Hua, L.K., 1956, *Mathematics Start with Yang Hui Triangle* (Beijing: People's Education Press) (in Chinese).

Hua, L.K., 2009, *An Introduction to Higher Mathematics,* Vols. 1–4 (Beijing: Higher Education Press) (in Chinese).

Hua, L.K., 2012, *An Introduction to Higher Mathematics,* translated by P. Shiu (Cambridge: Cambridge University Press).

Ince, E.L., 1956, *Ordinary Differential Equations* (New York: Dover).

Jackson, J.D., 1999, *Classical Electrodynamics*, 3rd ed. (New York: John Wiley).

James, I., 2002, *Remarkable Mathematicians: From Euler to von Neumann* (Cambridge: Cambridge University Press).

Jenkins-Jones, S., 1996, *The Hutchinson Dictionary of Scientists* (Bath: Helicon Publishing Ltd.).

John, F., 1981, *Partial Differential Equations,* 4th ed, Applied Mathematical Sciences Vol. 1 (New York: Springer-Verlag).

Kanigel, R., 1991, *The Man Who Knew Infinity: A Life of the Genius Ramanujan* (New York: Washington Square Press).

Kanok-Nukulchai, W. and Chau, K.T., 1990, Point sink fundamental solutions for subsidence prediction. *Journal of Engineering Mechanics ASCE*, **116**, pp. 1176–1182.

Kantorovich, L.V. and Krylov, V.I., 1964, *Approximate Methods of Higher Analysis,* translated by C.D. Benster (New York: Interscience).

Kelvin, Lord, 1887, On the waves produced by a single impulse in water of any depth, or in a dispersive medium. *Proceedings of the Royal Society of London,* **42**, pp. 80–83.

Kevorkian, J., 1990, *Partial Differential Equations: Analytical Solution Techniques* (Pacific Grove: Wadsworth & Brooks).

Kevorkian, J. and Cole, J.D., 1981, *Perturbation Methods in Applied Mathematics* (New York: Springer-Verlag).

Kline, M., 1959, *Mathematics and the Physical World* (New York: Dover).

Kline, M., 1972, *Mathematical Thought from Ancient to Modern Times*, Vols. 1–3 (New York: Oxford University Press).

Kline, M., 1977, *Why the Professor Can't Teach: Mathematics and The Dilemma of University Education* (New York: St. Martin's Press).

Kreyszig, E., 1979, *Advanced Engineering Mathematics*, 4th ed. (New York: Wiley).

Kuzmin, N.A., 1963, Representation of electrodynamic fields in curvilinear non-orthogonal space coordinates by means of two scalar functions and the inverse variational problem for the introduced functions. In *Electromagnetic Theory and Antennas,* edited by E.C. Jordan, Vol. 6, *International Series of Monographs on Electromagnetic Waves,* pp. 271–281 (Oxford: Pergamon).

Lagerstrom, P.A., 1988, *Matched Asymptotic Expansions: Ideas and Technique* (New York: Springer).

Lamb, H., 1932, *Hydrodynamics*, 6th ed. (New York: Dover).

Landau, L.D. and Lifshitz, E.M., 1987, *Fluid Mechanics* (Oxford: Pergamon).

Lebedev, N.N., 1972, *Special Functions & Their Applications,* translated and edited by R.A. Silverman (New York: Dover).

Lebedev, N.N., Sakalskaya, I.P. and Uflyand, Y.S., 1965, *Worked Problems in Applied Mathematics,* translated by R.A. Silverman and supplemented by E.L. Reiss (New York: Dover).

Le Veque, R.J., 2002, *Finite Volume Methods for Hyperbolic Problems* (New York: Cambridge University Press).

Lin, C.C. and Segel, L.A., 1988, *Mathematics Applied to Deterministic Problems in the Natural Sciences* (Philadelphia: Society for Industrial and Applied Mathematics).

Lipschutz, S., 1987, *Schaum's Outline Series: Theory and Problems of Linear Algebra*, SI ed. (New York: McGraw Hill).

Lopez, G., 2000, *Partial Differential Equations of First Order and Their Applications to Physics* (Singapore: World Scientific).

Lucy, L.B., 1977, A numerical approach to the testing of the fission hypothesis. *Astronomical Journal*, **82**, pp. 1013–1024.

Luschny, P., 2006, Is the gamma function misdefined? Hadamard versus Euler: Who found the better gamma function? http://www.luschny.de/math/factorial/hadamard/HadamardsGammaFunction.html.

Maor, E., 1994, *e: The Story of a Number* (Oxford: Princeton University Press).

Meleshko, V.V., 2003, Selected topics in the history of the two-dimensional biharmonic problem. *Applied Mechanics Reviews*, **56**(1), pp. 33–85.

Millar, D., Millar, I., Millar, J. and Millar, M., 2002, *The Cambridge Dictionary of Scientists,* 2nd ed. (Cambridge: Cambridge University Press).

Moulton, F.R., 1914, *An Introduction to Celestial Mechanics,* 2nd revised ed. (New York: Macmillan).

Mura, T., 1987, *Micromechanics of Defeats in Solids* (Dordrecht: Martinus Nijhoff).

Mura, T. and Koya, T., 1992, *Variational Methods in Mechanics* (New York: Oxford University Press).

Muskhelishvili, N.I., 1975, *Some Basic Problems of the Mathematical Theory of Elasticity*, translated by J.R.M. Radok, 2nd English ed. (Groningen: Noordhoff).

Myint-U, T., 1987, *Partial Differential Equations of Mathematical Physics* (New York: Elsevier).

Myint-U, T. and Debnath, L., 1987, *Partial Differential Equations for Scientists and Engineers* (Englewood Cliffs, N.J.: Prentice Hall).

Nahin, P.J., 2006, *Dr. Euler's Fabulous Formula: Cures Many Mathematical Ills* (Princeton: Princeton University Press).

Nasim, C., 1984, The Mehler-Fock transform of general order and arbitrary index and its inversion. *International Journal of Mathematics and Mathematical Science,* 7(1), pp. 171–180.

Nasim, C., 1989, Associated Weber integral transforms of arbitrary orders. *Indian Journal of Applied Mathematics,* **20**(11), pp. 1126–1138.

Nayfeh, A.H., 1973, *Perturbation Methods* (New York: John Wiley).

Nevel, D.E., 1959, *Tables of Kelvin Functions and Their Derivatives,* Technical Report 67 (Willmette, IL: U.S. Army Snow Ice and Permafrost Research Establishment).

Newton, I., 1687, *Principia,* translated into English by A. Motte (New York: Daniel Adee).

Oden, J.T., 1990, Historical Comments on Finite Elements. Originally published as "A History of Scientific Computing." Edited by Stephen G. Nash, pp. 128–152, 1990 by Association for Computing Machinery, Inc.

Olesiak, Z.S., 1990, Applications of Weber-Orr integral transforms in problems of thermoelasticity. In *Elasticity: Mathematical Methods and Applications*, The Ian N. Sneddon 70th Birthday Volume, edited by G. Eason and R.W. Ogden (New York: Ellis Horwood).

Olver, F.W.J., 1997, *Asymptotics and Special Functions* (Wellesley: A.K. Peters).

Olver, F.W.J., Lozier, D.W., Boisvert R.F. and Clark, C.W. 2010, *NIST Handbook of Mathematical Functions* (Cambridge: Cambridge University Press).

O'Malley, R.E., 2014, *Historical Development in Singular Perturbations* (Cham: Springer).

Petrovsky, I.G., 1991, *Lectures on Partial Differential Equations* (New York: Dover).

Piaggio, H.T.H., 1920, *An Elementary Treatise on Differential Equations and Their Applications* (London: Bell and Sons).

Polyanin, A.D., 2001, *Handbook of Linear Differential Equations for Engineers and Scientists* (Boca Raton: Chapman & Hall).

Polyanin, A.D. and Manzhirov, A.V., 2008, *Handbook of Integral Equations, 2nd edition* (Boca Raton: Chapman & Hall).

Polyanin, A.D. and Zaitsev, V.F., 2002, *Handbook of Exact Solutions for Ordinary Differential Equations* (Boca Raton: Chapman & Hall).

Polyanin, A.D. and Zaitsev, V.F., 2003, *Handbook of Nonlinear Differential Equations* (Boca Raton: Chapman & Hall).

Polyanin, A.D., Zaitsev, V.F. and Moussiaux, A., 2002, *Handbook of First Order Partial Differential Equations* (London: Taylor and Francis).

Poole, E.G.C., 1936, *Introduction to the Theory of Linear Differential Equations* (Oxford: Clarendon Press).

Press, W.H., Flannery, B.P., Teukolsky, S.A. and Vetterling, W.T., 1992, *Numerical Recipes (Fortran): The Art of Scientific Computing,* 2nd ed. (New York: Cambridge University Press).

Reddy, J.N., 2002, *Energy Principles and Variational Methods in Applied Mechanics* 2nd edition (Hoboken: John Wiley & Sons).

Reid, W.P., 1962, Free vibrations of a circular plate. *Journal of the Society for Industrial and Applied Mathematics*, **10**(4), 668–674.

Reiss, E.L., 1965, Variational and related methods. Supplement in *Worked Problems in Applied Mathematics,* by Lebedev, J.J., Sakalskaya, I.P. and Uflyand, Y.S. and translated by R.A. Silverman (New York: Dover).

Reiss, E.L., 1969, Column buckling: An elementary example of bifurcation. *Bifurcation Theory and Nonlinear Eigenvalue Problems,* ed. J.B. Keller and S. Antman (New York: Benjamin), pp. 1–16.

Reiss, E.L., 1977, Imperfect bifurcation. In *Applications of Bifurcation Theory,* ed. P.H. Rabinowitz (New York: Academic Press), pp. 37–71.

Reiss, E.L., 1980a, A modified two-time method for the dynamic transitions of bifurcation. *SIAM Journal of Applied Mathematics,* **38**(2), pp. 249–260.

Reiss, E.L., 1980b, A new asymptotic method for jump phenomena. *SIAM Journal of Applied Mathematics,* **39**(3), pp. 440–455.

Reiss, E.L., 1982, A nonlinear structural concept for compliant walls. *NASA Contractor Report 3628* (Washington DC: NASA).

Reiss, E.L., 1984, A nonlinear structural concept for drag-reducing compliant walls. *AIAA Journal,* **22**(3), pp. 399–402.

Reiss, E.L. and Matkowsky, B.J., 1971, Nonlinear dynamic buckling of a compressed elastic column. *Quarterly Journal of Applied Mathematics,* **29**, pp. 245–260.

Reyn, J.W., 1964, Classification and description of the singular points of a system of three linear differential equations. *Journal of Applied Mathematics and Physics* (*ZAMP*), **15**(5), pp. 540–557.

Rudnicki, J.W., 1986, Fluid mass sources and point forces in linear elastic diffusive solid. *Mechanics of Materials*, **5**, pp. 383–393.

Rudnicki, J.W., 1991, Boundary layer analysis of plane strain shear cracks propagating steadily on an impermeable plane in an elastic diffusive solid. *Journal of the Mechanics and Physics of Solids,* **39**(21), pp. 201–221.

Sabbagh, K., 2003, *The Riemann Hypothesis: The Great Unsolved Problem in Mathematics* (New York: Farrar, Straus & Giroux).

Saltykow, N., 1930, Etude bibliographique sur le memoire inedit de Charpit (Literature review of the unpublished memorandum Charpit). *Bulletin des Sciences Mathematiques* (2) **54**, pp. 255–264.

Saltykow, N., 1937, Etude bibliographique sur le la seconde partie du memoire inedit de Charpit (Literature review of the unpublished memorandum Charpit Part 2). *Bulletin des Sciences Mathematiques* (2) **61**, pp. 55–65.

Sedov, L.I., 1993, *Similarity and Dimensional Methods in Mechanics*, tenth edition (Boca Raton: CRC Press).

Segel, L.A., 1987, *Mathematics Applied to Continuum Mechanics* (New York: Dover).

Selvadurai, A.P.S., 2000a, *Partial Differential Equations in Mechanics*, Vol. 1, *Fundamentals, Laplace's Equation, Diffusion Equation, Wave Equation* (Berlin: Springer-Verlag).

Selvadurai, A.P.S., 2000b, *Partial Differential Equations in Mechanics*, Vol. 2, *The Biharmonic Equation, Poisson's Equation* (Berlin: Springer-Verlag).

Silverman, R.A., 1974, *Complex Analysis with Applications* (New York: Dover).

Simmons, G.F. and Krantz, S.G. 2007, *Differential Equations: Theory, Technique and Practice* (New York: McGraw Hill).

Singh, S., 1997, *Fermat's Last Theorem* (London: Fourth Estate).

Smirnov, M.M., 1978, *Equations of Mixed Type,* Translations of Mathematical Monographs, Vol. 51 (Providence: American Mathematical Society).

Sneddon, I.N., 1951, *Fourier Transforms* (New York: McGraw-Hill).

Sneddon, I.N., 1957, *Elements of Partial Differential Equations* (New York: Dover).

Sneddon, I.N., 1961, *Special Functions of Mathematical Physics and Chemistry* (New York: Interscience).

Sneddon, I.N., 1972, *The Use of Integral Transforms* (New York: McGraw-Hill).

Sommerfeld, A., 1949, *Partial Differential Equations in Physics.* Lectures on Theoretical Physics, Vol. VI, Translated by E. G. Straus (New York: Academic Press).

Spiegel, M.R., 1963, *Schaum's Outline Series: Theory and Problems of Advanced Calculus* (New York: McGraw-Hill).

Spiegel, M.R., 1964, *Schaum's Outline Series: Theory and Problems of Complex Variables* (New York: McGraw-Hill).

Spiegel, M.R., 1965, *Schaum's Outline Series: Theory and Problems of Laplace Transforms* (New York: McGraw-Hill).

Spiegel, M.R., 1968, *Schaum's Outline Series: Mathematical Handbook* (New York: McGraw-Hill).

Stakgold, I., 1967, *Boundary Value Problems of Mathematical Physics,* Vol. 1 (London: MacMillan).

Stakgold, I., 1968, *Boundary Value Problems of Mathematical Physics,* Vol. 2 (London: MacMillan).

Stakgold, I., 1979, *Green's Function Method and Boundary Value Problems* (New York: John Wiley & Sons).

Steele, C.R., 1976, Application of the WKB method in solid mechanics. In *Mechanics Today,* ed. S. Nemat-Nasser (New York: Pergamon), pp. 243–295.

Struik, D.J., 1987, *A Concise History of Mathematics,* 4th rev. ed. (New York: Dover).

Sulsky, D., Zhou, S.-J., and Schreyer, H. 1995, Application of particle-in-cell method to solid mechanics. *Computer Physics Communications*, **87**, pp. 236–252.

Temme, N.M., 1996, *Special Functions: An Introduction to the Classical Special Functions of Mathematical Physics* (New York: Wiley).

Tenenbaum, M. and Pollard, H., 1963, *Ordinary Differential Equations: An Elementary Textbook for Students of Mathematics, Engineering, and the Sciences* (New York: Dover).

Timoshenko, S.P., 1956, *Strength of Materials*, 3rd ed. (Princeton: Van Nostrand).

Timoshenko, S.P. and Goodier J.N., 1982, *Theory of Elasticity*, 3rd ed. (New York: McGraw-Hill).

Timoshenko, S.P. and Woinowsky-Krieger, S., 1959, *Theory of Plates and Shells*, 2nd ed. (New York: McGraw-Hill).

Tranter, C.J., 1957, *Techniques of Mathematical Analysis* (London: English Universities Press).

Tricomi, F.G., 1923, *Mixed Type Second Order Partial Differential Equations*, translated from Russian edition to Chinese by Qiu P. and Wang G. (Beijing: Science Press) (in Chinese).

Tricomi, F.G., 1957, *Integral Equations* (New York: Dover).

Trim, D.W., 1990, *Applied Partial Differential Equations* (Boston: PWS-KENT).

van Dyke, M., 1975, *Perturbation Methods in Fluid Mechanics* (Stanford: Parabolic Press).

Washizu, K., 1982. *Variational Methods in Elasticity & Plasticity*, 3rd ed. (Oxford: Pergamon).

Watson, G.N., 1914, *Complex Integration and Cauchy's Theorem* (Cambridge: Cambridge University Press).

Watson, G.N., 1918, The limits of applicability of the principle of stationary phase. *Proceedings of the Cambridge Philosophical Society*, **19**, pp. 49–55.

Watson, G.N., 1944, *A Treatise of the Theory of Bessel Functions*, 2nd ed. (New York: Dover).

Wazwaz, A.M., 2011, *Linear and Nonlinear Integral Equations* (Heidelberg: Springer).

Weber, O., Poranne, R. and Gotsman, C., 2012, Biharmonic coordinates. *Computer Graphics Forum*, **31**(8), pp. 2409–2422.

Weertman, J., 1996, *Dislocation Based Fracture Mechanics* (Singapore: World Scientific).

Weinstock, R., 1974, *Calculus of Variations with Applications to Physics and Engineering* (New York: Dover).

Whitham, G.B., 1974, *Linear and Nonlinear Waves* (New York: Wiley).

Whittaker, E. T. and Watson, G. N., 1927, *A Course of Modern Analysis,* 4th edition (London: Cambridge University Press).

Wong, C.W., 1991, *Introduction to Mathematical Physics: Methods and Concepts* (New York: Oxford University Press).

Wong, R., 2001, *Asymptotic Approximations of Integrals* (Philadelphia: Society for Industrial and Applied Mathematics).

Wylie, C.R., 1975, *Advanced Engineering Mathematics*, 4th ed. (New York: McGraw Hill).

Wyman, M., 1950, Deflections of an infinite plate. *Canadian Journal of Research,* A**28**, pp. 293–302.

Zachmanoglou, E.C. and Thoe, D.W., 1986, *Introduction to Partial Differential Equations with Applications* (New York: Dover).

Zangwill, A., 2013, *Modern Electrodynamics* (Cambridge: Cambridge University Press).

Zauderer, E., 1989, *Partial Differential Equations of Applied Mathematics,* 2nd ed. (New York: John Wiley & Sons).

Zemanian, A.H., 1965, *Distribution Theory and Transform Analysis* (New York: McGraw-Hill).

Zhu S., 1303, *Jade Mirror of the Four Unknowns* (in Chinese).

Zienkiewicz, O.C., 1977, *The Finite Element Method*, 3rd ed. (Maidenhead: McGraw-Hill).

Zienkiewicz, O.C. and Morgan, K., 1983, *Finite Elements and Approximation* (New York: John Wiley).

Zill, D.G., 1993, *A First Course of Differential Equations*, 5th ed. (Boston: PWS-KENT).

Zill, D.G. and Cullen, M.R., 2005, *Differential Equations with Boundary Value Problems*, 6th ed. (Thomson).

Zwillinger, D., 1997, *Handbook of Differential Equations,* 3rd ed. (Orlando: Academic Press).

Zwillinger, D., 2012, *Standard Mathematical Tables and Formulae* (Boca Raton: CRC Press).

Author Index

E

F

G

H

I

J

K

L

M

N

Subject Index

D

G

H

M

N

S

T